Decision variables	$(x_1, \ldots, x_n)$	$\mathbf{x}$
Solutions (or points) for iterations $t = 0, 1, \ldots$	$(x_1^{(0)}, \ldots, x_n^{(0)}), \ldots, (x_1^{(t)}, \ldots, x_n^{(t)}), \ldots$	$\mathbf{x}^{(0)}, \ldots, \mathbf{x}^{(t)}, \ldots$
Each new solution $(t+1)$ is obtained from current solution t by adding a step size λ times a change direction $\Delta \mathbf{x}$	$x_1^{(t+1)} \leftarrow x_1^{(t)} + \lambda \Delta x_1$ $\vdots$ $x_n^{(t+1)} \leftarrow x_n^{(t)} + \lambda \Delta x_n$	$\mathbf{x}^{(t+1)} \leftarrow \mathbf{x}^{(t)} + \lambda \Delta \mathbf{x}$
Each direction $\Delta \mathbf{x}$ should be improving at the current solution $\mathbf{x}^{(t)}$, i.e. it should improve the objective function value for sufficiently small steps λ.	(linear maximize) max $\sum_j c_j x_j$ requires $\sum_j c_j \Delta x_j > 0$ (linear minimize) min $\sum_j c_j x_j$ requires $\sum_j c_j \Delta x_j < 0$ (nonlinear maximize) max $f(x_1, \ldots, x_n)$ requires $\sum_j \frac{\partial f}{\partial x_j} \Delta x_j \geq 0$ but may not improve if $\sum_j \frac{\partial f}{\partial x_j} \Delta x_j = 0$ (nonlinear minimize) min $f(x_1, \ldots, x_n)$ requires $\sum_j \frac{\partial f}{\partial x_j} \Delta x_j \leq 0$ but may not improve if $\sum_j \frac{\partial f}{\partial x_j} \Delta x_j = 0$	(linear maximize) max $\mathbf{c} \cdot \mathbf{x}$ requires $\mathbf{c} \cdot \Delta \mathbf{x} > 0$ (linear minimize) min $\mathbf{c} \cdot \mathbf{x}$ requires $\mathbf{c} \cdot \Delta \mathbf{x} < 0$ (nonlinear maximize) max $f(\mathbf{x})$ requires $\nabla f(\mathbf{x}^{(t)}) \cdot \Delta \mathbf{x} \geq 0$ but may not improve if $\nabla f(\mathbf{x}^{(t)}) \cdot \Delta \mathbf{x} = 0$ (nonlinear minimize) min $f(\mathbf{x})$ requires $\nabla f(\mathbf{x}^{(t)}) \cdot \Delta \mathbf{x} \leq 0$ but may not improve if $\nabla f(\mathbf{x}^{(t)}) \cdot \Delta \mathbf{x} = 0$
Each direction $\Delta \mathbf{x}$ should also be feasible at the current solution $\mathbf{x}^{(t)}$, i.e. it should not cause violation of any constraints for sufficiently small steps λ.	(linear = constraints) $\sum_j a_j x_j = b$ requires $\sum_j a_j \Delta x_j = 0$ (linear $\geq$ constraints) $\sum_j a_j x_j \geq b$ requires $\sum_j a_j \Delta x_j \geq 0$ if $\sum_j a_j x_j^{(t)} = b$ (linear $\leq$ constraints) $\sum_j a_j x_j \leq b$ requires $\sum_j a_j \Delta x_j \leq 0$ if $\sum_j a_j x_j^{(t)} = b$ (lower bound constraints) $x_j \geq 0$ requires $\Delta x_j \geq 0$ if $x_j^{(t)} = 0$ (upper bound constraints) $x_j \leq u_j$ requires $\Delta x_j \leq 0$ if $x_j^{(t)} = u_j$	(linear = constraints) $\mathbf{A}\mathbf{x} = \mathbf{b}$ requires $\mathbf{A}\, \Delta \mathbf{x} = \mathbf{0}$ (linear $\geq$ constraints) $\mathbf{a} \cdot \mathbf{x} \geq b$ requires $\mathbf{a} \cdot \Delta \mathbf{x} \geq 0$ if $\mathbf{a} \cdot \mathbf{x}^{(t)} = b$ (linear $\leq$ constraints) $\mathbf{a} \cdot \mathbf{x} \leq b$ requires $\mathbf{a} \cdot \Delta \mathbf{x} \leq 0$ if $\mathbf{a} \cdot \mathbf{x}^{(t)} = b$ (lower bound constraints) $\mathbf{x} \geq \mathbf{0}$ requires $\Delta x_j \geq 0$ for all j with $x_j^{(t)} = 0$ (upper bound constraints) $\mathbf{x} \leq \mathbf{u}$ requires $\Delta x_j \leq 0$ for all j with $x_j^{(t)} = u_j$

| Each iteration applies the maximum step λ for which the selected direction $\Delta \mathbf{x}$ remains feasible and improving. If there is no limit, the model is unbounded. | (linear $\geq$ constraints)
 $\sum_j a_j x_j \geq b$ requires $\lambda \leq \left(\sum_j a_j x_j^{(t)} - b\right) / \left(-\sum_j a_j \Delta x_j\right)$ if $\left(\sum_j a_j \Delta x_j\right) < 0$
 (linear $\leq$ constraints)
 $\sum_j a_j x_j \leq b$ requires $\lambda \leq \left(b - \sum_j a_j x_j^{(t)}\right) / \left(\sum_j a_j \Delta x_j\right)$ if $\left(\sum_j a_j \Delta x_j\right) > 0$
 (lower bound constraints)
 $x_j \geq 0$ for all j or $\mathbf{x} \geq \mathbf{0}$ requires $\lambda \leq \min\left\{ \frac{x_j^{(t)}}{-\Delta x_j} : \Delta x_j < 0 \right\}$
 (upper bound constraints)
 $x_j \leq u_j$ for all j or $\mathbf{x} \leq \mathbf{u}$ requires $\lambda \leq \min\left\{ \frac{u_j - x_j^{(t)}}{\Delta x_j} : \Delta x_j > 0 \right\}$ |

Optimization in Operations Research

Optimization in Operations Research

RONALD L. RARDIN

Purdue University

Prentice Hall, Upper Saddle River, New Jersey 07458

Library of Congress Cataloging-in-Publication Data

Rardin, Ronald L.
    Optimization in operations research / Ronald L. Rardin.
       p.   cm.
    Includes bibliographic references and index.
    ISBN 0-02-398415-5
    1. Programming (Mathematics) 2. Operations research.
  3. Mathematical optimization.   I. Title.
T57.7.R37   1998
519.7--DC21   97–27030
                                            CIP

Associate editor: Alice Dworkin
Editor in chief: Marcia Horton
Production editor: Ann Marie Longobardo
Copy editor: Barbara Zeiders
Director of production and manufacturing: David W. Riccardi
Managing editor: Bayani Mendoza de Leon
Cover/Interior designer: Rosemarie Votta
Cover illustrator: David Bishop
Art director: Amy Rosen
Creative director: Paula Maylahn
Manufacturing buyer: Donna Sullivan
Editorial assistant: Dolores Mars

© 1998 by Prentice Hall, Inc.
Upper Saddle River, NJ 07458

For further information and free optimization software visit http://www.ecn.purdue.deu/~rardin/oorbook/

Printed in the United States of America

10 9 8 7 6 5 4 3

ISBN 0-02-398415-5

Reprinted with corrections April, 2000.

Prentice-Hall International (UK) Limited, *London*
Prentice-Hall of Australia Pty. Limited, *Sydney*
Prentice-Hall of Canada, Inc., *Toronto*
Prentice-Hall Hispanoamericana, S. A., *Mexico*
Prentice-Hall of India Private Limited, *New Delhi*
Prentice-Hall of Japan, Inc., *Tokyo*
Prentice-Hall Asia Pte. Ltd., *Singapore*
Editora Prentice-Hall do Brasil, Ltda., *Rio de Janeiro*

To my parents
Gene and Virginia,
my wife Blanca,
and my son Rob,
whose love and support
have made everything I do possible,
and to David Johnstone,
whose senseless murder diminished us all.

• • •

Contents

• •

Preface

• •

Operations Research really emerged as an academic discipline with the publication in the late 1960s of a series of pioneering introductory textbooks. A great deal has happened to the field in the three decades since, but most current introductory courses, and most new texts that have emerged along the way, still largely follow the treatment pioneered in those seminal books.

There is much good in that traditional development, but I undertook this new text on the deterministic (mathematical programming) half of OR because I believe students need something more. Emphasis should be on skills and intuitions they can carry away and apply in real settings or later courses. Modeling (or at least formulation) should be central. Students should be excited about the versatility and applicability of OR tools.

Most of the innovations of this book try to deliver on that vision. Numerous algorithms (including simplex) are developed in the context of a common improving search paradigm designed to prepare students for absorbing new material as they encounter it within the course and later. The full breadth of mathematical programming (including linear, integer, nonlinear, and multiobjective programming) is treated from the beginning to highlight the rich diversity of potential applications. Three entire chapters, and large portions of others, are devoted to describing and formulating examples, most of which are based on reports of real applications. The entire development is structured to facilitate its use in a variety of settings by making it as easy as possible to locate material and move around.

The book is intended for first undergraduate and graduate OR courses in engineering and mathematics, and undergraduate/graduate electives in management, as well as introductory courses in subtopics such as linear programming, integer and combinatorial optimization, network flows, and nonlinear programming. Its comprehensive and reentrant coverage should also make it a valuable reference for practitioners and researchers.

My rethinking of how to present mathematical programming begins with the conviction that you cannot do justice to the topic by teaching only linear programming and the simplex method. First of all, the overwhelming majority of optimization problems actually encountered in engineering and management practice cannot be modeled straightforwardly as linear programs. They involve discrete, nonlinear, and multiobjective elements that have to be confronted. Perhaps more importantly, OR practice has evolved to the point where integer, nonlinear, and multiobjective optimization models are dealt with every day. I was surprised myself, in researching the OR applications literature for this book, that accounts of discrete optimization significantly outnumber those using linear programming alone. Goal programs are everywhere. If students have never seen anything but classic LP, they will inevitably conclude OR is too narrow to deal with the complexities of real planning and design tasks.

Focusing heavily on the simplex algorithm also introduces a more subtle roadblock to learning. Regardless of whether simplex is taught as a combinatorial search for the right basic set or as a sequence of linear equation manipulations, decades of refinement on simplex development have almost totally obscured its relation to all the other tools of mathematical programming. A student who has learned only the simplex now carries away little insight around which to build any understanding of the wider field. Even the newer interior point algorithms for linear programming seem completely foreign.

I am not sugggesting that every introductory course must delve deeply into integer and nonlinear programming techniques, or that LP and the simplex should be discarded. What I believe we need is a development that treats LP as the best solved but far from the only form of mathematical program, and that equips students with the intuitions and paradigms to understand more advanced topics when they encounter them in later courses or practice. Where there is the time or the prerequisites to go more deeply into wider topics, that material should be there, developed as much as possible around the same concepts as those of basic LP.

This book tries to avoid equating linear programming with optimization by treating other cases from the very beginning. As in any introduction, the early priority is on formulating optimization problems in standard mathematical programming format. But by including integer, nonlinear, and multiobjective cases along with LPs, students are encouraged to get excited about how vast a range of applications can at least be modeled in this standard way. They can also begin early to recognize different model classes and to be alert to their relative tractability. My experience is that beginners will struggle with intracacies of subscripts, decision variables, constraints, and objectives—just as they do if we treat only LPs—but that the broader range of mathematical forms is well within their prior training. Binary variables are often the only new mathematical concept, and students find them quite intuitive.

Treatment of the improving search (hillclimbing) paradigm begins in Chapter 3 with basic notions of improving and feasible directions, step sizes, local and global solutions, and artificial starts. With the freedom to use nonlinear examples, this can be done in an intuitive and geometric way. Then the simplex algorithm is presented in Chapter 5 as an elegant specialization to linear programming. That improving search treatment of the simplex will seem unfamiliar even to many instructors, but very little is actually changed. Still, with the hillclimbing paradigm firmly established, it becomes much easier to develop interior point methods in Chapter 6, network algorithms in Chapter 10, nonlinear programming methods in Chapters 13-14, and even integer programming procedures in Chapter 12. Time probably will not permit covering all those topics in any single course, but my hope is that the paradigm prepares students for future learning, whether it comes in the current course, in later ones, or in professional practice.

Students can hardly be expected to get excited about using optimization methods if they have never seen them applied except on contrived examples small enough for hand computation. Also formulation skills, which I always find take the longest to develop, demand constant practice. These are why I have tried to make the book both an exposition of the rich diversity of OR applications and a continuing discourse on optimization modeling. Every algorithm and analytic principle is developed in the context of a brief story, and many computational exercises begin with a formulation step. The several whole or partial chapters devoted to formulation are full of realistic examples drawn from operations reseach practice, and exercises add many others. Time-short instructors will find both long enough to preserve a sense of the real application, yet sufficiently self-contained to be presented in 15 to 30 minutes of class time.

All serious mathematical programming computation requires computers, and I have tried to reflect that as well in my updated conceptualization of introductory OR. One way for students to judge whether their formulations are correct is to input them to optimization software and examine the results. That is why every formulation exercise with numerical constants also includes a computer solution step. Hand computation is still employed in numerous small exercises to develop student understanding of principles and algorithms, but here too the computer is exploited when possible. A few purely hand topics of traditional introductory courses have been dropped altogether because they provide neither tools for solving problems of practical size nor insights into the working of effective computer algorithms.

Realizing that few instructors will cover every topic in the book, and that even fewer students will read every page, I have tried the make the book as easy to move around in as possible. Dependencies between sections are minimized and clearly identified with explicit references. The entire development is broken into short and carefully labeled subsections addressing one topic each. Every major definition, principle and algorithm is set out for quick reference in an easy-to-spot box. One- or two-page "Primers" concisely review prerequisite material as it arises in the development. Computations and theoretical principles, which may be derived over several pages, are recapped in brief "Sample Exercises" also marked for easy identification. Icons in the exercises indicate which require computers ($\square$) or graphic calculators ($\boxplus$), and which have answers provided at the back of the book ($\boxtimes$).

Instructors who consider adopting the book may have to spend some time aligning their lectures and other materials with my innovations. Still, the cost is much less than it might seem at first glance because the changes have less to do with what is taught than the sequence and conceptual framework within which it is developed. I have tried to help with sample syllabi, notes, software and other aids posted through Internet site http://www.ecn.purdue.edu/~rardin/oorbook/.

I hope the effort to convert will prove justified for others. It has certainly been rewarded with improved learning and greater student motivation in my own classes at Purdue. I owe a great deal of thanks to the hundreds who have aided in its development as they studied from emerging drafts.

I also want to thank my wife Blanca and son Rob for their encouragement and patience through the long years this book has been underway, my two supportive department heads Ferd Leimkuhler and Marlin Thomas, and a sequence of helpful editors with Macmillan and Prentice-Hall. Special recognition is due editor David Johnstone, who launched me on this quest, and was senselessly murdered just as it began to take shape.

RONALD L. RARDIN, PROFESSOR
School of Industrial Engineering
Purdue University
West Lafayette, IN 47907-1287
rardin@ecn.purdue.edu
http://www.ecn.purdue.edu/~rardin/

About the Author

Ronald L. Rardin is professor of Industrial Engineering at Purdue University in West Lafayette, Indiana. Born and raised in Kansas, he received his B.A. and M.P.A. degrees from the University of Kansas. After working in city government, consulting and distribution for five years, he continued his education, earning a Ph.D. in Industrial Engineering from the Georgia Institute of Technology. He remained at Georgia Tech as a faculty member for nine years. Professor Rardin joined the faculty of Purdue University in 1982. His teaching and research interests center on optimization modeling and algorithms, particularly for large-scale integer and combinatorial problems. He is co-author of numerous research papers in that field and, with R. Gary Parker, author of a comprehensive graduate text *Discrete Optimization*. Professor Rardin is a recipient of Purdue University's Pritsker award for outstanding undergraduate teaching in industrial engineering. He is a member of IIE, INFORMS, and the Mathematical Programming Society.

C H A P T E R 1

Problem Solving with Mathematical Models

• •

Any student with the most elementary scientific training has encountered the idea of solving problems by analyzing mathematical equations that approximate the physical realities of the universe we inhabit. Countless questions about objects falling, beams shearing, gases diffusing, currents flowing, and so on, are reduced to simple computations upon skillful application of one of the natural laws passed to us by Newton, Ohm, Einstein, and others.

The applicable laws may be less enduring, but "operations" problems such as planning work shifts for large organizations, choosing investments for available funds, or designing facilities for customer service can also be posed in mathematical form. A **mathematical model** is the collection of variables and relationships needed to describe pertinent features of such a problem.

> **1.1** **Operations research (OR)** is the study of how to form mathematical models of complex engineering and management problems and how to analyze them to gain insight about possible solutions.

In this chapter some of the fundamental issues and vocabulary are introduced.

1.1 OR APPLICATION STORIES

Operations research techniques have proved useful in an enormous variety of application settings. One of the goals of this book is to expose students to as broad a sample as possible. All examples, many end-of-chapter exercises, several complete sections, and three full chapters present and analyze stories based on OR applications.

Whenever possible, these problems are drawn from reports of real operations research practice (identified in footnotes). Of course, they are necessarily reduced in size and complexity, and numerical details are almost always made up by the author. Other stories illustrate key elements of standard applications but greatly oversimplify, to facilitate quick learning.

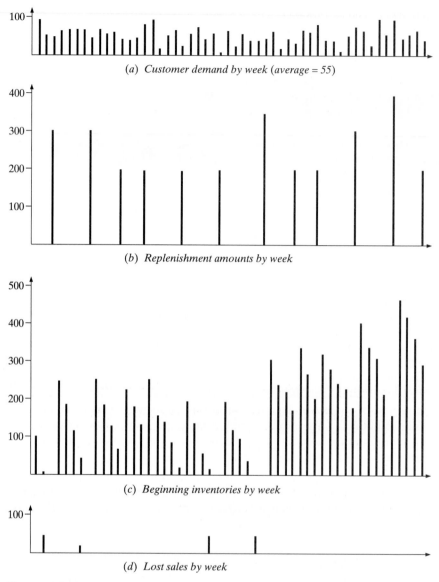

(a) Customer demand by week (average = 55)

(b) Replenishment amounts by week

(c) Beginning inventories by week

(d) Lost sales by week

FIGURE 1.1   Mortimer Middleman Example History

A handful of continuing examples are even smaller and more contrived. They still have a story, but convenience in illustrating methodological issues takes precedence over reality of application.

EXAMPLE 1.1: MORTIMER MIDDLEMAN

Our first story is of the totally made-up variety. Mortimer Middleman—friends call him MM—operates a modest wholesale diamond business. Several times each year MM travels to Antwerp, Belgium, to replenish his diamond supply on the international market. The wholesale price there averages approximately $700 per carat, but Antwerp market rules require him to buy at least 100 carats each trip. Mortimer and his staff then resell the diamonds to jewelers at a profit of $200 per carat. Each of the Antwerp trips requires 1 week, including the time for Mortimer to get ready, and costs approximately $2000.

Customer demand values in Figure 1.1(a) show that business has been good. Over the past year, customers have come in to order an average of 55 carats per week.

Part (c) of Figure 1.1 illustrates Mortimer's problem. Weekly levels of on-hand diamond inventory have varied widely, depending on the ups and downs in sales and the pattern of MM's replenishment trips [Figure 1.1(b)].

Sometimes Mortimer believes that he is holding too much inventory. The hundreds of carats of diamonds on hand during some weeks add to his insurance costs and tie up capital that he could otherwise invest. MM has estimated that these holding costs total 0.5% of wholesale value per week (i.e., $0.005 \times \$700 = \$3.50$ per carat per week).

At other times, diamond sales—and Mortimer's $200 per carat profit—have been lost because customer demand exceeded available stock [see Figure 1.1(d)]. When a customer calls, MM must either fill the order on the spot or lose the sale.

Adding this all up for the past year, MM estimates holding costs of $38,409, unrealized profits from lost sales of $31,600, and resupply travel costs of $24,000, making the annual total $94,009. Can he do better?

1.2 OPTIMIZATION AND THE OPERATIONS RESEARCH PROCESS

Operations research deals with **decision problems** like that of Mortimer Middleman by formulating and analyzing mathematical models—mathematical representations of pertinent problem features. Figure 1.2 illustrates this OR process.

The process begins with formulation or modeling. We define the variables and quantify the relationships needed to describe relevant system behavior.

Next comes analysis. We apply our mathematical skills and technology to see what conclusions the model suggests. Notice that these conclusions are drawn from

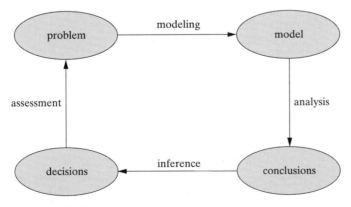

FIGURE 1.2   Operations Research Process

the model, not from the problem that it is intended to represent. To complete the process, we must engage in inference, that is, argue that conclusions drawn from the model are meaningful enough to infer decisions for the person or persons with the problem.

Often, an assessment of decisions inferred in this way shows them to be too inadequate or extreme for implementation. Further thought leads to revised modeling, and the loop continues.

Decisions, Constraints, and Objectives

We always begin modeling by focusing on three dimensions of the problem:

> __1.2__   The three fundamental concerns in forming operations research models are (a) the **decisions** open to decision makers, (b) the **constraints** limiting decision choices, and (c) the **objectives** making some decisions preferred to others.

In dealing with virtually any decision problem—engineering, management, or even personal—explicitly defining the decisions, constraints, and objectives helps to clarify the issues. Mortimer is obviously the decision maker in our diamond inventory management example. What decisions does he get to make?

Actually, MM makes hundreds of decisions each year about when to replenish his stock and how much to buy. However, it is common in inventory management circumstances such as Mortimer's to reduce the question to two policy decisions: What **reorder point** level of inventory should trigger a decision to buy new stock, and what **order quantity** should be purchased each time? These two variables constitute our decisions. We presume that each time on-hand inventory falls below the reorder point, Mortimer will head to Antwerp to buy a standard reorder quantity.

The next issue is constraints. What restrictions limit MM's decision choices? In this example there aren't very many. It is only necessary that both decisions be nonnegative numbers and that the order quantity conform to the 100-carat minimum of the Antwerp market.

The third element is objectives. What makes one decision better than another? In MM's case the objective is clearly to minimize cost. More precisely, we want to minimize the sum of holding, replenishment, and lost-sales costs.

Summarizing in a verbal model or word description, our goal is *to choose a nonnegative reorder point and a nonnegative reorder quantity to minimize the sum of holding, replenishment, and lost-sales costs subject to the reorder quantity being at least* 100.

Optimization and Mathematical Programming

Verbal models can help organize an analyst's thinking, but in this book we address a higher standard. We deal exclusively with optimization (also called mathematical programming).

> __1.3__   **Optimization models** (also called **mathematical programs**) represent problem choices as decision variables and seek values that maximize or min-

imize objective functions of the decision variables subject to constraints on variable values expressing the limits on possible decision choices.

With our Mortimer Middleman example, the decision variables are

$$q \triangleq \text{reorder quantity purchased at on each replenishment trip}$$

$$r \triangleq \text{reorder point signaling the need for replenishment}$$

(Here and throughout $\triangleq$ means "is defined to be.") Constraints require only that

$$q \geq 100$$
$$r \geq 0$$

The objective function,

$$c(q, r) \triangleq \text{total cost using a reorder quantity of } q \text{ and a reorder point } r$$

remains to be explicitly represented mathematically. We seek to minimize $c(q, r)$ over values of q and r satisfying all constraints.

Constant-Rate Demand Assumption

How we formulate constraints and objectives in terms of decision variables depends on what assumptions we are willing to make about the underlying system. We begin with a strong assumption regarding **constant-rate demand:** Assume that demand occurs at a constant rate of 55 carats per week. It is clear in Figure 1.1(a) that the demand rate is not exactly constant, but it does average 55 carats per week. Assuming that it is 55 carats in every week leads to some relatively simple analysis.

If the demand rate is constant, the pattern of on-hand inventory implied by a particular q and r will take one of the periodic "sawtooth" forms illustrated in Figure 1.3. Each time a shipment arrives, inventory will increase by order size q, then decline at the rate of 55 carats per week, producing regular cycles. Part (a) shows a case where inventory never runs out. A **safety stock** of (theoretically) untouched inventory protects against demand variability we have ignored. At the other extreme is part (c). Sales are lost because inventory runs out during the **lead time** between reaching the reorder point r and arrival of a new supply. Part (b) has neither safety stock nor lost sales. Stock runs out just as new supply arrives.

Back of Envelope Analysis

In cases where there are no lost sales [Figure 1.3(a) and (b)] it is easy to compute the length of each sawtooth cycle.

$$\frac{\text{order quantity}}{\text{demand rate}} = \frac{q}{55}$$

With lost sales [Figure 1.3(c)], each cycle is extended by a period when MM is out of stock that depends on both q and r.

Clearly, both modeling and analysis would be easier if we could ignore the lost-sales case. Can we afford to neglect lost sales? As in so many OR problems, a bit of crude "back of envelope" examination of the relevant costs will help us decide.

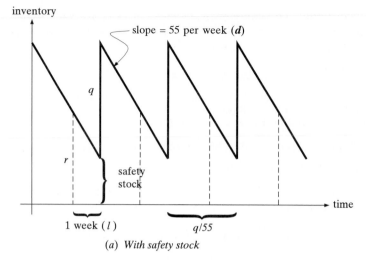

(a) With safety stock

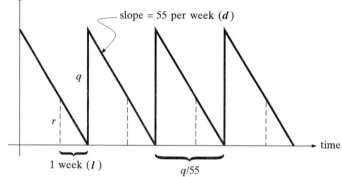

(b) No safety stock or lost sales

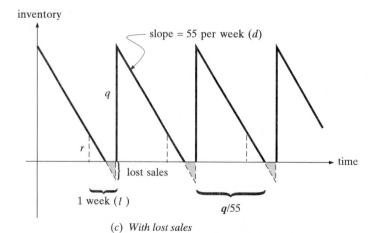

(c) With lost sales

FIGURE 1.3   Inventories Under Constant-Rate Demand

Lost sales may occur under the best of plans because of week-to-week variation in demand. Under our constant-rate demand assumption, however, there is no variation. Furthermore, MM can afford to add a unit to q and carry it for up to

$$\frac{\text{cost of lost sale}}{\text{weekly holding cost}} = \frac{\$200}{\$3.50} \approx 57.1 \text{ weeks}$$

rather than lose a carat of sales. Since the history in Figure 1.1 shows that inventory typically has been held no more than 4 to 6 weeks, it seems safe to make a second assumption regarding **no lost sales:** Assume that lost sales are not allowed.

Constant-Rate Demand Model

Since customers order a constant-rate 55 carats during the 1 week it takes Mortimer to carry out an Antwerp trip, both inventory at order arrival and lost sales can be computed by comparing 55 to r. If $r < 55$, we lose $(55 - r)$ carats of sales each cycle, something we have decided not to permit. Thus we may deduce the constraint

$$r \geq 55$$

With r restricted to be at least 55, $(r - 55)$ is the safety stock, and the cycle of rising and falling inventory repeats every $q/55$ weeks. Inventory on hand ranges from $(r - 55)$ at the low point of a cycle to $(r - 55) + q$ as a shipment arrives. The average will be the midpoint of these values, $(r - 55) + q/2$.

We are finally in a position to express all relevant costs. Holding cost per week is just the average inventory held times $3.50. Replenishment cost per week is $2000 divided by the cycle length or time between replenishments. Our first optimization model is

$$\begin{array}{ll} \text{minimize} & c = 3.50\left[(r - 55) + \dfrac{q}{2}\right] + \dfrac{2000}{q/55} \\ \text{subject to} & q \geq 100, \quad r \geq 55 \end{array} \tag{1.1}$$

Feasible and Optimal Solutions

Remember that our goal is to help Mortimer make decisions. Since the decisions are the variables in our model, we would like to characterize good values for **decision variables** q and r.

> 1.4   A **feasible solution** is a choice of values for the decision variables that satisfies all constraints. **Optimal solutions** are feasible solutions that achieve objective function value(s) as good as those of any other feasible solutions.

For example, $q = 200$, $r = 90$ is feasible in constant-rate demand model (1.1) because both constraints are satisfied: $200 \geq 100$ and $90 \geq 55$.

Here we can go farther and find an optimal solution. To begin, notice that if r deviates from demand 55, we incur extra holding cost and that no constraint prevents choosing r exactly 55. We conclude that

$$r^* = 55$$

will tell MM the perfect moment to start travel preparations. The asterisk (*) or **star** on a variable always denotes its optimal value.

Substituting this optimal choice of r of (1.1), the objective function reduces to

$$c(q, r) \triangleq 3.50 \left(\frac{q}{2}\right) + 2000 \left(\frac{55}{q}\right) \tag{1.2}$$

Elementary calculus will tell us how to finish (differentiate with respect to q and solve for a suitable point where the derivative is zero). To avoid being diverted by mathematical details in this introductory chapter, we leave the computation as an exercise for the reader.

The graphic presentation of cost function (1.2) in Figure 1.4 confirms the calculus result that the minimum average weekly cost occurs at

$$q^* = \pm\sqrt{\frac{2(2000)(55)}{3.50}} \approx 250.7$$

Since this value easily satisfies the $q \geq 100$ constraint, it is optimal.

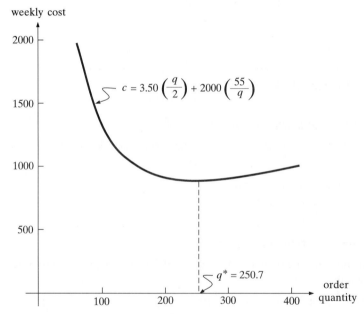

FIGURE 1.4   Optimal MM Order Quantity Under Constant-Rate Demand

To summarize, our assumptions of constant-rate demand and no lost sales have led us to advise Mortimer to go to Antwerp whenever inventory drops below $r^* = 55$ carats and to buy $q^* = 250.7$ carats of new diamonds each trip. Substituting these values in the objective function of (1.1), total cost should be about $877.50 per week or $45,630 per year—quite an improvement over Mortimer's real experience of $94,009.

1.3 SYSTEM BOUNDARIES, SENSITIVITY ANALYSIS, TRACTABILITY, AND VALIDITY

The modeling in Section 1.2 took as given many quantities, such as the demand per week and the cost per carat held, then computed optimal values for reorder point and reorder quantity. A line between those items taken as settled and those to be decided is called the **system boundary**. Figure 1.5 illustrates how **parameters**—quantities taken as given—define objective functions and constraints applicable to the decision model inside. Together, parameters and decision variables determine results measured as **output variables**.

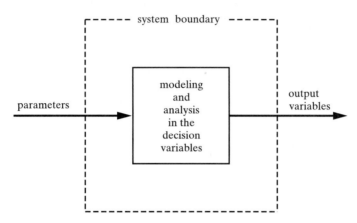

FIGURE 1.5   System Boundaries

EOQ Under Constant-Rate Demand

Only cost c is an output variable in our constant-rate demand model of Mortimer Middleman's problem. Enumerating the parameters, let

$d \triangleq$ weekly demand (55 carats)
$f \triangleq$ fixed cost of replenishment ($2000)
$h \triangleq$ cost per carat per week for holding inventory ($3.50)
$s \triangleq$ cost per carat of lost sales ($200)
$\ell \triangleq$ lead time between reaching the reorder point and receiving a new supply (1 week)
$m \triangleq$ minimum order size (100 carats)

A great attraction of our constant-rate demand analysis is that it can be done just as well in terms of these symbols. If lost sales are not allowed, repetition of the analysis (calculus) in terms of symbolic parameters will cause us to conclude that

$$
\boxed{1.5} \qquad
\begin{aligned}
\text{optimal reorder quantity } q^* &= \sqrt{\frac{2fd}{h}} \\
\text{optimal reorder point } r^* &= \ell d
\end{aligned}
$$

These results hold as long as $q^* \geq m$ (to be general, we also need to specify $q^* \geq \ell d$).

The square root expression for q^* now exhibits one of the oldest results in operations research: the classic **economic order quantity (EOQ)** formula for inventory management. Although we will soon see problems with the r^* part of the solution in MM's case, EOQ order quantities yield reliable inventory policies in a wide variety of settings.

System Boundaries and Sensitivity Analysis

To see the power of symbolic results such as equations $\boxed{1.5}$, we must recognize the inherent arbitrariness in system boundaries. If we took nothing as settled, models would mushroom in complexity and meaningful analysis would become impossible. Still, parameters we choose to regard as fixed at a system boundary often are known only vaguely. For example, MM's trips may cost approximately $f = \$2000$, but prices no doubt change with time, and more careful tabulation of past expenses might show that trips actually average $1000 or $3000. Figure 1.6 shows the significant implications for the optimal q^*. Our analysis with parameter $f = \$2000$ had $q^* = 250.7$. With $f = \$1000$, the optimal order quantity is 177.3; and with $f = \$3000$, it is 307.1.

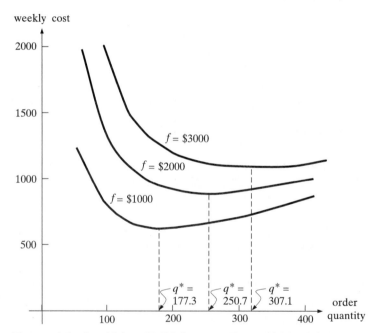

FIGURE 1.6   Sensitivity of MM Constant-Demand Results to Parameters

Such variations in input parameter values taken as fixed at a system boundary may dramatically affect the results of OR analysis.

$\boxed{1.6}$ **Sensitivity analysis** is an exploration of results from mathematical models to evaluate how they depend on the values chosen for parameters.

Any really complete operations research study includes an investigation of the sensitivity of results to parameter values.

Closed-Form Solutions

Solutions prescribing choices for decision variables by simple formulas in the input variables are called **closed-form** solutions. Expressions $\boxed{1.5}$ are an example.

The power of closed-form solutions lies with their providing results for many values of the parameters. Determining optimal q^* and r^* for the parameter values assumed in Section 1.2 was no small success. With closed-form solutions, however, we can determine sensitivities of optimal results to changes in input parameters. For example, we can see immediately that travel cost parameter f influences q^* in proportion to its square root and has no impact on r^*.

> **1.7** Closed-form solutions represent the ultimate in analysis of mathematical models because they provide both immediate results and rich sensitivity analysis.

Tractability versus Validity

> **1.8** **Tractability** in modeling means the degree to which the model admits convenient analysis—how much analysis is practical.

Our constant-rate demand model of Mortimer Middleman's problem has proved highly tractable because elementary calculus produced closed-form optimal solutions. Should we conclude that the goal of every OR study must be to formulate and analyze a model admitting closed-form solutions? Absolutely not!

Take another look at the operations research process depicted in Figure 1.2. Our purpose is not mathematical elegance but insights that will really help people like Mortimer Middleman deal with their problems. To assess the merit of operations research results, we must also consider another dimension.

> **1.9** The **validity** of a model is the degree to which inferences drawn from the model hold for the real system.

The complete analysis possible with our constant-rate demand model derived directly from the strong assumptions that we made to achieve a simple mathematical form. As we will see in the following subsections, those same assumptions bring the applicability of our closed-form results into real doubt.

Our study of OR will return to this dilemma again and again.

> **1.10** OR analysts almost always confront a tradeoff between validity of models and their tractability to analysis.

1.4 DESCRIPTIVE MODELS AND SIMULATION

Our constant-rate demand analysis of Mortimer Middleman's problem started by reducing all the information in Figure 1.1(a) to a single number, an average of 55 carats sold per week. Why not use more of what we know?

Simulation over MM's History

A **simulation model** is a computer program that simply steps through the behavior of a system of interest and reports experience. Of course, the behavior is tracked in computer variables and program logic rather than in a physical system.

To illustrate for Mortimer Middleman's inventory problem, suppose that we retain our verbal model (i.e., continue to focus on a reorder point r and a reorder quantity q). A straightforward computer program could then step through the 52 weeks of available history to try any given r and q. At each step, the program would:

1. Check whether MM is due to arrive with a new order of size q.
2. Determine if r has been reached so that a new trip is needed.
3. Reduce inventory by the demand actually experienced.

Holding cost for the week can be reported as $3.50 times the average number of carats in inventory. Replenishment cost is $2000 in each trip week. Lost-sales cost for the week is $200 times any excess of demand over the available inventory.

Table 1.1 details such a simulation from MM's demand history. Using the optimal $q^* = 251$ and $r^* = 55$ computed in our constant-rate demand model, this simulation reports beginning inventory for each week, customer demand taken from Figure 1.1(a), inventory management actions taken, and their consequences for holding cost, replenishment cost, and lost sales.

Total simulated cost for all 52 weeks is $108,621. Recall that Mortimer estimated actual cost at $94,009, and the constant demand analysis said that costs under q^* and r^* should decline to $45,630. The simulation model now is telling us that if MM adopts our supposedly optimal constant-rate demand policy, he will spend $108,621 - $94,009 = $14,621 more than if he simply keeps operating as he always has. This is hardly a help.

Simulation Model Validity

Can these new results be trusted? Only a clairvoyant could be certain, but it seems safe to conclude that the simulation results, based on an entire year's actual experience, should be taken more seriously than values derived from the constant-rate demand model.

> **1.11** Simulation models often possess high validity because they track true system behavior fairly accurately.

It should be noted, however, that the simulation was also based on some assumptions. For example, we have implicitly presumed that future demand will exactly mirror last year's experience.

Descriptive versus Prescriptive Models

Think now about tractability. How much did our simulation actually tell us? It estimated the holding, replenishment, and lost-sales costs of operating with $q = 251$ and $r = 55$ using fixed values for input parameters. Nothing more.

Models that evaluate fixed decision alternatives rather than indicating good choices may be termed **descriptive models**. Evaluating a few specific choices in this

TABLE 1.1   Deterministic Simulation of MM's Problem from Prior History

Week, t	Beginning Inventory	Customer Demand	Simulated Action	Holding Cost	Replenishment Cost	Lost Sales
1	100	94	Sell 94	$185.5	0	0
2	6	54	Below $r = 55$, so trip; sell 6	1.2	$2,000	$ 9,600
3	0	52	$q = 251$ arrive; sell 52	787.5	0	0
4	199	64	Sell 64	584.5	0	0
5	135	69	Sell 69	353.5	0	0
6	66	69	Sell 66	110.5	0	600
7	0	68	Below $r = 55$, so trip; sell 0	0.0	2,000	13,600
8	0	47	$q = 251$ arrive; sell 47	798.0	0	0
9	204	68	Sell 68	595.0	0	0
10	136	56	Sell 56	378.0	0	0
11	80	62	Sell 62	171.5	0	0
12	18	44	Below $r = 55$, so trip; sell 18	12.9	2,000	5,200
13	0	41	$q = 251$ arrive; sell 41	808.5	0	0
14	210	46	Sell 46	654.5	0	0
15	164	84	Sell 84	427.0	0	0
16	80	94	Sell 80	119.1	0	2,800
17	0	18	Below $r = 55$, so trip; sell 0	0.0	2,000	3,600
18	0	52	$q = 251$ arrive; sell 52	787.5	0	0
19	199	67	Sell 67	581.0	0	0
20	132	26	Sell 26	416.5	0	0
21	106	59	Sell 59	269.5	0	0
22	47	77	Below $r = 55$, so trip; sell 47	50.2	2,000	6,000
23	0	42	$q = 251$ arrive; sell 42	805.0	0	0
24	209	59	Sell 59	630.0	0	0
25	150	11	Sell 11	507.5	0	0
26	139	67	Sell 67	371.0	0	0
27	72	25	Sell 25	210.0	0	0
28	47	60	Below $r = 55$, so trip; sell 47	64.4	2,000	2,600
29	0	41	$q = 251$ arrive; sell 41	808.5	0	0
30	210	42	Sell 42	661.5	0	0
31	168	47	Sell 47	507.5	0	0
32	121	66	Sell 66	308.0	0	0
33	55	20	Below $r = 55$, so trip; sell 20	157.5	2,000	0
34	35	46	$q = 251$ arrive; sell 46	920.5	0	0
35	240	36	Sell 36	777.0	0	0
36	204	69	Sell 69	595.0	0	0
37	135	64	Sell 64	360.5	0	0
38	71	83	Sell 71	106.3	0	2,400
39	0	42	Below $r = 55$, so trip; sell 0	0.0	2,000	8,400
40	0	38	$q = 251$ arrive; sell 38	812.0	0	0
41	213	13	Sell 13	724.5	0	0
42	200	50	Sell 50	612.5	0	0
43	150	77	Sell 77	392.0	0	0
44	73	64	Sell 64	143.5	0	0
45	9	27	Below $r = 55$, so trip; sell 9	5.2	2,000	3,600
46	0	96	$q = 251$ arrive; sell 96	710.5	0	0
47	155	57	Sell 57	444.5	0	0
48	98	95	Sell 95	178.5	0	0
49	3	46	Below $r = 55$, so trip; sell 3	0.3	2,000	8,600
50	0	56	$q = 251$ arrive; sell 56	780.5	0	0
51	195	68	Sell 68	563.5	0	0
52	127	42	Sell 42	371.0	0	0

way sometimes tells a decision maker (here MM) as much as he or she needs to know. After all, many problems admit only a few practical solutions.

Still, results for a few cases are a very long way from what our **prescriptive** constant-rate demand optimization provided. Section 1.3's closed-form results both recommended optimal choices for q and r and offered insight about the sensitivity of results to changes in parameter values. The simulation did neither.

> **1.12** Descriptive models yield fewer analytic inferences than prescriptive, optimization models because they take both input parameters and decisions as fixed.

The tradeoff should now be clear. Resorting to a mathematically structureless form such as a simulation model can substantially improve validity. Still, that same lack of mathematical structure severely limits the possible analysis. More model validity almost always implies less tractability.

1.5 NUMERICAL SEARCH AND EXACT VERSUS HEURISTIC SOLUTIONS

The simulation in Section 1.4 provided a computer program for estimating the total inventory cost associated with a particular choice of reorder point and reorder quantity. Suppose we think of that computation as a function $c(q, r)$. That is, for any given q and r,

$c(q, r) \triangleq$ total cost computed by the simulation with reorder point fixed
  at r and reorder quantity at q

Mortimer Middleman's problem then reduces to the mathematical model

$$\begin{aligned} \text{minimize} \quad & c(q, r) \\ \text{subject to} \quad & q \geq 100, \quad r \geq 0 \end{aligned} \tag{1.3}$$

Numerical Search

Since we know very little about the properties of the mathematical function $c(q, r)$, restating our problem in this abstract form lends no immediate insight. Still, it does suggest a way to proceed.

Numerical search is the process of systematically trying different choices for the decision variables, keeping track of the feasible one with the best objective function value found so far. We call such a search numerical because it deals with specific values of the variables rather than with symbolic quantities such as those we were able to manipulate in analyzing the constant-rate demand model.

Here we search numerically over q and r. It seems reasonable to begin with the q and r recommended by our analysis of the constant-rate demand model. Using superscripts (note that these are *not exponents*) to identify specific choices of the decision variables, $q^{(0)} = 251$, $r^{(0)} = 55$, and we have already seen that $c(q^{(0)}, r^{(0)}) = \$108,621$.

Next, we need a systematic process for thinking of new q's and r's to try. Much of this book centers on how search processes should be structured. For now, we will try something naively simple: increasing and decreasing one variable at a time in steps of 10.

Table 1.1 showed considerable lost sales, so we start by increasing r to introduce a safety stock. Continuing until the objective function deteriorates yields.

$$q^{(0)} = 251 \qquad r^{(0)} = 55 \qquad c(q^{(0)}, r^{(0)}) = 108{,}621$$
$$q^{(1)} = 251 \qquad r^{(1)} = 65 \qquad c(q^{(1)}, r^{(1)}) = 108{,}421$$
$$q^{(2)} = 251 \qquad r^{(2)} = 75 \qquad c(q^{(2)}, r^{(2)}) = 63{,}254$$
$$q^{(3)} = 251 \qquad r^{(3)} = 85 \qquad c(q^{(3)}, r^{(3)}) = 63{,}054$$
$$q^{(4)} = 251 \qquad r^{(4)} = 95 \qquad c(q^{(4)}, r^{(4)}) = 64{,}242$$

All these $(q^{(t)}, r^{(t)})$ are feasible. Proceeding from one of the best, $r = 85$, we now try changing q. Increasing gives

$$q^{(5)} = 261 \qquad r^{(5)} = 85 \qquad c(q^{(5)}, r^{(5)}) = 95{,}193$$

and decreasing yields

$$q^{(6)} = 241 \qquad r^{(6)} = 85 \qquad c(q^{(6)}, r^{(6)}) = 72{,}781$$

Both are worse than $(q^{(3)}, r^{(3)})$, so we terminate the search.

A Different Start

Our numerical search has discovered a choice of q and r with simulated cost \$63,054, far better than either the \$108,621 of the constant-rate demand solution or MM's \$94,009 actual cost. Lacking any information on how much MM's costs might be reduced, our only way of learning more with numerical search is to try a new search from a different initial point. Consider the search sequence

$$q^{(0)} = 251 \qquad r^{(0)} = 145 \qquad c(q^{(0)}, r^{(0)}) = 56{,}904$$
$$q^{(1)} = 251 \qquad r^{(1)} = 155 \qquad c(q^{(1)}, r^{(1)}) = 59{,}539$$
$$q^{(2)} = 251 \qquad r^{(2)} = 135 \qquad c(q^{(2)}, r^{(2)}) = 56{,}900$$
$$q^{(3)} = 251 \qquad r^{(3)} = 125 \qquad c(q^{(3)}, r^{(3)}) = 59{,}732$$
$$q^{(4)} = 261 \qquad r^{(4)} = 135 \qquad c(q^{(4)}, r^{(4)}) = 54{,}193$$
$$q^{(5)} = 271 \qquad r^{(5)} = 135 \qquad c(q^{(5)}, r^{(5)}) = 58{,}467$$

This time, we have happened upon the better heuristic solution $q = 261, r = 135$ with simulated cost \$54,193. Certainly, the earlier best of $q = 251, r = 85$ was not optimal, but we still have no real reason to believe that our last result is the best achievable.

With only two variables we might try a more exhaustive search, forming a loop and evaluating $c(q, r)$ at an entire grid of points. The essential difficulty would remain, because results would depend on the size of the grid investigated.

| 1.13 | Inferences from numerical search are limited to the specific points explored unless mathematical structure in the model supports further deduction. |

Exact versus Heuristic Optimization

| 1.14 | An **exact optimal solution** is a feasible solution to an optimization model that is provably as good as any other in objective function value. A **heuristic** or |

> **approximate optimum** is a feasible solution derived from prescriptive analysis that is not guaranteed to yield an exact optimum.

Our numerical searches of MM's problem have produced only a heuristic optimum—a good feasible solution. Should we demand an exact optimum?

As usual, the answer is far from clear. The three "optimal" values of q depicted in Figure 1.6 are all mathematically exact, yet the decisions they recommend vary dramatically with the assumed value of input parameters. Furthermore, our simulation results have shown the true cost to be very different from any of the optimal objective function values computed assuming constant-rate demand.

> **1.15** | Losses from settling for heuristic instead of exact optimal solutions are often dwarfed by variations associated with questionable model assumptions and doubtful data.

Still, it would clearly make us more certain of whether to recommend the $q = 261, r = 135$ solution, which is the best uncovered in our numerical searches, if we knew something about how far from optimal it might be. Exact optima add a satisfying degree of certainty.

> **1.16** | The appeal of exact optimal solutions is that they provide both good feasible solutions and certainty about what can be achieved under a fixed set of model assumptions.

1.6 DETERMINISTIC VERSUS STOCHASTIC MODELS

All the searching in Section 1.5 was based on the assumption that Mortimer Middleman's future week-to-week demand will repeat identically the experience of the past year. This is almost certainly untrue. The best we can honestly claim regarding future events is that we know something about the probability of various outcomes.

> **1.17** | A mathematical model is termed **deterministic** if all parameter values are assumed to be known with certainty, and **probabilistic** or **stochastic** if it involves quantities known only in probability.

Random Variables and Realizations

Random variables represent quantities known only in terms of a probability in stochastic models. We distinguish random variables from ones with single values by using uppercase (capital) letters.

If we do not accept the notion that next year's demands in Mortimer's problem will exactly duplicate last year's, weekly demands are random variable parameters to his decision problem. Each weekly demand will eventually be a specific number. At the time we have to choose a reorder point and reorder quantity; however, we may know only something about their probability distributions.

If the magnitude of each week's random demand—denote it D_t—is independent of all others (an assumption), each of the values in Figure 1.1(a) is a **realization** of each D_t (i.e., a specific historical outcome). We can learn about the probability distribution of the D_t by counting how often different realizations appear in Figure 1.1(a). The solid line in Figure 1.7 presents such a **frequency histogram**. Demands ranged from 11 to 96, but Figure 1.7 clearly shows that demands in the range 40 to 70 are most common.

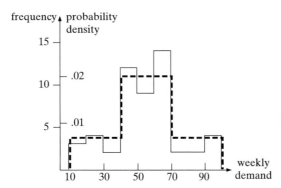

FIGURE 1.7   Weekly MM Demand Frequency Distribution

Eliminating some of the raggedness, the dashed lines in Figure 1.7 show how the true probability distribution for D probably looks. Outcomes in the range 40 to 70 are three times as probable as those in the regions 10 to 30 and 70 to 100, but demands are equally likely within regions.

Stochastic Simulation

The simulation model of Sections 1.4 and 1.5 is deterministic because all input parameters, including the week-by-week demands, are assumed known when computations such as Table 1.1 begin. We can develop a stochastic model of Mortimer's problem by improving on that simulation. Assuming that weekly demands $D_1, \ldots, D_{52}$ are independent random variables with known probability distribution such as the smoothed one (dashed lines) in Figure 1.7, we will investigate the distribution of the annual cost [now random variable $C(q, r)$] associated with any choice of reorder quantity q and reorder point r.

Stochastic simulation (sometimes called **Monte Carlo analysis**) provides the tool. It samples realizations from output variable distributions by:

1. Randomly generating a sequence of realizations for input parameters.
2. Simulating each realization against chosen values for the decision variables.

With a large enough (and random enough) sample, a frequency histogram of output realizations will approximate the true output distribution.

Figure 1.8 shows results of such stochastic simulation sampling in the Mortimer Middleman problem. A total of 200 different sets of realizations for demand variables $D_1, \ldots, D_{52}$ were randomly sampled from the distribution depicted in Figure 1.7. Then each was simulated with $q = 261$, $r = 135$ just as in Table 1.1, and the annual cost computed as one realization of $C(261, 135)$.

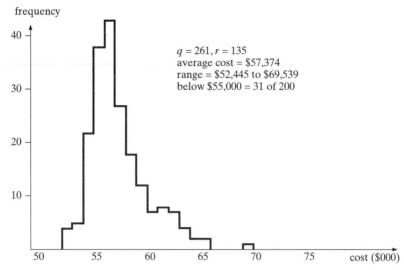

frequency

q = 261, r = 135
average cost = $57,374
range = $52,445 to $69,539
below $55,000 = 31 of 200

cost ($000)

FIGURE 1.8   MM Annual Cost Frequency Distribution

We can see from Figure 1.8 that for this best known choice of decision variables, annual cost has a distribution ranging from $52,445 to $69,539, with an average of about $57,374. Notice that this range of possible futures does include the single value $54,193 obtained in Section 1.5's numerical searches. Still, many other costs might result, depending on what demand pattern is actually realized.

Stochastic simulation is an effective and widely used OR modeling technique because the enormous flexibility of simulation models makes it possible to formulate validly a very wide range of problem situations. Still, the ragged shape of the frequency histograms in Figure 1.8 highlights the extra analytic challenge associated with computing distributions by simulation. Even with decision variables fixed, and after 200 runs, we know the $C(q, r)$ distributions in Figure 1.8 only approximately. Statistical estimation techniques would be required to determine how much confidence we can have in any conclusions.

> __1.18__   Besides providing only descriptive analysis, stochastic simulation models impose the extra analytic burden of having to estimate results statistically from a sample of system realizations.

Tradeoffs between Deterministic and Stochastic Models

Few, if any, parameters of real systems can actually be quantified with certainty, yet that is exactly what deterministic models assume. Distributional estimates such as Figure 1.8 usually provide a much more complete picture of what might actually happen with any chosen q and r.

Why not use stochastic models for every problem? It should be no surprise that the answer is: tractability. We have already seen how much work is required to estimate the output distribution for a single choice of q and r via stochastic simulation. Other stochastic methods can sometimes produce results with less effort, and validity sometimes demands that stochastic variation be modeled, but a major tractability difference remains.

> **1.19** The power and generality of available mathematical tools for analysis of stochastic models does not nearly match that available for deterministic models.

When optimization is the goal, deterministic modeling is often the only practical choice.

> **1.20** Most optimization models are deterministic—not because OR analysts really believe that all problem parameters are known with certainty, but because useful prescriptive results can often be obtained only if stochastic variation is ignored.

Of course, deterministic optimization models are also used because they work. We will see in illustrations throughout this book that very satisfactory results have been obtained in a wide variety of applications, often in circumstances that involve many thousands of decision variables.

1.7 PERSPECTIVES

Mortimer Middleman's inventory problem is only one of hundreds of operations design, planning, and control questions that we will encounter in this book. The purpose of treating it in so many different ways here is to introduce the possible approaches and to reveal some of their strengths and weaknesses.

OR analysts must daily decide whether a back-of-the-envelope computation is sufficient or a formal model is required; whether a more detailed and thus more valid model is preferred to a more tractable approximation. One goal of this book is to provide the vocabulary and sensitivity to issues that will help us make those choices.

Other Issues

The chapter's emphasis on the tractability versus validity tradeoff in operations research is not intended to suggest that it is the only issue. Models also differ in how easily they can be understood by the consumer of the analysis (Mortimer here), how fast analysis must be completed, how much mathematical and computer power is required to do the analysis, how many data must be collected, and in a variety of other ways.

More fundamentally, many situations do not lend themselves to the OR approach. Even from this chapter's brief introduction, it should be clear that it is not easy to formulate valid models that are tractable to analysis. Operations research is founded on the conviction, buttressed by a long history of successful practice, that formulation and analysis of mathematical decision models is often worth the trouble. Still, time and resources are required.

> **1.21** The model-based OR approach to problem solving works best on problems important enough to warrant the time and resources for a careful study.

The Rest of This Book

In the remainder of this book, the main classes of deterministic optimization models, and the solution techniques available to deal with them, are developed. Chapter 2 begins with formulation and classification of a series of real-world problems in terms of the mathematical properties of the models required. Later chapters address each class in turn, emphasizing formulation of important business and engineering applications and highlighting common elements of the sometimes bewildering variety of analysis techniques used to address various classes.

EXERCISES

1-1 A segment of automatically controlled highway is being equipped with sensors spaced equally along its length. The maximum error in estimating traffic volume, which occurs halfway between any two sensors, can be expressed as $(d/s)^2$, where d is the length of the segment and s is the number of sensors. Each sensor costs p dollars, and designers want to reduce the maximum error as much as possible within a budget for sensors of b dollars. Identify each of the following for this design problem.

(a) ☒ The decision variable
(b) ☒ The input parameters
(c) ☒ The objective function
(d) ☒ The constraints

1-2 Return to the problem of Exercise 1-1 and assume that $d = 10$, $p = 3.5$, and $b = 14$. Determine (if necessary by trying all the possibilties) whether each of the following is a feasible and/or optimal solution.

(a) ☒ $s = 4$
(b) ☒ $s = 6$
(c) ☒ $s = 2$

1-3 A factory has two production lines available to make a product. The first can produce one lot of the product in t_1 hours at cost c_1, and the second requires t_2 hours and cost c_2. The plant manager wishes to find the least costly way to produce b lots in a total of at most T hours. An integer number x_1 will be produced on line 1, and integer number x_2 on line 2. Identify each of the following for this design problem.

(a) The decision variables
(b) The input parameters
(c) The objective function
(d) The constraints

1-4 Return to the problem of Exercise 1-3 and assume that $t_1 = 10$, $t_2 = 20$, $c_1 = 500$, $c_2 = 100$, $b = 3$, and $T = 40$. Determine (if necessary by trying all the possibilties) whether each of the following is a feasible and/or optimal solution.

(a) $x_1 = 0, x_2 = 3$
(b) $x_1 = 2, x_2 = 1$
(c) $x_1 = 3, x_2 = 0$

1-5 A university wishes to purchase the maximum number of computer workstations that can be accommodated within the available laboratory floor space and budget. Determine whether each of the following outcomes is most likely the result of closed-form optimization, exact numerical optimization, heuristic optimization, or descriptive modeling.

(a) ☒ The maximum number of stations that can be fit within 2000 square feet and a budget of $500,000 is 110.
(b) The usual arrangement of 80 workstations would require 1600 square feet of floor space and cost $382,000.
(c) ☒ The maximum number of stations feasible with an area of f thousand square feet and a budget of b thousand dollars is min{$50f$, $b/5$}.
(d) The best of the usual layouts for an area of 2000 square feet and a budget of $500,000 can accommodate 85 workstations.

1-6 Explain why the first alternative of each pair constitutes more complete analysis of an optimization model than the second.

(a) Closed—form versus numerical optimization
(b) Exact optimization versus heuristic optimization
(c) Heuristic optimization versus simulation of some specific solutions

1-7 Explain why a model that admitted one of the preferred choices in Exercise 1-6 would not necessarily be more appropriate than one allowing only the alternative.

1-8 An engineer is working on a design problem with n parameters to be chosen, and each has 3 possible values. A highly valid descriptive model is available, and he is thinking of choosing the best design simply by using the model to evaluate the effect of every combination of design parameter values.

(a) Explain why this approach will require running the descriptive model on 3^n combinations of decision-variable values.

(b) ⊘ For $n = 10, 15, 20$, and 30, compute the time a computer would require to check all the combinations, assuming that it runs 24 hours per day, 365 days per year, and that it requires 1 second to apply the model for any particular choice of design parameters.

(c) Comment on the practical limits of this "try all combinations" analysis strategy in the light of your results from part (b).

1-9 Determine whether each of the following could probably be validly modeled only as a random variable or if a deterministic quantity would suffice.

(a) ⊘ The number of inches of rainfall in a city over the next 14 days

(b) The average 14-day rainfall in a city

(c) ⊘ The market price of a common stock 1 week ago

(d) The market price of a common stock 1 week from today

(e) ⊘ The seating capacity of a restaurant

(f) The number of customers who will arrive at a restaurant this evening

(g) ⊘ The production rate of an industrial robot subject to frequent breakdowns

(h) The production rate of a highly reliable industrial robot

(i) ⊘ Factory demand for a model of bulldozer over the next 7 days

(j) Factory demand for a model of bulldozer over the next 7 years

C H A P T E R 2

Deterministic Optimization Models in Operations Research

• •

With this chapter we begin our detailed study of deterministic models in operations research—models where it is reasonable to assume all problem data to be known with certainty. Few who have ever worked with real models can say the words *known with certainty* without breaking a smile. Input constants in OR models are almost always estimated, some rather crudely. We employ deterministic models because they often produce valid enough results to be useful and because deterministic models are almost always easier to analyze than are their stochastic counterparts.

The increased tractability of deterministic models permits us the luxury of dealing explicitly with optimization. Often, we achieve only an approximation, but the consistent goal is to prescribe the best possible solution.

Deterministic optimization models are also called **mathematical programs** because they decide how to plan or "program" activities. Our treatment is introduced in this chapter with a stream of examples. By formulating models for a variety of cases, we begin developing the modeling skills essential to every OR analyst. At the same time we illustrate the wide range of model forms available and introduce some terminology. Except for a tiny first example, all models presented are based on real applications by real organizations.

2.1 DECISION VARIABLES, CONSTRAINTS, AND OBJECTIVE FUNCTIONS

From the very early days of deterministic optimization, one of its heaviest users has been the petroleum refining industry. Refining operations are routinely planned by formal optimization, often on a daily or even hourly basis. We begin our survey of mathematical programming models with a made-up example from the refining setting, obviously much oversimplified.

EXAMPLE 2.1: TWO CRUDE PETROLEUM

Two Crude Petroleum runs a small refinery on the Texas coast. The refinery distills crude petroleum from two sources, Saudi Arabia and Venezuela, into three main products: gasoline, jet fuel, and lubricants.

The two crudes differ in chemical composition and thus yield different product mixes. Each barrel of Saudi crude yields 0.3 barrel of gasoline, 0.4 barrel of jet fuel, and 0.2 barrel of lubricants. On the other hand, each barrel of Venezuelan crude yields 0.4 barrel of gasoline but only 0.2 barrel of jet fuel and 0.3 barrel of lubricants. The remaining 10% of each barrel is lost to refining.

The crudes also differ in cost and availability. Two Crude can purchase up to 9000 barrels per day from Saudi Arabia at $20 per barrel. Up to 6000 barrels per day of Venezuelan petroleum are also available at the lower cost of $15 per barrel because of the shorter transportation distance.

Two Crude's contracts with independent distributors require it to produce 2000 barrels per day of gasoline, 1500 barrels per day of jet fuel, and 500 barrels per day of lubricants. How can these requirements be fulfilled most efficiently?

Decision Variables

The first step in formulating any optimization model is to identify the **decision variables**.

| 2.1 | Variables in optimization models represent the decisions to be taken. |

Numerous quantities are floating around in even the very simple Two Crude problem statement. Which do we get to decide? Cost, availabilities, yields, and requirements are all **input parameters**—quantities that we will take as fixed (see Section 1.3). What must be decided is how much of each crude to refine. Thus we define decision variables

$$
\begin{aligned}
x_1 &\triangleq \quad \text{barrels of Saudi crude refined per day (in thousands)} \\
x_2 &\triangleq \quad \text{barrels of Venezuelan crude refined per day (in thousands)}
\end{aligned}
\tag{2.1}
$$

Notice that like good modelers in all fields, we have specified both the meaning of each variable and the units in which it is denominated.

Variable-Type Constraints

The next issue in formulating an optimization model is **constraints**. What limits decisions?

The most elementary constraints declare variable type.

| 2.2 | **Variable-type constraints** specify the domain of definition for decision variables: the set of values for which the variables have meaning. |

For example, variables may be limited to nonnegative values or to nonnegative integer values, or they may be totally unrestricted.

Decision variables x_1 and x_2 in the Two Crude example represent quantities of petroleum refined. Thus they are subject to the most common variable-type constraint form: **nonnegativity**. Every meaningful choice of these decision variables must satisfy

$$x_1, x_2 \geq 0 \qquad (2.2)$$

That is, both quantities must be nonnegative real numbers.

It may seem a bit fastidious to specify constraints (2.2) when they are already implicit in definitions (2.1). Still, nothing can be taken for granted because our goal is to produce a form suitable for analysis by computer-based procedures. Formal methods for solving mathematical programs enforce only constraints explicitly expressed in the model formulation.

Main Constraints

The remainder of the limits on decision variable values constitute the main constraints.

> **2.3** | **Main constraints** of optimization models specify the restrictions and interactions, other than variable type, that limit decision variable values.

Even in as simple a case as Two Crude Petroleum, we have several main constraints. First consider output requirements. Petroleum volumes selected must meet contract requirements in the sense that

$$\underbrace{\sum (\text{yield per barrel}) (\text{barrels purchased})}_{\text{total output of a product}} \geq \text{product requirements}$$

Quantifying in terms of our decision variables produces one such constraint per product:

$$\begin{array}{ll} 0.3x_1 + 0.4x_2 \geq 2.0 & \text{(gasoline)} \\ 0.4x_1 + 0.2x_2 \geq 1.5 & \text{(jet fuel)} \\ 0.2x_1 + 0.3x_2 \geq 0.5 & \text{(lubricants)} \end{array} \qquad (2.3)$$

Each is denominated in thousands of barrels per day.

Availabilities yield the other class of main constraints. We can buy no more than 9000 barrels of Saudi crude per day, nor 6000 barrels of Venezuelan. Corresponding constraints are

$$\begin{array}{ll} x_1 \leq 9 & \text{(Saudi)} \\ x_2 \leq 6 & \text{(Venezuelan)} \end{array} \qquad (2.4)$$

Objective Functions

Objective or **criterion functions** tell us how to rate decisions.

> **2.4** | Objective functions in optimization models quantify the decision consequences to be maximized or minimized.

What makes one choice of decision variable values preferable to another in the Two Crude case? Cost. The best solution will seek to minimize total cost:

$$\min \quad \sum (\text{crude unit cost}) (\text{barrels purchased})$$

Quantifying this single objective in terms of the decision variables yields

$$\min \quad 20x_1 + 15x_2 \tag{2.5}$$

(in thousands of dollars per day).

Standard Model

Once decision variables have been defined, constraints detailed, and objectives quantified, the mathematical programming model we require is complete. Still, it is customary to follow variable and parameter definitions with a summary of the model in a standard format.

> **2.5** The standard statement of an optimization model has the form
>
> $$\text{min or max} \quad (\text{objective function(s)})$$
> $$\text{s.t.} \quad (\text{main constraints})$$
> $$\quad\quad\quad (\text{variable-type constraints})$$
>
> where "s.t." stands for "subject to."

Combining (2.2)–(2.5), we may formalize our Two Crude model as follows:

$$
\begin{aligned}
\min \quad & 20x_1 + 15x_2 & \text{(total cost)} \\
\text{s.t.} \quad & 0.3x_1 + 0.4x_2 \geq 2.0 & \text{(gasoline requirement)} \\
& 0.4x_1 + 0.2x_2 \geq 1.5 & \text{(jet fuel requirement)} \\
& 0.2x_1 + 0.3x_2 \geq 0.5 & \text{(lubricant requirement)} \\
& x_1 \quad\quad\quad\quad \leq 9 & \text{(Saudi availability)} \\
& x_2 \quad\quad\quad\quad \leq 6 & \text{(Venezuelan availability)} \\
& x_1, x_2 \geq 0 & \text{(nonnegativity)}
\end{aligned}
\tag{2.6}
$$

SAMPLE EXERCISE 2.1: FORMULATING FORMAL OPTIMIZATION MODELS

Suppose that we wish to enclose a rectangular equipment yard by at most 80 meters of fencing. Formulate an optimization model to find the design of maximum area.

Modeling: The decisions required are the dimensions of the rectangle. Thus define decision variables:

$$\ell \triangleq \text{length of the equipment yard (in meters)}$$

$$w \triangleq \text{width of the equipment yard (in meters)}$$

Both variables are nonnegative quantities, which implies variable-type constraints

$$\ell, w \geq 0$$

The only main constraint is that the perimeter of the equipment yard should not exceed 80 meters in length. In terms of decision variables, that limitation can be expressed as

$$2\ell + 2w \leq 80$$

Our objective is to maximize the enclosed area ℓw. Thus a complete model statement is as follows:

$$
\begin{array}{lll}
\max & \ell w & \text{(enclosed area)} \\
\text{s.t.} & 2\ell + 2w \leq 80 & \text{(fence length)} \\
& \ell, w \geq 0 & \text{(nonnegativity)}
\end{array}
$$

2.2 GRAPHIC SOLUTION AND OPTIMIZATION OUTCOMES

Methods for analyzing optimization models are the focus of many chapters to follow. However, very simple graphic techniques have enough power to deal with tiny models such as Two Crude formulation (2.6). They also provide "pictures" yielding helpful intuition about properties and solution methods for models of more realistic size.

In this section we develop the techniques of graphic solution. We also illustrate their intuitive power by exploring the unique optimal solution, alternative optimal solution, and infeasible and unbounded outcomes of optimization analysis.

Graphic Solution

Graphic solution solves 2 and 3-variable optimization models by plotting elements of the model in a coordinate system corresponding to the decision variables. For example, Two Crude model (2.6) involves decision variables x_1 and x_2. Every choice of values for these variables corresponds to a point (x_1, x_2) in a 2-dimensional plot.

Feasible Sets

The first issue in graphic solution is the feasible set (also called the feasible region or the feasible space).

> **2.6** The **feasible set** (or **region**) of an optimization model is the collection of choices for decision variables satisfying all model constraints.

Graphing Constraints and Feasible Sets

We want to draw a picture of the feasible set in a coordinate system defined by the decision variables. The process begins with the variable-type constraints (definition 2.2).

In the Two Crude example, both variables are nonnegative. Thus every feasible solution corresponds to a point in the shaded, nonnegative part of the following plot:

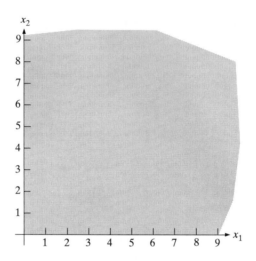

Next we introduce the main constraints (definition $\boxed{2.3}$) one at a time. Begin with the Two Crude gasoline requirement

$$0.3x_1 + 0.4x_2 \geq 2 \tag{2.7}$$

If it were of the equality form

$$0.3x_1 + 0.4x_2 = 2$$

feasible points would lie along a corresponding line.

However, constraint (2.7) is an inequality.

To identify feasible points, we first plot the boundary and then include whichever side of the line applies.

Over Two Crude's nonnegative (x_1, x_2), the result is

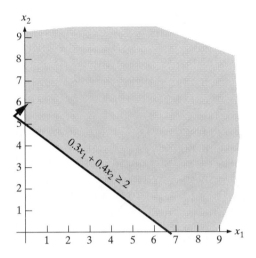

Points along the highlighted boundary satisfy the constraint with equality. To determine which side of the line depicts points satisfying the constraint as a strict inequality, we have only to substitute one point not on the line. For example, the origin $x_1 = x_2 = 0$ violates the constraint because

$$0.3x_1 + 0.4x_2 = 0.3(0) + 0.4(0) \not\geq 2$$

Feasible points must be on the other side of the equality line, just as the arrow on the constraint indicates.

Feasible (x_1, x_2) must satisfy all constraints simultaneously. To identify the feasible set fully, we merely repeat the foregoing process.

> **2.10**   The feasible set (or region) for an optimization model is plotted by introducing constraints one by one, keeping track of the region satisfying all at the same time.

Figure 2.1 shows the result for our Two Crude example. Each of the 5 main inequality constraints yields a line, where it is satisfied as an equality, plus a side of the line where it holds as an inequality. Together with variable-type constraints $x_1, x_2 \geq 0$, these define the shaded feasible set.

SAMPLE EXERCISE 2.2: GRAPHING CONSTRAINTS AND FEASIBLE SETS

Graph the feasible sets corresponding to each of the following systems of constraints.

(a) $x_1 + x_2 \qquad \leq 2$
$\quad\;\; 3x_1 + x_2 \quad\;\; \geq 3$
$\quad\;\; x_1, x_2 \geq 0$

(b) $x_1 + x_2 \qquad \leq 2$
$\quad\;\; 3x_1 + x_2 \quad\;\; = 3$
$\quad\;\; x_1, x_2 \geq 0$

(c) $(x_1)^2 + (x_2)^2 \quad \leq 4$
$\quad\;\; |x_1| - x_2 \qquad\;\; \leq 0$

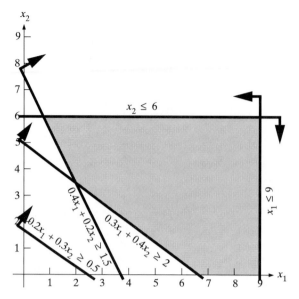

FIGURE 2.1   Feasible Set for the Crude Example

Analysis: Applying the process of 2.10 produces the following plots:

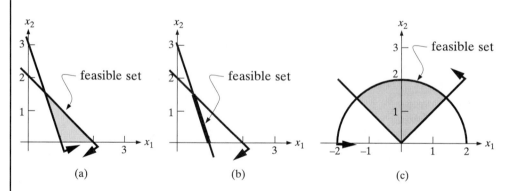

Notice that the feasible set for system (b) is just the highlighted line segment, because one constraint is an equality. Also, system (c) allows negative variable values because the variable type is unrestricted.

Graphing Objective Functions

To find the best feasible point, we must introduce the objective function into a plot like Figure 2.1. Observe that the objective in our Two Crude example, call it

$$c(x_1, x_2) \triangleq 20x_1 + 15x_2 \tag{2.8}$$

is a function of decision variables x_1 and x_2. Thus a plot requires a third dimension.

Figure 2.2(a) depicts surface (2.8) in an x_1 versus x_2 versus c coordinate system. For example, feasible solution $x_1 = x_2 = 4$ has value $c(4, 4) = 20(4) + 15(4) = 140$ on the objective function surface.

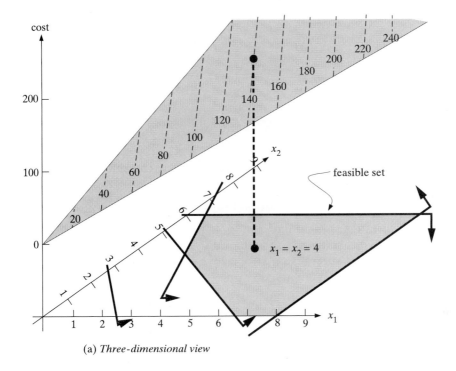

(a) *Three-dimensional view*

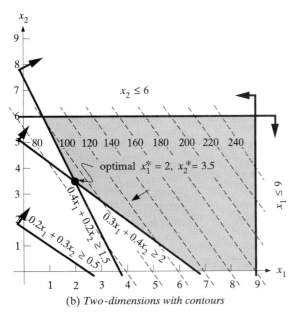

(b) *Two-dimensions with contours*

FIGURE 2.2 Graphic Solution of the Two Crude Model

Because most of us have trouble visualizing, much less drawing 3-dimensional plots, mathematical programmers customarily employ the 2-dimensional alternative of Figure 2.2(b). There, the third dimension, cost, is shown through contours.

> **2.11** Objective functions are normally plotted in the same coordinate system as the feasible set of an optimization model by introducing **contours**—lines or curves (usually dashed) through points having equal objective function value.

To introduce contours, we begin with any convenient point visible in the plot of the feasible space. For example, we might pick $x_1 = 9, x_2 = 0$ in the Two Crude model. Evaluating the objective function (2.8) at the point selected gives

$$20(9) + 15(0) = 180$$

thousand dollars. Thus all points on the contour where the objective function equals $180 thousand satisfy the equation

$$20x_1 + 15x_2 = 180$$

Plotting produces the straight-line contour marked "180" in Figure 2.2(b).

With one typical objective function value such as 180 known, we complete the contouring by trying several similar objective function values. Figure 2.2(b) uses increments of $20 thousand obtained by plotting equations

$$20x_1 + 15x_2 = 100$$
$$20x_1 + 15x_2 = 120$$
$$\vdots$$
$$20x_1 + 15x_2 = 240$$

A contour exists for every possible objective value, so many other levels could have been chosen. We only wish to convey the trend and shape of the objective function. A small arrow perpendicular to the contours shows the direction in which the objective function improves.

SAMPLE EXERCISE 2.3: PLOTTING OBJECTIVE FUNCTION CONTOURS

Show contours of each of the following objective functions over the feasible region defined by $y_1 + y_2 \leq 2, y_1, y_2 \geq 0$.

(a)   min $3y_1 + y_2$ (b)   max $3y_1 + y_2$ (c)   max $2(y_1)^2 + 2(y_2)^2$

Analysis: Contours for the three cases are as follows:

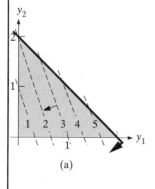

(a)

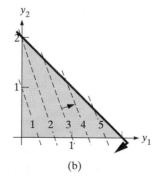

(b)

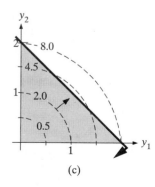

(c)

(a) After identifying the feasible region by plotting the main and variable-type constraints, we arbitrarily pick $y_1 = 0, y_2 = 1$ to begin introducing the objective function as in principle 2.11 . The corresponding contour is the line where

$$3y_1 + y_2 = 3(0) + (1) = 1$$

Then other contours are introduced by incrementing the level to $2, 3, \ldots, 5$.

(b) Contours for this maximize model are identical to those minimizing the same objective function in part (a). A small arrow on one of the contours reminds us of the direction in which the objective function improves.

(c) For this more complex objective function, contours will not plot as straight lines. We begin with $y_1 = 0$, $y_2 = 1$ and graph curve

$$2(y_1)^2 + 2(y_2)^2 = 2(0)^2 + 2(1)^2 = 2$$

Then other contours arise from trying nearby levels 0.5, 4.5, and 8.0.

Optimal Solutions

Our goal in graphic analysis of an optimization model is to identify an optimal solution if there is one.

2.12 | An **optimal solution** is a feasible choice for decision variables with objective function value at least equal to that of any other solution satisfying all constraints.

We have already determined that the feasible points in Figure 2.2(b) are the (x_1, x_2) in the shaded area. To identify an optimal solution, we examine the objective function contours.

2.13 | Optimal solutions show graphically as points lying on the best objective function contour that intersects the feasible region.

Our Two Crude model is of minimize form, so the best objective contour is the lowest. We can identify an optimal solution by finding a feasible point on the lowest possible contour.

Only a few contour lines are explicitly displayed in Figure 2.2(b), but infinitely many exist. It is clear from the pattern of those shown that the unique point satisfying condition 2.13 is $(x_1^*, x_2^*) = (2, 3.5)$. The asterisk (*) or "star" denotes an optimal solution.

This completes graphic solution of the Two Crude model. The optimal operating plan uses 2 thousand barrels per day of Saudi petroleum and 3.5 thousand barrels per day of Venezuelan. Total daily cost will be $20(2)+15(3.5) = 92.5$ thousand dollars.

SAMPLE EXERCISE 2.4: SOLVING OPTIMIZATION MODELS GRAPHICALLY

Return to the equipment yard model of Sample Exercise 2.1 and solve it graphically.

Analysis: Identifying the feasible set and introducing objective function contours produces the following plot:

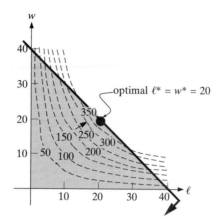

For this maximize problem we wish to be on the highest possible contour. Thus application of principle 2.13 yields optimal solution $\ell^* = w^* = 20$.

Optimal Values

Optimal solutions provide a best choice for decision variables and the optimal value is the corresponding objective function level.

> **2.14**   The **optimal value** in an optimization model is the objective function value of any optimal solutions.

For example, the Two Crude optimum of Figure 2.2(b) has optimal value $92,500.

Notice that two different objective values could not both be best.

> **2.15**   An optimization model can have only one optimal value.

We see this graphically because there can only be one best contour level in rule 2.13 .

Unique versus Alternative Optimal Solutions

The optimal solution $x_1^* = 2$, $x_2^* = 3.5$ in our Two Crude example is also unique because it is the only feasible solution in Figure 2.2(b) achieving the optimal value. This does not always happen. Many models have alternative optimal solutions.

> **2.16**   An optimization model may have a **unique optimal solution** or several **alternative optimal solutions**.

Still, all must have the same optimal value.

We can illustrate principle 2.16 very easily with another graphic solution. Suppose that crude prices change in the Two Crude example to produce objective function

$$\min 20x_1 + 10x_2$$

Figure 2.3 shows the impact. There is still only one optimal value, $75 thousand. But there are now an infinite number of alternative optimal solutions along the high-lighted boundary of the feasible set; all lie on the optimal $75 thousand contour.

> **2.17**   Unique optimal solutions show graphically by the optimal-value con-tour intersecting the feasible set at exactly one point. If the optimal-value contour intersects at more than one point, the model has alternative optimal solutions.

> **SAMPLE EXERCISE 2.5: IDENTIFYING UNIQUE AND ALTERNATIVE OPTIMAL SOLUTIONS**

Determine graphically which of the following optimization models has a unique optimal solution and which has alternative optima.

(a)   max $3w_1 + 3w_2$
     s.t. $w_1 + w_2 \leq 2$
          $w_1, w_2 \geq 0$

(b)   max $3w_1 + w_2$
     s.t. $w_1 + w_2 \leq 2$
          $w_1, w_2 \geq 0$

Analysis: Graphic solution of these models is as follows:

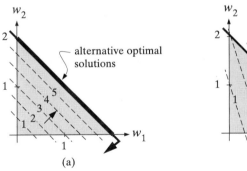
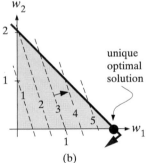

      (a)   (b)

Model (a) has alternative optimal solutions along the highlighted boundary, including $(w_1, w_2) = (0, 2)$, $(w_1, w_2) = (1, 1)$ and $(w_1, w_2) = (2, 0)$. Model (b) has unique optimal solution $(w_1, w_2) = (2, 0)$ because the optimal-value contour intersects the feasible space at only one point.

Infeasible Models

Infeasible models have no optimal solutions.

> **2.18**   An optimization model is **infeasible** if no choice of decision variables sat-isfies all constraints.

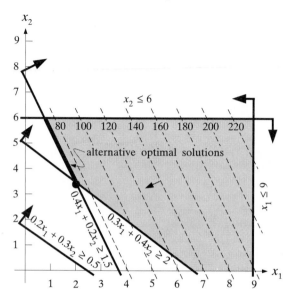

FIGURE 2.3   Variant of the Two Crude Model With Alternative Optima

Such models have no optimal solutions because they have no solutions at all.

Infeasibility is also easy to illustrate with models small enough to be analyzed graphically. For example, consider modifying our Two Crude case so that only 2 thousand barrels per day are available from each source. The resulting model is

$$
\begin{array}{lll}
\min & 20x_1 + 15x_2 & \text{(total cost)} \\
\text{s.t.} & 0.3x_1 + 0.4x_2 \geq 2.0 & \text{(gasoline requirement)} \\
& 0.4x_1 + 0.2x_2 \geq 1.5 & \text{(jet fuel requirement)} \\
& 0.2x_1 + 0.3x_2 \geq 0.5 & \text{(lubricant requirement)} \\
& x_1 \qquad\quad \leq 2 & \text{(Saudi availability)} \\
& x_2 \leq 2 & \text{(Venezuelan availability)} \\
& x_1, x_2 \geq 0 & \text{(nonnegativity)}
\end{array}
$$

An attempt to graph the feasible space produces Figure 2.4. As before, each constraint corresponds to a line where it holds as an equality and a side of the line satisfying it as an inequality. This time, however, there are no (x_1, x_2) satisfying all constraints simultaneously.

2.19 | An infeasible model shows graphically by no point falling within the feasible region for all constraints.

SAMPLE EXERCISE 2.6: IDENTIFYING INFEASIBLE MODELS GRAPHICALLY

Determine graphically which of the following optimization models is feasible and which is infeasible.

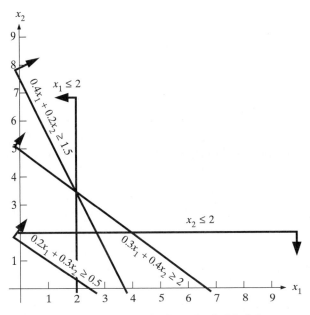

FIGURE 2.4 Infeasible of the Two Crude Model

(a) max $3w_1 + w_2$

    s.t. $w_1 + w_2 \leq 2$

          $w_1 + w_2 \geq 1$

          $w_1, w_2 \geq 0$

(b) max $3w_1 + w_2$

    s.t. $w_1 + w_2 \leq 2$

          $w_1 + w_2 \geq 3$

          $w_1, w_2 \geq 0$

Analysis: Graphic solution of these models is as follows:

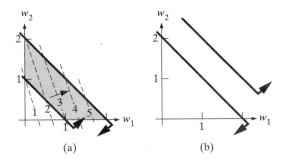

        (a)   (b)

    Model (a) has points in the shaded area that satisfy all constraints. Thus it is feasible. No points satisfy all constraints in model (b). That model is infeasible.

Unbounded Models

Another case where an optimization model has no feasible solution arises when it is unbounded.

2.20 | An optimization model is **unbounded** when feasible choices of the deci-
sion variables can produce arbitrarily good objective function values.

Unbounded models have no optimal solutions because any possibility can be improved.

We can illustrate this outcome graphically with still another variant of the Two
Crude case (although not a very realistic one). Suppose that Saudi Arabia decides to
subsidize oil prices heavily so that Two Crude is paid $2 for each barrel it consumes,
and further that Saudi Arabia will supply unlimited amounts of petroleum at this
negative price. The result is a revised model

$$
\begin{aligned}
\min \quad & -2x_1 + 15x_2 && \text{(total cost)} \\
\text{s.t.} \quad & 0.3x_1 + 0.4x_2 \geq 2.0 && \text{(gasoline requirement)} \\
& 0.4x_1 + 0.2x_2 \geq 1.5 && \text{(jet fuel requirement)} \\
& 0.2x_1 + 0.3x_2 \geq 0.5 && \text{(lubricant requirement)} \\
& x_2 \leq 6 && \text{(Venezuelan availability)} \\
& x_1, x_2 \geq 0 && \text{(nonnegativity)}
\end{aligned}
$$

Figure 2.5 shows a graphic solution. Notice that as x_1 (Saudi purchases) increases, we encounter feasible solutions with ever better objective function values.
No solution is optimal because a better one can always be found.

This is exactly what it means to be unbounded.

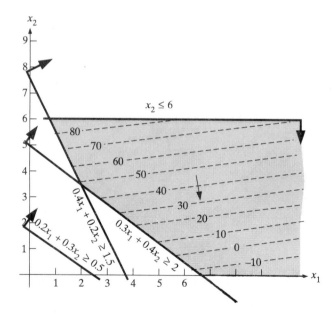

FIGURE 2.5   Unbounded Variant of the Two Crude
Example

| 2.21 | Unbounded models show graphically by there being points in the feasible set lying on ever-better objective function contours. |

SAMPLE EXERCISE 2.7: IDENTIFYING UNBOUNDED MODELS GRAPHICALLY

Determine graphically which of the following optimization models has an optimal solution and which is unbounded.

(a) max $-3w_1 + w_2$
     s.t. $-w_1 + w_2 \leq 1$
         $w_1, w_2 \geq 0$

(b) max $3w_1 + w_2$
     s.t. $-w_1 + w_2 \leq 1$
         $w_1, w_2 \geq 0$

Analysis: Graphic solution of these models is as follows:

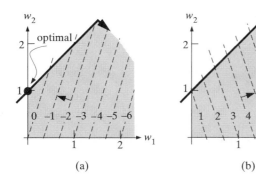

(a)   (b)

Model (a) has a unique optimal solution at $(w_1, w_2) = (0, 1)$. Model (b) is unbounded because we can find feasible solutions on arbitrarily good contours of the objective function.

2.3 LARGE-SCALE OPTIMIZATION MODELS AND INDEXING

Because it could be solved graphically, our tiny Two Crude Petroleum example facilitated introduction of a variety of mathematical programming concepts. Still, it seriously misrepresents problem size. In real applications optimization models quickly grow to thousands, even millions, of variables and constraints. We begin to see how in this section and introduce the indexed notational schemes that keep large models manageable.

EXAMPLE 2.2: Pi Hybrids

To illustrate, we introduce our first real application of operations research.[1] A large manufacturer of corn seed, which we will call Pi Hybrids, operates $\ell = 20$ facilities producing seeds of $m = 25$ hybrid corn varieties and distributes them to customers in $n = 30$ sales regions. They want to know how to carry out these production and distribution operations at minimum cost.

---

[1]Based on Mingjian Zuo, Way Kuo, and Keith L. McRoberts (1991), "Application of Mathematical Programming to a Large-Scale Agricultural Production and Distribution System," *Journal of the Operational Research Society, 42,* 639–648.

After some effort, a variety of parameters that we will take as constant have been estimated:

- The cost per bag of producing each hybrid at each facility
- The corn processing capacity of each facility in bushels
- The number of bushels of corn that must be processed to make a bag of each hybrid
- The number of bags of each hybrid demanded in each customer region
- The cost per bag of shipping each hybrid from each facility to each customer region

Our task is to produce a suitable optimization model.

Indexing

Indexes or **subscripts** permit representing collections of similar quantities with a single symbol. For example,

$$\{z_i : i = 1, \ldots, 100\}$$

represents 100 similar values with the same z name, distinguishing them with the index i.

Indexing is a powerful aid in keeping large-scale optimization models easy to think about and concise to write down. In fact, it often provides the initial model organization.

<u>2.22</u> The first step in formulating a large optimization model is to choose appropriate indexes for the different dimensions of the problem.

Our Pi Hybrids example has three main dimensions: facilities, hybrids, and sales regions. Thus we begin formulation of a model by assigning an index to each:

$$f \triangleq \text{production facility number } (f = 1, \ldots, \ell)$$
$$h \triangleq \text{hybrid variety number } (h = 1, \ldots, m)$$
$$r \triangleq \text{sales region number } (r = 1, \ldots, n)$$

Indexed Decision Variables

We are now ready to think about what decisions Pi Hybrids' problem requires. Clearly, they fall into two categories: how much to produce and how much to ship.

There is a production decision for each plant and each hybrid. We could sequentially number the corresponding decision variables $1, \ldots, km$, but it would then be difficult to keep plant and hybrid issues separate. It is much more convenient to employ decision variables with distinct subscripts for plant and hybrids:

$$x_{f,h} \triangleq \text{number of bags of hybrid } h \text{ produced at facility}$$
$$f (f = 1, \ldots, \ell; h = 1, \ldots, m)$$

Multiple subscripts are extremely common in large OR models.

<u>2.23</u> It is usually appropriate to use separate indexes for each problem dimension over which a decision variable or input parameter is defined.

Shipping at Pi Hybrid is distinguished by the production facility shipped from, the hybrid variety shipped, and the sales region shipped to. Following principle 2.23 , we complete the list of decision variables with a 3-subscript family:

$y_{f,h,r} \triangleq$ number of bags of hybrid h shipped from facility f to sales region r

$$(f = 1, \ldots, \ell; h = 1, \ldots, m; r = 1, \ldots, n)$$

Notice that indexing has made it possible to define quite a large number of variables with just two main symbols. There are $\ell m = 20(25) = 500$ x-variables and $\ell m n = 20(25)(30) = 15{,}000$ y-variables. The 15,500 total dwarfs the tiny examples of preceding sections, but it represents only a rather average size for applied models.

SAMPLE EXERCISE 2.8: COUNTING INDEXED DECISION VARIABLES

Suppose that an optimization model employs decision variables $w_{i,j,k,\ell}$, where i and k range over $1, \ldots, 100$, while j and ℓ index through $1, \ldots, 50$. Compute the total number of decision variables.

Analysis: The number of variables is

$$\text{(number of } i\text{) (number of } j\text{) (number of } k\text{) (number of } \ell\text{)}$$
$$= 100(50)(100)(50)$$
$$= 25{,}000{,}000$$

Indexed Symbolic Parameters

If Pi Hybrids made only 2 products at 2 facilities, we could easily express total production cost as

$$\begin{pmatrix} \text{cost of} \\ \text{hybrid} \\ \text{1 at} \\ \text{facility} \\ 1 \end{pmatrix} x_{1,1} + \begin{pmatrix} \text{cost of} \\ \text{hybrid} \\ \text{2 at} \\ \text{facility} \\ 1 \end{pmatrix} x_{1,2} + \begin{pmatrix} \text{cost of} \\ \text{hybrid} \\ \text{1 at} \\ \text{facility} \\ 2 \end{pmatrix} x_{2,1} + \begin{pmatrix} \text{cost of} \\ \text{hybrid} \\ \text{2 at} \\ \text{facility} \\ 2 \end{pmatrix} x_{2,2} \qquad (2.9)$$

using the actual cost values. But writing out the $\ell m = 20(25) = 500$ production cost terms actually required in our model would be bulky and almost impossible to read or explain.

The answer to this dilemma is more indexing, this time on input parameters.

2.24 | To describe large-scale optimization models compactly it is usually necessary to assign indexed symbolic names to most input parameters, even though they are being treated as constant.

For example, after defining

$$p_{f,h} \triangleq \text{cost per bag of producing hybrid } h \text{ at facility } f$$

we may employ **summation notation** to express form (2.9) for any number of facilities and hybrids as

$$\sum_{f=1}^{\ell}\sum_{h=1}^{m} p_{f,h} x_{f,h}$$

Moving indexes in the double sum capture terms for all combinations of facilities $f = 1, 2, \ldots, \ell$ and hybrids $h = 1, 2 \ldots, m$.

SAMPLE EXERCISE 2.9: USING SUMMATION NOTATION

(a) Write the following sum more compactly with summation notation:

$$2w_{1,5} + 2w_{2,5} + 2w_{3,5} + 2w_{4,5} + 2w_{5,5}$$

(b) Write out terms separately of the sum

$$\sum_{i=1}^{4} i w_i$$

Analysis:

(a) Introducing moving index i for the first subscript, the sum is

$$\sum_{i=1}^{5} 2w_{i,5} = 2\sum_{i=1}^{5} w_{i,5}$$

(b) The 4 terms of the sum are

$$1w_1 + 2w_2 + 3w_3 + 4w_4$$

Objective Function

For similar reasons of convenience, define the following symbolic names for other Pi Hybrids input parameters:

$u_f \triangleq$ corn processing capacity of facility f in bushels

$a_h \triangleq$ number of bushels of corn that must be processed to obtain a bag of hybrid h

$d_{h,r} \triangleq$ number of bags of hybrid h demanded in sales region r

$s_{f,h,r} \triangleq$ cost per bag of shipping hybrid h from facility f to sales region r

With the aid of our indexed decision variables and these indexed parameters, we are now ready to formulate the objective function of a Pi Hybrids model. It should minimize

$$\text{total cost} = \text{total production cost} + \text{total shipping cost}$$

or

$$\min \sum_{f=1}^{\ell}\sum_{h=1}^{m} p_{f,h} x_{f,h} + \sum_{f=1}^{\ell}\sum_{h=1}^{m}\sum_{r=1}^{n} s_{f,h,r} y_{f,h,r}$$

Indexed Families of Constraints

Turn now to constraints for the Pi Hybrids model. What restrictions must decision variables satisfy?

One family of constraints must enforce production capacities at the various facilities. Capacities u_f are measured in bushels processed, with a_h needed for each bag of hybrid h. Thus we may express the capacity requirement at each facility 1 by

$$\sum_{h=1}^{m} \begin{pmatrix} \text{bushels} \\ \text{per bag of} \\ h \end{pmatrix} \begin{pmatrix} \text{bags of } h \\ \text{produced} \\ \text{at 1} \end{pmatrix} \leq \text{ capacity at 1}$$

or

$$\sum_{h=1}^{m} a_h x_{1,h} \leq u_1$$

There are $\ell = 20$ such capacity constraints for different facilities. We could write each one down explicitly, but again, the model would become very bulky and hard to comprehend.

Standard mathematical programming notation deals with this difficulty by listing indexed families of constraints.

> __2.25__ Families of similar constraints distinguished by indexes may be expressed in a single-line format
>
> $$\text{(constraint for fixed indexes)   (ranges of indexes)}$$
>
> which implies one constraint for each combination of indexes in the ranges specified.

Written in this style, all capacity constraints of the Pi Hybrids model can be expressed in the single line

$$\sum_{h=1}^{m} a_h x_{f,h} \leq u_f \qquad f = 1, \ldots, \ell$$

Separate constraints are implied for $f = 1, 2, \ldots, \ell$. There are 20 in all. Equivalent forms are

$$\sum_{h=1}^{m} a_h x_{f,h} \leq u_f \qquad \text{for all } f$$

and

$$\sum_{h=1}^{m} a_h x_{f,h} \leq u_f \quad \forall f$$

because we know that f ranges from 1 through ℓ and the mathematical symbol $\forall$ means "for all".

SAMPLE EXERCISE 2.10: USING INDEXED FAMILIES OF CONSTRAINTS

An optimization model must decide how to allocate available supplies s_i at sources $i = 1, \ldots, p$ to meet requirements r_j at customers $j = 1, \ldots, q$. Using decision variables

$$w_{i,j} \triangleq \text{amount allocated from source } i \text{ to customer } j$$

formulate each of the following requirements in a single line.

(a) The amount allocated from source 32 cannot exceed the supply available at 32.

(b) The amount allocated from each source i cannot exceed the supply available at i.

(c) The amount allocated to customer n should equal the requirement at n.

(d) The amount allocated to each customer j should equal the requirement at j.

Analysis:

(a) Only one constraint is required:

$$\sum_{j=1}^{n} w_{32,j} \leq s_{32}$$

(b) Here we require constraints for all sources i. Using notation $\boxed{2.25}$, all m can be expressed:

$$\sum_{j=1}^{n} w_{i,j} \leq s_i \qquad i = 1, \ldots, m$$

(c) As with part (a) there is only one constraint:

$$\sum_{i=1}^{m} w_{i,n} = r_n$$

(d) This requirement implies constraints for each demand j. Using notation $\boxed{2.25}$, all n can be expressed

$$\sum_{i=1}^{m} w_{i,j} = r_j \qquad j = 1, \ldots, n$$

SAMPLE EXERCISE 2.11: COUNTING INDEXED CONSTRAINTS

Determine the number of constraints in the following systems.

(a) $\displaystyle\sum_{i=1}^{22} z_{i,3} \geq b_3$

(b) $\displaystyle\sum_{i=1}^{22} z_{i,p} \geq b_p, \quad p = 1, \ldots, 45$

(c) $\displaystyle\sum_{k=1}^{10} z_{i,j,k} \leq g_j, \quad i = 1, \ldots, 14; \quad j = 1, \ldots, 30$

Analysis:

(a) This expression represents only 1 constraint, associated with fixed index 3.

(b) This expression represents 45 constraints, one for each p.

(c) This expression represents $14(30) = 420$ constraints, one for each i and each j.

Pi Hybrids Example Model

In terms of the notation we have defined, a full model for Pi Hybrids' production–distribution problem is

$$\min \quad \sum_{f=1}^{\ell} \sum_{h=1}^{m} p_{f,h} x_{f,h} + \sum_{f=1}^{\ell} \sum_{h=1}^{m} \sum_{r=1}^{n} s_{f,h,r} y_{f,h,r} \qquad \text{(total cost)}$$

$$\text{s.t.} \quad \sum_{h=1}^{m} a_h x_{f,h} \le u_f \qquad f = 1, \ldots, \ell \qquad \text{(capacity)}$$

$$\sum_{f=1}^{\ell} y_{f,h,r} = d_{h,r} \qquad h = 1, \ldots, m; \quad r = 1, \ldots, n \quad \text{(demands)} \qquad (2.10)$$

$$\sum_{r=1}^{n} y_{f,h,r} = x_{f,h} \qquad f = 1, \ldots, \ell; \quad h = 1, \ldots, m \quad \text{(balance)}$$

$$x_{f,h} \ge 0 \qquad f = 1, \ldots, \ell; \quad h = 1, \ldots, m \quad \text{(nonnegativity)}$$

$$y_{f,h,r} \ge 0 \qquad f = 1, \ldots, \ell; \quad h = 1, \ldots, m; \quad r = 1, \ldots, n$$

Besides the objective function and capacity constraints derived earlier, model (2.10) includes four new systems of constraints. The first of these enforces demand for each hybrid in each sales region. Amounts shipped from various facilities are summed to compute the total applicable to each demand. A second system of constraints balances production and distribution. It requires that the total of any hybrid shipped from any facility should match the amount of that hybrid produced there. Finally, there are the variable-type constraints. The last two systems require all production and distribution quantities to be nonnegative.

How Models Become Large

The 500 x-variables and 15,000 y-variables of model (2.10) are subject to a total of $(20 + 750 + 500 + 15,500) = 16,770$ constraints:

$$\ell = 20 \qquad \qquad \text{capacity constraints}$$
$$mn = 25(30) = 750 \qquad \qquad \text{demand constraints}$$
$$\ell m = 20(25) = 500 \qquad \qquad \text{balance constraints}$$
$$\ell m + \ell mn = 20(25) + 20(25)(30) = 15,500 \quad \text{nonnegativity constraints}$$

Still, we have managed to write it compactly in only a few lines.

In part this compactness derives from the power of indexed notation. But there is another reason. Model (2.10) actually involves only a few simple notions: production cost, shipping cost, capacity, demand, balance, and nonnegativity. What makes it big is repetition of these notions over many combinations of facilities, hybrids, and sales regions.

This is typical of the way that OR models grow.

2.26 | Optimization models become large mainly by a relatively small number of objective function and constraint elements being repeated many times for different periods, locations, products, and so on.

With suitable indexing over such problem dimensions, we can express very large models in just a few lines.

2.4 LINEAR AND NONLINEAR PROGRAMS

Different classes of optimization models have enormously different tractability. This makes recognizing the major categories an important modeling skill. In this section we begin developing that ability by illustrating the fundamental distinction between linear programs and nonlinear programs.

General Mathematical Programming Format

The distinction begins with thinking of mathematical programs in terms of functions of the decision variables.

2.27 | The general form of a **mathematical program** or (single objective) optimization model is

$$\min \text{ or } \max \quad f(x_1, \ldots, x_n)$$

$$\text{s.t.} \quad g_i(x_1, \ldots, x_n) \left\{ \begin{array}{c} \leq \\ = \\ \geq \end{array} \right\} b_i \quad i = 1, \ldots, m$$

where $f, g_1, \ldots, g_m$ are given functions of decision variables $x_1, \ldots, x_n$, and $b_1, \ldots, b_m$ are specified constant parameters.

Individual constraints may be of $\leq$, $=$, or $\geq$ form.

To illustrate, return to the Two Crude petroleum model of Section 2.1:

$$\min \quad 20x_1 + 15x_2$$
$$\text{s.t.} \quad 0.3x_1 + 0.4x_2 \geq 2.0$$
$$0.4x_1 + 0.2x_2 \geq 1.5$$
$$0.2x_1 + 0.3x_2 \geq 0.5$$
$$x_1 \qquad\qquad \leq 9$$
$$x_2 \leq 6$$
$$x_1, x_2 \geq 0$$

Decision variables are x_1 and x_2, so that $n = 2$ and there are $m = 7$ constraints. In format 2.27 the implied functions are

$$f(x_1, x_2) \triangleq 20x_1 + 15x_2$$

$$g_1(x_1, x_2) \triangleq 0.3x_1 + 0.4x_2$$

$$g_2(x_1, x_2) \triangleq 0.4x_1 + 0.2x_2$$

$$g_3(x_1, x_2) \triangleq 0.2x_1 + 0.3x_2$$

$$g_4(x_1, x_2) \triangleq x_1 \tag{2.11}$$

$$g_5(x_1, x_2) \triangleq x_2$$

$$g_6(x_1, x_2) \triangleq x_1$$

$$g_7(x_1, x_2) \triangleq x_2$$

Notice that there are g-functions for both main and variable-type constraints.

Right-Hand Sides

Format $\boxed{2.27}$ collects everything involving the decision variables in the functions $f, g_1, \ldots, g_m$. Constraint limits $b_1, \ldots, b_m$ must be constants.

For obvious reasons these constraint constants b_i are called the **right-hand sides** (or **RHSs**) of the model. In our Two Crude example the right-hand sides are

$$b_1 = 2.0, \quad b_2 = 1.5, \quad b_3 = 0.5, \quad b_4 = 9, \quad b_5 = 6, \quad b_6 = 0, \quad \text{and} \quad b_7 = 0$$

SAMPLE EXERCISE 2.12: EXPRESSING MODELS IN FUNCTIONAL FORM

Assuming that the decision variables are w_1, w_2, and w_3, express the following optimization model in general functional format $\boxed{2.27}$ and identify all required functions and right-hand sides:

$$\begin{array}{ll} \max & (w_1)^2 + 8w_2 + (w_3)^2 \\ \text{s.t.} & w_1 + 6w_2 \leq 10 + w_2 \\ & (w_2)^2 = 7 \\ & w_1 \geq w_3 \\ & w_1, w_2 \geq 0 \end{array}$$

Modeling: In format $\boxed{2.27}$ the objective function is

$$f(w_1, w_2, w_3) \triangleq (w_1)^2 + 8w_2 + (w_3)^2$$

After collecting all terms involving the decision variables on the left-hand side, constraints have the form

$$g_1(w_1, w_2, w_3) \leq 10$$
$$g_2(w_1, w_2, w_3) = 7$$
$$g_3(w_1, w_2, w_3) \geq 0$$
$$g_4(w_1, w_2, w_3) \geq 0$$
$$g_5(w_1, w_2, w_3) \geq 0$$

where

$$g_1(w_1, w_2, w_3) \triangleq w_1 + 6w_2 - w_2 = w_1 + 5w_2$$
$$g_2(w_1, w_2, w_3) \triangleq (w_2)^2$$
$$g_3(w_1, w_2, w_3) \triangleq w_1 - w_3$$
$$g_4(w_1, w_2, w_3) \triangleq w_1$$
$$g_5(w_1, w_2, w_3) \triangleq w_2$$

Associated right-hand-side constants are

$$b_1 = 10, \quad b_2 = 7, \quad b_3 = 0, \quad b_4 = 0, \quad \text{and} \quad b_5 = 0$$

Linear Functions

We distinguish classes of mathematical programs according to whether functions $f, g_1, \ldots, g_m$ of format $\boxed{2.27}$ are linear or nonlinear in the decision variables.

> $\underline{2.28}$ A function is **linear** if it is a constant-weighted sum of decision variables. Otherwise, it is **nonlinear**.

Linear functions may involve only constants and terms with variables in the first power. For example, the objective function

$$f(x_1, x_2) \triangleq 20x_1 + 15x_2$$

of the Two Crude model is linear because it simply applies weights 20 and 15 in summing decision variables x_1 and x_2. On the other hand, the objective function in Sample Exercise 2.12,

$$f(w_1, w_2, w_3) \triangleq (w_1)^2 + 8w_2 + (w_3)^2$$

is nonlinear. It includes second powers of some decision variables.

> **SAMPLE EXERCISE 2.13: RECOGNIZING LINEAR FUNCTIONS**
>
> Assuming that x's are decision variables and all other symbols are constant, determine whether each of the following functions is linear or nonlinear.
>
> (a) $f(x_1, x_2, x_3) \triangleq 9x_1 - 17x_3$
>
> (b) $f(x_1, x_2, x_3) \triangleq \sum_{j=1}^{3} c_j x_j$
>
> (c) $f(x_1, x_2, x_3) \triangleq \dfrac{5}{x_1} + 3x_2 - 6x_3$
>
> (d) $f(x_1, x_2, x_3) \triangleq x_1 x_2 + (x_2)^3 + \ln(x_3)$
>
> (e) $f(x_1, x_2, x_3) \triangleq e^{\alpha} x_1 + \ln(\beta)x_3$
>
> (f) $f(x_1, x_2, x_3) \triangleq \dfrac{x_1 + x_2}{x_2 - x_3}$
>
> *Analysis:*
>
> (a) This function is linear because it merely sums the 3 decision variables with weights $9, 0$, and -17, respectively.
>
> (b) This function is also linear because the c_j are constants.

(c) This function is nonlinear because it involves negative powers of decision variable x_1.

(d) This function is nonlinear because it involves products, powers not 1, and logarithms of decision variables.

(e) This function is linear. With α and β constant, it is just a weighted sum of the decision variables.

(f) This function is nonlinear because it involves a quotient, even though both numerator and denominator are linear functions.

Linear and Nonlinear Programs Defined

The functional forms in format $\boxed{2.27}$ characterize linear programs and nonlinear programs.

$\boxed{2.29}$ An optimization model in functional form $\boxed{2.27}$ is a **linear program (LP)** if the (single) objective function f and all constraint functions $g_1, \ldots, g_m$ are linear in the decision variables. Also, decision variables should be able to take on whole-number or fractional values.

$\boxed{2.30}$ An optimization model in functional form $\boxed{2.27}$ is a **nonlinear program (NLP)** if the (single) objective function f or any of the constraint functions $g_1, \ldots, g_m$ is nonlinear in the decision variables. Also, decision variables should be able to take on whole-number or fractional values.

SAMPLE EXERCISE 2.14: RECOGNIZING LINEAR AND NONLINEAR PROGRAMS

Assuming that y's are decision variables and all other symbols are constant, determine whether each of the following mathematical programs is a linear program or a nonlinear program.

(a) min $\alpha(3y_1 + 11y_4)$

    s.t. $\sum_{j=1}^{5} d_j y_j \leq \beta$

        $y_j \geq 1 \quad j = 1, \ldots, 9$

(b) min $\alpha(3y_1 + 11y_4)^2$

    s.t. $\sum_{j=1}^{5} d_j y_j \leq \beta$

        $y_j \geq 1 \quad j = 1, \ldots, 9$

(c) max $\sum_{j=1}^{9} y_j$

    s.t. $y_1 y_2 \leq 100$

        $y_j \geq 1 \quad j = 1, \ldots, 9$

Analysis: We apply definitions 2.29 and 2.30 .

(a) This model is a linear program because the objective function and all constraints involve just weighted sums of the decision variables.

(b) This model has the same linear constraints as the model of part (a). However, it is a nonlinear program because its objective function is nonlinear.

(c) This model is a nonlinear program. Its objective function and last 9 constraints are linear, but the single nonlinear constraint

$$y_1 y_2 \leq 100$$

renders the entire model nonlinear.

Two Crude and Pi Hybrids Models Are LPs

Both the Two Crude Petroleum model of Section 2.1 and the Pi Hybrids model of Section 2.3 are linear programs. In the Two Crude case we exhibited the required functions in (2.11). Each obviously satisfies definition 2.28 . It is easier to be confused about the much larger Pi Hybrids model (2.10) because we assigned so many symbolic names to constants. However, a careful look will show that its objective function and every one of its constraints involve just a weighted sum of decision variables $x_{f,h}$ and $y_{f,h,r}$. It, too, is a linear program.

EXAMPLE 2.3: E-MART

For an example of a nonlinear program or NLP, consider the problem of budgeting advertising expenditures faced by a large European variety store chain we will call E-mart.[2] E-mart sells products in $m = 12$ major merchandise groups, such as children's wear, candy, music, toys, and electric. Advertising is organized into $n = 15$ campaign formats promoting specific merchandise groups through a particular medium (catalog, press, or television). For example, one variety of campaign advertises children's wear in catalogs, another promotes the same product line in newspapers and magazines, while a third sells toys with television. The profit margin (fraction) for each merchandise group is known, and E-mart wishes to maximize the profit gained from allocating its limited advertising budget across the campaign alternatives.

Indexing, Parameters, and Decision Variables for E-mart

We begin a model by introducing indexes for the two main dimensions of the problem:

$$g \triangleq \text{merchandise group number } (g = 1, \ldots, m)$$
$$c \triangleq \text{campaign type number } (c = 1, \ldots, n)$$

Then we may denote major input parameters by

$$p_g \triangleq \text{profit, as a fraction of sales, realized from merchandise group } g$$
$$b \triangleq \text{available advertising budget}$$

---

[2]Based on P. Doyle and J. Saunders (1990), "Multiproduct Advertising Budgeting," *Marketing Science, 9,* 97-113.

Decisions to be made concern how to spend E-mart's advertising budget. Thus we will employ decision variables

$$x_c \triangleq \text{amount spent on campaign type } c$$

Nonlinear Response

To complete a model, we must quantify how sales in each group g are affected by advertising expenditures on each campaign c. If the relationship is linear,

$$\begin{pmatrix} \text{sales increase} \\ \text{in group } g \text{ due} \\ \text{to campaign } c \end{pmatrix} = s_{g,c} x_c$$

where

$s_{g,c} \triangleq$ parameter relating advertising expenditures in campaign c to sales growth in merchandise group g

A linear form is attractive because it leads to easier analysis.

> **2.31** When there is an option, linear constraint and objective functions are preferred to nonlinear ones in optimization models because each nonlinearity of an optimization model usually reduces its tractability as compared to linear forms.

Unfortunately, marketing researchers often find the linear alternative inappropriate. The main difficulty is that linear functions produce what economists call **equal returns to scale**.

> **2.32** Linear functions implicitly assume that each unit increase in a decision variable has the same effect as the preceding increase: equal returns to scale.

The E-mart experience shows something different. Their advertising history exhibits **decreasing returns to scale**; that is, each dollar of advertising in a particular campaign yields less than did the preceding dollar.

Such unequal returns to scale imply nonlinearity. E-mart analysts chose the nonlinear form

$$\begin{pmatrix} \text{sales increase} \\ \text{in group } g \text{ due} \\ \text{to campaign } c \end{pmatrix} = s_{g,c} \log(x_c + 1) \qquad (2.12)$$

This sales response function has the required decreasing returns because logarithms grow at a declining rate as x_c becomes large. Adding $+1$ keeps the function nonnegative over $x_c \geq 0$.

E-mart Example Model

After estimating constants $s_{g,c}$ from history, we are ready to state the complete E-mart model:

$$\max \quad \sum_{g=1}^{m} p_g \sum_{c=1}^{n} s_{g,c} \log(x_c + 1) \qquad \text{(total profit)}$$

$$\text{s.t.} \quad \sum_{c=1}^{n} x_c \leq b \qquad \text{(budget limit)} \qquad\qquad (2.13)$$

$$x_c \geq 0 \quad c = 1, \ldots, n \qquad \text{(nonnegative expenditures)}$$

The objective function maximizes total profit by summing sales increase expressions (2.12) times corresponding profit factors. A single main constraint enforces the budget limit, and variable-type constraints keep all expenditures nonnegative. The model is a nonlinear program because its objective function is nonlinear.

2.5 DISCRETE OR INTEGER PROGRAMS

Variables in mathematical programs always encode decisions, but there are many types of decisions. In this section we introduce **discrete optimization** models, which include decisions of a logical character qualitatively different from those of linear and nonlinear) programs. Discrete optimization models are also called integer (linear or nonlinear) programs, mixed-integer (linear or nonlinear) programs, and combinatorial optimization problems.

EXAMPLE 2.4: BETHLEHEM INGOT MOLD

For an example, we turn to the problem confronted by Bethlehem Steel Corporation in choosing ingot sizes and molds.[3]. In their process for making steel products, molten output from main furnaces is poured into large molds to produce rectangular blocks called *ingots*. After the molds have been removed, the ingots are reheated and rolled into product shapes such as I-beams and flat sheets.

Bethlehem's mills using this process make approximately $n = 130$ different products. The dimensions of ingots directly affect efficiency. For example, ingots of one dimension may be easiest to roll into I-beams, but another produces sheet steel with less waste. Some ingot sizes cannot be used at all in making certain products.

A careful examination of the best mold dimensions for different products yielded $m = 600$ candidate designs. However, it is impractical to use more than a few because of the cost of handling and storage. We wish to select at most $p = 6$ and to minimize the waste associated with using them to produce all n products.

Indexes and Parameters of the Bethlehem Example

Our Bethlehem problem has two major index dimensions:

$$i \triangleq \text{mold design number } (i = 1, \ldots, m)$$
$$j \triangleq \text{product number } (j = 1, \ldots, n)$$

One set of input parameters are the

$$c_{i,j} \triangleq \text{amount of waste caused by using ingot mold } i \text{ on product } j$$

The other input required for a model is some indication of which products can use

[3]Based on F. J. Vasko, F. E. Wolf, K. S. Stott, and J. W. Scheirer (1989), "Selecting Optimal Ingot Sizes for Bethlehem Steel," *Interfaces, 19:1*, 68–84.

which molds. For this purpose we define index sets

$I_j \triangleq$ collection of indexes i corresponding to molds that
could be used for product j

If $i \in I_j$, mold i is feasible for product j.

Discrete versus Continuous Decision Variables

As usual, actual modeling begins with decision variables. Notice, however, that the decisions to be taken in our Bethlehem example are qualitatively different from those of earlier models. A mold design will either be selected or not. Once selected, it will either be used for a given product, or it will not.

Such decisions require a new, logical, discrete variable type.

> __2.33__   A variable is **discrete** if it is limited to a fixed or countable set of values. Often, the choices are only 0 and 1.

Frequently occurring discrete variable types include **binary** or **0–1 variables** limited to values 0 and 1, and nonnegative **integer variables** that may take any non-negative integer value.

We employ discrete variables mainly to model decisions of an all-or-nothing, either–or character such as those involved in our Bethlehem ingot mold example. In particular, we will use

$$y_i \triangleq \begin{cases} 1 & \text{if ingot mold } i \text{ is selected} \\ 0 & \text{if not} \end{cases}$$

Such y_i are discrete because they are limited to two values. We use value 1 to mean an event occurs, and value 0 when it does not. A value of 0.3 or $\frac{4}{5}$ has no physical meaning.

In a similar manner, we model decisions about which mold to use in making each product with decision variables

$$x_{i,j} \triangleq \begin{cases} 1 & \text{if ingot mold } i \text{ is used for product } j \\ 0 & \text{if not} \end{cases}$$

Again the variables are allowed only a countable set of values to reflect the logical character of the decisions.

Contrast these decision variables with the continuous variables we encountered in earlier models.

> __2.34__   A variable is **continuous** if it can take on any value in a specified interval.

For example, a nonnegative variable is continuous because it can assume any value in the interval $[0, +\infty)$.

Variables in the Two Crude model of Section 2.1 certainly have this continuous character. They are denominated in thousands of barrels of petroleum, and any nonnegative real value has a physical interpretation. Technically speaking, variables of the Pi Hybrids model in Section 2.3 should be limited to the integers (a discrete set) because they count bags of corn seed produced and shipped. Still, the numbers involved are likely to be so large that there is no loss of validity in allowing all nonnegative values. We will see in later chapters that there is a considerable gain in tractability.

> __2.35__   When there is an option, such as when optimal variable magnitudes are likely to be large enough that fractions have no practical importance, modeling with continuous variables is preferred to discrete because optimizations over continuous variables are generally more tractable than are ones over discrete variables.

SAMPLE EXERCISE 2.15: CHOOSING DISCRETE VERSUS CONTINUOUS VARIABLES

Decide whether a discrete or a continuous variable would be best employed to model each of the following quantities.

(a) The operating temperature of a chemical process

(b) The warehouse slot assigned a particular product

(c) Whether a capital project is selected for investment

(d) The amount of money converted from yen to dollars

(e) The number of aircraft produced on a defense contract

Modeling:

(a) A temperature can assume any value in a physical range. It is naturally continuous.

(b) The slot will be selected from a finite, and thus countable, list. It will probably require a discrete variable.

(c) Assuming that one cannot invest in just part of a project, there are only two possibilities: Take the project or reject it. A 0–1 discrete variable is appropriate.

(d) Amounts of money can assume any nonnegative value. They should be modeled with continuous variables.

(e) If the number of aircraft is likely to be large, principle 2.35 argues for a continuous variable even though the number of aircraft is clearly countable. If only a few expensive aircraft are planned, it may be necessary to model discreteness explicitly. Then a nonnegative integer variable is required.

Constraints with Discrete Variables

We are now ready to begin forming the constraints of our Bethlehem ingot mold example. One requirement is that at most p molds be selected:

$$\sum_{i=1}^{m} y_i \leq p$$

Notice that our convention of making the variable $= 1$ when something happens, and $= 0$ otherwise makes such requirements easy to express.

A similar requirement is that each product be assigned exactly one mold design. We may sum over the possible choice to express those constraints as

$$\sum_{i \in I_j} x_{i,j} = 1 \qquad j = 1, \dots, n$$

Notice again how easy it is to model "at least 1," "at most 1," and "exactly 1" restrictions with 0–1 discrete variables.

A last system of main constraints for the Bethlehem model must encode the dependency between x and y variables. Mold i cannot be assigned to product j unless it is one of the p selected. That is,

$$x_{i,j} \le y_i \qquad i = 1, \dots, m; \quad j = 1, \dots, n$$

SAMPLE EXERCISE 2.16: EXPRESSING CONSTRAINTS IN 0–1 VARIABLES

In choosing among a collection of 16 investment projects, variables

$$w_j \triangleq \begin{cases} 1 & \text{if project } j \text{ is selected} \\ 0 & \text{otherwise} \end{cases}$$

Express each of the following constraints in terms of these variables.

(a) At least one of the first eight projects must be selected.

(b) At most three of the last eight projects can be selected.

(c) Either project 4 or project 9 must be selected, but not both.

(d) Project 11 can be selected only if project 2 is also.

Modeling:

(a) $\sum_{j=1}^{8} w_j \ge 1$ (b) $\sum_{j=9}^{16} w_j \le 3$

(c) $w_4 + w_9 = 1$ (d) $w_{11} \le w_2$

Bethlehem Ingot Mold Example Model

We may complete a discrete optimization model of our Bethlehem ingot mold example as

$$\min \ \sum_{j=1}^{n} \sum_{i \in I_j} c_{i,j} x_{i,j} \qquad \text{(total waste)}$$

$$\text{s.t.} \ \sum_{i=1}^{m} y_i \le p \qquad \text{(select at most } p\text{)}$$

$$\sum_{i \in I_j} x_{i,j} = 1 \qquad j = 1, \dots, n \qquad \text{(one each product)} \qquad (2.14)$$

$$x_{i,j} \le y_i; \qquad j = 1, \dots, n; i \in I_j \qquad \text{(use only if selected)}$$

$$y_i = 0 \text{ or } 1 \qquad i = 1, \dots, m \qquad \text{(binary variables)}$$

$$x_{i,j} = 0 \text{ or } 1; \quad j = 1, \dots, n; i \in I_j$$

The objective function of this model merely totals the scrap waste associated with assigning molds to products. In addition to the main constraints formulated above, we have included m variable-type constraints on the y_j and mn such constraints on the $x_{i,j}$. The specification "$= 0$ or 1" signals that these variables are discrete and may take on only the values 0 and 1.

Integer and Mixed-Integer Programs

A mathematical program is a **discrete optimization model** if it includes any discrete variables at all. Otherwise, it is a **continuous optimization model**.

Discrete models are often called integer programs because we may think of discrete variables as being restricted to integers within an interval. For example,

$$y_j = 0 \text{ or } 1$$

is equivalent to

$$0 \le y_j \le 1$$
$$y_j \text{ integer}$$

Whenever the allowed values of a variable are countable, they can be aligned with the integers in a similar way.

> __2.36__ An optimization model is an **integer program (IP)** if any one of its decision variables is discrete. If all variables are discrete, the model is a **pure integer program**; otherwise, it is a **mixed-integer program**.

SAMPLE EXERCISE 2.17: RECOGNIZING INTEGER PROGRAMS

Determine whether an optimization model over each of the following systems of variables is an integer program, and if so, state whether it is pure or mixed.

(a) $w_j \ge 0, \quad j = 1, \ldots, q$

(b) $w_j = 0 \text{ or } 1, \quad j = 1, \ldots, p$
   $w_{p+1} \ge 0$ and integer

(c) $w_j \ge 0, \quad j = 1, \ldots, p$
   $w_{p+1} \ge 0$ and integer

Analysis: We apply definitions 2.36 .

(a) Here all variables are continuous, so the model is continuous, not discrete.

(b) The first p variables are limited to 0 and 1, and the last to any nonnegative integer. Thus the model is a pure integer program.

(c) The one integer variable makes this model discrete, and so an IP. It is a mixed-integer program because it also has continuous decision variables.

Integer Linear versus Integer Nonlinear Programs

Bethlehem ingot model (2.14) would fulfill definition $\boxed{2.29}$ of a linear program except for the binary type of its variables. Thus it is natural to classify it an integer linear program.

$\boxed{2.37}$ A discrete or integer programming model is an **integer linear program (ILP)** if its (single) objective function and all main constraints are linear.

The alternative is an integer nonlinear program.

$\boxed{2.38}$ A discrete or integer programming model is an **integer nonlinear program (INLP)** if its (single) objective function or any of its main constraints is nonlinear.

SAMPLE EXERCISE 2.18: RECOGNIZING ILPS AND INLPS

Assuming that all w_j are decision variables, determine whether each of the following mathematical programs is best described as a linear program (LP), a nonlinear program (NLP), an integer linear program (ILP), or an integer nonlinear program (INLP).

(a) max $\quad 3w_1 + 14w_2 - w_3$

s.t. $\quad w_1 \leq w_2$

$\qquad w_1 + w_2 + w_3 = 10$

$\qquad w_j = 0 \text{ or } 1 \qquad j = 1, \ldots, 3$

(b) min $\quad 3w_1 + 14w_2 - w_3$

s.t. $\quad w_1 w_2 \leq 1$

$\qquad w_1 + w_2 + w_3 = 10$

$\qquad w_j \geq 0 \qquad j = 1, \ldots, 3$

$\qquad w_1 \text{ integer}$

(c) min $\quad 3w_1 + 9\dfrac{\ln(w_2)}{w_3}$

s.t. $\quad w_1 \leq w_2$

$\qquad w_1 + w_2 + w_3 = 10$

$\qquad w_2, w_3 \geq 1$

$\qquad w_1 \geq 0$

(d) max $\quad 19w_1$

s.t. $\quad w_1 \leq w_2$

$\qquad w_1 + w_2 + w_3 = 10$

$\qquad w_2, w_3 \geq 1$

$\qquad w_1 \geq 0$

Analysis:

(a) Except for its discrete variable-type constraints, this model would be a linear program because the objective function and both main constraints are linear. Thus the model is an integer linear program (ILP).

(b) The product in its first main constraint makes this model nonlinear. However, it would not usually be called a nonlinear program because w_1 is discrete. The model is best classified as an integer nonlinear program (INLP).

(c) The logarithm and quotient terms in its objective function make this model nonlinear. Since all variables are continuous, it should be classified as a nonlinear program (NLP).

(d) The objective function and all main constraints of this model are linear and variable types are all continuous. Thus the model is a linear program (LP).

EXAMPLE 2.5: PURDUE FINAL EXAM SCHEDULING

We may illustrate integer nonlinear programming applications with a problem familiar to every college student: final exam scheduling. In a typical term Purdue University[4] picks one of $n = 30$ final exam time periods for each of over $m = 2000$ class units on its main campus. Most exams involve just one class section, but there are a substantial number of "unit exams" held at a single time for multiple sections.

The main issue in this exam scheduling is "conflicts," instances where a student has more than one exam scheduled during the same time period. Conflicts burden both students and instructors because a makeup exam will be required in at least one of the conflicting courses. Purdue's exam scheduling procedure begins by processing enrollment records to determine how many students are jointly enrolled in each pair of course units. Then an optimization scheme seeks to minimize total conflicts as it selects time periods for all class units.

Indexing, Parameters, and Decision Variables for Purdue Finals Example

As usual, we begin a model of this problem by introducing indexes for its main dimensions:

$$i \triangleq \text{class unit number } (i = 1, \ldots, m)$$

$$t \triangleq \text{exam time period number } (t = 1, \ldots, n)$$

Then discrete decision variables encode the schedule options:

$$x_{i,t} \triangleq \begin{cases} 1 & \text{if class } i \text{ is assigned to time period } t \\ 0 & \text{otherwise} \end{cases}$$

Also, we define joint enrollment input parameters:

$$e_{i,i'} \triangleq \text{number of students taking an exam in both class unit } i \text{ and class unit } i'$$

---

[4]Based on C. J. Horan and W. D. Coates (1990), "Using More Than ESP to Schedule Final Exams: Purdue's Examination Scheduling Procedure II (ESP II)," *College and University Computer Users Conference Proceedings, 35,* 133–142.

Nonlinear Objective Function

The main challenge in modeling Purdue's final exam scheduling problem with the foregoing notation is to represent total conflicts in an objective function. To begin, focus on any pair of courses i and i'. The product

$$x_{i,t}x_{i',t} = \begin{cases} 1 & \text{if } i \text{ and } i' \text{ are both scheduled at time period } t \\ 0 & \text{if not} \end{cases}$$

Thus we may sum over time periods and multiply by joint enrollment to express the conflict for any pair:

$$\text{conflicts between } i \text{ and } i' = e_{i,i'} \sum_{t=1}^{n} x_{i,t}x_{i',t}$$

It remains only to total such expressions over all course pairs i, i'. The result is the Purdue final exam scheduling objective function:

$$\min \quad \sum_{i=1}^{m-1} \sum_{i'=i+1}^{m} e_{i,i'} \sum_{t=1}^{n} x_{i,t}x_{i',t} \qquad \text{(total conflicts)} \qquad (2.15)$$

Notice that summations have been indexed so that $i' > i$, to avoid counting any pair more than once. The first sum considers all i except the last, which has no higher index, and the second adds in all pairs with $i' > i$.

Purdue Final Exam Scheduling Example Model

Beginning from objective function (2.15), we may model Purdue's problem:

$$
\begin{aligned}
\min \quad & \sum_{i=1}^{m-1} \sum_{i'=i+1}^{m} e_{i,i'} \sum_{t=1}^{n} x_{i,t}x_{i',t} && \text{(total conflicts)} \\
\text{s.t.} \quad & \sum_{t=1}^{n} x_{i,t} = 1 && i = 1, \ldots, m && \text{(class } i \text{ scheduled)} \\
& x_{i,t} = 0 \text{ or } 1 && i = 1, \ldots, m; \quad t = 1, \ldots, n
\end{aligned}
\qquad (2.16)
$$

Main constraints simply assure that each class unit i is assigned exactly one exam time period t.

Model (2.16) is an integer nonlinear program (INLP). Its main constraints are linear, but the objective function has nonlinear product terms and variable types discrete.

2.6 MULTIOBJECTIVE OPTIMIZATION MODELS

All the models considered so far have a clear, quantitative way to compare feasible solutions. That is, they have single objective functions. In many business and industrial applications, single objectives realistically model the true decision process. Such organizations often are really satisfied to maximize some measure of profit or minimize some approximation to cost, although other objectives may also be relevant.

Matters become much more confused when the problem arises in the government sector, or in a complex engineering design, or in circumstances where uncertainty cannot be ignored. In such applications, solutions may be evaluated quite differently by different participants in the decision process, or against different performance criteria. None can be discounted. A **multiobjective optimization** model is required to capture all the perspectives—one that maximizes or minimizes more than one objective function at the same time.

EXAMPLE 2.6: DuPAGE LAND USE PLANNING

Perhaps no public-sector problem involves more conflict between different interests and perspectives than land use planning. That is why a multiobjective approach was adopted when government officials in DuPage County, Illinois, which is a rapidly growing suburban area near Chicago, sought to construct a plan controlling use of its undeveloped land.[5]

Table 2.1 shows a simplified classification with $m = 7$ land use types. The problem was to decide how to allocate among these uses the undeveloped land in the county's $n = 147$ planning regions.

TABLE 2.1   Land Use Types
in DuPage Example

i	Land Use Type
1	Single-family residential
2	Multiple-family residential
3	Commercial
4	Offices
5	Manufacturing
6	Schools and other institutions
7	Open space

Multiple Objectives

No single criterion fully captures the appropriateness of assigning undeveloped acres to a given use. DuPage analysts employed five:

1. *Compatibility*: an index of the compatibility between each possible use in a region and the existing uses in and around the region.
2. *Transportation*: the time incurred in making trips generated by the land use to/from major transit and auto links.
3. *Tax load*: the ratio of added annual operating cost for government services associated with the use versus increase in the property tax assessment base.
4. *Environmental impact*: the relative degradation of the environment resulting from the land use.
5. *Facilities*: the capital costs of schools and other community facilities to support the land use.

A good plan should make the first of these objectives large and the others small. Assigning the indexes

$$i \triangleq \text{land use type } (i = 1, \ldots, m)$$

$$j \triangleq \text{planning region } (j = 1, \ldots, n)$$

the following symbolic constants parameterize the five objectives:

---

[5]Based on Deepak Bammi and Dalip Bammi (1979), "Development of a Comprehensive Land Use Plan by Means of a Multiple Objective Mathematical Programming Model," *Interfaces, 9:2*, part 2, 50–63.

$c_{i,j} \triangleq$ compatibility index per acre of land use i in planning region j

$t_{i,j} \triangleq$ transportation trip time generated per acre of land use i in planning region j

$r_{i,j} \triangleq$ property tax load ratio per acre of land use i in planning region j

$e_{i,j} \triangleq$ relative environmental degradation per acre of land use i in planning region j

$f_{i,j} \triangleq$ capital costs for community facilities per acre of land use i in planning region j

Then with nonnegative decision variables

$x_{i,j} \triangleq$ number of undeveloped acres assigned to land use i in planning region j

we have the following multiple objectives:

$$\max \quad \sum_{i=1}^{m} \sum_{j=1}^{n} c_{i,j} x_{i,j}$$

$$\min \quad \sum_{i=1}^{m} \sum_{j=1}^{n} t_{i,j} x_{i,j}$$

$$\min \quad \sum_{i=1}^{m} \sum_{j=1}^{n} r_{i,j} x_{i,j}$$

$$\min \quad \sum_{i=1}^{m} \sum_{j=1}^{n} e_{i,j} x_{i,j}$$

$$\min \quad \sum_{i=1}^{m} \sum_{j=1}^{n} f_{i,j} x_{i,j}$$

Constraints of the DuPage Land Use Example

Some of the main constraints enforced in the DuPage example are straightforward. Define symbolic parameters

$b_j \triangleq$ number of undeveloped acres in planning region j

$\ell_i \triangleq$ county-wide minimum number of acres allocated to land use type i

$u_i \triangleq$ county-wide maximum number of acres allocated to land use type i

$o_j \triangleq$ number of acres in planning region j consisting of undevelopable floodplains, rocky areas, etc.

Then a first system of main constraints assures that all undeveloped land in each planning region is allocated:

$$\sum_{i=1}^{m} x_{i,j} = b_j \qquad j = 1, \ldots, n$$

Two others enforce county-wide lower and upper limits on various uses:

$$\sum_{j=1}^{n} x_{i,j} \geq \ell_i \qquad i = 1, \ldots, m$$

$$\sum_{j=1}^{n} x_{i,j} \leq u_i \qquad i = 1, \ldots, m$$

Finally, we need to assure all undevelopable land is assigned to parks and other open space:

$$x_{7,j} \geq o_j \qquad j = 1, \ldots, n$$

The more complex constraints describe how land uses interact. In particular, single- and multiple-family residential development creates a demand for nearby commercial centers and open space, as well as new schools and other institutions. Using the symbolic parameters

$s_i \triangleq$ new acres of land use i implied by allocation of an acre of undeveloped land to single-family residential

$d_i \triangleq$ new acres of land use i implied by allocation of an acre of undeveloped land to multiple-family residential

we obtain

$$x_{i,j} \geq s_i x_{1,j} + d_i x_{2,j} \qquad i = 3, 6, 7; \quad j = 1, \ldots, n$$

Acres in $i = 3$, commercial, $i = 6$, institutions, and $i = 7$, open space, must meet the implied demands.

DuPage Land Use Example Model

Collecting all the above produces the following multiobjective optimization model of our DuPage example:

$$
\begin{array}{lll}
\max & \displaystyle\sum_{i=1}^{m} \sum_{j=1}^{n} c_{i,j} x_{i,j} & \text{(compatibility)} \\[2ex]
\min & \displaystyle\sum_{i=1}^{m} \sum_{j=1}^{n} t_{i,j} x_{i,j} & \text{(transportation)} \\[2ex]
\min & \displaystyle\sum_{i=1}^{m} \sum_{j=1}^{n} r_{i,j} x_{i,j} & \text{(tax load ratio)} \\[2ex]
\min & \displaystyle\sum_{i=1}^{m} \sum_{j=1}^{n} e_{i,j} x_{i,j} & \text{(environmental)} \\[2ex]
\min & \displaystyle\sum_{i=1}^{m} \sum_{j=1}^{n} f_{i,j} x_{i,j} & \text{(facilities)} \\[2ex]
\text{s.t.} & \displaystyle\sum_{i=1}^{m} x_{i,j} = b_j & j = 1, \ldots, n & \text{(all used)} \\[2ex]
& \displaystyle\sum_{j=1}^{n} x_{i,j} \geq l_i & i = 1, \ldots, m & \text{(use minimums)} \\[2ex]
& \displaystyle\sum_{j=1}^{n} x_{i,j} \leq u_i & i = 1, \ldots, m & \text{(use maximums)} \\[2ex]
& x_{7,j} \geq o_j & j = 1, \ldots, n & \text{(undevelopable)} \\[1ex]
& x_{i,j} \geq s_i x_{1,j} + d_i x_{2,j} & i = 3, 6, 7; \quad j = 1, \ldots, n & \text{(implied needs)} \\[1ex]
& x_{i,j} \geq 0 & i = 1, \ldots, m; \quad j = 1, \ldots, n & \text{(nonnegativity)}
\end{array}
$$

(2.17)

The only new element is nonnegativity type-constraints on the decision variables.

Conflict among Objectives

Like so many other applications, DuPage's land use planning could be modeled validly only with multiple objective functions. Still, we will see in later chapters that there is a price in tractability. It is almost certain that the objective functions will conflict about the best allocation of land uses. For example, a solution placing trip-generating manufacturing land uses near traffic arteries will score well on the transportation objective, but it may not be at all compatible with existing land use, and it may severely degrade sensitive environments.

Single objective models are easier to deal with because they avoid such conflicts. With multiple objectives it is not even clear how to define an "optimal" solution.

> **2.39** When there is an option, single-objective optimization models are preferred to multiobjective ones because conflicts among objectives usually make multiobjective models less tractable.

SAMPLE EXERCISE 2.19: UNDERSTANDING MULTIPLE-OBJECTIVE CONFLICT

Consider the multiobjective mathematical program

$$
\begin{aligned}
\max \quad & 3z_1 + z_2 \\
\min \quad & z_1 - z_2 \\
\text{s.t.} \quad & z_1 + z_2 \leq 3 \\
& z_1, z_2 \geq 0
\end{aligned}
$$

Graph the feasible region and show that the best solutions for the two objective functions conflict.

Analysis: The feasible region and contours of the two objective functions are as follows:

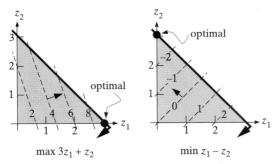

Feasible solution $z_1 = 3$, $z_2 = 0$ is optimal for the first objective function, but $z_1 = 0$, $z_2 = 3$ is optimal for the second. To find a solution that is overall "best," the analysis must somehow balance these conflicting objectives.

2.7 CLASSIFICATION SUMMARY

Over and over in the rest of this book we will see the importance of knowing whether a model is linear or nonlinear, continuous or discrete, single or multiobjective, when determining how tractable it will be and what methods are appropriate for trying to solve it. One of the most important skills for any practitioner of operations research is the ability to recognize these distinctions and know their implications.

To clarify the classification, Figure 2.6 provides a summary in a single display. The most tractable linear programming (LP) case has continuous variables, linear constraints, and a single linear objective function. If either constraints or the objective are nonlinear, it becomes a nonlinear program (NLP). Any discrete variables turn LPs into integer linear programs (ILPs) and NLPs into integer nonlinear programs (INLPs). All forms become more complex if there is more than one objective.

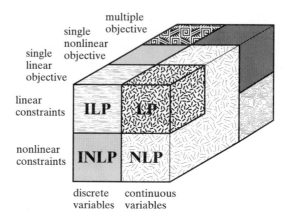

FIGURE 2.6 Classification of Optimization Models

EXERCISES

2-1 The Notip Table Company sells two models of its patented five-leg tables. The basic version uses a wood top, requires 0.6 hour to assemble, and sells for a profit of $200. The deluxe model takes 1.5 hours to assemble because of its glass top, and sells for a profit of $350. Over the next week the company has 300 legs, 50 wood tops, 35 glass tops, and 63 hours of assembly available. Notip wishes to determine a maximum profit production plan assuming that everything produced can be sold.

(a) ◻ Formulate a mathematical programming model with 4 main constraints to select an optimal production plan using decision variables $x_1 \triangleq$ number of basic models produced and $x_2 \triangleq$ number of deluxe models.

(b) ◻ ▱ Enter and solve your model with the class optimization software.

(c) Using a two-dimensional plot, solve your model graphically for an optimal product mix, and explain why it is unique.

(d) ◻ On a separate two-dimensional plot, show that that the model has alternative optimal solutions if profits are $120 and $300, respectively.

2-2 Wiley Wiz is a mutual fund manager trying to decide how to divide up to $12 million between domestic and foreign stocks. Domestic stocks have been returning 11% per year and foreign 17%. Naturally, Wiley would like to maximize the annual return from his investments. Still, he wants to exercise some caution. No more that $10 million of the fund should go into domestic stocks and no more than $7 million into foreign. Also, at least half as much should be invested in foreign as domestic, and at least half as much in domestic as foreign to maintain some balance.

(a) Formulate a mathematical programming model with 5 main constraints to decide Wiley's optimal investment plan using decision variables $x_1 \triangleq$ millions of dollars invested in domestic stocks and $x_2 \triangleq$ millions of dollars invested in foreign stocks.

(b) ◻ ▱ Enter and solve your model with the class optimization software.

(c) Using a two-dimensional plot, solve your model graphically for an optimal investment plan.

(d) On a separate two-dimensional plot, show graphically that the problem has alternative optimal solutions if rates of return are equal for domestic and foreign.

2-3 The Tall Tree lumber company owns 95,000 acres of forestland in the Pacific northwest, at least 50,000 of which must be aerially sprayed for insects this year. Up to 40,000 acres could be handled by planes based at Squawking Eagle, and up to 30,000 acres could be handled from a more distant airstrip at Crooked Creek. Flying time, pilots and materials together cost $3 per acre when spraying from Squawking Eagle and $5 per acre when handled from Crooked Creek. Tall Tree seeks a minimum cost spraying plan.

(a) Formulate a mathematical programming model to select an optimal spraying plan using decision variables $x_1 \triangleq$ thousands of acres sprayed from Squawking Eagle and $x_2 \triangleq$ thousands of acres sprayed from Crooked Creek.

(b) ⧉ ⧉ Enter and solve your model with the class optimization software.

(c) Using a two-dimensional plot, solve your model graphically for an optimal spraying plan, and explain why it is unique.

(d) On a separate two-dimensional plot, show graphically that the problem is unbounded if the Squawking Eagle capacity and both nonnegativity constraints are omitted.

(e) On a separate two-dimensional plot, show graphically that the problem becomes infeasible if the Crooked Creek facility is destroyed by fire.

2-4 The Fast Food Fantasy (Triple-F) hamburger chain is attempting to respond to customer demand for more healthy food by introducing a new birdburger made from a combination of beef and chicken. The new burger should weight at least 125 grams and have at most 350 calories, 15 grams of fat, and 360 milligrams of sodium. Each gram of beef used has 2.5 calories, 0.2 gram of fat, and 3.5 milligrams of sodium. Corresponding values for chicken are 1.8 calories, 0.1 gram, and 2.5 milligrams. Triple-F wants to find the mix that will meet all requirements and maximize beef content.

(a) Formulate a mathematical programming model to decide an optimal birdburger blend using decision variables $x_1 \triangleq$ grams of beef per burger and $x_2 \triangleq$ grams of chicken per burger.

(b) ⧉ ⧉ Enter and solve your model with the class optimization software.

(c) Using a two-dimensional plot, solve your model graphically for an optimal ingredient mix, and explain why it is unique.

(d) On a separate two-dimensional plot, show graphically that the model becomes infeasible if weight requirement is raised from 125 grams to 200 grams.

(e) On a separate two-dimensional plot, show graphically that the model is unbounded if the minimum weight and nonnegativity constraints are omitted.

2-5 Sun Agriculture (SunAg) operates a farm of 10,000 acres in the dry southwestern part of the United States. In the next season SunAg can plant acres in either vegetables, which return a profit of approximate $450 per acre, or cotton, which returns $200 per acre. As a precaution against bad weather, insects, and other factors, SunAg will plant no more than 70% of its total holdings in any one of these options. Also, irrigation water is limited. To grow vegetables requires 10 units (of water) per acre, and cotton requires 7, out of a government allocation of 70,000 units per season. SunAg wishes to develop a planting plan that maximumes profit.

(a) Formulate a mathematical programming model to solve SunAg's problem using 2 main constraints, 2 upper bound constraints, 2 variable-type constraints, and decision variables $v =$ acres in vegetables and $c =$ acres in cotton.

(b) ⧉ ⧉ Enter and solve your model with the class optimization software.

(c) Using a two-dimensional plot, find an optimal plan graphically.

(d) Show graphically that if we accidentally omitted the upper and lower bound constraints on the two variables, the resulting model would be unbounded.

(e) Now return to the correct model and imagine that SunAg is required by government regulation to plant all its acres. Show graphically that this case is infeasible.

2-6 The Kazak Film company needs to cut 15 long rolls and 10 short rolls of film from stock pieces. Each stock piece can be cut in one of two patterns. The first produces 5 long and 2 short rolls; the second yields 3 long and 5 short. Once any part of a piece of stock is cut, anything that remains is scrap. Also, neither pattern should be used more than 4 times because the jig used to cut it will become too inaccurate. Kazak wants to find the allowable combination of patterns that will minimize the number of stock pieces required.

(a) ◇ Formulate a mathematical model to decide what patterns to use. Use decision variables $x_1 \triangleq$ number of times pattern 1 is used and $x_2 \triangleq$ number of times pattern 2 is used.

(b) ◇ Both variables in your model should be restricted to take on only integer (whole number) values. Explain why.

(c) ◇ ▭ Enter and solve your model with class optimization software.

(d) Using a two-dimensional plot, solve your model graphically for an optimal cutting plan. Remember to consider only integer points.

(e) Explain how your plot shows that the model has alternative optimal solutions.

2-7 A factory is building a 500-square feet open rectangular cooling pool for water exhausted from its main process. The pool will be 8 feet deep, its length should be at least twice its width, and there is room for a width of at most 15 feet. We wish to choose the feasible design that minimizes cost by minimizing the concrete area of the pool walls.

(a) ◇ Formulate a mathematical programming model with 3 main constraints to choose an optimal design using decision variables $x_1 \triangleq$ length of the pool and $x_2 \triangleq$ width of the pool.

(b) ◇ ▭ Enter and solve your model with the class optimization software.

(c) Using a two-dimensional plot, solve your model graphically for an optimal design.

(d) ◇ On a separate two-dimensional plot, show graphically that the problem becomes infeasible if the pool can be at most 25 feet in length.

2-8 An architect is designing a cylindrical hotel that will have 150,000 square feet of floor space. She wishes to make the hotel have as many 10-feet-high floors as possible, but the height of the building should not exceed 4 times its diameter or it might be unstable. (Fractional numbers of floors are acceptable in this rough analysis.)

(a) Formulate a mathematical programming model with 2 main constraints to choose an optimal design using decision variables $x_1 \triangleq$ diameter of the hotel in feet and $x_2 \triangleq$ number of floors.

(b) ◇ ▭ Enter and solve your model with the class optimization software.

(c) Using a two-dimensional plot, solve your model graphically for an optimal design.

(d) On a separate two-dimensional plot, show graphically that the problem becomes infeasible if the diameter is limited to 50 feet.

2-9 Suppose that an optimization model in decision variables w_1 and w_2 has constraints
$$5w_1 + 2w_2 \le 10$$
$$w_2 \ge 0$$

(a) ◇ Devise a maximizing linear objective function for which the model has a unique optimal solution, and demonstrate that fact by solving the model graphically.

(b) ◇ Devise a maximizing linear objective function for which the model has multiple optimal solutions, and demonstrate that fact by solving the model graphically.

(c) ◇ Devise a maximizing linear objective function for which the model is unbounded, and demonstrate that fact by solving the model graphically.

2-10 Suppose that an optimization model in decision variables w_1 and w_2 has constraints
$$-w_1 + w_2 \le 7$$
$$w_1 \ge 0$$

(a) Devise a minimizing linear objective function for which the model has a unique optimal solution, and demonstrate that fact by solving the model graphically.

(b) Devise a minimizing linear objective function for which the model has multiple optimal solutions, and demonstrate that fact by solving the model graphically.

(c) Devise a minimizing linear objective function for which the model is unbounded, and demonstrate that fact by solving the model graphically.

2-11 Write each of the following as compactly as possible using summation and "for all" indexed notation.

(a) ◇ min $3y_{3,1} + 3y_{3,2} + 4y_{4,1} + 4y_{4,2}$

(b) max $1y_{3,1} + 2y_{3,2} + 3y_{3,3} + 4y_{3,4}$

(c) ◇ max $\alpha_1 y_{1,4} + \alpha_2 y_{2,4} + \cdots + \alpha_p y_{p,4}$

(d) min $\beta_1 y_1 + \beta_2 y_2 + \cdots + \beta_n y_n$

(e) ☒ $y_{1,1} + y_{1,2} + y_{1,3} + y_{1,4} = s_1$
$y_{2,1} + y_{2,2} + y_{2,3} + y_{2,4} = s_2$
$y_{3,1} + y_{3,2} + y_{3,3} + y_{3,4} = s_3$

(f) $a_{1,1}y_1 + a_{1,2}y_2 + a_{1,3}y_3 + a_{1,4}y_4 = b_1$
$a_{2,1}y_1 + a_{2,2}y_2 + a_{2,3}y_3 + a_{2,4}y_4 = b_2$
$a_{3,1}y_1 + a_{3,2}y_2 + a_{3,3}y_3 + a_{3,4}y_4 = b_3$

2-12 Suppose that the decision variables of a mathematical programming model are

$$x_{i,j,t} \triangleq \text{amount of product } i$$
$$\text{produced on manufacturing}$$
$$\text{line } j \text{ during week } t$$

where $i = 1, \ldots, 17; j = 1, \ldots, 5; t = 1, \ldots, 7$. Use summation and "for all" indexed notation to write expressions for each of the following systems of constraints in terms of these decision variables, and determine how many constraints belong to each system.

(a) ☒ Total production on any line in any week should not exceed 200.
(b) ☒ The total 7-week production of product 5 should not exceed 4000.
(c) ☒ At least 100 units of each product should be produced each week.

2-13 ☐ Repeat Exercise 2-12, this time coding the variables and constraints in the language of the class optimization software.

2-14 Suppose that the decision variables of a mathematical programming model are

$$x_{i,j,t} \triangleq \text{acres of land plot } i \text{ allocated}$$
$$\text{to crop } j \text{ in year } t$$

where $i = 1, \ldots, 47; j = 1, \ldots, 9; t = 1, \ldots, 10$. Use summation and "for all" indexed notation to write expressions for each of the following systems of constraints in terms of these decision variables, and determine how many constraints belong to each system.

(a) The acres allocated in each plot i cannot exceed the available acreage (call it a_i) in any year.
(b) At least 1000 total acres must be devoted to corn (crop $j = 4$) in each year.
(c) At least one-third of the total acreage planted over the 10 years must be in soybeans (crop $j = 2$).

2-15 ☐ Repeat Exercise 2-14, this time coding the variables and constraints in the language of the class optimization software.

2-16 Assuming that decision variables are $y_1, \ldots, y_3$, identify the objective function f, constraint functions g_i, and right-hand sides b_i of the general mathematical programming format 2.27 corresponding to each of the following optimization models.

(a) ☒ max $(y_1)^2 y_2 / y_3$
s.t. $y_1 + y_2 + y_3 + 7 = 20$
$2y_1 \geq y_2 - 9y_3$
$y_1, y_3 \geq 0$

(b) min $14y_1 y_2 y_3 + 100$
s.t. $y_1 + 2 \geq 3y_2 - y_3$
$y_1 + 8y_2 - 10 \leq 0$
$y_2, y_3 \geq 0$

2-17 Taking the x_j as variables, and all other symbols as given constants, determine whether each of the following is a linear or a nonlinear constraint, and briefly explain why.

(a) ☒ $3x_1 + 2x_2 - x_{17} = 9$
(b) $x_1 + x_2 - 4x_3 + x_{11} \leq 7$
(c) ☒ $\alpha/x_9 + 10x_{13} \leq 100$
(d) $x_4/\phi + 11x_{13} \geq 80$
(e) ☒ $\sum_{j=1}^{7} \beta_j (x_j)^2 \leq 10$
(f) $x_1 x_2 \ln(x_3) \leq 19$
(g) ☒ $\max\{x_1, 3x_1 + x_2\} \geq 111$
(h) $\sum_{j=1}^{7} e^{\beta_j} x_j = 234$

2-18 Assuming that the w_j are decision variables and all other symbols are constant, determine whether each of the following is a linear program (LP) or a nonlinear program (NLP), and briefly explain why.

(a) ☒ min $3w_1 + 8w_2 - 4w_3$
s.t. $\sum_{j=1}^{3} h_j w_j = 9$
$0 \leq w_j \leq 10, \quad j = 1, \ldots, 3$

(b) min $5/w_1 + 16w_2$
s.t. $6w_1 + w_2 \geq w_3$
$w_1, w_2, w_3 \geq 0$

(c) ☒ max $\sum_{j=1}^{10} \alpha_j w_j$
s.t. $(w_1)^2 + w_2 w_3 \geq 14$
$w_1, w_2, w_3 \leq 1$

(d) max $\sum_{j=1}^{200} w_j / (\sigma_j)^2$
s.t. $8 \sum_{j=1}^{100} w_j \geq \sum_{j=101}^{200} w_j$
$w_j \geq 0, \quad j = 1, \ldots, 200$

2-19 Determine whether a discrete or a continuous variable would be most appropriate to model each of the following quantities.

(a) ⊠ Amount of electricity consumed
(b) Whether a company wins a bidding
(c) ⊠ Process used to manufacture
(d) Height of a dam

2-20 The Forest Service can build a firewatch tower on any of 8 mountains. Using decision variables $x_j = 1$ if a tower is built on mountain j and $= 0$ otherwise, write constraint(s) enforcing each of the following requirements.

(a) ⊠ In all 3 sites will be selected.
(b) At least 2 of the first 5 mountains must be selected.
(c) ⊠ A tower should not be built on both mountains 3 and 8.
(d) A tower can be built on mountain 4 only if one is built on mountain 1.

2-21 The National Science Foundation (NSF) has received 4 proposals from professors to undertake new research in OR methods. Each proposal can be accepted for funding next year at the level (in thousands of dollars) shown in the following table or rejected. A total of $1 million is available for the year.

Proposal	1	2	3	4
Funding	700	400	300	600
Score	85	70	62	93

Scores represent the estimated value of doing each body of research that was assigned by NSF's advisory panel.

(a) ⊠ Formulate a discrete optimization model to decide what projects to accept to maximize total score within the available budget using decision variables $x_j = 1$ if proposal j is selected and $= 0$ otherwise.
(b) ⊠ ⌨ Enter and solve your model with the class optimization software.

2-22 The state Department of Labor is considering the establishment of area job training centers at up to 4 sites. The following table shows the land cost (in thousands of dollars) of the 4 sites and indicates with an × the sites that could provide adequate service to each of the five regions of the state.

Region	Site 1	2	3	4
Northwest	×	×	—	—
Southwest	×	×	×	—
Capital	×	—	×	×
Northeast	×	×	—	—
Southeast	×	—	—	×
Cost	200	40	55	75

The Department seeks a minimum total cost collection of sites that together could service all five regions.

(a) Formulate a discrete optimization model to decide what sites to build using decision variables $y_j = 1$ if site j is selected and $= 0$ otherwise.
(b) ⊠ ⌨ Enter and solve your model with the class optimization software.

2-23 Assuming that the z_j are decision variables, determine whether each of the following mathematical programs is best described as a linear program (LP), a nonlinear program (NLP), an integer linear program (ILP), or an integer nonlinear program (INLP), and briefly explain why.

(a) ⊠   max $3z_1 + 14z_2 + 7z_3$
        s.t. $10z_1 + 5z_2 + 18z_3 \leq 25$
                $z_j = 0$ or $1, \quad j = 1, \ldots, 3$

(b)   max $44z_1 + 2z_2 + 98z_3$
        s.t. $12z_1 + 92z_2 + 88z_3 \leq 95$
                $z_j \geq 0, \quad j = 1, \ldots, 3$

(c) ⊠   min $7z_1z_2 + 17z_2z_3 + 27z_1z_3$
        s.t. $\sum_{j=1}^{3} z_j = 2$
                $z_j = 0$ or $1, \quad j = 1, \ldots, 3$

(d)   min $7z_1 + \sin(z_2) + 8/z_3$
        s.t. $4z_1 + 16z_2 + z_3 \leq 29$
                $0 \leq z_j \leq 1, \quad j = 1, \ldots, 3$

(e) ⊠   max $12z_1 + 4z_2$
        s.t. $z_1z_2z_3 = 1$
                $z_1, z_2 \geq 0$
                $z_3 = 0$ or 1

(f)   min $8z_1 - z_2 + 14z_3$
        s.t. $z_1 \geq 3z_2$
                $z_j \geq 0, \quad j = 1, \ldots, 3$
                z_3 integer

2-24 Return to the mathematical programs of Exercise 2-23. Determine which of each of the following pairs of those models would normally be most tractable, and briefly explain why.

(a) ◇ Model (a) versus (b)
(b) Model (b) versus (d)
(c) ◇ Model (c) versus (d)
(d) Model (b) versus (f)

2-25 Consider the multiobjective optimization model

$$\text{max} \quad x_1$$
$$\text{max} \quad -3x_1 + x_2$$
$$\text{s.t.} \quad -x_1 + x_2 \leq 4$$
$$x_1 \leq 8$$
$$x_1, x_2 \geq 0$$

(a) Compute graphically a solution that is optimal if only the first objective is considered.
(b) Compute graphically a solution that is optimal if only the second objective is considered.
(c) Discuss the conflict inherent in trying to maximize both objectives at once.

2-26 Mexican Communications[6] is choosing cable for a new 16,000-meter telephone line. The following table shows the diameters available, along with the associated cost, resistance, and attenuation of each per meter.

Diameter (0.1 mm)	Cost ($/m)	Resistance (ohms/m)	Attenuation (db/m)
4	0.092	0.279	0.00175
5	0.112	0.160	0.00130
6	0.141	0.120	0.00161
9	0.420	0.065	0.00095
12	0.719	0.039	0.00048

The company wishes to choose the least cost combination of wires that will provide a new line with at most 1600 ohms resistance and 8.5 decibels attenuation.

(a) Formulate a mathematical programming model with three main constraints to choose an optimal combination of wires using decision variables

$(d = 4, 5, 6, 9, 12)$

$x_d \triangleq$ meters of diameter d wire used

Assume that resistance and attenuation grow linearly with the length of the wire used.

(b) ◇ 💻 Enter and solve your model with class optimization software.

2-27 The city of Lancaster's water distribution system[7] has 3 wells for water supply. There are 10 pumps at these 3 wells. It is estimated that a pumping rate of 10,000 gallons per minute is needed to satisfy the city's total water demand. There are limits on how much water can be pumped from each well: 3000 gal/min from well 1; 2500 gal/min from well 2; 7000 gal/min from well 3. There are also different costs of operating each pump and limits on the rate of each pump:

Pump	Maximum (gal/min)	Cost ($/gal/min)	From Well
1	1100	0.05	1
2	1100	0.05	2
3	1100	0.05	3
4	1500	0.07	1
5	1500	0.07	2
6	1500	0.07	3
7	2500	0.13	1
8	2500	0.13	2
9	2500	0.13	3
10	2500	0.13	3

Lancaster wishes to determine the least cost way to meet its pumping needs.

(a) Explain why appropriate decision variables for a model of this problem are $(j = 1, \ldots, 10)$

$x_j \triangleq$ pump rate per minute of pump j

(b) Assign suitable symbolic names to the constants of the cost and maximum rate values in the table above.
(c) Formulate an objective function to minimize the cost of the pumping plan selected.
(d) Formulate a system of 3 constraints enforcing well capacities.
(e) Formulate a system of 10 constraints enforcing pump capacities.
(f) Formulate a single constraint enforcing the overall pumping requirement.

[6]Based on L. F. Hernandez and B. Khoshnevis (1992), "Optimization of Telephone Wire Gauges for Transmission Standards," *European Journal of Operational Research*, 58, 389–392.

[7]Based on S. C. Sarin and W. El Benni (1982), "Determination of Optimal Pumping Policy of a Municipal Water Plant," *Interfaces*, 12:2, 43–48.

(g) Complete your model with an appropriate system of variable-type constraints.

(h) Is your model best classified as an LP, an NLP, an ILP, or an INLP, and is it single- or multiobjective? Explain.

(i) ◻ ▱ Enter and solve your model with class optimization software.

2-28 A small engineering consulting firm[8] is establishing its plan for the next year. The director and the three partners are to meet to decide which projects to pursue.

Preliminary research has been done on eight projects. The expected profit for each project is given in the following table together with the number of person-days of background preparation each will require and the computer processing unit (CPU) time (in hours) each will use.

Project	Profit	Person-Days	CPU
1	2.1	550	200
2	0.5	400	150
3	3.0	300	400
4	2.0	350	450
5	1.0	450	300
6	1.5	500	150
7	0.6	350	200
8	1.8	200	600

Excluding downtime, it is estimated that 1000 CPU hours will be available through the year. Presently there are 10 engineers (including the director and the partners); each works 240 days per year. At most three engineers could be let go, and management does not want to hire any new engineers for next year, due to market uncertainties. A minimum of 3 projects need to be selected, so each partner will be in charge of at least one project for the year. The director has four favorite projects (3, 4, 5, and 8), and the company needs to select at least one of these.

The firm wishes to formulate an optimization model to determine which projects to undertake, assuming that projects must be selected on an all-or-nothing basis.

(a) Justify why appropriate decision variables for the model are $(j = 1, \ldots, 8)$

$$x_j \triangleq \begin{cases} 1 & \text{if project } j \text{ is selected} \\ 0 & \text{otherwise} \end{cases}$$

(b) Assign suitable symbolic names to the constants in the foregoing table.

(c) Formulate an objective function to maximize total profit from projects selected.

(d) Formulate a pair of constraints to enforce the minimum and maximum engineer person-days available with different numbers laid off.

(e) Formulate 3 constraints to enforce the limit on computer time, meet the requirement to select at least three projects, and include at least one of the director's favorites.

(f) Complete your model with an appropriate system of variable-type constraints.

(g) Is your model best classified as an LP, an NLP, an ILP, or an INLP, and is it single- or multiobjective? Explain.

(h) ◻ ▱ Enter and solve your model with class optimization software.

2-29 A major expansion of the Brisbane airport[9] will require moving substantial quantities of earth from 4 sites where it is surplus to 7 locations where it is needed. The following table shows the haul distances (hundreds of meters) between points, as well as the quantity available (m^3) at each surplus site.

Need Site	Surplus Site			
	Apron	Term.	Cargo	Access
Extension	26	28	20	26
Dry pond	12	14	26	10
Roads	10	12	20	4
Parking	18	20	2	16
Fire station	11	13	6	24
Industrial park	8	10	22	14
Perimeter road	20	22	18	21
Quantity available	660	301	271	99

Quantities needed are 247 cubic meters at the extension, 394 at the dry pond, 265 along roads, 105 in the parking area, 90 at the fire station, 85 in the industrial park, and 145 along the perimeter road. The site engineer wishes to compute a minimum total distance times volume plan for accomplishing the required earth moving.

---

[8]Based on R. B. Gerdding and D. D. Morrison (1980), "Selecting Business Targets in a Competitive Environment," *Interfaces*, 10:4, 34–40.

[9]Based on C. Perry and M. Iliff (1983), "From the Shadows: Earthmoving on Construction Projects," *Interfaces*, 13:1, 79–84.

(a) Explain why appropriate decision variables for a model of this problem are $(i = 1, \ldots, 4, j = 1, \ldots, 7)$

$$x_{i,j} \triangleq \text{cubic meters moved from surplus } i \text{ to need } j$$

(b) Assign suitable symbolic names to the constants of the problem.

(c) Formulate an objective function to minimize total distance times volume movement.

(d) Formulate a system of 4 main constraints, assuring that the full available amount is moved from each surplus site.

(e) Formulate a system of 7 main constraints, assuring that the required amount is moved to each needing location.

(f) Complete your model with an appropriate system of variable-type constraints.

(g) Is your model best classified as an LP, an NLP, an ILP, or an INLP, and is it single- or multiobjective? Explain.

(h) ◇ ▣ Enter and solve your model with class optimization software.

2-30 The state highway department would like to have a formula for estimating the snow removal cost of each snow event as a function of the number of inches of snow to fall. A sample of n falls f_j and corresponding removal costs c_j, $j = 1, \ldots, n$, has been collected from history. Now the department would like to fit these data to an S-shaped curve of the form

$$c = \frac{k}{1 + e^{a + bf}}$$

in a way that minimizes the sum of squared errors. Here k, a, and b are empirical parameters of arbitrary sign.

(a) Explain why the decision variables in this optimization problem are k, a, and b.

(b) Formulate an unconstrained optimization model to perform the desired curve fit.

(c) Is your model best classified as an LP, an NLP, an ILP, or an INLP, and is it single- or multiobjective?

2-31 The Blue Hills Home Corporation (BHHC)[10] employs 22 remedial education teachers to service special needs students at 22 schools in the St. Louis area. BHHC assigns its teachers to schools for an entire year and is presently making decisions for the current year. One assignment consideration is cost; BHHC reimburses its teachers the cost $c_{i,j}$ of teacher i traveling to school j. However, the assignment must also consider 3 sets of preferences. Teachers express preferences scores $t_{i,j}$ of teacher i being assigned to school j, BHHC supervisors express preferences $s_{i,j}$, and school principals provide scores $p_{i,j}$. In all three cases a higher score indicates a greater preference.

(a) Explain why appropriate decision variables for a mathematical programming model of this problem are $(i, j = 1, \ldots, 22)$

$$x_{i,j} \begin{cases} 1 & \text{if teacher } i \text{ is assigned to} \\ & \text{school } j \\ 0 & \text{otherwise} \end{cases}$$

(b) Using these decision variables and the symbolic input constants above, express all BHHC assignment goals as separate objective functions.

(c) Write a system of 22 constraints expressing the requirement that each teacher be assigned to exactly one school.

(d) Write a system of 22 constraints expressing the requirement that each school be assigned exactly one teacher.

(e) Complete your model with an appropriate system of variable-type constraints.

(f) Is your model best classified as an LP, an NLP, an ILP, or an INLP, and is it single- or multiobjective? Explain.

2-32 Professor Proof is trying to decide which of 6 needed teaching assistant tasks he will assign to each of his 2 graduate assistants. Naturally, one assistant would probably be better at some tasks and the other assistant better at others. The following table shows his scoring of their potentials (high is good).

Assistant	Task					
	1	2	3	4	5	6
0	100	85	40	45	70	82
1	80	70	90	85	80	65

Professor Proof wants to assign three tasks to each assistant. However, tasks 5 and 6 are related and should be assigned to the same assistant.

[10]Based on Sang Lee and M. J. Schniederjans (1983), "A Multicriteria Assignment Problem: A Goal Programming Approach," *Interfaces, 13:4*, 75–81.

(a) Explain why appropriate decision variables for an optimization model of this problem are ($j = 1, \ldots, 6$)

$$x_j \triangleq \begin{cases} 0 & \text{if task } j \text{ is assigned to 0} \\ 1 & \text{if task } j \text{ is assigned to 1} \end{cases}$$

(b) Formulate an objective function to maximize the potential of the assignment chosen. (*Hint*: $1 - x_j = 1$ when $x_j = 0$.)

(c) Formulate a single main constraint to enforce requirements that at each assistant be assigned three tasks.

(d) Formulate a single main constraint to enforce the requirement that tasks 5 and 6 go to the same assistant.

(e) Complete your model with an appropriate system of variable-type constraints.

(f) Is your model best classified as an LP, an NLP, an ILP, or an INLP, and is it single- or multiobjective? Explain.

(g) ◇ 🖥 Enter and solve your model with class optimization software.

2-33 Fast Food Fantasy (Triple-F) cooks different types of hamburgers $j = 1, \ldots, 4$ in batches. A batch of burger j consists of at most u_j units and requires the entire cooking grill for t_j minutes. Assuming that the hourly demand for each burger is a known quantity d_j, Triple-F would like to decide the best batch size for each product. All required batches (and fractions of batches) must fit within the available grill time each hour, and the time required to sell out each burger should be minimized so that none will get too cold waiting to be sold.

(a) Explain why appropriate decision variables for a mathematical programming model of this problem are ($j = 1, \ldots, 4$)

$$x_j \triangleq \text{batch size of burger } j$$

(b) Formulate a system of 4 objective functions, minimizing the time to sell out batches of each burger assuming that demand is smooth over time.

(c) Formulate a single constraint assuring that all batches needed to meet demand each hour can be cooked.

(d) Complete your model with a suitable system of upper-bound and variable-type constraints.

(e) Is your model best classified as an LP, an NLP, an ILP, or an INLP, and is it single- or multiobjective? Explain.

2-34 The Kitty Railroad is in the process of planning relocations of freight cars among the 5 regions of the country to get ready for the fall harvest. The following table shows the cost of moving a car between each pair of regions, along with the number of cars in each at present and the number needed for harvest shipping.

From	Region				
	1	2	3	4	5
1	—	10	12	17	35
2	10	—	18	8	46
3	12	18	—	9	27
4	17	8	9	—	20
5	35	46	27	20	—
Present	115	385	410	480	610
Need	200	500	800	200	300

We want to choose a reallocation plan to get the required number of cars in each region at minimum total moving cost.

(a) Briefly justify why appropriate decision variables for this problem are ($i, j = 1, \ldots, 5, i \neq j$),

$$x_{i,j} \triangleq \text{number of cars moved from region } i \text{ to region } j$$

(b) The numbers of cars $x_{i,j}$ must physically be integer (whole numbers), but it is probably better to model them as continuous. Explain why.

(c) Assign symbolic names for the constants in the foregoing table.

(d) Write an objective function minimizing total movement cost.

(e) Write a system of 5 main constraints, assuring that the net number of cars in each region after the move will meet the need.

(f) Complete your model with an appropriate system of variable-type constraints.

(g) Is your model best classified as an LP, an NLP, an ILP, or an INLP, and is it single- or multiobjective? Explain.

(h) ◇ 🖥 Enter and solve your model with class optimization software.

2-35 A large copper company[11] has 23 plants, each of which can burn 4 different kinds of fuels to pro-

[11]Based on R. L. Bulfin and T. T. deMars (1983), "Fuel Allocation in Processing Copper Ore," *IIE Transactions, 15*, 217–222.

duce the energy needed in smelting. Energy require-ments at each plant p are known quantities r_p. We also know the energy output e_f of each ton of fuel f burned and the quantity of sulfur pollution s_f re-leased per ton of fuel f burned. Costs vary by loca-tion, but estimates c_{fp} are available of the cost per ton for fuel f at plant p. We want to choose mixes of fuels at plants to fulfill energy needs while minimiz-ing both cost and pollution.

(a) Briefly justify why appropriate decision variables for this problem are ($f = 1, \ldots, 4$; $p = 1, \ldots, 23$)

$$x_{fp} \triangleq \text{amount of fuel } f \text{ burned at plant } p$$

(b) Write an objective function minimizing total en-ergy cost.

(c) Write an objective function minimizing total sulfur pollution.

(d) Write a system of main constraints requiring that sufficient energy be produced at each plant. Also indicate how many constraints there are in this system.

(e) Complete your model with an appropriate sys-tem of variable-type constraints. Also indicate how many constraints there are in this system.

(f) Is your model best classified as an LP, an NLP, an ILP, or an INLP, and is it single- or multiob-jective? Explain.

2-36 Alabama Cabinet[12] runs a sawmill produc-ing wood panels called "blanks" for cabinetmaking. Some of the wood comes from logs sawed into boards at the company's mill, and the remainder derives from boards purchased green (undried). Lumber from both sources must be dried in the company's kilns before being cut into blanks.

Following are two tables, the first of which shows the purchase price per log, yield of green lum-ber per log, and availability for each of the three log diameters. Each board foot of green lumber sawed from logs gives 0.09 blank. The second table shows the price, yield in blanks, and availability of the two grades of purchased green lumber.

LOGS PURCHASED

Diameter	\$/Log	Bd-ft	Logs/Week
10	70	100	50
15	200	240	25
20	620	400	10

LUMBER PURCHASES

Grade	\$/bd-ft	Blanks/bd-ft	Bd-ft/Week
1	1.55	0.10	5000
2	1.30	0.08	Unlimited

We seek a minimum cost plan for producing at least 2350 blanks per week with the current mill's capacity to saw 1500 logs and dry 26,500 board feet of lumber each week.

(a) Explain why suitable decision variables for this model are

$x_d \triangleq$ number of logs of diameter d purchased ($d = 10, 15, 20$)

$y_g \triangleq$ board feet of green lumber grade g purchased ($g = 1, 2$)

(b) Numbers x_d must physically be integer (whole numbers), but it makes sense to model them as continuous. Explain why.

(c) Formulate an objective function minimizing to-tal material purchase cost. (We assume that other costs are essentially fixed.)

(d) Formulate a main constraint assuring that the required number of blanks will be produced.

(e) Formulate 2 main constraints enforcing sawing and drying capacities.

(f) Formulate 4 upper-bound constraints on deci-sion variables.

(g) Complete your model with a suitable system of variable-type constraints.

(h) Is your model best classified as an LP, an NLP, an ILP, or an INLP, and is it single- or multiob-jective? Explain.

(i) ⬨ 🖥 Enter and solve your model with class op-timization software.

2-37 The Bottles Film Festival draws thousands of people each year to view some of the latest motion pictures and award metals for the best. Planners are now selecting one of time slots $t = 1, \ldots, n$ for each of the $j = 1, \ldots, m$ films to be shown. From past experience, the festival can estimate numbers $a_{j,j'} \triangleq$ number of guests who would like to watch both film j and file j'. Now they wish to schedule the films so that no more than 4 are shown at any time and the minimum total number of guests are inconvenienced

---

[12]Based on H. F. Carino and C. H. LeNoir (1988), "Optimizing Wood Procurement in Cabinet Man-ufacturing," *Interfaces*, 18:2, 10–19.

by two movies they would like to see being scheduled at the same hour.

(a) Explain why appropriate decision variables for a mathematical programming model of this problem are $(j = 1, \ldots, m, t = 1, \ldots, n)$

$$x_{j,t} \triangleq \begin{cases} 1 & \text{if } j \text{ occurs at } t \\ 0 & \text{otherwise} \end{cases}$$

(b) Formulate an objective function to minimize the total number of guests inconvenienced by movies scheduled at the same time. (*Hint*: When $x_{j,t} x_{j',t} = 1$, films j and j' are scheduled at the same time t.)

(c) Formulate a system of m constraints assuring that each film is scheduled at some time.

(d) Formulate a system of n constraints assuring that no more than 4 movies are schedule at any time.

(e) Complete your model with an appropriate system of variable-type constraints.

(f) Is your model best classified as an LP, an NLP, an ILP, or an INLP, and is it single- or multiobjective? Explain.

2-38   To improve tax compliance[13] the Texas Comptroller's staff regularly audits at corporate home offices the records of out-of-state corporations doing business in Texas. Texas is considering the opening of a series of small offices near these corporate locations to reduce the travel costs now associated with such out-of-state audits. The following table shows the fixed cost (in thousands of dollars) of operating such offices at 5 sites i, the number of audits required in each of 5 states j, and the travel cost (in thousands of dollars) per audit performed in each state from a base at any of the proposed office sites.

Tax Site	Fixed Cost	Cost to Audit of Corporate Location:				
		1	2	3	4	5
1	160	0	0.4	0.8	0.4	0.8
2	49	0.7	0	0.8	0.4	0.4
3	246	0.6	0.4	0	0.5	0.4
4	86	0.6	0.4	0.9	0	0.4
5	100	0.9	0.4	0.7	0.4	0
Audits		200	100	300	100	200

We seek a minimum total cost auditing plan.

(a) Briefly explain why appropriate decision variables for an optimization model of this problem are

$$x_{i,j} \triangleq \text{fraction of audits at } j \text{ done from } i$$

$$y_i \triangleq \begin{cases} 1 & \text{if office } i \text{ is opened} \\ 0 & \text{otherwise} \end{cases}$$

(b) Explain why the y_i must be modeled as discrete.

(c) Assign suitable symbolic names to the constants in the foregoing table: the fixed cost of office i, the travel cost for audits done at j from i, and the number of audits at j.

(d) Formulate an objective function minimizing the sum of fixed office operating cost plus travel costs to audit sites. (*Hint*: The number of audits done at i from j is $x_{i,j}$ times the total number required at j.)

(e) Formulate a system of 5 main constraints requiring that 100% of audits at each j be performed.

(f) Formulate a system of 25 main constraints specifying that no part of the audits at any j can be done from i unless an office is opened at i.

(g) Complete your model with systems of variable-type constraints for the x and y decision variables.

(h) Is your model best classified as an LP, an NLP, an ILP, or an INLP, and is it single- or multiobjective? Explain.

(i) ⊘ ▭ Enter and solve your model with class optimization software.

2-39 The Speculators Fund is a stock mutual fund investing in categories $j = 1, \ldots, n$ of common stocks. At least a fraction ℓ_j and at most fraction u_j of the fund's capital is invested in any category j. The fund maintains estimates, v_j, of the expected annual return in capital gain and dividends for each dollar invested in category j. They also estimate the risk, r_j, per dollar invested in each category j. The goal is to maximize return at minimum risk.

(a) Explain why appropriate decision variables for a model of this problem are

$$x_j \triangleq \text{fraction of fund capital invested in category } j$$

(b) Formulate an objective function maximizing expected return per dollar invested.

---

[13]Based on J. A. Fitzsimmons and L. A. Allen (1983), "A Warehouse Location Model Helps Texas Comptroller Select Out-of-State Tax Offices," *Interfaces, 13:5*, 40–46.

(c) Formulate an objective function minimizing risk per dollar invested assuming that risks for different categories are independent of one another.

(d) Formulate a main constraint assuring that 100% of the fund's capital is invested somewhere.

(e) Formulate a system of n constraints enforcing lower bounds on the fraction of fund capital invested in each category.

(f) Formulate a system of n constraints enforcing upper bounds on the fraction of fund capital invested in each category.

(g) Is your model best classified as an LP, an NLP, an ILP, or an INLP, and is it single- or multiobjective? Explain.

2-40 Engineers[14] are designing the location for modules $i = 1, \ldots, m$ from among the $j = 1, \ldots, n$ available sites on a computer board. They already know

$$a_{i,i'} \triangleq \begin{cases} 1 & \text{if a wire is required from} \\ & \text{module } i \text{ to module } i' \\ 0 & \text{otherwise} \end{cases}$$

$d_{j,j'} \triangleq$ distance between sites j and j'

Using this information, they wish to choose a combination of locations that will minimize the total wire length required.

(a) Explain why appropriate decision variables for a model of this problem are $(i = 1, \ldots, m; j = 1, \ldots, n)$

$$x_{i,j} \triangleq \begin{cases} 1 & \text{if } i \text{ goes to site } j \\ 0 & \text{otherwise} \end{cases}$$

(b) Explain why the length of any wire required between modules i and i' can be expressed as

$$\sum_{j=1}^{n} \sum_{j'=1}^{n} d_{j,j'} x_{i,j} x_{i',j'}$$

(c) Use the expression of part (b) to formulate an objective function minimizing total wire length.

(d) Formulate a system of m constraints assuring that each module is assigned a location.

(e) Formulate a system of n constraints assuring that each location gets at most one module.

(f) Complete your model with an appropriate system of variable-type constraints.

(g) Is your model best classified as an LP, an NLP, an ILP, or an INLP, and is it single- or multiobjective? Explain.

---

[14]Based on L. Steinberg (1961), "The Backboard Wiring Problem: A Placement Algorithm," *SIAM Review*, 3, 37–50.

C H A P T E R 3

Improving Search

• •

To this point we have encountered a variety of types and sizes of deterministic optimization models but succeeded in analyzing only one or two. The time has come to begin looking seriously at solution methods.

Some optimization models admit closed-form solutions or similarly elegant analysis, but the overwhelming majority are approached by **numerical search**— repeatedly trying different values of the decision variables in a systematic way until a satisfactory one emerges. In fact, most optimization procedures can be thought of as variations on a single search theme: improving search.

Improving search tries to better a current solution by checking others nearby. If any proves superior, the search advances to such a solution, and the process repeats. Otherwise, we stop with the current solution. Synonyms for improving search include **local improvement**, **hillclimbing**, **local search**, and **neighborhood search**.

Improving search is one of the major themes of the book, and this chapter provides an elementary introduction. We explore the main strategies underlying improving search algorithms and identify special cases that prove particularly tractable. Familiarity with the model classifications of Chapter 2 is assumed.

Readers are strongly advised to absorb thoroughly the central ideas treated in this chapter before proceeding to the rest of the book. Much later development builds directly on Chapter 3.

3.1 IMPROVING SEARCH, LOCAL AND GLOBAL OPTIMA

Searching means hunting, and improving searches are hunts. We trek through a sequence of choices for decision variables, trying to find one good enough to justify stopping.

Solutions

Solutions are the points visited in a search.

> 3.1 | A **solution** is a choice of values for all decision variables.

For example, in the Two Crude model of Section 2.1, which had decision variables

$$x_1 \triangleq \text{thousands of barrels of Saudi crude refined per day}$$
$$x_2 \triangleq \text{thousands of barrels of Venezuelan crude refined per day}$$

a solution is a choice of two nonnegative quantities x_1 and x_2.

Notice that a solution need not be an "answer." Any pair of nonnegative real numbers forms a solution in the Two Crude case, but we were satisfied in our analysis only by an optimal solution—one that conforms to all constraints and minimizes cost.

Solutions as Vectors

If an optimization model has n decision variables, solutions are n-dimensional. It is convenient to deal with them as n-**vectors**—linear arrays of n **components**. For example, a Two Crude example solution refining 3 thousand barrels of Saudi crude per day and 2 thousand barrels of Venezuelan can be expressed as the vector $\mathbf{x} = (3, 2)$ with components $x_1 = 3$ and $x_2 = 2$.

Since improving search is about moving among such whole solutions, we use vector representations in most of our discussion. Primer 1 reviews vector notations and computations for those who feel rusty. We employ superscripts on solution vectors to indicate the order in which they are explored by a search.

> 3.2 | For a model with decision vector $\mathbf{x}$, the first solution visited by a search is denoted $\mathbf{x}^{(0)}$, the next $\mathbf{x}^{(1)}$, and so on.

SAMPLE EXERCISE 3.1: EXPRESSING SOLUTIONS AS VECTORS

The following table shows the sequence of solutions encountered by an improving search of an optimization model with 4 decision variables.

x_1	x_2	x_3	x_4
1	0	1	2
1	1	-2	4
2	1	-1	4
5	1	-1	6

(a) Express the solutions in standard vector notation $\boxed{3.2}$.

(b) Identify the value of $x_1^{(3)}$ and $x_3^{(1)}$ in part (a).

PRIMER 1: VECTORS

A **scalar** is a single real number such as 2, -0.25, $\frac{3}{7}$, $\sqrt{17}$, or π. **Scalar variables** such as x, y_6, Δp, and α are those that take on scalar values. Notice that we always show scalar variables in italic type.

Computations in operations research often involve procedures working on several quantities or variables at the same time. It is convenient to write such operations in terms of **vectors**—one-dimensional arrays of scalars. The vector may be displayed either vertically or horizontally. In this book it makes no difference. Representations

$$\begin{pmatrix} 3 \\ \frac{1}{3} \\ -2 \end{pmatrix} = (3, \tfrac{1}{3}, -2)$$

are just vertical and horizontal presentations of the same vector.

The number of scalars in a vector is its **dimension**. The example above is a 3-vector because it has dimension 3. We term the scalars making up a vector its **components**. Thus the second component of the example above is $\frac{1}{3}$.

Vector variables represent vectors of scalar quantities. In this book vector variables are typeset in boldface (e.g., $\mathbf{x}$, $\mathbf{a}$, $\Delta \mathbf{p}$), and their components are indicated by subscripts. Thus if $\mathbf{p}$ is a 5-vector (five-dimensional) variable, one possible value would be $\mathbf{p} = (0, -2, \frac{2}{5}, 0, 11)$ with components $p_2 = -2$ and $p_5 = 11$. In this book we distinguish vectors with superscripts in parentheses. That is, $\mathbf{y}^{(7)}$ and $\mathbf{y}^{(13)}$ represent distinct vectors having third components $y_3^{(7)}$ and $y_3^{(13)}$, respectively.

There are two closely related ways to conceptualize vectors geometrically. One scheme simply thinks of an n-vector as a point in n-dimensional space having coordinates equal to its components. For example, 2-vectors $\mathbf{x}^{(1)} = (2, -3)$ and $\mathbf{x}^{(2)} = (4, 1)$ correspond to points $(2, -3)$ and $(4, 1)$ in plot (a) in the following figure. The alternative conceptualization sees vectors as movements in n-space with components indicating displacements in different coordinates.

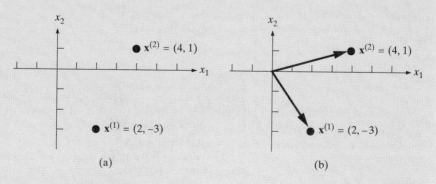

(a)   (b)

(*Continued*)

(Continued)

Arrows in part (b) illustrate that this is exactly the same thing if movement is assumed to begin at the origin.

The **length** or **norm** of n-vector **x**, denoted $||\mathbf{x}||$, is defined accordingly:

$$||\mathbf{x}|| \triangleq \sqrt{\sum_{j=1}^{n} (x_j)^2}$$

For example, $||\mathbf{x}^{(1)}||$ above is $\sqrt{(2)^2 + (-3)^2} = \sqrt{13}$.

Vectors of the same dimension are added and subtracted component by component. Thus for $\mathbf{x}^{(1)} = (4, 1)$ and $\mathbf{x}^{(2)} = (2, -3)$ above,

$$\mathbf{x}^{(1)} + \mathbf{x}^{(2)} = \begin{pmatrix} 2+4 \\ -3+1 \end{pmatrix} = \begin{pmatrix} 6 \\ -2 \end{pmatrix}, \qquad \mathbf{x}^{(1)} - \mathbf{x}^{(2)} = \begin{pmatrix} 2-4 \\ -3-1 \end{pmatrix} = \begin{pmatrix} -2 \\ -4 \end{pmatrix}$$

Similarly, scalar multiples of vectors are formed by simply multiplying each component by the scalar. Using the same $\mathbf{x}^{(1)}$ and $\mathbf{x}^{(2)}$ gives

$$0.3\mathbf{x}^{(1)} = (0.3(2), 0.3(-3)) = (0.6, -0.9)$$
$$\mathbf{x}^{(1)} + 0.3\mathbf{x}^{(2)} = (2 + 0.3(4), -3 + (0.3)(1)) = (3.2, -2.7)$$

The plots in the following figure demonstrate that these arithmetic operations also have a geometric interpretation. Adding vectors $\mathbf{x}^{(1)}$ and $\mathbf{x}^{(2)}$ has the effect of concatenating their associated movements in part (a). Similarly, in part (b), subtracting $\mathbf{x}^{(2)}$ from $\mathbf{x}^{(1)}$ extends $\mathbf{x}^{(1)}$'s movement by the negative of $\mathbf{x}^{(2)}$'s, and $\mathbf{x}^{(1)} + 0.3\mathbf{x}^{(2)}$ combines $\mathbf{x}^{(1)}$ with $\frac{3}{10}$ of $\mathbf{x}^{(2)}$.

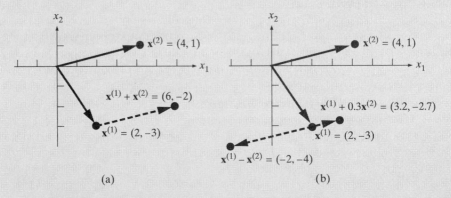

(a)   (b)

Vectors of the same dimension can also be multiplied. Although it seems natural to define the product of two vectors as the product of components, a less intuitive definition of multiplication is the one convenient in operations research. The **dot product** of two n-vectors **x** and **y** is the scalar quantity

$$\mathbf{x} \cdot \mathbf{y} \triangleq \mathbf{y} \cdot \mathbf{x} \triangleq \sum_{j=1}^{n} x_j y_j$$

(Continued)

> (*Continued*)
>
> For example, the 2-vectors $\mathbf{x}^{(1)}$ and $\mathbf{x}^{(2)}$ above yield $\mathbf{x}^{(1)} \cdot \mathbf{x}^{(2)} = \mathbf{x}^{(2)} \cdot \mathbf{x}^{(1)} = 2(4) + (-3)(1) = 3$. Notice that the dot product of vectors is indicated simply by writing the vectors with a multiplication dot between, and that it makes no difference which vector is mentioned first.
>
> We want multiplication of vectors to mean a dot product because we then have an easy way to denote weighted sums. For example, when we wish to show that the numbers $x_1, x_2, \ldots, x_6$ are summed with the weights $w_1, w_2, \ldots, w_6$, the result is exactly $\mathbf{w} \cdot \mathbf{x}$, where $\mathbf{w}$ is the vector of w_j and $\mathbf{x}$ the vector of x_j.

Analysis:

(a) In the notation of $\boxed{3.2}$ the four solutions are

$$\mathbf{x}^{(0)} = (1, 0, 1, 2)$$
$$\mathbf{x}^{(1)} = (1, 1, -2, 4)$$
$$\mathbf{x}^{(2)} = (2, 1, -1, 4)$$
$$\mathbf{x}^{(3)} = (5, 1, -1, 6)$$

(b) The first component of solution 3 is $x_1^{(3)} = 5$. The third component of solution 1 is $x_3^{(1)} = -2$.

EXAMPLE 3.1: DCLUB LOCATION

To illustrate some search ideas, consider the fictitious problem of choosing a location for the latest DClub discount department store. Dots on the map in Figure 3.1 show the three population centers of the area to be served. Population center 1 has approximately 60,000 persons, center 2 has 20,000, and center 3 has 30,000.

DClub wishes to locate one new store somewhere in the area in a way that maximizes business from the three populations. The obvious decision variables are x_1 and x_2, the coordinates of the chosen location.

The new store can be located anywhere except in the congested areas within $\frac{1}{2}$ mile of each population center. That is, constraints of the model are

$$[x_1 - (-1)]^2 + (x_2 - 3)^2 \geq \left(\tfrac{1}{2}\right)^2$$

$$(x_1 - 1)^2 + (x_2 - 3)^2 \geq \left(\tfrac{1}{2}\right)^2$$

$$(x_1 - 0)^2 + [x_2 - (-4)]^2 \geq \left(\tfrac{1}{2}\right)^2$$

Figure 3.1 shades the corresponding feasible set.

For an objective function, assume that experience shows that the business attracted from any population follows a "gravity" pattern—proportional to population (here in thousands) and inversely proportional to $1 +$ the square of its distance from the chosen location. Using this relationship and the coordinates of the three popu-

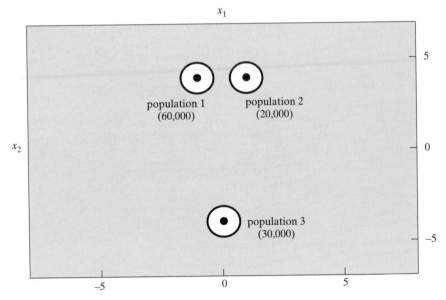

FIGURE 3.1   DClub Location Example

lation centers, our objective function becomes

$$\text{max } p(x_1, x_2) \triangleq \frac{60}{1 + (x_1 + 1)^2 + (x_2 - 3)^2} + \frac{20}{1 + (x_1 - 1)^2 + (x_2 - 3)^2}$$
$$+ \frac{30}{1 + (x_1)^2 + (x_2 + 4)^2} \tag{3.1}$$

Figure 3.2 provides a 3-dimensional view of the DClub example's nonlinear objective function. The peak occurs near population center 1. Figure 3.3 gives an easier-to-read contour view (as in Section 2.2) of the full model,

$$\text{max } p(x_1, x_2) \triangleq \frac{60}{1 + (x_1 + 1)^2 + (x_2 - 3)^2} + \frac{20}{1 + (x_1 - 1)^2 + (x_2 - 3)^2}$$
$$+ \frac{30}{1 + (x_1)^2 + (x_2 + 4)^2} \quad \text{(patronage)} \tag{3.2}$$

s.t. $(x_1 + 1)^2 + (x_2 - 3)^2 \geq \frac{1}{4}$ (avoid 1)

$(x_1 - 1)^2 + (x_2 - 3)^2 \geq \frac{1}{4}$ (avoid 2)

$(x_1 - 0)^2 + (x_2 + 4)^2 \geq \frac{1}{4}$ (avoid 3)

As usual, dashed lines connect points of equal objective value.

We want to maximize patronage $p(x_1, x_2)$ subject to avoiding congested circular areas around population centers. The point marked $\mathbf{x}^{(4)}$ in Figure 3.3 is (approximately) optimal because it is the feasible point falling on the highest contour (principle $\boxed{2.13}$).

$p(x_1, x_2)$

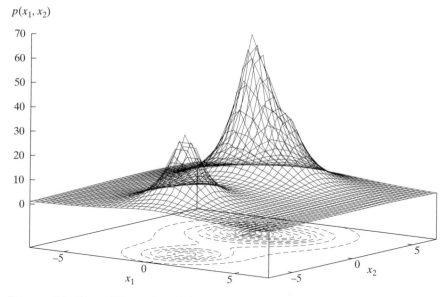

FIGURE 3.2   Three-Dimensional View of the DClub Patronage Function

Example of an Improving Search

Figure 3.3 also traces an improving search leading to optimal DClub solution $\mathbf{x}^{(4)}$. Beginning at

$$\mathbf{x}^{(0)} = (-5, 0) \qquad \text{with} \quad p(\mathbf{x}^{(0)}) \approx 3.5$$

it advances through solutions

$$\mathbf{x}^{(1)} = (-3, 4) \qquad \text{with} \quad p(\mathbf{x}^{(1)}) \approx 11.5$$

$$\mathbf{x}^{(2)} = (-1, 4.5) \qquad \text{with} \quad p(\mathbf{x}^{(2)}) \approx 21.6$$

$$\mathbf{x}^{(3)} = (0, 3.5) \qquad \text{with} \quad p(\mathbf{x}^{(3)}) \approx 36.1$$

to optimum

$$\mathbf{x}^{(4)} = (-0.5, 3) \qquad \text{with} \quad p(\mathbf{x}^{(4)}) \approx 54.8$$

Do not be concerned at the moment about where the various moves come from. Most of this chapter deals with principles for constructing such a search sequence.

For now, simply notice why we call it an improving search. The process begins with a feasible solution $\mathbf{x}^{(0)}$ and passes exclusively through feasible points. Furthermore, contours show that the objective function value constantly improves along the search path.

3.3   **Improving searches** are numerical algorithms that begin at a feasible solution to a given optimization model and advance along a search path of feasible points with ever-improving objective function value.

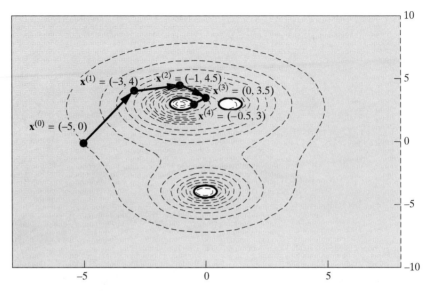

FIGURE 3.3   DClub Example Search Leading to an Optimal Solution

Neighborhood Perspective

For optimization models small enough to graph, as in Figure 3.3, it is easy to spot whether a search path maintains feasibility and constantly improves the objective function. Identifying optimal solutions is not much harder.

Unfortunately, such a global viewpoint is unavailable in typical models which have many decision variables. What we normally have to work with is illustrated by Figure 3.4. That plot zooms in on the region around $\mathbf{x}^{(4)} = (-0.5, 3)$ and blanks out the rest of the graph. We can tell something of the shape of the objective function near $\mathbf{x}^{(4)}$, and we can see the constraint limiting movement to the left. But we know nothing about other parts of the feasible region.

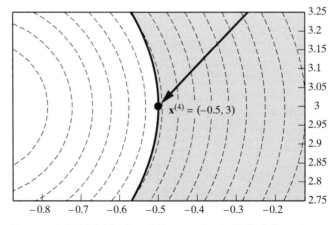

FIGURE 3.4   Neighborhood Perspective in DClub Search

Such a local perspective is typical of real searches. Lacking more complete insight, we must base search choices mostly on information about the neighborhood of the current solution.

> **3.4** | The **neighborhood** of a current solution $x^{(t)}$ consists of all nearby points; that is, all points within a small positive distance of $x^{(t)}$.

Local Optima

If Figure 3.4 were all we knew about our DClub model, we would be unable to tell whether points outside the visible range might prove superior to $x^{(4)}$. All we could say is that $(-0.5, 3)$ seems the best point in its neighborhood; that is, it is a local optimum.

> **3.5** | A solution is a **local optimum** (**local maximum** for a maximize problem or **local minimum** for a minimize problem) if it is feasible and if sufficiently small neighborhoods surrounding it contain no points that are both feasible and superior in objective value.

Local Optima and Improving Search

Observe that an improving search which has reached a local optimum can go no further.

> **3.6** | Improving searches stop if they encounter a local optimum.

Some solutions in the neighborhood of a local optimum may have better objective function value [e.g., $x = (-0.55, 3)$ in Figure 3.4]. Other neighbors may be feasible [e.g., $x = (-0.5, 3.05)$]. But no neighboring solution can continue the path of feasible points with ever-improving objective value (definition $\boxed{3.3}$) because none is both feasible and superior in objective value.

Local versus Global Optima

Truly optimal solutions to mathematical programs are feasible solutions with as good an objective function value as any other feasible point—neighbor or not. To distinguish this comprehensive notion of optimal, we employ the term global optimum.

> **3.7** | A solution is a **global optimum** (**global maximum** for a maximize problem or **global minimum** for a minimize problem) if it is feasible and no other feasible solution has superior objective value.

Notice that global optima cannot be improved in any neighborhood.

> **3.8** | Global optima are always local optima.

No matter how large a radius we consider around $\mathbf{x}^{(4)}$ of Figure 3.3, for example, the corresponding neighborhood contains no better point.

Unfortunately, the converse is not true.

| 3.9 | Local optima may not be global optima. |

Figure 3.5 illustrates with another improving search of the DClub example. This new search conforms to definition 3.3 in starting at a feasible point (in fact, the same one as Figure 3.3) and following an ever-improving path through feasible solutions. Still, it terminates at local maximum $\mathbf{x}^3 = (0, -3.5)$, where the patronage objective function $p(\mathbf{x}^{(3)}) \approx 25.8$, because no neighboring solution is both feasible and superior in the objective (principle 3.6). We already know from the earlier search of Figure 3.3 that solution $\mathbf{x}^* = (-0.5, 3)$ has a superior objective function value of approximately 54.8. The strictly worst local maximum of Figure 3.5 cannot be globally optimal.

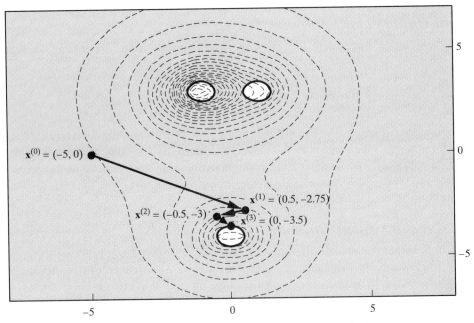

FIGURE 3.5 DClub Search Leading to a Local Maximum Not Global

---

SAMPLE EXERCISE 3.2: IDENTIFYING LOCAL AND GLOBAL OPTIMA

The figure that follows depicts constraints and contours of a minimizing optimization model.

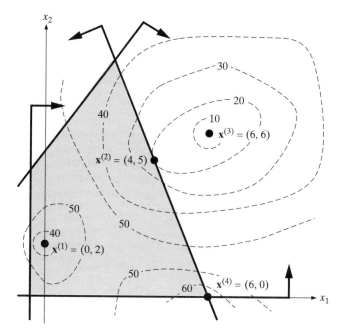

Determine whether each of the following points is apparently a global minimum, local minimum, or neither.

(a) $\mathbf{x}^{(1)} = (0, 2)$ (b) $\mathbf{x}^{(2)} = (4, 5)$

(c) $\mathbf{x}^{(3)} = (6, 6)$ (d) $\mathbf{x}^{(4)} = (6, 0)$

Analysis: We apply definitions 3.5 and 3.7 .

(a) Point $\mathbf{x}^{(1)}$ is apparently a local minimum because no neighboring point has better objective value even though all are feasible. Still, point $\mathbf{x}^{(2)}$ has superior objective function value, so $\mathbf{x}^{(1)}$ cannot be globally optimal.

(b) Point $\mathbf{x}^{(2)}$ appears to be a global and thus local minimum (principle 3.8). No other feasible point can match its objective value of 20.

(c) Even though point $\mathbf{x}^{(3)}$ has a very good objective function value, it is neither a local nor a global minimum because it is infeasible.

(d) Point $\mathbf{x}^{(4)}$ is neither a local nor a global minimum because every neighborhood contains feasible points with lower objective function value. One example is $\mathbf{x} = (5.9, 0.1)$.

Dealing with Local Optima

Principle 3.6 implies that improving searches can guarantee no more than a local optimum because they stop whenever one is encountered. But our DClub searches show that local optima may not always provide the globally best solutions we prefer (principle 3.9). Must we choose between the computational convenience of improving search and the analytical completeness of global optima?

Fortunately, the answer in many of the most frequently occurring cases is "no."

> $\boxed{3.10}$ The most tractable optimization models for improving search are those with mathematical forms assuring every local optimum is a global optimum.

We may pursue an improving search to a local optimum knowing in advance that it will also provide a global.

What can be done with the many models that fail tractability standard $\boxed{3.10}$? Sometimes we can still obtain a global optimum by switching to more complicated forms of search. Often, we must simply settle for less. After trying several improving searches—typically from different starting solutions—we keep the best of the results as an approximate or heuristic optimum.

> $\boxed{3.11}$ When models have local optima that are not global, the most satisfactory available analysis is often to run several independent improving searches and accept the best local optimum discovered as a **heuristic** or **approximate optimum**.

3.2 SEARCH WITH IMPROVING AND FEASIBLE DIRECTIONS

Having introduced improving search, we must now make it practical. Just how do we efficiently construct search paths satisfying the always feasible, constantly improving requirements of definition $\boxed{3.3}$? In this section we develop the short list of principles that point the way, and in Section 3.3 we translate them into algebraic conditions. Together, they comprise the foundation for nearly all practical implementations of improving search and much of this book.

Direction-Step Paradigm

Another look at the improving searches of Figures 3.3 and 3.5 will begin to reveal how practical search paths are constructed. Notice that the direction of search does not constantly change. Instead, we pursue a sequence of steps along straight-line move directions. Each begins at one of the numbered solutions $\mathbf{x}^{(t)}$. There, a move direction is chosen along with a step size specifying how far the direction should be pursued. Together they determine new point $\mathbf{x}^{(t+1)}$, and the search continues.

This **direction-step** paradigm lies at the heart of virtually all improving searches.

> $\boxed{3.12}$ Improving searches advance from current solution $\mathbf{x}^{(t)}$ to new solution $\mathbf{x}^{(t+1)}$ as
>
> $$\mathbf{x}^{(t+1)} \leftarrow \mathbf{x}^{(t)} + \lambda\, \Delta\mathbf{x}$$
>
> where vector $\Delta\mathbf{x}$ defines a **move direction** of solution change at $\mathbf{x}^{(t)}$, and **step size** multiplier $\lambda > 0$ determines how far to pursue the direction.

To illustrate, consider the first move of the search in Figure 3.5, which takes us from $\mathbf{x}^{(0)} = (-5, 0)$ to $\mathbf{x}^{(1)} = (0.5, -2.75)$. Most improving search algorithms would accomplish this move by first choosing a vector $\Delta\mathbf{x}$ of relative movement from $\mathbf{x}^{(0)}$ and then applying a suitable step size multiplier λ.

One vector sure to describe the direction chosen is the difference

$$\Delta\mathbf{x} = \mathbf{x}^{(1)} - \mathbf{x}^{(0)} = (0.5, -2.75) - (-5, 0) = (5.5, -2.75)$$

A step size of $\lambda = 1$ then yields the move

$$\mathbf{x}^{(1)} = \mathbf{x}^{(0)} + \lambda\,\Delta\mathbf{x} = (-5, 0) + 1(5.5, -2.75) = (0.5, -2.75)$$

However, vector $\Delta\mathbf{x}' = (2, -1)$ defines the same direction of movement. Application of step size $\lambda' = 2.75$ produces the identical move

$$\mathbf{x}^{(1)} = \mathbf{x}^{(0)} + \lambda'\,\Delta\mathbf{x}' = (-5, 0) + 2.75(2, -1) = (0.5, -2.75)$$

SAMPLE EXERCISE 3.3: DETERMINING SOLUTIONS FROM DIRECTIONS AND STEP SIZES

An improving search beginning at solution $\mathbf{w}^{(0)} = (5, 1, -1, 11)$ employs first move direction $\Delta\mathbf{w}^{(1)} = (0, 1, 1, 3)$ for step $\lambda_1 = \frac{1}{3}$, then $\Delta\mathbf{w}^{(2)} = (2, 0, \frac{1}{4}, -1)$ for step $\lambda_2 = 4$, and finally, $\Delta\mathbf{w}^{(3)} = (1, -\frac{1}{3}, 0, 2)$ for step $\lambda_3 = 1$. Determine the solutions visited.

Analysis: Applying $\boxed{3.12}$ gives

$$\mathbf{w}^{(1)} = \mathbf{w}^{(0)} + \lambda_1\,\Delta\mathbf{w}^{(1)} = (5, 1, -1, 11) + \tfrac{1}{3}(0, 1, 1, 3) \quad = (5, \tfrac{4}{3}, -\tfrac{2}{3}, 12)$$
$$\mathbf{w}^{(2)} = \mathbf{w}^{(1)} + \lambda_2\,\Delta\mathbf{w}^{(2)} = (5, \tfrac{4}{3}, -\tfrac{2}{3}, 12) + 4(2, 0, \tfrac{1}{4}, -1) \quad = (13, \tfrac{4}{3}, \tfrac{1}{3}, 8)$$
$$\mathbf{w}^{(3)} = \mathbf{w}^{(2)} + \lambda_3\,\Delta\mathbf{w}^{(3)} = (13, \tfrac{4}{3}, \tfrac{1}{3}, 8) + 1(1, -\tfrac{1}{3}, 0, 2) \quad = (14, 1, \tfrac{1}{3}, 10)$$

SAMPLE EXERCISE 3.4: DETERMINING MOVE DIRECTIONS FROM SOLUTIONS

The first four solutions visited by an improving search are $\mathbf{y}^{(0)} = (5, 11, 0)$, $\mathbf{y}^{(1)} = (4, 9, 3)$, $\mathbf{y}^{(2)} = (4, 9, 7)$, and $\mathbf{y}^{(3)} = (0, 8, 7)$. Determine the move directions employed assuming that all step sizes $\lambda = 1$.

Analysis: With all $\lambda = 1$, the sequence of move directions in computation $\boxed{3.12}$ must merely be the sequence of differences between successive solutions. First

$$\Delta\mathbf{y}^{(1)} = \mathbf{y}^{(1)} - \mathbf{y}^{(0)} = (4, 9, 3) - (5, 11, 0) = (-1, -2, 3)$$

so that

$$\mathbf{y}^{(1)} = \mathbf{y}^{(0)} + \lambda\,\Delta\mathbf{y}$$
$$= (5, 11, 0) + 1(-1, -2, 3) = (4, 9, 3)$$

Then

$$\Delta\mathbf{y}^{(2)} = \mathbf{y}^{(2)} - \mathbf{y}^{(1)} = (4, 9, 7) - (4, 9, 3) = (0, 0, 4)$$
$$\Delta\mathbf{y}^{(3)} = \mathbf{y}^{(3)} - \mathbf{y}^{(2)} = (0, 8, 7) - (4, 9, 7) = (4, -1, 0)$$

Improving Directions

We saw in Figure 3.4 that practical improving searches usually take a local perspective, limiting algorithmic decisions to information about the immediate neighborhood of current solution $\mathbf{x}^{(t)}$. Still, improving search definition $\boxed{3.3}$ demands that every move improve the objective function value.

How can we be sure of progress when we can "see" only a tiny neighborhood? We limit our choice to immediately improving directions.

> $\underline{3.13}$ Vector $\mathbf{\Delta x}$ is an **improving direction** at current solution $\mathbf{x}^{(t)}$ if the objective function value at $\mathbf{x}^{(t)} + \lambda\,\mathbf{\Delta x}$ is superior to that of $\mathbf{x}^{(t)}$ for all $\lambda > 0$ sufficiently small.

If a direction $\mathbf{\Delta x}$ does not immediately improve the objective in the neighborhood of current $\mathbf{x}^{(t)}$, it will not be pursued, regardless of how it affects the objective over larger steps.

Figure 3.6 illustrates for objective function (3.1) of our DClub location example. Constraints have been omitted since they have nothing to do with whether a direction improves.

Contours in part (a) show that direction $\mathbf{\Delta x} = (-3, 1)$ improves at $\mathbf{x}^{(3)} = (2, 0)$ because the (maximize) objective function increases. Notice that an improving direction is not required to yield progress forever. Multiple $\mathbf{\Delta x} = (-6, 2)$ (dashed line) remains improving even though a large enough λ produces a point with an objective value worse than that of $\mathbf{x}^{(3)}$.

Part (b) of Figure 3.6 demonstrates that an improving direction at one point need not improve everywhere. The same $\mathbf{\Delta x} = (-3, 1)$ that improved at $\mathbf{x}^{(3)} = (2, 0)$ fails for $\mathbf{x}^{(4)} = (-1.5, -1.5)$.

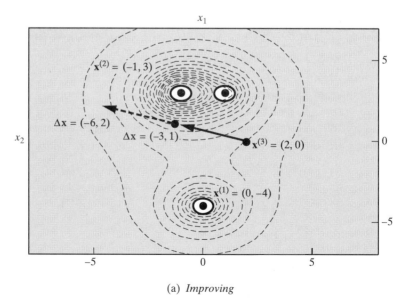

(a) *Improving*

FIGURE 3.6   Improving Directions of the DClub Objective

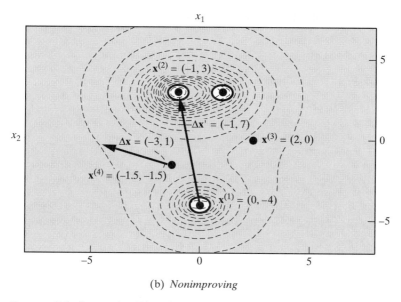

(b) *Nonimproving*

FIGURE 3.6   Improving Directions of the DClub Objective (Continued)

Finally, consider local maximum $\mathbf{x}^{(1)} = (0, -4)$ in Figure 3.6(b). We can see from contours that $\mathbf{x}^{(2)}$ has a better objective function value. Still, indicated direction

$$\Delta\mathbf{x}' = \mathbf{x}^{(2)} - \mathbf{x}^{(1)} = (-1, 3) - (0, -4) = (-1, 7)$$

is not an improving direction at $\mathbf{x}^{(1)}$. It fails definition $\boxed{3.13}$ because the improvement does not start immediately as we depart $\mathbf{x}^{(1)}$.

SAMPLE EXERCISE 3.5: RECOGNIZING IMPROVING DIRECTIONS GRAPHICALLY

The following figure plots contours of a minimizing objective function over decision variables y_1 and y_2.

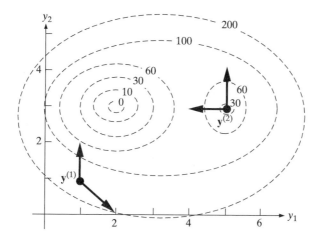

Determine graphically whether each of the following directions improves at the point indicated.

(a) $\Delta \mathbf{y} = (1, -1)$ at $\mathbf{y}^{(1)} = (1, 1)$

(b) $\Delta \mathbf{y} = (0, 1)$ at $\mathbf{y}^{(1)} = (1, 1)$

(c) $\Delta \mathbf{y} = (0, 1000)$ at $\mathbf{y}^{(1)} = (1, 1)$

(d) $\Delta \mathbf{y} = (0, 1000)$ at $\mathbf{y}^{(2)} = (5, 3)$

(e) $\Delta \mathbf{y} = (-1, 0)$ at $\mathbf{y}^{(2)} = (5, 3)$

Analysis: We apply definition 3.13 .

(a) At $\mathbf{y}^{(1)}$ a small movement in the indicated direction $\Delta \mathbf{y} = (1, -1)$ increases (degrades) the objective function value. Thus the direction is not improving.

(b) At the same $\mathbf{y}^{(1)}$ a small movement in the y_2-coordinate direction $\Delta \mathbf{y} = (0, 1)$ decreases (improves) the objective function value. Thus that direction improves at $\mathbf{y}^{(1)}$.

(c) The length of a move direction has no impact on whether it improves because definition 3.13 addresses only sufficiently small steps λ. Thus this case improves for the same reasons as part (b), albeit with smaller λ.

(d) This same direction of part (c) that improved at $\mathbf{y}^{(1)}$ fails to improve at $\mathbf{y}^{(2)}$ because a small step in the y_2-coordinate direction increases the objective value.

(e) Even though a move in direction $\Delta \mathbf{y} = (-1, 0)$ will eventually decrease the objective function from its value at $\mathbf{y}^{(2)} = (5, 3)$, the progress does not start immediately. Thus this $\Delta \mathbf{y}$ is not an improving direction at $\mathbf{y}^{(2)}$.

Feasible Directions

Moves in improving searches of constrained optimization models must both improve the objective function and maintain feasibility. For the latter, practical implementations parallel the foregoing discussion of improving directions by requiring immediately feasible directions.

> 3.14   Vector $\Delta \mathbf{x}$ is a **feasible direction** at current solution $\mathbf{x}^{(t)}$ if point $\mathbf{x}^{(t)} + \lambda \, \Delta \mathbf{x}$ violates no model constraint if $\lambda > 0$ is sufficiently small.

Just as with definition 3.13 , we evaluate directions $\Delta \mathbf{x}$ by considering only the immediate neighborhood of current solution $\mathbf{x}^{(t)}$. If feasibility is maintained for small enough steps, the direction is feasible. Otherwise, $\Delta \mathbf{x}$ will not be considered as our next search direction because we have no way to assess its impact outside the neighborhood.

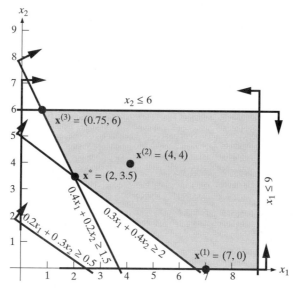

FIGURE 3.7 Constraints of the Two Crude Refinery Example

To illustrate, consider Figure 3.7, which depicts the variable-type and inequality constraints that define the feasible set of the Two Crude refinery model introduced in Sections 2.1 and 2.2:

$$\min \quad 20x_1 + 15x_2$$
$$\text{s.t.} \quad 0.3x_1 + 0.4x_2 \geq 2$$
$$0.4x_1 + 0.2x_2 \geq 1.5$$
$$0.2x_1 + 0.3x_2 \geq 0.5$$
$$0 \leq x_1 \leq 9$$
$$0 \leq x_2 \leq 6$$

Every direction is feasible at solution $x^{(2)} = (4, 4)$ because a short move in any direction leads to no violation of constraints.

Contrast with $x^{(1)} = (7, 0)$. There direction $\Delta x = (0, 1)$ is feasible because a small movement in the x_2-coordinate direction violates no constraints. As with improving directions, it does not matter that too big a step will produce a violation. The same direction is infeasible at $x^{(3)} = (0.75, 6)$ because even a tiny $\lambda > 0$ would lead to a violation of the bound constraint $x_2 \leq 6$.

SAMPLE EXERCISE 3.6: RECOGNIZING FEASIBLE DIRECTIONS GRAPHICALLY

The following figure shows the feasible region of a mathematical program over decision variables y_1 and y_2.

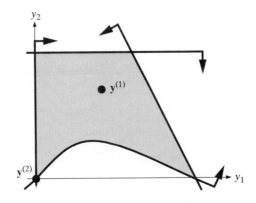

Determine graphically whether each of the following directions is feasible at the point indicated.

(a) $\Delta \mathbf{y} = (1, 0)$ at $\mathbf{y}^{(1)}$

(b) $\Delta \mathbf{y} = (1, 0)$ at $\mathbf{y}^{(2)}$

(c) $\Delta \mathbf{y} = (0, 1)$ at $\mathbf{y}^{(2)}$

(d) $\Delta \mathbf{y} = (0, 1000)$ at $\mathbf{y}^{(2)}$

Analysis: We apply definition $\boxed{3.14}$.

(a) We can move from $\mathbf{y}^{(1)}$ in any direction without (immediately) violating constraints. Thus all directions, including $\Delta \mathbf{y} = (1, 0)$, are feasible.

(b) At $\mathbf{y}^{(2)}$ a small step in the y_1-coordinate direction $\Delta \mathbf{y} = (1, 0)$ takes us outside the feasible region. Thus the direction is not feasible even though a long enough step would restore feasibility.

(c) A small step from $\mathbf{y}^{(2)}$ in the y_2-coordinate direction $\Delta \mathbf{y} = (0, 1)$ violates no constraint. Thus the direction is feasible.

(d) The length of a move direction has no impact on whether it is feasible because definition $\boxed{3.14}$ addresses only sufficiently small steps λ. Thus this direction is feasible for the same reason as in part (c), albeit with smaller λ.

Step Size: How Far?

Once an improving feasible move direction has been discovered at the current solution, how far should we follow it? That is, what step size λ should be applied?

We know from definitions $\boxed{3.13}$ and $\boxed{3.14}$ that we will improve the objection and retain feasibility for at least small steps. But having an improving feasible direction in hand, it is natural to pursue it as long as it remains so.

$\boxed{3.15}$ Improving searches normally apply the maximum step λ for which the selected move direction continues to retain feasibility and improve the objective function.

Notice that principle $\boxed{3.15}$ involves two issues: how long the direction improves the objective function and how long it remains feasible. We fix λ and choose a new direction when either the objective function stops improving or a constraint is violated.

SAMPLE EXERCISE 3.7: DETERMINING MAXIMUM STEP SIZE

Suppose that we are searching for an optimal solution to the mathematical program

$$\begin{array}{ll} \min & 10w_1 + 3w_2 \\ \text{s.t.} & w_1 + w_2 \leq 9 \\ & w_1, w_2 \geq 0 \end{array}$$

For current point $\mathbf{w}^{(19)} = (4, 5)$, determine the maximum step in improving feasible direction $\Delta\mathbf{w} = (-3, -8)$ consistent with principle $\boxed{3.15}$.

Analysis: Direction $\Delta\mathbf{w}$ reduces the objective function at every point because it decreases both decision variables, and they both have positive costs. Thus it remains improving for any $\lambda > 0$.

For feasibility, we first consider the main constraint. Any step $\lambda > 0$ will result in

$$(w_1 + \lambda \, \Delta w_1) + (w_2 + \lambda \, \Delta w_2) = (4 - 3\lambda) + (5 - 8\lambda) \leq 9$$

That is, the constraint remains satisfied.

We conclude that the maximum step size λ will be determined by feasibility in the nonnegativity constraints $w_1 \geq 0$ and $w_2 \geq 0$. Any $\lambda > \frac{4}{3}$ makes negative the first component of the new solution:

$$\mathbf{w}^{(20)} = \mathbf{w}^{(19)} + \lambda \, \Delta\mathbf{w} = \begin{pmatrix} 4 \\ 5 \end{pmatrix} + \lambda \begin{pmatrix} -3 \\ -8 \end{pmatrix} = \begin{pmatrix} 4 - 3\lambda \\ 5 - 8\lambda \end{pmatrix}$$

Similarly, any $\lambda > \frac{5}{8}$ will violate nonnegativity on w_2. Thus the maximum step retaining both improvement and feasibility in the objective function is

$$\lambda = \min\{\tfrac{4}{3}, \tfrac{5}{8}\} = \tfrac{5}{8}$$

Search of the DClub Example

Algorithm 3A collects principles $\boxed{3.12}$ to $\boxed{3.15}$ in a formal improving search procedure. We can illustrate with the DClub search shown in Figure 3.5. As usual, the search begins at a feasible solution vector, here $\mathbf{x}^{(0)} = (-5, 0)$. Its objective value is $p(-5, 0) \approx 3.5$.

Passing to step 1 of Algorithm 3A, we look for an improving feasible direction. The one adopted in Figure 3.5 is $\Delta\mathbf{x}^{(1)} = (2, -1)$. Although there are many other choices, this one clearly does conform to definitions $\boxed{3.13}$ and $\boxed{3.14}$ in improving the objective and maintaining feasibility near $\mathbf{x}^{(0)}$.

We must now pick a λ value at step 3. Any movement in the direction chosen leaves us at a feasible point. Thus the step size will be determined by where the objective function quits improving. In models with many variables, some work is required to find such a maximum λ. Here we can proceed graphically. Contours show stops after a step of approximately $\lambda_1 = 2.75$. Thus step 4's update yields

$$
\begin{aligned}
\mathbf{x}^{(1)} &\leftarrow \mathbf{x}^{(0)} + \lambda_1\,\Delta\mathbf{x}^{(1)} \\
&= (-5, 0) + 2.75(2, -1) \\
&= (0.5, -2.75)
\end{aligned}
$$

ALGORITHM 3A: CONTINUOUS IMPROVING SEARCH

Step 0: Initialization. Choose any starting feasible solution $\mathbf{x}^{(0)}$, and set solution index $t \leftarrow 0$.

Step 1: Local Optimum. If no improving feasible direction $\Delta\mathbf{x}$ exists at current solution $\mathbf{x}^{(t)}$, stop. Under mild assumptions about the form of the model, point $\mathbf{x}^{(t)}$ is a local optimum.

Step 2: Move Direction. Construct an improving feasible direction at $\mathbf{x}^{(t)}$ as $\Delta\mathbf{x}^{(t+1)}$.

Step 3: Step Size. If there is a limit on step sizes for which direction $\Delta\mathbf{x}^{(t+1)}$ continues to both improve the objective function and retain feasibility, choose the largest such step size as λ_{t+1}. If not, stop; the model is unbounded.

Step 4: Advance. Update

$$
\mathbf{x}^{(t+1)} \leftarrow \mathbf{x}^{(t)} + \lambda_{t+1}\,\Delta\mathbf{x}^{(t+1)}
$$

Then, increment $t \leftarrow t + 1$, and return to Step 1.

Our first iteration is now complete. Incrementing $t \leftarrow 1$, we return to step 1. Improving feasible directions still exist. This time the search selected $\Delta\mathbf{x}^{(2)} = (-4, -1)$. As before, the maximum appropriate step size is determined by progress in the objective function. Picking $\lambda_2 = 0.25$ leads to

$$
\begin{aligned}
\mathbf{x}^{(2)} &\leftarrow \mathbf{x}^{(1)} + \lambda_2\,\Delta\mathbf{x}^{(2)} \\
&= (0.5, -2.75) + 0.25(-4, -1) \\
&= (-0.5, -3)
\end{aligned}
$$

Returning again to step 1, we begin a third iteration. Improving feasible directions still exist, and the search chose $\Delta\mathbf{x}^{(3)} = (-1, 1)$. Unlike previous iterations, feasibility is now a consideration. A step of 0.5 in that direction brings us up against a constraint, even though further objective progress is possible. Thus $\lambda_3 = 0.5$, and

$$
\begin{aligned}
\mathbf{x}^{(3)} &\leftarrow \mathbf{x}^{(2)} + \lambda_3\,\Delta\mathbf{x}^{(3)} \\
&= (-0.5, -3) + 0.5(1, -1) \\
&= (0, -3.5)
\end{aligned}
$$

Upon still another return to step 1, no improving feasible direction is apparent. Algorithm 3A terminates with locally optimal solution $\mathbf{x} = (0, -3.5)$.

When Improving Search Stops

Algorithm 3A keeps going while improving feasible directions are available because local progress is still possible.

| 3.16 | No optimization model solution at which an improving feasible direction is available can be a local optimum. |

Since a direction improves and maintains feasibility for the smallest of steps $\lambda > 0$, every neighborhood includes a better point.

What if the algorithm stops at a point admitting no improving feasible direction? In most applications the result is a local optimum (definition 3.5).

| 3.17 | When a continuous improving search terminates at a solution admitting no improving feasible direction, and mild assumptions hold, the point is a local optimum. |

Figures 3.3 and 3.5 illustrate this typical case. Both searches lead us to points where no improving feasible direction is apparent. Both results are local maxima.

The "mild assumptions" caveat appears in principle 3.17 because it is possible to make up examples where there are no improving feasible directions at the current solution, yet it is not a local optimum. Figure 3.8 shows two. The $\mathbf{w}$ in part (a) is not a local minimum of the unconstrained model indicated because it is possible to decrease the unusually shaped objective function for arbitrarily small steps along the curved path shown. Still, every straight-line direction fails to improve. Thus Algorithm 3A would terminate at $\mathbf{w}$ because there are no improving directions.

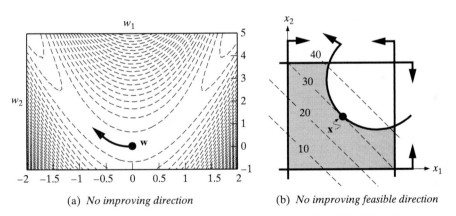

(a) *No improving direction*   (b) *No improving feasible direction*

FIGURE 3.8   Nonoptimal Points Admitting No Improving Feasible Directions

Even if the objective function has a very simple form, constraints can cause the same sorts of anomalies. Algorithm 3A would also terminate at the point $\mathbf{x}$ shown in part (b) because no (straight-line) feasible direction leads to higher objective function contours. Still, there are feasible solutions arbitrarily close to $\mathbf{x}$ with superior objective values. Stopping point $\mathbf{x}$ cannot be a local maximum.

Fortunately, such examples are rare for the standard models treated in this book. In the interest of efficiency, analysts are almost always willing to accept the result of an improving search as at least a local optimum.

Detecting Unboundedness

Most of the time, improving searches following Algorithm 3A will terminate at step 1 because there is no improving feasible direction. However, they can also stop in λ-choosing step 3.

An optimization model is **unbounded** if it admits feasible solutions with arbitrarily good objective value (definition 2.20).

> 3.18   If an improving search discovers an improving feasible direction for a model that can be pursued forever without ceasing to improve or losing feasibility, the model is unbounded.

Figure 3.9 shows such a problem and an application of Algorithm 3A starting at $\mathbf{y}^{(0)} = (1, 0)$. Contour lines illustrate how the objective value improves forever as y_1 becomes large.

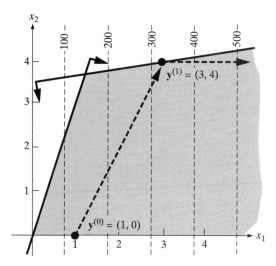

FIGURE 3.9   Improving Search of an Unbounded Model

The first iteration depicted in Figure 3.9 reached step 3 with chosen direction $\Delta\mathbf{y}^{(1)} = (1, 2)$. Progress in that direction is limited by constraints, so the step was fixed at the maximum $\lambda_1 = 2$ that retains feasibility (principle 3.15). On the next iteration, however, the chosen improving feasible direction is $\Delta\mathbf{y}^{(2)} = (1, 0)$. Step 3 terminates with a conclusion of unboundedness because there is no limit on step sizes improving the objective and retaining feasibility in that direction.

> **SAMPLE EXERCISE 3.8: DETECTING UNBOUNDEDNESS IN IMPROVING SEARCH**
>
> An improving search begins at $\mathbf{w}^{(0)} = (0, 0)$ in the model
>
> $$\min \quad w_2$$
> $$\text{s.t.} \quad 0 \leq w_1 \leq 1$$

(a) Explain why the model is unbounded.

(b) Identify an improving feasible direction $\Delta\mathbf{w}$ at $\mathbf{w}^{(0)}$ which demonstrates that the model is unbounded.

Analysis:

(a) Variable w_2 has no lower bound in the model. Thus we may leave w_1 at feasible value 0 and decrease w_2 to produce arbitrarily good objective function values.

(b) For the reasons given in part (a), arbitrarily good feasible solutions are obtained by leaving w_1 unchanged and decreasing w_2. The obvious move direction to accomplish this is $\Delta\mathbf{w} = (0, -1)$. There is no limit on the $\lambda > 0$ that may be applied to this direction while maintaining feasibility and improving the objective function.

3.3 ALGEBRAIC CONDITIONS FOR IMPROVING AND FEASIBLE DIRECTIONS

What distinguishes one implementation of improving search Algorithm 3A from another is the process employed to identify an improving feasible direction at step 2 (or to prove that none exists). We are now ready to develop the fundamental algebraic conditions at the heart of nearly every such construction.

Readers are advised to dwell on these simple conditions and to experiment with their own examples and plots until each idea has been absorbed completely. We will return to the algebra of this section many times as we develop the algorithms of subsequent chapters.

Gradients

To obtain algebraic characterizations of improving directions, we resort to a bit of differential calculus. If the objective function of our optimization model is **smooth** (i.e., differentiable with respect to all decision variables), simple conditions can easily be devised. Readers whose calculus is a bit inadequate or rusty should not panic. The brief synopsis in Primer 2 includes everything you will need.

When f is a function of n-vector $\mathbf{x} \triangleq (x_1, \ldots, x_n)$, it has n first partial derivatives. Our principal interest is the n-vector of such partial derivatives known as the **gradient**.

> 3.19 The gradient of $f(\mathbf{x}) \triangleq f(x_1, \ldots, x_n)$, denoted $\nabla f(\mathbf{x})$, is the vector of partial derivatives $\nabla f(\mathbf{x}) \triangleq (\partial f/\partial x_1, \ldots, \partial f/\partial x_n)$ evaluated at $\mathbf{x}$.

The gradient describes the shape of the objective function because each partial derivative of the objective function at any current solution quantifies the slope or rate of change per unit change in one of the coordinate directions.

To illustrate, we can return to patronage objective function (3.1) of Section 3.1's DClub example:

$$\max p(x_1, x_2) \triangleq \frac{60}{1 + (x_1 + 1)^2 + (x_2 - 3)^2} + \frac{20}{1 + (x_1 - 1)^2 + (x_2 - 3)^2}$$
$$+ \frac{30}{1 + (x_1)^2 + (x_2 + 4)^2}$$

PRIMER 2: DERIVATIVES AND PARTIAL DERIVATIVES

One of the major concepts of calculus is **derivatives**—rates of change in the value of a function with respect to small increases in its arguments. In this book we assume only an elementary understanding of derivatives, but we will need to compute them occasionally and to have an intuitive feel for derivatives as "slopes."

A function is **differentiable** or **smooth** at a point if its rate of change is unambiguous (i.e., if the function has no sudden rate changes). The prototypical example of a function that is not always differentiable is $f(x) \triangleq |x|$. At $x = 0$ the function makes a stark change from derivative -1 applicable for $x < 0$ to the $+1$ appropriate for $x > 0$. No derivative exists for x exactly zero.

For a function $f(x)$ of a single variable x, it is customary to denote the derivative of f with respect to x by df/dx or $f'(x)$. Thus constant function $f(x) \triangleq a$ has derivative $f'(x) = 0$ at every x because its value does not change with x.

Many familiar functions have easily expressed derivatives (a constant):

$f(x)$	$\dfrac{df}{dx}$	$f(x)$	$\dfrac{df}{dx}$	$f(x)$	$\dfrac{df}{dx}$
ax	ax	x^a	ax^{a-1}	$\sin(ax)$	$a\cos(ax)$
a^x	$a^x\ln(a)$	$\ln(ax)$	$\frac{1}{x}$	$\cos(ax)$	$-a\sin(ax)$

Also, a variety of computing formulas determine the derivative of a function $f(x)$ in terms of simpler functions g and h forming f:

$f(x)$	$\dfrac{df}{dx}$	$f(x)$	$\dfrac{df}{dx}$
$g(h(x))$	$\dfrac{dg}{dh} \cdot \dfrac{dh}{dx}$	$g(x) \cdot h(x)$	$g(x)\dfrac{dh}{dx} + h(x)\dfrac{dg}{dx}$
$g(x) \pm h(x)$	$\dfrac{dg}{dx} \pm \dfrac{dh}{dx}$	$\dfrac{g(x)}{h(x)}$	$\left[h(x)\dfrac{dg}{dx} - g(x)\dfrac{dh}{dx}\right] / h(x)^2$

For example, the derivative of $f(x) \triangleq (3x)^4$ can be computed as $dg/dh \cdot dh/dx = 4(-3x)^3(-3)$ by taking $g(h) \triangleq h^4$ and $h(x) \triangleq 3x$. Similarly, the derivative with respect to x of $f(x) \triangleq (4x)(e^x)$ can be computed as $(4x)(e^x)(1) + (e^x)(4)$ by thinking of f as the product of functions $g(x) \triangleq 4x$ and $h(x) \triangleq e^x$.

When a function has more than one argument, **partial derivatives** show rates of change with respect to single variables with all others held constant. The partial derivatives of $f(x_1, x_2, \ldots, x_n)$ are usually denoted $\partial f/\partial x_i, i = 1, 2, \ldots, n$. To illustrate, consider $f(x_1, x_2, x_3) \triangleq (x_1)^5(x_2)^7(x_3)$. Differentiating with x_2 and x_3 treated as constants produces the partial derivative $\partial f/\partial x_1 = 5(x_1)^4(x_2)^7(x_3)$. Similarly, $\partial f/\partial x_2 = 7(x_1)^5(x_2)^6(x_3)$ and $\partial f/\partial x_3 = (x_1)^5(x_2)^7$.

Figure 3.10 shows its now familiar contours. Differentiating yields

$$\nabla p(x_1, x_2) \triangleq \begin{pmatrix} \dfrac{\partial p}{\partial x_1} \\[2mm] \dfrac{\partial p}{\partial x_2} \end{pmatrix} =$$

(3.3)

$$\begin{pmatrix} -\dfrac{120(x_1 + 1)}{[1 + (x_1 + 1)^2 + (x_2 - 3)^2]^2} - \dfrac{40(x_1 - 1)}{[1 + (x_1 - 1)^2 + (x_2 - 3)^2]^2} - \dfrac{60(x_1)}{[1 + (x_1)^2 + (x_2 + 4)^2]^2} \\[4mm] -\dfrac{120(x_2 - 3)}{[1 + (x_1 + 1)^2 + (x_2 - 3)^2]^2} - \dfrac{40(x_2 - 3)}{[1 + (x_1 - 1)^2 + (x_2 - 3)^2])^2} - \dfrac{60(x_2 + 4)}{[1 + (x_1)^2 + (x_2 + 4)^2]^2} \end{pmatrix}$$

Thus at point $\mathbf{x} = (2, 0)$,

$$\nabla p(2, 0) \approx (-1.60, 1.45)$$

Thus in the neighborhood of $\mathbf{x} = (2, 0)$, function p declines at the rate of about -1.60 per unit increase in x_1 and grows at the rate of roughly 1.45 per unit increase in x_2. Of course, these rates may change if we move away from $(2, 0)$, but they provide fairly complete information at points in the immediate neighborhood.

Figure 3.10 also illustrates the geometry of gradient vectors.

3.20 | Gradients show graphically as vectors perpendicular to contours of the objective function and point in the direction of most rapid objective value increase.

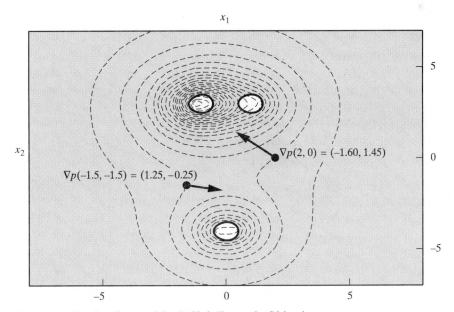

FIGURE 3.10   Gradients of the DClub Example Objective

For example, at $\mathbf{x} = (2, 0)$ the objective increases fastest by moving at right angles to the contour in direction $\mathbf{\Delta x} = (-1.60, 1.45)$, which is precisely $\nabla p(2, 0)$.

Gradient Conditions for Improving Directions

Suppose that our search of objective function f has arrived at current solution $\mathbf{x}$. Then the change associated with a step of size λ in direction $\mathbf{\Delta x}$ can be approximated as[1]

$$\text{objective change} \approx \sum_j \left(\frac{\partial f}{\partial x_j} \right) (\lambda \, \Delta x_j) = \lambda \, (\nabla f(\mathbf{x}) \, \cdot \, \mathbf{\Delta x})$$

That is, the rate of objective function change near the current $\mathbf{x}$ is roughly the dot product of $\nabla f(\mathbf{x})$ and $\mathbf{\Delta x}$ because it is approximated by the weighted sum of rates of change described by partial derivatives times move components in the various coordinate directions.

When dot product $\nabla f \cdot \mathbf{\Delta x} \neq 0$, this gradient information provides a simple algebraic test of whether a direction fulfills definition 3.13 as improving:

3.21 | Direction $\mathbf{\Delta x}$ is improving for maximize objective function f at point $\mathbf{x}$ if $\nabla f(\mathbf{x}) \cdot \mathbf{\Delta x} > 0$. On the other hand, if $\nabla f(\mathbf{x}) \cdot \mathbf{\Delta x} < 0$, $\mathbf{\Delta x}$ does not improve at $\mathbf{x}$.

3.22 | Direction $\mathbf{\Delta x}$ is improving for minimize objective function f at point $\mathbf{x}$ if $\nabla f(\mathbf{x}) \cdot \mathbf{\Delta x} < 0$. On the other hand, if $\nabla f(\mathbf{x}) \cdot \mathbf{\Delta x} > 0$, $\mathbf{\Delta x}$ does not improve at $\mathbf{x}$.

Cases where dot product $\nabla f(\mathbf{x}) \cdot \mathbf{\Delta x} = 0$ cannot be resolved without further information.

To illustrate, consider again the patronage function of Figure 3.10. We have already computed the gradient at $\mathbf{x} = (2, 0)$ as $\nabla p(2, 0) \approx (-1.60, 1.45)$. A glance at the figure shows that $\mathbf{\Delta x} = (-1, 1)$ is an improving direction because it leads us to higher contours. Confirming characterization 3.21 , we have

$$\nabla p(0, 2) \cdot \mathbf{\Delta x} \approx (-1.60, 1.45) \cdot (-1, 1) = 3.05 > 0$$

On the other hand, Figure 3.10 reports that evaluation of gradient expression (3.3) at $\mathbf{x} = (-1.5, -1.5)$ yields $\nabla p(-1.5, -1.5) \approx (1.25, -0.25)$. Testing the same direction $\mathbf{\Delta x} = (-1, 1)$, we see that

$$\nabla p(-1.5, -1.5) \cdot \mathbf{\Delta x} \approx (1.25, -0.25) \cdot (-1, 1) = -1.50 < 0$$

which verifies that direction $\mathbf{\Delta x} = (-1, 1)$ does not improve at $(-1.5, -1.5)$.

---

[1] This Taylor series approximation is developed more formally in Section 13.3.

SAMPLE EXERCISE 3.9: USING GRADIENTS TO DETERMINE IF DIRECTIONS IMPROVE

Either determine by an appropriate gradient test whether each of the following directions is improving for the specified objective function and point or show why further information is required.

(a) $\Delta\mathbf{w} = (1, 0, -2)$ for minimize $f(\mathbf{w}) \triangleq (w_1)^2 + 5w_2w_3$ at $\mathbf{w} = (2, 1, 0)$.

(b) $\Delta\mathbf{y} = (3, -6)$ for maximize $f(\mathbf{y}) \triangleq 9y_1 + 40y_2$ at $\mathbf{y} = (13, 2)$.

(c) $\Delta\mathbf{z} = (-6, 2)$ for minimize $f(\mathbf{z}) \triangleq 5(z_1)^2 - 3z_1z_2 + (z_2)^2$ at $\mathbf{z} = (1, 3)$.

Analysis:

(a) Computing the objective function gradient at the indicated $\mathbf{w}$ yields

$$\nabla f(\mathbf{w}) = \begin{pmatrix} \dfrac{\partial f}{\partial w_1} \\[2mm] \dfrac{\partial f}{\partial w_2} \\[2mm] \dfrac{\partial f}{\partial w_3} \end{pmatrix} = \begin{pmatrix} 2w_1 \\ 5w_3 \\ 5w_2 \end{pmatrix} = \begin{pmatrix} 2(2) \\ 5(0) \\ 5(1) \end{pmatrix} = \begin{pmatrix} 4 \\ 0 \\ 5 \end{pmatrix}$$

so that

$$\nabla f(\mathbf{w}) \cdot \Delta\mathbf{w} = (4, 0, 5) \cdot (1, 0, -2) = -6 < 0$$

By applying condition $\boxed{3.22}$, we conclude that $\Delta\mathbf{w}$ does improve at $\mathbf{w}$ for the minimize objective.

(b) Computing the objective function gradient at the $\mathbf{y}$ indicated gives

$$\nabla f(\mathbf{y}) = \begin{pmatrix} \dfrac{\partial f}{\partial y_1} \\[2mm] \dfrac{\partial f}{\partial y_2} \end{pmatrix} = \begin{pmatrix} 9 \\ 40 \end{pmatrix}$$

so that

$$\nabla f(\mathbf{y}) \cdot \Delta\mathbf{y} = (9, 40) \cdot (3, -6) = -213 < 0$$

Thus by condition $\boxed{3.21}$ the direction does not improve for the maximize objective.

(c) Computing partial derivatives of the objective function at the indicated $\mathbf{z}$, we have

$$\nabla f(\mathbf{z}) = \begin{pmatrix} \dfrac{\partial f}{z_1} \\[2mm] \dfrac{\partial f}{z_2} \end{pmatrix} = \begin{pmatrix} 10z_1 - 3z_2 \\ -3z_1 + 2z_2 \end{pmatrix} = \begin{pmatrix} 10(1) - 3(3) \\ -3(1) + 2(3) \end{pmatrix} = \begin{pmatrix} 1 \\ 3 \end{pmatrix}$$

and

$$\nabla f(\mathbf{z}) \cdot \Delta\mathbf{z} = (1, 3) \cdot (-6, 2) = 0$$

With the dot product $= 0$, conditions $\boxed{3.21}$ and $\boxed{3.22}$ are insufficient to determine whether this $\Delta\mathbf{z}$ improves.

Objective Function Gradients as Move Directions

One consequence of dot product tests $\boxed{3.21}$ and $\boxed{3.22}$ is that we can derive improving directions directly from any nonzero gradient (although we will see in Chapters 13 and 14 that gradients are not always the best directions to choose). Since the dot product of a nonzero gradient with itself is

$$\nabla f(\mathbf{x}) \cdot \nabla f(\mathbf{x}) = \sum_j \left(\frac{\partial f}{\partial x_j} \right)^2 > 0$$

we need only choose $\Delta \mathbf{x} = \pm \nabla f(\mathbf{x})$.

$\boxed{3.23}$ When objective function gradient $\nabla f(\mathbf{x}) \neq \mathbf{0}$, $\Delta \mathbf{x} = \nabla f(\mathbf{x})$ is an improving direction for a maximize objective f, and $\Delta \mathbf{x} = -\nabla f(\mathbf{x})$ is an improving direction for minimizing f.

Contours confirm that both of the gradients displayed for the maximize model in Figure 3.10 are indeed improving directions. We can verify the first algebraically by choosing

$$\Delta \mathbf{x} = \nabla p(0, 2) = (-1.60, 1.45)$$

then (principle $\boxed{3.21}$) the dot product

$$\nabla p(0, 2) \cdot \Delta \mathbf{x} = (-1.60, 1.45) \cdot (-1.60, 1.45) = (-1.60)^2 + (1.45)^2 > 0$$

establishes that $\Delta \mathbf{x}$ improves at $\mathbf{x} = (0, 2)$.

SAMPLE EXERCISE 3.10: CONSTRUCTING IMPROVING DIRECTIONS FROM GRADIENTS

Use the gradient of each of the following objective functions f to construct an improving direction at the indicated point.

(a) Minimize $f(\mathbf{w}) \triangleq (w_1)^2 \ln(w_2)$ at $\mathbf{w} = (5, 2)$.

(b) Maximize $f(\mathbf{y}) \triangleq 4y_1 + 5y_2 - 8y_3$ at $\mathbf{y} = (2, 0, 0.5)$.

Analysis:

(a) Computing the gradient at the indicated point yields

$$\nabla f(\mathbf{w}) \triangleq \begin{pmatrix} \dfrac{\partial f}{\partial w_1} \\ \dfrac{\partial f}{\partial w_2} \end{pmatrix} = \begin{pmatrix} 2w_1 \ln(w_2) \\ (w_1)^2 \\ \overline{w_2} \end{pmatrix} = \begin{pmatrix} 2(5) \ln(2) \\ \dfrac{(5)^2}{2} \end{pmatrix} \approx \begin{pmatrix} 6.93 \\ 12.5 \end{pmatrix}$$

Since this gradient is nonzero, we may apply construction $\boxed{3.23}$. Negative gradient

$$\Delta \mathbf{w} = -\nabla f(\mathbf{w}) = \begin{pmatrix} -6.93 \\ -12.5 \end{pmatrix}$$

must improve for the minimize objective.

(b) Computing the gradient at the indicated point yields

$$\nabla f(\mathbf{y}) \triangleq \begin{pmatrix} \dfrac{\partial f}{\partial y_1} \\[6pt] \dfrac{\partial f}{\partial y_2} \\[6pt] \dfrac{\partial f}{\partial y_3} \end{pmatrix} = \begin{pmatrix} 4 \\ 5 \\ -8 \end{pmatrix}$$

Since this gradient is nonzero, we may again apply construction $\boxed{3.23}$. Gradient

$$\Delta \mathbf{y} = \nabla f(\mathbf{y}) = \begin{pmatrix} 4 \\ 5 \\ -8 \end{pmatrix}$$

will improve for the maximize objective.

Active Constraints and Feasible Directions

Turning now to conditions for feasible directions (definition $\boxed{3.14}$), we focus our attention on constraints. Constraints define the boundary of the feasible region for an optimization model, so they will be the source of conditions for feasible directions.

Figure 3.11's repeat of our familiar Two Crude example demonstrates that not all the constraints of a model are relevant to whether a direction $\Delta\mathbf{x}$ is feasible at a particular solution $\mathbf{x}$. For example, at $\mathbf{x}^{(1)} = (7, 0)$ only

$$x_2 \geq 0$$

poses a threat to feasibility; small movements from $\mathbf{x}^{(1)}$ cannot violate other constraints. At $\mathbf{x}^{(3)} = (0.75, 6)$ the relevant constraints are different:

$$x_2 \leq 6 \quad \text{and} \quad 0.4x_1 + 0.2x_2 \geq 1.5$$

$\boxed{3.24}$ Whether a direction is feasible at a solution $\mathbf{x}$ depends on whether it would lead to immediate violation of any **active constraint** at $\mathbf{x}$, i.e., any constraint satisfied as equality at $\mathbf{x}$.

Other names for active constraints are **tight constraints** and **binding constraints**.

Equality constraints are active at every feasible point, because they are always satisfied as equalities. Inequalities are more complex. For example, the Two Crude constraint

$$0.4x_1 + 0.2x_2 \geq 1.5$$

is satisfied, but not active at $\mathbf{x}^{(1)} = (7, 0)$ of Figure 3.11 because

$$0.4(7) + 0.2(0) = 2.8 > 1.5$$

However, the same inequality is active at $\mathbf{x}^* = (2, 3.5)$. There

$$0.4(2) + 0.2(3.5) = 1.5$$

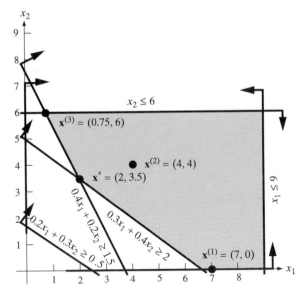

FIGURE 3.11   Active Constraints of the Two Crude Example

SAMPLE EXERCISE 3.11: RECOGNIZING ACTIVE CONSTRAINTS

Consider an optimization model with main constraints

$$(w_1 - 4)^2 + 3w_2 - 7w_3 \geq 31 \tag{3.4}$$

$$2w_1 + w_3 \leq 13 \tag{3.5}$$

$$w_1 + w_2 + w_3 = 25 \tag{3.6}$$

Determine which of these constraints are active at feasible solutions
(a) $\mathbf{w} = (8, 20, -3)$ (b) $\mathbf{w} = (5, 17, 3)$

Analysis: Equality (3.6) must be active at every feasible solution.

(a) Checking definition $\boxed{3.24}$ for inequalities (3.4) and (3.5) at $\mathbf{w} = (8, 20, -3)$, we have

$$(8 - 4)^2 + 3(20) - 7(-3) = 97 > 31$$

and

$$2(8) + (-3) = 13$$

Thus only (3.5) and (3.6) are active at $\mathbf{w} = (8, 20, -3)$.

(b) Checking definition $\boxed{3.24}$ for inequalities (3.4) and (3.5) at $\mathbf{w} = (5, 17, 3)$, we find that

$$(5 - 4)^2 + 3(17) - 7(3) = 31$$

and

$$2(5) + (3) = 13$$

Thus all three constraints are active at $\mathbf{w} = (5, 17, 3)$.

Linear Constraints

Constraints of the Two Crude example are all **linear**, that is, inequalities or equalities constraining a weighted sum of the decision variables by a right-hand-side constant (definition 2.28). We may denote general forms by

$$\mathbf{a} \cdot \mathbf{x} \triangleq \sum_{j=1}^{n} a_j x_j \geq b \tag{3.7}$$

$$\mathbf{a} \cdot \mathbf{x} \triangleq \sum_{j=1}^{n} a_j x_j \leq b \tag{3.8}$$

$$\mathbf{a} \cdot \mathbf{x} \triangleq \sum_{j=1}^{n} a_j x_j = b \tag{3.9}$$

where n denotes the number of decision variables, a_j is the constraint's coefficient for decision variable x_j, $\mathbf{a}$ is the vector of coefficients a_j, and b is the constraint's right-hand side.

One example is the first main Two Crude constraint,

$$0.3x_1 + 0.4x_2 \geq 2$$

In the notation of (3.7),

$$n = 2, \quad a_1 = 0.3, \quad a_2 = 0.4, \quad \mathbf{a} = (0.3, 0.4), \quad \text{and} \quad b = 2$$

Conditions for Feasible Directions with Linear Constraints

Conditions can be devised to characterize feasible directions in more complex cases (see Section 14.4), but we focus here on linear forms (3.7) to (3.9). Consider the Two Crude model constraint

$$0.4x_1 + 0.2x_2 \geq 1.5$$

At solution $\mathbf{x}^{(3)} = (0.75, 6)$ the constraint is active because

$$0.4(0.75) + 0.2(6) = 1.5$$

A step in move direction $\Delta \mathbf{x} \triangleq (\Delta x_1, \Delta x_2)$ will change the left-hand side as follows:

$$0.4(0.75 + \lambda \, \Delta x_1) + 0.2(6 + \lambda \, \Delta x_2) = 1.5 + \lambda(0.4 \, \Delta x_1 + 0.2 \, \Delta x_2)$$

Feasibility limit

$$1.5 + \lambda(0.4 \, \Delta x_1 + 0.2 \, \Delta x_2) \geq 1.5$$

can be maintained only if we restrict the sign "net change" coefficient of λ by requiring

$$\sum_{j=1}^{n} a_j \, \Delta x_j = 0.4 \, \Delta x_1 + 0.2 \, \Delta x_2 \geq 0$$

This sort of analysis leads immediately to general conditions for directions feasible to linear constraints:

3.25 | Direction $\Delta\mathbf{x} \triangleq (\Delta x_1, \ldots, \Delta x_n)$ is feasible for a linearly constrained optimization model at solution $\mathbf{x} \triangleq (x_1, \ldots, x_n)$ if and only if

$$\mathbf{a} \cdot \Delta\mathbf{x} \triangleq \sum_{i=1}^{n} a_j \Delta x_j \geq 0$$

for all active greater than or equal to constraints $\sum_j a_j x_j \geq b$;

$$\mathbf{a} \cdot \Delta\mathbf{x} \triangleq \sum_{i=1}^{n} a_j \Delta x_j \leq 0$$

for all active less than or equal to constraints $\sum_j a_j x_j \leq b$; and

$$\mathbf{a} \cdot \Delta\mathbf{x} \triangleq \sum_{i=1}^{n} a_j \Delta x_j = 0$$

for all equality constraints $\sum_j a_j x_j = b$.

To illustrate, return to Figure 3.11. At $\mathbf{x}^{(2)} = (4, 4)$, no constraints are active, and every direction is feasible.

At $\mathbf{x}^{(3)} = (0.75, 6)$, there are two active constraints:

$$0.4x_1 + 0.2x_2 \geq 1.5$$
$$x_2 \leq 6$$

If direction $\Delta\mathbf{x}$ is to be feasible, it must violate neither. Thus principle 3.25 yields one condition for an active $\geq$ form and one for an active $\leq$:

$$0.4\ \Delta x_1 + 0.2\ \Delta x_2 \geq 0$$
$$\Delta x_2 \leq 0$$

SAMPLE EXERCISE 3.12: FORMING CONDITIONS FOR A DIRECTION TO BE FEASIBLE

Consider an optimization model with linear constraints

$$3w_1 + w_3 \geq 26 \tag{3.10}$$
$$5w_1 - 2w_3 \leq 50 \tag{3.11}$$
$$2w_1 + w_2 + w_3 = 20 \tag{3.12}$$
$$w_1 \geq 0 \tag{3.13}$$
$$w_2 \geq 0 \tag{3.14}$$

State all conditions that must be satisfied for $\Delta\mathbf{w}$ to be a feasible move direction at $\mathbf{w} = (10, 0, 0)$.

Analysis: When we substitute the solution $\mathbf{w} = (10, 0, 0)$, the active constraints are (3.11), (3.12), and (3.14). Thus, applying conditions 3.25, the corresponding requirements for a feasible direction are

$$5\,\Delta w_1 - 2\,\Delta w_3 \leq 0$$
$$2\,\Delta w_1 + \Delta w_2 + \Delta w_3 = 0$$
$$\Delta w_2 \geq 0$$

> **SAMPLE EXERCISE 3.13: TESTING DIRECTIONS FOR FEASIBILITY**
>
> Return to the feasible region defined by constraints (3.10) to (3.14) and determine whether direction $\Delta\mathbf{w} = (0, -1, 1)$ is feasible at point $\mathbf{w} = (6, 0, 8)$.
>
> **Analysis:** At $\mathbf{w} = (6, 0, 8)$, the active constraints are (3.10), (3.12), and (3.14). Thus required conditions 3.25 for a feasible direction are
> $$3\,\Delta w_1 + \Delta w_3 \geq 0$$
> $$2\,\Delta w_1 + \Delta w_2 + \Delta w_3 = 0$$
> $$\Delta w_2 \geq 0$$
> Direction $\Delta\mathbf{w} = (0, -1, 1)$ meets the first two conditions because
> $$3\,\Delta w_1 + \Delta w_3 = 3(0) + (1) \geq 0$$
> $$2\,\Delta w_1 + \Delta w_2 + \Delta w_3 = 2(0) + (-1) + (1) = 0$$
> Still, the direction is not feasible because it violates the third condition,
> $$\Delta w_2 = (-1) \not\geq 0$$

3.4   UNIMODAL AND CONVEX MODEL FORMS TRACTABLE FOR IMPROVING SEARCH

Tractability in a model means convenience for analysis (definition 1.6). Sometimes difficult forms cannot be avoided, but in Chapter 1 (principle 1.7) we saw that modeling often involves tradeoffs. To obtain a model that is tractable enough to yield useful insights, we sometimes have the choice of making simplifying assumptions or other compromises.

What forms are best for improving search? In this subsection we introduce the classic unimodal objective function, convex feasible set cases we would always prefer when there is any option.

Tractability and Local Optima

Many issues might be raised in evaluating the tractability of model forms to improving search, but one stands out. The models considered most tractable to improving search are ones where every local optimum is necessarily global (principle 3.10).

Section 3.1 showed why. Improving searches stop upon encountering a local optimum because no better solution exists in the immediate neighborhood (principle 3.6). If the mathematical form of the model guarantees that every local optimum is a global optimum, we can pursue an improving search without being much concerned about stopping at less than the truly optimal solution we desire.

Unimodal Objective Functions

Local optima present difficulties only when they are not global (i.e., when there exist other solutions with strictly superior objective function values). Such cases cannot happen if every solution improvable in this way admits an improving direction under definition 3.13 . That is precisely the case when the objective function is unimodal.

> **3.26** An objective function $f(\mathbf{x})$ is **unimodal** if the straightline direction from every point in its domain to every better point is an improving direction. That is, for every $\mathbf{x}^{(1)}$ and every $\mathbf{x}^{(2)}$ with a better objective function value, direction $\Delta\mathbf{x} = (\mathbf{x}^{(2)} - \mathbf{x}^{(1)})$ should be improving at $\mathbf{x}^{(1)}$.

The DClub objective function of Figure 3.10 illustrates a case that fails the "one-hump" idea implicit in the name *unimodal*. To demonstrate that $p(x_1, x_2)$ is not unimodal, we must find two solutions with different objective function values such that the direct path from the first to the second does not satisfy the definition of an improving direction. The obvious choices are the local maxima labeled $\mathbf{x}^{(1)} = (0, -4)$ and $\mathbf{x}^{(2)} = (-1, 3)$. The straight-line move direction

$$\Delta\mathbf{x} = \mathbf{x}^{(2)} - \mathbf{x}^{(1)} = (-1, 3) - (0, -4) = (-1, 7)$$

is not improving at $\mathbf{x}^{(1)}$ because the objective function declines for small steps along $\Delta\mathbf{x}$.

Figure 3.12 graphs the more tractable objective function

$$\min\ f(y_1, y_2) \triangleq (y_1 - 2)^2 - (y_1 - 2)(y_2 - 3) + (y_2 - 3)^2 + 10$$

This case is unimodal because the direct path from any $\mathbf{y}^{(1)}$ to any $\mathbf{y}^{(2)}$ with a lower (better) objective value is an improving direction.

SAMPLE EXERCISE 3.14: RECOGNIZING UNIMODAL OBJECTIVE FUNCTIONS

Determine graphically whether each of the following 1-dimensional objective functions $f(w)$ is unimodal if f is maximized, and whether it is unimodal if f is minimized.

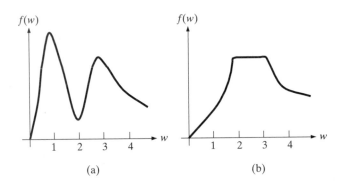

(a)   (b)

Analysis: With a single decision variable, the only search directions are movement to the left and movement to the right.

(a) This function is not unimodal for either a maximization or a minimization. We can demonstrate this for maximizing $f(w)$ by picking $w^{(1)} = 3$ and $w^{(2)} = 1$. Objective value $f(w^{(2)}) > f(w^{(1)})$, but the (leftward) move direction from $w^{(1)}$ to $w^{(2)}$ does not improve immediately at $w^{(1)}$. For minimizing $f(w)$ we choose $w^{(1)} = 2$ and $w^{(2)} = 0$. Again definition 3.26 is violated.

(b) Despite the fact that this $f(w)$ is constant from $w = 2$ to $w = 3$, it is unimodal for a maximization. The direct move left or right to a strictly better point always produces

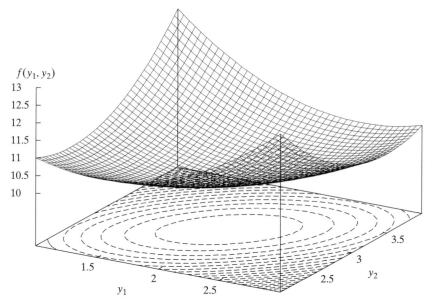

FIGURE 3.12 Example of a Unimodal Minimize Objective

immediate improvement. If we are minimizing $f(w)$, case (b) is not unimodal. Point pairs that fail definition $\boxed{3.26}$ include $w^{(1)} = 4$ and $w^{(2)} = 0$.

Linear Objective Functions

Recall that a **linear objective function** is one expressible as a weighted sum of the decision variables (definition $\boxed{2.28}$). The general form is

$$f(\mathbf{x}) \triangleq \sum_{j=1}^{n} c_j x_j = \mathbf{c} \cdot \mathbf{x} \tag{3.15}$$

with c_j the weight applied to decision variable x_j. For instance, linear objective

$$f(x_1, x_2, x_3) \triangleq 3.5x_1 - 2x_2 + x_3$$

has weights $c_1 = 3.5$, $c_2 = -2$, and $c_3 = 1$.

Models with linear objective functions are special; one reason is unimodality.

$\boxed{3.27}$ Linear objective functions are unimodal in both maximize and minimize optimization models.

To see why this is true, consider general form (3.15) and any two solutions $\mathbf{x}^{(1)}$ and $\mathbf{x}^{(2)}$.

$$f(\mathbf{x}^{(2)}) - f(\mathbf{x}^{(1)}) = \sum_{j=1}^{n} c_j x_j^{(2)} - \sum_{j=1}^{n} c_j x_j^{(1)}$$

$$= \sum_{j=1}^{n} c_j \left(x_j^{(2)} - x_j^{(1)} \right)$$

$$= \mathbf{c} \cdot \left(\mathbf{x}^{(2)} - \mathbf{x}^{(1)} \right)$$

$$= \nabla f(\mathbf{x}^{(1)}) \cdot \left(\mathbf{x}^{(2)} - \mathbf{x}^{(1)} \right)$$

[the last because the gradient of a linear function (3.15) is just its coefficient vector **c**].

Now we apply Section 3.3's gradient tests $\boxed{3.21}$ and $\boxed{3.22}$ for improving directions. In the maximize case, $f(\mathbf{x}^{(2)}) > f(\mathbf{x}^{(1)})$ implies from the algebra above that

$$f(\mathbf{x}^{(2)}) - f(\mathbf{x}^{(1)}) = \nabla f(\mathbf{x}^{(1)}) \left(\mathbf{x}^{(2)} - \mathbf{x}^{(1)} \right) > 0$$

This shows that $\Delta\mathbf{x} = (\mathbf{x}^{(2)} - \mathbf{x}^{(1)})$ improves at $\mathbf{x}^{(1)}$ as required in unimodal definition $\boxed{3.26}$ because gradient condition $\boxed{3.21}$ is satisfied. An identical argument holds for minimize linear objectives.

Unimodal Objective Functions and Unconstrained Local Optima

Unconstrained local optima are solutions for which no point in some surrounding neighborhood has a better objective function value. That is, they are local optima caused entirely by the objective function. An **unconstrained global optimum** is a solution yielding a better objective value than any other in the domain of the objective function. Point $\mathbf{y}^* = (2, 3)$ in Figure 3.12 is both an unconstrained local minimum and an unconstrained global minimum.

Unimodal objectives are tractable because they avoid unconstrained local optima that are not global.

$\boxed{3.28}$ If the objective function of an optimization model is unimodal, every unconstrained local optimum is an unconstrained global optimum.

This valuable property follows directly from unimodal definition $\boxed{3.26}$. For any $\mathbf{x}^{(1)}$ not an unconstrained global optimum there exists an $\mathbf{x}^{(2)}$ with better objective function value. Then unimodality implies that direction $(\mathbf{x}^{(2)} - \mathbf{x}^{(1)})$ improves at $\mathbf{x}^{(1)}$, which shows that $\mathbf{x}^{(1)}$ could not form an unconstrained local optimum (principle $\boxed{3.16}$). Every unconstrained local optimum of a unimodal objective function must be global.

Constraints and Local Optima

Objective functions are not the only parts of mathematical programming models that can induce local optima. For example, the only unconstrained local maximum in the part of the DClub feasible space displayed in Figure 3.13 occurs at unconstrained global optimum $\mathbf{x} = (-1, 3)$.

Solutions $\mathbf{x}^{(1)} = (-1.5, 3)$ and $\mathbf{x}^{(2)} = (1.5, 3)$ illustrate a new, constraint-induced form of local optimum. Both are (constrained) local maxima in the full model that do not achieve as high an objective value as global maximum $\mathbf{x}^* = (-0.5, 3)$. The difficulty is not the shape of the objective function, because improving directions $\Delta\mathbf{x} = (\mathbf{x}^* - \mathbf{x}^{(1)})$ at $\mathbf{x}^{(1)}$ and $\Delta\mathbf{x} = (\mathbf{x}^* - \mathbf{x}^{(2)})$ at $\mathbf{x}^{(2)}$ lead straight to the global optimum. The difficulty is constraints. No neighboring point at either $\mathbf{x}^{(1)}$ or $\mathbf{x}^{(2)}$ both improves the objective function and satisfies all constraints.

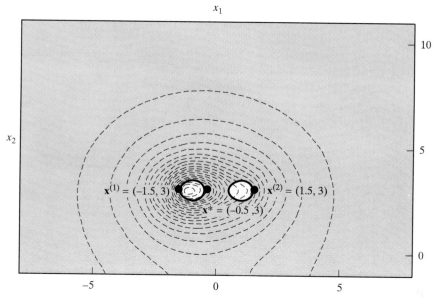

FIGURE 3.13 Constraint-Induced Local Optima for the DClub Example

Convex Feasible Sets

Convex feasible sets avoid such difficulties.

> **3.29** The feasible set of an optimization problem is **convex** if the line segment between every pair of feasible points falls entirely within the feasible region.

The idea is illustrated by the following feasible set:

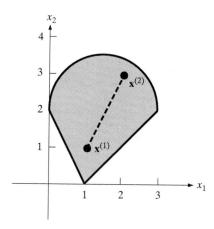

This set is convex because the line segment between every pair of feasible points lies entirely within the feasible set. The segment for $\mathbf{x}^{(1)} = (1, 1)$ and $\mathbf{x}^{(2)} = (2, 3)$ is shown.

Contrast with the DClub feasible region of Figure 3.13. That set is not convex because some pairs of points fail definition $\boxed{3.29}$. One of the many is $\mathbf{x}^{(1)} = (-1.5, 3)$ and $\mathbf{x}^{(2)} = (1.5, 3)$. The line segment joining them includes infeasible solutions in both of the forbidden boxes around population centers.

A similar situation occurs with discrete feasible sets much as

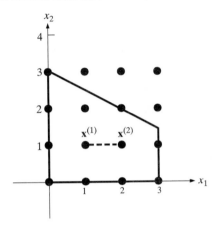

Here only the black integer points within the main constraints are feasible. Thus the line segment connecting any pair of feasible points passes primarily through noninteger points violating constraints.

$\boxed{3.30}$ Discrete feasible sets are never convex (except in the trivial case where there is only one feasible point).

SAMPLE EXERCISE 3.15: DEMONSTRATING NONCONVEXITY GRAPHICALLY

Show graphically that the following feasible region is not convex.

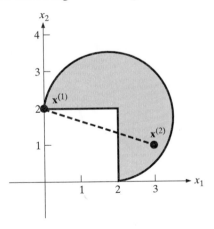

Analysis: To show that the set fails definition $\boxed{3.29}$, we need to find any pair of feasible solutions joined by a line segment that lies partly outside the feasible region.

Solutions $\mathbf{x}^{(1)} = (0, 2)$ and $\mathbf{x}^{(2)} = (3, 1)$ sketched in the figure suffice because the line segment between them clearly includes infeasible points.

Algebraic Description of Line Segments

What does it mean to have a line segment in more than two or three dimensions? There is an easy algebraic characterization. Since direction $\left(\mathbf{x}^{(2)} - \mathbf{x}^{(1)}\right)$ is the straight-line move from $\mathbf{x}^{(1)}$ to $\mathbf{x}^{(2)}$, the line segment between the two points consists exactly of those solutions obtained when we add a fraction of that direction to $\mathbf{x}^{(1)}$.

> **3.31** The line segment between vector solutions $\mathbf{x}^{(1)}$ and $\mathbf{x}^{(2)}$ consists of all points of the form $\mathbf{x}^{(1)} + \lambda \left(\mathbf{x}^{(2)} - \mathbf{x}^{(1)}\right)$ with $0 \le \lambda \le 1$.

As an example, suppose that $\mathbf{x}^{(1)} = (1, 5, 0)$ and $\mathbf{x}^{(2)} = (0, 1, 2)$ are both feasible. One end of the line segment connecting them is $\mathbf{x}^{(1)}$ with $\lambda = 0$ in $\boxed{3.31}$; the other end is $\mathbf{x}^{(2)}$ with $\lambda = 1$. In between lie all the points with $0 < \lambda < 1$. For instance, the one for $\lambda = 0.25$ is

$$\mathbf{x}^{(1)} + 0.25 \left(\mathbf{x}^{(2)} - \mathbf{x}^{(1)}\right) = \begin{pmatrix} 1 \\ 5 \\ 0 \end{pmatrix} + 0.25 \left[\begin{pmatrix} 0 \\ 1 \\ 2 \end{pmatrix} - \begin{pmatrix} 1 \\ 5 \\ 0 \end{pmatrix} \right] = \begin{pmatrix} 0.75 \\ 4 \\ 0.5 \end{pmatrix}$$

If the feasible set is to be convex, $(0.75, 4, 0.5)$ and all other solutions along the line segment must satisfy all constraints.

SAMPLE EXERCISE 3.16: REPRESENTING LINE SEGMENTS

Return to the line segment joining $\mathbf{x}^{(1)} = (0, 2)$ and $\mathbf{x}^{(2)} = (3, 1)$ in Sample Exercise 3.15.

(a) Represent the line segment algebraically.

(b) Show algebraically that one point on the line segment is $\mathbf{x} = (1, \frac{5}{3})$.

Analysis:

(a) Applying $\boxed{3.31}$, the vectors along that line segment are those representable as

$$\mathbf{x}^{(1)} + \lambda \left(\mathbf{x}^{(2)} - \mathbf{x}^{(1)}\right) = \begin{pmatrix} 0 \\ 2 \end{pmatrix} + \lambda \left[\begin{pmatrix} 3 \\ 1 \end{pmatrix} - \begin{pmatrix} 0 \\ 2 \end{pmatrix} \right] = \begin{pmatrix} 3\lambda \\ 2 - \lambda \end{pmatrix}$$

with $0 \le \lambda \le 1$.

(b) To represent $(1, \frac{5}{3})$, we must choose an appropriate λ. If first component $3\lambda = 1$, then $\lambda = \frac{1}{3}$, so that

$$(3\lambda, 2 - \lambda) = (3(\tfrac{1}{3}), 2 - (\tfrac{1}{3})) = (1, \tfrac{5}{3})$$

as required.

SAMPLE EXERCISE 3.17: DEMONSTRATING NONCONVEXITY ALGEBRAICALLY

Demonstrate algebraically that the feasible set excluding the unit circle with constraint

$$(w_1)^2 + (w_2)^2 \ge 1$$

is not convex.

Analysis: We must first pick a pair of feasible solutions joined by a line segment that passes through an infeasible point within the unit circle, say $\mathbf{w}^{(1)} = (-1, 0)$ and $\mathbf{w}^{(2)} = (1, 0)$. Under $\boxed{3.31}$, the line segment between these two points includes all vectors expressible as

$$\mathbf{w}^{(1)} + \lambda \left(\mathbf{w}^{(2)} - \mathbf{w}^{(1)} \right) = \begin{pmatrix} -1 \\ 0 \end{pmatrix} + \lambda \left[\begin{pmatrix} 1 \\ 0 \end{pmatrix} - \begin{pmatrix} -1 \\ 0 \end{pmatrix} \right] = \begin{pmatrix} -1 + 2\lambda \\ 0 \end{pmatrix}$$

with $0 \le \lambda \le 1$. Thus we have only to choose a λ corresponding to an infeasible point to complete the argument that the set is not convex. One is $\lambda = \frac{1}{2}$, which yields the vector $\mathbf{w} = (0, 0)$ in the expression above, which clearly violates the unit circle constraint.

Linear Constraints and Convexity

As with linear objective functions, linear constraints produce unusually tractable feasible sets.

$\boxed{3.32}$ If all constraints of an optimization model are linear (both main and variable-type), its feasible space is convex.

How can we be sure? Pick two solutions $\mathbf{x}^{(1)}$ and $\mathbf{x}^{(2)}$ feasible in an optimization model with linear constraints. For each model constraint of, say, the $\ge$ form (3.7), feasibility means

$$\sum_{j=1}^{n} a_j x_j^{(1)} \ge b \quad \text{and} \quad \sum_{j=1}^{n} a_j x_j^{(2)} \ge b \tag{3.16}$$

What about points along the line segment between $\mathbf{x}^{(1)}$ and $\mathbf{x}^{(2)}$? We know they are formed as in $\boxed{3.31}$ by some step λ with $0 < \lambda < 1$.

Choose any such λ. We may add $(1 - \lambda)$ times the first inequality of (3.16) to λ times the second, to obtain

$$(1 - \lambda) \sum_{j=1}^{n} a_j x_j^{(1)} + \lambda \sum_{j=1}^{n} a_j x_j^{(2)} \ge (1 - \lambda)b + \lambda b = b$$

Then a bit of regrouping gives

$$\sum_{j=1}^{n} a_j x_j^{(1)} - \lambda \sum_{j=1}^{n} a_j x_j^{(1)} + \lambda \sum_{j=1}^{n} a_j x_j^{(2)} \ge b$$

or

$$\sum_{j=1}^{n} a_j \left[x_j^{(1)} + \lambda(x_j^{(2)} - x_j^{(1)}) \right] \ge b$$

The last says that point $\mathbf{x}^{(1)} + \lambda(\mathbf{x}^{(2)} - \mathbf{x}^{(1)})$ satisfies the $\ge$ constraint. Since we could perform exactly the same computations for each of the other constraints, we may conclude that points along the line segment between $\mathbf{x}^{(1)}$ and $\mathbf{x}^{(2)}$ are all feasible—exactly what is required for the feasible set to be convex.

| **SAMPLE EXERCISE 3.18: SHOWING THAT LINEARLY CONSTRAINED SETS ARE CONVEX** |

Establish that the set of (w_1, w_2, w_3) satisfying

$$19w_1 + 3w_2 - w_3 \leq 14$$
$$w_1 \geq 0$$

is convex.

Analysis: Pick two arbitrary feasible solutions $\mathbf{w}^{(1)}$ and $\mathbf{w}^{(2)}$ satisfying both constraints. Then

$$19w_1^{(1)} + 3w_2^{(1)} - w_3^{(1)} \leq 14 \quad \text{and} \quad w_1^{(1)} \geq 0$$
$$19w_1^{(2)} + 3w_2^{(2)} - w_3^{(2)} \leq 14 \quad \text{and} \quad w_1^{(2)} \geq 0$$

We must show that any point along the line segment joining $\mathbf{w}^{(1)}$ and $\mathbf{w}^{(2)}$ also satisfies both constraints. Each such point corresponds to some λ strictly between 0 and 1 in representation $\boxed{3.31}$, so pick such a λ.

Multiplying the foregoing main inequality for $\mathbf{w}^{(1)}$ by $(1 - \lambda)$ and that of $\mathbf{w}^{(2)}$ by λ, we may conclude that

$$(1 - \lambda)\left(19w_1^{(1)} + 3w_2^{(1)} - w_3^{(1)}\right) + \lambda\left(19w_1^{(2)} + 3w_2^{(2)} - w_3^{(2)}\right)$$
$$\leq \quad (1 - \lambda)14 + (\lambda)14$$

Noting that the right-hand side is just 14 and regrouping the left, it follows that

$$19\left[w_1^{(1)} + \lambda(w_1^{(2)} - w_1^{(1)})\right] + 3\left[w_2^{(1)} + \lambda(w_2^{(2)} - w_2^{(1)})\right]$$
$$- \left[w_3^{(1)} + \lambda(w_3^{(2)} - w_3^{(1)})\right] \leq 14$$

As required, the point corresponding to λ on the line segment satisfies the constraint. For the second, nonnegativity constraint, the same computation yields

$$(1 - \lambda)w_1^{(1)} + \lambda(w_1^{(2)}) \geq (1 - \lambda)0 + (\lambda)0$$

or

$$w_1^{(1)} + \lambda\left(w_1^{(2)} - w_1^{(1)}\right) \geq 0$$

Thus the line segment point corresponding to λ also satisfies constraint $w_1 \geq 0$.

Convex Sets and Constraint Tractability

What attracts us to constraints producing convex feasible sets is the same thing that made unimodal objective functions desirable. Since the entire line segment between any pair of feasible points $\mathbf{x}^{(1)}$ and $\mathbf{x}^{(2)}$ lies within the feasible set, direct move $\Delta\mathbf{x} = (\mathbf{x}^{(2)} - \mathbf{x}^{(1)})$ must always be a feasible direction at $\mathbf{x}^{(1)}$.

| 3.33 | If the feasible set of an optimization model is convex, there is a feasible direction leading from any feasible solution to any other. |

Models for Which Local Optima Must Be Global

We are at last ready to characterize the most tractable optimization models for improving search—those that have no local optima that are not global. There can be no such local optima if improving search can move directly from any solution to any superior one; that is, if the straight-line path joining any solution and any better one must always be both an improving and a feasible direction. This is precisely what happens when optimizing unimodal objective functions over convex feasible sets.

> **3.34** | If the objective function of optimization model is unimodal and the constraints produce a convex feasible set, every local optimum of the model is a global optimum.

Optimization models satisfying these conditions can be truly well solved by improving search. Those that do not conform to 3.34 may be far from hopeless, but we can expect to have to settle for local optima that may not be global.

SAMPLE EXERCISE 3.19: RECOGNIZING MODELS WITH LOCAL OPTIMA GLOBAL

Use principle 3.34 to show that every local minimum of the following model is a global minimum.

$$\begin{aligned}
\min \quad & 8y_1 - 4y_3 \\
\text{s.t.} \quad & y_1 + 3y_2 + y_3 \geq 10 \\
& 0 \leq y_1 \leq 5 \\
& 0 \leq y_2 \leq 3 \\
& 0 \leq y_3 \leq 9
\end{aligned}$$

Analysis: The objective function is unimodal because it is linear (principle 3.27). Also, all constraints are linear, which makes the feasible set convex (principle 3.32). It follows from principle 3.34 that every local minimum is a global minimum.

3.5 SEARCHING FOR STARTING FEASIBLE SOLUTIONS

Up to now, all our discussion of improving search has envisioned moving from feasible solution to better feasible solution. But what if a first feasible solution is not readily available? Complex optimization problems, with thousands of constraints and variables, often have no obviously feasible solution. In fact, an early goal of analysis may be to determine whether any feasible solution exists. In this section we introduce the **two-phase** and **big-M** strategies, which deal with this startup issue.

Two Phase Method

Improving search can be adapted to deal with the absence of a starting feasible solution, simply by using improving search to find one. More precisely, optimization is effected in two phases. **Phase I** addresses an artificial problem with easier-to-satisfy constraints. Starting from some feasible solution for this artificial problem, we minimize how much the artificial solution violates constraints of the true problem. If the violation can be driven to $= 0$, a completely feasible solution is at hand; **Phase II** then performs a usual improving search starting from the Phase I result. If the (globally) optimal value in Phase I is not $\neq 0$, constraint violations cannot be completely eliminated; the true problem is infeasible. Algorithm 3B formalizes this two-phase method.

ALGORITHM 3B: TWO-PHASE IMPROVING SEARCH

Step 0: Artificial Model. Choose any convenient solution for the true model, and construct a corresponding Phase I model by adding (or subtracting) nonnegative artificial variables in each violated constraint.

Step 1: Phase I. Assign values to artificial variables to complete a starting feasible solution for the artificial model. Then begin at that solution and perform an improving search to minimize the sum of the artificial variables.

Step 2: Infeasibility. If Phase I search terminated with artificial sum $= 0$, proceed Step 3; the original model is feasible. If Phase I search terminated with a global minimum having artificial sum > 0, stop; the original model is infeasible. Otherwise, repeat Step 1 from a different starting solution.

Step 3: Phase II. Construct a starting feasible solution for the original model by deleting artificial components of the Phase I optimum. Then begin at that solution and perform an improving search to optimize the original objective function subject to original constraints.

Two Crude Model Example Revisited

The Two Crude model of Figure 3.11 provides a familiar example:

$$\begin{array}{ll}
\min & 20x_1 + 15x_2 \\
\text{s.t.} & 0.3x_1 + 0.4x_2 \geq 2 \\
& 0.4x_1 + 0.2x_2 \geq 1.5 \\
& 0.2x_1 + 0.3x_2 \geq 0.5 \\
& 0 \leq x_1 \leq 9, 0 \leq x_2 \leq 6
\end{array}$$

For this tiny case it is easy to find a starting feasible solution graphically or by trial and error. But for typical instances we employ a more formal procedure.

Algorithm 3B begins by picking arbitrarily some convenient values—feasible or not—for the true decision variables. Often, the choice is to make all decision variables $= 0$ as we will do here.

Our choice of $x_1 = x_2 = 0$ satisfies the bound constraints

$$0 \leq x_1 \leq 9 \quad \text{and} \quad 0 \leq x_2 \leq 6$$

Still, it violates all 3 main constraints.

Artificial Variables

We deal with constraints unsatisfied at our arbitrary starting solution by introducing artificial variables to absorb the infeasibility.

> **3.35** Phase I constraints are derived from those of the original model by considering each in relation to the starting solution chosen. Satisfied constraints simply become part of the Phase I model. Violated ones are augmented with a nonnegative **artificial variable** to permit artificial feasibility.

Including artificial variables x_3, x_4, and x_5 in the 3 constraints violated at $x_1 = x_2 = 0$ produces Phase I constraints

$$0.3x_1 + 0.4x_2 + x_3 \geq 2$$
$$0.4x_1 + 0.2x_2 + x_4 \geq 1.5$$
$$0.2x_1 + 0.3x_2 + x_5 \geq 0.5$$
$$0 \leq x_1 \leq 9, \ 0 \leq x_2 \leq 6$$
$$x_3, \ x_4, \ x_5 \geq 0$$

Notice that a different artificial variable was introduced in each constraint violated. It was added in each case because extra left-hand side is needed to satisfy the $\geq$ constraints at $x_1 = x_2 = 0$.

Occasionally, it is appropriate to subtract the artificial. For example, if the model had included a constraint

$$x_1 - x_2 = -10$$

the Phase I construction with artificial x_6 is

$$x_1 - x_2 - x_6 = -10$$

Then for $x_1 = x_2 = 0$, $x_6 = 10$ produces feasibility. Had we used a $+$ sign on x_6, the corresponding solution would require x_6 negative, a violation of nonnegativity.

Phase I Models

Artificial variables provide a way to get started in Phase I, but our goal is to drive out all the infeasibility, that is, to find a solution feasible with artificial variables all zero. This defines the Phase I objective function.

> **3.36** The **Phase I objective function** minimizes the sum of the artificial variables.

Minimizing the total minimizes each artificial because all are required to be nonnegative.

In our Two Crude example, the resulting Phase I model is

$$
\begin{aligned}
\min \quad & x_3 + x_4 + x_5 \\
\text{s.t.} \quad & 0.3x_1 + 0.4x_2 + x_3 \geq 2 \\
& 0.4x_1 + 0.2x_2 + x_4 \geq 1.5 \\
& 0.2x_1 + 0.3x_2 + x_5 \geq 0.5 \\
& 0 \leq x_1 \leq 9, \ 0 \leq x_2 \leq 6 \\
& x_3, \ x_4, \ x_5 \geq 0
\end{aligned}
\tag{3.17}
$$

SAMPLE EXERCISE 3.20: CONSTRUCTING A PHASE I MODEL

Consider the optimization model

$$\max \quad 14(w_1 - 10)^2 + (w_2 - 3)^2 + (w_3 + 5)^2$$

$$\text{s.t.} \quad 12w_1 + w_3 \qquad\qquad \geq 19 \qquad\qquad (3.18)$$

$$4w_1 + w_2 - 7w_3 \leq 10 \qquad\qquad (3.19)$$

$$-w_1 + w_2 - 6w_3 = -8 \qquad\qquad (3.20)$$

$$w_1, w_2, w_3 \geq 0 \qquad\qquad (3.21)$$

Construct an artificial model to begin Phase I improving search with $w_1 = w_2 = w_3 = 0$.

Analysis: The solution $w_1 = w_2 = w_3 = 0$ satisfies the main constraint (3.19) and nonnegativity constraints (3.21). In accord with principle $\boxed{3.35}$, an artificial variable w_4 must be added to deal with violated constraint (3.18), and another w_5 must be subtracted (because of the negative right-hand side -8) to satisfy the violated equality (3.20).

An objective function minimizing the sum of the two artificial variables (principle $\boxed{3.36}$) completes the Phase I model.

$$\min \quad w_4 + w_5$$

$$\text{s.t.} \quad 12w_1 + w_3 + w_4 \qquad\quad \geq 19$$

$$4w_1 + w_2 - 7w_3 \qquad\quad \leq 10$$

$$-w_1 + w_2 - 6w_3 - w_5 = -8$$

$$w_1, w_2, w_3, w_4, w_5 \geq 0$$

Starting Artificial Solution

Using separate artificials in each violated constraint makes it extremely easy to complete a starting feasible solution for the Phase I search.

$\boxed{3.37}$ After fixing original variables at their arbitrarily chosen values, each artificial variable is initialized at the smallest value still needed to achieve feasibility in the corresponding constraint.

For example, in Two Crude model (3.17), we have decided to initiate Phase I search at solution $\mathbf{x}^{(0)}$ with $x_1^{(0)} = x_2^{(0)} = 0$. For these values the first main constraint becomes

$$0.3(0) + 0.4(0) + x_3 \geq 2$$

It is violated without artificial variable x_3 present, but choosing $x_3^{(0)} = 2$ is just enough to provide feasibility in Phase I. Similarly,

$$0.4(0) + 0.2(0) + x_4 \geq 1.5$$

demands $x_4^{(0)} = 1.5$, and

$$0.2(0) + 0.3(0) + x_5 \geq 0.5$$

implies that $x_5^{(0)} = 0.5$. The result is $\mathbf{x}^{(0)} = (0, 0, 2, 1.5, 0.5)$, a starting feasible solution for the Phase I problem.

SAMPLE EXERCISE 3.21: CONSTRUCTING A PHASE I STARTING SOLUTION

Return to the artificial problem of Sample Exercise 3.20. Determine a starting Phase I solution having $w_1 = w_2 = w_3 = 0$.

Analysis: To start Phase I, we set $w_1^{(0)} = w_2^{(0)} = w_3^{(0)} = 0$. Then making $w_4^{(0)} = 19$ achieves feasibility in the first main constraint, and $w_5^{(0)} = 8$ satisfies the last. Phase I would start at $\mathbf{w}^{(0)} = (0, 0, 0, 19, 8)$.

Phase I Outcomes

How could the Phase I search end? Certainly it will not terminate with a negative objective function value. Artificial variables are restricted to be nonnegative, so their sum must be nonnegative. For the same reason, the Phase I problem cannot be unbounded. The objective value cannot ever fall below zero.

Three possibilities remain.

> **3.38** If Phase I terminates with a solution having (Phase I) objective function value = 0, the components of the Phase I solution corresponding to original variables provide a feasible solution for the original model.

> **3.39** If Phase I terminates with a global minimum having (Phase I) objective function value > 0, the original model is infeasible.

> **3.40** If Phase I terminates with a local minimum that may not be global but has (Phase I) objective function value > 0, we can conclude nothing. Phase I search should be repeated from a new starting solution.

Begin with the happiest case **3.38**. Here Phase I has been able to drive the sum of the artificial variables to zero. For example, in the Two Crude Phase I problem above, two iterations of improving search might produce the Phase I solution $\mathbf{x}^{(2)} = (4, 4, 0, 0, 0)$ with objective function value

$$x_3^{(2)} + x_4^{(2)} + x_5^{(2)} = 0 + 0 + 0 = 0$$

The only way nonnegative numbers can sum to zero is for all of them to equal zero. Thus, every artificial is necessarily zero in this final Phase I solution. Since artificials no longer have any effect on their constraints, the values for nonartificial variables must be feasible in the original model. We can simply drop all artificial variables and proceed with Phase II.

The starting Phase II solution will be that part of the Phase I result that involves variables of the real model. For example, our Two Crude Phase I solution $\mathbf{x}^{(2)} = (4, 4, 0, 0, 0)$ has components $x_1^{(2)} = x_2^{(2)} = 4$ on nonartificial variables. Phase II search can start from feasible solution $\mathbf{x}^{(0)} = (4, 4)$.

SAMPLE EXERCISE 3.22: VERIFYING FEASIBILITY WITH PHASE I

Verify that $w_1 = 2$, $w_2 = 0$, $w_3 = 1$ is a feasible solution to the original model in Sample Exercise 3.20. Then construct a corresponding optimal solution to exercise's Phase I model, and explain why it is optimal.

Analysis: The solution indicated is feasible because all components are nonnegative and

$$12(2) + 1 = 25 \geq 19$$
$$4(2) + (0) - 7(1) = 1 \leq 10$$
$$-(2) + (0) - 6(1) = -8$$

To construct a corresponding optimum for the Phase I problem, we set artificials $w_4 = w_5 = 0$. Full Phase I solution $\mathbf{w} = (2, 0, 1, 0, 0)$ is feasible because artificials are unneeded to satisfy constraints with $w_1 = 2$, $w_2 = 0$, and $w_3 = 1$. It is Phase I optimal because the sum of nonnegative quantities is minimum when all are zero.

Concluding Infeasibility from Phase I

Now consider the 3.39 and 3.40 cases, where Phase I improving search terminates with a positive objective function value. The final solution from Phase I improving search will probably approximate a local minimum (principle 3.17), but it may or may not be global.

If we have some way of being sure that the Phase I solution is a global optimum, our conclusion is clear. The original model is infeasible (principle 3.39), because every solution to the Phase I model has some artificial variables at positive values. Their sum just cannot be driven to zero.

To illustrate, let us modify our Two Crude model until it is infeasible. Specifically, reverse the direction of the last main inequality so that it reads

$$0.2x_1 + 0.3x_2 + x_5 \leq 0.5 \tag{3.22}$$

in the Phase I problem.

Improving search on this revised Phase I problem will terminate at some solution $\mathbf{x}^{(t)} = (2.5, 0, 1.25, 0.5, 0)$ with objective function value $1.25 + 0.5 + 0 = 1.75 > 0$. Since this Phase I problem has all linear constraints and a linear objective function, we know from Section 3.4 (principles 3.27 , 3.32 , and 3.34) that every Phase I local minimum is a global minimum. It follows that no feasible solution to our modified Phase I example can have artificial variable total less that $\mathbf{x}^{(t)}$'s 1.75. That is, artificials simply cannot all be driven out of the solution. We conclude (principle 3.39) that the model with last constraint reversed as in (3.22) is infeasible.

Notice how this analysis depended on our certainty that a local minimum in Phase I was a global minimum. When Phase I's solution may just be a local minimum with positive objective function value, there might still be a solution with artificials all zero. Like any improving search with local outcomes possible, we would have no choice but to repeat the analysis from a different Phase I starting solution (principle 3.40).

SAMPLE EXERCISE 3.23: PROCESSING PHASE I OUTCOMES

Suppose that a linearly constrained optimization model over variables z_1, z_2, and z_3 is converted for Phase I by adding nonnegative artificial variables z_4, z_5, and z_6. For each of the following original model objective functions and Phase I search results $\mathbf{z}$, indicate what we can conclude and how Algorithm 3B processing should proceed.

(a) Original model objective: maximize $14z_1 - z_3$; Phase I local optimum: $\mathbf{z} = (1, -1, 3, 0, 0, 0)$

(b) Original model objective: minimize $(z_1 z_2)^2 + \sin(z_3)$; Phase I local optimum: $\mathbf{z} = (1, 2, -3, 0, 1, 1)$

Analysis:

(a) All 3 artificial variables $= 0$ in the Phase I local maximum. Thus (principle $\boxed{3.38}$) $\mathbf{z}^{(0)} = (1, -1, 3)$ provides a feasible solution for the original model. Proceed to Phase II search, beginning at this $\mathbf{z}^{(0)}$.

(b) The corresponding Phase I model has a linear objective function and linear constraints. Thus $\mathbf{z}$ is a global optimum (principles $\boxed{3.27}$, $\boxed{3.32}$, and $\boxed{3.34}$) despite the highly non-unimodal nature of the true objective function. Applying $\boxed{3.39}$, we may conclude that the original model is infeasible because artificial sum $z_4 + z_5 + z_6 = 0 + 1 + 1 = 2 > 0$ in a Phase I global optimum.

Big-M Method

Two-Phase Algorithm 3B deals with feasibility and optimality separately. Phase I search tests feasibility. Phase II proceeds to an optimum.

The Big-M method combines these activities in a single search. Artificial variables are included in constraints exactly as in Phase I of Two-Phase search. However, the effort to drive their total to $=0$ is combined with a search for an optimal solution to the original model.

The key to combining feasibility and optimality considerations is a composite objective function.

$\boxed{3.41}$ The **Big-M method** uses a large positive multiplier M to combine feasibility and optimality in a single-objective function of the form

$$\max \ (\text{original objective}) - M(\text{artificial variable sum})$$

for an originally maximize problem, or

$$\min \ (\text{original objective}) + M(\text{artificial variable sum})$$

for a minimize problem.

Infeasibility reflected in positive values of artificial variables is forced toward zero by adding or subtracting that infeasibility in the objective with the large penalty multiplier M that gives the method its name.

We again illustrate with the Two Crude example. The big-M form of the problem is

$$\begin{aligned}
\min \quad & 20x_1 + 15x_2 + M(x_3 + x_4 + x_5) \\
\text{s.t.} \quad & 0.3x_1 + 0.4x_2 + x_3 \geq 2 \\
& 0.4x_1 + 0.2x_2 + x_4 \geq 1.5 \\
& 0.2x_1 + 0.3x_2 + x_5 \geq 0.5 \\
& 0 \leq x_1 \leq 9, \ 0 \leq x_2 \leq 6 \\
& x_3, \ x_4, \ x_5 \geq 0
\end{aligned} \tag{3.23}$$

Notice that the constraints are identical to those of Phase I model (3.17). That means that the $\mathbf{x}^{(0)} = (0, 0, 2, 1.5, 0.5)$ of construction 3.37 continues to provide a suitable starting solution.

What is new is the objective function. It combines the original objective to minimize $20x_1 + 15x_2$ with a large positive multiple of artificial variable sum $(x_3 + x_4 + x_5)$.

Pick, say, $M = 10,000$. Then a single search of big-M version (3.23) will lead to an optimal solution of the original model. Any solution with positive artificial variable total will be highly penalized and thus not optimal in (3.23). Among those that have artificial total $= 0$, (i.e., among the feasible solutions of the original model), one with lowest original objective value will be preferred.

SAMPLE EXERCISE 3.24: CONSTRUCTING A BIG-M MODEL

Return to the optimization of Sample Exercise 3.20 and construct the corresponding Big-M model.

Analysis: Constraints and artificial variables are exactly as in Sample Exercise 3.20. Including the Big-M objective of principle 3.41 results in the following big-M model:

$$\begin{aligned}
\max \quad & 14(w_1 - 10)^2 + (w_2 - 3)^2 + (w_3 + 5)^2 - M(w_4 + w_5) \\
\text{s.t.} \quad & 12w_1 + w_3 \qquad\quad + w_4 \geq 19 \\
& 4w_1 + w_2 - 7w_3 \qquad\quad \leq 10 \\
& -w_1 + w_2 - 6w_3 \ - w_5 = -8 \\
& w_1, w_2, w_3, w_4, w_5 \geq 0
\end{aligned}$$

The artificial sum is subtracted because the objective is maximized.

Big-M Outcomes

Possible outcomes from a Big-M search closely parallel those of the two-phase method detailed in principles 3.38 to 3.40 . When Big-M search terminates, we may have an optimum for the original model, a proof that the original model was infeasible, or just confusion.

Consider first the optimal case.

3.42 If a Big-M search terminates with a locally optimal solution having all artificial variables $= 0$, the components of the solution corresponding to original variables form a locally optimal solution for the original model.

For example, with $M = 10,000$, infeasibility is so expensive that big-M search of model (3.23) will compute global optimum $\mathbf{x} = (2, 3.5, 0, 0, 0)$. Components

$x_1 = 2$ and $x_2 = 3.5$, which were the original decision variables, constitute an optimal solution to model of interest.

When Big-M search terminates with artificial variables at positive values, matters are more murky.

3.43 | If M is sufficiently large and Big-M search terminates with a global optimum having some artificial variables > 0, the original model is infeasible.

3.44 | If Big-M search terminates with a local optimum having some artificial variables > 0, or the multiplier M may not be large enough, we can conclude nothing. The search should be repeated with a larger M and/or a new starting solution.

As with two-phase outcomes 3.39 and 3.40, we will not be able to reach any conclusions when big-M search stops with a positive artificial variable total unless we can be sure that the search has produced a global optimum. If the outcome is known only to be a local optimum, a feasible solution might yet be found.

However, notice in 3.43 and 3.44 that we encounter a new difficulty when we adopt the big-M approach. If artificial variables take nonzero values in the big-M optimum, it may only mean that we did not choose a large enough multiplier M. For example, solution of big-M form (3.23) with $M = 1$ will yield optimal solution $\mathbf{x} = (0, 0, 2.0, 1.5, 0.5)$ because the penalty of

$$M(x_3 + x_4 + x_5) = 1\,(2.0 + 1.5 + 0.5)$$

is insufficient to force out infeasibility. Only when M is sufficiently large—something that depends on details of each model—does a big-M optimum with positive artificial variables imply infeasibility.

SAMPLE EXERCISE 3.25: PROCESSING BIG-M OUTCOMES

Suppose that a linearly constrained optimization model over variables z_1, z_2, and z_3 is converted for big-M by including nonnegative artificial variables z_4, z_5, and z_6 and penalizing the original objective function (as in 3.41) by $M = 1000$ times their sum. For each of the following original model objective functions and big-M search results $\mathbf{z}$, indicate what we can conclude.

(a) Original model objective: maximize $14z_1 - z_3$; big-M local optimum: $\mathbf{z} = (1, -1, 3, 0, 0, 0)$

(b) Original model objective: maximize $z_2 + z_3$; big-M local optimum: $\mathbf{z} = (0, 0, 0, 1, 0, 2)$

(c) Original model objective: minimize $(z_1 z_2)^2 + \sin(z_3)$; big-M local optimum: $\mathbf{z} = (1, 1, 3, 0, 0, 0)$

(d) Original model objective: minimize $(z_1 z_2)^2 + \sin(z_3)$; big-M local optimum: $\mathbf{z} = (1, 2, -3, 0, 1, 1)$

Analysis:

(a) All three artificial variables $= 0$ in the big-M local optimum. Thus (principle $\boxed{3.42}$) $\mathbf{z} = (1, -1, 3)$ provides a local maximum for the original model. With both objective and constraints linear, it is also globally optimal.

(b) With both objective and constraints linear, solution $\mathbf{z}$ is a global optimum in the big-M model. However, we can conclude that the original model is infeasible (principle $\boxed{3.43}$) only if we know that $M = 1000$ imposes a large enough penalty to make any infeasible solution have a lower objective value than any feasible one.

(c) All three artificial variables $= 0$ in the big-M local optimum. Thus (principle $\boxed{3.42}$) $\mathbf{z} = (1, 1, 3)$ provides a local optimum for the original model. The highly non-unimodal nature of the objective makes it impossible to tell whether it is a global optimum.

(d) The highly non-unimodal nature of the big-M objective function makes it impossible to know whether the indicated $\mathbf{z}$ is a global minimum. Thus since some artificial variables have positive value, our only choices (principle $\boxed{3.44}$) are to increase M and/or repeat the search from a different starting point.

EXERCISES

3-1 In each of the following plots, determine whether the specified points are feasible, infeasible, local optimal, and/or global optimal in the depicted mathematical program over two continuous variables. Dashed lines indicate contours of the objective function, and solid lines show constraints.

(a) $\boxtimes$ Maximize problem; $\mathbf{x}^{(1)} = (5, 0)$, $\mathbf{x}^{(2)} = (2, -1)$, $\mathbf{x}^{(3)} = (3, 3)$, $\mathbf{x}^{(4)} = (1, 3)$

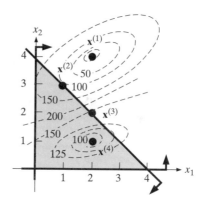

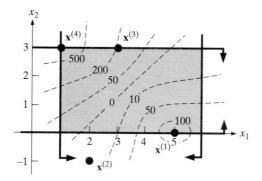

(b) Minimize problem; $\mathbf{x}^{(1)} = (2, 4)$, $\mathbf{x}^{(2)} = (1, 3)$, $\mathbf{x}^{(3)} = (2, 2)$, $\mathbf{x}^{(4)} = (2, 1)$

3-2 Each of the following shows the sequence of directions and steps employed by an improving search that began at $\mathbf{y}^{(0)} = (2, 0, 5)$. Compute the sequence of points visited by the search.

(a) $\boxtimes$ $\Delta \mathbf{y}^{(1)} = (3, -1, 0)$, $\lambda_1 = 2$, $\Delta \mathbf{y}^{(2)} = (-1, 2, 1)$, $\lambda_2 = 5$, $\Delta \mathbf{y}^{(3)} = (0, 6, 0)$, $\lambda_3 = \frac{1}{2}$

(b) $\Delta \mathbf{y}^{(1)} = (-1, 4, 1)$, $\lambda_1 = 3$, $\Delta \mathbf{y}^{(2)} = (0, 3, -6)$, $\lambda_2 = \frac{1}{3}$, $\Delta \mathbf{y}^{(3)} = (1, 1, 1)$, $\lambda_3 = 1$

3-3 Each of the following shows the sequence of points visited by an improving search. Compute the corresponding sequence of move directions assuming that all step sizes $\lambda = 1$.

(a) ◻ $\mathbf{w}^{(0)} = (0, 1, 1)$, $\mathbf{w}^{(1)} = (4, -1, 7)$, $\mathbf{w}^{(2)} = (4, -3, 19)$, $\mathbf{w}^{(3)} = (3, -3, 22)$

(b) $\mathbf{w}^{(0)} = (9, 0, 3)$, $\mathbf{w}^{(1)} = (2, 2, 7)$, $\mathbf{w}^{(2)} = (3, 6, 7)$, $\mathbf{w}^{(3)} = (3, 9, 8)$

3-4 Refer to the plots of Exercise 3-1 and determine graphically whether the following directions appear to be improving at the points indicated.

(a) ◻ $\Delta\mathbf{x} = (-3, 3)$ at $\mathbf{x}^{(1)}$ of 3-1(a)

(b) $\Delta\mathbf{x} = (0, 1)$ at $\mathbf{x}^{(2)}$ of 3-1(a)

(c) ◻ $\Delta\mathbf{x} = (-10, 1)$ at $\mathbf{x}^{(3)}$ of 3-1(a)

(d) $\Delta\mathbf{x} = (4, 0)$ at $\mathbf{x}^{(3)}$ of 3-1(b)

(e) ◻ $\Delta\mathbf{x} = (1, -2)$ at $\mathbf{x}^{(2)}$ of 3-1(b)

(f) $\Delta\mathbf{x} = (-3, 2)$ at $\mathbf{x}^{(4)}$ of 3-1(b)

3-5 Refer to the plots of Exercise 3-1 and determine graphically whether the following directions appear to be feasible at the points indicated.

(a) ◻ $\Delta\mathbf{x} = (-5, 5)$ at $\mathbf{x}^{(1)}$ of 3-1(a)

(b) $\Delta\mathbf{x} = (0, 1)$ at $\mathbf{x}^{(3)}$ of 3-1(a)

(c) ◻ $\Delta\mathbf{x} = (-10, 0)$ at $\mathbf{x}^{(3)}$ of 3-1(a)

(d) $\Delta\mathbf{x} = (-4, 4)$ at $\mathbf{x}^{(3)}$ of 3-1(b)

(e) ◻ $\Delta\mathbf{x} = (1, 2)$ at $\mathbf{x}^{(2)}$ of 3-1(b)

(f) $\Delta\mathbf{x} = (-3, 2)$ at $\mathbf{x}^{(4)}$ of 3-1(b)

3-6 Consider a mathematical program with constraints

$$x_1 - 2x_2 + 3x_3 \le 25$$
$$x_1, x_2, x_3 \ge 0$$

Determine the maximum step (possibly $+\infty$) that preserves feasibility in the direction indicated from the point specified. Also indicate whether that step indicates that the model is unbounded, assuming that directions improve everywhere.

(a) ◻ $\Delta\mathbf{x} = (-1, 3, -2)$ from $\mathbf{x} = (4, 0, 6)$

(b) $\Delta\mathbf{x} = (-3, -3, 9)$ from $\mathbf{x} = (9, 4, 6)$

(c) ◻ $\Delta\mathbf{x} = (1, 3, 1)$ from $\mathbf{x} = (0, 0, 4)$

(d) $\Delta\mathbf{x} = (-4, 0, 3)$ from $\mathbf{x} = (16, 2, 1)$

3-7 For each of the following combinations of objective function, point, and direction, determine whether conditions ⬚3.21 and ⬚3.22 show that the direction improves at the point, does not improve at the point, or that further information is required.

(a) ◻ max $4y_1 - 2y_3 + y_5$, $\mathbf{y} = (1, 0, 19, 4, 6)$, $\Delta\mathbf{y} = (2, -3, 4, 0, 6)$

(b) min $-y_2 + 8y_3 + 6y_4$, $\mathbf{y} = (2, 0, 8, 11, -3)$, $\Delta\mathbf{y} = (2, 0, 3, -1, 0)$

(c) ◻ min $y_1 y_2 + (y_1)^2 + 4y_2$, $\mathbf{y} = (3, -1)$, $\Delta\mathbf{y} = (-7, 5)$

(d) min $y_1 y_2 + (y_1)^2 + 4y_2$, $\mathbf{y} = (3, -1)$, $\Delta\mathbf{y} = (-1, 1)$

(e) ◻ max $(y_1 - 5)^2 + (y_2 + 1)^2$, $\mathbf{y} = (4, 1)$, $\Delta\mathbf{y} = (-1, 2)$

(f) min $(y_1 - 5)^2 + (y_2 + 1)^2$, $\mathbf{y} = (2, 3)$, $\Delta\mathbf{y} = (8, -6)$

3-8 Construct an improving direction from the gradient of each objective function at the point indicated.

(a) ◻ max $3w_1 - 2w_2 + w_4$ at $\mathbf{w} = (2, 0, 5, 1)$

(b) min $4w_1 + w_3 - 5w_4$ at $\mathbf{w} = (1, 1, 7, -1)$

(c) ◻ min $(w_1 + 2)^2 - w_1 w_2$ at $\mathbf{w} = (3, 2)$

(d) max $(w_1)^3 - 4w_1 + 6w_2$ at $\mathbf{w} = (2, 5)$

3-9 Determine which of the constraints

$$(z_1 - 2)^2 + (z_2 - 1)^2 \le 10 \quad [\text{i}]$$
$$2z_1 - z_2 = 8 \quad [\text{ii}]$$
$$z_1 \ge 0 \quad [\text{iii}]$$
$$z_2 \ge 0 \quad [\text{iv}]$$

are active at each of the following solutions.

(a) ◻ $\mathbf{z} = (4, 0)$

(b) $\mathbf{z} = (5, 2)$

3-10 Determine whether each of the directions specified is feasible at the solution indicated to the linear constraints

$$3y_1 - 2y_2 + 8y_3 = 14$$
$$6y_1 - 4y_2 - 1y_3 \le 11$$
$$y_1, y_2, y_3 \ge 0$$

(a) ◻ $\Delta\mathbf{y} = (0, 4, 1)$ at $\mathbf{y} = (2, 0, 1)$

(b) $\Delta\mathbf{y} = (0, -4, 1)$ at $\mathbf{y} = (2, 0, 1)$

(c) ◻ $\Delta\mathbf{y} = (2, 0, 1)$ at $\mathbf{y} = (0, 1, 2)$

(d) $\Delta\mathbf{y} = (0, 8, -2)$ at $\mathbf{y} = (0, 1, 2)$

3-11 State all conditions that must be satisfied by a feasible direction $\Delta\mathbf{w}$ at the solution indicated to each of the following systems of linear constraints.

(a) ◻
$$2w_1 + 3w_3 = 18$$
$$1w_1 + 1w_2 + 2w_3 = 14$$
$$w_1, w_2, w_3 \ge 0$$
at $\mathbf{w} = (0, 2, 6)$

(b) Same constraints as part (a) at $\mathbf{w} = (9, 5, 0)$.

(c) ◻
$$1w_1 + 1w_2 = 10$$
$$2w_1 - 1w_2 \ge 8$$
$$1w_1 - 8w_2 \le 1$$
at $\mathbf{w} = (6, 4)$

(d) Same constraints as part (c) at $\mathbf{w} = (9, 1)$.

3-12 Consider the mathematical program

$$\max \quad 4z_1 + 7z_2$$
$$\text{s.t.} \quad 2z_1 + z_2 \le 9$$
$$0 \le z_1 \le 4$$
$$0 \le z_2 \le 3$$

(a) ⬦ Show that directions $\Delta z^{(1)} = (2, 0)$ and $\Delta z^{(2)} = (-2, 4)$ are improving directions for this model at every z.

(b) ⬦ Beginning at $z^{(0)} = (0, 0)$, execute Improving Search Algorithm 3A on the model. Limit your search to the two directions of part (a), and continue until neither is both improving and feasible.

(c) Show in a two-dimensional plot the feasible space and objective function contours of the model. Then plot the path of your search in part (b).

3-13 Do Exercise 3-12 for mathematical program

$$\min \quad z_1 + z_2$$
$$\text{s.t.} \quad z_1 + 2z_2 \ge 4$$
$$0 \le z_1 \le 5$$
$$0 \le z_2 \le 3$$

directions $\Delta z^{(1)} = (0, -5)$, $\Delta z^{(2)} = (-2, 1)$, and initial point $z^{(0)} = (5, 3)$.

3-14 Determine whether each of the following objective functions is unimodal over the domain specified, and for those that are not, indicate points $x^{(1)}$ and $x^{(2)}$ that violate definition $\boxed{3.26}$.

(a) ⬦ The following maximize objective over x in the domain shown.

$f(x)$

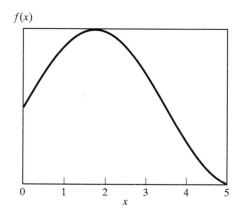

(b) The following minimize objective over x in the domain shown.

$f(x)$

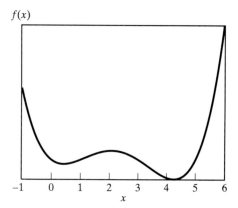

(c) ⬦ The following maximize objective over (x_1, x_2) in the domain shown.

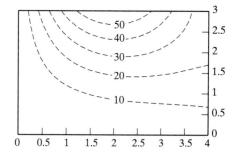

(d) The function of (c) if it is being minimized.

(e) ⬦ The maximize objective function plotted in Exercise 3-1(a) over (x_1, x_2) in the domain shown.

(f) The minimize objective function plotted in Exercise 3-1(b) over (x_1, x_2) in the domain shown.

(g) ⬦ $\min 3x_1 - 17x_2 + 7x_5$ over all $(x_1, \dots, x_5)$.

(h) $\max 37x_1 - 100x_2 - 13x_3 + 40x_4$ over all $(x_1, \dots, x_5)$.

3-15 Consider the line segment between each of the following solution pairs $z^{(1)}$ and $z^{(2)}$. Write an algebraic expression representing all points on the line segment, show that the given $z^{(3)}$ is on the line segment, and show that the $z^{(4)}$ specified is not.

(a) ⬦ $z^{(1)} = (3, 1, 0)$, $z^{(2)} = (0, 4, 9)$, $z^{(3)} = (2, 2, 3)$, $z^{(4)} = (3, 5, 9)$

(b) $z^{(1)} = (2, 1, 4)$, $z^{(2)} = (7, 0, -1)$, $z^{(3)} = (6, \frac{1}{5}, 0)$, $z^{(4)} = (0, 3, 1)$

3-16 Determine whether the feasible set for each of the following systems of constraints is convex, and if not, indicate points $\mathbf{x}^{(1)}$ and $\mathbf{x}^{(2)}$ that violate definition 3.29 .

(a) ▨ $(x_1)^2 + (x_2)^2 \geq 9$
$\quad x_1 + x_2 \leq 10$
$\quad x_2, x_2 \geq 0$

(b) $(x_1)^2 + (x_2)^2 \leq 25$
$\quad 2x_1 + 3x_2 \geq 6$
$\quad x_1, x_2 \geq 0$

(c) ▨ $x_1 - 2x_2 + x_3 = 2$
$\quad x_1 + 8x_2 - x_3 \leq 16$
$\quad x_1 + 4x_2 - x_3 \geq 5$
$\quad x_1, x_2, x_3 \geq 0$

(d) $\sum_{j=1}^{2} x_j = 1000$
$\quad 0 \leq x_j \leq 100, \; j = 1, \ldots, 20$

(e) ▨ $x_1 + 2x_2 + 3x_3 + x_4 \leq 24$
$\quad 0 \leq x_j \leq 10, \; j = 1, \ldots, 4$
$\quad x_j \text{ integer}, \; j = 1, \ldots, 4$

(f) $2x_1 + x_2 + x_3 - 5x_4 \geq 3$
$\quad x_j = 0 \text{ or } 1, \; j = 1, \ldots, 4$

3-17 Construct a Phase I model corresponding to each of the following, and indicate appropriate starting values for the artificial variables. Assume that all original decision variables start at $w_j = 0$.

(a) ▨ max $22w_1 - w_2 + 15w_3$
$\quad$ s.t. $40w_1 + 30w_2 + 10w_3 = 150$
$\quad\quad w_1 - w_2 \leq 0$
$\quad\quad 4w_2 + w_3 \geq 10$
$\quad\quad w_1, w_2, w_3 \geq 0$

(b) min $-11w_1 + 10w_2 + w_3$
$\quad$ s.t. $3w_1 + w_2 + 2w_3 \leq 9$
$\quad\quad 4w_1 + 4w_2 \geq 6$
$\quad\quad w_1 - w_2 - w_3 = -2$
$\quad\quad w_1, w_2, w_3 \geq 0$

(c) ▨ min $2w_1 + 3w_2$
$\quad$ s.t. $(w_1 - 3)^2 + (w_2 - 3)^2 \leq 4$
$\quad\quad 2w_1 + 2w_2 = 5$
$\quad\quad w_1 \geq 3$

(d) max $w_1 w_2$
$\quad$ s.t. $(w_1 - 5)^2 + (w_2)^2 \leq 25$
$\quad\quad 3w_1 - w_2 = 0$
$\quad\quad w_1 \geq 2$
$\quad\quad w_2 \geq 0$

3-18 Describe how a two-phase improving search of a model with original variables y_1, y_2, and y_3 would proceed if Phase I search terminated as follows:

(a) ▨ Global optimum $\mathbf{y} = (40, 7, 0, 9, 0)$
(b) Global optimum $\mathbf{y} = (22, 4, 3, 0, 0)$
(c) ▨ Local optimum $\mathbf{y} = (1, 3, 1, 0, 0)$
(d) Local optimum $\mathbf{y} = (0, 2, 6, 5, 1)$, which may not be a global optimum

3-19 Construct a Big-M starting model for each case in Exercise 3-17 and indicate appropriate starting values for the artificial variables. Assume that all original decision variables start at $w_j = 0$.

3-20 Describe how Big-M search of a model with original variables y_1, y_2, and y_3 would proceed if improving search of the Big-M version produced each of the outcomes in Exercise 3-18.

SUGGESTED READING

Bazarra, Mokhtar S., Hanif D. Sherali, and C. M. Shetty (1993), *Nonlinear Programming Theory and Algorithms*, Wiley, New York.

Luenberger, David G. (1984), *Linear and Nonlinear Programming*, Addison-Wesley, Reading, Mass.

CHAPTER 4

Linear Programming Models

● ●

Linear programs (or LP's) are mathematical programs combining all the characteristics that Chapter 3 showed lead to high tractability.

> **4.1** An optimization model is a **linear program** (or **LP**) if it has continuous variables, a single linear objective function, and all constraints are linear equalities or inequalities.

That is, the model's single-objective function and all constraints must consist of weighted sums of continuous decision variables.

Everything in life is not linear and continuous, but an enormous variety of applications can be validly modeled as LPs. Cases with thousands, millions, or even billions of variables can then be solved to global optimality.

We have already encountered linear programming models in Sections 2.1 to 2.4. Many others are found in network flow Chapter 10.

In this chapter we develop a more comprehensive picture by presenting several of the classic LP application forms. Most of the models described were actually used by real organizations. Only the details and numerical data are fictitious.

Linear programs are so central to the study of mathematical programming that four subsequent chapters also treat LP topics. In Chapter 5 we specialize the improving search notions of Chapter 3 to obtain a powerful algorithm called the simplex method; in Chapter 6 we present newer, interior point varieties of improving search for LPs; in Chapter 7 we develop the extensive duality and sensitivity analysis available in linear programming; and in Chapter 10 we address the special linear programs that model network flows.

4.1 ALLOCATION MODELS

One of the simplest forms of linear programs that occurs widely in application might be termed **allocation models**. The main issue is how to divide or allocate a valuable resource among competing needs. The resource may be land, capital, time, fuel, or anything else of limited availability. For example, the (nonlinear) E-mart model (2.13) of Section 2.4 allocated an advertising budget.

EXAMPLE 4.1: FOREST SERVICE ALLOCATION

The U.S. Forest Service has used just such an allocation model to address the sensitive task of managing 191 million acres of national forestland.[1] The Forest Service must tradeoff timber, grazing, recreational, environmental, national preservation, and other demands on forestland.

Models of a forest begin by dividing land into homogeneous *analysis areas*. Several *prescriptions* or land management policies are then proposed and evaluated for each. The optimization seeks the best possible allocation of land in the analysis areas to particular prescriptions, subject to forest-wide restrictions on land use.

Table 4.1 provides details of the fictional, 788 thousand acre Wagonho National Forest that we model. Wagonho is assumed to have 7 analysis areas, each subject to 3 different prescriptions. The first prescription encourages timbering, the second emphasizes grazing, and the third preserves the land as wilderness. Using index dimensions

$$i \triangleq \text{analysis area number } (i = 1, \ldots, 7)$$
$$j \triangleq \text{prescription number } (j = 1, \ldots, 3)$$

Table 4.1 provides values for all the following symbolic parameters:

$s_i \triangleq$ size of analysis area i (in thousands of acres)

$p_{i,j} \triangleq$ net present value (NPV) per acre of all uses in area i if managed under prescription j

$t_{i,j} \triangleq$ projected timber yield (in board feet per acre) of analysis area i if managed under prescription j

$g_{i,j} \triangleq$ projected grazing capability (in animal unit months per acre) of analysis area i if managed under prescription j

$w_{i,j} \triangleq$ wilderness index rating (0 to 100) of analysis area i if managed under prescription j

We wish to find an allocation that maximizes net present value, while producing 40 million board feet of timber, 5 thousand animal unit months of grazing, and keeping average wilderness index at least 70.

Allocation Decision Variables

As in all such models, our Forest Service example seeks an optimal allocation of a valuable resource. Corresponding decision variables define the allocation.

> 4.2 | Principal decision variables in allocation models specify how much of the critical resource is allocated to each use.

---

[1] Based on B. Kent, B. B. Bare, R. C. Field, and G. A. Bradley (1991), "Natural Resource Land Management Planning Using Large-Scale Linear Programs: The USDA Forest Service Experience with FORPLAN," *Operations Research, 39,* 13–27.

TABLE 4.1 Forest Service Example Data

Analysis Area, i	Acres, s_i (000)'s	Prescrip- tion, j	NPV, (per acre) $p_{i,j}$	Timber, (per acre) $t_{i,j}$	Grazing, (per acre) $g_{i,j}$	Wilderness Index, $w_{i,j}$
1	75	1	503	310	0.01	40
		2	140	50	0.04	80
		3	203	0	0	95
2	90	1	675	198	0.03	55
		2	100	46	0.06	60
		3	45	0	0	65
3	140	1	630	210	0.04	45
		2	105	57	0.07	55
		3	40	0	0	60
4	60	1	330	112	0.01	30
		2	40	30	0.02	35
		3	295	0	0	90
5	212	1	105	40	0.05	60
		2	460	32	0.08	60
		3	120	0	0	70
6	98	1	490	105	0.02	35
		2	55	25	0.03	50
		3	180	0	0	75
7	113	1	705	213	0.02	40
		2	60	40	0.04	45
		3	400	0	0	95

Our Forest Service case will employ nonnegative

$x_{i,j} \triangleq$ number of thousands of acres in analysis area i managed by prescription j

Forest Service Allocation Model

The Forest Service's objective is to maximize total net present value (NPV). In terms of the defined notation, this is

$$\max \sum_{i=1}^{7} \sum_{j=1}^{3} p_{i,j}x_{i,j} = \; 503x_{1,1} + 140x_{1,2} + 203x_{1,3} + 675x_{2,1} + 100x_{2,2} + 45x_{2,3}$$
$$+ \cdots + 705x_{7,1} + 60x_{7,2} + 400x_{7,3}$$

One system of constraints must assure that all acres of each analysis area are allocated. For example, in analysis area 1,

$$x_{1,1} + x_{1,2} + x_{1,3} = 75$$

Using symbolic constants, all 7 such constraints can be expressed by the system

$$\sum_{j=1}^{3} x_{i,j} = s_i \qquad i = 1, \ldots, 7$$

Finally, there are the performance requirements on timber, grazing, and wilderness index. For example, we want timber output

$$\sum_{i=1}^{7} \sum_{j=1}^{3} t_{i,j}x_{i,j} = 310x_{1,1} + 50x_{1,2} + 0x_{1,3} + 198x_{2,1} + 46x_{2,2} + 0x_{2,3}$$

$$+ \cdots + 213x_{7,1} + 40x_{7,2} + 0x_{7,3}$$

$$\geq 40,000$$

Combining produces the following allocation linear program:

$$\max \quad \sum_{i=1}^{7} \sum_{j=1}^{3} p_{i,j} x_{i,j} \qquad \text{(present value)}$$

$$\text{s.t.} \quad \sum_{j=1}^{3} x_{i,j} = s_i \quad i = 1, \dots, 7 \qquad \text{(allocation)}$$

$$\sum_{i=1}^{7} \sum_{j=1}^{3} t_{i,j} x_{i,j} \geq 40{,}000 \qquad \text{(timber)}$$

$$\sum_{i=1}^{7} \sum_{j=1}^{3} g_{i,j} x_{i,j} \geq 5 \qquad \text{(grazing)}$$

$$\frac{1}{788} \sum_{i=1}^{7} \sum_{j=1}^{3} w_{i,j} x_{i,j} \geq 70 \qquad \text{(wilderness)}$$

$$x_{i,j} \geq 0 \qquad i = 1, \dots, 7; \quad j = 1, \dots, 3$$

(4.1)

An optimal allocation makes

$$x_{1,1}^* = 0, \quad x_{1,2}^* = 0, \quad x_{1,3}^* = 75, \quad x_{2,1}^* = 90, \quad x_{2,2}^* = 0, \quad x_{2,3}^* = 0$$
$$x_{3,1}^* = 140, \quad x_{3,2}^* = 0, \quad x_{3,3}^* = 0, \quad x_{4,1}^* = 0, \quad x_{4,2}^* = 0, \quad x_{4,3}^* = 60$$
$$x_{5,1}^* = 0, \quad x_{5,2}^* = 154, \quad x_{5,3}^* = 58, \quad x_{6,1}^* = 0, \quad x_{6,2}^* = 0, \quad x_{6,3}^* = 98$$
$$x_{7,1}^* = 0, \quad x_{7,2}^* = 0, \quad x_{7,3}^* = 113$$

with total net present value $322,515,000.

SAMPLE EXERCISE 4.1: FORMULATING ALLOCATION LPS

Jill College is taking courses in operations research, engineering economics, statistics, and material science. She has 30 study hours to prepare for her finals and wishes to divide her time to improve her term grades as much as possible. Naturally, her favorite course is operations research, so she will spend as much on it as any other. Still, she believes up to 10 hours of study could be useful in any of the courses, with each hour on operations research increasing her grade by 2%, each on engineering economics yielding 3%, each on statistics producing 1%, and each on materials science adding 5%. Form an allocation linear program to help Jill optimize her study.

Modeling: Using decision variables

$$h_j \triangleq \text{hours spent on the } j\text{th course}$$

the required model is

$$\begin{aligned}
\max \quad & 2h_1 + 3h_2 + 1h_3 + 5h_4 & \text{(total gain)} \\
\text{s.t.} \quad & h_1 + h_2 + h_3 + h_4 = 30 & \text{(allocation)} \\
& h_1 \geq h_j \quad j = 2, \dots, 4 & \text{(OR most)} \\
& h_j \leq 10 \quad j = 1, \dots, 4 & \text{(maximum 10)} \\
& h_j \geq 0 \quad j = 1, \dots, 4 &
\end{aligned}$$

The objective maximizes score gain. Main constraints make the allocation total 30, keep OR study as great as any other, and limit the allocation for any class to 10 hours.

4.2 BLENDING MODELS

Allocation models split a resource. Blending models combine them. That is, **blending models** decide what mix of ingredients best fulfills specified output requirements. Various applications blend everything from chemicals, to diets, to metals, to animal food. For example, the Two Crude case of Section 2.1 is a blending model mixing crude petroleums to produce refinery products.

EXAMPLE 4.2: SWEDISH STEEL

The steel industry confronts another blending problem when it melts materials in high-temperature furnaces to manufacture new alloys from scrap. Fagersta AB of Fagersta, Sweden, is one of many companies that have used mathematical programming to plan this steel blending.[2]

An optimization arises each time a furnace is *charged*. Scrap in the available inventory is combined with pure additives to produce a blend having the required percentages of various chemical elements. It is critical to make maximum use of scrap because additives are much more expensive. Although there are some integer programming aspects discussed in Section 11.1, we deal here only with the simpler linear programming form.

Our fictitious version of Swedish steelmaking will produce a 1000-kilogram furnace charge. All steel consists primarily of iron. Table 4.2 shows the much smaller fractions of carbon, nickel, chromium, and molybdenum in the four available supplies of scrap, on which we can draw, along with the quantities held and their unit cost in Swedish kroner. It also shows the three higher-cost additives that can be used and the acceptable ranges for the resulting blend. For example, the 1000 kilograms of steel produced should contain between 0.65 and 0.75% carbon.

TABLE 4.2   Data for Swedish Steel Example

	Composition (%)				Available (kg)	Cost (kr/kg)
	Carbon	Nickel	Chromium	Molybdenum		
First scrap	0.80	18	12	—	75	16
Second scrap	0.70	3.2	1.1	0.1	250	10
Third scrap	0.85	—	—	—	Unlimited	8
Fourth scrap	0.40	—	—	—	Unlimited	9
Nickel	—	100	—	—	Unlimited	48
Chromium	—	—	100	—	Unlimited	60
Molybdenum	—	—	—	100	Unlimited	53
Minimum blend	0.65	3.0	1.0	1.1		
Maximum blend	0.75	3.5	1.2	1.3		

Ingredient Decision Variables

It is characteristic that we must make **ingredient** decisions.

[2]Based on C.-H. Westerberg, B. Bjorklund, and E. Hultman (1977), "An Application of Mixed Integer Programming in a Swedish Steel Mill," *Interfaces*, 7:2, 39-43.

> **4.3** Principal decision variables in blending models specify how much of each available ingredient to include in the mix.

In our Swedish Steel example we have seven such variables:

$$x_j \triangleq \text{number of kilograms of ingredient } j \text{ included in the charge}$$

where $j = 1, \ldots, 4$ refers to the four supplies of scrap, and $j = 5, \ldots, 7$ to the pure additives.

Composition Constraints

One requirement on any solution to our example is that the total charge sum to 1000 kilograms:

$$x_1 + x_2 + x_3 + x_4 + x_5 + x_6 + x_7 = 1000 \tag{4.2}$$

However, most of the main constraints restrict the composition of the mix.

> **4.4** **Composition** constraints in blending models enforce upper and/or lower limits on the properties of the resulting blend.

In our Swedish Steel example we have both upper and lower limits on the fraction of carbon, nickel, chromium, and molybdenum in the mix. Each such constraint will have the form

$$\sum_j \begin{pmatrix} \text{fraction in} \\ j\text{th} \\ \text{ingredient} \end{pmatrix} \cdot \begin{pmatrix} \text{amount of} \\ j\text{th} \\ \text{ingredient} \\ \text{used} \end{pmatrix} \begin{array}{c} \geq \\ \text{or} \\ \leq \end{array} \begin{pmatrix} \text{allowed} \\ \text{fraction in} \\ \text{the blend} \end{pmatrix} \cdot \begin{pmatrix} \text{blend} \\ \text{total} \end{pmatrix} \tag{4.3}$$

Specifically, the composition constraints required are

$$0.0080x_1 + 0.0070x_2 + 0.0085x_3 + 0.0040x_4 \geq 0.0065 \sum_{j=1}^{7} x_j$$

$$0.0080x_1 + 0.0070x_2 + 0.0085x_3 + 0.0040x_4 \leq 0.0075 \sum_{j=1}^{7} x_j$$

$$0.180x_1 + 0.032x_2 + 1.0x_5 \geq 0.030 \sum_{j=1}^{7} x_j$$

$$0.180x_1 + 0.032x_2 + 1.0x_5 \leq 0.035 \sum_{j=1}^{7} x_j$$

$$0.120x_1 + 0.011x_2 + 1.0x_6 \geq .010 \sum_{j=1}^{7} x_j$$

$$0.120x_1 + 0.011x_2 + 1.0x_6 \leq 0.012 \sum_{j=1}^{7} x_j$$

$$0.001x_2 + 1.0x_7 \geq 0.011 \sum_{j=1}^{7} x_j$$

$$0.001x_2 + 1.0x_7 \leq 0.013 \sum_{j=1}^{7} x_j$$

SAMPLE EXERCISE 4.2: FORMULATING COMPOSITION CONSTRAINTS

A food blending model with 3 ingredients employs nonnegative decision variables

$$x_j \triangleq \text{grams of ingredient } j \text{ used}$$

where ingredient 1 is 4% fiber and has 10 milligrams (mg) of sodium per gram, ingredient 2 is 9% fiber and has 15 mg of sodium per gram, and ingredient 3 is 3% fiber and has 5 mg of sodium per gram. Formulate linear constraints enforcing each of the following requirements.

(a) The blend must average at least 5% fiber.

(b) The blend must contain at most 100 mg of sodium.

Modeling:

(a) Following format (4.3), the needed constraint is

$$0.04x_1 + 0.09x_2 + 0.03x_3 \geq 0.05(x_1 + x_2 + x_3)$$

(b) Here the constraint deals with absolute amounts, not fractions. The required form is

$$10x_1 + 15x_2 + 5x_3 \leq 100$$

Swedish Steel Example Model

Collecting the elements derived so far yields the following LP model of our Swedish Steel example:

$$
\begin{array}{lll}
\min & 16x_1 + 10x_2 + 8x_3 + 9x_4 + 48x_5 + 60x_6 + 53x_7 & \text{(cost)} \\
\text{s.t.} & x_1 + x_2 + x_3 + x_4 + x_5 + x_6 + x_7 = 1000 & \text{(weight)} \\
& 0.0080x_1 + 0.0070x_2 + 0.0085x_3 & \text{(carbon)} \\
& \quad + 0.0040x_4 \geq 0.0065(1000) & \\
& 0.0080x_1 + 0.0070x_2 + 0.0085x_3 & \\
& \quad + 0.0040x_4 \leq 0.0075(1000) & \\
& 0.180x_1 + 0.032x_2 + 1.0x_5 \quad\geq 0.030(1000) & \text{(nickel)} \\
& 0.180x_1 + 0.032x_2 + 1.0x_5 \quad\leq 0.035(1000) & \quad\quad (4.4) \\
& 0.120x_1 + 0.011x_2 + 1.0x_6 \quad\geq 0.010(1000) & \text{(chromium)} \\
& 0.120x_1 + 0.011x_2 + 1.0x_6 \quad\leq 0.012(1000) & \\
& 0.001x_2 + 1.0x_7 \quad\geq 0.011(1000) & \text{(molybdenum)} \\
& 0.001x_2 + 1.0x_7 \quad\leq 0.013(1000) & \\
& x_1 \leq 75 & \text{(available)} \\
& x_2 \leq 250 & \\
& x_1, \ldots, x_7 \geq 0 & \text{(nonnegative)}
\end{array}
$$

The objective function merely sums the costs of the ingredients used. The only constraints not discussed above enforce supply limits on the first two types of scrap. Composition constraints have been simplified by taking advantage of the fact that total weight is fixed at 1000 kilograms.

The unique optimal solution to model (4.4) has

$$x_1^* = 75.00 \text{ kg}, \quad x_2^* = 90.91 \text{ kg}, \quad x_3^* = 672.28 \text{ kg}, \quad x_4^* = 137.31 \text{ kg}$$
$$x_5^* = 13.59 \text{ kg}, \quad x_6^* = 0.00 \text{ kg}, \quad x_7^* = 10.91 \text{ kg}$$

The total cost of an optimal charge is 9953.7 kroner.

Ratio Constraints

The composition constraints of our Swedish Steel example can be viewed as **ratio constraints** because they bound the fraction that one weighted sum of variables forms of another. For example, the lower limit on carbon is

$$0.0080x_1 + 0.0070x_2 + 0.0085x_3 + 0.0040x_4 \geq 0.0065 \sum_{j=1}^{7} x_j$$

or

$$\frac{0.0080x_1 + 0.0070x_2 + 0.0085x_3 + 0.0040x_4}{x_1 + x_2 + x_3 + x_4 + x_5 + x_6 + x_7} \geq 0.0065$$

In the latter form, the constraint does not even appear linear. Still, multiplying by the denominator, which we know will be nonnegative in every feasible solution, produces the linear version without reversing the direction of the inequality.

Many blending models have similar ratio constraints among ingredients. For example, if ingredients 3 and 4 must be in ratio no more than 2:3, we have the requirement that

$$\frac{x_3}{x_4} \leq \frac{2}{3}$$

Again, since we know the sign of denominator $x_4 \geq 0$, we may cross-multiply to produce the linear form

$$x_3 \leq \tfrac{2}{3}x_4$$

Contrast this with the ratio constraint

$$\frac{x_3}{x_1 - x_2} \leq \frac{2}{3}$$

Over nonnegative x_j the denominator of the latter constraint has an unpredictable sign. Without further information it must be considered nonlinear.

> __4.5__   **Ratio constraints**, which bound the quotient of linear functions by a constant, can often be converted to linear constraints by cross-multiplication. However, if the constraint is an inequality, the sign of the denominator function must be predictable over feasible solutions.

SAMPLE EXERCISE 4.3: FORMULATING RATIO CONSTRAINTS

Formulate linear constraints enforcing each of the following ratio requirements on nonnegative decision variables x_1, x_2, and x_3 determining the amount of three substances in a blend.

(a) The amounts of substance 1 and 2 should be in the ratio 4:7.

(b) The amount of substance 1 is at most half that of substance 3.

(c) The blend is at least 40% substance 1.

Modeling:

(a) The constraint required is

$$\frac{x_1}{x_2} = \frac{4}{7} \quad \text{or} \quad x_1 = \tfrac{4}{7}x_2$$

(b) The constraint needed is

$$x_1 \le \tfrac{1}{2}x_3$$

(c) The constraint specified is

$$\frac{x_1}{x_1 + x_2 + x_3} \ge 0.40$$

or (with the nonnegative denominator)

$$x_1 \ge 0.40(x_1 + x_2 + x_3)$$

4.3 OPERATIONS PLANNING MODELS

Another classic linear program form deals with **operations planning**. In organizations ranging from volunteer, to government, to manufacturing, to distribution, planners must decide what to do and when and where to do it. For example, Section 2.3's Pi Hybrids model (2.10) was used to plan the production and distribution of seed corn.

EXAMPLE 4.3: TUBULAR PRODUCTS OPERATIONS PLANNING

Sometimes a plan involves nothing more than allocation of work to operations. The Tubular Products Division (TP) of Babcock and Wilcox encountered just such a problem in investigating how work should be reallocated upon opening a new mill.[3] TP manufactured steel tubing in a variety of sizes and for many different uses, including electrical power generation. At the time of the study three mills handled production. The object was to consider how a fourth mill of different configuration would affect the optimal distribution of work (and associated costs) among the mills.

Table 4.3 shows fictional data for existing mills 1 to 3 and one design for new mill 4, versus an array of 16 products. The products comprise all combinations of standard or high-pressure tubing; $\tfrac{1}{2}$-, 1-, 2-, or 8-inch diameters; and thick or thin tube walls. The table includes the cost (in dollars) per 1000 pounds of each product according to which mill does the work, and the required processing time (in hours) per 1000 pounds produced. Missing values indicate products that cannot be manufactured feasibly at the mill indicated.

[3]Based on Wayne Drayer and Steve Seabury (1975), "Facilities Expansion Model," *Interfaces, 5:2*, part 2, 104–109.

Table 4.3 also shows the assumed division-wide demand for each of the 16 products in thousands of pounds per week. At present the three existing mills 1 to 3 have 800, 480, and 1280 hours per week of effective production capacity, respectively. New mill 4 is planned for 960 hours per week.

TABLE 4.3 Tubular Products Example Data

Product	Mill 1 Cost, $c_{p,1}$	Mill 1 Hours, $t_{p,1}$	Mill 2 Cost, $c_{p,2}$	Mill 2 Hours, $t_{p,2}$	Mill 3 Cost, $c_{p,3}$	Mill 3 Hours, $t_{p,3}$	Mill 4 Cost, $c_{p,4}$	Mill 4 Hours, $t_{p,4}$	Weekly Demand, d_p
Standard									
1: $\frac{1}{2}$ in. thick	90	0.8	75	0.7	70	0.5	63	0.6	100
2: $\frac{1}{2}$ in. thin	80	0.8	70	0.7	65	0.5	60	0.6	630
3: 1 in. thick	104	0.8	85	0.7	83	0.5	77	0.6	500
4: 1 in. thin	98	0.8	79	0.7	80	0.5	74	0.6	980
5: 2 in. thick	123	0.8	101	0.7	110	0.5	99	0.6	720
6: 2 in. thin	113	0.8	94	0.7	100	0.5	84	0.6	240
7: 8 in. thick	—	—	160	0.9	156	0.5	140	0.6	75
8: 8 in. thin	—	—	142	0.9	150	0.5	130	0.6	22
Pressure									
9: $\frac{1}{2}$ in. thick	140	1.5	110	0.9	—	—	122	1.2	50
10: $\frac{1}{2}$ in. thin	124	1.5	96	0.9	—	—	101	1.2	22
11: 1 in. thick	160	1.5	133	0.9	—	—	138	1.2	353
12: 1 in. thin	143	1.5	127	0.9	—	—	133	1.2	55
13: 2 in. thick	202	1.5	150	0.9	—	—	160	1.2	125
14: 2 in. thin	190	1.5	141	0.9	—	—	140	1.2	35
15: 8 in. thick	—	—	190	1.0	—	—	220	1.5	100
16: 8 in. thin	—	—	175	1.0	—	—	200	1.5	10

Tubular Products Operations Planning Model

In operations planning models, the decisions always revolve around what operations to undertake. Here the problem has two index dimensions:

$$p \triangleq \text{product number } (p = 1, \ldots, 16)$$

$$m \triangleq \text{mill number } (m = 1, \ldots, 4)$$

The corresponding decision variables are

$$x_{p,m} \triangleq \text{amount of product } p \text{ produced at mill } m$$
$$\text{(in thousands of pounds per week)}$$

In terms of these decision variables, it is straightforward to produce a linear programming model. For brevity, define the symbolic constants

$c_{p,m} \triangleq$ unit cost of producing product p at mill m shown in Table 4.3 [$= +\infty$ if this (p, m) combination is impossible]

$t_{p,m} \triangleq$ unit time to manufacture product p at mill m shown in Table 4.3 [$= 0$ if this (p, m) combination is impossible]

$d_p \triangleq$ weekly demand for product p shown in Table 4.3

$b_m \triangleq$ given production capacity at mill m

Then the model required is

$$\min \sum_{p=1}^{16} \sum_{m=1}^{4} c_{p,m} x_{p,m} \qquad \text{(total cost)}$$

$$\text{s.t.} \quad \sum_{m=1}^{4} x_{p,m} \ge d_p \qquad p = 1, \dots, 16 \qquad \text{(demands)}$$

$$\sum_{p=1}^{16} t_{p,m} x_{p,m} \le b_m \qquad m = 1, \dots, 4 \qquad \text{(capacities)}$$

$$x_{p,m} \ge 0 \qquad p = 1, \dots, 16; \quad m = 1, \dots, 4$$

$$(4.5)$$

The objective minimizes total production cost. One system of main constraints enforces product demands and the other mill capacities.

Table 4.4 shows an optimal solution $\mathbf{x}^*$. Old mill 1 should go virtually unused; mills 2 and 3 concentrate on pressure and standard tubing, respectively; and new mill 4 satisfies the remainder of demand for both. Total weekly cost is \$378,899.

TABLE 4.4   Tubular Products Example Optimum

Product		Mill 1, $x_{p,1}^*$	Mill 2, $x_{p,2}^*$	Mill 3, $x_{p,3}^*$	Mill 4, $x_{p,4}^*$
Standard					
1:	$\frac{1}{2}$ in. thick	0.0	0.0	0.0	100.0
2:	$\frac{1}{2}$ in. thin	0.0	0.0	630.0	0.0
3:	1 in. thick	0.0	0.0	404.8	95.2
4:	1 in. thin	0.0	0.0	980.0	0.0
5:	2 in. thick	0.0	0.0	0.0	720.0
6:	2 in. thin	0.0	0.0	0.0	240.0
7:	8 in. thick	—	0.0	0.0	75.0
8:	8 in. thin	—	0.0	0.0	22.0
Pressure					
9:	$\frac{1}{2}$ in. thick	0.0	50.0	—	0.0
10:	$\frac{1}{2}$ in. thin	0.0	22.0	—	0.0
11:	1 in. thick	0.0	214.1	—	138.9
12:	1 in. thin	55.0	0.0	—	0.0
13:	2 in. thick	0.0	125.0	—	0.0
14:	2 in. thin	0.0	0.0	—	35.0
15:	8 in. thick	—	100.0	—	0.0
16:	8 in. thin	—	10.0	—	0.0

EXAMPLE 4.4: CANADIAN FOREST PRODUCTS (CFPL) OPERATIONS PLANNING

Operations planning models become more complex when there are several stages of production. Activity at each stage consumes output of the preceding stage and creates input to the next stage.

Canadian Forest Products Limited (CFPL) employed such a model to plan their production of plywood.[4] Figure 4.1 shows the sequence of stages. Production

---

[4]Based on Dilip B. Kotak (1976), "Application of Linear Programming to Plywood Manufacture," *Interfaces, 7:1*, part 2, 56–68.

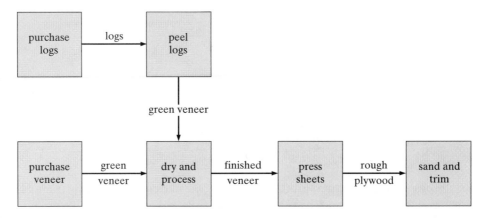

FIGURE 4.1   Plywood Processing Flow in CFPL Example

begins by purchasing logs and peeling them into strips of thin "green" veneer. Green veneer can also be purchased directly. All green veneer is next dried, classified by quality, and in some cases improved by patching knots and gluing thin strips together. After the veneer has been cut into sheet sizes, several layers are glued and pressed to produce plywood. A final production step sands completed plywood and trims it to exact size for sale.

The objective of CFPL's operations research analysis was to determine how to operate production facilities to maximize *contributed margin*: sales income less wood costs. Labor, maintenance, and other plant costs were assumed fixed. The principal constraint, other than limits on availability of wood and the market for various products, was the limited plant capacity to press plywood.

To have some numbers to work with, assume that logs are available from two vendors in "good" and "fair" qualities at the rate and price shown below. The table also shows the estimated yield in $\frac{1}{16}$- and $\frac{1}{8}$-inch green veneer of grades A, B, and C from peeling a log of the quality indicated.

	Veneer Yield (sq ft)			
	Vendor 1		Vendor 2	
	Good	Fair	Good	Fair
Available per month	200	300	100	1000
Cost per log ($ Canadian)	340	190	490	140
A $\frac{1}{16}$-inch green veneer (sq ft)	400	200	400	200
B $\frac{1}{16}$-inch green veneer (sq ft)	700	500	700	500
C $\frac{1}{16}$-inch green veneer (sq ft)	900	1300	900	1300
A $\frac{1}{8}$-inch green veneer (sq ft)	200	100	200	100
B $\frac{1}{8}$-inch green veneer (sq ft)	350	250	350	250
C $\frac{1}{8}$-inch green veneer (sq ft)	450	650	450	650

We can also purchase green veneer. Suppose that availabilities and purchase prices are as shown in the following table.

	$\frac{1}{16}$-Inch Green Veneer			$\frac{1}{8}$-Inch Green Veneer		
	A	**B**	**C**	**A**	**B**	**C**
Available (sq ft/month)	5000	25,000	40,000	10,000	40,000	50,000
Cost ($ Canadian/sq ft)	1.00	0.30	0.10	2.20	0.60	0.20

Our version of CFPL will make just 6 products—all 4- by 8-foot sheets of plywood for the U.S. market. A final table shows the composition of each product in veneer sheets, the available market per month, and the time required to glue and press each sheet of plywood out of a monthly capacity of 4500 hours.

	$\frac{1}{4}$-Inch Plywood Sheets			$\frac{1}{2}$-Inch Plywood Sheets		
	AB	**AC**	**BC**	**AB**	**AC**	**BC**
Front veneer	$\frac{1}{16}$A	$\frac{1}{16}$A	$\frac{1}{16}$B	$\frac{1}{16}$A	$\frac{1}{16}$A	$\frac{1}{16}$B
Core veneer	$\frac{1}{8}$C	$\frac{1}{8}$C	$\frac{1}{8}$C	$\frac{1}{8}$C $\frac{1}{8}$B $\frac{1}{8}$C	$\frac{1}{8}$C $\frac{1}{8}$B $\frac{1}{8}$C	$\frac{1}{8}$C $\frac{1}{8}$B $\frac{1}{8}$C
Back veneer	$\frac{1}{16}$B	$\frac{1}{16}$C	$\frac{1}{16}$C	$\frac{1}{16}$B	$\frac{1}{16}$C	$\frac{1}{16}$C
Market per month	1000	4000	8000	1000	5000	8000
Price ($ Canadian)	45.00	40.00	33.00	75.00	65.00	50.00
Pressing time (hours)	0.25	0.25	0.25	0.40	0.40	0.40

CFPL Decision Variables

As usual, we begin a model for the CFPL case by choosing variables deciding how much of what to do. Index dimensions include

$$q \triangleq \text{log quality } (q = \text{G for good, F for fair})$$
$$v \triangleq \text{log vendor number } (v = 1, 2)$$
$$t \triangleq \text{veneer thickness } (t = \tfrac{1}{16}, \tfrac{1}{8})$$
$$g \triangleq \text{veneer grade } (g = \text{A,B,C})$$

To formulate the problem as a linear program, we will use four classes of (continuous) decision variables over these index dimensions:

$w_{q,v,t} \triangleq$ number of logs of quality q bought from vendor v and peeled into green veneer of thickness t per month

$x_{t,g} \triangleq$ number of square feet of thickness t, grade g green veneer purchased directly per month

$y_{t,g,g'} \triangleq$ number of sheets of thickness t veneer used as grade g' after drying and processing from grade g green veneer per month

$z_{t,g,g'} \triangleq$ number of sheets of thickness t, front veneer grade g, back veneer grade g' plywood pressed and sold per month

Notice that these variables correspond to only 4 of the 6 processing boxes in Figure 4.1. This efficiency is possible because the way we have presented the problem offers no advantage for purchasing a log and not peeling it, or pressing a sheet of plywood and not sanding or selling it. If inventories had to be modeled, so that for example, purchased logs need not be immediately peeled, we would require many more variables.

Continuous Variables for Integer Quantities

Readers who are studying LP modeling for the first time may be perplexed about the fact that the CFPL decision variables are all treated as continuous. Don't quantities such as the number of logs and the number of sheets of plywood need to be integers? Indeed, how can CFPL's problem even be modeled as a linear program (which must have only continuous variables)?

Modeling physically integer quantities with continuous decision variables in this fashion is standard when optimal variable magnitudes are likely to be relatively large (principle $\boxed{1.11}$). If the LP-optimal number of plywood sheets sold of some type turns out to be, say, 953.2, there is little practical difficulty in rounding off to 953 sheets. After all, the costs, capacities, and other constants in the model are only estimates that contain a certain amount of error.

But we know that there is a big gain in tractability. Continuous optimization is almost always more efficient than discrete. To realize that gain without having much impact on the usability of optimal results, we choose to neglect integrality requirements.

> **4.6** | To gain tractability with little loss of validity, decision variables of relatively large magnitude are best modeled as continuous, even though they correspond to physically integer quantities.

Notice that this concession to tractability would be much more serious when decision variables were limited to, say, 0 and 1. If, for example, 0 means do not build a facility and 1 means build it, rounding continuous LP solutions could be much more problematic.

CFPL Objective Function

CFPL's maximum contributed margin objective is easily expressed in terms of the decision variables above. We compute

$$\max \ -(\text{log costs}) - (\text{purchased veneer costs}) + (\text{sales income})$$

That is,

$$
\begin{aligned}
\max \quad & -(340w_{G,1,1/16} + 190w_{F,1,1/16} + 490w_{G,2,1/16} + 140w_{F,2,1/16} \\
& + 340w_{G,1,1/8} + 190w_{F,1,1/8} + 490w_{G,2,1/8} + 140w_{F,2,1/8}) \\
& -(1.00x_{1/16,A} + 0.30x_{1/16,B} + 0.10x_{1/16,C} + 2.20x_{1/8,A} \\
& + 0.60x_{1/8,B} + 0.20x_{1/8,C}) + (45z_{1/4,A,B} + 40z_{1/4,A,C} \\
& + 33z_{1/4,B,C} + 75z_{1/2,A,B} + 65z_{1/2,A,C} + 50z_{1/2,B,C})
\end{aligned}
\tag{4.6}
$$

CFPL Constraints

Some constraints are equally easy. Log availability limits impose

$$
\begin{aligned}
w_{G,1,1/16} + w_{G,1,1/8} &\le 200, & w_{F,1,1/16} + w_{F,1,1/8} &\le 300 \\
w_{G,2,1/16} + w_{G,2,1/8} &\le 100, & w_{F,2,1/16} + w_{F,2,1/8} &\le 1000
\end{aligned}
\tag{4.7}
$$

purchased veneer availabilities imply that

$$x_{1/16,A} \le 5000, \quad x_{1/16,B} \le 25{,}000, \quad x_{1/16,C} \le 40{,}000$$
$$x_{1/8,A} \le 10{,}000, \quad x_{1/8,B} \le 40{,}000, \quad x_{1/8,C} \le 50{,}000 \tag{4.8}$$

and market sizes constrain

$$z_{1/4,A,B} \le 1000, \quad z_{1/4,A,C} \le 4000, \quad z_{1/4,B,C} \le 8000$$
$$z_{1/2,A,B} \le 1000, \quad z_{1/2,A,C} \le 5000, \quad z_{1/2,B,C} \le 8000 \tag{4.9}$$

Finally, the important pressing capacity limit yields the additional constraint

$$0.25(z_{1/4,A,B} + z_{1/4,A,C} + z_{1/4,B,C}) + 0.40(z_{1/2,A,B} + z_{1/2,A,C} + z_{1/2,B,C}) \le 4500 \tag{4.10}$$

Balance Constraints

So far we have done nothing to link log and veneer purchasing at the beginning of the process to sales at the end. In fact, we have not used the processing variables $y_{t,g,g'}$ at all.

What makes operations planning models with several processing stages special is the need to provide such links through balance constraints.

4.7 | A **balance constraint** assures that in-flows equal or exceed out-flows for materials and products created by one stage of production and consumed by others.

The first family of balance constraints needed in the CFPL model involves green veneer. Assume that with trim losses, 35 square feet of green veneer is required for each 4- by 8-foot sheet of finished veneer. We then have for each thickness and grade

(veneer from peeled logs) + (veneer purchased) $\ge$ 35(sheets of veneer finished)

Assuming that careful piecing and patching can permit green veneer of one grade to be used as the next higher, and veneer of any grade can be substituted for the next lower, we obtain the following six balance constraints for various grades and thicknesses of green veneer:

$$400w_{G,1,1/16} + 200w_{F,1,1/16} + 400w_{G,2,1/16} + 200w_{F,2,1/16} + x_{1/16,A}$$
$$\ge 35y_{1/16,A,A} + 35y_{1/16,A,B}$$

$$700w_{G,1,1/16} + 500w_{F,1,1/16} + 700w_{G,2,1/16} + 500w_{F,2,1/16} + x_{1/16,B}$$
$$\ge 35y_{1/16,B,A} + 35y_{1/16,B,B} + 35y_{1/16,B,C}$$

$$900w_{G,1,1/16} + 1300w_{F,1,1/16} + 900w_{G,2,1/16} + 1300w_{F,2,1/16} + x_{1/16,C}$$
$$\ge 35y_{1/16,C,B} + 35y_{1/16,C,C}$$

$$200w_{G,1,1/8} + 100w_{F,1,1/8} + 200w_{G,2,1/8} + 100w_{F,2,1/8} + x_{1/8,A} \tag{4.11}$$
$$\ge 35y_{1/8,A,A} + 35y_{1/8,A,B}$$

$$350w_{G,1,1/8} + 250w_{F,1,1/8} + 350w_{G,2,1/8} + 250w_{F,2,1/8} + x_{1/8,B}$$
$$\ge 35y_{1/8,B,A} + 35y_{1/8,B,B} + 35y_{1/8,B,C}$$

$$450w_{G,1,1/8} + 650w_{F,1,1/8} + 450w_{G,2,1/8} + 650w_{F,2,1/8} + x_{1/8,C}$$
$$\ge 35y_{1/8,C,B} + 35y_{1/8,C,C}$$

Six quite similar constraints enforce balance in sheets of finished veneer passing from the drying process to pressing:

$$\text{sheets finished for use at this grade} = \text{sheets consumed in pressing}$$

We can make the constraints equalities this time because no veneer would ever be finished unless it were going to be pressed. Again detailing for two thicknesses and three grades (other than the never-used $\frac{1}{8}$-inch, grade A finished veneer) gives

$$y_{1/16,A,A} + y_{1/16,B,A}$$
$$= z_{1/4,A,B} + z_{1/4,A,C} + z_{1/2,A,B} + z_{1/2,A,C}$$
$$y_{1/16,A,B} + y_{1/16,B,B} + y_{1/16,C,B}$$
$$= z_{1/4,A,B} + z_{1/4,B,C} + z_{1/2,A,B} + z_{1/2,B,C}$$
$$y_{1/16,B,C} + y_{1/16,C,C} \qquad\qquad (4.12)$$
$$= z_{1/4,A,C} + z_{1/4,B,C} + z_{1/2,A,C} + z_{1/2,B,C}$$
$$y_{1/8,A,B} + y_{1/8,B,B} + y_{1/8,C,B}$$
$$= z_{1/2,A,B} + z_{1/2,A,C} + z_{1/2,B,C}$$
$$y_{1/8,B,C} + y_{1/8,C,C}$$
$$= z_{1/4,A,B} + z_{1/4,A,C} + z_{1/4,B,C} + 2z_{1/2,A,B} + 2z_{1/2,A,C} + 2z_{1/2,B,C}$$

SAMPLE EXERCISE 4.4: FORMULATING BALANCE CONSTRAINTS

The following figure shows the assembly structure of two products:

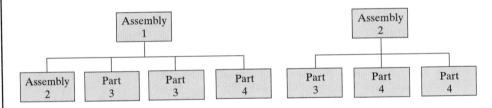

Use decision variables

$$x_j \triangleq \text{number of parts or assemblies } j \text{ produced}$$

to formulate balance constraints for assemblies/parts $j = 2, 3, 4$.

Modeling: Assembly 1 joins an assembly 2 with two part 3's and a part 4. Assembly 2 consists of one part 3 and two part 4's. Thus for $j = 2$, we require that the number of assembly 2's at least equal the number required for assembly 1's:

$$x_2 \geq x_1$$

Similarly, for parts $j = 3, 4$, we require production to at least meet requirements for the assemblies:

$$x_3 \geq 2x_1 + 1x_2$$
$$x_4 \geq 1x_1 + 2x_2$$

CFPL Example Model

Collecting (4.6)–(4.12) and adding variable-type constraints produces the full CFPL linear programming model detailed in Table 4.5. One optimal solution has the

TABLE 4.5   CFPL Example Linear Program Model

max $-(340w_{G,1,1/16} + 190w_{F,1,1/16} + 490w_{G,2,1/16} + 140w_{F,2,1/16}$	(logs)
$\quad + 340w_{G,1,1/8} + 190w_{F,1,1/8} + 490w_{G,2,1/8} + 140w_{F,2,1/8})$	
$\quad - (1.00x_{1/16,A} + 0.30x_{1/16,B} + 0.10x_{1/16,C}$	(veneer)
$\quad + 2.20x_{1/8,A} + 0.60x_{1/8,B} + 0.20x_{1/8,C})$	
$\quad + (45z_{1/4,A,B} + 40z_{1/4,A,C} + 33z_{1/4,B,C}$	(sales)
$\quad + 75z_{1/2,A,B} + 65z_{1/2,A,C} + 50z_{1/2,B,C})$	

s.t.
$$w_{G,1,1/16} + w_{G,1,1/8} \leq 200, \; w_{F,1,1/16} + w_{F,1,1/8} \leq 300 \quad \text{(log availability)}$$
$$w_{G,2,1/16} + w_{G,2,1/8} \leq 100, \; w_{F,2,1/16} + w_{F,2,1/8} \leq 1000$$
$$x_{1/16,A} \leq 5000, \; x_{1/16,B} \leq 25,000, \; x_{1/16,C} \leq 40,000 \quad \text{(veneer}$$
$$x_{1/8,A} \leq 10,000, \; x_{1/8,B} \leq 40,000, \; x_{1/8,C} \leq 50,000 \quad \text{availability)}$$
$$z_{1/4,A,B} \leq 1000, \; z_{1/4,A,C} \leq 4000, \; z_{1/4,B,C} \leq 8000 \quad \text{(market)}$$
$$z_{1/2,A,B} \leq 1000, \; z_{1/2,A,C} \leq 5000, \; z_{1/2,B,C} \leq 8000$$
$$0.25(z_{1/4,A,B} + z_{1/4,A,C} + z_{1/4,B,C}) + 0.40(z_{1/2,A,B} + z_{1/2,A,C} + z_{1/2,B,C}) \leq 4500 \quad \text{(pressing)}$$
$$400w_{G,1,1/16} + 200w_{F,1,1/16} + 400w_{G,2,1/16} + 200w_{F,2,1/16} + x_{1/16,A} \quad \text{(green}$$
$$\geq 35y_{1/16,A,A} + 35y_{1/16,A,B} \quad \text{veneer}$$
$$700w_{G,1,1/16} + 500w_{F,1,1/16} + 700w_{G,2,1/16} + 500w_{F,2,1/16} + x_{1/16,B} \quad \text{balance)}$$
$$\geq 35y_{1/16,B,A} + 35y_{1/16,B,B} + 35y_{1/16,B,C}$$
$$900w_{G,1,1/16} + 1300w_{F,1,1/16} + 900w_{G,2,1/16} + 1300w_{F,2,1/16} + x_{1/16,C}$$
$$\geq 35y_{1/16,C,B} + 35y_{1/16,C,C}$$
$$200w_{G,1,1/8} + 100w_{F,1,1/8} + 200w_{G,2,1/8} + 100w_{F,2,1/8} + x_{1/8,A}$$
$$\geq 35y_{1/8,A,A} + 35y_{1/8,A,B}$$
$$350w_{G,1,1/8} + 250w_{F,1,1/8} + 350w_{G,2,1/8} + 250w_{F,2,1/8} + x_{1/8,B}$$
$$\geq 35y_{1/8,B,A} + 35y_{1/8,B,B} + 35y_{1/8,B,C}$$
$$450w_{G,1,1/8} + 650w_{F,1,1/8} + 450w_{G,2,1/8} + 650w_{F,2,1/8} + x_{1/8,C}$$
$$\geq 35y_{1/8,C,B} + 35y_{1/8,C,C}$$
$$y_{1/16,A,A} + y_{1/16,B,A} = z_{1/4,A,B} + z_{1/4,A,C} + z_{1/2,A,B} + z_{1/2,A,C} \quad \text{(finished}$$
$$y_{1/16,A,B} + y_{1/16,B,B} + y_{1/16,C,B} = z_{1/4,A,B} + z_{1/4,B,C} \quad \text{veneer}$$
$$\quad + z_{1/2,A,B} + z_{1/2,B,C} \quad \text{balance)}$$
$$y_{1/16,B,C} + y_{1/16,C,C} = z_{1/4,A,C} + z_{1/4,B,C} + z_{1/2,A,C} + z_{1/2,B,C}$$
$$y_{1/8,A,B} + y_{1/8,B,B} + y_{1/8,C,B} = z_{1/2,A,B} + z_{1/2,A,C} + z_{1/2,B,C}$$
$$y_{1/8,B,C} + y_{1/8,C,C} = z_{1/4,A,B} + z_{1/4,A,C} + z_{1/4,B,C} + 2z_{1/2,A,B}$$
$$\quad + 2z_{1/2,A,C} + 2z_{1/2,B,C}$$
all variables $w_{q,v,t}, x_{t,g}, y_{t,g,g'}, z_{t,g,g'} \geq 0$

following variables nonzero:

$$w^*_{G,1,1/16} = 41.3, \quad w^*_{F,1,1/16} = 300.0, \quad w^*_{F,2,1/16} = 155.3$$
$$w^*_{F,2,1/8} = 844.7, \quad x^*_{1/16,C} = 40,000.0, \quad x^*_{1/8,C} = 50,000.0$$
$$y^*_{1/16,A,A} = 3073.2, \quad y^*_{1/16,B,A} = 7329.4, \quad y^*_{1/16,C,B} = 6355.8$$
$$y^*_{1/16,C,C} = 12,758.4, \quad y^*_{1/8,A,B} = 2413.5, \quad y^*_{1/8,B,C} = 6033.8 \quad (4.13)$$
$$y^*_{1/8,C,B} = 2989.1, \quad y^*_{1/8,C,C} = 14,127.3,$$
$$z^*_{1/4,A,B} = 1000.0, \quad z^*_{1/4,A,C} = 4000.0, \quad z^*_{1/4,B,C} = 4355.8$$
$$z^*_{1/2,A,B} = 1000.0, \quad z^*_{1/2,A,C} = 4402.6$$

The firm should enter all markets except the one for $\frac{1}{2}$-inch BC plywood. Total contributed margin is $484,878 Canadian per month.

Fractions in such variables as the number of sheets of plywood sold are physically impossible. Still, the advantage of globally solving this complex optimization efficiently with linear programming far outweighs the minute inaccuracy associated with rounding the LP optimum to obtain a plan.

SAMPLE EXERCISE 4.5: FORMULATING OPERATIONS PLANNING LPS

An orange juice company can sell up to 15,000 tons of juice to wholesalers at $1500 per ton. The juice is either squeezed from oranges purchased at $200 per ton or diluted from concentrate obtained at $1600 per ton. Approximately 15,000 tons of juice oranges are available and each yields 0.2 ton of juice. The supply of concentrate is essentially unlimited, and each ton dilutes into 2 tons of juice. Formulate a linear program to choose an operating plan that maximizes the company's net income (sales minus cost).

Modeling: We define decision variables for each of the 3 operations:

$$x_1 \triangleq \text{tons of oranges squeezed for juice}$$
$$x_2 \triangleq \text{tons of concentrated diluted for juice}$$
$$x_3 \triangleq \text{tons of juice sold}$$

Then the required model is

$$
\begin{array}{llll}
\max & -200x_1 - 1600x_2 + 1500x_3 & \text{(net income)} \\
\text{s.t.} & x_1 \leq 10,000 & \text{(orange availability)} \\
& x_3 \leq 15,000 & \text{(sales limit)} \\
& 0.2x_1 + 2x_2 = x_3 & \text{(balance)} \\
& x_1, x_2, x_3 \geq 0
\end{array}
$$

Its objective function maximizes the difference of purchase cost and sales revenue. The first main constraint enforces the limit on orange availability, the second limits sales to 15,000, and the third balances production with sales.

4.4 SHIFT SCHEDULING AND STAFF PLANNING MODELS

Operations planning models decide what work to undertake so that available resources are used efficiently. In **shift scheduling** or **staff planning models** the work is already fixed. We must now plan the resources to accomplish it. In particular, we must decide how many of what types of workers and shifts best cover all work requirements. Again, LP provides a powerful tool.

EXAMPLE 4.5: OHIO NATIONAL BANK (ONB) SHIFT SCHEDULING

The Ohio National Bank (ONB) confronted such a problem in staffing its check processing center.[5] Checks received by the bank already have account numbers and other identifying information encoded on them. Machine operators in the check processing center key the dollar amount of the check, which is then imprinted with the other information for computerized processing.

---

[5] Based on L. J. Krajewski, L. P. Ritzman, and Phil McKenzie (1980), "Shift Scheduling in Banking Operations: A Case Application," *Interfaces, 10:2,* 1–8.

Checks arrive through the business day in volumes peaking in the early evening. Our fictitious version will assume the following arrivals (in thousands):

Hour	Arrivals	Hour	Arrivals
11:00 (11 A.M.)	10	17:00 (5 P.M)	32
12:00 (noon)	11	18:00 (6 P.M)	50
13:00 (1 P.M.)	15	19:00 (7 P.M)	30
14:00 (2 P.M.)	20	20:00 (8 P.M)	20
15:00 (3 P.M.)	25	21:00 (9 P.M)	8
16:00 (4 P.M.)	28	—	—

Uncollected checks cost the bank money in lost interest. Thus it is essential that all checks be processed in time for collection on the next business day. ONB decided to enforce a requirement that all checks be completed by 22:00 (10 P.M.). Furthermore, the number unprocessed at any hour should not exceed 20 thousand.

Two types of employees can perform the check processing task. Full-time employees work an 8-hour shift with a 1-hour lunch break in the middle. Part-time employees work only 4 hours per day with no lunch. Both types of shifts can begin at any hour of the day, and full-time employees can be assigned an hour of overtime. Table 4.6 illustrates the possible shifts.

TABLE 4.6   Possible Shifts in ONB Example[a]

Start	Full-Time Shifts			Part-Time Shifts							
	11	12	13	11	12	13	14	15	16	17	18
11:00	R	—	—	R	—	—	—	—	—	—	—
12:00	R	R	—	R	R	—	—	—	—	—	—
13:00	R	R	R	R	R	R	—	—	—	—	—
14:00	R	R	R	R	R	R	R	—	—	—	—
15:00	—	R	R	—	R	R	R	R	—	—	—
16:00	R	—	R	—	—	R	R	R	R	—	—
17:00	R	R	—	—	—	—	R	R	R	R	—
18:00	RN	RN	RN	—	—	—	—	RN	RN	RN	RN
19:00	RN	RN	RN	—	—	—	—	—	RN	RN	RN
20:00	ON	RN	RN	—	—	—	—	—	—	RN	RN
21:00	—	ON	RN	—	—	—	—	—	—	—	RN

[a] R, regular duty; O, possible overtime; N, night differential.

In our analysis we assume that full-time employees receive $11 per hour in pay and benefits, plus an extra $1 per hour in "night differential" for time after 6 P.M. and 150% pay for daily overtime. Part-time employees are paid $7 per hour, plus $1 per hour night differential after 6 P.M. Also, to keep overtime under control, we require that no more than half the full-time employees on any shift work overtime and that the total number of scheduled overtime hours not exceed 20 per day.

Naturally, full-time employees work faster than part-timers. We will assume that full-time operators process 1000 checks per hour, and part-timers only 800.

One final complication is encoding stations. The number of machines available limits the number of employees who can work at any one time. Our center will have 35 machines.

ONB Decision Variables and Objective Function

The main decisions to be made in shift scheduling models are the number of employees to work various shifts. In the ONB case we have all the possibilities in Table 4.6. For example, the full-time shift starting at 11:00 works 4 hours, then takes a lunch break, then works 4 more hours. The final 2 hours come after 6 P.M., so a night differential applies. One additional hour may also be worked in overtime.

Using the index

$$h \triangleq (24\text{-hour clock}) \text{ shift start time}$$

we define the corresponding decision variables:

$x_h \triangleq$ number of full-time employees beginning a shift at hour h
$\quad (h = 11, \ldots, 13)$

$y_h \triangleq$ number of full-time employees with shift beginning at hour h who work
$\quad$ overtime $(h = 11, 12)$

$z_h \triangleq$ number of part-time employees beginning a shift at hour h
$\quad (h = 11, \ldots, 18)$

We need only add up the pay for each shift to obtain a minimum (daily) cost objective function:

$$\begin{align} \min \quad & 90x_{11} + 91x_{12} + 92x_{13} + 18y_{11} + 18y_{12} + 28z_{11} + 28z_{12} \\ & + 28z_{13} + 28z_{14} + 29z_{15} + 30z_{16} + 31z_{17} + 32z_{18} \end{align} \tag{4.14}$$

For example, the coefficient on x_{13} reflects 8 regular hours at \$11 per hour plus \$4 for the 4 hours after 6 P.M., or

$$8(\$11) + 4(\$1) = \$92$$

ONB Constraints

Table 4.6 also suggests how to model the requirement that no more than 35 operators be on duty at any time. We simply constrain the sum of full-time, overtime, and part-time employees on duty in each hour.

$$\begin{align} x_{11} + z_{11} & \leq 35 \quad \text{(11:00 machines)} \\ x_{11} + x_{12} + z_{11} + z_{12} & \leq 35 \quad \text{(12:00 machines)} \\ \vdots \qquad\qquad & \quad\vdots \qquad\qquad \vdots \\ y_{11} + x_{12} + x_{13} + z_{17} + z_{18} & \leq 35 \quad \text{(20:00 machines)} \\ y_{12} + x_{13} + z_{18} & \leq 35 \quad \text{(21:00 machines)} \end{align} \tag{4.15}$$

There are also overtime limits. Overtime cannot exceed half of any full-time shift or total more than 20 hours per day. These limits lead us to the constraints

$$\begin{align} y_{11} & \leq \tfrac{1}{2}x_{11} \quad \text{(11-shift overtime)} \\ y_{12} & \leq \tfrac{1}{2}x_{12} \quad \text{(12-shift overtime)} \\ y_{11} + y_{12} & \leq 20 \quad \text{(total overtime)} \end{align} \tag{4.16}$$

Covering Constraints

The main element in any staff planning model is a collection of covering constraints.

> **4.8** | **Covering constraints** in shift scheduling models assure that the shifts chosen provide enough worker output to cover requirements over each time period; that is,
>
> $$\sum_{\text{shifts}} (\text{output/worker})(\text{number on duty}) \geq \text{period requirement}$$

With the ONB case we have a slight complication in covering requirements. Work arrivals are specified on an hour-by-hour basis, but work completion is limited only by all checks being finished at 22:00 (10 P.M.). To model covering in such a case, we need some new decision variables reflecting the work carried over. Specifically, define

$$w_h \triangleq \text{uncompleted work backlog at (24-hour clock) hour } h \text{ (in thousands)}$$

Then our ONB covering constraints take the form

$$
\begin{array}{lll}
1x_{11} + 0.8z_{11} & \geq 10 - w_{12} & \text{(11:00 cover)} \\
1x_{11} + 1x_{12} + 0.8z_{11} + 0.8z_{12} & \geq 11 + w_{12} - w_{13} & \text{(12:00 cover)} \\
\qquad\qquad \vdots & \qquad \vdots & \qquad \vdots \qquad\quad (4.17) \\
1y_{11} + 1x_{12} + 1x_{13} + 0.8z_{17} + 0.8z_{18} & \geq 20 + w_{20} - w_{21} & \text{(20:00 cover)} \\
1y_{12} + 1x_{13} + 0.8z_{18} & \geq 8 + w_{21} & \text{(21:00 cover)}
\end{array}
$$

For example, the one for the 20:00 hour requires the total output of workers on duty from 20:00 to 21:00 to equal or exceed the 20 thousand checks arriving at that hour (see the table at the beginning of Example 4.5), plus checks held over from previous hours (w_{20}), less those passed on to later hours (w_{21}).

ONB Shift Scheduling Example Model

Combining (4.14)–(4.17) with suitable variable-type constraints and upper bounds of 20 on all backlog variables w_h produces the full ONB shift scheduling linear program shown in Table 4.7. An LP optimum makes the following variables nonzero:

$$
\begin{array}{lllll}
x_{12}^* = 8.57, & x_{13}^* = 12.86, & y_{12}^* = 4.29, & z_{14}^* = 13.57, & z_{16}^* = 5.36, \\
z_{17}^* = 7.50 & z_{18}^* = 0.71, & w_{12}^* = 10.00, & w_{13}^* = 12.43, \\
w_{14}^* = 6.00, & w_{18}^* = 2.29, & w_{19}^* = 20.00 & w_{20}^* = 17.71, & w_{21}^* = 9.71
\end{array}
$$

That is, full-time employees carry the load early in the day, with part-time beginning at 14:00 (2:00 P.M.) Total cost is $2836 per day.

Once again we have a fractional solution that certainly must be implemented in whole numbers of employees. Managers will need to round above LP-optimal values to obtain a satisfactory plan. Still, any loss of optimality resulting from rounding will fall well within the variability of hourly check arrivals and other data. Unless the numbers of persons working shifts are in the range of, say, 0 to 2, our LP model (4.7) is a valid approximation justified by its outstanding tractability.

TABLE 4.7   ONB Shift Scheduling Example LP Model

min	$90x_{11} + 91x_{12} + 92x_{13} + 18y_{11} + 18y_{12} + 28z_{11} + 28z_{12}$	(total pay)
	$+\, 28z_{13} + 28z_{14} + 29z_{15} + 30z_{16} + 31z_{17} + 32z_{18}$	
s.t.	$x_{11} + z_{11}$ ≤ 35	(11:00 machine)
	$x_{11} + x_{12} + z_{11} + z_{12}$ ≤ 35	(12:00 machine)
	$x_{11} + x_{12} + x_{13} + z_{11} + z_{12} + z_{13}$ ≤ 35	(13:00 machine)
	$x_{11} + x_{12} + x_{13} + z_{11} + z_{12} + z_{13} + z_{14}$ ≤ 35	(14:00 machine)
	$x_{12} + x_{13} + z_{12} + z_{13} + z_{14} + z_{15}$ ≤ 35	(15:00 machine)
	$x_{11} + x_{13} + z_{13} + z_{14} + z_{15} + z_{16}$ ≤ 35	(16:00 machine)
	$x_{11} + x_{12} + z_{14} + z_{15} + z_{16} + z_{17}$ ≤ 35	(17:00 machine)
	$x_{11} + x_{12} + x_{13} + z_{15} + z_{16} + z_{17} + z_{18}$ ≤ 35	(18:00 machine)
	$x_{11} + x_{12} + x_{13} + z_{16} + z_{17} + z_{18}$ ≤ 35	(19:00 machine)
	$y_{11} + x_{12} + x_{13} + z_{17} + z_{18}$ ≤ 35	(20:00 machine)
	$y_{12} + x_{13} + z_{18}$ ≤ 35	(21:00 machine)
	y_{11} $\le \frac{1}{2}x_{11}$	(11-shift overtime)
	y_{12} $\le \frac{1}{2}x_{12}$	(12-shift overtime)
	$y_{11} + y_{12}$ ≤ 20	(total overtime)
	$1x_{11} + 0.8z_{11}$ $\ge 10 - w_{12}$	(11:00 cover)
	$1x_{11} + 1x_{12} + 0.8z_{11} + 0.8z_{12}$ $\ge 11 + w_{12} - w_{13}$	(12:00 cover)
	$1x_{11} + 1x_{12} + 1x_{13} + 0.8z_{11} + 0.8z_{12} + 0.8z_{13}$ $\ge 15 + w_{13} - w_{14}$	(13:00 cover)
	$1x_{11} + 1x_{12} + 1x_{13} + 0.8z_{11} + 0.8z_{12} + 0.8z_{13} + 0.8z_{14}$ $\ge 20 + w_{14} - w_{15}$	(14:00 cover)
	$1x_{12} + 1x_{13} + 0.8z_{12} + 0.8z_{13} + 0.8z_{14} + 0.8z_{15}$ $\ge 25 + w_{15} - w_{16}$	(15:00 cover)
	$1x_{11} + 1x_{13} + 0.8z_{13} + 0.8z_{14} + 0.8z_{15} + 0.8z_{16}$ $\ge 28 + w_{16} - w_{17}$	(16:00 cover)
	$1x_{11} + 1x_{12} + 0.8z_{14} + 0.8z_{15} + 0.8z_{16} + 0.8z_{17}$ $\ge 32 + w_{17} - w_{18}$	(17:00 cover)
	$1x_{11} + 1x_{12} + 1x_{13} + 0.8z_{15} + 0.8z_{16} + 0.8z_{17} + 0.8z_{18}$ $\ge 50 + w_{18} - w_{19}$	(18:00 cover)
	$1x_{11} + 1x_{12} + 1x_{13} + 0.8z_{16} + 0.8z_{17} + 0.8z_{18}$ $\ge 30 + w_{19} - w_{20}$	(19:00 cover)
	$1y_{11} + 1x_{12} + 1x_{13} + 0.8z_{17} + 0.8z_{18}$ $\ge 20 + w_{20} - w_{21}$	(20:00 cover)
	$1y_{12} + 1x_{13} + 0.8z_{18}$ $\ge 8 + w_{21}$	(21:00 cover)
	all variables $w_h \le 20$	
	all variables $w_h, x_h, y_h, z_h \ge 0$	

SAMPLE EXERCISE 4.6: FORMULATING SHIFT SCHEDULING LPs

Clerical employees of a government agency are allowed to work four 10-hour days per week in any of the following patterns:

$j = 1$	Monday–Wednesday–Thursday–Friday
$j = 2$	Monday–Tuesday–Thursday–Friday
$j = 3$	Monday–Tuesday–Wednesday–Friday

Formulate a linear program to determine the minimum number of employees needed to have at least 10 on duty Mondays, 9 in the office on Fridays, and 7 working on Tuesdays through Thursdays.

Modeling: We employ decision variables

$$x_j \triangleq \text{number of employees working pattern } j$$

The required LP model is then

min	$x_1 + x_2 + x_3$		(total staff)
s.t.	$x_1 + x_2 + x_3$	≥ 10	(cover Monday)
	$x_2 + x_3$	≥ 7	(cover Tuesday)
	$x_1 + x_3$	≥ 7	(cover Wednesday)
	$x_1 + x_2$	≥ 7	(cover Thursday)
	$x_1 + x_2 + x_3$	≥ 9	(cover Friday)
	$x_1, x_2, x_3 \ge 0$		

The objective minimizes total staff, and the constraints enforce the specified coverage on all working days.

4.5 TIME-PHASED MODELS

So far in this chapter we have formulated only **static models**—those where all planning is for a single period of time. Many, perhaps most, linear programs are **dynamic** or **time phased** because they address circumstances that vary over time. In this section we introduce time-phased modeling.

EXAMPLE 4.6: INSTITUTIONAL FOOD SERVICES (IFS) CASH FLOW

LP models of almost any type may require time-phased decision making, but some of the most obviously time dependent involve **cash flow** management.[6] Every business must keep track of the coming and going of its cash accounts, borrowing where necessary and investing when wise.

We illustrate the modeling issues with a fictional Institutional Food Services company (IFS) that supplies food and other products to restaurants, schools, and similar institutions. Table 4.8 shows IFS's projections of some relevant accounts over the next 8 weeks (in thousands of dollars).

$s_t \triangleq$ projected revenue in week t from cash sales to small customers

$r_t \triangleq$ projected accounts receivable revenue received in week t from large customers who buy on credit

$p_t \triangleq$ projected accounts payable to IFS's suppliers in week t

$e_t \triangleq$ projected payroll, utility, and other expenses to be paid in week t

Cash sales and accounts receivable produce immediate income to IFS's checking account. Expenses are immediate deductions. Accounts payable amounts p_t are not actually due until week $t + 3$, but they are discounted by 2% if paid early in week t.

TABLE 4.8   IFS Cash Flow Example Data

Item	Projected Weekly Amount ($ 000's) for Week:							
	1	2	3	4	5	6	7	8
Cash Sales, s_t	600	750	1200	2100	2250	180	330	540
Accounts receivable, r_t	770	1260	1400	1750	2800	4900	5250	420
Accounts payable, p_t	3200	5600	6000	480	880	1440	1600	2000
Expenses, e_t	350	400	550	940	990	350	350	410

Values in Table 4.8 vary dramatically over the period as a holiday approaches. Besides the option on accounts payable, IFS's financial officer has two additional ways of dealing with the implied cash flow difficulties. First, the company's bank has extended a $4 million line of credit that may be drawn upon at 0.2% interest per week. However, the bank requires at least 20% of the current amount borrowed to be maintained (without earning interest) in IFS's checking account. The other

---

[6]Based on A. A. Robichek, D. Teichroew, and J. M. Jones (1965), "Optimal Short Term Financing Decision," *Management Science, 12*, 1–36.

option is investment of excess cash in short-term money markets. IFS can earn 0.1% interest per week on amounts invested in this way.

The financial officer wishes to minimize net total cost in interest and lost discounts while maintaining at least a $20,000 checking account safety balance. Our task is to help him decide how to exercise the available options.

Time-Phased Decision Variables

Time is always an index dimension in time-phased models because both input constants and decisions may be repeated in each time period. For our IFS example, time is the only index dimension. Decision variables for the three cash flow management options are (in thousands of dollars)

$g_t \triangleq$ amount borrowed in week t against the line of credit
$h_t \triangleq$ amount of line of credit debt paid off in week t
$w_t \triangleq$ amount of accounts payable in week t delayed until week $t + 3$ at a
      loss of discounts
$x_t \triangleq$ amount invested in short-term money markets during week t

For modeling convenience, we also define

$y_t \triangleq$ cumulative line of credit debt in week t
$z_t \triangleq$ cash on hand during week t

These variables could be eliminated by substituting suitable sums of the others, but many constraints are much easier to express when the extra variables are included.

Time-Phased Balance Constraints

Although separate decisions may be made in each period of a time-phased model, choices for different periods are rarely independent. Decisions in one period usually imply consequences that carry over into the next.

Such interactions among decisions for different time periods can often be modeled with balance constraints similar to those of definition 4.7 .

4.9   Time-phased models often link decisions in successive time periods with **balance constraints** of the form

$$\begin{pmatrix} \text{starting} \\ \text{level in} \\ \text{period } t \end{pmatrix} + \begin{pmatrix} \text{impacts of} \\ \text{period } t \\ \text{decisions} \end{pmatrix} = \begin{pmatrix} \text{starting} \\ \text{level in} \\ \text{period } t + 1 \end{pmatrix}$$

tracking commodities carried over from each period t to the next.

In our IFS example there are two main quantities carried over in this way: cash and debt. To develop the required balance constraints, we first enumerate the cash increments and decrements each week:

Cash Increments	Cash Decrements
Funds borrowed in week t	Borrowing paid off in week t
Investment principal from week $t-1$	Investment in week t
Interest on investment in week $t-1$	Interest on debt in week $t-1$
Cash sales from week t	Expenses paid in week t
Accounts receivable for week t	Accounts payable paid with discount for week t
	Accounts payable paid without discount for week $t-3$

Using the symbols defined above, these increments and decrements lead to the following system of balance constraints:

$$z_{t-1} + g_t - h_t + x_{t-1} - x_t + 0.001x_{t-1} - 0.002y_{t-1}$$
$$+ s_t - e_t + r_t - 0.98(p_t - w_t) - w_{t-3} = z_t \qquad t = 1, \ldots, 8 \qquad \text{(cash balance)}$$

(All symbols with subscripts outside the range $1, \ldots, 8$ are assumed $= 0$.)

A similar constraint system tracks cumulative debt. New borrowing increases, and paying off decreases:

$$y_{t-1} + g_t - h_t = y_t \qquad t = 1, \ldots, 8 \qquad \text{(debt balance)}$$

SAMPLE EXERCISE 4.7: FORMULATING BALANCE CONSTRAINTS OVER TIME

An LP model will decide

$x_q \triangleq$ thousands of snow shovels produced in quarter q

$i_q \triangleq$ thousands of snow shovels held in inventory at the end of quarter q

to meet customer demands for 11,000, 48,000, 64,000, and 15,000 shovels in quarters $q = 1, \ldots, 4$, respectively. Write balance constraints in shovels for the four quarters assuming inventory at the beginning of the first quarter $= 0$.

Modeling: Following principle 4.9 , the constraints will have the form

(beginning inventory) + (production) = (demand) + (ending inventory)

Now taking initial inventory $= 0$, we have

$$\begin{aligned}
0 + x_1 &= 11 + i_1 && \text{(quarter 1)} \\
i_1 + x_2 &= 48 + i_2 && \text{(quarter 2)} \\
i_2 + x_3 &= 64 + i_3 && \text{(quarter 3)} \\
i_3 + x_4 &= 15 + i_4 && \text{(quarter 4)}
\end{aligned}$$

IFS Cash Flow Model

We are now ready to state a full linear programming model for our IFS case:

$$\min \ 0.002 \sum_{t=1}^{8} y_t + 0.02 \sum_{t=1}^{8} w_t - 0.001 \sum_{t=1}^{8} x_t \qquad \text{(net interest)}$$

$$\begin{aligned}
\text{s.t.} \quad & z_{t-1} + g_t - h_t + x_{t-1} - x_t \\
& \quad + 0.001x_{t-1} - 0.002y_{t-1} + s_t - e_t \\
& \quad + r_t - 0.98(p_t - w_t) - w_{t-3} = z_t && t = 1, \ldots, 8 && \text{(cash balance)} \\
& y_{t-1} + g_t - h_t = y_t && t = 1, \ldots, 8 && \text{(debt balance)} \\
& y_t \le 4000 && t = 1, \ldots, 8 && \text{(credit limit)} \\
& z_t \ge 0.20y_t && t = 1, \ldots, 8 && \text{(bank rule)} \\
& w_t \le p_t && t = 1, \ldots, 8 && \text{(payables limit)} \\
& z_t \ge 20 && t = 1, \ldots, 8 && \text{(safety balance)} \\
& g_t, h_t, w_t, x_t, y_t, z_t \ge 0 && t = 1, \ldots, 8 && \text{(variable type)}
\end{aligned}$$

(4.18)

The objective function minimizes interest paid, plus discounts lost, less interest earned. Besides the two systems of balance requirements formulated above, constraints enforce the credit limit and the bank rule requiring that 20% of borrowed funds be kept as cash, keep delayed accounts payable within the value from Table 4.8, and ensure a continuing safety balance of $20,000. All symbols with subscripts outside the range $1, \ldots, 8$ are assumed $= 0$.

Table 4.9 presents an optimal solution. The corresponding optimal net interest and discounts total $158,492 for the 8 weeks.

TABLE 4.9   IFS Cash Flow Optimal Solution

Decision Variable	Optimal Weekly Amount ($ 000's) for Week:							
	1	**2**	**3**	**4**	**5**	**6**	**7**	**8**
Borrowing, g_t	100.0	505.7	3394.3	0.0	442.6	0.0	0.0	0.0
Debt payment, h_t	0.0	0.0	0.0	442.5	0.0	2715.3	1284.7	0.0
Payables delayed, w_t	2077.6	3544.5	1138.5	0.0	0.0	0.0	0.0	0.0
Short-term investments, x_t	0.0	0.0	0.0	0.0	0.0	0.0	2611.7	1204.3
Cumulative debt, y_t	100.0	605.7	4000.0	3557.4	4000.0	1284.7	0.0	0.0
Cumulative cash, z_t	20.0	121.1	800.0	711.5	800.0	256.9	20.0	20.0

Time Horizons

A **time horizon** establishes the range of time periods in a time-phased model. For example, our IFS cash flow example uses a **fixed time horizon** of $1, \ldots, 8$ because we model only 8 weeks.

Of course, IFS would have been operating before the current 8-week period and will continue operations after it. Thus our use of a fixed time horizon raises some special concerns about periods near the boundary.

In particular, model (4.18) assumes that all quantities outside the time horizon equal zero. Thus IFS begins week 1 with zero cash balance z_0, zero debt y_0, and zero short-term investments x_0. Optimal results in Table 4.9 could change dramatically if some of these boundary values proved to be nonsensical. To obtain a more valid model, it might be necessary to estimate typical values and include them as constants in balance equations for week 1.

At the other end of the time horizon we have similar issues. Although the optimum in Table 4.9 chose not to delay accounts payable in the last few weeks, it might have been severely tempted. Payables delayed in the last 3 weeks never have to be paid because the due date $(t + 3)$ falls beyond the time horizon.

Such issues do require particular attention if models with time horizons are to produce valid results.

4.10	Use of fixed time horizons, although necessary in most time-phased models, requires extra care in modeling and interpreting phenomena near both ends of the time epoch being modeled.

One way to avoid having to think about boundaries of a fixed time horizon is to employ an **infinite time horizon** model. Infinite horizon schemes "wrap around" output states of a last time period as input conditions for the first. The result is that time goes on infinitely even though only a few periods are modeled explicitly.

> **4.11** Infinite horizon modeling can avoid some of the boundary difficulties with finite horizons by treating the first explicitly modeled period as coming immediately after the last.

Infinite horizon modeling of our IFS example would treat $t = 1$ as the period immediately after week $t = 8$. Then, for example, the debt balance constraint for $t = 1$ would read

$$y_8 + g_1 - h_1 = y_1$$

SAMPLE EXERCISE 4.8: MODELING WITH TIME HORIZONS

Return to the snow shovel problem of Sample Exercise 4.7, and write balance constraints for the four quarters under each of the following assumptions about inventory at time horizon boundaries.

(a) The time horizon is a fixed four quarters, with beginning inventory in the first quarter of 9000 shovels.

(b) The time horizon is infinite, with quarter 1 following quarter 4.

Modeling:

(a) With initial inventory $= 9$, the required balance constraints are

$$\begin{aligned}
9 + x_1 &= 11 + i_1 && \text{(quarter 1)} \\
i_1 + x_2 &= 48 + i_2 && \text{(quarter 2)} \\
i_2 + x_3 &= 64 + i_3 && \text{(quarter 3)} \\
i_3 + x_4 &= 15 + i_4 && \text{(quarter 4)}
\end{aligned}$$

(b) With inventory wrapped around from the last to the first quarter, the balance constraints are

$$\begin{aligned}
i_4 + x_1 &= 11 + i_1 && \text{(quarter 1)} \\
i_1 + x_2 &= 48 + i_2 && \text{(quarter 2)} \\
i_2 + x_3 &= 64 + i_3 && \text{(quarter 3)} \\
i_3 + x_4 &= 15 + i_4 && \text{(quarter 4)}
\end{aligned}$$

4.6 MODELS WITH LINEARIZABLE NONLINEAR OBJECTIVES

Because LP models possess all the tractable features explored in Chapter 3, a linear programming model of a problem is almost always preferable to a nonlinear one of equal validity (principle 2.31 of Section 2.4). Nonlinearity is often unavoidable, but some frequently occurring nonlinear objective functions are exceptions. We introduce in this section those minimax, maximin, and min deviation objective functions which though nonlinear at first glance can be modeled with a linear objective function and linear constraints. Interested readers may also wish to refer to the related discussion of separable nonlinear programming in Section 14.8.

EXAMPLE 4.7: HIGHWAY PATROL

We begin with a real allocation problem addressed by the Highway Patrol of a southern state.[7] The Patrol wished to divide the effort of its on-duty officers among highway segments in each territory to maximize speeding reduction.

The first two lines of Table 4.10 illustrate the types of data available. Highway segments in our fictitious version are indexed by

$$j \triangleq \text{highway segment number } (j = 1, \ldots, 8)$$

with 25 officers per week to allocate. Analysts were able to estimate for each segment:

$u_j \triangleq$ upper bound on the number of officers assigned to segment
   j per week

$r_j \triangleq$ reduction potential for suppressing speeding on segment
   j per officer assigned

A high reduction potential indicates a segment where a patrol would be especially effective. In the real application, reduction potentials were obtained by directly measuring segment traffic speeds with and without an officer on patrol.

TABLE 4.10   Highway Patrol Example Data and Solutions

	Values by Highway Segment, j							
	1	2	3	4	5	6	7	8
Upper bound, u_j	4	8	5	7	6	5	6	4
Reduction potential, r_j	11	3	4	14	2	19	10	13
Maxisum optimum, x_j^*	4.00	0.00	0.00	7.00	0.00	5.00	5.00	4.00
Maximin optimum, x_j^*	1.09	4.00	3.00	0.86	6.00	4.85	1.20	4.00

Maxisum Highway Patrol Example Model

It is obvious that decision variables in our Highway Patrol allocation example should be

$$x_j \triangleq \text{number of officers per week assigned to patrol segment } j$$

Then a straightforward linear programming model is

$$\max \quad \sum_{j=1}^{8} r_j x_j \qquad \text{(total reduction)}$$

$$\text{s.t.} \quad \sum_{j=1}^{8} x_j \leq 25 \qquad \text{(officers available)} \qquad (4.19)$$

$$x_j \leq u_j \quad j = 1, \ldots, 8 \qquad \text{(upper bounds)}$$

$$x_j \geq 0 \quad j = 1, \ldots, 8 \qquad \text{(nonnegativity)}$$

The objective function is a **maxisum** because it seeks to maximize the sum of reductions in different segments (the analog in a minimize model is called **minisum**).

[7]Based on Don T. Phillips and Gary L. Hogg (1979), "The Algorithm That Converged Too Fast," *Interfaces, 9*:5, 90–93.

Main constraints restrict solutions to the 25 available officers and enforce upper bounds. The third line of Table 4.10 shows an optimal solution to maxisum model (4.19) that yields a total speed reduction of 339.

Minimax and Maximin Objective Functions

Notice that all but one of the maxisum optimal values in Table 4.10 are either zero or upper bound u_j. A little contemplation will reveal that this must always happen in a maxisum model with constraints as simple as those of (4.19).

Sometimes we would prefer a minimax or a maximin objective to spread the allocation more evenly.

> **4.12**   **Minimax** (minimize the maximum) or **maximin** (maximize the minimum) objective functions model cases where success is measured by worst rather than total performance.

Instead of optimizing total output, we focus on the model element with the least satisfactory result.

Nonlinear Maximin Highway Patrol Example Model

Adopting the maximin approach in our Highway Patrol example yields the model

$$\max \quad f(x_1, \ldots, x_8) \triangleq \min\{r_j x_j : j = 1, \ldots, 8\} \quad \text{(maximin reduction)}$$

$$\text{s.t.} \quad \sum_{j=1}^{8} x_j \leq 25 \quad \text{(officers available)} \tag{4.20}$$

$$x_j \leq u_j \quad j = 1, \ldots, 8 \quad \text{(upper bounds)}$$

$$x_j \geq 0 \quad j = 1, \ldots, 8 \quad \text{(nonnegativity)}$$

The objective now maximizes the least reduction among all highway segments.

Notice that (4.20) is a nonlinear program (definition $\boxed{2.14}$). Constraints remain linear, but the objective function is no longer a weighted sum of the decision variables. Still, this NLP may provide more valid results than maxisum LP (4.19) because speed reduction is addressed on every highway segment. The final line of Table 4.10 shows that the optimal allocation in this maximin model is much more uniform across segments. The specified optimum yields a reduction of at least 12 on every segment.

Linearizing Minimax and Maximin Objective Functions

With a model as simple as (4.20), nonlinearity may not produce much loss of tractability. Happily, we need not sacrifice tractability even in much more complicated cases. By a suitable modification of nonlinear form (4.20), we can formulate an exactly equivalent linear program.

The idea is simply to introduce a new continuous variable

$$f \triangleq \text{objective function value}$$

and maximize f subject to a system of linear constraints keeping f no more than any term of the minimum.

> **4.13** A minimax or maximin objective function can be modeled linearly by introducing a new decision variable f to represent the objective function value, and then minimizing f subject to $f \geq$ each max element in a minimax, or maximizing f subject to $f \leq$ each min element in a maximin.

Linearized Maximin Highway Patrol Example Model

Applying principle 4.13 to the maximin version of our Highway Patrol model yields linear program

$$
\begin{array}{llll}
\max & f & & \text{(maximin reduction)} \\
\text{s.t.} & f \leq r_j x_j & j = 1, \ldots, 8 & f \leq \text{each term} \\
\\
& \displaystyle\sum_{j=1}^{8} x_j \leq 25 & & \text{(officers available)} \\
& x_j \leq u_j & j = 1, \ldots, 8 & \text{(upper bounds)} \\
& x_j \geq 0 & j = 1, \ldots, 8 & \text{(nonnegativity)}
\end{array}
\tag{4.21}
$$

Unrestricted variable f is now the only term of the objective function, which makes the objective trivially linear. New linear constraints keep f less than or equal to all terms $r_j x_j$.

Transformation 4.13 works because any optimal solution in (4.21) must have

$$
f^* = \min\{r_j x_j^* : j = 1, \ldots, 8\}
$$

The new system of constraints keeps

$$
f^* \leq \min\{r_j x_j^* : j = 1, \ldots, 8\}
$$

and an f strictly less than the minimum $r_j x_j^*$ can be increased to improve the objective.

SAMPLE EXERCISE 4.9: MODELING MINIMAX OBJECTIVE FUNCTIONS

In terms of decision variables x_1, x_2, and x_3, the production times required on a company's two assembly lines are

$$
3x_1 + 2x_2 + 1x_3 \quad \text{and} \quad 1x_1 + 5x_2
$$

Assuming that all other constraints are linear, formulate an objective function and extra constraints needed to minimize the maximum time on the two lines as a linear program.

Modeling: Following principle 4.13 for a minimax case, we introduce new unrestricted variable f and employ objective function

$$
\min f
$$

To make this minimize the maximum, we also add constraints

$$
\begin{aligned}
f &\geq 3x_1 + 2x_2 + 1x_3 \\
f &\geq 1x_1 + 5x_2
\end{aligned}
$$

EXAMPLE 4.8: VIRGINIA PRESTRESS (VP) LOCATION

To see another common nonlinear objective that can be linearized, we consider the location problem confronted by a firm we will call Virginia Prestress (VP).[8] VP was planning for production of a new product: concrete utility poles. That production required a new concrete casting area to make poles and a storage area for finished products.

Figure 4.2 presents the implied facilities location problem for our fictitious case. The two new facilities will interact with each other and with three existing operations: the concrete batching facility, where premixed concrete is prepared, the steel area, where reinforcing steel is manufactured, and the shipping gate, where finished poles are processed out of the plant. A coordinate system quantifies the locations of all three existing facilities, and the adjoining table displays material handling costs of expected traffic between facilities. For example, each foot of distance between the pole storage area and the shipping gate adds $0.40 in crane activity. We must choose locations for the new facilities to minimize total material handling cost.

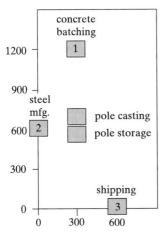

Material Handling Costs per Foot	Pole Casting $j = 1$	Pole Storage $j = 2$
1: Pole casting	—	—
2: Pole storage	4.00	—
3: Concrete batch	1.10	—
4: Steel manufacturing	0.70	0.65
5: Shipping	—	0.40

FIGURE 4.2   Virginia Prestress (VP) Location Example

Nonlinear VP Location Model

It should be clear that the main decisions to be made in this VP case are

$$x_j \triangleq x\text{-coordinate of new facility } j\text{'s location}$$
$$y_j \triangleq y\text{-coordinate of new facility } j\text{'s location}$$

We want to choose $x_1, y_1, x_2,$ and y_2 to minimize the sum of implied distances to other facilities times unit material handling costs. Using values and coordinates in Figure 4.2, a model is

$$\begin{aligned} \min \quad & 4.00d(x_1, y_1, x_2, y_2) + 1.10d(x_1, y_1, 300, 1200) \\ & + 0.70d(x_1, y_1, 0, 600) + 0.65d(x_2, y_2, 0, 600) \qquad \text{(handling cost)} \qquad (4.22) \\ & + 0.40d(x_2, y_2, 600, 0) \end{aligned}$$

where

$$d(x_j, y_j, x_k, y_k) \triangleq \text{distance from } (x_j, y_j) \text{ to } (x_k, y_k)$$

No constraints are required.

If we measure distance in the straight-line or **Euclidean** way depicted in Figure 4.3(a), model (4.22) is unavoidably nonlinear.[9] However, it is often more appropriate in facilities design to calculate distance in the **rectilinear** manner of Figure 4.3(b). Material movements tend to follow aisles and other paths aligned with either the x or the y axis. Thus travel distance is best modeled as

$$d(x_j, y_j, x_k, y_k) \triangleq |x_j - x_k| + |y_j - y_k| \tag{4.23}$$

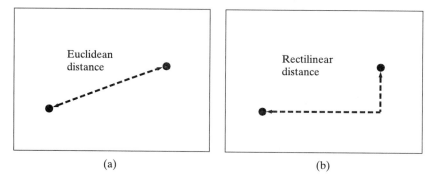

(a)   (b)

FIGURE 4.3   Euclidean versus Rectilinear Distance

Min Deviation Objective Functions

With rectilinear distance measure (4.23), mathematical program (4.22) takes a min deviation form.

> 4.14 | **Min deviation** objective functions model cases where the goal is to minimize positive-weighted sums of absolute differences between pairs of model quantities.

Here we seek to minimize the cost-weighted sum of location coordinate differences.

Linearizing Min Deviation Objective Functions

Any min deviation objective involving positive-weighted absolute differences of linear functions can be modeled linearly. We need only introduce new deviation variables expressing the required differences.

---

[9]See, for example, the DClub model of Section 3.1.

> 4.15 | Positive-weighted terms $|p(\mathbf{x}) - q(\mathbf{x})|$ of a min deviation objective function involving differences of linear functions $p(\mathbf{x})$ and $q(\mathbf{x})$ can be modeled linearly by (1) introducing new nonnegative **deviation variables** s^+ and s^-, (2) adding new constraints
>
> $$p(\mathbf{x}) - q(\mathbf{x}) = s^+ - s^-$$
>
> and (3) substituting $s^+ + s^-$ for the absolute differences in the objective function.

Linearized VP Location Model

With distance rectilinear, our VP location model (4.22) becomes

$$\begin{aligned}
\min \quad & 4.00|x_1 - x_2| + 4.00|y_1 - y_2| + 1.10|x_1 - 300| \\
& + 1.10|y_1 - 1200| + 0.70|x_1 - 0| + 0.70|y_1 - 600| \\
& + 0.65|x_2 - 0| + 0.65|y_2 - 600| + 0.40|x_2 - 600| \\
& + 0.40|y_2 - 0|
\end{aligned}$$

To apply principle 4.15, we introduce a pair of deviation variables for each objective function term $i = 1, \ldots, 10$:

$$s_i^+ \triangleq \text{positive difference in absolute value term } i$$
$$s_i^- \triangleq \text{negative difference in absolute value term } i$$

Then we may solve VP's location problem with linear program

$$\begin{aligned}
\min \quad & 4.00(s_1^+ + s_1^-) + 4.00(s_2^+ + s_2^-) + 1.10(s_3^+ + s_3^-) \\
& + 1.10(s_4^+ + s_4^-) + 0.70(s_5^+ + s_5^-) + 0.70(s_6^+ + s_6^-) \\
& + 0.65(s_7^+ + s_7^-) + 0.65(s_8^+ + s_8^-) + 0.40(s_9^+ + s_9^-) \\
& + 0.40(s_{10}^+ + s_{10}^-)
\end{aligned}$$

$$\begin{aligned}
\text{s.t.} \quad x_1 - x_2 &= s_1^+ - s_1^- && \text{(term 1)} \\
y_1 - y_2 &= s_2^+ - s_2^- && \text{(term 2)} \\
x_1 - 300 &= s_3^+ - s_3^- && \text{(term 3)} \\
y_1 - 1200 &= s_4^+ - s_4^- && \text{(term 4)} \\
x_1 - 0 &= s_5^+ - s_5^- && \text{(term 5)} \\
y_1 - 600 &= s_6^+ - s_6^- && \text{(term 6)} \\
x_2 - 0 &= s_7^+ - s_7^- && \text{(term 7)} \\
y_2 - 600 &= s_8^+ - s_8^- && \text{(term 8)} \\
x_2 - 600 &= s_9^+ - s_9^- && \text{(term 9)} \\
y_2 - 0 &= s_{10}^+ - s_{10}^- && \text{(term 10)} \\
s_i^+, s_i^- &\geq 0 \quad i = 1, \ldots, 10
\end{aligned}$$

(4.24)

Notice that new (linear) constraints express each absolute difference term of the min deviation objective function as the difference of corresponding deviation variables subject to nonnegativity constraints. Then the objective function minimizes the weighted sum (not the difference) of those variable pairs.

Here an optimal solution locates the new facilities next to each other at

$$x_1^* = x_2^* = 300$$
$$y_1^* = y_2^* = 600$$

Corresponding optimal values for deviation variables are

$$s_1^{+*} = 0, \quad s_2^{+*} = 0, \quad s_3^{+*} = 0, \quad s_4^{+*} = 0, \quad s_5^{+*} = 300$$
$$s_1^{-*} = 0, \quad s_2^{-*} = 0, \quad s_3^{-*} = 0, \quad s_4^{-*} = 600, \quad s_5^{-*} = 0$$
$$s_6^{+*} = 0, \quad s_7^{+*} = 300, \quad s_8^{+*} = 0, \quad s_9^{+*} = 0, \quad s_{10}^{+*} = 600$$
$$s_6^{-*} = 0, \quad s_7^{-*} = 0, \quad s_8^{-*} = 0, \quad s_9^{-*} = 300, \quad s_{10}^{-*} = 0$$

Why does transformation $\boxed{4.15}$ work? Because it can never be optimal for any term to have both $s_i^+ > 0$ and $s_i^- > 0$. For example, at optimality the fourth term of our VP model (4.24) has the corresponding constraint

$$y_1^* - 1200 = 600 - 1200 = -600 = s_4^+ - s_4^-$$

Many choices of s_4^+ and s_4^- will satisfy the constraint, but the objective function prefers the one with the smallest sum, that is,

$$s_4^{+*} = 0, \quad s_4^{-*} = 600$$

With at most one member of each deviation variable pair being positive in an optimal solution, their difference will always exactly equal the required absolute difference.

SAMPLE EXERCISE 4.10: MODELING MIN DEVIATION OBJECTIVE FUNCTIONS

In terms of design parameters x_1, x_2, and x_3 the speed and weight of a proposed vehicle can be expressed as

$$4x_1 - x_2 + 7x_3 \quad \text{and} \quad 9x_1 - 10x_2 + x_3$$

respectively. Assuming that all other constraints are linear, formulate an objective function and extra constraints needed in a linear program to find the design with speed as close as possible to 100, and weight as close as possible to 150.

Modeling: Following principle $\boxed{4.15}$, we define deviation variables s_1^+ and s_1^- for speed, together with s_2^+ and s_2^- for weight. Then we minimize the deviation of speed and weight from their desired values with objective function

$$\min \quad s_1^+ + s_1^- + s_2^+ + s_2^- \qquad \text{(total deviation)}$$

and special constraints

$$4x_1 - x_2 + 7x_3 - 100 \;=\; s_1^+ - s_1^- \qquad \text{(speed)}$$
$$9x_1 - 10x_2 + x_3 - 150 \;=\; s_2^+ - s_2^- \qquad \text{(weight)}$$
$$s_1^+, s_1^-, s_2^+, s_2^- \geq 0$$

EXERCISES

4-1 Bisco's new sugar-free, fat-free chocolate squares are so popular that the company cannot keep up with demand. Regional demands shown in the following table total 2000 cases per week, but Bisco can produce only 60% of that number.

	NE	SE	MW	W
Demand	620	490	510	380
Profit	1.60	1.40	1.90	1.20

The table also shows the different profit levels per case experienced in the regions due to competition and consumer tastes. Bisco wants to find a maximum profit plan that fulfills between 50 and 70% of each region's demand.

(a) ⊘ Formulate an allocation LP to choose an optimal distribution plan.
(b) ⊘ ▱ Enter and solve your model with the class optimization software.

4-2 A small engineering consulting firm has 3 senior designers available to work on the firm's 4 current projects over the next 2 weeks. Each designer has 80 hours to split among the projects, and the following table shows the manager's scoring (0 = nil to 100 = perfect) of the capability of each designer to contribute to each project, along with his estimate of the hours that each project will require.

	Project			
Designer	1	2	3	4
1	90	80	10	50
2	60	70	50	65
3	70	40	80	85
Required	70	50	85	35

The manager wants to assign designers to maximize total capability.

(a) Formulate an allocation LP to choose an optimal work assignment.
(b) ⊘ ▱ Enter and solve your model with the class optimization software.

4-3 Cattle feed can be mixed from oats, corn, alfalfa, and peanut hulls. The following table shows the current cost per ton (in dollars) of each of these ingredients, together with the percentage of recommended daily allowances for protein, fat, and fiber that a serving of it fulfills.

	Oats	Corn	Alfalfa	Hulls
% Protein	60	80	55	40
% Fat	50	70	40	100
% Fiber	90	30	60	80
Cost	200	150	100	75

We want to find a minimum cost way to produce feed that statisfies at least 60% of the daily allowance for protein and fiber while not exceeding 60% of the fat allowance.

(a) ⊘ Formulate a blending LP to choose an optimal feed mix.
(b) ⊘ Which of the constraints of your model are composition constraints? Explain.
(c) ⊘ ▱ Enter and solve your model with the class optimization software.

4-4 Several forms of gasoline are produced during the petroleum refining process, and a last step combines them to obtain market products with specified quality measures. Suppose 4 different gasolines are available, with values for the 2 indexes of quality being 99 and 210, 70 and 335, 78 and 280, and 91 and 265, respectively. Using corresponding costs per barrel of $48, $43, $58, and $46, we would like to choose a minimum cost blend with a first quality index between 85 and 90 and a second index between 270 and 280.

(a) Formulate a blending LP to choose an optimal gasoline blend.
(b) Which of the constraints of your model are composition constraints? Explain.
(c) ⊘ ▱ Enter and solve your model with the class optimization software.

4-5 Ronnie Runner distilleries blends $i = 1, \ldots, m$ scotch whiskeys to create its $j = 1, \ldots, n$ products with properties $k = 1, \ldots, p$. Unblended whiskey i measures $a_{i,k}$ on scale k. Express each of the following as linear constraint(s) in these parameters and the nonnegative decision variables $x_{i,j} \triangleq$ barrels of whiskey i used in product j. Assume that the properties combine in proportion to volume and that the total production of any blend is free to vary with the optimization.

(a) ⊘ Property $k = 11$ should fall between 45 and 48 in all products.
(b) Product $j = 14$ must have all properties $k = 5, \ldots, 9$ between 90 and 95.

(c) ⊗ Product $j = 26$ must measure at least 116 on property $k = 15$.

(d) No product should measure more than 87 on property $k = 8$.

(e) ⊗ Products 6 through 11 should combine input whiskey 1 with others in at most the ratio 3:7.

(f) Input whiskeys 4 and 7 should be in ratio 2:3 for all blends.

(g) ⊗ At least one-third of all output must come from inputs $i = 3, \ldots, 6$.

(h) No more than 5% of all output can come from input $i = 13$.

4-6 Problems are often modeled as linear programs even though some decision variables represent quantities such as the number of units processed or the number of times an alternative is used that must be integer in a physical implementation. Briefly justify this practice.

4-7 A metalworking shop needs to cut at least 37 large disks and 211 small ones from sheet metal rectangles of a standard size. Three cutting patterns are available. One yields 2 large disks with 34% waste, the second gives 5 small disks with 22% waste, and the last produces 1 large and 3 small disks with 27% waste. The shop seeks a minimum waste way to fulfill its requirements.

(a) ⊗ Formulate an operations management LP to choose an optimal cutting plan.

(b) ⊗ Enter and solve your model with the class optimization software.

4-8 Classic Candles handmakes three models of elegant Christmas candles. Santa models require 0.10 day of molding, 0.35 day of decorating, and 0.08 day of packaging and produce $16 of profit per unit sold. Corresponding values for Christmas trees are 0.10, 0.15, 0.03 and $9, while those of gingerbread houses are 0.25, 0.40, 0.05, and $27. Classic wants to maximize profit on what it makes over the next 20 working days with its 1 molder, 3 decorators, and 1 packager, assuming that everything made can be sold.

(a) Formulate an operations management LP to choose an optimal production plan.

(b) ⊗ ⌷ Enter and solve your model with the class optimization software.

4-9 Wobbly Office Equipment (WOE) makes two models of tables for libraries and other university facilities. Both models use the same tabletops, but model A has 4 short (18-inch) legs and model B has 4 longer ones (30-inch). It takes 0.10 labor hour to cut and shape a short leg from stock, 0.15 labor hour to do the same for a long leg, and 0.50 labor hour to produce a tabletop. An additional 0.30 labor hour is needed to attach the set of legs for either model after all parts are available. Estimated profit is $30 for each model A sold and $45 for each model B. Plenty of top material is on hand, but WOE wants to decide how to use the available 500 feet of leg stock and 80 labor hours to maximize profit, assuming that everything made can be sold.

(a) ⊗ Formulate an operations management LP to choose an optimal plan using the decision variables $x_1 \triangleq$ number of model A's assembled and sold, $x_2 \triangleq$ number of model B's assembled and sold, $x_3 \triangleq$ number of short legs manufactured, $x_4 \triangleq$ number of long legs manufactured, and $x_5 \triangleq$ number of tabletops manufactured.

(b) ⊗ Which of the constraints of your model are balance constraints? Explain.

(c) ⊗ ⌷ Enter and solve your model with the class optimization software.

4-10 Perfect Stack builds standard and extralong wooden palettes for a variety of manufacturers. Each model consists of 3 heavy separators of length equal to the palette. The standard model has 5 cross pieces above and 5 below the separators and requires 0.25 hour to assemble. The extralong version has 9 similar cross pieces on top and bottom and consumes 0.30 hour to assemble. The supply of wood is essentially unlimited, but it requires 0.005 hour to fabricate a standard separator, 0.007 hour to fabricate an extralong separator, and 0.002 hour to fabricate a cross piece. Assuming that it can sell as many standard models as can be made at $5 profit each and as many extralongs at $7 profit, Perfect wants to decide what to produce with the available 200 hours of assembly time and 40 hours of fabrication.

(a) Formulate an operations management LP to choose an optimal plan using the decision variables $x_1 \triangleq$ number of standard palettes assembled and sold, $x_2 \triangleq$ number of extralongs assembled and sold, $x_3 \triangleq$ number of standard separators manufactured, $x_4 \triangleq$ number of extralong separators manufactured, and $x_5 \triangleq$ number of cross pieces manufactured.

(b) Which of the constraints of your model are balance constraints? Explain.

(c) ⊗ ⌷ Enter and solve your model with the class optimization software.

4-11 Goings Engine produces diesel engines and assemblies $i = 1, \ldots, m$ at its plants $p = 1, \ldots, n$. There is some end demand $d_{i,p}$ for the various engines and assemblies, with the rest used in Goings production. The number of subassemblies i required to produce each assembly k is $a_{i,k}$.

(a) ⊗ Write a system of linear constraints specifying that the number of each engine and assembly must balance (with zero inventories) across the company using the parameters above and the nonnegative decision variable $x_{i,p} \triangleq$ number of assemblies i produced at plant p.

(b) Write a system of linear constraints specifying that the number of each engine and assembly must balance (with zero inventories) at each plant using the foregoing parameters and the nonnegative decision variable $x_{i,p,q} \triangleq$ number of assemblies i produced at plant p for use at plant q.

4-12 The River City Police Department uses work shifts in which officers work 5 of the 7 days of the week with 2 successive days off. For example, a shift might work Sunday through Thursday and then have Friday and Saturday off. A total of 6 officers must be on duty Monday, Tuesday, Wednesday, and Thursday; 10 are required on Friday and Saturday; and 8 are needed on Sunday. River City wants to meet these staffing needs with the minimum total number of officers.

(a) ⊗ Formulate a shift scheduling LP to select an optimal staffing plan.

(b) ⊗ Which of the constraints of your model are covering constraints? Explain.

(c) ⊗ ▭ Enter and solve your model with the class optimization software.

4-13 Mama's Kitchen serves from 5:30 A.M. each morning until 1:30 P.M. Tables are set and cleared by busers working 4-hour shifts beginning on the hour from 5 A.M. through 10 A.M. Most are college students who hate to get up in the morning, so Mama's pays $7 per hour for the 5, 6, and 7 A.M. shifts, and $6 per hour for all others. The manager seeks a minimum cost staffing plan that will have 2 busers on duty for the hour beginning at 5 A.M., plus 3, 5, 5, 3, 2, 4, 6 and 3 on duty for the hours to follow.

(a) Formulate a shift scheduling LP to select an optimal staffing plan.

(b) Which of the constraints of your model are covering constraints? Explain.

(c) ⊗ ▭ Enter and solve your model with the class optimization software.

4-14 The MacKensie's daughter will begin college 4 years from today. Her parents want to invest $10,000 at the beginning of each of the 4 years to accumulate a fund that can help pay the cost. Each year they expect to have available both certificates of deposit returning 5% after 1 year and ones returning 12% after 2 years. This year, they also have an opportunity to make a special investment that would return 21% after 4 years. The MacKensies want to choose investments to maximize their college fund assuming that all funds are reinvested at maturity.

(a) ⊗ Formulate a time-phased LP to choose an optimal investment plan.

(b) ⊗ Which of the constraints in your model are balance constraints? Explain.

(c) ⊗ What is the time horizon of your model? Explain.

(d) ⊗ ▭ Enter and solve your model with the class optimization software.

4-15 Global Minimum manufactures bikini swimming suits. Their business is highly seasonal, with expected demands being 2800, 500, 100, and 850 dozen suits over the four quarters of next year. The company can produce 1200 dozen suits per quarter, but inventories must be built up to meet larger demands at a holding cost of $15 per dozen per quarter. Global wants to meet demand while minimizing this inventory cost.

(a) Formulate a time-phased LP to choose an optimal production plan assuming an infinite time horizon.

(b) Which of the constraints in your model are balance constraints? Explain.

(c) Explain why your model has an infinite horizon even though it details only four quarters.

(d) ⊗ ▭ Enter and solve your model with the class optimization software.

4-16 A company manufactures parts $i = 1, \ldots, m$ in weeks $t = 1, \ldots, n$, where each unit of part i requires $a_{i,k}$ units of production resource $k = 1, \ldots, q$ and has value v_i. Production resource capacities b_k cannot be exceeded in any period and part demands $d_{i,t}$ must be met. Express each of the following as linear constraint(s) in these parameters and the nonnegative decision variables $x_{i,t} \triangleq$ number of units of i produced in week t and $z_{i,t} \triangleq$ inventory of product i held at the end of period t. Initial inventories all $= 0$.

(a) ⊠ No production capacity can ever be exceeded.
(b) The total value of held inventories should never exceed 200.
(c) ⊠ Quantities of each part i available after week 1 should balance with demand and accumulated inventory.
(d) Quantities of each part i available after weeks $2, \ldots, n - 1$ should balance with demand and accumulated inventory.

4-17 The following table shows observed electrical power consumption at several different levels of a factory's operation.

Level	2	3	5	7
Power	1	3	3	5

Engineers want to fit the estimating relationship

$$\text{power} = \beta_0 + \beta_1 \text{level}$$

to these data in a way that minimizes the sum of the absolute deviations between predicted and observed power requirements. Both parameters β_0 and β_1 should be nonnegative.

(a) ⊠ Formulate a linearized nonlinear LP to choose optimal parameter values. (*Hint*: β_0 and β_1 are among the decision variables.)
(b) ⊠ 🖳 Enter and solve your model with the class optimization software.
(c) ⊠ Determine from your solution in part (b) the absolute deviation at each observed point.

4-18 The following figure

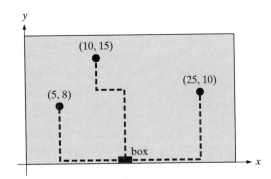

shows the ceiling locations of 3 sensors in a new factory relative to a coordinate system (in feet) with origin at the lower left. A control box will be located along the long (lower in the figure) wall with fiber-optic cables running rectilinearly to each sensor. Designers want to place the box to minimize the cable required.

(a) Formulate a linearized nonlinear LP to choose an optimal placement.
(b) ⊠ 🖳 Enter and solve your model with the class optimization software.
(c) How much cable would be required to implement your solution of part (b)?

4-19 ⊠ Repeat Exercise 4-17, this time choosing a fit that minimizes the maximum deviation between observed and predicted power requirements.

4-20 Repeat Exercise 4-18, this time minimizing the length of the longest cable.

4-21 The American Edwards Laboratories (AEL)[10] manufactures artificial human heart valves from pig hearts. One of the things making planning complex is that the size of pig hearts is highly variable, depending on breed, age when slaughtered, feed mix, and so on. The following (fictitious) table shows the fraction of hearts from suppliers $j = 1, \ldots, 5$ yielding each of the valve sizes $i = 1, \ldots, 7$, along with the maximum quantity available from each supplier per week and the unit cost of hearts obtained.

Size	Supplier j				
	1	2	3	4	5
1	0.4	0.1	—	—	—
2	0.4	0.2	—	—	—
3	0.2	0.3	0.4	0.2	—
4	—	0.2	0.3	0.2	—
5	—	0.2	0.3	0.2	0.2
6	—	—	—	0.2	0.3
7	—	—	—	0.2	0.5
Availability	500	330	150	650	300
Cost	2.5	3.2	3.0	2.1	3.9

AEL wants to decide how to purchase hearts to meet weekly requirements of 20 size 1, 30 size 2, 120 size 3, 200 size 4, 150 size 5, 60 size 6, and 45 size 7 valves at minimum total cost.

(a) ⊗ Formulate an LP model of this heart purchase planning problem using the decision variable ($j = 1, \ldots, 5$)

   $x_j \triangleq$ number of hearts purchased weekly from supplier j

(b) ⊗ ▯ Enter and solve your model with the class optimization software.

4-22 Midville Manufacturing assembles heavy duty materials handling carts to meet demand of 500 units in the first quarter of each year, 1200 in the second, 1000 in the third, and 300 in the fourth. Elementary components, which consist of wheels, steering yokes, and carrying platforms, are first assembled separately. Then each steering yoke is equipped with 4 wheels to form the front-end subassembly. Finally, front-end subassemblies are combined with a carrying platform and 8 additional wheels at the rear to complete the cart. Using $j = 1$ for steering yokes, $j = 2$ for wheels, $j = 3$ for platforms, $j = 4$ for front-end assemblies, and $j = 5$ for finished carts, the following table shows the estimated value of each element (in dollars) and the factory hours required to assemble it.

	Element j				
	1	2	3	4	5
Value	120	40	75	400	700
Time	0.06	0.07	0.04	0.12	0.32

Components, subassemblies, and finished carts produced in any quarter may be used or shipped in the same quarter or held over as inventory (including from the fourth quarter to the first) at 5% per quarter interest on the held value. Midville seeks a plan that minimizes these holding costs while conforming to the factory production capacity, 1150 hours per quarter.

(a) Formulate an LP model to choose a production plan using the decision variables ($j = 1, \ldots, 5$; $q = 1, \ldots, 4$)

   $x_{j,q} \triangleq$ number of units of element j produced in quarter q

   $h_{j,q} \triangleq$ number of units of element j held in inventory from quarter q to the next

Your model should include a system of main constraints for production capacity and an additional system for each element j to enforce material balance.

(b) ⊗ ▯ Enter and solve your model with the class optimization software.

4-23 A construction contractor has undertaken a job with 7 major tasks. Some of the tasks can begin at any time, but others have predecessors that must be completed first. The following table shows those predecessor task numbers, together with the minimum and maximum time (in days) allowed for each task, and the total cost that would be associated with accomplishing each task in its minimum and maximum times (more time usually saves expense).

j	Min. Time	Max. Time	Cost Min.	Cost Max.	Predecessor Tasks
1	6	12	1600	1000	None
2	8	16	2400	1800	None
3	16	24	2900	2000	2
4	14	20	1900	1300	1,2
5	4	16	3800	2000	3
6	12	16	2900	2200	3
7	2	12	1300	800	4

The contractor seeks a way to complete all work in 40 days at least total cost, assuming that the cost of each task is linearly interpolated for times between the minimum and maximum.

(a) Formulate an LP model of this time/cost planning problem using the decision variables ($j = 1, \ldots, 7$)

   $s_j \triangleq$ start time of task j (in days)
   $t_j \triangleq$ days to complete task j

Your model should have an objective function summing interpolated cost and main constraints to enforce precedence relationships and the time limit.

(b) ⊗ ▯ Enter and solve your model with the class optimization software.

4-24 Import Books, Incorporated (IBI)[11] stocks several thousand titles in its main warehouse. The titles can be categorized by sales volume, with $i = 1$ requiring a stored inventory of 0 to 20 books, $i = 2$ requiring 21 to 40, $i = 3$ requiring 41 to 100, and $i = 4$

---

[11]Based on R. J. Paul and R. C. Thomas (1977), "An Integrated Distribution, Warehousing and Inventory Control System for Imported Books," *Operational Research Quarterly, 28,* 629–640.

requiring 101 to 200. The number of titles in category i is b_i. Each title is stored in a separate bin, and each bin has at most one title. IBI has 500 bins with space for up to 100 books and 2000 larger ones with space for up to 200 books. The bins for 100 can also be subdivided to create either two bins for 40 books or three bins for 20. Costs $c_{i,j}$ for storing a category i title in a size j bin have been estimated by accounting for the material handling cost of accessing the bin and the wasted space if a bin is underutilized. Here $j = 1$ refers to bins for 20, $j = 2$ to bins for 40, $j = 3$ to bins for 100, and $j = 4$ to bins for 200. Formulate an LP model to find a minimum cost allocation of titles to bins using the decision variable $(i = 1, \ldots, 4; j = i, \ldots, 4)$

$$x_{i,j} \triangleq \text{number of titles of category}$$
$$i \text{ allocated to bins of size } j$$

4-25 Dairy cows[12] in most countries calve on a regular annual basis. Their milk output varies over the year accordingly, with a peak reached a few months after calving followed by a decline to almost zero in the tenth month. Knowing these facts, farmers in an agricultural cooperative are trying to plan calving months $c = 1, \ldots, 12$ to make it easier to meet seasonal milk demands r_d pounds in months $d = 1, \ldots, 12$. Any milk produced beyond these demands must be sold on the bulk market at b per pound below the regular price. The annual cost m_c of maintaining a cow calving in month c varies significantly over the year because low-cost grazing is available only in certain seasons. From scientific studies the farmers can estimate the yield $p_{d,c}$ pounds in demand month d per cow calving in month c. Formulate an LP model to determine a minimum total cost calving schedule using the decision variables $(c, d = 1, \ldots, 12)$

$$x_c \triangleq \text{number of cows calving in month } c$$
$$y_d \triangleq \text{pounds of excess milk}$$
$$\text{produced in demand month } d$$

4-26 Blue Bell[13] is planning its monthly production of a particular type of men's jeans. Demands

d_i are know for the $i = 1, \ldots, 75$ different fabric parts needed to make all the combinations of waist and inseam sizes being produced. Such parts are cut from fabric laid out on cutting tables in 60 to 70 layers. A predefined set of markers (cutting patterns) $m = 1, \ldots, 350$ define how various parts may be cut. Each use of marker m yields $a_{i,m}$ copies of part i per layer and wastes w_m square yards of fabric in areas between the usable parts. Formulate an LP model to choose a minimum total waste cutting plan using the decision variable $(m = 1, \ldots, 350; p = 60, \ldots, 70)$

$$x_{m,p} \triangleq \text{number of times marker pattern } m \text{ is cut}$$
$$\text{in a layup of } p \text{ layers}$$

4-27 To assess the impact on the U.S. coal market of different pollution control strategies, the Environmental Protection Agency (EPA)[14] wants to determine, for assumed control regimes, how much coal from supplies s_i in different mining regions $i = 1, \ldots, 24$ will be extracted, how much will then be processed into various deliverable coal types $m = 1, \ldots, 8$, and how much of each deliverable product will be sent to meet consumer demands $d_{m,j}$ in regions $j = 1, \ldots, 113$. Demands are expressed in Btu, with each ton of raw coal mined at i for processing into type m yielding $a_{i,m}$ Btu. Including the economic burden of pollution controls and transportation, the cost per ton of raw coal mined at i for processing into type m and use at j can be estimated at $c_{i,m,j}$. Formulate an LP model to determine how coal would be mined, processed, and distributed if the market seeks to minimize total cost. Use the decision variable $(i = 1, \ldots, 24; m = 1, \ldots, 8; j = 1, \ldots, 113)$

$$x_{i,m,j} \triangleq \text{tons of coal mined at } i \text{ for processing}$$
$$\text{into } m \text{ and use at } j$$

4-28 Quantas Airways Ltd.[15] must schedule its hundreds of reservation salesclerks around the clock to have at least r_t on duty during each 1-hour period starting at (24-hour) clock hour $t = 0, \ldots, 23$. A shift beginning at time t extends for 9 hours with 1 hour out for lunch in the fourth, fifth, or sixth hours of the shift. Shifts beginning at hour t cost the company c_t

[12]Based on L. Killen and M. Keane (1978), "A Linear Programming Model of Seasonality in Milk Production," *Journal of the Operational Research Society, 29,* 625–631.

[13]Based on J. R. Edwards, H. M Wagner, and W. P. Wood (1985), "Blue Bell Trims Its Inventory," *Interfaces, 15:1,* 34–52.

[14]Based on C. Bullard and R. Engelbrecht-Wiggans (1988), "Intelligent Data Compression in a Coal Model," *Operations Research, 38,* 521–531.

[15]Based on A. Gaballa and W. Pearce (1979), "Telephone Sales Manpower Planning at Quantas," *Interfaces, 9:3,* 1–9.

per day, including wages and night-hour premiums. Formulate an LP model to compute a minimum total cost daily shift schedule using the decision variables $(t = 0, \ldots, 23; i = t + 4, \ldots, t + 6)$

$x_t \triangleq$ number of clerks working a shift starting at hour t

$y_{t,i} \triangleq$ number of clerks working a shift starting at hour t who take lunch during hour i

4-29 An Indian reservation irrigation project[16] must decide how much water to release through the gate at the top of its main canal in each of the upcoming 4-hour periods $t = 1, \ldots, 18$. Ideal canal outflows, r_t, are known for each time period, and the total outflow over all 18 periods should equal or exceed the sum of these quantities. However, period-to-period deviations may be needed to avoid flooding. The initial canal storage is 120 units, and the net effect of releases and outflows should never leave more than u units stored after any period. Within these limits, managers would like to minimize the total absolute deviation between desired demands r_t and actual outflows. Formulate an LP model of this irrigation control problem using the decision variables $(t = 1, \ldots, 18)$

$x_t \triangleq$ gate release during period t

$s_t \triangleq$ amount of water stored in the canal at the end of period t

$w_t \triangleq$ canal outflow during period t

$d_t^+ \triangleq$ oversatisfaction of demand in period t

$d_t^- \triangleq$ undersatisfaction of demand in period t

4-30 Major shopping mall developer Homart[17] is selecting the tenant mix for its next facility. Stores of product types $i = 1, \ldots, 20$ are being considered for arrangement into the new mall's sectors $j = 1, \ldots, 5$. Each sector will have 150 thousand square feet, and an allowance of c_i per square foot will be set aside for finishing of areas allocated to stores of type i. From prior experience, Homart can estimate the present worth $p_{i,j}$ of revenues from a type i store located in sector j and the required floor space a_i (in thousands

of square feet). They seek a tenant mix that will maximize total present worth while having between $\underline{n}_i$ and $\bar{n}_i$ stores of each type i totaling between $\underline{f}_i$ and $\bar{f}_i$ thousand square feet, and not exceeding the budget b for finishing allowances. Formulate an LP model of this tenant mix problem using the decision variable

$x_{i,j} \triangleq$ number of stores of type i included in sector j

4-31 Once the configuration of molds is fixed, the planning of production of aluminum ingots[18] reduces to allocating the time of furnaces $j = 1, \ldots, n$ among alloys $i = 1, \ldots, m$ and ingot sizes $s = 1, \ldots, p$. Yields $a_{j,s}$ of ingots of size s producible from furnace j during the entire planning period can be estimated. Planning should meet demands $d_{i,s}$ for ingots of alloy i and size s, but this may require misapplying some ingots (i.e., trimming some larger ingots sizes s' greater than s to meet demands for size s of the same alloy). Misapplications result in a trim-loss cost $c_{i,s',s}$ for each ingot cut down from size s' to s. Managers want to find a feasible plan that minimizes the total cost of these misapplications. Formulate an LP model of this ingot production planning problem using decision variables $(i = 1, \ldots, m; j = 1, \ldots, n; s = 1, \ldots, p, s' > s)$

$x_{i,j,s} \triangleq$ fraction of time on furnace j dedicated to making ingots of alloy i, size s

$y_{i,s',s} \triangleq$ number of ingots of alloy i, size s', misapplied to meet demand for size s

4-32 S&S operates its large supermarkets[19] on a 24-hour per day basis using only part-time cashiers working shifts of 2 to 5 hours per day. All shifts start on the hour. The required number r_h of cashiers on duty at a given store is known for (24-clock) hours $h = 0, \ldots, 23$, and managers can also estimate the number of employees b_l willing to work shifts of lengths $l = 2, \ldots, 5$. They seek a shift schedule that meets requirements at minimum total cashier hours worked. Formulate an LP model of this shift scheduling problem using the decision

[16]Based on B. J. Boman and R. W. Hill (1989), "LP Operation Model for On-Demand Canal Systems," *Journal of Irrigation and Drainage Engineering, 115,* 687–700.

[17]Based on J. C. Bean, C. E. Noon, S. M. Ryan, and G. J. Salton (1988), "Selecting Tenants in a Shopping Mall," *Interfaces, 18:2,* 1–9.

[18]Based on M. R. Bowers, L. A. Kaplan, and T. L. Hooker (1995), "A Two-Phase Model for Planning the Production of Aluminum Ingots," *European Journal of Operational Research, 81,* 105–114.

[19]Based on E. Melachrinoudis and M. Olafsson (1992), "A Scheduling System for Supermarket Cashiers," *Computers and Industrial Engineering, 23,* 121–124.

variable ($h = 0, \ldots, 23; l = 2, \ldots, 5$)

$$x_{h,l} \triangleq \text{number of cashiers starting}$$
$$\text{an } l\text{-hour shift at hour } h$$

Neglect the fact that the numbers working each shift must physically be integers.

4-33 The transmitted gray-scale value $g_{i,j}$ of pixels $i = 1, \ldots, m, j = 1, \ldots, n$, in a digital space satellite photo[20] is distorted by both the usual random noise and a known problem with the video camera that effectively multiplies the value for pixel (i, j) by a blurring factor $b_{i,j}$. Engineers want to restore the image by estimating correct values for each pixel in a way that minimizes the total absolute deviation between predicted (after blurring) and observed gray-scale numbers. Formulate an LP model of this image restoration problem using decision variables ($i = 1, \ldots, m; j = 1, \ldots, n$)

$$x_{i,j} \triangleq \text{correct value for pixel } (i, j)$$

$$d_{i,j}^+ \triangleq \text{positive deviation of predicted over}$$
$$\text{observed value for pixel } (i, j)$$

$$d_{i,j}^- \triangleq \text{negative deviation of predicted below}$$
$$\text{observed value for pixel } (i, j)$$

4-34 The Hanshin expressway[21] serves the Osaka–Kobe area of Japan. Due to heavy congestion, the number of vehicles entering at each ramp $j = 1, \ldots, 38$ of the expressway is controlled by a system that reevaluates the situation every 5 minutes based on current queue lengths q_j at each ramp and estimated number of new entry-seeking arrivals d_j over the next 5 minutes. The system enforces end-of-period queue-length limits u_j and total traffic capacities b_i on 500-meter segments $i = 1, \ldots, 23$ of the highway. Traffic entering at j affects only downstream segments, and a part may exit before reaching any given i. Prior engineering studies have captured this behavior in estimated fractions $f_{i,j}$ of vehicles entering at j that persist to consume capacity at i. The system seeks the feasible control policy that permits the maximum total number of entry-seeking vehicles

into the expressway during the next time period. Formulate an LP model of this traffic control problem using the decision variables ($j = 1, \ldots, 38$)

$$x_j \triangleq \text{number of vehicles allowed to enter}$$
$$\text{at ramp } j \text{ during the time period}$$

4-35 Industrial engineers are planning the layout of cells $i = 1, \ldots, 18$ in a rectangular manufacturing facility of $x = 1000$ by $y = 200$ feet with a 6-foot-wide, two-way conveyor system along the $y = 0$ boundary[22]. It has already been decided that cells will be sequenced along the conveyor in the same order as their i subscripts, but the exact geometry of the cells remains to be fixed. Analysis of cell loadings has produced lower limits $\underline{x}_i$ and $\underline{y}_i$ on the x and y dimensions of each cell. Engineers have also specified minimum cell perimeters $\underline{p}_i$ as a surrogate for area (which would lead to nonlinear optimization). Conveyor traffic will enter and exit at input and output stations located at cell x-midpoints along the conveyor, and one-way traffic flows from cell i to cell j are estimated at $f_{i,j}$. The IEs seek a feasible design that minimizes total travel (flow times distance) on the conveyor. Formulate an LP model of this layout problem using the decision variables ($i, j = 1, \ldots, 18$)

$$x_j \triangleq \text{left } x\text{-coordinate of cell } j$$

$$y_j \triangleq y\text{-depth of cell } j$$

$$d_{i,j}^+ \triangleq \text{positive } x\text{-distance between I/O stations}$$
$$\text{of cells } i \text{ and } j$$

$$d_{i,j}^- \triangleq \text{negative } x\text{-distance between I/O stations}$$
$$\text{of cells } i \text{ and } j$$

4-36 Swift Chemical Company[23] mines phosphate rock, collects it in inventory piles $i = 1, \ldots, 8$, and blends it to meet contracts with customers $k = 1, \ldots, 25$ at profit p_{ik} per ton. The critical measure of phosphate content in rock is its BPL. Piles correspond to different average BPL contents b_i per ton, asset value a_i per ton, contract net profit $r_{i,k}$

[20]Based on R. V. Digumarthi, P. Payton, and E. Barrett (1991), "Linear Programming Solutions of Problems in Logical Inference and Space-Variant Image Restoration," *Image Understanding and the Man–Machine Interface III*, SPIE Vol. 1472, 128–136.

[21]Based on T. Yoshino, T. Sasaki, and T. Hasegawa (1995), "The Traffic-Control System on the Hanshin Expressway," *Interfaces, 25:1*, 94–108.

[22]Based on A. Langevin, B. Montreuil, and D. Riopel (1994), "Spine Layout Problem," *International Journal of Production Research, 32*, 429–442.

[23]Based on J. M. Reddy (1975), "A Model to Schedule Sales Optimally Blended from Scarce Resources," *Interfaces, 5:1*, 97–107.

per ton, starting inventory $\underline{h}_i$, and expected quantity q_i to arrive from mines vary accordingly. Each contract includes a minimum $\underline{s}_k$ and a maximum $\bar{s}_k$ number of tons to be shipped, along with a minimum $\underline{p}_k$ and a maximum $\bar{p}_k$ average BPL content. Managers want to schedule blending and sales to maximize total profit plus total ending inventory asset value. Formulate an LP model of this phosphate planning problem using the decision variables ($i = 1, \ldots, 8$; $k = 1, \ldots, 25$)

$$x_{i,k} \triangleq \text{tons of rock from pile } i \text{ included in shipment for contract } k$$

$$h_i \triangleq \text{ending inventory in pile } i$$

4-37 Any convex 3-dimensional object (i.e., a body such that the line segment between any two points in its volume falls entirely within the volume) with flat sides can be described as the set of points (x, y, z) satisfying a series of linear constraints[24]. For example, a 3- by 5- by 9-meter box with one corner at the origin can be modeled as

$$\{(x, y, z) : 0 \le x \le 3, \ 0 \le y \le 5, \ 0 \le z \le 9\}$$

Suppose that a stationary object is described in this way by constraints

$$a_i x + b_i y + c_i z \le d_i \qquad i = 1, \ldots, 19$$

and that a link of a robot arm is described at its initial position by the constraints

$$p_j x + q_j y + r_j z \le s_j \qquad j = 1, \ldots, 12$$

The object and the link do not intersect at that initial position, but the link is in motion. Its location is being translated from the initial location by growing a step $\alpha > 0$ in direction $(\Delta x, \Delta y, \Delta z)$. Formulate an LP in terms of decision variables x, y, z, and α to find the smallest step (if any) that will produce a collision between the object and the link, and indicate how the LP would detect the case where no collision will occur.

4-38 The principal export of Iceland[25] is fish which are very perishable and subject to high day-to-day variation in the size of the catch available for processing. Each day processing begins at any packing plant with estimates b_f of the kilograms of raw fish species $f = 1, \ldots, 10$ that will be available for processing for market $m = 1, \ldots, 20$. The marketable volume (in kilograms) of product m in any day is at most u_m. Each kilogram of raw fish f processed for market m yields a_{fm} kilograms of the final product and produces a gross profit of p_{fm} in sales minus costs other than labor. Processing fish f into product m requires $h_{f,m,i}$ hours of worker time at stations $i = 1$ (filleting), $i = 2$ (packing), and $i = 3$ (freezing). A total of q_i hours of workers can be obtained for workstation i at an average wage of c_i per hour. The daily plan should maximize total gross profit minus wage cost. Formulate an LP model to compute an optimal fish processing plan using the decision variables ($f = 1, \ldots, 10; m = 1, \ldots, 20; i = 1, \ldots, 3$)

$$x_{f,m} \triangleq \text{kilograms of raw fish } f \text{ processed for market } m$$

$$y_i \triangleq \text{worker hours at workstation } i$$

4-39 The U.S. Air Force (USAF)[26] must procure aircraft types $i = 1, \ldots, 10$ and associated munition types $j = 1, \ldots, 25$ to meet anticipated sortie requirements against target types $k = 1, \ldots, 15$ in weather conditions classes $\ell = 1, \ldots, 8$. Targets k are assigned a value r_k, and $t_{k,\ell}$ are anticipated under weather condition ℓ in the assumed war scenario. An aircraft type i using munitions type j under weather conditions ℓ has probability $p_{i,j,k,\ell}$ of killing a type k target each sortie, and a load of $b_{i,j,k,\ell}$ munitions j is required. A total of $s_{i,j,k,\ell}$ such sorties could be flown by each available type i aircraft during the assumed war. Currently, there are a_i aircraft of type i available, and new ones can be procured at \$$c_i$ billion per unit. Similarly, the current inventory of munitions type j is m_j and new ones can be procured at \$$d_j$ billion per unit. The USAF wants to buy planes and munitions to maximize the total expected value of targets it could kill if the assumed war scenario happened subject to a current-year procurement budget of \$100 billion. Formulate an LP model of this weapons procurement problem using the decision variables ($i = 1, \ldots, 10; j = 1, \ldots, 25; k = 1, \ldots, 15; \ell = 1, \ldots, 8$)

---

[24]Based on R. Gallerini and A. Sciomachen (1993), "On Using LP to Collision Detection between a Manipular Arm and Surrounding Obstacles," *European Journal of Operational Research, 63,* 343–350.

[25]Based on P. Jennson (1988), "Daily Production Planning in Fish Processing Firms," *European Journal of Operational Research, 36,* 410–415.

[26]Based on R. J. Might (1987), "Decision Support for Aircraft and Munitions Procurement," *Interfaces 17:5,* 55–63.

$x_{i,j,k,\ell} \triangleq$ number of sorties flown by aircraft type i with munitions type j against target type k under weather conditions class ℓ

$y_i \triangleq$ number of new type i aircraft procured

$z_j \triangleq$ number of new type j munitions procured

4-40 North American Van Lines[27] maintains a fleet of several thousand truck tractors, each of which is owned by one of its contract truckers. Tractors can be anywhere in the range $i = 0, \ldots, 9$ years old. Every 4-week planning period $t = 1, \ldots, 13$, tractors may be purchased new at price p, sold to contractors at price s_i, traded in to manufacturers at allowance a_i, and repurchased from contractors at price r_i. Only new units can be sold or traded as age $i = 0$, and no more total units of any age can be traded in any period than are purchased new in the same period. The number of tractors in the fleet inventory (i.e., with contractors) during each period t must fall between a minimum l_t and a maximum u_t to meet seasonal moving demands without accumulating excess capacity. Managers wish to find a plan that max-imizes their net total income from buying, selling, trading, and repurchasing trucks. Formulate an LP model of this fleet management problem using the decision variables $(i = 0, \ldots, 9; t = 1, \ldots, 3)$

$w_t \triangleq$ number of new tractors purchased in period t

$x_{i,t} \triangleq$ number of tractors of age i sold to contractors in period t

$y_{i,t} \triangleq$ number of tractors of age i traded in period t

$z_{i,t} \triangleq$ number of tractors of age i repurchased from contractors in period t

$f_{i,t} \triangleq$ number of tractors of age i in the fleet at the beginning of period t

Assume that vehicles enter and leave the fleet only through sales to and repurchases from contractors. Also assume that the fleet ages 1 year as inventory is passed from period 13 to period 1 and that no 9-year-old vehicles can be carried over to the new year.

SUGGESTED READING

Bazaraa, Mokhtar S., John J. Jarvis, and Hanif Sher-ali, (1990), *Linear Programming and Network Flows*, Wiley, New York.

Chvátal, Vašek (1983), *Linear Programming*, W.H. Freeman, San Francisco.

---

[27]Based on D. Avrmovich, T. M. Cook, G. D. Langston, and F. Sutherland (1982), "A Decision Support System for Fleet Management: A Linear Programming Approach," *Interfaces, 12:3*, 1–9.

C H A P T E R 5

Simplex Search for Linear Programming

Having sampled in Chapter 4 some of the enormous variety of linear programming models, it is time to focus on the powerful algorithms that make them tractable. In this chapter we develop a special form of improving search called the **simplex** method. Simplex is the most widely used of all optimization algorithms, although the newer interior-point methods of Chapter 6 are proving strong competitors. Both exploit the special properties of linear programs to produce highly efficient algorithms capable of globally optimizing huge models. Our treatment assumes reader familiarity with the search fundamentals in Chapter 3.

5.1 LP OPTIMAL SOLUTIONS AND STANDARD FORM

We begin our investigation of linear programming algorithms with some observations and conventions that enormously simplify the task of designing efficient computational procedures.

EXAMPLE 5.1: TOP BRASS TROPHY

As usual, it will help to have a tiny example at hand. We illustrate with the case of the (fictional) Top Brass Trophy Company, which makes large championship trophies for youth athletic leagues. At the moment they are planning production for fall sports: football and soccer. Each football trophy has a wood base, an engraved plaque, a large brass football on top, and returns $12 in profit. Soccer trophies are similar except that a brass soccer ball is on top, and the unit profit is only $9. Since the football has an asymmetric shape, its base requires 4 board feet of wood; the soccer base requires only 2 board feet. At the moment there are 1000 brass footballs in stock, 1500 soccer balls, 1750 plaques, and 4800 board feet of wood. What trophies should be produced from these supplies to maximize total profit assuming that all that are made can be sold?

The decisions to be made in this problem are

$$x_1 \triangleq \text{number of football trophies to produce}$$

$$x_2 \triangleq \text{number of soccer trophies to produce}$$

In terms of these decision variables, we can model the problem

$$
\begin{array}{lllll}
\max & 12x_1 + 9x_2 & & \text{(profit)} \\
\text{s.t.} & x_1 & \le 1000 & \text{(footballs)} \\
& x_2 & \le 1500 & \text{(soccer balls)} \\
& x_1 + x_2 & \le 1750 & \text{(plaques)} \\
& 4x_1 + 2x_2 & \le 4800 & \text{(wood)} \\
& x_1,\ x_2 \ge 0
\end{array}
\qquad (5.1)
$$

The objective seeks to maximize total profit, and the main constraints enforce limits on footballs, soccer balls, plaques, and wood, respectively.

Figure 5.1 solves the problem graphically. An optimal solution occurs at $x_1^* = 650$ and $x_2^* = 1100$ with a total profit of \$17,700.

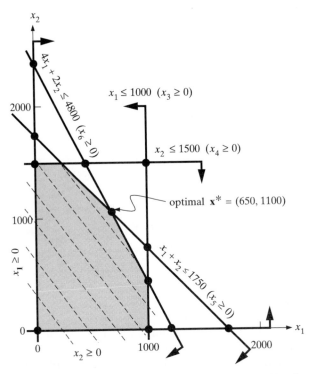

FIGURE 5.1 Graphical Solution of the Top Brass Trophy Example

Model (5.1) is a linear program (definition 4.1) because the single objective function and all constraints involve only weighted sums of the decision variables. Also, all variables are continuous.

Global Optima in Linear Programs

By applying some of the observations in Chapter 3, we can begin to appreciate the elegant model tractability implied by the characteristics required in definition 4.1. Linear objectives are unimodal (principle 3.27). Feasible sets defined by continuous variables and linear constraints are convex (principle $\boxed{3.32}$). Thus linear programs have all the properties most convenient for improving search, and principle $\boxed{3.34}$ applies.

$\underline{\boxed{5.1}}$ Every local optimum for a linear program is a global optimum.

Figure 5.1 illustrates graphically the Top Brass example. The unique local maximum is the global maximum. Notice how property $\boxed{5.1}$ simplifies the task of designing LP algorithms. Most implementations of improving search Algorithm 3A will produce local optima (principle $\boxed{3.17}$), and we now know that this is enough to obtain globally optimal solutions for linear programs. All our attention can be focused on making improving search efficient.

Interior, Boundary, and Extreme Points

One useful tool is a classification of the points in an LP feasible region. Not all have equal interest.

We first distinguish between points in the interior of the feasible region versus those on the boundary. Precise definitions follow from the fact that the boundary of an LP's feasible set is characterized by active inequality constraints (those satisfied as equality).

$\underline{\boxed{5.2}}$ A feasible solution to a linear program is a **boundary point** if at least one inequality constraint that can be strict for some feasible solutions is satisfied as equality at the given point, and an **interior point** if no such inequalities are active.

Figure 5.2 labels some specific points in the Top Brass Trophy feasible set. Each of the inequalities has some feasible points that satisfy it as a strict inequality, and solution $\mathbf{x}^{(7)}$ is interior because none of the inequalities is active. All of $\mathbf{x}^{(0)}$, $\mathbf{x}^{(1)}$, ..., $\mathbf{x}^{(6)}$ are boundary points because at least one inequality is active at each. For example, inequalities $x_1 \leq 1000$ and $x_2 \geq 0$ are satisfied as equalities at solution $\mathbf{x}^{(1)} = (1000, 0)$, and $x_1 \geq 0$ is active at $\mathbf{x}^{(6)}$. Solutions $\mathbf{x}^{(8)}$ and $\mathbf{x}^{(9)}$ constitute neither boundary nor interior points because both are infeasible.

Extreme points (also called corner points) such as $\mathbf{x}^{(0)}$, ..., $\mathbf{x}^{(5)}$ of Figure 5.2 are special boundary points so-named because they "stick out." To obtain a formal definition, we must recall that the feasible region of a linear program is a convex set and that convex means (definition $\boxed{3.29}$) that the line segment between any two feasible points lies entirely within the set. If a point is to form a "corner" of a feasible set, it cannot fall in the middle of any such line segment.

$\underline{\boxed{5.3}}$ **Extreme points** of convex sets are those that do not lie within the line segment between any two other points of the set.

Notice that points $\mathbf{x}^{(0)}$, ..., $\mathbf{x}^{(5)}$ in Figure 5.2 all fulfill this requirement. They can form endpoints of line segments within the set, but never midpoints. Contrast with

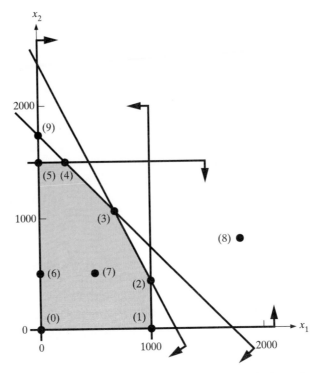

FIGURE 5.2   Interior, Boundary and Extreme Points of the Top Brass Example

nonextreme feasible solutions $\mathbf{x}^{(6)} = (0, 500)$ and $\mathbf{x}^{(7)} = (500, 500)$. The line segment from $\mathbf{x}^{(0)}$ to $\mathbf{x}^{(5)}$ passes through $\mathbf{x}^{(6)}$, and the segment from $\mathbf{x}^{(6)}$ to $\mathbf{x} = (501, 500)$ includes $\mathbf{x}^{(7)}$.

SAMPLE EXERCISE 5.1: CLASSIFYING FEASIBLE POINTS

Classify the labeled solutions as interior, boundary, and/or extreme points of the following LP feasible region:

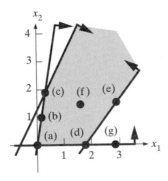

Analysis: Points (a), (c), and (d) are both extreme points and boundary points, because they do not lie along the line segment between any two other feasible points.

Points (b) and (e) are boundary points that are not extreme; each makes at least one constraint active, but these points do fall within a line segment joining others. Point (f) is interior because no constraint is active. Point (g) is neither interior nor boundary (nor extreme) because it is infeasible.

Optimal Points in Linear Programs

Linear programs achieve their optimal solutions on the boundary (except in trivial cases with all feasible solutions having the same objective value).

> **5.4**   Every optimal solution to a typical linear program will be a boundary point of its feasible region.

For example, the optimal solution to our Top Brass example (see Figure 5.1) occurs at boundary (and extreme-point) solution $\mathbf{x}^* = (650, 1100)$.

To gain some intuition about why LP optima usually occur on the boundary, consider the case where all constraints are inequalities. Observe that we can make at least a small move in any direction from an interior point without losing feasibility. Also, with a (nonzero) linear objective function, the coefficient vector $\Delta\mathbf{x} = \mathbf{c}$ is always an improving direction for a maximize problem because it is the gradient (principle $\boxed{3.23}$), and $\Delta\mathbf{x} = -\mathbf{c}$ improves for a minimize problem. Thus there is always an improving feasible move at interior points of a feasible set defined by inequalities unless $\mathbf{c} = \mathbf{0}$ and all solutions cost the same. Matters become more complex with equality constraints present, but the result is the same for every typical LP model. No interior point can be optimal.

What about **unique optimal solutions**? Intuition suggests that if a feasible point is to be the only optimal solution to a linear program, it must somehow "stick out" farther than other feasible points. This is true.

> **5.5**   If a linear program has a unique optimal solution, that optimum must occur at an extreme point of the feasible region.

To see exactly why, try to think of an objective function making non-extreme boundary solution $\mathbf{x}^{(6)}$ the unique optimum in Figure 5.2. The closest we can come is

$$\min\ x_1$$

which has $\mathbf{x}^{(6)}$ optimal but not uniquely so.

We can never make nonextreme boundary solutions uniquely optimal because such points always lie within a line segment joining two other feasible points (definition $\boxed{5.3}$). Thus they can be expressed as a weighted average of the endpoints of such a segment (principle $\boxed{3.31}$), and their objective value is the same weighted average of those for the two endpoints. Either all points of the segment have equal objective value, or one of the endpoints is superior to all the rest.

Continuing in this way, we can see that no collection of nonextreme boundary points can contain all the optimal solutions to a linear program. The collection could always be expanded, with the endpoints of line segments containing known members. We would inevitably have to come upon an extreme-point optimum.

> **5.6** If a linear program has any optimal solution, it has one at an extreme
> point of its feasible region.

Of course, the LP may be infeasible or unbounded, and there may be non-extreme
alternative optimal solutions along the boundary. But if a linear program has any
optimal solutions at all, at least one will occur at an extreme point.

Principle **5.6** is fundamental because of the flexibility it provides in designing
improving search algorithms for linear programming. Knowing that there will be an
optimal solution at an extreme point if there is any optimal solution at all, we are
free to restrict our search to extreme-point solutions. The simplex algorithm, which
is the main topic of this chapter, does exactly that.

SAMPLE EXERCISE 5.2: IDENTIFYING OPTIMAL POINTS

Indicate which of the labeled points in Sample Exercise 5.1 can be optimal or uniquely
optimal for any objective function.

Analysis: Following principle **5.5** , extreme points (a), (c), and (d) can be optimal
or uniquely optimal. Boundary points (b) and (e) can also be optimal. Still, neither can be uniquely optimal because any objective making either optimal will also
have extreme-point optimal solutions. Interior point (f) cannot be optimal for any
nonzero objective function; feasible improvement is always possible. Point (g) is not
even feasible.

LP Standard Form

A model can be a linear program even if variables are subject to a variety of (continuous) variable-type constraints, main constraints are mixtures of inequalities and
equalities, and expressions are nested through several levels of parentheses on both
sides of $\geq$, $\leq$, and $=$ signs. Still, it will be much easier to discuss solution methods if
we settle on an LP standard form.

> **5.7** Linear programs in **standard form** (1) have only equality main constraints;
> (2) have only nonnegative variables, and (3) have objective function and main
> constraints simplified so that variables appear at most once, on the left-hand
> side, and any constant term (possibly zero) appears on the right-hand side.

Converting Inequalities to Nonnegativities with Slack Variables

We will clearly have to do some rearranging to fit every "raw" linear program into
standard form. For example, Top Brass model (5.1) includes inequalities among
its main constraints. Standard form **5.7** allows inequalities only in the form of
nonnegativity variable-type constraints.

To accomplish the needed transformation we introduce new **slack variables**
in each main inequality that consumes the difference between left- and right-hand
sides.

> **5.8** Main inequality constraints of a given linear program can be convert-
> ed into nonnegativities by adding distinct, nonnegative, zero-cost slack vari-
> ables in every such $\leq$ inequality and subtracting such slack variables in every
> main $\geq$.

Applying construction **5.8** to the 4 main inequalities of Top Brass model (5.1),
we add slacks $x_3, \ldots, x_6$. The result is the standard-form model

$$
\begin{aligned}
\max \quad & 12x_1 + 9x_2 \\
\text{s.t.} \quad & x_1 && + x_3 && && = 1000 \\
& x_2 && && + x_4 && = 1500 \\
& x_1 + x_2 && && + x_5 && = 1750 \\
& 4x_1 + 2x_2 && && + x_6 && = 4800 \\
& x_1, x_2, x_3, x_4, x_5, x_6 \geq 0
\end{aligned}
\tag{5.2}
$$

Notice that a different slack was used in each constraint, nonnegativity limits apply
to each slack variable, and that the slacks do not appear in the objective function.
All slack variables carry plus signs in this example (compare with Sample Exercise
5.3 below) because all the modified inequalities were of the $\leq$ form.

Even though our standard form now has 6 variables versus an original 2, we
have not really changed the model. Labels in Figure 5.1 show that each original
inequality constraint corresponds to some slack's nonnegativity in standard form.
For example, the final main constraint

$$4x_1 + 2x_2 \leq 4800$$

becomes nonnegativity on slack variable x_6. Under standard-form equality

$$4x_1 + 2x_2 + x_6 = 4800$$

the inequality holds exactly when

$$x_6 = 4800 - 4x_1 - 2x_2 \geq 0$$

Also, with their coefficients in the objective function $=0$, slack variables have no
impact on cost.

Why bother with this modified version of a perfectly good linear program? To
see, we must look back to the development of feasible move directions in Section 3.5.
Most of the complexity in dealing with feasible directions relates to keeping track of
active inequality constraints (ones satisfied as equalities). Equality constraints are
always active, but inequalities may be active at one moment in a search and inactive
at the next. Introducing slack variables as in construction **5.8** does not eliminate
any inequalities, but it does convert them to the simplest possible form which will
simplify analysis.

SAMPLE EXERCISE 5.3: INTRODUCING SLACK VARIABLES

Introduce slack variables and simplify to place each of the following linear programs
in standard form.

(a) min $9w_1 + 6w_2$
    s.t. $2w_1 + w_2 \geq 10$
           $w_1 \leq 50$
           $w_1 + w_2 = 40$
           $100 \geq w_1 + 2w_2 \geq 15$
           $w_1, w_2 \geq 0$

(b) max $15(2x_1 + 8x_2) - 4x_3$
    s.t. $2(10 - x_1) + x_2 + 5(9 - x_3) \geq 10$
           $x_1 + 2x_3 \leq x_3$
           $2x_2 + 18x_3 = 50$
           $x_1, x_2, x_3 \geq 0$

Modeling:

(a) We introduce slack variables w_3, w_4, w_5, w_6 as in construction $\boxed{5.8}$ to convert the 4 main inequalities to nonnegativities. The result is the standard-form linear program

$$
\begin{aligned}
\min \quad & 9w_1 + 6w_2 \\
\text{s.t.} \quad & 2w_1 + w_2 \quad -w_3 = 10 \\
& w_1 \quad\quad\quad + w_4 = 50 \\
& w_1 + w_2 \quad\quad = 40 \\
& w_1 + 2w_2 \quad + w_5 = 100 \\
& w_1 + 2w_2 \quad -w_6 = 15 \\
& w_1, w_2, w_3, w_4, w_5, w_6 \geq 0
\end{aligned}
$$

Notice that the last two main constraints, which were written together in the original model, are separated in standard form. Slacks are added in $\leq$ inequalities and subtracted in $\geq$ forms. No slack is required in equality constraints.

(b) We begin by simplifying to collect variable terms on the left-hand side and constants on the right [(3) of definition $\boxed{5.7}$].

$$
\begin{aligned}
\max \quad & 30x_1 + 120x_2 - 4x_3 \\
\text{s.t.} \quad & -2x_1 + x_2 - 5x_3 \geq -55 \\
& x_1 + x_3 \leq 0 \\
& 2x_2 + 18x_3 = 50 \\
& x_1, x_2, x_3 \geq 0
\end{aligned}
$$

Now slacks are introduced according to construction $\boxed{5.8}$ to complete the standard form

$$
\begin{aligned}
\max \quad & 30x_1 + 120x_2 - 4x_3 \\
\text{s.t.} \quad & -2x_1 + x_2 - 5x_3 \quad -x_4 = -55 \\
& x_1 + x_3 \quad\quad + x_5 = 0 \\
& 2x_2 + 18x_3 \quad\quad = 50 \\
& x_1, x_2, x_3, x_4, x_5 \geq 0
\end{aligned}
$$

Converting Nonpositive and Unrestricted Variables to Nonegative

Like most applied linear programs, the Top Brass Trophy example employs only nonnegative decision variables. Condition (2) of standard format $\boxed{5.7}$ is fulfilled automatically.

Still, LPs do occasionally involve variables such as net income or temperature that can feasibly take on negative values. Such variables may be **nonpositive** (i.e., subject to ≤ 0 variable-type constraints) or **unrestricted** by sign. The latter are often designated **URS**, meaning "unrestricted sign."

We can convert linear programs with nonpositive and unrestricted variables to standard form simply by changes of variables. For nonpositive variables, the change substitutes the negative.

> **5.9** | Nonpositive variables in linear programs can be eliminated by substituting new variables equal to their negatives.

For example, in a model with original decision variables $x_1, \ldots, x_{10}$ and nonpositive-type restriction

$$x_7 \leq 0$$

we would introduce the new variable

$$x_{11} = -x_7$$

and substitute $-x_{11}$ everywhere that x_7 appears. In particular, the nonpositivity constraint

$$x_7 \leq 0 \quad \text{becomes} \quad -x_{11} \leq 0 \quad \text{or} \quad x_{11} \geq 0$$

The handling of unrestricted variables is slightly less obvious. How can nonnegative variables model a quantity that can take on any sign? The answer is to introduce two new nonnegative variables and consider their difference.

> **5.10** | Unrestricted (or URS) variables in linear programs can be eliminated by substituting the difference of two new nonnegative variables.

For example, a model with variables $y_1, \ldots, y_7$ and

$$y_1 \text{ URS}$$

can be placed in standard form by introducing two new variables $y_8, y_9 \geq 0$. Everywhere y_1 appears in the given model we then substitute

$$y_1 = y_8 - y_9$$

> **SAMPLE EXERCISE 5.4: CONVERTING NONPOSITIVE AND UNRESTRICTED**
> **VARIABLES**

Make suitable variable changes to convert the following linear program to standard form:

$$
\begin{aligned}
\min \quad & -9w_1 + 4w_2 + 16w_3 - 11w_4 \\
\text{s.t.} \quad & w_1 + w_2 + w_3 + w_4 = 100 \\
& 3w_1 - w_2 + 6w_3 - 2w_4 = 200 \\
& w_1 \geq 0, \ w_2 \leq 0
\end{aligned}
$$

Modeling: Variable w_1 is already nonnegative, but we must substitute for nonpositive variable w_2 and for URS variables w_3 and w_4 which are subject to no variable-type constraints. Following rules $\boxed{5.9}$ and $\boxed{5.10}$, we employ

$$w_2 = -w_5$$
$$w_3 = w_6 - w_7$$
$$w_4 = w_8 - w_9$$

After simplification the resulting standard-form linear program is

$$
\begin{aligned}
\min \quad & -9w_1 - 4w_5 + 16w_6 - 16w_7 - 11w_8 + 11w_9 \\
\text{s.t.} \quad & w_1 - w_5 + w_6 - w_7 + w_8 - w_9 = 100 \\
& 3w_1 + w_5 + 6w_6 - 6w_7 - 2w_8 + 2w_9 = 200 \\
& w_1, w_5, w_6, w_7, w_8, w_9 \geq 0
\end{aligned}
$$

Notice that w_2, w_3, and w_4 have been eliminated from the model completely.

Standard Notation for LPs

Once a linear program has been placed in standard form, its key elements are so neatly sorted out that we can begin to think of the model as a collection of coefficients. Widely used notation gives all these elements standard names.

$\boxed{5.11}$ Standard notation for linear programs is

$x_j \triangleq j$th decision variable
$c_j \triangleq$ cost or objective function coefficient of x_j
$a_{i,j} \triangleq$ constraint coefficient of x_j in the ith main constraint
$b_i \triangleq$ right-hand-side (or RHS) constant term of main constraint i
$m \triangleq$ number of main (equality) constraints
$n \triangleq$ number of decision variables

Then, LP standard form of every linear program has the generic format

$$\min \text{ (or max)} \quad \sum_{j=1}^{n} c_j x_j$$

$$\text{s.t.} \quad \sum_{j=1}^{n} a_{i,j} x_j = b_i \qquad \text{for all } i = 1, 2, \ldots, m$$

$$x_j \geq 0 \qquad \text{for all } j = 1, 2, \ldots, n$$

It will often be convenient to write standard-form linear programs in an even more compact way using notions of vectors and matrices. Primer 3 reviews some of the main ideas of matrix arithmetic for those who need it.

$\boxed{5.12}$ In the usual matrix notation, **LP standard form** is

$$\min \text{ (or max)} \quad \mathbf{c} \cdot \mathbf{x}$$
$$\text{s.t.} \quad \mathbf{Ax} = \mathbf{b}$$
$$\mathbf{x} \geq \mathbf{0}$$

PRIMER 3: MATRICES AND MATRIX ARITHMETIC

Matrices are 2-dimensional arrays of numbers, with the **dimension** of the array being described in terms of its number of rows and columns. For example,

$$Q = \begin{pmatrix} 2 & 0 & -\frac{7}{5} \\ 0 & -1.2 & 3 \end{pmatrix} \quad \text{and} \quad R = \begin{pmatrix} 12 & -2 & \frac{7}{5} \\ 1 & 0 & -2 \end{pmatrix}$$

are 2 by 3 matrices (2 rows and 3 columns). See Primer 1 for the related 1-dimensional notion of vectors.

In this book matrices are always denoted by uppercase (capital) boldface symbols (e.g., A, R, Σ). Their entries are indicated by corresponding lowercase italic symbols ($a_{i,j}, r_{2,6}, \sigma_{3,9}$) having one subscript for the row index and one for the column. Thus matrix Q above has $q_{1,2} = 0$ and $q_{2,3} = 3$.

Just as with vectors, matrices of like dimension are added, subtracted, and multiplied by a scalar in component by component fashion. Thus for Q and R above,

$$Q + R = \begin{pmatrix} 14 & -2 & 0 \\ 1 & -1.2 & 1 \end{pmatrix} \quad \text{and} \quad -.3R = \begin{pmatrix} -3.6 & 0.6 & -0.42 \\ -0.3 & 0 & 0.6 \end{pmatrix}$$

Also like vectors, multiplication of one matrix by another is defined in a somewhat nonintuitive way convenient for expressing linear combinations. Matrices P and A can be multiplied as $D = PA$ if the number of columns in P is the same as the number of rows in A. Then the i, j component of the result is defined as the dot product of row i of P and column j of A (i.e., $d_{i,j} = \sum_k p_{i,k} a_{k,j}$). For example, with

$$P = \begin{pmatrix} 1 & 3 \\ -1 & 2 \end{pmatrix} \quad \text{and} \quad A = \begin{pmatrix} 5 & -1 & 0 \\ 2 & 9 & 4 \end{pmatrix}, \quad PA = \begin{pmatrix} 11 & 26 & 12 \\ -1 & 19 & 8 \end{pmatrix}$$

Notice that the order of multiplication of matrices matters. Product AP is not defined for the P and A above because the number of columns of A does not equal the number of rows of P. Even if matrices admit multiplication in both orders, the results can be different.

Multiplication of a matrix by a vector is defined as if the vector were a row or column matrix, whichever is appropriate. Thus if $v = (-1, 4)$, $x = (2, 1, 2)$, and A is as above,

$$vA = (-1, 4) \begin{pmatrix} 5 & -1 & 0 \\ 2 & 9 & 4 \end{pmatrix} = (3, 37, 16)$$

$$Ax = \begin{pmatrix} 5 & -1 & 0 \\ 2 & 9 & 4 \end{pmatrix} \begin{pmatrix} 2 \\ 1 \\ 2 \end{pmatrix} = \begin{pmatrix} 9 \\ 21 \end{pmatrix}$$

In the first computation v was treated as a row matrix because it premultiplies matrix A. In the second computation x plays the role of a column matrix because it postmultiplies A.

Cost or **objective function vector c** presents all objective function coefficients c_j, **constraint matrix A** displays all main constraint coefficients $a_{i,j}$, and **right-hand-side**

(RHS) vector b shows constant terms b_i of the constraints. As usual, **x** is the vector of decision variables x_j.

We can illustrate with the Top Brass Trophy model and its $m = 4$ main constraints. In standard form the Top Brass decision vector has the form $\mathbf{x} = (x_1, x_2, \ldots, x_6)$, with $n = 6$ components. Corresponding coefficient arrays are n-vector **c**, m by n matrix **A**, and m-vector **b**:

$$\mathbf{c} = (\,12 \quad 9 \quad 0 \quad 0 \quad 0 \quad 0\,)$$

$$\mathbf{A} = \begin{pmatrix} 1 & 0 & 1 & 0 & 0 & 0 \\ 0 & 1 & 0 & 1 & 0 & 0 \\ 1 & 1 & 0 & 0 & 1 & 0 \\ 4 & 2 & 0 & 0 & 0 & 1 \end{pmatrix} \qquad \mathbf{b} = \begin{pmatrix} 1000 \\ 1500 \\ 1750 \\ 4800 \end{pmatrix}$$

SAMPLE EXERCISE 5.5: USING STANDARD LP MATRIX NOTATION

Return to the standard-form linear program of Sample Exercise 5.3(b). Identify the $m, n, \mathbf{A}, \mathbf{b}$, and **c** of standard matrix representation $\boxed{5.12}$.

Modeling: The standard form model of Sample Exercise 5.3(b) has $m = 3$ main constraints and $n = 5$ variables. Corresponding coefficient arrays are

$$\mathbf{c} = (\quad 30 \quad\quad 120 \quad -4 \quad\quad 0 \quad\quad 0)$$

$$\mathbf{A} = \begin{pmatrix} -2 & 1 & -5 & -1 & 0 \\ 1 & 0 & 1 & 0 & 1 \\ 0 & 2 & 18 & 0 & 0 \end{pmatrix} \qquad \mathbf{b} = \begin{pmatrix} -55 \\ 0 \\ 50 \end{pmatrix}$$

5.2 EXTREME-POINT SEARCH AND BASIC SOLUTIONS

One can fashion efficient algorithms for linear programs that pass through the interior of the feasible set (see Chapter 6). Still, since there is an extreme-point optimal solution if there is any at all (property $\boxed{5.6}$), we are free to limit our search to extreme points of an LP's feasible region.

Simplex is just such an extreme-point search algorithm. In this section we develop some underlying concepts that lead to the full algorithm in Section 5.3.

Determining Extreme Points with Active Constraints

To begin, we need a more convenient characterization of extreme points. Extreme points are boundary points, and every part of the boundary of an LP feasible space is formed by one or more active constraints. What makes extreme-point solutions special is that enough constraints are active to determine a point completely.

> $\boxed{5.13}$ Every extreme-point solution to a linear program is determined by a set of constraints that are simultaneously active only at that solution.

For example, extreme-point solution $\mathbf{x}^{(5)}$ in Top Brass Figure 5.2 is determined completely by the two constraints active there, $x_1 \geq 0$ and $x_2 \leq 1500$. Contrast this with point $\mathbf{x}^{(6)}$, where only $x_1 \geq 0$ is active. This is not enough to determine a point because many solutions have $x_1 = 0$.

To see more fully how active inequalities relate to extreme points, we need more dimensions. Figure 5.3 depicts the search of a 3-dimensional case maximizing x_3. The boundary area where each main inequality is active now forms one of the 2-dimensional faces labeled A–L. Extreme points are determined as in principle 5.13 by collections of three active constraints.

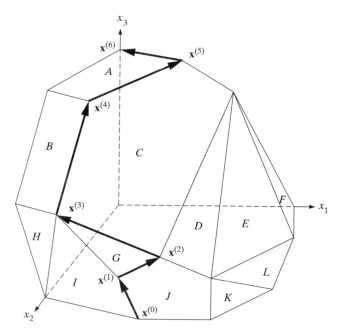

FIGURE 5.3   Adjacent Extreme Point Search to Max x_3 in Three Dimensions

Inequalities I and J, plus nonnegativity $x_3 \geq 0$, determine $\mathbf{x}^{(0)}$. Notice, however, that some extreme points can be determined in several ways. For example, any 3 of the inequalities for surfaces C, D, G, and J determine the extreme point labeled $\mathbf{x}^{(2)}$.

SAMPLE EXERCISE 5.6: DETERMINING EXTREME POINTS

List all sets of 3 active constraints determining points $\mathbf{x}^{(3)}$, $\mathbf{x}^{(4)}$ and $\mathbf{x}^{(5)}$ in Figure 5.3.

Analysis: Constraints B, C, G, H, and I are all active at $\mathbf{x}^{(3)}$. Any 3 of these 5 determine the point. Solution $\mathbf{x}^{(4)}$ is determined uniquely by constraints A, B, and C. Extreme point $\mathbf{x}^{(5)}$ is determined uniquely by inequalities A and C, plus $x_2 \geq 0$.

Adjacent Extreme Points and Edges

Improving search algorithms move from one solution to another nearby. We can define neighboring or adjacent extreme points in terms of their determining sets of active constraints.

> 5.14 | The extreme points of an LP-feasible space are **adjacent** if they are determined by active constraint sets differing in only one element.

Again Figure 5.3 illustrates. Extreme points $\mathbf{x}^{(1)}$ and $\mathbf{x}^{(2)}$ are adjacent because the first is determined by active inequalities G, I, and J, while D, G, and J provide a determining set for the second. These two lists have all but one member in common.

Contrast with $\mathbf{x}^{(2)}$ and $\mathbf{x}^{(4)}$, which are not adjacent. Point $\mathbf{x}^{(4)}$ is determined by active inequalities A, B, and C. Of these, only C is also active at $\mathbf{x}^{(2)}$; no choice of a determining set for $\mathbf{x}^{(2)}$ can satisfy definition 5.14.

Line segments joining adjacent extreme points in Figure 5.3 are called edges.

> 5.15 | An **edge** of the feasible region for a linear program is a 1-dimensional set of feasible points along a line determined by a collection of active constraints.

Adjacent extreme points are joined by an edge determined by the active constraints the extreme points have in common. For example, inequalities for surfaces G, I, and J determine the extreme point labeled $\mathbf{x}^{(1)}$. Keeping just G and J active produces the edge joining $\mathbf{x}^{(1)}$ and $\mathbf{x}^{(2)}$. Similarly, if I and G are kept active, we obtain the edge between adjacent extreme points $\mathbf{x}^{(1)}$ and $\mathbf{x}^{(3)}$.

SAMPLE EXERCISE 5.7: IDENTIFYING EDGES AND ADJACENT EXTREME POINTS

Consider the linear program with feasible set delimited by

$$-2x_1 + 3x_2 \le 6 \tag{5.3}$$
$$-x_1 + x_2 \le 1 \tag{5.4}$$
$$x_1 \ge 0 \tag{5.5}$$
$$x_2 \ge 0 \tag{5.6}$$

(a) Sketch the feasible region and identify points $\mathbf{x}^{(0)} = (0, 0)$, $\mathbf{x}^{(1)} = (0, 1)$, $\mathbf{x}^{(2)} = (3, 4)$, and $\mathbf{x}^{(3)} = (4, 0)$.

(b) Determine which pairs of those points are adjacent extreme points.

(c) Determine which pairs of the points are joined by an edge.

Analysis:

(a) The feasible region is as follows:

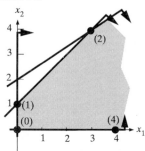

(b) Applying principle $\boxed{5.13}$, extreme points must be determined by active constraints. Solutions $\mathbf{x}^{(0)}$, $\mathbf{x}^{(1)}$, and $\mathbf{x}^{(2)}$ are thus extreme points determined by active constraints (5.5)–(5.6), (5.4)–(5.5), and (5.3)–(5.4), respectively. Point $\mathbf{x}^{(3)}$ is not an extreme point.

In two dimensions all-but-one condition $\boxed{5.14}$ makes extreme points adjacent if they have one active constraint in common. Thus $\mathbf{x}^{(0)}$ and $\mathbf{x}^{(1)}$ are adjacent, as are $\mathbf{x}^{(1)}$ and $\mathbf{x}^{(2)}$, but not $\mathbf{x}^{(0)}$ and $\mathbf{x}^{(2)}$.

(c) Each adjacent pair of part (b) is joined by an edge determined by their common active constraint. For example, $\mathbf{x}^{(1)}$ and $\mathbf{x}^{(2)}$ are joined by the edge along line

$$-x_1 + x_2 = 1$$

Although $\mathbf{x}^{(3)}$ is not an extreme point, the edge defined by line

$$x_2 = 0$$

joins it to $\mathbf{x}^{(0)}$.

Basic Solutions

We saw in Section 5.1 how LP standard form encodes every model as

$$\min \text{ (or max) } \sum_{j=1}^{n} c_j x_j$$

$$\text{s.t.} \quad \sum_{j=1}^{n} a_{i,j} x_j = b_i \quad \text{for all } i = 1, 2, \ldots, m$$

$$x_j \geq 0 \quad \text{for all } j = 1, 2, \ldots, n$$

One effect is to reduce all inequalities to nonnegativity constraints.

It follows from principle $\boxed{5.13}$ that extreme-point solutions to linear programs in standard form are determined by sets of active nonnegativity constraints. Enough variables must be fixed to $=0$ (making the corresponding nonnegativity constraints active) to uniquely determine all other components of a solution.

Basic solutions are produced in just this way.

$\boxed{5.16}$ A **basic solution** to a linear program in standard form is one obtained by fixing just enough variables to $=0$ that the model's equality constraints can be solved uniquely for the remaining variable values. Those variables fixed at zero are called **nonbasic** and the ones obtained by solving the equalities are termed **basic**.

We can illustrate with the standard form of our Top Brass example (Section 5.1):

$$
\begin{aligned}
\max \quad & 12x_1 && +9x_2 \\
\text{s.t.} \quad & +\ x_1 && +x_3 && && && = 1000 \\
& && +\ x_2 && +x_4 && && = 1500 \\
& +\ x_1 && +\ x_2 && && +x_5 && = 1750 \\
& +4x_1 && +2x_2 && && && +x_6 = 4800 \\
& x_1,\ x_2,\ x_3,\ x_4,\ x_5,\ x_6 \geq 0
\end{aligned}
\tag{5.7}
$$

Figure 5.4 again shows the feasible set, with constraints labeled by the applicable nonnegativity constraint of standard form.

One basic solution is obtained by choosing x_1, x_2, x_3, x_4 to be basic and x_5, x_6 to be nonbasic. Fixing $x_5 = x_6 = 0$ leaves the equality system

$$
\begin{array}{llllll}
+ \ x_1 & & + x_3 & & & = 1000 \\
& + \ x_2 & & + x_4 & & = 1500 \\
+ \ x_1 & + \ x_2 & & & + (0) & = 1750 \\
+ 4x_1 & + 2x_2 & & & + (0) & = 4800
\end{array}
\tag{5.8}
$$

The unique solution is $x_1 = 650$, $x_2 = 1100$, $x_3 = 350$, and $x_4 = 400$. Thus the full basic solution is $\mathbf{x} = (650, 1100, 350, 400, 0, 0)$.

Notice that this standard-form solution corresponds to extreme point $\mathbf{x}^{(3)}$ in Figure 5.4. This should come as no surprise, because $\mathbf{x}^{(3)}$ is defined by making active the inequalities corresponding to $x_5 \geq 0$ and $x_6 \geq 0$. That is, the point is defined by setting $x_5 = x_6 = 0$.

SAMPLE EXERCISE 5.8: COMPUTING BASIC SOLUTIONS

Suppose that a linear program in standard form has the constraints

$$
\begin{array}{llll}
4x_1 & - \ x_2 & + \ x_3 & = 1 \\
3x_1 & + 2x_2 & - 2x_3 & = 8 \\
x_1, x_2, x_3 \geq 0
\end{array}
$$

Compute the basic solution corresponding to x_1 and x_2 basic.

Analysis: The only nonbasic variable will be x_3. Setting it to zero as in construction 5.16 gives

$$
\begin{array}{lll}
4x_1 & - \ x_2 & + \ (0) = 1 \\
3x_1 & + 2x_2 & - 2(0) = 8
\end{array}
$$

The remaining 2 equations in 2 unknowns have a unique solution $x_1 = 1$, $x_2 = 3$. Thus the full basic solution is $\mathbf{x} = (1, 3, 0)$.

Existence of Basic Solutions

It is tempting to believe that we can form basic solutions by setting any collection of nonbasic variables to $=0$ (i.e., by making any collection of nonnegativity constraints active). Not so! For example, fixing only $x_4 = 0$ in the Top Brass standard-form model leaves the equation system

$$
\begin{array}{llllll}
+ \ x_1 & & + x_3 & & & = 1000 \\
& + \ x_2 & & + (0) & & = 1500 \\
+ \ x_1 & + \ x_2 & & & + x_5 & = 1750 \\
+ 4x_1 & + 2x_2 & & & + x_6 & = 4800
\end{array}
$$

There remain 5 unknowns in only 4 equations. Geometrically, this simply reflects the fact that only making active the inequality corresponding to $x_4 \geq 0$ in Figure 5.4 does not fully determine a point.

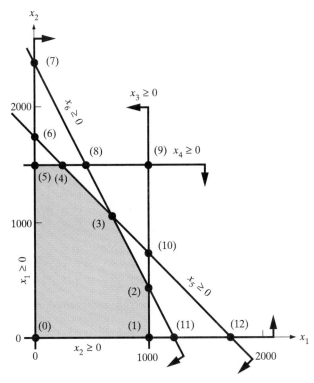

FIGURE 5.4   Basic Solutions of the Top Brass Example

We are no better off if we fix $x_2 = x_4 = 0$. The resulting system

$$
\begin{aligned}
+\; x_1 && +x_3 && &= 1000 \\
&& +(0) && +(0) && &= 1500 \\
+\; x_1 && +(0) && +x_5 && &= 1750 \\
+4x_1 && +2(0) && +x_6 && &= 4800
\end{aligned}
\tag{5.9}
$$

now has 4 equations in 4 unknowns, but it has no solutions as all (look carefully at the second equation). This establishes algebraically what is apparent in Figure 5.4. Making active the inequalities corresponding to $x_2 \geq 0$ and $x_4 \geq 0$ does not determine a point.

To characterize when basic solutions do exist, we must draw on the algebra of systems of simultaneous equations. Primer 4 provides a quick review. One condition assuring a system with a unique solution gives basic solutions their name:

> **5.17** │ A basic solution exists if and only if the columns of equality constraints corresponding to basic variables form a basis, that is, a largest possible linearly independent collection.

Systems (5.8) and (5.9) illustrate the possibilities. We can verify that the constraint columns for the basic variables x_1, x_2, x_3, and x_4 in (5.8) form a basis by checking that the corresponding matrix is nonsingular, that is,

$$\det \begin{pmatrix} 1 & 0 & 1 & 0 \\ 0 & 1 & 0 & 1 \\ 1 & 1 & 0 & 0 \\ 4 & 2 & 0 & 0 \end{pmatrix} = -2 \neq 0$$

It follows that the equations have a unique solution.

On the other hand, the columns for basic variables x_1, x_3, x_5, and x_6 in (5.9) do not produce a unique solution because they are linearly dependent; the column for x_1 is easily expressed as

$$\begin{pmatrix} 1 \\ 0 \\ 1 \\ 4 \end{pmatrix} = 1 \begin{pmatrix} 1 \\ 0 \\ 0 \\ 0 \end{pmatrix} + 1 \begin{pmatrix} 0 \\ 0 \\ 1 \\ 0 \end{pmatrix} + 4 \begin{pmatrix} 0 \\ 0 \\ 0 \\ 1 \end{pmatrix}$$

SAMPLE EXERCISE 5.9: CHECKING EXISTENCE OF BASIC SOLUTIONS

The following are the constraints of a linear program in standard form:

$$\begin{aligned} 4x_1 &- 8x_2 - x_3 = 15 \\ x_1 &- 2x_2 = 10 \\ x_1, x_2, x_3 &\geq 0 \end{aligned}$$

Check whether basic solutions exist for each of the following possible sets of basic variables: (a) x_1, x_2; (b) x_1, x_3; (c) x_1; (d) x_2, x_3; (e) x_1, x_2, x_3.

Analysis:

(a) Column vectors for x_1 and x_2 form a linearly dependent set because

$$\begin{pmatrix} 4 \\ 1 \end{pmatrix} = -\tfrac{1}{2} \begin{pmatrix} -8 \\ -2 \end{pmatrix} \quad \text{or equivalently,} \quad \det \begin{pmatrix} 4 & -8 \\ 1 & -2 \end{pmatrix} = 0$$

Thus basic solutions do not exist.

(b) Basic solutions do exist because column vectors for x_1 and x_3 form a basis. One way to check is to verify that the corresponding matrix is nonsingular, that is

$$\det \begin{pmatrix} 4 & -1 \\ 1 & 0 \end{pmatrix} = 1 \neq 0$$

(c) Column vector $(4, 1)$ for x_1 is linearly independent because it is nonzero. However, we have already seen in part (b) that a larger linearly independent set of columns is possible. Thus x_1 alone does not determine a basic solution.

(d) Basic solutions do exist because column vectors for x_2 and x_3 form a basis. One proof is

$$\det \begin{pmatrix} -8 & -1 \\ -2 & 0 \end{pmatrix} = -2 \neq 0$$

(e) Columns for this set of variables cannot form a basis because no more than two 2-vectors can be linearly independent. Thus basic solutions do not exist.

PRIMER 4: SIMULTANEOUS EQUATIONS, SINGULARITY AND BASES

A system of m **simultaneous linear equations** in m unknowns may have a unique solution, no solutions, or an infinite number of solutions. For example, the $m = 3$ instances

$$
\begin{array}{lll}
3x_1 + x_2 - 7x_3 = 17 & 2y_1 - y_2 - 5y_3 = 3 & 2z_1 - z_2 - 5z_3 = -3 \\
1x_1 + 5x_2 = 1 & -4y_1 + 8y_3 = 0 & -4z_1 + 8z_3 = 4 \\
-2x_1 + 11x_3 = -24 & -6y_1 - y_2 + 11y_3 = -2 & -6z_1 - z_2 + 11z_3 = 11
\end{array}
$$

have unique solution $\mathbf{x} = (1, 0, -2)$, no solution $\mathbf{y}$, and infinitely many solutions $\mathbf{z}$, respectively.

Whether a system falls within the first, unique-solution case depends entirely on the variable coefficient structure of its left-hand side. The x-system above continues to have a unique solution if the right-hand side $(17, 1, 24)$ is replaced by any other 3-vector. Notice that the same is not true of the two nonunique cases; the y and z systems above have identical coefficients on the left-hand side.

A square matrix is **singular** if its determinant $= 0$ and **nonsingular** otherwise. In these terms, systems of m linear equations in m unknowns have unique solutions if and only if the corresponding matrix of left-hand-side variable coefficients is nonsingular. Here the **determinant** of a square matrix $\mathbf{D}$ is the scalar quantity computed recursively as

$$
\det(\mathbf{D}) \triangleq \sum_j (-1)^{(j-1)} d_{1,j} \det(\mathbf{D}_j) \quad \text{with} \quad \det(d_{i,j}) \triangleq d_{i,j}
$$

and $\mathbf{D}_j$ the matrix obtained from $\mathbf{D}$ by deleting row 1 and column j. Thus the x-system above has a unique solution, and the others do not, because corresponding left-hand-side coefficient matrices

$$
\mathbf{N} \triangleq \begin{pmatrix} 3 & 1 & -7 \\ 1 & 5 & 0 \\ -2 & 0 & 11 \end{pmatrix} \quad \text{and} \quad \mathbf{S} \triangleq \begin{pmatrix} 2 & -1 & -5 \\ -4 & 0 & 8 \\ -6 & -1 & 11 \end{pmatrix}
$$

are nonsingular and singular, respectively. That is,

$$
\det(\mathbf{N}) = 3 \det \begin{pmatrix} 5 & 0 \\ 0 & 11 \end{pmatrix} - 1 \det \begin{pmatrix} 1 & 0 \\ -2 & 11 \end{pmatrix} - 7 \det \begin{pmatrix} 1 & 5 \\ -2 & 0 \end{pmatrix}
$$

$$
= 3(55 - 0) - 1(11 - 0) - 7(0 + 10) = 84 \neq 0
$$

and $\det(\mathbf{S}) = 0$.

It is often convenient to draw also on the completely equivalent characterization of when square systems of linear equations have unique solutions, which come from treating the columns of left-hand-side coefficients as vectors. For example, we could deal with coefficients of the y and z systems above as vectors

$$
\mathbf{s}^{(1)} \triangleq \begin{pmatrix} 2 \\ -4 \\ -6 \end{pmatrix}, \quad \mathbf{s}^{(2)} \triangleq \begin{pmatrix} -1 \\ 0 \\ -1 \end{pmatrix}, \quad \text{and} \quad \mathbf{s}^{(3)} \triangleq \begin{pmatrix} -5 \\ 8 \\ 11 \end{pmatrix}
$$

(Continued)

(Continued)

A **linear combination** of vectors is simply a weighted sum. Weights may be positive, negative, or zero. For example, weights $\frac{1}{2}$ and -3 applied to vectors $\mathbf{s}^{(1)}$ and $\mathbf{s}^{(2)}$ above produce the linear combination

$$\tfrac{1}{2}\mathbf{s}^{(1)} - 3\mathbf{s}^{(2)} = \tfrac{1}{2}(2, -4, -6) - 3(-1, 0, -1) = (4, -2, 0)$$

A collection of vectors is said to be **linearly independent** if all are nonzero, and none can be expressed as a linear combination of the others. Otherwise, the collection is **linearly dependent**. For example, the nonzero vectors $\mathbf{s}^{(1)}$ and $\mathbf{s}^{(2)}$ above are linearly independent because no multiple of one can produce the other. Still, the expanded collection $\{\mathbf{s}^{(1)}, \mathbf{s}^{(2)}, \mathbf{s}^{(3)}\}$ is linearly dependent because

$$-2\mathbf{s}^{(1)} + 1\mathbf{s}^{(2)} = -2(2, -4, -6) + 1(-1, 0, -1) = (-5, 8, 11) = \mathbf{s}^{(3)}$$

A **basis** is a largest or **maximal** collection of linearly independent vectors in the sense that members can be combined to produce any other vector. Such linear combinations are unique. Thus $\mathbf{e}^{(1)} = (1, 0)$ and $\mathbf{e}^{(2)} = (0, 1)$ form a basis of the 2-vectors because every (q_1, q_2) is expressed uniquely:

$$\begin{pmatrix} q_1 \\ q_2 \end{pmatrix} = \begin{pmatrix} 1 \\ 0 \end{pmatrix} q_1 + \begin{pmatrix} 0 \\ 1 \end{pmatrix} q_2 = \mathbf{e}^{(1)} q_1 + \mathbf{e}^{(2)} q_2$$

Any m linearly independent m-vectors form a basis, and vice versa. This property leads to the connection between systems of equations and bases, because solving a system for a right-hand side is the same as representing the right-hand side as a linear combination. To be precise, an m by m system of simultaneous linear equations has a unique solution if and only if the coefficient columns for the various variables form a basis (i.e., if and only if the coefficient columns are linearly independent). For example, the x-system above has a unique solution for every right-hand side (b_1, b_2, b_3), exactly because its coefficient columns form a basis, so that there always exist multipliers x_1, x_2, and x_3 satisfying

$$\begin{pmatrix} 3 \\ 1 \\ -2 \end{pmatrix} x_1 + \begin{pmatrix} 1 \\ 5 \\ 0 \end{pmatrix} x_2 + \begin{pmatrix} -7 \\ 0 \\ 11 \end{pmatrix} x_3 = \begin{pmatrix} b_1 \\ b_2 \\ b_3 \end{pmatrix}$$

On the other hand, corresponding y and z systems do not produce unique solutions because columns $\{\mathbf{s}^{(1)}, \mathbf{s}^{(2)}, \mathbf{s}^{(3)}\}$ are linearly dependent.

Basic Feasible Solutions and Extreme Points

Table 5.1 enumerates all possible choices of 2 nonbasic and 4 basic variables in the 4 equations of the Top Brass standard form. In two cases there is no basic solution because (principle $\boxed{5.17}$) the equality constraint columns corresponding to basic variables are linearly dependent.

All other combinations produce basic solutions. Notice, however, that nothing in the construction of basic solutions (definition $\boxed{5.16}$) guarantees feasibility. Some

TABLE 5.1 Basic Solutions of the Top Brass Example

Active Constraints	Basic Variables	Basic Solution	Solution Status	Point in Fig. 5.4
$x_1 \geq 0, x_2 \geq 0$	x_3, x_4, x_5, x_6	$\mathbf{x} = (0, 0, 1000, 1500, 1750, 4800)$	Feasible	$\mathbf{x}^{(0)}$
$x_1 \geq 0, x_3 \geq 0$	x_2, x_4, x_5, x_6	None	Dependent	—
$x_1 \geq 0, x_4 \geq 0$	x_2, x_3, x_5, x_6	$\mathbf{x} = (0, 1500, 1000, 0, 250, 1800)$	Feasible	$\mathbf{x}^{(5)}$
$x_1 \geq 0, x_5 \geq 0$	x_2, x_3, x_4, x_6	$\mathbf{x} = (0, 1750, 1000, -250, 0, 1300)$	Infeasible	$\mathbf{x}^{(6)}$
$x_1 \geq 0, x_6 \geq 0$	x_2, x_3, x_4, x_5	$\mathbf{x} = (0, 2400, 1000, -900, -650, 0)$	Infeasible	$\mathbf{x}^{(7)}$
$x_2 \geq 0, x_3 \geq 0$	x_1, x_4, x_5, x_6	$\mathbf{x} = (1000, 0, 0, 1500, 750, 800)$	Feasible	$\mathbf{x}^{(1)}$
$x_2 \geq 0, x_4 \geq 0$	x_1, x_3, x_5, x_6	None	Dependent	—
$x_2 \geq 0, x_5 \geq 0$	x_1, x_3, x_4, x_6	$\mathbf{x} = (1750, 0, -750, -1500, 0, -2200)$	Infeasible	$\mathbf{x}^{(12)}$
$x_2 \geq 0, x_6 \geq 0$	x_1, x_3, x_4, x_5	$\mathbf{x} = (1200, 0, -200, 1500, 550, 0)$	Infeasible	$\mathbf{x}^{(11)}$
$x_3 \geq 0, x_4 \geq 0$	x_1, x_2, x_5, x_6	$\mathbf{x} = (1000, 1500, 0, 0, -750, -2200)$	Infeasible	$\mathbf{x}^{(9)}$
$x_3 \geq 0, x_5 \geq 0$	x_1, x_2, x_4, x_6	$\mathbf{x} = (1000, 750, 0, 750, 0, -700)$	Infeasible	$\mathbf{x}^{(10)}$
$x_3 \geq 0, x_6 \geq 0$	x_1, x_2, x_4, x_5	$\mathbf{x} = (1000, 400, 0, 1100, 350, 0)$	Feasible	$\mathbf{x}^{(2)}$
$x_4 \geq 0, x_5 \geq 0$	x_1, x_2, x_3, x_6	$\mathbf{x} = (250, 1500, 750, 0, 0, 800)$	Feasible	$\mathbf{x}^{(4)}$
$x_4 \geq 0, x_6 \geq 0$	x_1, x_2, x_3, x_5	$\mathbf{x} = (450, 1500, 550, 0, -200, 0)$	Infeasible	$\mathbf{x}^{(8)}$
$x_5 \geq 0, x_6 \geq 0$	x_1, x_2, x_3, x_4	$\mathbf{x} = (650, 1100, 350, 400, 0, 0)$	Feasible	$\mathbf{x}^{(3)}$

of the solutions in Table 5.1 violate nonnegativity constraints. These correspond geometrically to points where active inequalities determine a solution falling outside the feasible region. For example, choosing x_3 and x_4 nonbasic yields point $\mathbf{x}^{(9)}$ in Figure 5.4, which has negative components for standard-form slack variables of the two constraints it violates.

Our interest is in basic feasible solutions.

> **5.18** A **basic feasible solution** to a linear program in standard form is a basic solution that satisfies all nonnegativity constraints.

Comparison of Table 5.1 and Figure 5.4 will show why. The six basic feasible solutions are exactly the six extreme points $\mathbf{x}^{(0)}, \ldots, \mathbf{x}^{(5)}$ of the feasible region.

This is no accident. We have seen how extreme points are feasible solutions determined by collections of active constraints (property 5.3). For linear programs in standard form, which have only nonnegativity-form inequalities, this means that extreme points are determined by basic solutions. It follows that the extreme-point solutions are the basic ones that are feasible.

> **5.19** The basic feasible solutions of a linear program in standard form are exactly the extreme-point solutions of its feasible region.

SAMPLE EXERCISE 5.10: IDENTIFYING BASIC FEASIBLE SOLUTIONS

The constraint set of a standard-form linear program is defined by the following constraints:

$$
\begin{aligned}
-x_1 &+ x_2 &- x_3 & & & = 0 \\
+x_1 & & &+ x_4 & & = 2 \\
&+ x_2 & & &+ x_5 &= 3 \\
x_1, &\ldots, x_5 &\geq 0
\end{aligned}
$$

(a) Assuming that variables x_3, x_4, and x_5 are slack variables added to produce standard form, graph the feasible region in the original variables x_1 and x_2.

(b) Compute the basic solutions corresponding to each of the following sets of basic variables, and determine which are basic feasible solutions: $B_1 = \{x_3, x_4, x_5\}$, $B_2 = \{x_2, x_4, x_5\}$, $B_3 = \{x_1, x_2, x_5\}$, $B_4 = \{x_1, x_2, x_4\}$, $B_5 = \{x_1, x_3, x_5\}$.

(c) Verify that each basic feasible solution of part (b) corresponds to an extreme point in the graph and that each infeasible basic solution corresponds to a point outside the feasible region determined by the intersection of constraints.

Analysis:

(a) Original constraints would have been

$$
\begin{array}{rrl}
-x_1 & +x_2 & \geq 0 \\
x_1 & & \leq 2 \\
& x_2 & \leq 3 \\
x_1, x_2 & \geq 0
\end{array}
$$

Thus the feasible region is as follows:

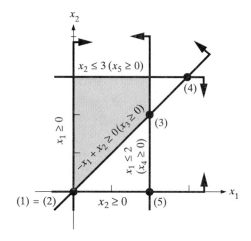

(b) Fixing nonbasics $x_1 = x_2 = 0$ for basic set B_1, the resulting equation system is

$$
\begin{array}{rrrrl}
-(0) & +(0) & -x_3 & & = 0 \\
+(0) & & & +x_4 & = 2 \\
& +(0) & & +x_5 & = 3
\end{array}
$$

Solving produces basic solution $\mathbf{x}^{(1)} = (0, 0, 0, 2, 3)$. Results for other basic sets are derived in the same way:

$$
\begin{array}{lll}
B_2 & \text{implies} & \mathbf{x}^{(2)} = (0, 0, 0, 2, 3) \\
B_3 & \text{implies} & \mathbf{x}^{(3)} = (2, 2, 0, 0, 1) \\
B_4 & \text{implies} & \mathbf{x}^{(4)} = (3, 3, 0, -1, 0) \\
B_5 & \text{implies} & \mathbf{x}^{(5)} = (2, 0, -2, 0, 3)
\end{array}
$$

Notice that different basic sets (here B_1 and B_2) can produce the same basic solution.

Under definition $\boxed{5.18}$, only $\mathbf{x}^{(1)}$, $\mathbf{x}^{(2)}$, and $\mathbf{x}^{(3)}$ are basic feasible. The others violate nonnegativity constraints.

(c) Points corresponding to each basic solution of part (b) are indicated on the plot of part (a). Confirming property $\boxed{5.19}$, basic feasible $\mathbf{x}^{(1)}$, $\mathbf{x}^{(2)}$, and $\mathbf{x}^{(3)}$ correspond to extreme points of the feasible region, while basic infeasible $\mathbf{x}^{(4)}$ and $\mathbf{x}^{(5)}$ correspond to infeasible points where active constraints intersect.

5.3 THE SIMPLEX ALGORITHM

The **simplex algorithm** is a variant of improving search elegantly adapted to exploit the special properties of linear programs in standard form. Like all improving searches, it moves from solution to solution, retaining feasibility and improving the objective function until a local optimum (global by $\boxed{5.1}$) is encountered.

What is unusual is that every step of the simplex leaves us at an extreme-point solution. Under property $\boxed{5.19}$ this means that we only need to think about moves among basic feasible solutions.

Standard Display

Definition $\boxed{5.12}$ showed how linear programs in standard form can be compactly described by a cost vector $\mathbf{c}$, a constraint coefficient matrix $\mathbf{A}$, and a right-hand-side vector $\mathbf{b}$. The evolution of a simplex search will produce many more vectors for current basic solutions, move directions, and so on.

In this book we display all these vectors simply by adding them as rows of a growing table. To see the idea, return to our Top Brass Trophy model (5.2). Standard-form data produce an initial table, with columns for each of the variables and the right-hand side:

	x_1	x_2	x_3	x_4	x_5	x_6	
max $\mathbf{c}$	12	9	0	0	0	0	$\mathbf{b}$
	1	0	1	0	0	0	1000
$\mathbf{A}$	0	1	0	1	0	0	1500
	1	1	0	0	1	0	1750
	4	2	0	0	0	1	4800

Initial Basic Solution

Any improving search begins by choosing a starting feasible solution, and simplex requires an extreme point.

> $\boxed{5.20}$ Simplex search begins at an extreme point of the feasible region (i.e., at a basic feasible solution to the model in standard form).

To illustrate for the Top Brass example, we choose (arbitrarily) to begin at extreme point $\mathbf{x}^{(0)} = (0, 0)$ in Figure 5.4. Table 5.1 will show that this solution is obtained as the basic feasible one for basic variables $B = \{x_3, x_4, x_5, x_6\}$ and nonbasic columns $N = \{x_1, x_2\}$. Adding this basis and the corresponding solution vector to our table gives

	x_1	x_2	x_3	x_4	x_5	x_6	
max **c**	12	9	0	0	0	0	**b**
	1	0	1	0	0	0	1000
A	0	1	0	1	0	0	1500
	1	1	0	0	1	0	1750
	4	2	0	0	0	1	4800
	N	N	B	B	B	B	
$\mathbf{x}^{(0)}$	0	0	1000	1500	1750	4800	

If we did not already know the solution from Table 5.1, it would have been computed by setting nonbasics $x_1 = x_2 = 0$ and solving as in definition $\boxed{5.16}$.

Simplex Directions

The next requirement is move directions. We want simplex to follow edge directions joining current extreme points to adjacent ones. Each such edge direction follows a line determined by all but one of the active constraints that fix our current extreme point because definition $\boxed{5.14}$ requires that adjacent extreme points share all but one of their determining active constraints. But the active constraints at a basis feasible solution are nothing more than the nonnegativity constraints on nonbasic variables.

We obtain the simplex directions by making the nonbasic nonnegativities inactive one at a time.

> $\boxed{5.21}$ **Simplex directions** are constructed by increasing a single nonbasic variable, leaving other nonbasics unchanged, and computing the (unique) corresponding changes in basic variables necessary to preserve equality constraints.

That is, $\Delta x_j = 1$ on the increasing nonbasic, $\Delta x_j = 0$ on other nonbasics, and basic components are obtained by solving conditions for a feasible direction in the equality constraints.

There is one simplex direction for each nonbasic variable. For example, at current solution $\mathbf{x}^{(0)}$ of the Top Brass example, we have one simplex direction increasing nonbasic x_1, and another increasing x_2.

Simplex directions always have $+1$ on the increasing nonbasic, and other nonbasic components $= 0$. To complete the directions, we must determine components for the basic variables.

Reaching all the way back to principle $\boxed{3.25}$ in Section 3.3, we can see what is required. A direction $\Delta\mathbf{x}$ follows the equality constraint

$$\sum_j a_j x_j = b$$

if and only if it satisfies the *net-change-zero condition*

$$\sum_j a_j \, \Delta x_j = 0$$

Thus for our entire system of equality constraints

$$\mathbf{Ax} = \mathbf{b}$$

every feasible direction must satisfy

$$\mathbf{A}\,\Delta\mathbf{x} = \mathbf{0} \tag{5.10}$$

At current Top Brass solution $\mathbf{x}^{(0)}$, condition (5.10) produces the following equation system for the simplex direction increasing x_1 when $\Delta\mathbf{x}_1$ is fixed $= 1$:

$$
\begin{aligned}
+1(1) &\quad +0(0) &\quad +1\,\Delta x_3 &\quad +0\,\Delta x_4 &\quad +0\,\Delta x_5 &\quad +0\,\Delta x_6 &= 0\\
+0(1) &\quad +1(0) &\quad +0\,\Delta x_3 &\quad +1\,\Delta x_4 &\quad +0\,\Delta x_5 &\quad +0\,\Delta x_6 &= 0\\
+1(1) &\quad +1(0) &\quad +0\,\Delta x_3 &\quad +0\,\Delta x_4 &\quad +1\,\Delta x_5 &\quad +0\,\Delta x_6 &= 0\\
+4(1) &\quad +2(0) &\quad +0\,\Delta x_3 &\quad +0\,\Delta x_4 &\quad +0\,\Delta x_5 &\quad +1\,\Delta x_6 &= 0
\end{aligned}
$$

Corresponding equations for the simplex direction increasing x_2 are

$$
\begin{aligned}
+1(0) &\quad +0(1) &\quad +1\,\Delta x_3 &\quad +0\,\Delta x_4 &\quad +0\,\Delta x_5 &\quad +0\,\Delta x_6 &= 0\\
+0(0) &\quad +1(1) &\quad +0\,\Delta x_3 &\quad +1\,\Delta x_4 &\quad +0\,\Delta x_5 &\quad +0\,\Delta x_6 &= 0\\
+1(0) &\quad +1(1) &\quad +0\,\Delta x_3 &\quad +0\,\Delta x_4 &\quad +1\,\Delta x_5 &\quad +0\,\Delta x_6 &= 0\\
+4(0) &\quad +2(1) &\quad +0\,\Delta x_3 &\quad +0\,\Delta x_4 &\quad +0\,\Delta x_5 &\quad +1\,\Delta x_6 &= 0
\end{aligned}
$$

Will these systems have a solution? Absolutely. A basis is a collection of column vectors that can represent every other vector of the same dimension, and we have been careful to leave undetermined only the components on basic variables. Each of the systems (5.10) amounts to finding multipliers Δx_j that weight columns of basic variables to produce the negative of the column for nonbasic k. Such multipliers have to exist. In fact, we know that they are unique (Primer 4).

Completing the two simplex directions for our current Top Brass solution yields the following updated table:

	x_1	x_2	x_3	x_4	x_5	x_6	
max $\mathbf{c}$	12	9	0	0	0	0	$\mathbf{b}$
	1	0	1	0	0	0	1000
$\mathbf{A}$	0	1	0	1	0	0	1500
	1	1	0	0	1	0	1750
	4	2	0	0	0	1	4800
	N	N	B	B	B	B	
$\mathbf{x}^{(0)}$	0	0	1000	1500	1750	4800	
$\Delta\mathbf{x}$ for x_1	1	0	−1	0	−1	−4	
$\Delta\mathbf{x}$ for x_2	0	1	0	−1	−1	−2	

SAMPLE EXERCISE 5.11: CONSTRUCTING SIMPLEX DIRECTIONS

A minimizing, standard-form linear program has the following coefficient data:

	x_1	x_2	x_3	x_4	
min $\mathbf{c}$	2	0	−3	18	$\mathbf{b}$
$\mathbf{A}$	1	−1	2	1	4
	1	1	0	3	2

Assume that x_1 and x_3 are basic, solve for the current basic feasible solution, and compute all corresponding simplex directions.

Analysis: Following definition $\boxed{5.16}$, the current basic solution is obtained by setting nonbasics $x_2 = x_4 = 0$ and solving equality constraints for basic variable values. Here,

$$
\begin{array}{rrrrr}
+1x_1 & -1(0) & +2x_3 & +1(0) & = 4 \\
+1x_1 & +1(0) & +0x_3 & +3(0) & = 2
\end{array}
$$

gives $\mathbf{x} = (2, 0, 1, 0)$.

There will be two simplex directions, one increasing nonbasic x_2 and another increasing nonbasic x_4. Following construction $\boxed{5.21}$, the direction increasing x_2 has $\Delta x_2 = 1$ and its components on all other nonbasics (here only x_4) $= 0$. We must solve for basic components Δx_1 and Δx_3 satisfying net-change-zero condition (5.10). The corresponding linear system

$$
\begin{array}{rrrrr}
+1\,\Delta x_1 & -1(1) & +2\,\Delta x_3 & +1(0) & = 0 \\
+1\,\Delta x_1 & +1(1) & +0\,\Delta x_3 & +3(0) & = 0
\end{array}
$$

has unique solution $\Delta x_1 = -1$, $\Delta x_3 = 1$. Thus the simplex direction increasing x_2 is $\Delta \mathbf{x} = (-1, 1, 1, 0)$.

For $\Delta \mathbf{x}^{(4)}$ the corresponding linear system is

$$
\begin{array}{rrrrr}
+1\,\Delta x_1 & -1(0) & +2\,\Delta x_3 & +1(1) & = 0 \\
+1\,\Delta x_1 & +1(0) & +0\,\Delta x_3 & +3(1) & = 0
\end{array}
$$

Solving for the two unknown components yields simplex direction $\Delta \mathbf{x} = (-3, 0, 1, 1)$.

Improving Simplex Directions and Reduced Costs

Having constructed the collection of simplex directions that can be pursued from our current basic solution without losing feasibility, our next task is to see if any of them improve the objective function

$$
f(\mathbf{x}) \triangleq \mathbf{c} \cdot \mathbf{x} \triangleq \sum_{j=1}^{n} c_j x_j
$$

We know from Section 3.3 that improvement can be checked by referring to the gradient $\nabla f(\mathbf{x})$ at our current solution.

For linear objective function $f(\mathbf{x})$, the gradient is just the vector of objective function coefficients. That is,

$$
\nabla f(\mathbf{x}) = \mathbf{c} \triangleq (c_1, c_2, \ldots, c_n) \qquad \text{for all } \mathbf{x}
$$

Thus gradient conditions $\boxed{3.21}$ and $\boxed{3.22}$ require checking quantities that we term reduced costs.

$\boxed{5.22}$ The **reduced cost** $\bar{c}_j$ associated with nonbasic variable x_j is

$$
\bar{c}_j = \mathbf{c} \cdot \Delta \mathbf{x}
$$

where $\Delta \mathbf{x}$ is the simplex direction increasing x_j.

$\boxed{5.23}$ The simplex direction increasing nonbasic x_j is improving for a maximize linear program if the corresponding reduced cost $\bar{c}_j > 0$, and for a minimize linear program if $\bar{c}_j < 0$.

These simple tests tell us immediately that both simplex directions at solution $\mathbf{x}^{(0)}$ of the Top Brass example improve the maximize objective. For example, the direction increasing x_1 gives

$$\bar{c}_1 = (12, 9, 0, 0, 0, 0) \cdot (1, 0, -1, 0, 0, -1, -4) = 12 > 0$$

and for the one increasing x_2,

$$\bar{c}_2 = (12, 9, 0, 0, 0, 0) \cdot (0, 1, 0, -1, -1, -2) = 9 > 0$$

More typically (see Sample Exercise 5.12), some simplex directions at a current basic solution will improve, and others will not.

SAMPLE EXERCISE 5.12: CHECKING IMPROVEMENT OF SIMPLEX DIRECTIONS

Determine which of the simplex directions computed in Sample Exercise 5.11 are improving for the specified minimizing objective function.

Analysis: We apply computations $\boxed{5.22}$ and $\boxed{5.23}$. For direction $\Delta\mathbf{x} = (-1, 1, 1, 0)$ increasing x_2,

$$\bar{c}_2 = (2, 0, -3, 18) \cdot (-1, 1, 1, 0) = -5 < 0$$

Thus the direction does improve. On the other hand, for $\Delta\mathbf{x} = (-3, 0, 1, 1)$, increasing x_4 does not improve because

$$\bar{c}_4 = (2, 0, -3, 18) \cdot (-3, 0, 1, 1) = 9 \not< 0$$

Step Size and the Minimum Ratio Rule

Simplex can adopt for the next move of improving search any simplex direction that improves the objective function. The next issue is "How far?", that is, what step size λ should be applied to chosen direction $\Delta\mathbf{x}$?

Following principle $\boxed{3.15}$ in Section 3.2, we would like to take the biggest step that preserves feasibility and improves the objective function value. With the constant gradient for linear programs, an improving simplex direction remains improving forever. Also, we constructed simplex directions to maintain all equality constraints $\mathbf{Ax} = \mathbf{b}$.

If there is any limit on step size λ, it must come from eventually violating a nonnegativity constraint. That is, some component of the revised solution must become negative. With all components feasible, and thus nonnegative at our current solution, infeasibility can occur only if some component of the chosen simplex direction is negative. The first solution component forced to zero fixes λ.

$\boxed{5.24}$ If any component is negative in improving simplex direction $\Delta\mathbf{x}$ at current basic solution $\mathbf{x}^{(t)}$, simplex search uses the maximum feasible step of **minimum ratio** computation

$$\lambda = \min\left\{ \frac{x_j^{(t)}}{-\Delta x_j} : \Delta x_j < 0 \right\}$$

> **5.25** If no component is negative in improving simplex direction $\Delta \mathbf{x}$ at current basic solution $\mathbf{x}^{(t)}$, the solution can be improved forever in direction $\Delta \mathbf{x}$. That is, the linear program is unbounded.

To illustrate, we arbitrarily set $\Delta \mathbf{x} = (1, 0, -1, 0, 0, -1, -4)$, increasing x_1 at $\mathbf{x}^{(0)}$ of our Top Brass example. The chosen direction $\Delta \mathbf{x}^{(1)}$ does have negative components, so there is no indication of unboundedness.

To decide the maximum step, we add a row to our table that computes the step size at which each component would drop to $=0$.

	x_1	x_2	x_3	x_4	x_5	x_6
	N	N	B	B	B	B
$\mathbf{x}^{(0)}$	0	0	1000	1500	1750	4800
$\Delta \mathbf{x}$	1	0	-1	0	-1	-4
	—	—	$\dfrac{1000}{-(-1)}$	—	$\dfrac{1750}{-(-1)}$	$\dfrac{4800}{-(-4)}$

The least of these ratios establishes λ in rule $\boxed{5.24}$:

$$\lambda = \min \left\{ \frac{1000}{1}, \frac{1750}{1}, \frac{4800}{4} \right\} = 1000$$

Thus our new solution is

$$\mathbf{x}^{(1)} \leftarrow \mathbf{x}^{(0)} + \lambda \Delta \mathbf{x}$$
$$= (0, 0, 1000, 1500, 1750, 4800) + 1000(1, 0, -1, 0, -1, -4)$$
$$= (1000, 0, 0, 1500, 750, 800)$$

SAMPLE EXERCISE 5.13: DETERMINING THE MAXIMUM SIMPLEX STEP

Let the current simplex solution to a linear program in standard form be $\mathbf{x}^{(17)} = (13, 0, 10, 2, 0, 0)$, with x_2 and x_5 nonbasic. Determine the maximum step and new solution (if any), assuming that each of the following is the improving simplex direction associated with increasing x_2.

(a) $\Delta \mathbf{x} = (12, 1, -5, -1, 0, 8)$

(b) $\Delta \mathbf{x} = (0, 1, 6, 3, 0, \frac{7}{2})$

(c) $\Delta \mathbf{x} = (-1, 1, -8, 0, 0, -5)$

Analysis:

(a) This improving simplex direction has negative components, so we apply rule $\boxed{5.24}$.

$$\lambda = \min \left\{ \frac{10}{-(-5)}, \frac{2}{-(-1)} \right\} = 2$$

Notice that values for x_1 were not included in this computation; with a positive directional component $\Delta x_1 = 12$, x_1 is increasing. The new simplex solution will be

$$\mathbf{x}^{(18)} \leftarrow \mathbf{x}^{(17)} + \lambda \, \Delta \mathbf{x}$$
$$= (13, 0, 10, 2, 0, 0) + 2(12, 1, -5, -1, 0, 8)$$
$$= (37, 2, 0, 0, 0, 16)$$

(b) This improving simplex direction has no negative components, so progress is unlimited. Under principle $\boxed{5.25}$, the model is unbounded.

(c) This improving simplex direction does have negative components. Applying rule $\boxed{5.24}$ yields

$$\lambda = \min \left\{ \frac{13}{-(-1)}, \frac{10}{-(-8)}, \frac{0}{-(-5)} \right\} = 0$$

and new solution

$$\mathbf{x}^{(18)} \leftarrow \mathbf{x}^{(17)} + 0 \, \Delta \mathbf{x} = \mathbf{x}^{(17)}$$

The zero step λ results from basic variable x_6 happening to take on the zero value more typical of nonbasics—a common occurrence with large-scale linear programs. In Section 5.7 we discuss this **degenerate** case further.

Updating the Basis

Step size rule $\boxed{5.24}$ tells us to continue from our present extreme-point solution, along the edge direction formed by increasing a single nonbasic, until an adjacent extreme point is formed by a newly active nonnegativity constraint (definitions $\boxed{5.14}$ and $\boxed{5.15}$). To continue the algorithm we need to find a new basis corresponding to this new extreme-point solution. Active nonnegativity constraints tell us how.

> $\underline{5.26}$ After each move of simplex search, the nonbasic variable generating the chosen simplex direction enters the basis, and any one of the (possibly several) basic variables fixing step size λ leaves the basis.

That is, we move to the new basis implied by the nonnegativity constraint on the increasing nonbasic variable becoming inactive, while a nonnegativity constraint on a blocking basic becomes active.

For our Top Brass example, the move from $\mathbf{x}^{(0)}$ increased nonbasic x_1, and x_3 is the first component to drop to $=0$. Thus x_1 enters and x_3 leaves, resulting in new basic set $\{x_1, x_4, x_5, x_6\}$.

Figure 5.5 interprets graphically. New solution $\mathbf{x}^{(1)}$ is indeed the adjacent extreme point obtained when we move along the edge direction for x_1 until the nonnegativity constraint for x_3 becomes active. The new basis makes x_1 basic and x_3 nonbasic.

SAMPLE EXERCISE 5.14: UPDATING THE BASIS

Return to Sample Exercise 5.13 and determine for bounded cases (a) and (c) what variables should enter and leave the basis.

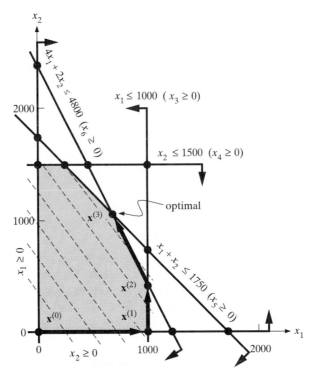

FIGURE 5.5   Simplex Search of the Top Brass Example

Analysis: We apply rule ⎣5.26⎦. Increasing nonbasic x_2 enters in both cases. For case (a) there is a choice of leaving variables because both x_3 and x_4 establish λ; either one could be selected to leave the basis. Variable x_6 alone fixes the value of λ in case (c); it must be chosen as the leaving basic.

Rudimentary Simplex Algorithm

We have now developed all the main ideas of simplex search. Simplex implements extreme-point search by moving from basic feasible solution to basic feasible solution until either the problem is shown to be unbounded or the current solution proves optimal. At every iteration a move is contemplated in each simplex direction. If any of the simplex directions is improving, one is selected and pursued as far as feasibility will permit. If no simplex direction improves, we stop and report an optimum. Algorithm 5A gives a formal statement.

ALGORITHM 5A: RUDIMENTARY SIMPLEX SEARCH FOR LINEAR PROGRAMS

Step 0: Initialization. Choose any starting feasible basis, construct the corresponding basic solution $\mathbf{x}^{(0)}$, and set solution index $t \leftarrow 0$.

Step 1: Simplex Directions. Construct the simplex direction $\Delta\mathbf{x}$ associated with increasing each nonbasic variable x_j, and compute the corresponding reduced cost $\bar{c}_j = \mathbf{c} \cdot \Delta\mathbf{x}$.

Step 2: Optimality. If no simplex direction is improving (no $\bar{c}_j > 0$ for a maximize problem, or no $\bar{c}_j < 0$ for a minimize), then stop; current solution $\mathbf{x}^{(t)}$ is globally optimal. Otherwise, choose any improving simplex direction as $\Delta\mathbf{x}^{(t+1)}$ and denote the associated entering basic variable x_p.

Step 3: Step Size. If there is no limit on feasible moves in simplex direction $\Delta\mathbf{x}^{(t+1)}$ (all components are nonnegative), then stop; the given model is unbounded. Otherwise, choose leaving variable x_r so that

$$\frac{x_r^{(t)}}{-\Delta x_r^{(t+1)}} = \min\left\{\frac{x_j^{(t)}}{-\Delta x_j^{(t+1)}} : \Delta x_j^{(t+1)} < 0\right\} \quad \text{and set} \quad \lambda \leftarrow \frac{x_r^{(t)}}{-\Delta x_r^{(t+1)}}$$

Step 4: New Point and Basis. Compute the new solution

$$\mathbf{x}^{(t+1)} \leftarrow \mathbf{x}^{(t)} + \lambda\,\Delta\mathbf{x}^{(t+1)}$$

and replace x_r in the basis by x_p. Then advance $t \leftarrow t+1$, and return to Step 1.

Rudimentary Simplex Solution of Top Brass Example

Table 5.2 details the full Algorithm 5A search of our Top Brass Trophy example in LP standard form. Figure 5.5 tracks progress graphically. As in our earlier discussion,

TABLE 5.2 Rudimentary Simplex Search of Top Brass Trophy Example

	x_1	x_2	x_3	x_4	x_5	x_6	
max **c**	12	9	0	0	0	0	**b**
	1	0	1	0	0	0	1000
A	0	1	0	1	0	0	1500
	1	1	0	0	1	0	1750
	4	2	0	0	0	1	4800
$t=0$	N	N	B	B	B	B	
$\mathbf{x}^{(0)}$	0	0	1000	1500	1750	4800	$\mathbf{c}\cdot\mathbf{x}^{(0)}=0$
$\Delta\mathbf{x}$ for x_1	1	0	−1	0	−1	−4	$\bar{c}_1 = \boxed{12}$
$\Delta\mathbf{x}$ for x_2	0	1	0	−1	−1	−2	$\bar{c}_2 = 9$
	—	—	$\dfrac{\boxed{1000}}{-(-1)}$	—	$\dfrac{1750}{-(-1)}$	$\dfrac{4800}{-(-4)}$	$\lambda = 1000$
$t=1$	B	N	N	B	B	B	
$\mathbf{x}^{(1)}$	1000	0	0	1500	750	800	$\mathbf{c}\cdot\mathbf{x}^{(1)}=12{,}000$
$\Delta\mathbf{x}$ for x_2	0	1	0	−1	−1	−2	$\bar{c}_2 = \boxed{9}$
$\Delta\mathbf{x}$ for x_3	−1	0	1	0	1	4	$\bar{c}_3 = -12$
	—	—	—	$\dfrac{1500}{-(-1)}$	$\dfrac{750}{-(-1)}$	$\dfrac{\boxed{800}}{-(-2)}$	$\lambda = 400$
$t=2$	B	B	N	B	B	N	
$\mathbf{x}^{(2)}$	1000	400	0	1100	350	0	$\mathbf{c}\cdot\mathbf{x}^{(2)}=15{,}600$
$\Delta\mathbf{x}$ for x_3	−1	2	1	−2	−1	0	$\bar{c}_3 = \boxed{6}$
$\Delta\mathbf{x}$ for x_6	0	−0.5	0	0.5	0.5	1	$\bar{c}_6 = -4.5$
	$\dfrac{1000}{-(-1)}$	—	—	$\dfrac{1100}{-(-2)}$	$\dfrac{350}{-(-1)}$	—	$\lambda = 350$
$t=3$	B	B	B	B	N	N	
$\mathbf{x}^{(3)}$	650	1100	350	400	0	0	$\mathbf{c}\cdot\mathbf{x}^{(3)}=17{,}700$
$\Delta\mathbf{x}$ for x_5	1	−2	−1	2	1	0	$\bar{c}_5 = -6$
$\Delta\mathbf{x}$ for x_6	0.5	0.5	0.5	−0.5	0	1	$\bar{c}_6 = -1.5$
							"optimal"

the search begins at basic feasible solution $\mathbf{x}^{(0)} = (0, 0, 1000, 1500, 1750, 4800)$. We have already detailed the first iteration $t = 0$. Simplex directions for nonbasics x_1 and x_2 both improve. Choosing $p = 1$, the maximum feasible step is determined by the component with subscript $r = 3$ at $\lambda = 1000$.

After the move, we have new solution $\mathbf{x}^{(1)} = (1000, 0, 0, 1500, 750, 800)$ with objective function value \$12,000. Increasing variable x_1 has replaced blocking x_3 in the basis. The process now repeats with $t = 1$. Simplex directions are available for x_2 and x_3, but only the first improves because $\bar{c}_2 = 9$ and $\bar{c}_3 = -12$. Thus $p = 2$, and we take a maximum step in the direction for x_2. Step size $\lambda = 400$ is fixed as basic x_6 decreases.

Taking this new step brings us to $\mathbf{x}^{(2)} = (1000, 400, 0, 1100, 350, 0)$ with objective value \$15,600. Figure 5.5 confirms that it too is an extreme-point solution.

The basis is now $\{x_1, x_2, x_4, x_5\}$. As usual, there are two simplex directions available, one for each nonbasic. However, only the one for x_3 improves. At $\lambda = 350$, the nonnegativity constraint on x_5 becomes active, producing new solution $\mathbf{x}^{(3)} = (650, 1100, 350, 400, 0, 0)$.

With $t = 3$ we repeat the sequence again. This time, however, neither simplex direction is improving. The search terminates with (global) optimal solution $\mathbf{x}^* = (650, 1100, 350, 400, 0, 0)$ having profit \$17,700.

Stopping and Global Optimality

If Algorithm 5A stops with a simplex direction along which we can improve forever without losing feasibility, the given model is clearly unbounded. But what if it stops with an optimum?

Simplex search considers only the simplex or edge directions as it looks for an improving feasible move. As long as one of the simplex directions improves, this approach is completely consistent with our previous development of improving search. Still, we stop when no simplex direction is improving and feasible.

Could some non-simplex feasible direction still improve? No.

5.27 | When simplex Algorithm 5A is applied to a linear program in standard form, it either stops with a correct indication of unboundedness or produces a globally optimal solution.

To see why, notice that any feasible direction at a current basic solution must increase nonbasics because the basic part of the solution is determined uniquely once nonbasics are fixed at $=0$. The key insight is that any non-simplex direction (i.e., any that increases more than one nonbasic) will just be a weighted sum of simplex directions. For example, at the last iteration of the Top Brass search, we might have considered a direction $\Delta\mathbf{x}$ increasing both x_5 and x_6. Say that $\Delta x_5 = 3$ and $\Delta x_6 = 7$. The only way this choice of nonbasic components can retain net change zero condition (5.10) is for $\Delta\mathbf{x}$ to have the form

$$\Delta\mathbf{x} = 3(\Delta\mathbf{x}) + 7(\Delta\mathbf{x}')$$

where $\Delta\mathbf{x}$ and $\Delta\mathbf{x}'$ are the simplex directions for x_5 and x_6, respectively. It follows that

$$\mathbf{c} \cdot \boldsymbol{\Delta}\mathbf{x} = \mathbf{c} \cdot \left(3\boldsymbol{\Delta}\mathbf{x} + 7\boldsymbol{\Delta}\mathbf{x}'\right)$$
$$= 3\left(\mathbf{c} \cdot \boldsymbol{\Delta}\mathbf{x}\right) + 7\left(\mathbf{c} \cdot \boldsymbol{\Delta}\mathbf{x}'\right)$$
$$= 3(-6) + 7(-1.5)$$
$$= -29.5$$

We see that no positive combination of simplex directions can meet the test for improvement unless at least one simplex direction does, too. A current basic feasible solution is optimal if and only if no simplex direction is improving.

5.4 DICTIONARY AND TABLEAU REPRESENTATIONS OF SIMPLEX

Our development of the simplex algorithm in Section 5.3 follows the improving search paradigm, which constitutes one of the main themes of this book. It also reflects the way professionals and researchers in linear programming think about simplex.

Still, readers who have encountered simplex in other introductory texts may find our form a bit difficult to recognize. In this section we make a brief connection to more traditional formats. Those not confused by our departures from tradition can skip the section without loss.

Simplex Dictionaries

Traditional introductory developments of the simplex algorithm view the process that we have conceived as a search in terms of manipulating the objective function and main standard-form equations

$$\sum_{j=1}^{n} a_{i,j}x_j = b_i \qquad \text{for all } i = 1, \ldots, m$$

At each iteration representations called simplex dictionaries express the basic variables and the objective function value in terms of the nonbasics.

> __5.28__   **Simplex dictionaries** express objective function value z and basic variables x_k, $k \in B$, in terms of nonbasic variables x_j, $j \in N$, as
>
> $$z = \bar{z} + \sum_{j\in N} \bar{c}_j x_j$$
> $$x_k = \bar{b}_k - \sum_{j\in N} \bar{a}_{k,j}x_j \qquad \text{for all } k \in B$$

Dictionary form is achieved by **Gaussian elimination**—solving for one basic variable at a time and substituting for it in other constraints and the objective. To see the idea, consider the basis corresponding to $t = 2$ in Table 5.2. There basic and nonbasic index sets are

$$B = \{1, 2, 4, 5\}$$
$$N = \{3, 6\}$$

We begin our derivation of the corresponding dictionary with the original objective and constraints:

$$z = 12x_1 + 9x_2$$
$$x_1 + x_3 = 1000$$
$$x_2 + x_4 = 1500$$
$$x_1 + x_2 + x_5 = 1750$$
$$4x_1 + 2x_2 + x_6 = 4800$$

The first constraint expresses basic variable x_1 as

$$x_1 = 1000 - (1x_3)$$

Substituting for x_1 leaves

$$x_1 = 1000 - (1x_3)$$
$$x_2 + x_4 = 1500$$
$$(1000 - x_3) + x_2 + x_5 = 1750$$
$$4(1000 - x_3) + 2x_2 + x_6 = 4800$$

We now continue with basic x_2. Solving the second equation and substituting gives

$$x_1 = 1000 - (1x_3)$$
$$x_2 = 1500 - (1x_4)$$
$$(1000 - x_3) + (1500 - x_4) + x_5 = 1750$$
$$+4(1000 - x_3) + 2(1500 - x_4) + x_6 = 4800$$

The third equation can now be solved for basic variable x_4. After substitution and solving the last equation for x_5, we complete the constraints as

$$
\begin{aligned}
x_1 &= 1000 &&- (+1x_3 \quad +0 \; x_6) \\
x_2 &= 400 &&- (-2x_3 \quad +0.5x_6) \\
x_4 &= 1100 &&- (+2x_3 \quad -0.5x_6) \\
x_5 &= 350 &&- (+1x_3 \quad -0.5x_6)
\end{aligned}
\tag{5.11}
$$

Finally, we substitute these expressions in the objective function to obtain

$$
\begin{aligned}
z &= 12(1000 - 1x_3) + 9(400 - 2x_3 - 0.5x_6) \\
&= 15{,}600 + 6x_3 - 4.5x_6
\end{aligned}
\tag{5.12}
$$

The dictionary (5.11)–(5.12) is now complete. In the notation of definition 5.28 , for example, $\bar{z} = 15{,}600$, $\bar{c}_3 = 6$, $\bar{b}_2 = 400$, and $\bar{a}_{4,6} = -0.5$.

SAMPLE EXERCISE 5.15: CONSTRUCTING SIMPLEX DICTIONARIES

The computed simplex directions and reduced costs for the linear program of Sample Exercises 5.11 and 5.12

$$
\begin{aligned}
\min \quad & z = 2x_1 - 3x_3 + 18x_4 \\
\text{s.t.} \quad & x_1 - x_2 + 2x_3 + x_4 = 4 \\
& x_1 + x_2 + 3x_4 = 2 \\
& x_1, x_2, x_3, x_4 \geq 0
\end{aligned}
$$

had x_1 and x_3 basic. Apply Gaussian elimination to derive the corresponding simplex dictionary.

Analysis: We want to solve for basic variables x_1 and x_3 in terms of nonbasics. From the first constraint,

$$x_1 = 4 - (-1x_2 + 2x_3 + 1x_4)$$

Substituting in the second constraint yields

$$(4 + x_2 - 2x_3 - 1x_4) + x_2 + 3x_4 = 2$$

Then, solving for x_3 and substituting produces the dictionary:

$$
\begin{aligned}
z &= 1 + & -5x_2 & +9x_4 \\
x_1 &= 2 - & (+1x_2 & +3x_4) \\
x_3 &= 1 - & (-1x_2 & -1x_4)
\end{aligned}
$$

Simplex Tableaux

Simplex tableaux are detached-coefficient displays of exactly the same information as simplex dictionaries.

> **5.29**   The **simplex tableau** associated with basic set $\{x_k : k \in B\}$ displays the coefficients $\bar{z}, \bar{c}_j, \bar{b}_i,$ and $\bar{a}_{i,j}$ of the corresponding simplex dictionary in detached-coefficient form with all variables translated to the left-hand side.

For example, the dictionary (5.11) corresponds to the simplex tableau

x_1	x_2	x_3	x_4	x_5	x_6	
0	0	–6	0	0	+4.5	15,600
1	0	1	0	0	0	1,000
0	1	–2	0	0	0.5	400
0	0	2	1	0	–0.5	1,100
0	0	1	0	1	–0.5	350

The only change is that nonbasic variables have been translated to the left-hand side, and coefficients have been extracted in a matrix.

> **SAMPLE EXERCISE 5.16: CONSTRUCTING SIMPLEX TABLEAUX**

Construct the simplex tableau corresponding to the dictionary of Sample Exercise 5.15.

Analysis: With all variables translated to the left-hand side, the coefficient tableau is

x_1	x_2	x_3	x_4	
0	5	0	–9	1
1	1	0	3	2
0	–1	1	–1	1

Simplex Algorithm with Dictionaries or Tableaux

In dictionary/tableau form, simplex still moves from basic feasible solution to basic feasible solution. Each simplex iteration begins by checking the sign of objective function coefficients $\bar{c}_j$ on nonbasic variables. If none is negative for a minimize problem (positive for a maximize), the current basic solution is optimal. Otherwise, the solution can be improved by increasing a nonbasic variable from its basic-solution value $=0$. We pick one such variable as the entering nonbasic. Its coefficient column in the dictionary or tableau tells us how basic variables will change as the nonbasic increases. If it can increase forever without losing feasibility, the problem is unbounded. Otherwise, the entering variable's increase eventually drives some basic variable to its lower limit of zero. This establishes a leaving basic variable. We update the dictionary or tableau to the new basis and repeat the process.

Correspondence to the Improving Search Paradigm

This brief synopsis of the simplex with dictionaries or tableaux should sound very familiar. It is, in fact, exactly Algorithm 5A. The only difference comes in how the arithmetic is accomplished.

Think first about the current basic solution. When nonbasic variables are fixed $=0$, dictionary format makes the corresponding values of basic variables obvious.

> **5.30** | Right-hand-side constants $\bar{b}_k$ of simplex dictionary/tableau format show current values for corresponding basic variables x_k. Similarly, $\bar{z}$ is the objective function value of the current basic solution.

For example, the dictionary (5.11) shows clearly that with nonbasics $x_3 = x_6 = 0$, the current basic solution is

$$(\bar{b}_1, \bar{b}_2, 0, \bar{b}_4, \bar{b}_5, 0) = (1000, 400, 0, 1100, 350, 0)$$

This is exactly the solution reported in Table 5.2 for $t = 2$. It objective function value is $\bar{z} = 15,600$.

Next consider the simplex directions. Definition 5.21 specifies components for nonbasic variables and leaves those for basic variables to be computed. Specifically, the changes in basics associated with increasing any nonbasic are unique solutions to equations $\mathbf{A}\Delta\mathbf{x} = \mathbf{0}$.

The simplex in dictionary/tableau form uses columns of the tableau to predict the same basic variable changes. Since the changes are unique, there must be a close connection to simplex directions.

> **5.31** | Simplex dictionary/tableau coefficients $\bar{a}_{k,j}$ for nonbasic variables x_j are exactly the negatives $-\Delta x_k$ of corresponding components in the simplex directions increasing x_j.

The dictionary (5.11) illustrates again. There

$$\bar{a}_{1,3} = +1 \qquad \bar{a}_{1,6} = 0$$
$$\bar{a}_{2,3} = -2 \qquad \bar{a}_{2,6} = +0.5$$
$$\bar{a}_{4,3} = +2 \qquad \bar{a}_{4,6} = -0.5$$
$$\bar{a}_{5,3} = +1 \qquad \bar{a}_{5,6} = -0.5$$

Table 5.2 for $t = 2$ shows that the simplex directions are

$$\begin{pmatrix} \Delta x_1 \\ \Delta x_2 \\ \Delta x_3 \\ \Delta x_4 \\ \Delta x_5 \\ \Delta x_6 \end{pmatrix} = \begin{pmatrix} -1 \\ +2 \\ +1 \\ -2 \\ -1 \\ 0 \end{pmatrix} \quad \text{for } x_3 \quad \text{and} \quad \begin{pmatrix} \Delta x_1 \\ \Delta x_2 \\ \Delta x_3 \\ \Delta x_4 \\ \Delta x_5 \\ \Delta x_6 \end{pmatrix} = \begin{pmatrix} 0 \\ -0.5 \\ 0 \\ 0.5 \\ 0.5 \\ 1 \end{pmatrix} \quad \text{for } x_6$$

Notice that tableau coefficients are exactly the negatives of corresponding simplex direction components for basic variables x_1, x_2, x_4, and x_5.

Finally, focus on the reduced costs which tell us whether a nonbasic variable can increase productively. We have employed the same $\bar{c}_j$ notation in both reduced cost definition $\boxed{5.22}$ and format $\boxed{5.28}$ because they refer to the same quantities.

> $\boxed{5.32}$ Simplex dictionary/tableau objective function coefficients $\bar{c}_j$ are exactly the reduced costs of definition $\boxed{5.22}$.

In both representations they show how objective function value z will change per unit increase in nonbasic variable x_j if basics are adjusted to preserve feasibility in main equality constraints. For example, the reduced cost of x_3 in both dictionary (5.11) and Table 5.2 (at $t = 2$) is $\bar{c}_3 = 6$.

Comparison of Formats

Observations $\boxed{5.30}$ to $\boxed{5.32}$ document how solving for current basic variables in dictionary format $\boxed{5.28}$ is just an alternative way to compute the basic solution, simplex directions, and reduced costs required in rudimentary simplex Algorithm 5A. Nothing is fundamentally different.

One can argue about which computation is easier for hand calculation. Still, it is important to realize that neither is ever employed in serious computer implementations of the simplex algorithm. Sections 5.7 and 5.8 will detail the much more efficient **revised simplex** version at the heart of all commercial simplex codes.

5.5 TWO PHASE SIMPLEX

As usual, we have developed the simplex version of improving search assuming that we know a starting basic feasible solution. In most real problems we will have to search for one.

In Section 3.6 we introduced the two-phase approach we will adopt (Algorithm 3B). A solution is chosen that satisfies at least part of the constraints; artificial variables are added to synthetically satisfy all other constraints; Phase I minimizes the sum of the artificials; Phase II proceeds from the result of Phase I to optimize the real objective function.

EXAMPLE 5.2: CLEVER CLYDE

We will illustrate two-phase simplex computation with the contrived but instructive case of Clever Clyde. Clyde is an entrepreneur of dubious integrity who is seeking to open a sports collectibles business. He has only $5000 of his own, and the business will require at least $100,000. Still, Clyde also has a real knack for convincing others to invest in his projects. One investor has already agreed to pay 50% of the initial cost of the firm in return for an equal partnership, and Clyde has an appointment tomorrow with another prospect. We want to consider several cases for the fraction of support, call it α, that Clyde may get from selling a second "equal" partnership. His goal is to maximize the size of the business.

To formulate Clyde's dilemma as a linear program, we will use the decision variables

$x_1 \triangleq$ amount invested by present investor 1 (thousands of dollars)
$x_2 \triangleq$ amount invested by new investor 2 (thousands of dollars)
$x_3 \triangleq$ amount Clyde invests of his own money (thousands of dollars)

Then clearly the objective function is to

$$\max x_1 + x_2 + x_3 \tag{5.13}$$

Some constraints are also easy:

$$\begin{aligned} x_1 + x_2 + x_3 &\geq 100 \\ x_1 \geq 0, \; x_2 \geq 0, \; 5 &\geq x_3 \geq 0 \end{aligned} \tag{5.14}$$

Total investment must be at least $100,000, all amounts are nonnegative, and Clyde's funds are limited to $5000.

Investment constraints are a bit more complex. If investor 1 is to provide half, the capital, and investor 2 a fraction α, we want

$$\frac{x_1}{x_1 + x_2 + x_3} = 0.5 \quad \text{and} \quad \frac{x_2}{x_1 + x_2 + x_3} = \alpha \tag{5.15}$$

These ratio constraints do not even look linear, but knowing $x_1 + x_2 + x_3$ will be positive in feasible solutions, we can clear denominators and collect terms (principle $\boxed{4.5}$) to obtain

$$\begin{aligned} -0.5x_1 + 0.5x_2 + 0.5x_3 &= 0 \\ \alpha x_1 + (\alpha - 1)x_2 + \alpha x_3 &= 0 \end{aligned} \tag{5.16}$$

Adding slacks x_4 and x_5 to place the LP in standard form, Clyde's dilemma is represented by the model

$$\begin{array}{rlllll}
\max & +1x_1 & +1x_2 & +1x_3 & & & \\
\text{s.t.} & +1x_1 & +1x_2 & +1x_3 & -1x_4 & & = 100 \\
& & & +1x_3 & & +1x_5 & = 5 \\
& -0.5x_1 & +0.5x_2 & +0.5x_3 & & & = 0 \\
& +\alpha x_1 & +(\alpha-1)x_2 & +\alpha x_3 & & & = 0 \\
& \multicolumn{6}{l}{x_1, \; x_2, \; x_3, \; x_4, \; x_5 \geq 0}
\end{array} \tag{5.17}$$

Starting Basis in the Two Phase Simplex

Two-phase computation begins by constructing a starting feasible solution to model (5.17) using artificial variables. There is one new issue in adapting the two-phase approach to simplex. Simplex search employs basic solutions. Our starting solution in Phase I must not just be feasible in the artificial model, but basic feasible.

One set of columns that is sure to qualify for a basis by being linearly independent is a collection where each column has only one nonzero component and no two of the columns are nonzero in the same component. That is, constraint columns with a single nonzero entry provide natural choices for a starting basis. Where none exists after the model has been placed in standard form, the column of an artificial variable will do the job. Either way, the single nonzero must have the same sign as the corresponding right-hand-side coefficient, so that the basis will produce a feasible basic solution.

5.33 | A starting basis for simplex can be obtained by making basic one variable for each constraint row, with that variable having its only nonzero coefficient in the row and coefficient sign matching that of the corresponding right-hand side. Where no standard-form variable meets these requirements, an artificial is introduced.

We can illustrate with Clever Clyde standard form (5.17). Our task is to find (or manufacture with artificial variables) a starting feasible basis. In the first constraint row we have a variable with its only nonzero coefficient there—slack x_4. However, the coefficient of x_4 is negative, while the right-hand-side 100 is positive. An artificial variable x_6 will be necessary.

Things go somewhat easier in the second row. There x_5 has its only nonzero coefficient with a sign that matches the right-hand side. Variable x_5 will be part of the initial basis. The last two constraints of standard form have no variable with its only nonzero coefficient in their rows. Artificials x_7 and x_8 will have to be added.

Summarizing, simplex will begin Phase I with the data

	x_1	x_2	x_3	x_4	x_5	x_6	x_7	x_8	
max **c**	1	1	1	0	0	0	0	0	
min **d**	0	0	0	0	0	1	1	1	**b**
A	1	1	1	−1	0	1	0	0	100
	0	0	1	0	1	0	0	0	5
	−0.5	0.5	0.5	0	0	0	1	0	0
	α	$\alpha - 1$	α	0	0	0	0	1	0
	N	N	N	N	B	B	B	B	
$\mathbf{x}^{(0)}$	0	0	0	0	5	100	0	0	

Initial solution $\mathbf{x}^{(0)}$ is the basic (artificially) feasible one obtained by assigning the value zero to all nonbasics and solving for the basics.

Notice that we now have two objective function rows. The vector **d** denotes the objective function vector for Phase I (sum of artificials). We first minimize $\mathbf{d} \cdot \mathbf{x}$, then return to maximize $\mathbf{c} \cdot \mathbf{x}$ in Phase II.

SAMPLE EXERCISE 5.17: CONSTRUCTING AN ARTIFICIAL BASIS

Introduce artificial variables as necessary to construct a Phase I artificial model and corresponding starting (artificially) feasible basis for the following standard-form linear program:

$$
\begin{aligned}
\min \quad & 14x_1 && -9x_3 && +x_4 \\
\text{s.t.} \quad & 8x_1 + x_2 && - x_3 && && = 74 \\
& && +4x_2 - 7x_3 && +x_4 && = -22 \\
& && + x_2 + x_3 && && = 11 \\
& x_1, x_2, x_3, x_4 \geq 0
\end{aligned}
$$

Analysis: We apply principle $\boxed{5.33}$. Variable x_1 appears only in the first main constraint and has the same sign as the right-hand-side 74. Thus it can be the basic variable for that constraint. Variable x_4 occurs only in the second main constraint, but its sign differs from the right-hand side; artificial variable x_5 will be needed. Another artificial x_6 is needed in the last constraint because no variable has its only nonzero coefficient there. Summarizing, the artificial model will be

$$
\begin{aligned}
\min \quad & x_5 + x_6 \\
\text{s.t.} \quad & 8x_1 + x_2 - x_3 && && = 74 \\
& +4x_2 - 7x_3 +x_4 && -x_5 && = -22 \\
& + x_2 + x_3 && +x_6 && = 11 \\
& x_1, x_2, x_3, x_4, x_5, x_6 \geq 0
\end{aligned}
$$

The starting (artificially) feasible basis is x_1, x_5, and x_6.

Three Possible Outcomes for Linear Programs

Recall that there are three possible outcomes for a linear programming model:

> $\boxed{5.34}$ A linear program may be **infeasible** (have no feasible solutions), be **unbounded** (have arbitrarily good feasible solutions), or have **finite optimal** solutions.

In the discussion of Section 3.6 we detailed how we detect each case, and Algorithm 5B provides details for the simplex. The problem is infeasible if Phase I reaches optimality without reducing the sum of artificials to $=0$ (principle $\boxed{3.39}$). Otherwise it has feasible solutions ($\boxed{3.38}$). Phase II then either detects unboundedness ($\boxed{5.25}$) or stops on an optimal solution ($\boxed{5.27}$).

Clever Clyde Infeasible Case

The Clever Clyde example of this section was contrived so that we can illustrate all three possibilities for LP outcomes and know by easy deduction that simplex is drawing the right conclusion. Think for a moment about what Clyde needs from investor 2. His own $5000 would provide 5% of the $100,000 minimum for the business, and investor 1 has guaranteed 50%. If $\alpha < 0.45$, there can be no feasible solution.

Take $\alpha = 0.4$. Table 5.3 shows the Phase I simplex computations leading to a conclusion of infeasibility. Beginning with x_5, x_6, x_7, and x_8 basic, the sum of artificial variables is $\mathbf{d} \cdot \mathbf{x}^{(0)} = 100$. There are 4 available simplex directions, and 3 improve the objective. (Remember that we are minimizing, so that $\bar{d}_j < 0$ implies improvement.) The direction increasing nonbasic x_3 is chosen.

ALGORITHM 5B: TWO-PHASE SIMPLEX SEARCH

Step 0: Artificial Model. If there are convenient starting feasible bases for the given standard-form linear program, identify one, and proceed to Step 3. Otherwise, construct an artificial model and artificially feasible basis by adding (or subtracting) artificial variables as in principle $\boxed{5.33}$.

Step 1: Phase I. Beginning from the artificially feasible basis of Step 0, apply simplex search to minimize the sum of the artificial variables subject to the constraints of the artificial model.

Step 2: Infeasibility. If Phase I search terminated with a minimum having artificial sum > 0, stop; the original model is infeasible. Otherwise, use the final Phase I basis in the artificial model to identify a starting feasible basis for the original model.

Step 3: Phase II. Beginning from the identified starting feasible basis, apply simplex search to compute an optimal solution to the original standard-form model or demonstrate that it is unbounded.

The next step is to determine how far we can pursue the Δx for x_3 without losing feasibility. Here, something new occurs. Basic variables x_7 and x_8, which were already $=0$ in solution $x^{(0)}$, are both decreased by Δx. As a consequence, the largest possible step size is $\lambda = 0$.

We will have more to say about this **degenerate** case in the next section. For the moment we merely act as if λ were small but positive. Variable x_3 enters the basis, and x_7 (one of the two variables establishing λ) leaves.

The new basic solution $x^{(1)}$ is identical to $x^{(0)}$, but the basis that determines it has changed. As a consequence, we have new simplex directions to consider. This time only the Δx for x_2 improves. It, too, leads to a degenerate step of $\lambda = 0$, but x_2 enters the basis and x_8 leaves.

In iteration $t = 2$ we finally see real progress. Nonbasic x_1 enters, and x_5 leaves after a step of $\lambda = 25$. The objective function value (sum of artificials) is halved to $d \cdot x^{(3)} = 50$. Iteration $t = 3$ is similar, reducing infeasibility to $d \cdot x^{(4)} = 5$.

Here the progress stops. As we consider the four available simplex directions at iteration $t = 4$, we see all fail the test for improvement. That is, $x^{(4)}$ is optimal in the Phase I problem. But the sum of artificials is still 5. No solution using only the original variables satisfies all constraints [i.e., the original model (with $\alpha = 0.4$) is infeasible].

SAMPLE EXERCISE 5.18: DETECTING INFEASIBILITY IN SIMPLEX PHASE I

Use two-phase simplex Algorithm 5B to establish that the linear program

$$\begin{array}{lll} \max & 8x_1 + 11x_2 & \\ \text{s.t.} & x_1 + x_2 & \leq 2 \\ & x_1 + x_2 & \geq 3 \\ & x_1, x_2 \geq 0 \end{array}$$

has no solution.

TABLE 5.3   Simplex Computation for Clever Clyde Infeasible Case

	x_1	x_2	x_3	x_4	x_5	x_6	x_7	x_8	
max **c**	1	1	1	0	0	0	0	0	
min **d**	0	0	0	0	0	1	1	1	**b**
A	1	1	-1	-1	0	1	0	0	100
	0	0	1	0	1	0	0	0	5
	-0.5	0.5	0.5	0	0	0	1	0	0
	0.4	-0.6	0.4	0	0	0	0	1	0
$t=0$	N	N	N	N	B	B	B	B	Phase I
$\mathbf{x}^{(0)}$	0	0	0	0	5	100	0	0	$\mathbf{d}\cdot\mathbf{x}^{(0)}=100$
$\Delta\mathbf{x}$ for x_1	1	0	0	0	0	-1	0.5	0.4	$\bar{d}_1=-0.9$
$\Delta\mathbf{x}$ for x_2	0	1	0	0	0	-1	-0.5	0.6	$\bar{d}_2=-0.9$
$\Delta\mathbf{x}$ for x_3	0	0	1	0	-1	-1	-0.5	-0.4	$\bar{d}_3=\boxed{-1.9}$
$\Delta\mathbf{x}$ for x_4	0	0	0	1	0	1	0	0	$\bar{d}_4=1.0$
	—	—	—	—	$\dfrac{5}{1}$	$\dfrac{100}{1}$	$\dfrac{0}{0.5}$	$\boxed{\dfrac{0}{0.4}}$	$\lambda=0.0$
$t=1$	N	N	B	N	B	B	B	N	Phase I
$\mathbf{x}^{(1)}$	0	0	0	0	5	100	0	0	$\mathbf{d}\cdot\mathbf{x}^{(1)}=100$
$\Delta\mathbf{x}$ for x_1	1	0	-1	0	1	0	1	0	$\bar{d}_1=1.0$
$\Delta\mathbf{x}$ for x_2	0	1	1.5	0	-1.5	-2.5	-1.25	0	$\bar{d}_2=\boxed{-3.75}$
$\Delta\mathbf{x}$ for x_4	0	0	0	1	0	1	0	0	$\bar{d}_4=1.0$
$\Delta\mathbf{x}$ for x_8	0	0	-2.5	0	2.5	2.5	1.25	1	$\bar{d}_8=4.75$
	—	—	—	—	$\dfrac{5}{1.5}$	$\dfrac{100}{2.5}$	$\boxed{\dfrac{0}{1.25}}$	—	$\lambda=0.0$
$t=2$	N	B	B	N	B	B	N	N	Phase I
$\mathbf{x}^{(2)}$	0	0	0	0	5	100	0	0	$\mathbf{d}\cdot\mathbf{x}^{(2)}=100$
$\Delta\mathbf{x}$ for x_1	1	0.8	0.2	0	-0.2	-2	0	0	$\bar{d}_1=\boxed{-2.0}$
$\Delta\mathbf{x}$ for x_4	0	0	0	1	0	1	0	0	$\bar{d}_4=1.0$
$\Delta\mathbf{x}$ for x_7	0	-0.8	-1.2	0	1.2	2	1	0	$\bar{d}_7=3.0$
$\Delta\mathbf{x}$ for x_8	0	1	-1	0	1	0	0	1	$\bar{d}_8=1.0$
	—	—	—	—	$\boxed{\dfrac{5}{0.2}}$	$\dfrac{100}{2}$	—	—	$\lambda=25$
$t=3$	B	B	B	N	N	B	N	N	Phase I
$\mathbf{x}^{(3)}$	25	20	5	0	0	50	0	0	$\mathbf{d}\cdot\mathbf{x}^{(3)}=50$
$\Delta\mathbf{x}$ for x_4	0	0	0	1	0	1	0	0	$\bar{d}_4=1.0$
$\Delta\mathbf{x}$ for x_5	-5	-4	-1	0	1	10	0	0	$\bar{d}_5=10.0$
$\Delta\mathbf{x}$ for x_7	6	4	0	0	0	-10	1	0	$\bar{d}_7=\boxed{-9.0}$
$\Delta\mathbf{x}$ for x_8	5	5	0	0	0	-10	0	1	$\bar{d}_8=-9.0$
	—	—	—	—	—	$\boxed{\dfrac{50}{10}}$	—	—	$\lambda=5$
$t=4$	B	B	B	N	N	N	B	N	Phase I
$\mathbf{x}^{(4)}$	55	40	5	0	0	0	5	0	$\mathbf{d}\cdot\mathbf{x}^{(4)}=5$
$\Delta\mathbf{x}$ for x_4	0.6	0.4	0	1	0	0	0.1	0	$\bar{d}_4=0.1$
$\Delta\mathbf{x}$ for x_5	1	0	-1	0	1	0	1	0	$\bar{d}_5=1.0$
$\Delta\mathbf{x}$ for x_6	-0.6	-0.4	0	0	0	1	-0.1	0	$\bar{d}_6=0.9$
$\Delta\mathbf{x}$ for x_8	-1	1	0	0	0	0	-1	1	$\bar{d}_8=0.0$
									"infeasible"

Analysis: We begin by including slack variables to obtain standard-form model

$$\max\quad 8x_1 + 11x_2$$
$$\begin{aligned}
\text{s.t.}\quad x_1 + x_2 &+ x_3 &&= 2\\
x_1 + x_2 && - x_4 &= 3\\
x_1, x_2, x_3, x_4 &\geq 0
\end{aligned}$$

Variable x_3 provides a starting basic variable in the first constraint, but artificial x_5 is needed in the second constraint. We are now ready to begin Phase I simplex with initial basis $\{x_3, x_5\}$.

	x_1	x_2	x_3	x_4	x_5	
max **c**	8	11	0	0	0	
min **d**	0	0	0	0	1	**b**
	1	1	1	0	0	2
A	1	1	0	−1	1	3
$t = 0$	N	N	B	N	B	Phase I
$\mathbf{x}^{(0)}$	0	0	2	0	3	$\mathbf{d} \cdot \mathbf{x} = 3$
$\mathbf{\Delta x}$ for x_1	1	0	−1	0	−1	$\bar{d}_1 = \boxed{-1}$
$\mathbf{\Delta x}$ for x_2	0	1	−1	0	−1	$\bar{d}_2 = -1$
$\mathbf{\Delta x}$ for x_4	0	0	0	1	1	$\bar{d}_4 = +1$
	—	—	$\boxed{\frac{2}{1}}$	—	$\frac{3}{1}$	$\lambda = 2$
$t = 1$	B	N	N	N	B	Phase I
$\mathbf{x}^{(1)}$	2	0	0	0	1	$\mathbf{d} \cdot \mathbf{x} = 1$
$\mathbf{\Delta x}$ for x_2	−1	1	0	0	0	$\bar{d}_2 = 0$
$\mathbf{\Delta x}$ for x_3	−1	0	1	0	1	$\bar{d}_3 = +1$
$\mathbf{\Delta x}$ for x_4	0	0	0	1	1	$\bar{d}_4 = +1$
						"infeasible"

The original model is infeasible because the Phase I optimum has artificial total 1.

Clever Clyde Optimal Case

Suppose now that investor 2 is willing to provide 49% of the startup capital Clever Clyde requires. There should be a finite optimal solution. Table 5.4 traces Phase I simplex computation on this $\alpha = 0.49$ case. The first two iterations closely parallel the infeasible variation of Table 5.3; two degenerate basis changes produce no actual progress.

Iteration $t = 2$ shows something new. The direction for entering nonbasic x_1 improves the objective, and the maximum feasible step of $\lambda = 50$ erases all infeasibility. The resulting solution $\mathbf{x}^{(3)}$ is 0 in all artificial components and thus corresponds to a feasible solution in the original model. Iteration $t = 3$ confirms that it is optimal in the Phase I problem, with objective function value $= 0$.

With a known basic feasible solution, we are ready to pass to Phase II. Table 5.5 details the computations. Now that only the standard-form variables are present, there is only one nonbasic. Its simplex direction $\mathbf{\Delta x}$ tests as improving for our maximize objective function, so it will enter the basis.

A check of ratios shows that feasible progress will stop at $\lambda = 400$ as x_5 leaves the basis. The new solution is $\mathbf{x}^{(1)} = (250, 245, 5, 400, 0)$. Now the only simplex direction is that of x_5 and it does not improve the objection function. We conclude that $\mathbf{x}^{(1)}$ is optimal. If investor 2 will provide 49% of the capital, a $500,000 business is achievable, taking $250,000 from investor 1, $245,000 from investor 2, and the remaining $5000 from Clyde.

TABLE 5.4 Phase I Simplex Computation for Clever Clyde Optimal Case

	x_1	x_2	x_3	x_4	x_5	x_6	x_7	x_8	
min **d**	0	0	0	0	0	1	1	1	**b**
A	1	1	1	−1	0	1	0	0	100
	0	1	0	0	1	0	0	0	5
	−0.50	0.50	0.50	0	0	0	1	0	0
	0.49	−0.51	0.49	0	0	0	0	1	0
$t=0$	N	N	N	N	B	B	B	B	Phase I
$\mathbf{x}^{(0)}$	0	0	0	0	5	100	0	0	$\mathbf{d}\cdot\mathbf{x}^{(0)}=100$
$\Delta\mathbf{x}$ for x_1	1	0	0	0	0	−1	0.50	−0.49	$\bar{d}_1=-0.99$
$\Delta\mathbf{x}$ for x_2	0	1	0	0	0	−1	−0.50	0.51	$\bar{d}_2=-0.99$
$\Delta\mathbf{x}$ for x_3	0	0	1	0	−1	−1	−0.50	−0.49	$\bar{d}_3=\boxed{-1.99}$
$\Delta\mathbf{x}$ for x_4	0	0	0	1	0	1	0	0	$\bar{d}_4=1.0$
	—	—	—	—	$\dfrac{5}{1}$	$\dfrac{100}{1}$	$\dfrac{0}{0.50}$	$\boxed{\dfrac{0}{0.49}}$	$\lambda=0.0$
$t=1$	N	N	B	N	B	B	B	N	Phase I
$\mathbf{x}^{(1)}$	0	0	0	0	5	100	0	0	$\mathbf{d}\cdot\mathbf{x}^{(1)}=100$
$\Delta\mathbf{x}$ for x_1	1	0	−1	0	1	0	1	0	$\bar{d}_1=1.0$
$\Delta\mathbf{x}$ for x_2	0	1	1.04	0	−1.04	−2.04	−1.02	0	$\bar{d}_2=\boxed{-3.06}$
$\Delta\mathbf{x}$ for x_4	0	0	0	1	0	1	0	0	$\bar{d}_4=1.0$
$\Delta\mathbf{x}$ for x_8	0	0	−2.04	0	2.04	2.04	1.02	1	$\bar{d}_8=4.06$
	—	—	—	—	$\dfrac{5}{1.04}$	$\dfrac{100}{2.04}$	$\boxed{\dfrac{0}{1.02}}$	—	$\lambda=0.0$
$t=2$	N	B	B	N	B	B	N	N	Phase I
$\mathbf{x}^{(2)}$	0	0	0	0	5	100	0	0	$\mathbf{d}\cdot\mathbf{x}^{(2)}=100$
$\Delta\mathbf{x}$ for x_1	1	0.98	0.02	0	−0.02	−2	0	0	$\bar{d}_1=\boxed{-2.0}$
$\Delta\mathbf{x}$ for x_4	0	0	0	1	0	1	0	0	$\bar{d}_4=1.0$
$\Delta\mathbf{x}$ for x_7	0	−0.98	−1.02	0	1.02	2	1	0	$\bar{d}_7=3.0$
$\Delta\mathbf{x}$ for x_8	0	1	−1	0	1	0	0	1	$\bar{d}_8=1.0$
	—	—	—	—	$\dfrac{5}{0.02}$	$\boxed{\dfrac{100}{2}}$	—	—	$\lambda=50$
$t=3$	B	B	B	N	B	N	N	N	Phase I
$\mathbf{x}^{(3)}$	50	49	1	0	4	0	0	0	$\mathbf{d}\cdot\mathbf{x}^{(3)}=0$
$\Delta\mathbf{x}$ for x_4	0.50	0.49	0.01	1	−0.01	0	0	0	$\bar{d}_4=0.0$
$\Delta\mathbf{x}$ for x_6	−0.50	−0.49	−0.01	0	0.01	1	0	0	$\bar{d}_6=1.0$
$\Delta\mathbf{x}$ for x_7	1	0	−1	0	1	0	1	0	$\bar{d}_7=1.0$
$\Delta\mathbf{x}$ for x_8	0	1	−1	0	1	0	0	1	$\bar{d}_8=1.0$
									"feasible"

SAMPLE EXERCISE 5.19: MOVING TO SIMPLEX PHASE II

A standard-form linear program has variables $x_1, \ldots, x_5$. To begin two-phase simplex Algorithm 5B, we introduce artificial variables $x_6, \ldots, x_8$. Explain for each of the following Phase I outcomes how the algorithm should proceed.

(a) Phase I optimum is $\mathbf{x} = (0, 0, 0, 0, 0, 0, 3, 6)$ with x_1, x_7, and x_8 basic.

(b) Phase I optimum is $\mathbf{x} = (0, 2, 0, 0, 9, 0, 0, 0)$ with x_1, x_2, and x_5 basic.

Analysis:

(a) Here the minimum artificial total is $(0 + 3 + 6) = 9$. The model is infeasible (principle $\boxed{3.39}$), so we terminate.

TABLE 5.5 Phase II Simplex Computation for Clever Clyde Optimal Case

	x_1	x_2	x_3	x_4	x_5	
max **c**	1	1	1	0	0	**b**
	1	1	1	−1	0	100
A	0	1	0	0	1	5
	−0.50	0.50	0.50	0	0	0
	0.49	−0.51	0.49	0	0	0
$t = 0$	B	B	B	N	B	Phase II
$\mathbf{x}^{(0)}$	50	49	1	0	4	$\mathbf{c} \cdot \mathbf{x}^{(0)} = 100$
Δ**x** for x_4	0.50	0.49	0.01	1	−0.01	$\bar{c}_4 = \boxed{1.0}$
	—	—	—	—	$\boxed{\dfrac{4}{0.01}}$	$\lambda = 400$
$t = 1$	B	B	B	B	N	Phase II
$\mathbf{x}^{(1)}$	250	245	5	400	0	$\mathbf{c} \cdot \mathbf{x}^{(1)} = 500$
Δ**x** for x_5	−50	−49	−1	−100	1	$\bar{c}_5 = -100$
						"optimal"

(b) With a minimum artificial total of $=0$, the model is feasible (principle $\boxed{3.38}$). We proceed to Phase II simplex, beginning with feasible basis $\{x_1, x_2, x_5\}$.

Clever Clyde Unbounded Case

If Clyde is particularly convincing, he might talk investor 2 into covering a full 50% of the investment. Clearly, the size of the business is then unbounded because Clyde's partners will be financing the entire cost.

Phase I simplex computation for the $\alpha = 0.50$ case is almost identical to that of Table 5.4. Only a few numbers differ. Table 5.6 shows Phase II. The feasible basis provided by Phase I consists of x_1, x_2, x_3, and x_5, with basic solution $\mathbf{x}^{(0)} = (50, 50, 0, 0, 5)$. The only nonbasic, x_4, yields an improving simplex direction.

TABLE 5.6 Phase II Simplex Computation for Clever Clyde Unbounded Case

	x_1	x_2	x_3	x_4	x_5	
max **c**	1	1	1	0	0	**b**
	1	1	1	−1	0	100
A	0	1	0	0	1	5
	−0.50	0.50	0.50	0	0	0
	0.50	−0.50	0.50	0	0	0
$t = 0$	B	B	B	N	B	Phase II
$\mathbf{x}^{(0)}$	50	50	0	0	5	$\mathbf{c} \cdot \mathbf{x}^{(0)} = 100$
Δ**x** for x_4	0.50	0.50	0.1	1	0.0	$\bar{c}_4 = \boxed{1.0}$
	—	—	—	—	—	$\lambda = \infty$
						"unbounded"

Notice that this improving simplex direction has no negative components (compare with Table 5.5). That is, the direction is not moving toward any nonnegativity constraint. It follows that there is no limit on how far we may follow the direction while retaining feasibility (principle $\boxed{5.25}$). Since every step in the direction improves the objective, the model must be unbounded.

5.6 DEGENERACY AND ZERO-LENGTH SIMPLEX STEPS

Our development of the simplex algorithm as a form of improving search has implicitly assumed that each iteration makes positive progress toward an optimal solution by advancing from a current extreme point solution to a superior one. If a better extreme point is encountered at each move, it is not hard to see that simplex must eventually produce an optimal solution or show that none exists (see Section 5.7).

Unfortunately, this is not always the case. It is much more typical for a simplex search to follow the pattern of Figure 5.6. Progress at some iterations is interspersed with periods of no advance. In this section we explore the degeneracy phenomenon in linear programming, which brings about simplex steps with no gain.

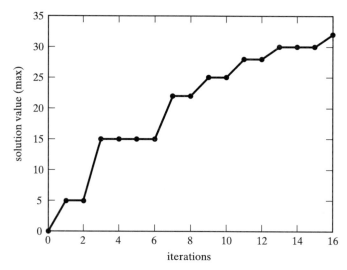

FIGURE 5.6   Typical Objective Function Progress of Simplex Search

Degenerate Solutions

Degeneracy happens in linear programming whenever more constraints are active than the minimum number needed to define a point.

5.35   A basic feasible solution to a standard-form linear program is **degenerate** if nonnegativity constraints for some basic variables are active (i.e., if some basic variables have value $=0$.

Basic solutions require nonnegativity constraints to be active for the nonbasics (definition 5.16), but when nonnegativities also happen to be active for some basics, there are alternative choices for the basic set.

5.36   In the presence of degeneracy, several bases can compute the same basic solution.

Figure 5.7 illustrates graphically. Any three of the five inequalities B, C, H, I, and G define extreme point $\mathbf{x}^{(1)}$. In standard form, each of those constraints will correspond to a nonnegativity constraint on a slack variable. This means that $(5 - 3) = 2$ basic variables will have value $= 0$. The solution is degenerate.

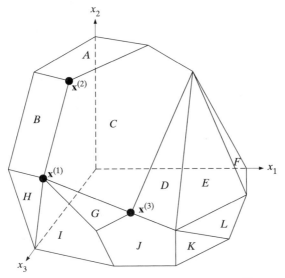

FIGURE 5.7   A Degenerate Extreme Point $\mathbf{x}^{(1)}$

SAMPLE EXERCISE 5.20: RECOGNIZING DEGENERATE SOLUTIONS

Determine for each of the following basic variable sets and basic solutions to a linear program in standard form whether the solution indicated is degenerate.

(a) $B = \{x_1, x_3, x_4\}$ for $\mathbf{x} = (3, 0, 7, 2, 0, 0)$

(b) $B = \{x_1, x_2, x_5\}$ for $\mathbf{x} = (0, 8, 0, 0, 1, 0)$

Analysis: We apply definition $\boxed{5.35}$.

(a) This basic solution is nondegenerate because none of the nonnegativity constraints on basic variables are active.

(b) This basic solution is degenerate because basic variable $x_1 = 0$.

Zero-Length Simplex Steps

We can see how degeneracy inhibits simplex algorithm progress by returning to the Clever Clyde example of Table 5.3. Table 5.7 recapitulates the first three iterations. Notice that the first three basic solutions encountered were the same:

$$\mathbf{x}^{(0)} = \mathbf{x}^{(1)} = \mathbf{x}^{(2)} = (0, 0, 0, 0, 5, 100, 0, 0)$$

But this basic solution was computed by three different basic sets:

$$\{x_5, x_6, x_7, x_8\} \quad \text{at} \quad t = 0$$
$$\{x_3, x_5, x_6, x_7\} \quad \text{at} \quad t = 1$$
$$\{x_2, x_3, x_5, x_6\} \quad \text{at} \quad t = 2$$

TABLE 5.7   Degenerate Iterations in Clever Clyde Example

	x_1	x_2	x_3	x_4	x_5	x_6	x_7	x_8	
$t=0$	N	N	N	N	B	B	B	B	Phase I (min)
$\mathbf{x}^{(0)}$	0	0	0	0	5	100	0	0	$\mathbf{d}\cdot\mathbf{x}^{(0)}=100$
$\Delta\mathbf{x}$ for x_1	1	0	0	0	0	−1	0.5	−0.4	$\bar{d}_1=-0.9$
$\Delta\mathbf{x}$ for x_2	0	1	0	0	0	−1	−0.5	0.6	$\bar{d}_2=-0.9$
$\Delta\mathbf{x}$ for x_3	0	0	1	0	−1	−1	−0.5	−0.4	$\bar{d}_3=\boxed{-1.9}$
$\Delta\mathbf{x}$ for x_4	0	0	0	1	0	1	0	0	$\bar{d}_4=1.0$
	—	—	—	—	$\frac{5}{1}$	$\frac{100}{1}$	$\frac{0}{0.5}$	$\boxed{\frac{0}{0.4}}$	$\lambda=0.0$
$t=1$	N	N	B	N	B	B	B	N	Phase I (min)
$\mathbf{x}^{(1)}$	0	0	0	0	5	100	0	0	$\mathbf{d}\cdot\mathbf{x}^{(1)}=100$
$\Delta\mathbf{x}$ for x_1	1	0	−1	0	1	0	1	0	$\bar{d}_1=1.0$
$\Delta\mathbf{x}$ for x_2	0	1	1.5	0	−1.5	−2.5	−1.25	0	$\bar{d}_2=\boxed{-3.75}$
$\Delta\mathbf{x}$ for x_4	0	0	1	1	0	1	0	0	$\bar{d}_4=1.0$
$\Delta\mathbf{x}$ for x_8	0	0	−2.5	0	2.5	2.5	1.25	1	$\bar{d}_8=4.75$
	—	—	—	—	$\frac{5}{1.5}$	$\frac{100}{2.5}$	$\boxed{\frac{0}{1.25}}$	—	$\lambda=0.0$
$t=2$	N	B	B	N	B	B	N	N	Phase I (min)
$\mathbf{x}^{(2)}$	0	0	0	0	5	100	0	0	$\mathbf{d}\cdot\mathbf{x}^{(2)}=100$
$\Delta\mathbf{x}$ for x_1	1	0.8	0.2	0	−0.2	−2	0	0	$\bar{d}_1=\boxed{-2.0}$
$\Delta\mathbf{x}$ for x_4	0	0	0	1	0	1	0	0	$\bar{d}_4=1.0$
$\Delta\mathbf{x}$ for x_7	0	−0.8	−1.2	0	1.2	2	1	0	$\bar{d}_7=3.0$
$\Delta\mathbf{x}$ for x_8	0	1	−1	0	1	0	0	1	$\bar{d}_8=1.0$
	—	—	—	—	$\boxed{\frac{5}{0.2}}$	$\frac{100}{2}$	—	—	$\lambda=25$
$t=3$	B	B	B	N	N	B	N	N	Phase I (min)
$\mathbf{x}^{(3)}$	25	20	5	0	0	50	0	0	$\mathbf{d}\cdot\mathbf{x}^{(3)}=50$

The two positive components were basic in every case, but there are several choices for the two remaining basics at value $=0$.

Simplex directions are structured to keep the equality constraints satisfied, and also nonnegativity constraints on nonbasic variables. But there is no guarantee that a basic component of the solution will not be decreased. If that component already happens to $=0$, as it does in degenerate cases, the result may be a step of $\lambda=0$.

> **5.37** | Simplex directions that decrease basic variables already $=0$ in a degenerate solution may produce moves with steps $\lambda=0$.

This is precisely what happens on the initial two moves of Table 5.7. In the first case it was decreasing $\Delta x_8=-0.4$, which produced a minimum ratio at $\lambda=0$. In the second, it was $\Delta x_7=-1.25$.

Progress through Changing of Bases

Computations continued in Table 5.7 as if there were nothing troublesome about $\lambda=0$. This is the normal way to proceed.

> **5.38** | When degenerate solutions cause the simplex algorithm to compute a step of $\lambda=0$, the basis should be changed according to rule 5.26 and computations continued as if a positive step had been taken.

At $t = 2$ in Table 5.7, such perseverance was rewarded. Solution $\mathbf{x}^{(2)}$ is just as degenerate as earlier $\mathbf{x}^{(0)}$ and $\mathbf{x}^{(1)}$, but real progress was made with a step of $\lambda = 25$.

In virtually all practical linear programs, eventual objective function progress will occur in this way (see Section 5.7 for exceptions). The reason is a much more subtle form of progress going on as we change bases at each iteration.

> **5.39**   Simplex computations will normally escape a sequence of degenerate moves because changing basic representations of the current solution, which also changes simplex directions, eventually produce a direction along which positive progress can be achieved.

The sequence of simplex directions for x_1 is illustrated in Table 5.7. At $t = 1$ that simplex direction was not even improving. But after a change of basis at $t = 2$, the direction for x_1 is completely different. As a result it now improves without decreasing a zero-valued basis variable, and a positive step is possible.

SAMPLE EXERCISE 5.21: DEALING WITH ZERO STEPS

Consider the standard-form linear program

$$\begin{aligned}
\max \quad & -x_1 + 2x_2 \\
\text{s.t.} \quad & x_1 - x_2 + x_3 = 0 \\
& x_1, x_2, x_3 \geq 0
\end{aligned}$$

(a) Begin with x_3 basic and apply rudimentary simplex Algorithm 5A to demonstrate that the first iteration produces a move with step $\lambda = 0$.

(b) Continue for another iteration to demonstrate that a positive (in fact, infinite) step λ is achieved at the second step.

(c) Explain how the change of basis the first iteration contributed to this progress.

Analysis: Simplex Algorithm 5A proceeds as follows:

	x_1	x_2	x_3	
max $\mathbf{c}$	-1	2	0	$\mathbf{b}$
$\mathbf{A}$	1	-1	-1	0
$t = 0$	N	N	B	
$\mathbf{x}^{(0)}$	0	0	0	$\mathbf{c} \cdot \mathbf{x}^{(0)} = 0$
$\Delta\mathbf{x}$ for x_1	1	0	1	$\bar{c}_1 = -1$
$\Delta\mathbf{x}$ for x_2	0	1	-1	$\bar{c}_2 = \boxed{2}$
	—	—	$\dfrac{0}{-(-1)}$	$\lambda = 0$
$t = 1$	N	B	N	
$\mathbf{x}^{(1)}$	0	0	0	$\mathbf{c} \cdot \mathbf{x}^{(1)} = 0$
$\Delta\mathbf{x}$ for x_1	1	1	0	$\bar{c}_1 = \boxed{1}$
$\Delta\mathbf{x}$ for x_3	0	-1	1	$\bar{c}_3 = -2$
	—	—	—	"unbounded"

(a) The only improving simplex direction at $t = 0$ decreases basic x_3, which is already at degenerate value $= 0$. A step $\lambda = 0$ results.

(b) After the change of basis at $t = 1$, the direction for x_1 now improves. It decreases no component of the solution, so $\lambda = +\infty$ is possible. The model is unbounded.

(c) Progress became possible because the change of basis produced a different simplex direction for x_1.

5.7 CONVERGENCE AND CYCLING WITH SIMPLEX

A numerical search algorithm is said to **converge** if iterations make steady progress toward a solution. It converges **finitely** if it is guaranteed to stop after a finite number of iterations.

Finite Convergence with Positive Steps

Does simplex converge? Yes, if there is progress at each iteration.

> **5.40** If each iteration of simplex search yields a positive step λ, the algorithm will stop after finitely many iterations with either an optimal solution or a conclusion of unboundedness.

With two phases we can also detect infeasibility.

To see why this is true, recall that every iteration of the simplex algorithm produces a basic feasible solution (i.e., one generated by a selected set of basic columns). If the model in standard form has constraint matrix $\mathbf{A}$ with m rows and n columns, there are only finitely many possible bases. Each chooses m columns from the list of n as we did in Table 5.1. This limit gives

$$\text{maximum number of bases} = \frac{n!}{m!(n-m)!}$$

where $k! \triangleq k(k-1), \ldots, (1)$. Certainly, this is a large number, but it is still finite.

When each step size λ is positive, no basis can ever repeat. Each time, we pursue a simplex direction that strictly improves the objective. Thus there is no way we could return to a previous one with poorer objective value.

> **SAMPLE EXERCISE 5.22: BOUNDING THE NUMBER OF SIMPLEX ITERATIONS**
>
> A standard-form linear program has 10 variables and 7 main constraints. Assuming step size $\lambda > 0$ at every step, compute a finite bound on the number of iterations that Algorithm 5A might require before terminating.
>
> *Analysis:* If there is positive progress at each iteration, we need only bound the number of possible bases. Each would have 7 basic variables selected from among the 10 available. Thus there can be at most
>
> $$\frac{n!}{m!\,(n-m)!} = \frac{10!}{7!\,3!} = \frac{3{,}628{,}800}{5040(6)} = 120$$
>
> iterations before simplex computation must stop.

Degeneracy and Cycling

Degenerate linear programs (definition 5.35) pose another threat to convergence of the simplex method. We saw in Section 5.6 (principle 5.37) that degeneracy can sometimes lead to simplex steps with $\lambda = 0$.

The concern raised by such degenerate moves of simplex search is that since no progress is being achieved in the objective function, a sequence of iterations might return to a basis we have already visited. Thereafter, the cycle would repeat, and the simplex would never stop.

Cycling can occur. Table 5.8 shows a carefully crafted example. For brevity, the table shows simplex directions only if they were actually used, but each is clearly an improving direction. Simplex passed through the degenerate basis sequence

$$x_1, \; x_2, \; x_3$$
$$x_2, \; x_3, \; x_4$$
$$x_3, \; x_4, \; x_5$$
$$x_3, \; x_5, \; x_6$$
$$x_3, \; x_6, \; x_7$$
$$x_1, \; x_3, \; x_7$$
$$x_1, \; x_2, \; x_3$$

starting and ending on the same basic set.

Despite examples like the one in Table 5.8, cycling is quite rare. In fact, small examples of LPs that cycle are so rare that the one in Table 5.8 is rather famous.

For the overwhelming majority of LP models it is safe to proceed just as set out in principle 5.38. If a step λ happens to $=0$, act as if it were small but positive, and continue the search.

> 5.41 It is usually safe to assume that cycling will not occur in applied linear programming models and thus that simplex search will converge finitely.

In those rare cases where cycling is a threat, careful choice of entering and leaving basic variables can restore simplex convergence. However, details of such **anticycling** rules are beyond the scope of this book.

5.8 DOING IT EFFICIENTLY: REVISED SIMPLEX

The approach we have taken to simplex search so far emphasizes the underlying logic of the algorithm. However, many of the steps can be executed much more efficiently with the aid of a bit of matrix algebra. In this section we outline the **revised simplex algorithm** which is at the heart of large-scale codes.

Computations with Basis Inverses

Rudimentary Algorithm 5A of Section 4.3 solves a great many systems of linear equations. We must solve one such system to find the first basic solution. Then at each iteration a linear system is solved for every simplex direction.

TABLE 5.8 Example of Cycling with Simplex

	x_1	x_2	x_3	x_4	x_5	x_6	x_7	
min **c**	0	0	0	−0.75	20	−0.50	6	**b**
A	1	0	0	0.25	−8	−1	9	0
	0	1	0	0.50	−12	−0.50	3	0
	0	0	1	0	0	1	0	1
$t=0$	B	B	B	N	N	N	N	
$\mathbf{x}^{(0)}$	0	0	1	0	0	0	0	$\mathbf{c}\cdot\mathbf{x}^{(0)}=0$
$\Delta\mathbf{x}$ for x_4	−0.25	−0.50	0	1	0	0	0	$\bar{c}_4=\boxed{-0.75}$
	$\boxed{\frac{0}{0.25}}$	$\frac{0}{0.50}$	—	—	—	—	—	$\lambda=0$
$t=1$	N	B	B	B	N	N	N	
$\mathbf{x}^{(1)}$	0	0	1	0	0	0	0	$\mathbf{c}\cdot\mathbf{x}^{(1)}=0$
$\Delta\mathbf{x}$ for x_5	0	−4	0	32	1	0	0	$\bar{c}_5=\boxed{-4.0}$
	—	$\boxed{\frac{0}{4}}$	—	—	—	—	—	$\lambda=0$
$t=2$	N	N	B	B	B	N	N	
$\mathbf{x}^{(2)}$	0	0	1	0	0	0	0	$\mathbf{c}\cdot\mathbf{x}^{(2)}=0$
$\Delta\mathbf{x}$ for x_6	0	0	−1	−8	−0.38	1	0	$\bar{c}_6=\boxed{-2.0}$
	—	—	$\frac{1}{1}$	$\boxed{\frac{0}{8}}$	$\frac{0}{0.38}$	—	—	$\lambda=0$
$t=3$	N	N	B	N	B	B	N	
$\mathbf{x}^{(3)}$	0	0	1	0	0	0	0	$\mathbf{c}\cdot\mathbf{x}^{(3)}=0$
$\Delta\mathbf{x}$ for x_7	0	0	−10.5	0	−0.19	10.5	1	$\bar{c}_7=\boxed{-2.0}$
	—	—	$\frac{1}{10.5}$	—	$\boxed{\frac{0}{0.19}}$	—	—	$\lambda=0$
$t=4$	N	N	B	N	N	B	N	
$\mathbf{x}^{(4)}$	0	0	1	0	0	0	0	$\mathbf{c}\cdot\mathbf{x}^{(4)}=0$
$\Delta\mathbf{x}$ for x_1	1	0	2	0	0	−2	−0.33	$\bar{c}_1=\boxed{-1.0}$
	—	—	—	—	—	$\boxed{\frac{0}{2}}$	$\frac{0}{0.33}$	$\lambda=0$
$t=5$	B	N	B	N	N	N	N	
$\mathbf{x}^{(5)}$	0	0	1	0	0	0	0	$\mathbf{c}\cdot\mathbf{x}^{(5)}=0$
$\Delta\mathbf{x}$ for x_2	3	1	0	0	0	0	−0.33	$\bar{c}_2=\boxed{-2.0}$
	—	—	—	—	—	—	$\boxed{\frac{0}{0.33}}$	$\lambda=0$
$t=6$	B	B	B	N	N	N	N	
$\mathbf{x}^{(6)}$	0	0	1	0	0	0	0	$\mathbf{c}\cdot\mathbf{x}^{(6)}=0$

The first insight leading to computational efficiency is to realize that the linear systems for any iteration have the same left-hand side. All involve solving for weights on basic columns to express another vector.

Consider, for example, iteration $t=1$ of the Top Brass computation detailed in Table 5.2. With all nonbasic variables fixed $=0$ (definition $\boxed{5.16}$), the current basic solution can be completed by solving the linear system

$$
\begin{aligned}
+1x_1 &+ 0x_4 &+ 0x_5 &+ 0x_6 &= 1000 \\
+0x_1 &+ 1x_4 &+ 0x_5 &+ 0x_6 &= 1500 \\
+1x_1 &+ 0x_4 &+ 1x_5 &+ 0x_6 &= 1750 \\
+4x_1 &+ 0x_4 &+ 0x_5 &+ 1x_6 &= 4800
\end{aligned}
$$

Simplex directions are similar. Definition 5.21 implicitly requires solving for a representation of the negative of the constraint column for x_j as we derive the corresponding simplex direction. For example, basic parts of such simplex directions $\Delta \mathbf{x}$ for Top Brass iteration $t = 1$ all come from linear equation systems

$$
\begin{array}{ccccc}
+1\,\Delta x_1 & +0\,\Delta x_4 & +0\,\Delta x_5 & +0\,\Delta x_6 & = & -a_{1,j} \\
+0\,\Delta x_1 & +1\,\Delta x_4 & +0\,\Delta x_5 & +0\,\Delta x_6 & = & -a_{2,j} \\
+1\,\Delta x_1 & +0\,\Delta x_4 & +1\,\Delta x_5 & +0\,\Delta x_6 & = & -a_{3,j} \\
+4\,\Delta x_1 & +0\,\Delta x_4 & +0\,\Delta x_5 & +1\,\Delta x_6 & = & -a_{4,j}
\end{array}
$$

Think of the collection of columns for basic variables as a **basis matrix**:

$$
\mathbf{B} \triangleq \begin{pmatrix} 1 & 0 & 0 & 0 \\ 0 & 1 & 0 & 0 \\ 1 & 0 & 1 & 0 \\ 4 & 0 & 0 & 1 \end{pmatrix}
$$

All needed linear systems for this $\mathbf{B}$ can be solved easily if we compute just once the **inverse** of matrix $\mathbf{B}$:

$$
\mathbf{B}^{-1} = \begin{pmatrix} 1 & 0 & 0 & 0 \\ 0 & 1 & 0 & 0 \\ -1 & 0 & 1 & 0 \\ -4 & 0 & 0 & 1 \end{pmatrix}
$$

Primer 5 reviews the key facts about matrix inverses. For our purposes the important observation is that multiplication by a basis inverse can solve a corresponding system of linear constraints.

5.42   Using a representation of the current basis matrix inverse $\mathbf{B}^{-1}$, basic components of the corresponding basic solution can be computed by matrix multiplication as $\mathbf{B}^{-1}\mathbf{b}$, and basic components of simplex directions for nonbasic variables x_j are $-\mathbf{B}^{-1}\mathbf{a}^{(j)}$, where $\mathbf{b}$ is the vector of right-hand-side coefficients, and $\mathbf{a}^{(j)}$ is the constraint column for x_j.

We can illustrate with the basis inverse above. Basic components of the corresponding basic solution can be computed (formula 5.42):

$$
\begin{pmatrix} x_1 \\ x_4 \\ x_5 \\ x_6 \end{pmatrix} = \mathbf{B}^{-1}\mathbf{b} = \begin{pmatrix} 1 & 0 & 0 & 0 \\ 0 & 1 & 0 & 0 \\ -1 & 0 & 1 & 0 \\ -4 & 0 & 0 & 1 \end{pmatrix} \begin{pmatrix} 1000 \\ 1500 \\ 1750 \\ 4800 \end{pmatrix} = \begin{pmatrix} 1000 \\ 1500 \\ 750 \\ 800 \end{pmatrix}
$$

Similarly, basic components of the simplex direction for x_2 are

$$
\begin{pmatrix} \Delta x_1 \\ \Delta x_4 \\ \Delta x_5 \\ \Delta x_6 \end{pmatrix} = -\mathbf{B}^{-1}\mathbf{a}^{(2)} = -\begin{pmatrix} 1 & 0 & 0 & 0 \\ 0 & 1 & 0 & 0 \\ -1 & 0 & 1 & 0 \\ -4 & 0 & 0 & 1 \end{pmatrix} \begin{pmatrix} 0 \\ 1 \\ 1 \\ 2 \end{pmatrix} = \begin{pmatrix} 0 \\ -1 \\ -1 \\ -2 \end{pmatrix}
$$

PRIMER 5: IDENTITY AND INVERSE MATRICES

Primer 3 reviewed general matrix arithmetic. One very special form of square matrix is an **identity** matrix, denoted $\mathbf{I}$, that leaves any matrix or vector unchanged after multiplication. That is,

$$\mathbf{IA} = \mathbf{AI} = \mathbf{A} \quad \text{and} \quad \mathbf{Ix} = \mathbf{xI} = \mathbf{x}$$

for any matrix $\mathbf{A}$ and any vector $\mathbf{x}$. The unique matrices that have this property are those of the form

$$\mathbf{I} \triangleq \begin{pmatrix} 1 & 0 & \cdots & 0 \\ 0 & 1 & \cdots & 0 \\ \vdots & & \ddots & \vdots \\ 0 & \cdots & 0 & 1 \end{pmatrix}$$

with 1's down the diagonal and 0's off-diagonal. The dimension of an $\mathbf{I}$ matrix is understood to be whatever is required for multiplication to be defined.

For every square nonsingular matrix $\mathbf{M}$, there is a unique square nonsingular matrix $\mathbf{M}^{-1}$ such that

$$\mathbf{MM}^{-1} = \mathbf{M}^{-1}\mathbf{M} = \mathbf{I}$$

Matrix $\mathbf{M}^{-1}$ is called the **inverse** of $\mathbf{M}$ and denoted by a -1 exponent. Matrices that are not square or are singular (see Primer 4) have no inverses.

As an example, the inverse of the matrix

$$\mathbf{M} \triangleq \begin{pmatrix} 5 & -1 & 3 \\ 0 & \frac{1}{2} & -\frac{1}{2} \\ 7 & 4 & 0 \end{pmatrix} \quad \text{is} \quad \mathbf{M}^{-1} = \begin{pmatrix} \frac{2}{3} & 4 & -\frac{1}{3} \\ -\frac{7}{6} & -7 & \frac{5}{6} \\ -\frac{7}{6} & -9 & \frac{5}{6} \end{pmatrix}$$

because

$$\mathbf{MM}^{-1} = \begin{pmatrix} 5 & -1 & 3 \\ 0 & \frac{1}{2} & -\frac{1}{2} \\ 7 & 4 & 0 \end{pmatrix} \begin{pmatrix} \frac{2}{3} & 4 & -\frac{1}{3} \\ -\frac{7}{6} & -7 & \frac{5}{6} \\ -\frac{7}{6} & -9 & \frac{5}{6} \end{pmatrix} = \begin{pmatrix} 1 & 0 & 0 \\ 0 & 1 & 0 \\ 0 & 0 & 1 \end{pmatrix}$$

For 2 by 2 nonsingular matrices, inverses have the simple form

$$\begin{pmatrix} p & q \\ r & s \end{pmatrix}^{-1} = \frac{1}{ps - qr}\begin{pmatrix} s & -q \\ -r & p \end{pmatrix}, \quad \text{so} \quad \begin{pmatrix} 2 & 3 \\ -4 & 5 \end{pmatrix}^{-1} = \begin{pmatrix} \frac{5}{22} & -\frac{3}{22} \\ \frac{4}{22} & \frac{2}{22} \end{pmatrix}$$

Many computational procedures are available to find matrix inverses in higher dimensions, but a calculator or computer is usually required.

Our principal use of matrix inverses is in computations involving systems of linear equations. In particular, if $\mathbf{Q}$ is a square nonsingular matrix, the unique solution to the equation system is

$$\mathbf{Qx} = \mathbf{r} \quad \text{is} \quad \mathbf{x} = \mathbf{Q}^{-1}\mathbf{r}$$

This follows because multiplication of the original system on the left by $\mathbf{Q}^{-1}$ gives $\mathbf{Q}^{-1}\mathbf{Qx} = \mathbf{Q}^{-1}\mathbf{r}$, and the left-hand side is by definition $\mathbf{Ix} = \mathbf{x}$. Similarly,

(Continued)

(Continued)

multiplication by Q^{-1} on the right shows that the unique solution to
$$vQ = h \quad \text{is} \quad v = hQ^{-1}$$

Illustrating with matrix M above and $r \triangleq (2, 1, 2)$, the unique solution x to the system $Mx = r$, or

$$
\begin{array}{rrrcl}
+5x_1 & -1x_2 & +3x_3 & = & 2 \\
+0x_1 & +\frac{1}{2}x_2 & -\frac{1}{2}x_3 & = & 1 \\
+7x_1 & +4x_2 & +0x_3 & = & 2
\end{array}
\quad \text{is} \quad x = M^{-1}r =
\begin{pmatrix} \frac{14}{3} \\ -\frac{23}{3} \\ -\frac{29}{3} \end{pmatrix}
$$

With $h = (-1, 1, 1)$ the unique solution v to the system $vM = h$, or

$$
\begin{array}{rrrcl}
+5v_1 & +0v_2 & +7v_3 & = & -1 \\
-1v_1 & +\frac{1}{2}v_2 & +4v_3 & = & 1 \\
+3v_1 & -\frac{1}{2}v_2 & +0v_3 & = & 1
\end{array}
\quad \text{is} \quad v = hM^{-1} =
\begin{pmatrix} -\frac{5}{3} \\ -20 \\ 2 \end{pmatrix}
$$

SAMPLE EXERCISE 5.23: COMPUTING WITH BASIS INVERSES

Consider the standard-form linear program

$$
\begin{array}{llll}
\min & 9x_1 & +3x_2 & & +1x_4 \\
\text{s.t.} & 2x_1 & +1x_2 & -1x_3 & & = 12 \\
& 1x_1 & & +9x_3 & +2x_4 & = 5 \\
& x_1, x_2, x_3, x_4 \geq 0
\end{array}
$$

and assume that x_1 and x_2 are basic.

(a) Identify the current basis matrix.

(b) Compute the current basis inverse.

(c) Use your basis inverse to compute the current basic solution.

(d) Use the basis matrix inverse to compute all simplex directions at the current solution.

Analysis:

(a) Columns for basic variables x_1 and x_2 make the basis matrix

$$B = \begin{pmatrix} 2 & 1 \\ 1 & 0 \end{pmatrix}$$

(b) Applying the formula in Primer 5 yields

$$B^{-1} = \frac{1}{(2)(0) - (1)(1)} \begin{pmatrix} 0 & -1 \\ -1 & 2 \end{pmatrix} = \begin{pmatrix} 0 & 1 \\ 1 & -2 \end{pmatrix}$$

(c) Using computation $\boxed{5.42}$, the components of the basic solution for basic variables are

$$\begin{pmatrix} x_1 \\ x_2 \end{pmatrix} = B^{-1}b = \begin{pmatrix} 0 & 1 \\ 1 & -2 \end{pmatrix}\begin{pmatrix} 12 \\ 5 \end{pmatrix} = \begin{pmatrix} 5 \\ 2 \end{pmatrix}$$

Thus the full basic solution is $x = (5, 2, 0, 0)$.

(d) Applying formula $\boxed{5.42}$, the basic components of the direction for x_3 can be computed:

$$\left(\begin{array}{c} \Delta x_1 \\ \Delta x_2 \end{array} \right) = -\mathbf{B}^{-1}\mathbf{a}^{(3)} = -\left(\begin{array}{cc} 0 & 1 \\ 1 & -2 \end{array} \right)\left(\begin{array}{c} -1 \\ 9 \end{array} \right) = \left(\begin{array}{c} -9 \\ 19 \end{array} \right)$$

Thus the full simplex direction is $\mathbf{x} = (-9, 19, 1, 0)$. Basic components of the corresponding direction for nonbasic x_4 are

$$\left(\begin{array}{c} \Delta x_1 \\ \Delta x_2 \end{array} \right) = -\mathbf{B}^{-1}\mathbf{a}^{(4)} = -\left(\begin{array}{cc} 0 & 1 \\ 1 & -2 \end{array} \right)\left(\begin{array}{c} 0 \\ 2 \end{array} \right) = \left(\begin{array}{c} -2 \\ 4 \end{array} \right)$$

which implies a full simplex direction $\mathbf{x} = (-2, 4, 0, 1)$.

Updating the Representation of B^{-1}

It would require much too much memory to keep handy a copy of the full basis inverse matrix for most real linear programs. Still, production-quality codes do employ a representation of $\mathbf{B}^{-1}$ in much the same way as we have just outlined to speed up the solution of required linear systems.

Recall that simplex changes the basis at each iteration. If a new basis matrix inverse (representation) had to be computed each time, some of the work we did in Sections 5.3 and 5.5 would be reduced, but an immense amount of calculation would remain.

The next insight for reducing simplex computation is that bases differ very little from iteration to iteration. Corresponding basis matrices are not very different either. Exactly one column of the old $\mathbf{B}$ (the one for the leaving basic variable) is replaced by one new one (the column of the entering nonbasic variable); other columns remain unchanged.

Fortunately, we can exploit this similarity of bases.

$\underline{5.43}$ The new basis inverse at each iteration can be computed from the old by $(\text{new } \mathbf{B}^{-1}) = \mathbf{E}\,(\text{old } \mathbf{B}^{-1})$, where $\mathbf{E}$ is an update matrix constructed from the chosen simplex direction.

That is, the inverse can be updated by a single matrix multiplication. The process is called a **pivot**.

The specific form of update matrices $\mathbf{E}$ is indicated by the schema

$$
\mathbf{E} =
\begin{pmatrix}
1 & 0 & \cdots & 0 & -\dfrac{\Delta x_{1\text{st}}}{\Delta x_{\text{leave}}} & 0 & \cdots & 0 \\
0 & 1 & \cdots & 0 & -\dfrac{\Delta x_{2\text{nd}}}{\Delta x_{\text{leave}}} & 0 & \cdots & 0 \\
\vdots & \vdots & \ddots & \vdots & \vdots & \vdots & & \vdots \\
\vdots & \vdots & & 1 & \vdots & 0 & \cdots & 0 \\
\vdots & \vdots & & \vdots & \dfrac{1}{\Delta x_{\text{leave}}} & & & \vdots \\
0 & 0 & \cdots & 0 & \vdots & 0 & \ddots & 0 \\
0 & 0 & \cdots & 0 & -\dfrac{\Delta x_{m\text{th}}}{\Delta x_{\text{leave}}} & 0 & \cdots & 1
\end{pmatrix}
\tag{5.18}
$$

leaving position $\searrow$

Here $\Delta x_{j\text{th}}$ denotes the simplex direction component for the jth basic variable, and Δx_{leave} indicates the component for the leaving basic. The matrix is almost an identity. Only the column position of the leaving basic variable is replaced. The jth row of that special column is $-\Delta x_{j\text{th}}/\Delta x_{\text{leave}}$, except on the diagonal, where it is $-1/\Delta x_{\text{leave}}$.

Again using iteration $t = 1$ of Table 5.2, the required pivot replaces x_6 in the fourth position of the old basis $\{x_1, x_4, x_5, x_6\}$ by x_2. The corresponding $\mathbf{E}$ matrix is

$$
\mathbf{E} = \begin{pmatrix}
1 & 0 & 0 & -\dfrac{0}{-2} \\[2mm]
0 & 1 & 0 & -\dfrac{-1}{-2} \\[2mm]
0 & 0 & 1 & -\dfrac{-1}{-2} \\[2mm]
0 & 0 & 0 & -\dfrac{1}{-2}
\end{pmatrix}
$$

The basis inverse after the iteration can then be computed from the old $\mathbf{B}^{-1}$ above as (pivot formula $\boxed{5.43}$)

$$
\text{new } \mathbf{B}^{-1} \triangleq \begin{pmatrix}
1 & 0 & 0 & 0 \\
0 & 1 & 0 & 1 \\
1 & 0 & 1 & 1 \\
4 & 0 & 0 & 2
\end{pmatrix}^{-1}
$$

$$
= \mathbf{E}(\text{old } \mathbf{B}^{-1})
$$

$$
= \begin{pmatrix}
1 & 0 & 0 & -\dfrac{0}{-2} \\[2mm]
0 & 1 & 0 & -\dfrac{-1}{-2} \\[2mm]
0 & 0 & 1 & -\dfrac{-1}{-2} \\[2mm]
0 & 0 & 0 & -\dfrac{1}{-2}
\end{pmatrix}
\begin{pmatrix}
1 & 0 & 0 & 0 \\
0 & 1 & 0 & 0 \\
-1 & 0 & 1 & 0 \\
-4 & 0 & 0 & 1
\end{pmatrix}
$$

$$
= \begin{pmatrix}
1 & 0 & 0 & 0 \\
2 & 1 & 0 & -0.5 \\
1 & 0 & 1 & -0.5 \\
-2 & 0 & 0 & 0.5
\end{pmatrix}
$$

Although it is beyond the scope of this book to provide details, update formula $\boxed{5.43}$ also hints at how production simplex computer codes actually represent basis inverses. Since each inverse is just $\mathbf{E}$ times the last, we could represent the current inverse simply by remembering its initial form along with the $\mathbf{E}$'s from iterations so far. Then the process of multiplying a vector by $\mathbf{B}^{-1}$, required, say, to compute a simplex direction, can be accomplished just as well by multiplying in turn by the initial form and all the $\mathbf{E}$'s.

SAMPLE EXERCISE 5.24: UPDATING BASIS INVERSES

Assume in the linear program of Sample Exercise 5.23 that nonbasic x_3 enters the basis and x_1 leaves. Use the simplex direction results of part (d) to compute update matrix $\mathbf{E}$, and apply the result to compute the new basis inverse.

Analysis: From Sample Exercise 5.23(d), the basic components of the simplex direction for x_3 are $\Delta x_1 = -9$ and $\Delta x_2 = 19$. With x_1 leaving the basic set $\{x_1, x_2\}$, schema (5.18) gives

$$
\mathbf{E} = \begin{pmatrix} -\dfrac{1}{\Delta x_1} & 0 \\[2mm] -\dfrac{\Delta x_2}{\Delta x_1} & 1 \end{pmatrix} = \begin{pmatrix} \frac{1}{9} & 0 \\[2mm] \frac{19}{9} & 1 \end{pmatrix}
$$

Using this $\mathbf{E}$ and the old basis inverse of Sample Exercise 5.23(b), updating **5.43** gives

$$
\text{new } \mathbf{B}^{-1} = \mathbf{E}(\text{old } \mathbf{B}^{-1}) = \begin{pmatrix} \frac{1}{9} & 0 \\[2mm] \frac{19}{9} & 1 \end{pmatrix}\begin{pmatrix} 0 & 1 \\ 1 & -2 \end{pmatrix} = \begin{pmatrix} 0 & \frac{1}{9} \\[2mm] 1 & \frac{1}{9} \end{pmatrix}
$$

Basic Variable Sequence in Revised Simplex

One slightly confusing aspect of equation (5.18) is its reference to basic variables by sequential position (1st, 2nd, etc.) rather than the original subscript. For example, the pivot above for Top Brass iteration $t = 1$ entered x_2 to replace x_6 as the 4th basic variable. The new basis thus has

$$
x_{1\text{st}} \triangleq x_1, \qquad x_{2\text{nd}} \triangleq x_4, \qquad x_{3\text{rd}} \triangleq x_5, \qquad x_{4\text{th}} \triangleq x_2
$$

and columns are arranged in that order in the foregoing expressions.

5.44 Variables in revised simplex enter the basis in the same sequential positions as the variables they replace.

This sequence-based numbering of basis variables recurs in several elements of revised simplex computation. Readers doing calculations by hand will find it somewhat tedious. Keep in mind, however, that revised simplex ideas are designed for efficient computer calculation, not ease of human understanding.

SAMPLE EXERCISE 5.25: TRACKING BASIC VARIABLE SEQUENCE

Assume that revised simplex computation on a linear program in standard form begins with basic variable sequence $\{x_1, x_2, x_3\}$. On subsequent pivots x_6 replaces x_1, x_5 replaces x_3, and x_4 replaces x_6. Show the basic variable sequences for those three iterations.

Analysis: The initial basis has

$$x_{1st} \triangleq x_1, \quad x_{2nd} \triangleq x_2, \quad x_{3rd} \triangleq x_3$$

As variables enter and leave, it changes to

$$x_{1st} \triangleq x_6, \quad x_{2nd} \triangleq x_2, \quad x_{3rd} \triangleq x_3$$
$$x_{1st} \triangleq x_6, \quad x_{2nd} \triangleq x_2, \quad x_{3rd} \triangleq x_5$$
$$x_{1st} \triangleq x_4, \quad x_{2nd} \triangleq x_2, \quad x_{3rd} \triangleq x_5$$

Computing Reduced Costs by Pricing

Simplex search chooses a move direction (or concludes optimality) by computing the reduced cost of definition $\boxed{5.22}$ for nonbasic variables x_j, that is,

$$\bar{c}_j \triangleq \mathbf{c} \cdot \Delta \mathbf{x} \triangleq \sum_{k=1}^{n} c_k \, \Delta x_k$$

where $\Delta \mathbf{x}$ is the simplex direction for nonbasic x_j.

The only reason for computing most of the simplex directions at any iteration is to determine reduced costs. We actually use just one simplex direction in the move. Obviously, a great deal of computation could be saved if we could obtain the $\bar{c}_j$ without generating full simplex directions.

To see exactly how to do that, recall that the simplex direction $\Delta \mathbf{x}$ for nonbasic x_j is $=1$ in the jth component and $=0$ on components for all other nonbasics (definition $\boxed{5.21}$). Thus we could write

$$\bar{c}_j = c_j + \sum_{k \in B} c_k \, \Delta x_k \tag{5.19}$$

where B denotes the set of basic variable subscripts. For example, in iteration $t = 1$ of the Top Brass example,

$$\bar{c}_3 = 0 + [12(-1) + 0(0) + 0(1) + 0(4)] = -12$$

We now know how to determine the basic directional components Δx_k in expression (5.19) using a representation of the basis inverse (formula $\boxed{5.42}$).

$$\begin{pmatrix} \Delta x_{1st} \\ \Delta x_{2nd} \\ \vdots \\ \Delta x_{mth} \end{pmatrix} = \mathbf{B}^{-1} \begin{pmatrix} -a_{1,k} \\ -a_{2,k} \\ \vdots \\ -a_{m,k} \end{pmatrix}$$

where again Δx_{jth} is the component of simplex direction $\Delta \mathbf{x}$ on the jth basic variable. Denoting corresponding basic objective function coefficients by c_{jth} and substituting produces the pricing vector for the current iteration.

$\boxed{5.45}$ The **pricing vector v** corresponding to the current basis $\{x_{1st}, x_{2nd}, \ldots, x_{mth}\}$ is

$$\mathbf{v} \triangleq (c_{1st}, c_{2nd}, \ldots, c_{mth}) \mathbf{B}^{-1}$$

Then, using formula $\boxed{5.42}$, equation (5.19) becomes

$$\bar{c}_j = c_j + (c_{1st}, c_{2nd}, \ldots, c_{mth}) \cdot \begin{pmatrix} \Delta \mathbf{x}_{1st} \\ \Delta \mathbf{x}_{2nd} \\ \vdots \\ \Delta \mathbf{x}_{mth} \end{pmatrix}$$

$$= c_j - (c_{1st}, c_{2nd}, \ldots, c_{mth}) \mathbf{B}^{-1} \begin{pmatrix} a_{1,j} \\ a_{2,j} \\ \vdots \\ a_{m,j} \end{pmatrix}$$

$$= c_j - \mathbf{v} \cdot \begin{pmatrix} a_{1,j} \\ a_{2,j} \\ \vdots \\ a_{m,j} \end{pmatrix}$$

We see that reduced costs $\bar{c}_j$ can be computed without explicitly generating simplex directions.

$\boxed{5.46}$ Reduced costs can be computed directly from original data as $\bar{c}_j = c_j - \mathbf{v} \cdot \mathbf{a}^{(j)}$, where $\mathbf{v}$ is the pricing vector of the current iteration and $\mathbf{a}^{(j)}$ is the original constraint matrix column for variable x_k.

Again we illustrate with iteration $t = 1$ of Table 5.2. The pricing vector for that iteration is

$$\mathbf{v} = (c_{1st}, c_{2nd}, \ldots, c_{mth}) \mathbf{B}^{-1} = (12, 0, 0, 0) \begin{pmatrix} 1 & 0 & 0 & 0 \\ 0 & 1 & 0 & 0 \\ -1 & 0 & 1 & 0 \\ -4 & 0 & 0 & 1 \end{pmatrix} = (12, 0, 0, 0)$$

Then reduced costs follow as

$$\bar{c}_2 = c_2 - \mathbf{v} \cdot \mathbf{a}^{(2)} \quad = 9 - (12, 0, 0, 0) \cdot \begin{pmatrix} 0 \\ 1 \\ 1 \\ 2 \end{pmatrix} = 9$$

$$\bar{c}_3 = c_3 - \mathbf{v} \cdot \mathbf{a}^{(3)} \quad = 0 - (12, 0, 0, 0) \cdot \begin{pmatrix} 1 \\ 0 \\ 0 \\ 0 \end{pmatrix} = -12$$

Formula $\boxed{5.46}$ changes absolutely nothing about the value of the reduced costs, but it decreases enormously the effort to compute them. We now need to compute only one simplex direction explicitly—the one that the nonbasic variable was chosen to enter.

SAMPLE EXERCISE 5.26: COMPUTING REDUCED COSTS BY PRICING

Consider the linear program

$$
\begin{array}{rllll}
\min & 3x_1 & +100x_2 & +12x_3 & -8x_4 \\
\text{s.t.} & 3x_1 & +1x_2 & -1x_3 & & = 90 \\
& -1x_1 & -1x_2 & & +1x_4 & = 22 \\
& x_1, x_2, x_3, x_4 \geq 0
\end{array}
$$

Assume that the current basic variable sequence is $\{x_3, x_1\}$.

(a) Compute the corresponding basis inverse.

(b) Compute the associated pricing vector.

(c) Compute the reduced costs on nonbasic variables without generating any simplex directions.

Analysis:

(a) In the sequence given, the basis inverse matrix is

$$
\mathbf{B}^{-1} = \begin{pmatrix} -1 & 3 \\ 0 & -1 \end{pmatrix}^{-1} = \begin{pmatrix} -1 & -3 \\ 0 & -1 \end{pmatrix}
$$

(b) Applying definition $\boxed{5.46}$, the associated pricing vector is

$$
\mathbf{v} = (c_{1\text{st}}, c_{2\text{nd}})\mathbf{B}^{-1} = (12, 3)\begin{pmatrix} -1 & -3 \\ 0 & -1 \end{pmatrix} = (-12, -39)
$$

(c) We can now compute all reduced costs via principle $\boxed{5.46}$.

$$
\begin{array}{rclclcl}
\bar{c}_2 &=& c_2 - \mathbf{v} \cdot \mathbf{a}^{(2)} &=& 100 - (-12, -39) \cdot (1, -1) &=& 73 \\
\bar{c}_4 &=& c_4 - \mathbf{v} \cdot \mathbf{a}^{(4)} &=& -8 - (-12, -39) \cdot (0, 1) &=& 31
\end{array}
$$

Revised Simplex Search of Top Brass Example

We now have all the pieces of the revised simplex procedure detailed in Algorithm 5C. Table 5.9 shows the complete revised simplex search of the Top Brass Trophy example done earlier in Table 5.2. It is important to note that the sequence of basic feasible solutions encountered is exactly the same in the two tables. Only the method of computation at intermediate steps has changed.

To avoid confusion about sequential numbering of basis variables, bases in Table 5.9 are marked with their positions. For example, the basis to start iteration $t = 3$ has x_1 as the first basic, x_2 as the fourth basic, x_3 as the third basic, and x_4 as the second basic. The variables x_5 and x_6 are nonbasic.

In Table 5.9 we maintain an updated representation of the basis matrix inverse at each iteration. That of iteration $t = 0$ is an identity matrix because the corresponding $\mathbf{B}$ is an identity, and it is easy to check that the inverse of an identity matrix is an identity matrix.

ALGORITHM 5C: REVISED SIMPLEX SEARCH FOR LINEAR PROGRAMS

Step 0: Initialization. Choose any starting feasible basis, and construct a representation of the corresponding basic column matrix inverse $\mathbf{B}^{-1}$. Then use that representation to solve $\mathbf{B}\mathbf{x}^B = \mathbf{b}$ for basic components of the initial solution $\mathbf{x}^{(0)}$, set all nonbasic $x_j^{(0)} \leftarrow 0$, and initialize solution index $t \leftarrow 0$.

Step 1: Pricing. Use the representation of the current basic column inverse $\mathbf{B}^{-1}$ to solve $\mathbf{v}\mathbf{B} = \mathbf{c}^B$ for pricing vector $\mathbf{v}$ where $\mathbf{c}^B$ is the vector of basic objective function coefficients. Then evaluate reduced costs $\bar{c}_j \leftarrow c_j - \mathbf{v} \cdot \mathbf{a}^{(j)}$ for each nonbasic x_j.

Step 2: Optimality. If no $\bar{c}_j > 0$ for a maximize problem (or no $\bar{c}_j < 0$ for a minimize), then stop; current solution $\mathbf{x}^{(t)}$ is globally optimal. Otherwise, choose an entering nonbasic variable x_p with improving $\bar{c}_p$.

Step 3: Simplex Direction. Construct the simplex direction $\Delta\mathbf{x}^{(t+1)}$ for nonbasic variable x_p using the representation of current basic column inverse $\mathbf{B}^{-1}$ to solve $\mathbf{B}\Delta\mathbf{x} = -\mathbf{a}^{(p)}$ for basic components.

Step 4: Step Size. If there is no limit on feasible moves in simplex direction $\Delta\mathbf{x}^{(t+1)}$ (all components are nonnegative), stop; the given model is unbounded. Otherwise, choose leaving variable x_r so that

$$\frac{x_r^{(t)}}{-\Delta x_r^{(t+1)}} = \min\left\{\frac{x_j^{(t)}}{-\Delta x_j^{(t+1)}} : \Delta x_j^{(t+1)} < 0\right\} \quad \text{and set} \quad \lambda \leftarrow \frac{x_r^{(t)}}{-\Delta x_r^{(t+1)}}$$

Step 5: New Point and Basis. Compute the new solution

$$\mathbf{x}^{(t+1)} \leftarrow \mathbf{x}^{(t)} + \lambda\Delta\mathbf{x}^{(t+1)}$$

and replace x_r in the basis by x_p. Also construct the associated pivot matrix $\mathbf{E}$ and update the basis inverse representation as $\mathbf{E}\mathbf{B}^{-1}$. Then advance $t \leftarrow t + 1$, and return to Step 1.

The next three iterations update $\mathbf{B}^{-1}$ via formula $\boxed{5.43}$. Required $\mathbf{E}$ matrices are, respectively,

$$
\begin{pmatrix}
-\dfrac{1}{-1} & 0 & 0 & 0 \\[2mm]
-\dfrac{0}{-1} & 1 & 0 & 0 \\[2mm]
-\dfrac{-1}{-1} & 0 & 1 & 0 \\[2mm]
-\dfrac{-4}{-1} & 0 & 0 & 1
\end{pmatrix},
\begin{pmatrix}
1 & 0 & 0 & -\dfrac{0}{-2} \\[2mm]
0 & 1 & 0 & -\dfrac{-1}{-2} \\[2mm]
0 & 0 & 1 & -\dfrac{-1}{-2} \\[2mm]
0 & 0 & 0 & -\dfrac{1}{-2}
\end{pmatrix},
\text{ and }
\begin{pmatrix}
1 & 0 & -\dfrac{-1}{-1} & 0 \\[2mm]
0 & 1 & -\dfrac{-2}{-1} & 0 \\[2mm]
0 & 0 & -\dfrac{1}{-1} & 0 \\[2mm]
0 & 0 & -\dfrac{2}{-1} & 1
\end{pmatrix}
$$

At each iteration of revised simplex, reduced costs $\bar{c}_j$ are computed directly using pricing vector $\mathbf{v}$. Table 5.9 shows all $\mathbf{v}$'s, and the resulting $\bar{c}_j$, boxing the one selected as improving at each iteration. The simplex direction associated with that

TABLE 5.9 Revised Simplex Search of Top Brass Trophy Example

	x_1	x_2	x_3	x_4	x_5	x_6	
max **c**	12	9	0	0	0	0	**b**
	1	0	1	0	0	0	1000
	0	1	0	1	0	0	1500
A	1	1	0	0	1	0	1750
	4	2	0	0	0	1	4800
$t = 0$	N	N	1st	2nd	3rd	4th	
$\mathbf{x}^{(0)}$	0	0	1000	1500	1750	4800	$\mathbf{c} \cdot \mathbf{x}^{(0)} = 0$

$$\mathbf{B}^{-1} = \begin{pmatrix} 1 & 0 & 0 & 0 \\ 0 & 1 & 0 & 0 \\ 0 & 0 & 1 & 0 \\ 0 & 0 & 0 & 1 \end{pmatrix}, \quad \mathbf{v} = \begin{pmatrix} 0 \\ 0 \\ 0 \\ 0 \end{pmatrix}$$

	x_1	x_2	x_3	x_4	x_5	x_6	
$\bar{c}_j$	$\boxed{12}$	9	0	0	0	0	
$\Delta \mathbf{x}$ for x_1	1	0	-1	0	-1	-4	
	—	—	$\boxed{\dfrac{1000}{-(-1)}}$	—	$\dfrac{1750}{-(-1)}$	$\dfrac{4800}{-(-1)}$	$\lambda = 1000$
$t = 1$	1st	N	N	2nd	3rd	4th	
$\mathbf{x}^{(1)}$	1000	0	0	1500	750	800	$\mathbf{c} \cdot \mathbf{x}^{(1)} = 12{,}000$

$$\mathbf{B}^{-1} = \begin{pmatrix} 1 & 0 & 0 & 0 \\ 0 & 1 & 0 & 0 \\ -1 & 0 & 1 & 0 \\ -4 & 0 & 0 & 1 \end{pmatrix}, \quad \mathbf{v} = \begin{pmatrix} 12 \\ 0 \\ 0 \\ 0 \end{pmatrix}$$

	x_1	x_2	x_3	x_4	x_5	x_6	
$\bar{c}_j$	0	$\boxed{9}$	-12	0	0	0	
$\Delta \mathbf{x}$ for x_2	0	1	0	-1	-1	-2	
	—	—	—	$\dfrac{1500}{-(-1)}$	$\dfrac{750}{-(-1)}$	$\boxed{\dfrac{800}{-(-2)}}$	$\lambda = 400$
$t = 2$	1st	4th	N	2nd	3rd	N	
$\mathbf{x}^{(2)}$	1000	400	0	1100	350	0	$\mathbf{c} \cdot \mathbf{x}^{(2)} = 15{,}600$

$$\mathbf{B}^{-1} = \begin{pmatrix} 1 & 0 & 0 & 0 \\ 2 & 1 & 0 & -0.5 \\ 1 & 0 & 1 & -0.5 \\ -2 & 0 & 0 & 0.5 \end{pmatrix}, \quad \mathbf{v} = \begin{pmatrix} -6 \\ 0 \\ 0 \\ 4.5 \end{pmatrix}$$

	x_1	x_2	x_3	x_4	x_5	x_6	
$\bar{c}_j$	0	0	$\boxed{6}$	0	0	-4.5	
$\Delta \mathbf{x}$ for x_3	-1	2	1	-2	-1	0	
	$\dfrac{1000}{-(-1)}$	—	—	$\dfrac{1100}{-(-2)}$	$\boxed{\dfrac{350}{-(-1)}}$	—	$\lambda = 350$
$t = 3$	1st	4th	3rd	2nd	N	N	
$\mathbf{x}^{(3)}$	650	1100	350	400	0	0	$\mathbf{c} \cdot \mathbf{x}^{(3)} = 17{,}700$

$$\mathbf{B}^{-1} = \begin{pmatrix} 0 & 0 & -1 & 0.5 \\ 0 & 1 & -2 & 0.5 \\ 1 & 0 & 1 & -0.5 \\ 0 & 0 & 2 & -0.5 \end{pmatrix}, \quad \mathbf{v} = \begin{pmatrix} 0 \\ 0 \\ 6 \\ 1.5 \end{pmatrix}$$

	x_1	x_2	x_3	x_4	x_5	x_6	
$\bar{c}_j$	0	0	0	0	-6	-1.5	"optimal"

improving reduced cost, which follows immediately in the table, is the only one explicitly generated at each iteration.

Once a move direction has been selected, the revised simplex makes the step size λ decision exactly as in the earlier simplex algorithm. A new basic solution

$\mathbf{x}^{(t+1)}$ is computed, a leaving variable is selected, the basis inverse representation is updated, and processing continues.

5.9 SIMPLEX WITH SIMPLE UPPER AND LOWER BOUNDS

In Section 5.1 we showed how LP standard form converts all inequalities of a problem to nonnegativity constraints $x_j \geq 0$. In later sections we demonstrated how easy this form makes construction of basic solutions, simplex directions, and so on.

Almost all those simplifications would follow just as well if we allowed inequalities to take the slightly more general **simple lower-bound** form

$$x_j \geq \ell_j$$

or **simple upper-bound** form

$$x_j \leq u_j$$

Here the ℓ_j and u_j are given model constants. Nonnegativities are just the special case of simple lower bounds with $\ell_j = 0$. In this section we explore briefly how the revised simplex method of Section 5.8 can be modified to encompass simple lower and upper bounds.

Lower and Upper-Bounded Standard Form

The standard form for linear programs with simple lower and upper bounds is

$$\text{min (or max)} \quad \sum_{j=1}^{n} c_j x_j$$

$$\text{s.t.} \quad \sum_{j=1}^{n} a_{i,j} x_j = b_i \qquad \text{for all } i = 1, 2, \ldots, m$$

$$u_j \geq x_j \geq \ell_j \qquad \text{for all } j = 1, 2, \ldots, n$$

Collecting lower and upper bounds in vectors ℓ and $\mathbf{u}$ produces the corresponding matrix form.

> **5.47**   In matrix notation, the **lower and upper-bounded standard form** for linear programs is
>
> $$\text{min (or max)} \quad \mathbf{c} \cdot \mathbf{x}$$
> $$\text{s.t.} \quad \mathbf{Ax} = \mathbf{b}$$
> $$\mathbf{u} \geq \mathbf{x} \geq \ell$$

Feasibility requires that $u_j \geq \ell_j$, and we allow the possibility that $u_j = \infty$, or $\ell_j = -\infty$, or both.

Often, allowing simple bounds in this way considerably reduces the number of rows m in main constraints $\mathbf{Ax} = \mathbf{b}$. The result is a savings in most of the complex steps of the simplex algorithm, such as generating simplex directions and updating $\mathbf{B}^{-1}$. For example, the Top Brass model of Figure 5.5 required that $m = 4$ rows in Tables 5.2 and 5.9; it needs only that $m = 2$ in lower or upper-bounded form.

	x_1	x_2	x_5	x_6	
max c	12	9	0	0	b
	1	1	1	0	1750
	4	2	0	1	4800
ℓ	0	0	0	0	
u	1000	1500	∞	∞	

Notice that only two slack variables are now required. We continue to call them x_5 and x_6, for consistency with Figure 5.5.

SAMPLE EXERCISE 5.27: CONSTRUCTING LOWER- AND UPPER-BOUNDED STANDARD FORM

Place the following linear program in lower- and upper-bounded standard form.

$$\min \quad 3x_1 - x_2 + x_3 + 11x_4$$
$$\text{s.t.} \quad x_1 + x_2 + x_3 + x_4 = 50$$
$$x_1 \leq 30$$
$$3x_1 + x_4 \leq 90$$
$$9x_3 + x_4 \geq 5$$
$$x_2 \leq 10$$
$$x_3 \geq -2$$
$$x_1, x_2 \geq 0$$

Analysis: Nonnegative slack variables x_5 and x_6 are required in the third and fourth constraints, but the rest of the model, including unrestricted variable x_4, can be accommodated by lower- and upper-bounded standard form $\boxed{5.47}$. A full coefficient array is

	x_1	x_2	x_3	x_4	x_5	x_6	
min c	3	−1	1	11	0	0	b
	1	1	1	1	0	0	50
	3	0	0	1	1	0	90
	0	0	9	1	0	−1	5
ℓ	0	0	−2	−∞	0	0	
u	30	10	∞	∞	∞	∞	

Basic Solutions with Lower and Upper Bounds

The main idea of basic solutions is to make active the inequalities for all nonbasic variables, then solve for the basics. With lower and upper bounds there may be choices for each nonbasic x_j, depending on which of ℓ_j and u_j are finite.

$\boxed{5.48}$ Basic solutions in lower- and upper-bounded simplex set nonbasics to either (finite) lower bound ℓ_j or (finite) upper bound u_j and solve $\mathbf{Ax} = \mathbf{b}$ for the basics.

The solution is **basic feasible** if computed basic values satisfy their lower and upper bound constraints

$$u_j \geq x_j \geq \ell_j \qquad \text{for all } j \text{ basic}$$

The linear system to be solved in computing basic variable values has the form

$$\sum_{j \in B} a_{i,j} x_j^{(t)} = b_i - \sum_{j \in L} a_{i,j} \ell_j - \sum_{j \in U} a_{i,j} u_j \qquad \text{for all } i = 1, \ldots, m \tag{5.20}$$

where L indexes the set of lower-bounded nonbasics and U the set of upper-bounded nonbasics. Except for the more complicated right-hand side, it is no more difficult to solve than that of the ordinary simplex.

SAMPLE EXERCISE 5.28: COMPUTING LOWER- AND UPPER-BOUNDED BASIC SOLUTIONS

Consider the linear program

$$
\begin{array}{lllll}
\min & -2x_1 & +4x_2 & +3x_3 \\
\text{s.t.} & 1x_1 & -1x_2 & & +3x_4 & = & 13 \\
& 2x_1 & +1x_2 & +2x_3 & -2x_4 & = & 6 \\
& 6 \geq x_1 \geq 4, & 10 \geq x_2 \geq -10, & 5 \geq x_3 \geq 1, & 3 \geq x_4 \geq 2
\end{array}
$$

Compute the basic solution having x_2 and x_3 basic, x_1 nonbasic lower-bounded, and x_4 nonbasic upper-bounded.

Analysis: Following definition 5.48 , nonbasic variables will be fixed:

$$x_1 = \ell_1 = 4 \quad \text{and} \quad x_4 = u_4 = 3$$

We solve for basic components in equations (5.20):

$$
\begin{array}{llllll}
-1x_2 & +0x_3 & = & 13 - (1)(4) - (3)(3) & = & 0 \\
+1x_2 & +2x_3 & = & 6 \ - (2)(4) - (-2)(3) & = & 4
\end{array}
$$

The unique solution is $x_2 = 0$, $x_3 = 2$, implying a full basic solution of $\mathbf{x} = (4, 0, 2, 3)$.

Unrestricted Variables with No Bounds

Sometimes models contain unrestricted variables that can take on any value. Examples include inventory levels where backordering is allowed, and temperatures. In such cases $\ell_j = -\infty$ and $u_j = \infty$. With neither finite, the variables cannot be nonbasic under definition 5.48 .

 5.49   Unrestricted variables must be basic in every lower- and upper-bounded basic solution.

Increasing and Decreasing Nonbasic Variable Values

Simplex directions $\Delta \mathbf{x}$ are constructed to be feasible when nonbasic x_j are increased. The jth directional component is $\Delta x_j = +1$.

This logic works fine for lower-bounded nonbasic variables because the only feasible move is an increase. Tests for improvement are as in principle $\boxed{5.23}$.

> $\boxed{5.50}$ The simplex direction $\Delta \mathbf{x}$ for lower-bounded nonbasic x_j improves in a maximize problem if $\bar{c}_j > 0$, and in a minimize problem if $\bar{c}_j < 0$.

The new element in lower- and upper-bounded simplex comprises upper-bounded nonbasic variables x_j. If they are to change values and stay feasible, they must decrease (i.e., the jth directional component of a move must be negative).

A little review of the derivations in Section 4.3 will show that the direction decreasing nonbasic x_j, changing no other nonbasic, and maintaining $\mathbf{Ax} = \mathbf{b}$ is precisely the negative of simplex direction increasing x_j. Improvement tests are similarly reversed.

> $\boxed{5.51}$ Negative simplex direction $-\Delta \mathbf{x}$ for upper-bounded nonbasic x_j improves in a maximize problem if $\bar{c}_j < 0$, and in a minimize problem if $\bar{c}_j > 0$.

> **SAMPLE EXERCISE 5.29: TESTING SIMPLEX DIRECTIONS IN LOWER AND UPPER FORMAT**

Return to the linear program of Sample Exercise 5.28. Compute reduced costs for all nonbasic variables, and determine which could enter to improve the objective value.

Analysis: Basis matrix

$$\mathbf{B} = \begin{pmatrix} -1 & 0 \\ 1 & 2 \end{pmatrix} \quad \text{has inverse} \quad \mathbf{B}^{-1} = \begin{pmatrix} -1 & 0 \\ \frac{1}{2} & \frac{1}{2} \end{pmatrix}$$

Thus, pricing vector

$$\mathbf{v} = \mathbf{c}^B \mathbf{B}^{-1} = (5, 3) \begin{pmatrix} -1 & 0 \\ \frac{1}{2} & \frac{1}{2} \end{pmatrix} = (-\tfrac{7}{2}, \tfrac{3}{2})$$

Then, applying computation $\boxed{5.46}$ yields

$$\begin{aligned}
\bar{c}_1 &= c_1 - \mathbf{v} \cdot \mathbf{a}^{(1)} = -2 - (-\tfrac{7}{2}, \tfrac{3}{2}) \cdot (1, 2) = -\tfrac{1}{2} \\
\bar{c}_4 &= c_4 - \mathbf{v} \cdot \mathbf{a}^{(4)} = 0 - (-\tfrac{7}{2}, \tfrac{3}{2}) \cdot (3, -2) = \tfrac{27}{2}
\end{aligned}$$

These reduced costs both qualify as improving under conditions $\boxed{5.50}$ and $\boxed{5.51}$.

Step Size with Increasing and Decreasing Values

Step size rule $\boxed{5.24}$ assumes that nonbasic variables always increase and that the lower bound of every basic variable is zero. With a lower- and upper-bounded simplex, many more possibilities exist. If a basic variable is decreasing from value x_k, the maximum feasible decrease is now $(x_k - \ell_k)$. If a basic variable is increasing from value x_k, the maximum feasible increase is $(u_k - x_k)$. Also, the changing nonbasic variable may itself fix λ because it can feasibly increase or decrease only $(u_k - \ell_k)$. These cases lead to a more complex step size rule.

5.52 | With upper and lower bounds the maximum feasible step size λ in direction $\delta\Delta\mathbf{x}$ ($\delta = \pm 1$) is $\lambda = \min\{\lambda^-, \lambda^+\}$, where

$$\lambda^- = \min\left\{\frac{x_j^{(t)} - \ell_j}{-\delta\,\Delta x_j} : \delta\,\Delta x_j < 0\right\} \quad (+\infty \text{ if none})$$

$$\lambda^+ = \min\left\{\frac{u_j - x_j^{(t)}}{\delta\,\Delta x_j} : \delta\,\Delta x_j > 0\right\} \quad (+\infty \text{ if none})$$

If $\lambda = \infty$, the problem is unbounded.

SAMPLE EXERCISE 5.30: DETERMINING STEP SIZE IN LOWER-
AND UPPER- BOUNDED SIMPLEX

Return again to the example of Sample Exercise 5.28. Sample Exercise 5.29 established that both nonbasic variables qualify to enter. Determine the corresponding stepsize and leaving variable for each.

Analysis: Using the results of Sample Exercise 5.29, basic components of the simplex directions for x_1 are

$$\begin{pmatrix} \Delta x_2 \\ \Delta x_3 \end{pmatrix} = -\mathbf{B}^{-1}\mathbf{a}^{(1)} = -\begin{pmatrix} -1 & 0 \\ \frac{1}{2} & \frac{1}{2} \end{pmatrix}\begin{pmatrix} 1 \\ 2 \end{pmatrix} = \begin{pmatrix} 1 \\ -\frac{3}{2} \end{pmatrix}$$

The full direction is $\Delta\mathbf{x} = (1, 1, -\frac{3}{2}, 0)$. With increasing orientation $\delta = +1$, rule 5.52 gives

$$\lambda^- \leftarrow \min\left\{\frac{2-1}{\frac{3}{2}}\right\} = \frac{2}{3}$$

$$\lambda^+ \leftarrow \min\left\{\frac{6-4}{1}, \frac{10-0}{1}\right\} = 2$$

$$\lambda \leftarrow \min\{\tfrac{2}{3}, 2\} = \tfrac{2}{3}$$

The leaving variable, which set the value of λ, is the basic x_3.

Basic components of the simplex directions for x_4 are

$$\begin{pmatrix} \Delta x_2 \\ \Delta x_3 \end{pmatrix} = -\mathbf{B}^{-1}\mathbf{a}^{(4)} = -\begin{pmatrix} -1 & 0 \\ \frac{1}{2} & \frac{1}{2} \end{pmatrix}\begin{pmatrix} 3 \\ -2 \end{pmatrix} = \begin{pmatrix} 3 \\ -\frac{1}{2} \end{pmatrix}$$

making the full direction $\Delta\mathbf{x} = (0, 3, -\frac{1}{2}, 1)$. With decreasing orientation $\delta = -1$, rule 5.52 gives

$$\lambda^- \leftarrow \min\left\{\frac{0-(-10)}{3}, \frac{3-2}{1}\right\} = 1$$

$$\lambda^+ \leftarrow \min\left\{\frac{5-2}{\frac{1}{2}}\right\} = 6$$

$$\lambda \leftarrow \min\{1, 6\} = 1$$

This time the leaving variable is the nonbasic x_4 itself.

Case with No Basis Change

The third possibility above, where λ is established by the changing nonbasic itself, presents another new issue. We have no leaving basic variable. In fact,

> **5.53** If a changing nonbasic establishes step size λ, it merely switches from lower-bounded to upper-bounded status, or vice versa. We need not change the basis or the associated representation of $\mathbf{B}^{-1}$.

Lower- and Upper-Bounded Simplex Algorithm

We are now ready to state the full lower- and upper-bounded revised simplex Algorithm 5D employed in most commercial computer codes. As in earlier sections, $\mathbf{a}^{(j)}$ denotes the (lower- and upper-bounded) standard-form constraint column for x_j, and $\mathbf{c}^B \triangleq (c_{1st}, c_{2nd}, \ldots, c_{mth})$ represents the vector of basic variable objective function coefficients.

Lower- and Upper-Bounded Simplex on Top Brass Example

Table 5.10 returns one last time to the Top Brass Trophy example of Figure 5.5. It details the lower- and upper-bounded simplex computations paralleling those in Tables 5.2 and 5.9.

 With lower and upper bounds, only 2 basic variables must be chosen. We select the two slacks and make both x_1 and x_2 nonbasic lower bounded. This produces starting basic feasible solution $\mathbf{x}^{(0)} = (0, 0, 1750, 4800)$ with objective function value $=0$. The corresponding basis inverse $\mathbf{B}^{-1}$ and pricing vector $\mathbf{v}$ are shown in Table 5.10.

 At iteration $t = 0$, both nonbasic lower-bounded variables have positive reduced costs for this $\mathbf{v}$. We choose the one for x_1, making $p = 1$ and orientation $\delta = +1$ to indicate increase.

 Next we compute step size λ by rules $\boxed{5.52}$. Each unit step in that direction decreases x_5 by 1 and x_6 by 4, so that

$$\lambda^- \leftarrow \min \left\{ \frac{1750 - 0}{1}, \frac{4800 - 0}{4} \right\} = 1200$$

The move also increases x_1 by 1 per unit step, making

$$\lambda^+ \leftarrow \min \left\{ \frac{1000 - 0}{1} \right\} = 1000$$

The maximum step feasible for both sorts of change is

$$\lambda = \min\{\lambda^-, \lambda^+\} = \min\{1200, 1000\} = 1000$$

with $r = 1$.

 A step of size λ in the chosen simplex direction moves us to new solution $\mathbf{x}^{(1)} = (1000, 0, 750, 800)$ at objective value $12,000$. Since it was the changing nonbasic x_1 that fixed λ (i.e., $p = r$), no modification of the basis is required (principle $\boxed{5.53}$). The variable x_1 merely switches to nonbasic upper bounded.

ALGORITHM 5D: LOWER- AND UPPER-BOUNDED REVISED SIMPLEX

Step 0: Initialization. Choose any starting feasible basis with initial nonbasic $j \in L$ at ℓ_j and nonbasic $j \in U$ at u_j, and construct a representation of the corresponding basis inverse $\mathbf{B}^{-1}$. Then use the representation of $\mathbf{B}^{-1}$ to solve

$$\sum_{j \in B} a_{i,j} x_j^{(0)} = b_i - \sum_{j \in L} a_{i,j} \ell_j - \sum_{j \in U} a_{i,j} u_j \qquad \text{for all } i = 1, \ldots, m$$

for basic components $j \in B$ of initial solution $\mathbf{x}^{(0)}$, and set solution index $t \leftarrow 0$.

Step 1: Pricing. Use the representation of the current basic column matrix inverse $\mathbf{B}^{-1}$ to solve $\mathbf{vB} = \mathbf{c}^B$ for pricing vector $\mathbf{v}$, where $\mathbf{c}^B$ is the vector of basic objective function coefficients. Then evaluate reduced costs $\bar{c}_j \leftarrow c_j - \mathbf{v} \cdot \mathbf{a}^{(j)}$ for each nonbasic x_j.

Step 2: Optimality. If no $\bar{c}_j$ indicates improvement (rules $\boxed{5.50}$ and $\boxed{5.51}$), stop; current solution $\mathbf{x}^{(t)}$ is globally optimal. Otherwise, choose an entering nonbasic variable x_p with improving $\bar{c}_p$ and set orientation $\delta = +1$ if the nonbasic is lower bounded and $\delta = -1$ if the nonbasic is upper bounded.

Step 3: Simplex Direction. Construct the simplex direction $\Delta \mathbf{x}^{(t+1)}$ for nonbasic variable x_p using the representation of current basic column inverse $\mathbf{B}^{-1}$ to solve $\mathbf{B} \Delta \mathbf{x} = -\delta \, \mathbf{a}^{(p)}$ for basic components.

Step 4: Step Size. Apply rule $\boxed{5.52}$ to determine the maximum feasible step λ in direction $\Delta \mathbf{x}^{(t+1)}$. If there is no limit ($\lambda = \infty$), stop; the given model is unbounded. Otherwise, choose as leaving x_r any blocking variable that established the value of λ in $\boxed{5.52}$.

Step 5: New Point and Basis. Compute the new solution

$$\mathbf{x}^{(t+1)} \leftarrow \mathbf{x}^{(t)} + \lambda \delta \, \Delta \mathbf{x}^{(t+1)}$$

If $p \neq r$, also replace x_r in the basis by x_p, construct the associated pivot matrix $\mathbf{E}$, and update the basis inverse representation as $\mathbf{EB}^{-1}$. Then advance $t \leftarrow t + 1$, and return to Step 1.

Iteration $t = 1$ proceeds in much the same way except that this time we consider both decreasing nonbasic x_1 and increasing x_2. Only the second satisfies tests for improvement, so the $\Delta \mathbf{x}$ for x_2 is pursued until the basic x_6 drops to $\ell_6 = 0$ at step $\lambda = 400$. The new solution is $\mathbf{x}^{(2)} = (1000, 400, 350, 0)$, with objective value 15,600.

At iteration $t = 2$ we encounter a nonbasic decreasing from its upper bound. With $\bar{c}_1 = -6 < 0$, the negative simplex direction for x_2 helps our maximize objective ($\boxed{5.51}$). Orientation indicator $\delta = -1$. We step distance $\lambda = 350$ in direction $\delta \Delta \mathbf{x}$ to reach new point $\mathbf{x}^{(3)} = (650, 1100, 0, 0)$ at value of 17,700. Iteration $t = 3$ completes the computation, verifying that no available simplex move can improve the objective function.

TABLE 5.10 Upper- and Lower-Bounded Simplex Search of Top Brass Trophy Example

	x_1	x_2	x_5	x_6	
max **c**	12	9	0	0	**b**
	1	1	1	0	1750
	4	2	0	1	4800
ℓ	0	0	0	0	
u	1000	1500	—	—	
$t = 0$	L	L	1st	2nd	
$\mathbf{x}^{(0)}$	0	0	1750	4800	$\mathbf{c} \cdot \mathbf{x}^{(0)} = 0$
	\multicolumn{4}{c}{$\mathbf{B}^{-1} = \begin{pmatrix} 1 & 0 \\ 0 & 1 \end{pmatrix}, \quad \mathbf{v} = \begin{pmatrix} 0 \\ 0 \end{pmatrix}$}	
$\bar{c}_j$	12	$\boxed{9}$	0	0	
$\Delta\mathbf{x}$ for x_1	1	0	−1	−4	
	$\boxed{\dfrac{1000}{1}}$	—	$\dfrac{1750}{1}$	$\dfrac{4800}{4}$	$\lambda = 1000$
$t = 1$	U	L	1st	2nd	
$\mathbf{x}^{(1)}$	1000	0	750	800	$\mathbf{c} \cdot \mathbf{x}^{(1)} = 12{,}000$
	\multicolumn{4}{c}{$\mathbf{B}^{-1} = \begin{pmatrix} 1 & 0 \\ 0 & 1 \end{pmatrix}, \quad \mathbf{v} = \begin{pmatrix} 0 \\ 0 \end{pmatrix}$}	
$\bar{c}_j$	12	$\boxed{9}$	0	0	
$\Delta\mathbf{x}$ for x_2	0	1	−1	−4	
	—	$\dfrac{1500}{1}$	$\dfrac{750}{1}$	$\boxed{\dfrac{800}{2}}$	$\lambda = 400$
$t = 2$	U	2nd	1st	L	
$\mathbf{x}^{(2)}$	1000	400	350	0	$\mathbf{c} \cdot \mathbf{x}^{(2)} = 15{,}600$
	\multicolumn{4}{c}{$\mathbf{B}^{-1} = \begin{pmatrix} 1 & -0.5 \\ 0 & 0.5 \end{pmatrix}, \quad \mathbf{v} = \begin{pmatrix} 0 \\ 4.5 \end{pmatrix}$}	
$\bar{c}_j$	$\boxed{-6}$	0	0	−4.5	
$\Delta\mathbf{x}$ for x_1	−1	2	−1	0	
	$\dfrac{1000}{1}$	$\dfrac{1100}{2}$	$\boxed{\dfrac{350}{1}}$	—	$\lambda = 350$
$t = 3$	1st	2nd	L	L	
$\mathbf{x}^{(3)}$	650	1100	0	0	$\mathbf{c} \cdot \mathbf{x}^{(3)} = 17{,}700$
	\multicolumn{4}{c}{$\mathbf{B}^{-1} = \begin{pmatrix} -1 & 0.5 \\ 2 & -0.5 \end{pmatrix}, \quad \mathbf{v} = \begin{pmatrix} 6 \\ 1.5 \end{pmatrix}$}	
$\bar{c}_j$	0	0	−6	−1.5	"optimal"

EXERCISES

5-1 Consider the linear constraints

$$-w_1 + w_2 \le 1$$
$$w_2 \le 3$$
$$w_1, w_2 \ge 0$$

(a) Sketch the feasible space in a 2-dimensional plot.

(b) ⊘ Determine geometrically whether each of the following solutions is infeasible, boundary, extreme, and/or interior: $\mathbf{w}^{(1)} = (2, 3)$, $\mathbf{w}^{(2)} = (0, 3)$, $\mathbf{w}^{(3)} = (2, 1)$, $\mathbf{w}^{(4)} = (3, 3)$.

(c) ⊘ For those points of part (b) that are feasible, demonstrate algebraically whether they are boundary or interior.

(d) ◻ Determine for each of the points in part (b) whether a suitable (nonconstant) objective function could make the point optimal or uniquely optimal. Explain.

5-2 Do Exercise 5-1 for the LP

$$3w_1 + 5w_2 \leq 15$$
$$5w_1 + 3w_2 \leq 15$$
$$w_1, w_2 \geq 0$$

and points $\mathbf{w}^{(1)} = (0,0)$, $\mathbf{w}^{(2)} = (1,1)$, $\mathbf{w}^{(3)} = (2,0)$, and $\mathbf{w}^{(4)} = (3,3)$.

5-3 Place each of the following LPs in standard form and identify the corresponding $\mathbf{A}$, $\mathbf{b}$, and $\mathbf{c}$ of definition $\boxed{5.12}$.

(a) ◻   min $4x_1 + 2x_2 - 33x_3$
     s.t. $x_1 - 4x_3 + x_3 \leq 12$
            $9x_1 + 6x_3 = 15$
            $-5x_1 + 9x_2 \geq 3$
            $x_1, x_2, x_3 \geq 0$

(b)   max $45x_1 + 15x_3$
     s.t. $4x_1 - 2x_2 + 9x_3 = 22$
            $-2x_1 + 5x_2 - x_3 \geq 1$
            $x_1 - x_2 \leq 5$
            $x_1, x_2, x_3 \geq 0$

(c) ◻   max $15(x_1 + 2x_2) + 11(x_2 - x_3)$
     s.t. $3x_1 \geq x_1 + x_2 + x_3$
            $0 \leq x_j \leq 3$ $j = 1, \ldots, 3$

(d)   min $-53x_1 + 33(x_1 + 3x_3)$
     s.t. $x_j + 1 \leq x_{j+1}$ $j = 1, 2$
            $\sum_{j=1}^{3} x_j = 10$
            $x_j \geq 0$ $j = 1, \ldots, 3$

(e) ◻   min $2x_1 + x_2 - 4x_3$
     s.t. $x_1 - x_2 - 5x_3 \leq 10$
            $3x_2 + 9x_1 = -6$
            $x_1 \geq 0, x_3 \leq 0$

(f)   max $4x_1 - x_2$
     s.t. $-4x_1 - x_2 + 7x_3 \geq 9$
            $-x_1 - x_2 + 3x_3 \leq 14$
            $x_1 \leq 0, x_3 \geq 0$

5-4 Consider the linear constraints

$$-y_1 + y_2 \leq 2$$
$$5y_1 \leq 10$$
$$y_1, y_2 \geq 0$$

(a) Sketch the feasible set in a 2-dimensional plot.
(b) ◻ Add slacks y_3 and y_4 to place constraints in LP standard form.

(c) ◻ Determine whether columns of standard form corresponding to each of the following sets of variables form a basis: $\{y_1, y_2\}$, $\{y_2, y_3\}$, $\{y_3, y_4\}$, $\{y_1, y_4\}$, $\{y_3\}$, $\{y_1, y_2, y_4\}$.

(d) ◻ For each set that does form a basis in part (c), determine the corresponding basic solution and classify it as feasible on infeasible.

(e) ◻ Identify each solution of part (d) on your plot of part (a), and comment on the connection between basic feasible solutions and extreme points.

5-5 Do Exercise 5-4 for the LP

$$y_1 + 2y_2 \leq 6$$
$$y_2 \leq 2$$
$$y_1, y_2 \geq 0$$

and possible basic sets $\{y_1, y_2\}$, $\{y_2, y_3\}$, $\{y_1\}$, $\{y_2, y_4\}$, $\{y_2, y_3, y_4\}$, $\{y_1, y_3\}$.

5-6 Write all conditions that a feasible direction $\Delta\mathbf{w}$ must satisfy, at the solution $\mathbf{w}$ indicated, to each of the following standard-form systems of LP constraints.

(a) ◻ $5w_1 + 1w_2 - 1w_3 = 9$ at $\mathbf{w} = (2, 0, 1)$
         $3w_1 - 4w_2 + 8w_3 = 14$
         $w_1, w_2, w_3 \geq 0$

(b) $4w_1 - 2w_2 + 5w_3 = 43$ at $\mathbf{w} = (0, 1, 9)$
       $4w_1 + 2w_2 - 3w_3 = -25$
       $w_1, w_2, w_3 \geq 0$

5-7 Following is a maximizing, standard-form linear program and a classification of variables as basic and nonbasic.

	x_1	x_2	x_3	x_4	
max $\mathbf{c}$	10	1	0	0	$\mathbf{b}$
	-1	1	4	21	13
	2	6	0	-2	2
	B	N	B	N	

(a) ◻ Compute the current basic solution.
(b) ◻ Compute all simplex directions available at the current basis.
(c) ◻ Verify that all your simplex directions are feasible at the current basic solution.
(d) ◻ Determine whether each of the simplex directions is improving.
(e) ◻ Regardless of whether it improves, determine the maximum step that preserves feasibility in each simplex direction and the new basis that would result after such a step.

5-8 Do Exercise 5-7 for minimizing the standard-form LP

	x_1	x_2	x_3	x_4	
min **c**	8	−5	0	1	**b**
	13	2	3	1	7
	−4	1	0	−1	−1
	N	N	B	B	

5-9 Consider the linear program

$$\max \quad 3z_1 + z_2$$
$$\text{s.t.} \quad -2z_1 + z_2 \le 2$$
$$z_1 + z_2 \le 6$$
$$z_1 \le 4$$
$$z_1, z_2 \ge 0$$

(a) ⊘ Solve the problem graphically.

(b) ⊘ Add slacks z_3, z_4, and z_5 to place the model in standard form.

(c) ⊘ Apply rudimentary simplex Algorithm 5A to compute an optimal solution to your standard form starting with all slacks basic.

(d) ⊘ Plot your progress in part (c) on the graph of part (a).

5-10 Do Exercise 5-9 for the LP

$$\max \quad 2z_1 + 5z_2$$
$$\text{s.t.} \quad 3z_1 + 2z_2 \le 18$$
$$z_1 \le 5$$
$$z_2 \le 3$$
$$z_1, z_2 \ge 0$$

5-11 The following plot shows several feasible points in a linear program and contours of its objective function.

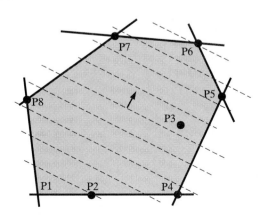

Determine whether each of the following sequences of solutions could have been one followed by the simplex algorithm applied to the corresponding LP standard form.

(a) ⊘ P1,P8,P7,P6

(b) P2,P4,P5,P6

(c) ⊘ P1,P3,P6

(d) P8,P7,P6

(e) ⊘ P1,P7,P6

(f) P4,P1,P8,P7,P6

5-12 Construct the simplex dictionary form 5.28 corresponding to each of the following.

(a) ⊘ The model and basis shown in Exercise 5-7

(b) The model and basis shown in Exercise 5-8

5-13 Rudimentary simplex Algorithm 5A is being applied to optimize a linear program with objective function

$$\min \quad 3w_1 + 11w_2 - 8w_3$$

Determine whether each of the following simplex directions for w_4 leads to a conclusion that the given LP in unbounded.

(a) ⊘ $\Delta \mathbf{w} = (1, 0, -4, 1)$

(b) $\Delta \mathbf{w} = (1, 3, 0, 1)$

(c) ⊘ $\Delta \mathbf{w} = (1, 0, 3, 1)$

(d) $\Delta \mathbf{w} = (-1, 1, -2, 1)$

5-14 Consider the linear program

$$\max \quad 4y_1 + 5y_2$$
$$\text{s.t.} \quad -y_1 + y_2 \le 4$$
$$y_1 - y_2 \le 10$$
$$y_1, y_2 \ge 0$$

(a) Show graphically that the model is unbounded.

(b) ⊘ Add slacks y_3 and y_4 to place the model in standard form.

(c) ⊘ Starting with all slacks basic, apply rudimentary simplex Algorithm 5A to establish that the original model is unbounded.

5-15 Do Exercise 5-14 for the LP

$$\min \quad -10y_1 + y_2$$
$$\text{s.t.} \quad -5y_1 + 3y_2 \le 15$$
$$3y_1 - 5y_2 \le 8$$
$$y_1, y_2 \ge 0$$

5-16 Setup each of the following to begin Phase I of two-phase simplex Algorithm 5B. Also indicate the basic variables of the initial Phase I solution.

(a) ⊠ max $2w_1 + w_2 + 9w_3$
    s.t. $w_1 + w_2 \leq 18$
          $-2w_1 + w_3 = -2$
          $3w_2 + 5w_3 \geq 15$
          $w_1, w_2, w_3 \geq 0$

(b) min $7w_1 - 9w_2 + 4w_3$
     s.t. $w_1 - 18w_2 \leq 9$
           $w_3 + w_4 \geq 14$
           $w_1 + w_2 - 2w_3 - 3w_4 = 1$
           $w_1, w_2, w_3, w_4 \geq 0$

5-17 Setup each of the models in Exercise 5-16 to begin a big-M solution using rudimentary simplex Algorithm 5A. Also indicate the basic variables of the initial solution.

5-18 Consider the linear program

       max $9y_1 + y_2$
       s.t. $-2y_1 + y_2 \geq 2$
             $y_2 \leq 1$
             $y_1, y_2 \geq 0$

(a) Show graphically that the model is infeasible.
(b) ⊠ Add slacks and artificials $y_3, \ldots, y_5$ to setup the model for Phase I of Algorithm 5B.
(c) ⊠ Apply rudimentary simplex Algorithm 5A to this Phase I problem to establish that the original model is infeasible.

5-19 Do Exercise 5-18 for the linear program

       min $2y_1 + 8y_2$
       s.t. $y_1 + y_2 \leq 2$
             $y_1 \geq 3$
             $y_1, y_2 \geq 0$

5-20 Assuming that step size $\lambda > 0$ at every step, compute a finite bound on the number of iterations of Algorithm 5A for each of the following standard-form linear programs.

(a) ⊠ The model in Exercise 5-7
(b) The model in Exercise 5-8
(c) ⊠ A model with 1150 main constraints and 2340 variables
(d) A model with 345 main constraints and 11,236 variables

5-21 Rudimentary simplex Algorithm 5A is being applied to a standard-form linear program with variables $x_1, \ldots, x_5$. Determine whether each of the following basic solutions is degenerate for the given basic variable set.

(a) ⊠ $B = \{x_1, x_2, x_3\}$, $\mathbf{x} = (1, 0, 5, 0, 0)$
(b) $B = \{x_3, x_4, x_5\}$, $\mathbf{x} = (0, 0, 1, 4, 9)$
(c) ⊠ $B = \{x_1, x_3, x_5\}$, $\mathbf{x} = (1, 0, 5, 0, 8)$
(d) $B = \{x_1, x_2, x_4\}$, $\mathbf{x} = (1, 0, 0, 0, 0)$

5-22 Consider the linear program

       max $x_1 + x_2$
       s.t. $x_1 + x_2 \leq 9$
             $-2x_1 + x_2 \leq 0$
             $x_1 - 2x_2 \leq 0$
             $x_1, x_2 \geq 0$

(a) ⊠ Solve the problem graphically.
(b) ⊠ Add slacks $x_3, \ldots, x_5$ to place the model in standard form.
(c) ⊠ Apply rudimentary simplex Algorithm 5A to compute an optimal solution to your standard form starting with all slacks basic.
(d) ⊠ Plot your progress in part (c) on the graph of part (a).
(e) How can the algorithm be making progress when $\lambda = 0$ if some moves of part (c) left the solution unchanged? Explain.

5-23 Do Exercise 5-22 for the LP

       max x_1
       s.t. $6x_1 + 3x_2 \leq 18$
             $12x_1 - 3x_2 \leq 0$
             $x_1, x_2 \geq 0$

5-24 Return to the LP of Exercise 5-7.

(a) ⊠ Compute the basis matrix inverse corresponding to the basic variables indicated.
(b) ⊠ Compute the corresponding pricing vector of $\boxed{5.45}$.
(c) ⊠ Without generating the implied simplex directions, use your pricing vector to determine whether each of them will be improving.
(d) ⊠ For each improving simplex direction in part (c), generate the corresponding matrix $\mathbf{E}$ of $\boxed{5.43}$, and compute the next basis matrix inverse.

5-25 Do Exercise 5-24 for the LP of Exercise 5-8.

5-26 Do Exercise 5-9 using revised simplex Algorithm 5C in part (c).

5-27 Do Exercise 5-10 using revised simplex Algorithm 5C in part (c).

5-28 Suppose lower- and upper-bounded simplex Algorithm 5D is being applied to a problem with objective function

$$\max \ 3x_1 - 4x_2 + x_3 - 4x_4 + 10x_5$$

3 main constraints, and bounds

$$0 \le x_j \le 5 \qquad j = 1, \dots, 5$$

For each of the following current basic solutions **x** and corresponding simplex directions $\Delta \mathbf{x}$, then determine whether the appropriate move of $\pm \Delta \mathbf{x}$ is improving. Also compute the maximum step λ that could be applied without losing feasibility and the basis status of variables that would result after such a step. Take the current basic variables to be those strictly between lower and upper bounds.

(a) ⊘ $\mathbf{x} = (2, 2, 4, 0, 5)$, $\Delta \mathbf{x} = (1, -1, 0, 0, 1)$ for x_5

(b) $\mathbf{x} = (5, 0, 2, 3, 2)$, $\Delta \mathbf{x} = (0, 1, \frac{1}{10}, -\frac{1}{5}, \frac{1}{3})$ for x_2

(c) ⊘ $\mathbf{x} = (0, 1, 0, 4, 2)$, $\Delta \mathbf{x} = (0, 0, 1, -\frac{2}{5}, \frac{2}{5})$ for x_3

(d) $\mathbf{x} = (5, 5, 1, 3, 1)$, $\Delta \mathbf{x} = (1, 0, 0, 4, 1)$ for x_1

5-29 Consider the linear program

$$\begin{aligned} \min \quad & 5z_1 + 6z_2 \\ \text{s.t.} \quad & z_1 + z_2 \ge 3 \\ & 3z_1 + 2z_2 \ge 8 \\ & 0 \le z_1 \le 6 \\ & 0 \le z_2 \le 5 \end{aligned}$$

(a) ⊘ Solve the problem graphically.

(b) ⊘ Add slacks z_3 and z_4 to place the model in standard form for a lower- and upper-bounded simplex.

(c) ⊘ Apply lower- and upper-bounded simplex Algorithm 5D to compute an optimal solution to your standard form starting with all slacks basic and original variables nonbasic at their upper bounds.

(d) ⊘ Plot your progress in part (c) on the graph of part (a).

5-30 Do Exercise 5-29 on the LP

$$\begin{aligned} \max \quad & 6z_1 + 8z_2 \\ \text{s.t.} \quad & z_1 + 3z_2 \le 10 \\ & z_1 + z_2 \le 5 \\ & 0 \le z_1 \le 4 \\ & 0 \le z_2 \le 3 \end{aligned}$$

Start with all original variables nonbasic lower-bounded.

SUGGESTED READING

Bazaraa, Mokhtar S., John J. Jarvis, and Hanif Sherali (1990), *Linear Programming and Network Flows*, Wiley, New York.

Chvátal, Vašek (1983), *Linear Programming*, W. H. Freeman, San Francisco.

CHAPTER 6

Interior Point Methods for Linear Programming

· ·

Although Chapter 5's simplex method remains the most widely used algorithm for solving linear programs, a very different strategy emerged in operations research practice at the end of the 1980s. New **interior point methods** still follow the improving search paradigm for linear programs, but they employ moves quite different from those in the simplex method. Instead of staying on the boundary of the feasible region and passing from extreme point to extreme point, interior point methods proceed directly across the interior.

Much more effort turns out to be required per move with interior point methods, but the number of moves decreases dramatically. In many large LPs, the result is a substantially shorter total solution time than any obtained so far with simplex.

The first commercial interior point method for linear programming was N. Karmarkar's **projective transformation** procedure, and developments have continued to this day. All methods are relatively complex mathematically, with many details beyond the scope of an introductory book.

In this chapter we seek to open up the exciting new interior point technology by concentrating on the broad strategies and intuitions behind the popular **affine scaling** and **log barrier** variants. In the final section we sketch some of the more advanced issues underlying commercial implementations. All development assumes reader familiarity with the search fundamentals of Chapter 3 and the LP conventions of Section 5.1.

6.1 SEARCHING THROUGH THE INTERIOR

Any linear programming search method allowed to pass through the interior of the feasible region poses a variety of new challenges. Before developing specific algorithms, we introduce some of the key issues.

EXAMPLE 6.1: FRANNIE'S FIREWOOD

To illustrate interior point computations, we need an example so small (and trivial) that it can be displayed graphically in a variety of ways. Frannie's Firewood problem will serve.

Each year Frannie sells up to 3 cords of firewood from her small woods. One potential customer has offered her $90 per half-cord and another, $150 per full cord. Our concern is how much Frannie should sell to each customer to maximize income, assuming that each will buy as much as he or she can.

To begin, we need a decision variable for each customer. Define

$$x_1 \triangleq \text{number of half-cords sold to customer 1}$$

$$x_2 \triangleq \text{number of cords sold to customer 2}$$

Then Frannie's problem can be modeled as the linear program

$$
\begin{aligned}
\max \quad & 90x_1 + 150x_2 \\
\text{s.t.} \quad & \tfrac{1}{2}x_1 + x_2 \le 3 \\
& x_1, x_2 \ge 0
\end{aligned}
\tag{6.1}
$$

Figure 6.1 solves model (6.1) graphically. The unique solution sells all 3 cords to the first customer, making $x_1^* = 6, x_2^* = 0$.

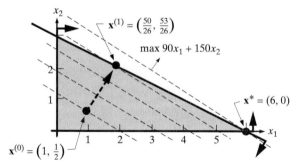

FIGURE 6.1   Graphic Solution of Frannie's Firewood Example

Interior Points

Recall (definition $\boxed{5.2}$, Section 5.1) that a feasible solution to a linear program is a **boundary point** if at least one inequality constraint of the model that can be strict for some feasible solutions is satisfied as an equality at the given point. A feasible solution is an **interior point** if no such inequalities are active. For example, solution $\mathbf{x}^{(0)} = (1, \tfrac{1}{2})$ of Figure 6.1 is an interior point because it satisfies all three constraints of model (6.1) as strict inequalities.

Objective as a Move Direction

There is one enormous convenience in searching from an interior point such as $\mathbf{x}^{(0)}$ of Figure 6.1 when all constraints are inequalities (we deal with equalities shortly). With no constraints active, every direction is **feasible**; that is, a small step in any direction retains feasibility (see Section 3.2). Thus our only consideration in picking the next move is to find a direction **improving** the objective function.

What direction improves the objective most rapidly? For linear programs it is easy to see that we should move with the gradient or objective coefficient vector as in principle $\boxed{3.23}$ (Section 3.3).

> $\boxed{6.1}$ The move direction of most rapid improvement for linear objective $\max \mathbf{c} \cdot \mathbf{x} = \sum_j c_j x_j$ is the objective function vector $\Delta \mathbf{x} = \mathbf{c}$. For a model to $\min \mathbf{c} \cdot \mathbf{x}$, it is $\Delta \mathbf{x} = -\mathbf{c}$.

For example, in the maximize Frannie model of Figure 6.1, we prefer $\Delta \mathbf{x} = \mathbf{c} = (90, 150)$ at $\mathbf{x}^{(0)}$. This direction runs exactly perpendicular to contours of the objective function. No alternative improves the objective faster.

SAMPLE EXERCISE 6.1: USING THE OBJECTIVE AS A DIRECTION

Each of the following is the objective function of a linear program with 3 decision variables. Determine for each the direction of most rapid objective improvement.

(a) $\min 4x_1 - 19x_2 + x_3$

(b) $\max -2x_1 - x_2 + 79x_3$

Analysis: We apply principle $\boxed{6.1}$.

(a) For a minimize objective the direction of steepest improvement is
$$\Delta \mathbf{x} = -\mathbf{c} = (-4, 19, -1)$$

(b) For a maximize objective the direction of steepest improvement is
$$\Delta \mathbf{x} = \mathbf{c} = (-2, -1, 79)$$

Boundary Strategy of Interior Point Methods

Figure 6.1 shows the effect of using cost direction $\Delta \mathbf{x} = (90, 150)$ at $\mathbf{x}^{(0)}$. A maximum feasible step of $\lambda = \frac{2}{195}$ would bring us to
$$\mathbf{x}^{(1)} = \mathbf{x}^{(0)} + \lambda \Delta \mathbf{x}^{(1)}$$
$$= (1, \tfrac{1}{2}) + \tfrac{2}{195}(90, 150)$$
$$= (\tfrac{50}{26}, \tfrac{53}{26})$$

Very rapid progress toward an optimum is achieved in a single move.

Notice that we would have to deal with the boundary at $\mathbf{x}^{(1)}$. Constraint
$$\tfrac{1}{2}x_1 + x_2 \leq 3$$

is now active, and we discovered in Chapter 3 (principle $\boxed{3.25}$, Section 3.3) that feasible directions $\Delta \mathbf{x}$ at $\mathbf{x}^{(1)}$ must preserve this inequality by satisfying
$$\tfrac{1}{2}\Delta x_1 + \Delta x_2 \leq 0$$

Such newly active constraints destroy the convenience of moves in the interior. That is why interior point algorithms stop short of the boundary as in the typical sequence plotted in Figure 6.2. Partial steps along suitable variations of the cost direction in principle $\boxed{6.1}$ continue progress while avoiding the boundary.

Of course, optimal solutions to linear programs lie along the boundary of the feasible region (principle $\boxed{5.4}$, Section 5.1). It cannot be avoided forever. The effectiveness of interior point methods depends on their keeping to the "middle" of the feasible region until an optimal solution is reached.

> $\boxed{6.2}$ Interior point algorithms begin at and move through a sequence of interior feasible solutions, converging to the boundary of the feasible region only at an optimal solution.

SAMPLE EXERCISE 6.2: IDENTIFYING INTERIOR POINT TRAJECTORIES

The following figure shows the Top Brass Trophy example from Section 5.1:

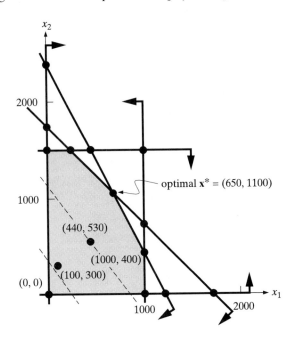

Determine whether each of the following sequences of solutions could have resulted from an interior point search.

(a) (0,0), (440,530), (650,1100)

(b) (100,300), (440,530), (650,1100)

(c) (100,300), (1000,400), (650,1100)

(d) (0,0), (1000,0), (1000,400), (650,1100)

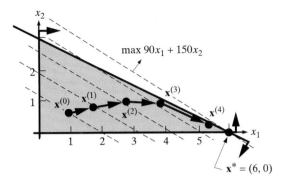

FIGURE 6.2 Typical Interior Point Search of Frannie's Firewood Example

Analysis: The sequence of solutions visited by an interior point search should conform to principle 6.2 .

(a) This sequence is not appropriate because it begins at boundary point (0,0).

(b) This could be the sequence of an interior point search. It starts at one interior solution, passes to another, and reaches the boundary only at the optimum.

(c) This sequence starts in the interior but goes to the boundary before optimality at (1000,400). It could not result from an interior point algorithm.

(d) This is the simplex algorithm extreme-point sequence of Figure 5.5. It never enters the interior.

Interior in LP Standard Form

In Section 5.1 we showed how every linear program can be placed in **standard form**:

$$\min \text{ (or max)} \quad \mathbf{c} \cdot \mathbf{x}$$
$$\text{s.t.} \quad \mathbf{Ax} = \mathbf{b} \qquad (6.2)$$
$$\mathbf{x} \geq \mathbf{0}$$

Any main inequalities are converted to equalities by adding slack variables, and original variables are transformed until each is subject to a nonnegativity constraint. For example, we can place the Frannie model (6.1) in standard form by adding slack variable x_3 in the main constraint to obtain

$$\max \quad 90x_1 + 150x_2$$
$$\text{s.t.} \quad \tfrac{1}{2}x_1 + x_2 + x_3 = 3 \qquad (6.3)$$
$$x_1, x_2, x_3 \geq 0$$

Chapter 5 placed LPs in form (6.2) to make it easier to perform simplex algorithm computations, but standard form is equally convenient for interior point search. With all inequalities reduced to nonnegativity constraints, it is easy to check whether a given solution lies in the interior of the feasible region.

6.3 | A feasible solution for a linear program in standard form is an interior point if every component of the solution that can be positive in any feasible solution is strictly positive in the given point.

For example, point $\mathbf{x}^{(0)}$ of Figure 6.2 corresponds to $\mathbf{x}^{(0)} = (1, \frac{1}{2}, 2)$ in standard form (6.3). In accord with conditions 6.3 , the vector is positive in every component.

SAMPLE EXERCISE 6.3: IDENTIFYING STANDARD-FORM INTERIOR POINTS

Consider the standard-form linear program

$$\min \quad 5x_1 - 2x_3 + 8x_4$$
$$\text{s.t.} \quad 2x_1 + 3x_2 - x_3 = 10$$
$$6x_1 - 2x_4 \quad = 12$$
$$x_1, x_2, x_3, x_4 \geq 0$$

which does have strictly positive feasible solutions. Determine whether each of the following solutions corresponds to an interior point.
(a) $\mathbf{x}^{(1)} = (8, 0, 6, 18)$ (b) $\mathbf{x}^{(2)} = (4, 1, 1, 6)$ (c) $\mathbf{x}^{(3)} = (3, 3, 1, 6)$

Analysis:

(a) Solution $\mathbf{x}^{(1)}$ cannot be interior, because a component $x_2^{(1)} = 0$ is not strictly positive.

(b) Solution $\mathbf{x}^{(2)}$ is positive in every component. Also,

$$2x_1^{(2)} + 3x_2^{(2)} - x_3^{(2)} = 2(4) + 3(1) - (1) = 10$$
$$6x_1^{(2)} - 2x_4^{(2)} \quad = 6(4) - 2(6) \quad = 12$$

which establishes the point is feasible. Thus $\mathbf{x}^{(2)}$ is an interior point.

(c) Solution $\mathbf{x}^{(3)}$ is also positive in every component. However,

$$2x_1^{(3)} + 3x_2^{(3)} - x_3^{(3)} = 2(3) + 3(3) - (1) = 14 \neq 10$$
$$6x_1^{(3)} - 2x_4^{(3)} \quad = 6(3) - 2(6) \quad = 6 \neq 12$$

Thus the point is infeasible and so not interior.

Projecting to Deal with Equality Constraints

The ease of identifying interior points in standard form comes at a price. Many equality constraints are added to the system $\mathbf{Ax} = \mathbf{b}$ as main inequalities are converted, and equality constraints are active at every solution. Straightforward moves such as the objective vector directions of principle 6.1 must now be modified to preserve the equalities.

Chapter 3 (principle 3.25 , Section 3.3) established that a direction $\mathbf{\Delta x}$ preserves equality constraints $\mathbf{Ax} = \mathbf{b}$ if and only if the net effect on every constraint is zero.

6.4 | A move direction Δx is feasible for equality constraints $Ax = b$ if it satisfies $A\Delta x = 0$.

How can we find a Δx direction satisfying conditions 6.4 that approximates as nearly as possible the direction d we would really like to follow? Interior point algorithms often use some form of projection.

6.5 | The **projection** of a move vector d on a given system of equalities is a direction preserving those constraints and minimizing the total squared difference between its components and those of d.

Readers familiar with statistics may recognize this idea as the one used in **least squares** curve fitting.

Figure 6.3 illustrates for Frannie's firewood optimization form (6.3). The shaded triangle in this 3-dimensional plot is the set of feasible (x_1, x_2, x_3). The search begins at $x^{(0)} = (1, \frac{1}{2}, 2)$.

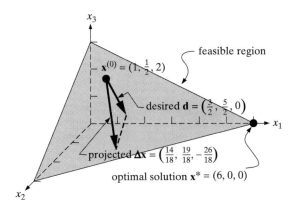

FIGURE 6.3 Projection in the Frannie's Firewood Example

To improve the objective function as quickly as possible, we would like to move parallel to standard-form cost vector $c = (90, 150, 0)$. Call the desired direction $d = \frac{1}{60}c = (\frac{3}{2}, \frac{5}{2}, 0)$ to keep it in the picture.

A feasible direction Δx must keep the next point in the feasible plane by satisfying 6.4, or

$$\tfrac{1}{2} \Delta x_1 + \Delta x_2 + \Delta x_3 = 0 \qquad (6.4)$$

But we would like it to be as much like d as possible. Figure 6.3 shows that the best choice is d's projection $\Delta x = (\frac{14}{18}, \frac{19}{18}, -\frac{26}{18})$. It is the closest to d in the sense that

$$(d_1 - \Delta x_1)^2 + (d_2 - \Delta x_2)^2 + (d_3 - \Delta x_3)^2$$

is minimized. It also satisfies feasibility condition (6.4), because

$$\tfrac{1}{2}\left(\tfrac{14}{18}\right) + \left(\tfrac{19}{18}\right) + \left(-\tfrac{26}{18}\right) = 0$$

Derivation of the projection computation that produced this move direction $\Delta \mathbf{x}$ is beyond the scope of this book. However, the formula for the required projection matrix $\mathbf{P}$ is well known:

> **6.6** The projection of direction $\mathbf{d}$ onto conditions $\mathbf{A}\Delta \mathbf{x} = \mathbf{0}$ preserving linear inequalities $\mathbf{A}\mathbf{x} = \mathbf{b}$ can be computed as
>
> $$\Delta \mathbf{x} = \mathbf{P}\mathbf{d}$$
>
> where **projection matrix**
>
> $$\mathbf{P} = \left(\mathbf{I} - \mathbf{A}^{\mathsf{T}}(\mathbf{A}\mathbf{A}^{\mathsf{T}})^{-1}\mathbf{A} \right)$$

Here $\mathbf{I}$ denotes an identity matrix and $\mathbf{A}^{\mathsf{T}}$ indicates the **transpose** of $\mathbf{A}$, which is obtained by swapping its rows for its columns. (Primer 6 reviews some properties for those who may require it.)

For our Frannie's Firewood example,

$$\mathbf{A} = (\tfrac{1}{2}, 1, 1), \qquad \mathbf{A}^{\mathsf{T}} = \begin{pmatrix} \tfrac{1}{2} \\ 1 \\ 1 \end{pmatrix}, \qquad \mathbf{A}\mathbf{A}^{\mathsf{T}} = \tfrac{9}{4}$$

Thus $\left(\mathbf{A}\mathbf{A}^{\mathsf{T}} \right)^{-1} = \tfrac{4}{9}$, and

$$\mathbf{A}^{\mathsf{T}} \left(\mathbf{A}\mathbf{A}^{\mathsf{T}} \right)^{-1} \mathbf{A} = \begin{pmatrix} \tfrac{1}{2} \\ 1 \\ 1 \end{pmatrix} (\tfrac{4}{9})(\tfrac{1}{2}, 1, 1) = \begin{pmatrix} \tfrac{1}{9} & \tfrac{2}{9} & \tfrac{2}{9} \\ \tfrac{2}{9} & \tfrac{4}{9} & \tfrac{4}{9} \\ \tfrac{2}{9} & \tfrac{4}{9} & \tfrac{4}{9} \end{pmatrix}$$

The needed projection matrix $\mathbf{P}$ then becomes

$$\mathbf{P} = \left(\mathbf{I} - \mathbf{A}^{\mathsf{T}}(\mathbf{A}\mathbf{A}^{\mathsf{T}})^{-1}\mathbf{A} \right)$$

$$= \begin{pmatrix} 1 & 0 & 0 \\ 0 & 1 & 0 \\ 0 & 0 & 1 \end{pmatrix} - \begin{pmatrix} \tfrac{1}{9} & \tfrac{2}{9} & \tfrac{2}{9} \\ \tfrac{2}{9} & \tfrac{4}{9} & \tfrac{4}{9} \\ \tfrac{2}{9} & \tfrac{4}{9} & \tfrac{4}{9} \end{pmatrix} = \begin{pmatrix} \tfrac{8}{9} & -\tfrac{2}{9} & -\tfrac{2}{9} \\ -\tfrac{2}{9} & \tfrac{5}{9} & -\tfrac{4}{9} \\ -\tfrac{2}{9} & -\tfrac{4}{9} & \tfrac{5}{9} \end{pmatrix}$$

Applying rule **6.6** for $\mathbf{d} = (\tfrac{3}{2}, \tfrac{5}{2}, 0)$ gives

$$\Delta \mathbf{x} = \mathbf{P}\mathbf{d}$$

$$= \begin{pmatrix} \tfrac{8}{9} & -\tfrac{2}{9} & -\tfrac{2}{9} \\ -\tfrac{2}{9} & \tfrac{5}{9} & -\tfrac{4}{9} \\ -\tfrac{2}{9} & -\tfrac{4}{9} & \tfrac{5}{9} \end{pmatrix} \begin{pmatrix} \tfrac{3}{2} \\ \tfrac{5}{2} \\ 0 \end{pmatrix} = \begin{pmatrix} \tfrac{14}{18} \\ \tfrac{19}{18} \\ -\tfrac{26}{18} \end{pmatrix} \tag{6.5}$$

PRIMER 6: MATRIX TRANSPOSES

The **transpose** of a matrix $\mathbf{M}$, which is denoted $\mathbf{M}^T$, has rows equal to the columns of $\mathbf{M}$. For example,

$$\mathbf{Q} = \begin{pmatrix} 2 & -\frac{1}{3} & 7 \\ 0 & 6 & 13 \end{pmatrix} \quad \text{has transpose} \quad \mathbf{Q}^T = \begin{pmatrix} 2 & 0 \\ -\frac{1}{3} & 6 \\ 7 & 13 \end{pmatrix}$$

A matrix is **symmetric** if transposing leaves it unchanged, and **nonsymmetric** otherwise. Thus with

$$\mathbf{R} = \begin{pmatrix} 3 & -1 \\ 6 & 0 \end{pmatrix} \quad \text{and} \quad \mathbf{S} = \begin{pmatrix} 3 & 1 & 8 \\ 1 & -11 & 0 \\ 8 & 0 & 25 \end{pmatrix}$$

$\mathbf{R}$ is nonsymmetric, because $\mathbf{R} \neq \mathbf{R}^T$, but $\mathbf{S} = \mathbf{S}^T$ is symmetric.

When matrices can be multiplied (see Primer 3, Section 5.1), the transpose of the product is the product of the transposes after the sequence of multiplication is reversed. For example, with the $\mathbf{Q}$ and $\mathbf{R}$ above,

$$(\mathbf{R}\mathbf{Q})^T = \left(\begin{pmatrix} 3 & -1 \\ 6 & 0 \end{pmatrix} \begin{pmatrix} 2 & -\frac{1}{3} & 7 \\ 0 & 6 & 13 \end{pmatrix} \right)^T = \begin{pmatrix} 6 & -7 & 8 \\ 12 & -2 & 42 \end{pmatrix}^T$$

$$= \begin{pmatrix} 6 & 12 \\ -7 & -2 \\ 8 & 42 \end{pmatrix} = \begin{pmatrix} 2 & 0 \\ -\frac{1}{3} & 6 \\ 7 & 13 \end{pmatrix} \begin{pmatrix} 3 & 6 \\ -1 & 0 \end{pmatrix} = \mathbf{Q}^T\mathbf{R}^T$$

When a matrix has an inverse (see Primer 5, Section 5.8), the inverse of its transpose is the transpose of its inverse. Thus

$$(\mathbf{R}^T)^{-1} = \begin{pmatrix} 3 & 6 \\ -1 & 0 \end{pmatrix}^{-1} = \begin{pmatrix} 0 & -1 \\ \frac{1}{6} & \frac{1}{2} \end{pmatrix} = \begin{pmatrix} 0 & \frac{1}{6} \\ -1 & \frac{1}{2} \end{pmatrix}^T = (\mathbf{R}^{-1})^T$$

SAMPLE EXERCISE 6.4: PROJECTING TO PRESERVE EQUALITY CONSTRAINTS

Consider the standard-form linear program

$$\begin{array}{llll} \max & 5x_1 & +7x_2 & +9x_3 \\ \text{s.t.} & 1x_1 & & -1x_3 & = -1 \\ & & +2x_2 & +1x_3 & = 5 \\ & x_1, x_2, x_3 \geq 0 \end{array}$$

(a) Determine the direction $\mathbf{d}$ of most rapid objective function improvement at interior point $\mathbf{x} = (2, 1, 3)$.

(b) Project that vector $\mathbf{d}$ to find the nearest direction $\Delta\mathbf{x}$ that preserves the main equality constraints of the LP.

(c) Verify that your $\Delta\mathbf{x}$ satisfies all requirements for a feasible move direction at $\mathbf{x}$.

Analysis:

(a) Applying principle 6.1 , the direction of steepest improvement is

$$\mathbf{d} = \text{objective function vector} = (5, 7, 9)$$

(b) To compute the projection, we apply formula 6.6 . Here

$$\mathbf{A} = \begin{pmatrix} 1 & 0 & -1 \\ 0 & 2 & 1 \end{pmatrix} \quad \text{and} \quad \mathbf{A}^\mathrm{T} = \begin{pmatrix} 1 & 0 \\ 0 & 2 \\ -1 & 1 \end{pmatrix}$$

Thus

$$\mathbf{A}\mathbf{A}^\mathrm{T} = \begin{pmatrix} 2 & -1 \\ -1 & 5 \end{pmatrix} \quad \text{with inverse} \quad (\mathbf{A}\mathbf{A}^\mathrm{T})^{-1} = \begin{pmatrix} \frac{5}{9} & \frac{1}{9} \\ \frac{1}{9} & \frac{2}{9} \end{pmatrix}$$

Continuing yields

$$\mathbf{A}^\mathrm{T}(\mathbf{A}\mathbf{A}^\mathrm{T})^{-1}\mathbf{A} = \begin{pmatrix} 1 & 0 \\ 0 & 2 \\ -1 & 1 \end{pmatrix} \begin{pmatrix} \frac{5}{9} & \frac{1}{9} \\ \frac{1}{9} & \frac{2}{9} \end{pmatrix} \begin{pmatrix} 1 & 0 & -1 \\ 0 & 2 & 1 \end{pmatrix}$$

$$= \begin{pmatrix} \frac{5}{9} & \frac{2}{9} & -\frac{4}{9} \\ \frac{2}{9} & \frac{8}{9} & \frac{2}{9} \\ -\frac{4}{9} & \frac{2}{9} & \frac{5}{9} \end{pmatrix}$$

and

$$\mathbf{P} = \left[\mathbf{I} - \mathbf{A}^\mathrm{T}(\mathbf{A}\mathbf{A}^\mathrm{T})^{-1}\mathbf{A} \right]$$

$$= \left[\begin{pmatrix} 1 & 0 & 0 \\ 0 & 1 & 0 \\ 0 & 0 & 1 \end{pmatrix} - \begin{pmatrix} \frac{5}{9} & \frac{2}{9} & -\frac{4}{9} \\ \frac{2}{9} & \frac{8}{9} & \frac{2}{9} \\ -\frac{4}{9} & \frac{2}{9} & \frac{5}{9} \end{pmatrix} \right] = \begin{pmatrix} \frac{4}{9} & -\frac{2}{9} & \frac{4}{9} \\ -\frac{2}{9} & \frac{1}{9} & -\frac{2}{9} \\ \frac{4}{9} & -\frac{2}{9} & \frac{4}{9} \end{pmatrix}$$

We conclude

$$\Delta\mathbf{x} = \mathbf{P}\mathbf{d} = \begin{pmatrix} \frac{4}{9} & -\frac{2}{9} & \frac{4}{9} \\ -\frac{2}{9} & \frac{1}{9} & -\frac{2}{9} \\ \frac{4}{9} & -\frac{2}{9} & \frac{4}{9} \end{pmatrix} \begin{pmatrix} 5 \\ 7 \\ 9 \end{pmatrix} = \begin{pmatrix} \frac{14}{3} \\ -\frac{7}{3} \\ \frac{14}{3} \end{pmatrix}$$

(c) Checking net change zero conditions 6.4 , we see that

$$\mathbf{A}\Delta\mathbf{x} = \begin{pmatrix} 1 & 0 & -1 \\ 0 & 2 & 1 \end{pmatrix} \begin{pmatrix} \frac{14}{3} \\ -\frac{7}{3} \\ \frac{14}{3} \end{pmatrix} = \begin{pmatrix} 0 \\ 0 \end{pmatrix}$$

With no inequalities (nonnegativities) active at **x**, this is all that is required for $\Delta\mathbf{x}$ to be feasible.

Improvement with Projected Directions

Projection principle 6.6 yields a feasible direction $\Delta\mathbf{x}$ at any standard-form interior point. But does the direction remain improving? For example, we know that the direction $\mathbf{d} = \left(\frac{3}{2}, \frac{5}{2}, 0\right)$ used in Frannie's Firewood computation (6.5) is improving because it parallels the maximize objective function vector $\mathbf{c} = (90, 150, 0)$. Projection would do little good if it achieved feasibility of $\Delta\mathbf{x}$ at the loss of this improving property.

Fortunately, we can show that improvement is always preserved in projecting objective function vectors.

> 6.7 The projection $\Delta\mathbf{x} = \mathbf{Pc}$ of (nonzero) objective function vector $\mathbf{c}$ onto equality constraints $\mathbf{Ax} = \mathbf{b}$ is an improving direction at every $\mathbf{x}$.

For example, the Frannie's Firewood direction $\Delta\mathbf{x}$ of (6.5) satisfies gradient improvement conditions 3.21 and 3.22 of Section 3.3 because

$$\mathbf{c} \cdot \Delta\mathbf{x} = (90, 150, 0) \cdot \left(\tfrac{14}{18}, \tfrac{19}{18}, -\tfrac{26}{18}\right)$$

$$= \tfrac{685}{3} > 0$$

To see that this will always be true, we need to make a simple observation about projections. Computation 6.6 finds the nearest vector $\Delta\mathbf{x}$ to direction $\mathbf{d}$ that satisfies $\mathbf{A}\Delta\mathbf{x} = \mathbf{0}$. Thus if $\mathbf{d}$ already satisfies $\mathbf{Ad} = \mathbf{0}$, the nearest such direction must be $\Delta\mathbf{x} = \mathbf{d}$ itself. That is, re-projection of a $\Delta\mathbf{x}$ that has already been projected must leave $\Delta\mathbf{x}$ unchanged, or in symbols,

$$\text{projection}(\mathbf{d}) = \mathbf{Pd}$$

$$= \mathbf{PPd} \tag{6.6}$$

$$= \text{projection}[\text{projection}(\mathbf{d})]$$

Projection matrices are also known to be symmetric ($\mathbf{P} = \mathbf{P}^{\mathsf{T}}$). Combining with property (6.6) applied to maximize objective function vector $\mathbf{c}$, we can see that the corresponding $\Delta\mathbf{x}$ of principle 6.6 has

$$\mathbf{c} \cdot \Delta\mathbf{x} = \mathbf{c}^{\mathsf{T}}\mathbf{Pc}$$

$$= \mathbf{c}^{\mathsf{T}}\mathbf{PPc}$$

$$= \mathbf{c}^{\mathsf{T}}\mathbf{P}^{\mathsf{T}}\mathbf{Pc}$$

$$= (\mathbf{Pc})^{\mathsf{T}}(\mathbf{Pc})$$

$$= \Delta\mathbf{x}^{\mathsf{T}}\Delta\mathbf{x}$$

$$> 0$$

This is exactly what is required for improvement, and a similar analysis holds for minimize models.

SAMPLE EXERCISE 6.5: VERIFYING PROJECTED OBJECTIVE IMPROVEMENT

Return to the linear program of Sample Exercise 6.4 and its projection matrix $\mathbf{P}$ derived in part (b). Assuming that the objective function was changed to each of the following, compute the $\Delta\mathbf{x}$ value obtained by projecting the corresponding steepest improvement direction of principle 6.1 , and verify that the resulting move direction is improving.

(a) max $x_1 - x_2 + x_3$

(b) min $2x_1 + x_3$

Analysis:

(a) The corresponding steepest improvement direction is $\mathbf{c} = (1, -1, 1)$, so that the projected move direction would be

$$\Delta\mathbf{x} = \mathbf{Pc} = \begin{pmatrix} \frac{4}{9} & -\frac{2}{9} & \frac{4}{9} \\ -\frac{2}{9} & \frac{1}{9} & -\frac{2}{9} \\ \frac{4}{9} & -\frac{2}{9} & \frac{4}{9} \end{pmatrix} \begin{pmatrix} 1 \\ -1 \\ 1 \end{pmatrix} = \begin{pmatrix} \frac{10}{9} \\ -\frac{5}{9} \\ \frac{10}{9} \end{pmatrix}$$

Checking improvement yields

$$\mathbf{c} \cdot \Delta\mathbf{x} = (1, -1, 1) \cdot (\tfrac{10}{9}, -\tfrac{5}{9}, \tfrac{10}{9}) = \tfrac{25}{9} > 0$$

(b) For this minimize objective the steepest improvement direction is $-\mathbf{c} = (-2, 0, -1)$. Projecting gives

$$\Delta\mathbf{x} = \begin{pmatrix} \frac{4}{9} & -\frac{2}{9} & \frac{4}{9} \\ -\frac{2}{9} & \frac{1}{9} & -\frac{2}{9} \\ \frac{4}{9} & -\frac{2}{9} & \frac{4}{9} \end{pmatrix} \begin{pmatrix} -2 \\ 0 \\ -1 \end{pmatrix} = \begin{pmatrix} -\frac{4}{3} \\ \frac{2}{3} \\ -\frac{4}{3} \end{pmatrix}$$

Checking improvement yields

$$\mathbf{c} \cdot \Delta\mathbf{x} = (2, 0, 1) \cdot (-\tfrac{4}{3}, \tfrac{2}{3}, -\tfrac{4}{3}) = -3 < 0$$

6.2 SCALING WITH THE CURRENT SOLUTION

We have already seen that it is important for interior point algorithms to avoid the boundary of the feasible set (until optimality). One of the tools used by virtually all such procedures to keep in the "middle" of the feasible region is **scaling**— revising the units in which decision variables are expressed to place all solution components a comfortable number of units from the boundary. In this section we introduce the most common **affine** type of rescaling used in all algorithms of this chapter.

Affine Scaling

Affine scaling adopts the simplest possible strategy for keeping away from the boundary. The model is rescaled so that the transformed version of current solution $\mathbf{x}^{(t)}$ is equidistant from all inequality constraints.

Figure 6.4 shows the idea for our Frannie's Firewood example model (6.3). Part (a) depicts the original feasible space, with a current interior point solution

$\mathbf{x}^{(t)} = (3, \frac{1}{2}, 1)$. The rescaled version in part (b) converts to new variables

$$y_1 \triangleq \frac{x_1}{x_1^{(t)}} = \frac{x_1}{3}$$

$$y_2 \triangleq \frac{x_2}{x_2^{(t)}} = \frac{x_2}{\frac{1}{2}}$$

$$y_3 \triangleq \frac{x_3}{x_3^{(t)}} = \frac{x_3}{1}$$

After dividing by current (positive because interior) x-values in this way, the corresponding current solution $\mathbf{y}^{(t)} = (1, 1, 1)$—equidistant from all inequality (nonnegativity) constraints.

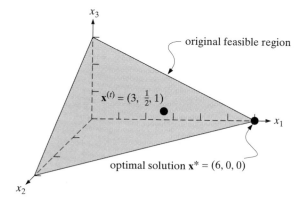

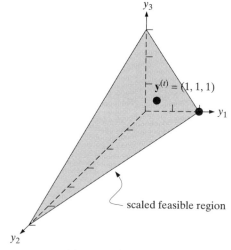

(a) *Original space*

(b) *Affine scaled space*

FIGURE 6.4   Affine Scaling of Frannie's Firewood Example

Diagonal Matrix Formalization of Affine Scaling

Although it may seem excessive at this stage, it will prove helpful to formalize affine scaling's simple notion of dividing components by the current solution in terms of **diagonal matrices**. In particular, we convert current solution vectors

$$\mathbf{x}^{(t)} = (x_1^{(t)}, x_2^{(t)}, \ldots, x_n^{(t)})$$

into the square matrix

$$\mathbf{X}_t = \begin{pmatrix} x_1^{(t)} & 0 & \cdots & 0 \\ 0 & x_2^{(t)} & \ddots & \vdots \\ \vdots & \ddots & \ddots & 0 \\ 0 & \cdots & 0 & x_n^{(t)} \end{pmatrix} \quad \text{with} \quad \mathbf{X}_t^{-1} = \begin{pmatrix} \dfrac{1}{x_1^{(t)}} & 0 & \cdots & 0 \\ 0 & \dfrac{1}{x_2^{(t)}} & \ddots & \vdots \\ \vdots & \ddots & \ddots & 0 \\ 0 & \cdots & 0 & \dfrac{1}{x_n^{(t)}} \end{pmatrix}$$

Affine scaling and unscaling can then be expressed in terms of multiplication by such matrices.

> **6.8** At current solution $\mathbf{x}^{(t)} > \mathbf{0}$, **affine scaling** transforms points $\mathbf{x}$ into $\mathbf{y}$ defined by
>
> $$\mathbf{y} = \mathbf{X}_t^{-1}\mathbf{x} \quad \text{or} \quad y_j = \frac{x_j}{x_j^{(t)}} \quad \text{for all } j$$
>
> where $\mathbf{X}_t$ denotes a square matrix with the components of $\mathbf{x}^{(t)}$ on its diagonal.

With $y_j = x_j/x_j^{(t)}$ the inverse affine scaling is equally easy to express.

> **6.9** At current solution $\mathbf{x}^{(t)} > \mathbf{0}$, the point $\mathbf{x}$ corresponding to affine-scaled solution $\mathbf{y}$ is
>
> $$\mathbf{x} = \mathbf{X}_t\mathbf{y} \quad \text{or} \quad x_j = (x_j^{(t)})y_j \quad \text{for all } j$$
>
> where $\mathbf{X}_t$ denotes a square matrix with the components of $\mathbf{x}^{(t)}$ on its diagonal.

For example, in the Frannie's firewood case of Figure 6.4,

$$\mathbf{x}^{(t)} = \begin{pmatrix} 3 \\ \frac{1}{2} \\ 1 \end{pmatrix} \quad \text{gives} \quad \mathbf{X}_t = \begin{pmatrix} 3 & 0 & 0 \\ 0 & \frac{1}{2} & 0 \\ 0 & 0 & 1 \end{pmatrix} \quad \text{with} \quad \mathbf{X}_t^{-1} = \begin{pmatrix} \frac{1}{3} & 0 & 0 \\ 0 & 2 & 0 \\ 0 & 0 & 1 \end{pmatrix}$$

Then under formula 6.8,

$$\mathbf{y} = \mathbf{X}_t^{-1}\mathbf{x}^{(t)} = \begin{pmatrix} \frac{1}{3} & 0 & 0 \\ 0 & 2 & 0 \\ 0 & 0 & 1 \end{pmatrix}\begin{pmatrix} 3 \\ \frac{1}{2} \\ 1 \end{pmatrix} = \begin{pmatrix} 1 \\ 1 \\ 1 \end{pmatrix}$$

Similarly, applying inverse formula 6.9 retrieves

$$\mathbf{x} = \mathbf{X}_t \mathbf{y} = \begin{pmatrix} 3 & 0 & 0 \\ 0 & \frac{1}{2} & 0 \\ 0 & 0 & 1 \end{pmatrix} \begin{pmatrix} 1 \\ 1 \\ 1 \end{pmatrix} = \begin{pmatrix} 3 \\ \frac{1}{2} \\ 1 \end{pmatrix}$$

SAMPLE EXERCISE 6.6: AFFINE SCALING WITH DIAGONAL MATRICES

Suppose that the first two solutions visited by a linear programming search algorithm are $\mathbf{x}^{(0)} = (12, 3, 2)$ and $\mathbf{x}^{(1)} = (1, 4, 7)$. Compute the affine scalings of $\mathbf{x} = (24, 12, 14)$ relative to each of these $\mathbf{x}^{(t)}$, and verify that applying reverse scaling to the resulting $\mathbf{y}$'s recovers $\mathbf{x}$.

Analysis: For $\mathbf{x}^{(0)}$ the diagonal matrix of computation 6.8 is

$$\mathbf{X}_0 = \begin{pmatrix} 12 & 0 & 0 \\ 0 & 3 & 0 \\ 0 & 0 & 2 \end{pmatrix} \quad \text{with} \quad \mathbf{X}_0^{-1} = \begin{pmatrix} \frac{1}{12} & 0 & 0 \\ 0 & \frac{1}{3} & 0 \\ 0 & 0 & \frac{1}{2} \end{pmatrix}$$

Thus

$$\mathbf{y} = \mathbf{X}_0^{-1} \mathbf{x} = \begin{pmatrix} \frac{1}{12} & 0 & 0 \\ 0 & \frac{1}{3} & 0 \\ 0 & 0 & \frac{1}{2} \end{pmatrix} \begin{pmatrix} 24 \\ 12 \\ 14 \end{pmatrix} = \begin{pmatrix} 2 \\ 4 \\ 7 \end{pmatrix}$$

Reversing the scaling with formula 6.9 recovers

$$\mathbf{x} = \mathbf{X}_0 \mathbf{y} = \begin{pmatrix} 12 & 0 & 0 \\ 0 & 3 & 0 \\ 0 & 0 & 2 \end{pmatrix} \begin{pmatrix} 2 \\ 4 \\ 7 \end{pmatrix} = \begin{pmatrix} 24 \\ 12 \\ 14 \end{pmatrix}$$

After the search advances to $\mathbf{x}^{(1)}$, the scaling changes. Now the diagonal matrix of computation 6.8 is

$$\mathbf{X}_1 = \begin{pmatrix} 1 & 0 & 0 \\ 0 & 4 & 0 \\ 0 & 0 & 7 \end{pmatrix} \quad \text{with} \quad \mathbf{X}_1^{-1} = \begin{pmatrix} 1 & 0 & 0 \\ 0 & \frac{1}{4} & 0 \\ 0 & 0 & \frac{1}{7} \end{pmatrix}$$

Thus solution $\mathbf{x} = (24, 12, 14)$ scales to

$$\mathbf{y} = \mathbf{X}_1^{-1} \mathbf{x} = \begin{pmatrix} 1 & 0 & 0 \\ 0 & \frac{1}{4} & 0 \\ 0 & 0 & \frac{1}{7} \end{pmatrix} \begin{pmatrix} 24 \\ 12 \\ 14 \end{pmatrix} = \begin{pmatrix} 24 \\ 3 \\ 2 \end{pmatrix}$$

Still, reversing the scaling with formula 6.9 recovers

$$\mathbf{x} = \mathbf{X}_1\mathbf{y} = \begin{pmatrix} 1 & 0 & 0 \\ 0 & 4 & 0 \\ 0 & 0 & 7 \end{pmatrix} \begin{pmatrix} 24 \\ 3 \\ 2 \end{pmatrix} = \begin{pmatrix} 24 \\ 12 \\ 14 \end{pmatrix}$$

Affine-Scaled Standard Form

Affine scaling in effect changes the objective function and constraint coefficients of the LP standard form:

$$\begin{aligned} \text{min or max} \quad & \mathbf{cx} \\ \text{s.t.} \quad & \mathbf{Ax} = \mathbf{b} \\ & \mathbf{x} \geq \mathbf{0} \end{aligned}$$

Substituting $\mathbf{x} = \mathbf{X}_t\mathbf{y}$ from formula 6.9 and collecting $\mathbf{c} \cdot \mathbf{x} = \mathbf{cX}_t\mathbf{y} \triangleq \mathbf{c}^{(t)}\mathbf{y}$ and $\mathbf{Ax} = \mathbf{AX}_t\mathbf{x} \triangleq \mathbf{A}_t\mathbf{y}$ produces a new affine-scaled standard form for each solution $\mathbf{x}^{(t)}$:

6.10   At current feasible solution $\mathbf{x}^{(t)} > \mathbf{0}$, the **affine-scaled version of a standard-form linear program** is

$$\begin{aligned} \text{min or max} \quad & \mathbf{c}^{(t)} \cdot \mathbf{y} \\ \text{s.t.} \quad & \mathbf{A}_t\mathbf{y} = \mathbf{b} \\ & \mathbf{y} \geq \mathbf{0} \end{aligned}$$

where $\mathbf{c}^{(t)} \triangleq \mathbf{cX}_t$, $\mathbf{A}_t \triangleq \mathbf{AX}_t$, and $\mathbf{X}_t$ is a square matrix with the components of $\mathbf{x}^{(t)}$ on its diagonal.

Our Frannie's Firewood example has

$$\begin{aligned} \mathbf{c} &= (90, 150, 0) \\ \mathbf{A} &= (\tfrac{1}{2}, 1, 1), \qquad \mathbf{b} = (3) \end{aligned}$$

Thus the affine-scaled form corresponding to the $\mathbf{x}^{(t)} = (3, \tfrac{1}{2}, 1)$ of Figure 6.4 is

$$\begin{aligned} \text{max} \quad & 270y_1 + 75y_2 \\ \text{s.t.} \quad & \tfrac{3}{2}y_1 + \tfrac{1}{2}y_2 + y_3 = 3 \\ & y_1, y_2, y_3 \geq 0 \end{aligned}$$

with

$$\mathbf{c}^{(t)} = \mathbf{cX}_t = (90, 150, 0) \begin{pmatrix} 3 & 0 & 0 \\ 0 & \tfrac{1}{2} & 0 \\ 0 & 0 & 1 \end{pmatrix} = (270, 75, 0)$$

$$\mathbf{A}_t = \mathbf{AX}_t = (\tfrac{1}{2}, 1, 1) \begin{pmatrix} 3 & 0 & 0 \\ 0 & \tfrac{1}{2} & 0 \\ 0 & 0 & 1 \end{pmatrix} = (\tfrac{3}{2}, \tfrac{1}{2}, 1)$$

SAMPLE EXERCISE 6.7: AFFINE SCALING STANDARD FORM

After 7 moves in an improving search of the linear program

$$\min \quad -3x_1 + 9x_3$$
$$\text{s.t.} \quad -x_1 + x_3 \quad\quad = 3$$
$$x_1 + 2x_2 \quad\quad = 4$$
$$x_1, x_2, x_3 \geq 0$$

$\mathbf{x}^{(7)} = (2, 1, 5)$ has been reached. Derive the corresponding affine-scaled standard-form model.

Analysis: Here

$$\mathbf{c} = (-3, 0, 9), \quad \mathbf{A} = \begin{pmatrix} -1 & 0 & 1 \\ 1 & 2 & 0 \end{pmatrix}, \quad \mathbf{b} = \begin{pmatrix} 3 \\ 4 \end{pmatrix}$$

Corresponding elements of affine-scaled standard form 6.10 are

$$\mathbf{c}^{(7)} = \mathbf{cX}_7 = (-3, 0, 9) \begin{pmatrix} 2 & 0 & 0 \\ 0 & 1 & 0 \\ 0 & 0 & 5 \end{pmatrix} = (-6, 0, 45)$$

$$\mathbf{A}_7 = \mathbf{AX}_7 = \begin{pmatrix} -1 & 0 & 1 \\ 1 & 2 & 0 \end{pmatrix} \begin{pmatrix} 2 & 0 & 0 \\ 0 & 1 & 0 \\ 0 & 0 & 5 \end{pmatrix} = \begin{pmatrix} -2 & 0 & 5 \\ 2 & 2 & 0 \end{pmatrix}$$

Thus the scaled model is

$$\min \quad -6y_1 + 45y_3$$
$$\text{s.t.} \quad -2y_1 + 5y_3 \quad = 3$$
$$2y_1 + 2y_2 \quad\quad = 4$$
$$y_1, y_2, y_3 \geq 0$$

Projecting on Affine-Scaled Equality Constraints

We have already seen in Section 6.1 (formula 6.6) that interior point algorithms often employ some form of projection to find directions that preserve the equality main constraints of standard form. We will soon see that most of those projections involve scaled problem coefficients $\mathbf{c}^{(t)}$ and $\mathbf{A}_t$ of affine-scaled standard form 6.10 .

6.11   The **projection** of direction $\mathbf{d}$ onto affine-scaled conditions $\mathbf{A}_t \Delta \mathbf{y} = \mathbf{0}$ preserving linear inequalities $\mathbf{A}_t \mathbf{y} = \mathbf{b}$ can be computed:

$$\Delta \mathbf{y} = \mathbf{P}_t \mathbf{d}$$

where **projection matrix**

$$\mathbf{P}_t = \left(\mathbf{I} - \mathbf{A}_t^T (\mathbf{A}_t \mathbf{A}_t^T)^{-1} \mathbf{A}_t \right)$$

and all other symbols are as in 6.10 .

SAMPLE EXERCISE 6.8: PROJECTING IN SCALED y-SPACE

Return to the linear program of Sample Exercise 6.7 at current solution $\mathbf{x}^{(7)} = (2, 1, 5)$, and compute the projection of direction $\mathbf{d} = (1, 0, -1)$ onto corresponding the standard-form equality constraints in affine-scaled y-space.

Analysis: We apply scaled projection formula $\boxed{6.11}$. From Sample Exercise 6.7 we have

$$\mathbf{A}_7 = \mathbf{A}\mathbf{X}_7 = \begin{pmatrix} -2 & 0 & 5 \\ 2 & 2 & 0 \end{pmatrix}$$

Thus

$$\mathbf{A}_7\mathbf{A}_7^{\mathsf{T}} = \begin{pmatrix} 29 & -4 \\ -4 & 8 \end{pmatrix} \quad \text{with} \quad (\mathbf{A}_7\mathbf{A}_7^{\mathsf{T}})^{-1} = \begin{pmatrix} \frac{8}{216} & \frac{4}{216} \\ \frac{4}{216} & \frac{29}{216} \end{pmatrix}$$

so

$$\mathbf{P}_7 = (\mathbf{I} - \mathbf{A}_7^{\mathsf{T}}(\mathbf{A}_7\mathbf{A}_7^{\mathsf{T}})^{-1}\mathbf{A}_7) = \begin{pmatrix} \frac{25}{54} & -\frac{25}{54} & \frac{10}{54} \\ -\frac{25}{54} & \frac{25}{54} & -\frac{10}{54} \\ \frac{10}{54} & -\frac{10}{54} & \frac{4}{54} \end{pmatrix}$$

Thus the projected direction

$$\Delta\mathbf{y} = \mathbf{P}_7\mathbf{d} = \begin{pmatrix} \frac{25}{54} & -\frac{25}{54} & \frac{10}{54} \\ -\frac{25}{54} & \frac{25}{54} & -\frac{10}{54} \\ \frac{10}{54} & -\frac{10}{54} & \frac{4}{54} \end{pmatrix}\begin{pmatrix} 1 \\ 0 \\ -1 \end{pmatrix} = \begin{pmatrix} \frac{5}{18} \\ -\frac{5}{18} \\ \frac{1}{9} \end{pmatrix}$$

Computational Effort in Interior Point Computations

The distinction between revised projection formulas of $\boxed{6.11}$ and the unscaled formulas of $\boxed{6.6}$ may seem rather trivial. Haven't we just changed from $\mathbf{A}$, $\mathbf{x}$, and $\mathbf{P}$ to $\mathbf{A}_t$, $\mathbf{y}$, and $\mathbf{P}_t$?

In fact, the difference is much more profound and accounts for most of the computational effort in interior point algorithms. The key insight is that the model constraint matrix $\mathbf{A}$ does not change as the algorithm proceeds. Thus its projection matrix $\mathbf{P}$ does not change either. If algorithms required only projection onto $\mathbf{A}\mathbf{x} = \mathbf{b}$, some representation of $\mathbf{P}$ could be computed once and stored. Thereafter, each move would involve only choosing a desired direction and multiplying by the representation of $\mathbf{P}$.

Contrast now with the fact that algorithms actually have to project onto scaled constraints $\mathbf{A}_t\mathbf{y} = \mathbf{b}$ (or something similar). Since $\mathbf{A}_t \triangleq \mathbf{A}\mathbf{X}_t$, $\mathbf{A}_t$ changes at every move with current solution $\mathbf{x}^{(t)}$, it follows that new projection computations have to be performed at every move. Clever techniques of numerical linear algebra can do much to make the computation more efficient, but the fact remains that a very considerable effort has to be expended at each move of the search.

6.12 | The bulk of the computational effort in most interior point algorithms is devoted to projection operations on scaled constraint matrices such as $\mathbf{A}_t$, which change with each solution visited.

6.3 AFFINE SCALING SEARCH

The affine problem scaling of Section 6.2 produces a new version of the model for which scaled current solution $\mathbf{y}^{(t)}$ is $= 1$ in every component. Thus the transformed point is neatly equidistant from all boundary (nonnegativity) constraints. It is natural now to think of computing a search move in this simpler scaled solution space and then converting it back to the true decision variables $\mathbf{x}$ via unscaling formula 6.9 . In this section we develop an affine scaling form of interior point search for linear programs that adopts just such a strategy.

Affine Scaling Move Directions

Principle 6.1 tells us that the preferred move direction in the scaled problem follows objective function vector $\mathbf{c}^{(t)}$. Projection formula 6.11 computes the closest $\Delta\mathbf{y}$ satisfying feasible direction requirement $\mathbf{A}_t\Delta\mathbf{y} = \mathbf{0}$. Combining gives the direction

$$\Delta\mathbf{y} = \mathbf{P}_t\mathbf{c}^{(t)} \tag{6.7}$$

for a maximize problem, and

$$\Delta\mathbf{y} = -\mathbf{P}_t\mathbf{c}^{(t)} \tag{6.8}$$

for a minimize problem.

To complete a move direction we have only to translate back into the original variables via inverse scaling formula 6.9 .

6.13 | An affine scaling search that has reached feasible solution $\mathbf{x}^{(t)} > \mathbf{0}$ moves next in the direction

$$\Delta\mathbf{x} = \pm\mathbf{X}_t\mathbf{P}_t\mathbf{c}^{(t)}$$

($+$ to maximize and $-$ to minimize) where $\mathbf{c}^{(t)}$, $\mathbf{X}_t$, and $\mathbf{P}_t$ are as in definitions 6.10 and 6.11 .

We can illustrate with the initial solution $\mathbf{x}^{(0)} = (1, \frac{1}{2}, 2)$ to our Frannie's Firewood example of Figure 6.4. Scaled standard-form definition 6.10 makes

$$\mathbf{c}^{(0)} = \mathbf{c}\mathbf{X}_0 = (90, 150, 0) \begin{pmatrix} 1 & 0 & 0 \\ 0 & \frac{1}{2} & 0 \\ 0 & 0 & 2 \end{pmatrix} = (90, 75, 0)$$

and

$$\mathbf{A}_0 = \mathbf{A}\mathbf{X}_0 = (\tfrac{1}{2}, 1, 1) \begin{pmatrix} 1 & 0 & 0 \\ 0 & \frac{1}{2} & 0 \\ 0 & 0 & 2 \end{pmatrix} = (\tfrac{1}{2}, \tfrac{1}{2}, 2)$$

Thus

$$\mathbf{P}_0 = \left[\mathbf{I} - \mathbf{A}_0^{\mathrm{T}}\left(\mathbf{A}_0\mathbf{A}_0^{\mathrm{T}}\right)^{-1}\mathbf{A}_0\right] = \begin{pmatrix} \frac{17}{18} & -\frac{1}{18} & -\frac{2}{9} \\ -\frac{1}{18} & \frac{17}{18} & -\frac{2}{9} \\ -\frac{2}{9} & -\frac{2}{9} & \frac{1}{9} \end{pmatrix}$$

We can now compute our maximizing [6.13] move direction as

$$\Delta\mathbf{x} = \mathbf{X}_0\mathbf{P}_0\mathbf{c}^{(0)}$$

$$= \begin{pmatrix} 1 & 0 & 0 \\ 0 & \frac{1}{2} & 0 \\ 0 & 0 & 2 \end{pmatrix} \begin{pmatrix} \frac{17}{18} & -\frac{1}{18} & -\frac{2}{9} \\ -\frac{1}{18} & \frac{17}{18} & -\frac{2}{9} \\ -\frac{2}{9} & -\frac{2}{9} & \frac{1}{9} \end{pmatrix} \begin{pmatrix} 90 \\ 75 \\ 0 \end{pmatrix} = \begin{pmatrix} 80\frac{5}{6} \\ 32\frac{11}{12} \\ -73\frac{1}{3} \end{pmatrix} \qquad (6.9)$$

SAMPLE EXERCISE 6.9: COMPUTING AN AFFINE-SCALED MOVE DIRECTION

Return to the linear program of Sample Exercises 6.7 and 6.8 at current solution $\mathbf{x}^{(7)} = (2, 1, 5)$. Compute the next affine scaling move direction.

Analysis: From Sample Exercise 6.7,

$$\mathbf{c}^{(7)} = (-6, 0, 45) \quad \text{and} \quad \mathbf{A}_7 = \begin{pmatrix} -2 & 0 & 5 \\ 2 & 2 & 0 \end{pmatrix}$$

and from Sample Exercise 6.8,

$$\mathbf{P}_7 = \begin{pmatrix} \frac{25}{54} & -\frac{25}{54} & \frac{10}{54} \\ -\frac{25}{54} & \frac{25}{54} & -\frac{10}{54} \\ \frac{10}{54} & -\frac{10}{54} & \frac{4}{54} \end{pmatrix}$$

Continuing with formula [6.13] for a minimizing problem gives

$$\Delta\mathbf{x}^{(8)} = -\begin{pmatrix} 2 & 0 & 0 \\ 0 & 1 & 0 \\ 0 & 0 & 5 \end{pmatrix} \begin{pmatrix} \frac{25}{54} & -\frac{25}{54} & \frac{10}{54} \\ -\frac{25}{54} & \frac{25}{54} & -\frac{10}{54} \\ \frac{10}{54} & -\frac{10}{54} & \frac{4}{54} \end{pmatrix} \begin{pmatrix} -6 \\ 0 \\ 45 \end{pmatrix} = \begin{pmatrix} -11\frac{1}{9} \\ 5\frac{5}{9} \\ -11\frac{1}{9} \end{pmatrix}$$

Feasibility and Improvement of Affine Scaling Directions

Alert readers will notice that our derivation of direction [6.13] is guided by rules to produce an improving and feasible move in scaled *y*-space. But we are using the direction in the original *x*-space. Will the familiar improving and feasible direction properties of improving search be preserved in that original problem setting? Fortunately, the answer is yes.

> **6.14** (Nonzero) affine scaling search directions derived by formula [6.13] are improving and feasible for the original model over **x** variables.

For feasibility, the $\mathbf{A}\Delta\mathbf{x} = \mathbf{0}$ required by condition $\boxed{6.4}$ is the same as $\mathbf{AX}_t\Delta\mathbf{y} = \mathbf{0}$, which must be true when projected on $\mathbf{A}_t\Delta\mathbf{y} = \mathbf{AX}_t\Delta\mathbf{y}$. The argument for improvement is much the same as the one given for unscaled projection $\boxed{6.7}$ in Section 6.1.

We may illustrate $\boxed{6.14}$ with Frannie's Firewood direction

$$\Delta\mathbf{x}^{(1)} = (80\tfrac{5}{6}, 32\tfrac{11}{12}, -73\tfrac{1}{3})$$

of expression (6.9). Since we are in the interior, feasiblity requires only $\frac{1}{2}\Delta x_1 + \Delta x_2 + \Delta x_3 = 0$ (condition $\boxed{6.4}$). Checking, we find that

$$\tfrac{1}{2}\left(80\tfrac{5}{6}\right) + \left(32\tfrac{11}{12}\right) + \left(-73\tfrac{1}{3}\right) = 0$$

Similarly, improvement for this maximize problem requires that $\mathbf{c} \cdot \Delta\mathbf{x} > 0$. Here

$$(90, 150, 0) \cdot (80\tfrac{5}{6}, 32\tfrac{11}{12}, -73\tfrac{1}{3}) \approx 12212 > 0$$

Affine Scaling Step Size

With directions of construction $\boxed{6.13}$ in hand, the next question involves how far to move. That is, we need to choose an appropriate improving search step size λ.

As usual, there may be no limit on progress in direction $\Delta\mathbf{x}$. For linear programs in standard form, the only inactive constraints at interior point solution $\mathbf{x}^{(t)} > \mathbf{0}$ are the nonnegativities. If $\Delta\mathbf{x}$ decreases no component of the solution, we can improve forever without encountering a nonnegativity constraint.

$\boxed{6.15}$ A linear program in standard form is unbounded if affine scaling construction $\boxed{6.13}$ ever produces a direction $\Delta\mathbf{x} \geq \mathbf{0}$.

When affine scaling produces a move direction with some negative components, we must limit the step so that

$$\mathbf{x}^{(t+1)} = \mathbf{x}^{(t)} + \lambda\Delta\mathbf{x} \geq \mathbf{0}$$

There is also a second consideration. Principle $\boxed{6.2}$ observed that interior point algorithms avoid the boundary until an optimal solution is obtained. That is, we should move all the way until a nonnegativity constraint is active only if such a move will produce an optimal solution.

There are many step size rules that fulfill both of these requirements. The one we will adopt is best understood in affine-scaled space over y-variables. Scaling makes the current solution there a vector of 1's. For example, Figure 6.5 displays current scaled solution $\mathbf{y}^{(0)} = (1, 1, 1)$. The highlighted unit sphere around $\mathbf{y}^{(0)}$ shows how a move of length 1 in any direction maintains all nonnegativity constraints. We can implement affine scaling by stepping to the limits of such a sphere.

$\boxed{6.16}$ At the current feasible solution $\mathbf{x}^{(t)} > \mathbf{0}$ of a linear program in standard form, if the $\Delta\mathbf{x}$ computed from $\boxed{6.13}$ is negative in some components, affine scaling search applies step size

$$\lambda = \frac{1}{||\Delta x X_t^{-1}||} = \frac{1}{||\Delta y||}$$

where $||d||$ denotes $\sqrt{\sum_j (d_j)^2}$, the length or **norm** of vector **d**.

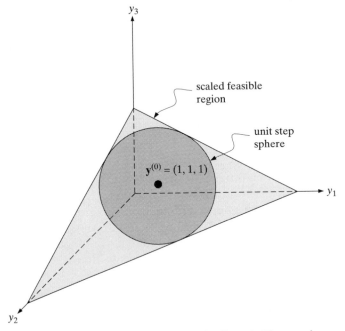

FIGURE 6.5   Feasible Unit Sphere in the Frannie Firewood Example

Figure 6.5 shows this move to the limit of the sphere along the direction computed at equation (6.9). The scaled direction is

$$\Delta y = X_0^{-1} \Delta x \begin{pmatrix} 1 & 0 & 0 \\ 0 & 2 & 0 \\ 0 & 0 & \frac{1}{2} \end{pmatrix} \begin{pmatrix} 80\frac{5}{6} \\ 32\frac{11}{12} \\ -73\frac{1}{3} \end{pmatrix} = \begin{pmatrix} 80\frac{5}{6} \\ 65\frac{5}{6} \\ -36\frac{2}{3} \end{pmatrix}$$

with length (norm)

$$||\Delta y|| = \sqrt{(80\tfrac{5}{6})^2 + (65\tfrac{5}{6})^2 + (-36\tfrac{2}{3})^2} \approx 110.5$$

Thus the computation 6.16 step of

$$\lambda = \frac{1}{||\Delta y||} = \frac{1}{110.5} \approx 0.00905 \qquad (6.10)$$

brings us exactly to the boundary of the unit sphere.

SAMPLE EXERCISE 6.10: COMPUTING AFFINE SCALING STEP SIZE

Determine the appropriate step size to apply to the move direction $\Delta \mathbf{x} = (-11\frac{1}{9}, 5\frac{5}{9},$ $-11\frac{1}{9})$ derived in Sample Exercise 6.9 at point $\mathbf{x}^{(7)} = (2, 1, 5)$.

Analysis: Applying principle $\boxed{6.16}$ yields

$$\Delta \mathbf{y} = \mathbf{X}_7^{-1} \Delta \mathbf{x} = \begin{pmatrix} \frac{1}{2} & 0 & 0 \\ 0 & 1 & 0 \\ 0 & 0 & \frac{1}{5} \end{pmatrix} \begin{pmatrix} -11\frac{1}{9} \\ 5\frac{5}{9} \\ -11\frac{1}{9} \end{pmatrix} = \begin{pmatrix} -5\frac{5}{9} \\ 5\frac{5}{9} \\ -2\frac{2}{9} \end{pmatrix}$$

This $\Delta \mathbf{y}$ has

$$\|\Delta \mathbf{y}\| = \sqrt{(-5\frac{5}{9})^2 + (5\frac{5}{9})^2 + (2\frac{2}{9})^2} \approx 8.165$$

Thus the appropriate step is

$$\lambda = \frac{1}{8.165} = 0.1225$$

Termination in Affine Scaling Search

Applying the step computed in expression (6.10) advances our search of the Frannie's firewood example to point

$$\begin{aligned} \mathbf{x}^{(1)} &= \mathbf{x}^{(0)} + \lambda \Delta \mathbf{x} \\ &= (0.5, 1, 1) + 0.00905(80.83, 32.92, -73.33) \\ &\approx (1.73, 0.80, 1.34) \end{aligned} \tag{6.11}$$

Notice that the new point remains positive in all components and thus in the interior of the feasible region.

With step rule $\boxed{6.16}$ constructed to keep updated (scaled) points within the unit sphere of feasible $\mathbf{y}$, the new solution will always be interior unless the search moves to one of the points where that sphere touches a nonnegativity constraint. Such moves are rare, but if they occur, the new solution can be shown to be *optimal*.

Much more typically, an affine scaling search will remain in the interior, stepping ever closer to an optimal solution on the boundary. Then a stopping rule is required to terminate computation when a solution becomes sufficiently close to an optimum.

But how do we know that we are near an optimal solution? Crude schemes simply stop when the solution value is no longer changing very much (as with Algorithm 6A). More precise rules derive bounds on the optimal solution value at each step so that we can know how much room remains for improvement. We look at the latter briefly in Section 6.5.

Affine Scaling Search of the Frannie's Firewood Example

Algorithm 6A collects all the insights of this section in an affine scaling algorithm for linear programs in standard form. Table 6.1 and Figure 6.6 detail its application to our Frannie's firewood example.

ALGORITHM 6A: AFFINE SCALING SEARCH FOR LINEAR PROGRAMS

Step 0: Initialization. Choose any starting feasible interior point solution, $\mathbf{x}^{(0)} > \mathbf{0}$, and set solution index $t \leftarrow 0$.

Step 1: Optimality. If any component of $\mathbf{x}^{(t)}$ is 0, or if recent algorithm steps have made no significant change in the solution value, stop. Current point $\mathbf{x}^{(t)}$ is either optimal in the given LP or very nearly so.

Step 2: Move Direction. Construct the next move direction by projecting in affine-scaled space as

$$\Delta \mathbf{x}^{(t+1)} \leftarrow \pm \mathbf{X}_t \mathbf{P}_t \mathbf{c}^{(t)}$$

(+ to maximize, − to minimize) where $\mathbf{X}_t$, $\mathbf{P}_t$, and $\mathbf{c}^{(t)}$ are the scaled values of Section 6.2.

Step 3: Step Size. If there is no limit on feasible moves in direction $\Delta \mathbf{x}^{(t+1)}$ (all components are nonnegative), stop; the given model is unbounded. Otherwise, construct step size

$$\lambda \leftarrow \frac{1}{||\Delta \mathbf{x}^{(t+1)} \mathbf{X}_t^{-1}||}$$

Step 4: Advance. Compute the new solution

$$\mathbf{x}^{(t+1)} \leftarrow \mathbf{x}^{(t)} + \lambda \Delta \mathbf{x}^{(t+1)}$$

Then advance $t \leftarrow t + 1$, and return to Step 1.

TABLE 6.1   Affine Scaling Search of the Frannie's Firewood Example

	x_1	x_2	x_3	
max c	90	150	0	**b**
A	0.5	1	1	3
$\mathbf{x}^{(0)}$	1.00	0.50	2.00	$\mathbf{c} \cdot \mathbf{x}^{(0)} = 165.00$
$\Delta\mathbf{x}^{(1)}$	80.83	32.92	−73.33	$\lambda = 0.00905$
$\mathbf{x}^{(1)}$	1.73	0.80	1.34	$\mathbf{c} \cdot \mathbf{x}^{(1)} = 275.51$
$\Delta\mathbf{x}^{(2)}$	160.94	49.25	−129.72	$\lambda = 0.00676$
$\mathbf{x}^{(2)}$	2.82	1.13	0.46	$\mathbf{c} \cdot \mathbf{x}^{(2)} = 423.40$
$\Delta\mathbf{x}^{(3)}$	87.26	−10.29	−33.34	$\lambda = 0.0126$
$\mathbf{x}^{(3)}$	3.92	1.00	0.04	$\mathbf{c} \cdot \mathbf{x}^{(3)} = 502.84$
$\Delta\mathbf{x}^{(4)}$	48.13	−23.79	−0.27	$\lambda = 0.0362$
$\mathbf{x}^{(4)}$	5.66	0.14	0.03	$\mathbf{c} \cdot \mathbf{x}^{(4)} = 530.45$
$\Delta\mathbf{x}^{(5)}$	1.49	−0.58	−0.16	$\lambda = 0.1472$
$\mathbf{x}^{(5)}$	5.88	0.05	0.01	$\mathbf{c} \cdot \mathbf{x}^{(5)} = 537.25$
$\Delta\mathbf{x}^{(6)}$	0.19	−0.09	−0.01	$\lambda = 0.5066$
$\mathbf{x}^{(6)}$	5.98	0.01	0.003	$\mathbf{c} \cdot \mathbf{x}^{(6)} = 539.22$
$\Delta\mathbf{x}^{(7)}$	0.008	−0.003	−0.001	$\lambda = 1.7689$
$\mathbf{x}^{(7)}$	5.99	0.005	+0.000	$\mathbf{c} \cdot \mathbf{x}^{(7)} = 539.79$
$\Delta\mathbf{x}^{(8)}$	0.001	−0.001	−0.000	$\lambda = 6.3305$
$\mathbf{x}^{(8)}$	6.00	0.001	+0.000	$\mathbf{c} \cdot \mathbf{x}^{(8)} = 539.95$
$\Delta\mathbf{x}^{(9)}$	+0.000	−0.000	−0.000	$\lambda = 23.854$
$\mathbf{x}^{(9)}$	6.00	+0.000	+0.000	$\mathbf{c} \cdot \mathbf{x}^{(9)} = 539.99$

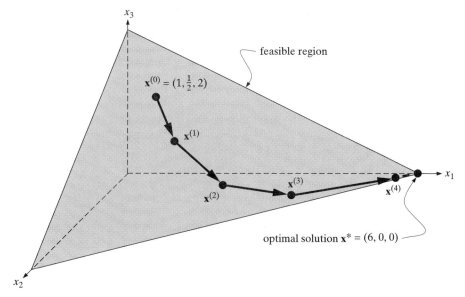

FIGURE 6.6 Affine Scaling Search of the Frannie's Firewood Example

Notice that the search makes very rapid progress while it is near the "middle" of the feasible region. Later iterations approach the boundary but never quite reach it. We terminate after the ninth iteration changes the objective function by less than 0.1.

6.4 LOG BARRIER METHODS FOR INTERIOR POINT SEARCH

Affine scaling is only one way to keep algorithms away from the boundary. In this section we develop a **log barrier** alternative.

Barrier Objective Functions

Remembering that the boundary of a standard-form linear program is defined by nonnegativity constraints, log barrier methods exploit the fact that $\ln(x_j) \to -\infty$ as $x_j \to 0$. A modified barrier objective function includes $\ln(x_j)$ terms to keep the x_j away from the boundary.

> **6.17** A maximize objective $\sum_j c_j x_j$ of a standard-form linear program is modified with a **logarithmic barrier function** as
>
> $$\max \sum_j c_j x_j + \mu \sum_j \ln(x_j)$$
>
> where $\mu > 0$ is a specified weighting constant. The corresponding form for a minimize objective is
>
> $$\min \sum_j c_j x_j - \mu \sum_j \ln(x_j)$$

We can illustrate with our familiar Frannie's firewood standard form

$$\max \quad 90x_1 + 150x_2$$

$$\text{s.t.} \quad \tfrac{1}{2}x_1 + x_2 + x_3 = 3$$

$$x_1, x_2, x_3 \geq 0$$

Adding a barrier term produces the modified model

$$\max \quad 90x_1 + 150x_2 + \mu \left[\ln(x_1) + \ln(x_2) + \ln(x_3)\right]$$

$$\text{s.t.} \quad \tfrac{1}{2}x_1 + x_2 + x_3 = 3 \qquad\qquad (6.12)$$

$$x_1, x_2, x_3 \geq 0$$

Suppose that we (arbitrarily) choose multiplier $\mu = 64$. Then the objective at feasible point $\mathbf{x} = (2, 1, 1)$, which is far from the boundary, evaluates to

$$90(2) + 150(1) + 0(1) + 64\left[\ln(2) + \ln(1) + \ln(1)\right] \approx 374.36$$

The log-barrier terms do have an influence, because the true objective value at this $\mathbf{x}$ is $90(2) + 150(1) = 330$. Still, the difference is modest.

Compare with near-boundary feasible point $\mathbf{x} = (0.010, 2.99, 0.005)$. There the barrier objective function evaluates as

$$90(0.010) + 150(2.99) + 0(0.005) + 64\left[\ln(0.010) + \ln(2.99) + \ln(0.005)\right]$$

$$= 449.4 + 64(-4.605 + 1.095 - 5.298)$$

$$\approx -114.31$$

The corresponding true objective value is $90(0.010) + 150(2.99) = 449.4$. We see that the negative logarithms of x_j near 0.0 have severely penalized this solution in the modified maximize objective.

It is in this penalization sense that modified objective functions of construction 6.17 erect a "barrier" to near-boundary solutions. As components x_j approach their boundary value of 0.0, penalties become larger and larger, thus guaranteeing that no improving search will choose to approach the boundary too closely.

SAMPLE EXERCISE 6.11: FORMING LOG BARRIER OBJECTIVES

Linear program

$$\min \quad 5x_1 + 3x_2$$

$$\text{s.t.} \quad x_1 + x_2 \geq 1$$

$$0 \leq x_1 \leq 2$$

$$0 \leq x_2 \leq 2$$

has feasible region as displayed in the following figure:

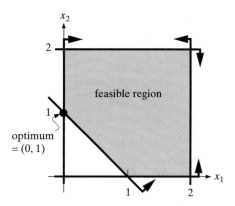

(a) Place the model in standard form.

(b) Show the modified objective function obtained when a logarithmic barrier is introduced to discourage approaching the boundary.

(c) Choose a solution near the middle of the feasible region, and compare the true and modified objective functions at that point using $\mu = 10$.

(d) Choose an interior point solution near the boundary of the feasible region, and compare the true and modified objective functions at that point using the same $\mu = 10$.

Analysis:

(a) Introducing slack variables x_3, x_4 and x_5, the standard-form model is

$$\begin{aligned}
\min \quad & 5x_1 + 3x_2 \\
\text{s.t.} \quad & x_1 + x_2 - x_3 = 1 \\
& x_1 \qquad\ \ + x_4 = 2 \\
& x_2 \qquad\quad + x_5 = 2 \\
& x_1, \ldots, x_5 \geq 0
\end{aligned}$$

(b) Following construction $\boxed{6.17}$, the barrier objective with multiplier μ is

$$\min \ 5x_1 + 3x_2 - \mu \left[\ln(x_1) + \ln(x_2) + \ln(x_3) + \ln(x_4) + \ln(x_5) \right]$$

(c) From the figure it is clear that $x_1 = x_2 = \frac{5}{4}$ is near the middle of the feasible region. There the true objective function value is $5(\frac{5}{4}) + 3(\frac{5}{4}) = 10$. Corresponding values for slack variables are $x_3 = \frac{5}{4} + \frac{5}{4} - 1 = \frac{3}{2}$ and $x_4 = x_5 = 2 - \frac{5}{4} = \frac{3}{4}$. Thus the log barrier objective with $\mu = 10$ evaluates

$$5(\tfrac{5}{4}) + 3(\tfrac{5}{4}) - 10 \left[\ln(\tfrac{5}{4}) + \ln(\tfrac{5}{4}) + \ln(\tfrac{3}{2}) + \ln(\tfrac{3}{4}) + \ln(\tfrac{3}{4}) \right] \approx 7.237$$

Modified cost is a bit lower that the true objective value.

(d) One point near the boundary is $x_1 = 1.999$, $x_2 = 0.001$. There the true objective function value is $5(1.999) + 3(0.001) = 9.998$. Corresponding values for slack variables are $x_3 = 1.999 + 0.001 - 1 = 1.000$, $x_4 = 2 - 1.999 = 0.001$, and $x_5 = 2 - 0.001 = 1.999$. Thus the log barrier objective with $\mu = 10$ evaluates

$$5(1.999) + 3(0.001) - 10[\ln(1.999) + \ln(0.001) + \ln(1.000)$$

$$+ \ln(1.999) + \ln(0.001)] \approx 134.3$$

This near-boundary point is penalized severely.

Problems with Gradient Directions

To make any progress on our underlying LP with barrier methods, we must find improving feasible directions for the now nonlinear standard form

$$\text{max or min} \quad f(\mathbf{x}) \triangleq \sum_j [c_j x_j \pm \mu \ln(x_j)] \tag{6.13}$$

$$\text{s.t.} \quad \mathbf{A}\mathbf{x} = \mathbf{b}$$

Nonnegativity constraints can be ignored because the barrier objective will keep us away from the boundary.

Although the gradient-based directions introduced in Section 3.3 have served us well so far, they usually perform poorly with nonlinear objective functions (see Section 13.5 for details). In essence they treat the objective as if it were approximated as

$$f(\mathbf{x}^{(t)} + \lambda \Delta \mathbf{x}) \quad \approx \quad f(\mathbf{x}^{(t)}) + \lambda \sum_j \frac{\partial f}{\partial x_j} \Delta x_j \tag{6.14}$$

where $\mathbf{x}^{(t)}$ is the current solution and $\Delta \mathbf{x}$ a proposed move direction. This approximation, known as the first-order **Taylor series** (see also Section 13.3), takes the function to be roughly its current value plus the net effect of partial derivative slopes times move components. Choosing gradient-based direction $\Delta \mathbf{x} = \pm \nabla f(\mathbf{x}^{(t)})$ produces the most rapid change in the approximation per unit λ.

With a linear objective function, partial derivatives are constant and approximation (6.14) is exact. But with nonlinear cases, the slope information in partial derivatives can decay rapidly as we move away from point $\mathbf{x}^{(t)}$. The result is that a direction based on rapid progress for the simple (6.14) approximation may prove very ineffective for the real nonlinear objective.

Newton Steps for Barrier Search

For those who remember their calculus (refer to Primers 2 and 7 if needed), it is natural to think of enhancing approximation (6.14) with second partial derivatives. The corresponding second-order Taylor form is

$$f(\mathbf{x}^{(t)} + \lambda \Delta \mathbf{x}) \quad \approx \quad f(\mathbf{x}^{(t)}) + \lambda \sum_j \frac{\partial f}{\partial x_j} \Delta x_j$$

$$+ \frac{\lambda^2}{2} \sum_j \sum_k \frac{\partial^2 f}{\partial x_j \partial x_k} \Delta x_j \Delta x_k \tag{6.15}$$

New terms account for changes in the rates of change $\partial f / \partial x_j$ as we move away from $\mathbf{x}^{(t)}$.

The partial derivatives needed in this new approximation have a particularly easy form for our log barrier objective $\boxed{6.17}$ (+ for maximize, − for minimize):

$$\frac{\partial f}{\partial x_j} = c_j \pm \frac{\mu}{x_j}$$

$$\frac{\partial^2 f}{\partial x_j \partial x_k} = \begin{cases} \mp \dfrac{\mu}{(x_j)^2} & \text{if } j=k \\ 0 & \text{otherwise} \end{cases} \tag{6.16}$$

Thus fixing $\lambda = 1$ for the moment and taking second-order approximation as exact, it makes sense to choose a move direction $\Delta\mathbf{x}$ at $\mathbf{x}^{(t)}$ that solves

$$\text{max or min} \quad f(\mathbf{x}^{(t)} + \Delta\mathbf{x}) \approx \sum_j \left[c_j \pm \frac{\mu}{x_j^{(t)}} \mp \tfrac{1}{2} \frac{\mu}{\left(x_j^{(t)}\right)^2} (\Delta x_j)^2 \right] \tag{6.17}$$

s.t. $\qquad\qquad \mathbf{A}\Delta\mathbf{x} = \mathbf{0}$

That is we choose the $\Delta\mathbf{x}$ that most improves our barrier objective function—as approximated by Taylor expression (6.15)—subject to the familiar requirements (conditions $\boxed{6.4}$) that the direction preserve all equality constraints of standard from (6.13).

Using Lagrange multiplier methods beyond the scope of this section (but developed in Section 14.3), it can be shown that the $\Delta\mathbf{x}$ which solves the direction-finding problem (6.17) is remarkably like the affine-scaled steps of Section 6.3. It takes the form

$$\Delta\mathbf{x} = \pm \frac{1}{\mu}\mathbf{X}_t\mathbf{P}_t \begin{pmatrix} c_1^{(t)} \pm \mu \\ \vdots \\ c_n^{(t)} \pm \mu \end{pmatrix} \tag{6.18}$$

where $\mathbf{X}_t$, $\mathbf{P}_t$, and $\mathbf{c}^{(t)}$ are the scaled problem data of $\boxed{6.10}$ and $\boxed{6.11}$ in Section 6.2. The only difference from affine scaling direction $\boxed{6.13}$ is the inclusion of barrier multiplier μ in forming the scaled direction to project.

This move is often called a **Newton step** because it is based on the same second-order Taylor approximation (6.15) that gives rise to the famous Newton method for equation solving and unconstrained optimization (see Section 13.6 for a full development). Allowing a step size λ to be applied, which means that we may drop the leading constant $1/\mu$, yields the directions to be employed in our barrier form of interior point LP search.

$\boxed{6.18}$ A Newton step barrier algorithm that has reached feasible solution $\mathbf{x}^{(t)} > \mathbf{0}$ with barrier multiplier $\mu > 0$ moves next in direction

$$\Delta\mathbf{x} = \pm\mathbf{X}_t\mathbf{P}_t \begin{pmatrix} c_1^{(t)} \pm \mu \\ \vdots \\ c_n^{(t)} \pm \mu \end{pmatrix}$$

(+ to maximize and − to minimize) where X_t, P_t, and $c^{(t)}$ are the affine-scaled problem data of definitions 6.10 and 6.11.

To illustrate construction 6.18, we return to our Frannie's Firewood example at $x^{(0)} = (1, \frac{1}{2}, 2)$ with $\mu = 16$. Scaling of Section 6.2 gives

$$c^{(0)} = cX_0 = (90, 150, 0) \begin{pmatrix} 1 & 0 & 0 \\ 0 & \frac{1}{2} & 0 \\ 0 & 0 & 2 \end{pmatrix} = (90, 75, 0)$$

and

$$A_0 = AX_0 = (\tfrac{1}{2}, 1, 1)$$

Thus

$$P_0 = \left(I - A_0^T(A_0A_0^T)^{-1}A_0\right) = \begin{pmatrix} \frac{17}{18} & -\frac{1}{18} & -\frac{2}{9} \\ -\frac{1}{18} & \frac{17}{18} & -\frac{2}{9} \\ -\frac{2}{9} & -\frac{2}{9} & \frac{1}{9} \end{pmatrix}$$

Now projecting to find the next move direction,

$$\Delta x^{(1)} = X_0 P_0 \begin{pmatrix} c_1^{(0)} + \mu \\ \vdots \\ c_n^{(0)} + \mu \end{pmatrix}$$

$$= \begin{pmatrix} 1 & 0 & 0 \\ 0 & \frac{1}{2} & 0 \\ 0 & 0 & 2 \end{pmatrix} \begin{pmatrix} \frac{17}{18} & -\frac{1}{18} & -\frac{2}{9} \\ -\frac{1}{18} & \frac{17}{18} & -\frac{2}{9} \\ -\frac{2}{9} & -\frac{2}{9} & \frac{1}{9} \end{pmatrix} \begin{pmatrix} 90 + 16 \\ 75 + 16 \\ 0 + 16 \end{pmatrix}$$

$$= (91\tfrac{1}{2}, 38\tfrac{1}{4}, -84)$$

SAMPLE EXERCISE 6.12: COMPUTING NEWTON STEP BARRIER SEARCH DIRECTIONS

Previous Sample Exercises 6.7 and 6.8 considered the model

$$\begin{aligned} \min \quad & -3x_1 + 9x_3 \\ \text{s.t.} \quad & -x_1 + x_3 = 3 \\ & x_1 + 2x_2 = 4 \\ & x_1, x_2, x_3 \geq 0 \end{aligned}$$

at current solution $x^{(7)} = (2, 1, 5)$. Compute the next Newton step barrier search direction assuming a barrier multiplier $\mu = 120$.

Analysis: From the earlier sample exercises we know that
$$\mathbf{c}^{(7)} = \mathbf{cX}_7 = (-6, 0, 45)$$
at the specified $\mathbf{x}^{(7)}$,
$$\mathbf{A}_7 = \mathbf{AX}_7 = \begin{pmatrix} -2 & 0 & 5 \\ 2 & 2 & 0 \end{pmatrix}$$
and
$$\mathbf{P}_7 = \left(\mathbf{I} - \mathbf{A}_7^{\mathsf{T}}(\mathbf{A}_7\mathbf{A}_7^{\mathsf{T}})^{-1}\mathbf{A}_7\right) = \begin{pmatrix} \frac{25}{54} & -\frac{25}{54} & \frac{10}{54} \\ -\frac{25}{54} & \frac{25}{54} & -\frac{10}{54} \\ \frac{10}{54} & -\frac{10}{54} & \frac{4}{54} \end{pmatrix}$$

Thus construction $\boxed{6.18}$ for this minimizing problem gives
$$\Delta\mathbf{x}^{(8)} = -\mathbf{X}_7\mathbf{P}_7 \begin{pmatrix} c_1^{(7)} - \mu \\ \vdots \\ c_n^{(7)} - \mu \end{pmatrix}$$
$$= -\begin{pmatrix} 2 & 0 & 0 \\ 0 & 1 & 0 \\ 0 & 0 & 5 \end{pmatrix} \begin{pmatrix} \frac{25}{54} & -\frac{25}{54} & \frac{10}{54} \\ -\frac{25}{54} & \frac{25}{54} & -\frac{10}{54} \\ \frac{10}{54} & -\frac{10}{54} & \frac{4}{54} \end{pmatrix} \begin{pmatrix} -6 - 120 \\ 0 - 120 \\ 45 - 120 \end{pmatrix} = \begin{pmatrix} 33\frac{1}{3} \\ -16\frac{2}{3} \\ 33\frac{1}{3} \end{pmatrix}$$

Newton Step Barrier Search Step Sizes

As usual, our next concern is how big a step to take in barrier search direction $\boxed{6.18}$. Any such rule must first assure that
$$\mathbf{x}^{(t+1)} = \mathbf{x}^{(t)} + \lambda\Delta\mathbf{x}^{(t+1)} > \mathbf{0}$$
so that the new point will also be in the interior.

This is easily accomplished by familiar minimum ratio computations. To remain interior, we keep λ less than or equal to say 90% of the maximum feasible step size, that is,
$$\lambda \leq 0.9\lambda_{\max} \tag{6.19}$$
where
$$\lambda_{\max} = \min\left\{ \frac{x_j^{(t)}}{-\Delta x_j^{(t+1)}} : \Delta x_j^{(t+1)} < 0 \right\}$$

With barrier objective $\boxed{6.17}$ nonlinear, there is also the possibility that improvement stops before we approach $\lambda_{\max}$. A direction that improves near current $\mathbf{x}^{(t)}$ may begin to degrade the barrier objective function value after a larger step.

Figure 6.7 illustrates both cases. At Fannie's firewood $\mathbf{x}^{(0)} = (1, \frac{1}{2}, 2)$, direction $\Delta\mathbf{x}^{(1)} = (91.5, 38.25, -84)$ decreases only x_3. Thus expression (6.19) yields
$$0.9\lambda_{\max} = 0.9\left(\frac{2}{84}\right) = 0.0214$$

Part (a) of the figure shows the barrier function improves throughout the range $0 \le \lambda \le 0.0214$.

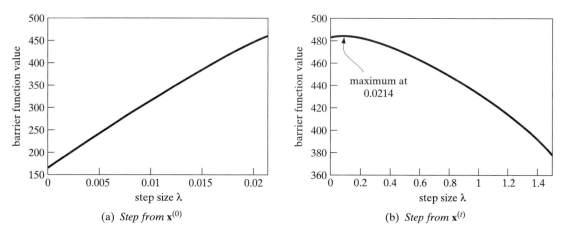

(a) *Step from* $\mathbf{x}^{(0)}$ (b) *Step from* $\mathbf{x}^{(t)}$

FIGURE 6.7   Frannie's Firewood Barrier Function Changes with Step Size

Contrast with the more typical step at $\mathbf{x}^{(t)} = (5.205, 0.336, 0.062)$. The corresponding $\Delta\mathbf{x}^{(t+1)} = (-3.100, 1.273, 0.278)$. Only x_1 decreases and the minimum ratio expression (6.19) gives

$$0.9\left(\frac{x_1^{(t)}}{-\Delta x_1^{(t+1)}}\right) = 0.9\left(\frac{5.205}{3.100}\right) = 1.511$$

However, Figure 6.7(b) shows that the barrier function reaches a maximum long before this limit is encountered.

We need an enhancement of the basic minimum ratio step size rule to account for this possibility of initial improvement followed by worsening of the barrier objective. Fortunately, one is readily available in the Newton step computations of expression (6.18). That best move for the second-order Taylor approximation consisted of $1/\mu$ times the barrier search direction of construction $\boxed{6.18}$.

Combining this idea with a minimum ratio gives a step size that both keeps to the interior and approximately optimizes the barrier objective:

$\boxed{6.19}$ At current feasible solution $\mathbf{x}^{(t)} > \mathbf{0}$ and barrier multiplier $\mu > 0$, a Newton step barrier search algorithm should apply step size

$$\lambda = \min\left\{\frac{1}{\mu}, 0.9\lambda_{max}\right\}$$

to the move direction of $\boxed{6.18}$, where

$$\lambda_{max} = \min\left\{\frac{x_j^{(t)}}{-\Delta x_j^{(t+1)}} : \Delta x_j^{(t+1)} < 0\right\}$$

SAMPLE EXERCISE 6.13: COMPUTING BARRIER SEARCH STEP SIZES

Return to the minimize problem of Sample Exercise 6.12.

(a) Determine the maximum step that can be applied to the direction $\Delta \mathbf{x}^{(8)}$ computed there before the boundary is reached.

(b) Sketch the barrier objective function for that model as a function of the step size λ applied to direction $\Delta \mathbf{x}^{(8)}$.

(c) Compute the step size that would be applied by a Newton step barrier search algorithm.

Analysis:

(a) In Sample Exercise 6.12 we computed the next move direction,

$$\Delta \mathbf{x}^{(8)} = \left(33\tfrac{1}{3}, -16\tfrac{2}{3}, 33\tfrac{1}{3}\right)$$

Applying the minimum ratio computation (6.19), the boundary would be reached at

$$\lambda_{\max} = \min \left\{ \frac{x_j^{(7)}}{-\Delta x_j^{(8)}} : \Delta x_j^{(t+1)} < 0 \right\}$$

$$= \min \left\{ \frac{1}{16\tfrac{2}{3}} \right\}$$

$$= 0.060$$

We should stop before, say, 90% or $\lambda = 0.9(0.060) = 0.054$ to stay interior.

(b) The barrier objective function for this exercise is

$$\min \ -3x_1 + 9x_3 - \mu \left[\ln(x_1) + \ln(x_2) + \ln(x_3) \right]$$

Plotting as a function of step size gives the following:

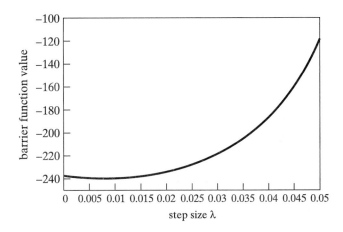

We seek the λ that yields a minimum value without leaving the interior.

(c) Applying rule $\boxed{6.19}$, Newton step barrier search would use step size

$$\lambda \;=\; \min\left\{\frac{1}{\mu}, 0.9\lambda_{max}\right\} \;=\; \min\left\{\tfrac{1}{120}, 0.054\right\} \;=\; \tfrac{1}{120} = 0.00833$$

Impact of the Barrier Multiplier μ

Having worked out all the details of a barrier search for any fixed μ in barrier forms $\boxed{6.17}$, it is time to consider how to manage that barrier multiplier. Parameter μ controls how much weight is assigned to keeping a search away from the boundary. For example, Table 6.2 details the effect for our Frannie's Firewood barrier form (6.12). The optimal barrier problem values of x_1, x_2, and x_3 are shown for a range of barrier multiplier values. For example, high weight $\mu = 2^{16} = 65,536$ on logarithmic barriers makes the optimal $\mathbf{x}$ in model (6.12) approximately $(2, 1, 1)$. As the multiplier decreases to $\mu = 2^{-5} = \frac{1}{32}$, the optimal $\mathbf{x}$ in (6.12) approaches the true optimum (without barriers) of $\mathbf{x}^* = (6, 0, 0)$.

TABLE 6.2   Impact of μ on the Frannie's Firewood Example

μ	x_1	x_2	x_3	μ	x_1	x_2	x_3
$2^{16} = 65,536$	2.002	1.001	0.998	$2^5 = 32$	4.250	0.711	0.164
$2^{15} = 32,768$	2.004	1.001	0.997	$2^4 = 16$	4.951	0.439	0.086
$2^{14} = 16,384$	2.009	1.002	0.993	$2^3 = 8$	5.426	0.243	0.044
$2^{13} = 8,192$	2.017	1.005	0.987	$2^2 = 4$	5.702	0.127	0.022
$2^{12} = 4,096$	2.034	1.009	0.974	$2^1 = 2$	5.847	0.065	0.011
$2^{11} = 2,048$	2.068	1.018	0.948	$2^0 = 1$	5.923	0.033	0.006
$2^{10} = 1,024$	2.134	1.035	0.898	$2^{-1} = \frac{1}{2}$	5.961	0.017	0.003
$2^9 = 512$	2.261	1.060	0.809	$2^{-2} = \frac{1}{4}$	5.980	0.009	0.001
$2^8 = 256$	2.494	1.088	0.665	$2^{-3} = \frac{1}{8}$	5.990	0.004	0.001
$2^7 = 128$	2.889	1.079	0.477	$2^{-4} = \frac{1}{16}$	5.995	0.002	0.000
$2^6 = 64$	3.489	0.960	0.295	$2^{-5} = \frac{1}{32}$	5.997	0.001	0.000

$\boxed{6.20}$ High values of barrier weight $\mu > 0$ severely penalize interior points near the boundary. Low values encourage the search to approach the boundary.

Barrier Algorithm Multiplier Strategy

The power of multiplier μ to control how close barrier optimal solutions come to the boundary suggests a strategy for using barrier methods to solve the underlying LP.

$\boxed{6.21}$ Barrier algorithms begin with multiplier $\mu > 0$ relatively high and slowly reduce it toward zero as the search proceeds.

Initial moves seek good solutions far from the boundary, with later ones coming ever closer to an optimum in the LP as barrier multiplier $\mu \to 0$.

Newton Step Barrier Algorithm

Algorithm 6B collects all the ideas developed so far in a Newton step barrier algorithm for linear programming. The search begins with an interior (feasible) point $\mathbf{x}^{(0)}$ and a relatively large barrier multiplier μ. For each μ, an inner loop seeks to approximately optimize the corresponding log barrier problem. That is, we take one or more steps to approach the barrier model optimum for that μ.

ALGORITHM 6B: NEWTON STEP BARRIER SEARCH FOR LINEAR PROGRAM

Step 0: Initialization. Choose any starting feasible interior point solution, $\mathbf{x}^{(0)} > \mathbf{0}$, and a relatively large initial barrier multiplier μ. Also set solution index $t \leftarrow 0$.

Step 1: Move Direction. Construct the next move direction by projecting in affine-scaled space as

$$\Delta \mathbf{x}^{(t+1)} \leftarrow \pm \mathbf{X}_t \mathbf{P}_t \begin{pmatrix} c_1^{(t)} \pm \mu \\ \vdots \\ c_n^{(t)} \pm \mu \end{pmatrix}$$

(+ to maximize and − to minimize) where $\mathbf{X}_t$, $\mathbf{P}_t$, and $\mathbf{c}^{(t)}$ are the scaled problem data of Section 6.2.

Step 2: Step Size. Choose a step size

$$\lambda \leftarrow \min \left\{ \frac{1}{\mu}, 0.9\lambda_{\max} \right\}$$

where

$$\lambda_{\max} \leftarrow \min \left\{ \frac{x_j^{(t)}}{-\Delta x_j^{(t+1)}} : \Delta x_j^{(t+1)} < 0 \right\}$$

Step 3: Advance. Compute the new solution

$$\mathbf{x}^{(t+1)} \leftarrow \mathbf{x}^{(t)} + \lambda \Delta \mathbf{x}^{(t+1)}$$

Step 4: Inner Loop. If progress indicates that $\mathbf{x}^{(t+1)}$ is far from optimal for the current μ, increment $t \leftarrow t + 1$, and return to Step 1.

Step 5: Outer Loop. If barrier multiplier μ is near zero, stop. The current point $\mathbf{x}^{(t+1)}$ is either optimal in the given LP or very nearly so. Otherwise, reduce the barrier multiplier μ, increment $t \leftarrow t + 1$, and return to Step 1.

Once slow progress indicates that we are near enough to the optimal value in the barrier model, an outer loop reduces μ and repeats the inner loop. The process terminates when μ nears zero, indicating that the current solution is very close to an optimum in the LP.

As with affine scaling Algorithm 6A, many details of stopping and convergence are left vague in the statement of Algorithm 6B because they involve mathematical issues beyond the scope of this chapter. However, in Section 6.5 we provide further insight regarding how such issues are handled in commercial-quality algorithms.

Newton Barrier Solution of Frannie's Firewood Example

Table 6.3 illustrates Algorithm 6B on our Frannie's firewood example. The table shows both the true objective function value at every step and the modified objective value, including log-barrier terms (in parentheses). Figure 6.8 tracks progress graphically.

TABLE 6.3   Newton Barrier Search of Frannie's Firewood Example

	x_1	x_2	x_3	
max c	90	150	0	**b**
A	0.5	1	1	3
$\mathbf{x}^{(0)}$	1.000	0.500	2.000	obj = 165.00 (165.00)
$\Delta\mathbf{x}^{(1)}$	91.500	38.250	−84.000	$\mu = 16.0, \lambda = 0.0214$
$\mathbf{x}^{(1)}$	2.961	1.320	0.200	obj = 464.41 (460.46)
$\Delta\mathbf{x}^{(2)}$	59.996	−26.113	−3.885	$\mu = 16.0, \lambda = 0.0455$
$\mathbf{x}^{(2)}$	5.689	0.132	0.023	obj = 531.84 (467.12)
$\Delta\mathbf{x}^{(3)}$	−3.519	1.487	0.272	$\mu = 16.0, \lambda = 0.0625$
$\mathbf{x}^{(3)}$	5.470	0.225	0.040	obj = 526.00 (477.93)
$\Delta\mathbf{x}^{(4)}$	−4.227	1.771	0.343	$\mu = 16.0, \lambda = 0.0625$
$\mathbf{x}^{(4)}$	5.205	0.336	0.062	obj = 518.82 (483.19)
$\Delta\mathbf{x}^{(5)}$	−3.100	1.273	0.278	$\mu = 16.0, \lambda = 0.0625$
$\mathbf{x}^{(5)}$	5.012	0.415	0.079	obj = 513.31 (484.44)
$\Delta\mathbf{x}^{(6)}$	−0.917	0.359	0.099	$\mu = 16.0, \lambda = 0.0625$
$\mathbf{x}^{(6)}$	4.954	0.438	0.085	obj = 511.52 (484.51)
$\Delta\mathbf{x}^{(7)}$	6.804	−2.756	−0.646	$\mu = 8.0, \lambda = 0.1188$
$\mathbf{x}^{(7)}$	5.762	0.110	0.009	obj = 535.16 (493.42)
$\Delta\mathbf{x}^{(8)}$	−1.076	0.483	0.055	$\mu = 8.0, \lambda = 0.1250$
$\mathbf{x}^{(8)}$	5.628	0.171	0.015	obj = 532.11 (498.40)
$\Delta\mathbf{x}^{(9)}$	0.426	−0.232	0.019	$\mu = 4.0, \lambda = 0.2500$
$\mathbf{x}^{(9)}$	5.734	0.113	0.020	obj = 533.01 (515.63)
$\Delta\mathbf{x}^{(10)}$	−0.118	0.052	0.007	$\mu = 4.0, \lambda = 0.2500$
$\mathbf{x}^{(10)}$	5.705	0.126	0.022	obj = 532.29 (515.67)
$\Delta\mathbf{x}^{(11)}$	0.552	−0.233	−0.043	$\mu = 2.0, \lambda = 0.4614$
$\mathbf{x}^{(11)}$	5.959	0.018	0.002	obj = 539.06 (522.36)
$\Delta\mathbf{x}^{(12)}$	−0.059	0.026	0.004	$\mu = 2.0, \lambda = 0.5000$
$\mathbf{x}^{(12)}$	5.930	0.031	0.004	obj = 538.35 (523.91)
$\Delta\mathbf{x}^{(13)}$	−0.006	0.002	0.001	$\mu = 1.0, \lambda = 1.0000$
$\mathbf{x}^{(13)}$	5.924	0.033	0.005	obj = 538.10 (531.18)
$\Delta\mathbf{x}^{(14)}$	−0.001	0.000	0.000	$\mu = 1.0, \lambda = 1.0000$
$\mathbf{x}^{(14)}$	5.923	0.033	0.006	obj = 538.02 (531.18)
$\Delta\mathbf{x}^{(15)}$	0.038	−0.016	−0.003	$\mu = 0.50, \lambda = 1.8$
$\mathbf{x}^{(15)}$	5.992	0.003	0.001	obj = 539.80 (534.10)
$\Delta\mathbf{x}^{(16)}$	−0.003	0.001	0.000	$\mu = 0.50, \lambda = 2.0$
$\mathbf{x}^{(16)}$	5.986	0.006	0.001	obj = 539.64 (534.53)

Initial point $\mathbf{x}^{(0)}$ is interior and satisfies the only equality constraint $\frac{1}{2}x_1 + x_2 + x_3 = 3$. We begin with barrier multiplier $\mu = 16$. After six moves, the search arrives at point $\mathbf{x}^{(6)}$, which very nearly optimizes barrier objective function

$$\text{max } 90x_1 + 150x_2 + 0x_3 + 16\left[\ln(x_1) + \ln(x_2) + \ln(x_3)\right]$$

Now we begin to reduce μ. In Table 6.3 each reduction updates $\mu \leftarrow \frac{1}{2}\mu$, although μ would normally not be reduced so rapidly. Two steps follow each change to bring the search near the best point for each new μ.

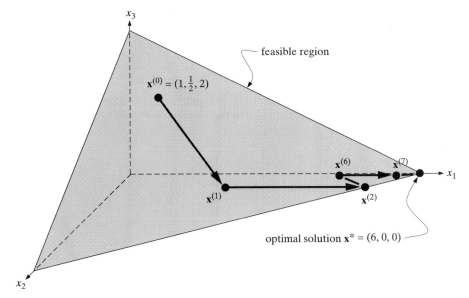

FIGURE 6.8   Barrier Search of Frannie's Firewood Example

Algorithm 6B terminates when μ approaches zero and no further progress is being made. Here we stop after $\mu = \frac{1}{2}$ with $\mathbf{x}^{(16)} = (5.986, 0.006, 0.001)$, which is very near the true LP optimum $\mathbf{x}^* = (6, 0, 0)$.

6.5 DUAL AND PRIMAL-DUAL EXTENSIONS

In the interest of reducing the background needed to understand the notions behind interior point search, the development of affine scaling and barrier algorithms in Sections 6.2 to 6.4 omitted many details of stopping and convergence rules needed in real implementations. To help complete the picture, in this section we sketch briefly some of the missing elements. A full understanding requires familiarity with the development of LP duality in Chapter 7; the presentation of Newton's method in Section 13.6, the treatment of Karush–Kuhn–Tucker optimality conditions in Section 14.4, and more. Still, readers who have absorbed the rest of this chapter should be able to follow the main concepts without further preparation.

Duals of Standard-Form Linear Programs

Now and for the remainder of this section, assume that we are dealing with the minimizing standard-form linear program

$$
\begin{aligned}
\min \quad & \mathbf{c} \cdot \mathbf{x} \\
\text{s.t.} \quad & \mathbf{A}\mathbf{x} = \mathbf{b} \\
& \mathbf{x} \geq \mathbf{0}
\end{aligned}
\qquad (6.20)
$$

Also assume that the rows of $\mathbf{A}$ are linearly independent and that the model is feasible and bounded, so that it has a finite optimal solution.

In Chapter 7 we explain in detail how a closely related **dual** linear program can be posed over the same constants $\mathbf{A}$, $\mathbf{b}$, and $\mathbf{c}$ as this given LP, now called the

primal. Specifically, the dual assigns variables $\mathbf{v}$ to the rows of the main constraints. Including slack variables $\mathbf{r} \geq \mathbf{0}$, the dual's standard form becomes

$$\begin{aligned} \max \quad & \mathbf{b} \cdot \mathbf{v} \\ \text{s.t.} \quad & \mathbf{v}\mathbf{A} + \mathbf{r} = \mathbf{c} \\ & \mathbf{v} \text{ URS}, \quad \mathbf{r} \geq \mathbf{0} \end{aligned} \tag{6.21}$$

This dual is a maximize problem because the primal minimizes. There is one dual constraint for each column or primal variable x_j. Right-hand sides $\mathbf{b}$ of the primal become objective function coefficients in the dual, and vice versa.

To illustrate, return to the minimizing standard-form LP used in Sample Exercises 6.7 to 6.13. Its primal is

$$\begin{aligned} \min \quad & -3x_1 \quad +0x_2 \quad +9x_3 \\ \text{s.t.} \quad & -1x_1 \quad +0x_2 \quad +1x_3 \quad = 3 \\ & +1x_1 \quad +2x_2 \quad +0x_3 \quad = 4 \\ & x_1, x_2, x_3 \geq 0 \end{aligned} \tag{6.22}$$

Assigning dual variables v_1 and v_2 to first and second main constraints, respectively, the corresponding dual is

$$\begin{aligned} \max \quad & +3v_1 \quad +4v_2 \\ \text{s.t.} \quad & -1v_1 \quad +1v_2 \quad +r_1 \quad\quad\quad\quad = -3 \\ & +0v_1 \quad +2v_2 \quad\quad\quad +r_2 \quad\quad = 0 \\ & +1v_1 \quad +0v_2 \quad\quad\quad\quad\quad +r_3 = 9 \\ & v_1, v_2 \text{ URS} \\ & r_1, r_2, r_3 \geq 0 \end{aligned} \tag{6.23}$$

A primal LP and its duals are related in far more ways than a surface rows-versus-columns symmetry. Most significantly, if either has a finite optimal solution, they both do, and the optimal solution values are the same. For example, primal (6.22) has optimal solution $\mathbf{x}^* = (0, 2, 3)$ with value 27, and its dual (6.23) has optimum $\mathbf{v}^* = (9, 0)$, $\mathbf{r}^* = (6, 0, 0)$ with the same optimal value.

For our present purposes the most important primal–dual relationship connects primal solutions $\mathbf{x}$ in (6.20) and dual solutions $(\mathbf{v}, \mathbf{r})$ in (6.21) that are not yet known to be optimal.

 6.22   If $\mathbf{x}$ is feasible in standard-form (minimizing) primal linear program (6.20), and $(\mathbf{v}, \mathbf{r})$ is feasible in corresponding standard-form dual linear program (6.21), the objectives satisfy

$$\mathbf{c} \cdot \mathbf{x} \geq \text{mutual optimal value} \geq \mathbf{b} \cdot \mathbf{v}$$

and the gap

$$\mathbf{c} \cdot \mathbf{x} - \mathbf{b} \cdot \mathbf{v} = \mathbf{x} \cdot \mathbf{r}$$

That is, any feasible primal solution gives an upper bound on the mutual optimal value, any feasible dual solution provides a lower bound, and the expression $\mathbf{x} \cdot \mathbf{r}$ measures the difference. For example, solution $\mathbf{x} = (2, 1, 5)$ is feasible in primal (6.22) with value 39, and $(\mathbf{v}, \mathbf{r}) = (4, -1, 2, 2, 5)$ is feasible in dual (6.22) with value 8. Their mutual optimal value 29 falls between, and the gap is $\mathbf{x} \cdot \mathbf{r} = (2, 1, 5) \cdot (2, 2, 5) = 31$.

Using Duals to Bound Affine Scaling Progress

Algorithm 6A of Section 6.3 can be called a primal search because it moves through a sequence of feasible solutions to primal LP (6.20). Still, a corresponding dual solution can be constructed at each iteration. If that dual solution happens to be feasible in (6.21), we can apply relationship $\boxed{6.22}$ to bound how far our current $\mathbf{x}^{(t)}$ is from optimal.

Recall (formula $\boxed{6.13}$) that Algorithm 6A employs (minimize problem) move directions

$$\Delta \mathbf{x} = -\mathbf{X}_t \mathbf{P}_t \mathbf{c}^{(t)} = -\mathbf{X}_t \left(\mathbf{I} - \mathbf{A}_t^{\mathrm{T}} (\mathbf{A}_t \mathbf{A}_t^{\mathrm{T}})^{-1} \mathbf{A}_t \right) \mathbf{c}^{(t)}$$

where $\mathbf{A}_t = \mathbf{A}\mathbf{X}_t$, $\mathbf{c}^{(t)} = \mathbf{c}\mathbf{X}_t$, and $\mathbf{X}_t$ is the diagonal matrix of current solution $\mathbf{x}^{(t)}$. In the process of carrying out this directional computation, it is straightforward to extract an estimated dual that will measure our progress toward optimality if it happens to be feasible.

$\boxed{6.23}$ As a by-product of direction forming in affine scaling Algorithm 6A for standard-form (minimizing) LP (6.20), estimated dual variable values can be computed:

$$\mathbf{v}^{(t)} = \left(\mathbf{A}_t \mathbf{A}_t^{\mathrm{T}} \right)^{-1} \mathbf{A}_t \mathbf{c}^{(t)}$$

$$\mathbf{r}^{(t)} = \mathbf{c} - \mathbf{v}_t \mathbf{A}$$

Then if $\mathbf{r}^{(t)} \geq \mathbf{0}$, the difference between current solution value $\mathbf{c} \cdot \mathbf{x}^{(t)}$ and an LP optimum is bounded by $\mathbf{x}^{(t)} \cdot \mathbf{r}^{(t)}$. If this does not lead to a stop, the next move direction is

$$\Delta \mathbf{x}^{(t+1)} = - (\mathbf{X}_t)^2 \mathbf{r}^{(t)}$$

For our example at $\mathbf{x}^{(7)} = (2, 1, 5)$, we have already computed in Sample Exercises 6.7 and 6.8 that

$$\mathbf{A}_7 = \begin{pmatrix} -2 & 0 & 5 \\ 2 & 2 & 0 \end{pmatrix}, \qquad (\mathbf{A}_7 \mathbf{A}_7^{\mathrm{T}})^{-1} = \begin{pmatrix} \frac{8}{216} & \frac{4}{216} \\ \frac{4}{216} & \frac{29}{216} \end{pmatrix}, \qquad \mathbf{c}^{(7)} = \begin{pmatrix} -6 \\ 0 \\ 45 \end{pmatrix}$$

These lead to estimated duals

$$\mathbf{v}^{(7)} = \left(\mathbf{A}_7 \mathbf{A}_7 \mathbf{A}_7^{\mathrm{T}} \right)^{-1} \mathbf{A}_7 \mathbf{c}^{(7)} = (8\tfrac{5}{9}, 2\tfrac{7}{9})$$

$$\mathbf{r}^{(7)} = \mathbf{c} - \mathbf{v}_7 \mathbf{A} = (2\tfrac{7}{9}, -5\tfrac{5}{9}, \tfrac{4}{9})$$

With $r_5^{(7)} = -5\tfrac{5}{9} < 0$, we do not obtain a bound, but the next move direction can still be computed:

$$\Delta \mathbf{x}^{(8)} = - (\mathbf{X}_7)^2 \mathbf{r}^{(7)} = (-11\tfrac{1}{9}, 5\tfrac{5}{9}, -11\tfrac{1}{9})$$

On the other hand, feasible (and indeed optimal) dual estimate $(\mathbf{v}^{(t)}, \mathbf{r}^{(t)}) = (9, 0, 6, 0, 0)$ results from computation $\boxed{6.23}$ if we use $\mathbf{x}^{(t)} = (0, 2, 3)$. Thus, as we move very close to that optimal primal solution, bounds from the estimated duals would alert us that further progress is impossible.

Using Duals to Bound Newton Barrier Search Progress

Newton step barrier Algorithm 6B is also a primal search, and the same sort of dual bounding strategy can be employed to know how close it has come to optimality. The only differences relate to the fact that direction $\boxed{6.18}$ (for minimize problems) projects a vector adjusted for the barrier multiplier μ:

$$\Delta \mathbf{x} = -\mathbf{X}_t \mathbf{P}_t \begin{pmatrix} c_1^{(t)} - \mu \\ \vdots \\ c_n^{(t)} - \mu \end{pmatrix}$$

$$= -\mathbf{X}_t \left(\mathbf{I} - \mathbf{A}_t^\mathsf{T} (\mathbf{A}_t \mathbf{A}_t^\mathsf{T})^{-1} \mathbf{A}_t \right) \begin{pmatrix} c_1^{(t)} - \mu \\ \vdots \\ c_n^{(t)} - \mu \end{pmatrix}$$

$\boxed{6.24}$ As a by-product of direction forming in Newton step barrier Algorithm 6B or standard-form (minimizing) LP (6.20), estimated dual variable values can be computed

$$\mathbf{v}^{(t)} = \left(\mathbf{A}_t \mathbf{A}_t^\mathsf{T} \right)^{-1} \mathbf{A}_t \begin{pmatrix} c_1^{(t)} - \mu \\ \vdots \\ c_n^{(t)} - \mu \end{pmatrix}$$

$$\mathbf{r}^{(t)} = \mathbf{c} - \mathbf{v}_t \mathbf{A}$$

Then if $\mathbf{r}^{(t)} \geq \mathbf{0}$, the difference between current solution value $\mathbf{c} \cdot \mathbf{x}^{(t)}$ and an LP optimum is bounded by $\mathbf{x}^{(t)} \cdot \mathbf{r}^{(t)}$. If this does not lead to a stop, the next move direction is

$$\Delta \mathbf{x}^{(t+1)} = -(\mathbf{X}_t)^2 \mathbf{r}^{(t)} + \mu \mathbf{x}^{(t)}$$

We may illustrate again with primal linear program (6.22). Taking $\mathbf{x}^{(7)} = (2, 1, 5)$ and $\mu = 120$ yields

$$\mathbf{v}^{(7)} = \left(\mathbf{A}_7 \mathbf{A}_7^\mathsf{T} \right)^{-1} \mathbf{A}_7 \begin{pmatrix} c_1^{(7)} - \mu \\ \vdots \\ c_n^{(7)} - \mu \end{pmatrix} = (-13\tfrac{2}{3}, -68\tfrac{1}{3})$$

and

$$\mathbf{r}^{(7)} = \mathbf{c} - \mathbf{v}_7 \mathbf{A} = \left(51\tfrac{2}{3}, 136\tfrac{2}{3}, 22\tfrac{2}{3} \right)$$

This $\mathbf{r}^{(t)} \geq \mathbf{0}$, so it provides a valid bound. We may conclude that the present $\mathbf{x}^{(t)}$ has objective value at most $\mathbf{x}^{(t)} \cdot \mathbf{r}^{(t)} = 353\tfrac{1}{3}$ above optimal.

Primal–Dual Interior Point Barrier Search

The most effective commercial interior point algorithms for linear programming adopt a more direct approach than either of the primal methods presented so far. They employ **primal–dual search**, which purposefully changes both primal and dual solutions at the same time.

Primal–dual methods depend on optimality conditions much like those discussed for the two primal methods. This time, however, they apply to the nonlinear barrier model

$$\min \quad \mathbf{c} \cdot \mathbf{x} - \mu \sum_{j} \ln(x_j)$$

$$\text{s.t.} \quad \mathbf{Ax} = \mathbf{b}$$

__6.25__   A vector $\mathbf{x} > \mathbf{0}$ is optimal in the barrier model with multiplier $\mu > 0$ if and only if there exist corresponding dual vectors $\mathbf{v}$ and $\mathbf{r} > \mathbf{0}$ such that

$$\mathbf{Ax} = \mathbf{b}$$
$$\mathbf{vA} + \mathbf{r} = \mathbf{c}$$
$$\sum_{j} \left(x_j r_j - \mu \right) = 0$$

Paralleling the primal and dual relationships above for the linear objective case, conditions __6.25__ now show that interior feasible solutions $\mathbf{x}$ and $(\mathbf{v}, \mathbf{r})$ relate through a final equation which measures the gap between their respective solution values.

Primal–Dual Move Directions

Primal–dual search procedures maintain a primal solution $\mathbf{x}^{(t)} > \mathbf{0}$ and a dual solution $(\mathbf{v}^{(t)}, \mathbf{r}^{(t)})$ that has $\mathbf{r}^{(t)} > \mathbf{0}$. Each move steps in a Newton-like direction $(\Delta\mathbf{x}^{(t+1)}, \Delta\mathbf{v}^{(t+1)}, \Delta\mathbf{r}^{(t+1)})$ constructed to improve satisfaction of conditions __6.25__.

__6.26__   At current interior feasible primal solution $\mathbf{x}^{(t)} > \mathbf{0}$ and interior feasible dual solution $(\mathbf{v}^{(t)}, \mathbf{r}^{(t)})$ with $\mathbf{r}^{(t)} > \mathbf{0}$, primal–dual interior point search adopts the Newton move direction

$$\Delta\mathbf{v}^{(t+1)} \leftarrow -\left(\mathbf{AX}_t\mathbf{R}_t^{-1}\mathbf{A}^\mathsf{T}\right)^{-1}\mathbf{AR}_t^{-1}\begin{pmatrix} \mu_t - x_1^{(t)}r_1^{(t)} \\ \vdots \\ \mu_t - x_n^{(t)}r_n^{(t)} \end{pmatrix}$$

$$\Delta\mathbf{r}^{(t+1)} \leftarrow -\Delta\mathbf{v}^{(t+1)}\mathbf{A}$$

$$\Delta\mathbf{x}^{(t+1)} \leftarrow \mathbf{R}_t^{-1}\left(\begin{pmatrix} \mu_t - x_1^{(t)}r_1^{(t)} \\ \vdots \\ \mu_t - x_n^{(t)}r_n^{(t)} \end{pmatrix} - \mathbf{X}_t\Delta\mathbf{r}^{(t+1)}\right)$$

where $\mathbf{X}_t$ and $\mathbf{R}_t$ are diagonal matrix forms of $\mathbf{x}^{(t)}$ and $\mathbf{r}^{(t)}$, respectively.

Notice that these move direction are of the same scaled projection format as earlier ones $\boxed{6.23}$ and $\boxed{6.24}$. As befits the primal–dual character of the search, however, matrix $\mathbf{A}$ is sometimes scaled by primal solution $\mathbf{X}_t$ and sometimes by dual $\mathbf{R}_t^{-1}$. Also, the vector being projected has components $(\mu_t - x_j^{(t)} r_j^{(t)})$, relating to the last condition of $\boxed{6.25}$ instead of the primal objective function $\mathbf{c}$.

We may illustrate computations $\boxed{6.26}$ for our LP of (6.22) and (6.23) from interior feasible solutions $\mathbf{x}^{(7)} = (2, 1, 5)$, $\mathbf{v}^{(7)} = (4, -1)$, and $\mathbf{r}^{(7)} = (2, 2, 5)$ with barrier multiplier $\mu_7 = 24$. There

$$\mathbf{X}_7 = \begin{pmatrix} 2 & 0 & 0 \\ 0 & 1 & 0 \\ 0 & 0 & 5 \end{pmatrix}, \quad \mathbf{R}_7^{-1} = \begin{pmatrix} \frac{1}{2} & 0 & 0 \\ 0 & \frac{1}{2} & 0 \\ 0 & 0 & \frac{1}{5} \end{pmatrix}$$

and

$$\begin{pmatrix} \mu_7 - x_1^{(7)} r_1^{(7)} \\ \vdots \\ \mu_7 - x_n^{(7)} r_n^{(7)} \end{pmatrix} = \begin{pmatrix} 24 - 2 \cdot 2 \\ 24 - 1 \cdot 2 \\ 24 - 5 \cdot 5 \end{pmatrix} = \begin{pmatrix} 20 \\ 22 \\ -1 \end{pmatrix}$$

Then

$$\Delta \mathbf{v}^{(8)} \leftarrow -\left(\mathbf{A} \mathbf{X}_7 \mathbf{R}_7^{-1} \mathbf{A}^{\mathsf{T}} \right)^{-1} \mathbf{A} \mathbf{R}_7^{-1} \begin{pmatrix} \mu_7 - x_1^{(7)} r_1^{(7)} \\ \vdots \\ \mu_7 - x_n^{(7)} r_n^{(7)} \end{pmatrix} = \begin{pmatrix} -\frac{7}{25} \\ -10\frac{19}{25} \end{pmatrix}$$

$$\Delta \mathbf{r}^{(8)} \leftarrow -\Delta \mathbf{v}^{(8)} \mathbf{A} = \begin{pmatrix} 10\frac{12}{25} \\ 21\frac{13}{25} \\ \frac{7}{25} \end{pmatrix}$$

$$\Delta \mathbf{x}^{(8)} \leftarrow \mathbf{R}_7^{-1} \left[\begin{pmatrix} \mu_7 - x_1^{(7)} r_1^{(7)} \\ \vdots \\ \mu_7 - x_n^{(7)} r_n^{(7)} \end{pmatrix} - \mathbf{X}_7 \Delta \mathbf{r}^{(8)} \right] = \begin{pmatrix} -\frac{12}{25} \\ \frac{6}{25} \\ -\frac{12}{25} \end{pmatrix}$$

Primal–Dual Algorithms

Algorithm 6C summarizes these observations in a primal–dual barrier search algorithm for linear programming. There are many variations, depending on the step sizes λ and barrier multipliers μ that are generated at Step 3. Experience has shown that the best implementations can challenge the simplex algorithm on a variety of LP models and significantly outperform it on many larger instances.

ALGORITHM 6C: PRIMAL–DUAL INTERIOR SEARCH FOR LINEAR PROGRAMS

Step 0: Initialization. Choose any starting primal feasible interior point solution, $\mathbf{x}^{(0)} > \mathbf{0}$, and dual feasilble $(\mathbf{v}^{(0)}, \mathbf{r}^{(0)})$ with $\mathbf{r}^{(0)} > \mathbf{0}$. Also select a realtively large initial barrier multiplier μ_0, and set solution index $t \leftarrow 0$.

Step 1: Optimality. If primal-to-dual solution gap $\mathbf{cx}^{(t)} - \mathbf{bv}^{(t)}$ is sufficiently small, stop. Current point $\mathbf{x}^{(t)}$ is either optimal in the given LP, or very nearly so.

Step 2: Move Direction. Construct the next move direction to improve satisfaction of optimality conditions $\boxed{6.25}$ as

$$\Delta \mathbf{v}^{(t+1)} \leftarrow -\left(\mathbf{A}\mathbf{X}_t\mathbf{R}_t^{-1}\mathbf{A}^{\mathsf{T}}\right)^{-1}\mathbf{A}\mathbf{R}_t^{-1}\begin{pmatrix} \mu_t - x_1^{(t)}r_1^{(t)} \\ \vdots \\ \mu_t - x_n^{(t)}r_n^{(t)} \end{pmatrix}$$

$$\Delta \mathbf{r}^{(t+1)} \leftarrow -\Delta \mathbf{v}^{(t+1)}\mathbf{A}$$

$$\Delta \mathbf{x}^{(t+1)} \leftarrow \mathbf{R}_t^{-1}\left(\begin{pmatrix} \mu_t - x_1^{(t)}r_1^{(t)} \\ \vdots \\ \mu_t - x_n^{(t)}r_n^{(t)} \end{pmatrix} - \mathbf{X}_t\Delta \mathbf{r}^{(t+1)}\right)$$

where $\mathbf{X}_t$ and $\mathbf{R}_t$ are diagonal matrix forms of $\mathbf{x}^{(t)}$ and $\mathbf{r}^{(t)}$, respectively.

Step 3: Step Size and Barrier Multiplier. Choose a compatible step size $\lambda_{t+1} > 0$ and next barrier multiplier $\mu_{t+1} \leq \mu_t$.

Step 4: Advance. Compute a new solution

$$\mathbf{x}^{(t+1)} \leftarrow \mathbf{x}^{(t)} + \lambda_{t+1}\Delta \mathbf{x}^{(t+1)}$$
$$\mathbf{v}^{(t+1)} \leftarrow \mathbf{v}^{(t)} + \lambda_{t+1}\Delta \mathbf{v}^{(t+1)}$$
$$\mathbf{r}^{(t+1)} \leftarrow \mathbf{r}^{(t)} + \lambda_{t+1}\Delta \mathbf{r}^{(t+1)}$$

Then advance $t \leftarrow t + 1$, and return to Step 1.

EXERCISES

6-1 Consider the linear program

$$\max \quad 2w_1 + 3w_2$$
$$\text{s.t.} \quad 4w_1 + 3w_2 \leq 12$$
$$w_2 \leq 2$$
$$w_1, w_2 \geq 0$$

(a) ⊠ Solve the problem graphically.

(b) ⊠ Determine the direction $\Delta\mathbf{w}$ of most rapid improvement in the objective function at any solution $\mathbf{w}$.

(c) Explain why the direction of part (b) is feasible at any interior point solution to the model.

(d) ⊠ Show that $\mathbf{w}^{(0)} = (1, 1)$ is an interior point solution.

(e) ⊠ Determine the maximum step $\lambda_{\max}$ from the point $\mathbf{w}^{(0)}$ that preserves feasibility in the direction of part (b).

(f) ⊠ Plot the move of part (e) and the resulting new point $\mathbf{w}^{(1)}$ in the graph of part (a).

(g) Explain why it is easier to find a good move direction at $\mathbf{w}^{(0)}$ than at $\mathbf{w}^{(1)}$.

6-2 Do Exercise 6-1 for the LP

$$\min \quad 9w_1 + 1w_2$$
$$\text{s.t.} \quad 3w_1 + 2w_2 \geq 6$$
$$2w_1 + 3w_2 \geq 6$$
$$w_1, w_2 \geq 0$$

and point $\mathbf{w}^{(0)} = (2, 3)$.

6-3 The following plot shows several feasible points in a linear program and contours of its objective function.

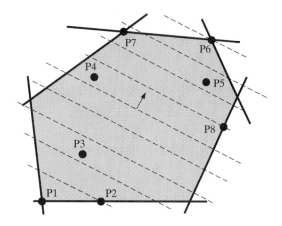

Determine whether each of the following sequences of solutions could have been one followed by an interior point algorithm applied to the corresponding standard-form LP.

(a) ⊠ P1,P5,P6
(b) P3,P5,P6
(c) ⊠ P3,P8,P6
(d) P3,P7,P6
(e) ⊠ P3,P4,P5,P6
(f) P2,P5,P6

6-4 Determine whether each of the following is an interior point solution to the standard-form LP constraints

$$4x_1 + 1x_3 = 13$$
$$5x_1 + 5x_2 = 15$$
$$x_1, x_2, x_3 \geq 0$$

(a) ⊠ $\mathbf{x} = (3, 0, 1)$
(b) $\mathbf{x} = (2, 1, 5)$
(c) ⊠ $\mathbf{x} = (1, 2, 9)$
(d) $\mathbf{x} = (5, 1, 1)$

(e) ⊠ $\mathbf{x} = (2, 2, 1)$
(f) $\mathbf{x} = (0, 3, 13)$

6-5 Write all conditions that a feasible direction $\Delta \mathbf{w}$ must satisfy at any interior point solution to each of the following standard-form systems of LP constraints.

(a) ⊠ $2w_1 + 3w_2 - 3w_3 = 5$
$$4w_1 - 1w_2 + 1w_3 = 3$$
$$w_1, w_2, w_3 \geq 0$$

(b) $7w_1 + 1w_2 - 1w_3 = 8$
$$2w_1 - 5w_2 + 3w_3 = -7$$
$$w_1, w_2, w_3 \geq 0$$

6-6 Table 6.4 shows several constraint matrices $\mathbf{A}$ (or $\mathbf{A}_t$) of standard-form LPs and the corresponding projection matrices $\mathbf{P}$ (or $\mathbf{P}_t$). Use these results to compute the feasible direction for the specifed equality constraints that is nearest to the given direction $\mathbf{d}$, and verify that the result satisfies feasible direction conditions at any interior point solution.

(a) ⊠ $x_1 + 2x_2 + x_3 = 4$ and $\mathbf{d} = (3, -6, 3)$
$$-2x_1 + x_2 = -1$$

(b) $3x_1 - x_2 + 4x_3 = 4$ and $\mathbf{d} = (1, -1, 7)$
$$x_2 - 2x_3 = 1$$

6-7 Consider the standard-form LP

$$\min \quad 14z_1 + 3z_2 + 5z_3$$
$$\text{s.t.} \quad 2z_1 - z_3 = 1$$
$$z_1 + z_2 = 1$$
$$z_1, z_2, z_3 \geq 0$$

(a) ⊠ Determine the direction of most rapid objective function improvement at any solution $\mathbf{z}$.
(b) ⊠ Compute the projection matrix $\mathbf{P}$ for the main equality constraints.
(c) ⊠ Apply your $\mathbf{P}$ to project the direction of (a).
(d) ⊠ Verify that the result of part (c) is improving and feasible at any interior point solution.
(e) Describe the sense in which the direction of part (c) is good for improving search at any interior point solution.

6-8 Do Exercise 6-7 for

$$\max \quad 4z_1 - z_2 + 7z_3$$
$$\text{s.t.} \quad z_1 + z_3 = 4$$
$$2z_2 = 12$$
$$z_1, z_2, z_3 \geq 0$$

TABLE 6.4 Projection Matrices

A or A_t			P or P_t		
1	−1	2	0.3333	−0.3333	−0.3333
0	1	−1	−0.3333	0.3333	0.3333
			−0.3333	0.3333	0.3333
1	10	3	0.2540	0.1016	−0.4232
−2	5	0	0.1016	0.0406	−0.1693
			−0.4232	−0.1693	0.7054
1	2	1	0.0333	0.0667	−0.1667
−2	1	0	0.0667	0.1333	−0.3333
			−0.1667	−0.3333	0.8333
3	−1	4	0.0816	−0.2449	−0.1224
0	1	−2	−0.2449	0.7347	0.3673
			−0.1224	0.3673	0.1837
12	−3	4	0.0819	−0.0755	−0.1132
0	3	−2	−0.0755	0.3019	0.4528
			−0.1132	0.4528	0.6793
4	−3	2	0.0533	−0.0710	−0.2130
0	3	−1	−0.0710	0.0947	0.2840
			−0.2130	0.2840	0.8521
−1	2	0	0.6667	0.3333	0.3333
1	−1	−1	0.3333	0.1667	0.1667
			0.3333	0.1667	0.1667

6-9 An interior point search using scaling has reached current solution $\mathbf{x}^{(7)} = (2, 5, 1, 9)$. Compute the affine scaled $\mathbf{y}$ that would correspond to each of the following $\mathbf{x}$'s.

(a) ⊠ $(1, 1, 1, 1)$
(b) $(4, 2, 3, 5)$
(c) ⊠ $(3, 5, 1, 6)$
(d) $(2, 5, 1, 9)$

6-10 Do Exercise 6-9 taking the listed vectors as scaled $\mathbf{y}$'s and computing the corresponding $\mathbf{x}$.

6-11 Consider the standard-form LP

$$\begin{array}{ll} \min & 2x_1 + 3x_2 + 5x_3 \\ \text{s.t.} & 2x_1 + 5x_2 + 3x_3 = 12 \\ & x_1, x_2, x_3 \geq 0 \end{array}$$

with current interior point solution $\mathbf{x}^{(3)} = (2, 1, 1)$.

(a) ⊠ Sketch the feasible space in a diagram like Figure 6.4(a) and identify both the current solution and an optimal extreme point.
(b) ⊠ Sketch the corresponding affine-scaled feasible space showing scaled equivalents of all points in part (a).
(c) ⊠ Derive the associated affine-scaled standard form 6.10 .

6-12 Do Exercise 6-11 using the LP

$$\begin{array}{ll} \max & 6x_1 + 1x_2 + 2x_3 \\ \text{s.t.} & x_1 + x_2 + 5x_3 = 18 \\ & x_1, x_2, x_3 \geq 0 \end{array}$$

and current solution $\mathbf{x}^{(9)} = (1, 7, 2)$.

6-13 Return to the LP of Exercise 6-11.

(a) ⊠ Compute in both $\mathbf{x}$ and $\mathbf{y}$ space the next move direction that would be pursued by affine scaling Algorithm 6A.
(b) ⊠ Verify that the $\Delta\mathbf{x}$ of part (a) is both improving and feasible.
(c) ⊠ Compute the step size λ that Algorithm 6A would apply to your move directions in part (a).
(d) ⊠ Plot the move of parts (a) and (c) in both original x-space and scaled y-space.

6-14 Do Exercise 6-13 on the LP of Exercise 6-12.

6-15 Consider the standard-form LP

$$\begin{array}{ll} \min & 10x_1 + 1x_2 \\ \text{s.t.} & x_1 - x_2 + 2x_3 = 3 \\ & x_2 - x_3 = 2 \\ & x_1, x_2, x_3 \geq 0 \end{array}$$

(a) ⊠ Show that $\mathbf{x}^{(0)} = (4, 3, 1)$ is an appropriate point to start affine scaling Algorithm 6A.
(b) ⊠ Derive the associated scaled standard form corresponding to solution $\mathbf{x}^{(0)}$.
(c) ⊠ Compute the move direction $\Delta\mathbf{x}$ that would be pursued from $\mathbf{x}^{(0)}$ by Algorithm 6A (refer to Table 6.4 for projection matrices).
(d) ⊠ Show that your direction is improving and feasible at $\mathbf{x}^{(0)}$.
(e) ⊠ Compute the step size λ that Algorithm 6A would apply to your direction, and determine the new point $\mathbf{x}^{(1)}$ that results.

6-16 Do Exercise 6-15 using the LP

$$\begin{array}{ll} \max & 6x_1 + 8x_2 + 10x_3 \\ \text{s.t.} & 9x_1 - 2x_2 + 4x_3 = 6 \\ & 2x_2 - 2x_3 = -1 \\ & x_1, x_2, x_3 \geq 0 \end{array}$$

and initial solution $\mathbf{x}^{(0)} = (\frac{1}{3}, \frac{1}{2}, 1)$.

6-17 Suppose that affine scaling Algorithm 6A has reached current solution $\mathbf{x}^{(11)} = (3, 1, 9)$. Determine whether each of the following next move directions would cause the algorithm to stop and why. If not, compute $\mathbf{x}^{(12)}$.

(a) ⊠ $\Delta\mathbf{x} = (6, 2, 2)$
(b) $\Delta\mathbf{x} = (0, 4, -9)$
(c) ⊠ $\Delta\mathbf{x} = (6, -6, 0)$
(d) $\Delta\mathbf{x} = (0, -2, 0)$

6-18 Consider the standard-form LP

$$\max \quad 13w_1 - 2w_2 + w_3$$
$$\text{s.t.} \quad 3w_1 + 6w_2 + 4w_3 = 12$$
$$w_1, w_2, w_3 \geq 0$$

(a) ⊠ Sketch the feasible space in a plot like Figure 6.6, identify an optimal extreme point, and show that $\mathbf{w}^{(1)} = (1.4, 0.7, 0.9)$ and $\mathbf{w}^{(2)} = (0.01, 0.01, 2.9775)$.
(b) ⊠ Form the corresponding log barrier problem with multiplier $\mu > 0$.
(c) ⊠ Using $\mu = 10$, evaluate the original and log barrier objective functions at $\mathbf{w}^{(1)}$ and $\mathbf{w}^{(2)}$. Then comment on the effect of barrier terms at points far from the boundary versus ones near the boundary.
(d) ▢ ⊠ Use the class optimization software to solve log barrier form (b) with multipliers $\mu = 100, 10,$ and 1.
(e) How does the trajectory of optimal solutions to part (b) evolve as $\mu \to 0$?

6-19 Do Exercise 6-18 for the LP

$$\min \quad 2w_1 + 5w_2 - w_3$$
$$\text{s.t.} \quad w_1 + 6w_2 + 2w_3 = 18$$
$$w_1, w_2, w_3 \geq 0$$

and points $\mathbf{w}^{(1)} = (6, 1, 3)$, $\mathbf{w}^{(2)} = (0.01, 0.01, 8.965)$.

6-20 Determine whether each of the following could be the first four barrier multiplier values μ used in the outer loop of a barrier search Algorithm 6B.

(a) ⊠ 100,80,64,51.2
(b) 100,200,400,800
(c) ⊠ 100,500,1,90
(d) 600,300,150,75

6-21 Assume that barrer Algorithm 6B has computed an appropriate move direction $\Delta\mathbf{x}$ for its standard-form LP and a maximum feasible step size λ_{max}. For each of the original objective functions below, determine which of the following curves best depicts how the corresponding barrier objective function will vary with $\lambda \in [0, \lambda_{max}]$.

barrier objective value after step λ

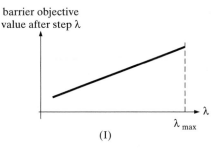

(I)

barrier objective value after step λ

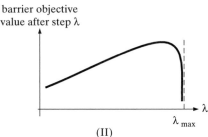

(II)

barrier objective value after step λ

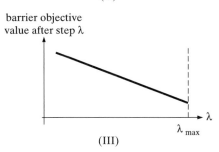

(III)

barrier objective value after step λ

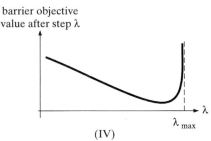

(IV)

(a) ⊠ $\max 34x_1 - 19x_2 - 23x_3 + 4x_4$
(b) $\min 444x_1 + 15x_2 + 1x_3 + 98x_4$

6-22 Consider the LP

$$\min \quad 4x_1 - x_2 + 2x_3$$
$$\text{s.t.} \quad 4x_1 - 3x_2 + 2x_3 = 13$$
$$3x_2 - x_3 = 1$$
$$x_1, x_2, x_3 \geq 0$$

(a) ⊘ Show that $\mathbf{x}^{(0)} = (3, 1, 2)$ is an appropriate point to start barrier Algorithm 6B.

(b) ⊘ Form the corresponding log barrier problem with multiplier $\mu = 10$.

(c) ⊘ Compute the move direction $\Delta\mathbf{x}$ that could be pursued from $\mathbf{x}^{(0)}$ by barrier Algorithm 6B. (See Table 6.4 for the projection matrix required.)

(d) ⊘ Verify that your direction of part (c) is improving and feasible in the barrier model of part (b) at $\mathbf{x}^{(0)}$.

(e) ⊘ Determine the maximum step λ_{max} from $\mathbf{x}^{(0)}$ that preserves feasibility in your direction of part (c) and the λ to be employed.

(f) ⊘ Will the barrier objective function first increase, then decrease, or first decrease, then increase, as the step λ applied to your direction of part (c) grows from 0 to λ_{max}? Explain.

(g) The next time that multiplier μ is changed, should it increase or decrease? Explain.

6-23 Do Exercise 6-22 for the LP

$$\max \quad -x_1 + 3x_2 + 8x_3$$
$$\text{s.t.} \quad x_1 + 2x_2 + x_3 = 14$$
$$-2x_1 + x_2 = 3$$
$$x_1, x_2, x_3 \geq 0$$

at $\mathbf{x}^{(0)} = (1, 5, 3)$.

6-24 Return to the LP and $\mathbf{x}^{(0)}$ of Exercise 6-15.

(a) ⊘ Show that $\mathbf{x}^{(0)} = (4, 3, 1)$ is an appropriate point to start barrier Algorithm 6B.

(b) ⊘ Compute the move direction $\Delta\mathbf{x}$ that would be pursued from $\mathbf{x}^{(0)}$ by Algorithm 6B with $\mu = 10$ (refer to Table 6.4 for projection matrices).

(c) ⊘ Show that your direction is improving and feasible at $\mathbf{x}^{(0)}$.

(d) ⊘ Compute the maximum step size λ_{max} that can be applied to your direction of part (b) at $\mathbf{x}^{(0)}$ without losing feasibility.

(e) ⊘ Compute the step size λ that Algorithm 6B would apply to your direction, and determine the new point $\mathbf{x}^{(1)}$ that results.

6-25 Do Exercise 6-24 for the LP and $\mathbf{x}^{(0)}$ of Exercise 6-16.

SUGGESTED READING

Bazaraa, Mokhtar S., John J. Jarvis, and Hanif Sherali (1990), *Linear Programming and Network Flows*, Wiley, New York.

Fang, Shu-Cherng, and Sarat Puthenpura (1993), *Linear Optimization and Extensions Theory and Algorithms*, Prentice Hall, Upper Saddle River, N.J.

C H A P T E R 7

Duality and Sensitivity in Linear Programming

• •

With the simplex and interior point methods of Chapters 5 and 6 we can compute mathematically optimal solutions to linear programming models. By comparison to nonlinear, discrete, and other more difficult optimization forms, this is no small achievement. Still, it rarely fulfills all an analyst's needs.

Mathematical optima do not suffice because the constant parameters from which they are derived—such items as costs, profits, yields, supplies, and demands—are almost never known with certainty at the time the model is solved. Often, they are not even within a factor of 10.

How much can we trust mathematically optimal answers to very imperfectly parameterized models? Maybe a cost or demand controls everything. Maybe it could change dramatically with absolutely no impact.

These are exactly the sorts of questions that **sensitivity analysis** tries to address. In Section 1.3 we explained that constants of OR models are really **input parameters**—values we agree to take as fixed at the system boundary in order to produce a tractable model. Optimal solutions simply provide a best choice of decision variables for one fixing of the inputs. Sensitivity analysis then tries to complete the picture by studying how results would vary with changes in parameter values.

Since linear programs have proved the most tractable of mathematical programs to numerical search, it should not surprise us that they also admit powerful sensitivity analysis. In this chapter we will see how post-optimality analyses of LP models can exploit the by-products of our search for an optimum to illuminate how the solution might change with variations in input variables. Remarkably in fact, we will see that there is an entirely separate **dual** linear program, defined on the same input constants as the main **primal** model, that has optimal solutions replete with sensitivity insights.

7.1 GENERIC ACTIVITIES VERSUS RESOURCES PERSPECTIVE

We have encountered linear programming models drawn from a wide range of application settings. How can we speak about the sensitivity of their optimal solutions to changes in input parameters when they model such different things?

299

We need a generic perspective—a standard intuition about the variables, constraints, and objective functions of LP models. Then, although the details will certainly change with the model, we can still speak in broad, common terms about the meaning of sensitivity results.

Objective Functions as Costs and Benefits

The easiest part of optimization models to interpret generically is their objective functions. Although the exact meaning varies greatly, virtually every model we will encounter can be thought about in terms of **costs** and **benefits**.

> **7.1**   Optimization model objective functions usually can be interpreted as minimizing some measure of cost or maximizing some measure of benefit.

Choosing a Direction for Inequality Constraints

Now consider inequality constraints. What does a $\geq$ inequality sign typically signify? Is a $\leq$ sign any different?

Every beginning algebra student knows that there is no mathematical distinction. For example, the gasoline demand constraint of the Two Crude refining model (2.3) can be written equivalently as either

$$0.3x_1 + 0.4x_2 \geq 2.0 \quad \text{or} \quad -0.3x_1 - 0.4x_2 \leq -2.0$$

Still, the first certainly depicts more clearly the idea that output must meet or exceed demand.

Most constraints have a "natural" format of this sort. No absolute principle tells us which direction is most intuitive, but a rule of thumb covers the foregoing and most other examples:

> **7.2**   The most natural expression of a constraint is usually the one making the right-hand-side constant nonnegative.

Inequalities as Resource Supplies and Demands

At the level of intuitive understanding, we see that $\leq$ inequalities do differ from $\geq$ inequalities. To assign a generic meaning, let us review some of the many $\leq$ inequalities that we have encountered:

- $x_1 \leq 9$ restricts the supply of Saudi petroleum in the Two Crude model (2.6) in Section 2.1.
- $4x_1 + 2x_2 \leq 4800$ in model (5.1) (Section 5.1) limits the wood supply in the Top Brass Trophy example.
- $0.25(z_{1/4,A,B} + z_{1/4,A,C} + z_{1/4,B,C}) + 0.40(z_{1/2,A,B} + z_{1/2,A,C} + z_{1/2,B,C}) \leq 4500$ enforces the supply limit on pressing capacity in the CFPL model of Table 4.5 in Section 4.3.
- $y_{11} + y_{12} \leq 20$ fixes the availability of overtime in the ONB example of Table 4.7 in Section 4.4.

Common elements are a resource and a limited supply of that resource.

| 7.3 | Optimization model constraints of the $\leq$ form usually can be interpreted as restricting the **supply** of some commodity or resource.

The corresponding interpretation of $\geq$ inequalities is as output demands.

- $0.4x_1 + 0.2x_2 \geq 1.5$ in model (2.6) (Section 2.1) requires jet fuel output to meet demand in the Two Crude refining example.
- $0.120x_1 + 0.011x_2 + 1.0x_6 \geq 10$ set the minimum acceptable level of chromium in the blend of the Swedish Steel model (4.4) (Section 4.2).
- $y_{1/8,A,B} + y_{1/8,B,B} + y_{1/8,C,B} - z_{1/2,A,B} - z_{1/2,A,C} - z_{1/2,B,C} \geq 0$ demands production of $\frac{1}{8}$-inch veneer to exceed consumption in the CFPL example of Table 4.5 (Section 4.3).
- $1y_{12} + 1x_{13} + 0.8z_{20} - w_{21} \geq 8$ enforces the demand that all checks be processed by 22:00 (10 P.M.) in the ONB model of Table 4.7 (Section 4.4).

As with $\leq$ inequalities, the common element is a commodity or resource, but the direction is different.

| 7.4 | Optimization model constraints of the $\geq$ form usually can be interpreted as requiring satisfaction of a **demand** for some commodity or resource.

Equality Constraints as Both Supplies and Demands

Equality constraints are hybrids. Consider, for example, the Swedish Steel model (4.4) constraint:

$$x_1 + x_2 + x_3 + x_4 + x_5 + x_6 + x_7 = 1000$$

enforcing the requirement that ingredients in the melting furnace charge should have a total weight of exactly 1000 kilograms. This (or any other) equality can just as well be expressed as two opposed inequalities:

$$x_1 + x_2 + x_3 + x_4 + x_5 + x_6 + x_7 \leq 1000$$

and

$$x_1 + x_2 + x_3 + x_4 + x_5 + x_6 + x_7 \geq 1000$$

Thinking of equalities in this way makes it easy to see the hybrid.

| 7.5 | Optimization model equality constraints usually can be interpreted as imposing both a supply restriction and a demand requirement on some commodity or resource.

The effect is a mixture of both limits on supply of a resource and demand for its output.

Variable-Type Constraints

Variable-type limitations such as nonnegativities are constraints in a linear program. We can even stretch principle | 7.4 | to think of nonnegativities as demands for some "positiveness" resource. Still, it usually makes more sense to keep variable-type constraints separate.

| 7.6 | Nonnegativity and other sign restriction constraints are usually best interpreted as declarations of variable type rather than supply or demand limits on resources. |

Variables as Activities

Turning now to the decision variables in an optimization model, let us again recall some examples:

- x_1 in the Two Crude model (2.6) (Section 2.1) chooses the amount of Saudi crude to be refined.
- $x_{p,m}$ in the TP model (4.5) (Section 4.3) decides the amount of product p produced at mill m.
- z_h in the ONB model of Table 4.7 (Section 4.4) sets the number of part-time employees starting at hour h.
- x_j in the Swedish Steel model (4.4) (Section 4.2) establishes the number of kilograms of ingredient j used in the blend.

The common element shared by these and other examples is a sense of **activity**.

| 7.7 | Decision variables in optimization models can usually be interpreted as choosing the level of some activity. |

LHS Coefficients as Activity Inputs and Outputs

Summarizing, our generic linear program chooses activity levels of appropriate sign to minimize cost or maximize benefit, subject to ≤ supply limits on input resources, ≥ demand requirements for output resources, and = constraints doing both. It follows immediately how we should think about the objective function and constraint coefficients on decision variables:

| 7.8 | Nonzero objective function and constraint coefficients on LP decision variables display the impacts per unit of the variable's activity on resources or commodities associated with the objective and constraints. |

Sometimes those impacts become clearer in block diagrams such as those of Figure 7.1. Each variable or activity forms a block. Inputs and outputs to the block indicate how a unit of the activity affects the constraints and objective. For example, each unit (thousand barrels) of Saudi petroleum refined in the Two Crude example (2.6) inputs 1 unit (thousand barrels) of Saudi availability and 20 units (thousand dollars) cost to output 0.3 unit (thousand barrels) of gasoline, 0.4 unit (thousand barrels) of jet fuel, and 0.2 unit (thousand barrels) of lubricants.

The second diagram in Figure 7.1 comes from the CFPL model of Table 4.5. Peeling "good"-quality logs from supplier 1 as $\frac{1}{16}$-inch green veneer consumes log availability and cost to obtain specified amounts of different veneer grades.

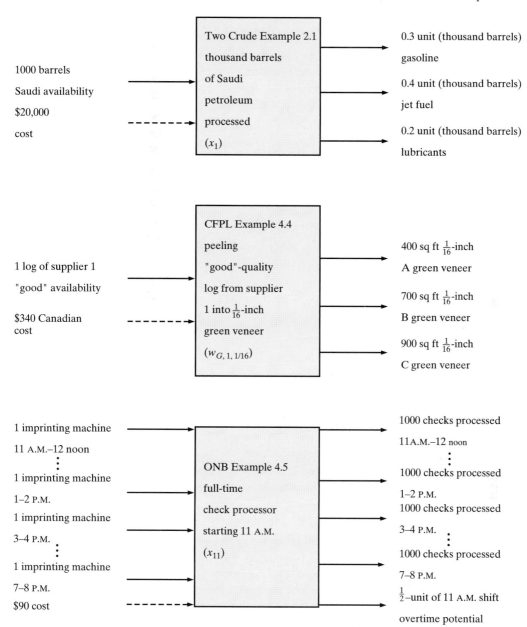

Figure 7.1   Inputs and Outputs for Activities in Various Models

Finally, we have an activity from the ONB model of Table 4.7. Using a full-time employee on the 11 A.M. shift in the ONB example inputs cost and equipment for each on-duty hour to output check processing through the shift and overtime potential. Each input or output produces a nonzero model coefficient.

SAMPLE EXERCISE 7.1: GENERICALLY INTERPRETING LINEAR PROGRAMS

Give a generic interpretation of the objective function, constraints variables, and coefficients in the following linear program:

$$
\begin{array}{lrrrrcr}
\max & +13x_1 & +24x_2 & +5x_3 & +50x_4 & & \\
\text{s.t.} & +1x_1 & +3x_2 & & & \geq & 89 \\
& & & -3x_3 & -5x_4 & \leq & -60 \\
& +10x_1 & +6x_2 & +8x_3 & +2x_4 & \leq & 608 \\
& & +1x_2 & & +1x_4 & \leq & 28 \\
& x_1, \ldots, x_4 \geq 0
\end{array}
$$

Analysis: We begin by reversing the direction of the second inequality to make its right-hand side nonnegative (principle $\boxed{7.2}$):

$$
\begin{array}{lrrrrcr}
\max & +13x_1 & +24x_2 & +5x_3 & +50x_4 & & \\
\text{s.t.} & +1x_1 & +3x_2 & & & \geq & 89 \\
& & & +3x_3 & +5x_4 & \geq & 60 \\
& +10x_1 & +6x_2 & +8x_3 & +2x_4 & \leq & 608 \\
& & +1x_2 & & +1x_4 & \leq & 28 \\
& x_1, \ldots, x_4 \geq 0
\end{array}
$$

We may now interpret the model as one of deciding the optimal level of 4 activities corresponding to the 4 decision variables (principle $\boxed{7.7}$). The objective function maximizes the benefit derived from these activities (principle $\boxed{7.1}$) with coefficients showing the benefit per unit activity (principle $\boxed{7.8}$).

The 4 main constraints deal with 4 commodities or resources. The initial 2, being of $\geq$ form, specify a demand of 89 for the first commodity and 60 for the second (principle $\boxed{7.4}$). The last 2, being of $\leq$ form, restrict the supply of commodities 3 and 4. Nonnegativity constraints merely enforce the variable type (principle $\boxed{7.6}$).

Coefficients on the left-hand side of main constraints show inputs and outputs per unit of each variable's activity (principle $\boxed{7.8}$). Specifically, each unit of activity 1 consumes 10 of commodity 3 to produce 1 of commodity 1; each unit of activity 2 consumes 6 of commodity 3 and 1 of commodity 4 to produce 3 of commodity 1; each unit of activity 3 consumes 8 of commodity 3 to produce 3 of commodity 2; and each unit of activity 4 consumes 2 of commodity 3 and 1 of commodity 4 to produce 5 of commodity 2.

7.2 QUALITATIVE SENSITIVITY TO CHANGES IN MODEL COEFFICIENTS

Armed with a generic way of thinking about LP models, we are ready to consider the sensitivity of optimization model results to changes in their input parameters or constants. Let us begin qualitatively. Much can be learned from looking just at the directions of change rather than their magnitudes.

Relaxing versus Tightening Constraints

Consider first relaxing versus tightening constraints. Figure 7.2 shows the idea graphically. A two-variable model in part (a) has the feasible region indicated by shading. Relaxing a constraint as in part (b) admits new feasible solutions. The corresponding optimal value must stay the same or improve. On the other hand, if the constraint is tightened as in part (c), fewer feasible solutions are available. The optimal value can only stay the same or worsen.

> **7.9** | **Relaxing** the constraints of an optimization model either leaves the optimal value unchanged or makes it better (higher for a maximize, lower for a minimize). **Tightening** the constraints either leaves the optimal value unchanged or makes it worse.

Principle 7.9 is the most broadly applicable of all sensitivity insights. It holds for any optimization model—LP or not—and any constraint type.

Swedish Steel Example Revisited

For a more concrete illustration, we revisit our Swedish Steel blending example of Section 4.2. The associated LP model is

$$
\begin{array}{lll}
\min & 16x_1 + 10x_2 + 8x_3 + 9x_4 + 48x_5 + 60x_6 + 53x_7 & \text{(cost)} \\
\text{s.t.} & x_1 + x_2 + x_3 + x_4 + x_5 + x_6 + x_7 = 1000 & \text{(weight)} \\
& 0.0080x_1 + 0.0070x_2 + 0.0085x_3 + 0.0040x_4 \geq 6.5 & \text{(carbon)} \\
& 0.0080x_1 + 0.0070x_2 + 0.0085x_3 + 0.0040x_4 \leq 7.5 & \\
& 0.180x_1 + 0.032x_2 + 1.0x_5 \geq 30.0 & \text{(nickel)} \\
& 0.180x_1 + 0.032x_2 + 1.0x_5 \leq 30.5 & \\
& 0.120x_1 + 0.011x_2 + 1.0x_6 \geq 10.0 & \text{(chromium)} \\
& 0.120x_1 + 0.011x_2 + 1.0x_6 \leq 12.0 & \\
& 0.001x_2 + 1.0x_7 \geq 11.0 & \text{(molybdenum)} \\
& 0.001x_2 + 1.0x_7 \leq 13.0 & \\
& x_1 \leq 75 & \text{(availability)} \\
& x_2 \leq 250 & \\
& x_1, \ldots, x_7 \geq 0 & \text{(nonnegativity)}
\end{array}
$$

$$(7.1)$$

Effects of Changes in Right-Hand Sides

Figure 7.3 plots the optimal value in the Swedish Steel model (7.1) versus two of its right-hand-side coefficients. Specifically, part (a) follows the impact of the right-hand side (RHS) for the scrap 1 availability constraint

$$x_1 \leq 75 \tag{7.2}$$

assuming that all other parameters are held constant. Similarly, part (b) tracks changes with the right-hand side of the minimum chromium content constraint:

$$0.120x_1 + 0.011x_2 + 1.0x_6 \geq 10.0 \tag{7.3}$$

Infeasible cases of this minimize cost problem are plotted as infinite costs.

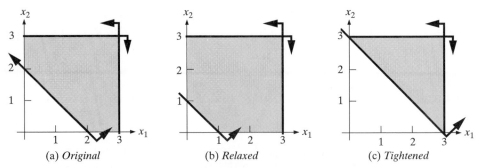

(a) *Original*   (b) *Relaxed*   (c) *Tightened*

FIGURE 7.2   Effects of Relaxing Versus Tightening Constraints

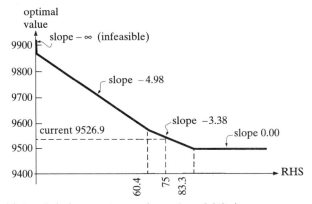

(a) *Supply* ($\leq$) *constraint case* (*scrap 1 availability*)

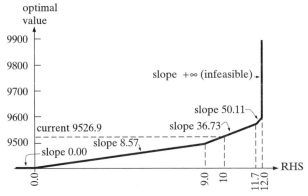

(b) *Demand* ($\geq$) *constraint case* (*minimum chromium content*)

FIGURE 7.3   Sensitivity of Swedish Steel Optimal Value to Right-Hand Sides

In Section 7.7 we discuss how to generate such plots. For the moment, just notice how the trends differ. Increases in the scrap 1 right-hand side help the minimize cost objective, and decreases hurt. Changes in the minimum chromium right-hand side produce exactly the opposite effect.

This apparently contradictory experience follows from one of the constraints being of $\leq$ supply form and the other a $\geq$ demand. An increase in the right-hand side for the two forms produces different effects on the feasible set.

| 7.10 | Changes in LP model right-hand-side coefficients affect the feasible space as follows:

Constraint Type	RHS Increase	RHS Decrease
Supply ($\leq$)	Relax	Tighten
Demand ($\geq$)	Tighten	Relax

The plot in Figure 7.3(a) declines because an increase in the right-hand side relaxes a $\leq$ constraint. Under principle | 7.9 | the optimal value must stay the same or decrease (improve). Increases in the RHS of part (b) tighten the corresponding $\geq$ constraint. The optimal value must either stay the same or increase (worsen).

SAMPLE EXERCISE 7.2: QUALITATIVELY ASSESSING RHS SENSITIVITY

Determine for each of the following objective function and constraint pairs the qualitative effect of increasing or decreasing the constraint right-hand side. Assume that the given constraint is not the only one.

(a) max $13w_1 - 11w_2 + w_3$
     s.t. $9w_1 + w_2 - w_3 \leq 50$

(b) max $13w_1 - 11w_2 + w_3$
     s.t. $9w_1 + w_2 - w_3 \geq 50$

(c) min $8z_1 - 4z_2 + 15z_3$
     s.t. $6z_1 - 3z_2 \leq -19$

(d) min $8z_1 - 4z_2 + 15z_3$
     s.t. $6z_1 - 3z_2 \geq -19$

Analysis: We apply principles | 7.10 | and | 7.9 |.

(a) A RHS increase relaxes this $\leq$ constraint, meaning that the optimal value will stay the same or increase (improve for a maximize objective). RHS decreases have the opposite effect.

(b) A RHS increase tightens this $\geq$ constraint, meaning that the optimal value will stay the same or decrease (worsen).

(c) A RHS increase relaxes this $\leq$ constraint, meaning that the optimal value will stay the same or decrease (improve for a minimize objective).

(d) A RHS increase tightens this $\geq$ constraint, meaning that the optimal value will stay the same or increase (worsen).

Effects of Changes in LHS Constraint Coefficients

With linear programs, principle | 7.9 | holds just as well for coefficient changes on the left-hand side (LHS) of a constraint as for RHS variations such as those in Figure 7.3. Since constraint functions are just weighted sums of the variables, minimum chromium content constraint (7.3) is relaxed if we increase the 0.120 yield coefficient on x_1 to, say, 0.500. It is tightened if, for instance, we reduce it to -0.400. A larger coefficient on the left-hand side of a $\geq$ constraint makes it easier (for nonnegative variables) to satisfy, and a smaller coefficient makes it harder to satisfy. Other cases are similar.

> 7.11 | Changes in LP model LHS constraint coefficients on nonnegative decision variables affect the feasible space as follows:
>
> | Constraint Type | Coefficient Increase | Coefficient Decrease |
> |---|---|---|
> | Supply ($\leq$) | Tighten | Relax |
> | Demand ($\geq$) | Relax | Tighten |

SAMPLE EXERCISE 7.3: QUALITATIVELY ASSESSING LHS SENSITIVITY

Return to the objective functions and constraints of Sample Exercise 7.3, and determine the qualitative impact of each of the following coefficient changes. Assume that all variables are required to be nonnegative.

(a) Change the coefficient of w_2 to 6 in Sample Exercise 7.2(a).

(b) Change the coefficient on w_3 to 0 in Sample Exercise 7.2(b).

(c) Change the coefficient on z_1 to 2 in Sample Exercise 7.2(c).

(d) Change the coefficient on z_3 to -1 in Sample Exercise 7.2(d).

Analysis: We apply principles 7.11 and 7.9 .

(a) The change from 1 to 6 is an increase, meaning that the $\leq$ constraint is tightened and the maximize optimal value will stay the same or decrease.

(b) The change from -1 to 0 is an increase, meaning that the $\geq$ constraint is relaxed and the maximize optimal value will stay the same or increase.

(c) The change from 6 to 2 is a decrease, meaning that the $\leq$ constraint is relaxed and the minimize optimal value will stay the same or decrease.

(d) The change from 0 to -1 is a decrease, meaning that the $\geq$ constraint is tightened and the minimize optimal value will stay the same or increase.

Effects of Adding or Dropping Constraints

Principle 7.9 can even be stretched to cases where we change the model more dramatically by adding or dropping constraints.

> 7.12 | Adding constraints to an optimization model tightens its feasible set, and dropping constraints relaxes its feasible set.

It follows that adding constraints can only make the optimal value worse. Dropping constraints can only make it better.

Figure 7.3(a) implicitly includes the dropping case. Elimination of the scrap 1 availability constraint is the same thing as making its right-hand side very large. Part (a) of the figure shows how the optimal value declines, an improvement for a minimize cost problem.

SAMPLE EXERCISE 7.4: QUALITATIVELY ASSESSING ADDED AND DROPPED CONSTRAINTS

Consider the linear program

$$\max \quad 6y_1 + 4y_2$$
$$\text{s.t.} \quad y_1 + y_2 \leq 3$$
$$y_1 \leq 2$$
$$y_2 \leq 2$$
$$y_1, y_2 \geq 0$$

Assess the qualitative impact of each of the following constraint changes.

(a) Dropping the first main constraint

(b) Adding the new constraint $y_2 \geq y_1$

Analysis: We apply principles 7.12 and 7.9 .

(a) Dropping a constraint relaxes the feasible set. Thus the maximize optimal value can only stay the same or increase.

(b) Adding a new constraint tightens the feasible set. Thus the maximize optimal value can only stay the same or decrease.

Effects of Unmodeled Constraints

Often in OR studies the possibility of adding constraints arises in the context of unmodeled phenomena. For example, managers in the Swedish Steel model (7.1) might prefer to use all or none of the 250 kilograms of scrap 2 instead of trying to measure amounts in between.

Such a requirement is discrete. (Do you see why?) Thus there are good reasons for modelers to neglect it in the interest of tractability. But how would results change if such a difficult-to-express limit on solutions had been modeled?

> 7.13   Explicitly including previously unmodeled constraints in an optimization model must leave the optimal value either unchanged or worsened.

Changing Rates of Constraint Coefficient Impact

Look again at Figure 7.3. Besides the fact that tightening constraints leaves the optimal value the same or worse, there is another pattern. The more we tighten, the greater the rate of damage. In part (a), for instance, tightening means reducing the

RHS coefficient. It has no impact at all for right-hand sides above 83.3. As we continue to squeeze the feasible set, however, the rates of change increase. Ultimately, the model becomes infeasible.

Relaxing constraints produces an opposite effect. For example, in Figure 7.3(b) relaxing means reducing the RHS. The figure shows a steep rate of decline at higher values of the right-hand side that dimishes as we make the coefficient smaller. Every linear program exhibits similar behavior.

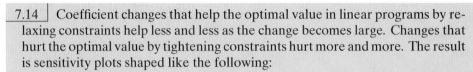

__7.14__ Coefficient changes that help the optimal value in linear programs by relaxing constraints help less and less as the change becomes large. Changes that hurt the optimal value by tightening constraints hurt more and more. The result is sensitivity plots shaped like the following:

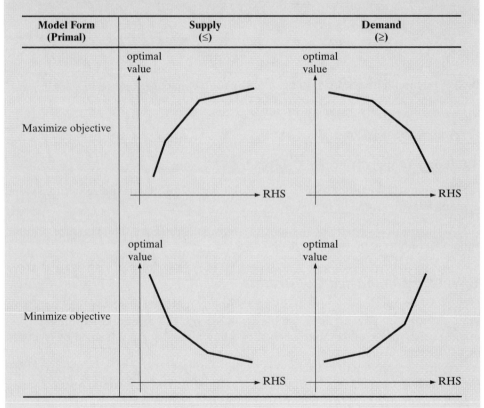

Note that principle ⎡7.14⎤ is limited to linear programs. Models with nonconvex feasible sets (see Section 3.4) may behave differently.

To get some intuition about ⎡7.14⎤, think of yourself as struggling heroically to force solutions to conform to your favorite supply or demand requirement. The more you tighten the constraint you control, the more it becomes the driving force behind all decisions; its impact becomes greater and greater. On the other hand, if you are driven back and forced to relax your constraint, there are many others to take up the cause; you make less and less difference.

SAMPLE EXERCISE 7.5: QUALITATIVELY ASSESSING RATES OF CONSTRAINT
CHANGE

Determine whether rates of optimal objective function value change would be steeper
or less steep with the magnitude of each of the following RHS changes.

(a) Increase the RHS of $4y_1 - 3y_2 \leq 19$ in a maximize linear program.

(b) Decrease the RHS of $3y_1 + 50y_2 \geq 40$ in a maximize linear program.

(c) Decrease the RHS of $14y_1 + 8y_2 \leq 90$ in a minimize linear program.

(d) Increase the RHS of $3y_1 - 2y_2 \geq 10$ in a minimize linear program.

Analysis:

(a) Under principle $\boxed{7.10}$, the optimal value will improve (become larger) with
increases in this RHS because the $\leq$ constraint is being relaxed. Principle $\boxed{7.14}$
implies that the rate of improvement should stay the same or diminish.

(b) Under principle $\boxed{7.10}$, the optimal value will improve (become larger) with
decreases in this RHS because the $\geq$ constraint is being relaxed. Principle $\boxed{7.14}$
implies that the rate of decline should stay the same or diminish.

(c) Under principle $\boxed{7.10}$, the optimal value will worsen (become larger) with increases in this RHS because the $\leq$ constraint is being relaxed. Principle $\boxed{7.14}$ implies
that the rate of improvement should stay the same or steepen.

(d) Under principle $\boxed{7.10}$, the optimal value will worsen (become larger) with increases in this RHS because the $\geq$ constraint is being tightened. Principle $\boxed{7.14}$
implies that the rate of decline should stay the same or steepen.

Effects of Objective Function Coefficient Changes

Objective function coefficients neither relax nor tighten constraints, but we may still
be very interested in their impact on results. Figure 7.4(a) plots the optimal value
in the (minimize cost) Swedish Steel model (7.1) as a function of the unit cost for
scrap 4 (coefficient of x_4). Part (b) tracks the optimal value in the (maximize margin)
CFPL model in Table 4.5 versus the selling price for sheets of $\frac{1}{2}$-inch AC plywood
(coefficient of $z_{1/2,a,c}$).

When an objective function coefficient multiplies a nonnegative variable, as
most of them do, the direction of change is easy to predict. For example, decreasing
the unit cost of an activity certainly makes lower-cost solutions easier to find; the
optimal value could only improve.

$\boxed{7.15}$ Changing the objective function coefficient of a nonnegative variable in
an optimization model affects the optimal value as follows:

Model Form (Primal)	Coefficient Increase	Coefficient Decrease
Maximize objective	Same or better	Same or worse
Minimize objective	Same or worse	Same or better

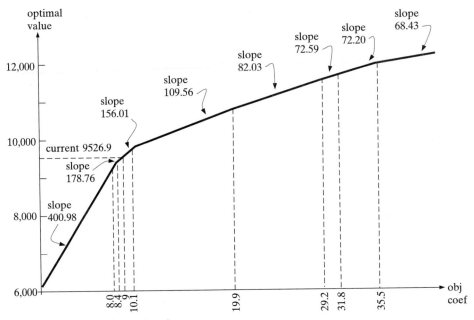

(a) *Min case (Swedish Steel scrap 4 cost)*

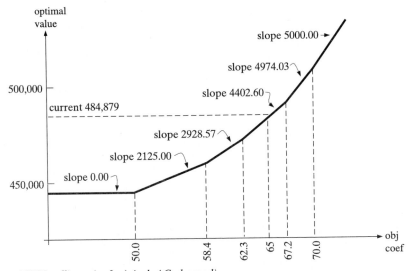

(b) *Max case (CFPL selling price for $\frac{1}{2}$-inch AC plywood)*

FIGURE 7.4   Sensitivity of Optimal Values to Changes in Objective Function Coefficients

SAMPLE EXERCISE 7.6: QUALITATIVELY ASSESSING OBJECTIVE CHANGES

For each of the following objective functions, determine the qualitative impact of the indicated change in objective function coefficient. Assume that all variables are nonnegative.

(a) Change the coefficient of w_1 to 7 in max $12w_1 - w_2$.

(b) Change the coefficient of w_2 to 9 in max $12w_1 - w_2$.

(c) Change the coefficient of w_1 to 60 in min $44w_1 + 3w_2$.

(d) Change the coefficient of w_2 to -9 in min $44w_1 + 3w_2$.

Analysis: We apply principle $\boxed{7.15}$.

(a) A change from 12 to 7 is a decrease, implying that the maximize optimal value will stay the same or worsen (decrease).

(b) A change from -1 to 9 is an increase, implying that the maximize optimal value will stay the same or improve (increase).

(c) A change from 44 to 60 is an increase, implying that the minimize optimal value will stay the same or worsen (increase).

(d) A change from 3 to -9 is a decrease, implying that the minimize optimal value will stay the same or improve (decrease).

Changing Rates of Objective Function Coefficient Impact

Readers who absorbed the discussion above about rates of optimal value change caused by variations in constraint coefficients should be surprised by the shapes of curves in Figure 7.4. Consider, for instance, the maximize case of part (b). A move to the right, which increases the coefficient of $z_{1/2,a,c}$, helps the objective function—just as we concluded in $\boxed{7.15}$.

What is different is that the more we increase the coefficient, the more rapidly the optimal value improves. In contrast to the decreasing returns we saw with constraint changes that helped the optimal value, favorable objective function changes produce increasing rates of return.

$\boxed{7.16}$ Objective function coefficient changes that help the optimal value in linear programs help more and more as the change becomes large. Changes that hurt the optimal value hurt less and less. The result is sensitivity plots shaped like the following:

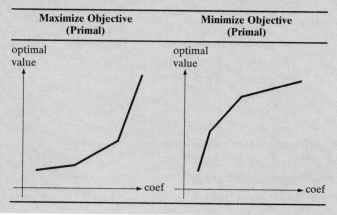

As with $\boxed{7.14}$, principle $\boxed{7.16}$ is limited to linear programs with their convex feasible sets. Nonconvex optimization models may perform differently.

A full explanation of the incongruity between rates of change in constraint and objective function coefficients must await the duality development of the next few sections. For now, return to your struggling hero role, this time by manipulating the unit cost or benefit of some activity. Changing your objective function coefficient in the direction that helps the optimal value draws more and more action to your activity. The more it dominates the solution, the more its objective function coefficient affects the optimal value. On the other hand, changes that hurt the optimal value are cushioned by transfer of responsibility to other activities. They have less and less impact as the change becomes large.

SAMPLE EXERCISE 7.7: QUALITATIVELY ASSESSING RATES OF OBJECTIVE CHANGE

Determine whether the objective function coefficient change indicated would increase or decrease the optimal objective function value in each of the following, and indicate whether the rate of change will steepen or become less steep with the magnitude of the change. Assume that all variables are nonnegative.

(a) Increase 42 on x_1 in max $42x_1 + 13x_2 - 9x_3$.

(b) Decrease -9 on x_3 in max $42x_1 + 13x_2 - 9x_3$.

(c) Increase -2 on x_2 in min $12x_1 - 2x_2 + x_4$.

(d) Decrease 0 on x_3 in min $12x_1 - 2x_2 + x_4$.

Analysis:

(a) The optimal value improves (increases) with any objective function coefficient increase in a maximize model (principle $\boxed{7.15}$). Thus, under principle $\boxed{7.16}$ the rate of increase should stay the same or steepen.

(b) A decrease would be to $c_3 < -9$. The optimal value worsens (decreases) with any objective function coefficient decrease in a maximize model (principle $\boxed{7.15}$). Thus, under principle $\boxed{7.16}$ the rate of increase should stay the same or lessen.

(c) The optimal value worsens (increases) with any objective function coefficient increase in a minimize model (principle $\boxed{7.15}$). Thus, under principle $\boxed{7.16}$ the rate of increase should stay the same or lessen.

(d) The optimal value improves (decreases) with any objective function coefficient decrease in a minimize model (principle $\boxed{7.15}$). Thus, under principle $\boxed{7.16}$ the rate of increase should stay the same or steepen.

Effects of Adding or Dropping Variables

The final sort of sensitivity analysis we need to consider is the adding or dropping of activities (which is implemented mathematically as adding or dropping of decision variables). For example, what happens if we decide to consider a new scrap source in the Swedish Steel model (7.1)? What happens if we drop scrap 4?

Adding activities offers new choices; dropping them reduces possibilities. The direction of optimal value change should then be apparent.

> 7.17 | Adding optimization model activities (variables) must leave the optimal value unchanged or improved. Dropping activities will leave the value unchanged or degraded.

SAMPLE EXERCISE 7.8: QUALITATIVELY ASSESSING ADDED AND DROPPED VARIABLES

Determine the qualitative impact of adding or dropping a variable in linear programs with each of the following objective functions.

(a) max $27y_1 - y_2 + 4y_3$

(b) min $33y_1 + 11y_2 + 39y_3$

Analysis: We apply principle $\boxed{7.17}$.

(a) Adding a variable in this maximize LP can only help, leaving the optimal value the same or higher. Dropping a variable would make it the same or lower.

(b) Adding a variable in the minimize LP can only help, leaving the optimal value the same or lower. Dropping a variable would make it the same or higher.

7.3 QUANTIFYING SENSITIVITY TO CHANGES IN LP MODEL COEFFICIENTS: A DUAL MODEL

All the LP sensitivities of Section 7.2 related to directions of changes rather than magnitudes. It is certainly helpful to know whether, say, a right-hand-side modification will help or hurt the optimal value. Still, we would gain much more insight if we could quantify the rates of change.

The OR approach to computing quantities we would like to know is to represent them as variables, formulate a model of how they interrelate, and solve the model to obtain their values. In this and Section 7.4 we pursue that approach. More specifically, we define new sensitivity variables and formulate the conditions that a quantitative sensitivity analysis should fulfill.

Primals and Duals Defined

With one model being used to analyze sensitivity of another, it is important to distinguish between primal and dual.

> 7.18 | The **primal** is the given optimization model, the one formulating the application of primary interest.

All the linear programs formulated in earlier chapters were primals.

7.19 | The **dual** is a subsidiary optimization model, defined over the same input parameters as the primal but characterizing the sensitivity of primal results to changes in inputs.

Dual Variables

To begin deriving the dual, take another look at the Figure 7.3 plots of the (primal) Swedish Steel model's optimal value as a function of right-hand side. Dual variables quantify slopes of such curves.

7.20 | There is one **dual variable** for each main primal constraint. Each reflects the rate of change in primal optimal value per unit increase from the given right-hand-side value of the corresponding constraint.

We often denote the dual variable on primal main constraint i by v_i.

The right-hand side of primal constraint (7.2) is 75. Figure 7.3(a) shows the rate of change in the primal optimal value, and thus the value of the dual variable corresponding to this availability constraint is -3.38 kroner/kilogram.

Of course, we would prefer to know such slopes for all possible values of right-hand sides. For instance, in Figure 7.3(a) we have

$$
\begin{array}{rlrll}
+\infty & \text{between} & -\infty & \text{and} & 0 \\
-4.98 & \text{between} & 0 & \text{and} & 60.42 \\
-3.38 & \text{between} & 60.42 & \text{and} & 83.33 \\
0 & \text{between} & 83.33 & \text{and} & +\infty
\end{array}
$$

However, several LPs have to be solved to derive such plots.

Since the dual only provides subsidiary analysis of the primal, which is the model of real interest, definition 7.20 adopts a more manageable standard. Optimal dual variable values for any single LP yield rates of change only at the original RHS value. For instance, only the -3.38 rate at RHS $= 75$ would be obtained for the constraint of Figure 7.3(a).

This definition is somewhat unclear if the RHS value happens to fall at exactly one of the points where slopes change. For example, if the given RHS in Figure 7.3(a) had been 83.33, the corresponding dual variable value could come out either -3.38 or 0.00 (or anything in between). The first applies to the left of 83.33 and the second to the right. At exactly 83.33 the slope is ambiguous.

SAMPLE EXERCISE 7.9: UNDERSTANDING DUAL VARIABLES

Refer to Figure 7.3(b)'s plot of sensitivity to minimum chromium constraint (7.3).

(a) What is the corresponding optimal dual variable value?

(b) What would it be if the constraint RHS had been 7.0?

(c) What would it be if the constraint RHS had been 9.0?

Analysis:

(a) By definition $\boxed{7.20}$ the dual variable corresponding to this minimum chromium constraints should equal the slope of the sensitivity curve at RHS = 10.0. That is, the dual variable has value 36.73 kroner/kilogram.

(b) For RHS = 7.0 the corresponding dual variable value or rate of change is 8.57 kroner/kilogram.

(c) At RHS = 9.0 the dual variable value is ambiguous. Computation might produce either the 8.57 to the left or the 36.73 to the right.

Dual Variable Types

We want to develop constraints and conditions that dual variables must satisfy to provide meaningful sensitivity information. Qualitative principles $\boxed{7.9}$ and $\boxed{7.10}$ already yield sign restrictions on rates of change in inequality right-hand sides. For example, an increase in the RHS of a $\leq$ constraint in a minimize model relaxes the constraint, which leaves the optimal value unchanged or reduced. The corresponding $v \leq 0$.

$\boxed{7.21}$ The linear programming dual variable on constraint i has type as follows:

Primal	i is $\leq$	i is $\geq$	i is =
Minimize objective	$v_i \leq 0$	$v_i \geq 0$	Unrestricted
Maximize objective	$v_i \geq 0$	$v_i \leq 0$	Unrestricted

Notice that principle $\boxed{7.21}$ specifies dual variables for = constraints to be unrestricted in sign (URS). Since an equality can act as either a $\leq$ or a $\geq$ (principle $\boxed{7.5}$), the corresponding dual rate of change can be either positive or negative.

Two Crude Example Again

To have an easy example to follow, let us return to the familiar Two Crude refining model of Section 2.1. That primal linear program was

$$
\begin{array}{llll}
\min & 20x_1 & +15x_2 \\
\text{s.t.} & 0.3x_1 & +0.4x_2 \geq 2 & : v_1 & \text{(gasoline demand)} \\
& 0.4x_1 & +0.2x_2 \geq 1.5 & : v_2 & \text{(jet fuel demand)} \\
& 0.2x_1 & +0.3x_2 \geq 0.5 & : v_3 & \text{(lubricant demand)} & \quad (7.4) \\
& 1x_1 & \leq 9 & : v_4 & \text{(Saudi availability)} \\
& & +1x_2 \leq 6 & : v_5 & \text{(Venezuelan availability)} \\
& x_1, \ x_2 \geq 0
\end{array}
$$

The unique primal optimal solution is $x_1^* = 2$, $x_2^* = 3.5$, with optimal value 92.5.
    Notice that we have assigned dual variables to each main constraint. Applying principle $\boxed{7.21}$, sign restriction constraints for the five Two Crude dual variables

become

$$v_1 \geq 0, \quad v_2 \geq 0, \quad v_3 \geq 0, \quad v_4 \leq 0, \quad v_5 \leq 0 \qquad (7.5)$$

Increasing $\geq$ right-hand sides in a minimize problem can only raise the optimal cost or leave it unchanged. Increasing those of $\leq$'s can only lower it or leave it unchanged.

SAMPLE EXERCISE 7.10: CHOOSING DUAL VARIABLE TYPE

For each of the following primal linear programs, determine whether the dual variables indicated should be nonnegative, nonpositive, or of URS type.

(a)
$$
\begin{array}{lllllll}
\min & +5x_1 & +1x_2 & +4x_3 & +5x_4 & & \\
\text{s.t.} & +1x_1 & +4x_2 & +2x_3 & & = 36 & : v_1 \\
& +3x_1 & +2x_2 & +8x_3 & +2x_4 & \leq 250 & : v_2 \\
& -5x_1 & -2x_2 & +1x_3 & +1x_4 & \leq 7 & : v_3 \\
& & & +1x_3 & +1x_4 & \geq 60 & : v_4 \\
\end{array}
$$

$$x_1, x_2, x_3, x_4 \geq 0$$

(b)
$$
\begin{array}{llllll}
\max & +13x_1 & +24x_2 & +5x_3 & +50x_4 & & \\
\text{s.t.} & +1x_1 & +3x_2 & & & \geq 89 & : v_1 \\
& & & +1x_3 & +5x_4 & \geq 60 & : v_2 \\
& +10x_1 & +6x_2 & +8x_3 & +2x_4 & \leq 608 & : v_3 \\
& & +1x_2 & & +1x_4 & = 28 & : v_4 \\
\end{array}
$$

$$x_1, x_2, x_3, x_4 \geq 0$$

Analysis: We apply rule $\boxed{7.21}$.

(a) v_1 URS, $v_2 \leq 0, v_3 \leq 0, v_4 \geq 0$

(b) $v_1 \leq 0, v_2 \leq 0, v_3 \geq 0, v_4$ URS

Dual Variables as Implicit Marginal Resource Prices

Because the dual variables tell us the rate of change in the optimal value for another unit of each RHS, they provide a sort of implicit price on the resource of each constraint model. To be more precise, they yield what economists call **marginal prices**.

7.22 | Dual variables provide **implicit prices** for the marginal unit of the resource modeled by each constraint as its right-hand-side limit is encountered.

In Two Crude model (7.4), for instance, the first constraint models gasoline demand. Variable v_1, which is in units of thousands of dollars per thousand barrels, will tell us the implicit price of gasoline at the margin when RHS $= 2000$ barrels are demanded, that is, how much the last 1000 barrels cost to produce in the optimal operating plan. Similarly, variable v_4 reflects the marginal impact of the Saudi availability constraint at its current level of 9000 barrels, which is the implicit value of another 1000 barrels of Saudi crude.

SAMPLE EXERCISE 7.11: INTERPRETING DUAL VARIABLES

The Top Brass Trophy example model (5.1) of Section 5.1 is

$$
\begin{array}{llll}
\max & 12x_1 + 9x_2 & & \text{[profit (dollars)]} \\
\text{s.t.} & x_1 & \leq 1000 & : v_1 & \text{(brass footballs)} \\
& x_2 & \leq 1500 & : v_2 & \text{(brass soccer balls)} \\
& x_1 + x_2 & \leq 1750 & : v_3 & \text{(brass plaques)} \\
& 4x_1 + 2x_2 & \leq 4800 & : v_4 & \text{[wood (board feet)]} \\
& x_1, \; x_2 \geq 0
\end{array}
$$

Interpret each of the 4 corresponding dual variables and determine their signs and units.

Analysis: Applying interpretation $\boxed{7.22}$, v_1 is the marginal value or contribution to profit of brass footballs at the current 1000 availability level, that is, how much we would pay for another one. Under principle $\boxed{7.21}$, $v_1 \geq 0$, and its units are dollars per brass football. Similarly, $v_2 \geq 0$ is the marginal value of brass soccer balls at the current 1500 availability level (in dollars per soccer ball), $v_3 \geq 0$ is the marginal value of brass plaques at the current 1750 availability (in dollars per plaque), and $v_4 \geq 0$ is the marginal value of wood at the current 4800 availability (in dollars per board foot).

Implicit Activity Pricing in Terms of Resources Produced and Consumed

The activity associated with any decision variable both consumes and creates constraints' resources (review Figure 7.1). We also know that the activity's nonzero constraint coefficients can be interpreted as amounts consumed or created per unit of the activity (principle $\boxed{7.8}$).

   By summing those coefficients times the implicit price v_i of the resources involved, we can obtain an implicit marginal value (minimize problems) or price (maximize problems) for the entire activity.

$\boxed{7.23}$ The implicit marginal value (minimize problems) or price (maximize problems) of a unit of LP activity (primal variable) j implied by dual variable values v_i is $\sum_i a_{i,j} v_i$, where $a_{i,j}$ denotes the coefficient of activity j in the left-hand side of constraint i.

   Venezuelan activity $j = 2$ in the Two Crude example illustrates. Given resource prices $v_1, \ldots, v_5$, each unit of Venezuelan crude activity is implicitly worth

$$
\sum_{i=1}^{5} a_{i,2} v_i = 0.4v_1 + 0.2v_2 + 0.3v_3 + 1v_5
$$

Noting sign restrictions (7.5), the first three terms measure the positive contribution of fulfilling demand constraints. The last deducts ($v_5 \leq 0$) the implicit worth of the availability resource consumed.

SAMPLE EXERCISE 7.12: IMPLICITLY PRICING ACTIVITIES

Develop and interpret expressions for the price per unit of each activity in the Top Brass Trophy model reviewed in Sample Exercise 7.11.

Analysis: In the Top Brass example, x_1 represents production of football trophies, and x_2, production of soccer trophies. Under principle $\boxed{7.23}$, the implicit price or marginal cost to produce a football trophy is the sum of coefficients on x_1 times the marginal value of the associated commodities, that is,

$$\sum_{i=1}^{4} a_{i,1} v_i = 1v_1 + 1v_3 + 4v_4$$

This expression simply sums the marginal cost of the 1 football, 1 plaque, and 4 board feet of wood needed to make a football trophy. Similarly, the marginal cost of producing a soccer trophy is

$$\sum_{i=1}^{4} a_{i,2} v_i = 1v_2 + 1v_3 + 2v_4$$

Main Dual Constraints to Enforce Activity Pricing

The $\boxed{7.23}$ notion of implicit activity worth per unit also leads to the main dual constraints. If dual variables v_i are really to reflect the value of constraint resources in the primal optimal value, the implied activity worths must be consistent with explicit unit costs or benefits in the primal objective function.

To be specific, dual variables in a minimize (cost) problem would overvalue an activity if they priced it above its true cost coefficient c_j. An activity implicitly worth more than it costs at optimality should be used in greater quantity. But then we would be improving on an optimal solution—an impossibility. Similarly for maximize (benefit) problems, we would want to reject any v_i's that priced the net value of resources that an activity uses below its real benefit c_j. Otherwise, it too would yield an improvement on an already optimal solution.

> $\boxed{7.24}$ For each nonnegative variable activity x_j in a minimize linear program, there is a corresponding **main dual constraint** $\sum_i a_{i,j} v_i \leq c_j$ requiring the net marginal value of the activity not to exceed its given cost. In a maximize problem main dual constraints for $x_j \geq 0$ are $\sum_i a_{i,j} v_i \geq c_j$, which keeps the net marginal cost of each activity at least equal to its given benefit.

Illustrating again with the Two Crude model (7.4), principle $\boxed{7.24}$ leads to one constraint for primal variable x_1 and another for x_2.

$$\begin{aligned}
0.3v_1 + 0.4v_2 + 0.2v_3 + 1v_4 &\leq 20 \\
0.4v_1 + 0.2v_2 + 0.3v_3 + 1v_5 &\leq 15
\end{aligned} \tag{7.6}$$

Without such limits, say if the worth evaluation on the left-hand side of the first inequality were allowed to reach 25, the v_i would be telling us that another unit of activity $j = 1$ would produce a net impact on the optimal value of $25 (thousand) yet costs only $20 (thousand). Such bargains cannot be allowed if the v_i are to mean what we want them to mean.

SAMPLE EXERCISE 7.13: FORMING MAIN DUAL CONSTRAINTS

Formulate and interpret the main dual constraints of the linear programs in Sample Exercise 7.10 .

Analysis: We apply principle 7.24 .

(a) For this minimize model, main dual constraints are

$$
\begin{aligned}
+1v_1 &+3v_2 &-5v_3 & & &\le 5 \\
+4v_1 &+2v_2 &-2v_3 & & &\le 1 \\
+2v_1 &+8v_2 &+1v_3 &+1v_4 & &\le 4 \\
       &+2v_2 &+1v_3 &+1v_4 & &\le 5
\end{aligned}
$$

They may be interpreted as requiring that the net marginal value of each activity under an optimal choice of dual variable values cannot exceed its given objective function cost.

(b) For this maximize model, main dual constraints are

$$
\begin{aligned}
+1v_1 & & &+10v_3 & & &\ge 13 \\
+3v_1 & & &+6v_3 &+1v_4 & &\ge 24 \\
       &+1v_2 &+8v_3 & & &\ge 5 \\
       &+5v_2 &+2v_3 &+1v_4 & &\ge 50
\end{aligned}
$$

They may be interpreted as requiring that the net marginal cost of each activity under an optimal choice of dual variable values cannot be less than its given objective function benefit.

Optimal Value Equality between Primal and Dual

If dual variables are to price the resources associated with constraints correctly, the supplies and demands for those resources should evaluate to exactly the primal optimal value.

> **7.25** If a primal linear program has an optimal solution, its optimal value $\sum_j c_j x_j^*$ equals the corresponding optimal dual implicit total value $\sum_i b_i v_i^*$ of all constraint resources.

In the Two Crude case, requirement 7.25 implies that

$$
20x_1^* + 15x_2^* = 2v_1^* + 1.5v_2^* + 0.5v_3^* + 9v_4^* + 6v_6^*
$$

The sum of optimal prices on all constraints times current limiting values should recover the optimal primal objective value.

SAMPLE EXERCISE 7.14: EXPRESSING PRIMAL–DUAL VALUE EQUALITY

Express and interpret optimal value equality requirement $\boxed{7.25}$ for each of the linear programs in Sample Exercise 7.10.

Analysis: We want the primal optimal value to equal exactly the implicit total value of all constraint resources.

(a) For this minimize model

$$5x_1 + 1x_2 + 4x_3 + 5x_4 = 36v_1 + 250v_2 + 7v_3 + 60v_4$$

(b) For this maximize model

$$13x_1 + 24x_2 + 5x_3 + 50x_4 = 89v_1 + 60v_2 + 608v_3 + 28v_4$$

Primal Complementary Slackness between Primal Constraints and Dual Variable Values

Recall that an inequality is **active** if it is satisfied as an equality by a given solution, and **inactive** otherwise. If an inequality is inactive at the primal optimal solution, small changes in its right-hand side would not affect the optimal value at all; the constraint has slack. We can immediately deduce the value of the corresponding dual variable v_i.

$\boxed{7.26}$ Either the primal optimal solution makes main inequality constraint i active or the corresponding dual variable $v_i = 0$.

Equalities are always active, so there is no corresponding condition.

Principle $\boxed{7.26}$ carries the somewhat bulky name **primal complementary slackness** because it asserts that either each primal inequality or its corresponding dual sign restriction ($v_i \geq 0$ or $v_i \leq 0$) will be slack (inactive) at optimality. To illustrate for the Two Crude model (7.4), we first substitute the primal optimum to determine which constraints are active.

$$
\begin{array}{llllll}
+0.3(2) & +0.4(3.5) & & = & 2.0 & \text{(active)} \\
+0.4(2) & +0.2(3.5) & & = & 1.5 & \text{(active)} \\
+0.2(2) & +0.3(3.5) & = 1.45 & > & 0.5 & \text{(inactive)} \\
+1(2) & & = 2 & < & 9 & \text{(inactive)} \\
& +1(3.5) & = 3.5 & < & 6 & \text{(inactive)}
\end{array}
$$

The last three have slack. Thus small changes in corresponding right-hand-side coefficients would have no impact on optimal value. In accord with principle $\boxed{7.26}$, associated dual variables must $= 0$.

SAMPLE EXERCISE 7.15: EXPRESSING PRIMAL COMPLEMENTARY SLACKNESS

Express and interpret primal complementary slackness conditions for the linear programs of Sample Exercise 7.10.

Analysis: We apply principle 7.26 .

(a) Primal complementary slackness conditions for this minimize model are

$$+3x_1 + 2x_2 + 8x_3 + 2x_4 = 250 \quad \text{or} \quad v_2 = 0$$
$$-5x_1 - 2x_2 + 1x_3 + 1x_4 = 7 \quad \text{or} \quad v_3 = 0$$
$$+1x_3 + 1x_4 = 60 \quad \text{or} \quad v_4 = 0$$

They specify that a main primal inequality's right-hand side can affect the optimal value only if the constraint is active. There is no condition for equality constraint 1 because it is always active.

(b) Primal complementary slackness conditions for this maximize model are

$$+1x_1 + 3x_2 = 89 \quad \text{or} \quad v_1 = 0$$
$$+1x_3 + 5x_4 = 60 \quad \text{or} \quad v_2 = 0$$
$$+10x_1 + 6x_2 + 8x_3 + 2x_4 = 608 \quad \text{or} \quad v_3 = 0$$

They, too, specify that changes in a main primal inequality's right-hand side can affect the optimal value only if the constraint is active. There is no such condition for equality constraint 4.

Dual Complementary Slackness between Dual Constraints and Primal Variable Values

Constraints 7.24 keep activity prices below true cost in minimize problems and above true benefit in maximize problems. At first glance it might seem reasonable to ask for more (i.e., demand that activity prices match exactly corresponding c_j).

There would be a practical mathematical difficulty with such a requirement in the common case where we have many more primal variables than main constraints. Relatively few v_i would simultaneously have to satisfy constraints

$$\sum_i a_{i,j} v_i = c_j$$

for too many activities j.

More important, we want the dual variables to measure resource value at optimality. The only (nonnegative) primal variables involved in an optimal solution are those with optimal $x_j^* > 0$. Limiting perfect valuation to this more limited list of activities produces our final set of requirements—**dual complementary slackness conditions**.

> 7.27 | Either a nonnegative primal variable has optimal value $x_j = 0$ or the corresponding dual prices v_i must make the jth dual constraint 7.24 active.

This time the complementarity is between the primal nonnegativity $x_j \geq 0$ and the corresponding dual inequality 7.24 .

For example, the tiny Two Crude optimum had both primal variables positive at optimality. Principle 7.27 now tells us that the dual values v_i we are trying to understand must make both constraints of (7.6) active.

SAMPLE EXERCISE 7.16: EXPRESSING DUAL COMPLEMENTARY SLACKNESS

Express and interpret dual complementary slackness conditions for the linear programs of Sample Exercise 7.10.

Analysis: We apply principle $\boxed{7.27}$ to the main dual constraints of Sample Exercise 7.13.

(a) For this minimize model, dual complementary slackness conditions are

$$x_1 = 0 \quad \text{or} \quad 1v_1 + 3v_2 - 5v_3 = 5$$
$$x_2 = 0 \quad \text{or} \quad 4v_1 + 2v_2 - 2v_3 = 1$$
$$x_3 = 0 \quad \text{or} \quad 2v_1 + 8v_2 + 1v_3 + 1v_4 = 4$$
$$x_4 = 0 \quad \text{or} \quad 2v_2 + 1v_3 + 1v_4 = 5$$

They may be interpreted to mean that marginal resource prices v_i must make the implicit value of each activity used in an optimal solution equal its given cost.

(b) For this maximize model, dual complemenary slackness conditions are

$$x_1 = 0 \quad \text{or} \quad 1v_1 + 10v_3 = 13$$
$$x_2 = 0 \quad \text{or} \quad 3v_1 + 6v_3 + 1v_4 = 24$$
$$x_3 = 0 \quad \text{or} \quad 1v_2 + 8v_3 = 5$$
$$x_4 = 0 \quad \text{or} \quad 5v_2 + 2v_3 + 1v_4 = 50$$

They may be interpreted to mean that marginal resource prices v_i must make the implicit cost of each activity used in an optimal solution equal to its given benefit.

7.4 FORMULATING LINEAR PROGRAMMING DUALS

The remarkable fact about linear programs is that all the quantitative sensitivity requirements of Section 7.3 can actually be achieved. Not only that, but they can be achieved with very minor computation as a by-product of the search for a primal optimum.

The secret behind these elegant results is deceptively simple: We think of dual variables v_i as decision variables of a new dual linear program defined on the same constants as the primal and determine their values by optimizing that dual. In this section we show how to formulate duals, and in the next section we verify the primal-to-dual relationships of Section 7.3 and more. As usual, x_j will always denote the jth primal variable, c_j its objective function coefficient, and $a_{i,j}$ its coefficient in the ith constraint; v_i is the dual variable for the ith constraint, and b_i is the right-hand side.

Form of the Dual for Nonnegative Primal Variables

To form a dual when primal decision variables are nonnegative, we optimize total resource value $\sum_i b_i v_i$ subject to the main dual constraints $\boxed{7.24}$ and dual variable-type restrictions $\boxed{7.21}$ derived in Section 7.3.

7.28 | The dual of a minimize primal over $x_j \geq 0$ is

$$\max \quad \sum_i b_i v_i$$

$$\text{s.t.} \quad \sum_i a_{i,j} v_i \leq c_j \qquad \text{for all primal activities } j$$

$$v_i \geq 0 \qquad\qquad\quad \text{for all primal } \geq\text{'s } i$$

$$v_i \leq 0 \qquad\qquad\quad \text{for all primal } \leq\text{'s } i$$

$$v_i \text{ URS} \qquad\qquad \text{for all primal } =\text{'s } i$$

7.29 | The dual of a maximize primal over $x_j \geq 0$ is

$$\min \quad \sum_i b_i v_i$$

$$\text{s.t.} \quad \sum_i a_{i,j} v_i \geq c_j \qquad \text{for all primal activities } j$$

$$v_i \geq 0 \qquad\qquad\quad \text{for all primal } \leq\text{'s } i$$

$$v_i \leq 0 \qquad\qquad\quad \text{for all primal } \geq\text{'s } i$$

$$v_i \text{ URS} \qquad\qquad \text{for all primal } =\text{'s } i$$

Notice some elegant symmetries:

- A minimize primal yields a maximize dual, and vice versa.
- There is one dual variable for every primal constraint and one dual constraint for every primal variable.
- Objective function coefficients of the primal become the right-hand sides of the dual, and right-hand sides of the primal provide the objective function coefficients of the dual.

Our familiar Two Crude model (7.4) is a specific example. That minimize primal has dual

$$\max \quad 2v_1 + 1.5v_2 + 0.5v_3 + 9v_4 + 6v_5$$

$$\text{s.t.} \quad 0.3v_1 + 0.4v_2 + 0.2v_3 + 1v_4 \quad \leq \quad 20$$

$$0.4v_1 + 0.2v_2 + 0.3v_3 + 1v_5 \quad \leq \quad 15 \qquad (7.7)$$

$$v_1, \ v_2, \ v_3 \geq 0$$

$$v_4, \ v_5 \leq 0$$

An optimal solution is $v_1^* = 20$, $v_2^* = 35$, $v_3^* = v_4^* = v_5^* = 0$, with dual objective function value 92.5.

SAMPLE EXERCISE 7.17: STATING DUALS WITH NONNEGATIVE x_j

State the duals of each of the following primal linear programs.

(a)
$$\min \quad +30x_1 \qquad\qquad +5x_3$$

$$\text{s.t.} \quad +1x_1 \quad -1x_2 \quad +1x_3 \quad \geq 1 \qquad : v_1$$

$$+3x_1 \quad +1x_2 \qquad\qquad = 4 \qquad : v_2$$

$$+4x_2 \quad +1x_3 \quad \leq 10 \qquad : v_3$$

$$x_1, x_2, x_3 \geq 0$$

(b) max $\quad +10x_1 \quad +9x_2 \quad -6x_3$

s.t. $\quad +2x_1 \quad +1x_2 \qquad\qquad \geq 3 \qquad : v_1$

$\qquad +5x_1 \quad +3x_2 \quad -1x_3 \quad \leq 15 \qquad : v_2$

$\qquad\qquad\qquad +1x_2 \quad +1x_3 \quad = 1 \qquad : v_3$

$\qquad x_1, x_2, x_3 \geq 0$

Analysis:

(a) From construction $\boxed{7.28}$, the dual of this minimize model is a maximize model with objective function

$$\text{max} \quad +1v_1 \quad +4v_2 \quad +10v_3$$

derived from the RHS of the primal. Constraints include one main constraint for each primal variable (principle $\boxed{7.24}$) and type restrictions from table $\boxed{7.21}$. Specifically,

s.t. $\quad +1v_1 \quad +3v_2 \qquad\qquad \leq 30$

$\qquad -1v_1 \quad +1v_2 \quad +4v_3 \quad \leq \;\; 0$

$\qquad +1v_1 \qquad\qquad +1v_3 \quad \leq \;\; 5$

$\qquad v_1 \geq 0,\, v_2 \text{ URS},\, v_3 \leq 0$

(b) From construction $\boxed{7.29}$, the dual of this maximize model is a minimize with objective function

$$\text{min} \quad +3v_1 \quad +15v_2 \quad +1v_3$$

derived from the RHS of the primal. Constraints include one main constraint for each primal variable (principle $\boxed{7.24}$) and type restrictions from table $\boxed{7.21}$. Specifically,

s.t. $\quad +2v_1 \quad +5v_2 \qquad\qquad \geq 10$

$\qquad +1v_1 \quad +3v_2 \quad +1v_3 \quad \geq \;\; 9$

$\qquad\qquad\qquad -1v_2 \quad +1v_3 \quad \geq -6$

$\qquad v_1 \leq 0,\, v_2 \geq 0,\, v_3 \text{ URS}$

Duals of LP Models with Nonpositive and Unrestricted Variables

So far, all our duality results have assumed nonnegative primal variables. Not much is lost, because the majority of LP models have only nonnegative variables. Still, to obtain full symmetry between primal and dual, both should be allowed to have nonpositive, nonnegative, and unrestricted variables.

Table 7.1 shows the complete picture. Nonpositive primal variables switch the direction of main dual inequalities. Unrestricted primal variables lead to dual equalities.

SAMPLE EXERCISE 7.18: FORMING DUALS OF ARBITRARY LPS

Form the dual of each of the following linear programs.

(a) $\qquad\qquad$ max $\quad +6x_1 \quad -1x_2 \quad +13x_3$

$\qquad\qquad$ s.t. $\quad +3x_1 \quad +1x_2 \quad +2x_3 \quad = 7$

$\qquad\qquad\qquad\quad +5x_1 \quad -1x_2 \qquad\qquad \leq 6$

$\qquad\qquad\qquad\qquad\qquad +1x_2 \quad +1x_3 \quad \geq 2$

$\qquad\qquad\qquad x_1 \geq 0,\, x_2 \leq 0$

TABLE 7.1   Corresponding Elements of Primal and Dual Linear Programs

	Primal Element	Corresponding Dual Element
max form	Objective $\max \sum_j c_j x_j$	Objective $\min \sum_i b_i v_i$
	Constraint $\sum_j a_{i,j} x_j \geq b_i$	Variable $v_i \leq 0$
	Constraint $\sum_j a_{i,j} x_j = b_i$	Variable v_i unrestricted
	Constraint $\sum_j a_{i,j} x_j \leq b_i$	Variable $v_i \geq 0$
	Variable $x_j \geq 0$	Constraint $\sum_i a_{i,j} v_i \geq c_j$
	Variable x_j URS	Constraint $\sum_i a_{i,j} v_i = c_j$
	Variable $x_j \leq 0$	Constraint $\sum_i a_{i,j} v_i \leq c_j$
min form	Objective $\min \sum_j c_j x_j$	Objective $\max \sum_i b_i v_i$
	Constraint $\sum_j a_{i,j} x_j \geq b_i$	Variable $v_i \geq 0$
	Constraint $\sum_i a_{i,j} x_j = b_i$	Variable v_i unrestricted
	Constraint $\sum_j a_{i,j} x_j \leq b_i$	Variable $v_i \leq 0$
	Variable $x_j \geq 0$	Constraint $\sum_i a_{i,j} v_i \leq c_j$
	Variable x_j URS	Constraint $\sum_j a_{i,j} v_i = c_j$
	Variable $x_j \leq 0$	Constraint $\sum_j a_{i,j} v_i \geq c_j$

(b)

$$\min \quad +7x_1 \qquad\qquad +44x_3$$
$$\text{s.t.} \quad -2x_1 \;-4x_2 \;+1x_3 \;\leq\; 15$$
$$+1x_1 \;+4x_2 \qquad\quad \geq\; 5$$
$$+5x_1 \;-1x_2 \;+3x_3 \;=-11$$
$$x_1 \leq 0, x_3 \geq 0$$

Analysis:

(a) Assigning dual variables v_1, v_2, v_3 to the three main constraints, we apply the "max form" part of Table 7.1 to produce the following dual:

$$\min \quad +7v_1 \;+6v_2 \;+2v_3$$
$$\text{s.t.} \quad +3v_1 \;+5v_2 \qquad\qquad \geq\; 6$$
$$+1v_1 \;-1v_2 \;+1v_3 \;\leq\; -1$$
$$+2v_1 \qquad\qquad +1v_3 \;=\; 13$$
$$v_1 \text{ URS}, v_2 \geq 0, v_3 \leq 0$$

Notice that unrestricted x_3 yields an equality dual constraint.

(b) Assigning dual variables v_1, v_2, v_3 to the three main constraints, we apply the "min form" part of Table 7.1 to produce the following dual:

$$\max \quad +15v_1 \;+5v_2 \;-11v_3$$
$$\text{s.t.} \quad -2v_1 \;+1v_2 \;+5v_3 \;\geq\; 7$$
$$-4v_1 \;+4v_2 \;-1v_3 \;=\; 0$$
$$+1v_1 \qquad\qquad +3v_3 \;\leq\; 44$$
$$v_1 \leq 0, v_2 \geq 0, v_3 \text{ URS}$$

Again, unrestricted x_2 yields an equality dual constraint.

Dual of the Dual Is the Primal

With this full symmetry of Table 7.1, we can also demonstrate one final form of primal-to-dual symmetry. Look back at the Two Crude model dual (7.7) and suppose that it were the primal:

$$\max \quad 2x_1 + 1.5x_2 + 0.5x_3 + 9x_4 + 6x_5$$

$$\text{s.t.} \quad 0.3x_1 + 0.4x_2 + 0.2x_3 + 1x_4 \leq 20$$

$$0.4x_1 + 0.2x_2 + 0.3x_3 + 1x_5 \leq 15$$

$$x_1, \ x_2, \ x_3 \geq 0$$

$$x_4, \ x_5 \leq 0$$

Like all linear programs, this primal must have a dual. Applying Table 7.1, we obtain

$$\min \quad +20v_1 \quad +15v_2$$

$$\text{s.t.} \quad +0.3v_1 \quad +0.4v_2 \quad \geq 2$$

$$+0.4v_1 \quad +0.2v_2 \quad \geq 1.5$$

$$+0.2v_1 \quad +0.3v_2 \quad \geq 0.5$$

$$+1v_1 \qquad\qquad \leq 9$$

$$+1v_2 \quad \leq 6$$

$$v_1, v_2 \geq 0$$

Except for variable naming, this dual exactly matches the original primal (7.4). This will always be true.

> **7.30** The dual of the dual of any linear program is the LP itself.

SAMPLE EXERCISE 7.19: FORMING THE DUAL OF THE DUAL

Show that the dual of the dual of the linear program in part (a) of Sample Exercise 7.18 is the primal.

Analysis: Assigning dual variables w_1, w_2, w_3 to the three main constraints of the dual part (a) of Sample Exercise 7.18, the dual of that dual is

$$\max \quad +6w_1 \quad -1w_2 \quad +13w_3$$

$$\text{s.t.} \quad +3w_1 \quad +1w_2 \quad +2w_3 \quad = 7$$

$$+5w_1 \quad -1w_2 \qquad \leq 6$$

$$+1w_2 \quad +1w_3 \quad \geq 2$$

$$w_1 \geq 0, w_2 \leq 0$$

In accord with principle 7.30 , this is exactly the primal (with different variable names).

7.5 PRIMAL-TO-DUAL RELATIONSHIPS

Having formed duals for all sorts of primal LPs in Section 7.4, we are ready to verify that they actually satisfy all the quantitative sensitivity requirements of Section 7.3.

Weak Duality between Objective Values

Dual forms $\boxed{7.28}$ and $\boxed{7.29}$ and Table 7.1 explicitly enforce type restrictions and main constraints of our Section 7.3 derivation. But what about value match $\boxed{7.25}$?

Given primal solution $\{x_j\}$ and dual solution $\{v_i\}$, a little algebraic manipulation shows that any objective function difference can be expressed as

$$\sum_j c_j x_j - \sum_i b_i v_i$$

$$= \sum_j c_j x_j \left(-\sum_i \sum_j v_i a_{i,j} x_j + \sum_i \sum_j v_i a_{i,j} x_j \right) - \sum_i b_i v_i$$

$$= \left(\sum_j c_j x_j - \sum_i \sum_j v_i a_{i,j} x_j \right) + \left(\sum_i \sum_j v_i a_{i,j} x_j - \sum_i b_i v_i \right) \quad (7.8)$$

$$= \left(\sum_j c_j x_j - \sum_j \sum_i a_{i,j} v_i x_j \right) + \left(\sum_i \sum_j a_{i,j} x_j v_i - \sum_i b_i v_i \right)$$

$$= \sum_j \left(c_j - \sum_i v_i a_{i,j} \right) x_j + \sum_i \left(\sum_j a_{i,j} x_j - b_i \right) v_i$$

Look closely at the last line of (7.8). Notice that each term in the first sum multiplies slack in a main dual constraint times the corresponding primal variable value. Similarly, each term in the second multiplies slack in a primal constraint times the corresponding dual variable value.

We can also predict the sign of all terms in that last line of expression (7.8) at feasible primal and dual variable values. For example, if the primal LP minimizes, the dual constraint $\boxed{7.24}$ corresponding to nonnegative variable x_j is

$$\sum_i a_{i,j} v_i \leq c_j \quad \text{so that} \quad \text{slack } (c_j - \sum_i a_{i,j} v_i) \geq 0$$

Thus if every x_j satisfies its nonnegativity and duals v_i satisfy all main dual constraints, every term in the first (7.8) sum is a product of nonnegative quantities. The entire sum would also have to be nonnegative. Similar reasoning accounts for nonpositive and URS variables and proves that the first sum will be nonpositive if our primal maximizes.

Now consider the second sum in the last line of (7.8). It deals with primal constraints, which we know may be of $\leq$, $=$, or $\geq$ form. Still, every term will be ≥ 0 if the primal is a minimize and ≤ 0 for a maximize. For example, in a minimize primal a $\leq$ constraint

$$\sum_j a_{i,j} x_j \leq b_i \quad \text{implies} \quad \text{slack} \left(\sum_j a_{i,j} x_j - b_i \right) \leq 0$$

at feasible x_j. But the corresponding dual variable is subject to type restriction $v_i \leq 0$, making the (7.8) term

$$\left(\sum_j a_{i,j} x_j - b_i \right) v_i \geq 0$$

at feasible primal and dual variable values. Other cases are similar.

A uniform sign for all terms in the last line of equation (7.8) means that primal and dual feasible solution values bound each other in what is known as **weak duality**.

> 7.31   The primal objective function evaluated at any feasible solution to a minimize primal is $\geq$ the objective function value of the corresponding dual evaluated at any dual feasible solution. For a maximize primal it is $\leq$.

We already know that 7.31 works if we use primal and dual optimal solutions in the Two Crude model:

$$20(2) + 15(3.5) = 2(20) + 1.5(35) + 0.5(0) + 9(0) + 6(0) = 92.5 \qquad (7.9)$$

Still, weak duality holds for any feasible primal and dual solutions. Suppose that we try $x_1 = 9$, $x_2 = 6$ in the primal, and $v_1 = v_2 = v_3 = 20$, $v_4 = v_5 = -5$ in the dual. Simple substitution will verify that these solutions satisfy all constraints of their respective problems. Checking 7.31 , we find

$$20(9) + 15(6) = 270 \quad \text{and} \quad 2(20) + 1.5(20) + 0.5(20) + 9(-5) + 6(-5) = 35$$

As required, the primal objective value is greater than or equal to the dual.

SAMPLE EXERCISE 7.20: VERIFYING WEAK DUALITY

For the two linear programs in Sample Exercise 7.17, verify that the following primal and dual solutions are feasible, and show that they conform to weak duality property 7.31 .

(a) Primal solution: $x_1 = 1$, $x_2 = 1$, $x_3 = 1$
Dual solution: $v_1 = 2$, $v_2 = 6$, $v_3 = -1$

(b) Primal solution: $x_1 = 2$, $x_2 = 1$, $x_3 = 0$
Dual solution: $v_1 = -2$, $v_2 = 4$, $v_3 = -1$
Analysis:

(a) To check primal feasiblity, we first note all specified $x_j \geq 0$. For main primal constraints,

$$\begin{aligned}
+1x_1 - 1x_2 - 1x_3 &= +1(1) - 1(1) + 1(1) = & 1 &\geq 1 \\
+3x_1 + 1x_2 & & = +3(1) + 1(1) & & = 4 \\
+4x_2 + 1x_3 & & = +4(1) + 1(1) & = 5 &\leq 10
\end{aligned}$$

Turning to dual feasibility, main constraints (see Sample Exercise 7.17) have

$$\begin{aligned}
+1v_1 + 3v_2 &= +1(2) + 3(6) & &= 14 \leq 30 \\
-1v_1 + 1v_2 + 4v_3 &= -1(2) + 1(6) + 4(-1) = 0 &\leq& 0 \\
+1v_1 + 1v_3 &= +1(2) + 1(-1) & &= 1 \leq 5
\end{aligned}$$

The given solution also satisfies type restrictions because

$$v_1 = 2 \geq 0 \quad \text{and} \quad v_3 = -1 \leq 0$$

Objective function values for the given solutions are

$$
\begin{aligned}
+30x_1 + 5x_3 &= +30(1) + 5(1) &&= 35 \\
+1v_1 + 4v_2 + 10v_3 &= +1(2) + 4(6) + 10(-1) &&= 16
\end{aligned}
$$

As implied by weak duality principle $\boxed{7.31}$, the primal value is $\geq$ the dual.

(b) To check primal feasiblity, we first note again that all specified $x_j \geq 0$. Verifying main primal constraints yields

$$
\begin{aligned}
+2x_1 + 1x_2 &= +2(2) + 1(1) &&= 5 &\geq 4 \\
+5x_1 + 3x_2 - 1x_3 &= 5(2) + 3(1) - 1(0) &&= 13 &\leq 15 \\
+1x_2 + 1x_3 &= +1(1) + 1(0) &&= 1
\end{aligned}
$$

Turning to dual feasibility, the main constraints have

$$
\begin{aligned}
+2v_1 + 5v_2 &= 2(-2) + 5(4) &&= 16 &\geq 10 \\
+1v_1 + 3v_2 + 1v_3 &= +1(-2) + 3(4) + 1(-1) &&= 9 &\geq 9 \\
-1v_2 + 1v_3 &= -1(4) + 1(-1) &&= -5 &\geq -6
\end{aligned}
$$

The given solution also satisfies type restrictions because

$$v_1 = -2 \leq 0 \quad \text{and} \quad v_2 = 4 \geq 0$$

Objective function values for the given solutions are

$$
\begin{aligned}
+10x_1 + 9x_2 - 6x_3 &= +10(2) + 9(1) - 6(0) &&= 29 \\
+3v_1 + 15v2 + 1v_3 &= +3(-2) + 15(4) + 1(-1) &&= 53
\end{aligned}
$$

Consistent with weak duality principle $\boxed{7.31}$, the primal value is $\leq$ the dual.

Strong Duality between Objective Values

The **strong duality** property of linear programs extends $\boxed{7.31}$ to exact equality when primal and dual objective functions are evaluated at optimal solutions:

$\boxed{7.32}$ If either a primal LP or its dual has an optimal solution, both do, and their optimal objective function values are equal.

For example, equation (7.9) has demonstrated strong duality for the Two Crude case.

SAMPLE EXERCISE 7.21: VERIFYING STRONG DUALITY

Consider the primal linear program

$$
\begin{aligned}
\max \quad & 6x_1 + 1x_2 \\
\text{s.t.} \quad & 1x_1 && \leq 1 \\
& 2x_1 + 1x_2 &&\leq 4 \\
& x_1, x_2 \geq 0
\end{aligned}
$$

(a) State the dual of this linear program.

(b) Solve both primal and dual graphically.

(c) Verify that their optimal values are equal.

Analysis: Assigning dual variables v_1 and v_2 to the two main primal constraints, the dual is

$$\min \quad 1v_1 \; + 4v_2$$
$$\text{s.t.} \quad 1v_1 \; + 2v_2 \geq 6$$
$$+ 1v_2 \geq 1$$
$$v_1, v_2 \geq 0$$

(a) The following figures solve primal and dual graphically:

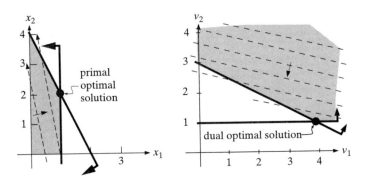

Optimal solutions are $x_1^* = 1, x_2^* = 2$ and $v_1^* = 4, v_2^* = 1$.

(b) Verifying the strong duality property $\boxed{7.32}$, we have

$$\text{primal optimal value} = 6x_1^* + 1x_2^* = 6(1) + 1(2)$$
$$= 8$$
$$= 1(2) + 4(1) = 1v_1^* + 4v_2^*$$
$$= \text{dual optimal value}$$

Dual Optimum as a By-product

We now wish to see why $\boxed{7.32}$ is true in general. For simplicity, assume that our primal LP is a minimize; the argument is almost identical for a maximize.

We proceed by looking back to the revised form of the simplex algorithm in Section 5.8. Recall that a key efficiency tool of the revised simplex is a **pricing vector** $\boxed{5.45}$, obtained by solving

$$\mathbf{vB} \triangleq (c_{1st}, c_{2nd}, \dots, c_{mth}) \tag{7.10}$$

where $c_{1st}, c_{2nd}, \dots, c_{mth}$ refer to the objective function coefficients of the 1st, 2nd, ..., mth basic variables, and $\mathbf{B}$ is the current basis column matrix of a (primal) LP in standard form (see Section 5.1). The revised simplex algorithm always finds an optimal solution if the primal model has one, and it recognizes that optimality has been reached in a minimize problem when **reduced costs** satisfy

$$\bar{c}_j \triangleq c_j - \sum_i a_{i,j} v_i \geq 0 \qquad \text{for all standard-form variables } j \qquad (7.11)$$

It is no accident that both the pricing vector of revised simplex and the dual variables of this chapter are denoted v. We will establish strong duality by showing that they are essentially the same thing.

| 7.33 | If the revised simplex algorithm stops with a primal optimal solution, the final pricing vector **v** yields an optimal solution in the corresponding dual.

That is, we obtain an optimal solution to the dual of a linear program as a by-product of revised simplex solution of its primal.

First, we want to establish that any pricing v_i's satisfying optimality conditions (7.11) are feasible in the dual $\boxed{7.28}$. To do so, separate conditions (7.11) for standard-form variables that were decision variables in the original primal model from those associated with slack variables added to achieve standard form. For decision variables, c_j and all $a_{i,j}$'s come directly from the given primal. Then by (7.11), pricing v_i's satisfy

$$c_j - \sum_i a_{i,j} v_i \geq 0 \quad \text{or} \quad \sum_i a_{i,j} v_i \leq c_j$$

This is exactly what is needed for feasibility in all main constraints of dual $\boxed{7.28}$.

Now focus on any slacks. We constructed standard form by choosing $c_j \triangleq 0$ for slack variable x_j in primal inequality i, and

$$a_{k,j} \triangleq \begin{cases} +1 & \text{if } k = i \text{ and } i \text{ is of } \leq \text{ form} \\ -1 & \text{if } k = i \text{ and } i \text{ is of } \geq \text{ form} \\ 0 & \text{otherwise} \end{cases}$$

Thus, with each slack column having only one nonzero entry, optimality test (7.11) on a slack in a $\leq$ constraint amounts to

$$(+1) v_i \geq 0$$

For a $\geq$ we have

$$(-1) v_i \leq 0 \quad \text{or} \quad v_i \leq 0$$

We conclude that pricing v_i's also satisfy the type restriction constraints in dual $\boxed{7.28}$. They are completely dual feasible.

Our next task is to show that the dual solution obtained from revised simplex pricing has the same objective function value as the primal optimum. The dual objective value for those v_i's is [using construction (7.10)]

$$\sum_i b_i v_i = \sum_i v_i b_i = \left[(c_{1\text{st}}, c_{2\text{nd}}, \ldots, c_{m\text{th}}) \mathbf{B}^{-1} \right] \cdot \mathbf{b}$$

Here **b** denotes the vector of b_i. But we also know that the only nonzero components of the primal optimal solution $\mathbf{x}^*$ are the basic ones,

$$(x_{1\text{st}}, x_{2\text{nd}}, \ldots, x_{m\text{th}}) = \mathbf{B}^{-1}\mathbf{b}$$

Thus the primal optimal value

$$\sum_j c_j x_j^* = c_{1st} x_{1st} + c_{2nd} x_{2nd} + \ldots + c_{mth} x_{mth}$$

$$= (c_{1st}, c_{2nd}, \ldots, c_{mth}) \cdot \left(\mathbf{B}^{-1} \mathbf{b} \right)$$

$$= \left[(c_{1st}, c_{2nd}, \ldots, c_{mth}) \mathbf{B}^{-1} \right] \cdot \mathbf{b}$$

That is, both primal and dual objective have the same value.

We already know that the primal solution $\mathbf{x}^*$ is optimal in its problem; revised simplex stopped. To see that pricing v_i's are also optimal in the dual, we exploit weak duality 7.31 . Weak duality holds for every primal feasible solution, so choose $\mathbf{x}^*$. No dual feasible solution can ever have a better dual objective function value than corresponding primal cost $\sum_j c_j x_j^*$. But we already know a dual feasible solution (our pricing v_i's) that can achieve that limit. We conclude that those v_i's form a feasible solution in the dual having the best possible objective function value, precisely what it means to be dual optimal.

SAMPLE EXERCISE 7.22: OBTAINING THE DUAL OPTIMUM AS A BY-PRODUCT

The LP

$$
\begin{array}{rlllll}
\max & +5x_1 & +7x_2 & +10x_3 \\
\text{s.t.} & +2x_1 & -1x_2 & +5x_3 & \leq 10 \\
& +1x_1 & +3x_2 & & \leq 15 \\
& x_1, x_2, x_3 \geq 0
\end{array}
$$

was converted to standard form by adding slack variables x_4 and x_5 in main constraints (principle 5.8). Then, revised simplex Algorithm 5C terminated with optimal basis matrix

$$
\begin{array}{cc}
& \begin{array}{cc} x_3 & x_2 \end{array} \\
\mathbf{B} = & \begin{pmatrix} 5 & -1 \\ 0 & 3 \end{pmatrix}
\end{array}
\quad \text{having inverse} \quad
\begin{array}{cc}
& \begin{array}{cc} x_3 & x_2 \end{array} \\
\mathbf{B}^{-1} = & \begin{pmatrix} \frac{3}{15} & \frac{1}{15} \\ 0 & \frac{5}{15} \end{pmatrix}
\end{array}
$$

(a) State the dual of the original linear program.

(b) Use $\mathbf{B}^{-1}$ to compute pricing vector $\mathbf{v}$ and verify that it is feasible in the dual.

(c) Use $\mathbf{B}^{-1}$ to compute the primal optimal solution, and verify that it has the same objective value as the primal.

Analysis:

(a) Following construction 7.29 , the dual is

$$
\begin{array}{rlll}
\min & +10v_1 & +15v_2 \\
\text{s.t.} & +2v_1 & +1v_2 & \geq 5 \\
& -1v_1 & +3v_2 & \geq 7 \\
& +5v_1 & & \geq 10 \\
& v_1, v_2 \geq 0
\end{array}
$$

(b) Here the first basic variable is x_3 and the second is x_2. Thus equation (7.10) gives

$$\mathbf{v} = (c_{1st}, c_{2nd})\mathbf{B}^{-1} = (10, 7)\begin{pmatrix} \frac{3}{15} & \frac{1}{15} \\ 0 & \frac{5}{15} \end{pmatrix} = (2, 3)$$

To check dual feasibility of $\mathbf{v}$, first note that both components are nonnegative as required by type restrictions. In main constraints,

$$\begin{aligned} 2v_1 + 1v_2 &= \ \ \ 2(2) + 1(3) \ = 7 \ \geq \ 5 \\ -1v_1 + 3v_2 &= -1(2) + 3(3) \ = 7 \ \geq \ 7 \\ 5v_1 \ \ \ \ \ \ &= \ \ \ \ 5(2) \ \ \ \ \ \ \ \ = 10 \geq 10 \end{aligned}$$

(c) The primal optimal basic solution has nonbasics $x_1^* = x_4^* = x_5^* = 0$ and basic components

$$\begin{pmatrix} x_3^* \\ x_2^* \end{pmatrix} = \mathbf{B}^{-1}\mathbf{b} = \begin{pmatrix} \frac{3}{15} & \frac{1}{15} \\ 0 & \frac{5}{15} \end{pmatrix}\begin{pmatrix} 10 \\ 15 \end{pmatrix} = \begin{pmatrix} 3 \\ 5 \end{pmatrix}$$

Thus the primal optimal value is $5(7) + 10(3) = 65$, which is the same as dual's $10(2) + 15(3)$.

Complementary Slackness between Primal and Dual Optima

Complementary slackness conditions 7.26 and 7.27 also hold when both primal and dual have an optimal solution (principle 7.32) as a consequence of strong duality. To see why, observe that one way to express dual complementarity conditions 7.27 is

$$(\text{dual slack})_j (\text{primal variable})_j = \left(c_j - \sum_i a_{i,j} v_i \right) x_j = 0 \qquad (7.12)$$

Either slack or variable must be $= 0$, possibly both. Similarly, primal complementarity conditions 7.26 amount to

$$(\text{primal slack})_i (\text{dual variable})_i = \left(\sum_j a_{i,j} x_j - b_i \right) v_i = 0 \qquad (7.13)$$

Now look back to weak duality computation (7.8). There we expressed the difference between primal and dual objective function values as sums of terms exactly like (7.12) and (7.13). In our derivation of weak duality we also concluded that each such term would be nonnegative in a minimize problem and nonpositive in a maximize problem. The only way that such terms of uniform sign can sum to the zero difference we know follows from strong duality is for every term to be zero. That is, strong duality's

$$\begin{aligned} 0 &= \sum_j c_j x_j^* - \sum_i b_i v_i^* \\ &= \sum_j \left(c_j - \sum_i v_i^* a_{i,j} \right) x_j^* + \sum_i \left(\sum_j a_{i,j} x_j^* - b_i \right) v_i^* \end{aligned}$$

implies that optimal primal and dual solutions satisfy every complementary slackness condition (7.12)–(7.13).

SAMPLE EXERCISE 7.23: VERIFYING COMPLEMENTARY SLACKNESS

Return to the linear program of Sample Exercise 7.22.

(a) Write all primal and dual complementary slackness conditions for model (7.12)–(7.13) in product form.

(b) Verify that the conditions are satisfied by optimal primal and dual solutions computed in Sample Exercise 7.22.

Analysis:

(a) Primal complementary slackness $\boxed{7.26}$ relates primal inequality slack to dual variable values:

$$(2x_1 - 1x_2 + 5x_3 - 10)v_1 = 0$$
$$(1x_1 + 3x_2 - 15)v_2 \quad\quad = 0$$

Dual complementary slackness $\boxed{7.27}$ relates dual inequality slack to primal variable values:

$$(5 - 2v_1 - 1v_2)x_1 = 0$$
$$(7 + 1v_1 - 3v_2)x_2 = 0$$
$$(10 - 5v_1)x_3 \quad\quad = 0$$

(b) From Sample Exercise 7.22, $\mathbf{x}^* = (0, 5, 3)$ and $\mathbf{v}^* = (2, 3)$. Checking primal complementary slackness conditions, we find that

$$(2x_1 - 1x_2 + 5x_3 - 10)v_1 = [2(0) - 1(5) + 5(3) - 10](2) = 0$$
$$(1x_1 + 3x_2 - 15)v_2 \quad\quad = [1(0) + 3(5) - 15](3) \quad\quad = 0$$

Similarly, in dual complementary slackness conditions,

$$(5 - 2v_1 - 1v_2)x_1 = [5 - 2(2) - 1(3)](0) = 0$$
$$(7 + 1v_1 - 3v_2)x_2 = [7 + 1(2) - 3(3)](5) = 0$$
$$(10 - 5v_1)x_3 \quad\quad = [10 - 5(2)](3) \quad\quad = 0$$

Unbounded and Infeasible Cases

Strong duality and complementary slackness conditions depend on the primal (and thus the dual) having optimal solutions. We would hardly undertake post-optimality analysis if there were no optimal solution. Still, other cases are sometimes of interest.

We learned very early that LPs can prove infeasible or unbounded. Weak duality condition $\boxed{7.31}$ provides the key tool in infeasible and unbounded instances. Suppose, for example, that a primal minimize problem is unbounded. Weak duality says that any dual feasible solution yields a lower bound on primal objective function values. But if the primal is infeasible, no such bound exists. The only possible explanation is that no dual feasible solutions exist.

We could make a similar argument for unbounded duals. Any primal feasible solution would limit dual objective function values, so none can exist.

| 7.34 | If either a primal LP model or its dual is unbounded, the other is infeasible.

After seeing so much elegant symmetry between primal and dual, one might guess that infeasibility in one problem also implies unboundedness in the other. This is false! Both primal and dual in the following are obviously infeasible:

$$\begin{array}{lll} & \min & -x_1 \\ (\text{primal}) \quad \text{s.t.} & x_1 - x_2 \geq 1 \\ & x_1 - x_2 \leq 0 \\ & x_1, x_2 \geq 0 \end{array} \qquad \begin{array}{lll} & \max & v_1 \\ (\text{dual}) \quad \text{s.t.} & v_1 + v_2 \leq -1 \\ & -v_1 - v_2 \leq 0 \\ & v_1 \geq 0, v_2 \leq 0 \end{array}$$

| 7.35 | The following shows which outcome pairs are possible for a primal linear program and its dual:

	Dual		
Primal	Optimal	Infeasible	Unbounded
Optimal	Possible	Never	Never
Infeasible	Never	Possible	Possible
Unbounded	Never	Possible	Never

SAMPLE EXERCISE 7.24: RELATING UNBOUNDEDNESS AND INFEASIBILITY

(a) Show graphically that the following linear program is unbounded and verify the implication for its dual.

$$\begin{array}{llll} \max & +1x_1 \\ \text{s.t.} & -1x_1 & +1x_2 & \geq 1 \\ & & +1x_2 & \geq 2 \\ & x_1, x_2 \geq 0 \end{array}$$

(b) Show graphically that the following linear program has an unbounded dual and verify the implication for the primal.

$$\begin{array}{llll} \min & +1x_1 & +1x_2 \\ \text{s.t.} & +2x_1 & +1x_2 & \geq 3 \\ & +5x_1 & +1x_2 & \leq 0 \\ & x_1, x_2 \geq 0 \end{array}$$

Analysis:

(a) The following plot clearly shows that the model is unbounded:

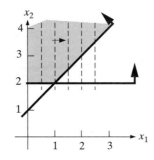

Its dual is

$$\begin{aligned}
\min \quad & +1v_1 \\
\text{s.t.} \quad & -1v_1 && \geq 1 \\
& +1v_1 && +1v_2 \geq 0 \\
& v_1, v_2 \leq 0
\end{aligned}$$

which is infeasible as implied by principle $\boxed{7.34}$.

(b) The dual of this linear program is

$$\begin{aligned}
\max \quad & +3v_1 \\
\text{s.t.} \quad & +2v_1 && +5v_2 \leq 1 \\
& +1v_1 && +1v_2 \leq 1 \\
& v_1 \geq 0, v_2 \leq 0
\end{aligned}$$

Solving graphically gives

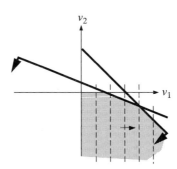

which is clearly unbounded.   It follows from principle $\boxed{7.34}$ that the primal is infeasible.

7.6 COMPUTER OUTPUTS AND WHAT IF CHANGES OF SINGLE PARAMETERS

All the machinery for quantitative sensitivity analysis is now in place. It is time to do some.

CFPL Example Primal and Dual

We illustrate with the CFPL model formulated in Section 4.3 to plan operations of a plywood factory. Table 7.2 displays the full primal in terms of the decision variables

$w_{q,v,t} \triangleq$ number of logs of quality q bought from vendor v and peeled into green veneer of thickness t per month

$x_{t,g}$ $\triangleq$ number of square feet of grade g green veneer of thickness t, purchased directly per month

$y_{t,g,g'} \triangleq$ number of sheets of thickness t used as grade g' veneer after drying and processing from grade g green veneer per month

$z_{t,g,g'} \triangleq$ number of sheets of front veneer grade g and back veneer grade g' plywood of thickness t, pressed and sold per month

A dual variable v_i has been assigned at the extreme right of each constraint. The optimal solution value is \$484,879 per month.

Table 7.3 provides the corresponding dual. Key points in its derivation include:

- The dual minimizes because the primal maximizes. Objective function coefficients come directly from the primal right-hand side.
- There is one main constraint for each of the 32 primal variables. All such constraints have the $\geq$ form because all primal variables are nonnegative. Dual right-hand sides come from the objective function coefficients of corresponding primal variables. Dual variables weighted on the left-hand side are those for constraints where the associated primal variable has nonzero coefficients.
- The first 17 dual variables are nonnegative because they correspond to $\leq$ constraints in a maximize primal. The next 6 are nonpositive because they relate to $\geq$'s, and the last 5 are unrestricted because they correspond to $=$'s.

Constraint Sensitivity Outputs

No two linear programming codes are identical, but all report the primal optimal solution, the corresponding dual solution and some related sensitivity information. Table 7.4 displays a typical output of constraint information. For each of the 28 CFPL (primal) constraints, the printout shows

Typ	Whether the constraint is of L = $\leq$, or G = $\geq$, or E = equality form
Optimal Dual	The optimal value of the dual variable for the constraint
RHS Coef	The specified right-hand side for the constraint
Slack	The amount of slack in the constraint at primal optimality
Lower Range	The lowest right-hand-side value for which the optimal dual solution must remain unchanged
Upper Range	The highest right-hand-side value for which the optimal dual solution must remain unchanged

TABLE 7.2   CFPL Example Primal

maximize:

(log costs)

$$-340w_{G, 1, 1/16} - 190w_{F, 1, 1/16} - 490w_{G, 2, 1/16} - 140w_{F, 2, 1/16}$$
$$-340w_{G, 1, 1/8} - 190w_{F, 1, 1/8} - 490w_{G, 2, 1/8} - 140w_{F, 2, 1/8}$$

(green veneer costs)

$$-1.00x_{1/16, A} - 0.30x_{1/16, B} - 0.10x_{1/16, C} - 2.20x_{1/8, A} - 0.60x_{1/8, B} - 0.20x_{1/8, C}$$

(finished plywood sales)

$$+45z_{1/4, A, B} + 40z_{1/4, A, C} + 33z_{1/4, B, C} + 75z_{1/2, A, B} + 65z_{1/2, A, C} + 50z_{1/2, B, C}$$

subject to:

(log availability)

$w_{G, 1, 1/16} + w_{G, 1, 1/8}$	≤ 200	v_1
$w_{F, 1, 1/16} + w_{F, 1, 1/8}$	≤ 300	v_2
$w_{G, 2, 1/16} + w_{G, 2, 1/8}$	≤ 100	v_3

(purchased green veneer availability)

$w_{F, 2, 1/16} + w_{F, 2, 1/8}$	≤ 1000	v_4
$x_{1/16, A}$	≤ 5000	v_5
$x_{1/16, B}$	$\leq 25,000$	v_6
$x_{1/16, C}$	$\leq 40,000$	v_7
$x_{1/8, A}$	$\leq 10,000$	v_8
$x_{1/8, B}$	$\leq 40,000$	v_9
$x_{1/8, C}$	$\leq 50,000$	v_{10}

(plywood market limits)

$z_{1/4, A, B}$	≤ 1000	v_{11}
$z_{1/4, A, C}$	≤ 4000	v_{12}
$z_{1/4, B, C}$	≤ 8000	v_{13}
$z_{1/2, A, B}$	≤ 1000	v_{14}
$z_{1/2, A, C}$	≤ 5000	v_{15}
$z_{1/2, B, C}$	≤ 8000	v_{16}

(plywood pressing capacity)

$0.25z_{1/4, A, B} + 0.25z_{1/4, A, C} + 0.25z_{1/4, B, C} + 0.40z_{1/2, A, B} + 0.40z_{1/2, A, C} + 0.40z_{1/2, B, C}$	≤ 4500	v_{17}

(green veneer balance)

$400w_{G, 1, 1/16} + 200w_{F, 1, 1/16} + 400w_{G, 2, 1/16} + 200w_{F, 2, 1/16} + x_{1/16, A}$ $-35y_{1/16, A, A} - 35y_{1/16, A, B}$	≥ 0	v_{18}
$700w_{G, 1, 1/16} + 500w_{F, 1, 1/16} + 700w_{G, 2, 1/16} + 500w_{F, 2, 1/16} + x_{1/16, B}$ $-35y_{1/16, B, A} - 35y_{1/16, B, B} + 35y_{1/16, B, C}$	≥ 0	v_{19}
$900w_{G, 1, 1/16} + 1300w_{F, 1, 1/16} + 900w_{G, 2, 1/16} + 1300w_{F, 2, 1/16} + x_{1/16, C}$ $-35y_{1/16, C, B} - 35y_{1/16, C, C}$	≥ 0	v_{20}
$200w_{G, 1, 1/8} + 100w_{F, 1, 1/8} + 200w_{G, 2, 1/8} + 100w_{F, 2, 1/8} + x_{1/8, A}$ $-35y_{1/8, A, A} - 35y_{1/8, A, B}$	≥ 0	v_{21}
$350w_{G, 1, 1/8} + 250w_{F, 1, 1/8} + 350w_{G, 2, 1/8} + 250w_{F, 2, 1/8} + x_{1/8, B}$ $-35y_{1/8, B, A} - 35y_{1/8, B, B} - 35y_{1/8, B, C}$	≥ 0	v_{22}
$450w_{G, 1, 1/8} + 650w_{F, 1, 1/8} + 450w_{G, 2, 1/8} + 650w_{F, 2, 1/8} + x_{1/8, C}$ $-35y_{1/8, C, B} - 35y_{1/8, C, C}$	≥ 0	v_{23}

(finished veneer sheet balance)

$y_{1/16, A, A} + y_{1/16, B, A} - z_{1/4, A, B} - z_{1/4, A, C} - z_{1/2, A, B} - z_{1/2, A, C}$	$= 0$	v_{24}
$y_{1/16, A, B} + y_{1/16, B, B} + z_{1/16, C, B} - z_{1/4, A, B} - z_{1/4, B, C} - z_{1/2, A, B} - z_{1/2, B, C}$	$= 0$	v_{25}
$y_{1/16, B, C} + y_{1/16, C, C} - z_{1/4, A, C} - z_{1/4, B, C} - z_{1/2, A, C} - z_{1/2, B, C}$	$= 0$	v_{26}
$y_{1/8, A, B} + y_{1/8, B, B} + z_{1/8, C, B} - z_{1/2, A, B} - z_{1/2, A, C} - z_{1/2, B, C}$	$= 0$	v_{27}
$y_{1/8, B, C} + y_{1/8, C, C} - z_{1/4, A, B} - z_{1/4, A, C} - z_{1/4, B, C} + 2z_{1/2, A, B} + 2z_{1/2, A, C} + 2z_{1/2, B, C}$	$= 0$	v_{28}

(nonnegativity)

all variables ≥ 0

TABLE 7.3   CFPL Example Dual

---

minimize:

$200v_1 + 300v_2 + 100v_3 + 1000v_4 + 5000v_5 + 25{,}000v_6$
$+ 40{,}000v_7 + 10{,}000v_8 + 40{,}000v_9 + 50{,}000v_{10} + 1000v_{11} + 4000v_{12}$
$+ 8000v_{13} + 1000v_{14} + 5000v_{15} + 8000v_{16} + 4500v_{17}$

subject to:

(*w*-variable columns)

$v_1 + 400v_{18} + 700v_{19} + 900v_{20}$	≥ -340
$v_2 + 200v_{18} + 500v_{19} + 1300v_{20}$	≥ -190
$v_3 + 400v_{18} + 700v_{19} + 900v_{20}$	≥ -490
$v_4 + 200v_{18} + 500v_{19} + 1300v_{20}$	≥ -140
$v_1 + 200v_{21} + 350v_{22} + 450v_{23}$	≥ -340
$v_2 + 100v_{21} + 250v_{22} + 650v_{23}$	≥ -190
$v_3 + 200v_{21} + 350v_{22} + 450v_{23}$	≥ -490
$v_4 + 100v_{21} + 250v_{22} + 650v_{23}$	≥ -140

(*x*-variable columns)

$v_5 + v_{18}$	≥ -1.00
$v_6 + v_{19}$	≥ -0.30
$v_7 + v_{20}$	≥ -0.10
$v_8 + v_{21}$	≥ -2.2
$v_9 + v_{22}$	≥ -0.60
$v_{10} + v_{23}$	≥ -0.20

(*y*-variable columns)

$v_{24} - 35v_{18}$	≥ 0
$v_{24} - 35v_{19}$	≥ 0
$v_{25} - 35v_{18}$	≥ 0
$v_{25} - 35v_{19}$	≥ 0
$v_{25} - 35v_{20}$	≥ 0
$v_{26} - 35v_{19}$	≥ 0
$v_{26} - 35v_{20}$	≥ 0
$v_{27} - 35v_{21}$	≥ 0
$v_{27} - 35v_{22}$	≥ 0
$v_{27} - 35v_{23}$	≥ 0
$v_{28} - 35v_{22}$	≥ 0
$v_{28} - 35v_{23}$	≥ 0

(*z*-variable columns)

$v_{11} + 0.25v_{17} - v_{24} - v_{25} - v_{28}$	≥ 45
$v_{12} + 0.25v_{17} - v_{24} - v_{26} - v_{28}$	≥ 40
$v_{13} + 0.25v_{17} - v_{25} - v_{26} - v_{28}$	≥ 33
$v_{14} + 0.40v_{17} - v_{24} - v_{25} - v_{27} - 2v_{28}$	≥ 75
$v_{15} + 0.40v_{17} - v_{24} - v_{26} - v_{27} - 2v_{28}$	≥ 65
$v_{16} + 0.40v_{17} - v_{25} - v_{26} - v_{27} - 2v_{28}$	≥ 50

(sign restrictions)

$v_1, v_2, \ldots, v_{17} \geq 0$
$v_{18}, v_{19}, \ldots, v_{23} \leq 0$
$v_{24}, v_{25}, \ldots, v_{28}$ unrestricted

Notice that slack and dual values conform to primal complementary slackness conditions 7.26 . Dual variables are positive only when the corresponding slack is zero.

Right-Hand-Side Ranges

The last two items for each constraint are new to our discussion. To understand their meaning, look back at Figure 7.3. We know that optimal dual variable values

TABLE 7.4   Typical Constraint Sensitivity Analysis Output for CFPL Model

Name	Typ	Optimal Dual	RHS Coef	Slack	Lower Range	Upper Range
c1	L	-0.000	200.000	158.727	41.273	+infin
c2	L	122.156	300.000	0.000	242.500	331.972
c3	L	-0.000	100.000	100.000	0.000	+infin
c4	L	172.156	1000.000	0.000	942.500	1031.972
c5	L	-0.000	5000.000	5000.000	0.000	+infin
c6	L	-0.000	25000.000	25000.000	0.000	+infin
c7	L	0.032	40000.000	0.000	0.000	85858.586
c8	L	-0.000	10000.000	10000.000	0.000	+infin
c9	L	-0.000	40000.000	40000.000	0.000	+infin
c10	L	0.112	50000.000	0.000	0.000	81971.831
c11	L	9.564	1000.000	0.000	342.857	2621.429
c12	L	4.564	4000.000	0.000	3342.857	5621.429
c13	L	-0.000	8000.000	3644.156	4355.844	+infin
c14	L	10.000	1000.000	0.000	402.597	5402.597
c15	L	-0.000	5000.000	597.403	4402.597	+infin
c16	L	-0.000	8000.000	8000.000	0.000	+infin
c17	L	51.418	4500.000	0.000	4404.622	4609.524
c18	G	-0.201	0.000	-0.000	-51111.111	131759.465
c19	G	-0.201	0.000	-0.000	-51111.111	194861.111
c20	G	-0.132	0.000	-0.000	-45858.586	41818.182
c21	G	-0.312	0.000	-0.000	-31971.831	57500.000
c22	G	-0.312	0.000	-0.000	-31971.831	57500.000
c23	G	-0.312	0.000	-0.000	-31971.831	57500.000
c24	E	-7.046	0.000	0.000	-1460.317	5567.460
c25	E	-4.610	0.000	0.000	-1310.245	1194.805
c26	E	-4.610	0.000	0.000	-1310.245	1194.805
c27	E	-10.925	0.000	0.000	-913.481	1642.857
c28	E	-10.925	0.000	0.000	-913.481	1642.857

quantify the slope or rate of change in such relations between the right-hand side of a constraint and the overall optimal value. In the CFPL output, for instance, $v_2^* = 122.156$ implies that the optimal value will improve by \$122.156 per log increase in RHS = 300 of the corresponding log availability constraint.

The RHS range question is how much we could change the right-hand side before that slope would no longer apply.

7.36   **Right-hand-side ranges** in LP sensitivity outputs show the interval within which the corresponding dual variable value provides the exact rate of change in optimal value per unit change in RHS (all other data held constant).

For example, output Table 7.4 tells us that slope v_2^* would remain unchanged for any RHS in the interval [242.5, 331.972]. Outside the range, we have only qualitative principle 7.14 .

The idea behind range computation is to identify the maximum change in any RHS for which the optimal basis remains (primal) feasible. In rows with slack, such as the first constraint in Table 7.4, feasibility persists exactly until the slack is eliminated. Thus with the original constraint one RHS of 200 permitting 158.727 units of slack, dual slope $v_1^* = 0$ applies for any RHS value at least

$$\text{RHS} - \text{slack} = 200 - 158.727 = 41.273$$

The upper range is $+\infty$ because increasing the RHS only relaxes an already slack constraint.

The computation is a bit more complex when a constraint is active, but the issue remains keeping the primal solution feasible. We omit details.

SAMPLE EXERCISE 7.25: INTERPRETING RHS RANGES

Suppose that solution of the LP

$$
\begin{array}{rllll}
\min & +5x_1 & +9x_2 & \\
\text{s.t.} & +1x_1 & +1x_2 & \geq 3 \\
& +1x_1 & -1x_2 & \leq 4 \\
& x_1, x_2 \geq 0
\end{array}
$$

produces the constraint sensitivity output

Name	Typ	Optimal Dual	RHS Coef	Slack	Lower Range	Upper Range
c1	G	5.000	3.000	-0.000	0.000	4.000
c2	L	0.000	4.000	1.000	3.000	+infin

Sketch what can be deduced from this output about how the optimal value varies with each RHS.

Analysis: What we can know about how optimal value varies with each RHS is summarized in the following two plots:

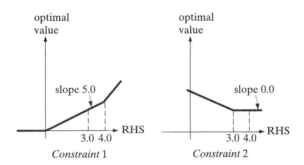

Constraint 1   *Constraint 2*

The printout's RHS range for constraint 1 shows that the rate of change in optimal value with its right-hand side is dual variable value 5.0 in the range 0.0 to 4.0 (principle 7.36). Above that range qualitative principle 7.14 implies that the rate can only increase because we are tightening a $\geq$ constraint. Below the range the rate can only decrease.

For constraint 2 the RHS range is $[3.0, +\infty)$. Within this range there will be no change in optimal value because the corresponding dual variable is 0.0. Below the range the tightening $\leq$ may produce a steeper rate of change (principle 7.14).

Constraint What If's

There are hundreds of possible "what if" questions about the CFPL model that can be asked and answered with the aid of results in Table 7.4. Much can be said if we consider changing only one coefficient while holding all others constant. We illustrate with a few examples.

- Question: How sensitive is the optimal value to our 8000 estimate of the market for $\frac{1}{2}$-inch BC plywood?

  Analysis: The market constraint for $\frac{1}{2}$-inch BC has optimal dual $v_{16}^* = 0$ and RHS range $[0, +\infty)$. Thus the optimal value will remain unchanged no matter what (nonnegative) market estimate we employ.

- Question: How sensitive is the optimal value to our 300 estimate of the supply of "fair" logs available from supplier 1?

  Analysis: The availability constraint for fair logs from supplier 1 has optimal dual $v_2^* = 122.156$ and the RHS range $[242.5, 331.972]$. Thus, if the estimate of 300 is too high, every log deducted will reduce the optimal value by at least \$122.156. If it is too low, every log added will increase the optimal value by at most the same amount.

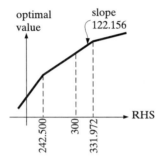

The \$122.156 rate is exact within the RHS range. Below the range, the rate of change will become steeper because we are tightening a constraint (principle 7.14). Above the range, the rate will diminish.

- Question: What is the marginal value of plywood pressing capacity?

  Analysis: The marginal value is the optimal dual on the pressing capacity constraint, $v_{17}^* = \$51.418$ per sheet.

- Question: How much should we be willing to spend in promotional costs to increase the market for $\frac{1}{4}$-inch AB plywood from 1000 to 2000 sheets per month?

  Analysis: Because 2000 is within the RHS range on the market constraint for $\frac{1}{4}$-inch AB, dual $v_{11}^* = \$9.564$ gives the exact rate of change. An increase from 1000 to 2000 would be worth

  $$\text{(new value} - \text{current RHS)} \, v_{11}^* = (2000 - 1000)(9.564)$$

  $$= \$9564 \text{ per month}$$

  Any promotional expense up to that amount would be justified.

- Question: How much should we be willing to pay to increase plant pressing capacity from 4500 to 6000 sheets per month?

  Analysis: New value 6000 falls well beyond the upper range limit of 4609.624 on the pressing capacity RHS, thus we can only bound the value of the proposed capacity increase.

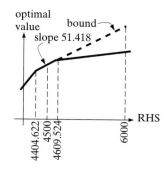

At a minimum, we know that the optimal value would improve:

$$(\text{range limit} - \text{current RHS})\, v_{17}^* = (4609.624 - 4500)(51.418)$$

$$\approx \$5637 \text{ per month}$$

The dual provides an exact rate of change through the upper range limit.

Above the range limit, the rate of change may decline (principle $\boxed{7.14}$); we are relaxing a constraint. Still, the gain from raising capacity to 6000 cannot exceed the estimate provided by the dual-variable value:

$$(\text{new value} - \text{current RHS})\, v_{17}^* = (6000 - 4500)(51.418)$$

$$= \$77{,}127 \text{ per month}$$

We conclude that any equivalent monthly capacity expansion cost up to \$5637 could be justified and any over \$77,127 could not. In between, we cannot be definitive.

- Question: How much loss would we incur if we stopped using "good" logs from supplier 2?

    Analysis: The dual variable on supplier 2 good log availability has $v_3^* = 0.0$ and RHS range $[0.0, +\infty)$. Thus no change in optimal value would result if we reduced availability all the way to zero.

- Question: How much loss would we experience if we stopped using fair logs from supplier 2?

    Analysis: The dual variable on supplier 2 fair log availability 1000 has $v_4^* = 172.156$ and RHS range $[942.5, 1031.972]$. Eliminating such log purchases amount to changing the RHS to zero. For such a value, which is well outside the RHS range, we can only bound the effect.

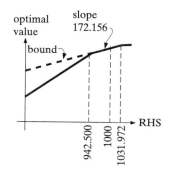

We are tightening the availability constraint, so principle $\boxed{7.14}$ shows that the true rate of change will be v_4^* (actually, its negative) or worse. Thus, closing out supplier 2 fair logs would produce a loss of at least

$$\text{(new value} - \text{current RHS)}\ v_3^* = (0 - 1000)(172.156)$$

$$= -\$172{,}156 \text{ per month}$$

There is no way to use Table 7.4 information to set an outer limit on the loss (except the full optimal value, \$484,878).

Variable Sensitivity Outputs

We saw in Section 7.1 that sensitivity analysis encompasses much more than just changes in right-hand-side values. Table 7.5 provides an example of the variable or column-oriented sensitivity information that is part of almost any LP optimizer output. The following are displayed for each primal variable:

`Name`	The name of the primal variable
`Optimal Value`	The optimal value for the variable
`Bas Sts`	Whether the variable is BAS = basic, NBL = nonbasic lower-bounded, or NBU = nonbasic upper-bounded in the optimal solution (see Section 5.9
`Lower Bound`	The specified lower bound on the variable
`Upper Bound`	The specified upper bound on the variable
`Object Coef`	The specified objective function coefficient for the variable
`Reduced Object`	The reduced objective function coefficient for the variable at optimality
`Lower Range`	The lowest objective function coefficient value for which the optimal primal solution must remain unchanged
`Upper Range`	The highest objective function coefficient value for which the optimal primal solution must remain unchanged

Most of these items are familiar from our previous discussion. Optimal values, objective function coefficients, and reduced objective function coefficients have usually been denoted x_j^*, c_j, and $\bar{c}_j$, respectively. Notice that given values satisfy dual complementary slackness conditions $\boxed{7.27}$. Whenever `Optimal Value> 0`, the corresponding dual constraint slack `Reduced Object = 0`.

Objective Coefficient Ranges

As with constraint information, the two completely new items are **ranges**, this time on the objective. Items `Lower Range` and `Upper Range` delimit changes in objective coefficient value that would leave the primal optimal solution unchanged.

Figure 7.5 illustrates how keeping the same optimal solution does not necessarily mean keeping the same optimal value. The optimal level for sales variable $z_{1/4,A,B}$ (zqAB in the computer output) is 1000. Thus the optimal objective value will increase at the rate of \$1000 per \$1 increase in the $z_{1/4,A,B}$ objective function coefficient as long as the primal optimal solution remains unchanged.

TABLE 7.5 Typical Variable Sensitivity Analysis Output for CFPL Model

Name	Optimal Value	Bas Sts	Lower Bound	Upper Bound	Object Coef	Reduced Object	Lower Range	Upper Range
wG1s	41.273	BAS	0.000	+infin	-340.000	0.000	-361.569	-304.762
wF1s	300.000	BAS	0.000	+infin	-190.000	0.000	-190.000	+infin
wG2s	0.000	NBL	0.000	+infin	-490.000	150.000	-infin	-340.000
wF2s	155.273	BAS	0.000	+infin	-140.000	0.000	-185.306	-140.000
wG1e	0.000	NBL	0.000	+infin	-340.000	27.844	-infin	-312.156
wF1e	0.000	NBL	0.000	+infin	-190.000	0.000	-infin	-190.000
wG2e	0.000	NBL	0.000	+infin	-490.000	177.844	-infin	-312.156
wF2e	844.727	BAS	0.000	+infin	-140.000	0.000	-140.000	-94.694
xsA	0.000	NBL	0.000	+infin	-1.000	0.799	-infin	-0.201
xsB	0.000	NBL	0.000	+infin	-0.300	0.099	-infin	-0.201
xsC	40000.000	BAS	0.000	+infin	-0.100	0.000	-0.132	+infin
xeA	0.000	NBL	0.000	+infin	-2.200	1.888	-infin	-0.312
xeB	0.000	NBL	0.000	+infin	-0.600	0.288	-infin	-0.312
xeC	50000.000	BAS	0.000	+infin	-0.200	0.000	-0.312	+infin
zqAB	1000.000	BAS	0.000	+infin	45.000	0.000	35.436	+infin
zqAC	4000.000	BAS	0.000	+infin	40.000	0.000	35.436	+infin
zqBC	4355.844	BAS	0.000	+infin	33.000	0.000	31.612	34.675
zhAB	1000.000	BAS	0.000	+infin	75.000	0.000	65.000	+infin
zhAC	4402.597	BAS	0.000	+infin	65.000	0.000	62.320	67.220
zhBC	0.000	NBL	0.000	+infin	50.000	12.564	-infin	62.564
yeBB	0.000	NBL	0.000	+infin	-0.000	0.000	-infin	0.000
ysAA	3073.247	BAS	0.000	+infin	0.000	0.000	-2.351	5.045
ysAB	0.000	NBL	0.000	+infin	0.000	2.436	-infin	2.436
ysBA	7329.351	BAS	0.000	+infin	0.000	0.000	-2.792	3.964
ysBB	0.000	NBL	0.000	+infin	0.000	2.436	-infin	2.436
ysBC	0.000	NBL	0.000	+infin	0.000	2.436	-infin	2.436
ysCB	6355.844	BAS	0.000	+infin	0.000	0.000	-1.388	1.675
ysCC	12758.442	BAS	0.000	+infin	0.000	0.000	-3.700	4.467
yeAB	2413.506	BAS	0.000	+infin	0.000	0.000	-12.867	15.857

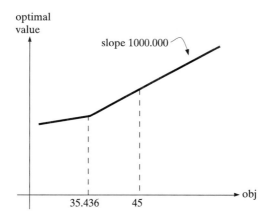

FIGURE 7.5 Optimal Value Impact of CFPL Objective Changes

7.37 Optimal primal LP variable values show the rate of change in optimal value per unit increase in the corresponding objective function coefficient.

7.38 **Objective coefficient ranges** in LP sensitivity outputs display the interval within which the current primal solution remains optimal and individual variable values continue to show exactly how optimal value changes with their objective coefficients (all other data held constant).

In Figure 7.5 the range is $[35.436, +\infty)$. Outside the range we must rely on qualitative principle **7.16** to impute rates of change. Since selling prices below \$35.436 help the objective less, the rate ≤ 1000.

Like the RHS case, objective coefficient range computations involve details inappropriate for this book. However, calculations center on keeping the dual solution feasible so that optimality conditions will continue to hold for the current primal optimum.

The computation is easy when dual slack $\bar{c}_j$ is nonzero. Certainly, dual feasibility will not be lost until all slack is eliminated. For instance, sales variable $z_{1/2,B,C}$ (zhBC), with reduced objective coefficient $-\$12.564$ on an original coefficient of \$50, has objective coefficient range $(-\infty, 62.564]$. Reductions from \$50 only relax an already slack dual constraint, and increases eliminate slack only at the coefficient value

$$\text{(current coefficient - dual slack)} = 50 - (-12.564)$$

$$= \$62.564 \text{ per month}$$

SAMPLE EXERCISE 7.26: INTERPRETING OBJECTIVE COEFFICIENT RANGES

Return to the LP of Sample Exercise 7.25, and assume that the variable part of sensitivity output is as follows:

Name	Optimal Value	Bas Sts	Object Coef	Reduced Object	Lower Range	Upper Range
x1	5.000	BAS	5.000	0.000	0.000	9.000
x2	0.000	NBL	9.000	4.000	5.000	+infin

Sketch what can be deduced from this output about how the optimal value will change with the two objective function coefficients.

Analysis: What we can know about how the optimal value varies with objective function coefficients is summarized in the following two plots:

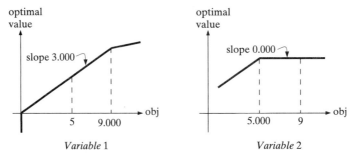

The range on the first coefficient is $[0.0, 9.0]$, meaning that the current primal solution will remain optimal for any c_1 in the range and that the optimal value varies within that range at rate $x_1^* = 3.0$ (principle 7.38). Increasing that coefficient hurts the optimal value, so the rate may decline for $c_1 > 9.0$ (principle 7.16). Below the range the rate becomes steeper.

Similarly, the output above shows that the range for c_2 is $[5.0, +\infty)$. Any cost in that range will leave $x_2^* = 0.0$ and the optimal value unchanged. Below the range the objective may decline because helping the objective will steepen the rate (principle 7.16).

Variable What If's

The value of output information in Table 7.5 is best illustrated by considering some of the "what if" questions that it might help to answer. Again we assume only one parameter changes with all else held constant.

- Question: How sensitive are the optimal solution and value to our $33 estimate of the selling price for $\frac{1}{4}$-inch BC plywood?

    Analysis: The objective range of $[31.612, 34.675]$ on $zqBC$ implies that the optimal operating plan would remain unchanged for any price between those values. However, the optimal value would vary.

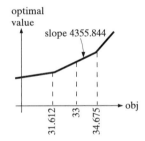

Every price increase of $1 would add $z_{1/4,B,C}^* = \$4355.844$ per week to the optimal value within the range and that much or more above (principle 7.16). Every $1 decrease would reduce the optimal value at the same rate through limit $31.612 and that much or less thereafter.

- Question: How sensitive are the optimal solution and value to our $1.00 per square foot (objective coefficient -1.00) estimate of the cost of $\frac{1}{16}$-inch A green veneer?

    Analysis: Coefficient range $(-\infty, -0.201]$ on xsA implies that the optimal solution would remain unchanged for any coefficient below -0.201 (cost above 0.201) per square foot. Within that range, optimal value would also remain unchanged because $x_{1/16,A}^* = 0$.

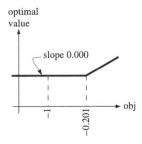

Coefficient values above -0.201 (costs below 0.201) might improve the optimal value because the rate of change cannot decrease.

- Question: How much difference in the optimal value would it make if we increased or decreased the $65 per sheet selling price of $\frac{1}{2}$-inch AC plywood by 20%?

  Analysis: A 20% increase would bring the price to $(1.2)(65) = \$78$. Because this value is well beyond the upper range limit of $67.220 for variable zhAC, we can only bound the optimal value impact. It will equal or exceed

$$(\text{new coefficient} - \text{original coefficient})\, z^*_{1/2,A,C} \;=\; (78 - 65)(4402.597)$$
$$\approx \quad \$57{,}234 \text{ per month}$$

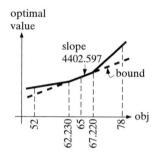

A 20% price decrease also takes us well outside the range. We can compute the least possible impact by considering only the range where we know the exact rate of change.

$$(\text{range limit} - \text{original coefficient})\, z^*_{1/2,A,C} = (62.320 - 65)(4402.597)$$
$$\approx \; -\$11{,}799 \text{ per month}$$

Extending the current rate of change all the way to $(0.8)(65) = \$52$ bounds the maximum impact:

$$(\text{new coefficient} - \text{original coefficient})\, z^*_{1/2,A,C} = (52 - 65)(4402.597)$$
$$\approx \; -\$57{,}234 \text{ per month}$$

- Question: The current optimal plan produces no $\frac{1}{2}$-inch BC plywood at our estimated $50 per sheet selling price. At what price would producing that product become profitable?

  Analysis: Coefficient ranges indicate $z^*_{1/2,B,C} = 0$ would remain the optimal choice for any price up to $62.564. At prices above that value a positive profit is possible because the rate of change can only increase as we move a coefficient to help the objective (7.16).

- Question: The current optimal plan buys $x^*_{1/16,C} = 40,000$ square feet of $\frac{1}{16}$-inch C green veneer per week at cost $0.10 (objective coefficient -0.10) per square foot. At what price would such purchases become unprofitable?

  Analysis: The coefficient range for $x_{1/16,C}$ is $[-0.132, +\infty)$. Thus we know that the 40,000-square foot value will remain optimal at coefficients above -0.132 (i.e., cost below $0.132 per square foot). Beyond $0.132 it may be optimal to buy less.

Dropping and Adding Constraint What If's

Another familiar form of "what if" questions involve dropping or adding constraints. Results in Figures 7.4 and 7.5 help here, too. The necessary insights are almost obvious:

7.39 Dropping a constraint can change the optimal solution only if the constraint is active at optimality.

7.40 Adding a constraint can change the optimal solution only if that optimum violates the constraint.

Some examples (assuming the rest of the model is held constant):

- Question: Would dropping the press capacity constraint change the optimal plan?
  Analysis: Yes. The constraint has no slack at the optimal solution.
- Question: Would dropping the supplier 1 good log availability constraint change the optimal plan?
  Analysis: No. At the optimal solution that constraint has a slack of 158.727 logs.
- Question: We have an informal commitment to buy at least 325 logs per month from supplier 1. Would including that constraint change the optimal solution?
  Analysis: No. Such a constraint would have the form

$$w_{G,1,1/16} + w_{G,1,1/8} + w_{F,1,1/16} + w_{F,1,1/8} \geq 325$$

Substituting the current optimal solution yields

$$
\begin{aligned}
w^*_{G,1,1/16} + w^*_{G,1,1/8} + w^*_{F,1,1/16}w^*_{F,1,1/8} &= 41.273 + 0 + 300 + 0 \\
&= 341.273 \\
&> 325
\end{aligned}
$$

The constraint would not be active if it were included.

- Question: Would a policy limiting green veneer purchase to $10,000 per month change the current optimal plan?
  Analysis: Yes. A new constraint enforcing this policy would have the form

$$1.00x_{1/16,A} + 0.30x_{1/16,B} + 0.10x_{1/16,C} + 2.20x_{1/8,A} + 0.60x_{1/8,B} + 0.20x_{1/8,C} \leq 10,000$$

Substituting the current plan yields

$$
\begin{aligned}
1.00x^*_{1/16,A} + 0.30x^*_{1/16,B} &+ 0.10x^*_{1/16,C} + 2.20x^*_{1/8,A} + 0.60x^*_{1/8,B} + 0.20x^*_{1/8,C} \\
&= 1.00(0) + 0.30(0) + 0.10(40,000) + 2.20(0) + 0.60(0) + 0.20(50,000) \\
&= 14,000 \\
&\not\leq 10,000
\end{aligned}
$$

Our current optimal solution violates the proposed constraint.

SAMPLE EXERCISE 7.27: ANALYZING DROPPED AND ADDED CONSTRAINTS

Suppose that a linear program has optimal solution $x^*_1 = 3, x^*_2 = 0, x^*_3 = 1$. Determine whether each of the following modifications in the model will change this optimal solution.

(a) Dropping constraint $6x_1 - x_2 + 2x_3 \geq 20$

(b) Dropping constraint $4x_1 - 3x_3 \leq 15$

(c) Adding constraint $x_1 + x_2 + x_3 \leq 2$

(d) Adding constraint $2x_1 + 7x_2 + x_3 \leq 7$

Analysis:

(a) Substituting the optimal solution gives us

$$6(3) - (0) + 2(1) = 20$$

so that the constraint is active. Under principle $\boxed{7.39}$, dropping the constraint may change the optimal solution.

(b) Substituting the optimal solution, we have

$$4(3) - 3(1) = 9 < 15$$

and the constraint is inactive. Dropping the constraint will not change the optimal solution (principle $\boxed{7.39}$).

(c) Substituting the optimal solution, we have

$$(3) + (0) + (1) = 4 \not\leq 2$$

Since the constraint is violated, adding it will change the optimal solution (principle $\boxed{7.40}$).

(d) Substituting the optimal solution, we have

$$2(3) + 7(0) + (1) = 7 \leq 7$$

Adding the constraint will not change the optimal solution because it is already satisfied (principle $\boxed{7.40}$).

Dropping and Adding Variable What If's

The final category of "what if" questions that we can answer using Figures 7.4 and 7.5 involves dropping and adding variables. The dropping case is trivial:

> **7.41**   An LP variable can be dropped without changing the optimal solution only if its optimal value is zero.

For example, Table 7.5 implies that the optimal plan would remain unchanged if we eliminated $\frac{1}{2}$-inch BC producing activity $z_{1/2,B,C}$. Its optimal value $z^*_{1/2,B,C} = 0$.

Adding is a bit more complex. We need to decide whether the optimal value for an added variable would be zero, implying that it is just as well not to include it. But this depends on whether current optimal dual prices v_i^* produce a reduced objective function coefficient with the proper sign (i.e., whether the main dual constraint associated with the new variable is satisfied).

> **7.42**   A new LP variable can change the current primal optimal solution only if its dual constraint is violated by the current dual optimum.

Some examples (assuming the rest of the model is held constant):

- A new $\frac{1}{4}$-inch AA plywood may be introduced that could be pressed in 0.25 hour using two $\frac{1}{16}$-inch sheets of A veneer and one $\frac{1}{8}$-inch sheet of B veneer. At what selling price would this product enter the optimal plan?

Analysis: The corresponding dual constraint would be

$$0.25v_{17} - 2v_{24} - v_{27} \geq c$$

Substituting the current dual solution yields

$$0.25v_{17}^* - 2v_{24}^* - v_{27}^* = 0.25(51.418) - 2(-7.046) - (-10.925)$$
$$\approx \$37.87$$

At any price above this value, the new product would be profitable.

- Question: A new supplier offers $\frac{1}{16}$-inch B green veneer at $0.40 per square foot. Would adding the possibility of purchasing his veneer change the optimal solution?
  Analysis: No. The only nonzero coefficient on this new activity would appear in the balance constraint for $\frac{1}{16}$-inch B veneer. Thus the constraint would be

$$v_{19} \geq -0.40$$

The present $v_{19}^* = -0.201$ satisfies this constraint, so it would be optimal to use none of the new activity.

SAMPLE EXERCISE 7.28: ANALYZING DROPPED AND ADDED VARIABLES

Suppose that a minimize LP over nonnegative variables has primal optimal solution $x_1 = 0$, $x_2 = 4$, $x_3 = 2$ and corresponding dual optimal solution $v_1 = 1$, $v_2 = 0$, $v_3 = 6$. Determine whether each of the following modifications in the model will change this optimal solution.

(a) Dropping variable x_1

(b) Dropping variable x_2

(c) Adding a variable with coefficients 2, -1, and 3 in the main constraints and cost 18

(d) Adding a variable with coefficients 1, 0, and 2 in the main constraints and cost 50

Analysis:

(a) In accord with principle 7.41 , dropping x_1 will not change the optimal solution because its optimal value $= 0$.

(b) Since x_2 has a nonzero optimal value, dropping it would change the optimal solution (principle 7.41).

(c) The main dual constraint for the new primal variable would be

$$2v_1 - 1v_2 + 3v_3 \leq 18$$

Substituting the optimal dual solution, we obtain

$$2(1) - 1(0) + 3(6) = 20 \not\leq 18$$

Adding this variable would change the primal optimum because the current dual solution violates the new constraint (principle 7.42).

(d) The main dual constraint for this new primal variable would be

$$1v_1 + 2v_3 \leq 50$$

Substituting the optimal dual solution, we obtain

$$1(1) + 2(6) = 13 \leq 50$$

Since the constraint is satisfied, adding the new variable would not change the primal optimum (principle $\boxed{7.42}$).

7.7 BIGGER MODEL CHANGES, REOPTIMIZATION, AND PARAMETRIC PROGRAMMING

Although standard LP sensitivity outputs lend insight on a host of minor "what if" questions, optimization by-products can go only so far. In this section we review briefly some of the practical limits and introduce more informative approaches requiring reoptimization.

Ambiguity at Limits of the RHS and Objective Coefficient Ranges

In Section 7.6 (principles $\boxed{7.36}$ and $\boxed{7.38}$) we explained that the optimal primal solution provides reliable quantitative sensitivity information only as long as an objective coefficient change stays within the range provided in LP outputs, and the optimal dual solution yields precise rates of change only if a right-hand-side change is restricted to the RHS range displayed. Beyond the ranges, we have only upper or lower bounds on rates of change from qualitative principles $\boxed{7.14}$ and $\boxed{7.16}$.

Figure 7.6 illustrates how the ambiguity about rates of change, which we first encountered in Section 7.2, occurs at the limits of RHS ranges. Three different versions of our venerable Two Crude refining model (7.4) are solved graphically using varying gasoline requirement RHS values $b_1 = 2.0$, 2.625, and 3.25, respectively.

Computer output for the 2.0 case shows optimal dual slope $v_1^* = 20$ holding within the RHS range $1.125 \leq b_1 \leq 2.625$. A similar output for the $b_1 = 3.25$ choice indicates that a rate of change $v_1^* = 66.667$ applies over the range $2.625 \leq b_1 \leq 5.100$.

One or the other of these dual and range outputs would print out if we solved with b_1 exactly 2.625, but which one it would be is unpredictable. Either would mislead. The correct rate for an increase from $b_1 = 2.625$ is 66.667, and for a decrease it is 20, but neither possible output provides both.

> $\boxed{7.43}$ At the limits of the RHS and objective function sensitivity ranges rates of optimal value change are ambiguous, with one value applying below the limit and another above. Computer outputs may show either value.

SAMPLE EXERCISE 7.29: INTERPRETING SENSITIVITY AT RANGE LIMITS

Return to the Swedish Steel sensitivity plot of Figure 7.3(a). Determine from the plot the RHS ranges and optimal dual values that might result if an LP optimization code were invoked with the given RHS (a) 75.0; (b) 60.4.

Analysis:

(a) RHS value 75.0 falls within an unambiguous range. Output would show dual variable -3.38 and range $[60.4, 83.3]$.

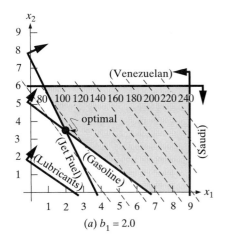

(a) $b_1 = 2.0$

Optimal Dual	Lower Range	Upper Range
20.000	1.125	2.625

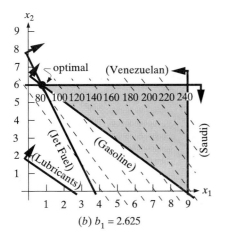

(b) $b_1 = 2.625$

either

Optimal Dual	Lower Range	Upper Range
20.000	1.125	2.625

or

Optimal Dual	Lower Range	Upper Range
66.667	2.625	5.100

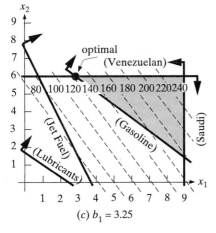

(c) $b_1 = 3.25$

Optimal Dual	Lower Range	Upper Range
66.667	2.625	5.100

FIGURE 7.6 Sensitivity Ranges for Varying Two Crude Gasoline Demand

(b) RHS value 60.4 forms the boundary between two ranges. Output might show dual variable -3.38 and range $[60.4, 83.3]$, or it might produce dual -4.98 with range $[0.0, 60.4]$.

Connection between Rate Changes and Degeneracy

A closer look at the three cases in Figure 7.6 shows what causes rates to change at $b_1 = 2.625$. Below that value, the optimal solution is defined by active gasoline and jet fuel requirement constraints

$$0.3x_1 + 0.4x_2 \geq b_1$$
$$0.4x_1 + 0.2x_2 \geq 1.5$$

The corresponding dual slope is $v_1^* = 20$. Above 2.625, the Venezuelan availability limit replaces jet fuel to give active set

$$0.3x_1 + 0.4x_2 \geq b_1$$
$$x_2 \leq 6$$

Rate $v_1^* = 66.667$.

> **7.44** | Rates of variation in optimal value with model constants change when the collection of active primal or dual constraints changes.

At exactly $b_1 = 2.625$ all three constraints are active, but any two define primal solution $x_1^* = 0.75$, $x_2^* = 6$. This is the degenerate case of more constraints being active than are needed to settle the primal solution that we introduced in Section 5.6. Rates of change displayed in sensitivity outputs will reflect whichever pair the optimization search happens to focus on.

Many, perhaps most, large linear programs will have degenerate optimal solutions. It follows that we will often find that the given value of a model parameter also forms one or the other end of its range. Even if the base-case value lies strictly within a range, that interval of reliable quantitative sensitivity information is likely to be rather narrow.

> **7.45** | Degeneracy, which is extremely common in large-scale LP models, limits the usefulness of sensitivity by-products from primal optimization because it leads to narrow RHS and objective coefficient ranges and ambiguity at the range limits.

Reoptimization to Make Sensitivity Exact

The obvious remedy for weaknesses in sensitivity analysis based on optimization by-products is reoptimization—repeating the optimization search with changed model constants. For example, if we want to know how the optimal plan at Two Crude would change if gasoline demand increases from 2.0 to 3.5, we could just run again with the revised RHS.

> **7.46** │ If the number of "what if" variations does not grow too big, **reoptimization** using different values of model input parameters often provides the most practical avenue to good sensitivity analysis.

Practical OR analyses rely heavily on running a variety of cases, and many linear programming optimization codes provide for inputting several alternative RHSs or objective functions at the same time. Still, the approach has its limits. All combinations of even 10 parameter variations leads to $2^{10} = 1024$ cases to try; 11 variations imply twice as many.

Parametric Variation of One Coefficient

To do sensitivity analysis by trying different cases, we must, of course, know what cases to try. For example, we may know we are interested in how sensitive optimal results are to changes in an RHS or objective coefficient, yet be unclear about exactly what values to consider.

Parametric studies track the optimal value as a function of model inputs. Figures 7.3 and 7.4 displayed just such parametric functions in our qualitative discussion of Section 7.2. Figure 7.7 adds another—the Two Crude optimal value as a function of the demand for gasoline.

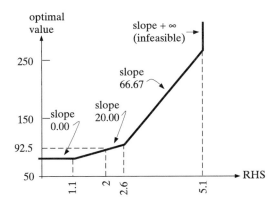

FIGURE 7.7   Parametric Variation of Optimal Value with Two Crude Gasoline Demand

Many LP optimization codes include automatic features to produce such parametric analysis, but we are now in a position to see how it can be done with the tools already at hand. Just four carefully chosen runs can construct the entire Figure 7.7 curve. In particular, we employ the following cases and sensitivity outputs:

Case	RHS	Dual	Lower Range	Upper Range
Base model	2.000	20.000	1.125	2.625
Variant 1	$2.625 + \epsilon$	66.667	2.626	5.100
Variant 2	$5.100 + \epsilon$	$+\infty$	5.100	$+\infty$
Variant 3	$1.125 - \epsilon$	0.000	$-\infty$	1.125

where ϵ is a small positive number.

Parametric analysis begins with the base model case of $b_1 = 2.000$ and its optimal value of 92.5. The first line of table values shows that the slope or rate of change in optimal value at that point is $v_1^* = 20.000$. Range information tells us that this slope holds from $b_1 = 1.125$ to 2.625. The result is the segment of the Figure 7.7 curve passing through $b_1 = 2.000$, optimal value = 92.5.

New rates of change arise only outside that range. Our first variation changes b_1 to $2.625 + \epsilon$, just beyond the upper limit. The result is new slope 66.667 and an indication that it holds through upper limit $b_1 = 5.100$. Repeating for $b_1 = 5.100 + \epsilon$ shows that the model is infeasible beyond 5.100. We complete the analysis by moving b_1 below the base-case range limit of 1.125. Reoptimization with $b_1 = 1.125 - \epsilon$ indicates that the slope $v_1^* = 0.000$ will hold for all further decreases.

> **7.47**   Parametric studies of optimal value as a function of a single-model RHS or objective function coefficient can be constructed by repeated optimization using new coefficient values just outside the previously applicable sensitivity range.

SAMPLE EXERCISE 7.30: PARAMETRICALLY ANALYZING ONE COEFFICIENT

Return to the parametric plots of Figures 7.3 and 7.4.

(a) Tabulate the separate optimizations that would have to be run to produce RHS parametric analysis Figure 7.3(a), beginning from the base the RHS value 75.0.

(b) Tabulate the separate optimizations that would have to be run to produce objective coefficient parametric analysis Figure 7.4(a), beginning from base cost 9.0.

Analysis:

(a) Proceeding from the figure according to principle 7.47 produces the following cases:

Case	RHS	Dual	Lower Range	Upper Range
Base model	75.0	−3.38	60.4	83.3
Variant 1	$83.3 + \epsilon$	0.00	83.3	$+\infty$
Variant 2	$60.4 + \epsilon$	−4.98	0.0	60.4
Variant 3	$0.0 - \epsilon$	$-\infty$	$-\infty$	0.0

(b) Proceeding from the figure according to principle 7.47 produces the following cases:

Case	Cost	Primal	Lower Range	Upper Range
Base model	9.0	156.1	8.4	10.1
Variant 1	$10.1 + \epsilon$	109.56	10.1	19.9
Variant 2	$19.9 + \epsilon$	82.03	19.9	29.2
Variant 3	$29.2 + \epsilon$	72.59	29.2	31.8
Variant 4	$31.8 + \epsilon$	72.21	31.8	35.5
Variant 5	$35.5 + \epsilon$	68.43	35.5	$+\infty$
Variant 6	$8.4 - \epsilon$	178.76	8.0	8.4
Variant 7	$8.0 - \epsilon$	400.98	$-\infty$	8.0

Assessing Effects of Multiple Parameter Changes

The greatest limitation of sensitivity studies we have explained so far is that all vary only one model input at a time. They increase or decrease a single RHS or objective coefficient, or add or drop a single variable or constraint. The implicit assumption is that all other data are held constant.

Computations behind sensitivity printouts like those of Figures 7.4 and 7.5 depend explicitly on this "single change" assumption.

7.48	Elementary LP sensitivity rates of change and ranges hold only for a single coefficient change, with all other data held constant.

Unfortunately, one-at-a-time analysis often does not suffice. Many "what if" questions that arise in operations research studies involve multiple changes, with several model constants varying simultaneously. For example, we may wish to know the impact on Two Crude refinery's optimal plan if demand for gasoline increases by some percent at the same time as demand for jet fuel increases at twice that percent because of an anticipated upturn in the economy; two RHSs would be changing at the same time.

As with single variations, we can deal with a modest number of such multiple-change questions simply by reoptimizing with new coefficients. For example, a fraction θ increase in both gasoline and jet fuel demand at Two Crude would be implemented by a new run with

$$b_1 = (1 + \theta)2.0$$
$$b_2 = (1 + 2\theta)1.5 \tag{7.14}$$

and all other data as before.

Parametric Multiple-RHS Change

To see how to track the parametric effect of such multiple changes over a variety of change magnitudes θ, we need to think of right-hand sides in the form

$$b_i^{new} = b_i^{base} + \theta \, \Delta b_i$$

where each Δb_i shows how much its RHS b_i is changing per unit variation, and θ defines the size of the step. In the example of (7.14), change components for the five model RHSs are

$$\Delta b_1 = 2.0, \quad \Delta b_2 = 2(1.5) = 3, \quad \Delta b_3 = \Delta b_4 = \Delta b_5 = 0$$

Thinking of θ as a decision variable, adding $(\theta \, \Delta b_i)$ on the right-hand side of each constraint translates as $-(\Delta b_i)\theta$ on the left. Therein lies the key insight.

7.49	The effect of a multiple change in right-hand sides with step θ can be analyzed parametrically by treating θ as a new decision variable with constraint coefficients $-\Delta b_i$ that detail the rates of change in RHSs b_i and a value fixed by a new equality constraint.

For example, the revised Two Crude model with varying rate of gasoline and jet fuel demand growth would be

$$
\begin{array}{lllll}
\min & +20x_1 & +15x_2 \\
\text{s.t.} & +0.3x_1 & +0.4x_2 & -2\theta & \ge 2 \\
& +0.4x_1 & +0.2x_2 & -3\theta & \ge 1.5 \\
& +0.2x_1 & +0.3x_2 & & \ge 0.5 \\
& +1x_1 & & & \le 9 \\
& & +1x_2 & & \le 6 \\
& & & +1\theta & = b_6
\end{array}
$$

$$x_1, x_2 \ge 0, \theta \text{ URS}$$

We can now proceed exactly as we did to construct Figure 7.7. New equality constraint

$$\theta = b_6$$

reduces the task of parametric analysis to one of seeing how optimal value changes with RHS b_6.

Using RHS range outputs for that constraint leads us to the following sequence of runs and the parametric curve of Figure 7.8.

Case	RHS	Dual	Lower Range	Upper Range
Base model	2.000	20.000	1.125	2.625
Variant 1	$2.625 + \epsilon$	66.667	2.626	5.100
Variant 2	$5.100 + \epsilon$	$+\infty$	5.100	$+\infty$
Variant 3	$1.125 - \epsilon$	0.000	$-\infty$	1.125

SAMPLE EXERCISE 7.31: ANALYZING MULTIPLE-RHS CHANGES

Consider the linear program

$$
\begin{array}{llll}
\max & +5x_1 & +2x_2 \\
\text{s.t.} & +1x_1 & +1x_2 & \le 3 \\
& +1x_1 & & \le 2 \\
& x_1, x_2 \ge 0
\end{array}
$$

Show how to modify the model to parametrically analyze the effect of simultaneously increasing the first RHS and decreasing the second by the same amount.

Analysis: The indicated change involves $\Delta b_1 = +1$, $\Delta b_2 = -1$. Thus, applying principle $\boxed{7.49}$, we may parametrically analyze the change by varying new RHS b_3 in the modified model:

$$
\begin{array}{lllll}
\max & +5x_1 & +2x_2 \\
\text{s.t.} & +1x_1 & +1x_2 & -1\theta & \le 3 \\
& +1x_1 & & +1\theta & \le 2 \\
& & & +1\theta & = b_3
\end{array}
$$

$$x_1, x_2 \ge 0, \theta \text{ URS}$$

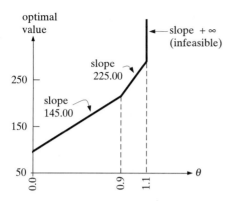

slope + ∞
(infeasible)

slope
225.00

slope
145.00

FIGURE 7.8   Parametric Multiple-RHS Change of the Two Crude Example

Parametric Change of Multiple Objective Function Coefficients

To see how to do the same sort of parametric multiple-change analysis when objective function coefficients are changing, we must think back to the relation between primal and dual. Parametric revision

$$c_j^{new} = c_j^{base} + \theta \, \Delta c_j$$

modifies the objective coefficients of the primal, which are the right-hand sides of the dual. Thus the analog of adding a new constraint $-\Delta \mathbf{b}$ column in the RHS case is the addition of a new constraint row $-\Delta \mathbf{c}$ to vary the c_j. To be more specific,

> **7.50**   The effect of a multiple change in objective function with step θ can be analyzed parametrically by treating objective rates of change $-\Delta c_j$ as coefficients in a new equality constraint having right-hand side zero and a new unrestricted variable with objective coefficient θ.

Again we illustrate using the Two Crude model (7.4). If we wish to explore parametrically the impact of a uniform crude price rise by a fraction θ,

$$c_1^{new} = 20 + \theta \, \Delta c_1 = 20 + \theta(20)$$
$$c_2^{new} = 15 + \theta \, \Delta c_2 = 15 + \theta(15)$$

The revised primal model to manipulate is

$$
\begin{array}{lrrrl}
\min & +20x_1 & +15x_2 & +\theta x_3 & \\
\text{s.t.} & +0.3x_1 & +0.4x_2 & & \geq 2 \\
 & +0.4x_1 & +0.2x_2 & & \geq 1.5 \\
 & +0.2x_1 & +0.3x_2 & & \geq 0.5 \\
 & +1x_1 & & & \leq 9 \\
 & & +1x_2 & & \leq 6 \\
 & -20x_1 & -15x_2 & +1x_3 & = 0 \\
 & \multicolumn{4}{l}{x_1, x_2 \geq 0, x_3 \text{ URS}}
\end{array}
$$

Negatives of rates Δc_j form coefficients for a new equality constraint with RHS $= 0.0$, and new variable x_3 appears in that constraint and the objective function.

We can now determine parametric effects of our multiple changes by analyzing the effect on optimal value of changes in the single objective coefficient θ. The process works because the main dual constraint corresponding to new variable x_3 is

$$v_6 = \theta$$

Thus θ affects all other main dual constraints by a term $-\Delta c_j v_6 = -\Delta c_j \theta$ on the left-hand side, which is equivalent to increasing at the same rate in the dual right-hand side $\mathbf{c}$. With primal and dual guaranteed to yield equal optimal values, manipulating the dual in this way does to the primal exactly what we want.

Using objective function range outputs for parameter θ leads us to the following sequence of runs and the parametric curve of Figure 7.9.

Case	Coefficient θ	Optimal x_3	Lower Range	Upper Range
Base model	2.000	20.000	1.125	2.625
Variant 1	$2.625 + \epsilon$	66.667	2.626	5.100
Variant 2	$5.100 + \epsilon$	$+\infty$ (infeasible)	5.100	$+\infty$
Variant 3	$1.125 - \epsilon$	0.000	$-\infty$	1.125

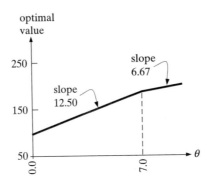

FIGURE 7.9   Parametric Multiple Objective Coefficient Change of the Two Crude Example

---

SAMPLE EXERCISE 7.32: ANALYZING MULTIPLE OBJECTIVE CHANGES

Show how to modify the model in Sample Exercise 7.31 to parametrically analyze the effect of simultaneously increasing the first objective coefficient and decreasing the second at twice the rate.

Analysis: The indicated change involves $\Delta c_1 = +1$, $\Delta c_2 = -2$. Thus, applying principle $\boxed{7.50}$, we may parametrically analyze the change by varying new objective coefficient θ in the modified model:

$$
\begin{array}{llllr}
\max & +5x_1 & +2x_2 & +\theta x_3 & \\
\text{s.t.} & +1x_1 & +1x_2 & & \leq 3 \\
& +1x_1 & & & \leq 2 \\
& -1x_1 & +2x_2 & +1x_3 & = 0 \\
& x_1, x_2 \geq 0, x_3 \text{ URS} & & &
\end{array}
$$

EXERCISES

7-1 As a result of a recent decision to stop production of toy guns that look too real, the Super Slayer Toy Company is planning to focus its production on two futuristic models: beta zappers and freeze phasers. Beta zappers produce $2.50 in profit for the company, and freeze phasers, $1.60. The company is contracted to sell 10 thousand beta zappers and 15 thousand freeze phasers in the next month, but all that are produced can be sold. Production of either model involves three crucial steps: extrusion, trimming, and assembly. Beta zappers use 5 hours of extrusion time per thousand units, 1 hour of trimming time, and 12 hours of assembly. Corresponding values per thousand units of freeze phasers are 9, 2, and 15. There are 320 hours of extrusion time, 300 hours of trimming time, and 480 hours of assembly time available over the next month. (An optimization output appears in Table 7.6.)

(a) Briefly explain how this problem can be modeled by the linear program:

$$\max \quad 2500x_1 + 1600x_2$$
$$\text{s.t.} \quad x_1 \geq 10$$
$$x_2 \geq 15$$
$$5x_1 + 9x_2 \leq 320$$
$$1x_1 + 2x_2 \leq 300$$
$$12x_1 + 15x_2 \leq 480$$
$$x_1, x_2 \geq 0$$

(b) ⊘ Identify the resource associated with the objective function and each main constraint in part (a).

(c) ⊘ Identify the activity associated with each decision variable in part (a).

(d) ⊘ Interpret the left-hand-side coefficients of each decision variable in part (a) as inputs and outputs of resources per unit activity.

7-2 Eli Orchid can manufacture its newest pharmacutical product in any of three processes. One costs $14,000 per batch, requires 3 tons of one major ingredient and 1 ton of the other, and yields 2 tons of output product. The second process costs $30,000 per batch, requires 2 and 7 tons of the ingredients, respectively, and yields 5 tons of product. The third process costs $11,000 per batch, requires 9 and 2 tons of the ingredients, respectively, and yields 1 ton of product. Orchid wants to find the least costly way to produce at least 50 tons of the new product given that there are 75 tons of ingredient 1 and 60 tons of ingredient 2 on hand. (An optimization output appears in Table 7.7.)

(a) Briefly explain how this problem can be modeled by the LP

$$\min \quad 14x_1 + 30x_2 + 11x_3$$
$$\text{s.t.} \quad 2x_1 + 5x_2 + 1x_3 \geq 50$$
$$3x_1 + 2x_2 + 9x_3 \leq 75$$
$$1x_1 + 7x_2 + 2x_3 \leq 60$$
$$x_1, x_2, x_3 \geq 0$$

(b) through (d) as in Exercise 7-1.

TABLE 7.6 Optimization Output for Super Slayer Exercise 7-1

```
Solution value (max) = 77125.000
VARIABLE SENSITIVITY ANALYSIS:
Name  Optimal  Bas  Lower   Upper   Object   Reduced   Lower     Upper
      Value    Sts  Bound   Bound   Coef     Object    Range     Range
----- -------- ---- ------  ------  -------  --------  ------    ------
x1    21.250   BAS  0.000   +infin  2500.00  0.000     1280.000  +infin
x2    15.000   BAS  0.000   +infin  1600.00  0.000     -infin    3125.000
CONSTRAINT SENSITIVITY ANALYSIS:
Name  Typ   Optimal    RHS      Slack    Lower     Upper
            Dual       Coef              Range     Range
----- ----  --------   ------   ------   ------    ------
c1    G       0.000    10.000   11.250   -infin    21.250
c2    G   -1525.000    15.000   -0.000    0.000    24.000
c3    L      -0.000   320.000   78.750   241.250   +infin
c4    L      -0.000   300.000   248.750  51.250    +infin
c5    L     208.333   480.000   0.000    345.000   669.000
```

TABLE 7.7   Optimization Output for Eli Orchid Exercise 7-2

```
Solution value (min) = 311.111
VARIABLE SENSITIVITY ANALYSIS:

Name   Optimal   Bas   Lower    Upper    Object   Reduced   Lower    Upper
       Value     Sts   Bound    Bound    Coef     Object    Range    Range
-----  -------   ---   -----    -----    ------   -------   ------   ------
 x1     5.556    BAS   0.000    +infin   14.000    0.000    12.000   +infin
 x2     7.778    BAS   0.000    +infin   30.000    0.000    -infin   35.000
 x3     0.000    NBL   0.000    +infin   11.000    5.667     5.333   +infin

CONSTRAINT SENSITIVITY ANALYSIS:

Name   Typ   Optimal   RHS       Slack     Lower     Upper
             Dual      Coef                Range     Range
-----  ---   -------   ------    ------    ------    ------
 c1     G     7.556    50.000    -0.000    42.857    70.263
 c2     L     0.000    75.000    42.778    32.222    +infin
 c3     L    -1.111    60.000     0.000    25.000    70.000
```

7-3 Professor Proof is trying to arrange for the implementation in a computer program of his latest operations research algorithm. He can contract with any mix of three sources for help: unlimited hours from undergraduates at $4 per hour, up to 500 hours of graduate students at $10 per hour, or unlimited hours of professional programmers at $25 per hour. The full project would take a professional at least 1000 hours, but grad students are only 0.3 as productive, and undergraduates, 0.2. Proof has only 164 hours of his own time to devote to the effort, and he knows from experience that undergraduate programmmers require more supervision than graduates, and graduates more than professionals. In particular, he estimates that he will have to invest 0.2 hour of his own time per hour of undergraduate programming, 0.15 hour of his time per hour of graduate programming, and 0.05 hour of his time per hour of professional programming. (An optimization output appears in Table 7.8.)

(a) Briefly explain how this problem can be modeled by the LP

$$\min \quad 4x_1 + 10x_2 + 25x_3$$
$$\text{s.t.} \quad 0.2x_1 + 0.3x_2 + x_3 \geq 1000$$
$$0.2x_1 + 0.15x + 0.15x_3 \leq 164$$
$$x_2 \leq 500$$
$$x_1, x_2, x_3 \geq 0$$

(b) ⊠ through (d) ⊠ as in Exercise 7-1.

7-4 The NCAA is making plans for distributing tickets to the upcoming regional basketball championships. The up to 10,000 available seats will be divided between the media, the competing universities, and the general public. Media people are admitted free, but the NCAA receives $45 per ticket from universities and $100 per ticket from the general public. At least 500 tickets must be reserved for the media, and at least half as many tickets should go

TABLE 7.8   Optimization Output for Professor Proof Exercise 7-3

```
Solution value (min) = 24917.647
VARIABLE SENSITIVITY ANALYSIS:

Name   Optimal   Bas   Lower    Upper    Object   Reduced   Lower    Upper
       Value     Sts   Bound    Bound    Coef     Object    Range    Range
-----  -------   ---   -----    -----    ------   -------   ------   ------
 x1     82.353   BAS   0.000    +infin    4.000    0.000    -4.444    5.000
 x2      0.000   NBL   0.000    +infin   10.000    2.235     7.765   +infin
 x3    983.529   BAS   0.000    +infin   25.000    0.000    20.000   31.333

CONSTRAINT SENSITIVITY ANALYSIS:

Name   Typ   Optimal    RHS         Slack     Lower      Upper
             Dual       Coef                  Range      Range
-----  ---   -------    ------      ------    ------     ------
 c1     G     25.882    1000.000    -0.000    164.000    1093.333
 c2     L     -5.882     164.000     0.000    150.000    1000.00
 c3     L      0.000     500.000   500.000      0.000    +infin
```

to the competing universities as to the general public. Within these restrictions, the NCAA wishes to find the allocation that raises the most money. An optimization output appears in Table 7.9.

(f) Minimize problem, $5w_1 - 2w_2 \leq 11$, increase -2.
(g) ◈ Minimize problem, $4w_1 - 3w_2 \geq 15$, decrease -3.
(h) Maximize problem, $3w_1 + 4w_2 \leq 17$, decrease 3.

TABLE 7.9 Optimization Output for NCAA Ticket Exercise 7-4

```
Solution value (max) = 775833.333
VARIABLE SENSITIVITY ANALYSIS:
Name   Optimal   Bas      Lower     Upper     Object     Reduced    Lower     Upper
       Value     Sts      Bound     Bound     Coef       Object     Range     Range
-----  --------  ----     ------    -------   --------   --------   --------  --------
x1      500.000  BAS      0.000     +infin       0.000     0.000     -infin    81.667
x2     3166.667  BAS      0.000     +infin      45.000     0.000   -200.000   100.000
x3     6333.333  BAS      0.000     +infin     100.000     0.000     45.000   +infin
CONSTRAINT SENSITIVITY ANALYSIS:
Name   Typ   Optimal      RHS         Slack      Lower       Upper
             Dual         Coef                   Range       Range
-----  ----  ---------    ----------  -------    ----------  ----------
c1      L      81.667    10000.000     0.000      500.000      +infin
c2      G     -36.667        0.000    -0.000    -4750.000    9500.000
c3      G     -81.667      500.000    -0.000        0.000   10000.000
```

(a) Briefly explain how this problem can be modeled by the LP

$$\begin{aligned} \max \quad & 45x_2 + 100x_3 \\ \text{s.t.} \quad & x_1 + x_2 + x_3 \leq 10,000 \\ & x_2 - \tfrac{1}{2}x_3 \geq 0 \\ & x_1 \geq 500 \\ & x_1, x_2, x_3 \geq 0 \end{aligned}$$

(b) through (d) as in Exercise 7-1.

7-5 For each of the following constraint coefficient changes, determine whether the change would tighten or relax the feasible set, whether any implied change in the optimal value would be an increase or a decrease, and whether the rate of any such optimal value effect would become more or less steep if it varied with the magnitude of coefficient change. Assume that the model is a linear program, that all variables are nonnegative, and that the constraint is not the only one.

(a) ◈ Maximize problem, $3w_1 + w_2 \geq 9$, increase 9.
(b) Minimize problem, $5w_1 - 2w_2 \leq 11$, decrease 11.
(c) ◈ Minimize problem, $4w_1 - 3w_2 \geq 15$, decrease 15.
(d) Maximize problem, $3w_1 + 4w_2 \leq 17$, increase 17.
(e) ◈ Maximize problem, $3w_1 + 1w_2 \geq 9$, increase 3.

7-6 Determine whether adding each of the following constraints to a mathematical program would tighten or relax the feasible set and whether any implied change in the optimal value would be an increase or a decrease. Assume that the constraint is not the only one.

(a) ◈ Maximize problem, $2w_1 + 4w_2 \geq 10$.
(b) Maximize problem, $14w_1 - w_2 \leq 20$.
(c) ◈ Minimize problem, $45w_1 + 34w_2 \leq 77$.
(d) Minimize problem, $32w_1 + 67w_2 \geq 49$.

7-7 For each of the following objective coefficient changes, determine whether any implied change in the optimal value would be an increase or a decrease and whether the rate of any such optimal value effect would become more or less steep if it varied with the magnitude of coefficient change. Assume that the model is a linear program and that all variables are nonnegative.

(a) ◈ max $13w_1 + 4w_2$, increase 13.
(b) max $5w_1 - 10w_2$, decrease -10.
(c) ◈ min $-5w_1 + 17w_2$, increase -5.
(d) min $29w_1 + 14w_2$, decrease 29.

7-8 Return to Super Slayer Exercise 7-1.

(a) ◈ Assign dual variables to each main constraint of the formulation in part (a), and define their meanings and units of measurement.

(b) ⊠ Show and justify the appropriate variable-type restrictions on all dual variables.

(c) ⊠ Formulate and interpret the main dual constraint corresponding to each primal variable.

(d) ⊠ Formulate and interpret an appropriate dual objective function.

(e) Use optimal solutions in Table 7.6 to verify that optimal primal and dual objective function values are equal.

(f) ⊠ Formulate and interpret all primal complementary slackness conditions for the model.

(g) ⊠ Formulate and interpret all dual complementary slackness conditions for the model.

(h) Verify that optimal primal and dual solutions in Table 7.6 satisfy the complementary slackness conditions of parts (f) and (g).

7-9 Do Exercise 7-8 for the problem of Exercise 7-2 using Table 7.7.

7-10 ⊠ Do Exercise 7-8 for the problem of Exercise 7-3 using Table 7.8.

7-11 Do Exercise 7-8 for the problem of Exercise 7-4 using Table 7.9.

7-12 State the dual of each of the following LPs.

(a) ⊠ min $17x_1 + 29x_2 + x_4$
  s.t. $2x_1 + 3x_2 + 2x_3 + 3x_4 \leq 40$
    $4x_1 + 4x_2 + x_4 \geq 10$
    $3x_3 - x_4 = 0$
    $x_1, \ldots, x_4 \geq 0$

(b) max $44x_1 - 3x_2 + 15x_3 + 56x_4$
  s.t. $x_1 + x_2 + x_3 + x_4 = 20$
    $x_1 - x_2 \leq 0$
    $9x_1 - 3x_2 + x_3 - x_4 \geq 25$
    $x_1, \ldots, x_4 \geq 0$

(c) ⊠ max $30x_1 - 2x_3 + 10x_4$
  s.t. $2x_1 - 3x_2 + 9x_4 \leq 10$
    $4x_2 - x_3 \geq 19$
    $x_1 + x_2 + x_3 = 5$
    $x_1 \geq 0, x_3 \leq 0$

(d) min $5x_1 + x_2 - 4x_3$
  s.t. $x_1 + x_2 + x_3 + x_4 = 19$
    $4x_2 + 8x_4 \leq 55$
    $x_1 + 6x_2 - x_3 \geq 7$
    $x_2, x_3 \geq 0, x_4 \leq 0$

(e) ⊠ max $2x_1 + 9x_2$
  s.t. $3w + 2x_1 - x_2 \geq 10$
    $w - y \leq 0$
    $x_1 + 3x_2 + y = 11$
    $x_1, x_2 \geq 0$

(f) max $19y_1 + 4y_2 - 8z_2$
  s.t. $11y_1 + y_2 + z_1 = 15$
    $z_1 + 5z_2 \leq 0$
    $y_1 - y_2 + z_2 \geq 4$
    $y_1, y_2 \geq 0$

(g) ⊠ min $32x_2 + 50x_3 - 19x_5$
  s.t. $\left(15 \sum_{j=1}^{3} x_j\right) + x_5 = 40$
    $12x_1 - 90x_2 + 14x_4 \geq 18$
    $x_4 \leq 11$
    $x_j \geq 0, \ j = 1, \ldots, 5$

(h) max $10(x_3 + x_4)$
  s.t. $\sum_{j=1}^{4} x_j = 80$
    $x_j - 2x_{j+1} \geq 0 \ j = 1, \ldots, 3$
    $x_1, x_2 \geq 0$

7-13 State (primal and dual) complementary slackness conditions for each LP in Exercise 7-12.

7-14 Each of the following LP has a finite optimal solution. State the corresponding dual, solve both primal and dual graphically, and verify that optimal objective function values are equal.

(a) ⊠ max $14x_1 + 7x_2$
  s.t. $2x_1 + 5x_2 \leq 14$
    $5x_1 + 2x_2 \leq 14$
    $x_1, x_2 \geq 0$

(b) min $4x_1 + 10x_2$
  s.t. $2x_1 + x_2 \geq 6$
    $x_1 \leq 10$
    $x_1, x_2 \geq 0$

(c) ⊠ min $8x_1 + 11x_2$
  s.t. $2x_1 + 9x_2 \geq 24$
    $3x_1 + x_2 \geq 11$
    $x_1, x_2 \geq 0$

(d)   max $7x_1 + 9x_2$

    s.t. $4x_1 + 2x_2 \leq 7$

        $3x_1 + 7x_2 \leq 19$

        $x_1, x_2 \geq 0$

(f)   max $-20x_1 + 15x_2$

    s.t. $3x_1 - 2x_2 \leq 6$

        $-3x_1 + 2x_2 \leq -12$

        $x_1, x_2 \geq 0$

7-15 Compute the dual solution corresponding to each of the following basic sets in the standard-form LP

    max $6x_1 + 1x_2 + 21x_3 - 54x_4 - 8x_5$

    s.t. $2x_1 + 5x_3 + 7x_5 = 70$

        $+3x_2 + 3x_3 - 9x_4 + 1x_5 = 1$

        $x_1, \ldots, x_5 \geq 0$

(a)   ⧄ $\{x_1, x_2\}$

(b) $\{x_1, x_3\}$

(c)   ⧄ $\{x_2, x_3\}$

(d) $\{x_4, x_5\}$

7-16 Each of the following is a linear program with no optimal solution. State the corresponding dual, solve both primal and dual graphically, and verify that whenever primal or dual is unbounded, the other is infeasible.

(a)   ⧄   max $4x_1 + x_2$

    s.t. $2x_1 + x_2 \geq 4$

        $3x_2 \leq 12$

        $x_1, x_2 \geq 0$

(b)   max $4x_1 + 8x_2$

    s.t. $4x_1 \geq 8$

        $x_1 + x_2 \leq 1$

        $x_1, x_2 \geq 0$

(c)   ⧄   min $10x_1 + 3x_2$

    s.t. $x_1 + x_2 \geq 2$

        $-x_2 \geq 5$

        $x_1, x_2 \geq 0$

(d)   min $x_1 - 2x_2$

    s.t. $-x_1 + x_2 \leq 6$

        $x_1 - 5x_2 \leq 5$

        $x_1, x_2 \geq 0$

(e)   ⧄   min $-3x_1 + 4x_2$

    s.t. $-x_1 + 2x_2 \geq 2$

        $x_1 - 2x_2 \geq 5$

        $x_1, x_1 \geq 0$

7-17 For each of the following LPs and solution vectors, demonstrate that the given solution is feasible, and compute the bound it provides on the optimal objective function value of the corresponding dual.

(a)   ⧄   min $30x_1 + 2x_2$ and $\mathbf{x} = (2, 5)$

    s.t. $4x_1 + x_2 \leq 15$

        $5x_1 - x_2 \geq 2$

        $15x_1 - 4x_2 = 10$

        $x_1, x_2 \geq 0$

(b)   max $10x_1 - 6x_2$ and $\mathbf{x} = (0, 5)$

    s.t. $12x_1 + 4x_2 \geq 8$

        $3x_1 - x_2 \leq 5$

        $2x_1 + 8x_2 = 40$

        $x_1, x_2 \geq 0$

7-18 Demonstrate for each linear program in Exercise 7-17 that the dual of its dual is the primal.

7-19 Return to Exercise 7-1. Answer each of the following as well as possible from the results in Table 7.6.

(a)   ⧄   Is the optimal solution sensitive to the exact value of the trimming hours available? At what number of hours capacity would it become relevant?

(b)   ⧄   How much should Super Slayer be willing to pay for an additional hour of extrusion time? For an additional hour of assembly time?

(c)   ⧄   What would be the profit effect of increasing assembly capacity to 580 hours? To 680 hours?

(d)   ⧄   What would be the profit effect of increasing the profit margin on beta zappers by $1500 per thousand? What would be the effect of a decrease in that amount?

(e)   ⧄   Suppose that the model of Exercise 7-1 ignores packaging capacity because it is hard to estimate, even though each thousand beta zappers requires 2 hours of packaging, and each thousand freeze phasers requires 3 hours. At what capacity would packaging affect the current optimal solution?

(f) ⊗ Suppose that Super Slayer also has a ninja nailer model it could manufacture that requires 2 hours of extrusion, 4 hours of trimming, and 3 hours of assembly per thousand units? At what profit per thousand would it be economic to produce?

7-20 Return to Eli Orchid Exercise 7-2. Answer each of the following as well as possible from the results in Table 7.7.

(a) What is the marginal cost of production (per ton of output)?

(b) How much would it cost to produce 70 tons of the new pharmaceutical product? To produce 100 tons?

(c) How much should Orchid be willing to pay to obtain 20 more tons of ingredient 1? How about ingredient 2?

(d) How cheap would the third process have to become before it might be used in an optimal solution?

(e) How much would the cost of the 50 tons of product increase if process 2 actually cost $32,000 per batch? If it cost $39,000?

(f) How much would the cost of the 50 tons of product decrease if process 1 cost $13,000 per batch? If it cost $10,000 per batch?

(g) Suppose that the engineering department is thinking about a new process that produces 6 tons of product using 3 tons of each of the two original ingredients. At what cost would this new process be economic to use?

(h) Suppose that the three processes actually use 0.1, 0.3, and 0.2 ton per batch of a third ingredient but we do not know exactly how much of it is available. Determine the minimum amount needed if the optimal primal solution in Table 7.7 is not to change.

7-21 Return to Exercise 7-3. Answer each of the following as well as possible from the results in Table 7.8.

(a) ⊗ What is the marginal cost per professional-equivalent hour of programming associated with the optimal solution in Table 7.8?

(b) ⊗ How much would cost increase if 1050 professional-equivalent hours of programming are required? How about 1100?

(c) ⊗ Does Professor Proof's availability limit the optimal solution? How much would cost change if Professor Proof could devote only 150 hours to supervision? How about 100 hours?

(d) ⊗ How much would the hourly rate of graduate student programmers have to be reduced before Professor Proof might optimally hire some?

(e) ⊗ How much would project cost increase if professional programmers cost $30 per hour? If they cost $45?

(f) ⊗ Suppose that Professor Proof decides to require at least half of the total programmer hours to go to students. Could this requirement change the optimal solution?

(g) ⊗ Suppose that Professor Proof decides to allow unlimited hours of graduate student programming. Could this revision change the optimal solution?

(h) ⊗ One of Professor Proof's colleagues has expressed interest in doing some of the programming to earn outside income. At what price per hour should Proof be interested if he estimates that the colleague would be 80% as efficient as a professional and require 0.10 hour of supervision per hour of work?

7-22 Return to NCAA ticket Exercise 7-4. Answer each of the following as well as possible from the results in Table 7.9.

(a) What is the marginal cost to the NCAA of each seat guaranteed the media?

(b) Suppose that there is an alternative arrangement of the dome where the games will be played that can provide 15,000 seats. How much additional revenue would be gained from the expanded seating? How much would it be for 20,000 seats?

(c) Since television revenue provides most of the income for NCAA events, another proposal would reduce the price of general public tickets to $50. How much revenue would be lost from this change? What if the price were $30?

(d) Media-hating coach Sobby Day wants the NCAA to restrict media seats to 20% of those allocated for universities. Could this policy change the optimal solution? How about 10%?

(e) To accommodate high demand from student supporters of participating universities, the NCAA is considering marketing a new "scrunch seat" that consumes only 80% of a regular bleacher seat but counts fully against the "university ≥ half public" rule. Could an optimal solution allocate any such seats at a ticket price of $35? At a price of $25?

7-23 Paper can be made from new wood pulp, from recycled office paper, or from recycled newsprint. New pulp costs $100 per ton, recycled office paper, $50 per ton, and recycled newsprint, $20 per ton. One available process uses 3 tons of pulp to make 1 ton of paper; a second uses 1 ton of pulp and 4 tons of recycled office paper; a third uses 1 ton of pulp and 12 tons of recycled newsprint; a fourth uses 8 tons of recycled office paper. At the moment only 80 tons of pulp is available. We wish to produce 100 tons of new paper at minimum total cost.

(a) Explain why this problem can be modeled as the LP

$$\begin{aligned}
\min \quad & 100x_1 + 50x_2 + 20x_3 \\
\text{s.t.} \quad & x_1 = 3y_1 + y_2 + y_3 \\
& x_2 = 4y_2 + 8y_4 \\
& x_3 = 12y_3 \\
& x_1 \leq 80 \\
& \sum_{j=1}^{4} y_j \geq 100 \\
& x_1, \ldots, x_3, y_1 \ldots, y_4 \geq 0
\end{aligned}$$

(b) State the dual of the given primal LP.

(c) ▣ Enter and solve the given LP with the class optimization software.

(d) ▣ Use your computer output to determine a corresponding optimal dual solution.

(e) ▣ Verify that your computer dual solution is feasible in the stated dual and that it has the same optimal solution value as the primal.

(f) ▣ Use your computer output to determine the marginal cost of paper production at optimality.

(g) ▣ Use your computer output to determine how much we should be willing to pay to obtain an additional ton of pulp.

(h) ▣ Use your computer output to determine or bound as well as possible how much optimal cost would change if the price of pulp increased to $150 per ton.

(i) ▣ Use your computer output to determine or bound as well as possible how much optimal cost would change if the price of recycled office paper decreased to $20 per ton.

(j) ▣ Use your computer output to determine or bound as well as possible how much optimal cost would change if the price of recycled office paper increased to $75 per ton.

(k) ▣ Use your computer output to determine or bound as well as possible how much optimal cost

would change if the number of tons of new paper needed decreased to 60.

(l) ▣ Use your computer output to determine or bound as well as possible how much optimal cost would change if the number of tons of new paper needed increased to 200.

(m) ▣ Use your computer output to determine how cheap recycled newsprint would have to become before the primal solution could change.

(n) ▣ An experimental new process will use 6 tons of newsprint and an undetermined number α tons of office paper. Use your computer output to determine how low α would have to be for the new process to be competitive with existing ones.

(o) ▣ Use your computer output to determine whether a limit of 400 tons on recycled office paper would change the primal optimal solution.

(p) ▣ Use your computer output to determine whether a limit of 400 tons on recycled newsprint would change the primal optimal solution.

7-24 Silva and Sons Ltd. (SSL)[1] is the largest coconut processor in Sri Lanka. SSL buys coconuts at 300 rupees per thousand to produce two grades (fancy and granule) of desiccated (dehydrated) coconut for candy manufacture, coconut shell flour used as a plastics filler, and charcoal. Nuts are first sorted into those good enough for desiccated coconut (90%) versus those good only for their shells. Those dedicated to desiccated coconut production go to hatcheting/pairing to remove the meat and then through a drying process. Their shells pass on for use in flour and charcoal. The 10% of nuts not suitable for desiccated coconut go directly to flour and charcoal. SSL has the capability to hatchet 300,000 nuts per month and dry 450 tons of desiccated coconut per month. Every 1000 nuts suitable for processing in this way yields 0.16 ton of desiccated coconut, 18% of which is fancy grade and the rest granulated. Shell flour is ground from coconut shells; 1000 shells yield 0.22 ton of flour. Charcoal also comes from shells; 1000 shells yield 0.50 ton of charcoal. SSL can sell fancy desiccated coconut at 3500 rupees per ton over hatcheting and drying cost, but the market is limited to 40 tons per month. A contract requires SSL delivery of at least 30 tons of granulated-quality desiccated coconut per month at 1350 rupees per ton over hatcheting and drying, but any larger amounts can be sold

[1] R. A. Cabraal (1981), "Production Planning in a Sri Lanka Coconut Mill Using Parametric Linear Programming," *Interfaces*, 11:3, 16–23.

at that price. The market for shell flour is limited to 50 tons per month at 450 rupees each. Unlimited amounts of charcoal can be sold at 250 rupees per ton.

(a) Explain how this coconut production planning problem can be modeled as the LP

$$\max \quad 3500s_1 + 1350s_2 + 450s_3$$
$$+ 250s_4 - 300p_1 - 300p_2$$

$$\text{s.t.} \quad 0.10p_1 - 0.90p_2 \quad\quad\quad = 0$$
$$0.82s_1 - 0.18s_2 \quad\quad\quad = 0$$
$$p_1 \quad\quad\quad \le 300$$
$$s_1 + s_2 \quad\quad\quad \le 450$$
$$s_1 \quad\quad\quad \le 40$$
$$s_2 \quad\quad\quad \ge 30$$
$$s_3 \quad\quad\quad \le 50$$
$$0.16p_1 - s_1 - s_2 \quad\quad\quad = 0$$
$$0.11p_1 + 0.11p_2 - 0.50s_3 - 0.22s_4 = 0$$
$$p_1, p_2, s_1, s_2, s_3, s_4 \quad\quad\quad \ge 0$$

(b) State the dual of your primal linear program.
(c) 🖳 Enter and solve the given primal LP with the class optimization software.
(d) 🖳 Use your computer output to determine a corresponding optimal dual solution.
(e) 🖳 Verify that your computer dual solution is feasible in the stated dual and that it has the same optimal solution value as the primal.
(f) 🖳 On the basis of your computer output, determine how much SSL should be willing to pay to increase hatcheting capacity by 1 unit (1000 nuts per month).
(g) 🖳 On the basis of your computer output, determine how much SSL should be willing to pay to increase drying capacity by 1 unit (1 ton per week).
(h) 🖳 On the basis of your computer output, determine or bound as well as possible the profit impact of a decrease in hatcheting capacity (thousands of nuts per month) to 250. Do the same for a capacity of 200.
(i) 🖳 On the basis of your computer output, determine or bound as well as possible the profit impact of an increase in hatcheting capacity (thousands of nuts per month) to 1000. Do the same for a capacity of 2000.

(j) 🖳 The company now has excess drying capacity. On the basis of your computer output, determine how low it could go before the optimal plan was affected.
(k) 🖳 The optimum now makes no shell flour. On the basis of your computer output, determine at what selling price per ton it would begin to be economical to make and sell flour.
(l) 🖳 On the basis of your computer output, determine or bound as well as possible the profit impact of a decrease in the selling price of granulated desiccated coconut to 800 rupees per ton. Do the same for a decrease to 600 rupees.
(m) 🖳 On the basis of your computer output, determine or bound as well as possible the profit impact of an increase to 400 rupees per ton in the price of charcoal. Do the same for an increase to 600 rupees.
(n) 🖳 On the basis of your computer output, determine whether the primal optimal solution would change if we dropped the constraint on drying capacity.
(o) 🖳 On the basis of your computer output, determine whether the primal optimal solution would change if we added a new limitation that the total number of nuts available per month cannot exceed 400,000. Do the same for a total not to exceed 200,000.

7-25 Tube Steel Incorporated (TSI) is optimizing production at its 4 hot mills. TSI makes 8 types of tubular products which are either solid or hollow and come in 4 diameters. The following two tables show production costs (in dollars) per tube of each product at each mill and the extrusion times (in minutes) for each allowed combination. Missing values indicate product–mill combinations that are not feasible.

Product	Unit Cost			
	Mill 1	Mill 2	Mill 3	Mill 4
0.5 in. solid	0.10	0.10	—	0.15
1 in. solid	0.15	0.18	—	0.20
2 in. solid	0.25	0.15	—	0.30
4 in. solid	0.55	0.50	—	—
0.5 in. hollow	—	0.20	0.13	0.25
1 in. hollow	—	0.30	0.18	0.35
2 in. hollow	—	0.50	0.28	0.55
4 in. hollow	—	1.0	0.60	—

	Unit Time			
Product	**Mill 1**	**Mill 2**	**Mill 3**	**Mill 4**
0.5 in. solid	0.50	0.50	—	0.10
1 in. solid	0.60	0.60	—	0.30
2 in. solid	0.80	0.60	—	0.60
4 in. solid	0.10	1.0	—	—
0.5 in. hollow	—	1.0	0.50	0.50
1 in. hollow	—	1.2	0.60	0.60
2 in. hollow	—	1.6	0.80	0.80
4 in. hollow	—	2.0	1.0	—

Yearly minimum requirements for the solid sizes (in thousands) are 250, 150, 150, and 80, respectively. For the hollow sizes they are 190, 190, 160, and 150. The mills can operate up to three 40-hour shifts per week, 50 weeks a year. Present policy is that each mill must operate at least one shift.

(a) Formulate a linear program to meet demand and shift requirements at minimum total cost using the decision variables

$x_{p,m}$ thousands of units of product p produced annually at mill m

Main constraints should have a system of 4 minimum time constraints, followed by a system of 4 maximum time constraints, followed by a system of 8 demand constraints.

(b) State the dual of your primal LP model.

(c) ⌨ Enter and solve your primal linear program with the class optimization software.

(d) ⌨ Use your computer output to determine a corresponding optimal dual solution.

(e) ⌨ Verify that your computer dual solution is feasible in the stated dual and that it has the same optimal solution value as the primal.

(f) ⌨ Use your computer results to determine the marginal cost of producing each of the eight products.

(g) ⌨ Use your computer results to explain why the policy of operating all mills at least one shift is costing the company money.

(h) ⌨ Two options being considered would open mills 3 or 4 on weekends (i.e., add up to 16 extra hours to each of 3 shifts over 50 weeks). Taking each option separately, determine or bound as well as possible from your computer results the impact these changes would have on total production cost.

(i) ⌨ Another option being considered is to hire young industrial engineers to find ways of reducing the unit costs of production at high-cost mill 4. For each of the 6 products there taken separately, use your computer results to determine to what level unit costs would have to be reduced before there could be any change in the optimal production plan.

(j) ⌨ A final pair of options being considered is to install equipment to produce 4-inch solid and 4-inch hollow tubes at mill 4. The new equipment would produce either product in 1 minute per unit. Taking each product separately, determine the unit production cost that would have to be achieved to make it economical to use the new facilities.

SUGGESTED READING

Bazaraa, Mokhtar S., John J. Jarvis, and Hanif Sherali (1990), *Linear Programming and Network Flows*, Wiley, New York.

Chvátal, Vašek (1983), *Linear Programming*, W. H. Freeman, San Francisco.

Luenberger, David G. (1984), *Linear and Nonlinear Programming*, Addison-Wesley, Reading, Mass.

CHAPTER 8

Multiobjective Optimization and Goal Programming

• •

Most of the methods of this book address optimization models with a **single objective** function—a lone criterion to be maximized or minimized. Although practical problems almost always involve more than one measure of solution merit, many can be modeled quite satisfactorily with a single cost or profit objective. Other criteria are either represented as constraints or weighted in a composite objective function to produce a model tractable enough for productive analysis.

Other applications—especially those in the public sector—must simply be treated as **multiobjective**. When goals cannot be reduced to a common scale of cost or benefit, trade-offs have be addressed. Only a model with multiple objective functions is satisfactory, even though analysis will almost certainly become more challenging.

This chapter introduces the key notions and approaches available when such multiobjective analysis is required. Emphasis is on **efficient solutions**, which are optimal in a certain multiobjective sense, and **goal programming**, which is the most commonly employed technique for dealing with multiobjective settings.

8.1 MULTIOBJECTIVE OPTIMIZATION MODELS

As usual, we begin our investigation of multiobjective optimization with some examples. In this section we formulate three cases illustrating the broad range of applications requiring multiobjective analysis. All are based on real contexts documented in published reports.

EXAMPLE 8.1: BANK THREE INVESTMENT

Every investor must tradeoff return versus risk in deciding how to allocate his or her available funds. The opportunities that promise the greatest profits are almost always the ones that present the most serious risks.

Commercial banks must be especially careful in balancing return and risk because legal and ethical obligations demand that they avoid undue hazards, yet their goal as a business enterprise is to maximize profit. This dilemma leads naturally to multiobjective optimization of investment that includes both profit and risk criteria.

Our investment example[1] adopts this multiobjective approach to a fictitious Bank Three. Bank Three has a modest $20 million capital, with $150 million in demand deposits (checking accounts) and $80 million in time deposits (savings accounts and certificates of deposit).

Table 8.1 displays the categories among which the bank must divide its capital and deposited funds. Rates of return are also provided for each category together with other information related to risk.

TABLE 8.1   Bank Three Investment Opportunities

Investment Category, j	Return Rate (%)	Liquid Part (%)	Required Capital (%)	Risk Asset?
1: Cash	0.0	100.0	0.0	No
2: Short term	4.0	99.5	0.5	No
3: Government: 1 to 5 years	4.5	96.0	4.0	No
4: Government: 5 to 10 years	5.5	90.0	5.0	No
5: Government: over 10 years	7.0	85.0	7.5	No
6: Installment loans	10.5	0.0	10.0	Yes
7: Mortgage loans	8.5	0.0	10.0	Yes
8: Commercial loans	9.2	0.0	10.0	Yes

We model Bank Three's investment decisions with a decision variable for each category of investment in Table 8.1:

$$x_j \triangleq \text{amount invested in category } j \text{ ($ million)} \qquad j = 1, \ldots, 8$$

Bank Three Example Objectives

The first goal of any private business is to maximize profit. Using rates of return from Table 8.1, this produces objective function

$$\begin{aligned} \max \quad & 0.040x_2 + 0.045x_3 + 0.055x_4 + 0.070x_5 \qquad \text{(profit)} \\ & + 0.105x_6 + 0.085x_7 + 0.092x_8 \end{aligned}$$

It is less clear how to quantify investment risk. We employ two common ratio measures.

One is the *capital-adequacy ratio*, expressed as the ratio of required capital for bank solvency to actual capital. A low value indicates minimum risk. The "required capital" rates of Table 8.1 approximate U.S. government formulas used to compute this ratio, and Bank Three's present capital is $20 million. Thus we will express a second objective as

$$\begin{aligned} \min \quad & \tfrac{1}{20}(0.005x_2 + 0.040x_3 + 0.050x_4 + 0.075x_5 \qquad \text{(capital-adequacy)} \\ & + 0.100x_6 + 0.100x_7 + 0.100x_8) \end{aligned}$$

Another measure of risk focuses on illiquid *risk assets*. A low risk asset/capital ratio indicates a financially secure institution. For our example, this third measure of success is expressed as

$$\min \quad \tfrac{1}{20}(x_6 + x_7 + x_8) \qquad \text{(risk-asset)}$$

---

[1] Based on J. L. Eatman and C. W. Sealey, Jr. (1979), "A Multiobjective Linear Programming Model for Commercial Bank Balance Sheet Management," *Journal of Bank Research, 9*, 227–236.

Bank Three Example Model

To complete a model of Bank Three's investment plans, we must describe the relevant constraints. Our example will assume five types:

1. Investments must sum to the available capital and deposit funds.
2. Cash reserves must be at least 14% of demand deposits plus 4% of time deposits.
3. The portion of investments considered liquid (see Table 8.1) should be at least 47% of demand deposits plus 36% of time deposits.
4. At least 5% of funds should be invested in each of the eight categories, for diversity.
5. At least 30% of funds should be invested in commercial loans, to maintain the bank's community status.

Combining the 3 objective functions above with these 5 systems of constraints completes a multiobjective linear programming model of Bank Three's investment problem:

$$
\begin{aligned}
\max \quad & 0.040x_2 + 0.045x_3 + 0.055x_4 + 0.070x_5 && \text{(profit)} \\
& + 0.105x_6 + 0.085x_7 + 0.092x_8 \\
\min \quad & \tfrac{1}{20}(0.005x_2 + 0.040x_3 + 0.050x_4 + 0.075x_5 && \text{(capital-adequacy)} \\
& + 0.100x_6 + 0.100x_7 + 0.100x_8) \\
\min \quad & \tfrac{1}{20}(x_6 + x_7 + x_8) && \text{(risk-asset)} \\
\text{s.t.} \quad & x_1 + \cdots + x_8 = (20 + 150 + 80) && \text{(invest all)} \\
& x_1 \geq 0.14(150) + 0.04(80) && \text{(cash reserve)} \\
& 1.00x_1 + 0.995x_2 + 0.960x_3 + 0.900x_4 && \text{(liquidity)} \\
& + 0.850x_5 \geq 0.47(150) + 0.36(80) \\
& x_j \geq 0.05(20 + 150 + 80) \quad \text{for all } j = 1,\ldots,8 && \text{(diversification)} \\
& x_8 \geq 0.30(20 + 150 + 80) && \text{(commercial)} \\
& x_1,\ldots,x_8 \geq 0
\end{aligned}
\tag{8.1}
$$

EXAMPLE 8.2: DYNAMOMETER RING DESIGN

Multiobjective optimization also occurs in the engineering design of almost any product or service. Competing performance measures must be balanced to choose a best design.

We will illustrate with the design of a simple mechanical part, an octagonal ring used in dynamometer instrumentation of machine tools.[2] Figure 8.1 depicts the critical design variables:

$$w \triangleq \text{width of the ring (in centimeters)}$$

$$t \triangleq \text{thickness of the outer surface (in centimeters)}$$

$$r \triangleq \text{radius of the two openings (in centimeters)}$$

It also shows upper and lower limits allowed for the three dimensions.

---

[2]Based on N. Singh and S. K. Agarwal (1983), "Optimum Design of an Extended Octagonal Ring by Goal Programming," *International Journal of Production Research*, 21, 891–898.

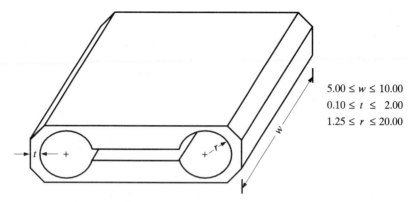

$5.00 \leq w \leq 10.00$
$0.10 \leq t \leq 2.00$
$1.25 \leq r \leq 20.00$

FIGURE 8.1   Dynamometer Ring Design Example

Dynamometer Ring Design Model

To model this machine part optimization, we must characterize performance in terms of decision variables w, t, and r. One consideration is sensitivity. We would like the instrument to be as sensitive as possible. Analysis of the relevant strains and deflections shows that sensitivity can be represented as

$$\max \quad \frac{0.7r}{Ewt^2} \quad \text{(sensitivity)}$$

where $E = 2.1 \times 10^6$, Young's modulus of elasticity for the ring material.

Another measure of effectiveness is rigidity. Increased rigidity produces greater accuracy. Again considering critical strains and deflections, rigidity can be modeled as

$$\max \quad \frac{Ewt^3}{r^3} \quad \text{(rigidity)}$$

Combining these two objectives with upper and lower bounds on the variables yields a multiobjective optimization model of our ring design problem:

$$\max \quad \frac{0.7r}{Ewt^2} \quad \text{(sensitivity)}$$
$$\max \quad \frac{Ewt^3}{r^3} \quad \text{(rigidity)}$$
$$5.0 \leq w \leq 10.0$$
$$0.1 \leq t \leq 2.0$$
$$1.25 \leq r \leq 20.0$$

(8.2)

Notice that this model is a multiobjective nonlinear program. It can, however, be handled by linear programming if we change variables to $\ln(w)$, $\ln(t)$, and $\ln(r)$ and maximize the logarithms of the two objective functions (See Section 14.9).

EXAMPLE 8.3: HAZARDOUS WASTE DISPOSAL

Many, perhaps most, government planning problems are multiobjective because they involve competing interests and objectives that are difficult to quantify in monetary terms. We illustrate with the planning of disposal sites for dangerous nuclear wastes.[3]

Figure 8.2 depicts the source points and potential disposal sites we assume in our fictitious version of the problem. Waste material is generated at the seven sources as nuclear power plants and other reactors use up fuel. The goal of the study is to choose 2 of the 3 possible disposal sites to receive the waste.

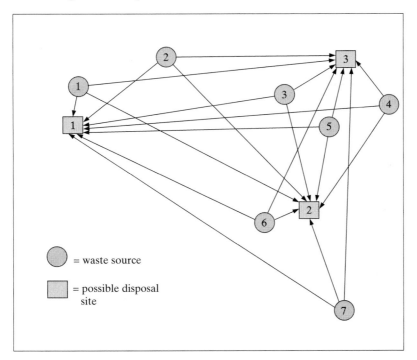

○ = waste source

□ = possible disposal site

FIGURE 8.2 Hazardous Waste Example Sources and Disposal Sites

When the system is ready, wastes will be trucked from sources to disposal sites over the public highway system. Conflicting objectives arise in routing materials to the sites. Costs will be reduced, and in-transit time of wastes minimized if we ship along shortest routes from source to disposal site. But population densities must also be considered. Routes through densely populated areas threaten more people and incur greater risk of traffic mishaps.

Table 8.2 shows parameters

$$s_i \triangleq \text{amount of waste expected to be produced at source } i$$

$$d_{i,j,k} \triangleq \text{distance from source } i \text{ to site } j \text{ along route } k$$

$$p_{i,j,k} \triangleq \text{population along route } k \text{ from source } i \text{ to site } j$$

for 2 possible routes from each source i to each possible disposal site. We wish to minimize some combination of distance and population to select the best sites and routes.

[3]Based on C. ReVelle, J. Cohon, and D. Shobrys (1991), "Simultaneous Siting and Routing in the Disposal of Hazardous Wastes," *Transportation Science, 25*, 138–145.

TABLE 8.2   Hazardous Waste Example

Source i		Site $j = 1$		Site $j = 2$		Site $j = 3$		Supply
		$k = 1$	$k = 2$	$k = 1$	$k = 2$	$k = 1$	$k = 2$	
1	Distance	200	280	850	1090	900	1100	1.2
	Population	50	15	300	80	400	190	
2	Distance	400	530	730	860	450	600	0.5
	Population	105	60	380	210	350	160	
3	Distance	600	735	550	600	210	240	0.3
	Population	300	130	520	220	270	140	
4	Distance	900	1060	450	570	180	360	0.7
	Population	620	410	700	430	800	280	
5	Distance	600	640	390	440	360	510	0.6
	Population	205	180	440	370	680	330	
6	Distance	900	1240	100	120	640	800	0.1
	Population	390	125	80	30	800	410	
7	Distance	1230	1410	400	460	1305	1500	0.2
	Population	465	310	180	105	1245	790	

Hazardous Waste Disposal Model

To model this shipping and site selection problem, we require the two sorts of decision variables typical of facility location models:

$$y_i \triangleq \begin{cases} 1 & \text{if site } i \text{ is opened} \\ 0 & \text{otherwise} \end{cases}$$

$x_{i,j,k} \triangleq$ amount shipped from source i to site j along route k

Then a multiobjective integer linear programming model is

$$\min \sum_{i=1}^{7} \sum_{j=1}^{3} \sum_{k=1}^{2} d_{i,j,k} x_{i,j,k} \qquad \text{(distance)}$$

$$\min \sum_{i=1}^{7} \sum_{j=1}^{3} \sum_{k=1}^{2} p_{i,j,k} x_{i,j,k} \qquad \text{(population)}$$

$$\text{s.t.} \quad \sum_{j=1}^{3} \sum_{k=1}^{2} x_{i,j,k} = s_i \qquad i = 1, \dots, 7 \qquad \text{(sources)} \qquad (8.3)$$

$$\sum_{j=1}^{3} y_j = 2 \qquad \text{(2 sites)}$$

$$x_{i,j,k} \leq s_i y_j \qquad i = 1, \dots, 7; \quad j = 1, \dots, 3; \quad k = 1, 2$$

$$x_{i,j,k} \geq 0 \qquad i = 1, \dots, 7; \quad j = 1, \dots, 3; \quad k = 1, 2$$

$$y_j = 0 \text{ or } 1 \qquad j = 1, \dots, 3$$

The first main constraints assure that all material arising at each source is shipped somewhere, and the second selects 2 sites. Switching constraints makes sure that a site is opened before any material can be shipped there.

8.2 EFFICIENT POINTS AND THE EFFICIENT FRONTIER

The familiar notion of an "optimal" solution becomes somewhat murky when an optimization model has more than one objective function. A solution that proves best by one criterion may rate among the poorest on another. In this section we develop the concept of **efficient points** and the **efficient frontier**, also known as **Pareto optima** and **nondominated points**, which help to characterize "best" feasible solutions in multiobjective models.

Efficient Points

A feasible solution to an optimization model cannot be best if there are others that equal or improve on it by every measure. Efficient points are ones that cannot be dominated in this way.

> **8.1** A feasible solution to a multiobjective optimization model is an **efficient point** if no other feasible solution scores at least as well in all objective functions and strictly better in one.

Efficient points are also termed Pareto optimal and nondominated.

We can illustrate with the dynamometer ring design model of Section 8.1:

$$\max \quad \frac{0.7r}{(2.1 \times 10^6)(wt^2)} \qquad \text{(sensitivity)}$$

$$\max \quad \frac{(2.1 \times 10^6)(wt^3)}{r^3} \qquad \text{(rigidity)}$$

$$5.0 \le w \le 10.0$$

$$0.1 \le t \le 2.0$$

$$1.25 \le r \le 20.0$$

(8.4)

Consider solution $(w^{(1)}, t^{(1)}, r^{(1)}) = (5, 0.1, 20)$. No feasible point has higher sensitivity, because the sole variable r in the numerator of the sensitivity objective is at its upper bound, and both w and t in the denominator are at lower bounds. But there are feasible points with higher rigidity; we need only decrease radius r. Even though it can be improved in one objective, the point is efficient. Any such change to improve rigidity results in a strict decrease in sensitivity.

Contrast with the solution $(w^{(2)}, t^{(2)}, r^{(2)}) = (7, 1, 3)$. Its values in the two objective functions are

$$\text{sensitivity} = \frac{0.7(3)}{(2.1 \times 10^6)(7)(1)^2} = 1.429 \times 10^{-7}$$

$$\text{rigidity} = \frac{(2.1 \times 10^6)(7)(1)^3}{(3)^3} = 5.444 \times 10^5$$

This feasible point cannot be efficient because it is dominated by $(w^{(3)}, t^{(3)}, r^{(3)}) = (7, 0.9, 2.7)$. The latter has the same rigidity and superior sensitivity 1.587×10^{-7}.

Identifying Efficient Points Graphically

When a multiobjective optimization model has only two decision variables, we can see graphically whether points are efficient. Consider, for example, the simple multiobjective linear program

$$\begin{aligned} \max \quad & +3x_1 + 1x_2 \\ \max \quad & -1x_1 + 2x_2 \\ \text{s.t.} \quad & x_1 + x_2 \le 4 \\ & 0 \le x_1 \le 3 \\ & 0 \le x_2 \le 3 \end{aligned}$$ (8.5)

plotted in Figure 8.3. Feasible solution $\mathbf{x} = (2, 2)$ of part (a) is efficient and the $\mathbf{x} = (3, 0)$ of part (b) is not.

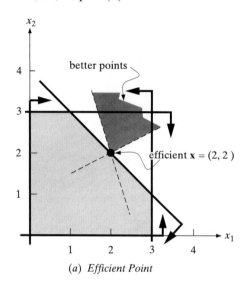

(a) Efficient Point

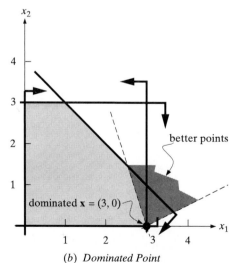

(b) Dominated Point

FIGURE 8.3   Graphical Characterization of Efficient Points

How can we be sure? In each case we have plotted the contours of the two objective functions that pass through the point. Other solutions with equal or better values of both objective functions must belong to the dark shaded areas bounded by those contours. In part (a) no such point is feasible. In part (b), others, such as $\mathbf{x}' = (3, 1)$, satisfy all constraints and thus dominate.

| 8.2 | Efficient points show graphically as ones for which no distinct feasible point lies in the region bounded by contours of the objective functions through the point, which contains every solution with equal or superior value of all objectives. |

SAMPLE EXERCISE 8.1: IDENTIFYING EFFICIENT POINTS

Determine whether each of the following points is efficient in model (8.5) and Figure 8.3.

(a) $\mathbf{x} = (1, 3)$

(b) $\mathbf{x} = (1, 1)$

Analysis: We apply principle $\boxed{8.2}$ in the following plots:

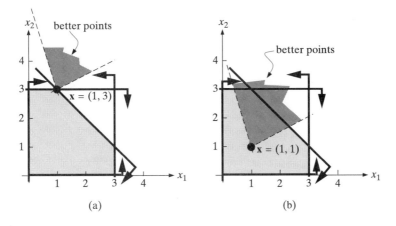

(a)   (b)

(a) In plot (a) the set of possibly dominant points, which is bounded by the contours, contains no other feasible point. Solution $\mathbf{x} = (1, 3)$ is efficient.

(b) In plot (b) the set of possibly dominant points, which is bounded by the contours, contains numerous feasible solutions. Solution $\mathbf{x} = (1, 1)$ cannot be efficient.

Efficient Frontier

When confronting a multiobjective optimization model, it seems natural to demand an efficient solution. But we have seen that very simple examples can have many efficient points. Some will score better by one criterion; others will evaluate superior by another.

To deal with such conflicts, we would often like to generate and consider a range of efficient solutions. The entire set is called the efficient frontier.

| 8.3 | The **efficient frontier** of a multiobjective optimization model is the collection of efficient points for the model. |

Plots in Objective Value Space

The term efficient frontier comes from an alternative way to plot solutions to multiobjective models. Instead of using axes corresponding to the decision variables, we plot in **objective value space** with axes for the different objective functions.

Figure 8.4 illustrates for our dynamometer ring design example (8.4). Every feasible solution for the model corresponds to a point in this plot, with horizontal dimension equal to its sensitivity, and vertical equal to its rigidity. For example, efficient point $(w^{(1)}, t^{(1)}, r^{(1)}) = (5, 0.1, 20)$ produces point $(1.333 \times 10^{-4}, 1.312)$ at the lower right. Dominated point $(w^{(2)}, t^{(2)}, r^{(2)}) = (7, 1, 3)$ graphs as $(1.429 \times 10^{-7}, 5.444 \times 10^{5})$.

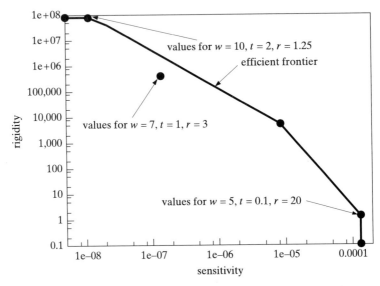

FIGURE 8.4   Efficient Frontier for Dynamometer Ring Example

The efficient frontier forms the boundary of the region defined by objective values for feasible points. Every efficient point lies along this boundary because no further progress is possible in one objective function without degrading another. On the other hand, dominated points plot in the interior; other feasible solutions along the boundary are equal in all objectives and superior in at least one.

Constructing the Efficient Frontier

When a multiobjective optimization model has only a few objectives, it is often practical to construct efficient frontier curves such as Figure 8.4. We have only to parametrically vary specified levels of all but one objective while optimizing the other.

> 8.4 | The set of points on the efficient frontier can be constructed by repeated optimization. New constraints enforce achievement levels for all but one objective, and the other is treated as a single objective.

The process parallels the parametric sensitivity analysis of Section 7.7.

Table 8.3 details the computations used to construct the efficient frontier of the dynamometer ring example in Figure 8.4. We begin by maximizing the two objectives separately. Maximizing sensitivity without regard to rigidity produces the first point $(1.333 \times 10^{-4}, 1.312)$ in the table. Then maximizing rigidity without considering sensitivity yields to the last point $(1.042 \times 10^{-8}, 8.6 \times 10^{7})$.

TABLE 8.3 Efficient Frontier of the Dynamometer Ring Example

Sensitivity Objective	Rigidity Objective	Efficient Point		
		w	t	r
1.333×10^{-4}	1.312	5.00	0.100	20.00
6.776×10^{-5}	10^{1}	5.00	0.100	10.16
3.145×10^{-5}	10^{2}	5.00	0.100	4.72
1.460×10^{-5}	10^{3}	5.00	0.100	2.19
5.510×10^{-6}	10^{4}	5.00	0.123	1.25
1.187×10^{-6}	10^{5}	5.00	0.265	1.25
2.557×10^{-7}	10^{6}	5.00	0.571	1.25
5.510×10^{-8}	10^{7}	5.00	1.230	1.25
1.042×10^{-8}	8.6×10^{7}	10.00	2.000	1.25

We now know the range of relevant rigidity values. The remaining points in Table 8.3 are obtained by maximizing sensitivity subject to a constraint on rigidity. For example, the values for rigidity 10^{3} result from solving

$$\max \quad \frac{0.7r}{(2.1 \times 10^{6})(wt^{2})} \qquad \text{(sensitivity)}$$

$$\text{s.t.} \quad \frac{(2.1 \times 10^{6})(wt^{3})}{r^{3}} \geq 10^{3} \qquad \text{(rigidity)}$$

$$5.0 \leq w \leq 10.0$$

$$0.1 \leq t \leq 2.0$$

$$1.25 \leq r \leq 20.0$$

SAMPLE EXERCISE 8.2: CONSTRUCTING THE EFFICIENT FRONTIER

Return to example model (8.5) of Figure 8.3 and construct an objective value space diagram of its efficient frontier like the one in Figure 8.4.

Analysis: We begin by separately maximizing the first and second objectives to obtain solutions with objective values $(10, -1)$ and $(3, 6)$, respectively. Now solving the LP

$$\max \quad +3x_1 + 1x_2$$
$$\text{s.t.} \quad -1x_1 + 2x_2 \geq \theta$$
$$x_1 + x_2 \leq 4$$
$$0 \leq x_1 \leq 3$$
$$0 \leq x_2 \leq 3$$

for several values of $\theta \in [-1, 6]$ produces the following plot:

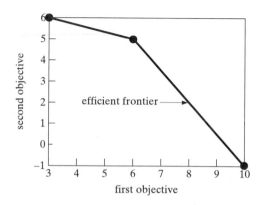

8.3 PREEMPTIVE OPTIMIZATION AND WEIGHTED SUMS OF OBJECTIVES

In a typical multiobjective model of realistic size, especially one with more than two objectives, the range of efficient solutions can be enormous. Explicit construction of an efficient frontier like Figure 8.4 is computationally impractical.

To obtain useful results, we must reduce the multiobjective model to a sequence of single objective optimizations. In this section we explore two of the most straightforward ways: **preemptive** (or **lexicographic**) **optimization** and **weighted sums**.

Preemptive Optimization

Although a model may have several objective criteria, they are rarely of equal importance. Preemptive optimization takes them in priority order.

> **8.5** **Preemptive** or **lexicographic optimization** performs multiobjective optimization by considering objectives one at a time. The most important is optimized; then the second most important is optimized subject to a requirement that the first achieve its optimal value; and so on.

Preemptive Optimization of the Bank Three Example

We can illustrate with the Bank Three example, model (8.1). That model has 3 objectives: profit, capital-adequacy ratio, and risk-asset ratio.

Suppose we decide that the third, risk-asset, objective is the single most important one. Preemptive optimization would then begin by minimizing risk-asset ratio in the single objective linear program

$$
\begin{array}{lll}
\min & \frac{1}{20}(x_6 + x_7 + x_8) & \text{(risk-asset)} \\
\text{s.t.} & x_1 + \cdots + x_8 = 20 + 150 + 80 & \text{(invest all)} \\
& x_1 \geq 0.14(150) + 0.04(80) & \text{(cash reserve)} \\
& 1.00x_1 + 0.995x_2 + 0.960x_3 + 0.900x_4 & \text{(liquidity)} \\
& \quad + 0.850x_5 \geq 0.47(150) + 0.36(80) &
\end{array}
$$

$$x_j \geq 0.05(20 + 150 + 80) \quad \text{for all } j = 1, \ldots, 8 \qquad \text{(diversification)}$$
$$x_8 \geq 0.30(20 + 150 + 80) \qquad \text{(commercial)}$$
$$x_1, \ldots, x_8 \geq 0$$

An optimal solution allocates funds (in millions of dollars)

$$x_1^* = 100.0, \quad x_2^* = 12.5, \quad x_3^* = 12.5, \quad x_4^* = 12.5$$
$$x_5^* = 12.5, \quad x_6^* = 12.5, \quad x_7^* = 12.5, \quad x_8^* = 75.0 \tag{8.6}$$

with optimal risk-asset ratio 5.0, profit \$11.9 million, and capital-adequacy ratio 0.606.

Next we introduce a constraint keeping risk-asset ratio at the optimal 5.0 and maximize the second priority, profit objective:

$$\text{max} \quad 0.040x_2 + 0.045x_3 + 0.055x_4 + 0.070x_5 \qquad \text{(profit)}$$
$$+ 0.105x_6 + 0.085x_7 + 0.092x_8$$

$$\text{s.t.} \quad \tfrac{1}{20}(x_6 + x_7 + x_8) \leq 5.0 \qquad \text{(risk-asset)}$$
$$x_1 + \cdots + x_8 = 20 + 150 + 80 \qquad \text{(invest all)}$$
$$x_1 \geq 0.14(150) + 0.04(80) \qquad \text{(cash reserve)}$$
$$1.00x_1 + 0.995x_2 + 0.960x_3 + 0.900x_4 \qquad \text{(liquidity)}$$
$$+ 0.850x_5 \geq 0.47(150) + 0.36(80)$$
$$x_j \geq 0.05(20 + 150 + 80) \quad \text{for all } j = 1, \ldots, 8 \qquad \text{(diversification)}$$
$$x_8 \geq 0.30(20 + 150 + 80) \qquad \text{(commercial)}$$
$$x_1, \ldots, x_8 \geq 0$$

The result is (millions of dollars)

$$x_1^* = 24.2, \quad x_2^* = 12.5, \quad x_3^* = 12.5, \quad x_4^* = 12.5$$
$$x_5^* = 88.3, \quad x_6^* = 12.5, \quad x_7^* = 12.5, \quad x_8^* = 75.0 \tag{8.7}$$

with optimal risk-asset ratio still 5.0, but profit now \$17.2 million, and capital-adequacy ratio 0.890.

Notice that the character of the optimum has changed significantly from solution (8.6). Considerable funds have been shifted from $x_1 =$ cash, to $x_5 =$ long-term government bonds in order to increase profit.

Finally, we come to the capital-adequacy objective function. Imposing a new constraint for profit, we solve

$$\text{min} \quad \tfrac{1}{20}(0.005x_2 + 0.040x_3 + 0.050x_4 + 0.075x_5 \qquad \text{(capital-adequacy)}$$
$$+ 0.100x_6 + 0.100x_7 + 0.100x_8)$$

$$\text{s.t.} \quad \tfrac{1}{20}(x_6 + x_7 + x_8) \leq 5.0 \qquad \text{(risk-asset)}$$
$$0.040x_2 + 0.045x_3 + 0.055x_4 + 0.070x_5 \qquad \text{(profit)}$$
$$+ 0.105x_6 + 0.085x_7 + 0.092x_8 \geq 17.2$$
$$x_1 + \cdots + x_8 = 20 + 150 + 80 \qquad \text{(invest all)}$$
$$x_1 \geq 0.14(150) + 0.04(80) \qquad \text{(cash reserve)}$$
$$1.00x_1 + 0.995x_2 + 0.960x_3 + 0.900x_4 \qquad \text{(liquidity)}$$
$$+ 0.850x_5 \geq 0.47(150) + 0.36(80)$$

$$x_j \geq 0.05(20 + 150 + 80) \quad \text{for all } j = 1, \ldots, 8 \quad \text{(diversification)}$$
$$x_8 \geq 0.30(20 + 150 + 80) \quad \text{(commercial)}$$
$$x_1, \ldots, x_8 \geq 0$$

This time solution (8.7) remains optimal. The capital-adequacy ratio cannot be decreased without worsening other objectives.

SAMPLE EXERCISE 8.3: SOLVING MULTIOBJECTIVE MODELS PREEMPTIVELY

Consider the multiobjective mathematical program

$$\begin{aligned}
\max \quad & w_1 \\
\max \quad & 2w_1 + 3w_2 \\
\text{s.t.} \quad & w_1 \leq 3 \\
& w_1 + w_2 \leq 5 \\
& w_1, w_2 \geq 0
\end{aligned}$$

Solve the model graphically by preemptive optimization, taking objectives in the sequence given.

Analysis: Graphic solution produces the following plot:

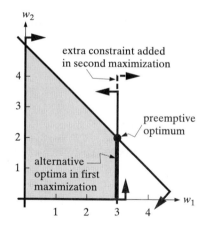

A first optimization maximizes w_1 subject to the given constraints. Any of the solutions along the line segment from $(3, 0)$ to $(3, 2)$ is optimal.

Next we impose an extra constraint

$$w_1 \geq 3$$

and maximize the second objective. The result is final preemptive solution $\mathbf{w} = (3, 2)$.

Preemptive Optimization and Efficient Points

One advantage of the preemptive approach to multiobjective optimization is that it results in solutions that cannot be improved in one objective without degrading another.

| 8.6 | If each stage of preemptive optimization yields a single-objective optimum, the final solution is an efficient point of the full multiobjective model. |

The preemptive process requires that we try objective functions in turn, trying to improve one without worsening the others. When we finish, no further improvement is possible. As usual, infeasible and unbounded cases can produce complications, but the typical outcome is an efficient point.

Preemptive Optimization and Alternative Optima

Although it usually will produce an efficient point, a moment's reflection on the preemptive optimization approach will reveal a major limitation:

| 8.7 | After one objective function has been optimized in preemptive processing of a multiobjective model, solutions obtained in subsequent stages must all be alternative optima in the first. |

That is, preemptive optimization places very great emphasis on the first objective addressed, with all later steps limited to alternative optima in the highest-priority objective.

The Bank Three computations above showed considerable change from initial optima (8.6) to final (8.7) because the first, risk-asset objective

$$\min \quad \frac{1}{20}(x_6 + x_7 + x_8)$$

admits many alternative optima among the unmentioned $x_1, \ldots, x_5$. But with cases where alternative optima are rare, the preemptive approach becomes essentially one of optimizing a priority objective while ignoring all the others.

Weighted Sums of Objectives

An alternative scheme for dealing with multiobjective models that permits more balanced handling of the objectives is simply to combine them in a **weighted sum**.

| 8.8 | Multiple objective functions can be combined into a single composite one to be maximized by summing objectives with positive weights on maximizes and negative weights on minimizes. If the composite is to be minimized, weights on maximize objectives should be negative, and those on minimizes should be positive. |

Signs orient all objectives in the same direction, and weights reflect their relative importance.

SAMPLE EXERCISE 8.4: FORMING WEIGHTED OBJECTIVES

Form a single weighted-sum composite objective from each of the following collections of objective functions. Indicate whether weights γ_i should be positive or negative and whether the composite objective should be maximized or minimized.

(a) min $+2w_1$ $+3w_2$ $-1w_3$
    max $+4w_1$ $-2w_2$
    max $+1w_2$ $+1w_3$

(b) min $+3w_1$ $-1w_2$
    min $+4w_1$ $+2w_2$ $+9w_3$

Analysis: We apply principle 8.8 .

(a) Using weights $\gamma_1, \ldots, \gamma_3$, the weighted objective is

$$\max (2\gamma_1 + 4\gamma_2)w_1 + (3\gamma_1 - 2\gamma_2 + 1\gamma_3)w_2 + (-1\gamma_1 + 1\gamma_3)w_3$$

This maximize composite form requires that $\gamma_1 < 0$, $\gamma_2 > 0$, and $\gamma_3 > 0$.

(b) Using weights γ_1 and γ_2, the weighted objective is

$$\min (3\gamma_1 + 4\gamma_2)w_1 + (-1\gamma_1 + 2\gamma_2)w_2 + (9\gamma_2)w_3$$

This minimize composite form requires that $\gamma_1 > 0$ and $\gamma_2 > 0$.

Weighted-Sum Optimization of the Hazardous Waste Example

Hazardous waste planning model (8.3):

$$\min \ \sum_{i=1}^{7}\sum_{j=1}^{3}\sum_{k=1}^{2} d_{i,j,k}x_{i,j,k} \qquad \text{(distance)}$$

$$\min \ \sum_{i=1}^{7}\sum_{j=1}^{3}\sum_{k=1}^{2} p_{i,j,k}x_{i,j,k} \qquad \text{(population)}$$

s.t. $\quad \sum_{j=1}^{3}\sum_{k=1}^{2} x_{i,j,k} = s_i \qquad i = 1, \ldots, 7 \qquad \text{(sources)}$

$\qquad \sum_{j=1}^{3} y_j = 2 \qquad\qquad\qquad\qquad \text{(2 sites)}$

$\qquad x_{i,j,k} \le s_i y_j \qquad\quad i = 1, \ldots, 7; \quad j = 1, \ldots, 3; \quad k = 1, 2$

$\qquad x_{i,j,k} \ge 0 \qquad\qquad\ i = 1, \ldots, 7; \quad j = 1, \ldots, 3; \quad k = 1, 2$

$\qquad y_j = 0 \text{ or } 1 \qquad\quad\ \ j = 1, \ldots, 3$

illustrates a setting where weighted-sum analysis of a multiobjective model can usefully be applied. The first objective minimizes the shipping distance to chosen disposal sites. The second minimizes the population exposed along the way. Each source is provided 2 alternative routes to each potential disposal site, one denoted $k = 1$, emphasizing short distance, and the other $k = 2$, avoiding population.

    Since both objective minimize, we may produce a single composite objective by applying weights $\gamma_1, \gamma_2 > 0$ and minimizing the result (principle 8.8):

$$\min \ \sum_{i=1}^{7}\sum_{j=1}^{3}\sum_{k=1}^{2} \left(\gamma_1 d_{i,j,k} + \gamma_2 p_{i,j,k}\right)x_{i,j,k} \qquad \text{(composite)}$$

Table 8.4 illustrates the impact for some different combinations of weights. With comparatively high weight γ_1 on distance, the optimization routes almost all

TABLE 8.4 Weighting Objectives in the Hazardous Waste Example

Weight		Ton-Miles		Total Ton-Miles	Total Ton-Population	Optimal Sites
γ_1	γ_2	$k = 1$	$k = 2$			
10	1	1155	0	1155	1334.5	1, 3
10	5	754	591	1345	782.5	1, 3
10	10	1046	404	1450	607.5	1, 2
5	10	440	1114	1554	533.0	1, 3
1	10	0	1715	1715	468.5	1, 3

along short routes $k = 1$. As population is given greater relative weight, activity shifts to the longer routes that avoid population except when the optimal sites change at $\gamma_1 = \gamma_2 = 10$. Eventually, all shipping is along the safer routes. Confronted with a range of such alternatives, decision makers could decide an appropriate balance.

Weighted-Sum Optimization and Efficient Points

Although offering more flexibility in trading off objectives, using weighted totals still assures an efficient solution.

> 8.9 If a single weighted-sum objective model derived from a multiobjective optimization as in 8.8 produces an optimal solution, the solution is an efficient point of the multiobjective model.

To see why this is true, we need only consider the nature of a weighted-sum objective:

$$\begin{pmatrix} \text{weight} \\ 1 \end{pmatrix} \begin{pmatrix} \text{objective} \\ 1 \text{ value} \end{pmatrix} + \begin{pmatrix} \text{weight} \\ 2 \end{pmatrix} \begin{pmatrix} \text{objective} \\ 2 \text{ value} \end{pmatrix} + \cdots$$
$$+ \begin{pmatrix} \text{weight} \\ p \end{pmatrix} \begin{pmatrix} \text{objective} \\ p \text{ value} \end{pmatrix}$$

With signs as in construction 8.8 , any solution that can improve in one objective without degrading the others would also score better in the weighted objective. Only an efficient point could be optimal.

8.4 GOAL PROGRAMMING

Multiobjective models of complex problems assume that we always want more of everything—lower values of objectives being minimized at the same time as higher values of criteria being maximized. Notice that this is independent of how much we may already have achieved on one or another objective. For example, a multiobjective model would keep the same priority or weight on a minimize cost objective whether or not we already have in hand a solution with extraordinarily low cost.

In this section we explore the **goal programming** alternative, which is constructed in terms of target levels to be achieved rather than quantities to be maximized or minimized. It is probably more realistic to assume that the importance of any criterion diminishes once a target level has been achieved. We will certainly see that it is easier to implement. That is why goal programming is by far the most popular approach to finding good solutions in multicriteria problem settings.

Goal or Target Levels

Goal program modeling of a multiobjective optimization begins by asking decision makers to specify new data: goal or target levels for each of the criteria used to evaluate solutions.

> **8.10** **Goal** or **target levels** specify the values of the criteria functions in an optimization model that decision makers consider sufficient or satisfactory.

Goal Form of Bank Three Example

To illustrate, return to our Bank Three example of Section 8.1.

$$
\begin{aligned}
\max \quad & 0.040x_2 + 0.045x_3 + 0.055x_4 + 0.070x_5 && \text{(profit)} \\
& + 0.105x_6 + 0.085x_7 + 0.092x_8 && \\
\min \quad & \tfrac{1}{20}(0.005x_2 + 0.040x_3 + 0.050x_4 + 0.075x_5 && \text{(capital-adequacy)} \\
& + 0.100x_6 + 0.100x_7 + 0.100x_8) && \\
\min \quad & \tfrac{1}{20}(x_6 + x_7 + x_8) && \text{(risk-asset)} \\
\text{s.t.} \quad & x_1 + \cdots + x_8 = (20 + 150 + 80) && \text{(invest all)} \\
& x_1 \geq 0.14(150) + 0.04(80) && \text{(cash reserve)} \\
& 1.00x_1 + 0.995x_2 + 0.960x_3 + 0.900x_4 && \text{(liquidity)} \\
& + 0.850x_5 \geq 0.47(150) + 0.36(80) && \\
& x_j \geq 0.05(20 + 150 + 80) \quad \text{for all } j = 1, \ldots, 8 && \text{(diversification)} \\
& x_8 \geq 0.30(20 + 150 + 80) && \text{(commercial)} \\
& x_1, \ldots, x_8 \geq 0 &&
\end{aligned}
$$

Here solutions are evaluated on three criteria: profit, capital-adequacy ratio, and risk-asset ratio. Assume that instead of seeking ever higher levels of the first criterion and lower values of the last two, we set some goals:

$$
\begin{aligned}
\text{profit} &\geq 18.5 \\
\text{capital-adequacy ratio} &\leq 0.8 \\
\text{risk-asset ratio} &\leq 7.0
\end{aligned}
\tag{8.8}
$$

Then the problem might be specified in goal format as

$$
\begin{aligned}
\text{goal} \quad & 0.040x_2 + 0.045x_3 + 0.055x_4 + 0.070x_5 && \text{(profit)} \\
& + 0.105x_6 + 0.085x_7 + 0.092x_8 \geq 18.5 && \\
\text{goal} \quad & \tfrac{1}{20}(0.005x_2 + 0.040x_3 + 0.050x_4 + 0.075x_5 && \text{(capital-adequacy)} \\
& + 0.100x_6 + 0.100x_7 + 0.100x_8) \leq 0.8 && \\
\text{goal} \quad & \tfrac{1}{20}(x_6 + x_7 + x_8) \geq 7.0 && \text{(risk-asset)} \\
\text{s.t.} \quad & x_1 + \cdots + x_8 = (20 + 150 + 80) && \text{(invest all)} \\
& x_1 \geq 0.14(150) + 0.04(80) && \text{(cash reserve)} \\
& 1.00x_1 + 0.995x_2 + 0.960x_3 + 0.900x_4 && \text{(liquidity)} \\
& + 0.850x_5 \geq 0.47(150) + 0.36(80) && \\
& x_j \geq 0.05(20 + 150 + 80) \quad \text{for all } j = 1, \ldots, 8 && \text{(diversification)} \\
& x_8 \geq 0.30(20 + 150 + 80) && \text{(commercial)} \\
& x_1, \ldots, x_8 \geq 0 &&
\end{aligned}
\tag{8.9}
$$

Soft Constraints

Goals in statement (8.9) may be thought of as soft constraints.

> **8.11**   **Soft constraints** such as the criteria targets of goal programming specify requirements that are desirable to satisfy but which may still be violated in feasible solutions.

The more familiar **hard constraints** still determine what solutions are feasible, leaving the soft ones to influence which solutions are preferred.

Deficiency Variables

Once target levels have been specified for soft constraints, we proceed to a more familiar mathematical programming formulation by adding constraints that enforce goal achievement. However, we cannot just impose the constraint that each objective meet its goal. There may be no solution that simultaneously achieves the desired levels of all soft constraints. Instead, we introduce new deficiency variables.

> **8.12**   Nonnegative **deficiency variables** are introduced to model the extent of violation in goal or other soft constraints that need not be rigidly enforced. With a $\geq$ target, the deficiency is the under achievement. With a $\leq$ target, it is the excess. With $=$ soft constraints, deficiency variables are included for both under- and over achievement.

In the 3-objective Bank Three example (8.9), we enforce goal levels (8.8) with deficiency variables

$$d_1 \triangleq \text{amount profit falls short of its goal}$$

$$d_2 \triangleq \text{amount capital-adequacy ratio exceeds its goal}$$

$$d_3 \triangleq \text{amount risk-asset ratio exceeds its goal}$$

Expressing Soft Constraints in Mathematical Programs

Goal and other soft constraints can now be expressed in the usual (hard) mathematical programming format with deficiency variables allowing violation.

> **8.13**   Goal or soft constraints use nonnegative deficiency variables to express a $\geq$ target
>
> $$\text{(criterion function)} + \text{(deficiency variable)} \geq \text{target value}$$
>
> and a $\leq$ target
>
> $$\text{(criterion function)} - \text{(deficiency variable)} \leq \text{target value}$$
>
> Equality-form soft constraints are modeled
>
> $$\text{(criterion function)} - \text{(oversatisfaction deficiency variable)}$$
> $$+ \text{(undersatisfaction deficiency variable)} = \text{target value}$$

For example, the three goals in Bank Three model (8.9) lead to the following main (and variable type) constraints:

$$0.040x_2 + 0.045x_3 + 0.055x_4 + 0.070x_5 \qquad \text{(profit)}$$
$$+ 0.105x_6 + 0.085x_7 + 0.092x_8 + d_1 \geq 18.5$$
$$\tfrac{1}{20}(0.005x_2 + 0.040x_3 + 0.050x_4 + 0.075x_5 \qquad \text{(capital-adequacy)}$$
$$+ 0.100x_6 + 0.100x_7 + 0.100x_8) - d_2 \leq 0.8$$
$$\tfrac{1}{20}(x_6 + x_7 + x_8) - d_3 \leq 7.0 \qquad \text{(risk-asset)}$$
$$d_1, d_2, d_3 \geq 0$$

The first keeps profit at least 18.5 or makes up the difference with deficiency variable d_1. The other two main constraints force the capital-adequacy and risk-asset ratios below our target value unless the corresponding deficiency variables are nonzero.

Notice that nonnegativity constraints are required. We want the deficiency to compute as zero if the target is achieved.

SAMPLE EXERCISE 8.5: FORMULATING GOAL CONSTRAINTS

Return to the multiobjective model

$$\begin{aligned} \max \quad & w_1 \\ \max \quad & 2w_1 + 3w_2 \\ \text{s.t.} \quad & w_1 \leq 3 \\ & w_1 + w_2 \leq 5 \\ & w_1, w_2 \geq 0 \end{aligned}$$

of Sample Exercise 8.3, and assume that instead of seeking to maximize the two criteria, we decide to seek a target level of 2.0 on the first and 14.0 on the second. Introduce deficiency variables and formulate new goal constraints to model these soft constraints as those of a linear program.

Analysis: We apply construction $\boxed{8.13}$ using deficiency variables

$d_1 \triangleq$ undersatisfaction of the first goal

$d_2 \triangleq$ undersatisfaction of the second goal

Then the new linear constraints are

$$\begin{aligned} w_1 \qquad\quad + d_1 \quad & \geq 2.0 \\ 2w_1 + 3w_2 + d_2 \quad & \geq 14.0 \\ d_1, d_2 \geq 0 \end{aligned}$$

Goal Program Objective Function: Minimizing (Weighted) Deficiency

Having modeled undersatisfaction of goals with deficiency variables, we complete a formulation by minimizing violation.

8.14 | The objective in a goal programming model expresses the desire to satisfy all goals as nearly as possible by minimizing a weighted sum of the deficiency variables.

Often, all deficiency is weighted equally.

Goal Linear Program Model of the Bank Three Example

Using equal goal weights in our Bank Three example (8.9) produces the goal linear program

$$
\begin{array}{lll}
\min & d_1 + d_2 + d_3 & \text{(total deficiency)} \\
\text{s.t.} & 0.040x_2 + 0.045x_3 + 0.055x_4 + 0.070x_5 & \text{(profit)} \\
& \quad + 0.105x_6 + 0.085x_7 + 0.092x_8 + d_1 \geq 18.5 \\
& \tfrac{1}{20}(0.005x_2 + 0.040x_3 + 0.050x_4 + 0.075x_5 & \text{(capital-adequacy)} \\
& \quad + 0.100x_6 + 0.100x_7 + 0.100x_8) - d_2 \leq 0.8 \\
& \tfrac{1}{20}(x_6 + x_7 + x_8) - d_3 \leq 7.0 & \text{(risk-asset)} \\
& x_1 + \cdots + x_8 = 20 + 150 + 80 & \text{(invest all)} \\
& x_1 \geq 0.14(150) + 0.04(80) & \text{(cash reserve)} \\
& 1.00x_1 + 0.995x_2 + 0.960x_3 + 0.900x_4 & \text{(liquidity)} \\
& \quad + 0.850x_5 \geq 0.47(150) + 0.36(80) \\
& x_j \geq 0.05(20 + 150 + 80) \quad \text{for all } j = 1, \ldots, 8 & \text{(diversification)} \\
& x_8 \geq 0.30(20 + 150 + 80) & \text{(commercial)} \\
& x_1, \ldots, x_8 \geq 0 \\
& d_1, d_2, d_3 \geq 0
\end{array}
$$
(8.10)

The goals have been expressed with new constraints involving deficiency variables. All original constraints are retained.

Alternative Deficiency Weights in the Objective

Table 8.5 shows an optimal solution to equal-weighted goal LP in column (1). Notice how it seeks only the $18.5 million goal for profit and the 7.0 goal for risk-asset ratio. Once corresponding deficiency variables d_1 and d_2 are driven to $= 0.0$, effort can be directed toward the remaining capital-adequacy goal.

There is no requirement that deficiency weights be equal. In our Bank Three example, the magnitude of the capital-adequacy ratio is much smaller than that of the other two objectives. Thus better results might be obtained by a different weighting that scales deficiencies more equally.

Column (2) of Table 8.5 shows the effect of multiplying the capital adequacy weight by 10. Now the profit slips below its $18.5 million goal (to $17.53 million), but the capital-adequacy ratio deficiency is reduced.

TABLE 8.5   Goal Programming Solution of the Bank Three Example

	(1) Equal Weights	(2) Unequal Weights	(3) Preempt Profit	(4) Preempt Profit, CA	(5) Preempt One Step
Profit goal weight	1	1	1	0	10,000
Cap-adequacy goal weight	1	10	0	1	100
Risk-asset goal weight	1	1	0	0	1
Extra constraints	—	—	—	$d_1 = 0$	—
Profit	18.50	17.53	18.50	18.50	18.50
Deficiency, d_1^*	0.00	0.97	0.00	0.00	0.00
Capital-adequacy ratio	0.928	0.815	0.943	0.919	0.919
Deficiency, d_2^*	0.128	0.015	0.143	0.119	0.119
Risk-asset ratio	7.000	7.000	7.097	7.158	7.158
Deficiency, d_3^*	0.000	0.000	0.097	0.158	0.158
Cash, x_1^*	24.20	24.20	24.20	24.20	24.20
Short term, x_2^*	16.03	48.30	12.50	19.73	19.73
Government: 1–5, x_3^*	12.50	12.50	12.50	12.50	12.50
Government: 5–10, x_4^*	12.50	12.50	12.50	12.50	12.50
Government: over 10, x_5^*	44.77	12.50	46.37	37.91	37.91
Installment, x_6^*	52.50	52.50	41.08	55.67	55.67
Mortgages, x_7^*	12.50	12.50	12.50	12.50	12.50
Commercial, x_8^*	75.00	75.00	88.36	75.00	75.00

SAMPLE EXERCISE 8.6: FORMULATING GOAL PROGRAMS

Assigning equal weights to violations of the two goals, formulate the goal linear program corresponding to the constraints and target values of Sample Exercise 8.5.

Analysis: Including the goal constraints of Sample Exercise 8.5 and minimizing total deficiency as in construction 8.14 produces the goal linear program

$$\begin{array}{ll} \min & d_1 + d_2 \\ \text{s.t.} & w_1 + d_1 \geq 2.0 \\ & 2w_1 + 3w_2 + d_2 \geq 14.0 \\ & w_1 \leq 3 \\ & w_1 + w_2 \leq 5 \\ & w_1, w_2 \geq 0 \\ & d_1, d_2 \geq 0 \end{array}$$

Notice that all original constraints have been retained.

Preemptive Goal Programming

The preemptive optimization of Section 8.4 (definition 8.5) takes objective functions in order: optimizing one, then a second subject to the first achieving its optimal value, and so on. A similar preemptive goal programming approach is often adopted after criteria have been modeled as goals.

8.15   **Preemptive** or **lexicographic goal programming** considers goals one at a time. Deficiency in the most important is minimized; then deficiency in the second most important is minimized subject to a requirement that the first achieve its minimum; and so on.

Preemptive Goal Programming of the Bank Three Example

Columns (3) and (4) of Table 8.5 illustrate this preemptive variant of goal programming. The first was obtained by focusing entirely on the profit goal. Using the objective function

$$\text{min} \quad d_1$$

we concern ourselves only with achieving the desired profit of \$18.5 million. Column (3) shows a distribution of investments that achieves that goal, albeit at the cost of violating both the others.

Now we turn to the capital-adequacy ratio goal. After adding the extra constraint

$$d_1 = 0$$

to keep the profit goal fully satisfied, we address capital adequacy with objective function

$$\text{min} \quad d_2$$

The result [column (4) of Table 8.5] shifts investments to improve the capital-adequacy ratio while continuing to achieve a profit of \$18.5 million.

To complete the preemptive goal programming solution of this example, we should now address the last risk-asset ratio objective by

$$\text{min} \quad d_3$$

subject to extra constraints

$$d_1 = 0.0, \quad d_2 = 0.119$$

The effect is to seek a solution coming closer to the risk-asset ratio goal without losing ground on either of the other two. For this example the resulting solution is the same as column (4).

SAMPLE EXERCISE 8.7: DOING PREEMPTIVE GOAL PROGRAMMING SEQUENTIALLY

Return to the weighted goal program of Sample Exercise 8.6.

(a) Formulate the first model to be solved if we wish to give highest priority to fulfilling the first goal.

(b) Assuming that the first goal can be completely achieved in the model of part (a), formulate a second model to seek satisfaction of the second goal subject to satisfying the first.

Analysis: We execute optimization sequence 8.15 .

(a) The first model to be solved emphasizes deficiency in goal 1:

$$\min \quad d_1$$
$$\text{s.t.} \quad w_1 + d_1 \geq 2.0$$
$$2w_1 + 3w_2 + d_2 \geq 14.0$$
$$w_1 \leq 3$$
$$w_1 + w_2 \leq 5$$
$$w_1, w_2 \geq 0$$
$$d_1, d_2 \geq 0$$

(b) Assuming that deficiency in the first goal can be completely eliminated, the second optimization to be solved is

$$\min \quad d_2$$
$$\text{s.t.} \quad w_1 + d_1 \geq 2.0$$
$$2w_1 + 3w_2 + d_2 \geq 14.0$$
$$w_1 \leq 3$$
$$w_1 + w_2 \leq 5$$
$$w_1, w_2 \geq 0$$
$$d_2 \geq 0$$
$$d_1 = 0$$

New constraint $d_1 = 0$ requires that the first goal be fully satisfied. Within that limitation, we minimize violation of the second goal.

Preemptive Goal Programming by Weighting the Objective

If a given model has a number of objective functions (or goals), preemptive goal programming in the successive optimization manner of principle $\boxed{8.15}$ can become rather tedious. Fortunately, the same effect can be achieved in a single optimization by choosing appropriate weights.

> $\underline{8.16}$ Preemptive goal programming can be accomplished in a single step by solving with an objective function that puts very great weight on the deficiency in the most important goal, less weight on the deficiency in the second goal, and so on.

We may implement the profit first, capital-adequacy ratio second, risk-asset ratio third prioritization of our Bank Three goal programming model (8.10) in this way by using the objective function

$$\min \quad 10{,}000d_1 + 100d_2 + 1d_3 \qquad \text{(preemptive-weighted deficiency)}$$

Column (5) of Table 8.5 shows that this weighting achieves the same solution as the one obtained sequentially in column (4).

SAMPLE EXERCISE 8.8: DOING PREEMPTIVE GOAL PROGRAMMING WITH WEIGHTS

Use appropriate weights to formulate a single goal program that implements the preemptive priority sequence in the example of Sample Exercise 8.7.

Analysis: We apply construction $\boxed{8.16}$. Using multiplier 100 on the first deficiency and 1 on the second assures that every effort will be made to fulfill the first goal before the second is considered. The result is the goal program

$$
\begin{aligned}
\min \quad & 100d_1 + d_2 \\
\text{s.t.} \quad & w_1 + d_1 \geq 2.0 \\
& 2w_1 + 3w_2 + d_2 \geq 14.0 \\
& w_1 \leq 3 \\
& w_1 + w_2 \leq 5 \\
& w_1, w_2 \geq 0 \\
& d_1, d_2 \geq 0
\end{aligned}
$$

Practical Advantage of Goal Programming in Multiobjective Problems

Although the same techniques can be used for any soft constraints, we have seen from our treatment of the Bank Three example how goal programming can provide a practical tool for dealing with multiobjective problem settings. If decision makers can be persuaded to specify target levels for their various objectives—and often that is not easy—modeling as a goal LP or ILP or NLP reduces the multiobjective analysis to standard mathematical programming form.

> $\boxed{8.17}$ Goal programming is the most popular approach to dealing with multiobjective optimization problems because it reduces complex multiobjective tradeoffs to a standard, single-objective mathematical program in a way that decision makers often find intuitive.

Goal Programming and Efficient Points

We have seen in Section 8.2 that efficient points (definition $\boxed{8.1}$) provide the closest analog to an optimal solution in multiobjective settings because they cannot be improved in one objective without degrading another. Unfortunately, goal programming reformulation of a multiobjective model cannot always guarantee a solution with this desirable property.

> $\boxed{8.18}$ If a goal program has alternative optimal solutions, some of them may not be efficient points of the corresponding multiobjective optimization model.

To see the difficulties that can occur, suppose that we modify Bank Three goal LP (8.10) by ignoring the capital-adequacy objective. The result is the two-goal form

$$\min \quad d_1 + d_3 \qquad\qquad\qquad\qquad \text{(total deficiency)}$$

$$\text{s.t.} \quad 0.040x_2 + 0.045x_3 + 0.055x_4 + 0.070x_5 \qquad \text{(profit)}$$
$$+ 0.105x_6 + 0.085x_7 + 0.092x_8 + d_1 \geq 18.5$$
$$+ 0.100x_6 + 0.100x_7 + 0.100x_8) - d_2 \leq 0.8$$
$$\tfrac{1}{20}(x_6 + x_7 + x_8) - d_3 \leq 7.0 \qquad\qquad \text{(risk-asset)}$$
$$x_1 + \cdots + x_8 = (20 + 150 + 80) \qquad\qquad \text{(invest all)}$$
$$x_1 \geq 0.14(150) + 0.04(80) \qquad\qquad \text{(cash reserve)} \qquad (8.11)$$
$$1.00x_1 + 0.995x_2 + 0.960x_3 + 0.900x_4 \qquad\qquad \text{(liquidity)}$$
$$+ 0.850x_5 \geq 0.47(150) + 0.36(80)$$
$$x_j \geq 0.05(20 + 150 + 80) \quad \text{for all } j = 1, \ldots, 8 \qquad \text{(diversification)}$$
$$x_8 \geq 0.30(20 + 150 + 80) \qquad\qquad \text{(commercial)}$$
$$x_1, \ldots, x_8 \geq 0$$
$$d_1, d_3 \geq 0$$

Solving gives the goal programming optimum

$$x_1^* = 24.2, \quad x_2^* = 12.5, \quad x_3^* = 12.5, \quad x_4^* = 12.5$$
$$x_5^* = 48.3, \quad x_6^* = 44.35, \quad x_7^* = 12.5, \quad x_8^* = 83.15 \tag{8.12}$$

with profit \$18.5 million and risk-asset ratio 7.00.

Both goals are satisfied completely. Still, solution (8.12) is not efficient because either of the objectives can be improved without worsening the other. For example, the efficient solution (an alternative goal program optimum)

$$x_1^* = 24.2, \quad x_2^* = 12.5, \quad x_3^* = 12.5, \quad x_4^* = 12.5$$
$$x_5^* = 51.33, \quad x_6^* = 49.47, \quad x_7^* = 12.5, \quad x_8^* = 75.0 \tag{8.13}$$

continues to produce \$18.5 million in profit but reduces the risk-asset ratio to 6.849.

What makes such nonefficient goal programming solutions occur? A moment's reflection will reveal that the problem is goals that are fully satisfied. Both objectives achieved their target values in (8.12). Once such a solution drives the corresponding deficiency variable $= 0$, there is nothing in the goal programming formulation to encourage further improvement. Progress stops without reaching an efficient point.

SAMPLE EXERCISE 8.9: UNDERSTANDING NONEFFICIENT GOAL PROGRAM SOLUTIONS

Consider the simple multiobjective optimization model

$$\min \quad w_1$$
$$\max \quad w_2$$
$$\text{s.t.} \quad 0 \leq w_1 \leq 2$$
$$\quad\quad 0 \leq w_2 \leq 1$$

(a) Introduce deficiency variables to formulate a goal program with target values 1 for the first objective function and 2 for the second.

(b) Show that $(w_1, w_2) = (1, 1)$ is optimal in the goal program of part (a) but not efficient in the original multiobjective form.

(c) Show that $(w_1, w_2) = (0, 1)$ is also optimal in the goal program of part (a) but that it is efficient in the original multiobjective form.

Analysis:

(a) Using deficiency variables d_1 in the first objective and d_2 in the second, the required goal program is

$$
\begin{aligned}
\min \quad & d_1 + d_2 \\
\text{s.t.} \quad & w_1 - d_1 \leq 1 \\
& w_2 + d_2 \geq 2 \\
& 0 \leq w_1 \leq 2 \\
& 0 \leq w_2 \leq 1 \\
& d_1, d_2 \geq 0
\end{aligned}
$$

(b) At solution $(w_1, w_2) = (1, 1)$, the first goal is satisfied $(d_1 = 0)$ and the second misses by $d_2 = 2 - 1 = 1$. However, no feasible solution can make $w_2 > 1$, so the solution is optimal in the goal program.

Even though this solution is optimal in the goal program, it is not efficient. We may feasibly make the first objective less that 1.0 while retaining 1.0 in the second.

(c) The solution $(w_1, w_2) = (0, 1)$ is feasible and achieves the same minimum total deficiency as the solution of (b). But this alternative optimum is efficient.

Modified Goal Program Formulation to Assure Efficient Points

In many applications, model constraints and objectives are so complex that goal programming's potential for producing non-efficient points never arises. Fortunately, there is an easy correction when it is a concern.

> 8.19 To assure that goal programming treatment of a multiobjective optimiza-
> tion yields an efficient point, we need only add a small positive multiple of each
> original minimize objective function to the goal program objective and subtract
> the same multiple of each original maximize.

We may illustrate with a modified Bank Three goal program (8.11). Efficient solution (8.13) results when we change the goal program objective to

$$
\begin{aligned}
\min \quad & d_1 + d_3 \\
& - 0.001(0.040x_2 + 0.045x_3 + 0.055x_4 + 0.070x_5 + 0.105x_6 \\
& + 0.085x_7 + 0.092x_8) + \frac{0.001}{20}(x_6 + x_7 + x_8)
\end{aligned}
$$

Multiple 0.001 of the maximize profit objective has been subtracted and the same multiple of the risk-asset ratio added to the standard deficiency objective.

For the same reason that weighted-sum approaches produce efficient solutions (principle 8.9), any result from this modified goal program can be optimal only if no original objective can be improved without degrading another. Still, the underlying goal program resolution of the multiple objectives will not be affected as long as the weight on original objectives is kept small.

SAMPLE EXERCISE 8.10: MAKING GOAL PROGRAMS PRODUCE EFFICIENT SOLUTIONS

Produce a revised version of the goal program in Sample Exercise 8.9(a) that must yield an efficient solution in the original multiobjective optimization.

Analysis: Following construction 8.19 , we introduce a small multiple of the original objective functions. Using multiplier 0.001, the result is

$$\min \quad d_1 + d_2 + 0.001(w_1) - 0.001(w_2)$$
$$\text{s.t.} \quad w_1 - d_1 \leq 1$$
$$w_2 + d_2 \geq 2$$
$$0 \leq w_1 \leq 2$$
$$0 \leq w_2 \leq 1$$
$$d_1, d_2 \geq 0$$

The multiplier is positive for the first, minimize objective, and negative on the maximize objective. The goal program objective of minimizing deficiency is still dominant, but new terms now assure that alternative optima will resolve to efficient points.

EXERCISES

8-1 The sketch that follows shows the District 88 river system in the southwestern United States.

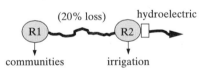

All water arises at mountain reservoir R1. The estimated flow is 294 million acre-feet. At least 24 million acre-feet of the water is contracted to go directly from R1 to nearby communities, but the communities would accept as much as could be supplied. The remainder flows through the desert to a second reservoir, R2, losing 20% to evaporation along the way. Some water at R2 can be allocated for irrigation of nearby farms. The remainder flows over a hydroelectric dam and passes downstream. To maintain the equipment, the flow over the dam must be at least 50 million acre-feet. District 88 sells water to communities at $0.50 per acre-foot and to irrigation farmers at $0.20 per acre foot. Water passing through

the hydroelectric dam earns $0.80 per acre-foot. The district would like to maximize both water supplied for irrigation and sales income.

(a) Briefly explain how this problem can be modeled by the multiobjective linear program

$$\max \quad x_2$$
$$\max \quad 0.50x_1 + 0.20x_2 + 0.80x_3$$
$$\text{s.t.} \quad x_1 + 1.25x_2 + 1.25x_3 = 294$$
$$x_1 \geq 24, x_2 \geq 0, x_3 \geq 50$$

(b) ⊗ ▣ Use the class optimization software to sketch the efficient frontier for the model of part (a) in an objective value graph like Figure 8.4.

(c) ⊗ ▣ Use the class optimization software to show that the optimal solutions taken separately for each objective do not coincide.

8-2 A semiconductor manufacturer has 3 different types of silicon wafers in stock to manufacture its 3 varieties of computer chips. Some wafer types cannot be used for some chips, but there are two alternatives for each chip. The table that follows shows the cost and on-hand supply of each wafer type, the number of each chip needed, and a score (0 to 10) of the appropriateness match of each wafer type for making each chip.

Wafer Type	Chip Match			Unit Cost	On Hand
	1	2	3		
1	7	8	—	15	500
2	10	—	6	25	630
3	—	10	10	30	710
Need	440	520	380		

The company would like to minimize the total cost of the wafers used while maximizing the total match score.

(a) Briefly explain how this problem can be modeled by the multiobjective LP

$$\text{min} \quad 15x_{1,1} + 15x_{1,2} + 25x_{2,1}$$
$$+ 25x_{2,3} + 30x_{3,2} + 30x_{3,3}$$
$$\text{max} \quad 7x_{1,1} + 8x_{1,2} + 10x_{2,1}$$
$$+ 6x_{2,3} + 10x_{3,2} + 10x_{3,3}$$
$$\text{s.t.} \quad x_{1,1} + x_{1,2} \leq 500$$
$$x_{2,1} + x_{2,3} \leq 630$$
$$x_{3,2} + x_{3,3} \leq 710$$
$$x_{1,1} + x_{2,1} = 440$$
$$x_{1,2} + x_{3,2} = 520$$
$$x_{2,3} + x_{3,3} = 380$$
$$\text{all } x_{i,j} \geq 0$$

(b) and (c) as in Exercise 8-1.

8-3 A national commission is deciding which military bases to close in order to save at least $85 million per year. The table that follows shows projected annual savings (millions of dollars) for each of five possible closings, together with the implied percent loss of military readiness and the number of civilian workers (thousands) who would lose their jobs.

	Base				
	1	2	3	4	5
Savings	24	29	45	34	80
Readiness	1.0	0.4	1.4	1.8	2.0
Workers	2.5	5.4	4.6	4.2	14.4

Each base must be left open or completely closed, and the commission wants to meet the required savings while minimizing both readiness loss and unemployment.

(a) Briefly explain how this base-closing problem can be modeled by the multiobjective ILP

$$\text{min} \quad 1.0x_1 + 0.4x_2 + 1.4x_3$$
$$+ 1.8x_4 + 2.0x_5$$
$$\text{min} \quad 2.5x_1 + 5.4x_2 + 4.6x_3$$
$$+ 4.2x_4 + 14.4x_5$$
$$\text{s.t.} \quad 24x_1 + 29x_2 + 45x_3$$
$$+ 34x_4 + 80x_5 \geq 85$$
$$x_j = 0 \text{ or } 1, \quad j = 1, \dots, 5$$

(b) and (c)⊘ as in Exercise 8-1.

8-4 A garden superstore sells four kinds of fertilizer from 20 pallet spaces aligned along a fence. Weekly demands for the four fertilizers are 20, 14, 9, and 5 pallets, respectively, and each pallet holds 10 bags. The sales counter is at the beginning of the fence, and every bag sold must be carried from its pallet to the counter. Whenever all pallets for a product are empty, the space allocated for that fertilizer is refilled completely using a fork truck from secondary storage. Store managers need to decide how many pallet spaces to assign each product to minimize both the number of times per week such refilling will be necessary and the total carrying distance of bags sold. Each fertilizer will have at least one pallet space.

(a) Assuming that fractional numbers of pallet spaces are allowed, briefly explain how this pallet allocation problem can be modeled by the multiobjective nonlinear program

$$\text{min} \quad 20/x_1 + 14/x_2 + 9/x_3 + 5/x_4$$
$$\text{min} \quad 200(x_1/2) + 140(x_1 + x_2/2)$$
$$+ 90(x_1 + x_2 + x_3/2)$$
$$+ 50(x_1 + x_2 + x_3 + x_4/2)$$
$$\text{s.t.} \quad \sum_{j=1}^{4} x_j = 20$$
$$x_j \geq 1, \quad j = 1, \dots, 4$$

(b) ⊘ and (c) ⊘ as in Exercise 8-1.

8-5 Consider the multiobjective LP

$$\text{max} \quad x_1 + 5x_2$$
$$\text{max} \quad x_1$$
$$\text{s.t.} \quad x_1 - 2x_2 \leq 2$$
$$x_1 + 2x_2 \leq 12$$
$$2x_1 + x_2 \leq 9$$
$$x_1, x_2 \geq 0$$

(a) ⊘ Show graphically that the optimal solutions taken separately for each objective do not coincide.

(b) ⊠ Determine graphically whether each of the following is an efficient point: (2,0), (4,7), (3,3), (2,5), (2,2), (0,6).

(c) ⊠ Use graphic solution to sketch the efficient frontier for this model in an objective value graph like Figure 8.4.

8-6 Do Exercise 8-5 for multiobjective LP

$$\begin{aligned}
\min \quad & 5x_1 - x_2 \\
\min \quad & x_1 + 4x_2 \\
\text{s.t.} \quad & -5x_1 + 2x_2 \leq 10 \\
& x_1 + x_2 \geq 3 \\
& x_1 + 2x_2 \geq 4 \\
& x_1, x_2 \geq 0
\end{aligned}$$

and points (4,0), (2,1), (3,3), (1,2), (5,0), (0,0).

8-7 Consider the multiobjective LP

$$\begin{aligned}
\max \quad & 6x_1 + 4x_2 \\
\max \quad & x_2 \\
\text{s.t.} \quad & 3x_1 + 2x_2 \leq 12 \\
& x_1 + 2x_2 \leq 10 \\
& x_1 \leq 3 \\
& x_1, x_2 \geq 0
\end{aligned}$$

(a) ⊠ State and solve graphically a sequence of linear programs to compute a preemptive solution giving priority to the first objective. Also verify that the result is an efficient point.

(b) ⊠ State and solve graphically a sequence of linear programs to compute a preemptive solution giving priority to the second objective. Also verify that the result is an efficient point.

8-8 Do Exercise 8-7 for the multiobjective LP

$$\begin{aligned}
\min \quad & x_1 + x_2 \\
\min \quad & x_1 \\
\text{s.t.} \quad & 2x_1 + x_2 \geq 4 \\
& 2x_1 + 2x_2 \geq 6 \\
& x_1 \leq 4 \\
& x_1, x_2 \geq 0
\end{aligned}$$

8-9 Combine each of the following sets of objective functions into a single weighted-sum objective of the specified maximize or minimize form, weighting individual objectives in the proportions indicated.

(a) ⊠   min $3x_1 + 5x_2 - 2x_3 + 19x_4$
　　　 max $17x_2 - 28x_4$
　　　 min $34x_2 + 34x_3$
Results form min, proportions 5:1:3.

(b)   max $20x_1 - 4x_2 + 10x_4$
　　 min $7x_2 + 9x_3 + 11x_4$
　　 min $23x_1$
Results form max, proportions 2:1:1.

8-10 Return to the multiobjective linear program of Exercise 8-7.

(a) ⊠ Solve graphically with a weighted-sum objective that weights the first objective twice as heavily as the second, and verify that the result is an efficient point.

(b) ⊠ Solve graphically with a weighted-sum objective that weights the second objective twelve times as heavily as the first, and verify that the result is an efficient point.

8-11 Do Exercise 8-10 on the multiobjective linear program of Exercise 8-8.

8-12 Convert each of the following multiobjective optimization models to a goal program seeking the specified target levels, and minimizing the unweighted sum of all goal violations.

(a) ⊠   min $3x_1 + 5x_2 - x_3$
　　 max $11x_2 + 23x_3$
　　 s.t. $8x_1 + 5x_2 + 3x_3 \leq 40$
　　　　　 $x_2 - x_3 \leq 0$
　　　　　 $x_1, x_2, x_3 \geq 0$
Targets 20, 100.

(b)   max $17x_1 - 27x_2$
　　 min $90x_2 + 97x_3$
　　 s.t. $x_1 + x_2 + x_3 = 100$
　　　　　 $40x_1 + 40x_2 - 20x_3 \geq 8$
　　　　　 $x_1, x_2, x_3 \geq 0$
Targets 500, 5000.

(c) ⊠   max $40x_1 + 23x_2$
　　 min $20x_1 - 20x_2$
　　 min $5x_2 + x_3$
　　 s.t. $x_1 + x_2 + 5x_3 \geq 17$
　　　　　 $40x_1 + 4x_2 + 33x_3 \leq 300$
　　　　　 $x_1, x_2, x_3 \geq 0$
Targets 700, 25, 65.

(d)   min $12x_1 + 34x_2 + 7x_3$
　　 max $x_2 - x_3$
　　 max $10x_1 + 7x_3$
　　 s.t. $5x_1 + 5x_2 + 15x_3 \leq 90$
　　　　　 $x_2 \leq 19$
　　　　　 $x_1, x_2, x_3 \geq 0$
Targets 600, 20, 180.

(e) ⬦ min $22x_1 + 8x_2 + 13x_3$
      max $3x_1 + 6x_2 + 4x_3$
      s.t. $5x_1 + 4x_2 + 2x_3 \leq 6$
            $x_1 + x_2 + x_3 \geq 1$
            $x_1, x_2, x_3 = 0$ or 1
    Targets 20, 12.

(f) max $4x_2 + \ln(x_2) + x_3 + \ln(x_3)$
     min $(x_1)^2 + 9(x_2)^2 - x_1 x_2$
     s.t. $x_1 + x_2 + x_3 \leq 10$
          $4x_2 + x_3 \geq 6$
          $x_1, x_2, x_3 \geq 0$
    Targets 20, 40.

8-13 Consider the multiobjective LP

$$\begin{aligned}
\max \quad & x_1 \\
\max \quad & 2x_1 + 2x_2 \\
\text{s.t.} \quad & 2x_1 + x_2 \leq 9 \\
& x_1 \leq 4 \\
& x_2 \leq 7 \\
& x_1, x_2 \geq 0
\end{aligned}$$

(a) Sketch the feasible space and contours where the first objective equals 3 and the second equals 14.

(b) ⬦ Formulate a corresponding goal program seeking target levels 3 and 14 on the two objectives and minimizing the unweighted sum of goal violations.

(c) ⬦ Explain why $\mathbf{x} = (2, 5)$ is optimal in the goal program of part (b) by reference to the plot of part (a).

8-14 Do Exercise 8-13 on the multiobjective LP

$$\begin{aligned}
\min \quad & x_2 \\
\max \quad & 5x_1 + 3x_2 \\
\text{s.t.} \quad & 2x_1 + 3x_2 \geq 6 \\
& x_1 \leq 5 \\
& -x_1 + x_2 \leq 2 \\
& x_1, x_2 \geq 0
\end{aligned}$$

using target levels 1 and 30 and solution $\mathbf{x} = (5, \frac{5}{3})$.

8-15 Return to the multiobjective LP of Exercise 8-13 with target levels 3 and 14 for the two objectives.

(a) ⬦ State and solve graphically [in (x_1, x_2) space] a sequence of linear programs to compute a preemptive goal programming solution, prioritizing objectives in the sequence given and using the given target levels.

(b) ⬦ Formulate a single goal program that implements the preemptive goal programming process of part (a) by suitable deficiency variable

weighting, and demonstrate graphically that it produces the same solution.

(c) ⬦ Determine whether your solution in part (a) is an efficient point of the original multiobjective model. Could the outcome have been different if the target for the first objective were revised to 1? Explain.

8-16 Return to the multiobjective LP of Exercise 8-14. Do Exercise 8-15 on the model using target levels of 1 and 30 and a revised objective 1 target of 3.

8-17 Return to the multiobjective District 88 water allocation problem of Exercise 8-1.

(a) ⬦ 🖳 State and solve with class optimization software a sequence of linear programs to compute a preemptive solution for the model of part (a) taking objectives in the order given.

(b) ⬦ 🖳 State and solve with class optimization software a weighted-sum objective LP that weights the first objective twice as heavily as the second.

(c) ⬦ Suppose that district management sets a goal of at least 100 million acre-feet for irrigation and \$144 million in income. Convert the model of Exercise 8-1(a) to a goal program seeking these target levels and minimizing the unweighted sum of goal violations.

(d) 🖳 Use class optimization software and your goal program of part (c) to show that all goals cannot be fulfilled simultaneously by any feasible solution.

(e) ⬦ 🖳 Solve the model of part (c) with class optimization software and compare to results for earlier methods.

(f) ⬦ 🖳 State and solve with class optimization software a sequence of linear programs to compute a preemptive goal programming solution for the model of part (c) taking objectives in the order given.

(g) ⬦ Formulate an alternative objective function in the goal program of part (c) that accomplishes the preemptive goal programming optimization of part (f) in a single step.

8-18 Do Exercise 8-17 on the silicon wafer problem of Exercise 8-2 using target levels of 30,000 for cost and 13,000 for match.

8-19 ⬦ Do Exercise 8-17 on the base closing problem of Exercise 8-3 using target levels of 3% for readiness loss and 12,000 for displaced workers.

8-20 Do Exercise 8-17 on the pallet allocation problem of Exercise 8-4 using target levels of 10 for refills and 2500 for walking.

8-21 An architect is trying to decide what mix of single, double, and luxury rooms to include in a new hotel. Only $10 million is available for the project, and single rooms cost $40,000 to build, double rooms $60,000, and luxury rooms $120,000 each. Business travelers (in groups of one or more) average 0.7 of single-room rentals, 0.4 of double-room rentals, and 0.9 of luxury room rentals. Family travelers occupy the remainder of each category. The architect would like her design to accommodate 100 total rooms for business travelers, but she would also like it to provide 120 total rooms for family travelers.

(a) ⊘ Formulate a weighted goal LP to decide how many of each room to include in the design. Weight the business traveler goal twice as heavily as the family traveler goal.

(b) ⊘ ⊑ Enter and solve your goal program using class optimization software.

8-22 Trustees of a major state university are trying to decide how much tuition[4] should be charged next year from four categories of full-time students: in-state undergraduates, in-state graduates, out-of-state undergraduates, and out-of-state graduates. The following table shows present tuition levels (thousands of dollars) for each category, together with projected enrollments (thousands) for next year and estimates of the true university cost to educate students.

	Undergrad	Grad
In-state tuition	4	10
Out-of-state tuition	12	15
In-state enrollment	20	1.5
Out-of-state enrollment	10	4
True cost	20	36

New tuitions must raise $292 million, but trustees would also like in-state tuitions to recover at least 25% of true cost and out-of-state, 50%. No tuition will be reduced from its present level, but the trustees would also like to assure that no tuition category increases by more that 10%.

(a) Formulate a goal LP to compute suitable tuitions giving priority to the 10% increase goals over cost recovery.

(b) ⊘ ⊑ Enter and solve your model with class optimization software.

8-23 A university library[5] must cut annual subscription expenses s_j to some or all scientific journals $j = 1, \ldots, 40$ to absorb a $5000 per year budget cut. One consideration will be the sum of published counts c_j of the number of times other journals cite papers in journal j, which is a measure of how seminal a journal is to research. Another is the sum of usefulness ratings r_j ($1 = $ low to $10 = $ high) solicited from university faculty. Finally, the library wants to consider the sum of ratings a_j of the relative availability ($1 = $ low to $8 = $ high) in nearby libraries, believing that journals readily available elsewhere need not be retained.

(a) Formulate a multiobjective ILP to choose journals to drop.

(b) Convert your multiobjective model to a goal program with targets at most C for total citations, at most R for total faculty ratings, and at least A for total availability ratings. Weight all goals equally.

8-24 The household sector is the largest consumer of energy in India,[6] requiring at least 10^8 kilowatthours per day in a large city such as Madras. This energy can be obtained from sources $j = 1$ kerosene, $j = 2$ biogas, $j = 3$ photovoltaic, $j = 4$ fuelwood-generated electricity, $j = 5$ biogas-generated electricity, $j = 6$ diesel-generated electricity, and $j = 7$ electricity from the national power grid, with efficiencies η_j expressing the number of kilowatthours per unit of original fuel input of source j. Biogas is limited 1.3×10^9 units. Energy planners are looking for the best mix of these sources to meet the 10^8 requirement in terms of a variety of conflicting objectives. One is low total cost at p_j per unit source

[4]Based on A. G. Greenwood and L. J. Moore (1987), "An Inter-temporal Multi-goal Linear Programming Model for Optimizing University Tuition and Fee Structures," *Journal of the Operational Research Society, 38*, 599–613.

[5]Based on M. J. Schniederjans and R. Santhanam (1989), "A Zero–One Goal Programming Approach for the Journal Selection and Cancellation Problem," *Computers and Operations Research, 16*, 557–565.

[6]Based on R. Ramanathan and L. S. Ganesh (1995), "Energy Alternatives for Lighting in Households: An Evaluation Using and Integrated Goal Programming-AHP Model," *Energy, 20*, 66–72.

j. Another is to maximize local employment, which is estimated to grow at e_j per unit of energy source *j*. Finally, there are three types of pollution to be minimized: carbon oxides produces at c_j per unit of source *j*, sulfides at s_j per unit of *j*, and nitrogen oxides at n_j per unit of *j*.

(a) Formulate a multiobjective LP to choose a best mix of sources.

(b) Revise your model as a goal program with maximum targets P, C, S, and N for cost, carbon, sulfur, and nitrogen, together with minimum target E for employment.

8-25 The table that follows lists tasks like those required of the U.S. Food and Drug Administration (FDA) in carrying out its responsibilities to regulate drug laboratories, along with an importance weighting of each tasks and the estimated minimum number of staff hours required to accomplish the task fully.

j	Task	Import	Hours
1	Laboratory quality control program	10	1500
2	Analytical protocol	1	100
3	Drug protocol	2	50
4	Quality assurance guidelines	5	55
5	Safety programs	10	135
6	Stability of testing	5	490
7	Calibration of equipment	4	2000
Total			4330

Unfortunately only 3600 of the 4330 total hours are available to execute all tasks. FDA management seeks an allocation of the 3600 hours that addresses the goal hours in the table on the basis of minimizing total importance of undersatisfaction. Formulate a goal LP to determine the most appropriate allocation.

8-26 The Lake Lucky dam[7] retains water in the reservoir lake of the same name to prevent flooding downstream and to assure a steady year-round supply of water for wildlife and nearby cities. The best available projections call for inflows i_t (cubic meters of water) to arrive at the reservoir on upcoming days $t = 1, \ldots, 120$. A minimum of at least $\underline{r}$ must be released through the dam every day to maintain downstream water quality, but the daily release can go as high as $\bar{r}$. The current storage in the reservoir is s_0 cubic meters, and storage levels must be kept between minimum and maximum limits $\underline{s}$ and $\bar{s}$ throughout the 120-day period. Reservoir managers seek a 120-day plan that releases, as close as possible to target, R cubic meters each day and maintains a target level of S cubic meters stored in the reservoir. Both below- and above-target outcomes are considered equally bad, and release deviations are weighted equally with those for storage. Formulate a goal LP to compute a reservoir operating plan. Assume that no water is lost (i.e. all water that enters the reservoir eventually flows over the dam).

8-27 A retail store has hired a marketing consultant to help it decide how to evaluate the quality[8] of its service. The store would like to evaluate customer perceptions about $i = 1$, employee attitude; $i = 2$, employee competence; $i = 3$, product quality; and $i = 4$, sensory (look, sound, smell) appeal of the store. The consultant is considering 6 survey instruments for the task: $j = 1$, point-of-purchase cards measuring employee service delivery; $j = 2$, point-of-purchase cards measuring product quality; $j = 3$, open-ended comment cards; $j = 4$, a focus group of known customers; and $j = 5$, a telephone survey of known customers. The instruments are not equally useful in measuring the desired quality attributes, so the consultant has developed (1 = poor to 10 = excellent) ratings $r_{i,j}$ scoring the value of instrument j in assessing attribute i. Instruments j would cost c_j dollars and require h_j hours of store employee time. Ideally, the instruments chosen should total at least 30 points on each rated attribute, but their total cost must not exceed $10,000, and the total employee hours invested must not be more than 500. Formulate a goal ILP to choose a suitable combination of instruments.

8-28 A steel workpiece is to be cut[9] in a lathe to a target depth of 0.04 inch. Empirical studies with this

[7]Based on K. K. Reznicek, S. P. Simonovic, and C. R. Bector (1991), "Optimization of Short-Term Operations of a Single Multipurpose Reservoir—A Goal Programming Approach," *Canadian Journal of Civil Engineering, 18*, 397–406.

[8]Based on M. J. Schniederjans and C. M. Karuppan (1995), "Designing a Quality Control System in a Service Organization: A Goal Programming Case Study," *European Journal of Operational Research 81*, 249–258.

[9]Based on R. M. Sundaram (1978), "An Application of Goal Programming Technique in Metal Cutting," *International Journal of Production Research, 16*, 375–382.

type of steel express the properties of the resulting product as

$$\text{finish} \triangleq 0.41 v^{3.97} f^{3.46} d^{0.91}$$
$$\text{power} \triangleq v f^{0.75} d^{0.90}$$
$$\text{time} \triangleq v f$$

where v is the cutting speed in revolutions per minute, f is the feed rate in inches per revolution, and d is the depth of cut in inches. Finish variation must be kept no greater than 150, required power no more that 4.0, cutting speed between 285 and 680, and feed rate between 0.0075 and 0.0104. Within these limits, the first priority goal is to make the depth of cut as close as possible to 0.04 inch, and the second is to complete the cut in at most time 1.5 (minutes).

(a) Formulate a preemptive goal NLP to choose cutting parameters.

(b) ◇ ⊒ Enter and (at least locally) solve your goal program using class optimization software.

8-29 Trees in available wild forest[10] stands $j = 1, \ldots, 300$ have been measured in terms of desirable traits $i = 1, \ldots, 12$ such as rate of growth, disease resistance, and wood density. Computations of averages and standard deviations in traits over the stands then produced numbers $z_{i,j}$ of standard deviations that stand j falls above or below the average for trait i. Corresponding scores for the offspring of a planned breeding program may be assumed to combine linearly (i.e. the result will be a sum of scores for the wild stands used, weighted by the proportion each stand makes up of the breeding population).

(a) Formulate a multiobjective LP to choose a breeding population that maximizes the offspring standard deviation score of each trait.

(b) Convert your model to a goal program with a target of breeding offspring 2 standard deviations above the mean on all traits. Weight the undersatisfaction of each goal equally.

8-30 Soar, a retailer catering to upscale suburbanites,[11] wishes to be represented in all the major malls

$j = 1, \ldots, 5$ in the Atlanta region. However, Soar wants to divide its investment budget b in proportion to three major measures of mall attractiveness: weekly patronage, p_j; average patron annual income, a_j; and number of large anchor stores, s_j. That is, they would like ratios of investment in j to p_j all to be equal, and similarly for corresponding ratios to the a_j and s_j. Formulate a goal LP model to choose an appropriate allocation, weighting all deviations equally.

8-31 Shipley Company[12] is designing a new photoresist, which is a chemical coating used in the photoengraving of silicon chips. Important characteristics of photoresists are $i = 1$, high flow (liquefication) temperature; $i = 2$, low minimum line-space resolution; $i = 3$, retention of unexposed areas when exposed circuits are developed away at least $\underline{b}_3$; $i = 4$, photospeed (minimum exposure time to create an image) between $\underline{b}_4$ and $\overline{b}_4$; and $i = 5$, exposure energy requirements between $\underline{b}_5$ and $\overline{b}_5$. Ingredients $j = 1, \ldots, 24$ produce highly nonlinear effects on these five important characteristics. However, designed experiments have demonstrated that the effects can be assumed linear within ingredient quantity ranges $[\ell_j, u_j]$ under consideration. Between these limits, each additional unit of ingredient j adds $a_{i,j}$ to characteristic i of the photoresist.

(a) Formulate a 2-objective LP model to choose the best composition for both flow temperature and line-space resolution.

(b) Explain the meaning of an efficient point solution to your model in part (a), and describe how one could be computed.

8-32 South African wildlife management officials[13] must decide annual quotas (in thousands of tons) for harvesting of pilchards, anchovies, and other fish of the pelagic family in the rich waters to the west of the Cape of Good Hope. One consideration is the maximization of fishing industry income, which is estimated at 110 rand/ton for pilchards, 30 rand/ton for anchovies, and 100 rand/ton for other pelagic. Still,

[10]Based on T. H. Mattheiss and S. B. Land (1984), "A Tree Breeding Strategy Based on Multiple Objective Linear Programming," *Interfaces, 14:5*, 96–104.

[11]Based on R. Khorramshahgol and A. A. Okoruwa (1994), "A Goal Programming Approach to Investment Decisions: A Case Study of Fund Allocation Among Different Shopping Malls," *European Journal of Operational Research, 73*, 17–22.

[12]Based on J. S. Schmidt and L. C. Meile (1989), "Taguchi Designs and Linear Programming Speed New Product Formulation," *Interfaces, 19:5*, 49–56.

[13]Based on T. J. Stewart (1988), "Experience with Prototype Multicriteria Decision Support Systems for Pelagic Fish Quota Determination," *Naval Research Logistics, 35*, 719–731.

the harvest should leave the year-end populations of the 3 types of fish as large as possible. Scientists estimate that initial biomasses of the 3 types are 140 thousand tons, 1,750 thousand tons, and 500 thousand tons, respectively. Due to differences in breeding characteristics, each thousand tons of pilchards harvested will deplete the population by 0.75 thousand tons, each of anchovies by 1.2 thousand tons, and each of other pilchards by 1.5 thousand tons. A final consideration is maintaining the ocean ecosystem. Important measures are the number of breeding pairs of penguins (thousands), which is estimated as $70 + 0.60$ (ending biomass of pilchards), and the number of such pairs of gannets (thousands), which is estimated at $5+0.20$ (ending biomass of pilchards) $+ 0.20$ (ending biomass of anchovies).

(a) Formulate a multiobjective LP to maximize all measures of success using the decision variables ($i = 1$, pilchards; $i = 2$, anchovies; $i = 3$, other pelagic)

$$x_i \overset{\Delta}{=} \text{quota for fish type } i \text{ (thousand tons)}$$

$$y_i \overset{\Delta}{=} \text{ending biomass of fish type } i$$
$$\text{(thousand tons)}$$

(b) Revise your formulation as a goal program with minimum targets for the 6 objectives being 38 million rands income, ending biomasses of 100, 1150 and 250 thousand tons, and ending breeding populations of 130 and 25 thousand pairs. Give preemptive priority to maintaining the populations of the 3 types of fish.

(c) ⊗ ▢ Enter and solve your goal program version using class optimization software.

8-33 The state of Calizona will soon be opening centers[14] in 12 of its county seats $i = 1, \ldots, 100$ to promote solid waste recycling programs. Each selected site will service a district of surrounding counties. The state wants to minimize the sum of distances $d_{i,j}$ populations p_i in counties i must travel to their district center at j. But they would also like to come as close as possible (in total deviation) to having 3 centers in each of the 4 Environment Department regions. Counties $i = 1, \ldots, 12$ make up region 1, $i = 13, \ldots, 47$ region 2, $i = 48, \ldots, 89$ region 3, and $i = 90, \ldots, 100$ region 4.

(a) Formulate a 2-objective ILP model of this facility location problem using the main decision variables ($i, j = 1, \ldots, 100$)

$$x_{i,j} \overset{\Delta}{=} \begin{cases} 1 & \text{if county } i \text{ is served by a center} \\ & \text{at } j \\ 0 & \text{otherwise} \end{cases}$$

$$x_j \overset{\Delta}{=} \begin{cases} 1 & \text{if county } j \text{ gets a center} \\ 0 & \text{otherwise} \end{cases}$$

(b) Explain the meaning of an efficient point solution to your model of part (a) and describe how one could be computed.

8-34 Every year the U.S. Navy must plan the reassignment of thousands of sailors finishing one tour of duty in specialties $i = 1, \ldots, 300$ and preparing for their next one, k. The reassignment may also require a move of the sailor and his or her dependents from present base location $j = 1, \ldots, 25$ to a new base ℓ. To plan the move, personnel staff estimate numbers $s_{i,j}$ of sailors at base j in specialty i who are ready for reassignment, $d_{k,\ell}$ of sailors needed at base ℓ in specialty k, as well as average costs $c_{j,\ell}$ of moving a sailor and dependents from base j to base ℓ. Training is required for the new position if it involves a different specialty than the present one, and capacity limits u_k at training schools limit the number that can be trained for any specialty k during the year. One Navy objective in planning reassignment is to keep total relocation costs as low as possible. However, other considerations arise from the fact that full staffing levels $d_{k,\ell}$ can almost never all be met. Naturally, the Atlantic fleet (bases $j, \ell = 1, \ldots, 15$) wishes to maximize the fraction of its slots actually filled, and the Pacific fleet (bases $j, \ell = 16, \ldots, 25$) prefers to maximize the fraction of its needs accommodated. Formulate a 3-objective LP model to aid the Navy in planning reassignments using the decision variables

$$x_{i,j,k,\ell} \overset{\Delta}{=} \text{number of sailors presently in specialty } i \text{ at}$$
$$\text{base } j \text{ reassigned to specialty } k \text{ at location } \ell$$

8-35 The sales manager of an office systems distributor[15] has sales representatives $i = 1, \ldots, 18$ to assign to both current accounts $j = 1, \ldots, 250$ and

---

[14]Based on R. A. Gerrard and R. L. Church (1994), "Analyzing Tradeoffs between Zonal Constraints and Accessibility in Facility Location," *Computers and Operations Research, 21*, 79–99.

[15]Based on A. Stam, E. A. Joachimsthaler, and L. A. Gardiner (1992), "Interactive Multiple Objective Decision Support for Sales Force Sizing and Deployment," *Decision Sciences, 23*, 445–466.

the development of new accounts. Each current account will be assigned at most one representative. From prior experience, the manager can estimate monotone-increasing nonlinear functions

$r_{i,j}(w_{i,j}) \triangleq$ current-period sales revenue that will result from allocating $w_{i,j}$ hours of representative i time to account j

$a_i(x_i) \triangleq$ present value of future sales revenue that will derive from assigning representative i to x_i hours of new account development

Each representative has 200 hours to divide between his or her assignments during the planning period, but the manager would like to maximize both total current revenue and total new account value.

(a) Formulate a 2-objective INLP to determine an allocation plan using the decision variables ($i = 1, \ldots, 18, j = 1, \ldots, 250$)

$w_{i,j} \triangleq$ hours spent by representative i on account j

$x_i \triangleq$ hours spent by representative i on new account development

$$y_{i,j} \triangleq \begin{cases} 1 & \text{if representative } i \text{ is assigned to} \\ & \text{account } j \\ 0 & \text{otherwise} \end{cases}$$

(b) Explain the meaning of an efficient point solution to your model of part (a) and describe how one could be computed.

SUGGESTED READING

Goicoechea, Ambrose, Don R. Hansen, and Lucien Duckstein (1982), *Multiobjective Decision Analysis* *with Engineering and Business Applications*, Wiley, New York.

Shortest Paths and Discrete Dynamic Programming

From our earliest discussions in Chapters 1 and 2, we have seen a tradeoff between generality of operations research models and tractability of their analysis. The more specialized the model form and the longer the list of underlying assumptions, the richer and more efficient the analysis that is possible.

In this chapter we extend that insight to a new extreme by introducing the most specialized and thus most efficiently solved of all broad classes of optimization models: **shortest paths and discrete dynamic programs**. Dynamic programming methods proceed by viewing the problem or problems we really want to solve as members of a family of closely related optimizations. When model forms are specialized enough to admit just the right sort of family, strong connections among optimal solutions to the various members can be exploited to organize a very efficient search.

9.1 SHORTEST PATH MODELS

Whether in urban traffic, or college hallways, or satellite communications, or the surface of a microchip, it makes sense to follow the shortest route. Such **shortest path** problems are the first we will attack.

EXAMPLE 9.1: LITTLEVILLE

As usual, it will help to begin with an example setting. Suppose that you are the city traffic engineer for the town of Littleville. Figure 9.1(a) depicts the arrangement of one- and two-way streets in a proposed improvement plan for Littleville's downtown, including the estimated average time in seconds that a car will require to transit each block.

From survey and other data we can estimate how many driver trips originate at various origin points in the town, and the destination for which each trip is bound. But such survey data cannot indicate what streets will be selected by motorists to move from origin to destination in a street network that does not yet exist.

One of the tasks of a traffic engineer is to project the route that drivers will elect, so that city leaders can evaluate whether flows will concentrate where they hope. A

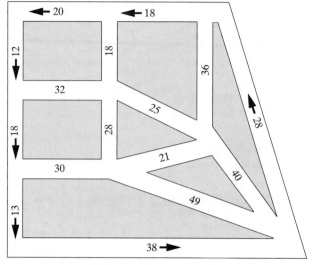

(a) *Proposed street network*

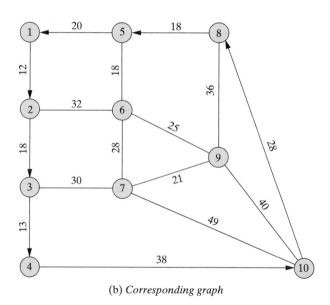

(b) *Corresponding graph*

FIGURE 9.1   Littleville Shortest Path Example

good starting point is to assume that drivers will do the most rational thing—follow the shortest time path from their origin to their destination. We need to compute all such shortest paths.

Nodes, Arcs, Edges, and Graphs

Figure 9.1(b) shows the first step. We abstract the given flow pattern—here a street system—into a graph or network.

These are not the newspaper sort of graphs comparing bars, or zigzag lines, or piles of little coins.

> **9.1** Mathematical **graphs** model travel, flow, and adjacency patterns in a **network**.

Such graphs begin with a collection of nodes (or vertices).

> **9.2** The **nodes** or **vertices** of a graph represent entities, intersections, and transfer points of the network.

Pairs of nodes may be linked in a graph by either **arcs** or **edges**.

> **9.3** The **arcs** of a graph model available directed (one-way) links between nodes. **Edges** represent undirected (two-way) links.

In Figure 9.1(b) nodes represent intersections. For convenience we have numbered the nodes 1 to 10. One-way streets between intersections of the Littleville network yield arcs in our graph; two-way streets produce edges.

We denote arcs or edges by naming their terminal nodes. Thus $(5, 1)$ and $(3, 4)$ are arcs in Figure 9.1(b), while $(7, 9)$ and $(5, 6)$ are edges. Order matters for arcs, so that it would be incorrect, for example, to call the arc from node 10 to node 8 arc $(8, 10)$. It must be $(10, 8)$. There is more flexibility in referencing edges. An edge between nodes i and j could be called either (i, j), or (j, i), or both, although it is more common to name the smallest node number first. For example, we could refer to the edge between nodes 2 and 6 of Figure 9.1(b) as either $(2, 6)$ or $(6, 2)$, but $(2, 6)$ would be more standard.

SAMPLE EXERCISE 9.1: IDENTIFYING ELEMENTS OF A GRAPH

Consider the following graph:

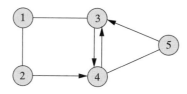

Identify its nodes, arcs, and edges.

Modeling: The node set of this graph is $V \triangleq \{1, 2, 3, 4, 5\}$. Arcs form $A \triangleq \{(2, 4), (3, 4),$ $(4, 3), (5, 3)\}$. Edges are $E \triangleq \{(1, 2), (1, 3), (4, 5)\}$.

Paths

Our interest in this chapter is paths.

9.4 | A **path** is a sequence of arcs or edges connecting two specified nodes in a graph. Each arc or edge must have exactly one node in common with its predecessor in the sequence, any arcs must be passed in the forward direction, and no node may be visited more than once.

Two of several paths from node 3 to node 8 in the Littleville example are illustrated in Figure 9.2(a). One proceeds 3–7–10–8, using edges (3, 7) and (7, 10), plus arc (10, 8). Another is 3–4–10–8, following arc sequence (3, 4), (4, 10), (10, 8). The pattern 3–7–6–5–8 of Figure 9.2(b) is not a path because it transits arc (8, 5) in the wrong direction. Sequence 3–7–6–9–7–10–8 also fails the definition of a path; it repeats node 7.

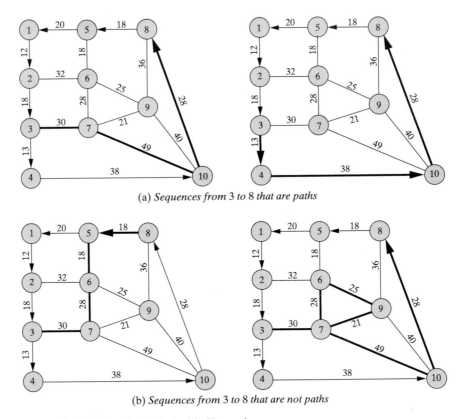

(a) *Sequences from 3 to 8 that are paths*

(b) *Sequences from 3 to 8 that are not paths*

FIGURE 9.2   Paths of the Littleville Example

SAMPLE EXERCISE 9.2: RECOGNIZING PATHS

Return to the graph of Sample Exercise 9.1. Identify all paths from node 1 to node 5.

Modeling: The two available paths are 1–2–4–5 and 1–3–4–5. Sequence 1–3–5 is not a path because it violates direction on arc (3, 5)

Shortest Path Problems

When an application includes costs or lengths on the arcs and edges of a graph, we are confronted with an optimization.

> 9.5 | **Shortest path problems** seek minimum total length paths between specified pairs of nodes in a graph.

In the Littleville example of Figure 9.1, lengths are travel times in seconds. The first path from node 3 to node 8 shown in Figure 9.2(a) totals $(30 + 49 + 28) = 107$ seconds in length. The shortest path from 3 to 8 is the second 3–4–10–8 sequence with length $(13 + 38 + 28) = 79$ seconds.

Classification of Shortest Path Models

Shortest path problems arise both as main decision questions and as steps in other computations. There are many variations, depending on the type of network and costs involved, and the pairs of nodes for which we need solutions.

We will see in the sections to come that shortest path algorithms are highly specialized to exploit particular properties. Thus it is important to distinguish cases.

Our Littleville example illustrates one combination. The graph has both arcs and edges, all link lengths are nonnegative, and our traffic engineering task requires shortest paths between all pairs of nodes. To summarize:

- **Name:**   Littleville
- **Graph:**   arcs and edges
- **Costs:**   nonnegative
- **Output:**   shortest paths
- **Pairs:**   all nodes to all others

Before passing to algorithms, we introduce some examples illustrating other possibilities.

EXAMPLE 9.2: TEXAS TRANSFER

Figure 9.3 displays a map of highway links between several major cities in Texas. Numbers on the edges in Figure 9.3 show standard driving distance in miles.

Texas Transfer, a major trucker in the southwest, ships goods from its hub warehouse in Ft. Worth to all the cities shown. Trucks leave the hub and proceed directly to their destination city, with no intermediate dropoffs or pickups.

Texas Transfer drivers are allowed to choose their own route from Ft. Worth to their destination. However, management's proposal in current labor negotiations calls for drivers to be paid on the basis of shortest standard mileage to the location. That is, they will be paid according to the length of the shortest path from Ft. Worth to their destination city in the network of Figure 9.3. To see the impact of this proposal, we need to compute such shortest path distances for all cities.

Notice that the character of this shortest path task differs significantly from the Littleville example. Here we need only optimal path lengths, not the paths themselves. Also, we need shortest path lengths for only one origin or **source**—the Ft. Worth hub.

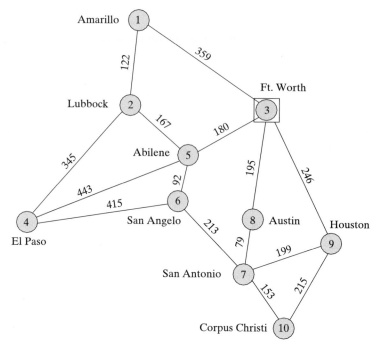

FIGURE 9.3   Network for Texas Transfer Example

- **Name:**   Texas Transfer
- **Graph:**   edges only
- **Costs:**   nonnegative
- **Output:**   shortest path lengths
- **Pairs:**   one node to all others

EXAMPLE 9.3: TWO RING CIRCUS

The Two Ring Circus is nearing the end of its season and planning a return to winter headquarters near Tallahassee, Florida. Present commitments will end in Lincoln, Nebraska, but there are still some opportunities for bookings in cities along the route home.

Figure 9.4(a) shows the travel routes available and the estimated costs (in thousands of dollars) of moving the circus over those routes. It also designates the cities where bookings have been offered and the anticipated net receipts (in thousands of dollars).

We want to compute the optimal path home for Two Ring. Notice that the cost of a path is the difference of travel costs and net receipts from performances along the way. But receipts occur at nodes of the network, not on edges as shortest path models require.

Undirected and Directed Graphs (Digraphs)

The Two Ring network of Figure 9.4(a) is an **undirected graph** because it has only edges (undirected links). The key to incorporating receipts at nodes is first to convert

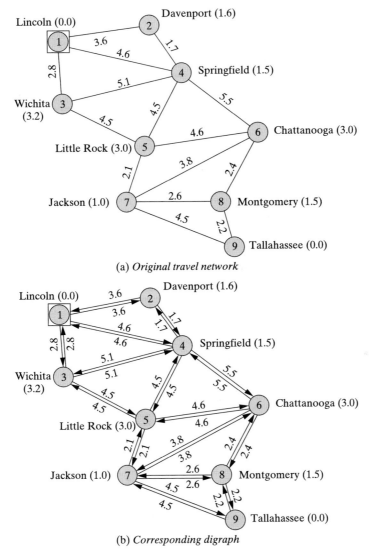

(a) *Original travel network*

(b) *Corresponding digraph*

FIGURE 9.4 Two Ring Circus Example Graphs

to an equivalent **directed graph** or **digraph**, that is, a graph having only arcs (directed links).

Such a **directing** of a graph is easy:

9.6 A shortest path problem including edges (i, j) of cost $c_{i,j}$ can be converted to an equivalent one on a digraph by replacing each edge with a pair of opposed arcs as

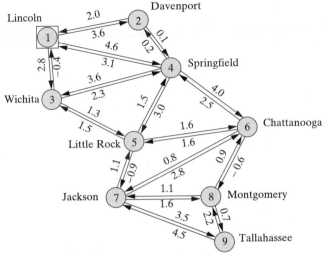

(c) *After incorporating node values*

FIGURE 9.4   Two Ring Circus Example Graphs (Continued)

The two parallel arcs replacing each edge merely make explicit the two directional options for using the edge in a path. Whether directed or not, a path will use the link no more than once because paths cannot repeat nodes.

SAMPLE EXERCISE 9.3: DIRECTING A GRAPH

Consider the following graph:

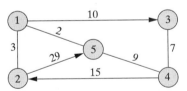

Show the equivalent digraph.

Modeling: Applying 9.6 , we replace all edges of the given graph with a pair of opposed arcs with the same length as the edge. The result is

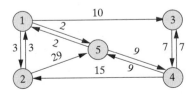

Notice that arcs are unchanged and that the two arcs replacing each edge both have the same length as the edge.

Two Ring Example Model

Figure 9.4(b) shows the result of directing the Two Ring network. For example, Wichita-to-Little Rock edge $(3, 5)$ becomes arcs $(3, 5)$ and $(5, 3)$, both with the same cost as the original edge.

Net receipt values remain on nodes in part (b) of the figure. Part (c) moves them to arcs by subtracting net receipts in each city from the cost of all outbound arcs there. For example, Little Rock-to-Jackson arc $(5, 7)$ now has cost

$$2.1 - 3.0 = -0.9$$

to account for net receipts if Two Ring plays Little Rock. Reverse arc $(7, 5)$ from Jackson to Little Rock has cost

$$2.1 - 1.0 = 1.1$$

to include Jackson receipts. If either of these arcs is used in an optimal path, the full effect of both travel costs and show revenues will be reflected.

It may seem wrong to subtract receipts from all outbound arcs from a city when show receipts can be realized only once. Remember, however, that a path can visit any node only once—entering on an inbound arc and leaving on an outbound. It follows that receipts for a town can be part of any path's length at most once.

The Two Ring digraph of Figure 9.4(c) also introduces another class of shortest path models. This time arc costs, being net dollar amounts, have unpredictable sign. Also, we require a shortest path for only one pair of nodes: Lincoln to Tallahassee.

- **Name:**   Two Ring Circus
- **Graph:**   directed
- **Costs:**   arbitrary
- **Output:**   shortest path
- **Pairs:**   one source to one destination

9.2 DYNAMIC PROGRAMMING APPROACH TO SHORTEST PATHS

Dynamic programming methods exploit the fact that it is sometimes easiest to solve one optimization problem by taking on an entire family. If strong enough relationships can be found among optimal solutions to the various members of the family, attacking all together can be the most efficient way of solving the one(s) we really care about.

Families of Shortest Path Models

We will see later in the chapter how we sometimes have to invent a family of problems to solve a particular one by dynamic programming. With shortest path models the appropriate family is already at hand. We need only exploit the fact that most applications require optimal paths or path lengths for more than one pair of nodes.

> **9.7**   Shortest path algorithms exploit relationships among optimal paths (and path lengths) for different pairs on nodes.

The Littleville and Texas Transfer examples of Section 9.1 already seek paths for multiple pairs of nodes. Littleville requires an optimal path between every pair of nodes, and Texas Transfer needs optimal path lengths from one node to every other.

Even if only one optimal path is required, our dynamic programming approach to shortest path problems embeds the case of application interest in such a family of problems for different origin–destination pairs. In particular, we will solve the Two Ring circus example, which needs an optimal path for only one origin–destination pair, by embedding it in the problem of finding shortest paths from the origin to all other nodes.

Functional Notation

With many optimizations going on at once, we will need some notation to keep score. Specifically, we denote optimal solutions and solution values as functions of the family member. Square brackets [...] enclose the parameters in such **functional notation**.

The Texas Transfer and Two Ring examples require optima for one source to all other nodes. Corresponding functional notation is

$$v[k] \triangleq \text{length of a shortest path from the source node to node } k$$
$$(= +\infty \text{ if no path exists})$$

$$x_{i,j}[k] \triangleq \begin{cases} 1 & \text{if arc/edge } (i, j) \text{ is part of the optimal} \\ & \text{path from the source node to node } k \\ 0 & \text{otherwise} \end{cases}$$

Notice that we adopt the convention that $v[k] = +\infty$ when there is no path to k.

Examples such as Littleville, which need optimal paths from all nodes to all others, have two parameters:

$$v[k, \ell] \triangleq \text{length of a shortest path from node } k \text{ to node } \ell$$
$$(= +\infty \text{ if no path exists})$$

$$x_{i,j}[k, \ell] \triangleq \begin{cases} 1 & \text{if arc/edge } (i, j) \text{ is part of the optimal path from node } k \text{ to node } \ell \\ 0 & \text{otherwise} \end{cases}$$

Again $v[k, \ell] = +\infty$ indicates that no path exists from k to ℓ.

SAMPLE EXERCISE 9.4: UNDERSTANDING FUNCTIONAL NOTATION

Consider the problem of finding the shortest path from source node 1 in the following graph to all other nodes.

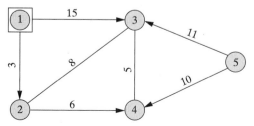

(a) Use inspection to determine all required shortest paths.

(b) Detail optimal solutions and solution values in functional notation.

Modeling:

(a) It is easy to check that optimal paths to nodes 2, 3, and 4 are 1–2, 1–2–3, and 1–2–4, respectively. There is no path from source 1 to node 5.

(b) In functional notation these optimal paths imply that

$$v[1] = 0 \qquad x_{1,2}[1] = x_{1,3}[1] = x_{2,3}[1] = x_{2,4}[1] = x_{3,4}[1] = 0$$
$$v[2] = 3 \qquad x_{1,2}[2] = 1, x_{1,3}[2] = x_{2,3}[2] = x_{2,4}[2] = x_{3,4}[2] = 0$$
$$v[3] = 11 \qquad x_{1,2}[3] = x_{2,3}[3] = 1, x_{1,3}[3] = x_{2,4}[3] = x_{3,4}[3] = 0$$
$$v[4] = 9 \qquad x_{1,2}[4] = x_{2,4}[4] = 1, x_{1,3}[4] = x_{2,3}[4] = x_{3,4}[4] = 0$$
$$v[5] = +\infty \qquad \text{(no path)}$$

Optimal Paths and Subpaths

To formulate a dynamic programming approach to our shortest path models, we must identify connections among optimal solutions to problems for different pairs of nodes. How is the optimal path for one pair of nodes related to the optimal path for another?

To begin to see, examine Figure 9.5. The highlighted path through Austin and San Antonio is the shortest from the origin at Ft. Worth to Corpus Christi. Think now about San Antonio. The figure's Ft. Worth–Austin–San Antonio path is one way to get to San Antonio, but could any other path be shorter? Certainly not. If there were a better way to get from Ft. Worth to San Antonio than the path indicated in

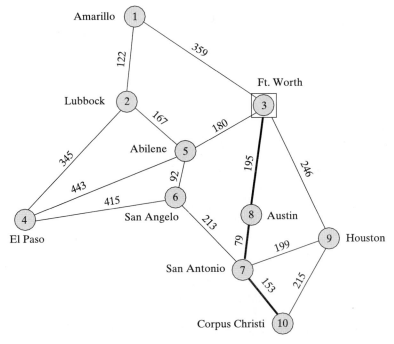

FIGURE 9.5   Optimal Path of the Texas Transfer Example

Figure 9.5, that better route would also yield an improvement on the Corpus Christi optimum. We need only follow it to San Antonio and finish with the link from San Antonio to Corpus Christi.

Negative Dicycles Exception

It seems safe to conclude from examples like Figure 9.5 that optimal paths must always have optimal subpaths. Unfortunately, there is a glitch.

Consider the example in Figure 9.6. The shortest path from source $s = 1$ to node 3 clearly proceeds 1–2–3 with $v[3] = 5$. Still, the subpath 1–2 is not optimal. Sequence 1–3–4–2 has lesser length $v[2] = -3$.

The logic of the preceding subsection suggests that there must be a contradiction. Simply extending the optimal path to node 2 with arc $(2, 3)$ should improve the optimum at node 3. But that extension produces the sequence 1–3–4–2–3, which is not a path because it repeats node 3. Eliminating the repetition gives path 1–3, but its length 10 is worse than 1–2–3.

The troublemaker is a negative dicycle.

> **9.8** | A **dicycle** is a path that begins and ends at the same node, and a **negative dicycle** is a dicycle of negative total length.

Figure 9.6 contains negative dicycle 3–4–2–3 with length

$$12 + (-25) + 3 = -10$$

Its presence permits nonoptimal subpath 1–2 because extending shortest 1 to 2 path 1–3–4–2 with arc $(2, 3)$ closes a dicycle that cannot be eliminated without increasing the length of the route it offers to node 3.

Negative dicycles do not occur often in applied shortest path models because they imply a sort of economic perpetual motion. Each time we transit a negative dicycle we end up closer (in total length) to any source that we were when we started. However, when negative dicycles are present, shortest path models become so dramatically more difficult that our whole solution approach must change.

> **9.9** | Shortest path models with negative dicycles are much less tractable than other cases because dynamic programming methods usually do not apply.

> **SAMPLE EXERCISE 9.5: IDENTIFYING NEGATIVE DICYCLES**
>
> Identify all dicycles and negative dicycles of the following graph:

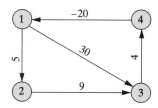

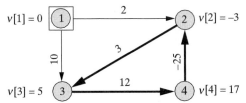

FIGURE 9.6   Example with a Negative Dicycle

Modeling: The dicycles of this example are 1–2–3–4–1 of length -2, and 1–3–4–1 of length 14. Only the first is a negative dicycle. Sequence 1–2–3–1 is not a dicycle because it violates direction.

Principle of Optimality

Fortunately, negative dicycles are the only bad case in shortest path analysis. Just as with the instance of Figure 9.6, a subpath of a shortest path can fail to be optimal only if a negative dicycle is present. Other forms satisfy the simple insight known as the **principle of optimality** for shortest path models:

> **9.10** | In a graph with no negative dicycles, optimal paths must have optimal subpaths.

Functional Equations

To see how principle | 9.10 | helps us to compute optimal paths, let us limit our attention to one node to all other circumstances of models such as that of Texas Transfer and the Two Ring Circus. If optimal paths must have optimal subpaths, it follows that a shortest path can be constructed by extending known shortest subpaths.
    Functional equations detail such recursive relationships.

> **9.11** | **Functional equations** of a dynamic program encode the recursive connections among optimal solution values that are implied by a principle of optimality.

Functional Equations for One Node to All Others

In one node to all other shortest path cases, functional equations relate shortest path lengths $v[k]$.

> **9.12** | The functional equations for shortest path problems from a single source s in a graph with no negative dicycles are
>
> $$v[s] = 0$$
> $$v[k] = \min\{v[i] + c_{i,k} : (i, k) \text{ exists}\} \quad \text{all } k \neq s$$

The minimization in the second expression is taken over all neighbors i leading to k in the graph. If none exist, we adopt the convention that $\min\{\text{nothing}\} \triangleq +\infty$.

The first part of the functional equations in 9.12 simply says that the length of the shortest path from s to itself should $= 0$. The rest of the functional equations fix the recursive relationships implied by principle 9.10. In words, they say that the shortest path length for node j must be the best single arc/edge extension of optimal paths to neighboring nodes i.

For an example, return again to San Antonio node $j = 7$ in Texas Transfer Figure 9.5. The corresponding equation 9.12 is

$$v[7] = \min\{v[6] + c_{6,7}, v[8] + c_{8,7}, v[9] + c_{9,7}, v[10] + c_{10,7}\}$$
$$= \min\{v[6] + 213, v[8] + 79, v[9] + 199, v[10] + 153\}$$

That is, the length of the shortest path to San Antonio node 7 must be the least of 1-edge extensions to optimal paths for neighboring nodes $i = 6$ (San Angelo), 8 (Austin), 9 (Houston), and 10 (Corpus Christi). One of those must constitute an optimal subpath of the shortest path from Ft. Worth to San Antonio.

SAMPLE EXERCISE 9.6: UNDERSTANDING FUNCTIONAL EQUATIONS

Return to Sample Exercise 9.4. Write all corresponding functional equations and verify that they are satisfied by shortest path lengths computed in Sample Exercise 9.4(b).

Analysis: Following 9.12, functional equations are

$$v[1] = 0$$
$$v[2] = \min\{v[1] + c_{1,2}, v[3] + c_{3,2}\}$$
$$v[3] = \min\{v[1] + c_{1,3}, v[2] + c_{2,3}, v[4] + c_{4,3}\}$$
$$v[4] = \min\{v[2] + c_{2,4}, v[3] + c_{3,4}\}$$
$$v[5] = \min\{\}$$

Now substituting optimal values yields

$$v[1] = 0$$
$$v[2] = \min\{0 + 3, 11 + 8\} = 3$$
$$v[3] = \min\{0 + 15, 3 + 8, 9 + 5\} = 11$$
$$v[4] = \min\{3 + 6, 11 + 5\} = 9$$
$$v[5] = \min\{\} = +\infty$$

Sufficiency of Functional Equations in the One to All Case

In the absence of negative dicycles, shortest path lengths satisfy functional equations 9.12 because they satisfy principle of optimality 9.10. Optimal paths must extend optimal subpaths.

The algorithms that make up most of this chapter depend on the fact that the result works in both directions.

> **9.13** | Node values $v[k]$ in a graph with no negative dicycles are lengths of shortest paths from a given source s if and only if they satisfy functional equations **9.12**.

That is, we can compute shortest path lengths $v[k]$ simply by computing values that satisfy the functional equations.

To see why, examine the $v[k]$ values of Texas Transport Figure 9.7. Certainly, the value a Ft. Worth source node $s = 3$ is correct. The length of a shortest path from Ft. Worth to Ft. Worth must $= 0$.

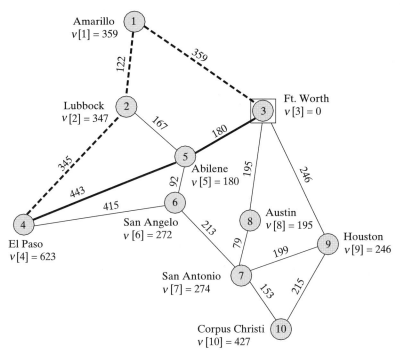

FIGURE 9.7   Sufficiency of Functional Equations

For other nodes, say El Paso node 4, we make two observations. First, functional equations **9.12** assure that $v[4]$ is the length of some path from s to node 4. This is true because there must be at least one neighbor i of node 4 that has

$$v[4] = v[i] + c_{i,4}$$

Here $i = 5$ (Abilene) works because it achieves the minimum in **9.12**. But Abilene node 5 must, in turn, have a neighbor j with

$$v[5] = v[j] + c_{j,5}$$

Here it is $j = 3$. We continue in this way until source node s is reached (as it just was). Then summing the relationships, simplifying, and using $v[s] = 0$ produces

$$v[4] + v[5] = v[5] + v[s] + c_{s,5} + c_{5,4}$$
$$v[4] = v[s] + c_{s,5} + c_{5,4}$$
$$v[4] = c_{s,5} + c_{5,4}$$

That is, $v[4]$ is the length of path $s - 5 - 4$.

The other half of the argument is to show that no other path can have shorter length than the one associated with $v[4]$. Consider the s–1–2–4 path highlighted with dashed lines in Figure 9.7. Because all values satisfy functional equations

$$v[1] \leq v[s] + c_{s,1}$$
$$v[2] \leq v[1] + c_{1,2}$$
$$v[4] \leq v[2] + c_{2,4}$$

Again summing, simplifying, and using $v[s] = 0$, we obtain

$$v[1] + v[2] + v[4] \leq v[s] + v[1] + v[2] + c_{s,1} + c_{1,2} + c_{2,4}$$
$$v[4] \leq v[s] + c_{s,1} + c_{1,2} + c_{2,4}$$
$$v[4] \leq c_{s,1} + c_{1,2} + c_{2,4}$$

The path indicated cannot have length less than $v[4]$.

SAMPLE EXERCISE 9.7: VERIFYING SUFFICIENCY OF FUNCTIONAL EQUATIONS

Consider the following graph and associated node values $v[k]$:

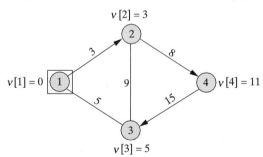

(a) Verify that given values $v[k]$ satisfy functional equations 9.12 .

(b) Identify the path associated with the $v[k]$ of each node k.

(c) Use functional equations to show why the length of path 1–3–2–4 must be $\geq v[4]$.

Analysis:

(a) $\quad v[1] = 0$
$$v[2] = \min\{v[1] + c_{1,2}, v[3] + c_{3,2}\} = \min\{0 + 3, 5 + 9\} = 3$$
$$v[3] = \min\{v[1] + c_{1,3}, v[2] + c_{2,3}\} = \min\{0 + 5, 3 + 9\} = 5$$
$$v[4] = \min\{v[2] + c_{2,4}, v[3] + c_{3,4}\} = \min\{3 + 8, 5 + 15\} = 11$$

(b) Label $v[s] = 0$ is the length of a null path from $s = 1$ to itself. For $k = 2$, we look for the neighbor achieving the minimum in the functional equation of $v[2]$. Here it is $s = 1$, so the path is 1–2 with length $v[2] = 3$. For $k = 3$, similar thinking yields path 1–3 with length $v[3] = 5$. Node $k = 4$ takes two steps. At node 4 the neighbor achieving the minimum is $i = 2$. Its minimum, in turn, occurred for $i' = 1$. Thus the path is 1–2–4 with length $v[4] = 11$.

(c) Functional equations for nodes 4, 2, and 3 imply that

$$v[4] \leq v[2] + c_{2,4}$$
$$v[2] \leq v[3] + c_{3,2}$$
$$v[3] \leq v[1] + c_{3,1}$$

Summing, simplifying, and using $v[1] = 0$, we have

$$v[4] + v[2] + v[3] \leq v[1] + v[2] +_v [3] + c_{1,3} + c_{3,2} + c_{2,4}$$
$$v[4] \leq v[1] + c_{1,3} + c_{3,2} + c_{2,4}$$
$$v[4] \leq c_{1,3} + c_{3,2} + c_{2,4}$$

Functional Equations for All Nodes to All Others

For shortest path problems requiring optimal paths between all pairs of nodes (Littleville is an example), almost everything said so far generalizes immediately. Only the functional equations change.

> **9.14**   The functional equations for shortest path problems from all nodes to all other nodes in a graph with no negative dicycles are
>
> $$v[k, k] = 0 \text{ all } k$$
> $$v[k, \ell] = \min \left\{ c_{k,l}, \{v[k, i] + v[i, \ell] : i \neq k, \ell\} \right\} \qquad \text{all } k \neq \ell$$

As with single-source cases, functional equations **9.14** merely detail the principle of optimality (**9.10**) insight that optimal paths must have optimal subpaths. Thus the shortest path from k to ℓ must consist either of arc/edge (k, ℓ), or an optimal path from k to some intermediate node i, plus an optimal path from i to ℓ. Sufficiency for computation follows exactly as it did for earlier model forms.

> **9.15**   Node pair values $v[k, \ell]$ in a graph with no negative dicycles are lengths of shortest paths from k to ℓ if and only if they satisfy functional equations **9.14** .

SAMPLE EXERCISE 9.8: VERIFYING FUNCTIONAL EQUATIONS FOR ALL PAIRS CASES

Consider the problem of finding shortest path between all pairs of nodes in the following graph.

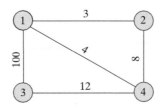

(a) Use inspection to identify optimal paths for all pairs.

(b) Write the functional equations for $(i, j) = (1, 4)$ and $(2, 3)$.

(c) Verify that your optimal path lengths of part (a) satisfy these equations.

Analysis:

(a) Optimal paths are detailed in the following table:

	$j = 1$		$j = 2$		$j = 3$		$j = 4$	
i	v	Path	v	Path	v	Path	v	Path
1	0	—	3	1–2	16	1–4–3	4	1–4
2	3	2–1	0	—	19	2–1–4–3	7	2–1–4
3	16	3–4–1	19	3–4–1–2	0	—	12	3–4
4	4	1–4	7	4–1–2	12	4–3	0	—

(b) Following 9.14 , the functional equations for $(i, j) = (1, 4)$ and $(2, 3)$ are

$$v[1, 4] = \min\{c_{1,4}, v[1, 2] + v[2, 4], v[1, 3] + v[3, 4]\}$$
$$v[2, 3] = \min\{c_{2,3}, v[2, 1] + v[1, 3], v[2, 4] + v[4, 3]\}$$

(c) Substitution of optimal values from the table in functional equations gives

$$v[1, 4] = \min\{4, 3 + 7, 16 + 12\} = 4$$
$$v[2, 3] = \min\{+\infty, 3 + 16, 7 + 12\} = 19$$

Solving Shortest Path Problems by Linear Programming

All the shortest path problems of this chapter can be formulated and solved by linear programming as long as the given graphs contain no negative dicycles. Details are provided in Section 10.6. Still, it is important to understand that the dynamic programming methods of the remainder of this chapter are far more efficient.

9.16 | Although shortest path problems over graphs with no negative dicycles can be solved by linear programming, dynamic programming methods based on the principle of optimality are far more efficient.

9.3 SHORTEST PATHS FROM ONE NODE TO ALL OTHERS: BELLMAN–FORD

We have seen that the task of computing shortest path lengths from single source s to all other nodes reduces to a search for quantities $v[k]$ satisfying functional equations $\boxed{9.12}$. Principle $\boxed{9.13}$ guarantees such $v[k]$ will provide shortest path lengths whenever the given graph contains no negative dicycles.

Solving the Functional Equations

The central issue in designing shortest path algorithms is how to compute functional equation solutions efficiently. Notice that the task is not as straightforward as, say, solving a system of linear equations. The min{...} operator of our functional equations makes them nonlinear.

The format of equations $\boxed{9.12}$ suggests another way. Each equation shows an expression for a single value $v[k]$. Why not just evaluate those expressions?

Sometimes we can employ such a one-pass evaluation strategy (see Sections 9.6 and 9.7). But if the graph contains a dicycle like the following one:

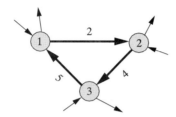

there is a difficulty. Notice that this dicycle has positive total length; the issue is not a negative dicycle. The difficulty comes from the form of the three corresponding functional equations:

$$v[1] = \min\{v[3] + 5, \ldots\}$$
$$v[2] = \min\{v[1] + 2, \ldots\}$$
$$v[3] = \min\{v[2] + 4, \ldots\}$$

Evaluating the expression for $v[2]$ requires $v[1]$; evaluating the expression for $v[3]$ requires $v[2]$; and evaluating the expression for $v[1]$ requires $v[3]$.

$\boxed{9.17}$ Dicycles introduce circular dependencies in functional equations that preclude their solution by one-pass evaluation, even if the length of all dicycles is nonnegative.

Repeated Evaluation Algorithm: Bellman–Ford

Algorithm 9A, which is attributed to R .E. Bellman and L. R. Ford, Jr., draws on the notion of evaluation of functional equations to produce an algorithm that does work for any graph with no negative dicycles. The key insight is repeated evaluation. Each major iteration of the search (i.e., each t) evaluates the functional equation for each $v[k]$ using results from the preceding iteration. We stop when no result changes.

Just as with all the other searches of this book, Algorithm 9A distinguishes search values at different iterations by attaching superscripts. That is,

$$v^{(t)}[k] \triangleq \text{value of } v[k] \text{ obtained on the } t\text{th iteration}$$

When the algorithm finishes, we may want to know both shortest path lengths $v[k]$ and the actual paths that achieve those optimal values. Labels $d[k]$ keep notes

ALGORITHM 9A: ONE TO ALL (NO NEGATIVE DICYCLES); BELLMAN–FORD

Step 0: Initialization. With s the source node, initialize optimal path lengths

$$v^{(0)}[k] \leftarrow \begin{cases} 0 & \text{if } k = s \\ +\infty & \text{otherwise} \end{cases}$$

and set iteration counter $t \leftarrow 1$.

Step 1: Evaluation. For each k evaluate

$$v^{(t)}[k] \leftarrow \min\{v^{(t-1)}[i] + c_{i,k} : (i, k) \text{ exists}\}$$

If $v^{(t)}[k] < v^{(t-1)}[k]$, also set $d[k] \leftarrow$ the number of a neighboring node i achieving the minimum $v^{(t)}[k]$.

Step 2: Stopping. Terminate if $v^{(t)}[k] = v^{(t-1)}[k]$ for every k, or if $t =$ the number of nodes in the graph. Values $v^{(t)}[k]$ then equal the required shortest path lengths unless some $v^{(t)}[k]$ changed at the last t, in which case the graph contains a negative dicycle.

Step 3: Advance. If some $v[k]$ changed and $t <$ the number of nodes, increment $t \leftarrow t + 1$ and return to Step 1.

that we will see allow us to recover the optimal paths. Specifically,

$$d[k] \triangleq \text{node preceding } k \text{ in the best known path from } s \text{ to } k$$

Bellman–Ford Solution of the Two Ring Circus Example

Before dealing with all the details of Bellman-Ford Algorithm 9A, let us apply it to the Two Ring Circus example [Figure 9.4(c)]. Table 9.1 provides details.

TABLE 9.1   Bellman–Ford Algorithm Solution of Two Ring Example

t	$v^{(t)}[1]$	$v^{(t)}[2]$	$v^{(t)}[3]$	$v^{(t)}[4]$	$v^{(t)}[5]$	$v^{(t)}[6]$	$v^{(t)}[7]$	$v^{(t)}[8]$	$v^{(t)}[9]$
0	0	$+\infty$	$+\infty$	$+\infty$	$+\infty$	$+\infty$	$+\infty$	$+\infty$	$+\infty$
1		3.6	2.8	4.6					
2				3.7	4.1				
3						5.7	3.2		
4								4.8	6.7
5									5.5
6									

t	$d[1]$	$d[2]$	$d[3]$	$d[4]$	$d[5]$	$d[6]$	$d[7]$	$d[8]$	$d[9]$
1		1	1	1					
2				2	3				
3						5	5		
4								7	7
5									8
6									

The process begins by initializing node values $v[k]$. Lacking better information, the algorithm initializes all except $v[s]$ at the worst possible value for a minimizing optimization, $+\infty$.

The first main iteration of the algorithm computes new $v^{(1)}[k]$ by evaluating the functional equations using the $v^{(0)}[k]$. For example,

$$v^{(1)}[2] = \min\{v^{(0)}[1] + c_{1,2}, v^{(0)}[4] + c_{4,2}\}$$
$$= \min\{0 + 3.6, \infty + 0.2\}$$
$$= 3.6$$

Values change for $v^{(1)}[2] = 3.6$, $v^{(1)}[3] = 2.8$, and $v^{(1)}[4] = 4.6$.

Whenever a new value is assigned to any path length $v[k]$, we also want to record the decision option that produced it. Here that means keeping track of the neighboring node i through which a new value for $v[k]$ was derived. For example, $d[2] \leftarrow 1$ because the minimum establishing $v^{(1)}[2]$ was achieved by extending the best known path to neighboring node $i = 1$. Similarly, $d[3] \leftarrow 1$ and $d[4] \leftarrow 1$.

Advancing to iteration $t = 2$, we repeat the process. This time, values change at nodes $k = 4$ and 5. The first of these illustrates the temporary nature of $v[k]$ values in the Bellman–Ford algorithm. At iteration 1 we set $v^{(1)} = 4.6$ because

$$v^{(1)}[4] = \min\{v^{(0)}[1] + c_{1,4}, v^{(0)}[2] + c_{2,4}, v^{(0)}[3] + c_{3,4}, v^{(0)}[5] + c_{5,4}, v^{(0)}[6] + c_{6,4}\}$$
$$= \min\{0 + 4.6, \infty + 0.1, \infty + 2.3, \infty + 1.5, \infty + 2.5\}$$
$$= 4.6$$

After iteration 2, however, updated values produce

$$v^{(2)}[4] = \min\{v^{(1)}[1] + c_{1,4}, v^{(1)}[2] + c_{2,4}, v^{(1)}[3] + c_{3,4}, v^{(1)}[5] + c_{5,4}, v^{(1)}[6] + c_{6,4}\}$$
$$= \min\{0 + 4.6, 3.6 + 0.1, 2.8 + 2.3, \infty + 1.5, \infty + 2.5\}$$
$$= 3.7$$

Processing continues in this way through iterations $t = 3, 4, 5$, with better and better values for the $v[k]$ being computed. However, evaluation of functional equations at $t = 6$ produces no changed values. This is the signal to stop and accept values $v^{(6)}[k]$ as optimal. Figure 9.8 shows corresponding optimal paths.

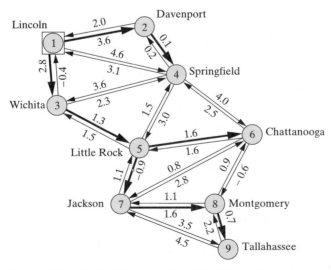

FIGURE 9.8 Optimal Paths in Two Ring Example

SAMPLE EXERCISE 9.9: APPLYING THE BELLMAN–FORD ALGORITHM

Apply Bellman–Ford Algorithm 9A to compute the lengths of shortest paths from source node $s = 1$ to all other nodes of the following graph.

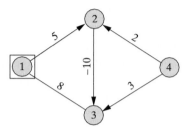

Analysis: Initialization Step 0 sets

$$v^{(0)}[1] = 0 \quad v^{(0)}[2] = v^{(0)}[3] = v^{(0)}[4] = \infty$$

On iteration $t = 1$, evaluation of functional equations gives

$$v^{(1)}[1] = 0$$

$$v^{(1)}[2] = \min\{v^{(0)}[1] + c_{1,2}, \, v^{(0)}[4] + c_{4,2}\}$$
$$= \min\{0 + 5, \, \infty + 2\} = 5 \text{ (making } d[2] = 1)$$

$$v^{(1)}[3] = \min\{v^{(0)}[1] + c_{1,3}, \, v^{(0)}[2] + c_{2,3}, \, v^{(0)}[4] + c_{4,3}\}$$
$$= \min\{0 + 8, \, \infty - 10, \, \infty + 3\} = 8 \text{ (making } d[3] = 1)$$

$$v^{(1)}[4] = \min\{\}$$

$$= \infty$$

Continuing in tabular form, we have

t	$v^{(t)}[1]$	$v^{(t)}[2]$	$v^{(t)}[3]$	$v^{(t)}[4]$
0	0	$+\infty$	$+\infty$	$+\infty$
1		5	8	
2			-5	
3				

t	$d[1]$	$d[2]$	$d[3]$	$d[4]$
1		1	1	
2			2	
3				

Iteration $t = 2$ revises $v^{(2)}[3] = -5$, which makes $d[3] = 2$. The algorithm terminates when all values repeat at iteration $t = 3$.

Extracting final values from the table, the shortest path to node 2 has length $v[2] = 5$, and the shortest path to node 3 has length $v[3] = -5$. Since the final $v[4] = \infty$, there is no path to node 4.

Justification of the Bellman–Ford Algorithm

Having seen how Algorithm 9A works, we now need to investigate why. What guarantees that the $v^{(t)}[k]$ correspond to optimal values when termination criteria are fulfilled?

Look again at the complex of optimal paths in Figure 9.8. Some nodes are one arc away from the source along optimal paths, some are two arcs away, some three, and so on. Notice, however, that none is more than

$$\text{number of nodes} - 1 = 8$$

arcs away. A path in a 9 node network can contain no more than $(9 - 1) = 8$ arcs. With any greater number, some node would have to repeat.

The value $v^{(t)}[1] = 0$ at the source is correct from $t = 0$. After iteration $t = 1$, all optimal lengths that directly depend upon it (i.e., all $v^{(1)}[k]$ with optimal paths of 1 arc) must also be correct. Similarly, after $t = 2$, those $v^{(2)}[k]$ with optimal paths of 2 arcs must also be final. In general, intermediate $v^{(t)}[k]$ reflect optimal paths of t or fewer steps. All must be final before $t = $ the number of nodes. Furthermore, if we go a whole iteration without changing any node's value, no more changes will occur. We might as well terminate computation.

Recovering Optimal Paths

Some shortest path models require only shortest path lengths, but many require the optimal paths that achieve those lengths.

To recover optimal paths, we exploit a consequence of principle of optimality 9.10 . When optimal paths must have optimal subpaths, the shortest path to any node is just a 1-arc/edge extension of the shortest path to a neighbor. This implies a record of that last link is enough to recover the full path. We need only move to the indicated neighbor and follow its optimal path.

The $d[k]$ labels we record during computation provide the needed information because they record the last link of the sequence establishing optimal path length $v[k]$.

> 9.18 At the completion of Algorithm 9A, a shortest path from source s to any other node k can be recovered by starting at k, backtracking to neighboring node $d[k]$, and continuing with an optimal path from s to the neighbor until source s is encountered.

To illustrate, consider the Lincoln-to-Tallahassee path we really need in Two Ring Circus Figure 9.8. Starting at Tallahassee node 9, final $d[9] = 8$ tells us the optimal path enters through node 8. Since the optimal path to node 9 must include an optimal subpath to node 8, we can continue the process by focusing on the shortest path to node 8. There $d[8] = 7$ tells us the path comes from node 7.

Continuing through $d[7] = 5$, $d[5] = 3$, and $d[3] = 1$ we finally reach the source. Thus the full optimal path is 1–3–5–7–8–9: Lincoln to Wichita to Little Rock to Jackson to Montgomery to Tallahassee.

SAMPLE EXERCISE 9.10: RECOVERING PATHS WITH BELLMAN–FORD

Return to the graph of Sample Exercise 9.9. Recover an optimal path to node 3 using the $d[k]$ labels assigned during Sample Exercise 9.9 processing.

Analysis: Applying principle $\boxed{9.18}$, we begin at destination node 3. Since $d[3] = 2$, the optimal path enters from node 2. The optimal path to node 2, in turn, enters through node $d[2] = 1$. Having reached the source, we conclude that a shortest path is 1–2–3.

Encountering Negative Dicycles with Bellman–Ford

We know that shortest path models with negative dicycles usually cannot be solved by the dynamic programming methods of Algorithm 9A (principle $\boxed{9.9}$). But if we confront a very large model, it could be quite difficult to know whether negative dicycles are present.

To see what happens, we could try applying the Bellman–Ford algorithm to our negative dicycle example of Figure 9.6:

t	$v^{(t)}[1]$	$v^{(t)}[2]$	$v^{(t)}[3]$	$v^{(t)}[4]$	$d[1]$	$d[2]$	$d[3]$	$d[4]$
0	0	$+\infty$	$+\infty$	$+\infty$				
1		2	10		1	1		
2			5	22			2	3
3		-3		17	4			
4		-8	0					

Initial values $v^{(0)}[k]$ are chosen just as in other examples, and each iteration updates $v[k]$ and $d[k]$ labels. The critical difference is that labels continued to change on iteration $t = 4$, even though the graph has only 4 nodes. Evaluation around negative dicycle drives $v^{(t)}$ lower and lower because

$$v^{(t)}[2] \leftarrow v^{(t-1)}[4] - 25$$
$$v^{(t)}[3] \leftarrow v^{(t-1)}[2] + 3$$
$$v^{(t)}[4] \leftarrow v^{(t-1)}[3] + 12$$

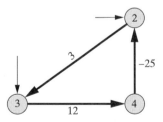

This behavior illustrates why the Bellman–Ford algorithm terminates if labels continue to change on iteration $t =$ the number of nodes.

$\boxed{9.19}$ If Algorithm 9A encounters negative dicycles, it will demonstrate their presence by continuing to change $v^{(t)}[k]$ on iteration $t =$ the number of nodes. Any node k with such a changing $v^{(t)}[k]$ belongs to a negative dicycle.

We can actually exhibit a negative dicycle by following the $d[k]$ labels of any k for which $v^{(t)}[k]$ changed on the final iteration. For example, we could begin at $k = 2$ because $v^{(4)}[2] = -8 \neq v^{(3)}[2] = -3$. Tracing backward through the $d[k]$ labels, we find that

$$d[2] = 4$$
$$d[4] = 3$$
$$d[3] = 2$$

As soon as node 2 repeats, we know we have completed negative dicycle 2–3–4–2.

9.4 SHORTEST PATHS FROM ALL NODES TO ALL OTHERS: FLOYD–WARSHALL

When shortest paths between all pairs of nodes are required in a graph with no negative dicycles (e.g., in our Littleville example of Figure 9.1), the task is to compute $v[k, \ell]$ efficiently, satisfying all-to-all functional equations 9.14 .

Floyd–Warshall Algorithm

Algorithm 9B, which is attributed to R. W. Floyd and S. Warshall, does just that. Search quantities

$$v^{(t)}[k, \ell] \triangleq \text{length of a shortest path for } k \text{ to } \ell \text{ using only}$$
$$\text{intermediate nodes numbered less than or equal to } t$$

converge to the required shortest path lengths $v[k, \ell]$. Corresponding decision labels

$$d[k, \ell] \triangleq \text{node just before } \ell \text{ on the current path from } k \text{ to } \ell$$

track the associated paths.

The key to Algorithm 9B's effectiveness is the clever sequence in which quantities are calculated. Initialization correctly sets the $v^{(0)}[k, \ell] = c_{k,\ell}$ because the only path from k to ℓ that has no intermediate nodes (i.e., none with positive node number ≤ 0) is an arc/edge (k, ℓ) with cost $c_{k,\ell}$.

Subsequent iterations consider concatenating previous results as indicated in the following sketch:

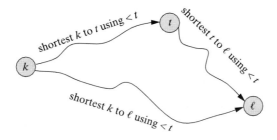

By iteration t we already know optimal paths using nodes numbered $< t$ as intermediaries. Thus $v^{(t)}[k, \ell]$, which is the shortest using nodes numbered $\leq t$, must be either the current best or one visiting t via an optimal path from k to t followed by

---

ALGORITHM 9B: ALL TO ALL (NO NEGATIVE DICYCLES); FLOYD–WARSHALL

Step 0: Initialization. All nodes should have consecutive positive numbers starting with 1. For all arcs and edges (k, ℓ) in the graph, initialize

$$v^{(0)}[k, \ell] \leftarrow c_{k,\ell}$$
$$d[k, \ell] \;\;\leftarrow k$$

For k, ℓ pairs with no arc/edge (k, ℓ), assign

$$v^{(0)}[k, \ell] \leftarrow \begin{cases} 0 & \text{if } k = \ell \\ +\infty & \text{otherwise} \end{cases}$$

Also set iteration counter $t \leftarrow 1$.

Step 1: Evaluation. For all $k, \ell \neq t$ update

$$v^{(t)}[k, \ell] \leftarrow \min\{v^{(t-1)}[k, \ell],\; v^{(t-1)}[k, t] + v^{(t-1)}[t, \ell]\}$$

If $v^{(t)}[k, \ell] < v^{(t-1)}[k, \ell]$, also set $d[k, \ell] \leftarrow d[t, \ell]$.

Step 2: Stopping. Terminate if $t = $ the number of nodes in the graph, or if $v^{(t)}[k, k] < 0$ for any node k. Values $v^{(t)}[k, \ell]$ then equal the required shortest path lengths unless some $v^{(t)}[k, k]$ is negative, in which case the graph contains a negative dicycle through k.

Step 3: Advance. If $t < $ the number of nodes and all $v^{(t)}[k, k] \geq 0$, increment $t \leftarrow t + 1$ and return to Step 1.

---

one from t to ℓ. The latter subpaths must involve intermediaries numbered $< t$, so they are already recorded in labels $v^{(t-1)}[k, t]$ and $v^{(t-1)}[t, \ell]$. At termination, all the possibilities of the minimization in functional equations $\boxed{9.14}$ have been checked, which implies that the final $v[k, \ell]$ are optimal.

Floyd–Warshall Solution of the Littleville Example

To illustrate Floyd–Warshall Algorithm 9B, we apply it to the Littleville example of Figure 9.1. Table 9.2 presents initial values, along with the results for iterations $t = 1$, 9, and 10.

The algorithm begins by initializing $v^{(0)}[k, \ell]$ for all arcs/edges (k, ℓ) at direct cost $c_{k,\ell}$, and fixing labels $d[k, \ell]$ accordingly. For example, $v^{(0)}[6, 7] \leftarrow c_{6,7} = 28$, and $d[6, 7] \leftarrow 6$ to indicate that the current (direct) path from node 6 to node 7 enters 7 via 6.

Values for which there is no arc/edge (k, ℓ) begin at $+\infty$ except for k-to-k path lengths $v^{(0)}[k, k]$, which are assigned value 0. Thus $v^{(0)}[1, 10] \leftarrow \infty$ in Table 9.2, because the Littleville network has no direct link from node 1 to node 10, and $v^{(0)}[9, 9] \leftarrow 0$ to indicate that the shortest path from node 9 to itself has length $= 0$.

Iteration $t = 1$ advances to Algorithm 9B, Step 1. For all node pairs k and ℓ, both different than $t = 1$, we update $v[k, \ell]$ to the minimum of what it was before and the concatenation the current path from k to t with the current path from t to ℓ.

Table 9.2 shows that the only new (boxed) value occurs at $k = 5$, $\ell = 2$. There is no direct path from node 5 to node 2 [refer back to Figure Example 9.1(b)], so

TABLE 9.2   Floyd–Warshall Algorithm Solution of Littleville Example

Initial Values

$v^{(0)}[k,\ell]$	$\ell=1$	$\ell=2$	$\ell=3$	$\ell=4$	$\ell=5$	$\ell=6$	$\ell=7$	$\ell=8$	$\ell=9$	$\ell=10$
$k=1$	0	12	∞	∞	∞	∞	∞	∞	∞	∞
$k=2$	∞	0	18	∞	∞	32	∞	∞	∞	∞
$k=3$	∞	∞	0	13	∞	∞	30	∞	∞	∞
$k=4$	∞	∞	∞	0	∞	∞	∞	∞	∞	38
$k=5$	20	∞	∞	∞	0	18	∞	∞	∞	∞
$k=6$	∞	32	∞	∞	18	0	28	∞	25	∞
$k=7$	∞	∞	30	∞	∞	28	0	∞	21	49
$k=8$	∞	∞	∞	∞	18	∞	∞	0	36	∞
$k=9$	∞	∞	∞	∞	∞	25	21	36	0	40
$k=10$	∞	∞	∞	∞	∞	∞	49	28	40	0

$d[k,\ell]$	$\ell=1$	$\ell=2$	$\ell=3$	$\ell=4$	$\ell=5$	$\ell=6$	$\ell=7$	$\ell=8$	$\ell=9$	$\ell=10$
$k=1$	—	1	—	—	—	—	—	—	—	—
$k=2$	—	—	2	—	—	2	—	—	—	—
$k=3$	—	—	—	3	—	—	3	—	—	—
$k=4$	—	—	—	—	—	—	—	—	—	4
$k=5$	5	—	—	—	5	—	—	—	—	—
$k=6$	—	6	—	—	6	—	6	—	6	—
$k=7$	—	—	7	—	—	7	—	—	7	7
$k=8$	—	—	—	—	8	—	—	—	8	—
$k=9$	—	—	—	—	—	9	9	9	—	9
$k=10$	—	—	—	—	—	—	10	10	10	—

After Iteration $t=1$

$v^{(1)}[k,\ell]$	$\ell=1$	$\ell=2$	$\ell=3$	$\ell=4$	$\ell=5$	$\ell=6$	$\ell=7$	$\ell=8$	$\ell=9$	$\ell=10$
$k=1$	0	12	∞	∞	∞	∞	∞	∞	∞	∞
$k=2$	∞	0	18	∞	∞	32	∞	∞	∞	∞
$k=3$	∞	∞	0	13	∞	∞	30	∞	∞	∞
$k=4$	∞	∞	∞	0	∞	∞	∞	∞	∞	38
$k=5$	20	[32]	∞	∞	0	18	∞	∞	∞	∞
$k=6$	∞	32	∞	∞	18	0	28	∞	25	∞
$k=7$	∞	∞	30	∞	∞	28	0	∞	21	49
$k=8$	∞	∞	∞	∞	18	∞	∞	0	36	∞
$k=9$	∞	∞	∞	∞	∞	25	21	36	0	40
$k=10$	∞	∞	∞	∞	∞	∞	49	28	40	0

$d[k,\ell]$	$\ell=1$	$\ell=2$	$\ell=3$	$\ell=4$	$\ell=5$	$\ell=6$	$\ell=7$	$\ell=8$	$\ell=9$	$\ell=10$
$k=1$	—	1	—	—	—	—	—	—	—	—
$k=2$	—	—	2	—	—	2	—	—	—	—
$k=3$	—	—	—	3	—	—	3	—	—	—
$k=4$	—	—	—	—	—	—	—	—	—	4
$k=5$	5	[1]	—	—	—	5	—	—	—	—
$k=6$	—	6	—	—	6	—	6	—	6	—
$k=7$	—	—	7	—	—	7	—	—	7	7
$k=8$	—	—	—	—	8	—	—	—	8	—
$k=9$	—	—	—	—	—	9	9	9	—	9
$k=10$	—	—	—	—	—	—	10	10	10	—

$v^{(0)}[5,2]=+\infty$. But iteration $t=1$ evaluates node 1 as an intermediary. The length of the path 5–1–2 is recorded by evaluating the corresponding functional equation
9.14 :

TABLE 9.2 Floyd–Warshall Algorithm Solution of Littleville Example (Continued)

After Iteration $t = 9$

$v^{(9)}[k,\ell]$	$\ell=1$	$\ell=2$	$\ell=3$	$\ell=4$	$\ell=5$	$\ell=6$	$\ell=7$	$\ell=8$	$\ell=9$	$\ell=10$
$k=1$	0	12	30	43	62	44	60	[105]	69	81
$k=2$	70	0	18	31	50	32	48	[93]	57	69
$k=3$	96	90	0	13	76	58	30	[87]	51	51
$k=4$	∞	∞	∞	0	∞	∞	∞	∞	∞	38
$k=5$	20	32	50	63	0	18	46	[79]	43	[83]
$k=6$	38	32	50	63	18	0	28	[61]	25	[65]
$k=7$	66	60	30	43	46	28	0	[57]	21	49
$k=8$	38	50	68	81	18	36	[57]	0	36	[76]
$k=9$	63	57	51	64	43	25	21	36	0	40
$k=10$	66	78	79	92	46	64	49	28	40	0

$d[k,\ell]$	$\ell=1$	$\ell=2$	$\ell=3$	$\ell=4$	$\ell=5$	$\ell=6$	$\ell=7$	$\ell=8$	$\ell=9$	$\ell=10$
$k=1$	—	1	2	3	6	2	3	[9]	6	4
$k=2$	5	—	2	3	6	2	3	[9]	6	4
$k=3$	5	6	—	3	6	7	3	[9]	7	4
$k=4$	—	—	—	—	—	—	—	—	—	4
$k=5$	5	1	2	3	—	5	6	[9]	6	[9]
$k=6$	5	6	2	3	6	—	6	[9]	6	[9]
$k=7$	5	6	7	3	6	7	—	[9]	7	7
$k=8$	5	1	2	3	8	5	[9]	—	8	[9]
$k=9$	5	6	7	3	6	9	9	9	—	9
$k=10$	5	1	7	3	8	5	10	10	10	—

After Iteration $t = 10$

$v^{(10)}[k,\ell]$	$\ell=1$	$\ell=2$	$\ell=3$	$\ell=4$	$\ell=5$	$\ell=6$	$\ell=7$	$\ell=8$	$\ell=9$	$\ell=10$
$k=1$	0	12	30	43	62	44	60	105	69	81
$k=2$	70	0	18	31	50	32	48	93	57	69
$k=3$	96	90	0	13	76	58	30	[79]	51	51
$k=4$	[104]	[116]	[117]	0	[84]	[102]	[87]	[66]	[78]	38
$k=5$	20	32	50	63	0	18	46	79	43	83
$k=6$	38	32	50	63	18	0	28	61	25	65
$k=7$	66	60	30	43	46	28	0	57	21	49
$k=8$	38	50	68	81	18	36	57	0	36	76
$k=9$	63	57	51	64	43	25	21	36	0	40
$k=10$	66	78	79	92	46	64	49	28	40	0

$d[k,\ell]$	$\ell=1$	$\ell=2$	$\ell=3$	$\ell=4$	$\ell=5$	$\ell=6$	$\ell=7$	$\ell=8$	$\ell=9$	$\ell=10$
$k=1$	—	1	2	3	6	2	3	9	6	4
$k=2$	5	—	2	3	6	2	3	9	6	4
$k=3$	5	6	—	3	6	7	3	[10]	7	4
$k=4$	[5]	[1]	[7]	—	[8]	[5]	[10]	[10]	[10]	4
$k=5$	5	1	2	3	—	5	6	9	6	9
$k=6$	5	6	2	3	6	—	6	9	6	9
$k=7$	5	6	7	3	6	7	—	9	7	7
$k=8$	5	1	2	3	8	5	9	—	8	9
$k=9$	5	6	7	3	6	9	9	9	—	9
$k=10$	5	1	7	3	8	5	10	10	10	—

$$v^{(1)}[5, 2] \leftarrow \min\{v^{(0)}[5, 2],\ v^{(0)}[5, 1] + v^{(0)}[1, 2]\}$$
$$= \min\{\infty,\ 20 + 12\}$$
$$= 32$$

Decision label $d[5, 2]$ is also corrected to indicate that the current path from $k = 5$ to $\ell = 2$ now enters through node $d[5, 2] = 1$.

Skipping ahead, Table 9.2 shows other values changing as nodes $t = 9$ and $t = 10$ are tried as intermediate nodes. One is

$$v^{(10)}[3, 8] \leftarrow \min\{v^{(9)}[3, 8], \ v^{(9)}[3, 10] + v^{(9)}[10, 8]\}$$
$$= \min\{87, \ 51 + 28\}$$
$$= 79$$

Another value changed at $t = 10$ is

$$v^{(10)}[4, 1] \leftarrow v^{(9)}[4, 10] + v^{(9)}[10, 1]$$
$$= 38 + 66 = 104$$

Look carefully at the corresponding $d[4, 1]$. Notice that it is set to $d[10, 1] = 5$, not $t = 10$. We want $d[4, 1]$ to be the next-to-last node of the newly recorded path, which is information available in $d[10, 1]$. This may or may not equal the intermediate t.

After iteration $t = 10$ every node has been tried as an intermediary; the Littleville network has only 10 nodes. Thus the algorithm terminates, and values $v^{(10)}[k, \ell]$ are optimal. That is, they provide the lengths of shortest paths from each k to each ℓ.

SAMPLE EXERCISE 9.11: APPLYING THE FLOYD–WARSHALL ALGORITHM

Consider the following digraph:

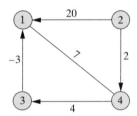

Apply Floyd–Warshall Algorithm 9B to compute the lengths of shortest paths between all pairs of nodes.

Analysis: Paralleling Table 9.2, we show tables of $v^{(t)}[k, \ell]$ and $d[k, \ell]$, boxing changed values.

	$v^{(0)}[k,\ell]$				$d[k,\ell]$			
k	$\ell = 1$	$\ell = 2$	$\ell = 3$	$\ell = 4$	$\ell = 1$	$\ell = 2$	$\ell = 3$	$\ell = 4$
1	0	∞	∞	7	—	—	—	1
2	20	0	∞	2	2	—	—	2
3	–3	∞	0	∞	3	—	—	—
4	7	∞	4	0	4	—	4	—

$v^{(1)}[k,\ell] = v^{(2)}[k,\ell]$				$d[k,\ell]$			
k	$\ell=1$	$\ell=2$	$\ell=3$	$\ell=4$	$\ell=1$	$\ell=2$	$\ell=3$	$\ell=4$
1	0	∞	∞	7	—	—	—	1
2	20	0	∞	2	2	—	—	2
3	−3	∞	0	[4]	3	—	—	[1]
4	7	∞	4	0	4	—	4	—

$v^{(3)}[k,\ell]$				$d[k,\ell]$			
k	$\ell=1$	$\ell=2$	$\ell=3$	$\ell=4$	$\ell=1$	$\ell=2$	$\ell=3$	$\ell=4$
1	0	∞	∞	7	—	—	—	1
2	20	0	∞	2	2	—	—	2
3	−3	∞	0	4	3	—	—	1
4	[1]	∞	4	0	[3]	—	4	—

$v^{(4)}[k,\ell]$				$d[k,\ell]$			
k	$\ell=1$	$\ell=2$	$\ell=3$	$\ell=4$	$\ell=1$	$\ell=2$	$\ell=3$	$\ell=4$
1	0	∞	[11]	7	—	—	[4]	1
2	3	0	[6]	2	[3]	—	[4]	2
3	−3	∞	0	4	3	—	—	1
4	1	∞	4	0	3	—	4	—

The table for $t = 1$ is identical to that of $t = 2$ because no paths use node 2 as an intermediary.

With only 4 nodes in the graph, computations are complete after iteration $t = 4$. Value $v^{(4)}[1, 2] = \infty$ indicates that there is no path from node 1 to node 2.

Recovering Optimal Paths

Just as with earlier cases (principle 9.18), final labels $d[k, \ell]$ allow us to recover an optimal path between any pair of nodes.

> 9.20   At the completion of Algorithm 9B, a shortest path from any node k to any other node ℓ can be recovered by starting at ℓ, backtracking to neighboring node $d[k, \ell]$, and continuing with an optimal path from k to the neighbor until k itself is encountered.

To illustrate, refer again to final $t = 10$ labels for the Littleville example displayed in Table 9.2. A shortest path from node $k = 3$ to node $\ell = 8$ is recovered by beginning at destination ℓ. Label $d[3, 8] = 10$ indicates that an optimal path enters from node 10. Now backtracking to node 10, we apply principle of optimality 9.10 and follow the best path from 3 to 10. Label $d[3, 10] = 4$ tells us that that path enters through node 4. Backtracking again, we discover label $d[3, 4] = 3$. Origin $k = 3$ has been reached, and the optimal path is 3–4–10–8.

SAMPLE EXERCISE 9.12: RECOVERING A FLOYD–WARSHALL PATH

Use final labels of Sample Exercise 9.11 to recover an optimal path from $k = 2$ to $\ell = 1$.

Analysis: Applying principle | 9.20 |, we backtrack from destination $\ell = 1$. Label $d[2, 1] = 3$ indicates the optimal path enters through node 3. Continuing, the optimal path from 2 to 3 enters through $d[2, 3] = 4$, and the one for 4 enters through $d[2, 4] = 2$, which brings us to origin $k = 2$. Thus the optimal path is 2–4–3–1.

Detecting Negative Dicycles with Floyd–Warshall

Just as with the Bellman–Ford algorithm for a single source, Floyd–Warshall Algorithm 9B guarantees optimal paths and path lengths only if the given graph contains no negative dicycles. Functional equations | 9.14 |, on which the algorithm is based, need not hold in the presence of negative dicycles.

What happens when Algorithm 9B is applied to a graph with negative dicycles? The secondary stopping criterion of algorithm Step 2 comes into play, and computation terminates with the conclusion that a negative dicycle exists. The actual condition flagging presence of a negative dicycle in the Floyd–Warshall procedure is $v^{(t)}[k, k] < 0$ (i.e., a shortest path from any k to itself turning negative).

> **9.21** If Algorithm 9B is applied to a graph with negative dicycles, it will demonstrate their presence by making some $v^{(t)}[k, k] < 0$ and terminating. Under those circumstances, the implied negative dicycle includes node k.

To illustrate, return to the negative dicycle example of Figure 9.6. Application of Algorithm 9B produces the following:

	$v^{(0)}[k,\ell] = v^{(1)}[k,\ell]$				$d[k,\ell]$			
k	$\ell=1$	$\ell=2$	$\ell=3$	$\ell=4$	$\ell=1$	$\ell=2$	$\ell=3$	$\ell=4$
1	0	2	10	∞	—	1	1	—
2	∞	0	3	∞	—	—	2	—
3	∞	∞	0	12	—	—	—	3
4	∞	-25	∞	0	—	4	—	—

	$v^{(2)}[k,\ell]$				$d[k,\ell]$			
k	$\ell=1$	$\ell=2$	$\ell=3$	$\ell=4$	$\ell=1$	$\ell=2$	$\ell=3$	$\ell=4$
1	0	2	$\boxed{5}$	∞	—	1	$\boxed{1}$	—
2	∞	0	3	∞	—	—	2	—
3	∞	∞	0	12	—	—	—	3
4	∞	-25	$\boxed{-22}$	0	—	4	$\boxed{2}$	—

	$v^{(3)}[k,\ell]$				$d[k,\ell]$			
k	$\ell=1$	$\ell=2$	$\ell=3$	$\ell=4$	$\ell=1$	$\ell=2$	$\ell=3$	$\ell=4$
1	0	2	5	$\boxed{17}$	—	1	2	$\boxed{3}$
2	∞	0	3	$\boxed{15}$	—	—	2	$\boxed{3}$
3	∞	∞	0	12	—	—	—	3
4	∞	-25	-22	$\boxed{-10}$	—	4	3	$\boxed{3}$

Notice that on iteration $t = 3$, value

$$v^{(3)}[4, 4] \leftarrow \min\{0, \; v^{(2)}[4, 3] + v^{(2)}[3, 4]\}$$
$$= \min\{0, \; -22 + 12\}$$
$$= -10$$

This would cause the algorithm to terminate at Step 2 with the conclusion that a negative dicycle involving node 4 is present in the graph.

As usual, final $d[k, \ell]$ labels allow us to retrieve the dicycle if needed. Starting at node 4 in the table for $t = 3$,

$$d[4, 4] = 3$$
$$d[4, 3] = 2$$
$$d[4, 2] = 4$$

indicating the negative dicycle 4–2–3–4.

9.5 SHORTEST PATH FROM ONE NODE TO ALL OTHERS WITH COSTS NONNEGATIVE: DIJKSTRA

The Bellman–Ford and Floyd–Warshall algorithms of Sections 9.3 and 9.4 can solve any of the shortest path models presented in Section 9.1 and any of the others we will encounter in Sections 9.6 and 9.7. They require only that the given graph contains no negative dicycle. Still, those two algorithms are not the most efficient option when given graphs satisfy further assumptions. In this section we develop the algorithm credited to E. W. Dijkstra, which is considerably more efficient for cases where we need shortest paths from one node to all others, and all costs are nonnegative ($c_{i,j} \geq 0$). Algorithm 9C provides a formal statement.

ALGORITHM 9C: ONE TO ALL (NONNEGATIVE COSTS); DIJKSTRA

Step 0: Initialization. With s the source node, initialize optimal path lengths

$$v[i] \leftarrow \begin{cases} 0 & \text{if } i = s \\ +\infty & \text{otherwise} \end{cases}$$

Then mark all nodes temporary, and choose $p \leftarrow s$ as the next permanently labeled node.

Step 1: Processing. Mark node p permanent, and for every arc/edge (p, i) leading from p to a temporary node, update

$$v[i] \leftarrow \min\{v[i], \ v[p] + c_{p,i}\}$$

If $v[i]$ changed in value, also set $d[i] \leftarrow p$.

Step 2: Stopping. If no temporary nodes remain, stop; values $v[i]$ now reflect the required shortest path lengths.

Step 3: Next Permanent. Choose as next permanently labeled node p a temporary node with least current value $v[i]$, that is,

$$v[p] = \min\{v[i] : i \text{ temporary}\}$$

Then return to Step 1.

Permanently and Temporarily Labeled Nodes

A graph with nonnegative arc/edge costs certainly contains no negative total length dicycles. Thus functional equations 9.12 apply.

What is new about Dijkstra Algorithm 9C is how values satisfying functional equations are computed. Bellman–Ford Algorithm 9A simply evaluates functional equations for all nodes on all iterations. Thus it processes all the inbound arcs and edges at a node many times while computing the minimum in $\boxed{9.12}$.

Dijkstra's method processes outbound arcs and edges instead of inbounds. Much more important, it processes each arc/edge only once. Each major iteration makes one new node p permanent.

> $\boxed{9.22}$ Once a node is classified **permanent** by Dijkstra Algorithm 9C, its $v[p]$ and $d[p]$ labels never change again. Nodes that are not yet permanently labeled are classified **temporary**.

As each new permanent node p is selected, we correct

$$v[i] \leftarrow \min\{v[i], \; v[p] + c_{p,i}\}$$

at all temporarily labeled neighbors i reachable from p. Knowing that the labels for p are final, we need never again consider arc/edge (p, i) in our search for the final value of $v[i]$. This finalizing of a node label at each step is what makes Algorithm 9C the best for its class.

> $\boxed{9.23}$ Dijkstra Algorithm 9C is the most efficient method available for computing shortest paths from one node to all others in (general) graphs and digraphs having all arc/edge costs nonnegative.

Of course, even faster schemes can be developed if the given graph or digraph has special features beyond nonnegative costs (see Section 9.6).

Least Temporary Criterion for Next Permanent Node

It should be obvious that the heart of Dijkstra's algorithm is a rule for selecting the next temporary node to make permanent. The one required is elegantly simply:

> $\boxed{9.24}$ Each iteration of Dijkstra Algorithm 9C selects as the new permanently labeled node p a temporary node of minimum $v[i]$.

All current $v[i]$ on temporary nodes are examined, and a p is selected with

$$v[p] = \min\{v[i] : i \text{ temporary}\}$$

Dijkstra Algorithm Solution of the Texas Transfer Example

Before investigating why criterion $\boxed{9.24}$ works, we will apply Dijkstra Algorithm 9C to the Texas Transfer example reproduced in Figure 9.9. Table 9.3 provides computational details.

Solution begins in exactly the same way as previous methods. Values $v[i] \leftarrow \infty$ except at the source node 3 with $v[3] \leftarrow 0$. All nodes start temporary. Source 3 is automatically the first permanent.

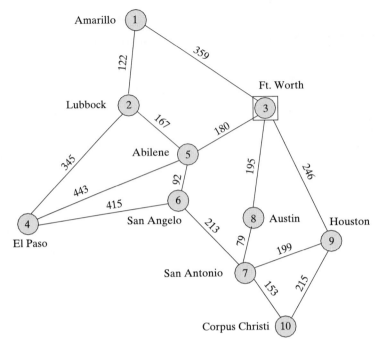

FIGURE 9.9   Texas Transfer Example Network

Processing a new permanent node p means checking the $v[i]$ values of all temporary nodes i reachable from p in one step. For $p = 3$ these include nodes 1, 5, 8, and 9. Thus

$$v[1] \leftarrow \min\{v[1], \; v[3] + c_{3,1}\}$$
$$= \min\{\infty, \; 0 + 359\}$$
$$= 359$$
$$v[5] \leftarrow \min\{v[5], \; v[3] + c_{3,5}\}$$
$$= \min\{\infty, \; 0 + 180\}$$
$$= 180$$
$$v[8] \leftarrow \min\{v[8], \; v[3] + c_{3,8}\}$$
$$= \min\{\infty, \; 0 + 195\}$$
$$= 195$$
$$v[9] \leftarrow \min\{v[9], \; v[3] + c_{3,9}\}$$
$$= \min\{\infty, \; 0 + 246\}$$
$$= 246$$

Corresponding labels $d[1] \leftarrow d[5] \leftarrow d[8] \leftarrow d[9] \leftarrow 3$ because all four $v[i]$ changed from 3.

It is now time to apply criterion $\boxed{9.24}$. The next permanent node p must be a temporary i of minimum $v[i]$. With all nodes temporary except 3,

TABLE 9.3 Dijkstra Algorithm Solution of Texas Transfer Example

p	$v[1]$	$v[2]$	$v[3]$	$v[4]$	$v[5]$	$v[6]$	$v[7]$	$v[8]$	$v[9]$	$v[10]$
(init)	∞	∞	0	∞	∞	∞	∞	∞	∞	∞
3	359		(perm)		180			195	246	
5		347		623	(perm)	272				
8							274	(perm)		
9									(perm)	461
6						(perm)				
7							(perm)			427
1	(perm)									
2		(perm)								
10										(perm)
4				(perm)						
(final)	359	347	0	623	180	272	274	195	246	427
p	$d[1]$	$d[2]$	$d[3]$	$d[4]$	$d[5]$	$d[6]$	$d[7]$	$d[8]$	$d[9]$	$d[10]$
3	3				3			3	3	
5		5		5		5				
8							8			
9										9
6										
7										7
1										
2										
10										
4										
(final)	3	5	—	5	3	5	8	3	3	7

$$\min\{v[1], v[2], v[4], v[5], v[6], v[7], v[8], v[9], v[10]\}$$
$$= \min\{359, \infty, \infty, 180, \infty, \infty, 195, 246, \infty\}$$
$$= 180$$

The smallest value occurs at node 5 and $p = 5$.

Correcting labels on outbound edges gives

$$v[2] \leftarrow \min\{v[2], \ v[5] + c_{5,2}\}$$
$$= \min\{\infty, \ 180 + 167\}$$
$$= 347$$
$$v[4] \leftarrow \min\{v[4], \ v[5] + c_{5,4}\}$$
$$= \min\{\infty, \ 180 + 443\}$$
$$= 623$$
$$v[6] \leftarrow \min\{v[6], \ v[5] + c_{5,6}\}$$
$$= \min\{\infty, \ 180 + 92\}$$
$$= 272$$

Corresponding $d[2] \leftarrow d[4] \leftarrow d[6] \leftarrow 5$.

To choose the next p, we examine the remaining temporary nodes:

$$\min\{v[1], v[2], v[4], v[6], v[7], v[8], v[9], v[10]\}$$
$$= \min\{359, 347, 623, 272, \infty, 195, 246, \infty\}$$
$$= 195$$

Here $p = 8$, leading to $v[7] \leftarrow 195 + 79 = 274$ and $d[7] \leftarrow 8$.
The next permanent node is $p = 9$ because

$$\min\{v[1], v[2], v[4], v[6], v[7], v[9], v[10]\}$$
$$= \min\{359, 347, 623, 272, 274, 246, \infty\}$$
$$= 246$$

Processing yields

$$v[7] \quad \leftarrow \min\{v[7], \ v[9] + c_{9,7}\}$$
$$= \min\{274, \ 246 + 199\}$$
$$= 274$$
$$v[10] \leftarrow \min\{v[10], \ v[9] + c_{9,10}\}$$
$$= \min\{\infty, \ 246 + 215\}$$
$$= 461$$

Notice that $v[7]$ did not change. Thus only label $d[10] \leftarrow 9$.

Remaining processing of Table 9.3 follows in a similar way. Successive iterations make permanent nodes 6, 7, 1, 2, 10, and 4. When all nodes are permanent, the algorithm terminates. Final $v[i]$ now represent the required lengths of shortest paths from source node 3.

SAMPLE EXERCISE 9.13: APPLYING DIJKSTRA'S ALGORITHM

Apply Dijkstra Algorithm 9C to compute the length of a shortest path from node $s = 1$ in the following graph to every other node.

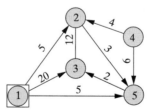

Analysis: Paralleling Table 9.3, we may summarize computations in tables for $v[i]$ and $d[i]$:

p	$v[1]$	$v[2]$	$v[3]$	$v[4]$	$v[5]$
(init)	0	∞	∞	∞	∞
1	(perm)	5	20		5
2		(perm)	17		
5			7		(perm)
3			(perm)		
4				(perm)	
(final)	0	5	7	∞	5

p	$d[1]$	$d[2]$	$d[3]$	$d[4]$	$d[5]$
1		1	1		1
2			2		
5			5		
3					
4					
(final)	—	1	5	—	1

One new feature of this example occurs with the selection of the second permanent node $p = 2$. Criterion $\boxed{9.24}$ requires us to choose a temporary node i of minimum $v[i]$, but there is a tie. Both $v[2] = 5$ and $v[5] = 5$. Either could be selected. The computations above arbitrarily chose $p = 2$.

One other different element is the presence of a node $i = 4$ to which there is no path. As with other shortest path algorithms, this condition is indicated by final $v[4] = \infty$.

Recovering Paths

The Texas Transfer example required only shortest path lengths, not the paths themselves. However, decision labels $d[i]$ make it possible to recover any needed paths using principle $\boxed{9.18}$. For example, the shortest path from source node $s = 3$ to $i = 10$ could be traced backward from $i = 10$ as

$$d[10] = 7$$
$$d[7] \ = 8$$
$$d[8] \ = 3$$

The resulting shortest path is 3–8–7–10.

Justification of the Dijkstra Algorithm

The Dijkstra algorithm designation of permanently labeled nodes certainly speeds computation. But how do nonnegative costs and selection criterion $\boxed{9.24}$ make it work?

The key insight is an interpretation of $v[i]$ values during computation.

$\boxed{9.25}$ After each major iteration, Dijkstra Algorithm 9C values $v[i]$ represent lengths of shortest paths from source s to i that use only permanently labeled nodes.

That is, interim results reflect optimal paths through nodes already classified permanent.

For an example, focus on $v[10]$ in Table 9.3. There is no path from $s = 3$ to $i = 10$ in the Texas Transfer network of Figure 9.9 that uses only nodes 3, 5, and 8. That is why $v[10] = \infty$ through iterations making those nodes permanent. When node 9 becomes permanent on the fourth iteration, an all-permanent path to node 10 does become available. Value $v[10] = 461$ to account for that path 3–9–10. Although the path is shortest among nodes classified permanent through 4 iterations, it is not the shortest overall. Optimal path 3–8–7–10 (length $v[10] = 427$) is discovered only 2 iterations later when node 7 is made permanent.

With interpretation $\boxed{9.25}$, we can easily see why the node p chosen by criterion $\boxed{9.24}$ is ready to become permanent.

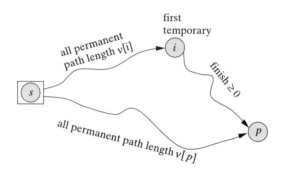

Label $v[p]$ corresponds to the shortest path to node p using only permanently labeled nodes. Noting that s becomes permanent on the very first iteration, any other path begins among permanent nodes, passes along a subpath to a first temporary i, and completes with an i to p subpath. The length of such a path is

$$(s\text{-to-}i \text{ subpath}) + (i\text{-to-}p \text{ subpath}) \ge v[i] + 0$$

because $v[i]$ is the length of a shortest all-permanent path to i, and arc/edges costs of the i-to-p subpath are nonnegative. But since p was chosen (criterion $\boxed{9.24}$) as the temporary node with smallest v-value,

$$v[i] + 0 = v[i] \ge v[p]$$

Thus no path can be shorter than the one for $v[p]$, and we may justly term it permanent.

9.6 SHORTEST PATHS FROM ONE NODE TO ALL OTHERS IN ACYCLIC DIGRAPHS

The key to Dijkstra Algorithm 9C's efficiency was careful selection of permanently labeled nodes, so that arcs needed to be processed only once. Computation becomes even more efficient if a sequence for permanent-node processing can be determined before we begin.

Acyclic Digraphs

A predetermined sequence is possible if the model arises on an **acyclic digraph**, which is a directed graph with no dicycles. Figure 9.10 illustrates the definition. Part (a) shows a graph that is acyclic because it has only arcs (no edges) and has no directed cycles (dicycles). Graphs of parts (b) and (c) fail the definition. The first is only partly directed; for example, it contains edge $(2, 3)$. The second is fully directed, but contains dicycles; one is 2–3–4–5–2.

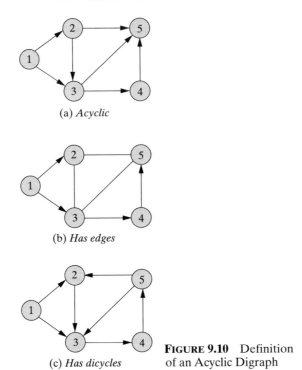

(a) *Acyclic*

(b) *Has edges*

(c) *Has dicycles*   **FIGURE 9.10**   Definition
                      of an Acyclic Digraph

We can determine whether small graphs such as those of Figure 9.10 are acyclic simply by inspection. For larger graphs we require an easy-to-check condition.

| 9.26 | A digraph (completely directed graph) is acyclic if and only if its nodes can be numbered so that every arc (i, j) has $i < j$. |

Figure 9.10(a) is already numbered in this way. Each arc goes from a lower-numbered to a higher-numbered node. Notice how this guarantees that the digraph contains no dicycles. Since each arc takes us to a higher-numbered node, no path could ever close a cycle by repeating a node that we have already visited.

Conversely, it is not hard to find a suitable numbering for a digraph that is acyclic.

| 9.27 | Any acyclic digraph can be numbered as required in principle 9.26 by transiting the graph in depth-first fashion, numbering nodes in decreasing sequence as soon as all their outbound arcs lead to nodes already visited. |

Here **depth-first** simply means that we pass to new nodes before backtracking to ones already visited.

We can illustrate by computing the node numbers of Figure 9.10(a). To avoid confusion, suppose that the graph was given to us with alphabetic node labels as follows:

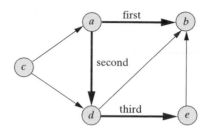

Beginning arbitrarily at node a, we pass first to node b. Since it has no outbound arcs, we assign it the highest possible number, $b = 5$, and backtrack to a. There, an outbound arc remains to node d, and it has an arc to e. Node e can now be numbered $e = 4$ because it leads only to already visited node $b = 5$. Backtracking to node d, we now assign it number $d = 3$. Similarly, node a is numbered $a = 2$. Having completed all arcs from our origin $a = 2$, we must start again at some still unnumbered node. Here the only choice is c, which immediately becomes node $c = 1$.

SAMPLE EXERCISE 9.14: DETERMINING WHETHER A DIGRAPH IS ACYCLIC

Determine whether each of the following digraphs is acyclic.

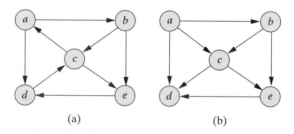

(a)   (b)

Analysis:

(a) This graph is not acyclic. For example, it contains dicycle a–b–c–a.

(b) This graph is acyclic. To demonstrate that it is, we apply principle $\boxed{9.27}$. Starting at node a, depth-first processing proceeds as follows:

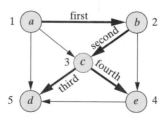

We advance to b, then c, then d before being blocked. Numbering $d = 5$ and back-tracking, we can now go to e. It leads nowhere, so we number $e = 4$ and backtrack to c. With no further unexplored neighbors, $c = 3$, and further backtracking yields $b = 2$ and $a = 1$. Node numbers now satisfy condition $\boxed{9.26}$.

Shortest Path Algorithm for Acyclic Digraphs

An acyclic digraph certainly has no negative dicycles, because it has no dicycles at all. Thus functional equations $\boxed{9.12}$ do apply.

Still, the most important convenience of acyclic digraphs in shortest path computation arises from the numbering of principle $\boxed{9.26}$. If we evaluate functional equations for nodes in that acyclic number sequence, each $v[k]$ is permanent as soon as it is computed. This is true because all terms in equations $\boxed{9.12}$ involve inbound arcs from lower-numbered nodes with previously fixed $v[i]$. Algorithm 9D provides details, and it should be no surprise that it is the most efficient possible.

ALGORITHM 9D: SHORTEST ONE TO ALL PATHS (ACYCLIC DIGRAPH)

Step 0: Initialization. Number nodes so that each arc (i, j) of the digraph has $i < j$. Then set source s optimal path length

$$v[s] \leftarrow 0$$

Step 1: Stopping. Terminate if all $v[k]$ have now been fixed. Otherwise, let p be the lowest number of an unprocessed node.

Step 2: Processing. If there are no inbound arcs at node p, set $v[p] \leftarrow +\infty$. Otherwise, compute

$$v[p] \leftarrow \min\{v[i] + c_{i,p} : (i, p) \text{ exists}\}$$

and let $d[p] \leftarrow$ the number of a node i achieving the minimum. Then return to Step 1.

$\boxed{9.28}$ Algorithm 9D is the most efficient possible for computing shortest paths from one node to all others in acyclic digraphs.

Acyclic Shortest Path Example

None of the models of Section 9.1 are posed on acyclic digraphs, so we will illustrate Algorithm 9D on the digraph of Figure 9.11. Table 9.4 provides results.

Notice that the digraph in Figure 9.11 is already numbered as in principle $\boxed{9.26}$, and we will assume that paths are to begin at source $s = 1$. Initialization sets $v[1] \leftarrow 0$.

Node processing now takes place in node number sequence. At node $p = 2$, values are updated as

$$\begin{aligned}
v[2] &\leftarrow \min\{v[1] + c_{1,2}\} \\
&= \min\{0 + 5\} \\
&= 5
\end{aligned}$$

with $d[2] \leftarrow 1$. Then at node $p = 3$, we compute

$$\begin{aligned}
v[3] &\leftarrow \min\{v[1] + c_{1,3}, v[2] + c_{2,3}\} \\
&= \min\{0 + 8, 5 - 10\} \\
&= -5
\end{aligned}$$

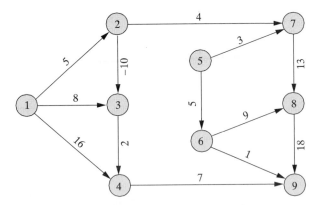

FIGURE 9.11   Acyclic Shortest Path Example

TABLE 9.4   Solution to Acyclic Shortest Path Example

p	$v[p]$	$d[p]$	p	$v[p]$	$d[p]$
1	0	—	6	∞	—
2	5	1	7	9	2
3	−5	2	8	22	7
4	7	3	9	14	4
5	∞	—			

with $d[3] \leftarrow 2$. Computation now continues to $p = 4$, $p = 5$, and so on, until we complete at $p = 9$. If required, shortest paths can be retrieved from final $d[k]$ by applying principle 9.18 .

9.7 CPM PROJECT SCHEDULING AND LONGEST PATHS

Section 9.1 introduced some of the more common shortest path model forms. However, many other models are closely related, even though they appear at first glance to have nothing to do with computing paths.

Project Management

One of the most commonly occurring such problems arises in the management of large work projects. For planning and control purposes, projects are usually subdivided into a collection of work **activities** that must all be completed to accomplish the project. Each activity has an estimated **duration**,

$$a_k \triangleq \text{time required to accomplish activity } k$$

and a list of predecessor activities.

> 9.29 | Activity j is a **predecessor** of activity k if activity j must be completed before activity k can begin.

The main issue is to compute an **early start schedule** for the project. That is, we want to know the earliest time that each activity can begin, subject to the requirement that all its predecessor activities have already completed.

EXAMPLE 9.4: WE BUILD CONSTRUCTION

For a contrived but suggestive example, consider We Build Construction's latest project. We Build is developing a 1-story medical office building on available land near a hospital.

Table 9.5 details the project's 9 work activities. For example, the table shows the estimated duration of the heating and air conditioning activity $k = 7$ is $a_7 = 13$ days, and that activity 7 cannot begin until predecessor activities $2 = $ rough plumbing and $4 = $ structural members have been completed.

TABLE 9.5 We Build Construction Example Tasks

k	Activity	Duration, a_k (days)	Predecessor Activities
1	Foundation	15	—
2	Rough plumbing	5	—
3	Concrete slab	4	1, 2
4	Structural members	3	3
5	Roof	7	4
6	Rough electrical	10	4
7	Heating and air conditioning	13	2, 4
8	Walls	18	4, 6, 7
9	Interior finish	20	5, 8

To plan materials deliveries and arrange subcontractors to do the various activities, We Build needs a schedule. In particular, they want to know the earliest time after project start that each activity can begin.

CPM Project Networks

To address project scheduling problems with shortest path technology, we require a network or graph. The **critical path method (CPM)** forms such project networks from precedence relationships.

> **9.30** | **CPM project networks** have special *start* and *finish* nodes, plus one node for each activity. Arcs of zero length connect *start* to all activities without predecessors. Other arcs of length a_k connect each activity node k to all activities of which it is a predecessor, or to *finish* if there are no such nodes.

Figure 9.12 illustrates for the We Build example of Table 9.5. The construction begins by creating a node for each of the 9 activities, plus special *start* and *finish* nodes to represent the beginning and ending of project activity. Each precedence relationship generates an arc with length equal to the predecessor's duration. For example, arc (8,9) represents the fact that activity 8 is a predecessor of activity 9. Its length is $a_8 = 18$ days.

Special arcs from *start* and to *finish* complete the digraph. Zero-length arcs lead from *start* to each activity having no predecessors. In the We Build example those are $1 = $ foundation and $2 = $ rough plumbing. Arcs of length a_k join activities that are the predecessors of no other activity to *finish*. Here only activity $9 = $ interior finish qualifies.

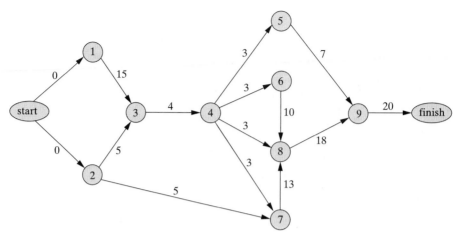

FIGURE 9.12   We Build Construction Project Network

SAMPLE EXERCISE 9.15: CONSTRUCTING A CPM NETWORK

The following table lists the activities required of a political advance team in arranging a campaign rally. Construct the corresponding CPM project network.

k	Activity	Duration, a_k (days)	Predecessor Activities
1	Contact local party	2	—
2	Find location	$1\frac{1}{2}$	1
3	Arrange date and time	1	1, 2
4	Notify news media	1	3
5	Arrange sound system	3	3
6	Coordinate police security	1	3
7	Install speaking platform	$1\frac{1}{2}$	3, 5
8	Decorate platform and site	1	7

Modeling: Following construction ⬛9.30⬛ , we obtain the following CPM project network:

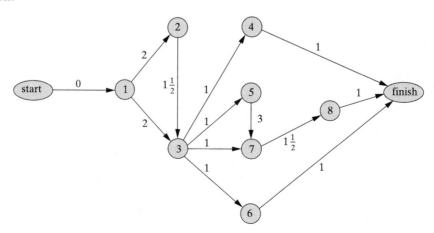

CPM Schedules and Longest Paths

Focus on activity 3 in Figure 9.12. When is the earliest time it can begin? One limit comes from predecessor activity 1 with duration 15. Another derives from predecessor activity 2 with duration 5. Both must be completed before activity 3 can begin. Since neither has a predecessor, both can begin at time 0. Thus the earliest start time for activity 3 is day

$$\max\{a_1, a_2\} = \max\{15, 5\} = 15$$

Notice that the constraint imposed by predecessor 1 corresponds to path *start–1–3* in Figure 9.12, and that of predecessor 2 corresponds to path *start–2–3*. The longest such path defines the **early start time** for activity 3. This is true in general.

> **9.31** The earliest start time of any activity k in a project equals the length of the longest path from *start* to node k in the corresponding project network.

The longest path to node k is the longest succession of activities that must complete before activity k can begin.

Critical Paths

Such longest paths are also called **critical paths**, from which the Critical Path Method derives its name, because delay in any activity along the critical path to node k will delay the start of activity k. The most important is the critical path from *start* to *finish*, which links activities that could delay completion of the entire project if they extend beyond their planned durations. Early start times and critical paths for all activities in the We Build project are shown in the following:

k	Activity	Early Start	Critical Path
1	Foundation	0	start–1
2	Rough plumbing	0	start–2
3	Concrete slab	15	start–1–3
4	Structural members	19	start–1–3–4
5	Roof	22	start–1–3–4–5
6	Rough electrical	22	start–1–3–4–6
7	Heating and air conditioning	22	start–1–3–4–7
8	Walls	35	start–1–3–4–7–8
9	Interior finish	53	start–1–3–4–7–8–9

In a similar way the least possible duration of a project corresponds to the longest *start*-to-*finish* path.

> **9.32** The minimum time to complete a project equals the length of the longest or critical path from *start* to *finish* in the corresponding project network.

For We Build, this project critical path runs

start–1–3–4–7–8–9–finish

with a total length of 73. Thus the minimum time to complete the building detailed in Table 9.5 is 73 days.

SAMPLE EXERCISE 9.16: IDENTIFYING CRITICAL PATHS

Use inspection to determine the earliest start times and critical paths for all activities of the political advance project in Sample Exercise 9.15. Also determine the minimum project duration and the activities which, if delayed, will delay completion of the whole project.

Analysis: Inspecting for the longest paths in the network of Sample Exercise 9.15, start times and critical paths are as follows:

k	Activity	Early Start	Critical Path
1	Contact local party	0	start–1
2	Find location	2	start–1–2
3	Arrange date and time	$3\frac{1}{2}$	start–1–2–3
4	Notify news media	$4\frac{1}{2}$	start–1–2–3–4
5	Arrange sound system	$4\frac{1}{2}$	start–1–2–3–5
6	Coordinate police security	$4\frac{1}{2}$	start–1–2–3–6
7	Install speaking platform	$7\frac{1}{2}$	start–1–2–3–5–7
8	Decorate platform and site	9	start–1–2–3–5–7–8

Minimum project duration will correspond to the longest *start*-to-*finish* path

$$start–1–2–3–5–7–8–finish$$

of length 10 days. Thus the activities that must complete as planned if the project is not to be delayed are activities 1, 2, 3, 5, 7, and 8.

Computing an Early Start Schedule for the We Build Construction Example

Algorithm 9E details a longest path scheme for computing early start times. Table 9.6 shows the results from application of Algorithm 9E to compute an early start schedule for our We Build example. Notation now signifies the following:

$$v[k] \triangleq \text{length of the longest path from source } start \text{ to node } k$$

$$d[k] \triangleq \text{immediate predecessor of node } k \text{ in a longest path}$$

TABLE 9.6   Early Start Schedule for the We Build Example

p	$v[p]$	$d[p]$	p	$v[p]$	$d[p]$
start	0	—	6	22	4
1	0	start	7	22	4
2	0	start	8	35	7
3	15	1	9	53	8
4	19	3	finish	73	9
5	22	4			

ALGORITHM 9E: CPM EARLY START SCHEDULING

Step 0: Initialization. Number activity nodes so that each precedence arc (i, j) of the CPM project network has $i < j$. Then initialize the schedule time of the project *start* node as

$$v[start] \leftarrow 0$$

Step 1: Stopping. Terminate if the early start time of the project *finish* node has been fixed. Otherwise, let p be the lowest number of an unprocessed node.

Step 2: Processing. Compute the activity p early start schedule time

$$v[p] \leftarrow \max\{v[i] + a_i : i \text{ is a predecessor of } p\}$$

and let $d[p] \leftarrow$ the number of a node i achieving the maximum. Then return to Step 1.

As before, computation begins by initializing source node time $v[start] \leftarrow 0$. Then values for other nodes are established in increasing number order. First,

$$v[1] \leftarrow \max\{v[start] + a_{start}\}$$
$$= \max\{0 + 0\} = 0$$

with $d[1] \leftarrow start$. In a similar way, $v[2] \leftarrow 0$ and $d[2] \leftarrow start$. Then, at $p = 3$

$$v[3] \leftarrow \max\{v[1] + a_1, v[2] + a_2\}$$
$$= \max\{0 + 15, 0 + 5\} = 15$$

making $d[3] \leftarrow 1$. The final results in Table 9.6 correspond exactly to the early start schedule detailed above.

Critical paths can be retrieved using optimal $d[k]$. For example, the entire project's critical path

$$start–1–3–4–7–8–9–finish$$

is recovered by backtracking (principle 9.18) from *finish* as

$$d[finish] = 9$$
$$d[9] = 8$$
$$d[8] = 7$$
$$d[7] = 4$$
$$d[4] = 3$$
$$d[3] = 1$$
$$d[1] = start$$

SAMPLE EXERCISE 9.17: COMPUTING A CPM EARLY START SCHEDULE

Return to the political advance project of Sample Exercise 9.15, and apply Algorithm 9E to compute an early start schedule and identify the critial path for the entire project *finish*.

Analysis: Following Algorithm 9E, $v[p]$ and $d[p]$ are set in increasing node number sequence. Results are summarized in the following table:

p	$v[p]$	$d[p]$	p	$v[p]$	$d[p]$
start	0	—	5	$4\frac{1}{2}$	3
1	0	start	6	$4\frac{1}{2}$	3
2	2	1	7	$7\frac{1}{2}$	5
3	$3\frac{1}{2}$	2	8	9	7
4	$4\frac{1}{2}$	3	finish	10	8

The critical path from *start* to *finish* is traced $d[finish] = 8$, $d[8] = 7$, $d[7] = 5$, $d[5] = 3$, $d[3] = 2$, $d[2] = 1$ and $d[1] = start$, yielding

$$start–1–2–3–5–7–8–finish$$

Late Start Schedules and Schedule Slack

Computations of early start Algorithm 9E implicitly assume that all activities can start at time 0. Early start schedules are then derived by finding the longest path of predecessor activity durations and summing (principle 9.31).

Suppose now that we have a **due date** for the project, that is, a time by which all activity must be completed. For example, our We Build project might be required to be complete on day 80. We can perform longest path computations in reverse to compute a **late start time** for each activity showing the latest it can begin if the due date is to be met.

> 9.33 | The latest start time of any activity k in a project equals the due date less the length of a longest path from k to the *finish* node in the corresponding CPM network.

To find this late start schedule, we simply reverse the sequence of computations in Algorithm 9E. Table 9.7 illustrates for the We Build example with a due date of 80.

Computation begins by fixing the late start schedule for the *finish* node at the due date. Then we advance to the highest-numbered activity, 9. Its late start time is the minimum of those for successor activities less its own duration, or $80 - 20 = 60$. Continuing in this way, the late start time for activity 4 is computed:

$$v[4] = \min\{v[5], v[6], v[7], v[8]\} - a_4$$
$$= \min\{53, 32, 29, 42\} - 3$$
$$= 26$$

and so on, until we complete the *start* node.

With both early and late schedules in hand, we can also subtract to determine the schedule slack for each activity.

> 9.34 | **Schedule slack** is the difference between the earliest and latest times that an activity can begin.

TABLE 9.7   Late Start Schedule for the We Build Example

p	$v[p]$	$d[p]$	p	$v[p]$	$d[p]$
finish	80	—	4	26	7
9	60	finish	3	22	4
8	42	9	2	17	3
7	29	8	1	7	3
6	32	8	start	7	1
5	53	9			

That is, schedule slack shows how much discretion we have in scheduling any single activity while meeting the overall project due date. Table 9.8 computes such slacks from the We Build example activities in Tables 9.6 and 9.7.

TABLE 9.8   Schedule Slack for the We Build Example

k	Activity	Early Start	Late Start	Slack
1	Foundation	0	7	7
2	Rough plumbing	0	17	17
3	Concrete slab	15	22	7
4	Structural members	19	26	7
5	Roof	22	53	31
6	Rough electrical	22	32	10
7	Heating and air conditioning	22	29	7
8	Walls	35	42	7
9	Interior finish	53	60	7

SAMPLE EXERCISE 9.18: COMPUTING LATE START SCHEDULES AND SLACK

Return to the political advance project of Sample Exercises 9.15 and 9.17. Compute the corresponding late start schedule for each activity and its schedule slack assuming that all work must be completed by day 10.

Analysis: As above, late start times are set in decreasing activity number sequence. Then differences are computed versus early start times to determine slack. The result is summarized in the following table:

k	Activity	Early Start	Late Start	Slack
1	Contact local party	0	0	0
2	Find location	2	2	0
3	Arrange date and time	$3\frac{1}{2}$	$3\frac{1}{2}$	0
4	Notify news media	$4\frac{1}{2}$	9	$4\frac{1}{2}$
5	Arrange sound system	$4\frac{1}{2}$	$4\frac{1}{2}$	0
6	Coordinate police security	$4\frac{1}{2}$	9	$4\frac{1}{2}$
7	Install speaking platform	$7\frac{1}{2}$	$7\frac{1}{2}$	0
8	Decorate platform and site	9	9	0

Difficulty of General Longest Path Problems

Longest path problems are the maximize analogs of the shortest path problems which were solved very efficiently by Algorithms 9A to 9D. All the functional equation and algorithmic technology of Sections 9.2 to 9.7 can be adapted to treat longest path cases simply by substituting *max* for *min* in update operations.

Notice, however, that the difficult case of negative dicycles in shortest path problems (principle 9.9) becomes the case of **positive dicycles** in longest path models—dicycles of positive total length.

Elementary dynamic programming methods usually cannot deal with longest path problems on graphs with positive dicycles. The fact that every one of the graphs of Section 9.1 contains positive dicycles explains why longest path problems are generally considered much less tractable than shortest path. Positive dicycles are far more common than their negative relatives.

Acyclic Character of Project Networks

Positive dicycles cannot occur in acyclic digraphs, which have no dicycles at all. We can do CPM scheduling and the associated longest path computations because CPM networks fit this acyclic assumption.

> 9.35   Every well-formed project network is acyclic.

To see why, recall that *start* nodes in construction 9.30 have only outbound arcs, and *finish* nodes have only inbound arcs. Thus any dicycle would have to run through other nodes (i.e., those for activities). But arcs connecting activity nodes correspond to precedence relationships. Any dicycle i–j–k–i

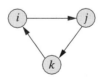

would mean that activity i must precede activity j, activity j must precede activity k, and activity k must precede activity i. This is physically impossible. A project network with a dicycle admits no feasible schedule (if durations $a_k > 0$).

9.8 DISCRETE DYNAMIC PROGRAMMING MODELS

Elementary discrete **dynamic programming** encompasses more than shortest paths, including a variety of problems that can still be modeled as shortest or longest path problems in a suitable digraph even though they have nothing to do with routes and distances. In this section we explore some of the possibilities.

Sequential Decision Problems

One common feature of problems amenable to dynamic programming is **sequential decision making**—decisions that can be arranged in a clear sequence. Then the decisions can be confronted one by one in sequence. That same sequence produces an acyclic digraph on which a shortest or longest path problem can be posed.

EXAMPLE 9.5: WAGNER–WHITIN LOT SIZING

A classic application, due to H. Wagner and T. Whitin, concerns **lot sizing**, that is, planning production in an environment where there is a substantial **setup cost** incurred with each production run. Table 9.9 details an example. For each time period $k = 1, \ldots, n$, we know

$r_k \triangleq$ quantity of product required at time k
$s_k \triangleq$ setup cost if production occurs at time k
$p_k \triangleq$ variable, per unit cost of production at time k
$h_k \triangleq$ unit cost of holding goods through period k

TABLE 9.9   Wagner–Whitin Lot Sizing Data

	Period, k					
	1	2	3	4	5	6
Requirement, r_k	10	40	20	5	5	15
Setup cost, s_k	50	50	50	50	50	50
Production cost, p_k	1	3	3	1	1	1
Holding cost, h_k	2	2	2	2	2	2

An optimal solution must carefully balance production and holding costs. Sometimes it will be better to pay the setup cost and produce just as goods are required. Other times we will produce in one period for the next several, incurring a holding cost of storage and lost investment income on goods not required immediately.

States in Dynamic Programming

For our Wagner–Whitin inventory example the passage of time sequences the required decisions. To form a dynamic programming model, we must characterize states of incomplete solution that capture the history on which each new decision should be based.

> **9.36**   **States** in dynamic programming characterize conditions of incomplete solution at which decisions should be considered.

If an optimal solution is known for any state, those of subsequent states can build upon it.

Development of a state description for Wagner–Whitin lot sizing begins with a simplifying insight: production needs to occur in an optimal solution only when no inventory is held through the preceding period. That is, we should produce or hold for any period k, but not both. Any solution that combined holding and production could at least be matched by shifting the full r_k to whichever of unit holding cost since the last production and variable production cost p_k was lower.

The result is a definition of states. We may say that the process has reached state k when requirements for periods $\ell < k$ have been fulfilled and there is no on-hand inventory.

Digraphs for Dynamic Programs

A suitable definition of states begins the construction of the digraph underlying any elementary dynamic program.

> 9.37 | Nodes of the digraph associated with any dynamic program correspond to states of incomplete solution.

The remaining element is **decisions**. What options are available upon reaching any given state, and what state will we reach if a decision is adopted?

> 9.38 | Arcs of the digraph associated with any dynamic program correspond to decisions. They link state nodes with a subsequent state to which the decision leads.

Figure 9.13 displays the digraph for our Wagner–Whitin lot-sizing example. Here decisions correspond to producing in some period to meet its requirements and, possibly, those of several subsequent periods. For example, the arc from state $k = 3$ to $\ell = 6$ reflects production at period 3 to meet requirements $r_3 = 20$, $r_4 = 5$, and $r_5 = 5$, arriving at period 6 with zero inventory. Its cost is

$$
\begin{aligned}
c_{3,6} &= \text{setup} + \text{production} + \text{holding} \\
&= s_3 + (r_3 + r_4 + r_5)\,p_3 + [(r_4 + r_5)h_3 + (r_5)h_4] \\
&= 50 + (20 + 5 + 5)3 + (5 + 5)2 + (5)2 \\
&= 170
\end{aligned}
$$

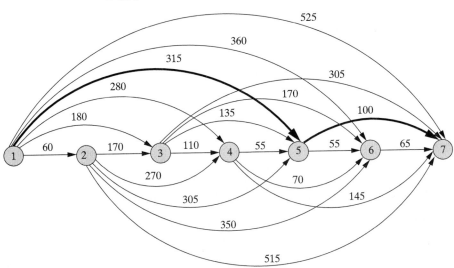

FIGURE 9.13 Digraph for Wagner–Whitin Example

Dynamic Programming Solutions as an Optimal Path

Notice now that each path from node 1 in the digraph of Figure 9.13 to node $(n+1) = 7$ constitutes a lot-sizing plan. For example, path 1–3–6–7 corresponds to production in period 1 for requirements r_1 and r_2, then in period 3 for requirements r_3, r_4, and r_5, then in period 6 for r_6. The shortest such path identifies an optimal solution.

> **9.39** Optimal solutions to elementary dynamic programs correspond to shortest or longest paths from a beginning to an ending state in the associated digraph.

They can be solved by applying the appropriate optimal path algorithm.

Like most dynamic programs, our lot-sizing problem is posed over an acyclic digraph because decision arcs at any state lead only to later states. Application of shortest acyclic Algorithm 9D produces the optimal solution highlighted in Figure 9.13. It corresponds to producing in periods 1 and 5 at total cost of 415.

SAMPLE EXERCISE 9.19: FORMULATING DYNAMIC PROGRAMS

The management team of a 5-year research project is planning a replacement policy for its personal computers. New models may be purchased for $3000 each. If sold after 1 year, they retain a salvage value of $1200. If sold after 2 years, the salvage value falls to $500, and after 3 years the units are obsolete and have no value. Maintenance costs also increase with age. They are estimated at $300 in the first year of service, $400 in the second year, and $500 in the third. Formulate a dynamic program and associated digraph to compute a minimum total cost replacement policy.

Modeling: States in this problem correspond to years completed, and decisions relate to keeping computers for 1, 2, or 3 years. Cost of any decision is

$$(\text{purchase cost}) - (\text{salvage value}) + (\text{maintenance cost})$$

The result is the following digraph:

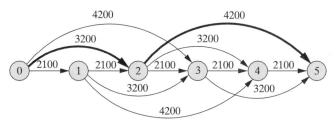

An optimal solution corresponds to a shortest path from time 0 through year 5. An optimum is highlighted in the figure with total cost $7400.

Dynamic Programming Functional Equations

Although it is usually instructive to draw the digraph associated with a dynamic program, it is common to proceed directly to the underlying functional equation recursions.

| 9.40 | Dynamic programming **functional equation** recursions encode connections among optimal values for different states of incomplete solution.

The process begins by interpreting the optimal value of any state. For example, with our Wagner–Whitin lot-sizing model,

$$v[k] \triangleq \text{minimum cost of fulfilling requirements for periods } \ell < k$$
$$\text{and arriving at } k \text{ with no inventory}$$

Dynamic programming solution of the problem is then encapsulated in functional equation recursions

$$v[1] = 0$$
$$v[k] = \min\{v[\ell] + c_{\ell,k} : 1 \le \ell < k\} \qquad k = 2, \ldots, n+1$$

where

$$c_{\ell,k} \triangleq s_\ell + (r_\ell + \cdots + r_{k-1})p_\ell + (r_{\ell+1} + \ldots + r_{k-1})h_\ell + \cdots + (r_{k-1})h_{k-2}$$

For example,

$$v[4] = \min\{v[1] + c_{1,4}, \ v[2] + c_{2,4}, \ v[3] + c_{3,4}\}$$
$$= \min\{v[1] + 280, \ v[2] + 270, \ v[3] + 110\}$$

These are precisely the shortest path functional equations of definition $\boxed{9.11}$ for the digraph of Figure 9.13.

SAMPLE EXERCISE 9.20: FORMULATING DYNAMIC PROGRAM RECURSIONS

Formulate functional equations for the dynamic programming model of Sample Exercise 9.19.

Modeling: For this model

$$v[k] \triangleq \text{minimum cost of providing computers through year } k$$

This leads to functional equations

$$v[0] = 0$$
$$v[k] = \min\{v[k-3] + c_3, \ v[k-2] + c_2, \ v[k-1] + c_1\} \qquad k = 1, \ldots, 5$$

where c_j is the total cost of keeping a computer for j years.

Dynamic Programming Models with Both Stages and States

A sequence of decisions based on states of incomplete solution lies at the heart of all discrete dynamic programs. However, it is often convenient to distinguish stages of decision making from states of solution.

| 9.41 | In dynamic programs with both stages and states, **stages** delineate the sequence of required decisions and **states** encode the conditions within which decisions can be considered.

EXAMPLE 9.6: PRESIDENT'S LIBRARY

We may illustrate with the design of shelving in the president's library.[1] The retiring president is collecting all his papers in a new presidential library. Materials for its archives are stored in covered cardboard boxes. All are 1.25 feet wide, but their height varies. Table 9.10 shows the estimated number of boxes at each height.

TABLE 9.10   President's Library Storage Requirements

i	1	2	3	4	5	6	7
Height in feet, h_i	0.25	0.40	0.80	1.00	1.50	2.00	3.0
Thousands of boxes, b_i	10	2	12	30	8	6	4

Boxes in the archives will be placed on metal shelves, with no box stacked on top of another. The issue is how much shelving will be required. Figure 9.14 shows that if, for example, two shelf spacings are employed, the total face area of shelving is

$$\begin{pmatrix} \text{linear feet of} \\ \text{boxes with} \\ \text{height} \le \text{smaller} \\ \text{shelf spacing} \end{pmatrix} \begin{pmatrix} \text{smaller} \\ \text{shelf} \\ \text{spacing} \end{pmatrix} + \begin{pmatrix} \text{linear feet of} \\ \text{boxes with} \\ \text{height} > \text{smaller} \\ \text{shelf spacing} \end{pmatrix} \begin{pmatrix} \text{largest} \\ \text{box} \\ \text{height} \end{pmatrix}$$

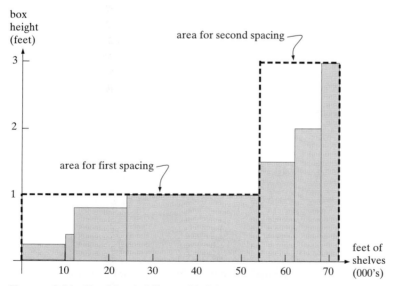

FIGURE 9.14   President's Library Shelving Face Areas

The most compact shelving obviously will result if different spacings are provided for each of the 7 sizes in Table 9.10. Still, designers would like to keep storage relatively uniform. Whether they choose to provide $1, 2, \ldots, 7$ spacings, they want to use the space as efficiently as possible.

---

[1]Based in part on F. F. Leimkuhler and J. G. Cox (1964), "Compact Book Storage in Libraries," *Operations Research, 12*, 419–427.

Dynamic Programming Modeling of the President's Library Example

To distinguish stages for a dynamic programming model of the president's library example, we must consider the sequence of decisions. Here decisions correspond to choices of shelf spacing. Thus we will have one stage for each distinct spacing. More precisely,

$$\text{stage } k \triangleq k\text{th-to-last choice of a spacing}$$

Next come states. On what preconditions should decisions be based? States are often the least intuitive part of forming a dynamic programming model. In our president's library example, the decision to allocate a new shelf spacing depends on what box sizes have already been provided for. Thus we define states

$$\text{state } i \triangleq \text{having provided shelves for box sizes } 0, 1, \ldots, i \ (0 \text{ indicates none})$$

Decisions, which constitute the third element of any dynamic programming model, are now easy to describe. The choice to provide a spacing for box sizes through j, given that those through i are already accommodated, means commitment of shelf face area

$$c_{i,j} \triangleq 1.25 \left(\sum_{i < s \leq j} b_s \right) h_j$$

For example, accommodating box sizes through $j = 6$, given that those up $i = 3$ have already been provided for, implies an area of

$$c_{3,6} = 1.25(b_4 + b_5 + b_6)h_6$$
$$= 1.25(30 + 8 + 6)(2.0) = 110.0$$

thousand square feet.

Figure 9.15 shows the digraph for this dynamic program. Notice that there is one group of nodes for each stage. In accord with principle 9.37 , nodes distinguish states. Arcs of the digraph reflect decisions (principle 9.38). For example, the arc from node 3 of stage 2 to node 6 of stage 1 corresponds to making the second-to-last spacing for boxes accommodate sizes 4 to 6. All arcs from the last stage lead to an artificial finish node that corresponds to having provided for all box sizes.

SAMPLE EXERCISE 9.21: MODELING WITH STAGES AND STATES

An entrepreneur with $8 million to invest is considering acquisition of four companies. She estimates the cost of acquiring the four at $5, $1, $2, and $7 million, respectively, and the present value of owning them at $8, $3, $4, and $10. Investments are on an "all or nothing" basis (i.e., she must buy entire companies). Model as a dynamic program the problem of deciding which investments she should choose.

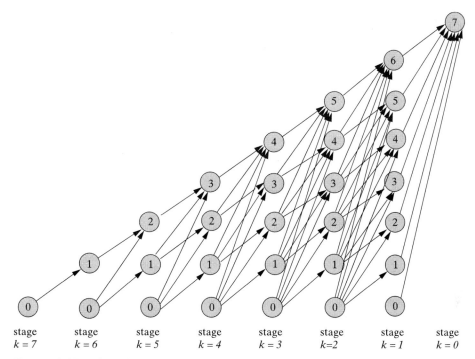

stage   stage   stage   stage   stage   stage   stage   stage
k = 7   k = 6   k = 5   k = 4   k = 3   k=2   k = 1   k = 0

FIGURE 9.15 Digraph for President's Library Example

Modeling: It is clear that decisions concern whether our entrepreneur should invest in each of the 4 companies. Thus we associate stages with companies $k = 1, 2, 3, 4$.

For states we must decide what information about prior decisions is needed to know whether a current one is possible. Here we need to know the amount of investment money remaining when the decision stage for a particular company is reached. The result is states $i = 0, 1, \ldots, 8$.

At each stage k and state i, two decisions are possible. The entrepreneur can choose not to buy the kth company and pass along amount i, or she can buy the kth company and advance to state $(i - a_k)$, where $a_k \triangleq$ purchase price of company k. Of course, the latter option is available only when $i \geq a_k$.

Denote

$$v_k[i] \triangleq \text{maximum total present worth obtainable from}$$
$$\text{companies } \ell \geq k \text{ given that } i \text{ million dollars is}$$
$$\text{available for investment}$$

Then functional equations are maximize versions of format (9.1):

$$v_k[i] = 0 \qquad\qquad\qquad k > 4 \text{ or } i < 0$$
$$v_k[i] = \max\{v_{k+1}[i], c_k + v_{k+1}[i - a_k]\} \qquad k = 1, \ldots, 4; \quad i = 0, \ldots, 8$$

where $c_k \triangleq$ present worth of company k. The associated digraph is as follows:

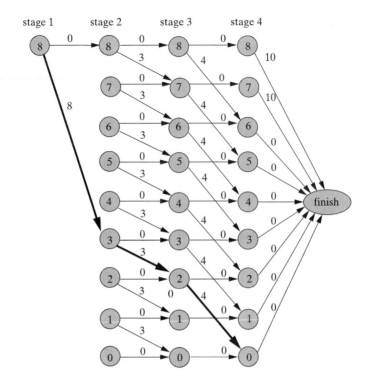

The highlighted optimal solution corresponds to a longest path from state 8 of stage 1 to *finish*.

Backward Solution of Dynamic Programs

Readers may wonder why we have chosen to number stages of the president's library example in reverse or **backward** sequence. This is to highlight the sequence of solution we will apply.

> **9.42**   Elementary dynamic programs are often most easily conceptualized in backward sequence, that is, proceeding from final to initial conditions.

Then optimal solutions correspond to shortest or longest paths to the final state in an associated digraph.

The president's library example is no exception to principle 9.42 . Its functional equations take the form

$$\begin{pmatrix} \text{optimal cost} \\ \text{to finish} \end{pmatrix} = \min \left\{ \begin{pmatrix} \text{immediate} \\ \text{decision} \\ \text{cost} \end{pmatrix} + \begin{pmatrix} \text{optimal cost} \\ \text{to finish} \\ \text{from the} \\ \text{resulting} \\ \text{state} \end{pmatrix} \right\} \tag{9.1}$$

where the minimization is over the available decisions. More precisely, define

$$v_k[i] \triangleq \text{optimal cost to finish from stage } k, \text{ state } i$$

Then

$$v_7[7] = 0$$
$$v_k[i] = \min\{c_{i,j} + v_{k-1}[j] : j > i\} \qquad k = 1, \ldots, 6; \quad i \le (7 - k)$$

An optimal solution now corresponds to a shortest path to *finish* rather than the usual shortest path from a single source. However, Algorithm 9D is easily adapted. We need only evaluate the $v_k[\ell]$ in reverse acyclic sequence.

Table 9.11 processing begins by setting the cost of the final state $v_{\text{finish}} \leftarrow 0$. Then all stage $k = 1$ states are solved in turn, each with only one out arc leading to *finish*.

A more typical step arises if fixing $v_2[3]$. This second-to-last shelf spacing (given boxes through size 3 are handled) could provide for size 4, or sizes 4 and 5, or sizes 4 to 6. Thus

$$v_2[3] \leftarrow \min\{c_{3,4} + v_1[4], \; c_{3,5} + v_1[5], \; c_{3,6} + v_1[6]\}$$
$$= \min\{37.5 + 67.5, \; 71.25 + 37.5, \; 110.0 + 15.0\}$$
$$= \min\{105.0, 108.75, 125.0\} = 105.0$$

thousand square feet, with the optimal decision leading to state 4.

TABLE 9.11 Backward Solution
of President's Library Example

Stage, k	State, i	Optimal Area, $v_k[i]$	Next State, $d_k[i]$
finish	—	0.000	0
1	0	270.000	7
	1	232.500	7
	2	225.000	7
	3	180.000	7
	4	67.500	7
	5	37.500	7
	6	15.000	7
2	0	135.000	4
	1	122.500	4
	2	120.000	4
	3	105.000	4
	4	50.000	6
	5	30.000	6
3	0	117.500	4
	1	105.000	4
	2	102.500	4
	3	87.500	4
	4	45.000	5
4	0	108.125	1
	1	100.000	4
	2	97.500	4
	3	82.500	4
5	0	103.125	1
	1	96.500	3
	2	94.500	3
6	0	99.625	1
	1	95.500	2
7	0	98.625	1

Multiple Problem Solutions Obtained Simultaneously

What solution is optimal in Table 9.11? There are several, depending on how many distinct shelf spacings are adopted. Optimal shelf areas are

$$
\begin{aligned}
v_1[0] &= 270.0 &\quad &\text{for } k = 1 \text{ spacing} \\
v_2[0] &= 135.0 &\quad &\text{for } k = 2 \text{ spacings} \\
v_3[0] &= 117.5 &\quad &\text{for } k = 3 \text{ spacings} \\
v_4[0] &= 108.125 &\quad &\text{for } k = 4 \text{ spacings} \\
v_5[0] &= 103.125 &\quad &\text{for } k = 5 \text{ spacings} \\
v_6[0] &= 99.625 &\quad &\text{for } k = 6 \text{ spacings} \\
v_7[0] &= 98.625 &\quad &\text{for } k = 7 \text{ spacings}
\end{aligned}
$$

with corresponding optimal solutions retrievable from corresponding $d_k[i]$ labels.

Solutions for multiple assumptions are a common benefit of dynamic programming solution.

> **9.43** The dynamic programming strategy of exploiting connections among optimal solutions for a number of related problems often implies that optima for multiple problem scenarios are produced by a single shortest or longest path computation.

EXERCISES

9-1 Consider the following graph.

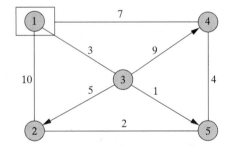

Numbers on arcs and edges represent lengths.

(a) ⊗ Identify the nodes, arcs, and edges of the graph.

(b) ⊗ Determine whether each of the following sequences is a path of the graph: 1–3–4–5, 2–5–3–4, 1–3–2–5–4, 1–3–4–1–2.

(c) ⊗ Direct the graph (i.e., exhibit a digraph with the same paths as those of the given graph).

9-2 Do Exercise 9-1 for the graph

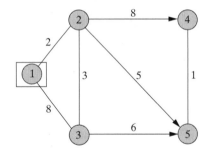

and sequences 3–2–1–3–5, 3–2–4, 1–3–2–4–5, 1–3–5–2.

9-3 Return to the problem of Exercise 9-1.

(a) ⊗ Find (by inspection) shortest paths from node 1 to all other nodes.

(b) ⊗ Verify that every subpath of the optimal 1 to 2 path in part (a) is itself optimal.

(c) ⊗ Detail your optimal solutions of part (a) in functional notation $v[k]$ and $x_{i,j}[k]$.

(d) ⊗ Write functional equations for the shortest path problems of part (a).

(e) Verify that the $v[k]$ of part (c) satisfy functional equations of part (d).

(f) ◻ Explain why functional equations are sufficient to characterize optimal values $v[k]$ for this instance.

9-4 Do Exercise 9-3 for the problem of Exercise 9-2, verifying the optimal 1 to 3 path in part (b).

9-5 Consider the following graph.

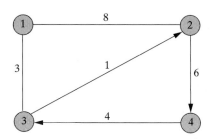

(a) ◻ Find (by inspection) shortest paths from all nodes to all other nodes.

(b) ◻ Verify that every subpath of the optimal 1 to 4 path in part (a) is itself optimal.

(c) ◻ Detail your optimal solutions of part (a) in functional notation $v[k][\ell]$ and $x_{i,j}[k][\ell]$.

(d) ◻ Write functional equations for the shortest path problems of part (a).

(e) Verify that the $v[k][\ell]$ of part (c) satisfy functional equations of part (d).

(f) ◻ Explain why functional equations are sufficient to characterize optimal values $v[k][\ell]$ for this instance.

9-6 Do Exercise 9-5 for the following graph, verifying the optimal 4 to 2 path in part (b).

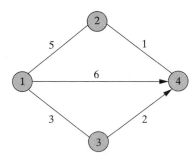

9-7 Consider the following digraph:

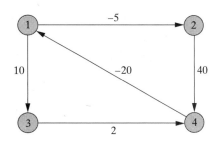

Numbers on arcs represent costs.

(a) ◻ Find (by inspection) shortest paths from node 1 to all other nodes.

(b) ◻ Enumerate the dicycles of the graph.

(c) ◻ Determine whether each dicycle is a negative dicycle.

(d) ◻ How does the presence of negative dicycles in this instance make it more difficult for an algorithm to compute shortest paths?

9-8 Do Exercise 9-7 for the digraph

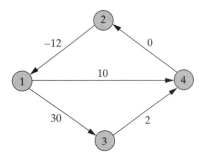

9-9 Return to the graph of Exercise 9-1, and suppose that we seek shortest paths from node 1 to all other nodes.

(a) ◻ Explain why Bellman-Form Algorithm 9A can be employed to compute the required shortest paths.

(b) ◻ Apply Algorithm 9A to compute the lengths of shortest paths from node 1 to all other nodes.

(c) ◻ Use $d[k]$ labels of your computations in part (b) to recover all optimal paths.

(d) ◻ Verify the interpretation of interim labels $v^{(t)}[k]$ as lengths of shortest paths using at most t arcs/edges by showing that values at the end of iteration $t = 2$ in your computations of part (b) correspond to lengths of shortest paths using 1 or 2 arcs/edges.

(e) ◻ Determine the maximum number of arcs/edges in any path of this graph, and explain how this bounds the computation required for Algorithm 9A.

9-10 Do Exercise 9-9 on the graph of Exercise 9-2.

9-11 Use Bellman-Ford Algorithm 9A to identify a negative dicycle in each of the following graphs.

(a) ◻ The digraph of Exercise 9-7
(b) The digraph of Exercise 9-8

9-12 Return to the graph of Exercise 9-5, and suppose that we seek shortest paths from all nodes to all other nodes.

(a) ◻ Explain why Floyd–Warshall Algorithm 9B can be employed to compute the required shortest paths.
(b) ◻ Apply Algorithm 9B to compute the length of shortest paths from all nodes to all other nodes.
(c) ◻ Use $d[k][\ell]$ labels of your computations in part (b) to recover all optimal paths.
(d) ◻ Verify the interpretation of interim labels $v^{(t)}[k][\ell]$ as lengths of shortest paths using only intermediate nodes up to t by showing that values at the end of iteration $t = 2$ in your computations of part (b) correspond to lengths of shortest paths using only nodes 1 and 2 as intermediates.

9-13 Do Exercise 9-12 on the graph of Exercise 9-6.

9-14 Use Floyd–Warshall Algorithm 9B to identify a negative dicycle in each of the following graphs.

(a) ◻ The digraph of Exercise 9-7
(b) The digraph of Exercise 9-8

9-15 Return to the graph of Exercise 9-1, and suppose that we seek shortest paths from node 1 to all other nodes.

(a) ◻ Explain why Dijkstra Algorithm 9C can be employed to compute the required shortest paths.
(b) ◻ Apply Algorithm 9C to compute the length of shortest paths from node 1 to all other nodes.
(c) ◻ Use $d[k]$ labels of your computations in part (b) to recover all optimal paths.
(d) ◻ Verify the interpretation of interim labels $v[k]$ as lengths of shortest paths using only permanently labeled nodes by showing that values

after two permanent nodes have been processed in your computations of part (b) correspond to lengths of shortest paths using only those nodes (and the destination).

9-16 Do Exercise 9-15 on the graph of Exercise 9-2.

9-17 For each of the following digraphs, determine which of Bellman-Ford Algorithms 9A, Floyd–Warshall Algorithm 9B, and Dijkstra Algorithm 9C could be applied to compute shortest paths from node 1 to all others. Then, if any is applicable, choose the most efficient and apply it to compute the required paths.

(a) ◻

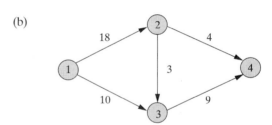

(b)

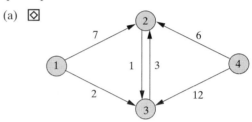

(c) ◻

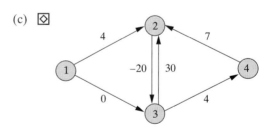

(d)

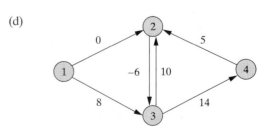

9-18 Demonstrate whether each of the following digraphs is acyclic by exhibiting a dicycle or numbering nodes so that every arc (i, j) has $i < j$.

(a)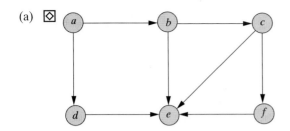

(b)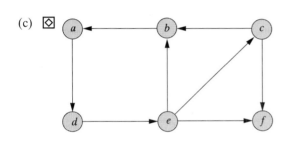

(c)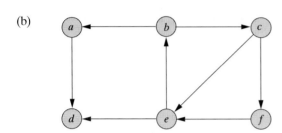

(d)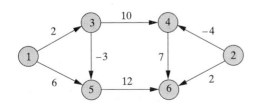

9-19 Consider the following digraph:

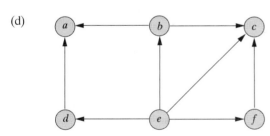

Suppose that labels on arcs are cost and that we seek a minimum total cost path from node 1 to all other nodes.

(a) ◻ Demonstrate that acyclic shortest path Algorithm 9D can be applied to compute the required paths.

(b) ◻ Use Algorithm 9D to compute the lengths of shortest paths from node 1 to all other nodes.

(c) ◻ Use $d[k]$ labels from your computations in part (b) to recover all optimal paths.

9-20 Do Exercise 9-19 for the digraph

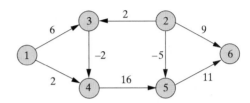

9-21 Determine whether acyclic shortest path Algorithm 9D could be applied to compute shortest paths from node 1 to all other nodes in each digraph of Exercise 9-17. If so, explain whether it would be more efficient than the Bellman–Ford, Floyd–Warshall, and Dijkstra alternatives, and why.

9-22 The table that follows lists the tasks that must be performed in preparing a simple breakfast. The table also shows the number of minutes each requires and the tasks that must be completed before each can begin.

k	Task	Time	Predecessors
1	Boil water	5	None
2	Get dishes	1	None
3	Make tea	3	1, 2
4	Pour cereal	1	2
5	Fruit on cereal	2	4
6	Milk on cereal	1	4
7	Make toast	4	None
8	Butter toast	3	7

(a) ◻ Construct the corresponding CPM project network 9.30 .

(b) ◻ Use given activity numbers to verify that your project network is acyclic.

(c) ◻ Apply CPM scheduling Algorithm 9E to compute early start times for each activity and an early finish time for the entire project.

(d) ◻ Use $d[k]$ labels of your scheduling computations to identify the activities along a critical path from project start to finish.

(e) ◻ Compute late start times for each activity assuming that the breakfast must be complete in

10 minutes, and combine with part (c) to determine schedule slacks.

9-23 The table that follows lists the activities that must be performed in organizing a soccer tournament. The table also shows the estimated number of days that each will require and the activities that must be completed before each can begin.

k	Activity	Time	Predecessors
1	Select dates	1	None
2	Recruit sponsors	4	1
3	Set fee	1	2
4	Buy souvenirs	5	1, 2
5	Mail invitations	1	2, 3
6	Wait responses	4	5
7	Plan pairings	1	6
8	Team packets	1	4, 7

(a) through (e) as in Exercise 9-22, assuming that the organization must be completed in 13 days.

9-24 Construction of a small two-story house involves the tasks listed in the following table. The table also shows the estimated duration of each task in days and the tasks that must complete before it can begin.

Task	Time	Predecessors
Foundation (FD)	8	None
Concrete slab (CS)	5	FD
First bearing walls (1B)	3	CS
First internal walls (1I)	4	CS
First finishing (1F)	12	1B, 1I, FL
Second floor (FL)	3	1B
Second bearing walls (2B)	4	2FL
Second internal walls (2I)	5	2FL
Second finishing (2F)	10	2B, 2I, R
Roof (RF)	2	2B

(a) ⊘ Construct the corresponding CPM project network 9.30 .

(b) ⊘ Number activity nodes in your project network so that every arc (i, j) has $i < j$.

(c) ⊘ Use CPM scheduling Algorithm 9E with your activity numbering to compute early start times for each activity and an early finish time for the entire project.

(d) ⊘ Use $d[k]$ labels of your scheduling computations to identify the activities along a critical path from project start to finish.

(e) ⊘ Compute late start times for each activity assuming that construction must be complete in 35 days, and combine with part (c) to determine schedule slacks.

9-25 The table that follows shows the activities required to construct a new computer laboratory, along with their estimated durations (in weeks) and predecessor activities.

Activity	Time	Predecessors
Order furniture (OF)	1	None
Order computers (OC)	1	None
Order software (OS)	1	OC
Furniture delivery (FD)	6	OF
Computer delivery (CD)	3	OC
Software delivery (SD)	2	OS
Assemble furniture (AF)	1	P, FD
Install computers (IC)	1	AF, CD
Install software (IS)	1	IC, SD
Wire room (W)	2	None
Paint room (P)	1	W

(a) through (e) as in Exercise 9-24, assuming that the lab must be completed in 15 weeks.

9-26 A pharmaceutical manufacturer must supply 30 batches of its new medication in the next quarter, then 25, 10, and 35 in successive quarters. Each quarter in which the company makes product requires a $100,000 setup, plus $3000 per batch produced. There is no limit on production capacity. Batches can be held in inventory, but the cost is a high $5000 per batch per quarter. The company seeks a minimum total cost production plan.

(a) Explain why this problem can be approached by dynamic programming, with states $k = 1, \ldots, 5$ representing the reaching of quarter k with all earlier demand fulfilled and no inventory on hand.

(b) ⊘ Sketch the digraph corresponding to the dynamic program structure of part (a). Include costs on all arcs.

(c) Explain why the feasible production plans correspond exactly to the paths from node $k = 1$ to node $k = 5$ in your digraph.

(d) ⊘ Solve a shortest path problem on your digraph to compute an optimal production plan.

(e) ⊘ Use your computations in part (d) to compute an optimal production plan for the first two quarters.

9-27 Do Exercise 9-26 with all parameters the same except a holding cost of $2000.

9-28 A copy machine repairman has four pieces of test equipment for which he estimates 25%, 30%, 55%, and 15% chances of using them at his next stop. However, the devices weigh 20, 30, 40, and 20 pounds, respectively, and he can carry no more than 60 pounds. The repairman seeks a maximum utility feasible collection of devices to carry with him.

(a) Explain why this problem can be approached by dynamic programming, with stages $k = 1, \ldots, 4$ representing the four devices and states $w = 0, 10, 20, 30, 40, 50, 60$ in each stage indicating reaching that decision stage with w units of weight limit remaining.

(b) ◻ Sketch the digraph corresponding to the dynamic program structure of part (a). Include objective function contributions on all arcs.

(c) Explain why the feasible production plans correspond exactly to the paths from stage $k = 1$, state $w = 60$ to the last stage in your digraph.

(d) ◻ Solve a longest path problem on your digraph to compute an optimal toolkit.

9-29 Do Exercise 9-28 with all parameters the same except a weight limit of 50 pounds.

9-30 The figure that follows shows a partially complete layout for a circuit board of a large computer peripheral. Lines show channels along which circuits can be placed, together with the lengths of the channels in centimeters. Several circuits can be placed in a single channel (on different layers).

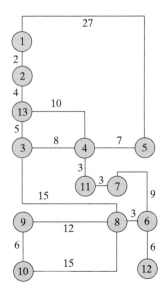

The last step in the design is to choose the routing from a component to be installed at point 1 to connections at points 8, 10, 11, and 12.

(a) Explain why this problem can be modeled as a shortest path problem.

(b) Explain why the most efficient procedure available from this chapter for computing optimal circuit routes is Dijkstra Algorithm 9C.

(c) ◻ Apply Algorithm 9C to compute optimal circuit routes to the 4 specified points.

9-31 The figure that follows shows the links of a proposed campus computer network. Each node is a computer, and links are fiber-optic cable.

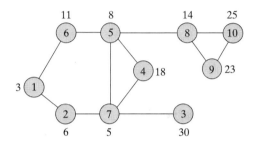

Designers now want to decide how E-mail, which will be broken into standard-length packets, should be routed from internet gateway node 1 to/from all other nodes. For example, E-mail for node 4 might be transmitted by 1 to 6, then repeated by 6 to 5, the repeated by 5 to 4. Numbers on nodes in the figure indicate minimum times (in nanoseconds) required by the corresponding computer to transmit or receive a message packet. The time to send a packet along any link of the network is the maximum of the times for the associated sending and receiving computers.

(a) Explain why this problem can be modeled as a shortest path problem, and sketch the graph and edge weights over which optimal paths are to be computed.

(b) Explain why the most efficient procedure available from this chapter for computing optimal E-mail routes is Dijkstra Algorithm 9C.

(c) ◻ Apply Algorithm 9C to compute optimal routes to all computers.

9-32 The campus shuttle bus begins running at 7:00 P.M. and continues until 2 A.M. Several drivers will be used, but only one should be on duty at any time. If a shift starts at or before 9:00 P.M., a regular driver can be obtained for a 4-hour shift at cost $50. Otherwise, part-time drivers will be used. Several would work 3-hour shifts at $40 and the rest are limited to 2-hour shifts at $30.

(a) Show that the problem of computing a minimum total cost nightly driver shift schedule can be modeled as a shortest path problem on a graph with nodes equal to the hours of the night from 7:00 P.M. through 2 A.M. Also sketch the corresponding digraph, and label it with arc lengths.

(b) ⊠ Is your digraph acyclic? Explain.

(c) Determine which of the algorithms of this chapter would solve your shortest path problem most efficiently, and justify your choice.

(d) ⊠ Apply your chosen algorithm to compute an optimal nightly shift schedule.

9-33 The machine shop depicted in the figure that follows has a heat treatment workstation at point 1, forges at points 2 and 3, machining centers at points 4, 5, and 6, and a grinding machine at 7. Each grid square indicated is the same size.

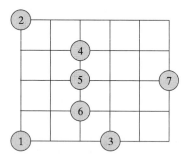

Processing of a camshaft product begins with heat treatment, then goes to any forge, then moves to any machining center, and finishes at the grinding machine. Movement between workstations is rectilinear (i.e., north/south displacement plus east/west).

(a) Show that the problem of computing an minimum total movement canshaft routing can be modeled as a shortest path problem. Also sketch the corresponding digraph, and label it with arc lengths.

(b) Is your digraph acyclic? Explain.

(c) Determine which of the algorithms of this chapter would solve your shortest path problem most efficiently, and justify your choice.

(d) ⊠ Apply your chosen algorithm to compute an optimal movement.

9-34 The figure that follows shows the trails of Littleville's Memorial Park.

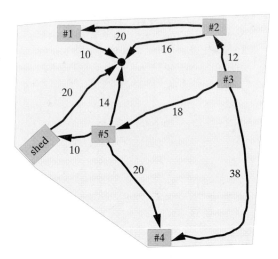

Whenever there is a major event there, heavy hot dog and soft-drink carts stored in the shed and rolled out along the trails to the five marked locations. Numbers on trail links indicate their length, and arrows show which way is uphill. The work to push a cart uphill on any link is proportional to that length. Carts can also be moved in the opposite direction, but the downhill effort is only half as much.

(a) Show how the problem of finding the least total-effort-push routes for all carts can be modeled as a shortest path problem. Also sketch the corresponding digraph, and label it with arc lengths.

(b) Is your digraph acyclic? Explain.

(c) Determine which of the algorithms of this chapter would solve your shortest path problem most efficiently, and justify your choice.

(d) ⊠ Apply your chosen algorithm to compute optimal move routes.

SUGGESTED READING

Ahuja, Ravindra K., Thomas L. Magnanti, and James B. Orlin (1993), *Network Flows: Theory, Algorithms and Applications*, Prentice Hall, Upper Saddle River, N.J.

Denardo, E. V. (1982), *Dynamic Programming: Models and Applications*, Prentice Hall, Upper Saddle River, N.J.

C H A P T E R 10

Network Flows

· ·

We have seen in Chapters 4 to 7 that linear program models admit some very elegant analysis. Global optima can be computed efficiently, and all sorts of "what if" sensitivity analysis can be performed on results.

Network flow problems are special, yet widely applicable cases of linear programs which prove even more tractable. Much larger models can be solved, because specialized algorithms apply. Most important, discrete cases, which we know are usually more difficult, can often be managed with no extra effort at all.

10.1 GRAPHS, NETWORKS, AND FLOWS

One of the things that make network flow models particularly tractable is that decisions and constraints have a form that we can easily represent in a diagram. More precisely, network flow models arise on structures called **directed graphs** or **digraphs**.

Digraphs, Nodes, and Arcs

We encountered digraphs in Section 9.1. They begin with a collection of **nodes** (or **vertices**), which we denote throughout this chapter by

$$V \triangleq \{\text{nodes or vertices of the network}\}$$

These indicate the facilities, or intersections, or transfer points of the network. Nodes are joined in digraphs by a collection of **arcs**, which we denote throughout this chapter by

$$A \triangleq \{\text{arcs of the network}\}$$

Arcs show possible flows or movements from one node to another. We indicate individual arcs simply by listing the pair of nodes they connect. For example, arc $(4, 7)$ would go from node 4 to node 7.

Digraphs are termed **directed** because the direction of flow matters. For example, an arc leading from node 7 to node 4, which would be denoted $(7, 4)$, is not the same as $(4, 7)$. They represent traffic in different directions.

EXAMPLE 10.1: OPTIMAL OVENS (OOI)

As usual, it will be much easier to absorb key notions with a small model in mind. Consider the (entirely ficticious) case of Optimal Ovens, Incorporated (OOI).

OOI makes home toaster ovens at plants in Wisconsin and Alabama. Completed ovens are shipped by rail to one of OOI's two warehouses in Memphis and Pittsburgh, and then distributed to customer facilities in Fresno, Peoria, and Newark. The two warehouses can also transfer small quantities of ovens between themselves, using company trucks.

Our task is to plan OOI's distribution of new model E27 ovens over the next month. Each plant can ship up to 1000 units during this period, and none are presently stored at warehouses. Fresno, Peoria, and Newark customers require 450, 500, and 610 ovens, respectively. Transfers between the warehouses are limited to 25 ovens, but no cost is charged. Unit costs (in dollars) of other possible flows are detailed in the following tables.

From/To	3: Memphis	4: Pittsburgh
1: Wisconsin	7	8
2: Alabama	4	7

From/To	5: Fresno	6: Peoria	7: Newark
3: Memphis	25	5	17
4: Pittsburgh	29	8	5

OOI Example Network

Figure 10.1 depicts OOI's network. The 2 plants, 2 warehouses and 3 customer sites make up the 7 nodes of this digraph. Arcs show the possible oven flows, with arrowheads indicating direction. For example, arc (3, 7) denotes ovens shipped from the Memphis warehouse to the Newark customer site. The absence of an arc (7, 3) means that ovens are not allowed to make the opposite move from Newark to Memphis. Two opposed arcs joining the warehouse nodes, indicate that traffic between them can flow in either direction.

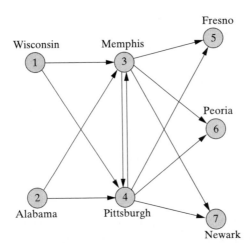

FIGURE 10.1 Network for Optimal Ovens, Incorporated (OOI) Example

Minimum Cost Flow Models

What linear program does a digraph like Figure 10.1 represent? We want it to describe flows, here of ovens starting at plants, passing through warehouses and terminating at customers.

As usual, we begin with decision variables.

| 10.1 | Decision variables $x_{i,j}$ in network flow models reflect the amount of flow in arcs (i, j). |

Letting $c_{i,j}$ denote the unit cost of flow on arc (i, j), the total cost to be minimized is simply

$$\sum_{(i,j)\in A} c_{i,j} x_{i,j}$$

Some constraints are equally easy. Flows must be nonnegative to make sense, and **capacities** or upper bounds $u_{i,j}$ may apply. These requirements lead to constraints:

$$0 \le x_{i,j} \le u_{i,j} \qquad \text{for all } (i, j) \in A \qquad (10.1)$$

The defining characteristic of network flow problems is the form of their main constraints.

| 10.2 | Main constraints of network flow problems enforce **balance** (or **conservation**) **of flow** at nodes. |

More precisely, we want

(total flow in) − (total flow out) = specified net demand

at every node. In symbols,

$$\sum_{(i,k)\in A} x_{i,k} - \sum_{(k,j)\in A} x_{k,j} = b_k \qquad \text{for all } k \in V \qquad (10.2)$$

where b_k denotes the specified net demand (required flow inbalance) at node k.

We are now ready to state the full minimum cost network flow model form:

| 10.3 | The **minimum cost network flow model** for a digraph on nodes $k \in V$ with net demands b_k, and arcs $(i, j) \in A$ with capacity $u_{i,j}$ and unit cost $c_{i,j}$ is |

$$\min \quad \sum_{(i,j)\in A} c_{i,j} x_{i,j}$$

$$\text{s.t.} \quad \sum_{(i,k)\in A} x_{i,k} - \sum_{(k,j)\in A} x_{k,j} = b_k \qquad \text{for all } k \in V$$

$$0 \le x_{i,j} \le u_{i,j} \qquad\qquad \text{for all } (i, j) \in A$$

Sources, Sinks, and Transshipment Nodes

Nodes come in three types. **Sink** or **demand** nodes such as the customer sites in the OOI example consume flow. **Source** or **supply** nodes such as the OOI plants create flow. **Transshipment** nodes such as OOI warehouses merely pass along flow. Net demands b_k have corresponding sign.

> **10.4** | **Net demand** b_k is positive at sink (demand) nodes, negative at source (supply) nodes, and zero at transshipment nodes.

OOI Example Model

Figure 10.2 includes net demands, costs and capacities on the digraph for our OOI example (plus an extra node 8 explained below). The corresponding minimum cost network flow model is

$$
\begin{aligned}
\min \quad & 7x_{1,3} + 8x_{1,4} + 4x_{2,3} + 7x_{2,4} + 25x_{3,5} + 5x_{3,6} \\
& + 17x_{3,7} + 29x_{4,5} + 8x_{4,6} + 5x_{4,7} && \text{(total cost)}
\end{aligned}
$$

$$
\begin{aligned}
\text{s.t.} \quad & -x_{1,3} - x_{1,4} - x_{1,8} && = -1000 && \text{(node 1)} \\
& -x_{2,3} - x_{2,4} - x_{2,8} && = -1000 && \text{(node 2)} \\
& +x_{1,3} + x_{2,3} + x_{4,3} - x_{3,4} - x_{3,5} - x_{3,6} - x_{3,7} = & 0 && \text{(node 3)} \\
& +x_{1,4} + x_{2,4} + x_{3,4} - x_{4,3} - x_{4,5} - x_{4,6} - x_{4,7} = & 0 && \text{(node 4)} \\
& +x_{3,5} + x_{4,5} && = & 450 && \text{(node 5)} \\
& +x_{3,6} + x_{4,6} && = & 500 && \text{(node 6)} \\
& +x_{3,7} + x_{4,7} && = & 610 && \text{(node 7)} \\
& +x_{1,8} + x_{2,8} && = & 440 && \text{(node 8)} \\
& x_{3,4} \le 25, \ x_{4,3} \le 25 && && \text{(capacities)} \\
& x_{i,j} \ge 0 \quad \text{for all } (i,j) \in A
\end{aligned}
$$

(10.3)

Heavy lines in Figure 10.3 show an optimal solution.

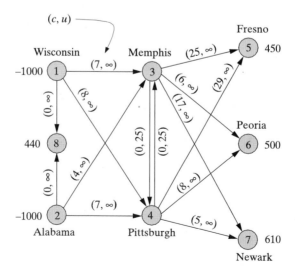

FIGURE 10.2 Minimum Cost Network Flow Problem for OOI Example

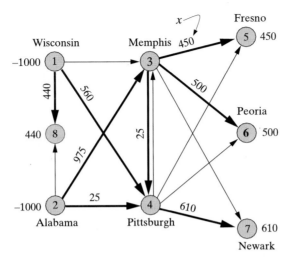

FIGURE 10.3 Optimal Flows for OOI Example

Plants are the source nodes of the OOI network. Thus their flow balance constraints (the first two) in model (10.3) have negative net demands. Warehouse nodes merely transship flow, so net demand is zero. The last four balance constraints of model (10.3) detail the sink nodes consuming flow; corresponding right-hand sides are positive.

The OOI example has capacities only on arcs connecting the two warehouses. No more than 25 ovens may be transferred in either direction. Bound constraints in model (10.3) reflect these two capacity limits and nonnegativity on all arcs.

SAMPLE EXERCISE 10.1: FORMULATING MINIMUM COST NETWORK FLOW MODELS

The figure that follows shows a network on four nodes. Numbers next to nodes are net demands b_k, and those on arcs are cost and capacity $(c_{i,j}, u_{i,j})$.

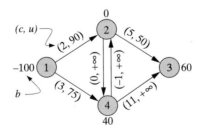

(a) Formulate the corresponding minimum cost network flow problem.

(b) Classify nodes of the problem as source, sink, or transshipment.

Modeling: Here

$$V = \{1, 2, 3, 4\}$$
$$A = \{(1, 2), (1, 4), (2, 3), (2, 4), (4, 2), (4, 3)\}$$

(a) Using variables $x_{1,2}$, $x_{1,4}$, $x_{2,3}$, $x_{2,4}$, $x_{4,2}$, and $x_{4,3}$ to represent flows on the six members of A, the formulation of principle $\boxed{10.3}$ is

$$\min \quad 2x_{1,2} + 3x_{1,4} + 5x_{2,3} - 1x_{4,2} + 11x_{4,3}$$

$$\begin{aligned}
\text{s.t.} \quad -x_{1,2} - x_{1,4} &= -100 \\
x_{1,2} + x_{4,2} - x_{2,3} - x_{2,4} &= 0 \\
x_{2,3} + x_{4,3} &= 60 \\
x_{1,4} + x_{2,4} - x_{4,2} - x_{4,3} &= 40
\end{aligned}$$

$$x_{1,2} \le 90, \ x_{1,4} \le 75, \ x_{2,3} \le 50$$

$$x_{i,j} \ge 0 \quad \text{for all } (i,j) \in A$$

(b) The only node with a negative net demand or supply is node 1. Thus it is the only source node. Nodes 3 and 4 have positive net demand, making them sinks. Remaining node 2, which has neither a demand nor a supply, is a transshipment node.

Total Supply = Total Demand

One element appears in Figure 10.2 and model (10.3) that was absent in original Figure 10.1. An extra sink node 8 has been added.

To see why, look again at flow balance constraints of $\boxed{10.3}$. Flow is created only at source nodes and consumed only at sink nodes. Thus there is no hope of a feasible flow unless

$$\text{total supply} = \text{total demand}$$

which means $\sum_k b_k = 0$.

When the given network of a minimum cost network flow problem does not come to us with total supply equal total demand, we must make some adjustments before proceeding.

$\boxed{10.5}$ If total supply is less that total demand in a given network flow problem, the problem is infeasible. If total supply exceeds total demand, a new sink node should be added to consume excess demand via zero-cost arcs from all sources.

Node 8 in Figure 10.2 was added because

$$\text{total supply} = 1000 + 1000 > 450 + 500 + 610 = \text{total demand}$$

in the OOI example. The excess of 440 determines node 8's demand. We want this excess supply to be able to reach node 8 without affecting any other part of the optimization. Zero-cost arcs from both source nodes do the job.

SAMPLE EXERCISE 10.2: BALANCING TOTAL SUPPLY AND TOTAL DEMAND

The digraph below shows net demand b_k next to each of its nodes and cost $c_{i,j}$ on its arcs.

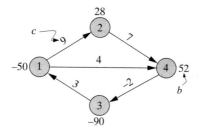

Modify the network as required to produce an equivalent one with total supply equal to total demand.

Modeling: Here total supply is $50 + 90 = 140$ and total demand is $28 + 52 = 80$. Thus total supply exceeds total demand by

$$140 - 80 = 60$$

To produce an equivalent model with total supply equal to total demand, we apply principle $\boxed{10.5}$ and add a "dummy" sink 5 to consume the excess.

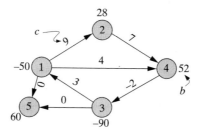

Zero-cost arcs $(1, 5)$ and $(3, 5)$ from the two sources assure that the extra supply can reach node 5 without affecting other costs.

Node–Arc Incidence Matrices and Matrix Standard Form

As in all our work with linear programs, we will often wish to think of minimum cost network flow model $\boxed{10.3}$ in matrix terms. To do so, we will abuse notation to treat the flow variables $x_{i,j}$ as a vector $\mathbf{x}$ even though components have two subscripts. Then, collecting costs $c_{i,j}$ and capacities $u_{i,j}$ in corresponding vectors $\mathbf{c}$ and $\mathbf{u}$, and arraying net demands b_k in vector $\mathbf{b}$ reduces the minimum cost network flow model $\boxed{10.3}$ to the familiar LP standard form

$$
\begin{aligned}
\min \quad & \mathbf{c} \cdot \mathbf{x} \\
\text{s.t.} \quad & \mathbf{A}\mathbf{x} = \mathbf{b} \\
& \mathbf{0} \leq \mathbf{x} \leq \mathbf{u}
\end{aligned}
\tag{10.4}
$$

Main constraint matrix $\mathbf{A}$ has a very special structure. Such matrices are called node–arc incidence matrices because they both encode the flow balance requirements and provide an algebraic description of the underlying digraph.

> **10.6** **Node–arc incidence matrices** represent both the flow balance require-
> ments and the graph structure of a network flow model with a row for every
> node and a column for every arc. The only nonzero entries in each column are
> a −1 in the row for the node the corresponding arc leaves and a +1 in the row
> for the node the arc enters.

Table 10.1 illustrates for the OOI example. The 8 rows correspond to the 8 nodes of the digraph in Figure 10.2. One column is present for each of the 14 arcs. The column for arc (3, 6) has a −1 in row 3 and a +1 in row 6 because arc (3, 6) leaves 3 and enters 6.

One of the conveniences of network flow problems is that we can depict them in network diagrams. Notice, however, that we could just as well start from the node–arc incidence matrix. If we had been given Table 10.1, it would be easy to sketch the corresponding digraph.

SAMPLE EXERCISE 10.3: CONSTRUCTING NODE–ARC INCIDENCE MATRICES

Construct the node–arc incidence matrix for the original digraph of Sample Exercise 10.2.

Modeling: Consistent with principle 10.6 , the node–arc incidence matrix will have a row for each of the 4 nodes and a column for each of the 5 arcs. The full matrix is as follows:

| | Arc | | | | |
Node	(1, 2)	(1, 4)	(2, 4)	(3, 1)	(4, 3)
1	−1	−1	0	+1	0
2	+1	0	−1	0	0
3	0	0	0	−1	+1
4	0	+1	+1	0	−1

SAMPLE EXERCISE 10.4: INTERPRETING NODE–ARC INCIDENCE MATRICES

Consider the following matrix:

$$\begin{pmatrix} -1 & +1 & 0 & +1 & 0 & 0 \\ 0 & -1 & -1 & 0 & 0 & +1 \\ 0 & 0 & +1 & -1 & -1 & 0 \\ +1 & 0 & 0 & 0 & +1 & -1 \end{pmatrix}$$

(a) Explain why it is a node–arc incidence matrix.

(b) Draw the corresponding digraph.

Modeling: Again we apply principle 10.6 .

(a) This is a node–arc incidence matrix because each column has only two nonzero entries, one −1 and one +1.

TABLE 10.1 Node-Arc Incidence Matrix of the OOI Example

Node	(1, 3)	(1, 4)	(1, 8)	(2, 3)	(2, 4)	(2, 8)	(3, 4)	(3, 5)	(3, 6)	(3, 7)	(4, 3)	(4, 5)	(4, 6)	(4, 7)
1	−1	−1	−1	0	0	0	0	0	0	0	0	0	0	0
2	0	0	0	−1	−1	−1	0	0	0	0	0	0	0	0
3	+1	0	0	+1	0	0	−1	−1	−1	−1	+1	0	0	0
4	0	+1	0	0	+1	0	+1	0	0	0	−1	−1	−1	−1
5	0	0	0	0	0	0	0	+1	0	0	0	+1	0	0
6	0	0	0	0	0	0	0	0	+1	0	0	0	+1	0
7	0	0	0	0	0	0	0	0	0	+1	0	0	0	+1
8	0	0	+1	0	0	+1	0	0	0	0	0	0	0	0

(b) Associating nodes 1 to 4 with the 4 rows of the matrix, we construct the digraph by inserting an arc for each column of the matrix. The arc leaves the node of the row where the column has a −1 and enters the node of the row where it has a +1.

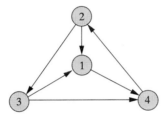

10.2 CYCLE DIRECTIONS FOR NETWORK FLOW SEARCH

Linear programming algorithms of Chapters 5 and 6 center on constructing improving feasible directions, directions of solution change that preserve feasibility and improve the objective function for suitably small steps. Knowing that network flow problems are linear programs with special properties, it should not surprise that unusually simple improving feasible directions are at the heart of their tractability.

Chains, Paths, Cycles, and Dicycles

Sections 9.1 introduced the notions of **paths** and **dicycles** in a graph. To derive directions for network flow problems, we need the additional concepts of chains and cycles.

> 10.7 | A **chain** is a sequence of arcs connecting two nodes. Each arc has exactly one node in common with its predecessor in the sequence, and no node is visited more than once.

> 10.8 | A **cycle** is a chain with the same beginning and ending node.

We describe a chain or cycle merely by listing its arcs in sequence. When no confusion will result, we may also use the corresponding sequence of nodes.

Figure 10.4(a) illustrates on the OOI digraph of Figure 10.2. The first example shows a chain $(1, 3)$, $(3, 6)$, $(4, 6)$, $(2, 4)$ connecting nodes 1 and 2. It could just as well be called by its node sequence 1–3–6–4–2 because there is only one arc that could provide each of the implied node-to-node connections. Figure 10.4(a)'s second example is chain $(1, 3)$, $(3, 4)$, $(4, 7)$. Here the node sequence would not be definitive because both this chain and $(1, 3)$, $(4, 3)$, $(4, 7)$ have node sequence 1–3–4–7.

Part (b) of Figure 10.4 shows some sequences that are not chains. The first is not connected, and the second repeats node 3.

Notice that chains need not observe direction on the arcs. This is how a chain differs from a path.

10.9 | Paths are chains that transmit all arcs in the forward direction.

Thus highlighted sequence 1–3–6–4–2 in Figure 10.4(a) is a chain but not a path; it violates direction on arcs $(4, 6)$ and $(2, 4)$. Second sequence 1–3–4–7 is both a chain and a path.

The only additional element with cycles is beginning and ending at the same node. For example, Figure 10.4(c) includes cycle 1–3–6–4–1 starting and ending at node 1 and cycle $(3, 4)$, $(4, 3)$ beginning and ending at node 3. Part (d) of the figure confirms that disconnected arc sequences or ones repeating a node cannot constitute cycles.

As with chains and paths, the distinction between cycles and dicycles involves direction.

10.10 | Dicycles are cycles that have all arcs oriented in the same direction.

Thus the first cycle 1–3–6–4–1 of Figure 10.4(c) is not a dicycle because it violates direction on arcs $(4, 6)$ and $(1, 4)$. Second cycle $(3, 4)$, $(4, 3)$ does meet the definition of a dicycle because it passes both arcs in the forward direction.

Cycle Directions

Cycles can pass arcs in either a **forward** (with direction) or the **reverse** (against direction) manner. Cycle directions derive from this visitation pattern.

10.11 | A **cycle direction** of a minimum cost network flow model increases flow on forward arcs and decreases flow on reverse arcs of a cycle in the given digraph; that is,

$$\Delta x_{i,j} \triangleq \begin{cases} +1 & \text{if arc } (i, j) \text{ is forward in the cycle} \\ -1 & \text{if arc } (i, j) \text{ is reverse in the cycle} \\ 0 & \text{if arc } (i, j) \text{ is not part of the cycle} \end{cases}$$

For example, consider the first cycle of Figure 10.4(c): 1–3–6–4–1. With the first two arcs forward in the cycle and the last two reverse, we obtain the cycle direction sketched below.

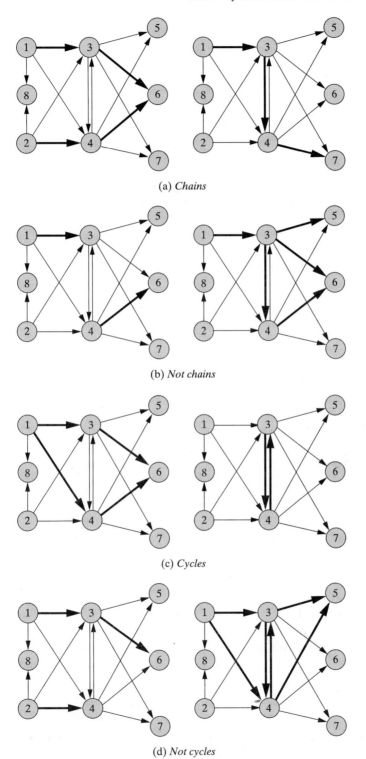

(a) *Chains*

(b) *Not chains*

(c) *Cycles*

(d) *Not cycles*

FIGURE 10.4 Chains and Cycles of the OOI Network

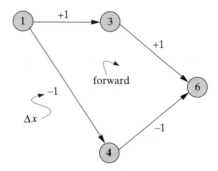

Forward arcs in the cycle have +1 components, reverse arcs have −1, and all other arcs have 0.

Notice that it matters which way we orient the cycle. If we used the same cycle as above but thought of visiting it in sequence 1–4–6–3–1, arcs (1, 4) and (4, 6) would be forward arcs with $\Delta x = +1$, and arcs (3, 6) and (1, 3) would be reverse with $\Delta x = -1$.

Maintaining Flow Balance with Cycle Directions

The main, flow conservation constraints of minimum cost network flow problems in matrix format (10.4) are equality constraints $\mathbf{Ax} = \mathbf{b}$ where $\mathbf{A}$ is the node–arc incidence matrix of the given digraph and $\mathbf{b}$ is the net demand vector. We know from our earliest investigation of improving search (principle $\boxed{3.29}$, Section 3.3) that any direction $\Delta \mathbf{x}$ preserving feasibility in such equality constraints must satisfy the condition $\mathbf{A}\Delta\mathbf{x}=\mathbf{0}$. We are interested in cycle directions because they satisfy this net-change-zero condition for node–arc incidence matrices $\mathbf{A}$.

$\boxed{10.12}$ Adjusting a feasible flow along a cycle direction of a network flow model leaves flow balance constraints satisfied.

To see that this is true, recall that each node of the cycle is touched by exactly two arcs, and think about the four ways two arcs can visit a node. Figure 10.5 illustrates how node–arc incidence matrix signs of the direction $\boxed{10.6}$ combine with those of the

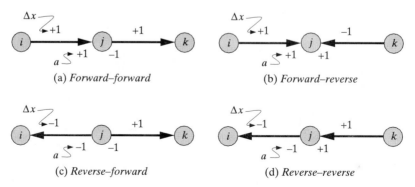

FIGURE 10.5 Possible Ways a Cycle Can Visit a Node

direction $\boxed{10.11}$ to produce a net change $= 0$ at each node. In the forward–reverse case, for example, we have

$$+1\,\Delta x_{i,j} + 1\,\Delta x_{k,j} = +1(+1) + 1(-1) = 0$$

Consider the following digraph:

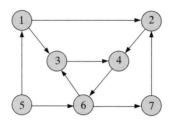

Construct the cycle direction for each of the following cycles, and verify that each retains flow balance at node 6.

(a) 1–2–7–6–5–1

(b) 3–4–6–3

Analysis:

(a) Applying definition $\boxed{10.11}$, the cycle direction has

$$\Delta x_{1,2} = \Delta x_{5,1} = +1$$
$$\Delta x_{7,2} = \Delta x_{6,7} = \Delta x_{5,6} = -1$$
$$\Delta x_{1,3} = \Delta x_{3,4} = \Delta x_{4,2} = \Delta x_{6,3} = \Delta x_{4,6} = 0$$

Checking principle $\boxed{10.12}$ at node 6,

$$-1\,\Delta x_{6,7} + 1\,\Delta x_{5,6} = -1(-1) + 1(-1) = 0$$

(b) Again applying definition $\boxed{10.11}$, the cycle direction has

$$\Delta x_{3,4} = \Delta x_{4,6} = \Delta x_{6,3} = 1$$
$$\Delta x_{1,2} = \Delta x_{1,3} = \Delta x_{2,4} = \Delta x_{7,2} = \Delta x_{6,7} = \Delta x_{5,6} = \Delta x_{5,1} = 0$$

At node 6

$$+1\,\Delta x_{4,6} - 1\,\Delta x_{6,3} = +1(+1) - 1(+1) = 0$$

Feasible Cycle Directions

Flow balance equalities are not the only constraints of a minimum cost network flow model $\boxed{10.3}$. We must also be concerned with nonnegativity constraints and arc capacities $u_{i,j}$. Section 3.3 principles $\boxed{3.27}$ and $\boxed{3.28}$ tell us the requirements that must be added for upper and lower bounds. No arc with zero flow can decrease, and no capacitated arc can increase if bound constraints are also to be maintained in a move from a current feasible solution $\mathbf{x}$.

10.13 | A cycle direction $\Delta\mathbf{x}$ is a feasible direction at current solution $\mathbf{x}$ if and only if $x_{i,j} > 0$ on all reverse arcs of the cycle, and $x_{i,j} < u_{i,j}$ on all forward arcs.

To illustrate, we need a feasible flow. Consider the OOI flow $\mathbf{x}^{(0)}$ depicted in Figure 10.6. That $\mathbf{x}^{(0)}$ satisfies net demand requirements at all nodes and conforms to all bounds.

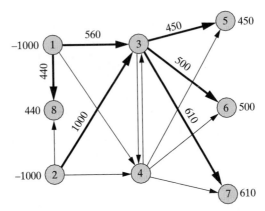

FIGURE 10.6   Initial Flow $\mathbf{x}^{(0)}$ in OOI Example

One cycle direction corresponds to arc sequence $(2, 3)$, $(3, 4)$, $(2, 4)$:

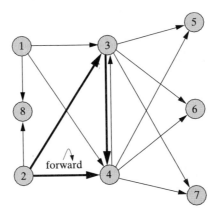

We increase flow on forward arcs $(2, 3)$ and $(3, 4)$, and decrease on reverse arc $(2, 4)$. Principle | 10.12 | assures that a move in this cycle direction will preserve flow balance at all nodes. Also, both forward arcs have infinite capacity, so that an increase will violate no upper bound. Still, reverse arc $(2, 4)$ has current flow zero. It cannot feasibly decrease, so this cycle direction is not a feasible direction.

Figure 10.7 shows a cycle 2–4–7–3–2 that does satisfy conditions | 10.13 |. Both reverse arcs have $x_{i,j}^{(0)} > 0$, and neither forward arc has a capacity.

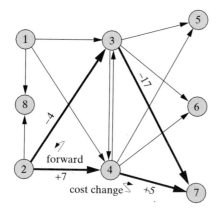

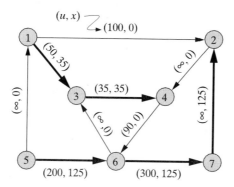

FIGURE 10.7 An Improving Feasible Cycle Direction at $x^{(0)}$ of Figure 10.6

SAMPLE EXERCISE 10.6: IDENTIFYING FEASIBLE CYCLE DIRECTIONS

The figure that follows shows a network flow problem with a feasible flow x on highlighted arcs. Labels on arcs indicate capacities and flows $(u_{i,j}, x_{i,j})$.

Determine whether the direction for each of the following cycles is a feasible cycle direction.

(a) 1–2–7–6–5–1

(b) 3–4–6–3

(c) 1–3–6–5–1

Analysis: We apply conditions $\boxed{10.13}$.

(a) This cycle direction is feasible because all reverse arcs currently have positive flow and no forward arc is at capacity.

(b) This cycle direction is not feasible. Forward arc $(3, 4)$ is at capacity and cannot increase.

(c) This cycle direction is not feasible. One of the reverse arcs, $(3, 6)$, has current flow 0 and cannot decrease.

Improving Cycle Directions

Feasible cycle directions aid a search for an optimal flow only if they improve the objective function. We know from principle $\boxed{3.18}$ (Section 3.3) that this will be true if $\bar{c} \triangleq \mathbf{c} \cdot \Delta\mathbf{x} < 0$ for our minimizing model. The simple $+1, -1, 0$ coefficient structure of cycle directions (principle $\boxed{10.11}$) make this test particularly easy to apply.

$$\bar{c} \triangleq \mathbf{c} \cdot \Delta\mathbf{x}$$
$$= \sum_{(i,j)\text{ in cycle}} c_{i,j}\,\Delta x_{i,j}$$
$$= \sum_{\text{forward }(i,j)} c_{i,j}(+1) + \sum_{\text{reverse }(i,j)} c_{i,j}(-1)$$
$$= (\text{total forward arc cost}) - (\text{total reverse arc cost})$$

$\boxed{10.14}$ A cycle direction improves for a minimum cost network flow model if the difference of total forward arc cost and total reverse arc cost < 0.

Cycle 2–4–7–3–2 of Figure 10.7 illustrates a direction that is both feasible and improving. Applying $\boxed{10.14}$ to check the latter gives

$$(\text{total forward}) - (\text{total reverse}) = (7 + 5) - (17 + 4) = -9 < 0$$

SAMPLE EXERCISE 10.7: IDENTIFYING IMPROVING CYCLE DIRECTIONS

The figure that follows adds costs to the network of Sample Exercise 10.6. Labels on arcs now show costs, capacities, and current flows $(c_{i,j}, u_{i,j}, x_{i,j})$.

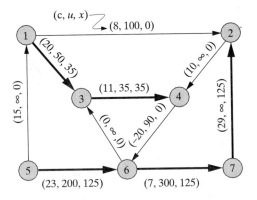

Determine whether each of the following cycles yields an improving cycle direction.

(a) 1–2–7–6–5–1

(b) 2–4–6–7–2

(c) 6–3–1–2–4–6

Analysis: We apply condition $\boxed{10.14}$.

(a) This cycle direction improves because

$$(\text{total forward}) - (\text{total reverse}) = (8 + 15) - (29 + 7 + 23) = -36$$

(b) This cycle direction does not improve because

$$(\text{total forward}) - (\text{total reverse}) = (10 - 20 + 7 + 29) - (0) = 26$$

(c) This cycle direction improves because

$$(\text{total forward}) - (\text{total reverse}) = (0 + 8 + 10 - 20) - (20) = -22$$

Step Size with Cycle Directions

If we can take an arbitrary step λ in improving feasible direction $\Delta\mathbf{x}$, we know that our model is unbounded. The objective can be improved forever without losing feasibility.

More often, upper and lower bounds on arc flows impose limits. Each unit step in a cycle direction increases forward arc flows by a unit and decreases reverse flows by a unit. Feasibility will be lost the first time these changes encounter a nonnegativity or capacity constraint. More specifically,

$\boxed{10.15}$ Steps from feasible flow $\mathbf{x}$ in cycle direction $\Delta\mathbf{x}$ retain feasibility for step sizes up to $\lambda = \min\{\lambda^+, \lambda^-\}$, where

$$\lambda^+ \triangleq \min\{(u_{i,j} - x_{i,j}) : (i, j) \text{ forward}\} \ (= +\infty \text{ if there are no forward arcs})$$
$$\lambda^- \triangleq \min\{x_{i,j} : (i, j) \text{ reverse}\} \ (= +\infty \text{ if there are no reverse arcs})$$

Return to the cycle direction 2–4–7–3–7 of Figure 10.7 and the flow $\mathbf{x}^{(0)}$ of Figure 10.6. For that example,

$$\lambda^+ = \min\{(\infty - 0), (\infty - 0)\} = \infty$$
$$\lambda^- = \min\{610, 1000\} = 610$$

Thus $\lambda = \min\{\infty, 610\} = 610$ is the largest step we can take in that cycle direction without losing feasibility.

SAMPLE EXERCISE 10.8: COMPUTING STEPS IN CYCLE DIRECTIONS

Return to the example of Sample Exercise 10.7. Whether or not directions corresponding to the three specified cycles improve the objective function, determine the maximum step λ in each direction that preserves feasibility.

Analysis: We apply principle $\boxed{10.15}$.

(a) For cycle 1–2–7–6–5–1,

$$\lambda^+ = \min\{(100 - 0), (\infty - 0)\} = 100$$
$$\lambda^- = \min\{125, 125, 125\} = 125$$
$$\lambda = \min\{\lambda^+, \lambda^-\} = \min\{100, 125\} = 100$$

(b) For cycle 2–4–6–7–2,

$$\lambda^+ = \min\{(\infty - 0), (90 - 0), (300 - 125), (\infty - 125)\} = 90$$
$$\lambda^- = +\infty$$
$$\lambda = \min\{\lambda^+, \lambda^-\} = \min\{90, \infty\} = 90$$

(c) For cycle 6–3–1–2–4–6,

$$\lambda^+ = \min\{(\infty - 0), (100 - 0), (\infty - 0), (90 - 0)\} = 90$$
$$\lambda^- = \min\{35\} = 35$$
$$\lambda = \min\{\lambda^+, \lambda^-\} = \min\{90, 35\} = 35$$

Sufficiency of Cycle Directions

Properties we long ago encountered for linear programs of any form, including network flow models of present interest, make a solution globally optimal if and only if it admits no improving feasible directions (principle $\boxed{5.1}$). We have seen how cycle directions can be improving and feasible. But what if none exist that satisfy both requirements? Certainly there are more complicated improving feasible directions for network flow models. Could any improve a solution when cycle directions fail? Happily, no.

$\boxed{10.16}$ A feasible flow in a minimum cost network flow problem is (globally) optimal if and only if it admits no improving feasible cycle direction.

A formal proof would be beyond the scope of this book, but an example will illustrate why $\boxed{10.16}$ must hold. The key insight is a decomposition:

$\boxed{10.17}$ Every direction $\Delta\mathbf{x}$ satisfying $\mathbf{A}\Delta\mathbf{x} = \mathbf{0}$ for node–arc incidence matrix $\mathbf{A}$ of a minimum cost network flow model can be decomposed into a weighted sum of cycle directions. Furthermore, if $\Delta\mathbf{x}$ is feasible and improving, at least one of the cycle directions will also be feasible and improving.

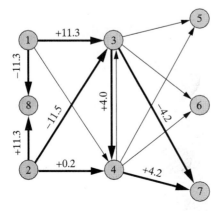

FIGURE 10.8 A Non-Cycle Improving Feasible Direction at $\mathbf{x}^{(0)}$ of Figure 10.6

Consider the direction shown in Figure 10.8. It certainly does not look like a cycle direction, but the reader can check that it satisfies conditions to be both feasible and improving. The objective changes by $\bar{c} = -\$15.9$ for every unit step.

This complex direction decomposes into a sum of the cycle directions for cycles 3–4–7–3, 1–3–2–8–1, and 2–4–7–3–2 as follows:

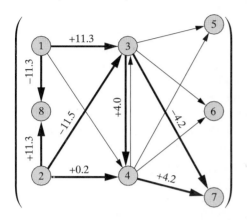

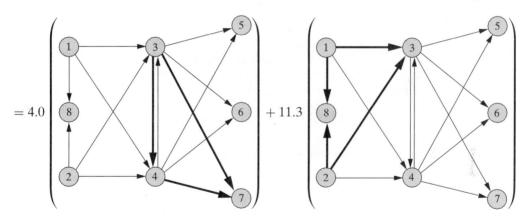

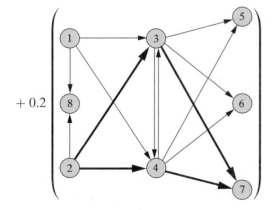

The three component cycle directions have reduced costs $\bar{c} = -12.0, +3.0$, and -1.8, respectively. We would never need the more complex direction because both the first and the third cycle directions are improving and feasible at $\mathbf{x}^{(0)}$ of Figure 10.6.

10.3 RUDIMENTARY CYCLE DIRECTION SEARCH ALGORITHMS FOR NETWORK FLOWS

Section 10.2 has developed all the building blocks. We are now ready to specify cycle direction-based algorithms for minimum cost network flow models.

Most of the known procedures for minimum cost network flow problems can be viewed as special forms of improving search using cycle directions. Every iteration identifies an improving feasible cycle direction and adjusts flows along it to reduce cost without losing feasibility. If no such cycle direction exists, the present solution is optimal (principle $\boxed{10.16}$).

Algorithm 10A provides a formal statement, but we omit for the moment many important details. In Sections 10.7 and 10.8 we specify schemes for determining at Step 1 whether any improving feasible cycle directions exist and for constructing them as needed at Step 2. For the moment these steps will be done by inspection.

Rudimentary Cycle Direction Search of the OOI Example

Figure 10.9 summarizes costs and capacities of the OOI example, along with the starting feasible flows of Figure 10.6, which have cost \$32,540. Figure 10.10 then details an application of Algorithm 10A to compute an optimal flow.

We first apply the improving feasible cycle direction of Figure 10.7. For that cycle $\lambda^+ = +\infty$ and $\lambda^- = \min\{610, 1000\} = 610$. A maximum feasible step of $\lambda = 610$ in that direction yields flow $\mathbf{x}^{(1)}$ at cost \$27,050.

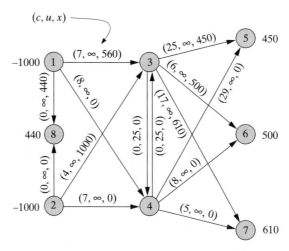

FIGURE 10.9 Data and Initial Flow for the OOI Example

ALGORITHM 10A: RUDIMENTARY CYCLE DIRECTION SEARCH

Step 0: Initialization. Choose any starting feasible flow $\mathbf{x}^{(0)}$, and set solution index $t \leftarrow 0$.

Step 1: Optimal. If no improving feasible cycle direction $\Delta \mathbf{x}$ exists at current solution $\mathbf{x}^{(t)}$ (conditions $\boxed{10.13}$ and $\boxed{10.14}$), then stop. Flow $\mathbf{x}^{(t)}$ is globally optimal.

Step 2: Cycle Direction. Choose an improving feasible cycle direction $\Delta \mathbf{x}$ at $\mathbf{x}^{(t)}$.

Step 3: Step Size. Compute the maximum feasible step λ in direction $\Delta \mathbf{x}$ (rule $\boxed{10.15}$):

$$\lambda^+ \leftarrow \min\{(u_{i,j} - x_{i,j}^{(t)}) : (i,j) \text{ forward}\} \ (+\infty \text{ if none})$$

$$\lambda^- \leftarrow \min\{x_{i,j}^{(t)} : (i,j) \text{ reverse}\} \ (+\infty \text{ if none})$$

$$\lambda \ \leftarrow \min\{\lambda^+, \lambda^-\}$$

If $\lambda = \infty$, stop; the model is unbounded.

Step 4: Advance. Update

$$\mathbf{x}^{(t+1)} \leftarrow \mathbf{x}^{(t)} + \lambda \Delta \mathbf{x}$$

by increasing flows on forward arcs of the cycle direction and decreasing those on reverse arcs by the amount λ. Then increment $t \leftarrow t+1$ and return to Step 1.

Next we employ cycle 2–3–1–4–2 at $\bar{c} = -2$ and $\lambda = 560$ to produce flow $\mathbf{x}^{(2)}$ at cost $25,930. A final step in the direction for the indicated cycle 2–3–4–2 completes recovery of the optimal flow in Figure 10.3 with value $25,855.

The reader can verify that each of the three directions employed was both improving and feasible for the solution to which it was applied. However, many other choices were available. Pending the development of Sections 10.7 and 10.8, readers should assume that choices were made arbitrarily, with appropriate cycles discovered by trial and error.

Many cycle directions remain in the optimal flow $\mathbf{x}^{(3)}$. Some are either improving or feasible. Still, it can be shown that no cycle direction is both improving and feasible. Algorithm 10A terminates with a conclusion of optimality.

SAMPLE EXERCISE 10.9: APPLYING RUDIMENTARY ALGORITHM 10A

Consider the network flow problem depicted below. Labels on nodes indicate net demands b_k. Labels on arcs show cost, capacities, and current flows ($c_{i,j}, u_{i,j}, x_{i,j}$).

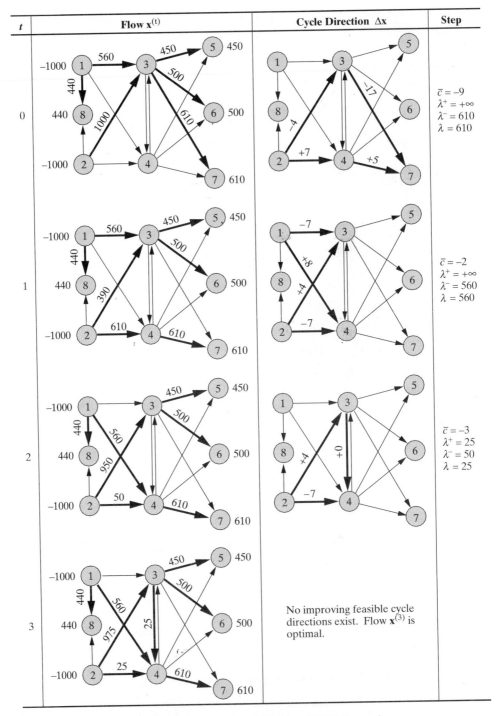

FIGURE 10.10 Rudimentary Cycle Direction Solution of OOI Example

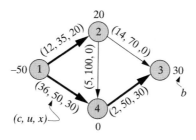

Verify that the direction for cycle 1–2–4–1 is improving and feasible. Then take the maximum feasible step in this direction and show the resulting improved flow.

Analysis: The direction is feasible under principle 10.13 because reverse arc $(1, 4)$ has positive flow, and neither forward arc $(1, 2)$ nor $(2, 4)$ is at capacity. The direction is improving under principle 10.14 because

$$\bar{c} = (12 + 5) - (36) = -19 < 0$$

Application of step size rule 10.15 gives

$$\lambda^+ = \min\{(35 - 20), (100 - 0)\} = 15$$
$$\lambda^- = \min\{30\} = 30$$
$$\lambda \ = \min\{15, 30\} = 15$$

Thus flows increase by $\lambda = 15$ units on forward arcs $(1, 2)$ and $(2, 4)$ of the cycle, while decreasing by 15 units on reverse arc $(1, 4)$. The result is

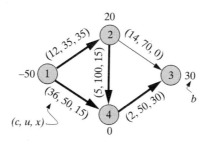

Finding Starting Feasible Solutions

As with all constrained optimization models, Algorithm 10A can be adapted to find a starting feasible flow or establish that none exists. We may employ either the two-phase or the big-M scheme of Section 3.6.

The key to both those methods is construction of an artificial model from which computation can begin. New artificial variables are introduced to force feasibility, and their sum is minimized. If that sum can be driven to $= 0$, what remains is a feasible flow for the original model. If not, the original model is infeasible.

Artificial Network Flow Model

The only new element in the network flow context is that we wish to create an artifical model that is itself a minimum cost network flow problem so that Algorithm 10A applies. In particular, we want to be able to interpret artificial variables as flows in arcs.

To obtain such a network flow artificial model, we simply put a zero flow on all arcs of the original model and add one artificial node. Supply and demand requirements are fulfilled by artificial arcs joining this special node to all others having net demand $b_k \neq 0$.

> 10.18 | An **artificial network model** and starting point for computing initial feasible solutions in minimum cost network flow problems can be constructed by (1) assigning 0 flow to all arcs of the original model, (2) introducing an artificial node, (3) creating artificial arcs from each supply node k ($b_k < 0$) to the artificial node with flow equal to the specified supply $|b_k|$, and (4) adding artificial arcs from the artificial node to each demand node k ($b_k > 0$) with flow equal to the required demand $|b_k|$.

Figure 10.11 illustrates for our OOI model. Artificial node 0 has been added to anchor artificial arcs. Then artificial arcs $(1, 0)$ and $(2, 0)$ balance flow at supply nodes 1 and 2 by carrying exactly the supply specified at those nodes. Similarly, artificial arcs $(0, 5)$, $(0, 6)$, $(0, 7)$, and $(0, 8)$ satisfy demand requirements by bringing the needed flow to each demand node. All original arcs have flow $= 0$.

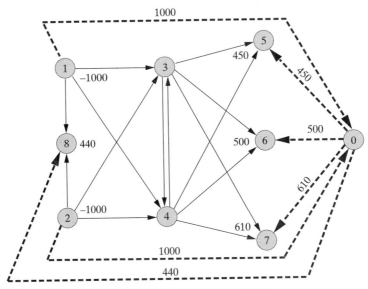

FIGURE 10.11   Starting Phase I Flow for the OOI Example

Zero flows on all original arcs satisfy their upper- and lower-bound constraints, and all specified artificial flows are nonnegative. Also, setting artificial flows to exactly the needed supply or demand has assured flow balance at all the original nodes.

But what about flow balance at the new, artificial node? To see that it too is balanced, observe that inflow will sum to total supply, and outflow will sum to total demand. Under total supply equals total demand requirement ⎡10.5⎤, these totals may be assumed to agree.

SAMPLE EXERCISE 10.10: CONSTRUCTING AN ARTIFICIAL NETWORK MODEL

Consider a network flow problem with net demands as depicted in the following figure:

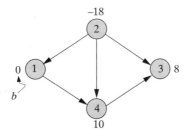

Construct both the corresponding artifical network model for phase I or big-M computation of a starting feasible solution and the associated artificial solution.

Modeling: Following construction ⎡10.18⎤, we introduce an artificial node 0, along with artificial arcs from supply node 2 and to demand nodes 3 and 4. Flows on original arcs are all zero, and those on artificials equal the specified supply or demand. The result is the following artificial model and artificial starting solution:

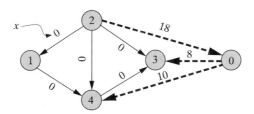

Network Flow Algorithms versus Linear Programming

Every network flow model is a linear program, so we could solve any of them with the simplex or interior point methods of Chapters 5 and 6. Still, implementations of cycle direction Algorithm 10A are usually faster.

⎡10.19⎤   Although any minimum cost network flow problem can be solved by the general-purpose linear programming methods of Chapter 5 and 6, implementations of cycle direction search Algorithm 10A usually prove much more efficient.

To see why, glance back at algorithm statements of Chapter 5 and 6. Even the most efficient revised simplex version 5C in Section 5.8 had to solve linear systems of equations to compute pricing vector (dual solution) **v** and simplex direction **Δx**. Interior point algorithms all require relatively complex projection computations (Sections 6.1 and 6.2).

Assuming that we can find improving feasible cycle directions without too much effort, which we will show how to do in Sections 10.7 and 10.8, the network computations of Algorithm 10A are trivial by comparison. We have only to develop a labeling scheme to mark the improving feasible cycle, pass around it to determine step size λ, and pass around again to add or subtract λ on forward and reverse arcs, respectively. Not one multiplication or division is required.

10.4 INTEGRALITY OF OPTIMAL NETWORK FLOWS

As far back as Chapter 2, we have seen that integer linear programs (ILPs) are generally much less tractable than linear programs (LPs). We are now prepared to show why network flows are often an exception. Under mild conditions, optimal network flows automatically take on integer (whole-number) values. Thus, if an ILP can be shown to have network form, we may solve it with methods based on Algorithm 10A which are even more efficient that general-purpose LP methods of Chapters 5 and 6 (principle 10.19).

When Optimal Network Flows Must Be Integer

Suppose that all constraint data of a given minimum cost network flow model—supplies, demands, and capacities—happen to have integer values as they do in the OOI example of Figure 10.9. Then we can think through some observations about the steps in a two-phase or big-M implementation of Algorithm 10A that have important consequences.

- The starting artificial flow of construction 10.18 will be integer because the only nonzero flows occur on artificial arcs, and all those flows equal supplies or demands that we have assumed integer.

- Regardless of whether we proceed by two-phase or big-M application of Algorithm 10A, an optimal solution can be computed using only cycle directions (principle 10.16).

- Each iteration of Algorithm 10A either increases arc flows by step size λ, decreases them by λ, or leaves the flows unchanged, because all components of cycle directions are $+1$, -1, or 0 (construction 10.11).

- Whenever the current flow $\mathbf{x}^{(t)}$ is integer, step size λ of 10.15 , which will equal either some $x_{i,j}^{(t)}$ or some $(u_{i,j} - x_{i,j}^{(t)})$, must also be integer under our assumption of integer $u_{i,j}$.

The consequence is a fundamental **integrality property**. When a network flow model has integer constraint data, we can start with an integer flow and execute every step of two-phase of big-M solution by adding or subtracting integer quantities λ to current flows. If the result is an optimal solution (i.e., not infeasible or unbounded), that solution must have integer flows on all arcs.

10.20 | If a minimum cost network flow model with integer constraint data (supplies, demands, and capacities) has any optimal solution, it has an integer optimal solution.

Notice one assumption not required in principle 10.20. Nothing was said about arc costs $c_{i,j}$. Integrality property 10.20 depends only on constraint data. What is important is that flows start integer and change by integer amounts at each step. Costs have no effect because only supplies, demands, and capacities are involved in flow adjustments.

SAMPLE EXERCISE 10.11: RECOGNIZING IF NETWORK OPTIMA WILL BE INTEGER

Each of the following details the constant data of a minimum cost network flow problem. Assuming that the models are neither infeasible nor unbounded, determine whether each must have an integer optimal solution.

(a) $\mathbf{b} = (100, 200, 0, -300)$, $\mathbf{u} = (90, 20, \infty, 220, 180)$, $\mathbf{c} = (8, 9, -4, 0, 6)$

(b) $\mathbf{b} = (-30, 40, -10, 0)$, $\mathbf{u} = (20, 12\frac{1}{2}, 23, 15, 92)$, $\mathbf{c} = (11, 0, 3, 81, 6)$

(c) $\mathbf{b} = (-13\frac{1}{3}, -20, 23\frac{1}{3}, 10)$, $\mathbf{u} = (10, 20, \infty, \infty, 40)$, $\mathbf{c} = (-4, 8, 0, 19, 31)$

(d) $\mathbf{b} = (25, 15, 0, -40)$, $\mathbf{u} = (20, \infty, 30, 45, 10)$, $\mathbf{c} = (3.5, 9.6, -2.1, 11.77, \sqrt{2})$

Analysis: We apply principle 10.20.

(a) This model has integer supplies, demands, and capacities. An integer optimal solution will exist.

(b) This model has a fractional capacity. An integer optimum is not assured.

(c) This model has a fractional supply and a fractional demand. An integer optimum is not assured.

(d) Even though cost data are highly noninteger, this model has integer constraint data. An integer optimal solution will exist.

Total Unimodularity of Node–Arc Incidence Matrices

The above "starts integer and stays integer" arguement is only one way of seeing that network flow optima will be integer when all constraint data are integer. Since we know that optimal solutions, at least unique ones, must be extreme points of the corresponding LP-feasible region (Section 5.1, principle 5.5), there must also be something special about the basic solution computations of extreme points (Section 5.2) in minimum cost network flow models.

The property is called total unimodularity:

10.21 | The constraint matrix **A** of a linear program is **totally unimodular** if each square submatrix has determinant $+1, 0$, or -1.

The main constraint matrices of network flow models are their node–arc incidence matrices, and they have this very special property:

> **10.22** Node–arc incidence matrices of network flow models are totally unimodular.

The relevance of total unimodularity comes in solving for basic (i.e., extreme-point) solutions to network LPs. The famous Cramer rule for solving systems of linear equations shows that the maximum denominator of the result is the determinant of the corresponding basis matrix. Under total unimodularity, that denominator will always be ±1, meaning that no fractions will be introduced.

It is beyond the scope of this book to establish principle 10.22 in general, but an example from the OOI node–arc matrix of Table 10.1 will illustrate. Extracting rows for nodes 2, 3, and 4, along with columns for arcs (2, 3), (2, 4), and (3, 4), gives

$$
\begin{pmatrix}
 & (2,3) & (2,4) & (3,4) \\
\hline
2 & -1 & -1 & 0 \\
3 & +1 & 0 & -1 \\
4 & 0 & -1 & +1
\end{pmatrix}
$$

All 1 by 1 submatrices are obviously $+1$, 0, or -1. The full 3 by 3 matrix has determinant $= 0$, its lower left corner yields

$$
\det \begin{pmatrix} +1 & 0 \\ 0 & -1 \end{pmatrix} = -1
$$

and all other 2 by 2 submatrices are similar.

10.5 TRANSPORTATION AND ASSIGNMENT MODELS

The OOI example of Section 10.1 contains all the major elements of network flow models, but it only begins to suggest the rich variety of problems that can be treated as networks. In this section we present the classic transportation and assignment special cases, and in Section 10.6 we introduce several more.

Transportation Problems

Among the simplest of classic flow models are transportation problems.

> **10.23** **Transportation problems** are special minimum cost network flow models for which every node is either a pure supply node (every arc points out) or a pure demand node (every arc points in).

That is, all flow goes immediately from some source node, where it is supplied, to a sink node where it is demanded. There are no intermediate steps and no transshipment nodes. The commodity flowing in a transportation problem may be people, water, oil, money, or almost anything else. It is only necesary that flows go direct from sources to sinks.

SAMPLE EXERCISE 10.12: IDENTIFYING TRANSPORTATION PROBLEMS

Each of the following digraphs shows a minimum cost network flow problem. Numbers on vertices are net demands, b_k, and those on arcs are costs.

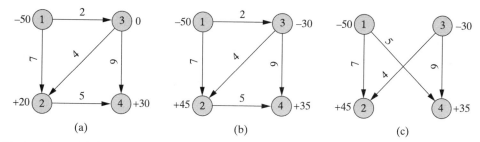

(a)   (b)   (c)

Determine which are transportation problems.

Modeling:

(a) This network flow problem is not a transportation problem because node 3 is a pure transshipment node, with neither supply nor demand.

(b) This network flow problem is also not a transportation problem because nodes 2 and 3 have both inbound and outbound arcs.

(c) This network flow problem is a transportation problem. Nodes 1 and 3 are pure source nodes; nodes 2 and 4 are pure demand nodes.

Standard Form for Transportation Problems

Using constants

$$s_i \triangleq \text{supply at node } i$$
$$d_j \triangleq \text{demand at node } j$$
$$c_{i,j} \triangleq \text{unit cost of flow from } i \text{ to } j$$

and decision variables

$$x_{i,j} \triangleq \text{flow from } i \text{ to } j$$

the network flow formulation ⎡10.3⎤ of a transportation problem simplifies to

$$\min \sum_i \sum_j c_{i,j} x_{i,j}$$

$$\text{s.t.} \quad -\sum_j x_{i,j} = -s_i \qquad \text{for all } i$$

$$\sum_i x_{i,j} = d_j \qquad \text{for all } j$$

$$x_{i,j} \geq 0 \qquad \text{for all } i, j$$

As usual, flows leaving supply nodes carry negative signs, and those entering demand points have positive signs.

The more common statement of transportation problems in operations research uses the standard form obtained when signs are reversed on the first set of constraints:

10.24 | For supplies s_i, demands d_j, and costs $c_{i,j}$, the **standard form** of **transportation problems** is

$$\min \quad \sum_i \sum_j c_{i,j} x_{i,j}$$

$$\text{s.t.} \quad \sum_j x_{i,j} = s_i \qquad \text{for all } i$$

$$\sum_i x_{i,j} = d_j \qquad \text{for all } j$$

$$x_{i,j} \geq 0 \qquad \text{for all } i, j$$

SAMPLE EXERCISE 10.13: FORMULATING TRANSPORTATION PROBLEMS

Write a standard-form transportation problem formulation for the diagraph of Sample Exercise 10.12(c).

Modeling: Following | 10.24 |, the model in standard form is

$$\min \quad 7x_{1,2} + 5x_{1,4} + 4x_{3,2} + 9x_{3,4}$$

$$\text{s.t.} \quad x_{1,2} + x_{1,4} \qquad\qquad\qquad = 50$$

$$x_{3,1} + x_{3,4} \qquad\qquad = 30$$

$$x_{1,2} + x_{3,2} \qquad\qquad = 45$$

$$x_{1,4} + x_{3,4} \qquad\qquad = 35$$

$$x_{1,2}, x_{1,4}, x_{3,1}, x_{3,4} \geq 0$$

EXAMPLE 10.2: MARINE MOBILIZATION TRANSPORTATION PROBLEM

A really massive transportation problem example arises in officer mobilization planning for the U.S. Marine Corps.[1] During any international emergency, thousands of officers must be mobilized from their regular duty or reserve positions into billets required for the emergency. However, not every officer is qualified by rank, training, or experience to fulfill every assignment. Using the very efficient network flow methods of this chapter, Marine planners are able to develop a mobilization scheme in a few minutes by solving a transportation problem with over 100,000 arcs.

Mobilization options can be depicted in a digraph like the fictitious one of Figure 10.12. Supply nodes exist for each group of like-qualified officers presently based in the same location. For example, Figure 10.12 shows one supply node for captains now in the 1st Division who are trained as intelligence officers. Another represents civil affairs–trained officers in Georgia units of the Marine Corps Reserve.

---

[1] Based on D. O. Bausch, G. G. Brown, D. R. Hundley, S. H. Rapp, and R. E. Rosenthal (1991), "Mobilizing Marine Corps Officers," *Interfaces, 21:4*, 26–38.

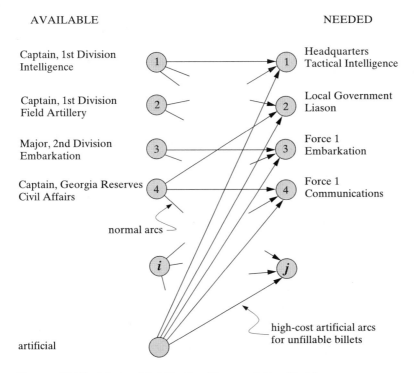

FIGURE 10.12   Marine Mobilization Transportation Problem

Demand nodes indicate needs for officers of particular qualifications at active locations of the emergency. For example, Figure 10.12 depicts the need for one or more local government liason officers in the forward area of a deployment.

Arcs exist whenever officers represented by a supply node would be qualified to fill billets of a demand node. Thus the Georgia civil affairs officers are linked to the local government liason billets but not to artillery or intelligence needs. There may be several feasible assignments for officers of any source node. For example, Figure 10.12 indicates that the same civil affairs officers could serve as communications officers with assault force 1.

The Marines' first priority is to fill all needed billets, but there are always some left unfilled. The transportation problem of Figure 10.12 models the possibility of unfilled billets with an artificial supply node connected to all demands. High cost on its arcs penalizes unfilled billets in the objective function.

Once as many requirements as possible have be filled, a secondary concern is to minimize turbulence. That is, the Marines try to assign officers to the same unit to which they were assigned before mobilization, or at least to assign them to a nearby unit by minimizing total travel cost.

Using the following notation:

s_i ≜ supply of officers available at source node i

d_j ≜ demand for officers at demand node j

$c_{i,j}$ ≜ distance officers at supply node i would have to travel to report for billet j

   (a large positive number if i is the artificial node)

$$I_j \triangleq \text{set of officer supply nodes } i \text{ suitable for billets } j$$

$$J_i \triangleq \text{set of billet demand nodes } j \text{ suitable for officers at supply node } i$$

$$x_{i,j} \triangleq \text{number of officers at node } i \text{ mobilized to billets at } j$$

this Marine mobilization problem reduces to the standard-form transportation problem

$$\min \sum_i \sum_j c_{i,j} x_{i,j}$$

$$\text{s.t.} \quad \sum_{j \in J_i} x_{i,j} = s_i \qquad \text{for all } i \qquad (\text{supply of } i)$$

$$\sum_{i \in I_j} x_{i,j} = d_j \qquad \text{for all } j \qquad (\text{demand for } j)$$

$$x_{i,j} \geq 0 \qquad \text{for all } i, j$$

Assignment Problems

Another important class of network flow models, known as (linear) assignment problems, do not seem at first glance to have anything to do with something flowing.

| 10.25 | **Assignment problems** deal with optimal pairing or matching of objects in two distinct sets. |

We might pair jobs to machines, male dating service clients to female, duties to employees, and so on.

Assignment problems are modeled using (discrete) decision variables

$$x_{i,j} \triangleq \begin{cases} 1 & \text{if } i \text{ is assigned to } j \\ 0 & \text{otherwise} \end{cases}$$

The first subscript refers to items in one set; the second subscript identifies items in the other.

Assignment problems may have quite complicated objective functions (see Section 11.4). We address here only the most common form with a linear objective. In that case we have known costs

$$c_{i,j} \triangleq \text{cost of assigning } i \text{ to } j$$

An optimal solution minimizes (or maximizes) total cost (or benefit).

Denoting by A the set of allowed assignments (i, j), we can express the linear assignment model as

$$\min \text{ or } \max \quad \sum_{(i,j) \in A} c_{i,j} x_{i,j} \qquad\qquad (\text{min or max total cost})$$

$$\text{s.t.} \qquad \sum_{j \text{ with } (i,j) \in A} x_{i,j} = 1 \qquad \text{for all } i \qquad (\text{every } i \text{ is assigned})$$

$$\sum_{i \text{ with } (i,j) \in A} x_{i,j} = 1 \qquad \text{for all } j \qquad (\text{every } j \text{ is assigned}) \qquad (10.5)$$

$$x_{i,j} = 0 \text{ or } 1 \qquad\qquad \text{for all } (i, j) \in A$$

The first system of constraints guarantees every i is assigned exactly once by summing over all possible assignments j. The second system does the same to assure that every j is assigned exactly once.

Notice that this formulation is an integer linear program (ILP). Decision variables are allowed only the discrete values 0 and 1. Still, we will soon see how the discreteness can be ignored in computing an optimum.

SAMPLE EXERCISE 10.14: FORMULATING LINEAR ASSIGNMENT MODELS

The following table shows a computerized dating service's compatibility ratings for male customer $i = 1, \ldots, 3$ with and female customers $j = 1, \ldots, 3$.

		j	
i	1	2	3
1	90	30	12
2	40	80	75
3	60	65	80

Formulate an assignment model to find the highest compatibility arrangement providing each customer with a single date of the opposite gender.

Modeling: Using variables $x_{i,j} = 1$ if male i is paired with female j, the model corresponding to system (10.5) is

$$\max \quad 90x_{1,1} + 30x_{1,2} + 12x_{1,3} + 40x_{2,1} + 80x_{2,2} + 75x_{2,3} + 60x_{3,1} + 65x_{3,2} + 80x_{3,3}$$

$$\begin{aligned}
\text{s.t.} \quad & x_{1,1} + x_{1,2} + x_{1,3} = 1 \\
& x_{2,1} + x_{2,2} + x_{2,3} = 1 \\
& x_{3,1} + x_{3,2} + x_{3,3} = 1 \\
& x_{1,1} + x_{2,1} + x_{3,1} = 1 \\
& x_{1,2} + x_{2,2} + x_{3,2} = 1 \\
& x_{1,3} + x_{2,3} + x_{3,3} = 1 \\
& \text{all } x_{i,j} = 0 \text{ or } 1
\end{aligned}$$

EXAMPLE 10.3: CAM ASSIGNMENT

For a more realistic assignment problem, consider a computer-aided manufacturing (CAM) system that automatically routes jobs through workstations of a computer-controlled factory. Each job consists of a sequence of required machining and assembly operations.

Often, there are several different workstations where the same operation can be performed. Thus the computer control system must make routing decisions. Each time that a job completes some operation, the system must select the next station to which the job should be routed from among the several that could perform the next operation required.

One method for accomplishing such control decisions in an approximately optimal way is to periodically solve an assignment model.[2] To illustrate, suppose that the 8 fictitious jobs i of Table 10.2 are either waiting for movement to their next

---

[2]Based on J. Chandra and J. Talavage (1991), "Optimization-Based Opportunistic Part Dispatching in Flexible Manufacturing Systems," School of Industrial Engineering, Purdue University, July.

TABLE 10.2   Transportation and Processing Times for CAM Assignment Example

Jobs, i	Next Workstations, j									
	1	2	3	4	5	6	7	8	9	10
1	8	—	23	—	—	—	—	—	5	—
2	—	4	—	12	15	—	—	—	—	—
3	—	—	20	—	13	6	—	8	—	—
4	—	—	—	—	19	10	—	—	—	—
5	—	—	—	8	—	—	12	—	—	16
6	14	—	—	—	—	—	8	—	3	—
7	—	6	—	—	—	—	—	27	—	12
8	—	5	15	—	—	—	—	32	—	—

workstation or will finish their current operation within the next 5 minutes. The 10 workstations j to which they might be routed are also shown. Entries in the table reflect

$$\begin{pmatrix} \text{transportation} \\ \text{time to the} \\ \text{station} \end{pmatrix} + \begin{pmatrix} \text{waiting time} \\ \text{until the station} \\ \text{becomes free} \end{pmatrix} + \begin{pmatrix} \text{operation} \\ \text{processing time} \\ \text{at the station} \end{pmatrix}$$

That is, they show the short-term time implications of assigning jobs i to stations j. Missing values in the table reflect assignments that are not possible because the next required operation cannot be performed at a workstation.

Balancing Unequal Sets with Dummy Elements

Letting $A \triangleq \{\text{feasible } (i, j) \text{ pairs}\}$, and $c_{i,j} \triangleq$ the times shown in Table 10.2, an optimal short-term control decision is to assign jobs to workstations in a manner minimizing total time. This is exactly what assignment model (10.5) will compute.

One small complication arises from the fact that there are more workstations than jobs. Model (10.6) assumes that the sets to be matched have equal numbers of objects.

This problem is easily solved with dummy members.

> **10.26** If the two sets to be paired in an assignment problem differ in size, the smaller can be augmented with **dummy members**. These dummy objects should be treated as assignable to all members of the other set at zero cost.

Principle 10.26 leads to dummy jobs $i = 9, 10$ in our CAM example.

Linear Assignment as Network Flows with Integer Solutions

Linear assignment model (10.6) is a discrete optimization problem—an ILP. Still, a quick review will show that the assignment model bears a striking resemblence to transportation problem standard form 10.24 .

In fact, they are the same. Thinking of i's as sources with supplies of 1, and j's as sinks with demands of 1, assignment is nothing more than finding an optimal source-to-sink transportation flow.

What about 0–1 integrality? Network flow integrality property $\boxed{10.20}$ of Section 10.4 completes the transformation. With supplies and demands all $= 1$ and arc capacities also $= 1$ (after relaxing to $0 \leq x_{i,j} \leq 1$) property $\boxed{10.20}$ assures that optimal solutions to the underlying network flow linear program will be integer valued. That is, the network structure of the assignment problem yields integer optimal solutions from a linear program.

In fact, we can ignore the capacities on arcs as well because each $x_{i,j}$ is constrained by a source constraint $\sum_i x_{i,j} = 1$. The result is the standard form for assignment models.

$\boxed{10.27}$ Every linear **assignment problem** can be formulated in **standard form** and solved as the network flow (or transportation problem) linear program

$$\text{min or max} \quad \sum_i \sum_j c_{i,j} x_{i,j}$$

$$\text{s.t.} \quad \sum_j x_{i,j} = 1 \quad \text{for all } i$$

$$\sum_i x_{i,j} = 1 \quad \text{for all } j$$

$$x_{i,j} \geq 0 \quad \text{for all } i \text{ and } j$$

where $c_{i,j}$ is the cost or benefit of assigning i to j, and $x_{i,j} = 1$ if i is assigned to j and $= 0$ otherwise.

In fact, assignment algorithms are known that are even more efficient that general network flow procedures (see, e.g., the **Hungarian algorithm** in references at the end of this chapter).

SAMPLE EXERCISE 10.15: FORMULATING ASSIGNMENT PROBLEMS AS NETWORK FLOWS

Formulate the assignment problem of Sample Exercise 10.14 as a minimum cost network flow problem, and explain why optimal flows must be integer.

Modeling: We must first reverse the sign of all objective function coefficients to convert the maximize problem of Sample Exercise 10.14 to minimize form. Then, treating $i = 1, 2, 3$ as sources with unit supply, and $j = 1, 2, 3$ as sinks with unit demand, gives the following network:

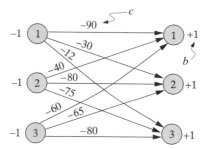

Optimal flows will be integer (binary) because supplies and demands are integer, and there are no capacities (property $\boxed{10.20}$).

CAM Assignment Example Model

We are now ready to state a full formulation of our CAM example in the standard format of 10.27 :

$$
\begin{aligned}
\min \quad & 8x_{1,1} + 23x_{1,3} + 5x_{1,9} + 4x_{2,2} + 12x_{2,4} + 15x_{2,5} \\
& + 20x_{3,3} + 13x_{3,5} + 6x_{3,6} + 8x_{3,8} + 19x_{4,5} + 10x_{4,6} \\
& + 8x_{5,4} + 12x_{5,7} + 16x_{5,10} + 14x_{6,1} + 8x_{6,7} + 3x_{6,9} \\
& + 6x_{7,2} + +27x_{7,8} + 12x_{7,10} + 5x_{8,2} + 15x_{8,3} + 32x_{8,8}
\end{aligned}
$$

$$
\begin{array}{lll}
\text{s.t.} & x_{1,1} + x_{1,3} + x_{1,9} = 1 & \text{(job 1)} \\
& x_{2,2} + x_{2,4} + x_{2,5} = 1 & \text{(job 2)} \\
& x_{3,3} + x_{3,5} + x_{3,6} + x_{3,8} = 1 & \text{(job 3)} \\
& x_{4,5} + x_{4,6} = 1 & \text{(job 4)} \\
& x_{5,4} + x_{5,7} + x_{5,10} = 1 & \text{(job 5)} \\
& x_{6,1} + x_{6,7} + x_{6,9} = 1 & \text{(job 6)} \\
& x_{7,2} + x_{7,8} + x_{7,10} = 1 & \text{(job 7)} \\
& x_{8,2} + x_{8,3} + x_{8,8} = 1 & \text{(job 8)} \\
& x_{9,1} + x_{9,2} + x_{9,3} + x_{9,4} + x_{9,5} & \\
& \quad + x_{9,6} + x_{9,7} + x_{9,8} + x_{9,9} + x_{9,10} = 1 & \text{(job 9)} \\
& x_{10,1} + x_{10,2} + x_{10,3} + x_{10,4} + x_{10,5} & \\
& \quad + x_{10,6} + x_{10,7} + x_{10,8} + x_{10,9} + x_{10,10} = 1 & \text{(job 10)} \\
& x_{1,1} + x_{6,1} + x_{9,1} + x_{10,1} = 1 & \text{(station 1)} \\
& x_{2,2} + x_{7,2} + x_{8,2} + x_{9,2} + x_{10,2} = 1 & \text{(station 2)} \\
& x_{1,3} + x_{3,3} + x_{8,3} + x_{9,3} + x_{10,3} = 1 & \text{(station 3)} \\
& x_{2,4} + x_{5,4} + x_{9,4} + x_{10,4} = 1 & \text{(station 4)} \\
& x_{2,5} + x_{3,5} + x_{4,5} + x_{9,5} + x_{10,5} = 1 & \text{(station 5)} \\
& x_{3,6} + x_{4,6} + x_{9,6} + x_{10,6} = 1 & \text{(station 6)} \\
& x_{5,7} + x_{6,7} + x_{9,7} + x_{10,7} = 1 & \text{(station 7)} \\
& x_{3,8} + x_{7,8} + x_{8,8} + x_{9,8} + x_{10,8} = 1 & \text{(station 8)} \\
& x_{1,9} + x_{6,9} + x_{9,9} + x_{10,9} = 1 & \text{(station 9)} \\
& x_{5,10} + x_{7,10} + x_{9,10} + x_{10,10} = 1 & \text{(station 10)} \\
& x_{i,j} = 0 \text{ or } 1 \text{ for all } i = 1, \ldots, 10;\, j = 1, \ldots, 10 &
\end{array}
$$

(10.6)

An optimal short-term routing of jobs to workstations has

$$
x^*_{1,1} = x^*_{2,2} = x^*_{3,8} = x^*_{4,6} = x^*_{5,4} = x^*_{6,9} = x^*_{7,10} = x^*_{8,3} = 1
$$

and all other $x^*_{i,j} = 0$.

Tableau Representation of Transportation and Assignment Problems

Especially when all (i, j) combinations are feasible, a digraph depiction of a transportation or assignment problem can become very confusing. That is why such problems are often represented in an alternative network tableau format.

10.28 | Transportation and assignment problems can be conveniently represented in **network tableau** form

with rows for sources i, columns for sinks j, supplies s_i along the right margin, demands d_j along the bottom, unit costs $c_{i,j}$ in the small upper-left box of each (i, j) cell, and flow $x_{i,j}$ in the larger part of the cells.

To see the idea, return to the transportation problem of Sample Exercise 10.13:

$$\begin{aligned} \min \quad & 7x_{1,2} + 5x_{1,4} + 4x_{3,2} + 9x_{3,4} \\ \text{s.t.} \quad & x_{1,2} + x_{1,4} = 50 \\ & x_{3,1} + x_{3,4} = 30 \\ & x_{1,2} + x_{3,2} = 45 \\ & x_{1,4} + x_{3,4} = 35 \\ & x_{1,2}, x_{1,4}, x_{3,1}, x_{3,4} \geq 0 \end{aligned}$$

with digraph depicted in Sample Exercise 10.12(c). The corresponding tableau would have the form

when $x_{1,2} = 15$, $x_{1,4} = 35$, $x_{3,2} = 30$, and $x_{3,4} = 0$.

10.6 OTHER SINGLE-COMMODITY NETWORK FLOW MODELS

The transportation and assignment models of Section 10.5 are the most famous special cases of network flow problems, but there are many others. In this section we develop several more that retain the **single-commodity** character of all models treated so far. That is, the underlying problems can be modeled adequately by simply tracking how much of some commodity enters or leaves the network at each node and how much of it flows through each arc. In Section 10.9 we introduce the more difficult cases where multiple commodities must be accounted for separately and where flow may be gained or lost as it transits the arcs of the network.

Network Flow Formulation of Shortest Path Problems

In Section 9.1 we introduced **shortest path problems**—models seeking the least cost path between specified pairs of nodes in a given graph. For example, the Two Ring Circus case reproduced in Figure 10.13 required a shortest path from Lincoln (node 1) to Tallahassee (node 9). Numbers on arcs reflect travel costs for the circus less revenue for appearing in cities along the way.

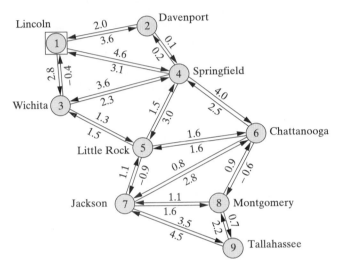

FIGURE 10.13   The Two Ring Circus Shortest Path Example

In Sections 9.3 to 9.7 we detailed very efficient algorithms for shortest path problems with no negative-length dicycles. When codes are available, those methods provide the most effective solution techniques.

We can, however, compute shortest paths as minimum cost network flows (and thus also as linear programs) by exploiting integrality property $\boxed{10.20}$. It is only necessary to define appropriate supplies and demands, then solve a flow problem over the given shortest path digraph.

$\boxed{10.29}$ Although it is usually more efficient to use dynamic programming-based methods of Chapter 9, the problem of finding a shortest path from source s to sink t in the absence of negative dicycles can be solved as a minimum cost flow problem over the same digraph and the same costs, with a supply of 1 at the source and a demand of 1 at the sink. All other nodes are transshipment.

On the Two Ring Circus example of Figure 10.13, construction $\boxed{10.29}$ creates a unit supply at Lincoln node $s = 1$ and a unit demand at Tallahassee sink $t = 9$. Certainly, flow introduced at Lincoln, and conserved at all nodes along the way, will eventually reach Tallahassee along one or more s-to-t paths.

Network integrality property $\boxed{10.20}$ assures that all flow will be along a single path because supplies and demands are integer. Those arcs with $x_{i,j} = 1$ will be on the optimal path, and those with $x_{i,j} = 0$ will not be traversed. This single path must be a shortest path because no flow following a path longer that the shortest could be a minimum cost flow.

SAMPLE EXERCISE 10.16: FORMULATING SHORTEST PATHS AS NETWORK FLOWS

Consider the following digraph:

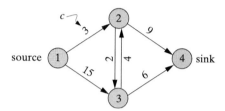

Numbers on arcs are lengths $c_{i,j}$.

(a) Formulate a minimum cost network flow model to compute a shortest path from node 1 to node 4.

(b) Solve you model by inspection and identify an optimal path.

Modeling:

(a) Applying construction $\boxed{10.29}$, the appropriate model $\boxed{10.3}$ is

$$\min \quad 3x_{1,2} + 15x_{1,3} + 2x_{2,3} + 9x_{2,4} + 4x_{3,2} + 6x_{3,4}$$
$$\text{s.t.} \quad -x_{1,2} - x_{1,3} \qquad\qquad = -1$$
$$x_{1,2} + x_{3,2} - x_{2,3} - x_{2,4} = 0$$
$$x_{1,3} + x_{2,3} - x_{3,2} - x_{3,4} = 0$$
$$x_{2,4} + x_{3,4} = 1$$
$$x_{1,2}, x_{1,3}, x_{2,3}, x_{2,4}, x_{3,2}, x_{3,4} \geq 0$$

(b) An optimal solution to the model of part (a) is $x_{1,2}^* = x_{2,3}^* = x_{3,4}^* = 1, x_{1,3}^* = x_{3,2}^* = x_{2,4}^* = 0$. The corresponding optimal path indicated by $x_{i,j}^* = 1$ is 1–2–3–4.

Maximum Flow Problems

Maximum flow problems are another very simple special case of network flows.

$\boxed{10.30}$ **Maximum flow problems** seek to find the largest possible flow between a specified source node and a specified sink node subject to conservation of flow at all other nodes and capacities on arcs.

The only issue is how much commodity can be moved from the single source to a single sink.

SAMPLE EXERCISE 10.17: IDENTIFYING MAXIMUM FLOWS

Determine by inspection the maximum flow from node 1 to node 4 in the following graph (numbers on arcs are capacities $u_{i,j}$):

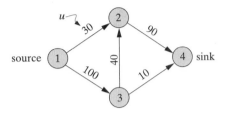

Analysis: Careful examination of the possibilities will establish that a maximum flow sends 80 units from 1 to 4 as follows:

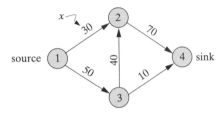

EXAMPLE 10.4: BUILDING EVACUATION MAXIMUM FLOW

Maximum flow problems arise most often as subproblems in more complex operations research studies. However, they occur naturally in evaluating the safety of proposed building designs.[3] Proper design requires adequate capacity for building evacuation in the event of an emergency.

Figure 10.14 shows a small example involving a proposed sports arena. Patrons in the arena would exit in an emergency through doors on all four sides that can accommodate 600 persons per minute. Those doors lead into an outer hallway that can move 350 persons per minute in each direction. Egress from the hallway is through four firestairs with capacity 400 persons per minute and a tunnel to the parking lot accommodating 800 persons per minute. Our interest is in the maximum rate of evacuation possible with this design.

Part (b) of Figure 10.14 shows how we reduce this safety analysis to a maximum flow model. Patron flows originate at source node 1. Outbound arcs model the four doorways. The flows around the outer hall lead to the four stairways and the tunnel. Persons exiting by any of those means pass to sink node 10. Capacities enforce the flow rates of the various facilities.

We wish to know the maximum flow from 1 to 10, subject to the capacities indicated. An optimal flow is provided in the arc labels of part (b). Patrons can escape at a total rate of 2100 per minute.

---

[3]Based in part on L. G. Chalmet, R. L. Frances, and P. B. Saunders (1982), "Network Models for Building Evacuation," *Management Science, 28*, 86–105. All numerical data and diagrams were made up by the author of this book.

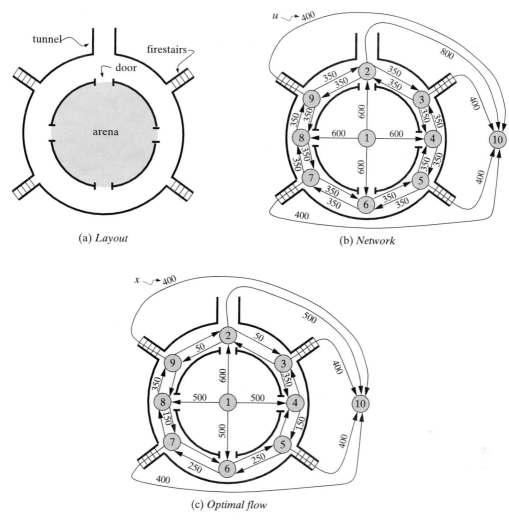

(a) *Layout*

(b) *Network*

(c) *Optimal flow*

FIGURE 10.14   Building Evacuation Maximum Flow Example

Return Arc Network Flow Formulation of Maximum Flow Problems

As so far presented, neither the tiny example of Sample Exercise 10.17 nor the larger one of Figure 10.14(b) is a minimum cost network flow problem. Flow conservation and capacity requirements are much like standard model 10.3 , but we have specified no costs, and flows do not balance at source and sink.

To create a true minimum cost flow problem, we add a return arc.

10.31   **Return arcs** balance unknown source-to-sink flows by feeding back that flow in an artificial arc from sink to source.

Adding a return arc to the maximum flow example of Figure 10.14(b) produces the following digraph:

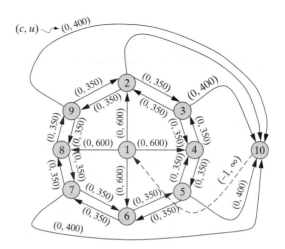

Artificial arc (10, 1) takes all flow reaching sink node 10 and returns it to source node 1, thus restoring flow balance (for net demand $b_k = 0$).

To finish a minimum cost network flow model, we need only introduce costs. Notice that the more the flow in return arc (10, 1), the greater the flow from source to sink. Thus we complete the model by placing a cost of -1 on flow in the return arc and taking all other costs as 0. Any minimum cost flow will necessarily maximize return arc, and thus source to sink flow.

| 10.32 | Maximum flow problems can be modeled as minimum cost flows by adding a return arc from sink to source with cost -1. All net demands and all costs on other arcs are zero. |

Although principle | 10.32 | assures that maximum flow problems can be solved by minimum cost network algorithms, even more efficient special-purpose procedures have been developed. See the references at the end of this chapter for details.

SAMPLE EXERCISE 10.18: MODELING MAXIMUM FLOWS AS NETWORKS

Develop a minimum cost network flow model corresponding to the maximum flow model of Sample Exercise 10.17.

Modeling: Following construction | 10.32 |, we add a return arc (4, 1) to obtain the following minimum cost flow problem:

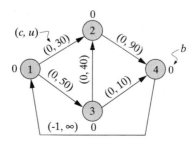

Time-Expanded Flow Models and Networks

As with the more general linear programs of Section 4.5, many applications of network flows involve time-expanded formulations to account for flows over time. This is especially true for those involving inventory management.

In the specially structured network flow case, such problem lead to time-expanded networks.

> 10.33 | **Time-expanded networks** model each node of a flow system as a series of nodes, one for each time interval. Arcs then reflect either flows between points in a particular time, or flows across time in a particular location.

EXAMPLE 10.5: AGRICO CHEMICAL TIME-EXPANDED NETWORK FLOW

To see the idea, consider the case of Agrico Chemical,[4] which is a large chemical fertilizer company. Figure 10.15 sketches our fictitious version of the network model used to plan Agrico's production, distribution, and inventory. Much in the spirit of OOI Example 10.1, products for this real company originate at 4 plants and are transshipped through 20 regional distribution centers before reaching customers in any of 500 service areas.

Highly seasonal demand makes the Agrico case more complex than OOI. Agrico produces fertilizers throughout the year, but much of the demand comes in the spring quarter. Production capacity cannot accommodate spring demand in just one quarter. Thus the company builds up inventories at distribution sites during off-seasons to ship in the spring. Of course, storage is not unlimited, and inventories result in holding costs.

Agrico's decision problem requires choosing the amount to produce in each of the four quarters of the year, the pattern of shipping and storing it at distribution sites, and a scheme for sending it on to customers. We want to do all this at minimum total cost.

Time-Expanded Modeling of Agrico Example

The Agrico digraph of Figure 10.15 illustrates a time-expanded network. Total annual flow originates at the source node, but separate arcs representing production connect it to the 8 nodes (Pi, t) modeling plants $P1$ through $P4$ in quarters $t = 1, \ldots, 4$. Capacities on those arcs would enforce the quarterly capacities of the correponding plants, and costs would reflect units costs of production.

Plants in any quarter, t, are connected by transportation arcs to distribution center nodes (Dj, t) in the same quarter. Distribution centers, in turn, are linked to customer demands of the corresponding quarter. Costs on these arcs would reflect the unit cost of transportation, perhaps differentiated by the quarter in which the shipment takes place.

Holding arcs between nodes for the same distribution center comprise the main new feature representable with time-expanded modeling. For example, the arc from $(D1, 2)$ to $(D1, 3)$ models holding of the product at distribution site number 1 from the second to the third quarter. Its cost would be the unit holding cost at distribution

---

[4]Based on F. Glover, G. Jones, D. Karney, D. Klingman, and J. Mote (1979), "An Integrated Production, Distribution, and Inventory Planning System," *Interfaces, 9:5,* 21–35.

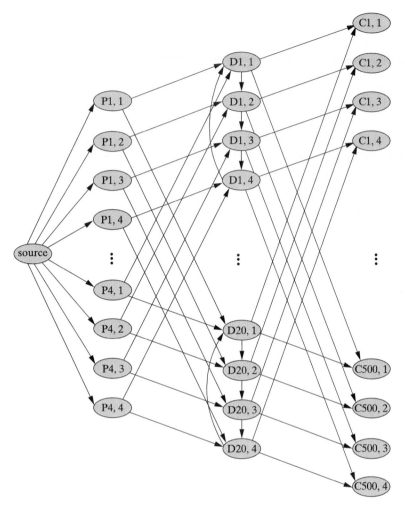

FIGURE 10.15   Agrico Time-Expanded Network Format

site 1, and its capacity the size of the available storage. It would be impossible to model both flows in time and flow among facilities without distinguishing nodes by both place and time.

SAMPLE EXERCISE 10.19: MODELING IN TIME-EXPANDED NETWORKS

A certain company can manufacture up to 15 thousand units of its product in any calendar quarter at a cost of $35 per thousand. The following table shows cost per thousand units shipped to each of the company's two customers, and the number (thousands) of units demanded by each in various quarters.

Customer	Shipping Cost	Demand by Quarter			
		1	2	3	4
1	11	5	9	2	1
2	17	3	14	6	4

Assuming that inventory can be maintained at the plant for $8 per thousand per quarter, develop a time-expanded network flow model to determine the company's best production, distribution, and inventory plan.

Modeling: Following $\boxed{10.33}$, we create 4 nodes for the plant in different quarters and 4 nodes for each of the 2 customers in different quarters. Commodities for all quarters and customers arises at a common source node. Production arcs link commodities to plant nodes by quarter, and holding arcs connect the plant nodes. Transportation arcs join the plant in each quarter to customer demands for that quarter. The result is the following time-expanded minimum cost network flow model:

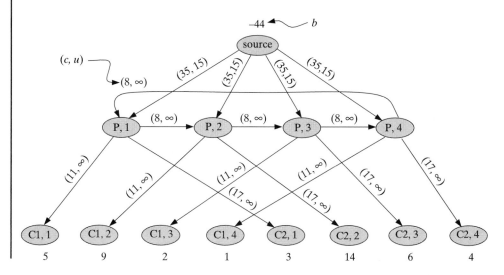

10.7   NETWORK SIMPLEX ALGORITHM FOR OPTIMAL FLOWS

All the minimum cost network flow models of Sections 10.1, 10.5, and 10.6 are linear programs, so the LP algorithms of Chapters 5 and 6 could be applied. Still, we have already seen in Section 10.3 how the elegant structure of cycle directions leads to much more efficient computations in the network setting (principle $\boxed{10.19}$).

In this section we fill in the details of rudimentary cycle direction Algorithm 10A to derive one such procedure. It is known as the **network simplex** because it specializes the usual simplex computations of Chapter 5, taking advantage of the fact that simplex directions turn out to be cycle directions in the network flow case. A very different, cycle canceling method is detailed in Section 10.8.

Linear Dependence in Node–Arc Matrices and Cycles

Simplex search centers on **bases**—maximal collections of linearly independent columns drawn from the constraint matrix of main (standard form) constraints $\mathbf{Ax} = \mathbf{b}$. Columns of the main constaints in network flow models are columns of the node–arc incidence matrix [format (10.4)]. They correspond to arcs. Thus, to understand simplex in the network context, we must first understand what collections of arcs correspond to linearly independent collections of columns.

Begin with cycles. For example, consider the OOI cycle (2, 3), (1, 3), (1, 4), (2, 4) illustrated in Figure 10.16. The figure displays both the cycle and the corresponding columns of the node–arc incidence matrix shown in Table 10.1.

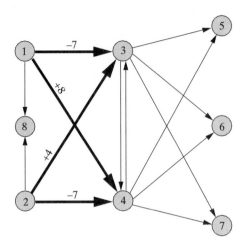

	Arc			
Node	(2, 3)	(1, 3)	(1, 4)	(2, 4)
1	0	−1	−1	0
2	−1	0	0	−1
3	+1	+1	0	0
4	0	0	+1	+1
5	0	0	0	0
6	0	0	0	0
7	0	0	0	0
8	0	0	0	0

FIGURE 10.16 Cycle and the Corresponding Node–Arc Incidence Columns of OOI Example

Suppose that we apply weights of $+1$ on all column vectors for all forward arcs in the cycle, and -1 on columns of reverse arcs. Then

$$
+1 \begin{pmatrix} 0 \\ -1 \\ +1 \\ 0 \\ 0 \\ 0 \\ 0 \\ 0 \end{pmatrix} - 1 \begin{pmatrix} -1 \\ 0 \\ +1 \\ 0 \\ 0 \\ 0 \\ 0 \\ 0 \end{pmatrix} + 1 \begin{pmatrix} -1 \\ 0 \\ 0 \\ +1 \\ 0 \\ 0 \\ 0 \\ 0 \end{pmatrix} - 1 \begin{pmatrix} 0 \\ -1 \\ 0 \\ +1 \\ 0 \\ 0 \\ 0 \\ 0 \end{pmatrix} = \begin{pmatrix} 0 \\ 0 \\ 0 \\ 0 \\ 0 \\ 0 \\ 0 \\ 0 \end{pmatrix}
$$

Our ± 1 weighted sum of the columns has produced the zero vector. This implies that the columns are linearly dependent because any one of the vectors can be expressed (by transposing) as a nonzero linear combination of the others.

The ± 1 weights we have chosen are exactly those of the corresponding cycle direction 10.11 . Thus for the same (Figure 10.5) reasons that cycle directions satisfy

$A\Delta x = 0$, such a ± 1 linear combination of node–arc incidence columns for the arcs of a cycle will always yield the zero vector.

> **10.34** Node–arc incidence matrix columns for arcs of a cycle form a linearly dependent set.

Since column vectors of a basis must be linearly independent, we have an immediate consequence:

> **10.35** Basic sets of arcs for minimum cost network flow models can contain no cycles.

Cycles are not the only way to produce linearly dependent sets in network flow models. For example, return to the complex direction of Figure 10.8. That weighting of node–arc incidence columns also produces the zero vector because the direction satisfies $A\Delta x = 0$. Thus the collection of arcs with nonzero weights is linearly dependent.

Still, we know from principle $\boxed{10.17}$ that complex directions like that of Figure 10.8 decompose into weighted sums of cycle directions. This implies that each includes at least one cycle.

> **10.36** Every linearly dependent arc set in a minimum cost network flow model contains a cycle.

SAMPLE EXERCISE 10.20: CHECKING LINEAR INDEPENDENCE OF ARCS

Consider the following digraph:

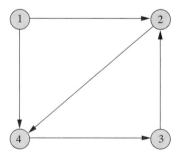

Determine which of the following arc sets could be part of a basis for the corresponding node–arc incidence matrix.

(a) $\{(1, 2), (4, 3)\}$

(b) $\{(2, 4), (4, 3), (3, 2)\}$

(c) $\{(1, 2), (1, 4), (4, 3), (2, 4)\}$

(d) $\{(1, 2), (2, 4), (3, 2)\}$

Analysis:

(a) This set is linearly independent under principle $\boxed{10.36}$ because it contains no cycle. It could be part of a basis.

(b) These arcs form a cycle. Thus, under principle $\boxed{10.35}$, they cannot be part of a basis.

(c) This set is linearly dependent because it contains cycle 1–2–4–1. Under principle $\boxed{10.35}$, it cannot be part of a basis.

(d) This set is linearly independent under principle $\boxed{10.36}$ because it contains no cycle. It could be part of a basis.

Spanning Trees of Networks

We say that a graph is **connected** if there is a chain between every pair of nodes. A graph is a **tree** if it is connected and contains no cycles. A tree is a **spanning tree** if it touches every node of a graph.

Figure 10.17 illustrates these definitions. Part (a) shows a spanning tree. It is connected, it contains no cycles, and it touches all nodes of the OOI network. Parts (b) to (d) demonstrate the various ways that graphs can fail the definition of a spanning tree.

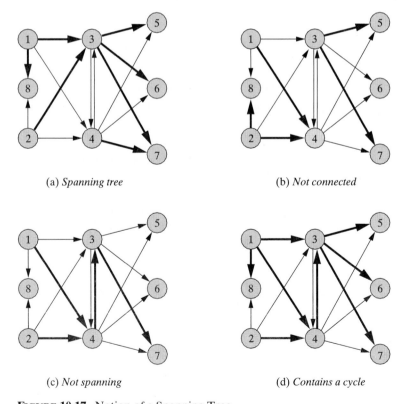

(a) *Spanning tree*   (b) *Not connected*

(c) *Not spanning*   (d) *Contains a cycle*

FIGURE 10.17 Notion of a Spanning Tree

The connected but cycle-free nature of trees leads to an important property:

> 10.37   Every pair of nodes in a tree is connected by a unique chain of the tree. If the tree spans a graph, every pair of nodes in the graph is connected by a unique chain of the tree.

For example, nodes 8 and 4 are connected in the spanning tree of Figure 10.17(a) by the chain 8–1–3–7–4 and by no other. If there were two such chains, they would form a cycle, and there must be one if the tree is connected.

Spanning Tree Bases for Network Flow Models

Returning to the issue of bases for node–arc incidence matrices, the example of Figure 10.17(d) has no hope of being a basis because it contains a cycle. Corresponding columns of the node–arc incidence matrix will be linearly dependent (principle 10.34). The graphs in parts (a) to (c) have no such cycles. Thus principle 10.36 assures us that corresponding columns are linearly independent. But are they a basis?

A basis must be a maximal linearly independent set (i.e., it should be impossible to enlarge it without creating a dependency). Notice that the examples in Figure 10.17(b) and (c) fail this test. We know from principle 10.36 that a dependency will be created only if we form a cycle. That cannot occur when we insert an arc between previously unconnected nodes. For example, adding arc (4, 3) in Figure 10.17(b) preserves linear independence. The same is true for adding arc (3, 5) in Figure 10.17(c).

The spanning tree of Figure 10.17(a) is different. Because there is already a unique chain between every pair of nodes (property 10.37), inserting any other arc closes a cycle. For example, adding arc (2, 4) closes cycle 2–4–7–3–2.

> 10.38   Adding an arc to a spanning tree produces a unique cycle.

It follows immediately that spanning trees correspond to maximal linearly independent sets and so bases. We have reached the key to understanding simplex computations in minimum cost network flow models:

> 10.39   A collection of columns in the node–arc incidence matrix of a minimum cost network flow problem form a basis if and only if corresponding arcs form a spanning tree of the associated digraph.

SAMPLE EXERCISE 10.21: IDENTIFYING NETWORK FLOW BASES

Determine which of the four arcs sets in Sample Exercise 10.20 form a basis for the corresponding network flow problem.

Analysis: We have already seen in Sample Exercise 10.20 that sets (b) and (c) are linearly dependent. Set (a) is linearly independent, but it is not a basis under principle 10.39 because the arcs do not form a spanning tree. Set (d) does form a spanning tree. Thus it is the only basis of the four.

Network Basic Solutions

Simplex algorithms proceed from basic feasible solution to basic feasible solution. For the network case, where capacities or upper bounds are present, we define basic solution in the upper-/lower-bound sense of Section 5.9 (principle $\boxed{5.48}$).

> $\boxed{10.40}$ In **basic solutions** for network flow problems, nonbasic arcs have flow equal to either 0 or capacity $u_{i,j}$. Basic arcs have the unique flow achieving flow balance for the nonbasic values specified. The flow is **basic feasible** of all basis flows are within bounds.

     Figure 10.18 illustrates for the OOI example of Figure 10.2. The corresponding basis is highlighted. There is positive flow on nonbasic arc (4, 3), but it equals capacity $u_{4,3} = 25$. Only basic arcs can have flow $x_{i,j}$ other than zero or $u_{i,j}$.

     To compute this basic solution, we first assign flow 0 to all nonbasics except (4, 3) and 25 to $x_{4,3}$. Only one choice of basic flows can then meet net demand requirements at each node. For example, with nonbasic flow $x_{3,7} = 0$, the only flow-balancing choice for basic flow $x_{4,7} = 610$. The solution of Figure 10.18 is basic feasible because all such basis flows are within bound limits.

> **SAMPLE EXERCISE 10.22: COMPUTING NETWORK BASIC SOLUTIONS**
>
> In the following network, numbers on nodes indicate net demands b_k and those on arcs show capacities $u_{i,j}$.
>
>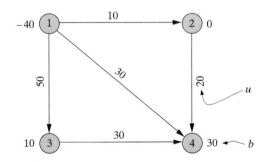
>
> For each of the following choices on nonbasic arcs, compute the corresponding basic solution and determine whether it is basic feasible.
>
> (a) (1, 4) nonbasic lower-bounded, (2, 4) nonbasic upper-bounded
>
> (b) (3, 4) nonbasic lower-bounded, (1, 2) nonbasic upper-bounded
>
> **Analysis:** We apply principle $\boxed{10.40}$.
>
> (a) Here the basis is $\{(1, 2), (1, 3), (3, 4)\}$. With $x_{1,4} = 0$ and $x_{2,4} = 20$, the unique choice of basic values that meets net demand requirements at all nodes is $x_{1,2} = x_{1,3} = 20, x_{3,4} = 10$. This basis solution is not basic feasible because flow $x_{1,2}$ exceeds capacity $u_{1,2} = 10$.

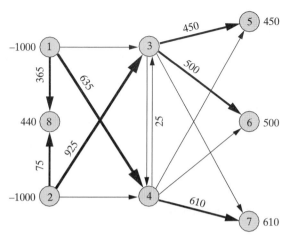

FIGURE 10.18 Initial Basic Solution for OOI Example

(b) Here the basis is $\{(1, 3), (1, 4), (2, 4)\}$. With $x_{3,4} = 0$ and $x_{1,2} = 10$, the unique choice of basic values that meets net demand requirements at all nodes is $x_{1,3} = x_{2,4} = 10$, $x_{1,4} = 20$. Since all basic flows are within bounds, this basic solution is basic feasible.

Simplex Cycle Directions

Simplex directions (construction $\boxed{5.21}$ of Section 5.3) in linear programs are formed by increasing one nonbasic variable from its lower bound (or decreasing a nonbasic from its upper bound) and changing basic variables in the unique way assuring that main constraints remain satisfied. For network flow cases, every basic set of columns (arcs) forms a spanning tree (principle $\boxed{10.39}$). Thus introducing any nonbasic arc creates a unique cycle (principle $\boxed{10.38}$), and the corresponding cycle direction must be the simplex direction.

$\boxed{10.41}$ The **network simplex direction** for increasing nonbasic arc at flow $= 0$ is the cycle direction of the unique cycle formed when the corresponding nonbasic arc is introduced into the current basis tree and the orientation is the same as that arc. The simplex direction for decreasing a nonbasic arc at capacity flow $u_{i,j}$ is the cycle direction obtained when the corresponding nonbasic arc is introduced and the orientation is opposite to that arc.

Simplex cycle directions need not be feasible because some decreasing basic flow may already equal zero or some increasing basic flow may be at capacity. As with other simplex methods, however, we may ignore such **degeneracy** without much practical effect except step sizes $\lambda = 0$ (principle $\boxed{5.38}$).

SAMPLE EXERCISE 10.23: CONSTRUCTING NETWORK SIMPLEX DIRECTIONS

Consider the following network. Numbers next to nodes are net demands b_k, and those on arcs are cost, capacities, and current flows $(c_{i,j}, u_{i,j}, x_{i,j})$.

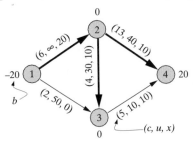

Taking highlighted arcs as the basis, construct the simplex directions for all nonbasic arcs.

Analysis: We apply construction $\boxed{10.41}$. Introducing nonbasic arc $(1, 3)$ produces unique cycle 1–3–2–1. Since nonbasic arc $(1, 3)$ is presently at flow 0, the corresponding simplex direction will be the cycle direction for 1–3–2–1 with $(1, 3)$ a forward arc. That is, $\Delta x_{1,3} = +1$, $\Delta x_{2,3} = \Delta x_{1,2} = -1$.

Nonbasic arc $(3, 4)$ is presently at capacity. To find its cycle/simplex direction we pass the unique cycle it forms in the basis so that $(3, 4)$ is a reverse arc. That is, we employ cycle direction 4–3–2–4 with $\Delta x_{3,4} = \Delta x_{2,3} = -1$, $\Delta x_{2,4} = +1$.

Network Simplex Algorithm

Algorithm 10B combines these ideas in a special simplex version of cycle direction-based network search, Algorithm 10A. Instead of considering all possible cycle directions at each iteration, we check only the simplex directions induced by the unique cycles for increasing or decreasing nonbasic arcs. Simplex theory for all linear programs (principle $\boxed{5.20}$ of Section 5.3) tells us that these directions are sufficient to find an optimal solution.

Comparison of this algorithm statement with the general simplex computations of Chapter 5 will highlight the convenience made possible by the special network flow structure. To find simplex directions and determine whether they improve the objective function, we need only trace our way through the basis tree.

Network Simplex Solution of OOI Example

Figure 10.19 details solution of our OOI example by Algorithm 10B, starting from the initial basic feasible solution in Figure 10.18. At the first iteration, the simplex directions for the 7 nonbasic arcs are as follows:

Nonbasic	Cycle	$\bar{c}$
Increase $(1, 3)$	1–3–2–8–1	3
Increase $(2, 4)$	2–4–1–8–2	−1
Increase $(3, 4)$	3–4–1–8–2–3 via $(3, 4)$	−4
Increase $(3, 7)$	3–7–4–1–8–2–3	8
Increase $(4, 3)$	3–4–1–8–2–3 via $(4, 3)$	−4
Increase $(4, 5)$	4–5–3–2–8–1–4	8
Increase $(4, 6)$	4–6–3–2–8–1–4	6

ALGORITHM 10B: NETWORK SIMPLEX SEARCH

Step 0: Initialization. Choose any starting basic feasible flow $\mathbf{x}^{(0)}$, identify the corresponding basis spanning tree, and set solution index $t \leftarrow 0$.

Step 1: Simplex Directions. For each nonbasic arc, examine the simplex direction associated with the unique cycle that arc forms in the basis spanning tree and apply test $\boxed{10.14}$ to determine whether the direction improves.

Step 2: Optimal. If no simplex cycle direction improves, stop; flow $\mathbf{x}^{(t)}$ is globally optimal. Otherwise, choose as $\Delta \mathbf{x}$ some improving simplex cycle direction, and let (p, q) denote the corresponding nonbasic arc.

Step 3: Step Size. Compute the maximum feasible step λ in direction $\Delta \mathbf{x}$ (rule $\boxed{10.15}$):

$$\lambda^+ \leftarrow \min\{(u_{i,j} - x_{i,j}^{(t)}) : (i, j) \text{ forward}\} \; (+\infty \text{ if none})$$

$$\lambda^- \leftarrow \min\{x_{i,j}^{(t)} : (i, j) \text{ reverse}\} \; (+\infty \text{ if none})$$

$$\lambda \; \leftarrow \min\{\lambda^+, \lambda^-\}$$

If $\lambda = +\infty$, stop; the model is unbounded.

Step 4: Advance. Update

$$\mathbf{x}^{(t+1)} \leftarrow \mathbf{x}^{(t)} + \lambda \Delta \mathbf{x}$$

by increasing flows on forward arcs of the cycle direction and decreasing those on reverse arcs by the amount λ.

Step 5: New Basis. If some arc other than (p, q) established the minimum λ in step 3, replace any such arc in the basis spanning tree by (p, q). Increment $t \leftarrow t + 1$ and return to step 1.

Any simplex direction with $\bar{c} < 0$ could provide a means of improving flow $\mathbf{x}^{(0)}$. Figure 10.19 employs the one associated with increasing nonbasic, lower-bounded arc $(2, 4)$. The unique cycle formed in the basis tree immediately identifies the rest of the associated cycle direction. Arcs $(2, 4)$ and $(1, 8)$ increase; $(1, 4)$ and $(2, 8)$ decrease. Thus $\lambda^+ = \infty$, $\lambda^- = \min\{75, 635\}$, and $\lambda = 75$. Arc $(2, 8)$ established the λ limit, so $(2, 4)$ replaces $(2, 8)$ in the basis. Updating produces flow $\mathbf{x}^{(1)}$ of Figure 10.19. Notice that the new basis is also a spanning tree.

The simplex direction decreasing nonbasic, upper-bounded arc $(4, 3)$ now improves for $\mathbf{x}^{(1)}$. The unique cycle formed by $(4, 3)$ in the current basis yields a cycle direction decreasing $(4, 3)$ and $(2, 4)$ while increasing $(2, 3)$. Move limit λ for this direction occurs when nonbasic $(4, 3)$ reaches its lower bound of 0. Thus flows are updated, but the basis remains unchanged.

Flow $\mathbf{x}^{(2)}$ in Figure 10.19 is the result. Increasing nonbasic $(3, 4)$ can improve this flow because $\bar{c} = -3$ on the corresponding simplex direction. Adjustment of $\lambda = 25$ around the implied cycle direction improves to flow $\mathbf{x}^{(3)}$. Again incoming arc $(3, 4)$ itself stops improvement, so the basis remains unchanged.

Comparison to Figure 10.3 will show that new flow $\mathbf{x}^{(3)}$ is now optimal. None of the 7 simplex cycle directions is improving.

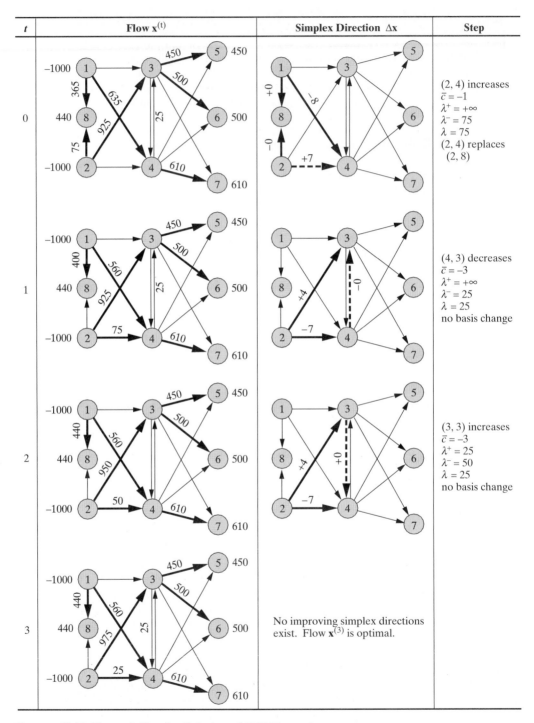

FIGURE 10.19 Network Simplex Solution of OOI Example

SAMPLE EXERCISE 10.24: APPLYING THE NETWORK SIMPLEX ALGORITHM

Return to the network flow example and starting basis of Sample Exercise 10.23, and compute an optimal solution with network simplex Algorithm 10B.

Analysis: At the first iteration, increasing $(1, 3)$ produces a simplex direction for cycle 1–3–2–1 having $\bar{c}_{1,3} = 2 - 4 - 6 = -8$. Decreasing nonbasic $(3, 4)$ yields the direction for cycle 4–3–2–4 with $\bar{c}_{3,4} = -5 - 4 + 13 = 4$. Thus we update along the former by $\lambda = 10$, and replace $(2, 3)$ in the basis by $(1, 2)$ because $(2, 3)$ established λ.

The new solution is

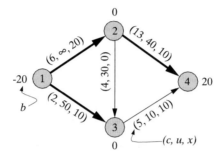

Now the cycle direction for increasing $(2, 3)$ has $\bar{c}_{2,3} = 8$ and that of decreasing $(3, 4)$ had $\bar{c}_{3,4} = 12$. Neither is improving, and the latest flow is optimal.

10.8 CYCLE CANCELING ALGORITHMS FOR OPTIMAL FLOWS

Algorithm 10A in Section 10.3 outlines the common computational logic of many network flow algorithms. The big question left unanswered is: How do we identify improving feasible cycle directions at each iteration, or prove that none exist? In Section 10.7 we presented one method based on the simplex algorithm. In this section we detail a quite different approach: **cycle canceling** by shortest path computation.

Residual Digraphs

The cycle canceling method (and many other network flow procedures) begins each iteration by constructing a residual digraph that details available options for improving feasible cycle directions. Each arc in the original digraph yields up to two in the residual one, depending on whether its present flow can feasibly increase or decrease or both.

> 10.42 | The **residual digraph** associated with current feasible flow $\mathbf{x}^{(t)}$ has the same nodes as the given network. It has one "increase arc" (i, j), with cost $c_{i,j}$, for each increasable arc flow $x_{i,j}^{(t)} < u_{i,j}$, and one (backward) "decrease arc" (j, i), with cost $-c_{i,j}$, for each decreasable arc flow $x_{i,j}^{(t)} > 0$.

Figure 10.20 illustrates for the OOI starting conditions depicted in Figure 10.9. Every flow that can both increase and decrease yields two opposed arcs in the residual digraph. For example, arc $(2, 3)$ with current flow $x_{2,3}^{(0)} = 1000$ produces both residual graph arc $(2, 3)$ and arc $(3, 2)$. The first, increase arc represents the fact that $x_{2,3}$ can feasibly become larger. Its cost is $c_{2,3} = 4$, the unit cost of such a flow increase. Decrease arc $(3, 2)$ shows that the flow can also become smaller without losing feasibility. It has cost $-c_{2,3} = -4$, the unit savings from a flow decrease.

Flows that cannot both increase and decrease produce just one arc in the residual digraph. For example, flow $x_{1,4}^{(0)} = 0$. It can only increase, so the residual graph has only arc $(1, 4)$ at cost $c_{1,4} = 8$.

Consider the network flow problem sketched below. Numbers on nodes show net demand b_k, and those on arcs show costs, capacities, and current flows $(c_{i,j}, u_{i,j}, x_{i,j})$.

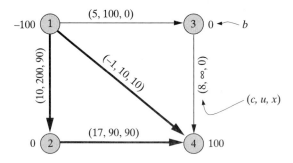

Construct the corresponding residual digraph.

Analysis: Applying construction $\boxed{10.42}$, the residual digraph is

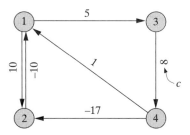

Feasible Cycle Directions and Dicycles of Residual Digraphs

The point of constructing a residual digraph is to make it easier to identify improving feasible cycle directions in the original graph. Notice first that any cycle direction meeting principle $\boxed{10.13}$'s requirements for feasibility—forward flows below capacity, reverse flows positive—now corresponds to a dicycle (definition $\boxed{10.10}$) in the residual graph. A backward, decrease arc has been provided for each reverse arc of the cycle.

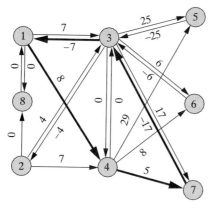

FIGURE 10.20 Residual Digraph for OOI Example Flow of Figure 10.9

10.43 | Cycle direction $\Delta\mathbf{x}$ is feasible for a current flow $\mathbf{x}^{(t)}$ if and only if the residual graph for $\mathbf{x}^{(t)}$ contains a dicycle with forward arcs of the $\Delta\mathbf{x}$ cycle corresponding to increase arcs in the residual dicycle, and reverse arcs of the cycle corresponding to decrease arcs.

One example is cycle 7–3–1–4–7 of the original digraph.

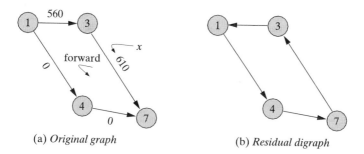

(a) *Original graph*   (b) *Residual digraph*

The corresponding cycle direction is feasible because forward arcs (1, 4) and (4, 7) are below capacity, and reverse arcs (3, 7) and (1, 3) have positive flow. It follows that each forward arc of (a) corresponds to an increase arc in (b), and each reverse arc of (a) yields a (backward) decrease arc in (b).

Improving Feasible Cycle Directions and Negative Dicycles of Residual Digraphs

We seek improving feasible cycle directions, those with negative reduced costs:

$$\bar{c} = (\text{total forward arc cost}) - (\text{total reverse arc cost})$$

But under construction 10.42 , forward arcs appear in the residual digraph as increase arcs with the same cost, and reverse arcs appear as backward, decrease arcs with the sign of their costs switched. It follows that the $\bar{c}$ of a feasible cycle direction is exactly the length of the corresponding dicycle in the residual digraph. Dicycles of negative length yield what we require.

10.44 | Feasible cycle direction $\Delta \mathbf{x}$ is improving if and only if the corresponding residual graph dicycle is a negative dicycle.

Again we can illustrate with the direction for OOI example cycle 7–3–1–4–7.

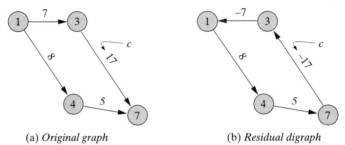

(a) *Original graph*   (b) *Residual digraph*

The original cycle is improving because the corresponding dicycle has negative total length.

SAMPLE EXERCISE 10.26: CONNECTING ORIGINAL AND RESIDUAL DIGRAPHS

Return to the example of Sample Exercise 10.25.

(a) Show that the direction for cycle 1–3–4–2–1 is both feasible and improving.

(b) Identify the corresponding dicycle in the residual digraph, and verify that it is a negative dicycle.

Analysis:

(a) The direction for cycle 1–3–4–2–1 is feasible because forward flows $(1, 3)$ and $(3, 4)$ are below capacity, and reverse flows $(2, 4)$ and $(1, 2)$ are positive. It is improving because

$$\bar{c} = (5 + 8) - (17 + 10) = -14$$

(b) The residual digraph of Sample Exercise 10.25 contains dicycle 1–3–4–2–1 with increase arcs $(1, 3)$ and $(3, 4)$, plus decrease arcs $(4, 2)$ and $(2, 1)$. Its total length is

$$5 + 8 - 17 - 10 = -14 = \bar{c}$$

Using Shortest Path Algorithms to Find Cycle Directions

Principles | 10.43 | and | 10.44 | reduce our search for improving feasible cycle directions to a hunt for negative dicycles in the residual digraph. The advantage is that we already know algorithms for identifying negative dicycles or for proving that none exists.

The Floyd–Warshall shortest path Algorithm 9B, which we derived in Section 9.4, is one. That procedure is designed primarily to compute shortest paths from all nodes of a graph to all other nodes. If the given graph contains a negative dicycle, shortest path computation fails, but a negative dicycle is returned. Interchanging roles to make negative dicycles the normal outcome and completion of shortest path calculations the exception produces exactly the improving feasible cycle direction subroutine that we require.

> 10.45   Application of Floyd–Warshall Algorithm 9B to the residual digraph for a current feasible flow either yields a negative dicycle, indicating that the direction for the corresponding cycle of the original graph is both improving and feasible, or completes a shortest path computation, proving that improving feasible cycle directions do not exist.

SAMPLE EXERCISE 10.27: FINDING DIRECTIONS WITH FLOYD–WARSHALL

Apply Floyd–Warshall Algorithm 9B to the residual digraph of Sample Exercise 10.25 either to produce an improving feasible cycle direction or to prove that none exists.

Analysis: After its third main iteration, Floyd–Warshall computation produces the following shortest path and inbound node results:

k	$v^{(0)}[k, \ell]$				$d[k, \ell]$			
	$\ell = 1$	$\ell = 2$	$\ell = 3$	$\ell = 4$	$\ell = 1$	$\ell = 2$	$\ell = 3$	$\ell = 4$
1	0	10	5	13	—	1	1	3
2	−10	0	−5	3	2	—	1	3
3	∞	∞	0	8	—	—	—	3
4	−27	−17	−22	−14	2	4	1	3

Computation halts with the conclusion that there is a negative dicycle because diagonal distance $v^{(3)}[4, 4] = -14$ is negative. Using $d[k, \ell]$ labels to recover the dicycle, $d[4, 4] = 3$, $d[4, 3] = 1$, $d[4, 1] = 2$, $d[4, 2] = 4$. Thus 4–2–1–3–4 is a negative dicycle of the residual graph. We already know from Sample Exercise 10.26 that the corresponding cycle direction is indeed improving and feasible.

Cycle Canceling Solution of the OOI Example

Algorithm 10C formalizes principle 10.45 in the cycle canceling method for network flow optimization. Figure 10.21 details its application to our OOI example of Figure 10.9.

As before, graphs on the left of Figure 10.21 show flows in the real network. Notice that the sequence of solutions differs from that of Table 10.10 because different improving feasible cycle directions were used. Still, both finished at the same optimal flow.

Graphs at the right in Figure 10.21 depict the residual digraphs for each flow encountered. Details of Floyd–Warshall computation have been omitted, but the negative dicycles obtained are highlighted. For example, the first application of Algorithm 9B identified negative dicycle 7–3–1–4–7 in the residual digraph for initial flow $\mathbf{x}^{(0)}$. Its total length is $-17 - 7 + 8 + 5 = -11$. Application of rule 10.15 around the corresponding cycle in the real OOI network yields $\lambda = 560$. Then increasing flows by 560 on forward arcs $(1, 4)$ and $(4, 7)$, while decreasing by the same amount on reverse arcs $(3, 7)$ and $(1, 3)$ produces the indicated flow $\mathbf{x}^{(1)}$.

ALGORITHM 10C: CYCLE CANCELING FOR NETWORK FLOWS

Step 0: Initialization. Choose any starting feasible flow $\mathbf{x}^{(0)}$, and set solution index $t \leftarrow 0$.

Step 1: Residual Digraph. Construct the residual digraph corresponding to the current flow $\mathbf{x}^{(t)}$ (principle $\boxed{10.42}$).

Step 2: Floyd–Warshall. Execute Algorithm 9B on the current residual graph. If Floyd–Warshall computation terminates with no indication of a negative dicycle, stop. No improving feasible cycle directions exist, and current flow $\mathbf{x}^{(t)}$ is globally optimal.

Step 3: Cycle Direction. Use Floyd–Warshall decision labels $d[k, \ell]$ to trace a negative dicycle in the residual graph, and construct the cycle direction $\Delta\mathbf{x}$ for the corresponding cycle in the original graph.

Step 4: Step Size. Compute the maximum feasible step λ in direction $\Delta\mathbf{x}$ (rule $\boxed{10.15}$):

$$\lambda^+ \leftarrow \min\{(u_{i,j} - x_{i,j}^{(t)}) : (i, j) \text{ forward}\} \ (+\infty \text{ if none})$$

$$\lambda^- \leftarrow \min\{x_{i,j}^{(t)} : (i, j) \text{ reverse}\} \ (+\infty \text{ if none})$$

$$\lambda \ \leftarrow \min\{\lambda^+, \lambda^-\}$$

If $\lambda = \infty$, stop; the model is unbounded.

Step 5: Advance. Update

$$\mathbf{x}^{(t+1)} \leftarrow \mathbf{x}^{(t)} + \lambda\Delta\mathbf{x}$$

by increasing flows on forward arcs of the cycle direction and decreasing those on reverse arcs by the amount λ. Then increment $t \leftarrow t + 1$ and return to Step 1.

Computation continues until $t = 4$. There Floyd–Warshall terminates on the residual digraph without discovering a negative dicycle. We may conclude (principle $\boxed{10.45}$) that no negative dicycle exists in the residual digraph, so that no improving feasible cycle direction exists for our current flow. The flow must be optimal.

SAMPLE EXERCISE 10.28: APPLYING THE CYCLE CANCELING ALGORITHM

Apply Cycle Canceling Algorithm 10C to the network flow problem of Sample Exercise 10.25.

Analysis: We have already seen in Sample Exercise 10.27 that Floyd–Warshall computation on the initial residual digraph yields negative dicycle 4–2–1–3–4. Forward arcs in the corresponding cycle imply that $\lambda^+ = \min\{(100 - 0), (\infty - 0)\} = 100$. Reverse arcs give $\lambda^- = \min\{90, 90\} = 90$. Thus step size $\lambda = \min\{100, 90\} = 90$.

Adjustment by this amount around cycle 4–2–1–3–4 produces the following new flow and residual digraph:

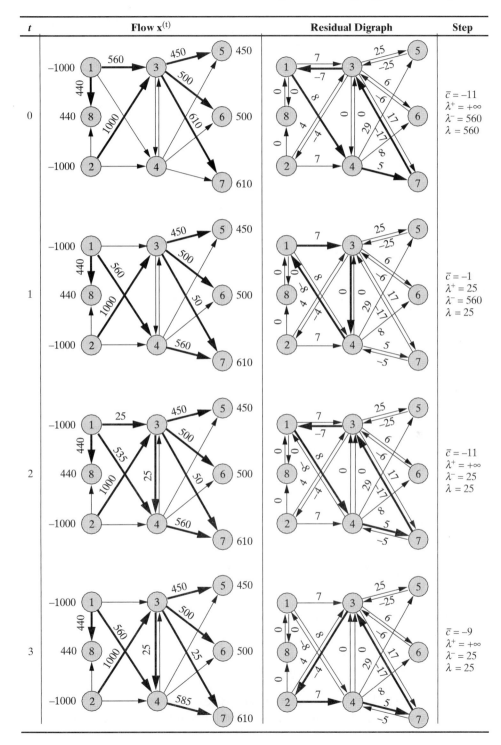

FIGURE 10.21 Cycle Canceling Solution of the OOI Example

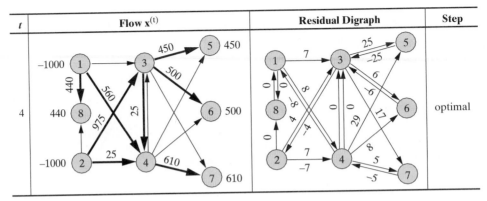

FIGURE 10.21 Cycle Cancelling Solution of the OOI Example (Continued)

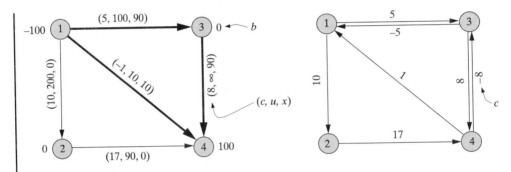

Floyd–Warshall computation on the new residual digraph will show that there is no negative dicycle. Thus the revised flow is optimal.

10.9 MULTICOMMODITY AND GAIN/LOSS FLOWS

There is almost always a tradeoff between the tractability of operations research models and the span of applications to which they can be adapted. Networks are no exception. In this final section we explore briefly some more broadly applicable network forms that retain part, but not all, of the tractability of minimum cost network flows.

Multicommodity Flows

An implicit assumption in our previous flow models has been that all flows are in a **single commodity**. In the OOI example of Section 10.1, it was toaster ovens. In other examples the commodity was people, chemicals, or manufacturing tasks. In every case we assumed that demands at sink nodes could be filled from any supply node with a path to the sink. That is, we assumed that flows were interchangeable or fungible.

Multicommodity flow problems arise when flows passing thorough a common network must be kept separate; flow is not fungible.

10.46 | **Multicommodity flow models** seek a minimum cost flow where separate commodities are moving through a common network.

EXAMPLE 10.6: BAY FERRY MULTICOMMODITY FLOW

As usual, it will help to think of a simple (fictitious) example. Figure 10.22 depicts traffic flows in communities around an ocean bay. Each morning, the population of the three residential communities travels to the two industrial and two commercial centers in the region. Table 10.3 details flows by origin and destination. For example, 1250 of the 6000 daily trips originating at residential node 4 have industrial park node 7 as their destination.

At present, geography limits each trip to a single path. Numbers on arcs of the network in Figure 10.22 show the distance in kilometers between various points. For instance, those originating at node 1 and bound for node 7 must drive all the way around the bay through nodes 2 to 6. The 21,100 trips arising daily in the three residential communities produce a total of 399,250 kilometers of driving along such routes.

Regional planners are considering various improvements to reduce air pollution by reducing the number of kilometers driven. One idea is the ferry indicated by

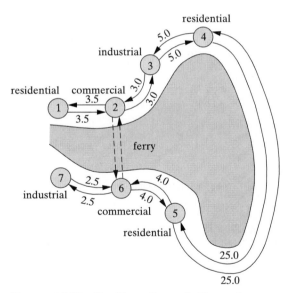

FIGURE 10.22   Bay Ferry Example Network

TABLE 10.3   Daily Trips in the Bay Ferry Example

Trip Origin Node	Total Trips	Trips by Destination						
		1	2	3	4	5	6	7
1	2,850	—	900	750	40	10	600	550
4	6,000	100	2000	1100	—	150	1400	1250
5	12,250	110	4000	2200	200	—	3300	2440

dashed arcs (2, 6) and (6, 2) in Figure 10.22. If a ferry were introduced, it could carry 2000 cars in each direction during the morning rush period. We want to know how many kilometers of driving might be saved.

Multicommodity Flow Models

Clearly, Figure 10.22 depicts a flow network. But what is flowing? If we treat all trips as equal, we have only a single commodity. Then, however, it would be feasible to fulfill the demand for 1250 trips from origin node 4 at sink node 7 with trips from any source. Naturally, a minimum distance solution would prefer ones from nearby source 5, leaving demands for node 5 trips at locations 2 and 3 to be filled from the closer source 1. Such a solution makes no sense for the application because trips are not fungible.

We must form separate commodity networks for trips from each of the three sources. That is, we must model in multiple commodities. Still, the commodities are not independent. All share the 2000-trip capacity of the proposed ferry. Such interdependencies are typical of multicommodity flows.

10.47 | Commodities of a multicommodity flow model cannot be analyzed separately because they interact through shared arc capacities.

Suppose that we employ constants

$$c_{q,i,j} \triangleq \text{unit cost of commodity } q \text{ flow in arc } (i, j)$$

$$u_{i,j} \triangleq \text{shared capacity of arc } (i, j)$$

$$b_{q,k} \triangleq \text{net demand for commodity } q \text{ at node } k$$

and the decision variables

$$x_{q,i,j} \triangleq \text{commodity } q \text{ flow in arc } (i, j)$$

10.48 | The multicommodity network flow model on a digraph with nodes $k \in V$ and arcs $(i, j) \in A$ is

$$\min \quad \sum_q \sum_{(i,j) \in A} c_{q,i,j} x_{q,i,j}$$

$$\text{s.t.} \quad \sum_{(i,k) \in A} x_{q,i,k} - \sum_{(k,j) \in A} x_{q,k,j} = b_{q,k} \qquad \text{for all } q, k \in V$$

$$\sum_q x_{q,i,j} \le u_{i,j} \qquad \text{for all } (i, j) \in A$$

$$x_{q,i,j} \ge 0 \qquad \text{for all } q, (i, j) \in A$$

Table 10.4 presents the corresponding formulation of our Bay Ferry example. There commodity 1 corresponds to flows from origin node 1, commodity 2 denotes flows from residential node 4, and commodity 3 relates to residential node 5. Notice that there are separate systems of flow conservation equations for each commodity, plus a common set of capacity constraints on the two arcs with flow limits.

TABLE 10.4   Bay Ferry Example Model

$$
\begin{aligned}
\min \quad & 3.5x_{1,1,2} + 3.5x_{1,2,1} + 3x_{1,2,3} + 3x_{1,3,2} + 5x_{1,3,4} + 5x_{1,4,3} && \text{(minimize driving)} \\
& + 15x_{1,4,5} + 15x_{1,5,4} + 4x_{1,5,6} + 4x_{1,6,5} + 2.5x_{1,6,7} + 2.5x_{1,7,6} \\
& + 3.5x_{2,1,2} + 3.5x_{2,2,1} + 3x_{2,2,3} + 3x_{2,3,2} + 5x_{2,3,4} + 5x_{2,4,3} \\
& + 15x_{2,4,5} + 15x_{2,5,4} + 4x_{2,5,6} + 4x_{2,6,5} + 2.5x_{2,6,7} + 2.5x_{2,7,6} \\
& + 3.5x_{3,1,2} + 3.5x_{3,2,1} + 3x_{3,2,3} + 3x_{3,3,2} + 5x_{3,3,4} + 5x_{3,4,3} \\
& + 15x_{3,4,5} + 15x_{3,5,4} + 4x_{3,5,6} + 4x_{3,6,5} + 2.5x_{3,6,7} + 2.5x_{3,7,6}
\end{aligned}
$$

$$
\begin{aligned}
\text{s.t.} \quad & x_{1,2,1} - x_{1,1,2} && = && -2{,}850 && \text{(commodity 1)} \\
& x_{1,1,2} + x_{1,3,2} + x_{1,6,2} - x_{1,2,1} - x_{1,2,3} - x_{1,2,6} && = && 900 \\
& x_{1,2,3} + x_{1,4,3} - x_{1,3,2} - x_{1,3,4} && = && 750 \\
& x_{1,3,4} + x_{1,5,4} - x_{1,4,3} - x_{1,4,5} && = && 40 \\
& x_{1,4,5} + x_{1,6,5} - x_{1,5,4} - x_{1,5,6} && = && 10 \\
& x_{1,2,6} + x_{1,5,6} + x_{1,7,6} + x_{1,6,2} - x_{1,6,5} - x_{1,6,7} && = && 600 \\
& x_{1,6,7} - x_{1,7,6} && = && 550 \\
& x_{2,2,1} - x_{2,1,2} && = && 100 && \text{(commodity 2)} \\
& x_{2,1,2} + x_{2,3,2} + x_{2,6,2} - x_{2,2,1} - x_{2,2,3} - x_{2,2,6} && = && 2{,}000 \\
& x_{2,2,3} + x_{2,4,3} - x_{2,3,2} - x_{2,3,4} && = && 1{,}100 \\
& x_{2,3,4} + x_{2,5,4} - x_{2,4,3} - x_{2,4,5} && = && -6{,}000 \\
& x_{2,4,5} + x_{2,6,5} - x_{2,5,4} - x_{2,5,6} && = && 150 \\
& x_{2,2,6} + x_{2,5,6} + x_{2,7,6} + x_{2,6,2} - x_{2,6,5} - x_{2,6,7} && = && 1{,}400 \\
& x_{2,6,7} - x_{2,7,6} && = && 1{,}250 \\
& x_{3,2,1} - x_{3,1,2} && = && 110 && \text{(commodity 3)} \\
& x_{3,1,2} + x_{3,3,2} + x_{3,6,2} - x_{3,2,1} - x_{3,2,3} - x_{3,2,6} && = && 4{,}000 \\
& x_{3,2,3} + x_{3,4,3} - x_{3,3,2} - x_{3,3,4} && = && 2{,}200 \\
& x_{3,3,4} + x_{3,5,4} - x_{3,4,3} - x_{3,4,5} && = && 200 \\
& x_{3,4,5} + x_{3,6,5} - x_{3,5,4} - x_{3,5,6} && = && -12{,}250 \\
& x_{3,2,6} + x_{3,5,6} + x_{3,7,6} - x_{3,6,2} - x_{3,6,5} - x_{3,6,7} && = && 3{,}300 \\
& x_{3,6,7} - x_{3,7,6} && = && 2{,}440 \\
& x_{1,2,6} + x_{2,2,6} + x_{3,2,6} && \le && 2{,}000 && \text{(capacities)} \\
& x_{1,6,2} + x_{2,6,2} + x_{3,6,2} && \le && 2{,}000 \\
& \text{all } x_{q,i,j} \ge 0
\end{aligned}
$$

An optimal solution reduces the total driving to 280,770 kilometers, a savings of 29.7%. This is accomplished by accommodating $x_{1,2,6} = 1160$ commodity 1 trips on the 2-to-6 ferry, and $x_{2,2,6} = 840$ commodity 2. In the reverse direction the ferry carries $x_{3,6,2} = 2000$ commodity 3 trips.

SAMPLE EXERCISE 10.29: FORMULATING MULTICOMMODITY FLOWS

Consider the following multicommodity flow problem:

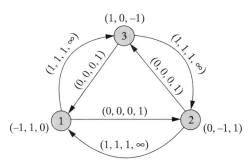

(a)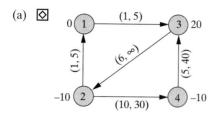

(b)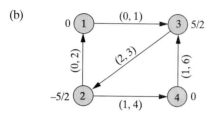

(c)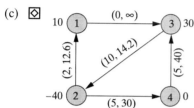

(d)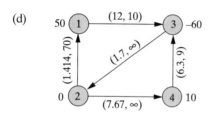

10-17 The Quick Chip gravel company has received a contract to supply two new construction projects in the towns of Brock and Wurst. A total of 60 truck-loads are needed at Brock in the next month and 90 at Wurst. Quick Chip has idle gravel pits in the towns of Nova, Scova, and Tova, each with a monthly production capacity of 50 truckloads. Travel distances from each pit to each project site are shown in the following table.

Pit	To Brock	To Wurst
Nova	23	77
Scova	8	94
Tova	53	41

The company wants to fulfill its contract at least total truck travel distance.

(a) ◙ Formulate a linear program to choose an optimal shipping plan.

(b) ◙ ▭ Use class optimization software to compute an optimal solution to your LP.
(c) ◙ Show that your LP can be represented as a minimum cost transportation problem by sketching the corresponding digraph and labeling as in Exercise 10-1.
(d) ◙ Sketch the transportation tableau representation of the problem with flows equal to your optimal solution in part (b).

10-18 Maize Mills has 800 thousand, 740 thousand and 460 thousand bushels of corn stored at its three rural elevators. Its three processing plants will soon require 220 thousand, 1060 thousand and 720 thousand respectively, to make cornstarch. The following table shows the cost per thousand bushels of shipping from each elevator to each plant.

	Plant		
Elevator	1	2	3
1	10	13	22
2	15	12	11
3	17	14	19

Maize wants to move its corn to plants at minimum total shipping cost.

(a) through (d) as in Exercise 10-17.

10-19 Senior design students are negotiating which of the four members of the team will take primary reponsibility for each of the four project tasks the team must complete. The following table shows the composite ratings (0 to 100) they have prepared to estimate the ability of each member to manage each task.

	Task Rating			
Member	1	2	3	4
1	90	78	45	69
2	11	71	50	89
3	88	90	85	93
4	40	80	65	39

The team wants to find a maximum total score plan that allocates exactly one task to each team member.

(a) ◙ Formulate a linear assignment problem (LP) to choose an optimal plan.
(b) ◙ ▭ Use class optimization software to compute an optimal assignment.
(c) ◙ Show that your assignment problem can be represented as a minimum total cost flow model by sketching the corresponding digraph and labeling as in Exercise 10-1.

(d) ◫ A feasible assignment must have decision variables $= 0$ or 1 in part (a), yet the model can be solved as a linear program. Explain how the underlying network nature of the problem makes this possible.

10-20 Paltry Properties has just acquired four rental homes. Paltry wishes to have the houses painted within the next week so that all can be available for the prime rental season. This means that each house will have to be painted by a different contractor. The following table shows the bids (thousands of dollars) received from four contractors on the four houses.

House	Painter Bid			
	1	**2**	**3**	**4**
1	2.5	1.3	3.6	1.8
2	2.9	1.4	5.0	2.2
3	2.2	1.6	3.2	2.4
4	3.1	1.8	4.0	2.5

Paltry want to decide which bids to accept in order to paint all houses at minimum total cost.

(a) through (d) as in Exercise 10-19

10-21 Formulate as a minimum cost network flow problem, and solve by inspection, the problem of finding a shortest path from the specified source to the specified sink in each of the following networks. Use costs specified on the original digraph.

(a) ◫ Source 3, sink 2 in the network of Exercise 10-11

(b) Source 1, sink 3 in the network of Exercise 10-12

(c) ◫ Source 1, sink 4 in the network of Exercise 10-30

(d) Source 3, sink 2 in the network of Exercise 10-31

10-22 Suppose that an analyst needs to compute the shortest path between two specified nodes of a 10,000-node 100,000-arc digraph with nonnegative arc costs. Would shortest path methods of Chapter 9 or a network flow formulation solved by a cycle direction algorithm be likely to provide the most efficient solution? Why?

10-23 A relief agency is urgently trying to get the maximum possible quantity of supplies from its base at Alto to the volcano-ravaged city of Epi. One available road goes via Billi. The agency estimates that the Alto-to-Billi part of that road can carry 500 tons per day, and the Billi-to-Epi segment, 320 tons. A second route goes via Chau and Domo, with the Alto-to-Chau part having capacity 650 tons, the

Chau-to-Domo section, 470, and the Domo-to-Epi segment, 800. There is also a small mountain road with capacity 80 tons that connects Billi to Domo.

(a) ◫ Formulate this problem as a maximum flow model by sketching the associated digraph. Indicate the source, the sink, and all capacities.

(b) ◫ Solve your maximum flow problem by inspection.

(c) ◫ Show how to modify your digraph of part (a) to represent the model as a minimum cost network flow problem.

10-24 Makers of the new Ditti Doll are urgently trying to get as many to market as possible because a craze has created almost unlimited demand. One plant can supply up to 8 thousand per week to its distribution center, but that center can then get only 3 thousand per week to east region customers, and 1 thousand to west. The other plant supplies up to 3 thousand per week to its (distinct) distribution center, which can ship 2 thousand per week to the east and the same number to the west.

(a) through (c) as in Exercise 10-23.

10-25 Formulate as a minimum cost network flow problem, and solve by inspection, the problem of finding a maximum flow from the specified source to the specified sink in each of the following networks. Use capacities specified on the original digraph.

(a) ◫ Source 3, sink 2 in the network of Exercise 10-11

(b) Source 1, sink 3 in the network of Exercise 10-12

(c) ◫ Source 1, sink 4 in the network of Exercise 10-30

(d) Source 3, sink 2 in the network of Exercise 10-31

10-26 Return to the Super Sleep problem of Exercise 10-7, and suppose now that we wish to plan over a 2-week time horizon. Customer demands remains 160 and 700 in the first week, but they are predicted to be 300 and 810 in the second. There is no initial inventory, but matresses may be held in the warehouse at $10 per week. All other parameters are the same in each week as those given in Exercise 10-7.

(a) ◫ Formulate a time-expanded linear program to determine an optimal shipping and holding plan.

(b) ◫ 🖳 Use class optimization software to compute an optimal solution to your LP.

(c) ⊠ Show that your LP can be represented as a minimum cost flow model (with total supply = total demand) by sketching the corresponding digraph and labeling as in Exercise 10-1.

10-27 Return to the Crazy Crude problem of Exercise 10-8, and suppose now that we wish to plan over a 2-day time horizon. Refinery demand remains 2000 on the first day but will be 3000 on the second. There are no initial inventories, but either tank farm can hold over petroleum at $0.05 per barrel per day. All other parameters are the same on each day as those given in Exercise 10-8.

(a) through (c) as in Exercise 10-26.

10-28 Demonstrate that columns of the node–arc incidence matrix corresponding to arcs in each of the following cycles of the digraph in Exercise 10-1 form a linearly dependent set.

(a) ⊠ (2, 5), (5, 2)
(b) (4, 2), (5, 2), (4, 5)
(c) ⊠ (1, 2), (4, 2), (1, 4)
(d) (1, 4), (4, 5), (5, 3), (3, 1)

10-29 The following depicts a network flow problem, with labels on nodes indicating net demand and those on arcs showing capacity.

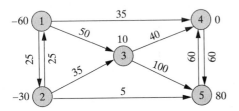

For each of the following lists of possible nonbasic arcs, either compute the corresponding basic solution and indicate whether it is basic feasible, or apply principle 10.39 to demonstrate that the unlisted arcs do not form a basis of the implied flow balance constraints.

(a) ⊠ (1, 2), (2, 1), (3, 4), (5, 4)   lower-bounded, (1, 3), (2, 5) upper-bounded
(b) (1, 2), (1, 4), (3, 4), (4, 5), (5, 4) lower-bounded, (1, 3), (3, 5) upper-bounded
(c) ⊠ (1, 3), (2, 1), (2, 5), (3, 4), (5, 4)   lower-bounded, (1, 4) upper-bounded
(d) (1, 2), (1, 4), (2, 3), (3, 4) lower-bounded, (2, 5) upper-bounded

(e) ⊠ (1, 2), (1, 4), (4, 5)   lower-bounded, (2, 5), (3, 4) upper-bounded
(f) (1, 2), (2, 1), (2, 5), (3, 4), (5, 4) lower-bounded, (1, 4) upper-bounded
(g) ⊠ (1, 2), (1, 4), (3, 4), (5, 4)   lower-bounded, (1, 3), (2, 5), (3, 5) upper-bounded
(h) (1, 2), (1, 4), (3, 4), (4, 5)   lower-bounded, (1, 3), (2, 5) upper-bounded

10-30 The following digraph depicts a partially solved minimum cost flow problem with labels on nodes indicating net demand and those on arcs showing unit cost, capacity, and current flow.

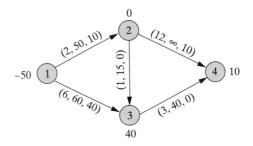

(a) ⊠ Verify that the given solution is a basic feasible solution for basis {(1, 2), (1, 3), (2, 4)}.
(b) ⊠ Compute all simplex directions available at this basis.
(c) ⊠ Determine whether each of the simplex directions is improving.
(d) ⊠ Regardless of whether they are improving, determine the maximum feasible step λ that could be applied to each of the simplex directions.

10-31 Do Exercise 10-30 on the problem

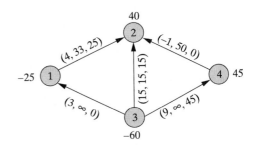

and basis {(1, 2), (3, 1), (3, 4)}.

10-32 Apply network simplex Algorithm 10B to compute an optimal flow in each of the following

networks. Start from the flow given in the figure, using the basis specified below.

(a) ⊘ The network of Exercise 10-11 with basis $\{(1, 2), (2, 4), (3, 4)\}$

(b) The network of Exercise 10-12 with basis $\{(1, 2), (2, 3), (4, 3)\}$

(c) ⊘ The network of Exercise 10-30 with basis $\{(1, 2), (1, 3), (2, 4)\}$

(d) The network of Exercise 10-31 with basis $\{(1, 2), (3, 1), (3, 4)\}$

10-33 Return to the partially solved minimum cost network flow problem of Exercise 10-30.

(a) ⊘ Verify that the given flow is feasible.

(b) ⊘ Construct the residual digraph corresponding to the current flow.

(c) ⊘ Use Floyd-Warshall Algorithm 9B on your residual digraph to identify an improving feasible cycle direction with respect to the given flow.

(d) ⊘ Compute the maximum feasible step λ that could be applied to your cycle direction.

10-34 Do Exercise 10-33 on the network of Exercise 10-31.

10-35 Apply cycle canceling Algorithm 10C to compute an optimal flow in each of the following networks. Start from the flow given in the figure, and identify negative dicycles in residual digraphs by inspection.

(a) ⊘ The network of Exercise 10-11

(b) The network of Exercise 10-12

(c) ⊘ The network of Exercise 10-30

(d) The network of Exercise 10-31

10-36 The following digraph depicts a partially solved minimum cost flow problem with labels on the nodes indicating net demand and those on the arcs showing unit cost, capacity, and current flow.

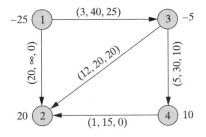

(a) ⊘ Verify that the given solution is a basic feasible solution for basis $\{(1, 2), (1, 3), (3, 4)\}$.

(b) ⊘ Identify a cycle for which the corresponding cycle direction could be pursued by cycle canceling Algorithm 10C but not by network simplex Algorithm 10B with the basis of part (a).

(c) ⊘ Identify a cycle for which the corresponding cycle direction could be pursued by network simplex Algorithm 10B with the basis of part (a) but not by cycle canceling Algorithm 10C.

10-37 Three workstations are located on the circular conveyor of a manufacturing facility. Most of the flow between them moves from one station to the next on the conveyor. However, 7 units per minute must move from each station to the station after the next. As much as possible of this 2-step flow should be carried within the 11 units per minute capacity of each link of the conveyor. The rest will be transported manually.

(a) ⊘ Formulate a linear program to determine an optimal way to carry the flows.

(b) ⊘ ▭ Use class optimization software to compute an optimal solution to your LP.

(c) ⊘ Show that your LP can be represented as a multicommodity flow problem by sketching the corresponding digraph and labeling with costs, capacities, and net demands.

(d) ⊘ Explain why your multicommodity model would give meaningless results if all flow were combined into a single commodity.

(e) ⊘ The optimal solution in this problem is fractional. Explain how this is compatible with the network nature of your multicommodity flow model.

10-38 The Wonder Waste disposal company has 5 truckloads of nuclear waste and 5 truckloads of hazardous chemical wastes that must be moved from its current cleanup site to nuclear and chemical disposal facilities, respectively. The following table shows that many of the available roads are restricted for one or the other type of waste.

Road		Nuclear OK	Chemical OK
From	To		
Site	NDisp	Yes	No
Site	CDisp	Yes	Yes
Site	Inter	No	Yes
NDisp	CDisp	No	Yes
CDisp	Inter	Yes	No
Inter	NDisp	Yes	Yes

Also, Wonder Waste wants to distribute any risk by allowing no more than half the 10 total truckloads on any road. One particular road, the link from crossing Inter to the nuclear disposal facility, is especially well suited to hazardous transfer because it runs through very remote areas. Wonder Waste seeks a feasible shipping plan that maximizes the use of that road.

(a) through (e) as in Exercise 10-37.

10-39 Maine Miracle's 2 restaurants sell lobsters obtained from 3 fisherman. A total of 350 lobsters per day are served at the first restaurant, and 275 at the second. Each fisherman can ship up to 300 per day, but not all arrive suitable for serving. The following table show the cost (including shipping) and the yield of servable lobsters for each combination of fisherman and restaurant.

	Restaurant			
	Cost		Yield (%)	
Fisherman	1	2	1	2
1	$7	$7	70	60
2	8	8	80	80
3	5	5	60	70

Maine Miracle seeks a minimum total cost way to meet its restaurant needs.

(a) ◻ Formulate a linear program to determine an optimal plan.

(b) ◻ ▢ Use class optimization software to compute an optimal solution to your LP.

(c) ◻ Show that your LP can be represented as a minimum cost flow model with gains or losses by sketching the corresponding digraph and labeling as in Sample Exercise 10.30.

(d) ◻ The optimal solution in this problem is fractional. Explain how this is compatible with the network nature of your flow with gains or losses model.

10-40 A new grocery store has 3 weeks to train its full staff of 39 employees. There are 5 employees now. At least 2 employees must work on preparing the store during the next week, 5 employees the week

after, and 10 in the final week before opening. Employees assigned to these duties earn $300 per week. Any other available employees, including those who were themselves trained in just the preceding week, can be assigned to train new workers. If an employee trains just one other, the two of them cost $500 for the week. If two are trained, the three employees cost $800 per week, including trainer overtime. Managements seeks a minimum total cost plan to meet all requirments.

(a) through (d) as in Exercise 10-39. You may use nonzero lower bounds on some variables.

10-41 Forest fire control organizations in Canada's provinces[6] must reposition the numbers of observation aircraft available at stations $i = 1, \ldots, 11$ on a daily basis to adjust for changing fire threats. The required number r_i and the present number p_i are known for all stations, along with the cost $c_{i,j}$ of moving an aircraft from any station i to station j. Explain how the problem of choosing a minimum total cost plan for repositioning aircraft can be modeled as a transportation problem by identifying the sources, sinks, supplies, demands, and arc costs.

10-42 A new highway[7] is being built through terrain points $i = 1, \ldots, 40$. The distance from i to $i + 1$ is d_i. To level the route, net earth deficits b_i truckloads must be corrected at all nodes ($b_i < 0$ if surplus, $\sum_i b_i = 0$). This will be done by moving truckloads of earth from surplus to deficit points along the route of the highway, but the same earth should not be handled more than once. Explain how the problem of choosing a minimum total truck travel distance leveling plan can be modeled as a transportation problem by identifying the sources, sinks, supplies, demands, and arc costs.

10-43 To estimate the impact of proposed tax changes, the U.S. Department of the Treasury[8] maintains two data files of records statistically characterizing the taxpayer population. Each of the $i = 1, \ldots, 10,000$ records in the first file represents a known number of families a_i and describes corresponding characteristics such as family size and age

---

[6]Based on P. Kourtz (1984), "A Network Approach to Least-Cost Daily Transfers of Forest Fire Control Resources," *INFOR, 22,* 283–290.

[7]Based on A. M. Farley (1980), "Levelling Terrain Trees: A Transshipment Problem," *Information Procession Letters, 10,* 189–192.

[8]Based on F. Glover and D. Klingman (1977), "Network Application in Industry and Government," *AIIE Transactions, 9,* 363–376.

distribution. Records $j = 1, \ldots, 40{,}000$ in the second file also represent a known number of families b_j and contain some of the same characteristics as the first ($\sum_i a_i = \sum_j b_j$). However, most of the entries in the second file relate to the sources of income for family class j. To do a better job of analyzing proposals, Treasury wants to merge these files into one with new records containing information drawn from both inputs. Each new record will represent a collection of families formed by matching some or all of those in population a_i with some of all in b_j. The quality of the similarity between classes i of the first file and j of the second can be described by a distance measure $d_{i,j}$, and the Treasury seeks a minimum total distance merge. Explain how this problem can be modeled as a transportation problem by identify sources, sinks, supplies, demands, and costs. Also, explain how an optimal flow can be understood as a merge.

10-44 Freight trains[9] run a regular weekly schedule in both the forward and reverse directions of a railroad's main line through section boundary points $i = 1, \ldots, 22$. Dividing the week into hourly time blocks $t = 1, \ldots, 168$ (with $t = 1$ following $t = 168$), a train leaving i bound for $j > i$ advances through both time and space as it transits sections $(i, i + 1)$, $(i + 1, i + 2)$, and so on. Summing requirements for all scheduled trains to pull the anticipated load over grades in the section, engine needs $f_{i,t}$ can be estimated for each forward section $(i, i+1)$, and each time t to $t + p_i$, where p_i is the number of hours required to transit the section. Reverse requirements $r_{i,t}$ provide the same information for trains moving i at t to $i - 1$ at $t + q_i$ with q_i the time to transit segment $(i, i - 1)$. We assume that all engine units are identical. Those located at any place and time where they are not immediately needed, may be held there, turned around to go the opposite way, or added as extras on a passing train. Cost c_i, d_i, and h reflect the cost of running one unit over segment $(i, i + 1)$, running one over $(i, i - 1)$, and holding for an hour at any site, respectively. Show that the problem of computing a minimum total cost engine schedule can be modeled as a network flow problem over

a suitable time-expanded network, by sketching a representative node and showing all adjacent nodes, arcs, costs, bounds, and node net demands. Some arcs will have nonzero lower bounds.

10-45 A substantial part of United Parcel Service's[10] freight traffic moves as trailer-on-flatcars (i.e., with truck trailers traveling most of the way on railroad flatcars). The required number of truckloads $d_{i,j}$ to be shipped between points $i, j = 1, \ldots, n$ in this way is known, but UPS can use either its own trailers at unit cost $c_{i,j}$ or rent trailers from the railroad at unit cost $r_{i,j}$. Rented trailers can be left anywhere, but UPS wishes to balance the number of its own trailers available at every point. That is, the number of company trailers inbound at any point should equal the number outbound. If necessary, trailers may be returned empty from i to j at unit cost $e_{i,j}$ to meet this requirement. Show that the problem of finding a minimum total cost shipping plan can be modeled as a (single-commodity) network flow problem by detailing the arcs that would join each node i to another j, including both their cost and any applicable capacity. Also indicate the net flow demand at each node. (*Hint:* Represent rented trailer flows indirectly from flows of loaded company-owned trailers with capacity $d_{i,j}$.)

10-46 KS brand tires[11] are shaped in changeable molds type $i = 1, \ldots, m$, installed in the company's 40 presses. Production plans dictate the minimum $\underline{r}_{i,t}$ and maximum $\bar{r}_{i,t}$ numbers of molds i that should be operational during time period $t = 1, \ldots, n$, and the planning period begins $t = 0$ with numbers of molds b_i installed ($\sum_i b_i = 40$). There is an adequate supply of molds, but changeovers from one mold to another are expensive, costing an amount c_i depending on the mold installed. All 40 presses should be in use in each time period. Demonstrate that the problem of finding a minimum total changeover cost mold schedule can be modeled as a network flow on a suitable time-expanded network by sketching the graph for a case with $m = 2$ and $n = 3$, and showing all nodes, arcs, costs, bounds, and node net demands. Some arcs will have nonzero lower bounds. (*Hint:* Also include supernodes balancing the total

[9]Based on M. Florian, G. Bushell, J. Ferland, G. Guerin, and L. Nastanshy (1976), "The Engine Scheduling Problem in a Railway Network," *INFOR, 14,* 121–138.

[10]Based on R. B. Dial (1994), "Minimizing Trailer-on-Flat-Car Costs: A Network Optimization Model," *Transportation Science, 28,* 24–35.

[11]Based on R. R. Love and R. R. Vemuganti (1978), "The Single-Plant Mold Allocation Problem with Capacity and Changeover Restrictions," *Operations Research, 26,* 159–165.

number of molds removed and inserted in each time period.)

10-47 Major league baseball umpires[12] work in crews that move among league cities to officiate series of 2–4 games. After every series, all crews move on to another that must involve different teams. To provide adequate travel time, any crew that works a series ending with a night game must also not go directly to one starting with a day game. Within these limits, league management would like to plan crew rotation to minimize the total of city i-to-city j travel costs $c_{i,j}$. Experience has shown that good results can be obtained by deciding each all crew move independently (i.e., without regard to where crews were before the most recent series or where they will be after the next). Explain how the problem of planning a move can be modeled as a linear assignment problem by describing the two sets being matched, the collection of feasible pairings, the associated linear costs, and whether the total is to be minimized or maximized.

10-48 One of the ways that airlines operating through major hubs can improve their service is to allow as many transferring passengers as possible who land at one of the scheduled arrival–departure peaks to carry on with their next flight on the same plane.[13] Such through-flight connections must involve flights scheduled for the same type of aircraft, and flights with sufficient arrival to departure time to complete the required servicing. The number of passengers $p_{i,j}$ arriving on flight i and continuing on flight j at any peak can be estimated in advance. Explain how this problem of optimizing through flights can be modeled as a linear assignment problem by describing the two sets being matched, the collection of feasible pairings, the associated linear costs, and whether the total is to be minimized or maximized.

10-49 As commercial airliner[14] makes stops $j = 1, \ldots, n$ of its daily routine and returns to where it started it takes on fuel for the next leg. Fuel is added at stop j to assure that the plane will arrive at stop $j + 1$ with at least the required safety reserve r_{j+1}. Fuel unit costs c_j (dollars per pound) vary considerably from stop to stop, so it is sometimes economical

to carry more fuel than the minimum required in order to buy less at high-cost stops. However, the take-off fuel load at any j must not exceed safety limit t_j. The amount of fuel at takeoff also affects the weight of the aircraft and thus its fuel consumption during flight. For each leg from stop j to $j + 1$, the fuel required can be estimated as a constant α_j plus a slope β_j times the onboard fuel at takeoff from j.

(a) Formulate this fuel management problem as a linear program in the decision variables ($j = 1, \ldots, n$)

$$x_j \triangleq \text{fuel added at stop } j$$
$$y_j \triangleq \text{onboard fuel at takeoff from } j$$

Assume that stop 1 is the successor of stop n, and use nonzero lower bounds if needed.

(b) Show how your model can be viewed as a flow with gains where the x_j correspond to 1-ended arcs, and the arcs associated with the y_j have both upper and (nonzero) lower bounds. Sketch the graph for a case with $n = 3$ and show all nodes, arcs, costs, bounds, and node net demands.

10-50 American Olean[15] makes product families $i = 1, \ldots, 10$ of tile at plants $p = 1, \ldots, 4$ to meet demands $d_{i,k}$ (square feet) at sales distribution points (SDPs) $k = 1, \ldots, 120$. Variable costs of production and transportation total $c_{i,p,k}$ per square foot to make tile of family i at plant p and ship to SDP k. Each plant can produce up to 100% of capacity, with $u_{i,p}$ being the capacity if plant p makes only group i. Management wants to find a minimum total cost way to meet demand.

(a) Formulate an LP model to compute an optimal plan in terms of the decision variables ($i = 1, \ldots, 10; p = 1, \ldots, 4; k = 1, \ldots, 120$)

$$x_{i,p,k} \triangleq \text{fraction of capacity at plant } p \text{ devoted}$$
$$\text{to making tile family } i \text{ for shipment}$$
$$\text{to SDP } k$$

---

[12]Based on J. R. Evans, (1988), "A Microcomputer-Based Decision Support System for Scheduling Umpires in the American Baseball League," *Interfaces, 18:6*, 42–51.

[13]Based on J. F. Bard and I. G. Cunningham (1987), "Improving Through-Flight Schedules," *IIE Transactions, 19*, 242–250.

[14]Based on J. S. Stroup and R. D. Wollmer (1992), "A Fuel Management Model for the Airline Industry," *Operations Research, 40*, 229–237.

[15]Based on M. J. Liberatore and T. Miller (1985), "A Hierarchial Production Planning System," *Interfaces 15:4*, 1–11.

(b) Show that your model can be viewed as a flow with gains of the transportation problem type by identifying sources, sinks, supplies, demands, arc gain multipliers, and arc costs.

10-51 The figure that follows shows a part of a distribution network like those used by Shell Oil Company[16] to supply the midwest with its three main classes of product: gasoline, kerosene/jet fuel, and fuel oil.

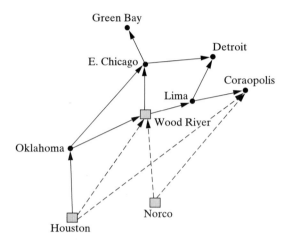

Solid links shown indicate available pipelines, but products can also be shipped by barge from the refineries at Houston and Norco to depots at Wood River and/or Coraopolis. Wood River is also a refinery. Flows must meet known demands $d_{i,p}$ for the various products p at all nodes i. The three refineries have known production capacities $b_{i,p}$ for production of product p at refinery i, and a pipeline from point i to point j can carry combined total of at most $u_{i,j}$ barrels of product. Barge capacity is essentially unlimited. Refining costs are assumed fixed, and cost per barrel $c_{i,j}$ to transport along links from i to j is the same for all products.

(a) Formulate an LP model to determine a minimum total cost distribution plan.

(b) Show that your model can be viewed as a multicommodity flow problem by sketching the corresponding network, labeling arcs with costs and capacities, and nodes with net demands.

SUGGESTED READING

Ahuja, Ravindra K., Thomas L. Magnanti, and James B. Orlin (1993), *Network Flows: Theory, Algorithms and Applications*, Prentice Hall, Upper Saddle River, N.J.

Bazaraa, Mokhtar S., John J. Jarvis, and Hanif Sherali (1990), *Linear Programming and Network Flows*, Wiley, New York.

---

[16]Based on T. K. Zierer, W. A. Mitchell, and T. R. White (1976), "Practical Applications of Linear Programming to Shell's Distribution Problems," *Interfaces, 6:4*, 13–26.

C H A P T E R

11

Discrete Optimization Models

• •

Most network flow, shortest path, and dynamic programming models of Chapters 9 and 10 solve elegantly, even if decision variables are treated as discrete. Fortunately, such special discrete models do occur in operations research practice. Unfortunately, they are hardly the norm. The overwhelming majority of integer and combinatorial optimization models prove much more challenging.

Before exploring methods for dealing with hard discrete models, we need to get some concept of their enormous range. In this chapter we formulate a series of classic types using real application contexts often drawn from published reports. In Chapter 12 we address integer programming methods.

11.1 LUMPY LINEAR PROGRAMS AND FIXED CHARGES

One broad class of discrete optimization problems add either/or side constraints or objective functions to what is otherwise a linear program. For want of a better term, we might call such models **lumpy linear programs**.

Swedish Steel Example with All-or-Nothing Constraints

The Swedish steel blending example of Section 4.2 provides an illustration. The model repeated below, which was formulated there, chooses a minimum cost mix of scrap metal and pure additives to produce a "charge" of steel. Constraints restrict the chemical content of the charge.

$$
\begin{array}{lll}
\min & 16x_1 + 10x_2 + 8x_3 + 9x_4 + 48x_5 + 60x_6 + 53x_7 & \text{(cost)} \\
\text{s.t.} & x_1 + x_2 + x_3 + x_4 + x_5 + x_6 + x_7 = 1000 & \text{(weight)} \\
& 0.0080x_1 + 0.0070x_2 + 0.0085x_3 + 0.0040x_4 \geq 6.5 & \text{(carbon)} \\
& 0.0080x_1 + 0.0070x_2 + 0.0085x_3 + 0.0040x_4 \leq 7.5 & \\
& 0.180x_1 + 0.032x_2 + 1.0x_5 \geq 30 & \text{(nickel)} \\
& 0.180x_1 + 0.032x_2 + 1.0x_5 \leq 35 & \\
& 0.120x_1 + 0.011x_2 + 1.0x_6 \geq 10 & \text{(chromium)} \\
& 0.120x_1 + 0.011x_2 + 1.0x_6 \leq 12 & \\
& 0.001x_2 + 1.0x_7 \geq 11 & \text{(molybdenum)} \\
& 0.001x_2 + 1.0x_7 \leq 13 & \\
& x_1 \leq 75 & \text{(availability)} \\
& x_2 \leq 250 & \\
& x_1, \ldots, x_7 \geq 0 &
\end{array}
\tag{11.1}
$$

In real application, steel blending often has an added complexity. Some of the scrap elements considered in the blend may be large blocks of reclaimed steel that cannot be subdivided. Either the entire block is used, or none of it is.

ILP Modeling of All-or-Nothing Requirements

Such indivisible input elements illustrate the need to model **all-or-nothing** phenomena. The usual approach to dealing with such cases is to rescale to new discrete variables.

> **11.1** All-or-nothing variable requirements of the form
> $$x_j = 0 \text{ or } u_j$$
> can be modeled by substituting $x_j = u_j y_j$, with new discrete variable
> $$y_j = 0 \text{ or } 1$$

The new y_j can be interpreted as the fraction of limit u_j chosen.

Swedish Steel Model with All-or-Nothing Constraints

Suppose in our example model (11.1) that the first two ingredients had this lumpy character. That is, we may use either none or all 75 kilograms of ingredient 1 and none or all 250 kilograms of ingredient 2. Then, instead of continuous decision variables x_1 and x_2 to reflect the quantities of each ingredient used, we simply employ discrete alternatives

$$
y_j \triangleq \begin{cases} 1 & \text{if scrap } j \text{ is part of the blend} \\ 0 & \text{otherwise} \end{cases}
$$

In terms of these new variables, the quantity of scraps 1 and 2 included in the mix are $75y_1$ and $250y_2$. Substituting produces the following integer linear program (definition 2.37):

$$\text{min} \quad 16(75)y_1 + 10(250)y_2 + 8x_3 + 9x_4 + 48x_5 + 60x_6 + 53x_7$$

$$\begin{array}{ll}
\text{s.t.} & 75y_1 + 250y_2 + x_3 + x_4 + x_5 + x_6 + x_7 & = 1000 \\
& 0.0080(75)y_1 + 0.0070(250)y_2 + 0.0085x_3 + 0.0040x_4 \geq 6.5 \\
& 0.0080(75)y_1 + 0.0070(250)y_2 + 0.0085x_3 + 0.0040x_4 \leq 7.5 \\
& 0.180(75)y_1 + 0.032(250)y_2 + 1.0x_5 & \geq 30 \\
& 0.180(75)y_1 + 0.032(250)y_2 + 1.0x_5 & \leq 35 \\
& 0.120(75)y_1 + 0.011(250)y_2 + 1.0x_6 & \geq 10 \\
& 0.120(75)y_1 + 0.011(250)y_2 + 1.0x_6 & \leq 12 \\
& 0.001(250)y_2 + 1.0x_7 & \geq 11 \\
& 0.001(250)y_2 + 1.0x_7 & \leq 13 \\
& x_3, \ldots, x_7 \geq 0 \\
& y_1, y_2 = 0 \text{ or } 1
\end{array} \tag{11.2}$$

An optimal solution now sets

$$y_1^* = 1, \qquad y_2^* = 0, \qquad x_3^* = 736.44, \quad x_4^* = 160.06$$
$$x_5^* = 16.50, \quad x_6^* = 1.00, \quad x_7^* = 11.00$$

Total cost increases to 9967.1 kroner, versus 9953.7 for the linear programming version (11.1), because of the all-or-nothing requirements.

SAMPLE EXERCISE 11.1: MODELING ALL-OR-NOTHING VARIABLES

Consider the linear program

$$\begin{array}{ll}
\text{max} & 18x_1 + 3x_2 + 9x_3 \\
\text{s.t.} & 2x_1 + x_2 + 7x_3 \leq 150 \\
& 0 \leq x_1 \leq 60 \\
& 0 \leq x_2 \leq 30 \\
& 0 \leq x_3 \leq 20
\end{array}$$

Revise the model so that each variable can be used only at zero or its upper bound.

Modeling: Following principle 11.1 , we introduce new 0–1 variables

$$y_j \triangleq \text{fraction of upper bound } u_j \text{ used}$$

Then substituting, the model becomes

$$\begin{array}{ll}
\text{max} & 18(60y_1) + 3(30y_2) + 9(20y_3) \\
\text{s.t.} & 2(60y_1) + (30y_2) + 7(20y_3) \leq 150 \\
& y_1, y_2, y_3 = 0 \text{ or } 1
\end{array}$$

or

$$\begin{array}{ll}
\text{max} & 1080y_1 + 90y_2 + 180y_3 \\
\text{s.t.} & 120y_1 + 30y_2 + 140y_3 \leq 150 \\
& y_1, y_2, y_3 = 0 \text{ or } 1
\end{array}$$

ILP Modeling of Fixed Charges

Another common source of lumpy phenomena in what are otherwise linear programs arises when the objective function involves **fixed charges**. For example, a nonnegative decision variable x may have cost

$$\theta(x) \triangleq \begin{cases} f + cx & \text{if } x > 0 \\ 0 & \text{otherwise} \end{cases}$$

That is, a fixed **initial**, or **construction**, or **setup** cost, f, must be paid before continuous decision variable x can be used at any nonzero level. Thereafter the usual **variable** cost c of linear programming applies.

If such fixed charges are nonnegative, which is almost always true, they can be modeled in mixed-integer linear programs by using new fixed charge variables.

___11.2___ Minimize objective functions with nonnegative fixed charges for making variables $x_j > 0$ can be modeled by introducing new fixed charge variables

$$y_j \triangleq \begin{cases} 1 & \text{if } x_j > 0 \\ 0 & \text{otherwise} \end{cases}$$

The objective coefficient of y_j is the fixed cost of x_j, and the coefficient of x_j is its variable cost.

New switching constraints are also required to link y_j with corresponding x_j.

___11.3___ **Switching constraints** model the requirement that continuous variable $x_j \geq 0$ can be used only if a corresponding binary variable $y_j = 1$ by

$$x_j \leq u_j y_j$$

where u_j is a given or derived upper bound on the value of x_j in any feasible solution.

If $y_j = 1$, x_j can assume any LP-feasible value. If $y_j = 0$, then $x_j = 0$ too.

Swedish Steel Example with Fixed Charges

To illustrate the modeling of fixed charges, return again to our original Swedish Steel model (11.1). This time, suppose that there are setup costs. Specifically, assume that ingredients 1 to 4 can be used in the furnace only after injection mechanisms are setup at a cost of 350 kroner each.

To model these fixed charges, we introduce new discrete variables

$$y_j \triangleq \begin{cases} 1 & \text{if setup for } j \text{ is performed} \\ 0 & \text{otherwise} \end{cases}$$

for $j = 1, \ldots, 4$.

We also require upper bounds on feasible values of the first four x_j. Bounds $u_1 = 75$ and $u_2 = 250$ are given in model statement (11.1). We must derive corresponding upper bounds for x_3 and x_4 in switching constraints $\boxed{11.3}$. Any value implied by constraints on these variables is valid, although we will see in Section 12.3 that it helps to employ the smallest possible. For simplicity, we look here only at the first main constraint of model (11.1). Since it sets the total weight of the charge at 1000, we know that $u_3 = u_4 = 1000$ are valid upper bounds.

Introducing these new variables and switching constraints of definition $\boxed{11.3}$ produces the fixed-charge version of our Swedish Steel model:

$$\begin{aligned}
\min \quad & 16x_1 + 10x_2 + 8x_3 + 9x_4 + 48x_5 + 60x_6 + 53x_7 \\
& + 350y_1 + 350y_2 + 350y_3 + 350y_4
\end{aligned}$$

$$\begin{aligned}
\text{s.t.} \quad & x_1 + x_2 + x_3 + x_4 + x_5 + x_6 + x_7 && = 1000 \\
& 0.0080x_1 + 0.0070x_2 + 0.0085x_3 + 0.0040x_4 && \geq 6.5 \\
& 0.0080x_1 + 0.0070x_2 + 0.0085x_3 + 0.0040x_4 && \leq 7.5 \\
& 0.180x_1 + 0.032x_2 + 1.0x_5 && \geq 30 \\
& 0.180x_1 + 0.032x_2 + 1.0x_5 && \leq 35 \\
& 0.120x_1 + 0.011x_2 + 1.0x_6 && \geq 10 \\
& 0.120x_1 + 0.011x_2 + 1.0x_6 && \leq 12 \\
& 0.001x_2 + 1.0x_7 && \geq 11 \\
& 0.001x_2 + 1.0x_7 && \leq 13 \\
& x_1 \leq 75y_1 \\
& x_2 \leq 250y_2 \\
& x_3 \leq 1000y_3 \\
& x_4 \leq 1000y_4 \\
& x_1, \ldots, x_7 \geq 0 \\
& y_1, \ldots, y_4 = 0 \text{ or } 1
\end{aligned}$$

An optimal solution is

$$\begin{aligned}
& x_1^* = 75, && x_2^* = 0, && x_3^* = 736.44, && x_4^* = 160.06 \\
& x_5^* = 16.5, && x_6^* = 1.00, && x_7^* = 11.00 \\
& y_1^* = 1, && y_2^* = 0, && y_3^* = 1, && y_4^* = 1
\end{aligned}$$

which sets up for ingredients 1, 3, and 4. Total cost is 11,017.1 kroner, versus 9953.7 in the linear version (11.1).

SAMPLE EXERCISE 11.2: MODELING FIXED CHARGES

Consider a fixed-charge objective function

$$\min \quad \theta_1(x_1) + \theta_2(x_2)$$

where

$$\theta_1(x_1) \triangleq \begin{cases} 150 + 7x_1 & \text{if } x_1 > 0 \\ 0 & \text{otherwise} \end{cases}$$

and

$$\theta_2(x_2) \triangleq \begin{cases} 110 + 9x_2 & \text{if } x_2 > 0 \\ 0 & \text{otherwise} \end{cases}$$

For each of the following systems of constraints on x_1 and x_2, form a corresponding mixed-integer linear programming model.

(a) $x_1 + x_2 \geq 8$
     $0 \leq x_1 \leq 3$
     $0 \leq x_2 \leq 8$

(b) $x_1 + x_2 \geq 8$
     $2x_1 + x_2 \leq 10$
     $x_1, x_2 \geq 0$

Modeling: We introduce y_j to carry fixed charges as in $\boxed{11.2}$ and switching constraints as in $\boxed{11.3}$.

(a) In this case upper bounds 3 and 8 are provided. Thus the mixed-integer formulation is

$$\begin{aligned} \min \quad & 7x_1 + 9x_2 + 150y_1 + 110y_2 \\ \text{s.t.} \quad & x_1 + x_2 \geq 8 \\ & x_1 \leq 3y_1 \\ & x_2 \leq 8y_2 \\ & x_1, x_2 \geq 0 \\ & y_1, y_2 = 0 \text{ or } 1 \end{aligned}$$

(b) Upper bounds on x_1 and x_2 are not explicit in these constraints. Still, we may infer from the second main constraint that any feasible solution has $x_1 \leq 5$ and $x_2 \leq 10$. Thus a mixed-integer formulation is

$$\begin{aligned} \min \quad & 7x_1 + 9x_2 + 150y_1 + 110y_2 \\ \text{s.t.} \quad & x_1 + x_2 \geq 8 \\ & 2x_1 + x_2 \leq 10 \\ & x_1 \leq 5y_1 \\ & x_2 \leq 10y_2 \\ & x_1, x_2 \geq 0 \\ & y_1, y_2 = 0 \text{ or } 1 \end{aligned}$$

11.2 KNAPSACK AND CAPITAL BUDGETING MODELS

In contrast to the cases of Section 11.1, which involve linear programs with some discrete side conditions, **knapsack** and **capital budgeting** problems are completely discrete. We must select an optimal collection of objects, or features, or projects, or investments subject to limits on budget resources. Each element is either all in or all out of the result, with no partial selections allowed.

Knapsack Problems

The simplest of these object choosing discrete problems, and indeed the simplest of all integer linear programs, are knapsack problems.

> 11.4 A **knapsack model** is a pure integer linear program with a single main constraint.

All knapsack decision variables are 0–1 in most applications.

The knapsack name derives from the problem confronted by a hiker packing a backpack. He or she must choose the most valuable collection of items to take subject to a volume or weight limit on the size of the pack.

EXAMPLE 11.1: INDY CAR KNAPSACK

We may illustrate more realistic forms of knapsack problems by considering the (fictitious) dilemma of mechanics in the Indy Car racing team. Six different features might still be added to this year's car to improve its top speed. Table 11.1 lists their estimated costs and speed enhancements.

TABLE 11.1 Indy Car Example Alternatives

	Proposed Feature, j					
	1	2	3	4	5	6
Cost ($ 000's)	10.2	6.0	23.0	11.1	9.8	31.6
Speed increase (mph)	8	3	15	7	10	12

Suppose first that Indy Car wants to maximize the performance gain without exceeding a budget of $35,000. Using decision variables

$$x_j \triangleq \begin{cases} 1 & \text{if feature } j \text{ is added} \\ 0 & \text{otherwise} \end{cases} \tag{11.3}$$

we can formulate the problem as the knapsack model

$$\begin{aligned} \max \quad & 8x_1 + 3x_2 + 15x_3 + 7x_4 + 10x_5 + 12x_6 && \text{(mph gain)} \\ \text{s.t.} \quad & 10.2x_1 + 6.0x_2 + 23.0x_3 + 11.1x_4 + 9.8x_5 + 31.6x_6 \le 35 && \text{(budget)} \tag{11.4} \\ & x_1, \ldots, x_6 = 0 \text{ or } 1 \end{aligned}$$

That is, we maximize total performance subject to a budget constraint. An optimal solution chooses features 1, 4, and 5 for a gain of 25 miles per hour.

Suppose now that the Indy Car team decides they simply must increase speed by 30 miles per hour to have any chance of winning the next race. Ignoring the budget, they wish to find the minimum cost way to achieve at least that much performance.

This scenario leads to an alternative, minimize knapsack form. With variables (11.3), we obtain

$$\text{min} \quad 10.2x_1 + 6.0x_2 + 23.0x_3 + 11.1x_4 + 9.8x_5 + 31.6x_6 \quad \text{(cost)}$$
$$\text{s.t.} \quad 8x_1 + 3x_2 + 15x_3 + 7x_4 + 10x_5 + 12x_6 \geq 30 \qquad \text{(mph required)} \qquad (11.5)$$
$$x_1, \ldots, x_6 = 0 \text{ or } 1$$

This model minimizes cost subject to a performance requirement. An optimal solution now chooses features 1, 3, and 5 at cost \$43,000.

SAMPLE EXERCISE 11.3: FORMULATING KNAPSACK MODELS

Readily available U.S. coins are denominated 1, 5, 10, and 25 cents. Formulate a knapsack model to minimize the number of coins needed to provide change amount q cents.

Modeling: We employ decision variables x_1, x_5, x_{10}, and x_{25} to represent the number of coins chosen from each denomination. Then the knapsack model is

$$\text{min} \quad x_1 + x_5 + x_{10} + x_{25} \qquad \text{(total coins)}$$
$$\text{s.t.} \quad x_1 + 5x_5 + 10x_{10} + 25x_{25} = q \qquad \text{(correct change)}$$
$$x_1, x_5, x_{10}, x_{25} \geq \text{ and integer}$$

Notice that these discrete variables are not limited to 0 and 1.

Capital Budgeting Models

The typical maximize form of a knapsack problem has its single main constraint enforcing a budget. When there are budget limits over more than one time period, or multiple limited resources, we obtain more general capital budgeting or multi-dimensional knapsack models.

> **11.5** **Capital budgeting models** (or **multidimensional knapsack**) select a maximum value collection of project, investments, and so on, subject to limitations on budgets or other resources consumed.

EXAMPLE 11.2: NASA CAPITAL BUDGETING

The U.S. space agency, NASA, must deal constantly with such decision problems in choosing how to divide its limited budgets among many competing missions proposed.[1] Table 11.2 shows a fictitious list of alternatives.

We must decide which of the 14 indicated missions to include in program plans for the 2000–2024 era. Thus it should be clear that the needed decision variables are

$$x_j \triangleq \begin{cases} 1 & \text{if mission } j \text{ is selected} \\ 0 & \text{otherwise} \end{cases} \qquad (11.6)$$

[1] Based on Gerald W. Evans and Robert Fairbairn (1989), "Selection and Scheduling of Advanced Missions for NASA Using 0–1 Integer Linear Programming," *Journal of the Operational Research Society,* *40,* 971–981.

TABLE 11.2 Proposed Missions in NASA Example

		Budget Requirements ($ billion)							
j	Mission	2000–2004	2005–2009	2010–2014	2015–2019	2020–2024	Value	Not With	Depends On
1	Communications satellite	6	—	—	—	—	200	—	—
2	Orbital microwave	2	3	—	—	—	3	—	—
3	Io lander	3	5	—	—	—	20	—	—
4	Uranus orbiter 2020	—	—	—	—	10	50	5	3
5	Uranus orbiter 2010	—	5	8	—	—	70	4	3
6	Mercury probe	—	—	1	8	4	20	—	3
7	Saturn probe	1	8	—	—	—	5	—	3
8	Infrared imaging	—	—	—	5	—	10	11	—
9	Ground-based SETI	4	5	—	—	—	200	14	—
10	Large orbital structures	—	8	4	—	—	150	—	—
11	Color imaging	—	—	2	7	—	18	8	2
12	Medical technology	5	7	—	—	—	8	—	—
13	Polar orbital platform	—	1	4	1	1	300	—	—
14	Geosynchronous SETI	—	4	5	3	3	185	9	—
	Budget	10	12	14	14	14			

Budget Constraints

The budget constraints that give capital budgeting problems their name limit project expenditures in particular time periods.

> **11.6** **Budget constraints** limit the total funds or other resources consumed by selected projects, investments, and so on, in each time period not to exceed the amount available.

Budget requirements in Table 11.2 span five time periods: 2000–2004, 2005–2009, 2010–2014, 2015–2019, and 2020–2024. We form budget constraints for each of these periods by summing project decision variables times their needs.

$$6x_1 + 2x_2 + 3x_3 + 1x_7 + 4x_9 + 5x_{12} \leq 10 \quad (2000\text{–}2004)$$

$$3x_2 + 5x_3 + 5x_5 + 8x_7 + 5x_9 + 8x_{10}$$
$$+ 7x_{12} + 1x_{13} + 4x_{14} \leq 12 \quad (2005\text{–}2009)$$

$$8x_5 + 1x_6 + 4x_{10} + 2x_{11} + 4x_{13} + 5x_{14} \leq 14 \quad (2010\text{–}2014)$$

$$8x_6 + 5x_8 + 7x_{11} + 1x_{13} + 3x_{14} \leq 14 \quad (2015\text{–}2019)$$

$$10x_4 + 4x_6 + 1x_{13} + 3x_{14} \leq 14 \quad (2020\text{–}2024)$$

SAMPLE EXERCISE 11.4: FORMULATING BUDGET CONSTRAINTS

A department store is considering 4 possible expansions into presently unoccupied space in a shopping mall. The following table shows how much (in millions of dollars) each expansion would cost in the next two fiscal years, and the required floor space (in thousands of square feet).

	Expansion, j			
	1	2	3	4
Year 1	1.5	5.0	7.3	1.9
Year 2	3.5	1.8	6.0	4.2
Space	2.2	9.1	5.3	8.6

Using decision variables

$$x_j \triangleq \begin{cases} 1 & \text{expansion } j \text{ is selected} \\ 0 & \text{otherwise} \end{cases}$$

formulate implied constraints on investment funds and floor space assuming that 10 million dollars are available in each of the two years and that the expansion cannot exceed 17 thousand square feet.

Modeling: The three required budget constraints are

$$1.5x_1 + 5.0x_2 + 7.3x_3 + 1.9x_4 \leq 10 \quad \text{(year 1 budget)}$$
$$3.5x_1 + 1.8x_2 + 6.0x_3 + 4.2x_4 \leq 10 \quad \text{(year 2 budget)}$$
$$2.2x_1 + 9.1x_2 + 5.3x_3 + 8.6x_4 \leq 17 \quad \text{(floor space)}$$

Modeling Mutually Exclusive Choices

Capital budgeting problems often come with other constraints besides simple budget limits. For example, two or more proposed projects may be **mutually exclusive**. That is, at most one of them can be included in a solution.

Table 11.2 indicates three such conflicts. Possibilities $j = 4$ and 5 represent alternative timing for the same mission. Technologies for missions $j = 8$ and 11 are incompatible. Numbers $j = 9$ and 14 involve two different ways to accomplish the SETI program.

Such incompatibilities are easily modeled with 0–1 decision variables (11.6).

> 11.7   **Mutually exclusiveness** conditions allowing at most one of a set of choices are modeled by $\sum x_j \leq 1$ constraints summing over each choice set.

For our NASA example, the result is

$$x_4 + x_5 \leq 1$$
$$x_8 + x_{11} \leq 1$$
$$x_9 + x_{14} \leq 1$$

SAMPLE EXERCISE 11.5: FORMULATING MUTUALLY EXCLUSIVE CONSTRAINTS

Suppose that a real estate development company is considering 5 investment decisions. Only 1 of the first 3 can be chosen because all require the same piece of land. Investment 4 is a new office building, and investment 5 is the same project delayed a year. Formulate suitable mutually exclusive constraints in terms of decision variables

$$x_j \triangleq \begin{cases} 1 & \text{if investment } j \text{ is selected} \\ 0 & \text{otherwise} \end{cases}$$

Modeling: The fact that only 1 of the first 3 can be selected produces constraint

$$x_1 + x_2 + x_3 \leq 1$$

Since x_4 and x_5 refer to the same investment, we also want

$$x_4 + x_5 \leq 1$$

Modeling Dependencies between Projects

Another characteristic relationship between projects arises when one project depends on another. We cannot choose such a dependent project unless we also include the option on which it depends.

We can model dependencies among projects as easily as mutual exclusiveness.

> 11.8 | **Dependence** of choice j on choice i can be enforced on corresponding binary variables by constraint $x_j \leq x_i$.

Variable x_j cannot $= 1$ unless x_i does too.

Table 11.2 shows that mission $j = 11$ depends on mission 2 in our NASA example, and also that mission $j = 3$ must be chosen if any of projects $4, \ldots, 7$ is. Implied constraints are

$$x_{11} \leq x_2$$
$$x_4 \leq x_3$$
$$x_5 \leq x_3$$
$$x_6 \leq x_3$$
$$x_7 \leq x_3$$

SAMPLE EXERCISE 11.6: MODELING PROJECT DEPENDENCIES

Phase 1 of a new city hall project will construct the first story of a building that could grow to two stories in phase 2, and to three in phase 3. Using the variables

$$x_j \triangleq \begin{cases} 1 & \text{if phase } j \text{ is constructed} \\ 0 & \text{otherwise} \end{cases}$$

write corresponding dependency constraints.

Modeling: Obviously, the second floor cannot be built without the first, and the third without the second. Thus we have depend project constraints

$$x_2 \leq x_1$$
$$x_3 \leq x_2$$

NASA Example Model

To complete formulation of our version of NASA's decision problem, we need an objective function. Like almost all public agencies, NASA has many. They may try to maximize the intellectual gains of selected missions, maximize the direct benefit to life on earth, and so on.

We will assume that a weighted sum of these different objective functions can be used to estimate a value for each mission. Resulting values are included in

Table 11.2. Combining with previous elements, we obtain the NASA example model:

$$
\begin{aligned}
\max \quad & 200x_1 + 3x_2 + 20x_3 + 50x_4 + 70x_5 && \text{(total value)} \\
& + 20x_6 + 5x_7 + 10x_8 + 200x_9 + 150x_{10} \\
& + 18x_{11} + 8x_{12} + 300x_{13} + 185x_{14} \\
\text{s.t.} \quad & 6x_1 + 2x_2 + 3x_3 + 1x_7 + 4x_9 + 5x_{12} && \leq 10 && \text{(2000–2004)} \\
& 3x_2 + 5x_3 + 5x_5 + 8x_7 + 5x_9 + 8x_{10} && && \text{(2005–2009)} \\
& \qquad + 7x_{12} + 1x_{13} + 4x_{14} && \leq 12 \\
& 8x_5 + 1x_6 + 4x_{10} + 2x_{11} + 4x_{13} + 5x_{14} && \leq 14 && \text{(2010–2014)} \\
& 8x_6 + 5x_8 + 7x_{11} + 1x_{13} + 3x_{14} && \leq 14 && \text{(2015–2019)} \\
& 10x_4 + 4x_6 + 1x_{13} + 3x_{14} && \leq 14 && \text{(2020–2024)} \\
& x_4 + x_5 \leq 1 && && \text{(mutually} \\
& x_8 + x_{11} \leq 1 && && \text{exclusive)} \\
& x_9 + x_{14} \leq 1 \\
& x_{11} \leq x_2 && && \text{(dependent} \\
& x_4 \leq x_3 && && \text{missions)} \\
& x_5 \leq x_3 \\
& x_6 \leq x_3 \\
& x_7 \leq x_3 \\
& x_j = 0 \text{ or } 1 \quad \text{for all } j = 1, \dots, 14
\end{aligned}
$$

(11.7)

11.3 SET PACKING, COVERING, AND PARTITIONING MODELS

Our capital budget models of Section 11.2 included mutual exclusiveness constraints involving subsets of decision variables, at most one of which can take part in a solution. **Set packing**, **covering**, and **partitioning models** feature such constraints. Using decision variables that $= 1$ if a object is part of a solution and $= 0$ otherwise, these models formulate problems where the core issue is membership in specified subsets.

EXAMPLE 11.3: EMS LOCATION PLANNING

As always, it will help to think of a specific example. A classic occurred when Austin, Texas undertook a study of the positioning of its emergency medical service (EMS) vehicles.[2] That city was divided into service districts needing EMS services, and vehicle stations selected from a list of alternatives so that as much of the population as possible would experience a quick response to calls for help.

Figure 11.1 shows the fictitious map we will assume for our numerical version. Our city is divided into 20 service districts that we wish to serve from some combination of the 10 indicated possibilities for EMS stations. Each station can provide service to all adjacent districts. For example, station 2 could service districts 1, 2, 6, and 7. Main decision variables are

$$
x_j \triangleq \begin{cases} 1 & \text{if location } j \text{ is selected} \\ 0 & \text{otherwise} \end{cases}
$$

[2] Based on David J. Eaton, Mark S. Daskin, Dennis Simmons, Bill Bulloch, and Glen Jansma (1985), "Determining Emergency Medical Service Vehicle Deployment in Austin, Texas," *Interfaces, 15:1*, 96–108.

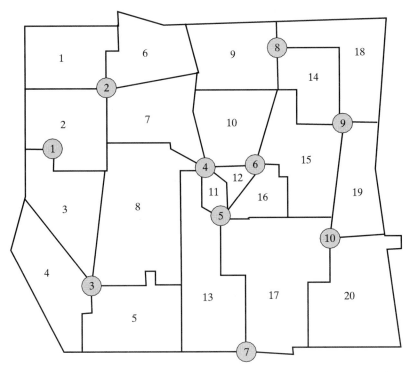

FIGURE 11.1   Service District and Candidate Locations for EMS Example

Set Packing, Covering, and Partitioning Constraints

The defining constraints of set packing, covering, and partitioning models deal with subcollections of problem objects. In our EMS example the subcollections are locations that can provide satisfactory response for a service district. For instance, Figure 11.1 shows that any of locations 4, 5, and 6 can protect downtown district 12.

Covering constraints demand that at least one member of each subcollection belongs to a solution, packing constraints allow at most one member, and partitioning constraints require exactly one member. Mathematical forms follow easily with 0–1 variables.

11.9 | **Set covering constraints** requiring that at least one member of subcollection J belongs to a solution are expressed

$$\sum_{j \in J} x_j \geq 1$$

11.10 | **Set packing constraints** requiring that at most one member of subcollection J belongs to a solution are expressed

$$\sum_{j \in J} x_j \leq 1$$

> **11.11**   **Set partitioning constraints** requiring that exactly one member of sub-collection J belongs to a solution are expressed
>
> $$\sum_{j \in J} x_j = 1$$

In our EMS example, we are covering, because each district should be protected by at least one location, but more is even better. In other settings, such as radio station location, we pack; at most one station on any particular frequency should reach a service area (see Exercise 11-7). The mutual exclusiveness constraints of capital budgeting also have the packing form. Partitioning applies when exactly one decision option can serve each sector. This is the case, for example, if decision variables relate to possible election districts, and constraints demand that each geographic unit belong to exactly one district (see Exercise 11-12).

> ### SAMPLE EXERCISE 11.7: FORMULATING SET PACKING, COVERING, AND PARTITIONING
>
> A university is acquiring mathematical programming software for use in operations research classes. The four codes available and the types of optimization algorithms they provide are indicated by $\times$'s in the following table.
>
> | Algorithm Type | Code, j | | | |
> |---|---|---|---|---|
> | | **1** | **2** | **3** | **4** |
> | LP | $\times$ | $\times$ | $\times$ | $\times$ |
> | IP | — | $\times$ | — | $\times$ |
> | NLP | — | — | $\times$ | $\times$ |
> | Objective | 3 | 4 | 6 | 14 |
>
> **(a)** Taking objective function coefficients as code costs, formulate a set covering model to acquire a minimum cost collection of codes providing LP, IP, and NLP capability.
>
> **(b)** Taking objective function coefficients as code costs, formulate a set partitioning model to acquire a minimum cost collection of codes with exactly one providing LP, one providing IP, and one providing NLP.
>
> **(c)** Taking objective function coefficients as indications of code quality, formulate a set packing model to acquire a maximum quality collection of codes with at most one providing LP, at most one providing IP, and at most one providing NLP.
>
> ***Modeling:*** In each case we use the decision variables
>
> $$x_j \triangleq \begin{cases} 1 & \text{if code } j \text{ is chosen} \\ 0 & \text{otherwise} \end{cases}$$

(a) Following form 11.9 , the required model is

$$
\begin{aligned}
\min \quad & 3x_1 + 4x_2 + 6x_3 + 14x_4 \\
\text{s.t.} \quad & x_1 + x_2 + x_3 + x_4 \geq 1 \quad \text{(LP)} \\
& x_2 + x_4 \qquad\qquad \geq 1 \quad \text{(IP)} \\
& x_3 + x_4 \qquad\qquad \geq 1 \quad \text{(NLP)} \\
& x_1, \ldots, x_4 = 0 \text{ or } 1
\end{aligned}
$$

(b) Following form 11.11 , the required model is

$$
\begin{aligned}
\min \quad & 3x_1 + 4x_2 + 6x_3 + 14x_4 \\
\text{s.t.} \quad & x_1 + x_2 + x_3 + x_4 = 1 \quad \text{(LP)} \\
& x_2 + x_4 \qquad\qquad = 1 \quad \text{(IP)} \\
& x_3 + x_4 \qquad\qquad = 1 \quad \text{(NLP)} \\
& x_1, \ldots, x_4 = 0 \text{ or } 1
\end{aligned}
$$

(c) Following form 11.10 , the required model is

$$
\begin{aligned}
\max \quad & 3x_1 + 4x_2 + 6x_3 + 14x_4 \\
\text{s.t.} \quad & x_1 + x_2 + x_3 + x_4 \leq 1 \quad \text{(LP)} \\
& x_2 + x_4 \qquad\qquad \leq 1 \quad \text{(IP)} \\
& x_3 + x_4 \qquad\qquad \leq 1 \quad \text{(NLP)} \\
& x_1, \ldots, x_4 = 0 \text{ or } 1
\end{aligned}
$$

Minimum Cover EMS Model

The most obvious approach to modeling our EMS example of Figure 11.1 is to minimize the number of locations needed to cover all districts. The following set covering model results.

$$
\begin{aligned}
\min \quad & \sum_{j=1}^{10} x_j && \text{(number of sites)} \\
\text{s.t.} \quad & x_2 && \geq 1 && \text{(district 1)} \\
& x_1 + x_2 && \geq 1 && \text{(district 2)} \\
& x_1 + x_3 && \geq 1 && \text{(district 3)} \\
& x_3 && \geq 1 && \text{(district 4)} \\
& x_3 && \geq 1 && \text{(district 5)} \\
& x_2 && \geq 1 && \text{(district 6)} && (11.8) \\
& x_2 + x_4 && \geq 1 && \text{(district 7)} \\
& x_3 + x_4 && \geq 1 && \text{(district 8)} \\
& x_8 && \geq 1 && \text{(district 9)} \\
& x_4 + x_6 && \geq 1 && \text{(district 10)} \\
& x_4 + x_5 && \geq 1 && \text{(district 11)} \\
& x_4 + x_5 + x_6 && \geq 1 && \text{(district 12)} \\
& x_4 + x_5 + x_7 && \geq 1 && \text{(district 13)}
\end{aligned}
$$

$$x_8 + x_9 \geq \quad 1 \quad \text{(district 14)}$$
$$x_6 + x_9 \geq \quad 1 \quad \text{(district 15)}$$
$$x_5 + x_6 \geq \quad 1 \quad \text{(district 16)}$$
$$x_5 + x_7 + x_{10} \geq 1 \quad \text{(district 17)}$$
$$x_8 + x_9 \geq \quad 1 \quad \text{(district 18)}$$
$$x_9 + x_{10} \geq \quad 1 \quad \text{(district 19)}$$
$$x_{10} \geq \quad 1 \quad \text{(district 20)}$$
$$x_1, \ldots, x_{10} = 0 \text{ or } 1$$

One optimal solution chooses the six sites 2, 3, 4, 6, 8, and 10. That is,

$$x_2^* = x_3^* = x_4^* = x_6^* = x_8^* = x_{10}^* = 1$$
$$x_1^* = x_5^* = x_7^* = x_9^* = 0$$

Maximum Coverage EMS Model

In the Austin case, as in many other real instances, the straightforward covering model (11.8) proves inadequate because it calls for too many sites. Suppose that we have funds for only 4 EMS locations. How can we find the collection of 4 that minimizes coverage insufficiency?

For this version of the model we need estimates of the demand or importance of covering each service district. We will assume the following values have been estimated by EMS staff:

District i	Value	District i	Value	District i	Value
1	5.2	8	12.2	15	15.5
2	4.4	9	7.6	16	25.6
3	7.1	10	20.3	17	11.0
4	9.0	11	30.4	18	5.3
5	6.1	12	30.9	19	7.9
6	5.7	13	12.0	20	9.9
7	10.0	14	9.3		

Next we introduce extra decision variables to model uncovered districts i.

11.12 | Set packing, covering, and partitioning models can be modified to penalize uncovered items i by introducing new variables

$$y_i \triangleq \begin{cases} 1 & \text{if item } i \text{ is uncovered in the solution} \\ 0 & \text{otherwise} \end{cases}$$

into each constraint i.

Including such variables, the EMS model becomes

$$\begin{aligned} \min \quad & 5.2y_1 + 4.4y_2 + 7.1y_3 + 9.0y_4 + 6.1y_5 \\ & + 5.7y_6 + 10.0y_7 + 12.2y_8 + 7.6y_9 + 20.3y_{10} \\ & + 30.4y_{11} + 30.9y_{12} + 12.0y_{13} + 9.3y_{14} + 15.5y_{15} \\ & + 25.6y_{16} + 11.0y_{17} + 5.3y_{18} + 7.9y_{19} + 9.9y_{20} \end{aligned}$$

(uncovered district importance)

$$
\begin{array}{llll}
\text{s.t.} & x_2 + y_1 & \geq 1 & \text{(district 1)} \\
& x_1 + x_2 + y_2 & \geq 1 & \text{(district 2)} \\
& x_1 + x_3 + y_3 & \geq 1 & \text{(district 3)} \\
& x_3 + y_4 & \geq 1 & \text{(district 4)} \\
& x_3 + y_5 & \geq 1 & \text{(district 5)} \\
& x_2 + y_6 & \geq 1 & \text{(district 6)} \\
& x_2 + x_4 + y_7 & \geq 1 & \text{(district 7)} \\
& x_3 + x_4 + y_8 & \geq 1 & \text{(district 8)} \\
& x_8 + y_9 & \geq 1 & \text{(district 9)} \\
& x_4 + x_6 + y_{10} & \geq 1 & \text{(district 10)} \\
& x_4 + x_5 + y_{11} & \geq 1 & \text{(district 11)} \\
& x_4 + x_5 + x_6 + y_{12} & \geq 1 & \text{(district 12)} \\
& x_4 + x_5 + x_7 + y_{13} & \geq 1 & \text{(district 13)} \\
& x_8 + x_9 + y_{14} & \geq 1 & \text{(district 14)} \\
& x_6 + x_9 + y_{15} & \geq 1 & \text{(district 15)} \\
& x_5 + x_6 + y_{16} & \geq 1 & \text{(district 16)} \\
& x_5 + x_7 + x_{10} + y_{17} & \geq 1 & \text{(district 17)} \\
& x_8 + x_9 + y_{18} & \geq 1 & \text{(district 18)} \\
& x_9 + x_{10} + y_{19} & \geq 1 & \text{(district 19)} \\
& x_{10} + y_{20} & \geq 1 & \text{(district 20)} \\
& \displaystyle\sum_{j=1}^{10} x_j & \leq 4 & \text{(at most four)} \\
& x_1, \ldots, x_{10} = 0 \text{ or } 1 \\
& y_1, \ldots, y_{20} = 0 \text{ or } 1
\end{array}
\tag{11.9}
$$

Here the objective function minimizes the total importance of uncovered districts. The last (noncovering) main constraint limits solutions to four EMS sites.

An optimal solution for this more realistic version chooses

$$
x_3^* = x_4^* = x_5^* = x_9^* = 1
$$
$$
y_1^* = y_2^* = y_6^* = y_9^* = y_{20}^* = 1
$$

with all other decision variables $= 0$. That is, sites 3, 4, 5, and 9 are chosen, leaving districts 1, 2, 6, 9, and 20 uncovered. The total importance of those districts, which is the optimal objective value, equals 32.8.

SAMPLE EXERCISE 11.8: FORMULATING MAXIMUM COVERING MODELS

Return to the set covering case of Sample Exercise 11.7(a). Revise the model to maximize the number of algorithms available within a budget of 12.

Modeling: We introduce new variables y_{LP}, y_{IP}, and y_{NLP} that $= 1$ if the indicated algorithm is not provided, and $= 0$ otherwise. Then the required model is

$$\begin{array}{rll} \min & y_{LP} + y_{IP} + y_{NLP} & \\ \text{s.t.} & x_1 + x_2 + x_3 + x_4 + y_{LP} \geq 1 & \text{(LP)} \\ & x_2 + x_4 + y_{IP} \geq 1 & \text{(IP)} \\ & x_3 + x_4 + y_{NLP} \geq 1 & \text{(NLP)} \\ & 3x_1 + 4x_2 + 6x_3 + 14x_4 \leq 12 & \text{(budget)} \\ & x_1, \dots, x_4 = 0 \text{ or } 1 & \\ & y_{LP}, y_{IP}, y_{NLP} = 0 \text{ or } 1 & \end{array}$$

Column Generation Models

Another common family of set packing, covering, and partitioning models are derived from problems involving a combinatorially large array of possibilities too complex to be modeled concisely. Column generation adopt a two-part strategy for such problems.

> **11.13** | **Column generation** approaches deal with complex combinatorial problems by first enumerating a sequence of columns representing viable solutions to parts of the problem, and then solving a set partitioning (or covering or packing) model to select an optimal collection of these alternatives fulfilling all problem requirements.

The convenience of this approach comes from its flexibility. Any appropriate scheme—however ad hoc—can be employed to generate a rich family of columns meeting complex and difficult-to-model constraints. Optimization is reserved for the second part of the strategy, when a well-formed model over the columns can be addressed by standard ILP technology.

EXAMPLE 11.4: AA CREW SCHEDULING

A classic application of column generation arises in the enormously complex problem of scheduling crews for airlines. For example, American Airlines[3] reports spending over \$1.3 billion per year on salaries, benefits, and travel expenses of air crews. Careful scheduling, or **crew pairing** as it is called, can produce enormous savings.

Figure 11.2 illustrates a (tiny) sequence of flights to be crewed in our fictitious case. For example, flight 101 originates in Miami and arrives in Chicago some hours later.

Each pairing is a sequence of flights to be covered by a single crew over a 2- to 3-day period. It must begin and end in the base city where the crew resides.

Table 11.3 enumerates possible pairings of flights in Figure 11.2. For example, pairing $j = 1$ begins at Miami with flight 101. After a layover in Chicago, the crew then covers flight 203 to Dallas–Ft. Worth and then flight 406 to Charlotte. Finally, flight 308 returns them to Miami.

In real applications, intricate government and union rules regulate exactly which sequences of flights constitute a reasonable pairing. Complex software is employed to generate a list such as that of Table 11.3. Here we simply allow all closed sequences of 3 or 4 flights in Figure 11.2.

---

[3]Based on R. Anbil, E. Gelman, B. Patty, and R. Tanga (1991), "Recent Advances in Crew-Pairing Optimization at American Airlines," *Interfaces, 21:1*, 62–74.

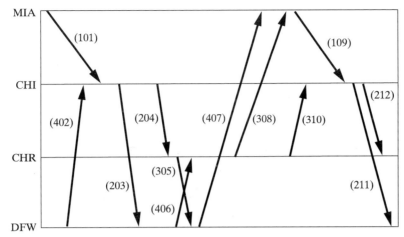

FIGURE 11.2   Flight Schedule for AA Example

TABLE 11.3   Possible Pairings for AA Example

j	Flight Sequence	Cost	j	Flight Sequence	Cost
1	101–203–406–308	2900	9	305–407–109–212	2600
2	101–203–407	2700	10	308–109–212	2050
3	101–204–305–407	2600	11	402–204–305	2400
4	101–204–308	3000	12	402–204–310–211	3600
5	203–406–310	2600	13	406–308–109–211	2550
6	203–407–109	3150	14	406–310–211	2650
7	204–305–407–109	2550	15	407–109–211	2350
8	204–308–109	2500			

Pairing costs are equally complex. On duty crews are guaranteed pay for minimum periods, regardless of the part of that time they are actually flying. If the pairing requires overnight stays away from home, hotel and other expenses are added. Values for our illustration are shown in Table 11.3.

Column Generation Model for AA Example

Having enumerated a list of alternatives like Table 11.3, the remaining task is to find a minimum total cost collection of columns staffing each flight exactly once. Define decision variables

$$x_j \triangleq \begin{cases} 1 & \text{if pairing } j \text{ is chosen} \\ 0 & \text{otherwise} \end{cases}$$

Then the following set partitioning model does the job:

$$
\begin{aligned}
\min \quad & 2900x_1 + 2700x_2 + 2600x_3 + 3000x_4 + 2600x_5 \\
& + 3150x_6 + 2550x_7 + 2500x_8 + 2600x_9 + 2050x_{10} \\
& + 2400x_{11} + 3600x_{12} + 2550x_{13} + 2650x_{14} + 2350x_{15}
\end{aligned}
$$

(11.10)

$$
\begin{aligned}
\text{s.t.} \quad & x_1 + x_2 + x_3 + x_4 && = 1 && \text{(flight 101)} \\
& x_6 + x_7 + x_8 + x_9 + x_{10} + x_{13} + x_{15} && = 1 && \text{(flight 109)} \\
& x_1 + x_2 + x_5 + x_6 && = 1 && \text{(flight 203)}
\end{aligned}
$$

$$x_3 + x_4 + x_7 + x_8 + x_{11} + x_{12} = 1 \qquad \text{(flight 204)}$$
$$x_{12} + x_{13} + x_{14} + x_{15} \qquad\qquad = 1 \qquad \text{(flight 211)}$$
$$x_9 + x_{10} \qquad\qquad\qquad\qquad = 1 \qquad \text{(flight 212)}$$
$$x_3 + x_7 + x_9 + x_{11} \qquad\qquad = 1 \qquad \text{(flight 305)}$$
$$x_1 + x_4 + x_8 + x_{10} + x_{13} \qquad = 1 \qquad \text{(flight 308)}$$
$$x_5 + x_{12} + x_{14} \qquad\qquad\qquad = 1 \qquad \text{(flight 310)}$$
$$x_{11} + x_{12} \qquad\qquad\qquad\qquad = 1 \qquad \text{(flight 402)}$$
$$x_1 + x_5 + x_{13} + x_{14} \qquad\qquad = 1 \qquad \text{(flight 406)}$$
$$x_2 + x_3 + x_6 + x_7 + x_9 + x_{15} = 1 \qquad \text{(flight 407)}$$
$$x_1, \ldots, x_{15} = 0 \text{ or } 1$$

An optimal solution makes

$$x_1^* = x_9^* = x_{12}^* = 1$$

and all other $x_j^* = 0$ at total cost \$9100.

SAMPLE EXERCISE 11.9: FORMING COLUMN GENERATION MODELS

A moving and storage company is allocating 5 long-distance moving loads to trucks. One feasible combination covers loads 1 and 3 in a 4525-mile route; a second combines loads 2, 3, and 4 in 2960 miles; a third hauls loads 2, 4, and 5 in 3170 miles; and a fourth covers load 1, 4, and 5 in 5230 miles. Form a set partitioning model to decide a minimum distance combination of routes covering each load exactly once.

Modeling: We use the decision variables

$$x_j \triangleq \begin{cases} 1 & \text{if route } j \text{ is chosen} \\ 0 & \text{otherwise} \end{cases}$$

Then the required model is

$$\begin{aligned}
\min \quad & 4525x_1 + 2960x_2 + 3170x_3 + 5230x_4 && \text{(total miles)} \\
\text{s.t.} \quad & x_1 + x_4 && = 1 && \text{(load 1)} \\
& x_2 + x_3 && = 1 && \text{(load 2)} \\
& x_1 + x_2 && = 1 && \text{(load 3)} \\
& x_2 + x_3 + x_4 && = 1 && \text{(load 4)} \\
& x_3 + x_4 && = 1 && \text{(load 5)} \\
& x_1, \ldots, x_4 = 0 \text{ or } 1
\end{aligned}$$

11.4 ASSIGNMENT AND MATCHING MODELS

We have already encountered **assignment problems** in network flow Section 10.5. The issue is optimal matching or pairing of objects of two distinct types—jobs to machines, sales personnel to customers, and so on. In this section we elaborate on a number of variations that cannot be solved by network flow methods.

Assignment Constraints

It is standard to model all assignment forms with the decision variables

$$x_{i,j} \triangleq \begin{cases} 1 & \text{if } i \text{ of the first set is matched with } j \text{ of the second} \\ 0 & \text{otherwise} \end{cases}$$

Then corresponding assignment constraints merely require that each object of each set be paired exactly once.

> **11.14** Over decision variables $x_{i,j} = 1$ if i is assigned to j and $= 0$ otherwise, **assignment constraints** take the form
>
> $$\sum_j x_{i,j} = 1 \qquad \text{for all } i$$
>
> $$\sum_i x_{i,j} = 1 \qquad \text{for all } j$$
>
> $$x_{i,j} = 0 \text{ or } 1 \qquad \text{for all } i, j$$
>
> where all sums are limited to (i, j) combinations allowed in the problem.

The first system of constraints forces every i to be assigned. The second does the same for every j.

CAM Linear Assignment Example Revisited

Section 10.5's computer-aided manufacturing (CAM) model illustrates one assignment case. Table 11.4 repeats total transportation, queueing, and processing times for 8 pending jobs on 10 workstations to which they might next be routed. Each workstation can accommodate only one job at a time. We want to find a minimum total time routing.

TABLE 11.4 Transportation and Processing Times for CAM Example

Job, i	\multicolumn{10}{c}{Next Workstation, j}

Job, i	1	2	3	4	5	6	7	8	9	10
1	8	—	23	—	—	—	—	—	5	—
2	—	4	—	12	15	—	—	—	—	—
3	—	—	20	—	13	6	—	8	—	—
4	—	—	—	—	19	10	—	—	—	—
5	—	—	—	8	—	—	12	—	—	16
6	14	—	—	—	—	—	8	—	3	—
7	—	6	—	—	—	—	—	27	—	12
8	—	5	15	—	—	—	—	32	—	—

After introducing dummy jobs 9 and 10, so that we have the same number of jobs as machines, this problem clearly has the assignment form. A complete model is given in (10.6) of Section 10.5. Our interest here is in the objective function

$$\begin{aligned}
\min \quad & 8x_{1,1} + 23x_{1,3} + 5x_{1,9} + 4x_{2,2} + 12x_{2,4} + 15x_{2,5} \\
& + 20x_{3,3} + 13x_{3,5} + 6x_{3,6} + 8x_{3,8} + 19x_{4,5} + 10x_{4,6} \\
& + 8x_{5,4} + 12x_{5,7} + 16x_{5,10} + 14x_{6,1} + 8x_{6,7} + 3x_{6,9} \\
& + 6x_{7,2} + +27x_{7,8} + 12x_{7,10} + 5x_{8,2} + 15x_{8,3} + 32x_{8,8}
\end{aligned} \qquad (11.11)$$

Linear Assignment Models

CAM model (10.6) is a linear assignment model because its objective function (11.11) is linear.

> **11.15** | **Linear assignment models** minimize or maximize a linear objective function of the form
>
> $$\sum_i \sum_j c_{i,j} x_{i,j}$$
>
> subject to assignment constraints | 11.14 |, where $c_{i,j}$ is the cost (or benefit) of assigning i to j.

Costs are sums of single assignment decisions.

SAMPLE EXERCISE 11.10: FORMULATING LINEAR ASSIGNMENT MODELS

A swimming coach is choosing his team for a medley relay. One swimmer will swim the back stroke leg $j = 1$, one the breast stroke leg $j = 2$, one the butterfly leg $j = 3$, and one the free-style leg $j = 4$. From previous experience the coach can estimate the time, $t_{i,j}$, that swimmer i could achieve on leg j. Formulate a linear assignment model to choose the fastest medley team.

Modeling: Using the decision variables

$$x_{i,j} \triangleq \begin{cases} 1 & \text{if swimmer } i \text{ swims leg } j \\ 0 & \text{otherwise} \end{cases}$$

the required model is

$$\min \quad \sum_{i=1}^{4} \sum_{j=1}^{4} t_{i,j} x_{i,j} \qquad \text{(team time)}$$

$$\text{s.t.} \quad \sum_{j=1}^{4} x_{i,j} = 1 \qquad i = 1, \ldots, 4 \qquad \text{(one leg per swimmer)}$$

$$\sum_{i=1}^{4} x_{i,j} = 1 \qquad j = 1, \ldots, 4 \qquad \text{(one swimmer per leg)}$$

$$x_{i,j} = 0 \text{ or } 1 \qquad i = 1, \ldots, 4; \quad j = 1, \ldots, 4$$

Swimmers are assigned to relay legs to minimize total team time.

Quadratic Assignment Models

What produces the linearity in definition | 11.15 | is that the objective function weights single decisions to pair i in one set with j in the other. For example, in objective (11.11), a decision to assign job 4 to workstation 5 adds 19 to the total time of the solution, regardless of how other decisions are resolved.

Many assignment problem circumstances do not fit the linear case because the objective function depends on combinations of decisions. That is, the impact of one decision cannot be assessed until we know how others are resolved. Such circumstances often lead to quadratic assignment models.

| 11.16 | **Quadratic assignment models** minimize or maximize a quadratic objective function of the form

$$\sum_i \sum_j \sum_{k>i} \sum_{\ell \neq j} c_{i,j,k,\ell}\, x_{i,j}\, x_{k,\ell}$$

subject to assignment constraints | 11.14 |, where $c_{i,j,k,\ell}$ is the cost (or benefit) of assigning i to j and k to ℓ.

Notice that each objective function term

$$c_{i,j,k,\ell} \cdot x_{i,j} \cdot x_{k,\ell}$$

involves two assignment decisions. Cost $c_{i,j,k,\ell}$ is realized only if both $x_{i,j} = 1$ and $x_{k,\ell} = 1$. That is, $c_{i,j,k,\ell}$ applies only if i is assigned to j and k is assigned to ℓ.

EXAMPLE 11.5: MALL LAYOUT QUADRATIC ASSIGNMENT

Some of the most common cases producing quadratic assignment models arise in **facility layout**. We are given a collection of machines, offices, departments, stores, and so on, to arrange within a facility, and a set of locations within which they must fit. The problem is to decide which unit to assign to each location.

Figure 11.3 illustrates with 4 possible locations for stores in a shopping mall. Walking distances (in feet) between the shop locations are displayed in the adjacent table. The 4 prospective tenants for the shop locations are listed in Table 11.5. The table also shows the number of customers each week (in thousands) who might wish to visit various pairs of shops. For example, a projected 5 thousand customers per week will visit both 1 (Clothes Are) and 2 (Computers Aye).

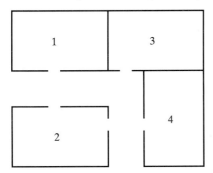

| Distance (feet) | | | |
j \ ℓ	**1**	**2**	**3**	**4**
1	—	80	150	170
2	80	—	130	100
3	150	130	—	120
4	150	100	120	—

FIGURE 11.3 Mall Layout Example Locations

Mall managers want to arrange the stores in the 4 locations to minimize customer inconvenience. One very common measure is **flow-distance**, the product of flow volumes between facilities and the distances between their assigned locations. For example, if shop 1 (Clothes Are) is located in space 1, and shop 4 (Book Bazaar) is located in space 2, their 7 thousand common customers will have to walk the 80 feet between the locations. This adds $7 \cdot 80 = 560$ thousand customer-feet to the flow-distance.

TABLE 11.5   Mall Layout Example Tenants

Store, i	Common Customers with k (000's)			
	1	2	3	4
1: Clothes Are	—	5	2	7
2: Computers Aye	5	—	3	8
3: Toy Parade	2	3	—	3
4: Book Bazaar	7	8	3	—

Mall Layout Example Model

Notice that the flow-distance for any pair of shops cannot be computed until we know where both are assigned. This is the assignment combinations characteristic that yields quadratic assignment models.

Using the decision variables

$$x_{i,j} \triangleq \begin{cases} 1 & \text{if shop } i \text{ is assigned to location } j \\ 0 & \text{otherwise} \end{cases}$$

the required quadratic assignment model is

$$
\begin{aligned}
\min \quad & 5(80x_{1,1}x_{2,2} + 150x_{1,1}x_{2,3} + 170x_{1,1}x_{2,4} && \text{(shops 1 and 2)} \\
& + 80x_{1,2}x_{2,1} + 130x_{1,2}x_{2,3} + 100x_{1,2}x_{2,4} \\
& + 150x_{1,3}x_{2,1} + 130x_{1,3}x_{2,2} + 120x_{1,3}x_{2,4} \\
& + 170x_{1,4}x_{2,1} + 100x_{1,4}x_{2,2} + 120x_{1,4}x_{2,3}) \\
& 2(80x_{1,1}x_{3,2} + 150x_{1,1}x_{3,3} + 170x_{1,1}x_{3,4} && \text{(shops 1 and 3)} \\
& + 80x_{1,2}x_{3,1} + 130x_{1,2}x_{3,3} + 100x_{1,2}x_{3,4} \\
& + 150x_{1,3}x_{3,1} + 130x_{1,3}x_{3,2} + 120x_{1,3}x_{3,4} \\
& + 170x_{1,4}x_{3,1} + 100x_{1,4}x_{3,2} + 120x_{1,4}x_{3,3}) \\
& 7(80x_{1,1}x_{4,2} + 150x_{1,1}x_{4,3} + 170x_{1,1}x_{4,4} && \text{(shops 1 and 4)} \\
& + 80x_{1,2}x_{4,1} + 130x_{1,2}x_{4,3} + 100x_{1,2}x_{4,4} \\
& + 150x_{1,3}x_{4,1} + 130x_{1,3}x_{4,2} + 120x_{1,3}x_{4,4} && \text{(11.12)} \\
& + 170x_{1,4}x_{4,1} + 100x_{1,4}x_{4,2} + 120x_{1,4}x_{4,3}) \\
& 3(80x_{2,1}x_{3,2} + 150x_{2,1}x_{3,3} + 170x_{2,1}x_{3,4} && \text{(shops 2 and 3)} \\
& + 80x_{2,2}x_{3,1} + 130x_{2,2}x_{3,3} + 100x_{2,2}x_{3,4} \\
& + 150x_{2,3}x_{3,1} + 130x_{2,3}x_{3,2} + 120x_{2,3}x_{3,4} \\
& + 170x_{2,4}x_{3,1} + 100x_{2,4}x_{3,2} + 120x_{2,4}x_{3,3}) \\
& 8(80x_{2,1}x_{4,2} + 150x_{2,1}x_{4,3} + 170x_{2,1}x_{4,4} && \text{(shops 2 and 4)} \\
& + 80x_{2,2}x_{4,1} + 130x_{2,2}x_{4,3} + 100x_{2,2}x_{4,4} \\
& + 150x_{2,3}x_{4,1} + 130x_{2,3}x_{4,2} + 120x_{2,3}x_{4,4} \\
& + 170x_{2,4}x_{4,1} + 100x_{2,4}x_{4,2} + 120x_{2,4}x_{4,3})
\end{aligned}
$$

$$3(80x_{3,1}x_{4,2} + 150x_{3,1}x_{4,3} + 170x_{3,1}x_{4,4} \qquad \text{(shops 3 and 4)}$$
$$+ 80x_{3,2}x_{4,1} + 130x_{3,2}x_{4,3} + 100x_{3,2}x_{4,4}$$
$$+ 150x_{3,3}x_{4,1} + 130x_{3,3}x_{4,2} + 120x_{3,3}x_{4,4}$$
$$+ 170x_{3,4}x_{4,1} + 100x_{3,4}x_{4,2} + 120x_{3,4}x_{4,3})$$

s.t.
$$x_{1,1} + x_{1,2} + x_{1,3} + x_{1,4} = 1 \qquad \text{(1, Clothes Are)}$$
$$x_{2,1} + x_{2,2} + x_{2,3} + x_{2,4} = 1 \qquad \text{(2, Computers Aye)}$$
$$x_{3,1} + x_{3,2} + x_{3,3} + x_{3,4} = 1 \qquad \text{(3, Toy Parade)}$$
$$x_{4,1} + x_{4,2} + x_{4,3} + x_{4,4} = 1 \qquad \text{(4, Book Bazaar)}$$
$$x_{1,1} + x_{2,1} + x_{3,1} + x_{4,1} = 1 \qquad \text{(location 1)}$$
$$x_{1,2} + x_{2,2} + x_{3,2} + x_{4,2} = 1 \qquad \text{(location 2)}$$
$$x_{1,3} + x_{2,3} + x_{3,3} + x_{4,3} = 1 \qquad \text{(location 3)}$$
$$x_{1,4} + x_{2,4} + x_{3,4} + x_{4,4} = 1 \qquad \text{(location 4)}$$
$$x_{i,j} = 0 \text{ or } 1 \quad i = 1, \ldots, 4; j = 1, \ldots, 4$$

The objective function computes total flow distance for all pairs of shops and all possible assigned locations. Assignment constraints assure that one shop goes to each location and each locations gets one shop. An optimal assignment places shop 1 in location 1, shop 2 in location 4, shop 3 in location 3, and shop 4 in location 2, for a total flow distance of 3260 thousand customer-feet.

SAMPLE EXERCISE 11.11: FORMULATING QUADRATIC ASSIGNMENT MODELS

An industrial engineer has divided a proposed machine shop's floor area into 12 grid squares, g, each of which will be the location of a single machine m. He has also estimated the distance, $d_{g,g'}$, between all pairs of grid squares and the number of units, $f_{m,m'}$, that will have to travel between machines m and m' (in both directions) during each week of operation. Formulate a quadratic assignment model to layout the shop in a way that will minimize material handling cost (i.e., minimize the product of between machine flows and the distance between their locations). Assume $d_{g,g'} = d_{g',g}$.

Modeling: Using the decision variables

$$x_{m,g} \triangleq \begin{cases} 1 & \text{if machine } m \text{ is located at grid square } g \\ 0 & \text{otherwise} \end{cases}$$

the required model is

$$\min \sum_{m=1}^{12} \sum_{g=1}^{12} \sum_{\substack{m'>m}}^{12} \sum_{\substack{g'=1 \\ g'\neq g}}^{12} f_{m,m'} d_{g,g'} x_{m,g} x_{m',g'} \qquad \text{(flow distance)}$$

s.t.
$$\sum_{g=1}^{12} x_{m,g} = 1 \qquad m = 1, \ldots, 12 \qquad \text{(square per machine)}$$

$$\sum_{m=1}^{12} x_{m,g} = 1 \qquad g = 1, \ldots, 12 \qquad \text{(machine per square)}$$

$$x_{i,j} = 0 \text{ or } 1 \qquad m = 1, \ldots, 12; \quad g = 1, \ldots, 12$$

Machines are assigned to grid squares to minimize the total flow distance which measures the material handling implication of a layout. Index ranges in the objective function assure that each pair of machines and locations is reflected just once.

Generalized Assignment Models

Main assignment constraints of definition $\boxed{11.14}$ require, respectively, that each object i of one set is assigned to exactly one j of the other, and that each j receives one i. Suppose, instead, that each object i must be assigned to some j, but that j's may receive several i. Specifically, define

$b_j \triangleq$ capacity of j

$s_{i,j} \triangleq$ size, space, or similar amount of j's capacity consumed if i is assigned to j

$c_{i,j} \triangleq$ cost (or benefit) of assigning i to j

Then finding the best way to assign all i without violating capacities is called a generalized assignment model.

$\boxed{11.17}$ **Generalized assignment models**, which encompass cases where allocation of i to j requires fixed size or space $s_{i,j}$ within j capacity b_j, have the form

$$\text{min or max} \quad \sum_i \sum_j c_{i,j} x_{i,j}$$

$$\text{s.t.} \quad \sum_j x_{i,j} = 1 \qquad \text{for all } i$$

$$\sum_i s_{i,j} x_{i,j} \leq b_j \qquad \text{for all } j$$

$$x_{i,j} = 0 \text{ or } 1 \qquad \text{for all } i, j$$

Here $c_{i,j}$ is the cost (or benefit) of assigning i to j and all sums are limited to (i, j) combinations allowed in the problem.

EXAMPLE 11.6: CDOT GENERALIZED ASSIGNMENT

The Canadian Department of Transportation encountered a problem of the generalized assignment form when reviewing their allocation of coast guard ships on Canada's Pacific coast.[4] The ships maintain such navigational aids as lighthouses and buoys. Each of the districts along the coast is assigned to one of a smaller number of coast guard ships. Since the ships have different home bases and different equipment and operating costs, the time and cost for assigning any district varies considerably among the ships. The task is to find a minimum cost assignment.

Table 11.6 shows data for our (fictitious) version of the problem. Three ships—the Estevan, the Mackenzie, and the Skidegate—are available to serve 6 districts. Entries in the table show the number of weeks each ship would require to maintain aides in each district, together with the annual cost (in thousands of Canadian dollars). Each ship is available 50 weeks per year.

---

[4]Based on Joseph G. Debanne and Jean-Noel Lavier (1979), "Management Science in the Public Sector—the Estevan Case," *Interfaces,* 9:2, part 2, 66–77.

TABLE 11.6   Costs and Times for the CDOT Example

Ship, *j*		**District, *i***					
		1	**2**	**3**	**4**	**5**	**6**
1: Estevan	Cost	130	30	510	30	340	20
	Time	30	50	10	11	13	9
2: Mackenzie	Cost	460	150	20	40	30	450
	Time	10	20	60	10	10	17
3: Skidegate	Cost	40	370	120	390	40	30
	Time	70	10	10	15	8	12

CDOT Example Model

Using the decision variables

$$x_{i,j} \triangleq \begin{cases} 1 & \text{if district } i \text{ is assigned to ship } j \\ 0 & \text{otherwise} \end{cases}$$

this CDOT example can be formulated

$$\begin{aligned}
\min \quad & 130x_{1,1} + 460x_{1,2} + 40x_{1,3} + 30x_{2,1} + 150x_{2,2} + 370x_{2,3} \\
& + 510x_{3,1} + 20x_{3,2} + 120x_{3,3} + 30x_{4,1} + 40x_{4,2} + 390x_{4,3} \\
& + 340x_{5,1} + 30x_{5,2} + 40x_{5,2} + 20x_{6,1} + 450x_{6,2} + 30x_{6,3}
\end{aligned}$$

$$\begin{aligned}
\text{s.t.} \quad & x_{1,1} + x_{1,2} + x_{1,3} = 1 && \text{(district 1)} \\
& x_{2,1} + x_{2,2} + x_{2,3} = 1 && \text{(district 2)} \\
& x_{3,1} + x_{3,2} + x_{3,3} = 1 && \text{(district 3)} \\
& x_{4,1} + x_{4,2} + x_{4,3} = 1 && \text{(district 4)} \\
& x_{5,1} + x_{5,2} + x_{5,3} = 1 && \text{(district 5)} \\
& x_{6,1} + x_{6,2} + x_{6,3} = 1 && \text{(district 6)} \\
& 30x_{1,1} + 50x_{2,1} + 10x_{3,1} && \text{(Estevan)} \\
& \quad + 11x_{4,1} + 13x_{5,1} + 9x_{6,1} \leq 50 \\
& 10x_{1,2} + 20x_{2,2} + 60x_{3,2} && \text{(Mackenzie)} \\
& \quad + 10x_{4,2} + 10x_{5,2} + 17x_{6,2} \leq 50 \\
& 70x_{1,3} + 10x_{2,3} + 10x_{3,3} && \text{(Skidegate)} \\
& \quad + 15x_{4,3} + 8x_{5,3} + 12x_{6,3} \leq 50 \\
& x_{i,j} = 0 \text{ or } 1 \quad i = 1, \dots, 6; j = 1, \dots, 3
\end{aligned}$$

(11.13)

The objective function minimizes total cost. The first 6 constraints assure that every district is assigned to one ship, and the last 3 keep work assigned to each ship within the 50 weeks available. An optimal solution assigns districts 1, 4, and 6 to the Estevan, districts 2 and 5 to the Mackenzie, and district 3 to the Skidegate at a total cost of $480,000.

SAMPLE EXERCISE 11.12: FORMULATING GENERALIZED ASSIGNMENT MODELS

Objects $i = 1, \dots, 100$ of volume c_i cubic meters are being stored in an automated warehouse. Storage locations $j = 1, \dots, 20$ are located d_j meters from the system's input/output station, and all have capacity b cubic meters. Formulate a generalized assignment model to store all items at minimum total travel distance assuming that as many objects can be placed in any location as volume permits.

Modeling: Using the decision variables

$$x_{i,j} \triangleq \begin{cases} 1 & \text{if object } i \text{ is stored in location } j \\ 0 & \text{otherwise} \end{cases}$$

the required generalized assignment model is

$$\min \sum_{i=1}^{100} \sum_{j=1}^{20} d_j x_{i,j} \qquad \text{(total distance)}$$

$$\sum_{j=1}^{20} x_{i,j} = 1 \qquad i = 1, \ldots, 100 \qquad \text{(each } i \text{ stored)}$$

$$\sum_{j=1}^{100} c_i x_{i,j} \leq b \qquad j = 1, \ldots, 20 \qquad (j \text{ capacity)}$$

$$x_{i,j} = 0 \text{ or } 1 \qquad i = 1, \ldots, 100; \quad j = 1, \ldots, 20$$

The objective function totals move distance to assigned locations, the first system of main constraints assures that every object is stored, and the second enforces capacities.

Matching Models

Assignment problems considered so far always pair objects in two distinct sets. One final variation eliminates the distinction between the sets. Decision variables of such matching models are

$$x_{i,i'} \triangleq \begin{cases} 1 & \text{if } i \text{ is paired with } i' \\ 0 & \text{otherwise} \end{cases}$$

where by convention index $i' > i$ to avoid double counting.

11.18 **Matching models**, which seek an optimal pairing of like objects i, have the form

$$\text{min or max} \quad \sum_{i} \sum_{i' > i} c_{i,i'} x_{i,i'}$$

$$\text{s.t.} \quad \sum_{i' < i} x_{i',i} + \sum_{i' > i} x_{i,i'} = 1 \qquad \text{for all } i$$

$$x_{i,i'} = 0 \text{ or } 1 \qquad \text{for all } i, i' > i$$

Here $c_{i,i'}$ is the cost (or benefit) of pairing i with i' and all sums are limited to (i, i') combinations allowed in the problem.

The two sums in each main constraint of formulation 11.18 are required because any particular i will be the higher index in some pairs and the lower index in others.

Matching models include linear assignment cases 11.15 if allowed pairings are restricted to those matching an object in one class with an object in another. However, the matching term is usually reserved for the more general case where all objects come from a single class.

EXAMPLE 11.7: SUPERFI SPEAKER MATCHING

We may illustrate matching with the task faced by fictitious high-fidelity speaker manufacturer Superfi. Superfi sells its speakers in pairs. Even though the manufacturing process maintains the most rigid quality standards, any two speakers produced will still interfere slightly with each other when connected to the same stereo system.

To improve its product quality even more, Superfi has measured the distortion $d_{i,i'}$ for each pair of speakers in the current lot. They wish to determine how to pair the speakers so that total distortion is minimized.

Notice that any two speakers may be paired. There is no distinction between large and small, or left and right.

Superfi Example Model

We may model this problem with the decision variables

$$
x_{i,i'} \triangleq \begin{cases} 1 & \text{if speakers } i \text{ and } i' \text{ are paired} \\ 0 & \text{otherwise} \end{cases}
$$

There is one for each pair (i, i'), $i < i'$. The corresponding matching model 11.18 is

$$
\min \quad \sum_i \sum_{i' > i} d_{i,i'} x_{i,i'} \qquad \text{(distortion)}
$$

$$
\text{s.t.} \quad \sum_{i' < i} x_{i',i} + \sum_{i' > i} x_{i,i'} = 1 \quad \text{for all } i \qquad \text{(each speaker paired)} \qquad (11.14)
$$

$$
x_{i,i'} = 0 \text{ or } 1 \qquad \text{for all } i, i' > i
$$

The objective function sums the distortion of all selected pairs, and main constraints assure that each speaker is part of exactly one pair.

SAMPLE EXERCISE 11.13: FORMULATING MATCHING MODELS

The instructor in an operations research class is assigning his students to 2-person teams for a term project. Each student s has scored his or her preference $p_{s,s'}$ for working with each other student s'. Formulate a matching model to form teams in a way that maximizes total preference.

Modeling: Using the decision variables

$$
x_{i,i'} \triangleq \begin{cases} 1 & \text{if } i \text{ is teamed with } i' \\ 0 & \text{otherwise} \end{cases}
$$

the required matching model is

$$\max \quad \sum_{i} \sum_{i' > i} (p_{i,i'} + p_{i',i}) x_{i,i'} \qquad\qquad \text{(preference)}$$

$$\text{s.t.} \quad \sum_{i' < i} x_{i',i} + \sum_{i' > i} x_{i,i'} = 1 \qquad \text{for all } i \qquad \text{(each student paired)}$$

$$x_{i,i'} = 0 \text{ or } 1 \qquad\qquad \text{for all } i, i' > i$$

Each term of the objective captures the two-way preference gain of teaming i with i', and the main constraints assure that each student is assigned to one team.

Tractability of Assignment and Matching Models

The various models of the assignment family presented in this section provide a good illustration of the tremendous variation in tractability of discrete optimization models.

> **11.19** Linear assignment models are highly tractable because they can be viewed as network flow problems, and thus as special cases of linear programming. Even more efficient special-purpose algorithms exist.

> **11.20** It is very difficult to compute global optima in quadratic assignment models because the nonlinear objective function precludes even the ILP methods of Sections 12.2 to 12.5. Improving search heuristics like those of Sections 12.6 and 12.7 are usually applied.

> **11.21** Generalized assignment models, which are ILPs, can be solved in moderate size by the methods of Sections 12.2 to 12.5. Still, they are far less tractable than the linear assignment case.

> **11.22** Matching models are more difficult to solve than linear assignment cases, but efficient special-purpose algorithms are known that can solve quite large instances.

11.5 TRAVELING SALESMAN AND ROUTING MODELS

Among the most common discrete optimization problems are those that organize a collection of customer locations, jobs, cities, points, and so on, into sequences or routes. Sometimes such **routing models** form all points into a single sequence. Other times, several routes are required.

Traveling Salesman Problem

The simplest and most famous of routing problems is known to researchers as the traveling salesman problem.

> **11.23** The **traveling salesman problem (TSP)** seeks a minimum-total-length route visiting every point in a given set exactly once.

The name derives from a mythical salesperson who must make a **tour** of the **cities** in his or her territory while traveling the least possible distance. TSPs actually occur in a much wider expanse of applications. Any task of sequencing objects in minimum total cost, length, or time can be viewed as a traveling salesman problem.

EXAMPLE 11.8: NCB CIRCUIT BOARD TSP

A practical scenario for thinking about traveling salesman problems occurs in the manufacture of printed circuit boards for electronics.[5] Circuit boards have many small holes through which chips and other components are wired. In a typical example, several hundred holes may have to be drilled in up to 10 different sizes. Efficient manufacture requires that these holes be completed as rapidly as possible by a moving drill. Thus for any single size, the question of finding the most efficient drilling sequence is a traveling salesman routing problem.

Figure 11.4 shows the tiny example that we will investigate for fictional board manufacturer NCB. We seek an optimal route through the 10 hole locations indicated. Table 11.7 reports straight-line distances $d_{i,j}$ between hole locations i and j. Lines in Figure 11.4 show a fair quality solution with total length 92.8 inches. The best route is 11 inches shorter (see Section 12.6).

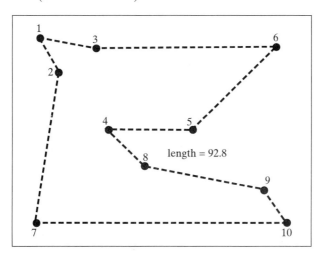

FIGURE 11.4   Board Drilling Locations for NCB Example

Symmetric versus Asymmetric Cases of the TSP

An important distinction among traveling salesman problems concerns whether distances between points are symmetric or asymmetric.

| 11.24 | A traveling salesman problem is **symmetric** if the distance or cost of passing from any point i to any other point j is the same as the distance from j to i. Otherwise, the problem is **asymmetric**. |

---

[5]Based in part on Surya Danusaputro, Chung-Yee Lee, and Louis A. Martin-Vega (1990), "An Efficient Algorithm for Drilling Printed Circuit Boards," *Computers and Industrial Engineering, 18,* 145–151.

TABLE 11.7   Distances between Holes in NCB Example

i \ j	1	2	3	4	5	6	7	8	9	10
1	—	3.6	5.1	10.0	15.3	20.0	16.0	14.2	23.0	26.4
2	3.6	—	3.6	6.4	12.1	18.1	13.2	10.6	19.7	23.0
3	5.1	3.6	—	7.1	10.6	15.0	15.8	10.8	18.4	21.9
4	10.0	6.4	7.1	—	7.0	15.7	10.0	4.2	13.9	17.0
5	15.3	12.1	10.6	7.0	—	9.9	15.3	5.0	7.3	11.3
6	20.0	18.1	15.0	15.7	9.9	—	25.0	14.9	12.0	15.0
7	16.0	13.2	15.8	10.0	15.3	25.0	—	10.3	19.2	21.0
8	14.2	10.6	10.8	4.2	5.0	14.9	10.3	—	10.2	13.0
9	23.0	19.7	18.4	13.9	7.8	12.0	19.2	10.2	—	3.6
10	26.4	23.0	21.9	17.0	11.3	15.0	21.0	13.0	3.6	—

Our NCB example is symmetric because $d_{i,j} = d_{j,i}$ in Table 11.7. In other cases, distances are not symmetric in this way because travel from i to j is with the traffic, and travel from j to i against, or because of a host of similar asymmetric circumstances.

Formulating the Symmetric TSP

One thing that intrigues researchers about traveling salesman problems is that there are many different formulations—none of them straightforward. Most ILP models of the symmetric case employ decision variables, $i < j$,

$$x_{i,j} \triangleq \begin{cases} 1 & \text{if the route includes a leg between } i \text{ and } j \\ 0 & \text{otherwise} \end{cases}$$

Notice that we define $x_{i,j}$ only for $i < j$. This numbering convention avoids the duplication that could result because a leg between i and j implies one between j and i, and costs are the same.

In terms of these new decision variables, the total route length now has the easy linear form

$$\min \sum_i \sum_{j>i} d_{i,j} x_{i,j} \qquad (11.15)$$

What makes ILP traveling salesman models complex is their constraints. A little contemplation will make one system clear. In the symmetric case, exactly two x-variables relating to any point i can be $= 1$ in a feasible solution. One links i to the city before it in the route, and the other links i to the city after. We can express this requirement mathematically with constraints

$$\sum_{j<i} x_{j,i} + \sum_{j>i} x_{i,j} = 2 \qquad \text{for all } i \qquad (11.16)$$

A specific instance for $i = 5$ in the NCB example of Figure 11.4 is

$$x_{1,5} + x_{2,5} + x_{3,5} + x_{4,5} + x_{5,6} + x_{5,7} + x_{5,8} + x_{5,9} + x_{5,10} = 2$$

SAMPLE EXERCISE 11.14: MODELING TSPs AS ILPs

The following graph show the available links joining 6 points, with numbers on edges indicating transit times.

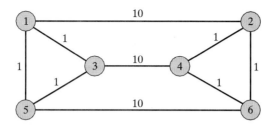

We wish to find the shortest time route visiting all nodes exactly once and using only links shown in the graph.

(a) Explain why this problem can be viewed as a symmetric traveling salesman problem.

(b) Formulate integer linear program objective (11.15) for this instance.

(c) Formulate constraints (11.16) for this instance.

Modeling:

(a) The problem is a TSP because it requires a closed route visiting each point. It is symmetric because the time is the same whether a link is passed in the *i*-to-*j* or the *j*-to-*i* direction.

(b) The linear objective required is

$$\min \quad 10x_{1,2} + 1x_{1,3} + 1x_{1,5} + 1x_{2,4} + 1x_{2,6}$$
$$+ 10x_{3,4} + 1x_{3,5} + 1x_{4,6} + 10x_{5,6}$$

which minimizes total tour length.

(c) Needed constraints (11.16) are

$$
\begin{aligned}
x_{1,2} + x_{1,3} + x_{1,5} &= 2 && \text{(node 1)} \\
x_{1,2} + x_{2,4} + x_{2,6} &= 2 && \text{(node 2)} \\
x_{1,3} + x_{3,4} + x_{3,5} &= 2 && \text{(node 3)} \\
x_{2,4} + x_{3,4} + x_{4,6} &= 2 && \text{(node 4)} \\
x_{1,5} + x_{3,5} + x_{5,6} &= 2 && \text{(node 5)} \\
x_{2,6} + x_{4,6} + x_{5,6} &= 2 && \text{(node 6)}
\end{aligned}
$$

Subtours

Figure 11.5 illustrates why constraints (11.16) are not usually enough. The solution shown does have two links at each point. But it divides the 10 hole locations among three **subtours** or miniroutes. We seek a single route through all the points.

No one has any difficulty understanding subtours, but constraints to prevent them are less obvious. One form of **subtour elimination constraints** is obtained from any

$$S \triangleq \text{proper subset of the points/cities to be routed}$$

Every tour must cross between points in S and points outside at least twice. This

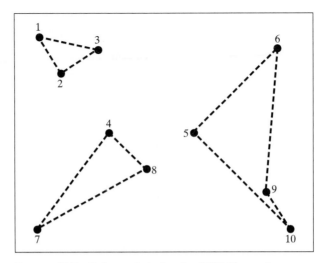

FIGURE 11.5   Subtour Solution for NCB Example

leads to constraints of the form

$$\begin{pmatrix} \text{number of legs} \\ \text{between points in } S \\ \text{and points not in } S \end{pmatrix} = \sum_{i \in S} \sum_{j \notin S} x_{i,j} + \sum_{i \notin S} \sum_{j \in S} x_{i,j} \geq 2 \qquad (11.17)$$

There is one such constraint for every proper subset S of at least 3 cities.

The subtour solution of Figure 11.5 violates several of these constraints. For example, pick $S \triangleq \{5, 6, 9, 10\}$. The corresponding subtour elimination constraint lists all possible tour legs passing into or out of set S:

$$x_{1,5} + x_{1,6} + x_{1,9} + x_{1,10} + x_{2,5}$$
$$+ x_{2,6} + x_{2,9} + x_{2,10} + x_{3,5} + x_{3,6}$$
$$+ x_{3,9} + x_{3,10} + x_{4,5} + x_{4,6} + x_{4,9}$$
$$+ x_{4,10} + x_{5,7} + x_{6,7} + x_{7,9} + x_{7,10}$$
$$+ x_{5,8} + x_{6,8} + x_{8,9} + x_{8,10}$$
$$\geq 2$$

SAMPLE EXERCISE 11.15: FORMING SUBTOUR ELIMINATION CONSTRAINTS

Return to the TSP of Sample Exercise 11.14.

(a) Show by inspection that the least cost binary solution over $= 2$ constraints of Sample Exercise 11.14(b) has subtours.

(b) Formulate a subtour elimination constraint (11.17) not satisfied by the subtour solution of part (a).

Analysis:

(a) It is obvious that the least cost binary solution touching each node with exactly two links has

$$x_{1,3} = x_{1,5} = x_{3,5} = 1$$
$$x_{2,4} = x_{2,6} = x_{4,6} = 1$$
$$x_{1,2} = x_{3,4} = x_{5,6} = 0$$

It contains subtours 1–3–5–1 and 2–4–6–2.

(b) One violated subtour elimination constraint (11.17) arises from

$$S \triangleq \{1, 3, 5\}$$

Summing all edges joining a node in S to one outside gives

$$x_{1,2} + x_{3,4} + x_{5,6} \geq 2$$

ILP Model of the Symmetric TSP

Combining expressions (11.15)–(11.17) produces a full integer linear programming formulation of the symmetric traveling salesman problem:

> **11.25**   An integer linear programming formulation of the symmetric traveling salesman problem is
>
> $$\min \quad \sum_{i} \sum_{j>i} d_{i,j} x_{i,j}$$
>
> $$\text{s.t.} \quad \sum_{j<i} x_{j,i} + \sum_{j>i} x_{i,j} = 2 \qquad \text{for all } i$$
>
> $$\sum_{i \in S} \sum_{j \notin S, j>i} x_{i,j} + \sum_{i \notin S} \sum_{j \in S, j>i} x_{i,j} \geq 2 \qquad \begin{array}{l}\text{for all proper point} \\ \text{subsets } S, |S| \geq 3\end{array}$$
>
> $$x_{i,j} = 0 \text{ or } 1 \qquad \text{for all } i; \quad j > i$$
>
> where $x_{i,j} = 1$ if link (i, j) is part of the solution, and $d_{i,j} \triangleq$ the distance from point i to point j.

ILP Model of the Asymmetric TSP

How must formulation 11.25 be modified to address the asymmetric case? Several insights are required.

- In the asymmetric case we use decision variables

$$x_{i,j} \triangleq \begin{cases} 1 & \text{if the tour passes } i \text{ to } j \\ 0 & \text{otherwise} \end{cases}$$

for all combinations of i and j. With costs asymmetric it matters whether the route goes i to j of j to i.

- Instead of just meeting each point twice, any asymmetric TSP route must enter each point once and leave each point once. Thus $= 2$ constraints of the symmetric formulation become assignment constraints (system 11.14) in the asymmetric case:

$$\sum_{j} x_{j,i} = 1 \qquad \text{for all i} \qquad (\text{enter } i)$$

$$\sum_{j} x_{i,j} = 1 \qquad \text{for all i} \qquad (\text{leave } i)$$

- Each tour must enter and leave every subset S of points. Thus we may accomplish subtour elimination by requiring the tour to leave every S at least once:

$$\sum_{i \in S} \sum_{j \notin S} x_{i,j} \geq 1 \qquad \text{all proper point subsets } S$$

Combing produces a full asymmetric formulation.

__11.26__ The asymmetric traveling salesman problem can be formulated as the integer linear program

$$\min \sum_{i} \sum_{j \neq i} d_{i,j} x_{i,j}$$

$$\begin{aligned}
\text{s.t.} \quad & \sum_{j} x_{j,i} = 1 && \text{for all } i \\
& \sum_{j} x_{i,j} = 1 && \text{for all } i \\
& \sum_{i \in S} \sum_{j \notin S} x_{i,j} \geq 1 && \text{for all proper point subsets } S, \\
& && |S| \geq 2 \\
& x_{i,j} = 0 \text{ or } 1 && \text{for all } i, j
\end{aligned}$$

where $x_{i,j} = 1$ if the tour passes from i to j and $d_{i,j} \triangleq$ distance from i to j.

SAMPLE EXERCISE 11.16: FORMULATING ASYMMETRIC TSPS

Return to the TSP of Sample Exercises 11.14 and 11.15, and assume that a 2-unit cost penalty is required when the tour passes from a higher to a lower node number. That is, $d_{1,2} = 10$, but $d_{2,1} = 10 + 2 = 12$, $d_{1,3} = 1$, but $d_{3,1} = 1 + 2 = 3$, and so on.

(a) Explain why the problem is now an asymmetric TSP.

(b) Formulate an objective function for the corresponding model 11.26 .

(c) Formulate constraints requiring the tour to enter and leave each node exactly once.

(d) Formulate a subtour elimination constraint requiring the tour to leave node subset $S = \{1, 3, 5\}$ at least once.

Analysis:

(a) The penalty of 2 makes $d_{i,j} \neq d_{j,i}$, which turns the problem asymmetric.

(b) Including variables for all one-way passages, the objective function of format 11.26 is

$$\begin{aligned}
\min \quad & 10x_{1,2} + 1x_{1,3} + 1x_{1,5} \\
& + 12x_{2,1} + 1x_{2,4} + 1x_{2,6} \\
& + 3x_{3,1} + 10x_{3,4} + 1x_{3,5} \\
& + 3x_{4,2} + 12x_{4,3} + 1x_{4,6} \\
& + 3x_{5,1} + 3x_{5,3} + 10x_{5,6} \\
& + 3x_{6,2} + 3x_{6,4} + 12x_{6,5}
\end{aligned}$$

(c) These constraints have the assignment form:

$$x_{2,1} + x_{3,1} + x_{5,1} = 1$$
$$x_{1,2} + x_{4,2} + x_{6,2} = 1$$
$$x_{1,3} + x_{4,3} + x_{5,3} = 1$$
$$x_{2,4} + x_{3,4} + x_{6,4} = 1$$
$$x_{1,5} + x_{3,5} + x_{6,5} = 1$$
$$x_{2,6} + x_{4,6} + x_{5,6} = 1$$

plus

$$x_{1,2} + x_{1,3} + x_{1,5} = 1$$
$$x_{2,1} + x_{2,4} + x_{2,6} = 1$$
$$x_{3,1} + x_{3,4} + x_{3,5} = 1$$
$$x_{4,2} + x_{4,3} + x_{4,6} = 1$$
$$x_{5,1} + x_{5,3} + x_{5,6} = 1$$
$$x_{6,2} + x_{6,4} + x_{6,5} = 1$$

(d) To avoid any subtour among nodes in $S = \{1, 3, 5\}$, we add subtour elimination constraint.

$$x_{1,2} + x_{3,4} + x_{5,6} \geq 1$$

Quadratic Assignment Formulation of the TSP

There are an enormous number of subtour elmination constraints for a TSP on even a modest number of points. That is why it is sometimes easier—especially in developing heuristic procedures of Section 12.6 and 12.7—to deal with a TSP formulation having simpler constraints. We can accomplish this, at the cost of making the objective function nonlinear, by formulating the TSP as a quadratic assignment (QAP) model 11.16 .

Think of the tour as a sequence or permutation of the points to be visited. Decision variables in the QAP form assign sequence positions k to points i; that is,

$$y_{k,i} \triangleq \begin{cases} 1 & \text{if } k\text{th point visited is } i \\ 0 & \text{otherwise} \end{cases}$$

For example, the illustrative route of Figure 11.4 has

$$y_{1,1} = y_{2,3} = y_{3,6} = y_{4,5} = y_{5,4} = y_{6,8}$$
$$= y_{7,9} = y_{8,10} = y_{9,7} = y_{10,2}$$
$$= 1$$

when viewed as starting at hole 1. In terms of these variables, either the symmetric or the asymmetric case can be formulated as an INLP.

11.27 | The traveling salesman problem can be formulated as the quadratic assignment model

$$\min \quad \sum_i \sum_j d_{i,j} \sum_k y_{k,i} y_{k+1,j} \qquad \text{(total length)}$$

$$\sum_i y_{k,i} = 1 \qquad \text{for all } k \qquad \text{(each position occupied)}$$

$$\sum_k y_{k,i} = 1 \qquad \text{for all } i \qquad \text{(each point visited)}$$

$$y_{k,i} = 0 \text{ or } 1 \qquad \text{for all } k, i$$

where $y_{k,i} = 1$ if the kth visited point is i, and $d_{i,j} \triangleq$ distance from point i to point j.

Constraints in | 11.27 | have the usual assignment format | 11.14 |, but the objective function is less transparent. Terms

$$d_{i,j} \sum_k y_{k,i} y_{k+1,j}$$

add in distance $d_{i,j}$ exactly when j follows i somewhere in the chosen tour sequence. (Here we take $k + 1$ to mean 1 when k is the highest-numbered point.) In Figure 11.4, for example, one of the nonzero terms would be

$$d_{3,6} \cdot y_{2,3} \cdot y_{3,6}$$

because holes 3 and 6 could be second and third in sequence. Summing over all i and j recovers total distance.

Problems Requiring Multiple Routes

For many trucking and other distribution organizations, the problem is to design many routes, not just one. The task begins with a collection of stops to be serviced. Decisions first subdivide the stops into several routes and then choose the visitation sequence for each route.

EXAMPLE 11.9: KI TRUCK ROUTING

Kraft Incorporated confronts such multiple-route design problems in planning truck delivery of its food products to over 100,000 commercial, industrial, and military customers in North America.[6] Known customer requirements must be grouped into truckloads and then routes planned.

Our tiny fictitious version of the KI case is illustrated in Figure 11.6. The 20 stops indicated are to be serviced from a single depot. Table 11.8 displays the requirements

---

[6]Based on Hong K. Chung and John B. Norback (1991), "A Clustering and Insertion Heuristic Applied to a Large Routing Problem in Food Distribution," *Journal of the Operational Research Society*, 42, 555–564.

at each stop, expressed in fractions, f_i, of a truckload. The 7 routes depicted in Figure 11.6 show one feasible solution.

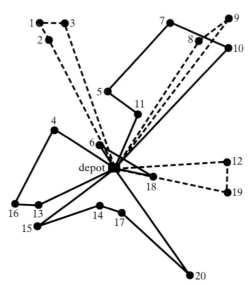

FIGURE 11.6 Depot and Delivery Locations in the KI Example

TABLE 11.8 Fractions of Truckloads to Be Delivered in KI Example

Stop, i	Fraction, f_i	Stop, i	Fraction, f_i	Stop, i	Fraction, f_i	Stop, i	Fraction, f_i
1	0.25	6	0.70	11	0.21	16	0.38
2	0.33	7	0.28	12	0.68	17	0.26
3	0.39	8	0.43	13	0.16	18	0.29
4	0.40	9	0.50	14	0.19	19	0.17
5	0.27	10	0.22	15	0.22	20	0.31

KI Truck Routing Example Model

With complex problems like KI's it is not easy to see where to begin a model.

11.28 Routing problems are characteristically difficult to represent concisely in optimization models.

We will start with the assignment of stops to routes. The decision variables

$$z_{i,j} \triangleq \begin{cases} 1 & \text{if stop } i \text{ is assigned to route } j \\ 0 & \text{otherwise} \end{cases}$$

Generalized assignment constraints (definition $\boxed{11.17}$) then manage the allocation of the 20 stops to seven routes:

$$\sum_{j=1}^{7} z_{i,j} = 1 \qquad \text{for all } i = 1, \ldots, 20 \qquad (\text{each } i \text{ to some } j)$$

$$\sum_{i=1}^{20} f_i z_{i,j} \le 1 \qquad \text{for all } j = 1, \ldots, 7 \qquad (\text{truck capacities}) \tag{11.18}$$

$$z_{i,j} = 0 \text{ or } 1 \qquad \text{for all } i = 1, \ldots, 20; \quad j = 1, \ldots, 7$$

The first set makes sure every stop goes to some route, and the second keeps loads to assigned trucks within capacity.

The difficulty of concisely formulating routing problems becomes apparent when we try to express an objective function to go with (11.18). Certainly, it is something like

$$\sum_{j=1}^{7} \theta_j(\mathbf{z}) \tag{11.19}$$

where $\mathbf{z}$ is the vector of $z_{i,j}$ and

$$\theta_j(\mathbf{z}) \triangleq \text{length of the best route through stops assigned} \\ \text{to truck } j \text{ by decision vector } \mathbf{z}$$

However, functions like $\theta_j(\mathbf{z})$ are highly nonlinear. In fact, $\theta_j(\mathbf{z})$ is the optimal solution value of a traveling salesman problem over the stops allocated to route j.

No concise expression of such functions exists. Still, we know what they mean in the problem context. Section 12.8 will show that good heuristic solutions can be obtained with rough approximations.

11.6 FACILITY LOCATION AND NETWORK DESIGN MODELS

Principles $\boxed{11.2}$ and $\boxed{11.3}$ of Section 11.1 demonstrated how **fixed charges** can be modeled with new binary variables and **switching constraints**. Virtually any linear program with fixed charges can be modeled in that way, but some forms are especially common. In this section we introduce the classic **facility location** and **network design** cases.

Facility Location Models

Facility location models, also called warehouse location models and plant location models, are perhaps the most frequently solved of all forms involving fixed charges.

> $\boxed{11.29}$ **Facility/plant/warehouse location models** choose which of a proposed list of facilities to open in order to service specified customer demands at minimum total cost.

Costs include both the variable cost of servicing customers from chosen facilities and the fixed cost of opening the facilities.

EXAMPLE 11.10: TMARK FACILITIES LOCATION

AT&T has confronted many facility location problems in recommending sites for the toll-free call-in centers of its telemarketing customers.[7] Such centers handle telephone reservations and orders arising in many geographic zones. Since telephone rates vary dramatically depending on the zone of call origin and the location of the receiving center, site selection is extremely important. A well-designed system should minimize the total of call charges and center setup costs.

Our version of this scenario will involve fictional firm Tmark. Figure 11.7 shows the 8 sites under consideration for Tmark's catalog order centers embedded in a map of the 14 calling zones. Table 11.9 shows corresponding unit calling charges, $r_{i,j}$, from each zone j to various centers i, and the zone's anticipated call load, d_j.

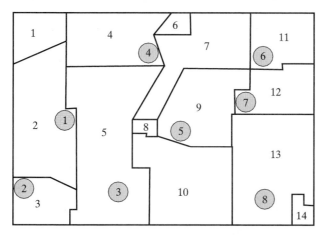

i	Fixed Cost
1	2400
2	7000
3	3600
4	1600
5	3000
6	4600
7	9000
8	2000

FIGURE 11.7 Customer Zones and Possible Facility Locations for Tmark Example

TABLE 11.9 Unit Call Charges and Demands for Tmark Example

Zone, j	Possible Center Location, i								Call Demand
	1	2	3	4	5	6	7	8	
1	1.25	1.40	1.10	0.90	1.50	1.90	2.00	2.10	250
2	0.80	0.90	0.90	1.30	1.40	2.20	2.10	1.80	150
3	0.70	0.40	0.80	1.70	1.60	2.50	2.05	1.60	1000
4	0.90	1.20	1.40	0.50	1.55	1.70	1.80	1.40	80
5	0.80	0.70	0.60	0.70	1.45	1.80	1.70	1.30	50
6	1.10	1.70	1.10	0.60	0.90	1.30	1.30	1.40	800
7	1.40	1.40	1.25	0.80	0.80	1.00	1.00	1.10	325
8	1.30	1.50	1.00	1.10	0.70	1.50	1.50	1.00	100
9	1.50	1.90	1.70	1.30	0.40	0.80	0.70	0.80	475
10	1.35	1.60	1.30	1.50	1.00	1.20	1.10	0.70	220
11	2.10	2.90	2.40	1.90	1.10	2.00	0.80	1.20	900
12	1.80	2.60	2.20	1.80	0.95	0.50	2.00	1.00	1500
13	1.60	2.00	1.90	1.90	1.40	1.00	0.90	0.80	430
14	2.00	2.40	2.00	2.20	1.50	1.20	1.10	0.80	200

[7]Based on Thomas Spencer III, Anthony J. Brigandi, Dennis R. Dargon, and Michael J. Sheehan (1990), "AT&T's Telemarketing Site Selection System Offers Customer Support," *Interfaces, 20:1*, 83–96.

Any Tmark center selected can handle between 1500 and 5000 call units per day. However, their fixed costs of operation vary significantly because of differences in labor and real estate prices. Estimated daily fixed costs, f_i, for the 8 centers are displayed in Figure 11.7.

ILP Model of Facilities Location

Clearly, there are two kinds of decisions to make in facilities location: which facilities to open, and how chosen facilities should serve customer demands. We employ the decision variables

$$y_i \triangleq \begin{cases} 1 & \text{if facility } i \text{ is opened} \\ 0 & \text{otherwise} \end{cases}$$

to determine which sites are selected. A second set,

$$x_{i,j} \triangleq \text{fraction of customer } j \text{ demand satisfied from facility } i$$

decides how service is to be allocated. Combining produces a standard model of facilities location:

> **11.30** Basic facilites location problems can be formulated as the integer linear program
>
> $$\min \quad \sum_i \sum_j c_{i,j} d_j x_{i,j} + \sum_i f_i y_i \qquad \text{(total cost)}$$
>
> $$\text{s.t.} \quad \sum_i x_{i,j} = 1 \qquad \text{for all } j \qquad \text{(fulfill } j \text{ demand)}$$
>
> $$\sum_j d_j x_{i,j} \leq u_i y_i \qquad \text{for all } i \qquad (i \text{ capacity switching)}$$
>
> $$x_{i,j} \geq 0 \qquad \text{for all } i, j$$
>
> $$y_1 = 0 \text{ or } 1 \qquad \text{for all } i$$
>
> where $x_{i,j} \triangleq$ fraction of demand j fulfilled from i, $y_i = 1$ if facility i is opened, d_j is the demand at j, $c_{i,j}$ is the unit cost of fulfilling j demand from i, f_i is the nonnegative fixed cost of opening facility i, and u_i is the capacity of facility i.

This objective function sums all variable and fixed costs. The first system of constraints assures that 100% of each customer demand is fulfilled. The second set switches "on" facility capacity when the corresponding $y_i = 1$. If the problem has no true capacities on facilities, any sufficiently large value can provide a u_i—perhaps total demand.

Sometimes customers must be 100% serviced from a single facility, so that constraints

$$x_{i,j} = 0 \text{ or } 1 \qquad \text{for all } i, j$$

are also enforced. Other times, demand can be split among several facilities, which leaves $x_{i,j}$ continuous.

Tmark Facilities Location Example Model

Following format 11.30 , we can express a Tmark's facilities location problem in the mixed-integer linear program

$$
\min \quad \sum_{i=1}^{8} \sum_{j=1}^{14} (r_{i,j} d_j) x_{i,j} + \sum_{i=1}^{8} f_i y_i \qquad \text{(total cost)}
$$

$$
\text{s.t.} \quad \sum_{i=1}^{8} x_{i,j} = 1 \qquad \text{for all } j = 1, \ldots, 14 \qquad \text{(carry } j \text{ load)}
$$

$$
1500 y_i \le \sum_{j=1}^{14} d_j x_{i,j} \qquad \text{for all } i = 1, \ldots, 8 \qquad \text{(minimum at } i) \qquad (11.20)
$$

$$
\sum_{j=1}^{14} d_j x_{i,j} \le 5000 y_i \qquad \text{for all } i = 1, \ldots, 8 \qquad \text{(maximum at } i)
$$

$$
x_{i,j} \ge 0 \qquad \text{for all } i = 1, \ldots, 8; \quad j = 1, \ldots, 14
$$

$$
y_i = 0 \text{ or } 1 \qquad \text{for all } i = 1, \ldots, 8
$$

Here the objective function sums calling and setup costs. The total cost of call load serviced by i for zone j is

$$
\text{(demand at } j)(i, j \text{ telephone rate})(\text{fraction of } j \text{ serviced from } i) = d_j r_{i,j} x_{i,j}
$$

Most constraints of model (11.20) are identical to pattern 11.30 . Variables $x_{i,j}$ are continuous because it is assumed that the load from a zone can be split among several centers.

The one new element is switching constraints dealing with minimum operating levels. Constraints

$$
1500 y_i \le \sum_{j=1}^{14} d_j x_{i,j}
$$

enforce the requirement that any open center i must handle at least 1500 call units per day.

Solution of model (11.20) produces

$$
y_4^* = y_8^* = 1
$$
$$
y_1^* = y_2^* = y_3^* = y_5^* = y_6^* = y_7^* = 0
$$

with a total cost of \$10,153 per day. Zones 1, 2, 4, 5, 6, and 7 are serviced from center 4, and the rest from center 8.

SAMPLE EXERCISE 11.17: FORMULATING FACILITIES LOCATION MODELS

Environmental protection authorities have identified 14 possible offices around the country from which inspectors would make annual visits to each of 111 sites at high risk for oil spills. They have also measured the travel cost $c_{i,j}$ from location i to every potential spill site j. Each spill site should be under the supervision of a single inspection office.

(a) Formulate a facilities location model to choose a minimum total cost collection of offices and inspection assignments assuming that there is an annual fixed cost of f for operating any office.

(b) Formulate a facilities location model to choose a minimum total cost collection of offices and inspection assignments assuming that fixed costs of operating office are unknown but that it has been decided to open at most 9.

Modeling: We follow integer linear programming format $\boxed{11.30}$.

(a) For this case the model required is

$$\min \quad \sum_{i=1}^{14}\sum_{j=1}^{111} c_{i,j}x_{i,j} + f\sum_{i=1}^{14} y_i \qquad \text{(total cost)}$$

$$\text{s.t.} \quad \sum_{i=1}^{14} x_{i,j} = 1 \qquad j = 1, \ldots, 111 \qquad \text{(each } j \text{ inspected)}$$

$$\sum_{j=1}^{111} x_{i,j} \le 111 y_i \qquad i = 1, \ldots, 14 \qquad (i \text{ switching})$$

$$x_{i,j} = 0 \text{ or } 1 \qquad i = 1, \ldots, 14; \quad j = 1, \ldots, 111$$

$$y_i = 0 \text{ or } 1 \qquad i = 1, \ldots, 14$$

Since no capacities are specified for the offices, switching constraints use $u_i = 111$. Every spill site can be covered by any open office. Variables $x_{i,j}$ are binary because spill sites must be inspected from a single office.

(b) For this case the model required is

$$\min \quad \sum_{i=1}^{14}\sum_{j=1}^{111} c_{i,j}x_{i,j} \qquad \text{(travel cost)}$$

$$\text{s.t.} \quad \sum_{i=1}^{14} x_{i,j} = 1 \qquad j = 1, \ldots, 111 \qquad \text{(each } j \text{ inspected)}$$

$$\sum_{j=1}^{111} x_{i,j} \le 111 y_i \qquad i = 1, \ldots, 14 \qquad (i \text{ switching})$$

$$\sum_{i=1}^{14} y_i \le 9 \qquad \text{(max nine chosen)}$$

$$x_{i,j} = 0 \text{ or } 1 \qquad i = 1, \ldots, 14; \quad j = 1, \ldots, 111$$

$$y_i = 0 \text{ or } 1 \qquad i = 1, \ldots, 14$$

Fixed charges have been omitted and a new constraint added that limits the number of offices to 9.

Network Design Models

Facility location models decide which nodes of a network to open. Network design or fixed-charge network flow models decide which arcs. A continuous network flow in variables $x_{i,j}$ is augmented with discrete variables $y_{i,j}$ implementing fixed charges for opening/constructing/setting up arcs.

11.31 The **fixed-charge network flow** or **network design model** on a digraph on nodes $k \in V$ with net demand b_k, and arcs $(i, j) \in A$ with capacity $u_{i,j}$, unit cost $c_{i,j}$, and nonnegative fixed cost $f_{i,j}$ is

$$\min \quad \sum_{(i,j) \in A} c_{i,j} x_{i,j} + \sum_{(i,j) \in A} f_{i,j} y_{i,j}$$

$$\text{s.t.} \quad \sum_{(i,k) \in A} x_{i,k} - \sum_{(k,j) \in A} x_{k,j} = b_k \qquad \text{for all } k \in V$$

$$0 \leq x_{i,j} \leq u_{i,j} y_{i,j} \qquad\qquad \text{for all } (i, j) \in A$$

$$y_{i,j} = 0 \text{ or } 1 \qquad\qquad\qquad \text{for all } (i, j) \in A$$

Main constraints in $x_{i,j}$ mirror network flow models $\boxed{10.3}$ (Section 10.1) in conserving flow at each node. Switching constraints (definition $\boxed{11.3}$) turn on arc capacities $u_{i,j}$ if the fixed charge is paid. As usual, capacities must be derived from other constraints if not given explicitly—possibly by taking $u_{i,j} = $ the largest feasible flow on arc (i, j).

EXAMPLE 11.11: WASTEWATER NETWORK DESIGN

Network design applications may involve telecommunications, electricity, water, gas, coal slurry, or any other substance that flows in a network. We illustrate with an application involving regional wastewater (sewer) networks.[8]

As new areas develop around major cities, entire networks of collector sewers and treatment plants must be constructed to service growing population. Figure 11.8 displays our particular (fictional) instance.

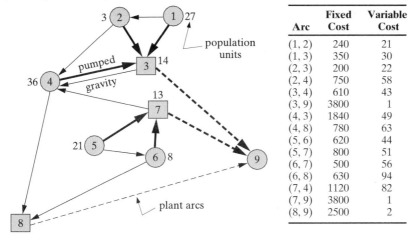

Arc	Fixed Cost	Variable Cost
(1, 2)	240	21
(1, 3)	350	30
(2, 3)	200	22
(2, 4)	750	58
(3, 4)	610	43
(3, 9)	3800	1
(4, 3)	1840	49
(4, 8)	780	63
(5, 6)	620	44
(5, 7)	800	51
(6, 7)	500	56
(6, 8)	630	94
(7, 4)	1120	82
(7, 9)	3800	1
(8, 9)	2500	2

FIGURE 11.8   Wastewater Network Design Example

---

[8]Based on J. J. Jarvis, R. L. Rardin, V. E. Unger, R. W. Moore, and C. C. Schimpeler (1978), "Optimal Design of Regional Wastewater Systems: A Fixed Charge Network Flow Model," *Operations Research,* 26, 538–550.

Nodes 1 to 8 of the network represent population centers where smaller sewers feed into the main regional network, and locations where treatment plants might be built. Wastewater loads are roughly proportional to population, so the inflows indicated at nodes represent population units (in thousands).

Arcs joining nodes 1 to 8 show possible routes for main collector sewers. Most follow the topology in gravity flow, but one pumped line (4, 3) is included. A large part of the construction cost for either type of line is fixed: right-of-way acquisition, trenching, and so on. Still, the cost of a line also grows with the number of population units carried, because greater flows imply larger-diameter pipes. The table in Figure 11.8 shows the fixed and variable cost for each arc in thousand of dollars.

Treatment plant costs actually occur at nodes—here nodes 3, 7, and 8. Figure 11.8 illustrates, however, that such costs can be modeled on arcs by introducing an artificial "supersink" node 9. Costs shown for arcs $(3, 9)$, $(7, 9)$, and $(8, 9)$ capture the fixed and variable expense of plant construction as flows depart the network.

Wastewater Network Design Example Model

To place our wastewater network design problem in format 11.31 , we need capacities $u_{i,j}$ on the arcs. None are explicitly provided, but it is not difficult to determine the maximum possible flow on each arc. For example, we may take

$$u_{2,3} = 27 + 3 = 30$$

because no solution would send more than the 27 thousand population units at node 1 and the 3 at node 2 along arc $(2, 3)$.

With such derived capacities, we may model the wastewater system design problem of Figure 11.8 as the following fixed-charge network flow problem:

$$
\begin{aligned}
\min \quad & 21x_{1,2} + 30x_{1,3} + 22x_{2,3} + 58x_{2,4} + 43x_{3,4} && \text{(total cost)} \\
& + 1x_{3,9} + 49x_{4,3} + 63x_{4,8} + 44x_{5,6} + 51x_{5,7} \\
& + 56x_{6,7} + 94x_{6,8} + 82x_{7,4} + 1x_{7,9} + 2x_{8,9} \\
& + 240y_{1,2} + 350y_{1,3} + 200y_{2,3} + 750y_{2,4} + 610y_{3,4} \\
& + 3800y_{3,9} + 1840y_{4,3} + 780y_{4,8} + 620y_{5,6} + 800y_{5,7} \\
& + 500y_{6,7} + 630y_{6,8} + 1120y_{7,4} + 3800y_{7,9} + 2500y_{8,9}
\end{aligned}
$$

$$
\begin{aligned}
\text{s.t.} \quad & {-x_{1,2} - x_{1,3}} & = -27 && \text{(node 1)} \\
& x_{1,2} - x_{2,3} - x_{2,4} & = -3 && \text{(node 2)} && (11.21) \\
& x_{1,3} + x_{2,3} + x_{4,3} - x_{3,4} - x_{3,9} & = -14 && \text{(node 3)} \\
& x_{2,4} + x_{3,4} + x_{7,4} - x_{4,3} - x_{4,8} & = -36 && \text{(node 4)} \\
& {-x_{5,6} - x_{5,7}} & = -21 && \text{(node 5)} \\
& x_{5,6} - x_{6,7} - x_{6,8} & = -8 && \text{(node 6)} \\
& x_{5,7} + x_{6,7} - x_{7,4} - x_{7,9} & = -13 && \text{(node 7)} \\
& x_{4,8} + x_{6,8} - x_{8,9} & = 0 && \text{(node 8)} \\
& x_{3,9} + x_{7,9} + x_{8,9} & = 122 && \text{(node 9)}
\end{aligned}
$$

$$0 \le x_{1,2} \le 27 y_{1,2}, \quad 0 \le x_{1,3} \le 27 y_{1,3}, \quad 0 \le x_{2,3} \le 30 y_{2,3} \qquad \text{(switching)}$$

$$0 \le x_{2,4} \le 30 y_{2,4}, \quad 0 \le x_{3,4} \le 44 y_{3,4}, \quad 0 \le x_{3,9} \le 122 y_{3,9}$$

$$0 \le x_{4,3} \le 108 y_{4,3}, \quad 0 \le x_{4,8} \le 122 y_{4,8}, \quad 0 \le x_{5,6} \le 21 y_{5,6}$$

$$0 \le x_{5,7} \le 21 y_{5,7}, \quad 0 \le x_{6,7} \le 29 y_{6,7}, \quad 0 \le x_{6,8} \le 29 y_{6,8}$$

$$0 \le x_{7,4} \le 42 y_{7,4}, \quad 0 \le x_{7,9} \le 42 y_{7,9}, \quad 0 \le x_{8,9} \le 122 y_{8,9}$$

$$y_{i,j} = 0 \text{ or } 1 \qquad \text{all arcs } (i, j)$$

Heavier lines in Figure 11.8 indicate an optimal design. Lines $(1, 3)$, $(2, 3)$, $(4, 3)$, $(5, 7)$, and $(6, 7)$ should be constructed, along with the plant at node 7. Total cost is $\$15{,}571{,}000$.

The following digraph shows possible routes for cable television lines from broadcasting center node 1 to towns at nodes 3 and 4. Node 2 is a connection box that may or may not be included in the ultimate system. Numbers on arcs are the fixed cost of the corresponding cable line.

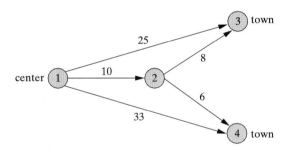

Formulate a fixed-charge network flow model to choose a least cost design providing service to both towns.

Modeling: The problem provides no explicit supplies, demands, or capacities, and there are no variable costs. Thus we may assume that demand $= 1$ at both cities, with 2 units of flow supplied at node 1. Capacity is then 2 on arc $(1,2)$, which might carry both flows, and 1 on all other arcs.

Substituting these values in form $\boxed{11.31}$ produces the integer linear programming model

$$
\begin{array}{lll}
\min & 10 y_{1,2} + 25 y_{1,3} + 33 y_{1,4} + 8 y_{2,3} + 6 y_{2,4} & \text{(cost)} \\
\text{s.t.} & -x_{1,2} - x_{1,3} - x_{1,4} = -2 & \text{(node 1)} \\
& x_{1,2} - x_{2,3} - x_{2,4} = 0 & \text{(node 2)} \\
& x_{1,3} + x_{2,3} = 1 & \text{(node 3)} \\
& x_{1,4} + x_{2,4} = 1 & \text{(node 4)}
\end{array}
$$

$$0 \le x_{1,2} \le 2 y_{1,2}, \quad 0 \le x_{1,3} \le y_{1,3}, \quad 0 \le x_{1,4} \le y_{1,4} \qquad \text{(switching)}$$

$$0 \le x_{2,3} \le y_{2,3}, \quad 0 \le x_{2,4} \le y_{2,4}$$

$$y_{1,2}, \; y_{1,3}, \; y_{1,4}, \; y_{2,3}, \; y_{2,4} = 0 \text{ or } 1$$

11.7 PROCESSOR SCHEDULING AND SEQUENCING MODELS

Scheduling is the allocation of resources over time. The enormous range of applications encompasses staffing activities like the ONB shift planning in Section 5.4, AA air crew scheduling in Section 11.3, Purdue final exam timetabling in Section 2.5, and construction project management in Section 9.7. In this section we introduce another very broad class: **processor scheduling** models that **sequence** a collection of jobs through a given set of processing devices.

EXAMPLE 11.12: NIFTY NOTES SINGLE-MACHINE SCHEDULING

We begin with the binder scheduling problem confronting a fictitious Nifty Notes copy shop. Just before the start of each semester, professors at the nearby university supply Nifty Notes with a single original of their class handouts, a projection of the class enrollment, and a due date by which copies should be available. Then the Nifty Notes staff must rush to print and bind the required number of copies before each class begins.

During the busy period each semester, Nifty Notes operates its single binding station 24 hours per day. Table 11.10 shows **process times**, **release times**, and **due dates** for the jobs $j = 1, \ldots, 6$ now pending at the binder:

$p_j \triangleq$ estimated process time (in hours) job j will require to bind

$r_j \triangleq$ release time (hour) at which job j has/will become available for processing (relative to time 0 = now)

$d_j \triangleq$ due date (hour) by which job j should be completed (relative to time 0 = now)

Notice that two jobs are already late ($d_j < 0$).

TABLE 11.10 Nifty Notes Scheduling Example Data

	Binder Job, j					
	1	**2**	**3**	**4**	**5**	**6**
Process time, p_j	12	8	3	10	4	18
Release time, r_j	–20	–15	–12	–10	–3	2
Due date, d_j	10	2	72	–8	–6	60

We wish to chose an optimal sequence in which to accomplish these jobs. No more than one can be in process at a time; and once started, a job must be completed before another can begin.

Single-Processor Scheduling Problems

Our Nifty Notes example is a very simple case of single-processor (or single-machine) scheduling.

> **11.32** **Single-processor** (or **single-machine**) **scheduling problems** seek an optimal sequence in which to complete a given collection of jobs on a single processor that can accommodate only one job at a time.

Usually (as in the Nifty Notes case), we also assume that **preemption** is not allowed. That is, one job cannot be interrupted to work on another.

Time Decision Variables

Since scheduling means assignment of resources over time, it is natural that one set of decision variables found in most models chooses job **start** or **completion times**.

> 11.33 │ A set of (continuous) decision variables in processor scheduling models usually determines either the start or the completion time(s) of each job on the processor(s) it requires.

In our Nifty Notes example we employ

$$x_j \triangleq \text{time binding starts for job } j \text{ (relative to time } 0 = \text{now)}$$

Then one set of constraints requires each job to start after the later of now (time $= 0$) and the hour it will be released for processing:

$$x_j \geq \max\{0, r_j\} \quad j = 1, \dots, 6 \quad \text{(earliest start)} \quad (11.22)$$

We could just as well have modeled in terms of completion times. Still, both start and completion times need not be decided because one can be computed from the other through

$$(\text{start time}) + (\text{process time}) = (\text{completion time})$$

(Recall that process times are given constants.)

Conflict Constraints and Disjunctive Variables

The central issue in processor scheduling is that only one job should be in progress on any processor at any time. If we know just start (or finish) times, it is not difficult to check for a violation or **conflict**. Still, it is not easy to write standard mathematical programming constraints that prevent conflicts.

For any pair of jobs j and j' that might conflict on a processor, the appropriate **conflict constraint** is either

$$(\text{start time of } j) + (\text{process time of } j) \leq \text{start time of } j'$$

or

$$(11.23)$$

$$(\text{start time of } j') + (\text{process time of } j') \leq \text{start time of } j$$

Either j must finish before j' begins, or vice versa. But only one possibility can hold, and to determine which, we must know whether j or j' starts first on the processor.

Often, operations research analysts deal with conflict prevention (11.23) outside the usual mathematical program format. Nonetheless, conflict avoidance requirements can be modeled explicitly with the aid of additional disjunctive variables.

> 11.34 │ A set of (discrete) **disjunctive variables** in processor scheduling models usually determines the sequence in which jobs are started on processors by specifying whether each job j is scheduled before or after each other j' with which it might conflict.

Then we can enforce conflict prevention (11.23) with pairs of linear constraints.

| 11.35 | A processor scheduling model with job start times x_j and process times p_j can prevent conflicts between jobs j and j' with **disjunctive constraint pairs**

$$x_j + p_j \leq x_{j'} + M(1 - y_{j,j'})$$
$$x_{j'} + p_{j'} \leq x_j + My_{j,j'}$$

where M is a large positive constant, and binary disjunctive variable $y_{j,j'} = 1$ when j is scheduled before j' on the processor and $= 0$ if j' is first.

To illustrate for our Nifty Notes example, define

$$y_{j,j'} \triangleq \begin{cases} 1 & \text{if } j \text{ binding comes before } j' \\ 0 & \text{if } j' \text{ binding comes before } j \end{cases}$$

Construction | 11.35 | yields constraints

$$\left. \begin{array}{l} x_j + p_j \leq x_{j'} + M(1 - y_{j,j'}) \\ x_{j'} + p_{j'} \leq x_j + My_{j,j'} \end{array} \right\} \quad j = 1, \ldots, 6; \quad j' > j \qquad (11.24)$$

As in many other cases, we consider only $j' > j$, to avoid listing the same pair twice.

To see how the constraints prevent conflicts, consider the specific case of $j = 2$, $j' = 6$. Using processing times from Table 11.10, the corresponding pair (11.24) is

$$x_2 + 8 \leq x_6 + M(1 - y_{2,6})$$
$$x_6 + 18 \leq x_2 + My_{2,6}$$

If job 2 is started before job 6, $y_{2,6} = 1$, and the first constraint keeps the start time for 6 after the finish of 2. The second constraint is also enforced. However, the big-M term $My_{2,6}$ makes it true for any x_2, x_6. On the other hand, if job 6 is started before job 2, so that $y_{2,6} = 0$, the first constraint of the pair is discounted and the second enforced.

SAMPLE EXERCISE 11.19: FORMULATING CONFLICT CONSTRAINTS

Formulate integer linear programming constraints for feasible schedules on a single processor with jobs $j = 1, \ldots, 3$ having process times 14, 3, and 7, respectively.

Modeling: We employ the decision variables

$$x_j \triangleq \text{start time of job } j$$

$$y_{j,j'} \triangleq \begin{cases} 1 & \text{if } j \text{ is scheduled before } j' \\ 0 & \text{otherwise} \end{cases}$$

Then conflict constraints | 11.35 | assure that only one job is processed at a time are

$$x_1 + 14 \leq x_2 + M(1 - y_{1,2})$$
$$x_2 + 3 \leq x_1 + My_{1,2}$$
$$x_1 + 14 \leq x_3 + M(1 - y_{1,3})$$

$$x_3 + 7 \leq x_1 + My_{1,3}$$
$$x_2 + 3 \leq x_3 + M(1 - y_{2,3})$$
$$x_3 + 7 \leq x_2 + My_{2,3}$$

Variable-type restrictions

$$x_1, x_2, x_3 \quad \geq 0$$
$$y_{1,2}, y_{1,3}, y_{2,3} = 0 \text{ or } 1$$

complete the constraints required.

Handling of Due Dates

Readers will note that we have not formulated any constraints enforcing due dates for various jobs. This could easily be done by adding conditions of the form

$$x_j + p_j \leq d_j$$

However, it is not standard to enforce such requirements because there may very well be no feasible schedule that meets all due dates. In Nifty Notes data of Table 11.10, for example, some due dates have already passed. it is far more customary to reflect due dates in the objective function as explained in the next subsection.

| 11.36 | **Due dates** in processor scheduling models are usually handled as goals to be reflected in the objective function rather than as explicit constraints. Dates that must be met are termed **deadlines** to distinguish. |

Processor Scheduling Objective Functions

One of the intriguing features of scheduling models is the wide variety of objective functions that may be appropriate.

| 11.37 | Denoting the start time of job $j = 1, \ldots, n$ by x_j, the process time by p_j, the release time by r_j and the due date by d_j, processor scheduling objective functions often minimize one of the following: |

Maximum completion time	$\max_j \{x_j + p_j\}$
Mean completion time	$\frac{1}{n} \Sigma_j (x_j + p_j)$
Maximum flow time	$\max_j \{x_j + p_j - r_j\}$
Mean flow time	$\frac{1}{n} \Sigma_j (x_j + p_j - r_j)\}$
Maximum lateness	$\max_j \{x_j + p_j - d_j\}$
Mean lateness	$\frac{1}{n} \Sigma_j (x_j + p_j - d_j)$
Maximum tardiness	$\max_j \{\max\{0, x_j + p_j - d_j\}\}$
Mean tardiness	$\frac{1}{n} \Sigma_j (\max\{0, x_j + p_j - d_j\})$

Maximum completion time is also known as **makespan**.

Total completion time, flow time, lateness, or tardiness may also be of interest, but optimization over each of these is equivalent to optimization over the corresponding mean measure because the total is constant n times the mean.

The completion time measures in definition 11.37 emphasize getting all jobs done as soon as possible. For example, the mean completion time version of our Nifty Notes example would have objective function

$$\min \quad \tfrac{1}{6}[(x_1 + 12) + (x_2 + 8) + (x_3 + 3) + (x_4 + 10) + (x_5 + 4) + (x_6 + 18)]$$

A corresponding optimal schedule has Nifty Notes binding jobs in sequence 3–5–2–4–1–6 with start times

$$x_1^* = 25, \quad x_2^* = 7, \quad x_3^* = 0, \quad x_4^* = 15, \quad x_5^* = 3, \quad x_6^* = 37 \qquad (11.25)$$

and mean completion time 23.67.

Completion time measures are particularly appropriate when there are a fixed number of jobs to complete and no more expected. Where the model relates to a more continuing operation, flow time may be more suitable. Flow time tracks the length of time that a job is in the system:

$$\text{flow time} \triangleq (\text{completion time}) - (\text{release time})$$

The idea is to minimize **work in process** so that inventory costs for partially finished goods are reduced.

The third category of measures becomes important when due dates are critical. Lateness counts both early and late jobs:

$$\text{lateness} \triangleq (\text{completion time}) - (\text{due date})$$

Tardiness considers only the late ones (positive lateness):

$$\text{tardiness} \triangleq \max\{0, (\text{lateness})\}$$

For example, minimizing maximum lateness in the Nifty Notes case produces the objective function

$$\min \quad \max \{(x_1 + 12 - 10), (x_2 + 8 - 2), (x_3 + 3 - 72), (x_4 + 10 + 8),$$
$$(x_5 + 4 + 6), (x_6 + 18 - 60)\} \qquad (11.26)$$

The corresponding optimal schedule has Nifty Notes binding jobs in sequence 4–2–5–3–1–6 with start times

$$x_1^* = 22, \quad x_2^* = 14, \quad x_3^* = 52, \quad x_4^* = 0, \quad x_5^* = 10, \quad x_6^* = 34 \qquad (11.27)$$

and maximum lateness $(22 + 12 - 10) = 24$ on job 1. Notice that this schedule differs significantly from the mean completion time schedule of (11.25). To decrease lateness, mean completion time has increased from 23.67 to 31.17.

SAMPLE EXERCISE 11.20: UNDERSTANDING PROCESSOR SCHEDULING OBJECTIVES

The following table shows the process times, release times, due dates, and scheduled start times for three jobs.

	Job 1	Job 2	Job 3
Process time	15	6	9
Release time	5	10	0
Due date	20	25	36
Scheduled start	9	24	0

Compute the corresponding value of each of the eight objective functions in definition 11.37 .

Analysis: Completion times (start + process) are

$$9 + 15 = 24, \quad 24 + 6 = 30, \quad \text{and } 0 + 9 = 9$$

Thus maximum completion time is max{24, 30, 9} = 30 and mean completion time is

$$\tfrac{1}{3}(24 + 30 + 9) = 21$$

Corresponding flow times (completion–release) are

$$24 - 5 = 19, \quad 30 - 10 = 20, \quad \text{and} \quad 9 - 0 = 9$$

Thus maximum flow time is max{19, 20, 9} = 20 and mean flow time is $\tfrac{1}{3}(19+20+9) =$ 16.

Lateness of the three jobs (completion–due date) is

$$24 - 20 = 4, \; 30 - 25 = 5, \quad \text{and} \quad 9 - 36 = -27$$

Thus maximum lateness is max{4, 5, −27} = 5 and mean lateness is $\tfrac{1}{3}(4+5-27) = -6$.
Finally, tardiness of the jobs (max{0, lateness}) is

$$\max\{0, 4\} = 4, \quad \max\{0, 5\} = 5, \quad \text{and} \quad \max\{0, -27\} = 0$$

Thus maximum tardiness is max{4, 5, 0} = 5 and mean tardiness is $\tfrac{1}{3}(4+5+0) = 3$.

ILP Formulation of Minmax Scheduling Objectives

Disjunctive constraints of formulation 11.35 can be combined with any of the mean objective forms in list 11.36 except tardiness to obtain an integer linear programming formulation of a processor scheduling problem. When tardiness or any of the minmax objectives are being optimized, however, the problem is an integer nonlinear program (INLP).

We can convert any of these INLP forms to the more tractable ILP by using the techniques of Section 4.6 (construction 4.13).

> 11.38   Any of the min max objective of list 11.37 can be linearized by introducing a new decision variable f to represent the objective function value, then minimizing f subject to new constraints of the form $f \geq$ each element in the maximize set. A similar construction can model tardiness by introducing new nonnegative tardiness variables for each job and adding constraints keeping each tardiness variable $\geq$ the corresponding lateness.

To illustrate, return to the Nifty Notes lateness objective of expression (11.26):

$$\min \quad \max \{(x_1 + 12 - 10), (x_2 + 8 - 2), (x_3 + 3 - 72), (x_4 + 10 + 8),$$
$$(x_5 + 4 + 6), (x_6 + 18 - 60)\}$$

We may convert to ILP form by introducing a new variable f and then solving

$$\min \quad f$$
$$\text{s.t.} \quad f \geq x_1 + 2$$
$$f \geq x_2 + 6$$
$$f \geq x_3 - 69$$
$$f \geq x_4 + 18$$
$$f \geq x_5 + 10$$
$$f \geq x_6 - 42$$
(all original constraints)

SAMPLE EXERCISE 11.21: LINEARIZING SCHEDULING OBJECTIVES

Using x_j to represent the scheduled start time of each job j in Sample Exercise 11.20, show how each of the following objective functions can be expressed in ILP format.

(a) Maximum completion

(b) Mean tardiness

Modeling: We apply construction 11.38 .

(a) To linearize maximum completion time, we introduce a new decision variable f and enforce new constraints to keep it as great as any completion time. Specifically, the formulation is

$$\min \quad f$$
$$\text{s.t.} \quad f \geq x_1 + 15$$
$$f \geq x_2 + 6$$
$$f \geq x_3 + 9$$
(all original constraints)

(b) To model tardiness, we introduce new decision variables $t_j \geq 0$ for each job j and force it to be $\geq$ lateness. Then the mean tardiness model is

$$\min \quad \tfrac{1}{3}(t_1 + t_2 + + t_3)$$
$$\text{s.t.} \quad t_1 \geq x_1 + 15 - 20$$
$$t_2 \geq x_2 + 6 - 25$$
$$t_3 \geq x_3 + 9 - 36$$
$$t_1, t_2, t_3 \geq 0$$
(all original constraints)

If job j is late, the corresponding new main constraint makes $t_j = $ lateness. If not, nonnegativity constraints $t_j \geq 0$ force tardiness $= 0$.

Equivalences among Scheduling Objective Functions

Different objective functions from list 11.37 do not always imply different optimal schedules.

 11.39   The mean completion time, mean flow time, and mean lateness scheduling objective functions are equivalent in the sense that an optimal schedule for one is also optimal for the others.

| 11.40 | An optimal schedule for the maximum lateness objective function is also optimal for maximum tardiness.

For example, the optimal mean completion time schedule (11.25) of our Nifty Notes example is also optimal for the mean flow time and mean lateness objective functions (principle 11.39). Schedule (11.27), which minimized maximum lateness, also minimizes maximum tardiness (principle 11.40).

To see why mean completion time and mean flow time are equivalent, we need only rearrange defining sums:

$$
\text{mean flow time} = \frac{1}{n} \sum_{j=1}^{n} (x_j + p_j - r_j)
$$

$$
= \frac{1}{n} \sum_{j=1}^{n} (x_j + p_j) - \frac{1}{n} \sum_{j=1}^{n} r_j
$$

$$
= (\text{mean completion time}) - \frac{1}{n} \sum_{j=1}^{n} r_j
$$

Expressed this way, it is apparent that the objective functions differ only by the constant last term. Adding or subtracting such a constant to the objective function cannot change what solutions are optimal. Similar arguments equate mean completion time and mean lateness.

Connection 11.40 between maximum lateness and maximum tardiness is also straightforward. If at least one job must be late in every schedule, maximum lateness = maximum tardiness. If no job has to be late, all schedules are optimal for the maximum tardiness objective, including any optimal for maximum lateness.

Job Shop Scheduling

In contrast to the single-processor case of definition 11.32 , job shop scheduling involves jobs that must be processed on several different machines.

| 11.41 | **Job shop scheduling problems** seek an optimal schedule for a given collection of jobs, each of which requires a known sequence of processors that can accommodate only one job at a time.

EXAMPLE 11.13: CUSTOM METALWORKING JOB SHOP

We illustrate job shop scheduling with a fictitious Custom Metalworking company which fabricates prototype metal parts for a nearby engine manufacturer. Figure 11.9 provides details on the 3 jobs waiting to be scheduled. First is a die requiring work on a sequence of 5 workstations: 1 (forging), then 2 (machining), then 3 (grinding), then 4 (polishing), and finally, 6 (electric discharge cutting). Job 2 is a cam shaft requiring 4 stations, and job 3 a fuel injector requiring 5 steps. Numbers in boxes indicate process times

$$
p_{j,k} \triangleq \text{process time (in minutes) of job } j \text{ on processor } k
$$

For example, job 1 requires 45 minutes at polishing workstation 4.

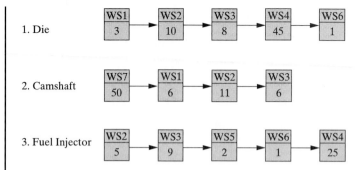

k	Workstations
1	Forging
2	Machining center
3	Grinding
4	Polishing
5	Drilling
6	Electric discharge
7	Heat treatment

FIGURE 11.9   Custom Metalworking Example Jobs

Any of the objective function forms in list $\boxed{11.37}$ could be appropriate for Custom Metalworking's scheduling. We will assume that the company wants to complete all 3 jobs as soon as possible (minimize maximum completion), so that workers can leave for a holiday.

Custom Metalworking Example Decision Variables and Objective

Job shop scheduling involves deciding when to start each step of each job on its processor. Thus start time decision variables $\boxed{11.33}$ are now indexed by both job and processor:

$$x_{j,k} \triangleq \text{start time of job } j \text{ on processor } k$$

Our assumed makespan scheduling objective can then be expressed as

$$\min \max\{x_{1,6} + 1, x_{2,3} + 6, x_{3,4} + 25\}$$

Notice that only the last step of each job is reflected. Completion of a multiprocessor job means completion of all steps.

Precedence Constraints

Steps of the various jobs being scheduled in a job shop must take place in the sequence given. That is, start times are subject to **precedence constraints**.

> $\boxed{11.42}$ The precedence requirement that job j must complete on processor k before activity on k' begins can be expressed as
>
> $$x_{j,k} + p_{j,k} \le x_{j,k'}$$
>
> where $x_{j,k}$ denotes the start time of job j on processor k, $p_{j,k}$ is the process time of j on k, and $x_{j,k'}$ is the start time of job j on processor k'.

Job shop models have precedence constraints $\boxed{11.42}$ between each step and its successor in each job. For instance, job 1 in Figure 11.9 implies precedence constraints

$$x_{1,1} + 3 \le x_{1,2}$$
$$x_{1,2} + 10 \le x_{1,3}$$
$$x_{1,3} + 8 \le x_{1,4}$$
$$x_{1,4} + 45 \le x_{1,6}$$

to maintain the required processing sequence.

SAMPLE EXERCISE 11.22: FORMULATING JOB SHOP PRECEDENCE CONSTRAINTS

A job shop must schedule product 1, which requires 12 minutes on machine 1 followed by 30 minutes on machine 2, and product 2, which requires 17 minutes on machine 1 followed by 29 minutes on machine 3. Formulate the implied precedence constraints in terms of the decision variables

$$x_{j,k} \triangleq \text{start time of product } j \text{ on machine } k$$

Modeling: There is one precedence constraint for each job because each has only two steps. In accord with $\boxed{11.42}$, those constraints are

$$x_{1,1} + 12 \le x_{1,2}$$
$$x_{2,1} + 17 \le x_{2,3}$$

Conflict Constraints in Job Shops

As in one machine case such as Nifty Notes, job shop models must also confront the possibility of conflicts—jobs scheduled simultaneously on the same processor. For example, in the Custom Metalworking example of Figure 11.9, jobs 1 and 2 may conflict at workstation 1, which both require. One must complete before the other can begin.

Paralleling $\boxed{11.34}$ and $\boxed{11.35}$, we may model conflicts by introducing the new discrete decision variables

$$y_{j,j',k} \triangleq \begin{cases} 1 & \text{if } j \text{ is scheduled before job } j' \text{ on processor } k \\ 0 & \text{otherwise} \end{cases}$$

$\boxed{11.43}$ Job shop models can prevent conflicts between jobs by introducing new disjunctive variables $y_{j,j',k}$ and constraint pair

$$x_{j,k} + p_{j,k} \le x_{j',k} + M(1 - y_{j,j',k})$$
$$x_{j',k} + p_{j',k} \le x_{j,k} + My_{j,j',k}$$

for each j, j' that both require any processor k. Here $x_{j,k}$ denotes the start time of job j on processor k, $p_{j,k}$ its process time, M is a large positive constant, and binary $y_{j,j',k} = 1$ when j is scheduled before j' on k and $= 0$ if j' is first.

For instance, the possible conflict between Custom Metalworking jobs 1 and 2 at workstation 1 produces constraint pair

$$x_{1,1} + 6 \le x_{2,1} + M(1 - y_{1,2,1})$$
$$x_{2,1} + 3 \le x_{1,1} + My_{1,2,1}$$

If job 1 uses the processor first, $y_{1,2,1} = 1$, and the first constraint is enforced. If job 2 comes first, $y_{1,2,1} = 0$, and the second constraint controls.

SAMPLE EXERCISE 11.23: FORMULATING JOB SHOP CONFLICT CONSTRAINTS

Return to the job shop of Sample Exercise 11.22 and formulate all constraints required to prevent conflicts.

Modeling: Conflicts can occur only on machine 1, which is the only one required for both products. Thus we require only one binary variable,

$$y_{1,2,1} \triangleq \begin{cases} 1 & \text{if product 1 is first on machine 1} \\ 0 & \text{if product 2 is first on machine 1} \end{cases}$$

and the needed constraints $\boxed{11.43}$ are

$$x_{1,1} + 12 \leq x_{2,1} + M(1 - y_{1,2,1})$$
$$x_{2,1} + 17 \leq x_{1,1} + My_{1,2,1}$$

Here $M = 12 + 17 = 29$ would be large enough to have the desired effect of discounting whichever constraint should not really apply.

Custom Metalworking Example Model

Combining our maximum completion time objective function with all required precedence and conflict constraints produces the following complete model of the Custom Metalworking example in Figure 11.9:

$$
\begin{array}{lll}
\min & \max\{x_{1,6} + 1, x_{2,3} + 6, x_{3,4} + 25\} & \text{(maximum completion)} \\
\text{s.t.} & x_{1,1} + 3 \leq x_{1,2} & \text{(job 1 precedence)} \\
& x_{1,2} + 10 \leq x_{1,3} & \\
& x_{1,3} + 8 \leq x_{1,4} & \\
& x_{1,4} + 45 \leq x_{1,6} & \\
& x_{2,7} + 50 \leq x_{2,1} & \text{(job 2 precedence)} \\
& x_{2,1} + 6 \leq x_{2,2} & \\
& x_{2,2} + 11 \leq x_{2,3} & \\
& x_{3,2} + 5 \leq x_{3,3} & \text{(job 3 precedence)} \\
& x_{3,3} + 9 \leq x_{3,5} & \\
& x_{3,5} + 2 \leq x_{3,6} & \\
& x_{3,6} + 1 \leq x_{3,4} & \\
& x_{1,1} + 6 \leq x_{2,1} + M(1 - y_{1,2,1}) & \text{(workstation 1 conflicts)} \\
& x_{2,1} + 3 \leq x_{1,1} + My_{1,2,1} & \\
& x_{1,2} + 10 \leq x_{2,2} + M(1 - y_{1,2,2}) & \text{(workstation 2 conflicts)} \\
& x_{2,2} + 11 \leq x_{1,2} + My_{1,2,2} & \\
& x_{1,2} + 10 \leq x_{3,2} + M(1 - y_{1,3,2}) & \\
& x_{3,2} + 5 \leq x_{1,2} + My_{1,3,2} & \\
& x_{2,2} + 11 \leq x_{3,2} + M(1 - y_{2,3,2}) & \\
& x_{3,2} + 5 \leq x_{2,2} + My_{2,3,2} & \\
& x_{1,3} + 8 \leq x_{2,3} + M(1 - y_{1,2,3}) & \text{(workstation 3 conflicts)}
\end{array}
$$

$$(11.28)$$

$$x_{2,3} + 6 \ \le x_{1,3} + My_{1,2,3}$$
$$x_{1,3} + 8 \ \le x_{3,3} + M(1 - y_{1,3,3})$$
$$x_{3,3} + 9 \ \le x_{1,3} + My_{1,3,3}$$
$$x_{2,3} + 6 \ \le x_{3,3} + M(1 - y_{2,3,3})$$
$$x_{3,3} + 9 \ \le x_{2,3} + My_{2,3,3}$$
$$x_{1,4} + 45 \le x_{3,4} + M(1 - y_{1,3,4}) \qquad \text{(workstation 4 conflicts)}$$
$$x_{3,4} + 25 \le x_{1,4} + My_{1,3,4}$$
$$x_{1,6} + 1 \ \le x_{3,6} + M(1 - y_{1,3,6}) \qquad \text{(workstation 6 conflicts)}$$
$$x_{3,6} + 1 \ \le x_{1,6} + My_{1,3,6}$$
$$\text{all } x_{j,k} \ge 0$$
$$\text{all } y_{j,j',k} = 0 \text{ or } 1$$

An optimal solution uses start times

$$x_{1,1}^* = 2, \quad x_{1,2}^* = 5, \quad x_{1,3}^* = 15, \quad x_{1,4}^* = 42, \quad x_{1,6}^* = 87$$
$$x_{2,7}^* = 0, \quad x_{2,1}^* = 50, \quad x_{2,2}^* = 56, \quad x_{2,3}^* = 67$$
$$x_{3,2}^* = 0, \quad x_{3,3}^* = 5, \quad x_{3,5}^* = 14, \quad x_{3,6}^* = 16, \quad x_{3,4}^* = 17$$

to complete all jobs in 88 minutes.

EXERCISES

11-1 A fertilizer plant can make a product by any of 3 processes. Using decision variables $x_j \triangleq$ number of units produced by process j, the following linear program computes a minimum cost way to produce 150 units with available resources:

$$\begin{array}{ll} \min & 15x_1 + 11x_2 + 18x_3 \\ \text{s.t.} & x_1 + x_2 + x_3 = 150 \\ & 2x_1 + 4x_2 + 2x_3 \le 310 \\ & 4x_1 + 3x_2 + x_3 \le 450 \\ & x_1, x_2, x_3 \ge 0 \end{array}$$

(a) ◒ Explain why this LP implicitly assumes that objective function coefficients are variable costs.

(b) ◒ ▭ Use class optimization software to solve the given LP.

(c) ◒ Formulate a revised ILP model implementing a requirement that only one of the 3 activities may be used.

(d) ◒ ▭ Use class optimization software to solve the ILP of part (c).

(e) ◒ Formulate a different revised ILP implementing a fixed setup cost of 400 charged for each activity used at all.

(f) ◒ ▭ Use class optimization software to solve the ILP of part (e).

(g) ◒ Formulate another revised ILP implementing a requirement that an activity can be used only if a minimum of 50 units are produced.

(h) ◒ ▭ Use class optimization software to solve the ILP of part (g).

11-2 A computer distributor can purchase workstations from any of 3 suppliers. Using decision variables $x_j \triangleq$ number of units purchased from supplier j, the following linear program computes a minimum cost way to purchase 300 workstations within applicable limits:

$$\begin{array}{ll} \min & 5x_1 + 7x_2 + 6.5x_3 \\ \text{s.t.} & x_1 + x_2 + x_3 = 300 \\ & 3x_1 + 5x_2 + 4x_3 \le 1500 \\ & 0 \le x_1 \le 200 \\ & 0 \le x_2 \le 300 \\ & 0 \le x_3 \le 200 \end{array}$$

(a) through (h) as in Exercise 11-1 using a setup cost of 100 and a minimum purchase of 125.

11-3 A retired executive has up to $8 million that he wishes to invest in apartment buildings. The following table shows the purchase price and the expected 10-year return (in millions of dollars) of the 4 buildings that he is considering.

	Building			
	1	**2**	**3**	**4**
Price	4.0	3.8	6.0	7.2
Return	4.5	4.1	8.0	7.0

The executive wishes to choose investments that maximize his total return. Assume that every option is available only on an all-or-nothing basis.

(a) ⊠ Formulate a knapsack ILP to choose an optimal investment plan.
(b) ⊠ Solve your knapsack by inspection.

11-4 The River City redevelopment authority wants to add a minimum of one thousand new parking spaces in the downtown area. The following table shows the estimated cost (in millions of dollars) of the 4 proposed projects and the number of spaces each would yield (in hundreds).

	Project			
	1	**2**	**3**	**4**
Cost	16	9	11	13
Spaces	8	3	6	6

The authority wants to meet its goal at minimum total cost. Assume that every project is available only on an all-or-nothing basis.

(a) Formulate a knapsack ILP to choose an optimal parking program.
(b) Solve your knapsack by inspection.

11-5 Silo State's School of Engineering is preparing a 5-year plan for building construction and expansion to accommodate new offices, laboratories, and classrooms. The Electrical Engineering faculty has proposed projects for all 3 available parcels of land: a digital circuits lab on the northwest parcel at $48 million, a faculty office annex on the southeast parcel at $20.8 million, and a computer vision lab on the northeast parcel at $32 million. The Mechanical Engineering faculty has 3 alternative proposals for the same northwest parcel: a large lecture room building at $28 million, a heat transfer lab at $44 million, and a computer-aided design expansion at $17.2 million. The Industrial Engineering faculty has only 2 proposals: a manufacturing research center on the southeast parcel at $36.8 million, and a tunnel from their current building to the new center for an additional $1.2 million. The Dean of Engineering scores

the impact of these projects as 9, 2, and 10 for the EE proposals, 2, 5, and 8 for the ME alternatives, 10 and 1 for the IE ideas. He wishes to allocate his available $100 million to maximize the total impact, selecting projects on an all-or-nothing basis with at most one per land parcel.

(a) ⊠ Describe verbally the budget limits, mutual exclusiveness constraints, and project dependency requirements of this capital budgeting problem.
(b) ⊠ Formulate a capital budgeting ILP to select an optimal combination of proposals.
(c) ⊠ ⌨ Use class optimization software to solve your ILP model.

11-6 A small pharmaceutical research laboratory must decide which product research activities to undertake with its $25 million available in each of the next two 5-year periods. The products optamine and feasibine are ready for field testing in the first 5-year period at a cost of $13 million and $14 million, respectively. Discretol and zeronex are at an earlier, development stage. Proposed development activities would require $4 million in the first 5 years for discretol and $3 million in the second. Corresponding values for zeronex are $2 million and $6 million. Field testing of the two products may also be chosen in the second 5-year period for $10 million and $15 million, respectively, if the corresponding development activity was undertaken in the first 5 years. The company wishes to maximize future profits from field-tested products, which it estimates at $510 million for optamine, $640 million for feasibine, $580 million for discretol, and $469 million for zeronex. All projects must be adopted on an all-or-nothing basis, and no more than one of the two development projects may be selected.

(a) through (c) as in Exercise 11-5.

11-7 The map that follows shows the locations of 8 applicants for low-power radio station licences and the approximate range of their signals.

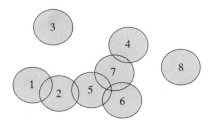

Regulators have scored the quality of applications on a scale of 0 to 100 as 45, 30, 84, 73, 80, 70, 61, and 91, respectively. They wish to select the highest-quality combination of applications that has no overlap in signal ranges.

(a) ⬦ Formulate this problem as a set packing ILP.
(b) ⬦ ▢ Use class optimization software to solve your ILP.

11-8 Time Sink Incorporated is about to begin selling its computer game software to college students in the midwest. The following table shows the states that could be covered by salespersons based at 4 possible locations, together with the annual sales (in thousands of dollars) expected if all those states were covered from that base.

	Base			
	Ames	**Beloit**	**Normal**	**Avon**
MN	×	×	—	—
IA	×	×	×	—
MO	×	—	×	—
WI	—	×	×	—
IN	—	—	×	×
KY	—	—	—	×
Sales	115	90	150	126

Time Sink wants to choose a collection of bases that maximizes total sales without assigning the same state to more than one base.

(a) and (b) as in Exercise 11-7.

11-9 The following map shows the 8 intersections at which automatic traffic monitoring devices might be installed. A station at any particular node can monitor all the road links meeting that intersection. Numbers next to nodes reflect the monthly cost (in thousands of dollars) of operating a station at that location.

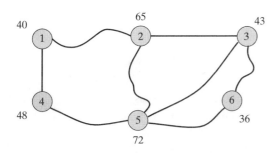

(a) ⬦ Formulate the problem of providing full coverage at minimum total cost as a set covering ILP.
(b) ⬦ ▢ Use class optimization software to solve your ILP of part (a).
(c) Revise your formulation of part (a) to obtain an ILP minimizing the number of uncovered road links while using at most 2 stations.
(d) ⬦ ▢ Use class optimization software to solve your ILP of part (c).

11-10 Top Tool Company wishes to hire part-time models to pass out literature about its machine tools at an upcoming trade show. Each of the 6 available models can work 2 of the 5 show days: Monday–Tuesday, Monday–Wednesday, Monday–Friday, Tuesday–Wednesday, Tuesday–Friday, and Thursday–Friday. If there is no more than one day separating their duty days, models will be paid $300, but the charge is $500 if two or more days intervene. Top Tool seeks a minimum cost way to have a least one model on all 5 days.

(a) through (d) as in Exercise 11-9 with the revised model minimizing the number of days uncovered using just 2 models.

11-11 Air Anton is a small commuter airline running 6 flights per day from New York City to surrounding resort areas. Flight crews are all based in New York, working flights to various locations and then returning on the next flight home. Taking into account complex work rules and pay incentives, Air Anton schedulers have constructed the 8 possible work patterns detailed in the following table. Each row of the table marks the flights that could be covered in a particular pattern and the daily cost per crew (in thousands of dollars).

Work Pattern	**Flight**						**Cost**
	1	**2**	**3**	**4**	**5**	**6**	
1	—	×	—	×	—	—	1.40
2	×	—	—	—	—	×	0.96
3	—	×	—	×	×	—	1.52
4	—	×	—	—	×	×	1.60
5	×	—	×	—	—	×	1.32
6	—	—	×	—	×	—	1.12
7	—	—	—	×	—	×	0.84
8	×	—	×	×	—	—	1.54

The company wants to choose a minimum total cost collection of work patterns that covers all flights exactly once.

(a) ⊗ Formulate this problem as a set partitioning ILP.

(b) ⊗ 💻 Use class optimization software to solve your ILP.

11-12 A special court commission appointed to resolve a bitter fight over legislative redistricting has proposed 6 combinations of the 5 disputed counties that could form new districts. The following table marks the counties composing each proposed district and shows the district's deviation from the equal-population norm.

County	District					
	1	2	3	4	5	6
1	×	—	—	—	×	×
2	×	—	×	—	—	×
3	—	—	×	×	×	—
4	—	×	—	×	—	—
5	—	×	—	—	—	×
Deviation	0.5	0.5	0.6	1.3	0.7	1.2

The court wants to select a minimum total deviation collection of districts that includes each county exactly once.

(a) and (b) as in Exercise 11-11.

11-13 Mogul Motors is planning a major overhaul of its automobile assembly plants as it introduces 4 new models. Exactly one exisiting plant must be converted to assemble each model. The following table shows for each model and plant the cost (in millions of dollars) of modifications at the plant to produce the model. Those marked with a dash reflect plants not large enough to accommodate the needed activity.

Model	Plant			
	1	2	3	4
1	18	26	—	31
2	—	50	22	—
3	40	29	52	39
4	—	—	43	46

Mogul seek a minimum total cost way to make the conversion.

(a) ⊗ Formulate this problem as a linear assignment LP.

(b) ⊗ Explain why a binary optimal solution is guaranteed even though your model is a linear program.

(c) ⊗ 💻 Use class optimization software to solve your assignment LP.

11-14 The sister communities program pairs cities in Russia with cities in the United States that have a similar size and economic base. Visits are then exchanged between the sister communities to improve international understanding. The following table shows the program's compatibility scores (0 to 100) for the 4 U.S. and 4 Russian cities about to join the program.

U.S.	Russian			
	1	2	3	4
1	80	65	83	77
2	54	87	61	66
3	92	45	53	59
4	70	61	81	76

Sister communities seeks a maximum total compatibility pairing.

(a) through (c) as in Exercise 11-13.

11-15 Sandbox State University is rearranging the locations of 3 equal-sized academic departments to provide for better faculty communication. The following tables shows the estimated number of person-to-person contacts per month between members of the various faculties, and the distances (in thousands of feet) between available office locations.

	Interaction	
	English	Math
History	20	12
English	—	14

	Distance	
	2	3
1	3	6
2	—	1

Sandbox State wants to place one department in each location in a way that minimizes total distance traveled for faculty interactions.

(a) ⊗ Formulate this problem as a quadratic assignment INLP.

(b) ⊗ Explain why the objective function in this problem must be quadratic rather than linear.

(c) ⊗ Compute an optimal assignment in your INLP by inspection.

11-16 The River City Operations Research society is planning a meeting that will have 2 morning and 2 afternoon sessions running at the same time. The 4 sessions will be on LP, NLP, ILP, and INLP, respectively, but times have not yet been fixed. The following table shows the estimated number of attendees who would like to be able to attend both of each combination of sessions.

	NLP	ILP	INLP
LP	10	30	14
NLP	—	5	8
ILP	—	—	18

The society would like to arrange sessions to minimize the number of persons who cannot attend a desired pair of sessions because they occur simultaneously.

(a) through (c) as in Exercise 11-15.

11-17 A warehouse facility has packing stations at both its front and back entrances. The following table shows the number of ton-feet (in thousands) of materials handling that would be required to move each of the 6 pending jobs to either of the 2 stations, along with the number of hours packing that would be required at whichever station does the work.

			Job			
	1	2	3	4	5	6
Front	21	17	10	30	40	22
Back	13	18	29	24	33	29
Time	44	60	51	80	73	67

Schedulers seek a minimum handling plan that completes all packing within the 200 hours available at the front station and 190 hours available at the back. Assume that jobs must go entirely to a single packing station.

(a) ◇ Formulate this problem as a generalized assignment ILP.
(b) ◇ Explain why this is a generalized rather than an ordinary assignment problem.
(c) ◇ ▭ Use class optimization software to solve your ILP.

11-18 Three professional baseball teams are trying to find places for 6 available players within their remaining salary limits of $35 million, $20 million, and

$26 million, respectively. The following table shows how valuable each player would be to each team on a scale of 0 to 10, and the player's current annual salary (in millions of dollars).

		Value		
Player	1	2	3	Salary
1	8	7	10	10
2	7	8	6	13
3	5	4	6	8
4	6	3	3	6
5	8	7	6	15
6	10	9	10	22

We want to find a maximum total score allocation of players to teams that fits with in salary limits.

(a) through (c) as in Exercise 11-17.

11-19 Military commanders are planning the command structure for 6 new radar stations. Three commanders will each be in charge of two of the stations. The following table shows the projected cost (in millions of dollars) of building the necessary communication links to connect jointly commanded locations.

	2	3	4	5	6
1	42	65	29	31	55
2	—	20	39	40	21
3	—	—	68	55	22
4	—	—	—	30	39
5	—	—	—	—	47

Planners seek a minimum cost way to organize the command.

(a) ◇ Formulate this problem as a matching ILP.
(b) ◇ Explain why this is not an assignment problem.
(c) ◇ ▭ Use class optimization software to solve your ILP.

11-20 Awesome Advertising manages the television promotion of a variety of products. In the next few months they are planning to cross-advertise six of their items by running interlocking television ads that mention both products. The following table shows Awesome's estimate of the number of viewers (in millions) who might be interested jointly in each pair of products.

	2	3	4	5	6
1	7	8	6	14	15
2	—	18	20	5	8
3	—	—	19	9	10
4	—	—	—	6	11
5	—	—	—	—	16

Awesome wants to find a product pairing that maximizes the appeal, with each product in exactly one pair.

(a) through (c) as in Exercise 11-19.

11-21 Engineers are designing a fixed route to be followed by automatic guided vehicles in a large manufacturing plant. The following table shows the east–west and north–south coordinates of the 6 stations to be served by vehicles moving continuously around the same route.

	1	2	3	4	5	6
E/W	20	40	180	130	160	50
N/S	90	70	20	40	10	80

Since traffic must move along east–west or north–south aisles, designers seek a route of shortest total rectilinear length (see Section 4.6).

(a) ⊘ Explain why this problem can be viewed as a traveling salesman problem.
(b) ⊘ Explain why distances in this problem are symmetric, and compute a matrix of rectilinear distances between all pairs of points.
(c) ⊘ Formulate this problem (incompletely) as an ILP with main constraints requiring only that every point be touched by 2 links of the route.
(d) ⊘ ▭ Use class optimization software to show that your ILP of part (c) produces a subtour 1–2–6–1.
(e) ⊘ Formulate a subtour elimination constraint that precludes the solution of part (d).
(f) ⊘ ▭ Use class optimization software to show that an optimal route results when your subtour elimination constraint is added to the formulation of part (c).
(g) ⊘ Formulate this problem as a quadratic assignment INLP.

11-22 An oil company currently has 5 platforms drilling off the Gulf coast of the United States. The following table shows the east–west and north–south coordinates of their shore base at point 0 and all the platform locations.

	0	1	2	3	4	5
E/W	80	10	60	30	85	15
N/S	95	15	70	10	75	30

Each day a helicopter delivers supplies by flying from the base to all platforms and then returning to base. Supervisors seek the route of shortest total length.

(a) through (f) as in Exercise 11-21 using straight-line (Euclidean) distance and subtour 0–2–4–0.

11-23 Every week Mighty Mo Manufacturing makes one production run of each of its 4 different kinds of metal cookware. Setup times to make any particular product vary depending on what was produced most recently. The following table shows the time (in hours) required to convert from any product to any other.

	1	2	3	4
1	—	4.2	1.5	6.5
2	5.0	—	8.5	1.0
3	1.2	7.7	—	8.0
4	5.5	1.8	6.0	—

Mighty Mo would like to find the production sequence that minimizes total setup time.

(a) ⊘ Explain why this problem can be viewed as an asymmetric traveling salesman problem.
(b) ⊘ Formulate this problem (incompletely) as a linear assignment problem to choose a successor for each product.
(c) ⊘ ▭ Use class optimization software to show that your ILP of part (b) produces a subtour 1–3–1.
(d) ⊘ Formulate a subtour elimination constraint that precludes the solution of part (c).
(e) ⊘ ▭ Use class optimization software to show that an optimal route results when your subtour elimination constraint is added to the formulation of part (b).
(f) ⊘ Formulate this problem as a quadratic assignment INLP.

11-24 Every weekday afternoon, at the height of the rush hour, a bank messenger drives from the a bank's central office to its 3 branches and returns with noncash records of the day's activity. The following figure shows the freeway routes that are not hopelessly clogged by traffic at that hour and the estimated driving time for each (in minutes).

The bank would like to find a route that minimizes total travel time.

(a) through (e) as in Exercise 11-23 with subtour 0–3–0.

11-25 Gotit Grocery Company is considering 3 locations for new distribution centers to serve it customers in 4 nearby cities. The following table shows the fixed cost (in millions of dollars) of opening each potential center, the number (in thousands) of truckloads forecasted to be demanded at each city over the next 5 years, and the transportation cost (in millions of dollars) per thousand truckloads moved from each center location to each city.

Center	Fixed Cost	City 1	2	3	4
1	200	6	5	9	3
2	400	4	3	5	6
3	225	5	8	2	4
Demand	—	11	18	15	25

Gotit seeks a minimum cost distribution system assuming any distribution center can meet any or all demands.

(a) ⬧ Formulate this problem as a facilities location ILP.
(b) ⬧ 💻 Use class optimization software to solve your ILP.

11-26 Basic Box Company is considering 5 new box designs of different sizes to package 4 upcoming lines of computer monitors. The following table shows the wasted space that each box would have if used to package each monitor. Missing values indicate a box that cannot be used for a particular monitor.

Box	Monitor 1	2	3	4
1	5	—	10	—
2	20	—	—	25
3	40	—	40	30
4	—	10	70	—
5	—	40	80	—

Basic wants to choose the smallest number of box designs needed to pack all products and to decide which box design to use for each monitor, to minimize waste.

(a) and (b) as in Exercise 11-25. (*Hint:* Use a large positive constant for fixed charges.)

11-27 The figure that follows shows 5 pipelines under consideration by a natural gas company to move gas from its 2 fields to its 2 storage areas. The numbers on the arcs show the number of miles of line that would have to be constructed at $100,000 per mile.

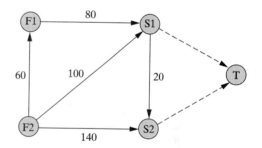

The figure also shows that storage facilities are both already connected to the company's main terminal through existing lines. An estimated 800 million cubic feet must be shipped each year from field 1 to the terminal, and 600 million from field 2. Variable shipping cost is $2000 per million cubic feet on each link of the network, and all links have an annual capacity of 1 billion cubic feet. The company wants a minimum total annual cost system for the required shipping.

(a) ⬧ Formulate this problem as a network design ILP.
(b) ⬧ 💻 Use class optimization software to solve you ILP.

11-28 Dandy Diesel manufacturing company assembles diesel engines for heavy construction equipment. Over the next 4 quarters the company expects to ship 40, 20, 60, and 15 units, respectively, but no more than 50 can be assembled in any quarter. There is a fixed cost of $2000 each time the line is setup for production, plus $200 per unit assembled. Engines may be held over in inventory at the plant for $100 per unit per month. Dandy seeks a minimum total cost production plan for the 4 quarters, assuming that there in no beginning or ending inventory.

(a) and (b) as in Exercise 11-27. (*Hint:* Create nodes for each quarter and a common source node for production arcs.)

11-29 Top-T shirt company imprints T-shirts with cartoons and celebrity photographs. For each of their 4 pending contracts, the following table shows the number of days of production required, the earliest day the order can begin, and the day the order is due.

	1	2	3	4
Production	10	3	16	8
Earliest	0	20	1	12
Due Date	12	30	20	21

The company wants to design an optimal schedule assuming that contracts can be processed in any sequence but that production cannot be interrupted once a job has started.

(a) ⊗ Ignoring objective functions for the moment, formulate constraints of a single-machine ILP to select an optimal start time for each contract.

(b) ⊗ Evaluate each of the 8 objective function in principle 11.37 for the schedule with start times 2, 20, 23, and 12 for the four contracts, respectively.

(c) ⊗ Extend your constraints of part (a) to formulate an ILP to compute a minimum mean completion time schedule.

(d) ⊗ ▭ Use class optimization software to solve your ILP of part (c).

(e) ⊗ Without actually solving, list the other objective functions of 11.37 for which your schedule of part (d) must be optimal.

(f) ⊗ Extend your constraints of part (a) to formulate an ILP to compute a minimum maximum lateness schedule.

(g) ⊗ ▭ Use class optimization software to solve your ILP of part (f).

(h) ⊗ Without actually solving, list the other objective functions of 11.37 for which your schedule of part (g) must be optimal.

11-30 Sarah is a graduate student who must make 4 large experimental runs on her personal computer as part of her thesis research. The jobs require virtually all the computer's resources, so only one can be processed at a time and none can be interrupted once it has begun. The following table shows the number of days of computing each job will require, the earliest that all data will be available, and the day Sarah has promised the result to her thesis advisor.

	1	2	3	4
Time	15	8	20	6
Earliest	0	0	10	10
Promise	20	20	30	20

Before beginning any work, Sarah wants to compute an optimal schedule. Assume promised times are only targets.

(a) through (h) as in Exercise 11-29, evaluating the schedule with start times 8, 0, 23, and 43, respectively.

11-31 Three new jobs have just arrived at Fancy Finishing's main furniture restoration shop. The following table shows the sequence that each must follow through the company's 3 finish removal processes and the time required for each.

Job	Sequence	Process Time 1	2	3
1	1–2–3	10	3	14
2	1–3–2	2	4	1
3	2–1–3	12	6	8

Once a process is begun, it cannot be interrupted. Although the shop was empty when the new jobs arrived, Fancy expects more in the next few days. To maintain efficiency, they seek a schedule that minimizes the average time that a job is in the shop.

(a) ⊗ Ignoring objective functions for the moment, formulate constraints of a job shop ILP to select an optimal start time for each job on each machine.

(b) ⊗ Which of the 8 objective functions in 11.37 is appropriate for this problem?

(c) ⊗ Complete an ILP model by introducing that objective.

(d) ⊗ ▭ Use class optimization software to solve your ILP model for an optimal schedule.

11-32 A team of auditors has divided itself into 3 groups, each to examine one category of records. Each group will review their speciality area for all 3 subsidiaries of the client being audited, but the required sequence and times differ, as shown in the following table.

Subsidiary	Sequence	Group Time		
		1	2	3
1	1–3–2	4	5	12
2	2–1–3	6	18	3
3	3–2–1	5	7	3

Once a group starts on a subsidiary, it should finish all work either before it moves to another or before a different group begins on theirs. The team seeks a schedule that will complete all work at the earliest possible time.

(a) through (d) as in Exercise 11-31.

11-33 With the addition of a new plant, Monsanto[9] now has more capacity than it needs to manufacture its main chemical product. Numerous reactors $i = 1, \ldots, m$ can be operated at a variety $j = 1, \ldots, n$ of discrete combinations of settings for feed rate, reactor velocity, and reactor pressure. Both the production yield $p_{i,j}$ and the operating cost $c_{i,j}$ vary with reactor and setting. Formulate an ILP model to find the least cost way to fulfill total production target b in terms of the decision variables ($i = 1, \ldots, m$; $j = 1, \ldots, n$)

$$x_{i,j} \triangleq \begin{cases} 1 & \text{if reactor } i \text{ operates at setting } j \\ 0 & \text{otherwise} \end{cases}$$

11-34 W.R. Grace[10] strip mines phosphates in strata numbered from $i = 1$ at the top to $i = n$ at the deepest level. Each stratum must be removed before the next can be mined, but only some of the layers contain enough suitable minerals to justify processing into the company's three products: pebble, concentrate, and flotation feed ($j = 1, 2, 3$). The company can estimate from drill samples the quantity $a_{i,j}$ of product j available in each stratum i, the fraction $b_{i,j}$ of BPL (a measure of phosphate content) in the part of i suitable for j, and the corresponding fraction $p_{i,j}$ of pollutant chemicals. They wish to choose a mining plan that maximizes the product output while keeping the average fraction BPL of material processed for each product j at least b_j and the average pollution fraction at most p_j. Formulate an ILP model of this mining problem using the decision variables ($i = 1, \ldots, n$)

$$x_i \triangleq \begin{cases} 1 & \text{if stratum } i \text{ is removed} \\ 0 & \text{otherwise} \end{cases}$$

$$y_i \triangleq \begin{cases} 1 & \text{if stratum } i \text{ is processed} \\ 0 & \text{otherwise} \end{cases}$$

11-35 Ault Food Limited[11] is planning the production and distribution system for its new line of food products. Plants may be opened at any of sites $i = 1, \ldots, 7$, and warehouses at locations $j = 1, \ldots, 13$, to meet demands d_k at customer regions $k = 1, \ldots, 219$. Each plant costs $50 million to open and produces up to 30 thousand cases per year. Warehouses cost $12 million to open and handle up to 10 thousand cases per year. Transportation costs are $r_{i,j}$ per case for rail shipment from plant i to warehouse j, and $t_{j,k}$ per case for trucking from warehouse j to customer k. No direct shipments from the plants are allowed. Formulate ILP model to decide which facilities to open and how to service customers using the decision variables ($i = 1, \ldots, 7$; $j = 1, \ldots, 13$; $k = 1, \ldots, 219$)

$$x_{i,j,k} \triangleq \text{thousand of cases produced at plant } i \text{ and shipped to customer } k \text{ via warehouse } j$$

$$y_i \triangleq \begin{cases} 1 & \text{if plant } i \text{ is opened} \\ 0 & \text{otherwise} \end{cases}$$

$$w_j \triangleq \begin{cases} 1 & \text{if warehouse } j \text{ is opened} \\ 0 & \text{otherwise} \end{cases}$$

[9]Based on R. R. Boykin (1985), "Optimizing Chemical Production at Monsanto," *Interfaces, 15:1*, 88–95.

[10]Based on D. Klingman and N. Phillips (1988), "Integer Programming for Optimal Phosphate-Mining Strategies," *Journal of the Operational Research Society, 9*, 805–809.

[11]Based on J. Pooley (1994), "Integrated Production and Distribution Facility Planning for Ault Foods," *Interfaces, 24:4*, 113–121.

11-36 Space structures[12] designed for zero gravity have no structural weight to support and nothing to which a foundation can be attached. The structure needs only to withstand vibrations in space. This is accomplished by replacing truss members with dampers at a given number p sites among $j = 1, \ldots, n$ candidate locations throughout the structure. Engineering analysis can identify the principal strain modes $i = 1, \ldots, m$, and estimate fractions $d_{i,j}$ of total modal strain energy imparted in mode i to truss site j. The best design places dampers to absorb as much energy as possible. Specifically we want to maximize the minimum total of $d_{i,j}$ reaching chosen sites over all modes i. Formulate an ILP model to choose an optimal design in terms of p, the $d_{i,j}$, and the decision variables ($j = 1, \ldots, n$)

$$x_j \triangleq \begin{cases} 1 & \text{if a damper is placed at } j \\ 0 & \text{otherwise} \end{cases}$$

$$z \triangleq \text{smallest modal } d\text{-total}$$

(*Hint:* Maximize z.)

11-37 The National Cancer Institute[13] has received proposals from 22 states to participate in its newest smoking intervention study. The first $j = 1, \ldots, 5$ are from the Northeast region, the next $j = 6, \ldots, 11$ from the Southeast, numbers $j = 12, \ldots, 17$ from the Midwest, and the last $j = 18, \ldots, 22$ from the West. At least 3 are to be selected from each region. Each proposal has been evaluated and rated with a merit score r_j based on rankings by a panel of experts. Selected project budgets b_j (in millions of dollars) must total no more than the $15 million available for the study, and the number of smokers s_j (in millions) living in chosen states must sum to at least 11 million. Proposals $j = 2, 7, 11, 19$ come from states in the highest quartile with respect to the fraction of the population that smokes; proposals $j = 1, 4, 13, 14, 21$ come from the lowest quartile. At least 2 states must be selected from each of these outlier groups. Formulate an ILP model to choose a maximum merit feasible set of proposals to fund

using the decision variables ($j = 1, \ldots, 22$)

$$x_j \triangleq \begin{cases} 1 & \text{if proposal } j \text{ is selected} \\ 0 & \text{otherwise} \end{cases}$$

11-38 Every year the city of Montreal[14] must remove large quantities of snow from sectors $i = 1, \ldots, 60$ of the city. Each sector is assigned to one of sites $j = 1, \ldots, 20$ as its primary disposal point. From prior-year history, planners have been able to estimate the expected volume of snowfall f_i (in cubic meters) in each sector and the capacity u_j (in cubic meters) of each disposal site. They also know the travel distance $d_{i,j}$ from each sector to each disposal site. Removal rates also have to be considered. The total of hourly removal rates r_i (in m³/hr) associated with sectors assigned to any disposal site must not exceed its receiving rate b_j (in m³/hr). Formulate an ILP model to select an assignment that minimizes total distance times volume moved using the decision variables ($i = 1, \ldots, 60; j = 1, \ldots, 20$)

$$x_{i,j} \triangleq \begin{cases} 1 & \text{if sector } i \text{ is assigned to disposal } j \\ 0 & \text{otherwise} \end{cases}$$

11-39 Highway maintenance in Australia[15] is performed by crews operating out of maintenance depots. The Victoria region is planning a major realignments of its depots to provide more effective service to the $i = 1, \ldots, 276$ highway segments in the region. A total of 14 sites must be selected from among the possible depot locations $j = 1, \ldots, 36$. Review of the possibilities has determined the indicators

$$a_{i,j} \triangleq \begin{cases} 1 & \text{if a depot at } j \text{ is close enough} \\ & \text{to provide good service to segment } i \\ 0 & \text{otherwise} \end{cases}$$

Not all segments will be able to receive this good service, so planners wish to select 14 locations to minimize to sum of service requirements s_i at segments with inadequate service. Formulate an ILP model to select an optimal collection of depots using the

[12]Based on R. K. Kincaid and R. T. Berger (1993), "Damper Placement for the CSI-Phase I Evolutionary Model," 34th AIAA/ASME/ASCE/AHS/ASC Structures, Structural Dynamics and Materials Conference, part 6, 3086–3095.

[13]Based on N. G. Hall, J. C. Hershey, L. G. Kessler, and R. C. Spotts (1992), "A Model for Making Project Funding Decisions at the National Cancer Institute," *Operations Research, 40,* 1040–1052.

[14]Based on J. F. Campbell and A. Langevin (1995), "The Snow Disposal Assignment Problem," *Journal of the Operational Research Society, 46,* 919–929.

[15]Based on G. Rose, D. W. Bennett, and A. T. Evans (1992), "Locating and Sizing Road Maintenance Depots," *European Journal of Operational Research, 63,* 151–163.

decision variables $(i = 1, \ldots, 276; j = 1, \ldots, 36)$

$$x_j \triangleq \begin{cases} 1 & \text{if depot } j \text{ is selected} \\ 0 & \text{otherwise} \end{cases}$$

$$y_i \triangleq \begin{cases} 1 & \text{if segment } i \text{ goes inadequately} \\ & \text{covered} \\ 0 & \text{otherwise} \end{cases}$$

11-40 Mobil Oil Corporation[16] serves 600,000 customers with 430 tanktrucks operating out of 120 bulk terminals. As illustrated in the following figure, tanktrucks have several compartments $c = 1, \ldots, n$ of varying capacity u_c.

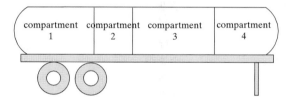

The final stage of distribution planning is to allocate outgoing gasoline products $p = 1, \ldots, m$ to compartments. The various products ordered are placed in one or more compartments, but each compartment can contain only one product. To avoid overfilling, the total gallons of any product loaded can be reduced from the ordered volume v_p by up to b_p gallons. The loading procedure seeks to minimize the sum of these underloadings while meeting all other requirements. Formulate an ILP model of this loading problem using the decision variables $(p = 1, \ldots, m; c = 1, \ldots, n)$

$$x_{p,c} \triangleq \text{gallons of product } p \text{ loaded in} \\ \text{compartment } c$$

$$y_{p,c} \triangleq \begin{cases} 1 & \text{if product } p \text{ is loaded in} \\ & \text{compartment } c \\ 0 & \text{otherwise} \end{cases}$$

$$z_p \triangleq \text{gallons underloading of product } p$$

11-41 Each of the 11 nurses[17] at Rosey Retirement Home works a total of 10 days within each 2-week period, alternating between a compatible pair of weekly schedules taken from those depicted in the following table $(1 = \text{work}, 0 = \text{off})$.

Day	Shift					
	1	2	3	4	5	6
1	1	1	0	1	1	0
2	1	0	1	1	1	1
3	1	1	1	1	1	1
4	1	1	1	1	0	1
5	0	1	1	0	1	1
6	0	0	0	1	1	1
7	0	0	0	1	1	1

Compatible weekly schedule pairs are $C \triangleq \{(1, 4), (1, 5), (1, 6), (2, 4), (2, 5), (2, 6), (3, 4), (3, 5), (3, 6), (4, 1), (4, 2), (5, 2), (5, 3), (6, 3)\}$, and the number working each shift should be constant from week to week. At least r_d nurses must be on duty on days $d = 1, \ldots, 7$, but excesses are allowed if mandated by the shift patterns. Rosey's objective is to minimize the sum of any such overstaffing. Formulate an ILP model to decide an optimal cyclic schedule for the staff of 11 using the decision variables $((i, j) \in C, d = 1, \ldots, 7)$

$$x_{i,j} \triangleq \text{number of nurses working shift } i \text{ then shift } j$$

$$z_d \triangleq \text{excess number of nurses scheduled on day } d$$

11-42 Regional Bell telephone operating companies,[18] which buy many products from a much smaller number of suppliers, often solicit discounts based on the total dollar volume of business awarded to particular suppliers. For example, suppliers $i = 1, \ldots, 25$ might be required to quote base prices $p_{i,j}$ for needed products $j = 1, \ldots, 200$, together with upper limits $b_{i,k}, k = 1, \ldots, 5$ for ranges on total dollar volume and corresponding discount fractions $d_{i,k}$ that increase with k. Then the actual cost to the telephone company of supplier i's goods will be $(1 - d_{i,1})$ times the total base price of those goods if that total dollar value falls within the interval $[u_{i,0}, u_{i,1}], (1 - d_{i,2})$ if the total dollar volume falls within $[u_{i,1}, u_{i,2}]$, and so on (assume that $u_{i,0} \triangleq 0, u_{i,5} \geq$ any feasible dollar volume). The company wants to choose bids to equal or exceed all required product quantities r_j at least total discounted cost. Formulate an ILP model

[16]Based on G. G. Brown, C. J. Ellis, G. W. Graves, and D. Ronen (1987), "Real-Time, Wide Area Dispatch of Mobil Tank Trucks," *Interfaces, 17:1*, 107–120.

[17]Based on E. S. Rosenbloom and N. F. Goertzen (1987), "Cyclic Nurse Scheduling," *European Journal of Operational Research, 31*, 19–23.

[18]Based on P. Katz, A. Sadrian, and P. Tendick (1994), "Telephone Companies Analyze Price Quotations with Bellcore's PDSS Software," *Interfaces, 24:1*, 50–63.

of this volume discount problem using the decision variables ($i = 1, \ldots, 25; j = 1, \ldots, 200; k = 1, \ldots, 5$)

$x_{i,j} \triangleq$ quantity of product j purchased from supplier i

$w_{i,k} \triangleq$ dollar volume of goods from supplier i when discount range k applies

$$y_{i,k} \triangleq \begin{cases} 1 & \text{if discount rang } k \text{ applies for supplier } i \\ 0 & \text{otherwise} \end{cases}$$

11-43 The small Adele[19] textile company knits products $p = 1, \ldots, 79$ on a variety of machines $m = 1, \ldots, 48$ to meet know output quotas q_p pounds for the next week. The variable cost per pound of making product p on machine m is known $c_{m,p}$. Machines operate with changeable cylinder types $j = 1, \ldots, 14$, which have different combinations of knitting needles, and thus yield different quantities $a_{m,j,p}$ (pounds) of product p per hour on machine m. A total of 100 hours is available on each machine over the next week, but a setup time $s_{m,j}$ must be deducted for each cylinder type j used on each machine m. Adele wants to find a minimum total variable cost schedule that conforms to all constraints. Formulate an ILP model of this production scheduling problem using decision variables ($m = 1, \ldots, 48$; $j = 1, \ldots, 14; p = 1, \ldots, 79$)

$x_{m,j,p} \triangleq$ pounds of product p made on machine m with cylinder type j

$$y_{m,j} \triangleq \begin{cases} 1 & \text{if cylinder type } j \text{ is used on machine } m \\ 0 & \text{otherwise} \end{cases}$$

11-44 Mail Order Mart (MOM)[20] will ship q_j pounds of small-order novelty goods over the next week to regions $j = 1, \ldots, 27$ of the United States. MOM's distribution facility is located in New England region 1. Orders can be direct shipped from the distribution center by small parcel carriers at cost $p_{1,j}$ per pound. An often cheaper alternative is to drop-ship (i.e., group) the week's orders for a region j into a bulk quantity that can be sent by common carrier freight to an intermediate point in region i for c_i per

pound ($c_1 \triangleq 0$), and then be direct-shipped from i on to j at small-parcel cost $p_{i,j}$ per pound. However, common carriers require a minimum of 1000 pounds per shipment. MOM wants to identify the least total cost way to meet this week's shipping needs. Formulate an ILP model of this shipping problem using the decision variables ($i, j = 1, \ldots, 27$)

$$x_{i,j} \triangleq \begin{cases} 1 & \text{if goods bound for region } j \text{ are drop shipped via region } i \\ 0 & \text{otherwise} \end{cases}$$

($x_{1,j} = 1$ implies direct shipping to j.)

11-45 Gas turbine engines[21] have the following radial assembly of nozzle guides

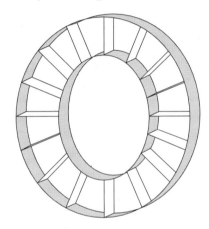

located immediately upstream from each rotor. The purpose of the vanes is to spread flow uniformly over the rotor, which improves its efficiency materially. During engine maintenance, the 55 old vanes are removed and replaced by new and refurbished ones $i = 1, \ldots, 55$ marked with previously assessed performance measures a_i and b_i of the two faces. The effect of each nozzle slot is greatly impacted by the sum of the a-value for the vane placed on one side and the b-value for the vane installed on the other. Maintenance personnel want to chosen a (counter-clockwise) arrangement of vanes around the assembly to balance this performance by minimizing the total of squared deviations between each resulting $a + b$ sum and known target value t. Show that this

[19]Based on U. Akinc (1993), "A Practical Approach to Lot and Setup Scheduling at a Textile Firm," *IIE Transactions, 25,* 54–64.

[20]Based on L. S. Franz and J. Woodmansee (1993), "Zone Skipping vs. Direct Shipment of Small Orders: Integrating Order Processing and Optimization," *Computers and Operations Research, 20,* 467–475.

[21]Based on R. D. Plante, T. J. Lowe, and R. Chandrasekaran (1987), "The Product Matrix Traveling Salesman Problem: An Application and Solution Heuristic," *Operations Research, 35,* 772–783.

vane arrangement task can be viewed as a traveling salesman problem and write an expression for the corresponding point i to point j costs.

11-46 A new freight airline[22] is designing a hub-and-spoke system for its operations. From a total of 34 airports to be served, 3 will be selected as hubs. Then (one-way) airport-to-airport freight quantities $f_{i,j}$ will be routed via the hubs ($f_{i,i} = 0$ for all i). That is, flow from i to j will begin at i, go to the unique hub k for i, then pass to the (possibly same) hub ℓ for j before being shipped on to j. The goal is to minimize the total of flows times unit transportation costs $c_{i,j}$ taking into account a 30% savings for flows between hubs that results from economies of scale ($c_{i,i} = 0$ for all i).

(a) Explain why appropriate decision variables for an integer programming model of this hub design problem are ($i, k = 1, \ldots, 34$)

$$x_{i,k} \triangleq \begin{cases} 1 & \text{if airport } i \text{ is assigned to a hub} \\ & \text{at } k \\ 0 & \text{otherwise} \end{cases}$$

$$y_k \triangleq \begin{cases} 1 & \text{if a hub opens at } k \\ 0 & \text{otherwise} \end{cases}$$

(b) Use only the $x_{i,k}$ (i.e., disregard whether hubs are open) to formulate a quadratic objective function summing origin-to-hub, hub-to-hub, and hub-to-destination transportation costs for each pair of nodes.

(c) Complete an INLP model of the problem by adding linear main constraints and appropriate variable-type constraints.

11-47 In an area with many suburban communities, telephone listings are usually grouped into several different books.[23] Patrons in each community $i = 1, \ldots, n$ are covered in exactly one directory $k = 1, \ldots, m$. Numbers of patrons p_i are known for each community, as well as (one-way) calling traffic levels $t_{i,j}$ between communities. Engineers seek to design books that maximize the traffic among patrons sharing a common telephone book without listing more than capacity q patrons in any single directory. Formulate an INLP model to select an optimal collection of telephone books in terms of the decision variables ($i = 1, \ldots, n, k = 1, \ldots, m$)

$$x_{i,k} \triangleq \begin{cases} 1 & \text{if community } i \text{ goes in book } k \\ 0 & \text{otherwise} \end{cases}$$

SUGGESTED READING

Nemhauser, George L., and Laurence A. Wolsey (1988), *Integer and Combinatorial Optimization*, Wiley, New York.

Parker, R. Gary, and Ronald L. Rardin (1988), *Discrete Optimization*, Academic Press, San Diego, Calif.

---

[22] Based on M. E. O'Kelly (1987), "A Quadratic Integer Program for the Location of Interacting Hub Facilities," *European Journal of Operational Research, 32*, 393–404.

[23] Based on S. Chen and C. J. McCallum (1977), "The Application of Management Science to the Design of Telephone Directories," *Interfaces, 8:1*, 58–69.

C H A P T E R 12

Discrete Optimization Methods

• •

In Chapter 11 we illustrated the wide range of integer and combinatorial optimization models encountered in operations research practice. Some are linear programs with a few discrete side constraints; others are still linear but involve only combinatorial decision variables; still others are both nonlinear and combinatorial. Every one includes logical decisions that just cannot be modeled validly as continuous, so most lack the elegant tractability of the LP and network models studied in earlier chapters.

Diminished tractability does not imply dimished importance. Discrete optimization models such as those presented in Chapter 11 all represent critical decision problems in engineering and management that must somehow be confronted. Even partial analysis can prove enormously valuable.

It should not surprise that discrete optimization methods span a range as wide as the models they address. In contrast to, say, linear programming, where a few prominent algorithms have proved adequate for the overwhelming majority of models, success in discrete optimization often requires methods cleverly specialized to an individual application. Still, there are common themes. In this chapter we introduce the best known.

12.1 SOLVING BY TOTAL ENUMERATION

Beginning students often find counterintuitive the idea that discrete optimization problems are more difficult than their continuous analogs. The algebra of LP algorithms in Chapters 5 and 6 is rather daunting. By comparison, a discrete model, which has only a finite number of choices for decision variables, can seem refreshingly easy. Why not just try them all and keep the best feasible solution as optimal?

Although naive, this point of view contains a kernel of wisdom.

> **12.1** If model has only a few discrete decision variables, the most effective method of analysis is often the most direct: enumeration of all the possibilities.

Total Enumeration

To be more specific, total or complete enumeration requires checking all possibilities implied by discrete variable values.

> **12.2**   **Total enumeration** solves a discrete optimization by trying all possible combinations of discrete variable values, computing for each the best corresponding choice of any continuous variables. Among combinations yielding a feasible solution, those with the best objective function value are optimal.

Swedish Steel All-or-Nothing Example

We can illustrate with the discrete version of our Swedish Steel example formulated in model (11.2) (Section 11.1):

$$
\begin{aligned}
\min \quad & 16(75)y_1 + 10(250)y_2 + 8x_3 + 9x_4 + 48x_5 + 60x_6 + 53x_7 \\
\text{s.t.} \quad & 75y_1 + 250y_2 + x_3 + x_4 + x_5 + x_6 + x_7 = 1000 \\
& 0.0080(75)y_1 + 0.0070(250)y_2 + 0.0085x_3 + 0.0040x_4 \geq 0.0065(1000) \\
& 0.0080(75)y_1 + 0.0070(250)y_2 + 0.0085x_3 + 0.0040x_4 \leq 0.0075(1000) \\
& 0.180(75)y_1 + 0.032(250)y_2 + 1.0x_5 \geq 0.030(1000) \\
& 0.180(75)y_1 + 0.032(250)y_2 + 1.0x_5 \leq 0.035(1000) \\
& 0.120(75)y_1 + 0.011(250)y_2 + 1.0x_6 \geq 0.010(1000) \\
& 0.120(75)y_1 + 0.011(250)y_2 + 1.0x_6 \leq 0.012(1000) \\
& 0.001(250)y_2 + 1.0x_7 \geq 0.011(1000) \\
& 0.001(250)y_2 + 1.0x_7 \leq 0.013(1000) \\
& x_3, \ldots, x_7 \geq 0 \\
& y_1, y_2 = 0 \text{ or } 1
\end{aligned}
\qquad (12.1)
$$

In this version the first two sources of scrap iron have to be entered on an all-or-nothing basis modeled with discrete variables. The other five sources can be employed in any nonnegative amount.

There are 2 possible values for y_1 and 2 for y_2, or a total of $2 \cdot 2 = 4$ combinations to enumerate. Table 12.1 provides details. Third option $y_1 = 1$, $y_2 = 0$ yields the optimal solution with objective value 9540.3.

TABLE 12.1   Enumeration of the Swedish Steel All-or-Nothing Model

Discrete Combination	Corresponding Continuous Solution					Objective Value
$y_1 = 0$, $y_2 = 0$	$x_3 = 814.3$,	$x_4 = 114.6$,	$x_5 = 30.0$,	$x_6 = 10.0$,	$x_7 = 1.1$	9914.1
$y_1 = 0$, $y_2 = 1$	$x_3 = 637.9$,	$x_4 = 82.0$,	$x_5 = 22.0$,	$x_6 = 7.3$,	$x_7 = 0.9$	9877.3
$y_1 = 1$, $y_2 = 0$	$x_3 = 727.6$,	$x_4 = 178.8$,	$x_5 = 16.5$,	$x_6 = 1.0$,	$x_7 = 1.1$	9540.3
$y_1 = 1$, $y_2 = 1$	$x_3 = 552.8$,	$x_4 = 112.9$,	$x_5 = 8.5$,	$x_6 = 0.0$,	$x_7 = 0.9$	9591.1

Since this model has both discrete and continuous variables, each case enumerated requires solving a continuous optimization over variables $x_3, \ldots, x_7$ to find the best continuous values to go with the choice of discrete variables selected. For example, fixing $y_1 = y_2 = 0$ in model (12.1) leaves the linear program

$$\begin{aligned}
\min \quad & 16(75)(0) + 10(250)(0) + 8x_3 + 9x_4 + 48x_5 + 60x_6 + 53x_7 \\
\text{s.t.} \quad & 75(0) + 250(0) + x_3 + x_4 + x_5 + x_6 + x_7 && = 1000 \\
& 0.0080(75)(0) + 0.0070(250)(0) + 0.0085x_3 + 0.0040x_4 && \geq 0.0065(1000) \\
& 0.0080(75)(0) + 0.0070(250)(0) + 0.0085x_3 + 0.0040x_4 && \leq 0.0075(1000) \\
& 0.180(75)(0) + 0.032(250)(0) + 1.0x_5 && \geq 0.030(1000) \\
& 0.180(75)(0) + 0.032(250)(0) + 1.0x_5 && \leq 0.035(1000) \\
& 0.120(75)(0) + 0.011(250)(0) + 1.0x_6 && \geq 0.010(1000) \\
& 0.120(75)(0) + 0.011(250)(0) + 1.0x_6 && \leq 0.012(1000) \\
& 0.001(250)(0) + 1.0x_7 && \geq 0.011(1000) \\
& 0.001(250)(0) + 1.0x_7 && \leq 0.013(1000) \\
& x_3, \ldots, x_7 \geq 0
\end{aligned}$$

Optimal solution $x_3 = 814.3$, $x_4 = 144.6$, $x_5 = 30.0$, $x_6 = 10.0$, $x_7 = 1.1$, completes the first case in Table 12.1.

SAMPLE EXERCISE 12.1: SOLVING BY TOTAL ENUMERATION

Solve the following discrete optimization model by total enumeration $\boxed{12.2}$.

$$\begin{aligned}
\max \quad & 7x_1 + 4x_2 + 19x_3 \\
\text{s.t.} \quad & x_1 + x_3 \leq 1 \\
& x_2 + x_3 \leq 1 \\
& x_1, x_2, x_3 = 0 \text{ or } 1
\end{aligned}$$

Analysis: Checking the $2^3 = 8$ combinations produces the following table:

Case	Objective	Case	Objective
$\mathbf{x} = (0, 0, 0)$	0	$\mathbf{x} = (1, 0, 0)$	7
$\mathbf{x} = (0, 0, 1)$	19	$\mathbf{x} = (1, 0, 1)$	Infeasible
$\mathbf{x} = (0, 1, 0)$	4	$\mathbf{x} = (1, 1, 0)$	11
$\mathbf{x} = (0, 1, 1)$	Infeasible	$\mathbf{x} = (1, 1, 1)$	Infeasible

Solution $\mathbf{x} = (0, 0, 1)$ is the feasible one with best objective value 19, so it is optimal.

Exponential Growth of Cases to Enumerate

Our Swedish Steel example has two discrete decision variables, each with two possible values 0 and 1. A total of

$$2 \cdot 2 = 2^2 = 4$$

combinations result.

Similar thinking shows that a model with k binary decision variables would have

$$\underbrace{2 \cdot 2 \cdots 2}_{k \text{ times}} = 2^k$$

cases to enumerate. This is **exponential growth**, with every additional 0–1 variable doubling the number of combinations.

An analyst can easily run $2^2 = 4$ or $2^4 = 16$ cases—perhaps even $2^{10} = 1024$ with the aid of a computer. But $2^{100} \approx 10^{30}$, and we know that a discrete model with $k = 100$ binary variables is not particularly large.

The fastest current computers perform a few billion (10^9) arithmetic operations in a second. Future computers might well solve the entire linear program associated with each choice of discrete variables in that same time. Assume even more, that a trillion (10^{12}) cases could be checked in a single second. Enumeration of a 100-variable model would still require

$$\frac{2^{100}}{10^{12}} \approx 1.27 \times 10^{18} \text{ seconds} \approx 402 \text{ million centuries}$$

too long for the most patient of decision makers to wait.

> **12.3** | Exponential growth makes total enumeration impractical with models having more than a handful of discrete decision variables.

SAMPLE EXERCISE 12.2: UNDERSTANDING EXPONENTIAL GROWTH

Suppose that your personal computer can enumerate one combination of discrete values each second of a given mixed-integer program, including solving the implied optimization for corresponding continuous variable values. Determine how long it would take to totally enumerate instances with 10, 20, 30, and 40 binary variables.

Analysis: For 10 variables, enumeration would require

$$2^{10} = 1024 \text{ seconds} \approx 17.1 \text{ minutes}$$

Each increment of 10 binary variables multiplies the number of combinations by $2^{10} = 1024$. Thus a case with 20 variables would require

$$1024 \cdot 17.1 \approx 17,500 \text{ minutes} \approx 12.1 \text{ days}$$

An instance with 30 binary variables would need about $1024 \cdot 12.1 = 12,390$ days, and one with 40 variables would require nearly 12.7 million days, more than 347 centuries.

Nonlinearities

The practicality of enumeration in mixed cases is also limited by the continuous problem that remains when discrete variables are enumerated. Our Swedish Steel case was an integer linear program (ILP), so that each case involved only solving an LP. If the remaining continuous model had been nonlinear, even evaluating the cases could have been a difficult task.

12.2 RELAXATIONS OF DISCRETE OPTIMIZATION MODELS AND THEIR USES

Because analysis of discrete optimization models is usually hard, it is natural to look for related but easier formulations that can aid in the analysis. **Relaxations** are auxiliary optimization models of this sort formed by weakening either the constraints or the objective function of the given discrete model.

EXAMPLE 12.1: BISON BOOSTERS

Before considering relaxation in the more realistic circumstances of models in Chapter 11, it will help to develop a more compact (albeit highly artificial) example. Consider the dilemma of the Bison Boosters club supporting the local atheletic team.

The Boosters are trying to decide what fundraising projects to undertake at the next country fair. One option is customized T-shirts, which will sell for $20 each; the other is sweatshirts selling for $30. History shows that everything offered for sale will be sold before the fair is over.

Materials to make the shirts are all donated by local merchants, but the Boosters must rent the equipment for customization. Different processes are involved, with the T-shirt equipment renting at $550 for the period up to the fair, and the sweatshirt equipment for $720. Display space presents another consideration. The Boosters have only 300 square feet of display wall area at the fair, and T-shirts will consume 1.5 square feet each, sweatshirts 4 square feet each. What plan will net the most income?

Certainly this problem centers on making shirts, so decision variables will include

$$x_1 \triangleq \text{number of T-shirts made and sold}$$
$$x_2 \triangleq \text{number of sweatshirts made and sold}$$

However, the Boosters also confront discrete decisions on whether to rent equipment:

$$y_1 \triangleq 1 \text{ if T-shirt equipment is rented and} = 0 \text{ otherwise}$$
$$y_2 \triangleq 1 \text{ if sweatshirt equipment is rented and} = 0 \text{ otherwise}$$

Using these decision variables, the Boosters' dilemma can be modeled:

$$
\begin{array}{lll}
\max & 20x_1 + 30x_2 - 550y_1 - 720y_2 & \text{(net income)} \\
\text{s.t.} & 1.5x_1 + 4x_2 \le 300 & \text{(display space)} \\
& x_1 \le 200y_1 & \text{(T-shirts if equipment)} \\
& x_2 \le 75y_2 & \text{(sweatshirts if equipment)} \\
& x_1, x_2 \ge 0 & \\
& y_1, y_2 = 0 \text{ or } 1 &
\end{array}
\qquad (12.2)
$$

The objective function maximizes net income, and the first main constraint enforces the display space limit. The next two constraints provide the switching we have seen in other models. Any sufficiently large big-M could be used as the y_j coefficient in these constraints. Values in (12.2) derive from the greatest production possible within the 300 square feet display limit. Coefficients $300/1.5 = 200$ for T-shirts and $300/4 = 75$ for sweatshirts introduce no limitation if y's equal 1, yet switch off all production if y's equal 0.

Enumeration of the 4 combinations of y_1 and y_2 values easily establishes that the Boosters should make only T-shirts. The unique optimal solution is $x_1^* = 200$, $x_2^* = 0$, $y_1^* = 1$, $y_2^* = 0$, with net income $3450.

Constraint Relaxations

Relaxations may weaken either the objective function or the constraints, but the elementary ones we explore in this book nearly all focus on constraints. A constraint relaxation produces an easier model by dropping or easing some constraints.

> 12.4 │ Model $(\tilde{P})$ is a **constraint relaxation** of model (P) if every feasible solution to (P) is also feasible in $(\tilde{P})$ and both models have the same objective function.

New feasible solutions may be allowed, but none should be lost.

Table 12.2 shows several constraint relaxations of the tiny Bison Boosters model (12.2). The first simply doubles capacities. The result is certainly a relaxation, because every solution fitting within the true capacity of 300 square feet will also fit within twice as much area. Still, this relaxation gains us little.

TABLE 12.2   Constraint Relaxations of Bison Boosters Model

Revised Constraints	Discussion
$1.5x_1 + 4x_2 \leq 600$ $x_1 \leq 400y_1$ $x_2 \leq 150y_2$ $x_1, x_2 \geq 0$ $y_1, y_2 = 0$ or 1	Doubled capacities. Relaxation optimum: $\tilde{x}_1 = 400$, $\tilde{x}_2 = 0$, $\tilde{y}_1 = 1$, $\tilde{y}_2 = 0$, net income \$7450
$x_1 \leq 200y_1$ $x_2 \leq 75y_2$ $x_1, x_2 \geq 0$ $y_1, y_2 = 0$ or 1	Dropped first constraint. Relaxation optimum: $\tilde{x}_1 = 200$, $\tilde{x}_2 = 75$, $\tilde{y}_1 = 1$, $\tilde{y}_2 = 1$, net income \$4980
$1.5x_1 + 4x_2 \leq 300$ $x_1 \leq 200y_1$ $x_2 \leq 75y_2$ $x_1, x_2 \geq 0$ $0 \leq y_1 \leq 1$ $0 \leq y_2 \leq 1$	Linear programming relaxation with discrete variables treated as continuous. Relaxation optimum: $\tilde{x}_1 = 200$, $\tilde{x}_2 = 0$, $\tilde{y}_1 = 1$, $\tilde{y}_2 = 0$, net income \$3450

> 12.5 │ Relaxations should be significantly more tractable than the models they relax, so that deeper analysis is practical.

Doubling capacities fails this requirement because the character of the model is unchanged.

The second relaxation of Table 12.2 is more on track. Dropping the first constraint delinks decisions about the two types of shirts. It then becomes much easier to compute a (relaxation) optimal solution. We need only decide one by one whether the maximum production now allowed each x_j when its $y_j = 1$ justifies the fixed cost of equipment rental. Both do.

SAMPLE EXERCISE 12.3: RECOGNIZING RELAXATIONS

Determine whether or not each of the following mixed-integer programs is a constraint relaxation of

$$\min \quad 3x_1 + 6x_2 + 7x_3 + x_4$$
$$\text{s.t.} \quad 2x_1 + x_2 + x_3 + 10x_4 \geq 100$$
$$x_1 + x_2 + x_3 \leq 1$$
$$x_1, x_2, x_3 = 0 \text{ or } 1$$
$$x_4 \geq 0$$

(a)
$$\min \quad 3x_1 + 6x_2 + 7x_3 + x_4$$
$$\text{s.t.} \quad 2x_1 + x_2 + x_3 + 10x_4 \geq 100$$
$$x_1, x_2, x_3 = 0 \text{ or } 1$$
$$x_4 \geq 0$$

(b)
$$\min \quad 3x_1 + 6x_2 + 7x_3 + x_4$$
$$\text{s.t.} \quad 2x_1 + x_2 + x_3 + 10x_4 \geq 200$$
$$x_1 + x_2 + x_3 \leq 1$$
$$x_1, x_2, x_3 = 0 \text{ or } 1$$
$$x_4 \geq 0$$

(c)
$$\min \quad 3x_1 + 6x_2 + 7x_3 + x_4$$
$$\text{s.t.} \quad 2x_1 + x_2 + x_3 + 10x_4 \geq 100$$
$$x_1 + x_2 + x_3 \leq 1$$
$$x_1, x_2, x_3, x_4 \geq 0$$

(d)
$$\min \quad 3x_1 + 6x_2 + 7x_3 + x_4$$
$$\text{s.t.} \quad 2x_1 + x_2 + x_3 + 10x_4 \geq 100$$
$$x_1 + x_2 + x_3 \leq 1$$
$$1 \geq x_1 \geq 0, \ 1 \geq x_2 \geq 0,$$
$$1 \geq x_3 \geq 0, \ x_4 \geq 0$$

Analysis: We apply definition 12.4 .

(a) This model is a constraint relaxation because it is formed by dropping the second main constraint. Certainly, every solution feasible in the original model remains so with fewer constraints.

(b) This model is not a relaxation. The only change, which is increasing the right-hand side by 100, to 200, eliminates previously feasible solutions. One example is $\mathbf{x} = (0, 0, 0, 10)$.

(c) This model is a relaxation. Allowing x_1, x_2, and x_3 to take on any nonnegative value—rather than just 0 or 1—cannot eliminate previously feasible solutions.

(d) This model is also a relaxation. Allowing x_1, x_2, and x_3 to take on any values in the interval [0, 1] precludes none of their truly feasible values.

Linear Programming Relaxations

The third case in Table 12.2 illustrates the best known and most used of all constraint relaxation forms: linear programming, or more generally, continuous relaxations.

12.6 | **Continuous relaxations (linear programming relaxations** if the given model is an integer linear program) are formed by treating any discrete variables as continuous while retaining all other constraints.

In the real Bison Boosters model, each y_j must equal 0 or 1. In the continuous relaxation we also admit fractions, replacing each

$$y_j = 0 \text{ or } 1 \quad \text{by} \quad 1 \geq y_j \geq 0$$

Certainly, no feasible solutions are lost by allowing both fractional and integer choices for discrete variables, so the process does produce a valid relaxation. More important, the relaxed model usually proves significantly more tractable.

Our Bison Boosters model is an integer linear program (ILP), linear in all aspects except the discreteness of y_1 and y_2. Thus, relaxing discrete variables to continuous leaves a linear program to solve—the linear programming relaxation; we have already expended several chapters of this book showing how effectively linear programs can be analyzed.

> **12.7** | LP relaxations of integer linear programs are by far the most used relaxation forms because they bring all the power of linear programming to bear on analysis of the given discrete models.

SAMPLE EXERCISE 12.4: FORMING LINEAR PROGRAMMING RELAXATIONS

Form the linear programming relaxation of the following mixed-integer program:

$$\begin{aligned}
\min \quad & 15x_1 + 2x_2 - 4x_3 + 10x_4 \\
\text{s.t.} \quad & x_3 - x_4 \leq 0 \\
& x_1 + 2x_2 + 4x_3 + 8x_4 = 20 \\
& x_2 + x_4 \leq 1 \\
& x_1 \geq 0 \\
& x_2, x_3, x_4 = 0 \text{ or } 1
\end{aligned}$$

Analysis: Following definition $\boxed{12.6}$, we replace 0–1 constraints $x_j = 0$ or 1 by $x_j \in [0, 1]$ to obtain the LP relaxation

$$\begin{aligned}
\min \quad & 15x_1 + 2x_2 - 4x_3 + 10x_4 \\
\text{s.t.} \quad & x_3 - x_4 \leq 0 \\
& x_1 + 2x_2 + 4x_3 + 8x_4 = 20 \\
& x_2 + x_4 \leq 1 \\
& x_1 \geq 0 \\
& 1 \geq x_2 \geq 0, \ 1 \geq x_3 \geq 0, \ 1 \geq x_4 \geq 0
\end{aligned}$$

Proving Infeasibility with Relaxations

Exactly what do relaxations add to our analysis of discrete optimization models? One thing is prove infeasibility.

Suppose that a constraint relaxation comes out infeasible. Then it has no solutions at all. Since every solution to the full model must also be feasible in the relaxation, it follows that the original model was also infeasible. By analyzing the relaxation we have learned a critical fact about the model of real interest.

> **12.8** | If a constraint relaxation is infeasible, so is the full model it relaxes.

SAMPLE EXERCISE 12.5: PROVING INFEASIBILITY WITH RELAXATIONS

Use linear programming relaxation to establish that the following discrete optimization model is infeasible:

$$\min \quad 8x_1 + 2x_2$$
$$\text{s.t.} \quad x_1 - x_2 \geq 2$$
$$-x_1 + x_2 \geq -1$$
$$x_1, x_2 \geq 0 \text{ and integer}$$

Analysis: The linear programming relaxation of this model is

$$\min \quad 8x_1 + 2x_2$$
$$\text{s.t.} \quad x_1 - x_2 \geq 2$$
$$-x_1 + x_2 \geq -1$$
$$x_1, x_2 \geq 0$$

It is clearly infeasible, because the two main constraints can be written

$$x_1 - x_2 \geq 2$$
$$x_1 - x_2 \leq 1$$

Thus by principle $\boxed{12.8}$, the given integer program is also infeasible. Any solutions satisfying all constraints would also have to be feasible in the relaxation.

Solution Value Bounds from Relaxations

Figure 12.1 illustrates how relaxations also give us bounds on optimal solution values. Constraint relaxations expand the feasible set, allowing more candidates for relaxation optimum. The relaxation optimal value, which is the best over the expanded set of solutions, must then equal or improve on the best feasible solution value to the true model.

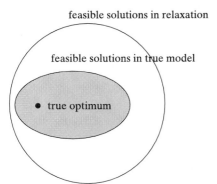

FIGURE 12.1   Relaxations and Optimality

$\boxed{12.9}$ The optimal value of any relaxation of a maximize model yields an upper bound on the optimal value of the full model. The optimal value of any relaxation of a minimize model yields a lower bound.

All three relaxations in Table 12.2 illustrate the maximize case. The optimal solution value of the Bison Boosters model (12.2) is $3450. One of the cases in Table 12.2 yields exactly this value. The others produce higher estimates of net income. All provide the upper bound guaranteed in principle $\boxed{12.9}$.

With the Bison Boosters model, which is so small that it is easily solved optimally, relaxation bounds offer little new insight. A better sense of their value comes from considering the somewhat larger EMS model (11.8) of Section 11.3 that minimizes the number of stations to cover the 20 metropolitan districts:

$$
\begin{array}{lll}
\min & \displaystyle\sum_{j=1}^{10} x_j & \text{(number of sites)} \\
\text{s.t.} & x_2 \geq 1 & \text{(district 1)} \\
& x_1 + x_2 \geq 1 & \text{(district 2)} \\
& x_1 + x_3 \geq 1 & \text{(district 3)} \\
& x_3 \geq 1 & \text{(district 4)} \\
& x_3 \geq 1 & \text{(district 5)} \\
& x_2 \geq 1 & \text{(district 6)} \\
& x_2 + x_4 \geq 1 & \text{(district 7)} \\
& x_3 + x_4 \geq 1 & \text{(district 8)} \\
& x_8 \geq 1 & \text{(district 9)} \\
& x_4 + x_6 \geq 1 & \text{(district 10)} \\
& x_4 + x_5 \geq 1 & \text{(district 11)} \\
& x_4 + x_5 + x_6 \geq 1 & \text{(district 12)} \\
& x_4 + x_5 + x_7 \geq 1 & \text{(district 13)} \\
& x_8 + x_9 \geq 1 & \text{(district 14)} \\
& x_6 + x_9 \geq 1 & \text{(district 15)} \\
& x_5 + x_6 \geq 1 & \text{(district 16)} \\
& x_5 + x_7 + x_{10} \geq 1 & \text{(district 17)} \\
& x_8 + x_9 \geq 1 & \text{(district 18)} \\
& x_9 + x_{10} \geq 1 & \text{(district 19)} \\
& x_{10} \geq 1 & \text{(district 20)} \\
& x_1, \ldots, x_{10} = 0 \text{ or } 1 &
\end{array}
\tag{12.3}
$$

How many stations does this model imply? Even with just 10 discrete variables, the answer is hardly obvious. But if we replace each $x_j = 0$ or 1 constraint by $0 \leq x_j \leq 1$, the resulting linear programming relaxation can be solved quickly with say, the simplex algorithm. An optimal solution is

$$
\begin{aligned}
\tilde{x}_1 &= \tilde{x}_7 = 0 \\
\tilde{x}_2 &= \tilde{x}_3 = \tilde{x}_8 = \tilde{x}_{10} = 1 \\
\tilde{x}_4 &= \tilde{x}_5 = \tilde{x}_6 = \tilde{x}_9 = \tfrac{1}{2}
\end{aligned}
\tag{12.4}
$$

with optimal value 6.0. Without looking any further into the discrete model, we can conclude that at least 6 EMS sites will be required because this LP relaxation value provides a lower bound (principle $\boxed{12.9}$).

SAMPLE EXERCISE 12.6: COMPUTING BOUNDS FROM RELAXATIONS

Compute (by inspection) the optimal solution value and the LP relaxation bound for each of the following integer programs.

(a) max $x_1 + x_2 + x_3$
    s.t. $x_1 + x_2 \leq 1$
         $x_1 + x_3 \leq 1$
         $x_2 + x_3 \leq 1$
         $x_1, x_2, x_3 = 0$ or 1

(b) min $20x_1 + 9x_2 + 7x_3$
    s.t. $10x_1 + 4x_2 + 3x_3 \geq 7$
         $x_1, x_2, x_3 = 0$ or 1

Analysis:

(a) Clearly, only one of the variables in this model can $= 1$, so the optimal solution value is 1. Corresponding linear programming relaxation

$$\begin{aligned} \max \quad & x_1 + x_2 + x_3 \\ \text{s.t.} \quad & x_1 + x_2 \leq 1 \\ & x_1 + x_3 \leq 1 \\ & x_2 + x_3 \leq 1 \\ & 1 \geq x_1, x_2, x_3 \geq 0 \end{aligned}$$

yields optimal solution $\tilde{\mathbf{x}} = (\frac{1}{2}, \frac{1}{2}, \frac{1}{2})$ with objective value $\frac{3}{2}$. In accord with principle 12.9 , relaxation value $\frac{3}{2}$ is an upper bound on the true optimal value 1 of this maximize model.

(b) Total enumeration shows that an optimal solution to this minimizing ILP is $\mathbf{x} = (0, 1, 1)$ with value 16. Its linear programming relaxation is

$$\begin{aligned} \min \quad & 20x_1 + 9x_2 + 7x3 \\ \text{s.t.} \quad & 10x_1 + 4x_2 + 3x_3 \geq 7 \\ & 1 \geq x_1, x_2, x_3 \geq 0 \end{aligned}$$

with optimal solution $\tilde{\mathbf{x}} = (\frac{7}{10}, 0, 0)$ and value 14. Demonstrating principle 12.9 , relaxation value 14 provides a lower bound on true optimal value 16.

Optimal Solutions from Relaxations

Sometimes relaxations not only bound the optimal value of the corresponding discrete model but produce an optimal solution.

12.10 If an optimal solution to a constraint relaxation is also feasible in the model it relaxes, the solution is optimal in that original model.

Another look at Figure 12.1 will show why. All (shaded-area) feasible solutions to the original discrete model must also belong to the larger relaxation feasible set. If the relaxation optimum happens to be one of them, it has as good an objective function value as any feasible solution to the relaxation. In particular, it has as good an objective function value as any feasible solution in the original model. It must be optimal in the full model.

The third, linear programming relaxation of the Bison Boosters model in Table 12.2 illustrates. Even though integrality was not required of y-components in the relaxation optimal solution

$$\tilde{x}_1 = 200, \quad \tilde{x}_2 = 0, \quad \tilde{y}_1 = 1, \quad \tilde{y}_2 = 0$$

it happened anyway. This relaxation optimum is feasible in the full discrete model and so optimal there.

SAMPLE EXERCISE 12.7: OBTAINING OPTIMAL SOLUTIONS FROM RELAXATIONS

Compute (by inspection) optimal solutions to each of the following relaxations, and determine whether we can conclude that the relaxation optimum is optimal in the original model.

(a) The linear programming relaxation of

$$\begin{aligned}
\max \quad & 20x_1 + 8x_2 + 2x_3 \\
\text{s.t.} \quad & x_1 + x_2 + x_3 \leq 1 \\
& x_1, x_2, x_3 = 0 \text{ or } 1
\end{aligned}$$

(b) The linear programming relaxation of

$$\begin{aligned}
\max \quad & x_1 + x_2 + x_3 \\
\text{s.t.} \quad & x_1 + x_2 \leq 1 \\
& x_1 + x_3 \leq 1 \\
& x_2 + x_3 \leq 1 \\
& x_1, x_2, x_3 = 0 \text{ or } 1
\end{aligned}$$

(c) The relaxation obtained by dropping the first main constraint of

$$\begin{aligned}
\min \quad & 2x_1 + 4x_2 + 8x_3 \\
\text{s.t.} \quad & x_1 + x_2 + x_3 \leq 2 \\
& 10x_1 + 3x_2 + x_3 \geq 8 \\
& x_1, x_2, x_3 = 0 \text{ or } 1
\end{aligned}$$

Analysis:

(a) The linear programming relaxation of this model is

$$\begin{aligned}
\max \quad & 20x_1 + 8x_2 + 2x_3 \\
\text{s.t.} \quad & x_1 + x_2 + x_3 \leq 1 \\
& 1 \geq x_1, x_2, x_3 \geq 0
\end{aligned}$$

with obvious optimal solution $\tilde{\mathbf{x}} = (1, 0, 0)$. Since this solution is also feasible in the original model, it follows from principle $\boxed{12.10}$ that it is optimal there.

(b) The linear programming relaxation of this model is

$$\begin{aligned}
\max \quad & x_1 + x_2 + x_3 \\
\text{s.t.} \quad & x_1 + x_2 \leq 1 \\
& x_2 + x_3 \leq 1 \\
& x_1 + x_3 \leq 1 \\
& 1 \geq x_1, x_2, x_3 \geq 0
\end{aligned}$$

with optimal solution $\tilde{\mathbf{x}} = (\frac{1}{2}, \frac{1}{2}, \frac{1}{2})$. Since this solution violates integrality requirements in the original model, it is infeasible there. It could not be optimal.

(c) The indicated relaxation is

$$\begin{aligned}
\min \quad & 2x_1 + 4x_2 + 8x_3 \\
\text{s.t.} \quad & 10x_1 + 3x_2 + x_3 \geq 8 \\
& x_1, x_2, x_3 = 0 \text{ or } 1
\end{aligned}$$

with obvious optimal solution $\tilde{\mathbf{x}} = (1, 0, 0)$. This relaxation optimum satisfies relaxed constraint

$$x_1 + x_2 + x_3 \leq 2$$

and so is feasible in the original model. It follows from principle $\boxed{12.10}$ that it is optimal in the full integer program.

Rounded Solutions from Relaxations

When principle $\boxed{12.10}$ applies, relaxation completely solves a hard discrete optimization model. More commonly, things are not that simple. As with the EMS solution (12.4) above, relaxation optima usually violate some constraints of the true model.

All is hardly lost. First, we have the bound of principle $\boxed{12.9}$. We may also have a starting point for constructing a good heuristic solution to the full discrete model.

> $\boxed{12.11}$ Many relaxations produce optimal solutions that are easily "rounded" to good feasible solutions for the full model.

Consider, for example, the EMS solution (12.4). The nature of model constraints (12.3), $\geq$ form with nonnegative coefficients on the left-hand side, means that feasibility of a solution is not lost if we increase some of its components. Beginning from the LP relaxation optimum and rounding up produces the approximate optimal solution

$$\begin{aligned}
\hat{x}_1 &= \lceil \tilde{x}_1 \rceil = \lceil 0 \rceil = 0 \\
\hat{x}_2 &= \lceil \tilde{x}_2 \rceil = \lceil 1 \rceil = 1 \\
\hat{x}_3 &= \lceil \tilde{x}_3 \rceil = \lceil 1 \rceil = 1 \\
\hat{x}_4 &= \lceil \tilde{x}_4 \rceil = \lceil \tfrac{1}{2} \rceil = 1 \\
\hat{x}_5 &= \lceil \tilde{x}_7 \rceil = \lceil \tfrac{1}{2} \rceil = 1 \\
\hat{x}_6 &= \lceil \tilde{x}_6 \rceil = \lceil \tfrac{1}{2} \rceil = 1 \\
\hat{x}_7 &= \lceil \tilde{x}_7 \rceil = \lceil 0 \rceil = 0 \\
\hat{x}_8 &= \lceil \tilde{x}_8 \rceil = \lceil 1 \rceil = 1 \\
\hat{x}_9 &= \lceil \tilde{x}_9 \rceil = \lceil \tfrac{1}{2} \rceil = 1 \\
\hat{x}_{10} &= \lceil \tilde{x}_{10} \rceil = \lceil 1 \rceil = 1
\end{aligned} \qquad (12.5)$$

with value $\sum_{j=1}^{10} \hat{x}_j = 8$. Here **ceiling** notation

$$\lceil x \rceil \triangleq \text{least integer greater than or equal to } x$$

The corresponding **floor** notation

$$\lfloor x \rfloor \triangleq \text{greatest integer less that or equal to } x$$

Heuristic optimum $\hat{\mathbf{x}}$ may not be truly optimal, but it does satisfy all constraints. Where time permits no deeper analysis, this rounded relaxation solution might well suffice. Also, feasible solutions provide bounds to complement those obtained from the optimal relaxation solution value (principle 12.9).

12.12   The objective function value of any (integer) feasible solution to a maximizing discrete optimization problem provides a lower bound on the integer optimal value, and any (integer) feasible solution to a minimizing discrete optimization problem provides an upper bound.

Set covering relaxation optima like (12.4) are particularly easy to round, because of the unusually simple form of the constraints. Many other forms admit similar rounding. Some round infeasible relaxation solutions up, some round down, and some do other straightforward patching. Details vary with model form.

Unfortunately, there are some discrete models that just do not round. For an example, return to our AA airline crew scheduling model (11.10) (Section 11.3). Its set partitioning form closely resembles the set covering case we just rounded easily. But set partitioning involves equality constraints. Each time we round some infeasible $\tilde{x}_j$ up to 1 or down to 0, other variables sharing constraints with that x_j will also have to be adjusted if feasibility is to be preserved. Much more complex rounding schemes are required, and success cannot be guaranteed.

SAMPLE EXERCISE 12.8: ROUNDING RELAXATION OPTIMA

In each of the following integer linear programs, develop and apply a scheme for rounding the indicated LP relaxation optimum to an approximate solution for the full model. Also, indicate the best lower and upper bounds on the optimal integer solution value available from relaxation and rounding.

(a) min $10x_1 + 8x_2 + 18x_3$ with LP relaxation optimum $\tilde{\mathbf{x}} = (0, 1, \frac{1}{7})$

    s.t. $2x_1 + 4x_2 + 7x_3 \geq 5$

          $x_1 + x_2 + x_3 \geq 1$

          $x_1, x_2, x_3 = 0 \text{ or } 1$

(b) max $40x_1 + 2x_2 + 18x_3$ with LP relaxation optimum $\tilde{\mathbf{x}} = (1, 0, \frac{3}{7})$

    s.t. $2x_1 + 11x_2 + 7x_3 \leq 5$

          $x_1 + x_2 + x_3 \leq 2$

          $x_1, x_2, x_3 = 0 \text{ or } 1$

(c) min $3x_1 + 5x_2 + 20x_3 + 14x_4$ with LP relaxation optimum $\tilde{\mathbf{x}} = (\frac{16}{3}, \frac{17}{3}, \frac{16}{33}, \frac{17}{33})$

   s.t. $x_1 + x_2 = 11$

$3x_1 + 6x_2 = 50$

$x_1 \leq 11x_3$

$x_2 \leq 11x_4$

$x_1, x_2 \geq 0$

$x_3, x_4 = 0 \text{ or } 1$

Analysis:

(a) All main constraints of this model are $\geq$ form, and coefficients on the left-hand side are nonnegative. Thus increasing feasible variable values cannot cause a violation. We may round up to integer-feasible solution

$$\lceil \tilde{\mathbf{x}} \rceil = (\lceil 0 \rceil, \lceil 1 \rceil, \lceil \tfrac{1}{7} \rceil) = (0, 1, 1)$$

Substituting this solution in the objective function gives an upper bound (principle 12.12) of 26 on the optimal value. The corresponding lower bound, which is obtained by substituting the relaxation optimal solution (principle 12.9), is 10.57.

(b) All main constraints of this model are $\leq$ form, and coefficients on the left-hand side are nonnegative. Thus decreasing feasible variable values cannot cause a violation. We may round down to integer-feasible solution

$$\lfloor \tilde{\mathbf{x}} \rfloor = (\lfloor 1 \rfloor, \lfloor 0 \rfloor, \lfloor \tfrac{3}{7} \rfloor) = (1, 0, 0)$$

Substituting this solution in the objective function gives a lower bound (principle 12.12) of 40 on the optimal value. The corresponding upper bound, which is obtained by substituting the relaxation optimal solution (principle 12.9), is 47.71.

(c) Each of the discrete variables in this mixed-integer linear program occurs in only one $\leq$ constraint on the right-hand side. Thus increasing x_3 and x_4 from their relaxation values cannot lose feasiblity. We may round up to

$$\left(\tfrac{16}{3}, \tfrac{17}{3}, \lceil\tfrac{16}{33}\rceil, \lceil\tfrac{17}{33}\rceil\right) = \left(\tfrac{16}{3}, \tfrac{17}{3}, 1, 1\right)$$

Notice that continuous variable values were not changed.

   Substituting this solution in the objective function gives an upper bound (principle 12.12) of 78.33 on the optimal value. The corresponding lower bound, which is obtained by substituting the relaxation optimal solution (principle 12.9), is 61.24.

12.3 STRONGER LP RELAXATIONS, VALID INEQUALITIES, AND LAGRANGIAN RELAXATIONS

It should be obvious that we will detect infeasibility quicker (principle 12.8), obtain sharper bounds (principle 12.9), have a better chance of discovering an optimal solution (principle 12.10), and find rounding much easier (principle 12.11) if the relaxations we employ closely approximate the full model of interest. Strong relaxations do just that.

> **12.13** | A relaxation is **strong** or **sharp** if its optimal value closely bounds that of the true model, and its optimal solution closely approximates an optimum in the full model.

Analysts dealing with hard discrete models via relaxations will almost always find it worthwhile to look for means to strengthen the relaxations without losing too much tractability.

Stronger LP Relaxations

To begin, let us will focus on the standard linear programming relaxations of ILP models. How can we make such LP relaxations strong? The key insight sometimes surprises:

> **12.14** | Equally correct integer linear programming formulations of a discrete problem may have dramatically different linear programming relaxation optima.

SAMPLE EXERCISE 12.9: UNDERSTANDING STRONGER LP RELAXATIONS

Show (by inspection) that even though the two following integer linear programming models have the same feasible solutions, the second yields a stronger linear programming relaxation.

$$\begin{array}{lll} \max & x_1 + x_2 + x_3 \\ \text{s.t.} & x_1 + x_2 \leq 1 \\ & x_1 + x_3 \leq 1 \\ & x_2 + x_3 \leq 1 \\ & x_1, x_2, x_3 = 0 \text{ or } 1 \end{array} \qquad \begin{array}{lll} \max & x_1 + x_2 + x_3 \\ \text{s.t.} & x_1 + x_2 \leq 1 \\ & x_1 + x_3 \leq 1 \\ & x_2 + x_3 \leq 1 \\ & x_1 + x_2 + x_3 \leq 1 \\ & x_1, x_2, x_3 = 0 \text{ or } 1 \end{array}$$

Analysis: Both ILPs have the same feasible solutions,

$$\mathbf{x}^{(1)} = (1, 0, 0)$$
$$\mathbf{x}^{(2)} = (0, 1, 0)$$
$$\mathbf{x}^{(3)} = (0, 0, 1)$$

Thus they are both valid models of the same problem. Still, the first has LP relaxation optimum $\tilde{\mathbf{x}} = (\frac{1}{2}, \frac{1}{2}, \frac{1}{2})$, and the second has relaxation optimum $\tilde{\mathbf{x}} = (1, 0, 0)$ (among others). The corresponding relaxation bounds are $\frac{3}{2}$ and 1, making the second relaxation stronger. In this simple case, in fact, it yields a discrete optimum (via principle 12.10).

Choosing Big-M Constants

The "sufficiently large" big-M constants needed in so many models offer one easy family where details of ILP modeling affect the LP relaxation. Return, for instance, to the tiny Bison Boosters model of (12.2) and Table 12.2. In formulating switching constraints $x_1 \leq 400y_1$ and $x_2 \leq 75y_2$, we constructed values 400 and 75 with a back-of-envelope computation. Any sufficiently large M would yield a correct integer linear programming model.

Suppose that we had used 10,000 for both. The new model is

$$
\begin{array}{llll}
\max & 20x_1 + 30x_2 - 550y_1 - 720y_2 & \text{(net income)} & \\
\text{s.t.} & 1.5x_1 + 4x_2 \leq 300 & \text{(display space)} & \\
& x_1 \leq 10,000y_1 & \text{(T-shirts if equipment)} & \\
& x_2 \leq 10,000y_2 & \text{(sweatshirts if equipment)} & (12.6) \\
& x_1, x_2 \geq 0 & & \\
& y_1, y_2 = 0 \text{ or } 1 & &
\end{array}
$$

Recall that the original model (12.2) had relaxation optimum

$$\tilde{x}_1 = 200, \quad \tilde{x}_2 = 0, \quad \tilde{y}_1 = 1, \quad \tilde{y}_2 = 0$$

matching perfectly the discrete optimal solution with value $3450. Its LP relaxation was indeed strong.

Revision (12.6) is every bit as correct as the original (12.2) in the sense that it has exactly the same (discrete) feasible set. However, the LP relaxation of (12.6) yields optimum

$$\tilde{x}_1 = 200, \quad \tilde{x}_2 = 0, \quad \tilde{y}_1 = 0.02, \quad \tilde{y}_2 = 0 \tag{12.7}$$

with value $3989. The value bound $3989 now differs significantly from the true optimal value $3450. Also, the relaxation optimal solution has component $\tilde{y}_1$ at a tiny fractional value. With only (12.7) at hand, it would be hard to tell whether to rent or not the T-shirt equipment.

This contrast between LP relaxations of integer-equivalent models (12.2) and (12.6) highlights an important and easy-to-implement principle for strengthening relaxations:

> **12.15** Whenever a discrete model requires sufficiently large big-M's, the strongest relaxations will result from models employing the smallest valid choice of those constants.

SAMPLE EXERCISE 12.10: CHOOSING SMALLEST BIG-M'S

We wish to decide which combination of 2 pharmaceutical facilities should be used to produce 80 units of a needed product. One costs $5000 to setup and has variable cost $20 unit. The other cost $7000 to setup and has variable cost $15. Both have capacity of 200 units.

(a) Formulate a mixed-integer linear programming model using capacities for needed big-M's.

(b) Strengthen the linear programming relaxation of your model in part (a) by reducing big-M's to their smallest valid value.

Modeling:

(a) Using decision variables x_1 and x_2 for the amount produced in each facility, and switching variables x_3 and x_4 to track setups, a valid formulation is

$$\min \quad 20x_1 + 15x_2 + 5000x_3 + 7000x_4$$
$$\text{s.t.} \quad x_1 + x_2 = 80$$
$$x_1 \leq 200x_3$$
$$x_2 \leq 200x_4$$
$$x_1, x_2 \geq 0$$
$$x_3, x_4 = 0 \text{ or } 1$$

Full capacity is available whenever setup cost is paid.

(b) Although capacities are 200, the problem calls for only 80 units to be produced. Thus neither x_1 nor x_2 will ever exceed 80 in an optimal solution. We may strengthen the model by reducing big-M constants from 200 to 80, to produce

$$\min \quad 20x_1 + 15x_2 + 5000x_3 + 7000x_4$$
$$\text{s.t.} \quad x_1 + x_2 = 80$$
$$x_1 \leq 80x_3$$
$$x_2 \leq 80x_4$$
$$x_1, x_2 \geq 0$$
$$x_3, x_4 = 0 \text{ or } 1$$

The reader can verify that this new formulation has relaxation optimum $\tilde{\mathbf{x}} = (80, 0, 1, 0)$ with value \$6600, versus the original model's $\tilde{\mathbf{x}} = (80, 0, .4, 0)$ at value \$3600.

Valid Inequalities

Sharpening big-M coefficients is only one of many ways to strengthen LP relaxations. We can also add new valid inequality constraints.

> 12.16   A linear inequality is a **valid inequality** for a given discrete optimization model if it holds for all (integer) feasible solutions to the model.

Relaxations can often be strenghtened dramatically by including valid inequalities that are not needed for a correct discrete model.

Not every valid inequality strengthens a relaxation. For example, all inequality constraints of the original formulation are trivially valid because they are satisfied by every feasible solution.

> 12.17   To strengthen a relaxation, a valid inequality must cut off (render infeasible) some feasible solutions to the current LP relaxation that are not feasible in the full ILP model.

This need to cut off noninteger relaxation solutions is why valid inequalities are sometimes called **cutting planes**.

The Tmark facilities location model of Section 11.6 illustrates a classic case. The model formulated there is:

$$\min \quad \sum_{i=1}^{8}\sum_{j=1}^{14}(d_j r_{i,j})x_{i,j}+\sum_{i=1}^{8}f_i y_i \qquad \text{(total fixed cost)}$$

$$\text{s.t.} \quad \sum_{i=1}^{8}x_{i,j}=1 \qquad \text{for all } j=1,\ldots,14 \quad \text{(carry } j \text{ load)}$$

$$1500y_i \le \sum_{j=1}^{14}d_j x_{i,j} \qquad \text{for all } i=1,\ldots,8 \quad \text{(minimum at } i) \qquad (12.8)$$

$$\sum_{j=1}^{14}d_j x_{i,j}\le 5000y_i \qquad \text{for all } i=1,\ldots,8 \quad \text{(maximum at } i)$$

$$x_{i,j}\ge 0 \qquad \text{for all } i=1,\ldots,8; \quad j=1,\ldots,14$$

$$y_i=0 \text{ or } 1 \qquad \text{for all } i=1,\ldots,8$$

where $x_{i,j}$ if the fraction of region j's call traffic handled by center i, y_i decides whether or not center i is opened, d_j is the anticipated call demand from region j, $r_{i,j}$ is the unit cost of calls from region j to center i, and f_i is the fixed cost of opening center i.

Focus on the third, maximum capacity set of constraints. Each forces discrete variable y_i to take on a value in the relaxation satisfying

$$y_i \ge \frac{\sum_{j=1}^{14}d_j x_{i,j}}{5000} \overset{\Delta}{=} \frac{\text{capacity used}}{\text{total available}}$$

For discrete modeling, these constraints do fine. Each y_i must equal 1 if corresponding x-variables are to use facility i at any level. In the LP relaxation, however, if x-variables use only a small part of the capacity, the corresponding y_i will take on a small fractional value.

The numerical values of Section 11.6 confirm this behavior. The LP relaxation of formulation (12.8) has

$$\tilde{y}_1 = 0.230, \quad \tilde{y}_2 = 0.000, \quad \tilde{y}_3 = 0.000, \quad \tilde{y}_4 = 0.301$$
$$\tilde{y}_5 = 0.115, \quad \tilde{y}_6 = 0.000, \quad \tilde{y}_7 = 0.000, \quad \tilde{y}_8 = 0.650 \qquad (12.9)$$
$$\text{total cost} = \$8036.60$$

with many of the $\tilde{y}_j$ small.

Compare the optimal mixed-integer solution

$$y_1^* = 0, \quad y_2^* = 0, \qquad y_3^* = 0, \quad y_4^* = 1$$
$$y_5^* = 0, \quad y_6^* = 0, \qquad y_7^* = 0, \quad y_8^* = 1 \qquad (12.10)$$
$$\text{total cost} = \$10,153$$

Bound \$8036 of (12.9) is only 79% of true optimal value \$10,153. Also, (12.9) suggests that 4 centers may be needed, while the optimum opens only 2.

Even when a center is used only fractionally, it may fulfill the whole demand for some single district. Such thinking suggests valid inequalities

$$x_{i,j}\le y_i \qquad \text{for all } i=1,\ldots,8; \quad j=1,\ldots,14 \qquad (12.11)$$

which require that the fraction a center is opened be as great as the fraction of any region's demand satisfied from the center.

Certainly, these inequalities satisfy validity definition $\boxed{12.16}$ because each is satisfied by every integer-feasible solution to model (12.8). Also, it is easy to fulfill requirement $\boxed{12.17}$ by finding solutions to the relaxation of (12.8) that violate (12.11).

Adding these valid inequalities improves the LP relaxation dramatically. The strengthened model has optimal solution

$$\tilde{y}_1 = 0.000, \quad \tilde{y}_2 = 0.000, \quad \tilde{y}_3 = 0.000, \quad \tilde{y}_4 = 0.537$$
$$\tilde{y}_5 = 0.000, \quad \tilde{y}_6 = 0.000, \quad \tilde{y}_7 = 0.000, \quad \tilde{y}_8 = 1.000$$
$$\text{total cost} = \$10,033.68$$

Its bound $10,033 is almost 99% of optimal value $10,153, and only one discrete variable comes out fractional. Addition of the valid inequalities (12.11) has produced a much stronger relaxation, which provides much better information about the form of a discrete optimum.

SAMPLE EXERCISE 12.11: RECOGNIZING USEFUL VALID INEQUALITIES

Consider the ILP

$$\begin{aligned} \max \quad & 3x_1 + 14x_2 + 18x_3 \\ \text{s.t.} \quad & 3x_1 + 5x_2 + 6x_3 \le 10 \\ & x_1, x_2, x_3 = 0 \text{ or } 1 \end{aligned}$$

with LP relaxation optimum $\tilde{\mathbf{x}} = (0, \frac{4}{5}, 1)$. Determine (by inspection) whether each of the following inequalities is valid for this model, and if so, whether adding it would strengthen the LP relaxation.

(a) $x_2 + x_3 \le 1$

(b) $x_1 + x_2 + x_3 \le 1$

(c) $3x_1 + 5x_2 \le 10$

Analysis: We apply definition $\boxed{12.16}$ and principle $\boxed{12.17}$.

(a) It is obvious from the main constraint that no feasible solution can have both $x_2 = 1$ and $x_3 = 1$. Thus the constraint is valid. Also, the current LP relaxation optimum is one LP-feasible solution that violates the inequality because

$$\tilde{x}_2 + \tilde{x}_3 = \tfrac{4}{5} + 1 \not\le 1$$

It follows that the constraint will strengthen the relaxation.

(b) This constraint is not valid. For example, $\mathbf{x} = (1, 0, 1)$ violates the constraint even though it is integer-feasible in the given model.

(c) This constraint is valid, because any integer-feasible solution satisfying main constraint $3x_1 + 5x_2 + 6x_3 \le 10$ certainly has $3x_1 + 5x_2 \le 10$. Still, this will also be true of all feasible solutions in the LP relaxation. Adding the inequality cannot improve the relaxation.

Lagrangian Relaxations

Even when the given model is an ILP, the strongest practical relaxation may not be the LP form obtained when integrality constraints are dropped. **Lagrangian relaxations**, which prove stronger for some model forms, adopt a completely different strategy. Instead of dropping integrality requirements, they relax some of the main linear constraints of the model. However, the relaxed constraints are not totally dropped. Instead, they are **dualized** or weighted in the objective function with suitable **Lagrange multipliers** to discourage violations.

> __12.18__ | **Lagrangian relaxations** partially relax some of the main, linear constraints of an ILP by moving them to the objective function as terms
>
> $$\cdots + v_i \left(b_i - \sum_j a_{i,j} x_j \right) \cdots$$
>
> Here v_i is a Lagrange multiplier chosen as the relaxation is formed. If the relaxed constraint has form $\sum_j a_{i,j} x_j \geq b_i$, multiplier $v_i \leq 0$ for a maximize model and $v_i \geq 0$ in a minimize. If the relaxed constraint is $\sum_j a_{i,j} x_j \leq b_i$, multiplier $v_i \geq 0$ for a maximize model and $v_i \leq 0$ for a minimize model. Equality constraints $\sum_j a_{i,j} x_j = b_i$ have unrestricted multipliers v_i.

Lagrangian Relaxation of the CDOT Example

We can illustrate with the CDOT generalized assignment model (11.13) of Section 11.4:

$$
\begin{aligned}
\min \quad & 130x_{1,1} + 460x_{1,2} + 40x_{1,3} + 30x_{2,1} + 150x_{2,2} + 370x_{2,3} \\
& + 510x_{3,1} + 20x_{3,2} + 120x_{3,3} + 30x_{4,1} + 40x_{4,2} + 390x_{4,3} \\
& + 340x_{5,1} + 30x_{5,2} + 40x_{5,2} + 20x_{6,1} + 450x_{6,2} + 30x_{6,3}
\end{aligned}
$$

s.t.	$x_{1,1} + x_{1,2} + x_{1,3} = 1$	(district 1)
	$x_{2,1} + x_{2,2} + x_{2,3} = 1$	(district 2)
	$x_{3,1} + x_{3,2} + x_{3,3} = 1$	(district 3)
	$x_{4,1} + x_{4,2} + x_{4,3} = 1$	(district 4)
	$x_{5,1} + x_{5,2} + x_{5,3} = 1$	(district 5)
	$x_{6,1} + x_{6,2} + x_{6,3} = 1$	(district 6)

$$(12.12)$$

$$
\begin{aligned}
& 30x_{1,1} + 50x_{2,1} + 10x_{3,1} \\
& \quad + 11x_{4,1} + 13x_{5,1} + 9x_{6,1} \quad \leq 50 \qquad \text{Estevan} \\
& 10x_{1,2} + 20x_{2,2} + 60x_{3,2} \\
& \quad + 10x_{4,2} + 10x_{5,2} + 17x_{6,2} \quad \leq 50 \qquad \text{(Mackenzie)} \\
& 70x_{1,3} + 10x_{2,3} + 10x_{3,3} \\
& \quad + 15x_{4,3} + 8x_{5,3} + 12x_{6,3} \quad \leq 50 \qquad \text{(Skidegate)} \\
& x_{i,j} = 0 \text{ or } 1 \qquad i = 1, 6; j = 1, 3
\end{aligned}
$$

where

$$x_{i,j} \triangleq \begin{cases} 1 & \text{if district } i \text{ is assigned to ship } j \\ 0 & \text{otherwise} \end{cases}$$

One strong Lagrangian relaxation dualizes the first 6 main constraints with weights

$$v_i \triangleq \text{ the Lagrange multiplier on the constraint for district } i$$

The result is

$$
\begin{aligned}
\min \quad & 130x_{1,1} + 460x_{1,2} + 40x_{1,3} + 30x_{2,1} + 150x_{2,2} + 370x_{2,3} \\
& + 510x_{3,1} + 20x_{3,2} + 120x_{3,3} + 30x_{4,1} + 40x_{4,2} + 390x_{4,3} \\
& + 340x_{5,1} + 30x_{5,2} + 40x_{5,2} + 20x_{6,1} + 450x_{6,2} + 30x_{6,3} \\
& + v_1(1 - x_{1,1} - x_{1,2} - x_{1,3}) + v_2(1 - x_{2,1} - x_{2,2} - x_{2,3}) \\
& + v_3(1 - x_{3,1} - x_{3,2} - x_{3,3}) + v_4(1 - x_{4,1} - x_{4,2} - x_{4,3}) \\
& + v_5(1 - x_{5,1} - x_{5,2} - x_{5,3}) + v_6(1 - x_{6,1} - x_{6,2} - x_{6,3})
\end{aligned}
\tag{12.13}
$$

$$
\begin{aligned}
\text{s.t.} \quad & 30x_{1,1} + 50x_{2,1} + 10x_{3,1} + 11x_{4,1} + 13x_{5,1} + 9x_{6,1} && \leq 50 \\
& 10x_{1,2} + 20x_{2,2} + 60x_{3,2} + 10x_{4,2} + 10x_{5,2} + 17x_{6,2} && \leq 50 \\
& 70x_{1,3} + 10x_{2,3} + 10x_{3,3} + 15x_{4,3} + 8x_{5,3} + 12x_{6,3} && \leq 50 \\
& x_{i,j} = 0 \text{ or } 1 \quad i = 1, 6; \ j = 1, 3
\end{aligned}
$$

Notice that the 6 equality constraints of full model (12.12) have not been completely dropped. Instead, they have been rolled into the objective function as in construction $\boxed{12.18}$. For example, the model (12.13) objective now includes the term

$$\cdots + v_3(1 - x_{3,1} - x_{3,2} - x_{3,3}) \cdots \tag{12.14}$$

Feasible solutions in this Lagrangian relaxation may very well have

$$x_{3,1} + x_{3,2} + x_{3,3} \neq 1 \quad \text{or} \quad (1 - x_{3,1} - x_{3,2} - x_{3,3}) \neq 0$$

But if weight $v_3 \neq 0$, violations will at least affect the objective value through (12.14).

SAMPLE EXERCISE 12.12: FORMING LAGRANGIAN RELAXATIONS

Return to Bison Boosters example model (12.2):

$$
\begin{aligned}
\max \quad & 20x_1 + 30x_2 - 550y_1 - 720y_2 \\
\text{s.t.} \quad & 1.5x_1 + 4x_2 \leq 300 \\
& x_1 - 200y_1 \leq 0 \\
& x_2 - 75y_2 \leq 0 \\
& x_1, x_2 \geq 0 \\
& y_1, y_2 = 0 \text{ or } 1
\end{aligned}
$$

(a) Use multipliers v_1 and v_2 to form a Lagrangian relaxation dualizing the last two, switching constraints.

(b) Indicate any required sign restrictions on multipliers v_1 and v_2.

Analysis: We apply construction 12.18 .

(a) The Lagrangian relaxation is formed by moving the two switching constraints to the objective function as

$$\max \quad 20x_1 + 30x_2 - 550y_1 - 720y_2 + v_1(0 - x_1 + 200y_1) + v_2(0 - x_2 + 75y_2)$$
$$\text{s.t.} \quad 1.5x_1 + 4x_2 \leq 300$$
$$x_1, x_2 \geq 0$$
$$y_1, y_2 = 0 \text{ or } 1$$

(b) For $\leq$ constraints in a maximize model, multipliers should satisfy $v_1, v_2 \geq 0$.

Tractable (Integer) Lagrangian Relaxations

Notice that the Lagrangian relaxation (12.13) keeps variables $x_{i,j}$ discrete. Integrality requirements of the original (12.12) have not been dropped.

Lagrangian relaxations fulfill principle 12.5 's mandate for increased tractability by dualizing enough linear constraints that the remaining discrete problem is manageable.

> 12.19 │ Constraints chosen for dualization in Lagrangian relaxations should leave a still-integer linear program with enough special structure to be relatively tractable.

A close look at the remaining constraints in Lagrangian relaxation (12.13) will reveal how it conforms to requirement 12.19 . After dualization, each $x_{i,j}$ occurs in exactly one main constraint. Thus these relaxations can be solved as a series of single-constraint, knapsack ILPs (definition 11.4 , Section 11.2), which are the simplest of integer programs. There is one for each of the 3 ships.

Lagrangian Relaxation Bounds

Dropping linear constraints in a Lagrangian relaxation 12.18 cannot eliminate any solutions. That is, Lagrangian relaxations parallel property 12.4 in having every solution feasible in the full model still feasible in the relaxation. But Lagrangian forms are more complex than constraint relaxations because they modify both constraints and objective. Fortunately, they still yield bounds.

> 12.20 │ The optimal value of any Lagrangian relaxation of a maximize model using multipliers conforming to 12.18 yields an upper bound on the optimal value of the full model. The optimal value of any valid Lagrangian relaxation of a minimize model yields a lower bound.

Sign rules of definition 12.18 assure that every feasible solution in the original ILP achieves no less in the objective function of the Lagrangian relaxation when we maximize and no more when we minimize. Thus the relaxation optimum, which either equals or improves on all these results for truly feasible solutions, must yield a bound.

Choosing Lagrange Multipliers

How strong the bounds of $\boxed{12.20}$ prove to be in any Lagrangian relaxation depends on the multiplier values chosen. Some choices of v_i will produce a very weak Lagrangian relaxation, and others can make it quite strong.

$\boxed{12.21}$ A search is usually required to identify Lagrange multiplier values v_i defining a strong Lagrangian relaxation.

Methods for determining good Lagrange multipliers are beyond the scope of this book, but we can illustrate their potential power with

$$v_1 = 300, \quad v_2 = 200, \quad v_3 = 200, \quad v_4 = 45, \quad v_5 = 45, \quad v_6 = 30$$

in our CDOT example. The corresponding optimal value in Lagrangian relaxation (12.13) is \$470,000, which is much closer to the integer optimal value of \$480,000 than is the LP relaxation bound \$326,100.

SAMPLE EXERCISE 12.13: UNDERSTANDING LAGRANGE MULTIPLIER IMPACTS

Return to the Bison Boosters Lagrangian relaxation of Sample Exercise 12.12. Solve (by inspection) the relaxation for each of the following choices of Lagrange multipliers, and comment on the strength of the results.

(a) $v_1 = 0, v_2 = 1$

(b) $v_1 = 3, v_2 = 3$

Analysis:

(a) For $v_1 = 0, v_2 = 1$, the Lagrangian relaxation reduces to

$$\begin{aligned} \max \quad & 20x_1 + 29x_2 - 550y_1 - 645y_2 \\ \text{s.t.} \quad & 1.5x_1 + 4x_2 \le 300 \\ & x_1, x_2 \ge 0 \\ & y_1, y_2 = 0 \text{ or } 1 \end{aligned}$$

With only 0–1 constraints enforced on the y_j, setting $\tilde{y}_1 = \tilde{y}_2 = 0$ produces a relaxation optimum. Corresponding optimal choices for the continuous variables are $\tilde{x}_1 = 200$ and $\tilde{x}_2 = 0$. The resulting relaxation bound of \$4000 is weak because the mixed-integer optimal value is \$3450.

(b) For $v_1 = v_2 = 3$, the Lagrangian relaxation is

$$\begin{aligned} \max \quad & 17x_1 + 27x_2 + 50y_1 - 495y_2 \\ \text{s.t.} \quad & 1.5x_1 + 4x_2 \le 300 \\ & x_1, x_2 \ge 0 \\ & y_1, y_2 = 0 \text{ or } 1 \end{aligned}$$

With only 0–1 constraints enforced on the y_j, setting $\tilde{y}_1 = 1, \tilde{y}_2 = 0$ produces a relaxation optimum. Corresponding optimal choices for the continuous variables are again $\tilde{x}_1 = 200, \tilde{x}_2 = 0$. This time, the resulting relaxation bound of \$3450 is very strong because it exactly matches the true mixed-integer value.

12.4 BRANCH AND BOUND SEARCH

Total enumerations of Section 12.1 are impractical for all but the simplest models because every one of an explosively growing number of discrete solutions must be considered explicitly. The process would become much more manageable if we could deal with those solutions in large classes, determining for each whole class whether it is likely to contain optimal solutions, and doing so without explicit enumeration of all its members. Only the most promising classes would have to be searched in detail.

Branch and bound algorithms combine such a partial or subset enumeration strategy with the relaxations of Sections 12.2 and 12.3. They systematically form classes of solutions and investigate whether the classes can contain optimal solutions by analyzing associated relaxations. More detailed enumeration ensues only if the relaxations fail to be definitive.

EXAMPLE 12.2: RIVER POWER

As with so many other topics, an artificially small example will aid in our development of branch and bound ideas. Here we consider an operations problems at River Power Company.

River Power has 4 generators currently available for production and wishes to decide which to put on line to meet the expected 700-megawatt peak demand over the next several hours. The following table shows the cost to operate each generator (in thousands of dollars per hour) and their outputs (in megawatts).

	Generator, j			
	1	**2**	**3**	**4**
Operating cost	7	12	5	14
Ouput power	300	600	500	1600

Units must be completely on or completely off.

We can formulate River Power's problem as a knapsack problem like those of Section 11.2. Decision variables

$$x_j \triangleq \begin{cases} 1 & \text{if generator } j \text{ is turned on} \\ 0 & \text{otherwise} \end{cases}$$

Then a model is

$$
\begin{array}{lll}
\min & 7x_1 + 12x_2 + 5x_3 + 14x_4 & \text{(total cost)} \\
\text{s.t.} & 300x_1 + 600x_2 + 500x_3 + 1600x_4 \geq 700 & \text{(demand)} \\
& x_1, x_2, x_3, x_4 = 0 \text{ or } 1
\end{array}
\qquad (12.15)
$$

The objective function minimizes total operating costs, and the main constraint assures that the chosen combination of generators will fulfill demand. Total enumeration establishes that an optimal solution use generators 1 and 3 and cost $12,000.

Partial Solutions

Much like the improving searches of most of this book, branch and bound searches iterate through a sequence of solutions until we are ready to conclude optimality or stop with the best fully feasible solution found so far. What is new is that branch and bound searches through partial solutions.

> 12.22 | A **partial solution** has some decision variables fixed, with others left **free** or undetermined. We denote free components of a partial solution by the symbol #.

For example, in the River Power model (12.15), $x = (1, \#, 0, \#)$ specifies a partial solution with $x_1 = 1$, $x_3 = 0$, while x_2 and x_4 remain free.

Completions of Partial Solutions

Each partial solution implicitly defines a class of full solutions called its completions.

> 12.23 | The **completions** of a partial solution to a given model are the possible full solutions agreeing with the partial solution on all fixed components.

For instance, the completions of partial solution $x = (1, \#, \#, 0)$ in our River Power model are

$$(1, 0, 0, 0), \quad (1, 0, 1, 0), \quad (1, 1, 0, 0), \quad \text{and} \quad (1, 1, 1, 0)$$

Every solution with $x_1 = 1$ and $x_4 = 0$ is among these 4. The last 3 are **feasible completions** because they satisfy all constraints of model (12.15).

> **SAMPLE EXERCISE 12.14: UNDERSTANDING PARTIAL SOLUTIONS AND COMPLETIONS**
>
> Suppose an integer program as decision variables $x_1, x_2, x_3 = 0$ or 1. List all completions of each of the following partial solutions.
>
> (a) $(1, \#, \#)$
>
> (b) $(1, \#, 0)$
>
> *Analysis:* We apply definitions 12.22 and 12.23 .
>
> (a) Completions of this partial solution consist of all full solutions with $x_1 = 1$ [i.e., $(1, 0, 0)$, $(1, 0, 1)$, $(1, 1, 0)$, and $(1, 1, 1)$].
>
> (b) Completions of this partial solution consist of all full solutions with $x_1 = 1$ and $x_3 = 0$ [i.e., $(1, 0, 0)$ and $(1, 1, 0)$].

Tree Search

Branch and bound investigates classes of solutions corresponding to completions of partial solutions in a treelike fashion that gives it the "branch" part of its name. Figure 12.2 provides a full example for River Power model (12.15). Nodes of this **branch and bound tree** represent partial solutions, with numbers indicating the sequence in which they are investigated. Edges or links of the tree specify how variables are fixed in partial solutions. For example, partial solution $x^{(6)}$ in the sequence has $x_4 = 0$ and $x_2 = 0$.

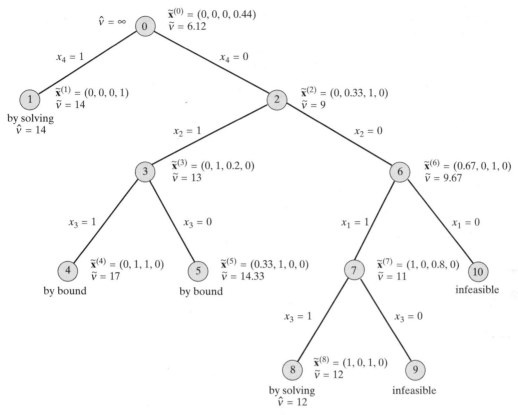

$\tilde{\mathbf{x}}^{(0)} = (0, 0, 0, 0.44)$
$\tilde{v} = 6.12$

$\hat{v} = \infty$ ⓪

$x_4 = 1$ $x_4 = 0$

① $\tilde{\mathbf{x}}^{(1)} = (0, 0, 0, 1)$
$\tilde{v} = 14$
by solving
$\hat{v} = 14$

② $\tilde{\mathbf{x}}^{(2)} = (0, 0.33, 1, 0)$
$\tilde{v} = 9$

$x_2 = 1$ $x_2 = 0$

③ $\tilde{\mathbf{x}}^{(3)} = (0, 1, 0.2, 0)$
$\tilde{v} = 13$

⑥ $\tilde{\mathbf{x}}^{(6)} = (0.67, 0, 1, 0)$
$\tilde{v} = 9.67$

$x_3 = 1$ $x_3 = 0$ $x_1 = 1$ $x_1 = 0$

④ $\tilde{\mathbf{x}}^{(4)} = (0, 1, 1, 0)$
$\tilde{v} = 17$
by bound

⑤ $\tilde{\mathbf{x}}^{(5)} = (0.33, 1, 0, 0)$
$\tilde{v} = 14.33$
by bound

⑦ $\tilde{\mathbf{x}}^{(7)} = (1, 0, 0.8, 0)$
$\tilde{v} = 11$

⑩ infeasible

$x_3 = 1$ $x_3 = 0$

⑧ $\tilde{\mathbf{x}}^{(8)} = (1, 0, 1, 0)$
$\tilde{v} = 12$
by solving
$\hat{v} = 12$

⑨ infeasible

FIGURE 12.2   Branch and Bound Tree for River Power Example

The process begins at root node 0.

12.24 | Branch and bound search begins at initial or **root** partial solution $\mathbf{x}^{(0)} = (\#, \ldots, \#)$ with all variables free.

This provides the first **active** or unanalyzed partial solution.

At any stage of the search one or more active nodes (distinguished by not yet having a number) remain in the tree. Analysis of each node or partial solution attempts to decide which, if any, completions warrants consideration as an overall optimal solution. Sometimes we can either find a best completion or conclude that none is worth further investigation. Then we **terminate** or **fathom** the entire class of (completion) solutions represented by the node. That is, we give it no further consideration.

12.25 | Branch and bound searches terminate or fathom a partial solution when they either identify a best completion or prove that none can produce an optimal solution in the overall model.

Node 1 in Figure 12.2 illustrates termination. It has no subsidiary nodes because analysis of partial solution $\mathbf{x}^{(1)} = (\#, \#, \#, 1)$, which we detail below, established that the best possible completion is $\mathbf{x} = (0, 0, 0, 1)$. No further investigation of any solution with $x_4 = 1$ is required.

Observe that this termination dealt, in a single step, with fully half the possible solutions to the River Power model. That is, we enumerated the half of all solutions with $x_4 = 1$ as a class. In this way, the exponentially growing effort to totally enumerate every member was avoided.

Unfortunately, it often happens that analysis is not definitive. In such cases the node or partial solution must be branched.

> **12.26** When a partial solution cannot be terminated in a branch-and-bound search of a 0–1 discrete optimization model, it is **branched** by creating 2 subsidiary partial solutions derived by fixing a previously free binary variable. One of these partial solutions matches the current except that the variable chosen is fixed $= 1$, and the other is identical except that the variable is fixed $= 0$.

Node 2 of Figure 12.2 illustrates the need for branching. Analysis (detailed below) was unable either to find the best completion of partial solution $\mathbf{x}^{(2)} = (\#, \#, \#, 0)$ or to prove that none could be optimal. Thus the node was branched into those numbered 3 and 6. Both have $x_4 = 0$, as in the $\mathbf{x}^{(2)}$. However, previously free variable x_2 has now been fixed. In partial solution 3, it is fixed $= 1$. In partial solution 6, it is fixed $= 0$.

Notice that this branching process loses no solutions. Every completion of node 2 has either $x_2 = 0$ or $x_2 = 1$. We have simply constructed 2 smaller classes of solutions in the hope that our analysis will now be strong enough to permit termination.

Since no solutions are lost, the enumeration is complete when all partial solutions have been resolved definitively.

> **12.27** Branch and bound search stops when every partial solution in the tree has been either branched or terminated.

As long as partial solutions do remain, branch and bound search must select an active one to explore next. The simplest such scheme is known as depth first.

> **12.28** **Depth first** search selects at each iteration an active partial solution with the most components fixed (i.e., one deepest in the search tree).

The River Power enumeration of Figure 12.2 employs this depth first rule. For example, after node 3 had been explored, the partial solutions corresponding to nodes 4, 5, and 6 were active in the tree. In accord with depth first rule **12.28**, one of the deeper nodes 4 and 5 was selected to investigate next.

SAMPLE EXERCISE 12.15: UNDERSTANDING BRANCH AND BOUND TREES

The following is the branch and bound tree for a discrete optimization model with decision variables $x_1, x_2, x_3 = 0$ or 1.

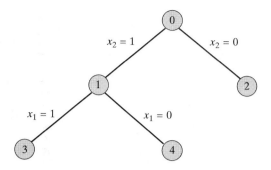

(a) List the sequence of partial solutions explored.

(b) Identify which of the partial solutions in part (a) were terminated and which branched.

(c) Determine which nodes would have been active just after processing of node 1, and explain which could have been the next explored under the depth-first enumeration rule.

(d) Demonstrate that all possible solutions were implicitly enumerated by specifying for each the node at which it was terminated.

Analysis:

(a) Under principle $\boxed{12.24}$, the first partial solution $\mathbf{x}^{(0)}$ of a branch and bound search is always the all-free (#, #, #). From variable restrictions on branches, we can see subsequent partial solutions visited were $\mathbf{x}^{(1)} = (\#, 1, \#)$, $\mathbf{x}^{(2)} = (\#, 0, \#)$, $\mathbf{x}^{(3)} = (1, 1, \#)$, and $\mathbf{x}^{(4)} = (0, 1, \#)$.

(b) A lack of subsidiary nodes shows that partial solutions 2, 3, and 4 were terminated. Remaining nodes 0 and 1 were branched.

(c) After processing node 1, 3 active partial solutions remained in the tree. None would then have had a number, but they eventually became node 2 created when partial 0 was branched, together with nodes 3 and 4 formed at 1. Depth first enumeration would have taken the search to either of the deeper nodes 3 and 4. However, the search in this example proceeded, instead, to node 2.

(d) A full solution is implicitly enumerated when a completion class to which it belongs is terminated. The following table shows the node at which each of the 8 solutions was terminated in this case.

Solution	Node	Solution	Node	Solution	Node	Solution	Node
(0, 0, 0)	2	(0, 1, 0)	4	(1, 0, 0)	2	(1, 1, 0)	3
(0, 0, 1)	2	(0, 1, 1)	4	(1, 0, 1)	2	(1, 1, 1)	3

Incumbent Solutions

Whether explicit or implicit, the goal of any enumeration is to identify an optimal (or at least a good feasible) solution to some optimization model. To that end, it is essential to keep track of the best known or incumbent solution.

12.29 │ The **incumbent solution** at any stage in a search of a discrete model is the best (in terms of objective value) feasible solution known so far. We denote the incumbent solution $\hat{\mathbf{x}}$ and its objective function value $\hat{v}$.

The incumbent solution may derive from experience prior to the search, or it may have been discovered as the search evolved.

When the search stops, the last incumbent solution is its output. Assuming that the given model has an optimal solution, the final incumbent at least provides an approximate optimum, $\hat{\mathbf{x}}$, with corresponding incumbent solution value $\hat{v}$. If the search was fully carried out, the optimum is exact.

12.30 │ If a branch and bound search stops as in │ 12.27 │, with all partial solutions having been either branched or terminated, the final incumbent solution is a global optimum if one exists. Otherwise, the model is infeasible.

SAMPLE EXERCISE 12.16: UNDERSTANDING INCUMBENT SOLUTIONS

Return to the branch and bound tree of Sample Exercise 12.15 and assume (i) that we were maximizing, (ii) that from prior experience we knew a feasible solution with objective value 10, and that (iii) analysis leading to terminations at nodes 2, 3, and 4 showed the best feasible completions of the corresponding partial solutions have objective values 8, 14, and 12, respectively.

(a) Show the sequence of incumbent solution objective values.

(b) Determine the optimal solution value.

Analysis:

(a) From assumption (ii) the initial incumbent solution value would have been $\hat{v} = 10$. This value held until node 3 because node 1 was branched and the best completion of node 2 did not improve on 10. At node 3, the search uncovered a feasible solution with better value 14, so this became the incumbent solution value to $\hat{v} = 14$. Node 4 produced no change.

(b) This search implicitly enumerated all possible solutions. Thus, by the principle │ 12.30 │, the final incumbent solution value $\hat{v} = 14$ is the optimal value.

Candidate Problems

Having introduced the tree search underlying branch and bound, we are now ready to see how relaxations of Section 12.2 and 12.3 can make it efficient (and justify the "bound" part of its name). Candidate problems provide the linkage.

12.31 │ The **candidate problem** associated with any partial solution to an optimization model is the restricted version of the model obtained when variables are fixed as in the partial solution.

We may illustrate with partial solution $\mathbf{x}^{(3)} = (\#, 1, \#, 0)$ in River Power Figure 12.2. The corresponding candidate problem is

$$\begin{array}{ll} \min & 7x_1 + 12x_2 + 5x_3 + 14x_4 \\ \text{s.t.} & 300x_1 + 600x_2 + 500x_3 + 1600x_4 \geq 700 \\ & x_1, x_3 = 0 \text{ or } 1 \\ & x_2 = 1, x_4 = 0 \end{array}$$

It derives from original model (12.15) by restricting x_2 and x_4 to their partial solution values.

Thinking about candidate problems aids branch and bound search because of the close connection with completions of the corresponding partial solution.

> **12.32** The feasible completions of any partial solution are exactly the feasible solutions to the corresponding candidate problem, and thus the objective value of the best feasible completion is the optimal objective value of the candidate problem.

That is, we can search for a best completion of any partial solution, or at least learn something about a solution's objective value, by trying to optimize the corresponding candidate problem.

SAMPLE EXERCISE 12.17: UNDERSTANDING CANDIDATE PROBLEMS

Consider the ILP

$$\begin{array}{ll} \max & 10w_1 + 3w_2 + 9w_3 \\ \text{s.t.} & 6w_1 + 4w_2 + 3w_3 \leq 10 \\ & w_2 - w_3 \geq 0 \\ & w_1, w_2, w_3 = 0 \text{ or } 1 \end{array}$$

(a) State the candidate problem corresponding to partial solution $\mathbf{w} = (1, \#, 0)$.

(b) State the LP relaxation of the candidate problem corresponding to partial solution $\mathbf{w} = (1, \#, 0)$.

Analysis:

(a) Following definition 12.31 , the required candidate problem is the restricted version with $w_1 = 1, w_3 = 0$:

$$\begin{array}{ll} \max & 10w_1 + 3w_2 + 9w_3 \\ \text{s.t.} & 6w_1 + 4w_2 + 3w_3 \leq 10 \\ & w_2 - w_3 \geq 0 \\ & w_1 = 1, \ w_3 = 0 \\ & w_2 = 0 \text{ or } 1 \end{array}$$

An optimal solution provides a best completion of the partial $\mathbf{w} = (1, \#, 0)$ (principle 12.32).

(b) Applying definition 12.6 , the LP relaxation of the candidate problem in part (a) is

$$\begin{aligned}
\max \quad & 10w_1 + 3w_2 + 9w_3 \\
\text{s.t.} \quad & 6w_1 + 4w_2 + 3w_3 \le 10 \\
& w_2 - w_3 \ge 0 \\
& w_1 = 1, \ w_3 = 0 \\
& 0 \le w_2 \le 1
\end{aligned}$$

Notice that it is a relaxation of the candidate problem in part (a), not a relaxation of the full model.

Terminating Partial Solutions with Relaxations

Now we are ready to take advantage of the relaxation principles in Section 12.2. We analyze nodes in a branch and bound tree by solving relaxations of the corresponding candidate problems. For example, relaxation solutions and solution values shown next to the nodes in River Power Figure 12.2 (denoted $\tilde{\mathbf{x}}$) come from the linear programming relaxation of the corresponding candidate problem.

Begin with infeasibility principle $\boxed{12.8}$. If any relaxation of a candidate problem is infeasible, so is the full candidate.

> $\underline{12.33}$ If any relaxation of a candidate problem proves infeasible, the associated partial solution can be terminated because it has no feasible completions.

Node 10 of Figure 12.2 illustrates. The LP relaxation of the corresponding candidate problem is infeasible. It follows that partial solution $\mathbf{x}^{(10)} = (0, 0, \#, 0)$ has no feasible completions, and we may terminate infeasible.

Now consider relaxation bound principle $\boxed{12.9}$. Relaxation optimal values bound those of the problem relaxed. Thus in the context of candidate problems, relaxation optimal values bound the objective value of the best possible completion. Comparison to the incumbent solution value (definition $\boxed{12.29}$) can lead to termination.

> $\underline{12.34}$ If any relaxation of a candidate problem has optimal objective value no better than the current incumbent solution value, the associated partial solution can be terminated because no feasible completion can improve on the incumbent.

Node 5 of Figure 12.2 illustrates such termination by bound. When the search reached partial solution $\mathbf{x}^{(5)} = (\#, 1, 0, 0)$, incumbent solution $\hat{\mathbf{x}} = (0, 0, 0, 1)$ with objective value $\hat{v} = 14$ was already in hand (from node 1). The linear programming relaxation of the candidate problem for node 5 had optimal value 14.33, which means that no feasible completion can do better than that value. It follows that none can improve on the incumbent solution in this minimizing problem, and we terminate by bound.

The third way to terminate with relaxations derives from optimality principle $\boxed{12.10}$.

> **12.35** If an optimal solution to any constraint relaxation of a candidate problem is feasible in the full candidate, it is a best feasible completion of the associated partial solution. After checking whether a new incumbent has been discovered, the partial solution can be terminated.

Consider node 1 of River Power Figure 12.2. The corresponding LP relaxation fixed $x_4 = 1$ but allowed free variables to take on any value between 0 and 1. Still, relaxation optimum $\tilde{\mathbf{x}} = (0, 0, 0, 1)$ meets integrality requirements on all components. It is optimal in the corresponding candidate problem and the best possible completion of partial solution $\mathbf{x}^{(1)} = (\#, \#, \#, 1)$. It is also the first fully feasible solution encountered in the search, so it provides the first incumbent solution. After saving it as the incumbent, node 1 was terminated by solving.

SAMPLE EXERCISE 12.18: TERMINATING PARTIAL SOLUTIONS WITH RELAXATIONS

Suppose that a maximizing branch and bound search over $y_1, \ldots, y_4 = 0$ or 1 reaches partial solution $\mathbf{y}^{(3)} = (\#, 0, \#, \#)$ with incumbent solution value $\hat{v} = 100$. Explain how the search should proceed assuming each of the following outcomes from an attempt to solve the linear programming relaxation of the corresponding candidate problem.

(a) Relaxation optimum $\tilde{\mathbf{y}} = (\frac{1}{3}, 0, 1, 0)$ with objective value $\tilde{v} = 85$.

(b) Relaxation optimum $\tilde{\mathbf{y}} = (1, 0, \frac{1}{2}, 0)$ with objective value $\tilde{v} = 100$.

(c) Relaxation optimum $\tilde{\mathbf{y}} = (0, 0, 1, 1)$ with objective value $\tilde{v} = 120$.

(d) Relaxation infeasible.

(e) Relaxation optimum $\tilde{\mathbf{y}} = (0, \frac{1}{4}, 1, 0)$ with objective value $\tilde{v} = 111$.

Analysis:

(a) The relaxation bound demonstrates no feasible completion can do better than 85 in objective value, which is worse than the known incumbent. The partial solution should be terminated as in principle 12.34 .

(b) The relaxation bound demonstrates that no feasible completion can do better than 100 in objective value, which is the same as the incumbent. The partial solution should be terminated as in principle 12.34 (unless alternative optimal solutions are of interest).

(c) This relaxation optimum is feasible in the full candidate problem and thus optimal. Having found the best possible completion, we terminate by principle 12.35 after updating the incumbent solution value to improved value $\hat{v} \leftarrow 120$.

(d) There are no feasible completions. The partial solution should be terminated as in principle 12.33 .

(e) Here none of principles $\boxed{12.33}$ to $\boxed{12.35}$ lead to termination because the relaxation optimum is better in objective value than the incumbent but still fractional in some components. The partial solution must be branched as in principle $\boxed{12.26}$.

LP-Based Branch and Bound

The discrete models most frequently solved by branch and bound are ILPs with 0–1 variables. Linear programming relaxations of candidate problems usually provide the basis for analysis.

Algorithm 12A details a formal algorithm for this **LP-based branch and bound** case. For simplicity we assume that the model is not unbounded.

ALGORITHM 12A: LP-BASED BRANCH AND BOUND (0–1 ILPS)

Step 0: Initialization. Make the only active partial solution the one with all discrete variables free, and initialize solution index $t \leftarrow 0$. If any feasible solutions are known for the model, also choose the best as incumbent solution $\hat{x}$ with objective value $\hat{v}$. Otherwise, set $\hat{v} \leftarrow -\infty$ if the model maximizes and $\hat{v} \leftarrow +\infty$ if it minimizes.

Step 1: Stopping. If active partial solutions remain, select one as $x^{(t)}$, and proceed to Step 2. Otherwise, stop. If there is an incumbent solution $\hat{x}$, it is optimal, and if not, the model is infeasible.

Step 2: Relaxation. Attempt to solve the linear programming relaxation of the candidate problem corresponding to $x^{(t)}$.

Step 3: Termination by Infeasibility. If the LP relaxation proved infeasible, there are no feasible completions of partial solution $x^{(t)}$. Terminate $x^{(t)}$, increment $t \leftarrow t + 1$, and return to Step 1.

Step 4: Termination by Bound. If the model maximizes and LP relaxation optimal value $\tilde{v}$ satisfies $\tilde{v} \leq \hat{v}$, or it minimizes and $\tilde{v} \geq \hat{v}$, the best feasible completion of partial solution $x^{(t)}$ cannot improve on the incumbent. Terminate $x^{(t)}$, increment $t \leftarrow t + 1$, and return to Step 1.

Step 5: Termination by Solving. If the LP relaxation optimum $\tilde{x}^{(t)}$ satisfies all binary constraints of the model, it provides the best feasible completion of partial solution $x^{(t)}$. After saving it as new incumbent solution

$$\hat{x} \leftarrow \tilde{x}^{(t)}$$
$$\hat{v} \leftarrow \tilde{v}$$

terminate $x^{(t)}$, increment $t \leftarrow t + 1$, and return to Step 1.

Step 6: Branching. Choose some free binary-restricted component x_p that was fractional in the LP relaxation optimum, and branch $x^{(t)}$ by creating two new actives. One is identical to $x^{(t)}$ except that x_p is fixed $= 0$, and the other the same except that x_p is fixed $= 1$. Then increment $t \leftarrow t + 1$ and return to Step 1.

Initialization begins as in principle $\boxed{12.24}$ with all binary variables free. If no incumbent solution $\hat{x}$ is known, the worst possible value $\hat{v} \leftarrow \pm\infty$ is assumed.

Each main iteration begins by selecting some active partial solution to pursue. Any one can be selected, although sequence does make a difference (see Section 12.5).

Processing begins with an attempt to solve the LP relaxation. Then we check to see if any of termination rules $\boxed{12.33}$ to $\boxed{12.35}$ apply. If so, the current partial is terminated and the process repeats. Partial solutions that cannot be terminated must be branched (principle $\boxed{12.26}$).

Branching Rules for LP-Based Branch and Bound

One peculiarity of the LP-based form of branch and bound is that this is done on a fractional free variable.

$\boxed{12.36}$ LP-based branch and bound algorithms always branch by fixing an integer-restricted decision variable that had a fractional value in the associated candidate problem relaxation.

For example, at node 0 of in Figure 12.2, branching occurred by fixing x_4, which had fractional value $\tilde{x}_4 = 0.44$ in the LP relaxation.

The motivation behind this fractional variable branching rule is avoiding duplicate computation. The alternative of branching on a free variable that came out integer in the LP relaxation produces one new candidate problem guaranteed to have the same relaxation optimum as the one just solved. For example, branching on x_1 at node 0 of Figure 12.2 would have created new partial solutions

$$\mathbf{x}^{(1)} = (1, \#, \#, \#) \quad \text{and} \quad \mathbf{x}^{(2)} = (0, \#, \#, \#)$$

But the relaxation optimum for the second would be exactly the same as $\tilde{\mathbf{x}}^{(0)} = (0, 0, 0, 0.438)$ because added constraint $x_1 = 0$ can affect solution value only if it is violated by the previous optimum. We prefer to get new information from both candidates.

Somewhat similar thinking applies when more than one integer variable is fractional in the relaxation optimum.

$\boxed{12.37}$ When more than one integer-restricted variable is fractional in the relaxation optimum, LP-based branch and bound algorithms often branch by fixing the one closest to an integer value.

For example, if the relaxation optimum were $\tilde{\mathbf{x}} = (0.3, 1, 0.5, 0.9)$, and all components were supposed to be binary, rule $\boxed{12.37}$ would next fix x_4 because it is the fractional variable closest to an integer.

The motivation behind rule $\boxed{12.37}$ is to take the most obvious decision, hoping that the farthest away of the two resulting partial solutions can be terminated before it has to be explored. Still, many other schemes are possible. Commercial codes often give the modeler an opportunity to explicitly designate priority variables for branching.

LP-Based Branch and Bound Solution of the River Power Example

We are now ready to fully trace Figure 12.2's branch and bound solution of River Power model (12.15).

The process begins with all-free partial solution $\mathbf{x}^{(0)} = (\#, \#, \#, \#)$ and $\hat{v} = +\infty$. The corresponding LP relaxation optimum is fractional on x_4, so the node is branched (rule $\boxed{12.36}$) as

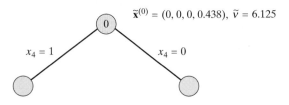

$\tilde{\mathbf{x}}^{(0)} = (0, 0, 0, 0.438)$, $\tilde{v} = 6.125$

$x_4 = 1$ $x_4 = 0$

There are now 2 active partial solutions, and we must choose one to process next. Any active partial solution could be chosen, but all Figure 12.2 computation employs the depth first rule $\boxed{12.28}$. That is, it chooses an active partial solution deepest in the tree.

At the moment both active candidates have equal depth because both have 1 fixed variable. Figure 12.2 adopts the simple tie-breaking rule of always selecting the partial solution with the branching variable fixed $= 1$. (See rule $\boxed{12.45}$ of Section 12.5 for another possibility.)

That tie-breaking rules leads us to $\mathbf{x}^{(1)} = (\#, \#, \#, 1)$. Here the relaxation solution is binary on all components. Thus we have found the best completion, and we terminate (rule $\boxed{12.35}$) after saving incumbent solution

$$\hat{\mathbf{x}} \leftarrow (0, 0, 0, 1) \quad \text{with} \quad \hat{v} \leftarrow 14$$

The only remaining active is now partial solution $\mathbf{x}^{(2)} = (\#, \#, \#, 0)$. Relaxation optimum $\tilde{\mathbf{x}}^{(2)} = (0, 0.333, 1, 0)$ fails all termination tests $\boxed{12.33}$ to $\boxed{12.35}$ because it is feasible, fractional, and has solution value 9 strictly better than $\hat{v} = 14$. We must branch on fractional x_2.

Processing continues in this way until we reach $\mathbf{x}^{(4)} = (\#, 1, 1, 0)$. There the relaxation bound $\tilde{v} = 17 \geq \hat{v} = 14$, so we terminate (rule $\boxed{12.34}$). Something similar happens at node 5.

After several more steps, we encounter a new incumbent solution at node 8,

$$\hat{\mathbf{x}} \leftarrow (1, 0, 1, 0) \quad \text{with} \quad \hat{v} \leftarrow 12$$

Termination of nodes 9 and 10 by infeasibility (rule $\boxed{12.33}$) completes the search. This final incumbent solution is optimal (principle $\boxed{12.32}$).

SAMPLE EXERCISE 12.19: PERFORMING LP-BASED BRANCH AND BOUND

The following table shows candidate problem LP relaxation optima for all possible combinations of fixed and free values in a maximizing mixed-integer linear program over $x_1, x_2, x_3 = 0$ or 1, $x_4 \geq 0$.

x_1	x_2	x_3	$\tilde{\mathbf{x}}$	$\tilde{v}$	x_1	x_2	x_3	$\tilde{\mathbf{x}}$	$\tilde{v}$
#	#	#	$(0.2, 1, 0, 0)$	82.80	0	0	1	Infeasible	—
#	#	0	$(0.2, 1, 0, 0)$	82.80	0	1	#	$(0, 1, 0.67, 0)$	80.67
#	#	1	$(0, 0.8, 1, 0)$	79.40	0	1	0	$(0, 1, 0, 2)$	28.00
#	0	#	$(0.7, 0, 0, 0)$	81.80	0	1	1	$(0, 1, 1, 0.5)$	77.00
#	0	0	$(0.7, 0, 0, 0)$	81.80	1	#	#	$(1, 0, 0, 0)$	74.00
#	0	1	$(0.4, 0, 1, 0)$	78.60	1	#	0	$(1, 0, 0, 0)$	74.00
#	1	#	$(0.2, 1, 0, 0)$	82.80	1	#	1	$(1, 0, 1, 0)$	63.00
#	1	0	$(0.2, 1, 0, 0)$	82.80	1	0	#	$(1, 0, 0, 0)$	74.00
#	1	1	$(0, 1, 1, 0.5)$	77.00	1	0	0	$(1, 0, 0, 0)$	74.00
0	#	#	$(0, 1, 0.67, 0)$	80.67	1	0	1	$(1, 0, 1, 0)$	63.00
0	#	0	$(0, 1, 0, 2)$	28.00	1	1	#	$(1, 1, 0, 0)$	62.00
0	#	1	$(0, 0.8, 1, 0)$	79.40	1	1	0	$(1, 1, 0, 0)$	62.00
0	0	#	Infeasible	—	1	1	1	$(1, 1, 1, 0)$	51.00
0	0	0	Infeasible	—					

Solve the model by LP-based branch and bound Algorithm 12A, applying the same depth first rule for selecting among actives (ties broken in favor of $x_j = 1$) that was implemented in the River Power example of Figure 12.2.

Analysis: Applying Algorithm 12A produces the following branch and bound tree:

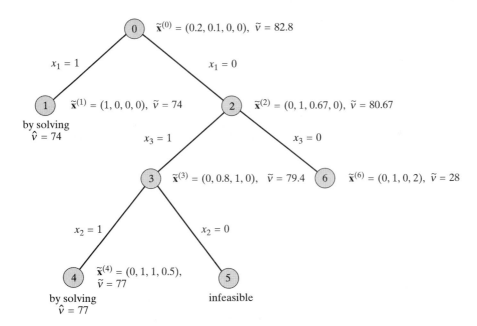

Processing stops with optimal solution $\mathbf{x}^* = (0, 1, 1, .5)$ at objective value 77.

Most of the processing is similar to our River Power example. One exception is that this model maximizes. For example, we terminate by bound at node 6 because

$$28 = \tilde{v} \le \hat{v} = 77$$

The other new element is that this model has continuous variable x_4 as well as 3 binary ones. This makes relaxation optimum $\tilde{\mathbf{x}}^{(4)} = (0, 1, 1, 0.5)$ feasible (and so optimal) in the full candidate, even though its last component is fractional. We terminate by having solved the candidate problem.

12.5 ROUNDING, PARENT BOUNDS, ENUMERATIONS SEQUENCES, AND STOPPING EARLY IN BRANCH AND BOUND

Algorithm 12A contains all the main elements of at least LP-based branch and bound, but it omits many details. In this section we briefly introduce some refinements.

Branch and Bound Solution of NASA Capital Budgeting Example

It will help to have a more serious example to illustrate. For this purpose we employ the NASA mission selection model formulated in Section 11.2. There, the decision variables

$$x_j \triangleq \begin{cases} 1 & \text{if mission } j \text{ is selected} \\ 0 & \text{otherwise} \end{cases}$$

The complete formulation of our version was

$$
\begin{array}{lll}
\max & 200x_1 + 3x_2 + 20x_3 + 50x_4 + 70x_5 & \text{(total value)} \\
& + 20x_6 + 5x_7 + 10x_8 + 200x_9 + 150x_{10} \\
& + 18x_{11} + 8x_{12} + 300x_{13} + 185x_{14} \\
\text{s.t.} & 6x_1 + 2x_2 + 3x_3 + 1x_7 + 4x_9 + 5x_{12} & \le 10 \quad \text{(2000–2004)} \\
& 3x_2 + 5x_3 + 5x_5 + 8x_7 + 5x_9 + 8x_{10} & \text{(2005-2009)} \\
& \quad + 7x_{12} + 1x_{13} + 4x_{14} & \le 12 \\
& 8x_5 + 1x_6 + 4x_{10} + 2x_{11} + 4x_{13} + 5x_{14} & \le 14 \quad \text{(2010–2014)} \\
& 8x_6 + 5x_8 + 7x_{11} + 1x_{13} + 3x_{14} & \le 14 \quad \text{(2015–2019)} \\
& 10x_4 + 4x_6 + 1x_{13} + 3x_{14} & \le 14 \quad \text{(2020–2024)} \\
& x_4 + x_5 \le 1 & \text{(mutually} \\
& x_8 + x_{11} \le 1 & \text{exclusives)} \\
& x_9 + x_{14} \le 1 \\
& x_{11} \le x_2 & \text{(dependent} \\
& x_4 \le x_3 & \text{missions)} \\
& x_5 \le x_3 \\
& x_6 \le x_3 \\
& x_7 \le x_3 \\
& x_j = 0 \text{ or } 1 \quad \text{for all } j = 1, \ldots, 14
\end{array}
\tag{12.16}
$$

Figure 12.3 shows the tree of an LP-based branch and bound search, and Table 12.3 details computations.

Rounding for Incumbent Solutions

One refinement of branch and bound exhibited in these NASA example computations is **rounding** relaxation optima as in principle $\boxed{12.11}$ to speed the process of finding good incumbent solutions.

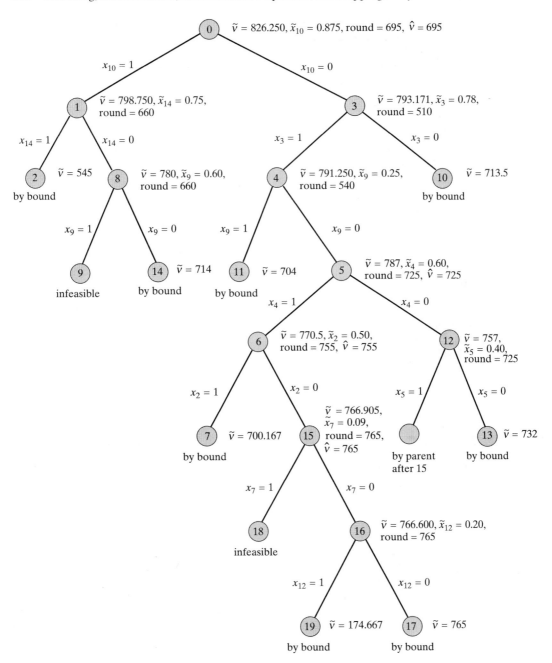

FIGURE 12.3   Branch and Bound Search of NASA Example

12.38 | If convenient rounding schemes are available, the relaxation optimum for every partial solution that cannot be terminated in a branch and bound search is usually rounded to a feasible solution for the full model prior to branching. The feasible solution provides a new incumbent if it is better than any known.

TABLE 12.3 Branch and Bound Search of NASA Example

t	Relax Value	Relaxation Solution[a]	Round Value	Action
0	826.250	$(1, 0, 0, 0, 0, 0, 0, 1, 0, 0.875, 0, 0, 1, 1)$	695	First incumbent $\hat{\mathbf{x}} \leftarrow$ $(1, 0, 0, 0, 0, 0, 0, 1, 0, 0, 0, 0, 1, 1)$ branch on x_{10}
1	798.750	$(1, 0, 0, 0, 0, 0, 0, 1, 0, 1, 0, \underline{1}, 0, 0, 1, 0.750)$	660	Branch on x_{14}
2	545.000	$(1, 0, 0, 0, 0, 0, 0, 1, 0, \underline{1}, 0, 0, 0, \underline{1})$	—	Terminate by bound
3	793.171	$(1, 0, 0.780, 0.463, 0.537, 0.780, 0, 1, 0.415, \underline{0}, 0, 0, 1, 0.585)$	510	Branch on x_3
4	791.250	$(1, 0, \underline{1}, 0.650, 350, 1, 0, 0.550, 0.250, \underline{0}, 0, 1, 0.750)$	540	Branch on x_9
5	787.000	$(1, 0, \underline{1}, 0.600, 0.400, 1, 0, 0.400, \underline{0}, \underline{0}, 0, 0, 1, 1)$	725	New incumbent $\hat{\mathbf{x}} \leftarrow$ $(1, 0, 1, 0, 0, 1, 0, 0, 0, 0, 0, 0, 1, 1)$ branch on x_4
6	770.500	$(1, 0.500, \underline{1}, \underline{1}, 0, 0, 0, 0.500, \underline{0}, \underline{0}, 0.500, 0, 1, 1)$	755	New incumbent $\hat{\mathbf{x}} \leftarrow$ $(1, 0, 1, 1, 0, 0, 0, 0, 0, 0, 0, 0, 1, 1)$ branch on x_2
7	700.167	$(0.833, \underline{1}, \underline{1}, \underline{1}, 0.188, 0, 0, \underline{0}, \underline{0}, 1, 0, 1, 0.750)$	—	Terminate by bound
8	780.000	$(1, 0, 0, 0, 0, 0, 0, 1, 0.600, \underline{1}, 0, 0, 1, \underline{0})$	660	Branch on x_9
9	Infeasible	None	—	Terminate by infeasible
10	713.500	$(1, 1, \underline{0}, 0, 0, 0, 0, 0, 0.500, \underline{0}, 1, 0, 1, 0.500)$	—	Terminate by bound
11	704.000	$(0.500, 0, \underline{1}, 0.800, 0.200, 1, 0, 1, \underline{1}, \underline{0}, 0, 0, 1, 0)$	—	Terminate by bound
12	757.000	$(1, 0, \underline{1}, \underline{0}, 0.400, 1, 0, 0.400, \underline{0}, \underline{0}, 0, 0, 1, 1)$	725	Branch on x_5
13	732.000	$(1, 0.462, \underline{1}, \underline{0}, \underline{0}, 0.846, 0, 0.077, 0, \underline{0}, \underline{0}, 0.462, 0, 1, 1)$	—	Terminate by bound
14	714.000	$(1, 0, 0.600, 0.600, 0, 0.600, 0, 0, 1, \underline{0}, \underline{1}, 0, 0, 1, \underline{0})$	—	Terminate by bound
15	766.909	$(1, \underline{0}, \underline{1}, \underline{1}, 0, 0, 0.091, 1, \underline{0}, \underline{0}, 0, 0.182, 1, 1)$	765	New incumbent $\hat{\mathbf{x}} \leftarrow$ $(1, 0, 1, 1, 0, 0, 0, 1, 0, 0, 0, 0, 1, 1)$ branch on x_7
16	766.600	$(1, \underline{0}, \underline{1}, \underline{1}, 0, 0, \underline{0}, 1, \underline{0}, \underline{0}, 0, 0.200, 1, 1)$	765	Branch on x_{12}
17	765.000	$(1, \underline{0}, \underline{1}, \underline{1}, 0, 0, \underline{0}, 1, \underline{0}, \underline{0}, 0, \underline{0}, 1, 1)$	—	Terminate by bound
18	Infeasible	None	—	Terminate by infeasible
19	174.667	$(0.333, \underline{0}, \underline{1}, \underline{1}, 0, 1, \underline{0}, 1, \underline{0}, \underline{0}, 0, \underline{1}, 0, 0)$	—	Terminate by bound

[a] Underlined values are fixed in the partial solution.

Computations in NASA Table 12.3 round down; that is, fractional components in relaxation optima are set $= 0$. Every partial solution that cannot be terminated is rounded before branching, and the resulting feasible solution considered as an incumbent. For instance, at node 5, the candidate relaxation produced optimum

$$\tilde{\mathbf{x}}^{(5)} = (1, 0, 1, 0.6, 0.4, 1, 0, 0.4, 0, 0, 0, 0, 1, 1), \qquad \tilde{v} = 787$$

The solution is fractional, and bound 787 is insufficient to terminate. Instead of branching immediately, however, the relaxation is first rounded by setting each $\hat{x}_j = \lfloor \tilde{x}_j \rfloor$. The result is new incumbent solution

$$\hat{\mathbf{x}} = (1, 0, 1, 0, 0, 1, 0, 0, 0, 0, 0, 0, 1, 1), \qquad \hat{v} = 725$$

Rounding is not guaranteed always to produce a new incumbent. For example, at node 1 of the NASA example, rounding produced a solution of value 660, which was no better that the existing incumbent $\hat{v} = 695$.

It is valuable to have good incumbent solutions early in a branch and bound search, for two reasons. First, time may not permit the search to be run until every node has been terminated or branched. Then the final incumbent solution provides an approximate optimum. Clearly, we would like it to be as good as possible.

The other advantage of early incumbents comes with termination by bound (principle $\boxed{12.34}$). In the maximize case, for example, we terminate if

$$\text{relaxation bound} \triangleq \tilde{v} \leq \hat{v} \triangleq \text{incumbent value}$$

We can certainly terminate more nodes with the same bound values if a strong incumbent value is discovered quickly.

The following tree shows Algorithm 12A computation on a minimizing ILP over $x_1, \ldots, x_3 = 0$ or 1. (Rounded solution values are shown but were not used in the computation.)

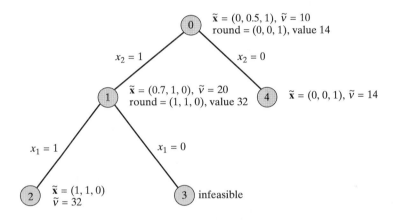

(a) Determine the earliest moment in the search at which ultimate optimal solution value 14 was known.

(b) Repeat the search using rounding to produce earlier incumbent solutions.

(c) Describe the savings that resulted in part (b) from using rounded solutions.

Analysis:

(a) The final incumbent solution, which proved optimal, was encountered when the relaxation optimum at partial solution 4 solved its candidate problem.

(b) At node 0 the relaxation optimum has value 10, and it rounds to a feasible solution with value 14. This becomes the first incumbent solution value. We then branch on x_2 as before. At node 1 relaxation bound 20 permits termination because it does not improve on the incumbent (principle $\boxed{12.34}$). Then the relaxation at node 4 the verifies optimality of the incumbent solution.

(c) In the search of part (b) the incumbent was uncovered at node 0, much earlier than node 4 in calculations depicted in the tree. This would be an advantage if we had to stop the search before all nodes had been terminated or branched. Also, the early incumbent avoided some candidate problem computations at nodes 2 and 3. The bound at 1 is now enough to terminate.

Branch and Bound Family Tree Terminology

To discuss issues connected with managing and controlling branch and bound trees, we need some terminology. It is standard to make analogies with family trees.

Any node created directly from another by branching is called a **child**, and the branched node is its **parent**. For example, in Figure 12.3, nodes 11 and 5 are the children of node 4. Its parent is node 3.

SAMPLE EXERCISE 12.21: UNDERSTANDING TREE TERMINOLOGY

Identify each of the following in the branch and bound tree of Sample Exercise 12.20.

(a) The parent of node 3

(b) The children of node 0

Analysis:

(a) Node 1 is the parent of node 3.

(b) Nodes 1 and 4 are the children of node 0.

Parent Bounds

Another way of refining LP-based branch and bound Algorithm 12A is to take advantage of an easy **parent bound** readily available from prior computations.

> **12.39**   The relaxation optimal value for the parent of any partial solution to a minimize model provides a lower bound on the objective value of any completion of its children. The relaxation optimal value for the parent in a maximize model provides an upper bound.

To see that these bounds are valid, recall that the candidate problem associated with any partial solution is just our original model augmented by constraints for variables fixed in the partial solution (definition 12.31). Extending the solution with a new fix in a child can only worsen the optimal objective function value. For example, in Figure 12.3, relaxation bound 826.250 for node 0 holds for its children nodes 1 and 3. Both have an additional constraint, so the solution value can only worsen. In fact, when relaxation bounds $\tilde{v}^{(1)} = 798.750$ and $\tilde{v}^{(3)} = 793.171$ were later computed, they proved strictly worse (for this maximize model).

SAMPLE EXERCISE 12.22: UNDERSTANDING PARENT BOUNDS

Identify for node 5 in NASA branch and bound Figure 12.3 the best (upper) bound known for the optimal value of the corresponding candidate problem *before* the candidate's LP relaxation was solved.

Analysis: Prior to solving the LP relaxation of the the candidate problem for node 5, the best bound available on its optimal value comes from its parent. Parent bound $\tilde{v} = 791.250$ is valid for child node 5 because the is identical except for the extra $x_9 = 0$ constraint.

Terminating with Parent Bounds

One way a sophisticated branch and bound algorithm can exploit parent bounds 12.39 arises whenever a new incumbent solution is discovered.

> 12.40   Whenever a branch and bound search discovers a new incumbent solution, any active partial solution with parent bound no better than the new incumbent solution value can immediately be terminated.

Processing of partial solution 15 in Table 12.3 illustrates. An incumbent solution was discovered there with value $\hat{v} = 765$. Earlier, partial solution 12 had been branched because its bound of $\tilde{v} = 757$ was insufficient to terminate versus the incumbent at that time. Now we may terminate the still-active child with $x_5 = 1$. The 757 bound available from the parent is no better than the incumbent discovered at node 15.

SAMPLE EXERCISE 12.23: TERMINATING WITH PARENT BOUNDS

The following tree shows a minimizing branch and bound search that has processed 4 partial solutions and just discovered its first incumbent solution at node 3. Nodes marked α, β, and γ are active but not yet explored.

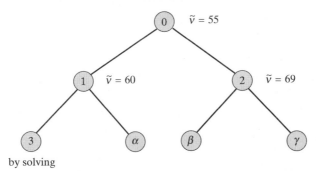

Determine which if any active partial solution can be terminated if the incumbent solution discovered at node 3 has value (a) $\hat{v} = 80$; (b) $\hat{v} = 63$.

Analysis: We apply principle 12.40 , comparing with parent bounds 60 at node α and 69 at nodes β and γ.

(a) Incumbent value $\hat{v} = 80$ is not as good as either parent bound for this minimize model. No active solution can be terminated.

(b) Incumbent value $\hat{v} = 63$ is not as good as 60, but superior to 69. Thus active partial solutions β and γ should be terminated.

Stopping Early: Branch and Bound as a Heuristic

Table 12.4 provides a historical summary of the NASA example in Figure 12.3 and Table 12.3. Incumbent solutions reported in the first column show a typical pattern. Good incumbents were uncovered fairly early in the search, with the final 50 to 80% of computational effort consumed in squeezing out the last improvements and proving global optimality. For example, nearly optimal $\hat{v} = 755$ was discovered after only 7 of the 20 partial solutions analyzed.

TABLE 12.4   Incumbent and Best Parent Bound History in NASA Example, Branch Bound Figure 12.3

After Node	Incumbent Value	Best Parent	After Node	Incumbent Value	Best Parent
0	695	826.250	10	755	791.250
1	695	826.250	11	755	780.000
2	695	826.250	12	755	780.000
3	695	798.750	13	755	780.000
4	695	798.750	14	755	770.500
5	725	798.750	15	765	766.909
6	755	798.750	16	765	766.909
7	755	798.750	17	765	766.909
8	755	793.171	18	765	765.000
9	755	793.171			

Often, we are willing to settle for less that global optimality to avoid the long final phase. That is, we want to **stop early**, accepting the last incumbent as a heuristic optimum.

Bounds on the Error of Stopping with the Incumbent Solution

Of course, we know that such incumbent solutions are feasible for the full problem, and better than any other feasible solution that we have encountered. But we can use the best of current parent bounds 12.39 to conclude much more.

> **12.41**   The least relaxation optimal value for parents of the active partial solutions in a minimizing branch and bound search always provides a lower bound on the optimal solution value of the full model. The greatest relaxation optimal value for parents in a maximizing search provides an upper bound.

Since every full solution that might still improve on the incumbent is a completion of some active partial solution, we can bound every such solution by finding the best of the corresponding parent bounds.

The second column of Table 12.4 tracks this overall bound on remaining active partials in our NASA example. For example, at the moment after partial solution $\mathbf{x}^{(6)}$ is explored and branched, the tree then contains active children of nodes 1, 3, 4, 5, and 6. The best of their relaxation bounds

$$\max\{798.750, 793.171, 791.250, 787.000, 770.500\} = 798.750$$

is as much as any future incumbent could ever achieve. Any unexplored solution has to be a feasible completion in one of those active partial solutions.

Suppose now that we decide to terminate the search after node 6. Table 12.4 shows that the incumbent solution value was 755 at that point, and we just computed the best parent bound of 798.750. Thus we can compute that our approximate optimum (the incumbent) is at most

$$\frac{(\text{best possible}) - (\text{best known})}{(\text{best known})} = \frac{798.750 - 755}{755} = 5.8\%$$

below optimal. That is, we can bound the error in our approximation.

> **12.42** At any stage of a branch and bound search, the difference between the incumbent solution value and the best parent bound of any active partial solution shows the maximum error in accepting the incumbent as an approximate optimum.

We use the lower bound in the denominator of percent computations because the optimal solution value could be that small.

SAMPLE EXERCISE 12.24: STOPPING BRANCH AND BOUND EARLY

Return to the minimizing branch and bound tree of Sample Exercise 12.23 and assume the incumbent solution value discovered at node 3 was $\hat{v} = 71$. Determine the maximum error and percent error in stopping the search at that point.

Analysis: After node 3 the best parent bound 12.41 is

$$\min\{60, 69\} = 60$$

Thus the maximum error for this minimize model is

$$(\text{best known}) - (\text{best possible}) = 71 - 60 = 11$$

As a percent the maximum error is

$$\frac{(\text{best known}) - (\text{best possible})}{(\text{best possible})} = \frac{71 - 60}{60} = 18.3\%$$

Depth First, Best First, and Depth Forward Best Back Sequences

Another important implementation question in branch and bound is how to select a partial solution to pursue among the many that may be active. We have already introduced the simple **depth first** rule 12.28 , which selects at each iteration an active partial solution with the most components fixed (i.e., one deepest in the search tree). Parent bounds 12.39 permit other alternatives.

> **12.43** **Best first** search selects at each iteration an active partial solution with best parent bound.

> **12.44** **Depth forward best back** search selects a deepest active partial solution after branching a node, but one with best parent bound after a termination.

All these rules require tie-breaking refinements when several partials have maximum depth or best parent bound. One alternative builds on fractional variable rule $\boxed{12.36}$ by preferring the nearest child.

> $\boxed{12.45}$ When several active partial solutions tie for deepest or best parent bound, the **nearest child** rule chooses the one with last fixed variable value nearest the corresponding component of the parent LP relaxation.

Assuming that the parent relaxation optimum tells us something about good values for the branching variable, this nearest child is more likely to lead to early discovery of good incumbent solutions.

Figure 12.4 illustrates all 3 rules $\boxed{12.28}$ – $\boxed{12.44}$ on NASA model (12.16). Branch and bound trees are shown after the first 10 partial solutions. Each uses nearest child tie-breaking rule $\boxed{12.45}$.

Part (a) applies depth first search, always selecting an active partial solution with the most variables fixed. The deepest active partials in the current tree are the children of node 9. Nearest child rule $\boxed{12.45}$ would chose the one with $x_4 = 1$ as $\mathbf{x}^{(10)}$ because the parent LP relaxation had $\tilde{x}_4 = 0.6$, which is closer to 1 than 0.

Notice that depth first search will automatically choose a child of the last partial solution analyzed if that partial was branched. It was a deepest solution before branching, and the children have an additional variable fixed. This means that depth first search tends to move rapidly to fix enough variables that a feasible solution to the full model is uncovered. When rounding is not easy, that may be the best way to produce early incumbent solutions.

Selecting a child of the last node investigated can also have important computational savings. For example, at node 3 of Figure 12.4(a) the associated candidate problem differs by only the $x_{14} = 0$ constraint from that of parent node 1. Often, this similarity can be exploited to solve the relaxation at a child very quickly by starting at the LP optimum of the parent.

Figure 12.4(b) illustrates best first search, always advancing to an active partial solution with best parent bound. At the moment depicted there are active children of nodes 4, 7 and 9. We would next select a child of node 7 because its parent bound

$$787 = \max\{780, 787, 757\}$$

is best for this maximizing model. Partial solution $\mathbf{x}^{(10)}$ would be the child of 7 with $x_4 = 1$.

Here the idea is always to pursue a partial solution that could lead to the best possible completion. We select one with best parent bound because that is the most accurate information at hand about how good completions might be.

Notice, however, that best first search tends to skip rapidly around the branch and bound tree, with selected partial solutions often rather different than the one just before. This tendency means that depth first's efficiency when advancing to a child is often lost.

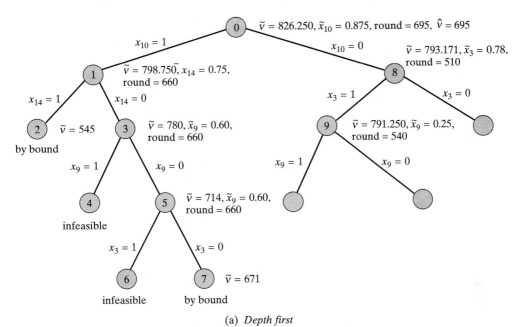

(a) *Depth first*

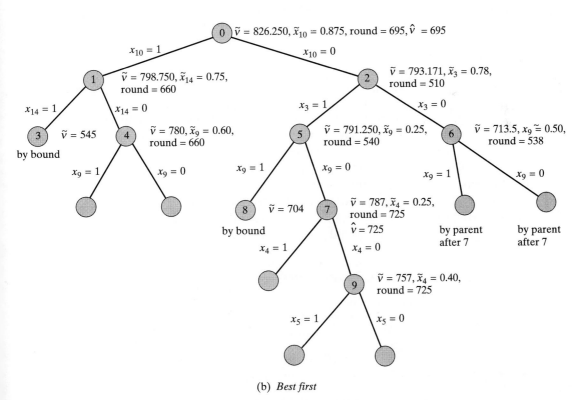

(b) *Best first*

FIGURE 12.4   Alternative Partial Solution Selection in NASA Example

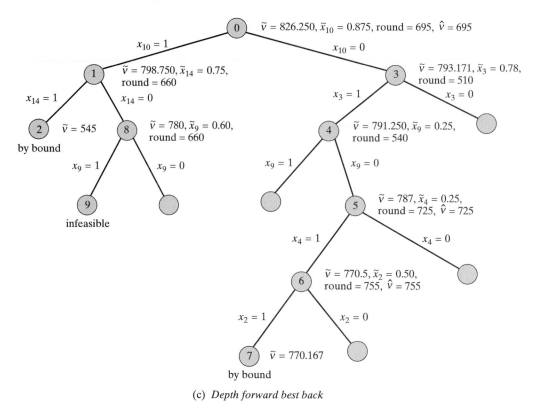

(c) *Depth forward best back*

FIGURE 12.4   Alternative Partial Solution Selection in NASA Example (Continued)

The depth forward best back rule of Figure 12.4(c) (and full search tree, Figure 12.3) provides a compromise. As long as nodes are branched, so that their children are a possible next choice, this rule follows depth first in selecting one of the children. When a node is terminated, however, so that some disruption is unavoidable, rule 12.44 pursues the more ambitious best-first policy.

In Figure 12.4(c), partial solution $\mathbf{x}^{(9)}$ was terminated. Thus $\mathbf{x}^{(10)}$ would be an active partial with best parent bound. Here that would be the child of node 3 with $x_3 = 0$. Had node 9 been branched, $\mathbf{x}^{(10)}$ would have been one of its children.

SAMPLE EXERCISE 12.25: SELECTING AMONG ACTIVE PARTIAL SOLUTIONS

Return to the maximizing branch and bound problem of Sample Exercise 12.19 and show the branch and bounds trees for searches guided by (a) depth first selection, (b) best first selection; (c) depth forward best back selection. Use nearest child tie-breaking rule 12.45 in each case.

Analysis:

(a) The branch and bound tree for depth first rule 12.28 is

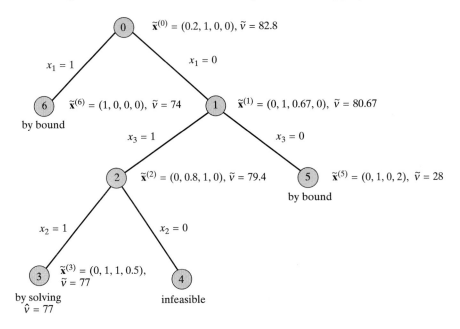

$\tilde{x}^{(0)} = (0.2, 1, 0, 0)$, $\tilde{v} = 82.8$

$x_1 = 1$

$x_1 = 0$

$\tilde{x}^{(6)} = (1, 0, 0, 0)$, $\tilde{v} = 74$

by bound

$\tilde{x}^{(1)} = (0, 1, 0.67, 0)$, $\tilde{v} = 80.67$

$x_3 = 1$

$x_3 = 0$

$\tilde{x}^{(2)} = (0, 0.8, 1, 0)$, $\tilde{v} = 79.4$

$\tilde{x}^{(5)} = (0, 1, 0, 2)$, $\tilde{v} = 28$

by bound

$x_2 = 1$

$x_2 = 0$

$\tilde{x}^{(3)} = (0, 1, 1, 0.5)$, $\tilde{v} = 77$

by solving $\hat{v} = 77$

infeasible

(b) The branch and bound tree for best first rule $\boxed{12.43}$ is

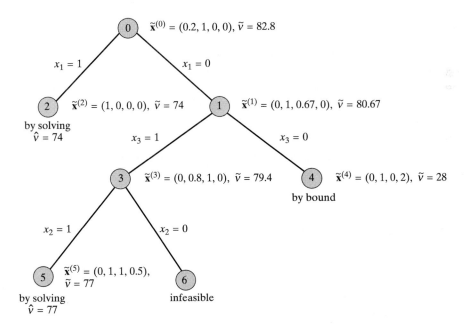

$\tilde{x}^{(0)} = (0.2, 1, 0, 0)$, $\tilde{v} = 82.8$

$x_1 = 1$

$x_1 = 0$

$\tilde{x}^{(2)} = (1, 0, 0, 0)$, $\tilde{v} = 74$

by solving $\hat{v} = 74$

$\tilde{x}^{(1)} = (0, 1, 0.67, 0)$, $\tilde{v} = 80.67$

$x_3 = 1$

$x_3 = 0$

$\tilde{x}^{(3)} = (0, 0.8, 1, 0)$, $\tilde{v} = 79.4$

$\tilde{x}^{(4)} = (0, 1, 0, 2)$, $\tilde{v} = 28$

by bound

$x_2 = 1$

$x_2 = 0$

$\tilde{x}^{(5)} = (0, 1, 1, 0.5)$, $\tilde{v} = 77$

by solving $\hat{v} = 77$

infeasible

(c) The branch and bound tree for depth forward best back rule $\boxed{12.44}$ is

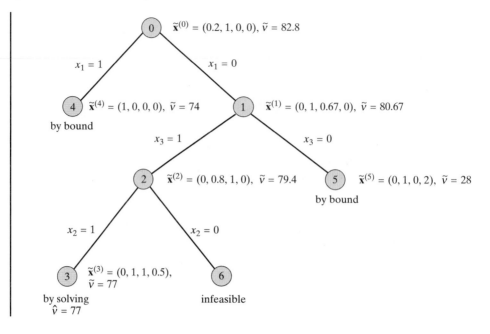

Branch and Cut Search

Section 12.3 showed that relaxations can often be strengthened by adding new, **valid inequality** constraints. A variation of branch and bound termed branch and cut takes advantage of such inequalities to speed computation.

> __12.46__ **Branch and cut** algorithms modify the basic branch and bound strategy of Algorithm 12A by attempting to strengthen relaxations with new inequalities before branching a partial solution. Added constraints should hold for all feasible solutions to the full discrete model, but they should cut off (render infeasible) the last relaxation optimum.

Branch and Cut Solution of the River Power Example

Algorithm 12B provides a formal statement of the branch and cut process. To see the idea, return to our River Power example of Section 12.4. Decisions there relate to which generators to activate, and the model is

$$\min \quad 7x_1 + 12x_2 + 5x_3 + 14x_4 \qquad \text{(total cost)}$$
$$\text{s.t.} \quad 300x_1 + 600x_2 + 500x_3 + 1600x_4 \geq 700 \qquad \text{(demand)} \qquad (12.17)$$
$$x_1, x_2, x_3, x_4 = 0 \text{ or } 1$$

Solution of the first, all-free candidate problem produces relaxation optimum

$$\tilde{\mathbf{x}}^{(0)} = (0, 0, 0, 0.438), \qquad \tilde{v} = 6.125$$

In normal branch and bound Figure 12.2 we immediately branched on fractional variable x_4.

ALGORITHM 12B: BRANCH AND CUT (0-1 ILP'S)

Step 0: Initialization. Make the only active partial solution the one with all discrete variables free, and initialize solution index $t \leftarrow 0$. If any feasible solutions are known for the model, also choose the best as incumbent solution $\hat{\mathbf{x}}$ with objective value $\hat{v}$. Otherwise, set $\hat{v} \leftarrow -\infty$ if the model maximizes and $\hat{v} \leftarrow +\infty$ if it minimizes.

Step 1: Stopping. If active partial solutions remain, select one as $\mathbf{x}^{(t)}$, and proceed to Step 2. Otherwise, stop. If there is an incumbent solution $\hat{\mathbf{x}}$, it is optimal, and if not, the model is infeasible.

Step 2: Relaxation. Attempt to solve the linear programming relaxation of the candidate problem corresponding to $\mathbf{x}^{(t)}$.

Step 3: Termination by Infeasibility. If the LP relaxation proved infeasible, there are no feasible completions of partial solution $\mathbf{x}^{(t)}$. Terminate $\mathbf{x}^{(t)}$, increment $t \leftarrow t+1$, and return to Step 1.

Step 4: Termination by Bound. If the model maximizes and LP relaxation optimal value $\tilde{v}$ satisfies $\tilde{v} \leq \hat{v}$, or it minimizes and $\tilde{v} \geq \hat{v}$, the best feasible completion of partial solution $\mathbf{x}^{(t)}$ cannot improve on the incumbent. Terminate $\mathbf{x}^{(t)}$, increment $t \leftarrow t+1$, and return to Step 1.

Step 5: Termination by Solving. If the LP relaxation optimum $\tilde{\mathbf{x}}^{(t)}$ satisfies all binary constraints of the model, it provides the best feasible completion of partial solution $\mathbf{x}^{(t)}$. After saving it as new incumbent solution by $\hat{\mathbf{x}} \leftarrow \tilde{\mathbf{x}}^{(t)}$ and $\hat{v} \leftarrow \tilde{v}$, terminate $\mathbf{x}^{(t)}$, increment $t \leftarrow t+1$, and return to Step 1.

Step 6: Valid Inequality. Attempt to identify a valid inequality for the full ILP model that is violated by the current relaxation optimum $\tilde{\mathbf{x}}^{(t)}$. If successful, make the constraint a part of the full model increment $t \leftarrow t+1$, and return to Step 2.

Step 7: Branching. Choose some free binary-restricted component x_p that was fractional in the last LP relaxation optimum, and branch $\mathbf{x}^{(t)}$ by creating two new actives. One is identical to $\mathbf{x}^{(t)}$ except that x_p is fixed $=0$, and the other the same except that x_p is fixed $=1$. Then increment $t \leftarrow t+1$ and return to Step 1.

Before turning the current partial solution into two in that way, branch and cut Algorithm 12B would try to improve the relaxation. The idea is to find an inequality satisfied by every binary solution to the full model but violated by $\tilde{\mathbf{x}}^{(0)}$.

Methods used to find such cutting inequalities vary enormously from one model to another. In this example we simply observe that any feasible solution in (12.17) must have a least one generator turned on. Thus

$$x_1 + x_2 + x_3 + x_4 \geq 1 \tag{12.18}$$

is valid. Also, constraint (12.18) cuts off previous relaxation optimum $\tilde{\mathbf{x}}^{(0)}$ because

$$0 + 0 + 0 + 0.438 \not\geq 1$$

Figure 12.5 shows that branch and cut advances by adding inequality (12.18) to improve the relaxation. With the same all-free partial solution, we now obtain stronger results:

$$\tilde{\mathbf{x}}^{(1)} = (0, 0, 0.818, 0.182), \qquad \tilde{v} = 6.636$$

Suppose now that a hunt for further cuts meets with no success. The search branches as usual on fractional variable x_3. Depth first rule $\boxed{12.28}$ with nearest child tie-breaker $\boxed{12.45}$ makes the next partial solution $\mathbf{x}^{(2)} = (\#, \#, 1, \#)$.

At node 2 the relaxation again proves inadequate for termination. This time analysis of possible cuts discovers violated inequality

$$x_1 + x_2 + x_3 + 2x_4 \geq 2 \tag{12.19}$$

which recognizes that either generator 4 or two others are required to meet the 700-megawatt output requirement. The improved relaxation at node 3 has $\tilde{v} = 12$, and the search continues.

Notice that both inequalities (12.18) and (12.19) are valid for the original model (12.17). That is, they do not depend on variables fixed in partial solutions. As a consequence, they may be retained when candidate problem relaxations are solved at subsequent nodes 4 to 6. Each new cut discovered strengthens all subsequent relaxations.

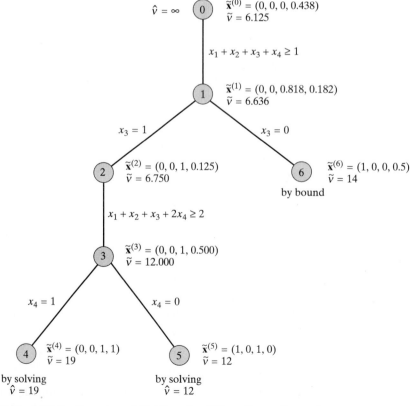

FIGURE 12.5   Branch and Cut Search of River Power Example

The following is a branch and cut tree detailing application of Algorithm 12B to a maximizing integer linear program over $x_1, x_2, x_3 = 0$ or 1 and $x_4 \geq 0$. The best first selection rule $\boxed{12.43}$ was employed with nearest child tie-breaker.

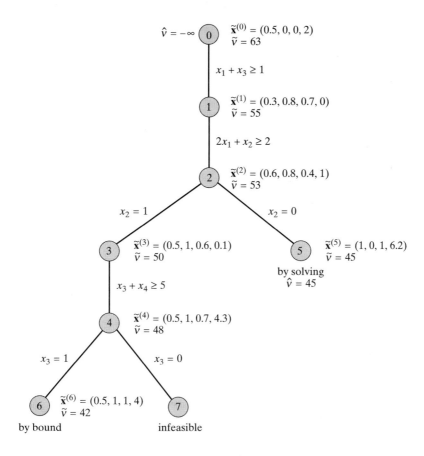

Assuming that all cutting inequalities shown are valid, trace the computation and justify each step.

Analysis: There is no initial incumbent solution. The first relaxation at node 0 proves feasible and fractional, but cut $x_1 + x_3 \geq 1$ could be generated. We can usefully add it because

$$x_1^{(0)} + x_3^{(0)} = 0.5 + 0 = 0.5 \not\geq 1$$

The improved relaxation at node 1 still cannot be terminated, but valid inequality $2x_1 + x_2 \geq 2$ can be added because

$$2x_1^{(1)} + x_2^{(1)} = 2(0.3) + (0.8) = 1.4 \not\geq 2$$

The resulting node 2 relaxation remains feasible and fractional, and no further cuts are added. Now we branch on fractional variable x_2, choosing the child with $x_2 = 1$ as partial solution 3 under the nearest child rule $\boxed{12.45}$. Its LP relaxation also fails to justify termination, but another valid inequality $x_3 + x_4 \geq 5$ is added with

$$x_3^{(3)} + x_4^{(4)} = 0.6 + 0.1 = 0.7 \not\geq 5$$

After branching on fractional variable x_3, there are active children of nodes 2 and 4. Best first enumeration rule $\boxed{12.43}$ takes us next to partial solution $\mathbf{x}^{(5)} = (\#, 0, \#, \#)$ because parent bound

$$53 = \max\{53, 48\}$$

The resulting relaxation optimum is feasible in the original model and so optimal in the corresponding candidate problem. It provides incumbent solution

$$\hat{\mathbf{x}} = (1, 0, 1, 6.2), \qquad \hat{v} = 45$$

Now returning to the children of node 4, the search goes first to the child with $x_3 = 1$ (rule $\boxed{12.45}$). Relaxation bound $\tilde{v} = 42$ is now sufficient to terminate because $42 \leq 45$. The search finishes by discovering that the relaxation of node 7 is infeasible. Final incumbent solution $\hat{\mathbf{x}}$ is optimal.

12.6 IMPROVING SEARCH HEURISTICS FOR DISCRETE OPTIMIZATION INLPs

Many large combinatorial optimization models, especially INLPs with nonlinear objective functions, are too large for enumeration and lack strong relaxations that are tractable. Still, much can be done. Suitable adaptations of **improving search** methods introduced in Chapter 3 can often yield very effective **heuristic** algorithms. That is, we can still find good feasible solutions even though we will not be able to guarantee their optimality or even be sure about how close they come to optimal.

Rudimentary Improving Search Algorithm

Algorithm 12C shows a rudimentary adaptation of improving search to discrete models. Like the continuous cases of Chapter 3, the process begins with an initial feasible solution $\mathbf{x}^{(0)}$. Each iteration t considers neighbors of current solution $\mathbf{x}^{(t)}$ and tries to advance to one that is feasible and superior in objective value. If no feasible neighbor is improving, the process stops with **local optimum** and **heuristic optimum** $\mathbf{x}^{(t)}$.

Discrete Neighborhoods and Move Sets

What is new about the discrete form of improving search is that we must explicitly define the **neighborhood** of a current solution. Unlike the continuous case, where there are infinitely many points near a current solution, discrete search must advance to a binary or integer point. Explicit **move sets** (denoted $\mathcal{M}$) control what solutions are considered neighbors of current $\mathbf{x}^{(t)}$.

$\boxed{12.47}$ Improving searches over discrete variables define neighborhoods by specifying a move set $\mathcal{M}$ of moves allowed. The current solution and all reachable from it in a single move $\Delta \mathbf{x} \in \mathcal{M}$ comprise its neighborhood.

ALGORITHM 12C: DISCRETE IMPROVING SEARCH

Step 0: Initialization. Choose any starting feasible solution $\mathbf{x}^{(0)}$, and set solution index $t \leftarrow 0$.

Step 1: Local Optimum. If no move $\Delta \mathbf{x}$ in move set $\mathcal{M}$ is both improving and feasible at current solution $\mathbf{x}^{(t)}$, stop. Point $\mathbf{x}^{(t)}$ is a local optimum.

Step 2: Move. Choose some improving feasible move $\Delta \mathbf{x} \in \mathcal{M}$ as $\Delta \mathbf{x}^{(t+1)}$.

Step 3: Step. Update

$$\mathbf{x}^{(t+1)} \leftarrow \mathbf{x}^{(t)} + \Delta \mathbf{x}^{(t+1)}$$

Step 4: Increment. Increment $t \leftarrow t+1$, and return to Step 1.

SAMPLE EXERCISE 12.27: DEFINING MOVE SETS

Consider the discrete optimization model

$$\begin{aligned}
\max \quad & 20x_1 - 4x_2 + 14x_3 \\
\text{s.t.} \quad & 2x_1 + x_2 + 4x_3 \leq 5 \\
& x_1, x_2, x_3 = 0 \text{ or } 1
\end{aligned}$$

and assume that an improving search begins at $\mathbf{x}^{(0)} = (1, 1, 0)$.

(a) List all neighbors of $\mathbf{x}^{(0)}$ under move set

$$\mathcal{M} \triangleq \left\{ \begin{pmatrix} 1 \\ 0 \\ 0 \end{pmatrix}, \begin{pmatrix} -1 \\ 0 \\ 0 \end{pmatrix}, \begin{pmatrix} 0 \\ 1 \\ 0 \end{pmatrix}, \begin{pmatrix} 0 \\ -1 \\ 0 \end{pmatrix}, \begin{pmatrix} 0 \\ 0 \\ 1 \end{pmatrix}, \begin{pmatrix} 0 \\ 0 \\ -1 \end{pmatrix} \right\}$$

(b) Determine which members of the neighborhood are both improving and feasible.

Analysis:

(a) Following definition $\boxed{12.47}$, the neighbors of $\mathbf{x}^{(0)}$ under the specified move set are

$$\begin{aligned}
(1, 1, 0) + (1, 0, 0) &= (2, 1, 0) \\
(1, 1, 0) + (-1, 0, 0) &= (0, 1, 0) \\
(1, 1, 0) + (0, 1, 0) &= (1, 2, 0) \\
(1, 1, 0) + (0, -1, 0) &= (1, 0, 0) \\
(1, 1, 0) + (0, 0, 1) &= (1, 1, 1) \\
(1, 1, 0) + (0, 0, -1) &= (1, 1, -1)
\end{aligned}$$

(b) Of the neighbors in part (a), only $\mathbf{x} = (0, 1, 0)$, which has objective value -4, and $\mathbf{x} = (1, 0, 0)$, which has objective value 20, are feasible for all constraints of the model. Current point $\mathbf{x}^{(0)} = (1, 1, 0)$ has objective value 16. Thus $\mathbf{x} = (1, 0, 0)$ is the only neighbor that is both improving and feasible (i.e., the one to which improving search would advance).

NCB Example Revisited

We will use the NCB example of Section 11.5 to illustrate improving search in discrete optimization. Recall that we seek a shortest-distance routing through 10 points in a printed circuit board that must be drilled. Table 12.5 details hole-to-hole travel distances.

For improving search it will be most convenient to employ the quadratic assignment formulation $\boxed{11.27}$ of Section 11.5:

$$\min \quad \sum_{k=1}^{10}\sum_{i=1}^{10}\sum_{j=1}^{10} d_{i,j} y_{k,i} y_{k+1,j} \qquad \text{(total distance)}$$

$$\sum_{i=1}^{10} y_{k,i} = 1 \qquad \text{for all } k = 1, \ldots, 10 \qquad \text{(some hole each } k\text{)} \qquad (12.20)$$

$$\sum_{k=1}^{10} y_{k,i} = 1 \qquad \text{for all } i = 1, \ldots, 10 \qquad \text{(each } i \text{ assigned)}$$

$$y_{k,i} = 0 \text{ or } 1 \qquad \text{for all } k = 1, \ldots, 10; \quad i = 1, \ldots, 10$$

where

$$y_{k,i} \triangleq \begin{cases} 1 & \text{if } k\text{th hole drilled is } i \\ 0 & \text{otherwise} \end{cases}$$

and $y_{10+1,j}$ is understood to mean $y_{1,j}$ in objective function summations.

We will abuse notation in the usual way to think of solutions as vectors $\mathbf{y}$ even though components have two subscripts. The NCB optimal solution $\mathbf{y}^*$, which is depicted in Figure 12.6, has length 81.8 inches and nonzero components

$$y_{1,1}^* = y_{2,3}^* = y_{3,6}^* = y_{4,10}^* = y_{5,9}^* = y_{6,5}^* = y_{7,8}^* = y_{8,7}^* = y_{9,4}^* = y_{10,2}^* = 1$$

We begin our improving searches with initial feasible solution

$$\mathbf{y}^{(0)} = (1, 0, \ldots, 0; 0, 1, 0, \ldots, 0; \ldots; 0, \ldots, 0, 1) \qquad (12.21)$$

corresponding to $y_{1,1} = y_{2,2} = \cdots = y_{10,10} = 1$. That is, hole 1 is drilled first, then hole 2, and so on. The total length is

$$d_{1,2} + d_{2,3} + d_{3,4} + d_{4,5} + d_{5,6} + d_{6,7} + d_{7,8} + d_{8,9} + d_{9,10} + d_{10,1}$$
$$= 3.6 + 3.6 + 7.1 + 7.0 + 9.9 + 25.0 + 10.3 + 10.2 + 3.6 + 26.4$$
$$= 106.7 \text{ inches}$$

TABLE 12.5   Distances between Holes in NCB Example

i \ j	1	2	3	4	5	6	7	8	9	10
1	—	3.6	5.1	10.0	15.3	20.0	16.0	14.2	23.0	26.4
2	3.6	—	3.6	6.4	12.1	18.1	13.2	10.6	19.7	23.0
3	5.1	3.6	—	7.1	10.6	15.0	15.8	10.8	18.4	21.9
4	10.0	6.4	7.1	—	7.0	15.7	10.0	4.2	13.9	17.0
5	15.3	12.1	10.6	7.0	—	9.9	15.3	5.0	7.3	11.3
6	20.0	18.1	15.0	15.7	9.9	—	25.0	14.9	12.0	15.0
7	16.0	13.2	15.8	10.0	15.3	25.0	—	10.3	19.2	21.0
8	14.2	10.6	10.8	4.2	5.0	14.9	10.3	—	10.2	13.0
9	23.0	19.7	18.4	13.9	7.8	12.0	19.2	10.2	—	3.6
10	26.4	23.0	21.9	17.0	11.3	15.0	21.0	13.0	3.6	—

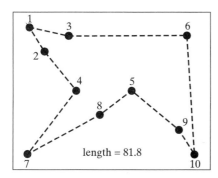

length = 81.8

FIGURE 12.6   Optimal Drill Path in the NCB Example

Choosing a Move Set

The critical element of a discrete improving search heuristic is its move set. If it were possible, we would make every solution a neighbor of every other. Then the search would yield global optima because a stop implies that no solution at all is feasible and superior in objective value to the current.

In a practical search, however, we must accept much less.

> **12.48**   The move set $\mathcal{M}$ of a discrete improving search must be compact enough to be checked at each iteration for improving feasible neighbors.

On the other hand, we would not want too limited a move set.

> **12.49**   The solution produced by a discrete improving search depends on the move set (or neighborhood) employed, with larger move sets generally resulting in superior local optima.

If the move set $\mathcal{M}$ of a discrete improving search is too restrictive, very few solutions will be considered at each iteration, and poor-quality local optima will result.

We will adopt one of the simplest move sets for our NCB example. Specifically, our $\mathcal{M}$ will consist of **pairwise interchanges** swapping one position k with another ℓ. Corresponding move vectors $\Delta \mathbf{y}$ have two -1 components at deleted assignments, two $+1$ components at revised ones, and all other components 0. For example, if the hole in route position $k = 3$ is now number 7, and the hole in position $\ell = 5$ in now number 1, the corresponding interchange move direction has

$$\Delta y_{3,7} = -1, \quad \Delta y_{5,1} = -1, \quad \Delta y_{3,1} = +1, \quad \Delta y_{5,7} = +1$$

changing hole 1 to position 3 and hole 7 to position 5.

In all, this pairwise interchange $\mathcal{M}$ contains a $\Delta \mathbf{y}$ for $(10 \cdot 9)/2 = 45$ choices of k and ℓ, each with $10 \cdot 9 = 90$ possible current hole assignment pairs—a total of $45 \cdot 90 = 4050$ moves. However, at any particular solution $\mathbf{y}$, only 45 moves interchanging its specific assignments lead to a feasible neighbor. In all searches in this section we adopt the one such move most improving the objective function.

SAMPLE EXERCISE 12.28: COMPARING MOVE SETS

Return to the discrete model of Sample Exercise 12.27 at initial point $\mathbf{x}^{(0)} = (1, 1, 0)$.

(a) Show that $\mathbf{x}^{(0)}$ is not locally optimal under the move set of Sample Exercise 12.27.

(b) Show that $\mathbf{x}^{(0)}$ is locally optimal over smaller move set

$$\mathcal{M} \triangleq \{(1, 0, 0), (0, 1, 0), (0, 0, 1)\}$$

Analysis:

(a) Sample Exercise 12.27(b) established that $\mathbf{x} = (1, 0, 0)$ is a feasible neighbor of $\mathbf{x}^{(0)}$ with superior objective value. Thus $\mathbf{x}^{(0)}$ is not best in its neighborhood, and so not locally optimal. Algorithm 12C would advance to $\mathbf{x}^{(1)} = (1, 0, 0)$ and repeat.

(b) Over this more restricted move set, neighbors are

$$(1, 1, 0) + (1, 0, 0) = (2, 1, 0)$$
$$(1, 1, 0) + (0, 1, 0) = (1, 2, 0)$$
$$(1, 1, 0) + (0, 0, 1) = (1, 1, 1)$$

None is feasible, so $\mathbf{x}^{(0)}$ is locally optimal.

Rudimentary Improving Search of the NCB Example

For initial solution (12.21) in our NCB example, the objective function impact of the 45 feasible pairwise interchanges is as follows:

$k \backslash \ell$	2	3	4	5	6	7	8	9	10
1	–1.9	–1.6	–0.3	6.5	–0.5	3.1	12.1	20.1	38.8
2		0.8	11.5	26.3	18.2	18.0	30.2	54.0	34.4
3			6.4	13.6	14.6	3.0	17.8	41.8	22.8
4				9.3	–7.1	1.6	5.1	19.5	13.0
5					–1.0	–2.3	4.8	11.5	8.2
6						10.0	–3.1	8.2	–0.6
7							–1.1	4.4	–2.1
8								18.3	–1.5
9									–0.6

For example, the best swap, position $k = 4$ for $\ell = 6$, implies savings of

$$d_{3,4} + d_{4,5} + d_{5,6} + d_{6,7} = 7.1 + 7.0 + 9.9 + 25.0 = 49 \text{ inches}$$

for delinking holes 4 and 6 from their current fourth and sixth tour positions, plus costs

$$d_{3,6} + d_{6,5} + d_{5,4} + d_{4,7} = 15.0 + 9.9 + 7.0 + 10.0 = 41.9 \text{ inches}$$

for relinking in their new positions. The net change is $41.9 - 49 = -7.1$ inches.

Table 12.6 displays the tour sequence resulting from this best interchange. It also details, swaps, move directions, and solutions visited to reach local optimality at $t = 5$. Local optimum $\hat{\mathbf{y}}$ has length 92.8 inches with

$$\hat{y}_{1,2} = \hat{y}_{2,1} = \hat{y}_{3,3} = \hat{y}_{4,6} = \hat{y}_{5,5} = \hat{y}_{6,4} = \hat{y}_{7,8} = \hat{y}_{8,9} = \hat{y}_{9,10} = \hat{y}_{10,7} = 1$$

TABLE 12.6 Rudimentary Improving Search of the NCB Example

t	Drill Sequence	Length	Interchange	Nonzero Move Components
0	1–2–3–4–5–6–7–8–9–10	106.7	4th for 6th	$\Delta y_{4,4} = \Delta y_{6,6} = -1, \quad \Delta y_{4,6} = \Delta y_{6,4} = 1$
1	1–2–3–6–5–4–7–8–9–10	99.6	1st for 2nd	$\Delta y_{1,1} = \Delta y_{2,2} = -1, \quad \Delta y_{1,2} = \Delta y_{2,1} = 1$
2	2–1–3–6–5–4–7–8–9–10	97.7	8th for 10th	$\Delta y_{8,8} = \Delta y_{10,10} = -1, \quad \Delta y_{8,10} = \Delta y_{10,8} = 1$
3	2–1–3–6–5–4–7–10–9–8	96.0	7th for 10th	$\Delta y_{7,7} = \Delta y_{10,8} = -1, \quad \Delta y_{7,8} = \Delta y_{10,7} = 1$
4	2–1–3–6–5–4–8–10–9–7	93.8	8th for 9th	$\Delta y_{8,10} = \Delta y_{9,9} = -1, \quad \Delta y_{8,9} = \Delta y_{9,10} = 1$
5	2–1–3–6–5–4–8–9–10–7	92.8	Local optimum	

Initial length 106.7 inches has been reduced by 13%, but locally optimal value 92.8 leaves us well above globally shortest tour length 81.8 inches.

SAMPLE EXERCISE 12.29: PERFORMING DISCRETE IMPROVING SEARCH

Consider the knapsack model (Section 11.2)

$$\max \quad 18x_1 + 25x_2 + 11x_3 + 14x_4$$
$$\text{s.t.} \quad 2x_1 + 2x_2 + x_3 + x_4 \leq 3$$
$$x_1, \ldots, x_4 = 0 \text{ or } 1$$

under the **single complement** move set admitting moves that change one 0-component to 1 or one 1-component to 0. Compute an approximate optimal solution by discrete improving search Algorithm 12C beginning at $\mathbf{x}^{(0)} = (1, 0, 0, 0)$. If more than one neighbor is feasible and improving at any move, choose the one that improves the objective the most.

Analysis: Feasible neighbors of the given $\mathbf{x}^{(0)}$ are $(0, 0, 0, 0)$, $(1, 0, 1, 0)$ and $(1, 0, 0, 1)$ with objective values 0, 29, and 32, respectively. The last improves the objective function the most, so the search advances to $\mathbf{x}^{(1)} = (1, 0, 0, 1)$.

At $\mathbf{x}^{(1)}$, feasible neighbors are $(0, 0, 0, 1)$ and $(1, 0, 0, 0)$. Since neither improves the objective function, the search stops with local optimum $\hat{\mathbf{x}} = \mathbf{x}^{(1)} = (1, 0, 0, 1)$ at objective value $\hat{v} = 32$.

Multistart Search

For our tiny NCB example we were able to know how far heuristic optimum sequence 2–1–3–6–5–4–8–9–10–7 is from globally optimal. In a larger example it would be difficult to tell. Thus it is natural to at least try for further improvement.

One obvious approach is multistart—repeating the search from several different initial solutions $\mathbf{y}^{(0)}$.

> **12.50** **Multistart** or keeping the best of several local optima obtained by searches from different starting solutions is one way to improve the heuristic solutions produced by improving search.

Table 12.7 details searches of the NCB example from 3 different starts. The first is the search of Table 12.6 with local minimum 92.8. Search 2 yields an improved local minimum with length 84.7 inches after only one iteration. Search 3 terminates

TABLE 12.7   Multistart Search of the NCB Example

t	Drill Sequence	Length	Interchange	Nonzero Move Components
			Search 1	
0	1–2–3–4–5–6–7–8–9–10	106.7	4th for 6th	$\Delta y_{4,4} = \Delta y_{6,6} = -1$, $\Delta y_{4,6} = \Delta y_{6,4} = 1$
1	1–2–3–6–5–4–7–8–9–10	99.6	1st for 2nd	$\Delta y_{1,1} = \Delta y_{2,2} = -1$, $\Delta y_{1,2} = \Delta y_{2,1} = 1$
2	2–1–3–6–5–4–7–8–9–10	97.7	8th for 10th	$\Delta y_{8,8} = \Delta y_{10,10} = -1$, $\Delta y_{8,10} = \Delta y_{10,8} = 1$
3	2–1–3–6–5–4–7–10–9–8	96.0	7th for 10th	$\Delta y_{7,7} = \Delta y_{10,8} = -1$, $\Delta y_{7,8} = \Delta y_{10,7} = 1$
4	2–1–3–6–5–4–8–10–9–7	93.8	8th for 9th	$\Delta y_{8,10} = \Delta y_{9,9} = -1$, $\Delta y_{8,9} = \Delta y_{9,10} = 1$
5	2–1–3–6–5–4–8–9–10–7	92.8	Local optimum	Incumbent value = 92.8
			Search 2	
0	1–2–7–3–4–8–5–6–9–10	100.8	1st for 3rd	$\Delta y_{1,1} = \Delta y_{3,7} = -1$, $\Delta y_{1,7} = \Delta y_{3,1} = 1$
1	7–2–1–3–4–8–5–6–9–10	84.7	Local optimum	Incumbent value = 84.7
			Search 3	
0	1–10–2–9–3–8–4–7–5–6	157.7	1st for 4th	$\Delta y_{1,1} = \Delta y_{4,9} = -1$, $\Delta y_{1,9} = \Delta y_{4,1} = 1$
1	9–10–2–1–3–8–4–7–5–6	97.5	6th for 8th	$\Delta y_{6,8} = \Delta y_{8,7} = -1$, $\Delta y_{6,7} = \Delta y_{8,8} = 1$
2	9–10–2–1–3–7–4–8–5–6	92.2	3rd for 6th	$\Delta y_{3,2} = \Delta y_{6,7} = -1$, $\Delta y_{3,7} = \Delta y_{6,2} = 1$
3	9–10–7–1–3–2–4–8–5–6	86.8	5th for 6th	$\Delta y_{5,3} = \Delta y_{6,2} = -1$, $\Delta y_{5,2} = \Delta y_{6,3} = 1$
4	9–10–7–1–2–3–4–8–5–6	86.0	4th for 5th	$\Delta y_{4,1} = \Delta y_{5,2} = -1$, $\Delta y_{4,2} = \Delta y_{5,1} = 1$
5	9–10–7–2–1–3–4–8–5–6	84.7	Local optimum	Incumbent value = 84.7

with the same solution after a longer sequence. Multistart would report the best of these as an approximate optimum.

SAMPLE EXERCISE 12.30: PERFORMING MULTISTART SEARCH

Return to the knapsack problem of Sample Exercise 12.29:

$$\begin{aligned} \max\quad & 18x_1 + 25x_2 + 11x_3 + 14x_4 \\ \text{s.t.}\quad & 2x_1 + 2x_2 + x_3 + x_4 \le 3 \\ & x_1, \ldots, x_4 = 0 \text{ or } 1 \end{aligned}$$

and its single complement search neighborhood. Perform a multistart discrete improving search starting from initial solutions $(1, 0, 0, 0)$, $(0, 1, 0, 0)$, and $(0, 0, 1, 0)$.

Analysis: Sample Exercise 12.29 already established that $\mathbf{x}^{(0)} = (1, 0, 0, 0)$ leads to local maximum $\hat{\mathbf{x}} = (1, 0, 0, 1)$ with value $\hat{v} = 32$. Now restarting with $\mathbf{x}^{(0)} = (0, 1, 0, 0)$, the best feasible neighbor is $\mathbf{x}^{(1)} = (0, 1, 0, 1)$ with value 39.

This solution is locally optimal because no neighbor is both feasible and improving. Since it also improves on the current best (or incumbent) solution, our approximate optimum becomes $\hat{\mathbf{x}} = (0, 1, 0, 1)$, $\hat{v} = 39$.

The third search starts with $\mathbf{x}^{(0)} = (0, 0, 1, 0)$. Its feasible neighbor that most improves the objective function is $\mathbf{x}^{(1)} = (0, 1, 1, 0)$ with value 36. Again, Algorithm 12C stops at a local maximum. However, this one is not better than incumbent value $\hat{v} = 39$, so it is not retained.

12.7 TABU, SIMULATED ANNEALING, AND GENETIC ALGORITHM EXTENSIONS OF IMPROVING SEARCH

Although improving search Algorithm 12C can be quite effective on many models, expecially if applied repeatedly as in multistart, the improving search idea can be extended in a variety of ways to produce more robust algorithms. In this section we briefly introduce three of the most popular: tabu search, simulated annealing, and genetic algorithms.

Difficulty with Allowing Nonimproving Moves

An alternative to restarting improving search when no improving feasible move remains is to escape a local optimum by allowing nonimproving feasible moves. A few such retrograde moves might very well take the search to a region where progress can resume.

Unfortunately, this appealing strategy has a fatal flaw (unless new technology is introduced). Consider, for example, the locally optimal point where the first search of Table 12.7 terminated. No improving move was available at final drilling sequence 2–1–3–6–5–4–8–9–10–7 with value 92.8 inches. However, the best nonimproving move, which swaps the 9 in the eighth position for the 10 in the ninth, increases the objective by only 1.0 inch.

Why not take it and hope it leads to a better local optimum? Try. Sequence 2–1–3–6–5–4–8–10–9–7 results with length 93.8 inches. There is now a feasible interchange that improves (perhaps only one): swap back the 10 and 9 in the eighth and ninth tour positions to reduce length to 92.8 inches. Adopting that move returns us to exactly where we began, and the search will cycle forever.

> **12.51** Nonimproving moves will lead to infinite cycling of improving search unless some provision is added to prevent repeating solutions.

Tabu Search

Several schemes for incorporating nonimproving moves in improving search without undo cycling have proved effective on a variety of discrete optimization problems. One is called **tabu search** because it proceeds by classifying some moves "tabu" or forbidden. To be more specific,

> **12.52** Tabu search deals with cycling by temporarily forbidding moves that would return to a solution recently visited.

The effect is to prevent short-term cycling, although solutions can repeat over a longer period.

Algorithm 12D gives a formal statement of this tabu modification to improving search Algorithm 12C. A **tabu list** records forbidden moves, and each iteration chooses a non-tabu feasible move. After each step, a collection of moves that includes any returning immediately to the previous point is added to the tabu list. No such

---

ALGORITHM 12D: TABU SEARCH

Step 0: Initialization. Choose any starting feasible solution $\mathbf{x}^{(0)}$ and an iteration limit $t_{\max}$. Then set incumbent solution $\hat{\mathbf{x}} \leftarrow \mathbf{x}^{(0)}$ and solution index $t \leftarrow 0$. No moves are tabu.

Step 1: Stopping. If no non-tabu move $\Delta \mathbf{x}$ in move set $\mathcal{M}$ leads to a feasible neighbor of current solution $\mathbf{x}^{(t)}$, or if $t = t_{\max}$, then stop. Incumbent solution $\hat{\mathbf{x}}$ is an approximate optimum.

Step 2: Move. Choose some non-tabu feasible move $\Delta \mathbf{x} \in \mathcal{M}$ as $\Delta \mathbf{x}^{(t+1)}$.

Step 3: Step. Update

$$\mathbf{x}^{(t+1)} \leftarrow \mathbf{x}^{(t)} + \Delta \mathbf{x}^{(t+1)}$$

Step 4: Incumbent Solution. If the objective function value of $\mathbf{x}^{(t+1)}$ is superior to that of incumbent solution $\hat{\mathbf{x}}$, replace $\hat{\mathbf{x}} \leftarrow \mathbf{x}^{(t+1)}$.

Step 5: Tabu List. Remove from the list of tabu of forbidden moves any that have been on it for a sufficient number of iterations, and add a collection of moves that includes any returning immediately from $\mathbf{x}^{(t+1)}$ to $\mathbf{x}^{(t)}$.

Step 6: Increment. Increment $t \leftarrow t + 1$, and return to Step 1.

---

move is allowed for a few iterations, but eventually all are removed from the tabu list and again available.

Since steps may either improve or degrade the objective function value, an incumbent solution $\hat{\mathbf{x}}$ tracks the best feasible point found so far. When the search stops, which is usually when user-supplied iteration limit $t_{\max}$ is reached, incumbent $\hat{\mathbf{x}}$ is reported as an approximate optimum.

Tabu Search of the NCB Example

Table 12.8 illustrates an implementation of tabu search on the NCB example. Figure 12.7 tracks the objective function values of points encountered.

The initial solution and move set of this tabu search are identical to those of the ordinary improving search in Table 12.6. This time, however, $t_{\max} = 50$ points were visited, with search proceeding to the feasible, non-tabu neighbor with best objective function value, whether or not it improves.

The design of tabu searches requires some judgment in deciding what moves to make tabu at each iteration. Too few will lead to cycling; too many inordinately restricts the search.

Table 12.8's search of the NCB example fixed the first of each two positions interchanged for a period of 6 iterations. For example, after the first interchange of fourth and sixth positions, no move again changing the fourth position was allowed for 6 steps. Tabu positions are underlined. Such a policy maintains a relatively rich set of available moves, yet prevents immediate reverses.

Figure 12.7 shows clearly how tabu search improves as long as improving moves are available, then begins controlled wandering. In the beginning the incumbent solution $\hat{\mathbf{y}}$ improves rapidly. (Incumbent values are recorded in Table 12.8.) Later progress slows. Still, the global optimum was discovered on iteration $t = 44$. Naturally, it would be the final incumbent reported when iteration limit $t_{\max} = 50$ was reached.

TABLE 12.8   Tabu Search of NCB Example

t	Drill Sequence[a]	Length	Incumbent	Interchange	Δ Obj.
0	1–2–3–4–5–6–7–8–9–10	106.7	106.7	4th for 6th	7.1
1	1–2–3–6–5–4–7–8–9–10	99.6	99.6	1st for 2nd	1.9
2	2–1–3–6–5–4–7–8–9–10	97.7	97.7	8th for 10th	1.7
3	2–1–3–6–5–4–7–10–9–8	96.0	96.0	7th for 10th	2.2
4	2–1–3–6–5–4–8–10–9–7	93.8	93.8	6th for 10th	–2.3
5	2–1–3–6–5–7–8–10–9–4	96.1	93.8	5th for 10th	.1
6	2–1–3–6–4–7–8–10–9–5	96.2	93.8	4th for 10th	2.9
7	2–1–3–5–4–7–8–10–9–6	93.3	93.3	1st for 3rd	1.6
8	3–1–2–5–4–7–8–10–9–6	91.7	91.7	8th for 9th	–0.2
9	3–1–2–5–4–7–8–9–10–6	91.9	91.7	2nd for 3rd	–1.7
10	3–2–1–5–4–7–8–9–10–6	93.6	91.7	6th for 7th	–3.2
11	3–2–1–5–4–8–7–9–10–6	96.8	91.7	5th for 7th	–3.0
12	3–2–1–5–7–8–4–9–10–6	99.8	91.7	4th for 7th	15.9
13	3–2–1–4–7–8–5–9–10–6	83.9	83.9	1st for 3rd	–2.1
14	1–2–3–4–7–8–5–9–10–6	86.0	83.9	8th for 9th	–0.5
15	1–2–3–4–7–8–5–10–9–6	86.5	83.9	2nd for 3rd	–0.8
⋮	⋮	⋮	⋮	⋮	⋮
40	1–2–7–8–5–9–10–4–6–3	96.3	83.9	1st for 2nd	–1.3
41	2–1–7–8–5–9–10–4–6–3	97.6	83.9	8th for 9th	9.9
42	2–1–7–8–5–9–10–6–4–3	87.7	83.9	2nd for 9th	0.9
43	2–4–7–8–5–9–10–6–1–3	86.8	83.9	9th for 10th	5.0
44	2–4–7–8–5–9–10–6–3–1	81.8	81.8	6th for 7th	–0.5
45	2–4–7–8–5–10–9–6–3–1	82.3	81.8	5th for 7th	–3.1
46	2–4–7–8–9–10–5–6–3–1	85.4	81.8	1st for 10th	–2.1
47	1–4–7–8–9–10–5–6–3–2	87.5	81.8	7th for 8th	0.7
48	1–4–7–8–9–10–6–5–3–2	86.8	81.8	2nd for 3rd	0.1
49	1–7–4–8–9–10–6–5–3–2	86.7	81.8	9th for 10th	–3.0
50	1–7–4–8–9–10–6–5–2–3	89.7	81.8	Stop	

[a] Underlining indicates components not allowed to change.

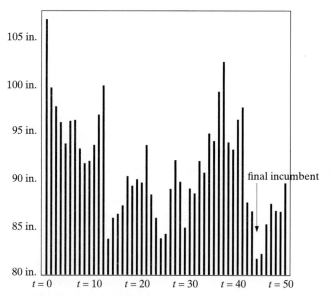

FIGURE 12.7   Solution Values in Tabu Search of NCB Example

Results will vary with models and details of tabu policy, but there is good reason to believe that such performance is typical. Suitable implementations of the tabu variation on improving search can greatly enhance the quality of heuristic solutions obtained.

SAMPLE EXERCISE 12.31: APPLYING TABU SEARCH

Return to the knapsack problem

$$\max \quad 18x_1 + 25x_2 + 11x_3 + 14x_4$$
$$\text{s.t.} \quad 2x_1 + 2x_2 + x_3 + x_4 \le 3$$
$$x_1, \dots, x_4 = 0 \text{ or } 1$$

of Sample Exercises 12.29 and 12.30 and assume that we will use the same single-complement move set.

(a) Explain how making it tabu to complement any component x_j for the next 2 iterations after it is changed by the search prevents short-term cycling.

(b) Begin from solution $\mathbf{x}^{(0)} = (1, 0, 0, 0)$ and use this tabu rule in executing Algorithm 12D through $t_{max} = 5$ steps.

Analysis:

(a) Once a component is changed for 0 to 1, or vice versa, the only way to return to the immediately preceding solution is to complement the same component again.

(b) The required search is summarized in the following table:

t	$\mathbf{x}^{(t)}$	Value	Incumbent Value	Complemented	Δ obj.
0	$(1, 0, 0, 0)$	18	18	$j = 4$	14
1	$(1, 0, 0, \underline{1})$	32	32	$j = 1$	−18
2	$(\underline{0}, 0, 0, 1)$	14	32	$j = 2$	25
3	$(\underline{0,1}, 0, 1)$	39	39	$j = 4$	−14
4	$(0, \underline{1}, 0, \underline{0})$	25	39	$j = 3$	11
5	$(0, 1, \underline{1}, \underline{0})$	36	39	Stop	

Each iteration begins by selecting an x_j to complement that is not among those marked tabu (underlined) and does preserve feasibility. The best such move produces the next solution. The result is also saved as a new incumbent if it is superior to any feasible solution encountered so far. Computation stops with approximate (here exact) optimum $\hat{\mathbf{x}} = (0, 1, 0, 1)$ at $t = t_{max} = 5$.

Simulated Annealing Search

Another method of introducing nonimproving moves into improving search is termed **simulated annealing** because of its analogy to the annealing process of slowly cooling metals to improve strength.

12.53 Simulated annealing algorithms control cycling by accepting nonimproving moves according to probabilities tested with computer-generated random numbers.

Improving and accepted nonimproving moves are pursued; rejected ones are not.

Algorithm 12E provides details. The move selection process at each iteration begins with random choice of a provisional feasible move, totally ignoring its objective function impact. Next, the net objective function improvement Δobj (nonpositive for nonimproving moves) is computed for the chosen move. The move is always accepted if it improves (Δ obj > 0), and otherwise

$$\text{probability of acceptance} = e^{\Delta \text{obj}/q} \qquad (12.22)$$

That is, all improving moves and some nonimproving ones are accepted. The probability of accepting a nonimproving move declines as net objective improvement Δobj becomes more negative.

ALGORITHM 12E: SIMULATED ANNEALING SEARCH

Step 0: Initialization. Choose any starting feasible solution $\mathbf{x}^{(0)}$, an iteration limit $t_{\max}$, and a relatively large initial temperature $q > 0$. Then set incumbent solution $\hat{\mathbf{x}} \leftarrow \mathbf{x}^{(0)}$ and solution index $t \leftarrow 0$.

Step 1: Stopping. If no move $\Delta \mathbf{x}$ in move set $\mathcal{M}$ leads to a feasible neighbor of current solution $\mathbf{x}^{(t)}$, or if $t = t_{\max}$, then stop. Incumbent solution $\hat{\mathbf{x}}$ is an approximate optimum.

Step 2: Provisional Move. Randomly choose a feasible move $\Delta \mathbf{x} \in \mathcal{M}$ as a provisional $\Delta \mathbf{x}^{(t+1)}$, and compute the (possibly negative) net objective function improvement Δobj for moving from $\mathbf{x}^{(t)}$ to $\left(\mathbf{x}^{(t)} + \Delta \mathbf{x}^{(t+1)} \right)$ (increase for a maximize, decrease for a minimize).

Step 3: Acceptance. If $\Delta \mathbf{x}^{(t+1)}$ improves, or with probability $e^{\Delta \text{obj}/q}$ if Δobj ≤ 0, accept $\Delta \mathbf{x}^{(t+1)}$ and update

$$\mathbf{x}^{(t+1)} \leftarrow \mathbf{x}^{(t)} + \Delta \mathbf{x}^{(t+1)}$$

Otherwise, return to Step 2.

Step 4: Incumbent Solution. If the objective function value of $\mathbf{x}^{(t+1)}$ is superior to that of incumbent solution $\hat{\mathbf{x}}$, replace $\hat{\mathbf{x}} \leftarrow \mathbf{x}^{(t+1)}$.

Step 5: Temperature Reduction. If a sufficient number of iterations have passed since the last temperature change, reduce temperature q.

Step 6: Increment. Increment $t \leftarrow t + 1$, and return to Step 1.

Parameter q in (12.22) is a **temperature** controlling the randomness of the search. If q is large, the exponent in (12.22) approaches 0, implying that the probability of accepting nonimproving moves approximates $e^0 = 1$. If q is small, the probability of accepting very bad moves decreases dramatically. Simulated annealing searches usually begin with q relatively large and decrease it every few iterations.

As with tabu and other searches that can make nonimproving moves, an incumbent solution $\hat{\mathbf{x}}$ must be maintained to keep track of the best feasible solution found so far. When computation stops, $\hat{\mathbf{x}}$ is output as an approximately optimal solution.

Simulated Annealing Search of NCB Example

Table 12.9 provides an abridged summary of a simulated annealing search of our NCB example. Figure 12.8 shows the complete history of accepted solutions through iteration limit $t_{max} = 50$. Like all other searches of this section, move set $\mathcal{M}$ included all single interchanges, and initial solution $\mathbf{y}^{(0)}$ is the one in (12.21).

TABLE 12.9   Simulated Annealing Search of NCB Example

t	Drill Sequence	Length	Incumbent	Temp q	Interchange	Δ Obj.	Outcome
0	1–2–3–4–5–6–7–8–9–10	106.7	106.7	5.00	7th for 10th	2.1	Accepted
1	1–2–3–4–5–6–10–8–9–7	104.6	104.6	5.00	1st for 9th	−20.1	Rejected
					1st for 4th	−3.1	Accepted
2	4–2–3–1–5–6–10–8–9–7	107.7	104.6	5.00	7th for 10th	1.3	Accepted
3	4–2–3–1–5–6–7–8–9–10	106.4	104.6	5.00	5th for 6th	5.0	Accepted
4	4–2–3–1–6–5–7–8–9–10	101.4	101.4	5.00	9th for 10th	0.3	Accepted
5	4–2–3–1–6–5–7–8–10–9	101.1	101.1	5.00	9th for 10th	−0.3	Accepted
6	4–2–3–1–6–5–7–8–9–10	101.4	101.1	5.00	2nd for 7th	−12.9	Rejected
					1st for 7th	3.6	Accepted
7	7–2–3–1–6–5–4–8–9–10	97.8	97.8	5.00	6th for 7th	−6.6	Rejected
					7th for 8th	−1.7	Accepted
8	7–2–3–1–6–5–8–4–9–10	99.5	97.8	5.00	2nd for 5th	−9.0	Rejected
					2nd for 4th	−0.9	Accepted
⋮	⋮	⋮	⋮	⋮	⋮	⋮	⋮
40	3–1–2–4–7–8–5–9–10–6	81.8	81.8	2.05	6th for 10th	−13.4	Rejected
					5th for 6th	−4.5	Rejected
					2nd for 5th	−24.2	Rejected
					9th for 10th	−15.3	Rejected
					9th for 10th	−15.3	Rejected
					8th for 9th	−0.5	Accepted
⋮	⋮	⋮	⋮	⋮	⋮	⋮	⋮
46	3–1–2–4–7–8–9–10–5–6	85.4	81.8	2.05	5th for 10th	−16.5	Rejected
					8th for 10th	−15.3	Rejected
					8th for 10th	−15.3	Rejected
					3rd for 6th	−20.8	Rejected
					2nd for 9th	−39.2	Rejected
					2nd for 9th	−39.2	Rejected
					4th for 9th	−22.5	Rejected
					3rd for 9th	−32.2	Rejected
					5th for 7th	−21.3	Rejected
					1st for 3rd	−3.8	Rejected
					2nd for 8th	−59.6	Rejected
					9th for 10th	0.7	Accepted
47	3–1–2–4–7–8–9–10–6–5	84.7	81.8	2.05	3rd for 8th	−52.6	Rejected
					8th for 9th	−9.8	Rejected
					4th for 5th	−0.7	Rejected
					7th for 10th	−12.4	Rejected
					7th for 10th	−12.4	Rejected
					5th for 10th	−12.0	Rejected
					1st for 7th	−34.0	Rejected
					4th for 10th	−13.3	Rejected
					2nd for 10th	−18.6	Rejected
					6th for 10th	−7.8	Rejected
					6th for 7th	−18.3	Rejected
					3rd for 5th	−12.7	Rejected
					9th for 10th	−0.7	Accepted
48	3–1–2–4–7–8–9–10–5–6	85.4	81.8	2.05	9th for 10th	0.7	Accepted
49	3–1–2–4–7–8–9–10–6–5	84.7	81.8	2.05	7th for 8th	0.2	Accepted
50	3–1–2–4–7–8–10–9–6–5	84.5	81.8	1.64	Stop		

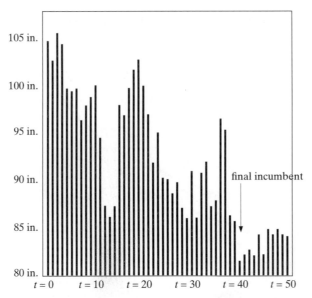

FIGURE 12.8   Solution Values in Simulated Annealing
Search of NCB Example

Temperatures in this simulated annealing example began at $q = 5.0$. Every 10
iterations they were reduced by a factor of 0.8, so that

$$q = 5.0 \qquad\qquad \text{for iterations } t = 0, \ldots, 9$$
$$q = 0.8(5.0) = 4.0 \qquad \text{for iterations } t = 10, \ldots, 19$$
$$q = 0.8(4.0) = 3.2 \qquad \text{for iterations } t = 20, \ldots, 29$$
$$\vdots$$

The first few iterations of Table 12.9 show most randomly chosen moves being
accepted. Interchange of the seventh and tenth tour positions improves on solution
$\mathbf{y}^{(0)}$ and is accepted immediately to produce $\mathbf{y}^{(1)}$. The first move generated from
solution $\mathbf{y}^{(1)}$ increases length by 20.1 inches. Thus its probability of acceptance was

$$e^{-20.1/5} \approx 0.018$$

Not surprisingly, it was rejected.

The next try produced a move that increased length by only 3.1 inches. With
better probability

$$e^{-3.1/5} \approx 0.538$$

it was adopted.

The latter part of Table 12.9 shows the impact of reducing temperature q as
the search proceeds. More and more nonimproving moves are rejected, because the
probability of acceptance has declined with q.

Results in Figure 12.8 show the wide-ranging evolution of the full simulated
annealing search. As with tabu, the final incumbent solution happens to be the
global optimum, but this is not guaranteed.

Again this behavior is typical of reported simulated annealing applications, although many more iterations would normally be required. Suitable implementations

of the simulated annealing variation on improving search can greatly enhance the quality of heuristic solutions obtained.

SAMPLE EXERCISE 12.32: APPLYING SIMULATED ANNEALING

Return again to the knapsack model of Sample Exercises 12.29 to 12.31 with its single-complement move set and initial solution $x^{(0)} = (1, 0, 0, 0)$. Using temperature $q = 10$, apply simulated annealing Algorithm 12E through $t_{max} = 3$ steps. Where random decisions are required, use the following random numbers (uniform between 0 and 1): 0.72, 0.83, 0.33, 0.41, 0.09, 0.94.

Analysis: Required computations are summarized in the following table.

t	$x^{(t)}$	Value	Incumbent Value	q	Complement	Δ obj	Outcome
0	$(1, 0, 0, 0)$	18	18	10	$j = 4$	14	Accepted
1	$(1, 0, 0, 1)$	32	32	10	$j = 4$	−14	Rejected
					$j = 1$	−18	Accepted
2	$(0, 0, 0, 1)$	14	32	10	$j = 3$	11	Accepted
3	$(0, 0, 1, 1)$	25	32	10	Stop		

The process begins by randomly selecting between feasible moves complementing $j = 3$ and $j = 4$ at $x^{(0)} = (1, 0, 0, 0)$. Since the first random number 0.72 is in the upper half of interval [0,1], $j = 4$ is provisionally selected. The corresponding move is improving, so the search advances to $x^{(1)} = (1, 0, 0, 1)$ with value 32.

Feasible complements at $x^{(1)}$ are $j = 1$ and $j = 4$. The next random number 0.83 selects $j = 4$. The corresponding move has Δobj $= -14$, so we test whether random number 0.33 is at most probability

$$e^{\Delta obj/q} = e^{-14/10} \approx 0.247$$

It is not, and the provisional move is rejected.

The next randomly generated feasible move complements $j = 1$ with Δobj $= -18$. This time

$$e^{\Delta obj/q} = e^{-18/10} \approx 0.165 \geq 0.09$$

We accept the move and advance to $x^{(2)} = (0, 0, 0, 1)$.

Three moves are feasible at $x^{(2)}$ and complementation of $j = 3$ is selected by random number 0.94. The corresponding improving move advances to $x^{(3)} = (0, 0, 1, 1)$. We now stop at $t = t_{max} = 3$ and report incumbent solution $\hat{x} = (1, 0, 0, 1)$ with value 32.

Genetic Algorithms

A third popular method for avoiding local optima in improving search is known as **genetic algorithms** because it attempts to parallel the process of biological evolution to find better and better solutions.

12.54 | Genetic algorithms evolve good heuristic optima by operations combining members of an improving **population** of individual solutions.

The best single solution encountered so far will always be part of the population (in the variant discussed here), but each **generation** will also include a spectrum of other solutions. Ideally, all will be feasible, and some may be nearly as good in the objective function as the best. Others may have quite poor solution values.

New solutions are created by combining pairs of individuals in the population. Local optima are less frequent because this combining process does not center entirely on the best current solution.

Crossover Operations in Genetic Algorithms

The standard genetic algorithm method for combining solutions of the population is known as crossover.

12.55 | **Crossover** combines a pair of "parent" solutions to produce a pair of "children" by breaking both parent vectors at the same point and reassembling the first part of one parent solution with the second part of the other, and vice versa.

We can illustrate with two binary solution vectors $\mathbf{x}^{(1)}$ and $\mathbf{x}^{(2)}$:

$$\mathbf{x}^{(1)} = (1, 0, 1, 1, 0, \,\big|\, 0, 1, 0, 0)$$
$$\mathbf{x}^{(2)} = (0, 1, 1, 0, 1, \,\big|\, 1, 0, 0, 1)$$

Crossover after component $j = 5$ leads to children

$$\mathbf{x}^{(3)} = (1, 0, 1, 1, 0, \,\big|\, 1, 0, 0, 1)$$
$$\mathbf{x}^{(4)} = (0, 1, 1, 0, 1, \,\big|\, 0, 1, 0, 0)$$

One child $\mathbf{x}^{(3)}$ combines the initial part of $\mathbf{x}^{(1)}$ with the final part of $\mathbf{x}^{(2)}$. Child $\mathbf{x}^{(4)}$ does just the opposite. Both become members of the new population, and the search continues.

There is no guarantee that crossover's rather arbitrary manipulation of parent solutions will yield improvement. Still, it does lead to fundamentally new solutions that preserve significant parts of their parents. Experience shows that this is often enough to produce very good results.

Managing Genetic Algorithms with Elites, Immigrants, and Crossovers

Many variations on the basic genetic algorithm strategy have been employed successfully in particular applications, including many alternatives to standard crossover operations | **12.55** |. Principal differences in the various implementations concern how to select pairs of current solutions to produce new ones via crossover, how to decide which new and/or old solutions will survive in the next population, and how to maintain diversity in the population as the search advances from generation to generation The only requirement is that better solutions have greater chance to breed.

In this brief introduction we consider only a single elitest method of population management. Each new generation will be composed of a combination of elite, immigrant, and crossover solutions.

> 12.56 | The **elitest** strategy for implementation of genetic algorithms forms each new generation as a mixture of **elite** (best) solutions held over from the previous generation, **immigrant** solutions added arbitrarily to increase diversity, and children of crossover operations on nonoverlapping pairs of solutions in the previous population.

Maintenance of the elite solutions from the preceding generation assures that the best solutions known so far will remain in the population and have more opportunities to produce offspring. Addition of new immigrant solutions will help to maintain diversity as solutions are combined. The bulk of the new solutions will be the product of crossovers, with elites in the preceding population allowed to serve as parents. Algorithm 12F details a full procedure.

ALGORITHM 12F: GENETIC ALGORITHM SEARCH

Step 0: Initialization. Choose a population size p, initial starting feasible solutions $\mathbf{x}^{(1)}, \ldots, \mathbf{x}^{(p)}$, a generation limit t_{max}, and population subdivisions p_e for elites, p_i for immigrants, and p_c for crossovers. Also set generation index $t \leftarrow 0$.

Step 1: Stopping. If $t = t_{max}$, stop and report the best solution of the current population as an approximate optimum.

Step 2: Elite. Initialize the population of generation $t + 1$ with copies of the p_e best solutions in the current generation.

Step 3: Immigrants. Arbitrarily choose p_i new immigrant feasible solutions, and include them in the $t + 1$ population.

Step 4: Crossovers. Choose $p_c/2$ nonoverlapping pairs of solutions from the generation t population, and execute crossover on each pair at an independently chosen random cut point to complete the generation $t + 1$ population.

Step 5: Increment. Increment $t \leftarrow t + 1$, and return to Step 1.

Solution Encoding for Genetic Algorithm Search

Just as design of an ordinary improving search requires careful construction of a move set (principles | 12.48 | and | 12.49 |), implementations of genetic algorithms require judicious choice of a scheme for encoding solutions in a vector. To see the difficulty, return to our NCB drilling example.

If solutions are encoded merely by displaying the drilling sequence, two that might come together as crossover parents would be

$$(3, 1, 2, 4, 7, 8, 5, 9, 10, 6) \quad \text{and} \quad (7, 2, 3, 1, 6, 5, 8, 4, 9, 10)$$

Crossing over the solutions after component $j = 6$ would yield children

$$(3, 1, 2, 4, 7, 8, 8, 4, 9, 10) \quad \text{and} \quad (7, 2, 3, 1, 6, 5, 5, 9, 10, 6)$$

Neither is a feasible drilling sequence because some holes are visited more than once and some never. A poor choice of solution encoding has made it almost impossible for crossover to produce useful new solutions.

> **12.57** Effective genetic algorithm search requires a choice for encoding problem solutions that often, if not always, preserves solution feasibility after crossover.

We can obtain a better encoding in the NCB example by a technique known as **random keys**. Sequences are encoded indirectly under random keys as a vector of random numbers such as

$$(0.32, 0.56, 0.91, 0.44, 0.21, 0.68, 0.51, 0.07, 0.12, 0.39)$$

The drilling sequence implied is the one obtained by aligning hole 1 with the lowest random component, hole 2 at the next lowest, and so on. That is, the given random vector encodes the drilling sequence

$$(4, 8, 10, 6, 3, 9, 7, 1, 2, 5)$$

But notice that crossover on two vectors of random numbers produces two others. Thus every crossover operation of these random key encodings will yield two new feasible solutions. Also, it is very easy to generate arbitrary new solutions for the initial population and immigration simply by generating random vectors of random numbers.

Genetic Algorithm Search of NCB Example

Figure 12.9 shows objective function values in a 30-generation Algorithm 12F search of our NCB example. The population had size 20, with the initial population generated randomly. Following principle 12.56, each new generation contained the 6 best (elite) solutions of the preceding generation, 4 randomly generated immigrant solutions, and 10 solutions obtained from crossover of 5 pairs of parents in the preceding generation. All solutions were encoded by random keys.

Bars in Figure 12.9 extend from the lowest to the highest route length of solutions in each population. Notice that the lowest ones converge systematically to 81.8, which we know is the optimal value of this example. Still, the population always contains a variety of solutions, some with rather poor objective values. This diversity makes it possible for the search to range into distinctly new parts of the feasible space as it seeks improved solutions.

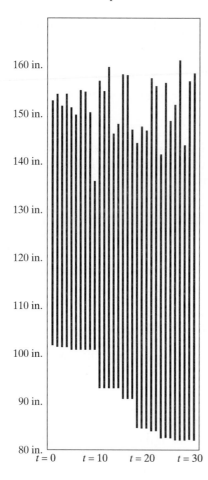

FIGURE 12.9   Genetic Search Population Ranges for NCB Example

12.8 CONSTRUCTIVE HEURISTICS

Improving search heuristics of Section 12.6 and 12.7 move from complete solution to complete solution. The **constructive search** alternative, which is the subject of this section, follows a strategy more like the branch and bound searches of Section 12.4 and 12.5. They proceed through partial solutions (definition 12.22), choosing values for decision variables one at a time and (often) stopping upon completion of a first feasible solution.

Rudimentary Constructive Search Algorithm

Constructive searches typically begin with every discrete component of the decision vector free. At each iteration, one previously free variable is fixed at a value feasible with decisions fixed so far. That is, the chosen value for the new component should not produce constraint violation when previously fixed values are substituted and optimistic assumptions about free variable values are adopted.

In the simplest case, the process terminates when no free variables remain. Algorithm 12G gives a more formal statement.

ALGORITHM 12G: RUDIMENTARY CONSTRUCTIVE SEARCH

Step 0: Initialization. Start with all-free initial partial solution $\mathbf{x}^{(0)} = (\#, \ldots, \#)$ and set solution index $t \leftarrow 0$.

Step 1: Stopping. If all components of current solution $\mathbf{x}^{(t)}$ are fixed, stop and output $\hat{\mathbf{x}} \leftarrow \mathbf{x}^{(t)}$ as an approximate optimum.

Step 2: Step. Choose a free component x_p of partial solution $\mathbf{x}^{(t)}$ and a value for it that plausibly leads to good feasible completions. Then, advance to partial solution $\mathbf{x}^{(t+1)}$ identical to $\mathbf{x}^{(t)}$ except that x_p is fixed at the chosen value.

Step 3: Increment. Increment $t \leftarrow t + 1$, and return to Step 1.

Greedy Choices of Variables to Fix

Obviously, the bulk of the effort in constructive searches goes to choosing the next free variable to fix and picking its value. Most common procedures accomplish these tasks in a **greedy** or **myopic** fashion.

12.58 | Greedy constructive heuristics elect the next variable to fix and its value that does least damage to feasibility and most helps the objective function, based on what has already been fixed in the current partial solution.

That is, greedy rules choose the fix that seems most likely, on the basis of what is presently known, to lead to a good feasible completions.

Just as with improving search, there is risk in looking only locally for information about the next choice. Quite possibly a decision that appears very good with only a few variables fixed will actually end up forcing the search into a very poor part of the feasible space. Still, if the procedure is to be computationally efficient, compromises have to be made.

Greedy Rule for NASA Example

We can illustrate the idea of constructive search with the NASA capital budgeting model (12.16). There, the decision variables

$$x_j \triangleq \begin{cases} 1 & \text{if mission } j \text{ is selected} \\ 0 & \text{otherwise} \end{cases}$$

It is natural to construct a solution for capital budgeting models like this one by successively adding missions until constraints block further inclusions. That is, we will seek to fix a previously free x_j at value 1 as long as any free mission can be selected without violating a constraint.

To implement this natural constructive search we need a greedy selection criterion of the type described in | 12.58 | . Certainly, we should prefer missions with high objective function "value" coefficients. But we also want to consider constraints. One high-value mission might consume so much of the various budgets that it would block all further decisions. Also, precedence constraints change the implicit value of a mission; choosing the mission gains its value and makes successors feasible.

Our search will tradeoff these objective and constraint considerations in a common way, by comparing missions according to the ratios

$$r_j \triangleq \frac{\left(\begin{array}{c}\text{project} \\ j \text{ value}\end{array}\right) + \left(\begin{array}{c}\text{allowance for} \\ \text{enabled successor} \\ \text{values}\end{array}\right)}{\sum_{i=1}^{8} \left(\begin{array}{c}\text{fraction of remaining} \\ \text{constraint } i \text{ right-hand} \\ \text{side consumed by } j\end{array}\right)} = \frac{c_j + \sum_{\text{free } k \text{ preceded by } j} \left(\frac{c_k}{2}\right)}{\sum_{i=1}^{8} \left(\frac{a_{i,j}}{b_i^{(t)}}\right)} \qquad (12.23)$$

where

$c_j \triangleq$ objective function coefficient for mission j

$a_{i,j} \triangleq$ coefficient for mission j in the ith main constraint

$b_i^{(t)} \triangleq$ right-hand side remaining in the ith main constraint after fixing variable values as in partial solution $\mathbf{x}^{(t)}$

Among the free projects for which all predecessors have been scheduled in partial solution $\mathbf{x}^{(t)}$, we will fix the x_j with maximum ratio r_j. If there remains room for j in all applicable budgets, it is fixed $= 1$. Otherwise, we set $x_j = 0$.

Like many such greedy indices, ratio (12.23) seems rather complicated at first glance. The numerator tries to account for both the immediate value of selecting a mission, and the potential it opens up to select missions of which it is a predecessor. Half the value of all enabled successors is arbitrarily added to the mission's direct value. The denominator of (12.23) sums the fractions of remaining row "resources" that a mission would consume if selected. Thus we favor missions using relatively little of now-scarce resources.

Ratioing value to resource use combines objective and constraint considerations. The highest r_j will correspond to a mission j high in value, or low in resource consumption, or both. Selecting that mission may not be the best long-term decision, but it does reflect about all we can know without looking more than one step into the future.

SAMPLE EXERCISE 12.33: DEVISING GREEDY HEURISTIC RULES

Recall from Section 11.3 that set cover models seek a minimum cost collection of columns or subsets that together include or cover every element of a given set. One example is

$$\begin{array}{llllll}
\min & 15x_1 & +18x_2 & +6x_3 & +20x_4 & \\
\text{s.t.} & +\ x_1 & & & +\ x_4 & \geq 1 \\
& +\ x_1 & +\ x_2 & & +\ x_4 & \geq 1 \\
& & +\ x_2 & +x_3 & +\ x_4 & \geq 1 \\
& x_1, \ldots, x_4 = 0 \text{ or } 1 & & & &
\end{array}$$

Explain why it would make sense to choose free x_j to fix $= 1$ by picking one with least ratio

$$r_j \triangleq \frac{\text{cost coefficient of column } j}{\text{number of uncovered elements that } j \text{ covers}}$$

Analysis: The proposed ratio explicitly seeks minimum cost by including the objective function coefficient in its numerator. Still, it also considers feasibility in dividing by the number of still uncovered rows or elements each free j could resolve. The effect is to seek the most efficient next choice of x_j to fix $= 1$, the best in the short-term or myopic sense.

Constructive Heuristic Solution of NASA Example

Starting from the completely free partial solution

$$\mathbf{x}^{(0)} = (\#, \#, \#, \#, \#, \#, \#, \#, \#, \#, \#, \#, \#, \#)$$

all $b_i^{(0)}$ equal initial right-hand sides. Ratios for the first two j's are

$$r_1 = \frac{200}{6/10} = 333.33$$

$$r_2 = \frac{3 + (18/2)}{(2/10) + (3/12)} = 26.67$$

and similar arithmetic yields

$$r_3 = 129.07, \quad r_4 = 29.17, \quad r_5 = 35.21, \quad r_6 = 21.54$$

$$r_7 = 6.52, \quad r_8 = 7.37, \quad r_9 = 110.09, \quad r_{10} = 157.50$$

$$r_{11} = 10.96, \quad r_{12} = 7.38, \quad r_{13} = 586.05, \quad r_{14} = 87.30$$

The highest of these ratios is 586.05 for mission 13. Since this mission fits within remaining right-hand sides $b_i^{(t)}$ and has no predecessors, we fix $x_{13} = 1$ to produce

$$\mathbf{x}^{(1)} = (\#, \#, \#, \#, \#, \#, \#, \#, \#, \#, \#, \#, 1, \#)$$

Table 12.10 provides an abridged summary of the rest of the search. Processing of $t = 1$ parallels the first iteration, selecting and fixing $= 1$ additional mission 1.

Something different occurs at $t = 3$. The mission with the maximum r_j there is number 9. But mission 9 requires \$5 billion in the 2000–2004 budget period, and projects already chosen use all but $b_2^{(3)} = \$3$ billion. We have to fix $x_9 = 0$ to maintain feasibility.

Another peculiarity arises at iteration $t = 9$. There the mission with the best ratio is $j = 11$. However, mission 11 cannot be selected before predecessor mission 2. Thus we pass to the second best ratio, which happens to be $j = 2$, and fix $x_2 = 1$.

Our constructive search terminates when all 14 components of the decision vector have been fixed. The heuristic optimal solution produced is

$$\hat{\mathbf{x}} \triangleq \mathbf{x}^{(14)} = (1, 1, 0, 0, 0, 0, 0, 0, 0, 1, 1, 0, 1, 0)$$

flying missions 1, 2, 10, 11, and 13 at a total value of 671. (Compare with the branch and bound results in Table 12.3.)

TABLE 12.10   Constructive Search of NASA Example

t	Computation	Choice
0	$\mathbf{x}^{(0)} = (\#, \#, \#, \#, \#, \#, \#, \#, \#, \#, \#, \#, \#, \#)$	
	$b_1^{(0)} = 10, \quad b_2^{(0)} = 12, \quad b_3^{(0)} = 14, \quad b_4^{(0)} = 14,$ $b_5^{(0)} = 14, \quad b_6^{(0)} = 1, \quad b_7^{(0)} = 1, \quad b_8^{(0)} = 1$	
	$r_1 = 333.33, \quad r_2 = 26.67, \quad r_3 = 129.07, \quad r_4 = 29.17,$ $r_5 = 35.21, \quad r_6 = 21.54, \quad r_7 = 6.52, \quad r_8 = 7.37,$ $r_9 = 110.09, \quad r_{10} = 157.50, \quad r_{11} = 10.96, \quad r_{12} = 7.38,$ $r_{13} = 586.05, \quad r_{14} = 87.30$	Select $j = 13$ and fix $x_{13} = 1$
1	$\mathbf{x}^{(1)} = (\#, \#, \#, \#, \#, \#, \#, \#, \#, \#, \#, \#, 1, \#)$	
	$b_1^{(1)} = 10, \quad b_2^{(1)} = 11, \quad b_3^{(1)} = 10, \quad b_4^{(1)} = 13,$ $b_5^{(1)} = 13, \quad b_6^{(1)} = 1, \quad b_7^{(1)} = 1, \quad b_8^{(1)} = 1$	
	$r_1 = 333.33, \quad r_2 = 25.38, \quad r_3 = 122.59, \quad r_4 = 28.26,$ $r_5 = 31.05, \quad r_6 = 19.55, \quad r_7 = 6.05, \quad r_8 = 7.22,$ $r_9 = 107.84, \quad r_{10} = 133.06, \quad r_{11} = 10.35, \quad r_{12} = 7.04,$ $r_{13} = \text{N/A}, \quad r_{14} = 79.56$	Select $j = 1$ and fix $x_1 = 1$
$\vdots$	$\vdots$	$\vdots$
3	$\mathbf{x}^{(3)} = (1, \#, \#, \#, \#, \#, \#, \#, \#, 1, \#, \#, 1, \#)$	
	$b_1^{(3)} = 4, \quad b_2^{(3)} = 3, \quad b_3^{(3)} = 6, \quad b_4^{(3)} = 13,$ $b_5^{(3)} = 13, \quad b_6^{(3)} = 1, \quad b_7^{(3)} = 1, \quad b_8^{(3)} = 1$	
	$r_1 = \text{N/A}, \quad r_2 = 8.00, \quad r_3 = 38.28, \quad r_4 = 28.26,$ $r_5 = 17.50, \quad r_6 = 18.35, \quad r_7 = 1.71, \quad r_8 = 7.22,$ $r_9 = 54.54, \quad r_{10} = \text{N/A}, \quad r_{11} = 9.61, \quad r_{12} = 2.23,$ $r_{13} = \text{N/A}, \quad r_{14} = 50.99$	Select $j = 9$ and fix $x_9 = 0$ because violates contraint 2
$\vdots$	$\vdots$	$\vdots$
9	$\mathbf{x}^{(9)} = (1, \#, 0, 0, 0, 0, \#, \#, 0, 1, \#, \#, 1, 0)$	
	$b_1^{(9)} = 4, \quad b_2^{(9)} = 3, \quad b_3^{(9)} = 6, \quad b_4^{(9)} = 13,$ $b_5^{(9)} = 13, \quad b_6^{(9)} = 1, \quad b_7^{(9)} = 1, \quad b_8^{(9)} = 1$	
	$r_1 = \text{N/A}, \quad r_2 = 8.00, \quad r_3 = \text{N/A}, \quad r_4 = \text{N/A},$ $r_5 = \text{N/A}, \quad r_6 = \text{N/A}, \quad r_7 = 1.71, \quad r_8 = 7.22,$ $r_9 = \text{N/A}, \quad r_{10} = \text{N/A}, \quad r_{11} = 9.61, \quad r_{12} = 2.23,$ $r_{13} = \text{N/A}, \quad r_{14} = \text{N/A}$	Select second best $j = 2$ and fix $x_2 = 1$ because $j = 11$ has free predecessor
10	$\mathbf{x}^{(10)} = (1, 1, 0, 0, 0, 0, \#, \#, 0, 1, \#, \#, 1, 0)$	
	$b_1^{(10)} = 2, \quad b_2^{(10)} = 0, \quad b_3^{(10)} = 6, \quad b_4^{(10)} = 13,$ $b_5^{(10)} = 13, \quad b_6^{(10)} = 1, \quad b_7^{(10)} = 1, \quad b_8^{(10)} = 1$	
	$r_1 = \text{N/A}, \quad r_2 = \text{N/A}, \quad r_3 = \text{N/A}, \quad r_4 = \text{N/A},$ $r_5 = \text{N/A}, \quad r_6 = \text{N/A}, \quad r_7 = 0.0000, \quad r_8 = 0.0001,$ $r_9 = \text{N/A}, \quad r_{10} = \text{N/A}, \quad r_{11} = 0.002, \quad r_{12} = 0.0000,$ $r_{13} = \text{N/A}, \quad r_{14} = \text{N/A}$	Select $j = 11$ and fix $x_{11} = 1$
$\vdots$	$\vdots$	$\vdots$
14	$\mathbf{x}^{(14)} = (1, 1, 0, 0, 0, 0, 0, 0, 0, 1, 1, 0, 1, 0)$	

SAMPLE EXERCISE 12.34: EXECUTING CONSTRUCTIVE HEURISTICS

Return to the set covering problem of Sample Exercise 12.33 and its suggested greedy ratio r_j. Use this ratio to apply Algorithm 12G, fixing selected $x_j = 1$ as long as rows remain uncovered and $= 0$ thereafter.

Analysis: The search begins with all-free partial solution $\mathbf{x}^{(0)} = (\#, \#, \#, \#)$. All rows of the set covering model are uncovered, so ratios compute

$$r_1 = \tfrac{15}{2}, \quad r_2 = \tfrac{18}{2}, \quad r_3 = \tfrac{6}{1}, \quad r_4 = \tfrac{20}{3}$$

The least of these values occurs at $j = 3$, so we fix x_3 to obtain $\mathbf{x}^{(1)} = (\#, \#, 1, \#)$.

The third row of the model is now satisfied. This leads to revised ratios

$$r_1 = \tfrac{15}{2}, \quad r_2 = \tfrac{18}{1}, \quad r_4 = \tfrac{20}{2}$$

Choosing the least fixes x_1 in $\mathbf{x}^{(2)} = (1, \#, 1, \#)$.

All rows are covered by partial solution $\mathbf{x}^{(2)}$. Thus the least cost choice for remaining components is zero. We stop with heuristic optimum $\hat{\mathbf{x}} = (1, 0, 1, 0)$ at cost $6 + 15 = 21$. Notice that this solution is not as good as optimal $\mathbf{x}^* = (0, 0, 0, 1)$ with cost 20.

Constructive Search as the Method of Last Resort

Many project selection and capital budgeting models are approached by greedy constructive heuristics exactly like the one just illustrated. Still, we have seen in Section 12.5 that more exact, branch and bound methods can also be effective.

The real need for constructive search methods becomes clear only with large, often nonlinear, highly combinatorial discrete models such as the KI truck routing problem of Section 11.5 or cases where we need an answer fast.

> **12.59** In large, especially nonlinear, discrete models, or when time is limited, constructive search is often the only effective optimization-based approach to finding good solutions.

If tractable and strong relaxations are available, branch and bound is preferred. When natural neighborhoods exist, improving search can be effective. If neither applies, constructive heuristics provide the method of last resort.

Constructive Search of KI Truck Routing Example

To illustrate constructive search in such highly combinatorial cases, we will develop an algorithm for the KI routing example (Section 11.5). Recall that stops $i = 1, \ldots, 20$ are to be organized into the smallest possible list of routes j originating and terminating at a single central depot. Each route is then sequenced by an improving search to minimize travel distance. Figure 12.10 shows stop locations, and Table 12.11 provides the fractions of a load to be delivered at each stop.

Our constructive search for KI begins each route with a "seed" stop. We will choose the free stop farthest from the depot—number $i = 9$ in the first route. Figure 12.10 shows that the idea is to create a starting, out-and-back route with a general direction anchored by the seed location.

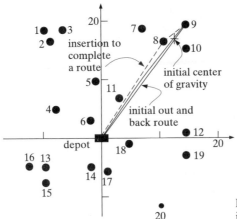

FIGURE 12.10   Locations and First Route in KI Example

As long as capacity remains in the truck for a route, we will next insert new stops. As usual, our approach is greedy. A "center of gravity" is computed for stops so far fixed into the route, and the closest stop to that center still fitting on the truck is added to the route.

Figure 12.10 shows that a 1-stop route's center of gravity is set arbitrarily 80% of the way from the depot to the seed location. This initial center of gravity for the route started by stop 9 has coordinates

$$0.8(x_9, y_9) = 0.8(15, 20) = (12, 16)$$

We hope to grow a cluster of stops near that point to form a compact route.

Stop 9 already uses $f_9 = 0.50$ truck. The nearest stop to the center of gravity is $i = 8$, with load $f_8 = 0.43$ within the remaining capacity. It becomes the first insertion.

After the route has more than one stop, the new selection is averaged into its center of gravity as

$$\tfrac{1}{2}[0.8(15, 20) + (11, 17)] = (11.5, 16.5)$$

However, the capacity fixed on this route already sums to

$$f_8 + f_9 = 0.50 + 0.43 = 0.93$$

and no remaining load will fit in the residual 0.07 truckload capacity. Route $j = 1$ is complete.

The next seed location is the farthest free stop from the depot—number $i = 10$. In turn, stop $i = 7$ is fixed in the route, then stop $i = 11$, and finally $i = 5$. These four

TABLE 12.11   Fractions of Truckloads to Be Delivered in KI Example

Stop, i	Fraction, f_i	Stop, i	Fraction, f_i	Stop, i	Fraction, f_i	Stop, i	Fraction, f_i
1	0.25	6	0.70	11	0.21	16	0.38
2	0.33	7	0.28	12	0.68	17	0.26
3	0.39	8	0.43	13	0.16	18	0.29
4	0.40	9	0.50	14	0.19	19	0.17
5	0.27	10	0.22	15	0.22	20	0.31

locations exhaust 0.98 of capacity, leaving no room for more.

Continuing in this way produces the full route structure of Figure 12.11. A total of 7 routes were constructed, seeded by $i = 9,\ 10,\ 1,\ 20,\ 19,\ 16,$ and $18,$ respectively.

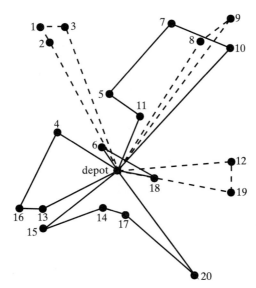

FIGURE 12.11   Final Routes in KI Example

EXERCISES

12-1 Solve each of the following discrete optimization models by total enumeration.

(a) ⊘   min $2x_1 + x_2 + 4x_3 + 10x_4$
    s.t. $x_1 + x_2 + x_3 \leq 2$
            $3x_1 + 7x_2 + 19x_3 + x_4 \geq 20$
            $x_1, x_2, x_3 = 0 \text{ or } 1$
            $x_4 \geq 0$

(b)   max $30x_1 + 12x_2 + 24x_3 + 55x_4$
    s.t. $30x_1 + 20x_2 + 40x_3 + 35x_4 \leq 45$
            $x_2 + 2x_3 + x_4 \geq 2$
            $x_1 \geq 0$
            $x_2, x_3, x_4 = 0 \text{ or } 1$

12-2 Suppose that you have been asked to solve a mixed-integer ILP with 10,000 continuous and n binary decision variables. Determine the largest n's for which the problem could be totally enumerated in one 24-hour day and in one 24-hour, 30-day month by each of the following computer environments.

(a) ⊘ An engineering workstation that can enumerate one choice of binary variables each second, including solving the resulting LP.

(b) A parallel processing computer that can evaluate 32,768 choices simultaneously every second, including solving the resulting LPs.

12-3 Consider the ILP

    max $14x_1 + 2x_2 - 11x_3 + 17x_4$
    s.t. $2x_1 + x_2 + 4x_3 + 5x_4 \leq 12$
            $x_1 - 3x_2 - 3x_3 - 3x_4 \leq 0$
            $x_1 \geq 0$
            $x_2, x_3, x_4 = 0 \text{ or } 1$

Determine whether each of the following is a constraint relaxation.

(a) ⊘   max $14x_1 + 2x_2 - 11x_3 + 17x_4$
    s.t. $2x_1 + x_2 + 4x_3 + 5x_4 \leq 12$
            $x_1 - 3x_2 - 3x_3 - 3x_4 \leq 0$
            $x_j \geq 0, \quad j = 1, \ldots, 4$

(b)   max $14x_1 + 2x_2 - 11x_3 + 17x_4$
    s.t. $2x_1 + x_2 + 4x_3 + 5x_4 \leq 12$
            $x_1 - 3x_2 - 3x_3 - 3x_4 \leq 0$
            $x_1 \geq 0 \text{ and integer}$
            $x_2, x_3, x_4 = 0 \text{ or } 1$

(c) ◪   max $14x_1 + 2x_2 - 11x_3 + 17x_4$
    s.t. $2x_1 + x_2 + 4x_3 + 5x_4 \leq 5$
            $x_1 - 3x_2 - 3x_3 - 3x_4 \leq 0$
            $x_1 \geq 0$
            $x_2, x_3, x_4 = 0$ or 1

(d)   max $14x_1 + 2x_2 - 11x_3 + 17x_4$
    s.t. $x_1 - 3x_2 - 3x_3 - 3x_4 \leq 0$
            $x_1 \geq 0$
            $x_2, x_3, x_4 = 0$ or 1

12-4 Form the linear programming relaxation of each of the following ILPs.

(a) ◪   min $12x_1 + 45x_2 + 67x_3 + 1x_4$
    s.t. $4x_1 + 2x_2 - x_4 \leq 10$
            $6x_1 + 19x_3 \geq 5$
            $x_2, x_3, x_4 \geq 0$
            $x_1 = 0$ or 1
            x_3 integer

(b)   max $9x_1 + 3x_2 + 4x_3 + 8x_4$
    s.t. $2x_1 + 2x_2 + 2x_3 + 3x_4 \leq 20$
            $29x_1 + 14x_2 + 78x_3 + 20x_4 \leq 200$
            $x_1, x_2, x_3 = 0$ or 1
            $x_4 \geq 0$

12-5 Each of the following ILPs has no feasible solutions. Solve the corresponding LP relaxation graphically and indicate whether your relaxation results are sufficient to show that the ILP is infeasible.

(a) ◪   min $10x_1 + 15x_2$
    s.t. $x_1 + x_2 \geq 2$
            $-2x_1 + 2x_2 \geq 1$
            $x_1, x_2 = 0$ or 1

(b)   max $11x_1 + 21x_2$
    s.t. $2x_1 + x_2 \geq 2$
            $2x_1 - x_2 \leq 0$
            $x_1, x_2 = 0$ or 1

(c) ◪   min $2x_1 + x_2$
    s.t. $x_1 + 4x_2 \leq 2$
            $-4x_1 + 4x_2 \geq 1$
            $x_1 \geq 0, \ x_2 = 0$ or 1

(d)   max $47x_1 + 58x_2$
    s.t. $x_1 + x_2 \geq 4$
            $x_1 = 0$ or 1
            $0 \leq x_2 \leq 2$

12-6 Determine the best bound on the optimal solution value of an ILP with each of the following objective functions that is available from the specified LP relaxation optima $\tilde{\mathbf{x}}$.

(a) ◪   max $24x_1 + 13x_2 + 3x_3$
        $\tilde{\mathbf{x}} = \left(2, \frac{1}{2}, 0\right)$
(b)   min $x_1 - 4x_2 + 70x_3$
        $\tilde{\mathbf{x}} = \left(1, 0, \frac{2}{7}\right)$
(c) ◪   min $60x_1 - 16x_2 + 10x_3$
        $\tilde{\mathbf{x}} = \left(\frac{1}{2}, 1, \frac{1}{2}\right)$
(d)   max $90x_1 + 11x_2 + 30x_3$
        $\tilde{\mathbf{x}} = \left(0, \frac{1}{2}, 5\right)$

12-7 Determine whether each of the following LP relaxation optima $\tilde{\mathbf{x}}$ is optimal in the corresponding ILP over the specified variable type constraints.

(a) ◪ $x_j = 0$ or $1, j = 1, \ldots, 4$
        $\tilde{\mathbf{x}} = \left(1, 0, \frac{1}{3}, \frac{2}{3}\right)$
(b) $x_1, x_2 = 0$ or $1, x_3, x_4 \geq 0$
        $\tilde{\mathbf{x}} = \left(0, 1, \frac{1}{2}, \frac{1}{2}\right)$
(c) ◪ $x_1, x_2, x_3 = 0$ or $1, x_4 \geq 0$
        $\tilde{\mathbf{x}} = \left(1, 0, 1, \frac{23}{7}\right)$
(d) $x_j \geq 0$ and integer, $j = 1, \ldots, 4$
        $\tilde{\mathbf{x}} = \left(0, 3, \frac{9}{2}, 1\right)$

12-8 The ILP

    max $3x_1 + 6x_2 + 4x_3 + 10x_4 + 3x_5$
    s.t. $2x_1 + 4x_2 + x_3 + 3x_4 + 7x_5 \leq 10$
            $x_1 + x_3 + x_4 \leq 2$
            $4x_2 + 4x_4 + 4x_5 \leq 7$
            $x_1, \ldots, x_5 = 0$ or 1

has LP relaxation optimal solution $\tilde{\mathbf{x}} = (0, 0.75, 1, 1, 0)$.

(a) ◪   Determine the best bound on the ILP optimal solution value available from relaxation results.
(b) ◪   Determine whether the relaxation optimum solves the full ILP. If not, round to an ILP-feasible solution either by moving all binary variables at fractional values in the relaxation up to 1 or by moving all down to 0.
(c) ◪   Combine parts (a) and (b) to determine the best upper and lower bounds on the ILP optimal solution value available from the combination of relaxation and rounding.
(d) ◪ ▢   Verify your bounds of part (c) by solving the full ILP with class optimization software.

12-9 Do Exercise 12-8 for the ILP

    min $4x_1 + 5x_2 + 12x_3 + 7x_4 + 6x_5$
    s.t. $6x_1 + 8x_2 + 21x_3 + 6x_4 + 5x_5 \geq 11$
            $x_1 + x_2 + 2x_3 + x_4 \geq 1$
            $2x_2 + 5x_3 + x_5 \geq 2$
            $x_1, \ldots, x_5 = 0$ or 1

and LP relaxation optimum $\tilde{\mathbf{x}} = (0, 0, 0.533, 0, 0)$.

12-10 ◇ Do Exercise 12-8 for the ILP

$$\begin{aligned}
\min\quad & 17x_1 + 12x_2 + 24x_3 + 2x_4 + 8x_5 \\
\text{s.t.}\quad & 3x_1 + 5x_3 + 7x_4 + 9x_5 \geq 13 \\
& 7x_2 + 4x_4 + 11x_5 \geq 5 \\
& 2x_1 + 3x_2 + 2x_3 + 3x_4 \geq 7 \\
& x_2, x_3, x_4 = 0 \text{ or } 1 \\
& x_1, x_5 \geq 0
\end{aligned}$$

and LP relaxation optimum $\tilde{\mathbf{x}} = (0.5, 1, 0, 1, 0.5)$.

12-11 Do Exercise 12-8 for the ILP

$$\begin{aligned}
\min\quad & 40x_3 + 500x_4 + 800x_5 + 900x_6 \\
\text{s.t.}\quad & 10x_1 + 6x_2 + 2x_3 = 45 \\
& 2x_1 + 3x_2 + x_3 \geq 12 \\
& 0 \leq x_1 \leq 5x_4 \\
& 0 \leq x_2 \leq 5x_5 \\
& 0 \leq x_3 \leq 5x_6 \\
& x_4, x_5, x_6 = 0 \text{ or } 1
\end{aligned}$$

and LP relaxation optimum $\tilde{\mathbf{x}} = (3.500, 1.667, 0, 0.70, 0.333, 0)$.

12-12 Consider the ILP

$$\begin{aligned}
\min\quad & 10x_1 + 20x_2 + 40x_3 + 80x_4 - 144y \\
\text{s.t.}\quad & x_1 + x_2 + x_3 + x_4 \geq 4y \\
& x_1, \ldots, x_4, y = 0 \text{ or } 1
\end{aligned}$$

(a) ◇ Solve the full ILP model by inspection.
(b) Verify by inspection that its LP relaxation has optimal solution $\tilde{\mathbf{x}} = (1, 1, 0, 0), \tilde{y} = \frac{1}{2}$.
(c) ◇ Show that an equivalent ILP would result if the main constraint were replaced by

$$x_j \geq y \qquad j = 1, \ldots, 4$$

(d) ◇ Verify that the revised formulation of part (c) has a stronger LP relaxation than the original of part (b).

12-13 Do Exercise 12-12 for ILP

$$\begin{aligned}
\min\quad & 14x_1 + 16x_2 + 15x_3 \\
\text{s.t.}\quad & x_1 + x_2 \geq 1 \\
& x_2 + x_3 \geq 1 \\
& x_1 + x_3 \geq 1 \\
& x_1, \ldots, x_3 = 0 \text{ or } 1
\end{aligned}$$

LP relaxation optimum $\tilde{\mathbf{x}} = \left(\frac{1}{2}, \frac{1}{2}, \frac{1}{2}\right)$ and revised main constraint

$$x_1 + x_2 + x_3 \geq 2$$

12-14 Return to the ILP of Exercise 12-12 with LP relaxation optimum $\tilde{\mathbf{x}} = (1, 1, 0, 0), \tilde{y} = \frac{1}{2}$. Determine whether each of the following is a valid inequality for the ILP, and if so, whether it would strengthen the original LP relaxation to add the inequality as a constraint.

(a) ◇ $x_2 + x_3 + x_4 \geq 3y$
(b) ◇ $x_1 + x_2 + x_3 + x_4 \geq 4y$
(c) ◇ $x_1 + x_2 \geq 1$
(d) ◇ $x_3 \geq y$

12-15 Return to the ILP of Exercise 12-13 with LP relaxation optimum $\tilde{\mathbf{x}} = \left(\frac{1}{2}, \frac{1}{2}, \frac{1}{2}\right)$. Determine whether each of the following is a valid inequality for the ILP, and if so, whether it would strengthen the original LP relaxation to add the inequality as a constraint.

(a) $10x_1 + 10x_2 + 10x_3 \geq 25$
(b) $x_1 + x_2 + x_3 \geq 1$
(c) $x_1 + x_2 + x_3 \geq 2$
(d) $14x_1 + 20x_2 + 16x_3 \geq 28$

12-16 The ILP

$$\begin{aligned}
\max\quad & 40x_1 + 5x_2 + 60x_3 + 8x_4 \\
\text{s.t.}\quad & 18x_1 + 3x_2 + 20x_3 + 5x_4 \leq 25 \\
& x_1, \ldots, x_4 = 0 \text{ or } 1
\end{aligned}$$

has LP relaxation optimum $\tilde{\mathbf{x}} = \left(\frac{5}{18}, 0, 1, 0\right)$. Determine whether each of the following is a valid inequality for the ILP, and if so, whether it would strengthen the LP relaxation to add the inequality as a constraint.

(a) ◇ $x_1 + x_3 \leq 1$
(b) $x_1 + x_2 + x_3 + x_4 \leq 4$
(c) ◇ $x_2 + x_4 \geq 1$
(d) $18x_1 + 20x_3 \leq 20$

12-17 The fixed-charge ILP

$$\begin{aligned}
\min\quad & 60x_1 + 78x_2 + 200y_1 + 400y_2 \\
\text{s.t.}\quad & 12x_1 + 20x_2 \geq 64 \\
& 15x_1 + 10x_2 \leq 60 \\
& x_1 + x_2 \leq 10 \\
& 0 \leq x_1 \leq 100y_1 \\
& 0 \leq x_2 \leq 100y_2 \\
& y_1, y_2 = 0 \text{ or } 1
\end{aligned}$$

has LP relaxation optimum $\tilde{\mathbf{x}} = (0, 3.2), \tilde{\mathbf{y}} = (0, 0.032)$.

(a) ◇ Compute the smallest replacements for big-M values of 100 in this formulation that can be inferred simply by examining constraints of the model.
(b) ◇ Show that the LP relaxation optimum will change if the lower big-M's of part (a) are employed.
(c) ◇ ▢ Verify part (b) by solving the model having smaller big-M's with class optimization software.

12-18 Do Exercise 12-17 for the average-completion-time, single-machine scheduling ILP

$$\min \quad 0.5(x_1 + 12 + x_2 + 8)$$
$$\text{s.t.} \quad x_1 + 12 \le x_2 + 100(1 - y)$$
$$x_2 + 8 \le x_1 + 100y$$
$$x_1, x_2 \ge 0$$
$$y = 0 \text{ or } 1$$

with LP relaxation optimum $\tilde{\mathbf{x}} = (0,0)$, $\tilde{y} = 0.08$. [*Hint*: Consider the sum of the process times in part (a).]

12-19 Consider the ILP

$$\max \quad 30x_1 + 55x_2 + 20x_3$$
$$\text{s.t.} \quad 40x_1 - 12x_2 + 11x_3 \le 55$$
$$19x_1 + 60x_2 + 3x_3 \ge 20$$
$$3x_1 + 2x_2 + 2x_3 = 5$$
$$x_1, x_2, x_3 = 0 \text{ or } 1$$

Form the Lagrangian relaxations obtained by dualizing each of the following collections of main constraints, and show all sign restrictions that apply to Lagrange multipliers.

(a) ⊠ Dualize the first and second main constraints.

(b) Dualize the second and third main constraints.

12-20 Consider the facilities location ILP

$$\min \quad 3x_{1,1} + 6x_{1,2} + 5x_{2,1} + 2x_{2,2}$$
$$+ 250y_1 + 300y_2$$
$$\text{s.t.} \quad 30x_{1,1} + 20x_{1,2} \le 50y_1$$
$$30x_{2,1} + 20x_{2,2} \le 50y_2$$
$$x_{1,1} + x_{2,1} = 1$$
$$x_{1,2} + x_{2,2} = 1$$
$$0 \le x_{1,1}, x_{1,2}, x_{2,1}, x_{2,2} \le 1$$
$$y_i = 0 \text{ or } 1$$

(a) ⊠ Use total enumeration to compute an optimal solution.

(b) ⊠ Form a Lagrangian relaxation dualizing the third and fourth constraints with Lagrange multipliers v_1 and v_2.

(c) ⊠ Explain how the dualization in part (b) leaves a relaxation that is easier to solve than the full ILP.

(d) ⊠ Use total enumeration to solve the Lagrangian relaxation of part (b) with $v_1 = v_2 = 0$, and verify that the relaxation optimal value provides a lower bound on the true optimal value computed in part (a).

(e) ⊠ Repeat part (d) with $v_1 = v_2 = 100$.

(f) ⊠ Repeat part (d) with $v_1 = 1000$, $v_2 = 500$.

12-21 Do Exercise 12-20 for the generalized assignment model

$$\min \quad 15x_{1,1} + 10x_{1,2} + 30x_{2,1} + 20x_{2,2}$$
$$\text{s.t.} \quad x_{1,1} + x_{1,2} = 1$$
$$x_{2,1} + x_{2,2} = 1$$
$$30x_{1,1} + 50x_{2,1} \le 80$$
$$30x_{1,2} + 50x_{2,2} \le 60$$
$$x_{1,1}, x_{1,2}, x_{2,1}, x_{2,2} = 0 \text{ or } 1$$

Dualize the first two main constraints and solve with $\mathbf{v} = (0, 0)$, $\mathbf{v} = (10, 12)$, and $\mathbf{v} = (200, 100)$.

12-22 Suppose that an integer linear program has decision variables $x_1, x_2, x_3 = 0$ or 1. List all completions of the following partial solutions.

(a) ⊠ (#, 0, #)

(b) (#, 1, 0)

12-23 The following is the complete branch-and-bound tree for an ILP over decision variables $x_1, \ldots, x_4 = 0$ or 1.

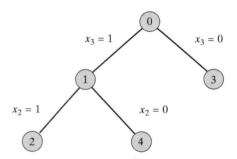

(a) ⊠ List the partial solutions associated with each node of the tree.

(b) ⊠ Which nodes were branched and which terminated?

(c) ⊠ Identify the nodes of the tree that have $\mathbf{x} = (0, 1, 0, 1)$ as a feasible completion.

12-24 Do Exercise 12-23 for the branch and bound tree

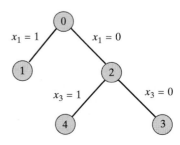

12-25 Suppose that the ILP of Exercise 12-8 is being solved by branch and bound. State the candidate problem associated with each of the following partial solutions.

(a) ⊠ (#, 1, #, #, 0)
(b) (0, 1, #, 1, 0)

12-26 Suppose that a minimizing ILP is being solved by LP-based branch and bound Algorithm 12A over decision variables $x_1, x_2, x_3 = 0$ or 1, $x_4 \geq 0$. Show how the search should process the node with $x_2 = 1$ and other variables free if the corresponding LP relaxation has each of the following outcomes. Assume that the incumbent solution value is 100.

(a) ⊠ $\tilde{x} = (.9, 1, 0, 6)$, value $\tilde{v} = 97$
(b) $\tilde{x} = (0.2, 1, 0.77, 4.5)$, value $\tilde{v} = 116$
(c) ⊠ $\tilde{x} = (1, 1, 0, 4.2)$, value $\tilde{v} = 75$
(d) LP relaxation infeasible
(e) ⊠ $\tilde{x} = (1, 1, 0.6, 0)$, value $\tilde{v} = 100$
(f) $\tilde{x} = (0.4, 1, 0.1, 5.9)$, value $\tilde{v} = 83$

12-27 ⊠ The following table shows the LP relaxation outcomes for all possible combinations of fixed and free variables in branch and bound solution of a minimizing integer linear program over decision variables $x_1, x_2, x_3 = 0$ or 1, $x_4 \geq 0$. Solve the problem by LP-based Algorithm 12A and record your results in a branch and bound tree. Apply the depth first rule for selecting among active nodes and pick whichever of $= 0$ and $= 1$ is closest to the preceding relaxation value when nodes have equal depth. Branch on the integer-restricted variable with fractional relaxation value nearest to integer.

x_1	x_2	x_3	$\tilde{x}$	$\tilde{v}$
#	#	#	(0, 0.60, 0.14, 0)	60.9
#	#	0	(0.20, 0.60, 0, 0)	61.0
#	#	1	(0.60, 0, 1, 0)	69.0
#	0	#	(0.60, 0, 1, 0)	69.0
#	0	0	Infeasible	—
#	0	1	(0.60, 0, 1, 0)	69.0
#	1	#	(0, 1, 0, 400)	4090.0
#	1	0	(0, 1, 0, 400)	4090.0
#	1	1	Infeasible	—
0	#	#	(0, 0.60, 0.14, 0)	60.9
0	#	0	(0, 0.60, 0, 1.9)	73.6
0	#	1	(0, 0, 1, 6)	108.0
0	0	#	(0, 0, 1, 6)	108.0
0	0	0	Infeasible	—
0	0	1	(0, 0, 1, 6)	108.0

(continued top of next column)

(continued)

x_1	x_2	x_3	$\tilde{x}$	$\tilde{v}$
0	1	#	(0, 1, 0, 400)	4090.0
0	1	0	(0, 1, 0, 400)	4090.0
0	1	1	Infeasible	—
1	#	#	(1, 0.33, 0, 0)	65.0
1	#	0	(1, 0.33, 0, 0)	65.0
1	#	1	(1, 0, 1, 0)	83.0
1	0	#	(1, 0, 0.71, 0)	69.3
1	0	0	Infeasible	—
1	0	1	(1, 0, 1, 0)	83.0
1	1	#	(1, 1, 0, 400)	4125.0
1	1	0	(1, 1, 0, 400)	4125.0
1	1	1	Infeasible	—

12-28 Do Exercise 12-27 for a maximizing ILP over $x_1 \geq 0$, $x_2, x_3, x_4 = 0$ or 1 with the following LP relaxation outcomes.

x_2	x_3	x_4	$\tilde{x}$	$\tilde{v}$
#	#	#	(48.1, 0.20, 0.77, 1)	78.0
#	#	0	(0, 1, 1, 0)	59.0
#	#	1	(48.1, 0.20, 0.77, 1)	78.0
#	0	#	(29.6, 0.20, 0, 1)	77.2
#	0	0	(0, 1, 0, 0)	34.0
#	0	1	(29.6, 0.20, 0, 1)	77.2
#	1	#	(41.6, 0.43, 1, 0.77)	75.0
#	1	0	(0, 1, 1, 0)	59.0
#	1	1	Infeasible	—
0	#	#	(42.5, 0, 0.77, 1)	76.8
0	#	0	(0, 0, 1, 0)	25.0
0	#	1	(42.5, 0, 0.77, 1)	76.8
0	0	#	(24, 0, 0, 1)	76.0
0	0	0	(0, 0, 0, 0)	0.0
0	0	1	(24, 0, 0, 1)	76.0
0	1	#	(29.6, 0, 1, 0.77)	72.4
0	1	0	(0, 0, 1, 0)	25.0
0	1	1	Infeasible	—
1	#	#	(12, 1, 1, 0.20)	67.0
1	#	0	(0, 1, 1, 0)	59.0
1	#	1	Infeasible	—
1	0	#	(0, 1, 0, 0.20)	54.0
1	0	0	(0, 1, 0, 0)	34.0
1	0	1	Infeasible	—
1	1	#	(12, 1, 1, 0.20)	67.0
1	1	0	(0, 1, 1, 0)	59.0
1	1	1	Infeasible	—

12-29 Students often mistakenly believe ILPs are more tractable than LPs because the straightforward rules of Algorithm 12A seem less complex than the simplex and interior point methods of Chapters 5 and 6.

(a) Explain why solution of any ILP by LP-based branch and bound always takes at least as much work as a linear program of comparable size and coefficients.

(b) Justify why the number of LP relaxations that might have to be solved in branch-and-bound enumeration of an ILP with n binary variables is $2^{(n+1)} - 1$.

(c) Use part (b) to compute the number of linear programs that could have to be solved in branch and bound search of ILP models with 100, 500, and 1000 binary variables respectively, and determine how long each such search could take at the rate of one LP per second.

(d) How practical is it be to solve LPs of 100, 500, and 1000 variables in reasonable amounts of time?

(e) Comment on the implications of your analysis in parts (a) to (d) for tractability of LPs versus ILPs of comparable size.

12-30 In most applications, LP based branch and bound Algorithm 12A actually investigates only a tiny fraction of the possible partial solutions. Still, this is not always the case. Consider the family of ILPs of the form

$$\min \quad y$$
$$\text{s.t.} \quad 2\sum_{j=1}^{n} x_j + y = n$$
$$x_j = 0 \text{ or } 1 \qquad j = 1, \dots, n$$
$$y = 0 \text{ or } 1$$

where n is odd.

(a) ▣ Enter and solve versions for $n = 7$, $n = 11$, and $n = 15$ with class branch and bound software, and record the number of branch and bound nodes explored. (*Warning:* Program limits must be set big enough to allow up to 20,000 branch and bound nodes.)

(b) Express your results in part (a) as fractions as the total number of nodes that might have to be investigated. [*Hint:* Use the formula in Exercise 12-29(b).]

(c) Comment on the implications for tractability of ILP's via branch and bound if fractions like those of part (b) were typical.

12-31 ▣ The branch and bound tree that follows records solution of the knapsack model

$$\min \quad 90x_1 + 50x_2 + 54x_3$$
$$\text{s.t.} \quad 60x_1 + 110x_2 + 150x_3 \geq 50$$
$$x_1, x_2, x_3 = 0 \text{ or } 1$$

by LP-based Algorithm 12A under rules of Exercise 12-27.

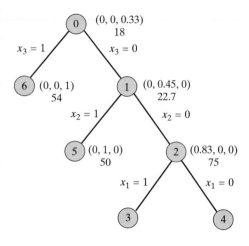

Briefly describe the processing, including how and why nodes were branched or terminated, when incumbent solutions were discovered, and what solution proved optimal. Assume that there was no initial incumbent solution.

12-32 Do Exercise 12-31 for

$$\max \quad 51x_1 + 72x_2 + 41x_3$$
$$\text{s.t.} \quad 17x_1 + 10x_2 + 14x_3 \leq 19$$
$$x_1, x_2, x_3 = 0 \text{ or } 1$$

and tree   *figure top of next column*

12-33 Return to the knapsack problem of Exercise 12-31.

(a) Explain why LP relaxation optimal solutions can be rounded to integer-feasible solutions by setting $\hat{x}_j \leftarrow \lceil \tilde{x}_j \rceil$.

(b) ▣ Repeat the branch and bound computations of Exercise 12-31, this time rounding up each relaxation solution in this way to produce earlier incumbent solutions.

(c) Comment on the computational savings with rounding.

12-34 Do Exercise 12-33 on the knapsack model of Exercise 12-32, this time rounding solutions $\hat{x}_j \leftarrow \lfloor \tilde{x}_j \rfloor$.

12-35 Determine the best lower and upper bounds on the ultimate ILP optimal available from parent

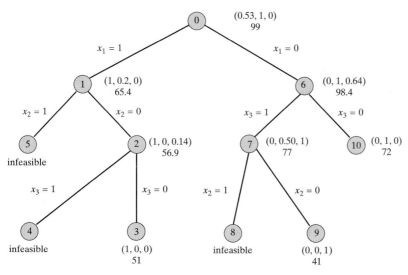

FIGURE 12.12   Branch and Bound Tree for Exercise 12-32

bounds and incumbent solutions after processing each node in the following:

(a) ⊘ The branch and bound tree of Exercise 12-31.

(b) The branch and bound tree of Figure 12.12.

12-36 The branch and bound tree that follows shows the incomplete solution of a maximizing ILP by LP-based Algorithm 12A, with numbers next to nodes indicating LP relaxation solution values.

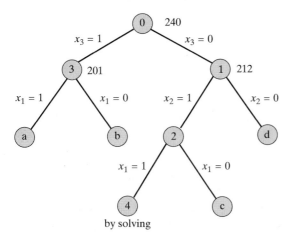

Node 4 has just produced the first incumbent solution, and nodes *a* to *d* remain unexplored.

(a) ⊘ Show which unexplored nodes could be immediately terminated by parent bound if the incumbent at node 4 had objective function value 205. How about 210?

(b) ⊘ Determine the best upper bound on the ultimate ILP optimal value that is available after processing of node 4.

(c) ⊘ Assuming the incumbent at node 4 has objective value 195, compute the maximum absolute and percent objective value error in accepting the incumbent as an approximate optimum.

12-37 The branch-and-bound tree that follows shows the incomplete solution of a minimizing ILP by LP-based Algorithm 12A, with numbers next to nodes indicating LP relaxation solution values.

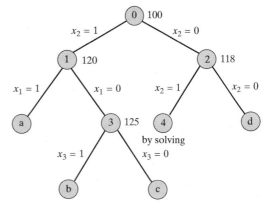

Node 4 has just produced the first incumbent solution, and nodes a to d remain unexplored.

(a) Show which unexplored nodes could be immediately terminated by the parent bound if the incumbent at node 4 had objective function value 120. How about 118?

(b) Determine the best lower bound on the ultimate ILP optimal value that is available after processing of node 4.

(c) Assuming that the incumbent at node 4 has objective value 130, compute the maximum absolute and percent objective value error in accepting the incumbent as an approximate optimum.

12-38 Repeat Exercise 12-27, following the same rules except:

(a) ⊘ Use the best-first enumeration sequence and allow termination by parent bounds.

(b) ⊘ Use the depth-forward best-back enumeration sequence and allow termination by parent bounds.

12-39 Do Exercise 12-38 for the branch and bound of Exercise 12-28.

12-40 Three company trucks must be assigned to pickup 7 miscellaneous loads on the way back from their regular deliveries. Truck capacities and load sizes (cubic yards) are shown in the following table, together with the extra distance (in miles) that each truck would have to travel if it is to deviate to pick up any of the loads.

Load	Distance for Truck: 1	2	3	Load Size
1	23	45	50	4
2	25	72	23	8
3	29	13	41	13
4	12	23	40	31
5	49	7	42	11
6	37	39	59	9
7	2	9	20	21
Capacity	30	40	50	

(a) Formulate this problem as a generalized assignment ILP using the decision variables ($i = 1, \ldots, 7; j = 1, \ldots, 3$)

$$x_{i,j} \triangleq \begin{cases} 1 & \text{if load } i \text{ goes to truck } j \\ 0 & \text{otherwise} \end{cases}$$

(b) ▭ Enter and use class optimization software to compute an optimal solution.

(c) ▭ Use class optimization software to solve the corresponding LP relaxation and verify that

the relaxation optimal value provides a lower bound.

(d) ▭ Using class optimization software to solve relaxations, verify your ILP optimal solution by executing LP-based branch and bound Algorithm 12A including parent bounds. Apply the depth-first rule for selecting among active nodes and pick whichever of $= 0$ and $= 1$ is closest to the preceding relaxation value when nodes have equal depth. Branch on the integer-restricted variable with fractional relaxation value nearest to integer picking the one with least subscript if there are ties. Record incumbent solutions only when an LP relaxation comes out integer (i.e., do not round).

(e) Determine when your search of part (d) could have been stopped if we were willing to accept an incumbent solution no worse than 25% above optimal.

(f) ▭ Do the same branch and bound computation as part (d) except with the best first enumeration rule using LP relaxation values as parent bounds.

(g) ▭ Do the same branch and bound computation as part (d) except with the depth forward best back enumeration rule using LP relaxation values as parent bounds.

(h) Compare your results in parts (d), (f) and (g).

12-41 ⊘ The following tree records solution of a maximizing ILP over $x_1, x_2, x_3 = 0$ or 1 by branch and cut Algorithm 12B. LP relaxations solutions show next to each node.

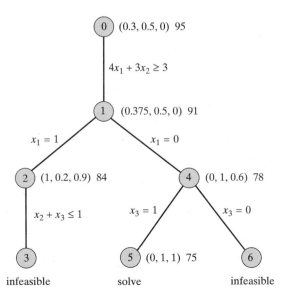

Briefly describe the processing, including how and why nodes were branched tightened or terminated, when incumbent solutions were discovered, and what solution proved optimal. Assume that all added inequalities are valid for the original ILP.

12-42 Do Exercise 12-41 for the following tree of a minimizing ILP over $x_1, x_2, x_3 = 0$ or 1.

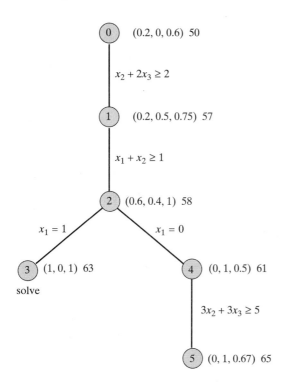

12-43 Consider solving (approximately) the ILP

$$\begin{aligned} \max \quad & 5x_1 + 7x_2 - 2x_3 \\ \text{s.t.} \quad & x_2 + x_3 \le 1 \\ & x_1, x_2, x_3 = 0 \text{ or } 1 \end{aligned}$$

by a version of discrete improving search Algorithm 12C that employs move set $\mathcal{M} = \{(1, 0, 0), (0, 1, 0), (0, 0, 1)\}$ and always advances to the feasible neighbor with best objective value.

(a) ⊠ Identify a global optimal solution by inspection.
(b) ⊠ List all points in the neighborhood of feasible solution $\mathbf{x}^{(0)} = (0, 0, 1)$.
(c) ⊠ Compute a local optimal solution by applying Algorithm 12C starting from $\mathbf{x}^{(0)} = (0, 0, 1)$.
(d) ⊠ Repeat part (c), this time starting at $\mathbf{x}^{(0)} = (1, 0, 0)$.

(e) Compare results in parts (c) and (d), and comment on the effect of starting solutions.
(f) ⊠ Repeat part (c), this time using the single-complement move set that allows any one $x_j = 1$ to be switched to $= 0$, or vice versa.
(g) Compare results in parts (c) and (f), and comment on the effect of move set.

12-44 Repeat Exercise 12-43 for the ILP

$$\begin{aligned} \min \quad & 2x_1 - 23x_2 + 14x_3 \\ \text{s.t.} \quad & x_1 + x_2 + x_3 \ge 1 \\ & x_1, x_2, x_3 = 0 \text{ or } 1 \end{aligned}$$

12-45 Consider solving (approximately) the ILP

$$\begin{aligned} \max \quad & 12x_1 + 7x_2 + 9x_3 + 8x_4 \\ \text{s.t.} \quad & 3x_1 + x_2 + x_3 + x_4 \le 3 \\ & x_3 + x_4 \le 1 \\ & x_1, \dots, x_4 = 0 \text{ or } 1 \end{aligned}$$

by a version of discrete improving search Algorithm 12C that always advances to the feasible neighbor with best objective value and uses the single-complement neighborhood permitting any one $x_j = 1$ to be switched to $= 0$, or vice versa.

(a) ⊠ Identify a global optimal solution by inspection.
(b) ⊠ Use Algorithm 12C to compute a local optimum starting from $\mathbf{x}^{(0)} = (0, 0, 0, 0)$.
(c) ⊠ Apply the multistart extension of improving search to compute a local optimum by trying starts at $\mathbf{x} = (0,0,0,0)$, $(0,1,0,0)$, and $(0,0,0,1)$.

12-46 Do Exercise 12-45 for the ILP

$$\begin{aligned} \min \quad & 50x_1 + 30x_2 + 20x_3 + 15x_4 \\ \text{s.t.} \quad & x_1 + x_2 \ge 1 \\ & x_1 + x_4 \ge 1 \\ & x_1, \dots, x_4 = 0 \text{ or } 1 \end{aligned}$$

Start Algorithm 12C at $\mathbf{x} = (1, 1, 1, 1)$, and multistart at $\mathbf{x} = (1, 1, 1, 1)$, $(1, 0, 1, 1)$, and $(1, 1, 1, 0)$.

12-47 Return to the improving search problem of Exercise 12-45.

(a) Show that $\mathbf{x} = (1, 0, 0, 0)$ is a local optimum.
(b) Show that if a nonimproving move is allowed at $\mathbf{x} = (1, 0, 0, 0)$, the next iteration will return the search to this same point.

12-48 Do Exercise 12-47 for the model of Exercise 12-46.

12-49 ⊠ Return to the improving search problem of Exercise 12-45, starting from $\mathbf{x}^{(0)} = (1, 0, 0, 0)$. Compute an approximate optimum by tabu search Algorithm 12D, forbidding complementation of a variable for one step after its value changes, and limiting the search to $t_{max} = 5$ moves.

12-50 Do Exercise 12-49 for the model of Exercise 12-46. Forbid complementation of a variable for two steps after its value changes.

12-51 ⊘ Return to the improving search problem of Exercise 12-45, starting from $\mathbf{x}^{(0)} = (0, 0, 0, 1)$. Compute an approximate optimum by simulated annealing Algorithm 12E, using a temperature of $q = 20$, limiting the search to $t_{max} = 4$ moves, and resolving probabilistic decisions with (uniform [0,1]) random numbers 0.65, 0.10, 0.40, 0.53, 0.33, 0.98, 0.88, 0.37.

12-52 Do Exercise 12-51 for the model of Exercise 12-46. Use random numbers 0.60, 0.87, 0.77, 0.43, 0.18, 0.13, 0.21, 0.48, 0.71, 0.83, 0.29. Start at $\mathbf{x}^{(0)} = (1, 0, 0, 0)$.

12-53 ⊘ Return to the improving search problem of Exercise 12-45.

(a)   ⊘ Show that the solutions $\mathbf{x}^{(1)} = (0, 0, 1, 0)$ and $\mathbf{x}^{(2)} = (0, 0, 0, 1)$ are eligible to belong to a genetic algorithm population for the problem.

(b)   ⊘ Construct all possible crossover results (all cut points) for the $\mathbf{x}^{(1)}$ and $\mathbf{x}^{(2)}$ of part (a).

(c)   ⊘ Determine whether all your resulting solutions in part (b) are feasible, and if not, explain what difficulty this presents for effective application of genetic algorithm search.

12-54 Do Exercise 12-53 on the model of Exercise 12-46 using $\mathbf{x}^{(1)} = (0, 1, 1, 1)$ and $\mathbf{x}^{(2)} = (1, 0, 1, 1)$.

12-55 ⊘ Return again to the model of Exercise 12-45, and consider employing genetic Algorithm 12F with initial population

$\{(0, 0, 1, 0), (0, 0, 0, 1), (0, 1, 1, 0), (1, 0, 0, 0)\}$, $p_e = p_i = 1$, and $p_c = 2$. Construct and evaluate each member of the next generation population, with crossover after component 2 of the best and worst current solutions. Use a large negative M as the objective value of any infeasible solutions produced by crossover.

12-56 Do Exercise 12-55 on the model of Exercise 12-46 with initial population $\{(0, 1, 1, 1), (1, 0, 1, 1), (0, 1, 0, 1), (1, 0, 0, 0)\}$.

12-57 Consider solving (approximately) the following knapsack problem by constructive search Algorithm 12G.

$$\begin{aligned} \max \quad & 11x_1 + 1x_2 + 9x_3 + 17x_4 \\ \text{s.t.} \quad & 9x_1 + 2x_2 + 7x_3 + 13x_4 \le 17 \\ & x_1, \ldots, x_4 = 0 \text{ or } 1 \end{aligned}$$

(a)   ⊘ Determine a global optimum by inspection.

(b)   ⊘ Explain why it is reasonable to fix variables in order of ratio

$$\frac{\text{objective coefficient}}{\text{constraint coefficient}}$$

(c)   ⊘ Apply constructive search Algorithm 12G to construct an approximate solution choosing variables to fix in this ratio sequence.

12-58 Do Exercise 12-57 for the knapsack model

$$\begin{aligned} \min \quad & 80x_1 + 150x_2 + 54x_3 + 180x_4 \\ \text{s.t.} \quad & 25x_1 + 30x_2 + 18x_3 + 45x_4 \ge 40 \\ & x_1, \ldots, x_4 = 0 \text{ or } 1 \end{aligned}$$

SUGGESTED READING

Nemhauser, George L., and Laurence A. Wolsey (1988), *Integer and Combinatorial Optimization*, Wiley, New York.

Parker, R. Gary, and Ronald L. Rardin (1988), *Discrete Optimization*, Academic Press, San Diego, Calif.

Reeves, Colin R. (1993), *Modern Heuristic Techniques for Combinatorial Problems*, Halsted Press, New York.

13

Unconstrained Nonlinear Programming

• •

A major theme of this book has been the power and elegance of linear programming models, which are models with continuous decision variables, linear constraints, and a linear objective function. **Nonlinear programming** (**NLP**) encompasses all the rest of single-objective optimization over continuous decision variables.

Being defined only by what it is not—linear—leaves nonlinear programming with a host of quite different forms and algorithms. Some models have constraints. Others have only an objective function. Calculus yields readily exploitable derivatives in many models. Derivatives do not even exist in others. In some cases, both objective function and constraints are nonlinear. In others it is only the objective function. Even single-variable optimization is a nontrivial topic when the objective is nonlinear.

This chapter begins our treatment of nonlinear programming with the **unconstrained** case where no constraints apply. Chapter 14 follows with the more complicated models having constraints that cannot be ignored. There are important unconstrained applications, but most real models have at least a few constraints. We treat unconstrained cases first because many of the underlying notions of nonlinear programming are easier to understand without the encumbrance of constraints, and because most methods for constrained optimization use unconstrained algorithms as building blocks. Familiarity with definitions of Section 2.4 and improving search concepts of Chapter 3 is assumed throughout.

13.1 UNCONSTRAINED NONLINEAR PROGRAMMING MODELS

One of the major differences between linear and nonlinear programs is that **unconstrained** NLPs—ones with no constraints—can still be meaningful.

> 13.1 Unconstrained optimization over a linear objective function is always unbounded (except in the trivial case where the objective is constant), but unconstrained nonlinear programs can have finite optimal solutions.

We begin our discussion of unconstrained NLPs with some typical examples.

EXAMPLE 13.1: USPS SINGLE VARIABLE

Even models with a single decision variable can be challenging when the objective function is nonlinear. For a real single-variable nonlinear program we turn to the U.S. Postal Service (USPS).[1] Service "territories" for the USPS typically consist of a city and its suburbs. Mail delivery is provided by a number of postal carriers driving, or sometimes walking, specified "delivery regions." Carriers are based at "delivery units" distributed throughout the territory, beginning and ending their routes there each workday. Often, a delivery unit is also a local post office that sells stamps, and so on, to the general public.

Determining the most efficient number of delivery units for a territory involves a tradeoff between the fixed overhead costs of operating delivery units and the travel savings from carriers being based nearer their delivery regions. More delivery units increases overhead, but it reduces the number of carriers required by saving travel as the units are dispersed closer to customers.

Increasing automation of mail handling has significantly changed the relative economics of such decisions. To adjust, the USPS has developed and applied a rough decision model computing the approximate number of delivery units appropriate for any given territory. Input parameters are

$a \triangleq$ land area of the territory

$m \triangleq$ number of customers in the territory

$t \triangleq$ average time for a carrier to service any customer site

$d \triangleq$ length of the carrier work day

$c \triangleq$ annual cost per carrier

$u \triangleq$ annual overhead cost of operating a delivery unit

We want to determine the decision variable

$x \triangleq$ number of delivery units

To develop a model, we employ approximations derived from an assumption that customers are spread evenly over the territory. Under that assumption it can be shown that

$$\text{average travel time per carrier to/from regions} \approx k_1 \sqrt{\frac{a}{x}}$$

$$\text{travel time between route stops for all routes} \approx k_2 \sqrt{am}$$

where k_1 and k_2 are constants of proportionality. Then the total number of routes is

$$\frac{(\text{total time at stops}) + (\text{total time between stops})}{\text{effective work time per carrier}} = \frac{tm + k_2\sqrt{am}}{d - k_1\sqrt{a/x}}$$

and total cost is

$$\text{overhead} + \text{operations} = ux + c\left(\frac{tm + k_2\sqrt{am}}{d - k_1\sqrt{a/x}}\right) \tag{13.1}$$

---

[1]Based on D. B. Rosenfield, I. Engelstein, and D. Feigenbaum (1992), "An Application of Sizing Service Territories," *European Journal of Operational Research*, 63, 164–177.

USPS Single-Variable Example Model

To have a specific example with which to deal, pick

$$a = 400, \quad m = 200{,}000, \quad d = 8, \quad t = 0.05$$
$$c = 0.10, \quad u = 0.75, \quad k_1 = 0.2, \quad k_2 = 0.1$$

in expression (13.1). Then our single-variable USPS nonlinear program is

$$\min f(x) \triangleq (0.75)x + (0.10)\left[\frac{(0.05)(200{,}000) + (0.1)\sqrt{400(200{,}000)}}{(8) - (0.2)\sqrt{400/x}}\right] \tag{13.2}$$

$$\approx 0.75x + \frac{1089.4}{8 - 0.2\sqrt{400/x}}$$

Figure 13.1 shows that a unique optimal solution occurs at $x^* \approx 15.3$.

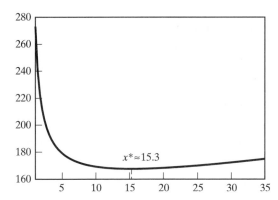

FIGURE 13.1   USPS Example Objective Function

Neglecting Constraints to Use Unconstrained Methods

Strictly speaking, model (13.2) is incomplete. A meaningful number of deliver units, x, should satisfy the constraint

$$x \geq 0$$

and perhaps also

$$x \text{ integer}$$

Still, we know (e.g., principle 7.40) that adding a constraint can change an optimal solution only if it is violated. With $x^* = 15.2 > 0$, the neglected nonnegativity constraint would have no impact even if we modeled it explicitly. Integrality is violated at x^*, but the rough planning nature of the model suggests that we would be quite justified in rounding to $x = 15$.

Many, perhaps most, problems modeled as unconstrained nonlinear programs actually have a few constraints that are neglected in this way.

13.2 │ The relative tractability of unconstrained models often justifies neglecting simple constraints that should apply until an optimum has been computed, checking only afterward that the solution is feasible.

Of course, if the unconstrained optimum violates an important constraint, we must resort to constrained methods.

Curve Fitting and Regression Problems

Perhaps the most common of unconstrained NLPs involve **curve fitting** or **regression**. We seek to choose coefficients for a functional form to make it fit closely some observed data.

> **13.3** | Decision variables in a regression problem are the coefficients of the fitted functional form, and the objective function measures the accuracy of the fit.

EXAMPLE 13.2: CUSTOM COMPUTER CURVE FITTING

For a simple example of curve fitting, we consider the problem of (fictitious) Custom Computer Company, which builds specialized computer workstations for engineers. Although there may be many commonalties, stations produced for each order are specially modified to meet customer specifications.

Table 13.1 shows the number of units and unit cost (in thousands of dollars) of $m = 12$ recent orders. Figure 13.2 plots this experience. Obviously, unit cost declines dramatically with the size of an order. Custom wants to fit an estimating function like the one depicted in Figure 13.2 to facilitate preparation of bids on future work.

TABLE 13.1   Cost Data for Custom Computer Example

i	Number, p_i	Cost, q_i	i	Number, p_i	Cost, q_i	i	Number, p_i	Cost, q_i
1	19	7.9	5	5	19.5	9	14	9.2
2	2	25.0	6	6	13.0	10	17	6.3
3	9	13.1	7	3	17.8	11	1	42.0
4	4	17.4	8	11	8.0	12	20	6.6

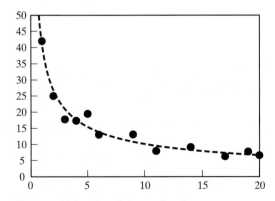

FIGURE 13.2   Fitted Curve for Custom Computer Example

Linear versus Nonlinear Regression

Our first task in dealing with Custom Computer's problem is choosing a regression form to fit. That is, we want to choose a function $r(p)$ with unknown coefficients to approximate

$$r(p_i) \approx q_i \qquad \text{for all } i = 1, \dots, m$$

Nonlinear optimization will choose the best coefficient values.

The fact that functional coefficients constitute the decision variables (principle 13.3) in curve fitting leads to considerable confusion. For example, regression analysts often denote a form something like

$$r(x) \triangleq a + bx \tag{13.3}$$

with x being data and $\{a, b\}$ the coefficients to determine. This choice of notation is exactly the reverse of the one familiar in optimization.

We will follow the mathematical programming tradition of reserving x for decision variables. Thus form (13.3) might be expressed as

$$r(p) \triangleq x_1 + x_2 p \tag{13.4}$$

with p the data and $\{x_1, x_2\}$ the undetermined coefficients.

A similar confusion arises in distinguishing **linear** versus **nonlinear regression**.

13.4 | A regression problem is termed linear if the functional form being fitted is linear in the unknown coefficients (decision variables) and nonlinear otherwise.

For example, the choice of

$$r(p) \triangleq x_1 + \frac{x_2}{p} \tag{13.5}$$

in our Custom Computer case would be a linear regression. The corresponding curve in Figure 13.2 would not be a straight line, but expression (13.5) is linear in the unknown coefficients x_1 and x_2.

The distinction between linear and nonlinear regression is important because there is often a closed-form solution for fitting linear forms, but nonlinear ones usually require search. Computations throughout this chapter will illustrate for nonlinear regression form

$$r(p) \triangleq x_1 p^{x_2} \tag{13.6}$$

on our Custom Computer example. The curve depicted in Figure 13.2 is

$$r(p) \triangleq 40.69 p^{-0.6024}$$

which provides an optimal fit.

SAMPLE EXERCISE 13.1: DISTINGUISHING LINEAR AND NONLINEAR REGRESSION

Taking variables x_j to be the unknown coefficients and all other symbols as given data, determine whether fitting each of the following forms is linear or nonlinear regression.

(a) $r(p) \triangleq x_1 + x_2 \sin(p)$

(b) $r(p) \triangleq x_1 + \sin(x_2)p$

(c) $r(p_1, p_2) \triangleq x_1 p_1^2 + x_2 e^{p_2}$

Analysis: We apply definition $\boxed{13.4}$.

(a) This regression form is linear because it is linear in the decision variables x_1 and x_2 for given data p.

(b) This regression form is nonlinear. Decision variable x_2 appears in the nonlinear expression $\sin(x_2)$.

(c) This regression form is linear. It fits a nonlinear function of two inputs p_1 and p_2, but decision variables x_1 and x_2 occur linearly.

Regression Objective Functions

To complete formulation of curve fitting as a nonlinear optimization, we require an objective function measuring fit. The error or **residual** associated with any data point is the difference between the fitted function and the actually value observed. For example, in our Custom Computer case, residuals under functional form (13.6) are

$$q_i - r(p_i) = q_i - x_1(p_i)^{x_2} \qquad \text{for all } i = 1, \ldots, m$$

Regression objective functions minimize some nondecreasing function of the magnitudes of residuals. Many possibilities have been employed, but the most common is the sum of residual squares or **least squares**. This objective possesses a number of desirable statistical properties, and it also has the intuitive appeal that small deviations cost little but large ones are heavily penalized.

Custom Computer Curve Fitting Example Model

For our Custom Computer example, the least squares objective produces unconstrained NLP model

$$\min \quad f(x_1, x_2) \triangleq \sum_{i=1}^{m} [q_i - x_1(p_i)^{x_2}]^2 \tag{13.7}$$

Figure 13.3 displays the objective graphically, with global minimum at

$$x_1^* \approx 40.69, \quad x_2^* \approx -0.6024$$

yielding the best fit.

SAMPLE EXERCISE 13.2: FORMULATING NONLINEAR REGRESSION MODELS

Three observations from a function believed to have the form

$$z = \alpha^u \beta^v$$

are $(u_1, v_1, z_1) = (1, 8, 3)$, $(u_2, v_2, z_2) = (4, 15, 2)$, and $(u_3, v_3, z_3) = (2, 29, 71)$. Formulate an unconstrained nonlinear program to choose the α and β yielding a least squares fit.

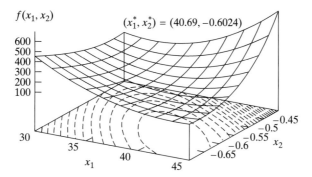

FIGURE 13.3 Objective Function in Custom Computer Example

Modeling: Residuals are $(z_1 - \alpha^{u_i}\beta^{v_i})$. Thus a least squares fit will be obtained at α and β solving

$$\min\ f(\alpha, \beta) \triangleq \left(3 - \alpha^1\beta^8\right)^2 + \left(2 - \alpha^4\beta^{15}\right)^2 + \left(71 - \alpha^2\beta^{29}\right)^2$$

Maximum Likelihood Estimation Problems

Another common application of unconstrained NLP arises in fitting continuous probability distributions to observed data. A **probability density function**, $d(p)$, characterizes any such distribution by showing how the probability is spread over different values of a random variable, P. For example, the $d(p)$ depicted in Figure 13.4 indicates relatively higher probability of values of P near 0.7 than near 0.2.

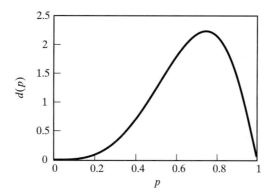

FIGURE 13.4 Density Function for PERT Maximum Likelihood Example

When say m independent random variables $P_1, P_2, \ldots, P_m$ have the same probability density $d(p)$, the **joint probability density function** or **likelihood** is

$$d(p_1, p_2, \ldots, p_m) = d(p_1)d(p_2)\cdots d(p_m) \qquad (13.8)$$

That is, the density for any particular independent combination of values is the product of the densities for the values separately.

Estimation involves choosing values for unknown coefficients of a density functional form $d(p)$. The best estimates by many criteria are **maximum likelihood** ones that maximize the joint density of a known sample of random variable values.

> **13.5** | Decision variables in maximum likelihood estimation are the coefficients of the fitted probability density, and the objective function is the corresponding joint density or likelihood evaluated at the observed data.

EXAMPLE 13.3: PERT MAXIMUM LIKELIHOOD

Maximum likelihood estimates for parameters of many standard probability densities can be obtained in closed form. However, some require a numerical optimization to maximize the likelihood function.

A specific example occurs in fitting the **beta distribution** often used in **project evaluation and review technique (PERT)** project management. PERT is an extension of the **CPM** project scheduling method introduced in Section 9.7. As with CPM, a project is divided into a series of work activities with specified time durations. The new element with PERT is that durations are taken as random variables (i.e., they are assumed to be known only in probability distribution at the time that planning takes place).

Beta random variables assign probability density over the interval $0 \leq p \leq 1$, so that in PERT they reflect the fraction p that some activity's duration forms of an allowed maximum. The beta probability density function is

$$d(p) \triangleq \frac{\Gamma(x_1 + x_2)}{\Gamma(x_1)\Gamma(x_2)}(p)^{x_1-1}(1-p)^{x_2-1} \tag{13.9}$$

where $x_1 > 0$ and $x_2 > 0$ are parameters controlling its shape.[2] For instance, $x_1 = 4.50$, and $x_2 = 2.20$ yields density shown in Figure 13.4. In expression (13.9), $\Gamma(x)$ is the standard Γ-function equal to the area under the curve

$$\gamma(x) \triangleq (h)^{x-1}(e)^{-h}$$

over $0 \leq h \leq +\infty$. $\Gamma(x)$ has no closed form.

Table 13.2 shows the data we will assume for $m = 10$ previous times a project activity was undertaken. Values p_i represent actual duration as a fraction of the maximum ever expected.

TABLE 13.2   Realized Data for PERT Maximum Likelihood Example

	Value		Value		Value		Value		Value
p_1	0.65	p_3	0.52	p_5	0.74	p_7	0.79	p_9	0.92
p_2	0.57	p_4	0.72	p_6	0.30	p_8	0.89	p_{10}	0.42

---

[2]Beta parameters are more commonly called α and β, but we will employ x_1 and x_2 to be consistent with the convention that x_j refers to a decision variable.

PERT Maximum Likelihood Example Model

The beta probability density for observation $p_1 = 0.65$ is

$$d(p_1) = d(0.65) = \frac{\Gamma(x_1 + x_2)}{\Gamma(x_1)\Gamma(x_2)}(0.65)^{x_1-1}(1 - 0.65)^{x_2-1}$$

and for the first two observations together it is [applying expression (13.8)]

$$d(p_1)d(p_2) = \left[\frac{\Gamma(x_1 + x_2)}{\Gamma(x_1)\Gamma(x_2)}(0.65)^{x_1-1}(1 - 0.65)^{x_2-1}\right]$$
$$\cdot \left[\frac{\Gamma(x_1 + x_2)}{\Gamma(x_1)\Gamma(x_2)}(0.57)^{x_1-1}(1 - 0.57)^{x_2-1}\right]$$

Continuing in this way our nonlinear program to maximize the likelihood of the full m-value sample over coefficients x_1 and x_2 is

$$\max \quad f(x_1, x_2) \triangleq \prod_{i=1}^{m}\left[\frac{\Gamma(x_1 + x_2)}{\Gamma(x_1)\Gamma(x_2)}(p_i)^{x_1-1}(1 - p_i)^{x_2-1}\right] \tag{13.10}$$

Figure 13.5 plots this objective for various x_1 and x_2. The unique global maximum is at $x_1^* \approx 4.50$ and $x_2^* \approx 2.20$ of the density in Figure 13.4.

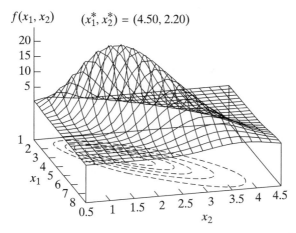

FIGURE 13.5 PERT Maximum Likelihood Example Objective Function

---

SAMPLE EXERCISE 13.3: FORMULATING MAXIMUM LIKELIHOOD MODELS

Exponential probability distributions have density function

$$d(p) \triangleq \alpha e^{-\alpha p}$$

Formulate an unconstrained nonlinear program to determine the maximum likelihood value of parameter α consistant with realizations $p_1 = 4$, $p_2 = 9$, and $p_3 = 8$.

Modeling: The likelihood is the product [expression (13.8)] of the densities for the three realizations. Thus the model required is

$$\max f(\alpha) \triangleq \left(\alpha e^{-4\alpha}\right)\left(\alpha e^{-9\alpha}\right)\left(\alpha e^{-8\alpha}\right)$$

Smooth versus Nonsmooth Functions and Derivatives

It is useful to classify nonlinear programs according to whether their objective functions are smooth or nonsmooth.

> **13.6**   A function $f(\mathbf{x})$ is said to be **smooth** if it is continuous and differentiable at all relevant $\mathbf{x}$. Otherwise, it is **nonsmooth**.

Figure 13.6 illustrates this for some functions of a single x. (Refer, if needed, to Primer 2 in Section 3.3 for a quick review of differential calculus.)

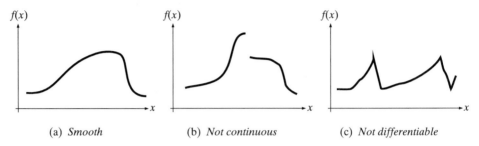

| (a) *Smooth* | (b) *Not continuous* | (c) *Not differentiable* |

FIGURE 13.6   Examples of Smooth and Nonsmooth Functions

The smooth/nonsmooth distinction is useful because the more erratic nature of nonsmooth functions usually implies a more difficult search.

> **13.7**   Nonlinear programs over smooth functions are generally more tractable than those over nonsmooth ones.

SAMPLE EXERCISE 13.4: RECOGNIZING SMOOTH FUNCTIONS

Determine whether each of the following single variable functions is smooth or nonsmooth over the specified domain.

(a) $f(x) \triangleq x^3$ for $x \in (-\infty, +\infty)$

(b) $f(x) \triangleq |x - 1|$ for $x \in (-\infty, +\infty)$

(c) $f(x) \triangleq \dfrac{1}{x}$ for $x > 0$

Analysis: We apply definition 13.6 .

(a) This function is differentiable at every point in the domain, so the function is smooth.

(b) This function is not differentiable at point $x = 1$, which is within the range. Thus the function is nonsmooth.

(c) This function is discontinuous at $x = 0$. However, that point is not in the domain specified. Thus the function is smooth for $x > 0$.

Usable Derivatives

All three objective functions for the examples of this section (Figures 13.1, 13.3, and 13.5) are smooth. Still, the existence of (partial) derivatives at every relevant $\mathbf{x}$ does not necessarily imply that derivatives are readily available to aid search algorithms.

For the USPS example

$$f(x) \triangleq 0.75x + \frac{1089.4}{8 - 0.2\sqrt{400/x}}$$

and

$$\frac{df}{dx} = 0.75 - \frac{1089.4}{\left(8 - 0.2\sqrt{400/x}\right)^2} \left(\frac{0.1}{\sqrt{400/x}}\right) \left(\frac{400}{x^2}\right) \qquad (13.11)$$

For the Custom Computer case

$$f(x_1, x_2) \triangleq \sum_{i=1}^{m} [q_i - x_1(p_i)^{x_2}]^2$$

with

$$\frac{\partial f}{\partial x_1} = -2 \sum_{i=1}^{m} [q_i - x_1(p_i)^{x_2}] (p_i)^{x_2}$$

$$\frac{\partial f}{\partial x_2} = -2 \sum_{i=1}^{m} [q_i - x_1(p_i)^{x_2}] [x_1(p_i)^{x_2}] \ln(p_i) \qquad (13.12)$$

Neither (13.11) nor (13.12) is a particularly simple expression, but both show derivatives that can be computed with reasonable effort to speed a search.

In elementary calculus, unconstrained optima are computed by solving the system of equations resulting from setting derivatives $= 0$. With complex derivative expressions such as (13.11) and (13.12), solving such a system of equations is often as difficult as solving the underlying nonlinear program. Nevertheless, practically available derivatives can be a significant aid to a numerical search for optimal variable values.

> **13.8** | Nonlinear programs over smooth functions with convenient-to-compute derivatives are usually more tractable than those without, because derivatives can be exploited to produce a much more efficient search.

Contrast with the PERT maximum likelihood objective

$$f(x_1, x_2) \triangleq \prod_{i=1}^{m} \left[\frac{\Gamma(x_1 + x_2)}{\Gamma(x_1)\Gamma(x_2)} (p_i)^{x_1 - 1} (1 - p_i)^{x_2 - 1} \right]$$

The Γ-function itself has no closed form, so derivatives are certainly not readily available, even though they do exist in theory. To compute an optimum for this case we require a search method that does not depend on derivatives.

> **13.9** | Although derivatives always exist for smooth functions, they are sometimes so difficult to compute that the search must proceed without using derivatives.

SAMPLE EXERCISE 13.5: ASSESSING PRACTICALITY OF DERIVATIVES

Return to the least squares objective function of Sample Exercise 13.2:

$$\min \; f(\alpha, \beta) \triangleq \left(3 - \alpha^1 \beta^8\right)^2 + \left(2 - \alpha^4 \beta^{15}\right)^2 + \left(71 - \alpha^2 \beta^{29}\right)^2$$

(a) Express partial derivatives with respect to α and β.

(b) Discuss the usefulness of those partial derivatives in computing an optimal α and β.

Analysis:

(a) Partial derivatives are

$$\frac{\partial f}{\partial \alpha} = 2(3 - \alpha^1 \beta^8)(-\beta^8) + 2(2 - \alpha^4 \beta^{15})(-4\alpha^3 \beta^{15})$$

$$+ 2(71 - \alpha^2 \beta^{29})(-2\alpha \beta^{29})$$

$$\frac{\partial f}{\partial \beta} = 2(3 - \alpha^1 \beta^8)(-8\alpha^1 \beta^7) + 2(2 - \alpha^4 \beta^{15})(-15\alpha^4 \beta^{14})$$

$$+ 2(71 - \alpha^2 \beta^{29})(-29\alpha^2 \beta^{28})$$

(b) Although derivative expressions of part (a) are somewhat complicated, they can be evaluated efficiently. Simply setting them = 0 is impractical because it would leave a pair of difficult nonlinear equations to solve. However, the derivatives can be employed to speed an improving search (principle $\boxed{13.8}$).

13.2 ONE-DIMENSIONAL SEARCH

The easiest case on unconstrained nonlinear programming is single-variable or **1-dimensional search**. One-dimensional NLPs occur both directly, as in our USPS example of Section 13.1, and as **line search** subroutines choosing step sizes to apply to move directions of more general algorithms.

Unimodal Objective Functions

Figure 13.7 illustrates for

$$f(x) \triangleq (x - 4)(x - 6)^3(x - 1)^2$$

how 1-dimensional optimization can be quite challenging.

- Points $x^{(1)}$ and $x^{(3)}$ are both **local minima** because small changes from either x do not decrease the objective function (definition $\boxed{3.5}$). Only $x^{(3)}$ is an overall, **global minimum** (definition $\boxed{3.7}$) for the displayed interval.
- Point $x^{(2)}$ is a local maximum because small changes from $x^{(2)}$ do not increase the objective function. Being the only local maximum, it is also a global maximum for the displayed interval.
- Point $x^{(4)}$ is neither a maximum nor a minimum, despite the fact that the slope of $f(x)$ (derivative df/dx) at the point = 0.

Fortunately, most of the 1-dimensional searches we encounter in application are somewhat better behaved. Recall from definition $\boxed{3.26}$ in Section 3.4 that an

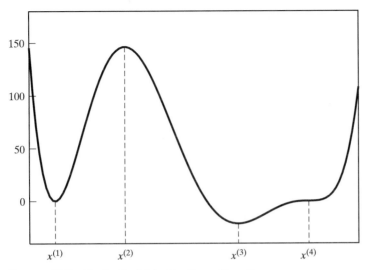

FIGURE 13.7 Single-Variable Nonlinear Function

objective function $f(\mathbf{x})$ is **unimodal** if the straight-line direction $\Delta \mathbf{x}$ to any $\mathbf{x}$ with a better objective value is an improving direction; that is, it helps the objective for even very small positive steps (definition $\boxed{3.13}$). Over unimodal objective functions, every unconstrained local optimum is a global optimum (property $\boxed{3.28}$).

The "one-hump" or "single-mode" character that gives unimodal functions their name is particularly easy to grasp in 1-dimensional cases. For example, the minimizing USPS objective of Figure 13.1 is unimodal because the objective is decreasing (improving) at every x to the left of $x^* \approx 15.2$ and increasing (degrading) at every x to the right. Since a tiny move toward x^* always helps, the local minimum there must be global.

Notice that an objective's being unimodal depends on whether we are maximizing or minimizing. The USPS objective of Figure 13.1 is not unimodal for a maximize problem because, for example, a move toward $x^{(2)} = 30$ does not immediately improve the objective function at $x^{(1)} = 10$, even though $f(30) > f(10)$.

Golden Section Search

Although derivatives can sometimes be of assistance, many 1-dimensional optimizations employ simpler methods not requiring derivatives. Among the most clever is **golden section search**, which deals with an unimodal objective by rapidly narrowing an interval guaranteed to contain an optimum.

Figure 13.8 illustrates the idea for a minimize problem. We iteratively consider the functional value at four carefully spaced points. Leftmost $x^{(lo)}$ is always a lower bound on the optimal x^*, and $x^{(hi)}$ is an upper bound, so that an optimum is certain to lie within the interval $[x^{(lo)}, x^{(hi)}]$. Points $x^{(1)}$ and $x^{(2)}$ fall in between.

Each iteration begins by determining whether the objective is better at $x^{(1)}$ or $x^{(2)}$. If $x^{(1)}$ proves superior [part (a)], we may conclude the optimum lies within the smaller interval $[x^{(lo)}, x^{(2)}]$. If $x^{(2)}$ has a better objective value [part (b)], we restrict further attention to the interval $[x^{(1)}, x^{(hi)}]$.

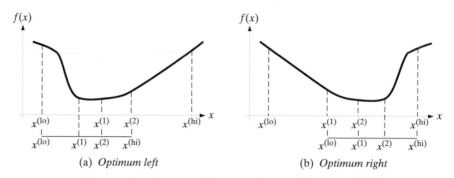

FIGURE 13.8   Interval Reduction in a Minimizing Golden Section Search

Each narrowing of the interval certain to contain an optimum leaves us with 2 endpoints and 1 interior point at which we already know the objective value. The efficiency of golden section search comes from how we choose 1 new interior point to evaluate. Any new point would allow the search to continue, but we want consistent progress regardless of whether the next interval is $[x^{(lo)}, x^{(2)}]$ or $[x^{(1)}, x^{(hi)}]$.

Golden section search proceeds by keeping both these possible intervals equal in length.

> **13.10**   The two middle points of golden section search are spaced
> $$x^{(1)} = x^{(hi)} - \alpha(x^{(hi)} - x^{(lo)})$$
> $$x^{(2)} = x^{(lo)} + \alpha(x^{(hi)} - x^{(lo)})$$
> where $\alpha \approx 0.618$ is a fraction known as the **golden ratio**.

Whichever of $[x^{(lo)}, x^{(2)}]$ or $[x^{(1)}, x^{(hi)}]$ provides the next interval, its length will be α times the current.

The golden ratio value of

$$\alpha = \frac{-1 + \sqrt{5}}{2} \approx 0.618 \tag{13.13}$$

arises from the need to maintain the spacing of principle 13.10 as the algorithm proceeds. For example, suppose that the chosen next interval is $[x^{(lo)}, x^{(2)}]$. As indicated in Figure 13.8(a), we want current $x^{(1)}$ to play the role of $x^{(2)}$ in the next interval. Applying formulas 13.10 yields

$$\text{current } x^{(1)} = x^{(hi)} - \alpha(x^{(hi)} - x^{(lo)})$$

and

$$\text{next } x^{(2)} = x^{(lo)} + \alpha(x^{(2)} - x^{(lo)})$$
$$= x^{(lo)} + \alpha\left(x^{(lo)} + \alpha(x^{(hi)} - x^{(lo)}) - x^{(lo)}\right)$$

Equating and regrouping produces

$$0 = \alpha^2(x^{(hi)} - x^{(lo)}) + \alpha(x^{(hi)} - x^{(lo)}) - (x^{(hi)} - x^{(lo)})$$

which further simplifies with $x^{(hi)} \neq x^{(lo)}$ to

$$0 = \alpha^2 + \alpha - 1$$

The unique positive root of this quadratic equation is $\alpha =$ the golden ratio of expression (13.13).

Golden Section Solution of USPS Example

Algorithm 13A formalizes these ideas of golden section search. Table 13.3 details its application to our minimizing USPS model (13.2).

ALGORITHM 13A: GOLDEN SECTION SEARCH

Step 0: Initialization. Choose lower bound $x^{(lo)}$ and upper bound $x^{(hi)}$ on an optimal solution x^* along with stopping tolerance $\epsilon > 0$, compute

$$x^{(1)} \leftarrow x^{(hi)} - \alpha(x^{(hi)} - x^{(lo)})$$
$$x^{(2)} \leftarrow x^{(lo)} + \alpha(x^{(hi)} - x^{(lo)})$$

for golden ratio α of (13.13), evaluate objective function $f(x)$ at all four points, and initialize iteration counter $t \leftarrow 0$.

Step 1: Stopping. If $(x^{(hi)} - x^{(lo)}) \leq \epsilon$, stop and report as an approximate optimal solution

$$x^* \leftarrow \tfrac{1}{2}(x^{(lo)} + x^{(hi)})$$

the midpoint of the remaining interval. Otherwise, proceed to Step 2 if $f(x^{(1)})$ is superior to $f(x^{(2)})$ (less for a minimize model, greater for a maximize), and to Step 3 if it is not.

Step 2: Left. Narrow the search to the left part of the interval by updating

$$x^{(hi)} \leftarrow x^{(2)}$$
$$x^{(2)} \leftarrow x^{(1)}$$
$$x^{(1)} \leftarrow x^{(hi)} - \alpha(x^{(hi)} - x^{(lo)})$$

and evaluate the objective at new point $x^{(1)}$. Then advance $t \leftarrow t + 1$, and return to Step 1.

Step 3: Right. Narrow the search to the right part of the interval by updating

$$x^{(lo)} \leftarrow x^{(1)}$$
$$x^{(1)} \leftarrow x^{(2)}$$
$$x^{(2)} \leftarrow x^{(lo)} + \alpha(x^{(hi)} - x^{(lo)})$$

and evaluate the objective at new point $x^{(2)}$. Then advance $t \leftarrow t + 1$, and return to Step 1.

Computation in Table 13.3 starts arbitrarily with interval

$$[x^{(lo)}, x^{(hi)}] = [8, 32]$$

TABLE 13.3 Golden Section Search of USPS Example

t	$x^{(lo)}$	$x^{(1)}$	$x^{(2)}$	$x^{(hi)}$	$f(x^{(lo)})$	$f(x^{(1)})$	$f(x^{(2)})$	$f(x^{(hi)})$	$x^{(hi)} - x^{(lo)}$
0	8.00	17.17	22.83	32.00	171.42	167.74	169.22	173.38	24.00
1	8.00	13.67	17.17	22.83	171.42	167.73	167.74	169.22	14.83
2	8.00	11.50	13.67	17.17	171.42	168.36	167.73	167.74	9.17
3	11.50	13.67	15.00	17.17	168.36	167.73	167.62	167.74	5.67
4	13.67	15.00	15.83	17.17	167.73	167.62	167.63	167.74	3.50
5	13.67	14.49	15.00	15.83	167.73	167.64	167.62	167.63	2.16
6	14.49	15.00	15.32	15.83	167.64	167.62	167.61	167.63	1.34
7	15.00	15.32	15.51	15.83	167.62	167.61	167.62	167.63	0.83
8	15.00	15.20	15.32	15.51	167.62	167.61	167.61	167.62	0.51
9	15.20	15.32	15.39	15.51	167.61	167.61	167.61	167.62	0.32

It is only necessary that the interval contain the optimum. Intermediate points $x^{(1)}$ and $x^{(2)}$ are then computed by principle 13.10 as

$$x^{(1)} \leftarrow x^{(hi)} - \alpha(x^{(hi)} - x^{(lo)})$$
$$= 32 - 0.618(32 - 8)$$
$$\approx 17.17$$
$$x^{(2)} \leftarrow x^{(lo)} + \alpha(x^{(hi)} - x^{(lo)})$$
$$= 8 + 0.618(32 - 8)$$
$$\approx 22.83$$

At iteration $t = 0$, objective value

$$f(x^{(1)}) = 167.74 < f(x^{(2)}) = 169.22$$

implying that the optimum lies in the left part of the current interval. Following Algorithm 13A, we update

$$x^{(hi)} \leftarrow x^{(2)} = 22.83$$
$$x^{(2)} \leftarrow x^{(1)} = 17.17$$

and compute new

$$x^{(1)} \leftarrow x^{(hi)} - \alpha(x^{(hi)} - x^{(lo)})$$
$$= 22.83 - 0.618(22.83 - 8.00)$$
$$\approx 13.67$$

The process now repeats for iteration $t = 1$.

Computation continues until interval $[x^{(lo)}, x^{(hi)}]$ has sufficiently small length. In Table 13.3 stopping was set to occur when

$$x^{(hi)} - x^{(lo)} < \epsilon = 0.5$$

which happened at $t = 9$. Then our estimate of the optimal solution is midpoint

$$x^* \leftarrow \tfrac{1}{2}(x^{(lo)} + x^{(hi)}) = \tfrac{1}{2}(15.20 + 15.51) = 15.36$$

If greater accuracy were desired, we would need to continue through more iterations.

SAMPLE EXERCISE 13.6: APPLYING GOLDEN SECTION SEARCH

Beginning with interval $[0, 40]$, apply golden section search to the unconstrained nonlinear program

$$\max \quad f(x) \triangleq 2x - \frac{(x-20)^4}{500}$$

plotted below.

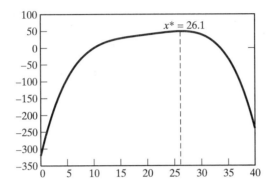

Continue until the interval containing an optimum has length at most 10.

Analysis: The algorithm proceeds exactly as in Table 13.3 except that this model maximizes. The following table provides details:

t	$x^{(lo)}$	$x^{(1)}$	$x^{(2)}$	$x^{(hi)}$	$f(x^{(lo)})$	$f(x^{(1)})$	$f(x^{(2)})$	$f(x^{(hi)})$	$x^{(hi)} - x^{(lo)}$
0	0.00	15.28	24.72	40.00	−320.00	29.57	48.45	−240.00	40.00
1	15.28	24.72	30.54	40.00	29.57	48.45	36.27	−240.00	24.72
2	15.28	21.12	24.72	30.56	29.57	42.23	48.45	36.27	15.28
3	21.12	24.72	26.95	30.56	42.23	48.45	49.23	36.27	9.44

Termination occurs at $t = 3$ with $x^{(hi)} - x^{(lo)} = 9.44 < \epsilon = 10$.

Bracketing and 3-Point Patterns

Golden section search begins solving a 1-variable model with an interval $[x^{(lo)}, x^{(hi)}]$ known to contain an optimum. But how do we determine such initial intervals; that is, how do we **bracket** an optimal solution before the main search begins?

Sometimes the initial bracket is given because the model includes implicit upper and lower bounds on the decision variable. Much more commonly, one endpoint is known and the other must be determined. For example, in line searches, where the single variable is the step size λ to apply to a chosen move direction, λ must be positive. Thus we begin with $x^{(lo)} = 0$.

To locate the corresponding $x^{(hi)}$ bracketing the optimum of a unimodal objective function requires a search for a **3-point pattern**.

| 13.11 | In 1-dimensional optimization, a 3-point pattern is a collection of 3 decision variable values $x^{(lo)} < x^{(mid)} < x^{(hi)}$ with the objective value at $x^{(mid)}$ superior to that of the other two (greater for a maximize, lesser for a minimize). |

Figure 13.9 illustrates for our USPS model (13.2). Points

$$x^{(lo)} = 8, \quad x^{(mid)} = 16, \quad x^{(hi)} = 32$$

surround the minimum in a 3-point pattern with

$$f(x^{(lo)}) = f(8) \approx 171.42 > f(x^{(mid)}) = f(16) \approx 167.63$$
$$f(x^{(hi)}) = f(32) \approx 173.38 > f(x^{(mid)}) = f(16) \approx 167.63$$

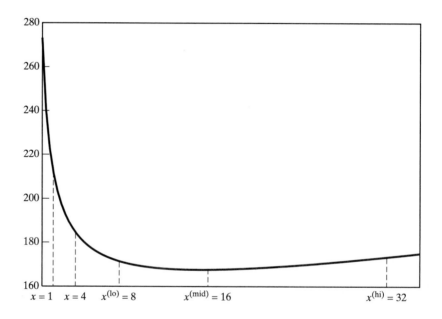

FIGURE 13.9   Bracketing the USPS Optimum with a 3-Point Pattern

A 3-point pattern provides the bracket we seek if the objective function is unimodal. Midpoint value $f(x^{(mid)})$ superior to $f(x^{(lo)})$ means that the function improves to the right of $x^{(lo)}$. Similarly, with $f(x^{(mid)})$ better than $f(x^{(hi)})$, the function improves to the left of $x^{(hi)}$. An optimum must lie in between.

| 13.12 | If $\{x^{(lo)}, x^{(mid)}, x^{(hi)}\}$ is a 3-point pattern for unimodal objective function $f(x)$, there is an optimal x^* in the interval $[x^{(lo)}, x^{(hi)}]$. |

Finding a 3-Point Pattern

Algorithm 13B details the most common scheme for quickly finding a 3-point pattern when we are given only an initial lower endpoint $x^{(lo)}$. Values of x are modified by exponentially changing step δ until the last three form a 3-point pattern.

ALGORITHM 13B: THREE-POINT PATTERN

Step 0: Initialization. Choose lower bound $x^{(lo)}$ on optimal solution x^* and initial step $\delta > 0$.

Step 1: Right or Left. If $f(x^{(lo)} + \delta)$ is superior to $f(x^{(lo)})$ (less for a minimize, greater for a maximize), set

$$x^{(mid)} \leftarrow x^{(lo)} + \delta$$

and go to Step 2 to search right. Otherwise, an optimum lies to the left; set

$$x^{(hi)} \leftarrow x^{(lo)} + \delta$$

and go to Step 3.

Step 2: Expand. Increase $\delta \leftarrow 2\delta$. If now $f(x^{(mid)})$ is superior to $f(x^{(mid)} + \delta)$, set

$$x^{(hi)} \leftarrow x^{(mid)} + \delta$$

and stop; $\{x^{(lo)}, x^{(mid)}, x^{(hi)}\}$ forms a 3-point pattern. Otherwise, update

$$\begin{aligned} x^{(lo)} &\leftarrow x^{(mid)} \\ x^{(mid)} &\leftarrow x^{(mid)} + \delta \end{aligned}$$

and repeat Step 2.

Step 3: Reduce. Decrease $\delta \leftarrow \frac{1}{2}\delta$. If $f(x^{(lo)} + \delta)$ is now superior to $f(x^{(lo)})$, set

$$x^{(mid)} \leftarrow x^{(lo)} + \delta$$

and stop; $\{x^{(lo)}, x^{(mid)}, x^{(hi)}\}$ forms a 3-point pattern. Otherwise, update

$$x^{(hi)} \leftarrow x^{(lo)} + \delta$$

and repeat Step 3.

Values in Figure 13.9 illustrate the idea. Computation starts with $x^{(lo)} = \delta = 1$. Since

$$f(x^{(lo)} + \delta) = f(1 + 1) = f(2) \approx 212.16 < f(x^{(lo)}) = f(1) \approx 273.11$$

we must expand to the right to find a 3-point pattern bracketing the optimum. Setting $x^{(mid)} = 2$, we double δ and consider

$$x^{(mid)} + \delta = 2 + 2 = 4$$

The function improves again, so

$$\begin{aligned} x^{(lo)} &\leftarrow x^{(mid)} &= 2 \\ x^{(mid)} &\leftarrow x^{(mid)} + \delta = 4 \end{aligned}$$

and the process continues.

Eventually, we have $x^{(lo)} = 8$, $x^{(mid)} = 16$, and $\delta = 16$. Then

$$f(x^{(mid)}) = f(16) \approx 167.63 < f(x^{(mid)} + \delta) = f(32) \approx 173.63$$

and the algorithm stops after completing the 3-point pattern with

$$x^{(\text{hi})} = x^{(\text{mid})} + \delta = 32$$

SAMPLE EXERCISE 13.7: FINDING A 3-POINT PATTERN

Return to the model

$$\max \quad f(x) \triangleq 2x - \frac{(x-20)^4}{500}$$

of Sample Exercise 13.6, and apply Algorithm 13B to compute 3-point patterns with initial $x^{(\text{lo})} = 0$ and (a) $\delta = 10$; (b) $\delta = 50$.

Analysis:

(a) Initial $f(x^{(\text{lo})}) = f(0) = -320$, and $f(x^{(\text{lo})} + \delta) = f(10) = 0$ improves. Thus $x^{(\text{mid})} \leftarrow 10$. Doubling δ and trying $f(x^{(\text{mid})} + \delta) = f(30) = 40$ produces further improvement. Thus $x^{(\text{lo})} \leftarrow x^{(\text{mid})} = 10$, $x^{(\text{mid})} \leftarrow 30$. Doubling δ again yields $f(x^{(\text{mid})} + \delta) = f(30 + 40) = -12,360$. Thus we stop with $x^{(\text{hi})} \leftarrow 70$.

(b) Initial $f(x^{(\text{lo})}) = f(0) = -320$, and $f(x^{(\text{lo})} + \delta) = f(50) = -1520$ is worse. Setting $x^{(\text{hi})} \leftarrow 50$, we could stop if our only purpose is to bracket the maximum. To complete a 3-point pattern, however, we must reduce δ. Halving to $\delta = 25$ produces $f(x^{(\text{lo})} + \delta) = f(25) = 48.75$, which does improve on $f(x^{(\text{lo})})$. Choosing $x^{(\text{mid})} \leftarrow 25$ completes the 3-point pattern.

Quadratic Fit Search

Golden search Algorithm 13A is reliable, but its slow and steady narrowing of the optimum-containing interval can require considerable computation before an optimum is identified with sufficient accuracy. **Quadratic fit search** closes in much more rapidly by taking full advantage of a current 3-point pattern.

Given a 3-point pattern, we can fit a quadratic function through corresponding functional values that has a unique maximum or minimum, $x^{(\text{qu})}$, whichever we are seeking for the given objective $f(x)$. Quadratic fit uses this approximation to improve the current 3-point pattern by replacing one of its points with approximate optimum $x^{(\text{qu})}$.

Figure 13.10 illustrates for USPS model (13.2) with initial 3-point pattern

$$x^{(\text{lo})} = 8, \quad x^{(\text{mid})} = 20, \quad x^{(\text{hi})} = 32$$

The main curve plots actual objective function $f(x)$. A second, dashed line shows the unique quadratic function fitting through the 3 pattern points.

That quadratic approximation has a minimum at

$$x^{(\text{qu})} \approx 18.56 \quad \text{with} \quad f(x^{(\text{qu})}) \approx 167.98 \tag{13.14}$$

Together with the current $x^{(\text{lo})}$ and $x^{(\text{mid})}$ it now forms a new 3-point pattern

$$x^{(\text{lo})} = 8, \quad x^{(\text{mid})} = 18.56, \quad x^{(\text{hi})} = 20$$

with a smaller interval [8, 20]. Repeating in this way isolates an optimum for $f(x)$ in an ever-narrowing range.

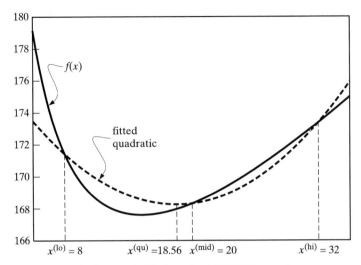

FIGURE 13.10 Quadratic Fit Search of USPS Example

The computation to determine $x^{(qu)}$ requires only some tedious algebra.

| 13.13 | The unique optimum of a quadratic function agreeing with $f(x)$ at 3-point pattern $\{x^{(lo)}, x^{(mid)}, x^{(hi)}\}$ occurs at

$$x^{(qu)} \triangleq \frac{1}{2} \frac{f^{(lo)}\left[s^{(mid)} - s^{(hi)}\right] + f^{(mid)}\left[s^{(hi)} - s^{(lo)}\right] + f^{(hi)}\left[s^{(lo)} - s^{(mid)}\right]}{f^{(lo)}\left[x^{(mid)} - x^{(hi)}\right] + f^{(mid)}\left[x^{(hi)} - x^{(lo)}\right] + f^{(hi)}\left[x^{(lo)} - x^{(mid)}\right]}$$

where $f^{(lo)} \triangleq f(x^{(lo)})$, $f^{(mid)} \triangleq f(x^{(mid)})$, $f^{(hi)} \triangleq f(x^{(hi)})$, $s^{(lo)} \triangleq (x^{(lo)})^2$, $s^{(mid)} \triangleq (x^{(mid)})^2$, and $s^{(hi)} \triangleq (x^{(hi)})^2$.

For example, the approximate minimum of expression (13.14) is

$$x^{(qu)} = \frac{1}{2} \frac{171.42\left[(20)^2 - (32)^2\right] + 168.32\left[(32)^2 - (8)^2\right] + 173.38\left[(8)^2 - (20)^2\right]}{171.42\left[20 - 32\right] + 168.32\left[32 - 8\right] + 173.38\left[8 - 20\right]}$$

$$\approx 18.56$$

Quadratic Fit Solution of USPS Example

Algorithm 13C details a quadratic fit procedure for 1-dimensional search, and Table 13.4 tracks progress for our USPS example. The reader can verify that each iteration produces a new 3-point pattern and that interval $[x^{(lo)}, x^{(hi)}]$ narrows constantly. Computation for Table 13.4 was stopped when that interval had length at most $\epsilon = 0.50$.

One new element arises when the computed $x^{(qu)}$ happens to nearly coincide with the current $x^{(mid)}$. If nothing were done, the algorithm would loop forever. Step 3 of Algorithm 13C addresses this difficulty by preturbing $x^{(qu)}$ by $\epsilon/2$ toward the most distant endpoint.

The one value changed in this way in Table 13.4 is marked with an asterisk (*) at $t = 7$. There formula $\boxed{13.13}$ produced $x^{(qu)} = 15.34$, which was too close to

---

ALGORITHM 13C: QUADRATIC FIT SEARCH

Step 0: Initialization. Choose starting 3-point pattern $\{x^{(lo)}, x^{(mid)}, x^{(hi)}\}$ along with a stopping tolerance $\epsilon > 0$, and initialize iteration counter $t \leftarrow 0$.

Step 1: Stopping. If $(x^{(hi)} - x^{(lo)}) \leq \epsilon$, stop and report approximate optimal solution $x^{(mid)}$.

Step 2: Quadratic Fit. Compute quadratic fit optimum $x^{(qu)}$ according to formula $\boxed{13.13}$. Then if $x^{(qu)} \approx x^{(mid)}$, go to Step 3; if $x^{(qu)} < x^{(mid)}$, go to Step 4; and if $x^{(qu)} > x^{(mid)}$, go to Step 5.

Step 3: Coincide. New $x^{(qu)}$ coincides essentially with current $x^{(mid)}$. If $x^{(mid)}$ if farther from $x^{(lo)}$ than from $x^{(hi)}$, preturb left

$$x^{(qu)} \leftarrow x^{(mid)} - \frac{\epsilon}{2}$$

and proceed to Step 4. Otherwise, adjust right

$$x^{(qu)} \leftarrow x^{(mid)} + \frac{\epsilon}{2}$$

and proceed to Step 5.

Step 4: Left. If $f(x^{(mid)})$ is superior to $f(x^{(qu)})$ (less for a minimize, greater for a maximize), then update

$$x^{(lo)} \leftarrow x^{(qu)}$$

Otherwise, replace

$$x^{(hi)} \leftarrow x^{(mid)}$$
$$x^{(mid)} \leftarrow x^{(qu)}$$

Either way, advance $t \leftarrow t + 1$, and return to Step 1.

Step 5: Right. If $f(x^{(mid)})$ is superior to $f(x^{(qu)})$ (less for a minimize, greater for a maximize), then update

$$x^{(hi)} \leftarrow x^{(qu)}$$

Otherwise, replace

$$x^{(lo)} \leftarrow x^{(mid)}$$
$$x^{(mid)} \leftarrow x^{(qu)}$$

Either way advance $t \leftarrow t + 1$, and return to Step 1.

TABLE 13.4 Quadratic Fit Solution of USPS Example

t	$x^{(lo)}$	$x^{(mid)}$	$x^{(hi)}$	$f(x^{(lo)})$	$f(x^{(mid)})$	$f(x^{(hi)})$	$x^{(hi)} - x^{(lo)}$	$x^{(qu)}$	$f(x^{(qu)})$
0	8.00	20.00	32.00	171.42	168.32	173.38	24.00	18.56	167.98
1	8.00	18.56	20.00	171.42	167.98	168.32	12.00	16.75	167.70
2	8.00	16.75	18.56	171.42	167.70	167.98	10.56	16.23	167.65
3	8.00	16.23	16.75	171.42	167.65	167.70	8.75	15.78	167.62
4	8.00	15.78	16.23	171.42	167.62	167.65	8.23	15.59	167.62
5	8.00	15.59	15.78	171.42	167.62	167.62	7.78	15.45	167.62
6	8.00	15.45	15.59	171.42	167.62	167.62	7.59	15.38	167.61
7	8.00	15.38	15.45	171.42	167.61	167.62	7.45	*15.13	167.62
8	15.13	$\boxed{15.38}$	15.45	167.62	167.61	167.62	0.32	—	—

$x^{(\text{mid})} = 15.38$. With $x^{(\text{lo})} = 8.00$ farther from this value than $x^{(\text{hi})} = 15.45$, Step 3 preturbed the computed value to

$$x^{(\text{qu})} = x^{(\text{mid})} - \frac{\epsilon}{2} = 15.38 - 0.25 = 15.13$$

SAMPLE EXERCISE 13.8: APPLYING QUADRATIC FIT SEARCH

Return to the unconstrained nonlinear program of Sample Exercise 13.6:

$$\max \quad f(x) \triangleq 2x - \frac{(x-20)^4}{500}$$

Using initial 3-point pattern $x^{(\text{lo})} = 0$, $x^{(\text{mid})} = 32$, $x^{(\text{hi})} = 40$, apply quadratic fit Algorithm 13C to identify an optimal solution within an interval $[x^{(\text{lo})}, x^{(\text{hi})}]$ of length at most 10.

Analysis: Computation parallels Table 13.4 except that this model maximizes. Details are contained in the following table:

t	$x^{(\text{lo})}$	$x^{(\text{mid})}$	$x^{(\text{hi})}$	$f(x^{(\text{lo})})$	$f(x^{(\text{mid})})$	$f(x^{(\text{hi})})$	$x^{(\text{hi})} - x^{(\text{lo})}$	$x^{(\text{qu})}$	$f(x^{(\text{qu})})$
0	0.00	32.00	40.00	−320.00	22.53	−240.00	40.00	20.92	41.84
1	0.00	20.92	32.00	−320.00	41.84	22.53	32.00	25.00	48.75
2	20.92	25.00	32.00	41.84	48.75	22.53	11.80	*30.00	40.03
3	20.92	[25.00]	30.00	41.84	48.75	40.03	9.08	—	—

13.3   DERIVATIVES, TAYLOR SERIES, AND CONDITIONS FOR LOCAL OPTIMA

Unconstrained nonlinear optimization is certainly possible without derivatives. Still, where derivatives are readily available, they can tell us a great deal about a model and substantially accelerate search algorithm progress (principle 13.8). This section develops some of the most important insights to be gained.

Improving Search Paradigm

In Sections 3.1 and 3.2 we introduced the principle of **improving search** (Algorithm 3A, Section 3.2) on which almost all nonlinear algorithms are based. We begin with a (vector) solution $\mathbf{x}^{(0)}$ satisfying all model constraints. In the unconstrained context of this chapter, any $\mathbf{x}^{(0)}$ will do. Iterations t advance current solution $\mathbf{x}^{(t)}$ to

$$\mathbf{x}^{(t+1)} \leftarrow \mathbf{x}^{(t)} + \lambda \Delta \mathbf{x}$$

where $\Delta \mathbf{x}$ is a **move direction** and λ a positive **step size**. Each $\Delta \mathbf{x}$ should be an **improving direction**; that is, it should produce immediate objective function improvement (definition 3.13). (It should also retain feasibility when constraints are present.) The process continues until a point is reached where no directions lead to such immediate improvement. There we stop (principle 3.17) with what is usually a **local optimum**—a point as good in objective value as any nearby (definition 3.5). [See Figure 3.8(a) for an exception that is not a local optimum even though it admits no improving direction.]

Figure 13.11 illustrates for our (minimizing) Custom Computers model (13.7). At initial point $\mathbf{x}^{(0)} = (32, -0.4)$ that search took a step of $\lambda = \frac{1}{2}$ in direction $\Delta\mathbf{x} = (2, -0.2)$ to produce

$$\mathbf{x}^{(1)} = \begin{pmatrix} 32 \\ -0.4 \end{pmatrix} + \tfrac{1}{2} \begin{pmatrix} 2 \\ -0.2 \end{pmatrix} = \begin{pmatrix} 33 \\ -0.5 \end{pmatrix}$$

Dashed lines in the figure, which show **contours** of the objective function plotted in Figure 13.3, demonstrate that the move is improving. Even very small steps from $\mathbf{x}^{(0)}$ in direction $\Delta\mathbf{x}$ advance the search to lower contours of the objective function.

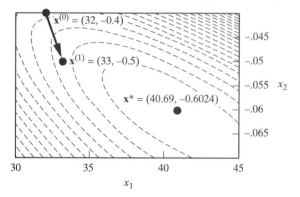

FIGURE 13.11   Improving Search of the Custom Computers Example

Local Information and Neighborhoods

What move direction should the search of Figure 13.11 adopt next? The best choice would make

$$\Delta\mathbf{x} = \mathbf{x}^* - \mathbf{x}^{(1)} = \begin{pmatrix} 40.69 \\ -0.6024 \end{pmatrix} - \begin{pmatrix} 33 \\ -0.5 \end{pmatrix} = \begin{pmatrix} 7.69 \\ -0.1024 \end{pmatrix}$$

which leads directly to the optimal solution.

Unfortunately, a search in progress does not have the global viewpoint available in Figure 13.11. The next move must be chosen using only experience with points already visited (here $\mathbf{x}^{(0)}$ and $\mathbf{x}^{(1)}$) plus local information about the shape of the objective function in the immediate **neighborhood** (definition $\boxed{3.4}$) of current $\mathbf{x}^{(1)}$.

First Derivatives and Gradients

We know from elementary calculus (see also Section 3.3 and Primer 2) that **first derivatives** or **gradients** provide information about how an objective function changes near a current solution $\mathbf{x}^{(t)}$.

$\boxed{13.14}$ The first derivative $f'(x)$ of a single-variable objective function $f(x)$, or the gradient vector $\nabla f(\mathbf{x})$ of first partial derivatives $\partial f/\partial x_1, \ldots, \partial f/\partial x_n$ with n variables, describes the slope or rate of change in f with small increments in current decision variable values.

For instance, at $\mathbf{x}^{(1)} = (33, -0.5)$ in minimizing Figure 13.11, we may apply partial derivative expressions (13.12) to compute

$$\frac{\partial f}{\partial x_1} \approx -23.07, \quad \frac{\partial f}{\partial x_2} \approx -174.23, \quad \text{so that} \quad \nabla f(\mathbf{x}^{(1)}) \approx (-23.07, -174.23)$$

Thus small increments from either $x_1 = 33$ or $x_2 = -0.5$ decrease $f(x_1, x_2)$, but the rate of change is much more rapid with increments in x_2.

Second Derivatives and Hessian Matrices

When an objective function is twice differentiable, which is typical for the smooth objectives most often occurring in applications, **second derivatives** can tell us still more about the shape of the function in the neighborhood of current solution $\mathbf{x}^{(t)}$. Primer 7 reviews some of the fundamentals.

The second derivative of a single-variable objective f is a scalar function $f''(x)$. For an n-variable objective, there is a whole **Hessian matrix** of second partial derivatives with row i, column j entry $\partial^2 f / \partial x_i \partial x_j$. For example, our Custom Computer objective [model (13.7)]

$$f(x_1, x_2) \triangleq \sum_{i=1}^{m} \left(q_i - x_1 p_i^{x_2}\right)^2$$

has first partial derivatives [expression (13.12)]

$$\frac{\partial f}{\partial x_1} = -2 \sum_{i=1}^{m} \left(q_i - x_1 p_i^{x_2}\right) p_i^{x_2}$$

$$\frac{\partial f}{\partial x_2} = -2 \sum_{i=1}^{m} \left(q_i - x_1 p_i^{x_2}\right) \left(x_1 p_i^{x_2}\right) \ln(p_i)$$

Thus second partials are

$$\frac{\partial^2 f}{\partial x_1^2} = 2 \sum_{i=1}^{m} p_i^{2x_2}$$

$$\frac{\partial^2 f}{\partial x_1 \partial x_2} = \frac{\partial^2 f}{\partial x_2 \partial x_1}$$

$$= -2 \sum_{i=1}^{m} \left[(q_i - x_1 p_i^{x_2})(p_i^{x_2}) \ln(p_i) - (p_i^{x_2})(x_1 p_i^{x_2}) \ln(p_i) \right]$$

$$\frac{\partial^2 f}{\partial x_2^2} = -2 \sum_{i=1}^{m} \ln^2(p_i) \left[(q_i - x_1 p_i^{x_2})(x_1 p_i^{x_2}) - (x_1 p_i^{x_2})^2 \right]$$

(13.15)

At the $\mathbf{x}^{(1)} = (33, -0.5)$ of Figure 13.11 constants p_i and q_i of Table 13.1 yield the Hessian matrix

$$\mathbf{H}(33, -0.5) \triangleq \begin{pmatrix} \dfrac{\partial^2 f}{\partial x_1^2} & \dfrac{\partial^2 f}{\partial x_1 \partial x_2} \\[2mm] \dfrac{\partial^2 f}{\partial x_2 \partial x_1} & \dfrac{\partial^2 f}{\partial x_2^2} \end{pmatrix} \approx \begin{pmatrix} 5.77 & 179.65 \\ 179.65 & 11{,}003.12 \end{pmatrix}$$

PRIMER 7: SECOND DERIVATIVES AND HESSIAN MATRICES

Primer 2 (Section 3.3) provides a brief overview of first derivatives df/dx [or $f'(x)$] and first partial derivatives $\partial f/\partial x_j$, which measure the rate of change function f with respect to increases in its arguments. The vector of partial derivatives for n-variable function $f(\mathbf{x})$ is its **gradient** $\nabla f(\mathbf{x})$.

First derivatives or partial derivatives of function f are themselves functions of its arguments. If such derivative functions are also differentiable, f is said to be **twice differentiable**, and we may determine **second derivatives**. Second derivatives describe the rate of change in slopes (i.e., the curvature of f).

Second derivatives of a single-variable $f(x)$ are customarily denoted d^2f/dx^2 or $f''(x)$. For example, $f(x) \triangleq 3x^4$ has first derivative $f'(x) = 12x^3$ and second $f''(x) = 36x^2$. At $x = 2$, $df/dx = 12(2)^3 = 96$, while $d^2f/dx^2 = 36(2)^2 = 144$.

Twice differentiable functions $f(\mathbf{x}) \triangleq f(x_1, \ldots, x_n)$ of n variables have **second partial derivatives** for each pair of variables x_i and x_j. Such second partial derivatives are customarily denoted $\partial^2 f/\partial x_i \partial x_j$ when $i \neq j$ and $\partial^2 f/\partial x_i^2$ if $i = j$. The order in which variables are listed indicates the sequence of differentiation. That is,

$$\frac{\partial^2 f}{\partial x_i \partial x_j} \triangleq \frac{\partial}{\partial x_j}\left(\frac{\partial f}{\partial x_i}\right)$$

To illustrate, consider $f(x_1, x_2) \triangleq 5x_1(x_2)^3$. First partial derivatives are $\partial f/\partial x_1 = 5(x_2)^3$ and $\partial f/\partial x_2 = 15x_1(x_2)^2$. Thus

$$\frac{\partial^2 f}{\partial x_1^2} = 0, \quad \frac{\partial^2 f}{\partial x_1 \partial x_2} = 15(x_2)^2, \quad \frac{\partial^2 f}{\partial x_2 \partial x_1} = 15(x_2)^2, \quad \frac{\partial^2 f}{\partial x_2^2} = 30x_1x_2$$

Notice in this example that $\partial^2 f/\partial x_1 \partial x_2 = \partial^2 f/\partial x_2 \partial x_1$. It is always true that

$$\frac{\partial^2 f}{\partial x_i \partial x_j} = \frac{\partial^2 f}{\partial x_j \partial x_i}$$

when f and all its first partial derivatives are continuous functions.

It is often convenient to deal with second partial derivatives in a **Hessian matrix** denoted $\mathbf{H}(\mathbf{x})$ and defined

$$\mathbf{H}(x_1, \cdots, x_n) \triangleq \begin{pmatrix} \dfrac{\partial^2 f}{\partial x_1^2} & \cdots & \dfrac{\partial^2 f}{\partial x_1 \partial x_n} \\ \vdots & \ddots & \vdots \\ \dfrac{\partial^2 f}{\partial x_n \partial x_1} & \cdots & \dfrac{\partial^2 f}{\partial x_n^2} \end{pmatrix}$$

For instance, at $x_1 = -3$, $x_2 = 2$, the $f(x_1, x_2)$ above has Hessian matrix

$$\mathbf{H}(-3, 2) = \begin{pmatrix} 0 & 15(2)^2 \\ 15(2)^2 & 30(-3)(2) \end{pmatrix} = \begin{pmatrix} 0 & 60 \\ 60 & -180 \end{pmatrix}$$

What second derivatives offer a search algorithm is information about the curvature of objective function f near current solution $\mathbf{x}^{(t)}$.

> **13.15** The second derivative $f''(x)$ of a single-variable objective function $f(x)$, or the Hessian matrix $\mathbf{H}(\mathbf{x})$ of second partial derivatives of $\partial^2 f / \partial x_i \, \partial x_j$ for n variables, describes the change in slope or curvature of f in the neighborhood of current decision variable values.

For example, the $\partial^2 f / \partial x_2{}^2 = 11{,}003.12$ confirms what we can see in Figure 13.11—that small changes in x_2 dramatically affect the slope of f near $\mathbf{x}^{(1)}$. The much smaller $\partial^2 f / \partial x_1{}^2 = 5.77$ indicates the function is flatter in the x_1 dimension.

Taylor Series Approximations with One Variable

A more concise description of what derivatives tell us about an objective function follows from classic **Taylor series**. For 1-dimensional function $f(x)$, Taylor's approximation represents the impact of a change λ from current $x^{(t)}$ as

$$f(x^{(t)} + \lambda) \approx f(x^{(t)}) + \frac{\lambda}{1!} f'(x^{(t)}) + \frac{\lambda^2}{2!} f''(x^{(t)}) + \frac{\lambda^3}{3!} f'''(x^{(t)}) + \cdots \qquad (13.16)$$

where $f'(x)$ is the first derivative of f, $f''(x)$ is the second derivative, and so on.

To illustrate, consider

$$f(x) \triangleq e^{3x-6} \qquad (13.17)$$

for which $f'(x) = 3e^{3x-6}$, $f''(x) = 9e^{3x-6}$, and $f'''(x) = 27e^{3x-6}$. Near $x^{(t)} = 2$, derivatives approximate the impact of a change λ as

$$f(2 + \lambda) \approx f(2) + \frac{\lambda}{1!} f'(2) + \frac{\lambda^2}{2!} f''(2) + \frac{\lambda^3}{3!} f'''(2) + \cdots$$

$$= 1 + 3\lambda + \tfrac{9}{2}\lambda^2 + \tfrac{27}{6}\lambda^3 + \cdots$$

Notice that as $|\lambda| \to 0$, higher powers of λ approach zero the most rapidly. That is why we may approximate a function with just the first few terms of expression (13.16) if our interest centers on the immediate neighborhood of current $x^{(t)}$. The results are the first-order or linear, and second-order or quadratic approximation to a function of a single variable.

> **13.16** The **first-order** or **linear**, and **second-order** or **quadratic Taylor series approximations** to single-variable function $f(x)$ near $x = x^{(t)}$ are, respectively,
>
> $$f_1(x^{(t)} + \lambda) \triangleq f(x^{(t)}) + \lambda f'(x^{(t)})$$
>
> and
>
> $$f_2(x^{(t)} + \lambda) \triangleq f(x^{(t)}) + \lambda f'(x^{(t)}) + \tfrac{1}{2}\lambda^2 f''(x^{(t)})$$
>
> where λ is the amount of change, f' is the first derivative of f, and f'' is the second.

Figure 13.12 illustrates for the $f(x) \triangleq e^{3x-6}$ of expression (13.17). Part (a) plots $f(x)$ and the first-order approximation for current $x^{(t)} = 2$

$$f_1(2 + \lambda) = f(2) + \lambda f'(2) = 1 + 3\lambda$$

and part (b) shows $f(x)$ versus second-order approximation

$$f_2(2 + \lambda) = f(2) + \lambda f'(2) + \tfrac{1}{2}\lambda^2 f''(2) = 1 + 3\lambda + \tfrac{9}{2}\lambda^2$$

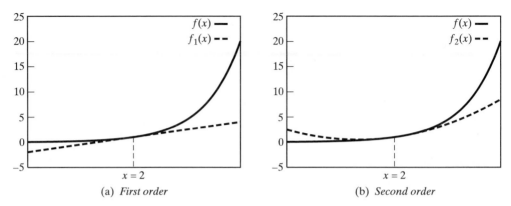

(a) *First order* (b) *Second order*

FIGURE 13.12   First- and Second-Order Taylor Series Approximations

Notice that the first-order approximation is a linear function of change λ. It assumes that the slope at $x^{(t)} = 2$ remains constant. Both approximations are fairly accurate near $\lambda = 0$ and deteriorate as λ becomes larger. Still, the second-order approximation in part (b) comes somewhat closer to the real f because it incorporates the curvature information in second derivative $f''(x^{(t)})$.

Taylor Series Approximations with Multiple Variables

We may extend Taylor series approximations to functions of more than one variable by using first and second partial derivatives.

> 13.17 │ The first-order or linear, and second-order or quadratic Taylor series approximations to n-variable function $f(\mathbf{x}) \triangleq f(x_1, \ldots, x_n)$ at point $\mathbf{x}^{(t)}$ are, respectively,
>
> $$f_1(\mathbf{x}^{(t)} + \lambda \Delta \mathbf{x}) \triangleq f(\mathbf{x}^{(t)}) + \lambda \nabla f(\mathbf{x}^{(t)}) \cdot \Delta \mathbf{x}$$
>
> $$\triangleq f(\mathbf{x}^{(t)}) + \lambda \sum_{j=1}^{n} \left(\frac{\partial f}{\partial x_j} \right) \Delta x_j$$
>
> and
>
> $$f_2(\mathbf{x}^{(t)} + \lambda \Delta \mathbf{x}) \triangleq f(\mathbf{x}^{(t)}) + \lambda \nabla f(\mathbf{x}^{(t)}) \cdot \Delta \mathbf{x} + \frac{\lambda^2}{2} \Delta \mathbf{x}\, \mathbf{H}(\mathbf{x}^{(t)}) \Delta \mathbf{x}$$
>
> $$\triangleq f(\mathbf{x}^{(t)}) + \lambda \sum_{j=1}^{n} \left(\frac{\partial f}{\partial x_j} \right) \Delta x_j + \frac{\lambda^2}{2} \sum_{i=1}^{n} \sum_{j=1}^{n} \left(\frac{\partial^2 f}{\partial x_i\, \partial x_j} \right) \Delta x_i\, \Delta x_j$$
>
> where $\Delta \mathbf{x} \triangleq (\Delta x_1, \ldots, \Delta x_n)$ is a direction of change, λ is the applied step size, $\nabla f(\mathbf{x}^{(t)})$ is the gradient of f at $\mathbf{x}^{(t)}$, and $\mathbf{H}(\mathbf{x}^{(t)})$ is the corresponding Hessian matrix.

As with 1-dimensional series (13.16), there are higher-order terms in the full Taylor series expansion of an n-variate function, but they become insignificant as $|\lambda| \rightarrow 0$.

To illustrate $\boxed{13.17}$, consider

$$f(x_1, x_2) \triangleq x_1 \ln(x_2) + 2$$

at $\mathbf{x}^{(t)} = (-3, 1)$. There $f(-3, 1) = 2$, and gradient

$$\nabla f(-3, 1) \triangleq \begin{pmatrix} \dfrac{\partial f}{\partial x_1} \\[2mm] \dfrac{\partial f}{\partial x_2} \end{pmatrix} = \begin{pmatrix} \ln(x_2) \\[2mm] \dfrac{x_1}{x_2} \end{pmatrix} = \begin{pmatrix} 0 \\ -3 \end{pmatrix}$$

Thus the first-order approximation to $f(x_1, x_2)$ near $\mathbf{x}^{(t)} = (-3, 1)$ in direction $\Delta\mathbf{x} \triangleq$ (Δx_1, Δx_2) is

$$\begin{aligned} f_1(\mathbf{x}^{(t)} + \lambda\Delta\mathbf{x}) &\triangleq f(\mathbf{x}^{(t)}) + \lambda \nabla f(\mathbf{x}^{(t)}) \cdot \Delta\mathbf{x} \\ &= 2 + \lambda(0, -3) \cdot (\Delta x_1, \ \Delta x_2) \\ &= 2 - 3\lambda \, \Delta x_2 \end{aligned}$$

To improve the approximation with second-order terms, we compute Hessian

$$\mathbf{H}(-3, 1) = \begin{pmatrix} 0 & \dfrac{1}{x_2} \\[2mm] \dfrac{1}{x_2} & \dfrac{-x_1}{(x_2)^2} \end{pmatrix} = \begin{pmatrix} 0 & 1 \\ 1 & 3 \end{pmatrix}$$

Then

$$\begin{aligned} f_2(\mathbf{x}^{(t)} + \lambda\Delta\mathbf{x}) &\triangleq f(\mathbf{x}^{(t)}) + \lambda \nabla f(\mathbf{x}^{(t)}) \cdot \Delta\mathbf{x} + \frac{\lambda^2}{2}\Delta\mathbf{x}\,\mathbf{H}(\mathbf{x}^{(t)})\Delta\mathbf{x} \\ &= 2 + \lambda(0, -3)\begin{pmatrix} \Delta x_1 \\ \Delta x_2 \end{pmatrix} + \frac{\lambda^2}{2}(\Delta x_1, \ \Delta x_2)\begin{pmatrix} 0 & 1 \\ 1 & 3 \end{pmatrix}\begin{pmatrix} \Delta x_1 \\ \Delta x_2 \end{pmatrix} \\ &= 2 - 3\lambda\,\Delta x_2 + \lambda^2\,\Delta x_1\,\Delta x_2 + \tfrac{3}{2}\lambda^2(\Delta x_2)^2 \end{aligned}$$

Stationary Points and Local Optima

First and second derivatives tell us a great deal about whether a solution is a local optimum. Begin with stationary points.

$\boxed{13.18}$ Solution $\mathbf{x}$ is a **stationary point** of smooth function f if $\nabla f(\mathbf{x}) = \mathbf{0}$.

That is, stationary points are solutions where all first (partial) derivatives equal zero.
   Figure 13.13 illustrates for

$$f(x_1, x_2) \triangleq 40 + (x_1)^3(x_1 - 4) + 3(x_2 - 5)^2 \tag{13.18}$$

Partial derivatives are

$$\frac{\partial f}{\partial x_1} = (x_1)^2(4x_1 - 12)$$

$$\frac{\partial f}{\partial x_2} = 6(x_2 - 5) \tag{13.19}$$

It is easy to check that they become zero at two stationary points:

$$\mathbf{x}^{(1)} = (3, 5) \quad \text{and} \quad \mathbf{x}^{(2)} = (0, 5) \tag{13.20}$$

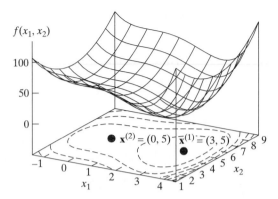

FIGURE 13.13 Stationary Points of a Minimize Objective

We can see in Figure 13.13 that one of these, $\mathbf{x}^{(1)}$, is a local (and here also global) minimum of f. This suggests our first (so called **first-order necessary**) condition for an unconstrained local optima.

> 13.19 | Every unconstrained local optimum of a smooth objective function must be a stationary point.

The reason that condition 13.19 must hold in every case is that a nonzero gradient $\nabla f(\mathbf{x}^{(t)})$ itself provides an improving direction at $\mathbf{x}^{(t)}$. Following principle 3.23 , we may adopt $\Delta \mathbf{x} = \pm \nabla f(\mathbf{x}^{(t)})$ with $+$ for maximize problems and $-$ to minimize. Then the first-order Taylor series approximation 13.17 gives

$$
\begin{aligned}
f(\mathbf{x}^{(t)} + \lambda \Delta \mathbf{x}) &\approx f(\mathbf{x}^{(t)}) + \lambda \nabla f(\mathbf{x}^{(t)}) \cdot \Delta \mathbf{x} \\
&= f(\mathbf{x}^{(t)}) \pm \lambda \nabla f(\mathbf{x}^{(t)}) \cdot \nabla f(\mathbf{x}^{(t)}) \\
&= f(\mathbf{x}^{(t)}) \pm \lambda \sum_{j=1}^{n} \left(\frac{\partial f}{\partial x_j} \right)^2
\end{aligned}
\tag{13.21}
$$

This is an improvement in the objective value unless all partial derivatives $= 0$, and we know that for λ sufficiently small the first-order part of the Taylor series expansion dominates all other terms.

SAMPLE EXERCISE 13.9: VERIFYING LOCAL OPTIMA AS STATIONARY POINTS

Consider the single-variable function

$$
f(x) = x^3 - 9x^2 + 24x - 14
$$

Plot the function for $1 \le x \le 5$ and verify that local maximum $x^{(1)} = 2$ and local minimum $x^{(2)} = 4$ are both stationary points.

Analysis: A plot of the function is as follows:

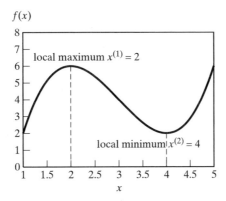

Its first derivative is

$$f'(x) = 3x^2 - 18x + 24$$

Both $f'(x^{(1)}) = f'(2) = 0$ and $f'(x^{(2)}) = f'(4) = 0$, confirming that both are stationary points.

Saddle Points

Look again at Figure 13.13. Stationary point $\mathbf{x}^{(1)} = (3, 5)$ is a local minimum, but $\mathbf{x}^{(2)} = (0, 5)$ is not. Increasing x_1 reduces the objective at the latter. Point $\mathbf{x}^{(2)}$ is also not a local maximum. Increasing x_2 makes the objective value bigger. Figure 13.14 shows that the remaining possibility is a saddle point.

> **13.20** A **saddle point** is a stationary point that is neither a local maximum nor a local minimum.

Every stationary point is either a local maximum, a local minimum, or a saddle point.

Saddle points get their name from the saddlelike possibility of the 2-dimensional case in Figure 13.14(c). The same stationary point is a local maximum in one dimension and a local minimum in another, yet neither a local maximum nor a local minimum when both directions are considered together.

Hessian Matrices and Local Optima

To distinguish better among the 3 types of stationary points in Figure 13.14 we must look at second (partial) derivatives. At stationary points, which have $\nabla f(\mathbf{x}^{(t)}) = \mathbf{0}$, second-order Taylor approximation $\boxed{13.17}$ simplifies as

$$f(\mathbf{x}^{(t)} + \lambda \Delta \mathbf{x}) \approx f(\mathbf{x}^{(t)}) + \lambda \nabla f(\mathbf{x}^{(t)}) \cdot \Delta \mathbf{x} + \frac{\lambda^2}{2} \Delta \mathbf{x}\, \mathbf{H}(\mathbf{x}^{(t)}) \Delta \mathbf{x}$$

$$= f(\mathbf{x}^{(t)}) + 0 + \frac{\lambda^2}{2} \Delta \mathbf{x}\, \mathbf{H}(\mathbf{x}^{(t)}) \Delta \mathbf{x}$$

(13.22)

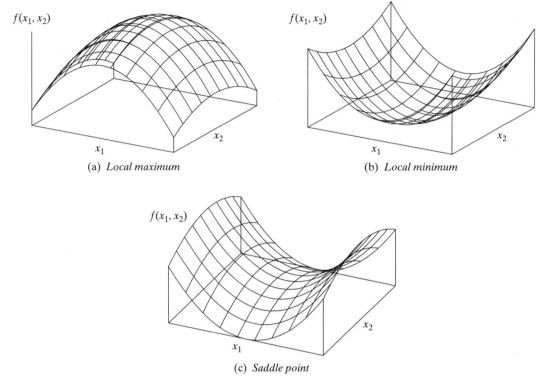

(a) *Local maximum*   (b) *Local minimum*

(c) *Saddle point*

FIGURE 13.14   Three Forms of Stationary Points

Thus (nonzero) Hessian-based quadratic forms $\Delta x H(x^{(t)})\Delta x$ critically influence whether improving directions Δx exist at stationary points (i.e., whether such points have any chance of being local optima).

Consider, for example, a direction Δx with $\Delta x H(x^{(t)})\Delta x < 0$ at stationary point $x^{(t)}$. Quadratic approximation (13.22) implies that

$$f(x^{(t)}) + \lambda \Delta x) \approx f(x^{(t)}) + \frac{\lambda^2}{2}\Delta x \, H(x^{(t)})\Delta x$$

$$< f(x^{(t)})$$

We may conclude that Δx is an improving direction for minimize problems at $x^{(t)}$ because moves in direction Δx strictly reduce the objective value if λ is small enough for this quadratic Taylor approximation to dominate higher-order terms. With a descent direction at hand, stationary point $x^{(t)}$ could not be a local minimum (principle $\boxed{3.16}$).

The ponderously named **positive** and **negative (semi)definite** properties of square matrices, which are reviewed briefly in Primer 8, address just such sign issues in quadratic forms. Combining with Taylor expression (13.22), we may use these properties to distinguish among stationary points. Semidefinite forms provide **second-order necessary optimality conditions**:

PRIMER 8: POSITIVE AND NEGATIVE (SEMI) DEFINITE MATRICES

Single-variable **quadratic form** $dad = ad^2$ is positive for all $d \neq 0$ if constant $a > 0$ and negative for all $d \neq 0$ if $a < 0$. In a similar way, whether n-variable quadratic form $\mathbf{dMd} \triangleq \sum_{i=1}^{n} \sum_{j=1}^{n} m_{i,j} d_i d_j$ is positive or negative depends on properties of the matrix $\mathbf{M}$.

Square matrix $\mathbf{M}$ is said to be **positive definite** if $\mathbf{dMd} > 0$ for all $\mathbf{d} \neq \mathbf{0}$, and **positive semidefinite** if $\mathbf{dMd} \geq 0$ for all $\mathbf{d}$. Similarly, $\mathbf{M}$ is **negative definite** if $\mathbf{dMd} < 0$ for all $\mathbf{d} \neq \mathbf{0}$, and **negative semidefinite** if $\mathbf{dMd} \leq 0$ for all $\mathbf{d}$.

To illustrate, consider

$$\mathbf{A} \triangleq \begin{pmatrix} 3 & 0 \\ 0 & 8 \end{pmatrix}, \quad \mathbf{B} \triangleq \begin{pmatrix} -1 & 2 \\ 2 & -4 \end{pmatrix}, \quad \mathbf{C} \triangleq \begin{pmatrix} 3 & 0 \\ 0 & -8 \end{pmatrix}$$

Matrix $\mathbf{A}$ is positive definite because

$$\mathbf{dAd} = 3(d_1)^2 + 8(d_2)^2$$

which is positive for every nonzero $\mathbf{d}$. Similarly, $\mathbf{B}$ is negative semi-definite because

$$\mathbf{dBd} = -(d_1)^2 + 4d_1 d_2 - 4(d_2)^2 = -(d_1 - 2d_2)^2 \leq 0$$

But with $\mathbf{dBd} = 0$ for $d_1 = 2d_2$, $\mathbf{B}$ is not negative definite. Matrix $\mathbf{C}$ is neither positive nor negative definite or semidefinite.

Obviously, $\mathbf{M}$ being positive (or negative) definite implies that $\mathbf{M}$ is positive (negative) semidefinite, so that example $\mathbf{A}$ is positive semidefinite. Conversely, a positive (or negative) semidefinite matrix that is also symmetric (Primer 6) and nonsingular (Primer 4) is positive (negative) definite. Thus example $\mathbf{A}$ being positive semidefinite, symmetric, and nonsingular proves that it is positive definite. Also, if $\mathbf{M}$ is positive (semi)definite, then $-\mathbf{M}$ is negative (semi)definite, and vice versa.

One way to test whether a symmetric matrix $\mathbf{M}$ satisfies any of these definitions is to check the determinants (Primer 4) of its **principal submatrices** [i.e., the submatrices made up of its first k rows and columns ($k = 1, \ldots, n$]. Symmetric matrix $\mathbf{M}$ is positive definite if all such principal determinants are positive, and positive semidefinite if they are all nonnegative. Similarly, symmetric $\mathbf{M}$ is negative definite if the principal determinants are nonzero and alternating in sign with the first negative; negative semidefinite allows zeros.

For example, one $\mathbf{D}$ and its principal submatrix determinants are

$$\mathbf{D} \triangleq \begin{pmatrix} 5 & -2 & 0 \\ -2 & 3 & 0 \\ 0 & 0 & 8 \end{pmatrix}, \quad \begin{aligned} \det(5) &= 5, \\ \det\begin{pmatrix} 5 & -2 \\ -2 & 3 \end{pmatrix} &= 11, \end{aligned} \quad \det\begin{pmatrix} 5 & -2 & 0 \\ -2 & 3 & 0 \\ 0 & 0 & 8 \end{pmatrix} = 88$$

Since all determinants are positive, $\mathbf{D}$ is positive definite. On the other hand, principal determinants of example $\mathbf{B}$ above are -1 and 0. Alternating signs with the first nonpositive confirm that $\mathbf{B}$ is negative semidefinite.

13.21   The Hessian matrix of a smooth function f is negative semidefinite at every unconstrained local maximum and positive semidefinite at every unconstrained local minimum.

The stronger definite forms give **sufficient conditions**:

13.22   A stationary point of a smooth function f is an unconstrained local maximum if the Hessian matrix at the point is negative definite. A stationary point is an unconstrained local minimum if the Hessian matrix is positive definite.

We may illustrate principles 13.21 and 13.22 by testing the two stationary points of the example in Figure 13.13 and function (13.18). Using first partial derivative expressions (13.19), the Hessian at $\mathbf{x}^{(1)}$ computes as

$$\mathbf{H}(\mathbf{x}^{(1)}) = \mathbf{H}(3, 5) = \begin{pmatrix} 12(x_1)^2 - 24x_1 & 0 \\ 0 & 6 \end{pmatrix} = \begin{pmatrix} 24 & 0 \\ 0 & 6 \end{pmatrix}$$

The matrix is positive definite because

$$\mathbf{\Delta x} \begin{pmatrix} 24 & 0 \\ 0 & 6 \end{pmatrix} \mathbf{\Delta x} = 24(\Delta x_1)^2 + 6(\Delta x_2)^2 > 0 \qquad \text{for all } \mathbf{\Delta x} \neq \mathbf{0}$$

Confirming principal 13.21 , local minimum $\mathbf{x}^{(1)}$ has a positive definite, and thus positive semidefinite Hessian. Conversely, we can establish that stationary point $\mathbf{x}^{(1)}$ is a local minimum by applying principle 13.22 with the Hessian positive definite.

The second stationary point $\mathbf{x}^{(2)} = (0, 5)$ shows that properties 13.21 and 13.22 are not always conclusive. There

$$\mathbf{H}(\mathbf{x}^{(2)}) = \mathbf{H}(0, 5) = \begin{pmatrix} 12(x_1)^2 - 24x_1 & 0 \\ 0 & 6 \end{pmatrix} = \begin{pmatrix} 0 & 0 \\ 0 & 6 \end{pmatrix}$$

and quadratic form

$$\mathbf{\Delta x} \begin{pmatrix} 0 & 0 \\ 0 & 6 \end{pmatrix} \mathbf{\Delta x} = 6(\Delta x_2)^2 \geq 0$$

We can apply principle 13.21 to rule out the possibility of a local maximum because this Hessian is not negative semidefinite. Still, principle 13.22 cannot be applied to assure a local minimum with the Hessian only positive semidefinite. Without extending to third derivatives, we cannot distinguish between a local minimum and a saddle point.

SAMPLE EXERCISE 13.10: VERIFYING LOCAL OPTIMA

Verify that function

$$f(x_1, x_2, x_3) \triangleq (x_1)^2 + x_1 x_2 + 5(x_2)^2 + 9(x_3 - 2)^2$$

has a local minimum at $\mathbf{x} = (0, 0, 2)$.

Analysis: We apply sufficient conditions $\boxed{13.22}$. First, **x** must be a stationary point. All three partial derivatives

$$\frac{\partial f}{\partial x_1} = 2x_1 + x_2, \quad \frac{\partial f}{\partial x_2} = x_1 + 10x_2, \quad \frac{\partial f}{\partial x_3} = 18(x_3 - 2)$$

$= 0$ at $\mathbf{x} = (0, 0, 2)$ as required.

Next we consider the Hessian

$$\mathbf{H}(\mathbf{x}) = \begin{pmatrix} 2 & 1 & 0 \\ 1 & 10 & 0 \\ 0 & 0 & 18 \end{pmatrix}$$

We may verify that this matrix is positive definite, and thus **x** a local minimum, by checking that all determinants of principal submatrices are positive (refer to Primer 8 if needed):

$$\det(2) = 2 > 0, \quad \det \begin{pmatrix} 2 & 1 \\ 1 & 10 \end{pmatrix} = 18 > 0, \quad \det \begin{pmatrix} 2 & 1 & 0 \\ 1 & 10 & 0 \\ 0 & 0 & 18 \end{pmatrix} = 324 > 0$$

SAMPLE EXERCISE 13.11: VERIFYING SADDLE POINTS

Verify that function

$$f(x_1, x_2) \triangleq (x_1)^2 - 2x_1 - (x_2)^2$$

has a saddle point at $\mathbf{x} = (1, 0)$.

Analysis: To fulfill definition $\boxed{13.20}$, a saddle point must first be a stationary point. Checking yields

$$\frac{\partial f}{\partial x_1} = 2x_1 - 2 = 2(1) - 2 = 0 \quad \text{and} \quad \frac{\partial f}{\partial x_2} = -2x_2 = -2(0) = 0$$

Now computing the Hessian gives

$$\mathbf{H}(1, 0) = \begin{pmatrix} 2 & 0 \\ 0 & -2 \end{pmatrix}$$

With first principal determinant $= 2$, and second $= -4$, this matrix is neither positive semidefinite nor negative semidefinite. Thus **x** violates requirements $\boxed{13.21}$ for both a local minimum and a local maximum. The remaining possibility is a saddle point.

13.4 CONVEX/CONCAVE FUNCTIONS AND GLOBAL OPTIMALITY

Improving search Algorithm 3A (Section 3.2), which provides the paradigm for nearly all unconstrained nonlinear programming algorithms, stops if it encounters a locally optimal solution (principle $\boxed{3.6}$). What then? We would clearly prefer an overall, or **global optimum**.

In this section we investigate objective functions having special **convex**, **concave**, and **unimodal** forms that allow us to prove that a local optimum must also be global (see also Section 3.4). With other objectives we must either accept the improving search stopping point or try for another by restarting the search from a different initial $\mathbf{x}^{(0)}$.

Convex and Concave Functions Defined

Convex and concave functions can be defined in terms of how $f(\mathbf{x})$ changes as we move from $\mathbf{x}^{(1)}$ to $\mathbf{x}^{(2)}$ along straight-line path $\Delta\mathbf{x} \triangleq (\mathbf{x}^{(2)} - \mathbf{x}^{(1)})$.

| 13.23 | A function $f(\mathbf{x})$ is **convex** if |

$$f(\mathbf{x}^{(1)} + \lambda(\mathbf{x}^{(2)} - \mathbf{x}^{(1)})) \leq f(\mathbf{x}^{(1)}) + \lambda\left(f(\mathbf{x}^{(2)}) - f(\mathbf{x}^{(1)})\right)$$

for every $\mathbf{x}^{(1)}$ and $\mathbf{x}^{(2)}$ in its domain and every step $\lambda \in [0, 1]$. Similarly, $f(\mathbf{x})$ is **concave** if

$$f(\mathbf{x}^{(1)} + \lambda(\mathbf{x}^{(2)} - \mathbf{x}^{(1)})) \geq f(\mathbf{x}^{(1)}) + \lambda\left(f(\mathbf{x}^{(2)}) - f(\mathbf{x}^{(1)})\right)$$

for all $\mathbf{x}^{(1)}$, $\mathbf{x}^{(2)}$ and $\lambda \in [0, 1]$.

Interpolation of f values along the line segment from $\mathbf{x}^{(1)}$ to $\mathbf{x}^{(2)}$ should neither underestimate for a convex function nor overestimate for a concave one.

Figure 13.15 illustrates for functions of 2-vectors $\mathbf{x} \triangleq (x_1, x_2)$. The indicated moves start at $\mathbf{x}^{(1)}$ and advance toward $\mathbf{x}^{(2)}$ along direction $(\mathbf{x}^{(2)} - \mathbf{x}^{(1)})$. Each point

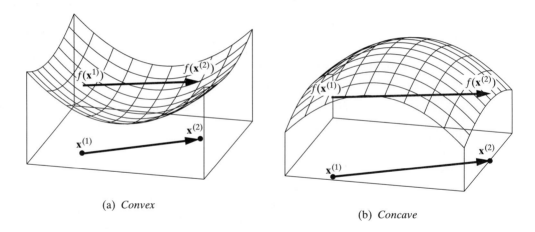

(a) *Convex*

(b) *Concave*

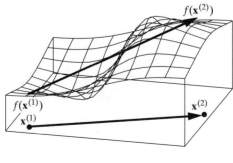

(c) *Neither*

FIGURE 13.15   Convex and Concave Functions

x in that trajectory has a representation

$$\mathbf{x} = \mathbf{x}^{(1)} + \lambda(\mathbf{x}^{(2)} - \mathbf{x}^{(1)})$$

for some $\lambda \in [0, 1]$ (property 3.31). For example, $\mathbf{x}^{(1)}$ corresponds to $\lambda = 0$, and $\mathbf{x}^{(2)}$ to $\lambda = 1$.

The issue in definition 13.23 is what happens when we interpolate an estimated value for f somewhere along the trajectory. For convex functions [Figure 13.15(a)] the corresponding interpolated values

$$f(\mathbf{x}^{(1)}) + \lambda\left(f(\mathbf{x}^{(2)}) - f(\mathbf{x}^{(1)})\right)$$

should always equal or exceed the true $f(\mathbf{x}^{(1)} + \lambda(\mathbf{x}^{(2)} - \mathbf{x}^{(1)}))$. For concave functions [part (b)] it should fall equal or below.

The property must hold for every pair of points $\mathbf{x}^{(1)}$ and $\mathbf{x}^{(2)}$. For example, some pairs would meet the test for convexity in Figure 13.15(c), and others would satisfy the definition of concave. Still, the function is neither convex nor concave because the indicated pair violates both definitions.

SAMPLE EXERCISE 13.12: RECOGNIZING CONVEX AND CONCAVE FUNCTIONS

Determine graphically whether each of the following single-variable functions are convex, concave, or neither over $x \in [0, 5]$.

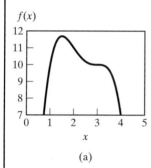

(a)

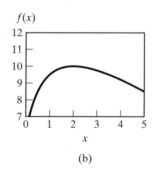

(b)

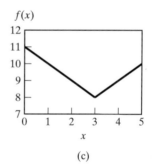

(c)

Analysis: We apply definition 13.23 .

(a) This function is neither convex nor concave. To demonstrate that it is not convex, take $x^{(1)} = 1$, $x^{(2)} = 2$, and $\lambda = \frac{1}{2}$.

$$f(x^{(1)} + \lambda(x^{(2)} - x^{(1)})) = f(1 + \tfrac{1}{2}(2 - 1)) = f(1.5) \approx 11.7$$

$$\nleq f(x^{(1)}) + \lambda\left(f(x^{(2)}) - f(x^{(1)})\right) = 10 + \tfrac{1}{2}(11 - 10) = 10.5$$

Similarly choosing $x^{(1)} = 3$, $x^{(2)} = 2$ and $\lambda = \frac{1}{2}$ establishes that the function is not concave because

$$f(x^{(1)} + \lambda(x^{(2)} - x^{(1)})) = f(3 + \tfrac{1}{2}(2 - 3)) = f(2.5) \approx 10.2$$

$$\ngeq f(x^{(1)}) + \lambda\left(f(x^{(2)}) - f(x^{(1)})\right) = 10 + \tfrac{1}{2}(11 - 10) = 10.5$$

(b) This function is apparently concave because definition $\boxed{13.23}$ holds for all pairs of points displayed.

(c) This function is apparently convex because definition $\boxed{13.23}$ holds for all pairs of points displayed. Notice that convex (and concave) functions need not be differentiable.

Sufficient Conditions for Unconstrained Global Optima

The importance of convex and concave objective functions lies with their unusual tractability for improving search.

> $\boxed{13.24}$ If $f(\mathbf{x})$ is a convex function, every unconstrained local minimum of f is an unconstrained global minimum. If $f(\mathbf{x})$ is concave, every unconstrained local maximum is an unconstrained global maximum.

That is, a search must only achieve a local optimum to produce with no additional effort a global minimum of a convex objective function or a global maximum of a concave one.

To see why principle $\boxed{13.24}$ must be true, consider a convex objective function $f(x)$, a global minimum x^*, and any $x^{(1)}$ that is not globally optimal:

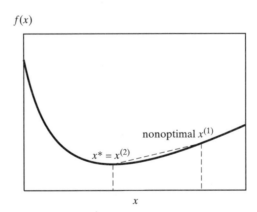

Then

$$f(x^*) < f(x^{(1)}) \quad \text{or} \quad \lambda \left(f(x^*) - f(x^{(1)}) \right) < 0 \tag{13.23}$$

for all $\lambda > 0$. Combining with the convexity definition $\boxed{13.23}$ yields

$$f(x^{(1)} + \lambda(x^* - x^{(1)})) \leq f(x^{(1)}) + \lambda \left(f(x^*) - f(x^{(1)}) \right) < f(x^{(1)}) \tag{13.24}$$

for all $\lambda \in (0, 1]$. That is, direction $\Delta x = x^* - x^{(1)}$ is an improving direction at every $x^{(1)}$ that is not globally optimal. A local optimum, which permits no improving directions, can exist only if it is also a global optimum.

Convex/Concave Functions and Stationary Points

Property 13.24 shows that we need only compute a local minimum of a convex function or a local maximum of a concave one to obtain an unconstrained global optimum. In fact, the requirement is even weaker when the objective function is differentiable.

> 13.25   Every stationary point of a smooth convex function is an unconstrained global minimum, and every stationary point of a smooth concave function is an unconstrained global maximum.

We require only an $\mathbf{x}$ with $\nabla f(\mathbf{x}) = \mathbf{0}$.

For an idea of why principle 13.25 holds, let f be a smooth convex function, and $\nabla f(\mathbf{x}^*) = \mathbf{0}$. Convexity definition 13.23 assures that

$$f(\mathbf{x}^* + \lambda(\mathbf{x} - \mathbf{x}^*)) \le f(\mathbf{x}^*) + \lambda(f(\mathbf{x}) - f(\mathbf{x}^*))$$

for any $\mathbf{x}$ and any $\lambda \in (0, 1]$. Furthermore, first-order Taylor approximation 13.16 gives

$$f(\mathbf{x}^* + \lambda(\mathbf{x} - \mathbf{x}^*)) \approx f(\mathbf{x}^*) + \lambda \nabla f(\mathbf{x}^*)(\mathbf{x} - \mathbf{x}^*)$$

Subtracting, simplifying, and dividing by $\lambda \to 0$ yields

$$f(\mathbf{x}) - f(\mathbf{x}^*) \ge \nabla f(\mathbf{x}^*)(\mathbf{x} - \mathbf{x}^*)$$

It follows that $\mathbf{x}^*$ is a global minimum when $\nabla f(\mathbf{x}^*) = \mathbf{0}$ because $f(\mathbf{x}) - f(\mathbf{x}^*) \ge 0$ for all $\mathbf{x}$.

SAMPLE EXERCISE 13.13: VERIFYING GLOBAL OPTIMA WITH CONVEXITY

The function

$$f(x) \triangleq 20 - x^2 + 6x$$

is concave. Use this fact to establish that it has an unconstrained global maximum at $x = 3$.

Analysis: Differentiating yields

$$f'(x) = -2x + 6$$

so that $f'(3) = 0$ Being a stationary point of a concave function, $x = 3$ must be an unconstrained global maximum (principle 13.25).

Tests for Convex and Concave Functions

Many familiar functions are either convex or concave, but it is often tedious to verify definitions 13.23 . Fortunately, when the function's domain is all real n-vectors or all positive n-vectors, or any other open convex set (definition 3.29), some important properties are available to simplify the analysis:

> 13.26   If $f(\mathbf{x})$ is convex, $-f(\mathbf{x})$ is concave, and vice versa.

13.27 An $f(\mathbf{x})$ with continuous second (partial) derivatives is convex if and only if Hessian $\mathbf{H}(\mathbf{x})$ is positive semidefinite at all $\mathbf{x}$ in its (open convex set) domain. It is concave if and only if $\mathbf{H}(\mathbf{x})$ is negative semidefinite at all $\mathbf{x}$ in the domain.

13.28 Linear functions are both convex and concave.

13.29 Any $f(\mathbf{x})$ formed as the nonnegative-weighted ($\alpha_i \geq 0$) sum

$$f(\mathbf{x}) \triangleq \sum_{i=1}^{k} \alpha_i g_i(\mathbf{x})$$

of convex functions $g_i(\mathbf{x})$, $i = 1, \ldots, k$, is itself convex. The nonnegative-weighted sum of concave functions is concave.

13.30 Any $f(\mathbf{x})$ formed as the maximum

$$f(\mathbf{x}) \triangleq \max\{g_i(\mathbf{x}) : i = 1, \ldots, k\}$$

of convex functions $g_i(\mathbf{x})$, $i = 1, \ldots, k$, is itself convex. The minimum of concave functions is concave.

13.31 If $g(y)$ is a nondecreasing, single-variable convex function, and $h(\mathbf{x})$ is convex, $f(\mathbf{x}) \triangleq g(h(\mathbf{x}))$ is convex. If $g(y)$ is a nondecreasing, single-variable concave function, and $h(\mathbf{x})$ is concave, $f(\mathbf{x}) \triangleq g(h(\mathbf{x}))$ is concave.

13.32 If $g(\mathbf{x})$ is a concave function, $f(\mathbf{x}) \triangleq 1/g(\mathbf{x})$ is convex over $\mathbf{x}$ with $g(\mathbf{x}) > 0$. If $g(\mathbf{x})$ is a convex function, $f(\mathbf{x}) \triangleq 1/g(\mathbf{x})$ is concave over $\mathbf{x}$ with $g(\mathbf{x}) < 0$.

To see the power of rules 13.26 to 13.32 , examine the curve-fitting objective in linear regression form (13.4):

$$\min \; f(x_1, x_2) \triangleq \sum_{i=1}^{m} [q_i - (x_1 + x_2 p_i)]^2$$

[The nonlinear case with $(q_i - x_1(p_i)^{x_2}$ is not convex.] Recall that the p_i and q_i are given constants.

To show this linear regression f is convex, notice first that it is the (unweighted) sum of functions

$$g_i(x_1, x_2) \triangleq [q_i - (x_1 + x_2 p_i)]^2$$

Under rule 13.29 , f will be convex if each of the g_i is convex.

Now, dropping the i subscripts, we examine

$$g(x_1, x_2) \triangleq [q - (x_1 + x_2 p)]^2$$
$$= [|q - (x_1 + x_2 p|)]^2$$
$$= [\max\{(q - x_1 - x_2 p), -(q - x_1 - x_2 p)\}]^2$$

(the last equality holds because $|z| = \max\{z, -z\}$). Expressions $(q - x_1 - x_2p)$ and $-(q - x_1 - x_2p)$ are both linear and thus convex by property $\boxed{13.28}$. Therefore,

$$h(x_1, x_2) \triangleq |q - (x_1 + x_2p)| = \max\{q - x_1 - x_2p, -(q - x_1 - x_2p)\}$$

is also convex; it is the maximum of convex functions (rule $\boxed{13.30}$). Finally, consider $s(y) \triangleq y^2$. Second derivative $s''(y) = 2$ proves $s(y)$ is convex because $s''(y)$ is the 1 by 1 Hessian matrix and positive definite (property $\boxed{13.27}$). Over domain $y \geq 0, s(y) \triangleq y^2$ is also nondecreasing. Thus we may apply composition rule $\boxed{13.31}$ to conclude that

$$g(x_1, x_2) \triangleq (q - (x_1 + x_2p))^2 = s(h(x_1, x_2))$$

is convex. This completes the argument for convexity f.

SAMPLE EXERCISE 13.14: VERIFYING CONVEXITY AND CONCAVITY

Apply properties $\boxed{13.26}$ to $\boxed{13.32}$ to establish that the first two of the following functions are convex and the last two are concave over the specified domains.

(a) $f(x_1, x_2) \triangleq (x_1 + 1)^4 + x_1x_2 + (x_2 + 1)^4$ over all $x_1, x_2 > 0$

(b) $f(x_1, x_2) \triangleq e^{-3x_1 + x_2}$ over all x_1, x_2

(c) $f(x_1, x_2, x_3) \triangleq -4(x_1)^2 + 5x_1x_2 - 2(x_2)^2 + 18x_3$ over all x_1, x_2

(d) $f(x_1, x_2) \triangleq \dfrac{1}{-7x_1} - e^{-3x_1 + x_2}$ over all $x_1, x_2 > 0$

Analysis:

(a) Here the Hessian matrix is

$$\mathbf{H}(x_1, x_2) = \begin{pmatrix} 12(x_1 + 1)^2 & 1 \\ 1 & 12(x_2 + 1)^2 \end{pmatrix}$$

Determinants of its principal submatrices are $12(x_1+1)^2$ and $144(x_1+1)^2(x_2+1)^2 - 1$, which are both positive for all $x_1, x_2 > 0$. Thus the Hessian is positive definite, and f is convex by rule $\boxed{13.27}$.

(b) Function $h(x_1, x_2) \triangleq -3x_1 + x_2$ is convex because it is linear (property $\boxed{13.28}$). Also, $g(y) \triangleq e^y$ in nondecreasing and convex because $g''(y) = e^y > 0$. Thus composition rule $\boxed{13.31}$ proves that $f(x_1, x_2) = g(h(x_1, x_2))$ is convex.

(c) For this function the Hessian matrix is

$$\mathbf{H}(x_1, x_2, x_3) = \begin{pmatrix} -8 & 5 & 0 \\ 5 & -4 & 0 \\ 0 & 0 & 0 \end{pmatrix}$$

Principal submatrix determinants are -8, $(32 - 25) = 7$, and 0, which imply that the Hessian is negative semidefinite and f is concave (rule $\boxed{13.27}$).

(d) Over $x_1, x_2 > 0$, first term $g_1(x_1, x_2) \triangleq 1/(-7x_1)$ is the reciprocal of negative-valued, linear, and thus convex function $h(x_1, x_2) \triangleq -7x_1$. It follows that this $g_1(x_1, x_2)$

is concave (property $\boxed{13.32}$). Part (b) already established that $g_2(x_1, x_2) \triangleq e^{-3x_1+x_2}$ is convex, meaning (rule $\boxed{13.26}$) that its negative is concave. Thus f is the sum of concave functions and so concave (property $\boxed{13.29}$).

Unimodal versus Convex/Concave Objectives

In Section 3.4 we introduced the notion of unimodal objective functions (definition $\boxed{3.26}$). A function $f(\mathbf{x})$ is **unimodal** if $f(\mathbf{x}^{(2)})$ superior to $f(\mathbf{x}^{(1)})$ (greater for a maximize, lesser for a minimize) implies that direct move $\Delta\mathbf{x} = \mathbf{x}^{(2)} - \mathbf{x}^{(1)}$ is an improving direction at $\mathbf{x}^{(1)}$. That is, heading straight toward any better point should produce immediate gain. It follows (principle $\boxed{3.28}$) that every unconstrained local optimum of a unimodal objective function is a global optimum because improving directions exist at every point that can be bettered.

Since both unimodal and convex/concave objective functions imply that unconstrained local optima are global (principles $\boxed{3.28}$ and $\boxed{13.24}$, respectively), it should no surprise that there is a connection.

$\boxed{13.33}$ Both convex objective functions in minimize problems and concave objective functions in maximize problems are unimodal.

Expressions (13.23) and (13.24) have already shown why. Improving directions exist at all solutions not globally optimal in a convex minimization or concave maximization.

Unimodality is a weaker requirement than convexity or concavity.

$\boxed{13.34}$ A unimodal objective function need not be either convex or concave.

For example, the following is unimodal for a maximize problem, but we showed in Sample Exercise 13.12(a) that it is not concave.

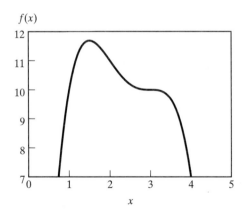

Other such examples are our Custom Computer objective in Figure 13.3 and PERT example in Figure 13.5.

Unfortunately, convenient combination rules such as 13.26 – 13.32 do not generally hold for arbitrary unimodal objectives. Thus in practice we must often establish the more restrictive convex or concave properties to be sure that an objective is unimodal. When the functions are not concave for a maximize or convex for a minimize, which often happens in applied models, we usually must accept the risk of local optima that are not global.

13.5 GRADIENT SEARCH

In Section 3.3, principle 3.23 established that a nonzero gradient $\nabla f(\mathbf{x}^{(t)})$ provides an improving direction at solution $\mathbf{x}^{(t)}$. First-order Taylor series computations of expression (13.21) guarantee improvement with sufficiently small steps in direction $\Delta \mathbf{x} = \nabla f(\mathbf{x}^{(t)})$ for a maximize problem or $\Delta \mathbf{x} = -\nabla f(\mathbf{x}^{(t)})$ for a minimize. In this section we develop the simple **gradient search** algorithm that adopts such gradient-based move directions.

Gradient Search Algorithm

Algorithm 13D provides details. The move direction for each iteration is derived from the gradient at the current point.

13.35   At any current point $\mathbf{x}^{(t)}$ with gradient $\nabla f(\mathbf{x}^{(t)}) \neq \mathbf{0}$, gradient search pursues move direction

$$\Delta \mathbf{x} \triangleq \pm \nabla f(\mathbf{x}^{(t)})$$

(+ for a maximize, − for a minimize).

ALGORITHM 13D: GRADIENT SEARCH

Step 0: Initialization. Choose any starting solution $\mathbf{x}^{(0)}$, pick stopping tolerance $\epsilon > 0$, and set solution index $t \leftarrow 0$.

Step 1: Gradient. Compute objective function gradient $\nabla f(\mathbf{x}^{(t)})$ at current point $\mathbf{x}^{(t)}$.

Step 2: Stationary Point. If gradient norm $\|\nabla f(\mathbf{x}^{(t)})\| < \epsilon$, stop. Point $\mathbf{x}^{(t)}$ is sufficiently close to a stationary point.

Step 3: Direction. Choose gradient move direction
$$\Delta \mathbf{x}^{(t+1)} \leftarrow \pm \nabla f(\mathbf{x}^{(t)})$$
(+ for maximize and − for minimize).

Step 4: Line Search. Solve (at least approximately) corresponding one-dimensional line search
$$\max \text{ or } \min f(\mathbf{x}^{(t)} + \lambda \Delta \mathbf{x}^{(t+1)})$$
to compute λ_{t+1}.

Step 5: New Point. Update
$$\mathbf{x}^{(t+1)} \leftarrow \mathbf{x}^{(t)} + \lambda_{t+1} \Delta \mathbf{x}^{(t+1)}$$

Step 6: Advance. Increment $t \leftarrow t + 1$, and return to Step 1.

Gradient norm

$$\|\nabla f(\mathbf{x}^{(t)})\| \triangleq \sqrt{\sum_j \left(\frac{\partial f}{\partial x_j}\right)^2}$$

provides a stopping rule at Step 1 of Algorithm 13D. If the gradient at $\mathbf{x}^{(t)}$ is very small in length (less than stopping tolerance ϵ), all its components must be very nearly zero. Thus the search has essentially reached a stationary point (definition $\boxed{13.18}$).

Of course, we know from Section 13.3 that a stationary point may be a saddle point (definition $\boxed{13.20}$), not the local optimum we seek. Still, a pure gradient algorithm can guarantee no more. When $\nabla f(\mathbf{x}^{(t)}) = \mathbf{0}$, construction $\boxed{13.35}$ provides no move direction to pursue.

Gradient Search of Custom Computer Example

Table 13.5 details application of Algorithm 13D to the Custom Computer regression model (13.7) (Section 13.1). Figure 13.16 plots the first few steps on a contour map.

TABLE 13.5 Gradient Search of Custom Computer Example

t	$\mathbf{x}^{(t)}$	$f(\mathbf{x}^{(t)})$	$\nabla f(\mathbf{x}^{(t)})$	$\|\nabla f(\mathbf{x}^{(t)})\|$	λ_{t+1}
0	(32.00, −0.4000)	174.746	(−6.24, 1053.37)	1053.39	0.00007
1	(32.00, −0.4687)	141.138	(−23.06, −0.14)	23.06	0.10558
2	(34.44, −0.4540)	112.599	(−4.57, 759.60)	759.61	0.00007
3	(34.44, −0.5078)	93.297	(−16.28, −0.10)	16.28	0.11303
4	(36.28, −0.4962)	78.123	(−3.34, 530.01)	530.02	0.00008
5	(36.28, −0.5365)	67.897	(−11.34, −0.07)	11.34	0.11970
6	(37.63, −0.5278)	60.133	(−2.33, 365.74)	365.75	0.00008
7	(37.63, −0.5571)	54.932	(−7.78, −0.05)	7.78	0.12905
8	(38.64, −0.5512)	51.006	(−1.48, 251.17)	251.17	0.00008
9	(38.64, −0.5722)	48.428	(−5.19, −0.03)	5.19	0.13926
10	(39.36, −0.5684)	46.548	(0.88, 168.22)	168.22	0.00009
11	(39.36, −0.5829)	45.348	(−3.34, −0.02)	3.34	0.14737
12	(39.85, −0.5806)	44.522	(−0.51, 108.66)	108.66	0.00009
13	(39.85, −0.5902)	44.007	(−2.10, −0.01)	2.10	0.15220
14	(40.17, −0.5888)	43.672	(−0.30, 68.07)	68.07	0.00009
15	(40.17, −0.5949)	43.466	(−1.29, 0.00)	1.29	0.16435
16	(40.38, −0.5941)	43.329	(−0.14, 42. 34)	42.34	0.00009
17	(40.38, −0.5980)	43.248	(−0.76, 0.00)	0.76	0.15652
18	(40.50, −0.5975)	43.203	(−0.10, 24.60)	24.60	0.00009
19	(40.50, −0.5997)	43.175	(−0.46, 0.00)	0.46	0.14805
20	(40.57, −0.5994)	43.160	(−0.08, 14.77)	14.77	0.00009
21	(40.57, −0.6007)	43.150	(−0.29, 0.00)	0.29	0.15527
22	(40.62, −0.6005)	43.143	(−0.04, 9.37)	9.37	0.00009
23	(40.62, −0.6014)	43.139	(−0.18, 0.00)	0.18	0.16376
24	(40.65, −0.6013)	43.137	(−0.02, 5.78)	5.78	0.00009
25	(40.65, −0.6018)	43.135	(−0.11, 0.00)	0.11	0.15266
26	(40.66, −0.6017)	43.134	(−0.02, 3.37)	3.37	0.00009
27	(40.66, −0.6020)	43.134	(−0.06, 0.00)	0.06	Stop

The search begins at $\mathbf{x}^{(0)} = (32, -0.4)$. There the move direction (construction $\boxed{13.35}$) is

$$\Delta\mathbf{x} = -\nabla f(\mathbf{x}^{(0)}) = -(-6.24, 1053.37) = (6.24, -1053.37)$$

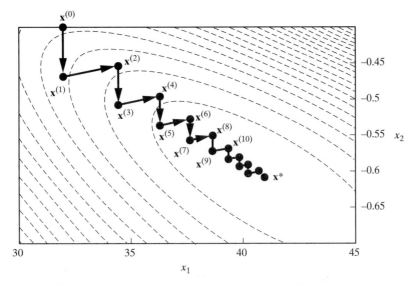

FIGURE 13.16 Gradient Search of Custom Computer Example

This move direction defines the first line search for step size λ. We identify the largest step for which $\Delta\mathbf{x}$ continues to improve by solving the 1-dimensional problem

$$\min f(\mathbf{x}^{(0)} + \lambda\Delta\mathbf{x}) \triangleq f(32 + 6.24\lambda, -0.4 - 1053.37\lambda)$$

Plotting shows that a minimum occurs at approximately $\lambda_1 = 0.00007$.

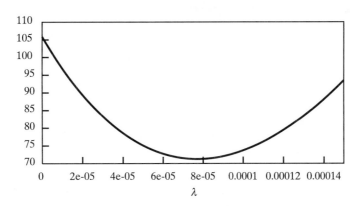

The result is new point

$$\mathbf{x}^{(1)} \leftarrow \mathbf{x}^{(0)} + \lambda_1\Delta\mathbf{x} \approx (32, -0.4687)$$

Computations in Table 13.5 employ stopping tolerance $\epsilon = 0.1$. Thus the algorithm continues until norm $||\nabla f(\mathbf{x}^{(t)})|| < 0.1$ at $t = 27$. The resulting (approximate) stationary point is $\mathbf{x}^{(27)} = (40.66, -0.6020)$ which we know from earlier analysis approximates a local (and here also global) minimum.

SAMPLE EXERCISE 13.15: EXECUTING GRADIENT SEARCH

Consider the unconstrained nonlinear program

$$\max \quad f(x_1, x_2) \triangleq \frac{x_1}{1 + e^{0.1x_1}} - (x_2 - 5)^2$$

(a) Compute the move direction that would be pursued by gradient search Algorithm 13D at $\mathbf{x}^{(0)} = (30, 2)$.

(b) State the line search problem implied by your direction of part (a).

Analysis:

(a) At the specified $\mathbf{x}^{(0)}$, the gradient is

$$\nabla f(30, 2) = \begin{pmatrix} \dfrac{1 + e^{0.1x_1} - 0.1x_1 e^{0.1x_1}}{(1 + e^{0.1x_1})^2} \\ -2(x_2 - 5) \end{pmatrix} = \begin{pmatrix} -0.088 \\ 6 \end{pmatrix}$$

Thus with a maximize problem we use direction

$$\Delta \mathbf{x} = +\nabla f(\mathbf{x}^{(0)}) = (-0.088, 6)$$

(b) The line search problem implied by the direction of part (a) is

$$\max f(30 - 0.088\lambda, 2 + 6\lambda) \triangleq \frac{(30 - 0.088\lambda)}{1 + e^{0.1(30 - 0.088\lambda)}} - [(2 + 6\lambda) - 5]^2$$

over $\lambda > 0$.

Steepest Ascent/Descent Property

Gradient search is sometimes called the **method of steepest ascent (steepest descent** for minimize problems) because the direction of construction 13.35 produces the most rapid rate of objective improvement near the current solution.

> 13.36   At any $\mathbf{x}^{(t)}$ with $\nabla f(\mathbf{x}^{(t)}) \neq \mathbf{0}$, direction $\Delta \mathbf{x} \triangleq \nabla f(\mathbf{x}^{(t)})$ produces the locally steepest rate of objective function ascent, and $\Delta \mathbf{x} \triangleq -\nabla f(\mathbf{x}^{(t)})$ yields the locally steepest rate of descent.

The search of Figure 13.16 illustrates graphically. The rate of objective improvement at any point depends on the angle between the move direction and nearby objective function contours. Gradient-based directions, which move perpendicular to the contours (principle 3.20), produce the most rapid local progress.

SAMPLE EXERCISE 13.16: COMPUTING DIRECTIONS OF STEEPEST ASCENT/DESCENT

Return to the nonlinear program of Sample Exercise 13.15 and compute the steepest descent direction of length 1 at point $\mathbf{x}^{(0)} = (30, 2)$.

Analysis: In accord with principle $\boxed{13.36}$, the steepest descent direction at $(30, 2)$ will be the negative of the improving gradient direction computed in Sample Exercise 13.15(a). Thus the steepest descent direction is $\Delta \mathbf{x} = (0.088, -6)$.

Zigzagging and Poor Convergence of Gradient Search

Gradient search is appealingly straightforward, but it is not very effective in most applications. To see why, look again at the search in Figure 13.16 and Table 13.5. The first move was in direction $\Delta \mathbf{x} = (6.24, -1053.37)$, the second pursued $\Delta \mathbf{x} = (23.06, 0.14)$, and the third adopted $\Delta \mathbf{x} = (4.57, -759.60)$. This third direction is almost exactly parallel to the first, and the fourth will parallel the second. Later iterations continued this **zigzagging** alternation of almost perpendicular move directions.

We would prefer to approach the optimum more directly, but these gradient-based directions still produced good progress in early iterations. The objective fell from 174.746 to 46.548 in 10 steps.

The difficulty arises later in the search. Near an optimal solution, the shape of the objective function changes rapidly with very small step sizes. Thus although gradient-based directions produce the locally steepest rate of improvement, they can be followed only a very short distance before the best direction changes dramatically. The resulting zigzagging consumed the last 17 iterations in Table 13.5 to reduce the objective from 46.548 to 43.134.

Unfortunately, this poor convergence is typical of gradient methods.

> $\boxed{13.37}$ Although gradient search may produce good initial progress, zigzagging as it approaches a stationary point makes the method too slow and unreliable to provide satisfactory results in many unconstrained nonlinear applications.

Zigzagging is not the only convergence problem with gradient search. With small solution changes having a big objective function impact, numerical errors also can hopelessly bog down the procedure far from an optimal solution. More sophistication is required to obtain a really satisfactory improving search algorithm.

13.6 NEWTON'S METHOD

Gradient search can be viewed as pursuing the move direction suggested by the first-order Taylor series approximation (definition $\boxed{13.17}$)

$$f_1(\mathbf{x}^{(t)} + \lambda \Delta \mathbf{x}) \triangleq f(\mathbf{x}^{(t)}) + \lambda \nabla f(\mathbf{x}^{(t)}) \cdot \Delta \mathbf{x}$$

Aligning $\Delta \mathbf{x}$ with gradient $\nabla f(\mathbf{x}^{(t)})$ produces the most rapid improvement in this first-order approximation to $f(\mathbf{x})$.

To improve on the slow, zigzagging progress characteristic of gradient search (principle $\boxed{13.37}$) requires more information. An obvious possibility is extending to the second-order Taylor approximation

$$f_2(\mathbf{x}^{(t)} + \lambda \Delta \mathbf{x}) \triangleq f(\mathbf{x}^{(t)}) + \lambda \nabla f(\mathbf{x}^{(t)}) \cdot \Delta \mathbf{x} + \frac{\lambda^2}{2} \Delta \mathbf{x} \mathbf{H}(\mathbf{x}^{(t)}) \Delta \mathbf{x}$$

This section explores the famous **Newton's method**, which does exactly that.

Newton Step

Unlike the first-order Taylor approximation, which is linear in directional components Δx_j, the quadratic, second-order version may have a local maximum or minimum. To determine the move $\lambda \Delta \mathbf{x}$ taking us to such a local optimum of the second-order approximation, we may fix $\lambda = 1$ and differentiate f_2 with respect to components of $\Delta \mathbf{x}$. With $\lambda = 1$ the scalar-notation form of f_2 is

$$f_2(\mathbf{x}^{(t)} + \Delta \mathbf{x}) \triangleq f(\mathbf{x}^{(t)}) + \sum_{i=1}^{n} \left(\frac{\partial f}{\partial x_i} \right) \Delta x_i + \frac{1}{2} \sum_{i=1}^{n} \sum_{j=1}^{n} \left(\frac{\partial^2 f}{\partial x_i \, \partial x_j} \right) \Delta x_i \, \Delta x_j$$

Then partial derivatives with respect to move components are

$$\frac{\partial f_2}{\partial \Delta x_i} = \left(\frac{\partial f}{\partial x_i} \right) + \sum_{j=1}^{n} \left(\frac{\partial^2 f}{\partial x_i \partial x_j} \right) \Delta x_j, \qquad i = 1, \ldots, n$$

or in matrix format,

$$\nabla f_2(\Delta \mathbf{x}) = \nabla f(\mathbf{x}^{(t)}) + \mathbf{H}(\mathbf{x}^{(t)}) \Delta \mathbf{x}$$

Either way, setting $\nabla f_2(\Delta \mathbf{x}) = \mathbf{0}$ to find a stationary point produces the famous **Newton step**.

> **13.38** Newton steps $\Delta \mathbf{x}$, which move to a stationary point (if there is one), of the second-order Taylor series approximation to $f(\mathbf{x})$ at current point $\mathbf{x}^{(t)}$ are obtained by solving the linear equation system
> $$\mathbf{H}(\mathbf{x}^{(t)}) \Delta \mathbf{x} = -\nabla f(\mathbf{x}^{(t)})$$

Figure 13.17 illustrates for our Custom Computer curve-fitting model (13.7) (Section 13.1). First and second partial derivatives at initial point $\mathbf{x}^{(0)} = (32, -0.4)$

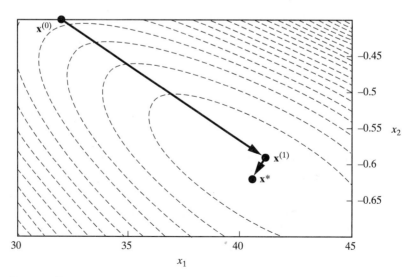

FIGURE 13.17   Newton's Method on the Custom Computer Example

are

$$\nabla f(\mathbf{x}^{(0)}) = \begin{pmatrix} -6.240 \\ 1053 \end{pmatrix} \quad \text{and} \quad \mathbf{H}(\mathbf{x}^{(0)}) = \begin{pmatrix} 7.13 & 293.99 \\ 293.99 & 18{,}817 \end{pmatrix}$$

Solving system

$$\begin{pmatrix} 7.13 & 293.99 \\ 293.99 & 18{,}817 \end{pmatrix} \Delta \mathbf{x} = - \begin{pmatrix} -6.240 \\ 1053 \end{pmatrix}$$

produces Newton step $\Delta \mathbf{x} = (8.956, -0.1959)$, which takes us to second-order Taylor approximation minimum

$$\mathbf{x}^{(1)} = \mathbf{x}^{(0)} + \Delta \mathbf{x} = (32, -0.4) + (8.956, -0.1959) = (40.96, -0.5959) \qquad (13.25)$$

SAMPLE EXERCISE 13.17: COMPUTING NEWTON STEPS

Compute the Newton step corresponding to current point $\mathbf{x}^{(0)} = (0, 1)$ in a search of unconstrained NLP

$$\min \; f(x_1, x_2) \triangleq (x_1 + 1)^4 + x_1 x_2 + (x_2 + 1)^4$$

Analysis: To develop linear system $\boxed{13.38}$, we compute partial derivatives

$$\nabla f(0, 1) = \begin{pmatrix} 4(x_1 + 1)^3 + x_2 \\ x_1 + 4(x_2 + 1)^3 \end{pmatrix} = \begin{pmatrix} 5 \\ 32 \end{pmatrix}$$

and

$$\mathbf{H}(0, 1) = \begin{pmatrix} 12(x_1 + 1)^2 & 1 \\ 1 & 12(x_2 + 1)^2 \end{pmatrix} = \begin{pmatrix} 12 & 1 \\ 1 & 48 \end{pmatrix}$$

Then Newton step $\Delta \mathbf{x}$ is the solution to the system

$$\begin{pmatrix} 12 & 1 \\ 1 & 48 \end{pmatrix} \begin{pmatrix} \Delta x_1 \\ \Delta x_2 \end{pmatrix} = - \begin{pmatrix} 5 \\ 32 \end{pmatrix}$$

which is approximately $\Delta x_1 = -0.3617$, $\Delta x_2 = -0.6591$.

Newton's Method

Newton's method proceeds by repeating the process above. That is, it uses first and second partial derivatives at the current point to compute a Newton step, updates the solution with that step, and repeats the process. Algorithm 13E provides details.

Newton's Method on the Custom Computer Example

Table 13.6 and Figure 13.17 apply Algorithm 13E to our Custom Computer example using stopping tolerance $\epsilon = 0.1$. Equation (13.25) already derived the first move to $\mathbf{x}^{(1)} = (40.96, -0.5959)$. There derivatives are recomputed, and new Newton step $\Delta \mathbf{x} = (-0.2733, -0.0061)$ brings us to

$$\mathbf{x}^{(2)} = \mathbf{x}^{(1)} + \Delta \mathbf{x} = (40.96, -0.5959) + (-0.2733, -0.0061) = (40.68, -0.6020)$$

Notice that a step size of $\lambda = 1$ is assumed because the Newton step represents a full move rather than just a direction.

ALGORITHM 13E: NEWTON'S METHOD

Step 0: Initialization. Choose any starting solution $\mathbf{x}^{(0)}$, pick stopping tolerance $\epsilon > 0$, and set solution index $t \leftarrow 0$.

Step 1: Derivatives. Compute objective function gradient $\nabla f(\mathbf{x}^{(t)})$ and Hessian matrix $\mathbf{H}(\mathbf{x}^{(t)})$ at current point $\mathbf{x}^{(t)}$.

Step 2: Stationary Point. If $||\nabla f(\mathbf{x}^{(t)})|| < \epsilon$, stop. Point $\mathbf{x}^{(t)}$ is sufficiently close to a stationary point.

Step 3: Newton Move. Solve the linear system

$$\mathbf{H}(\mathbf{x}^{(t)})\Delta\mathbf{x} = -\nabla f(\mathbf{x}^{(t)})$$

for Newton move $\Delta\mathbf{x}^{(t+1)}$.

Step 4: New Point. Update

$$\mathbf{x}^{(t+1)} \leftarrow \mathbf{x}^{(t)} + \Delta\mathbf{x}^{(t+1)}$$

Step 5: Advance. Increment $t \leftarrow t+1$, and return to Step 1.

TABLE 13.6   Newton's method on the Custom Computer Example

t	$\mathbf{x}^{(t)}$	$f(\mathbf{x}^{(t)})$	$\nabla f(\mathbf{x}^{(t)})$	$\mathbf{H}(\mathbf{x}^{(t)})$	$||\nabla f(\mathbf{x}^{(t)})||$	$\Delta\mathbf{x}^{(t+1)}$
0	(32.00, −0.4000)	174.746	(−6.240, 1053)	$\begin{pmatrix} 7.13 & 293.99 \\ 293.99 & 18{,}817 \end{pmatrix}$	1053.4	(8.956, −0.1959)
1	(40.96, −0.5959)	43.820	(2.347, 116.3)	$\begin{pmatrix} 4.86 & 166.21 \\ 166.21 & 11{,}564 \end{pmatrix}$	116.36	(−0.2733, −0.0061)
2	(40.68, −0.6020)	43.133	(0.0289, 2.989)	$\begin{pmatrix} 4.81 & 158.97 \\ 158.97 & 10{,}918 \end{pmatrix}$	2.9895	(0.0058, −0.0004)
3	(40.69, −0.6024)	43.133	(0.0000, 0.0027)	$\begin{pmatrix} 4.81 & 158.73 \\ 158.73 & 10{,}899 \end{pmatrix}$	0.00272	Stop

Algorithm 13E stops after one additional move to $\mathbf{x}^{(3)} = (40.68, -0.6024)$. The gradient norm 0.00272 at that point is less than stopping tolerance $\epsilon = 0.1$, which implies that we are sufficiently close to a stationary point of the full objective function.

SAMPLE EXERCISE 13.18: EXECUTING NEWTON'S METHOD

Return to the model of Sample Exercise 13.17, and execute 2 iterations Newton's method Algorithm 13E starting with $\mathbf{x}^{(0)} = (0, 1)$.

Analysis: Sample Exercise 13.17 already computed partial derivative expressions

$$\nabla f(x_1, x_2) = \begin{pmatrix} 4(x_1 + 1)^3 + x_2 \\ x_1 + 4(x_2 + 1)^3 \end{pmatrix}$$

and

$$\mathbf{H}(x_1, x_2) = \begin{pmatrix} 12(x_1 + 1)^2 & 1 \\ 1 & 12(x_2 + 1)^2 \end{pmatrix}$$

along with first Newton step $\Delta\mathbf{x}^{(1)} = (-0.3617, -0.6591)$. Thus the first iteration produces

$$\mathbf{x}^{(1)} = (0, 1) + (-0.3617, -0.6591) = (-0.3617, 0.3409)$$

Notice that no step size λ is applied (or equivalently, $\lambda = 1$).

Substituting this $\mathbf{x}^{(1)}$ in gradient and Hessian expressions produces the next
13.38 | linear system

$$\begin{pmatrix} 4.889 & 1 \\ 1 & 28.93 \end{pmatrix} \begin{pmatrix} \Delta x_1 \\ \Delta x_2 \end{pmatrix} = - \begin{pmatrix} 1.381 \\ 9.282 \end{pmatrix}$$

There the solution is $\Delta\mathbf{x}^{(2)} = (-0.2184, -0.3133)$, and we complete the second iteration with

$$\mathbf{x}^{(2)} = (-0.3617, -0.6591) + (-0.2184, -0.3133) = (-0.5801, 0.0276)$$

Rapid Convergence Rate of Newton's Method

Comparison of Tables 13.5 and 13.6 shows dramatically improved convergence with Newton's method versus gradient search. The gradient algorithm required 27 moves to reach an optimum. Newton's method took only 3.

Although the mathematical theory to fully explain this gain is beyond the scope of this book, it is typical of comparative experience with the methods.

13.39 | If Newton's method converges to a local optimum, it usually does so in many fewer steps than first-order procedures such as gradient search.

Computational Tradeoffs between Gradient and Newton Search

Of course, the number of iterations is not the only consideration in comparing the efficiency of algorithms. We must also take into account the effort per iteration.

There Newton's method has both advantages and disadvantages. On the positive side is the absence of line searches (although some extensions add them to Algorithm 13E). Once each direction $\Delta\mathbf{x}^{(t+1)}$ is computed, we may update immediately, with no need for a relatively costly search for the best step size λ.

The extra burdens of Newton's method come with its use of the second-order Taylor approximation. Each directional computation on an n-vector $\mathbf{x}^{(t)}$ requires evaluating n expressions for the various first partial derivatives, and (using symmetry) another $\frac{1}{2}n(n+1)$ expressions for the Hessian. This is roughly the same amount of work as evaluating the original objective $n + \frac{1}{2}n(n+1)$ times versus only n for gradient search. In addition, we must solve an n by n system of linear equations to find the next Newton move. This too represents a substantial computational burden at every iteration.

13.40 | Computing both first and second partial derivatives plus solving a linear system of equations at each iteration makes Newton's method computationally burdensome as the dimension of the decision vector becomes large.

Starting Close with Newton's Method

Perhaps the greatest disadvantage of Newton's method is that it may not converge at all.

> 13.41 Newton's method is assured of converging to local optimum only if it starts relatively close to a local optimum.

There are two main reasons convergence can fail. First is the quadratic Taylor approximation itself. Far from an optimal solution, the second-order approximation can give such poor information that the computed Newton step does not even improve the objective function. For instance, suppose that the Newton search of Sample Exercise 13.18 had begun at $\mathbf{x}^{(0)} = (-1, 1)$ instead of $(0, 1)$. Then the linear system of construction 13.38 would have been

$$\begin{pmatrix} 0 & 1 \\ 1 & 48 \end{pmatrix} \begin{pmatrix} \Delta x_1 \\ \Delta x_2 \end{pmatrix} = - \begin{pmatrix} 1 \\ 31 \end{pmatrix}$$

The implied Newton step $\mathbf{\Delta x}^{(1)} = (-1, 17)$ moves the minimizing objective value from $f(-1, 1) = 15$ to $f(-2, 18) = 130{,}286$. Hardly an improvement!

Another potential difficulty arises with the linear system that must be solved at each Newton iteration. How do we know that it can be solved efficiently; that is (Primer 4), what assures that Hessian matrix $\mathbf{H}(\mathbf{x}^{(t)})$ in 13.38 is nonsingular?

Near a strict local optimum, second-order sufficient conditions 13.22 suggest that the Hessian may be positive or negative definite. Either implies nonsingularity (Primer 8). But if we are farther away from an optimum, there is no guarantee whatever.

13.7 QUASI-NEWTON METHODS AND BFGS SEARCH

In Section 13.5 we saw that gradient search requires only first partial derivatives but often gives poor numerical performance. Newton's method of Section 13.6 yields much improved convergence but requires second derivatives and solving a system of linear equations at each iteration. It is natural to look for a blend of the two that preserves their advantages while ameliorating their worst defects. That is precisely the idea behind **quasi-Newton methods**, which provide the most effective known algorithms for many unconstrained nonlinear programs.

Deflection Matrices

The Newton step of construction 13.38 solves

$$\mathbf{H}(\mathbf{x}^{(t)})\mathbf{\Delta x} = -\nabla f(\mathbf{x}^{(t)})$$

for move $\mathbf{\Delta x}$. Assuming that the Hessian is nonsingular, we may left-multiply by its matrix inverse to express the move as

$$\mathbf{\Delta x} = -\mathbf{H}^{-1}(\mathbf{x}^{(t)})\nabla f(\mathbf{x}^{(t)})$$

That is, directions are computed by applying a suitable deflection matrix $\mathbf{D}_t \triangleq \mathbf{H}^{-1}(\mathbf{x}^{(t)})$ to the current gradient.

> **13.42** **Deflection matrices $\mathbf{D}_t$ produce modified gradient search directions**
> $$\Delta\mathbf{x}^{(t+1)} = -\mathbf{D}_t\nabla f(\mathbf{x}^{(t)})$$

By stretching a point, we can also think of gradient search as a deflection matrix method. For example, the maximize case of Algorithm 13D employs directions

$$\Delta\mathbf{x} = \nabla f(\mathbf{x}^{(t)}) = -(-\mathbf{I})\nabla f(\mathbf{x}^{(t)})$$

which can be viewed as adopting negative identity deflection matrix $\mathbf{D}_t = -\mathbf{I}$. The corresponding minimize case uses $\mathbf{D}_t = +\mathbf{I}$.

Quasi-Newton Approach

Quasi-Newton methods work with a deflection matrix that approximates the Hessian inverse $\mathbf{H}^{-1}(\mathbf{x}^{(t)})$ of Newton's method. Unlike the full Newton's method, however, this $\mathbf{D}_t$ is built up from prior search results using only first derivatives.

The key to this approach is identifying properties that a deflection matrix should possess if it is to do the job of an inverse Hessian. Principal among these is the idea that the Hessian $\mathbf{H}(\mathbf{x}^{(t)})$ reflects the rates of change in first derivatives $\nabla f(\mathbf{x}^{(t)})$. As we move from $\mathbf{x}^{(t)}$ to $\mathbf{x}^{(t+1)}$, it follows that

$$\nabla f(\mathbf{x}^{(t+1)}) - \nabla f(\mathbf{x}^{(t)}) \approx \mathbf{H}(\mathbf{x}^{(t)})(\mathbf{x}^{(t+1)} - \mathbf{x}^{(t)})$$

or

$$\mathbf{H}^{-1}(\mathbf{x}^{(t)})\left(\nabla f(\mathbf{x}^{(t+1)}) - \nabla f(\mathbf{x}^{(t)})\right) \approx \mathbf{x}^{(t+1)} - \mathbf{x}^{(t)}$$

The analogous requirement on $\mathbf{D}_t$ is known as the **quasi-Newton condition**.

> **13.43** Deflection matrices of quasi-Newton algorithms approximate the gradient change behavior of inverse Hessian matrices by satisfying the quasi-Newton condition
> $$\mathbf{D}_{t+1}\mathbf{g} = \mathbf{d}$$
> at every iteration, where $\mathbf{d} \triangleq \mathbf{x}^{(t+1)} - \mathbf{x}^{(t)}$ and $\mathbf{g} \triangleq \nabla f(\mathbf{x}^{(t+1)}) - \nabla f(\mathbf{x}^{(t)})$.

Another characteristic property of Hessian matrices is their symmetry. For most common functions both $\mathbf{H}(\mathbf{x}^{(t)})$ and $\mathbf{H}^{-1}(\mathbf{x}^{(t)})$ are symmetric matrices. If quasi-Newton deflection matrices are to have any hope of approximating such inverse Hessians, they must also preserve this property.

> **13.44** Deflection matrices of quasi-Newton algorithms should parallel inverse Hessians by being symmetric.

Guaranteeing Directions Improve

One of the difficulties that we encountered with Newton's method in Section 13.6 is that it gives unpredictable results far from an optimal solution. It may not even produce an improving step.

We would like our quasi-Newton algorithms to avoid this difficulty. Recall from earliest principles $\boxed{3.21}$ and $\boxed{3.22}$ that direction $\Delta \mathbf{x}$ improves for a maximize problem at $\mathbf{x}^{(t)}$ if $\nabla f(\mathbf{x}^{(t)}) \cdot \Delta \mathbf{x} > 0$ and for a minimize if $\nabla f(\mathbf{x}^{(t)}) \cdot \Delta \mathbf{x} < 0$. With directions from deflection matrix construction $\boxed{13.42}$, these conditions become

$$\nabla f(\mathbf{x}^{(t)})(-\mathbf{D}_t \nabla f(\mathbf{x}^{(t)})) = \nabla f(\mathbf{x}^{(t)})(-\mathbf{D}_t)\nabla f(\mathbf{x}^{(t)}) > 0$$

and

$$\nabla f(\mathbf{x}^{(t)})(-\mathbf{D}_t \nabla f(\mathbf{x}^{(t)})) = \nabla f(\mathbf{x}^{(t)})(-\mathbf{D}_t)\nabla f(\mathbf{x}^{(t)}) < 0$$

Notice that the maximize case will be satisfied for any gradient if every $\mathbf{D}_t$ is negative definite, so that $-\mathbf{D}_t$ is positive definite (Primer 8). Similarly, the minimize case requires $\mathbf{D}_t$ positive definite. These concerns motivate another specification.

$\boxed{13.45}$ Deflection matrices in quasi-Newton algorithms should assure improving directions by keeping $\mathbf{D}_t$ negative definite for maximize problems and positive definite for minimizes.

BFGS Formula

It turns out that a variety of deflection matrix update formulas can meet quasi-Newton requirements $\boxed{13.43}$ to $\boxed{13.45}$. Still, one has proved more effective than all the others. Developed through the combined work of C. Broyden, R. Fletcher, D. Goldfarb, and D. Shanno, it is known as the BFGS formula.

$\boxed{13.46}$ The **BFGS formula** updates deflection matrices by

$$\mathbf{D}_{t+1} \leftarrow \mathbf{D}_t + \left(1 + \frac{\mathbf{g}\mathbf{D}_t\mathbf{g}}{\mathbf{d} \cdot \mathbf{g}}\right)\frac{\mathbf{d}\mathbf{d}^\mathsf{T}}{\mathbf{d} \cdot \mathbf{g}} - \frac{\mathbf{D}_t\mathbf{g}\mathbf{d}^\mathsf{T} + \mathbf{d}\mathbf{g}^\mathsf{T}\mathbf{D}_t}{\mathbf{d} \cdot \mathbf{g}}$$

where $\mathbf{d} \triangleq \mathbf{x}^{(t+1)} - \mathbf{x}^{(t)}$ and $\mathbf{g} \triangleq \nabla f(\mathbf{x}^{(t+1)}) - \nabla f(\mathbf{x}^{(t)})$.

Although it appears rather imposing, BFGS update $\boxed{13.46}$ actually changes deflection matrices rather modestly at each iteration. The update has the form

$$\mathbf{D}_t + \phi\mathbf{C}_1 - \left[(\mathbf{D}_t\mathbf{C}_2) + (\mathbf{D}_t\mathbf{C}_2)^\mathsf{T}\right]$$

where weight

$$\phi \triangleq 1 + \frac{\mathbf{g}\mathbf{D}_t\mathbf{g}}{\mathbf{d} \cdot \mathbf{g}} \tag{13.26}$$

is applied to combine simple matrices

$$
\mathbf{C}_1 \triangleq \frac{\mathbf{dd}^{\mathrm{T}}}{\mathbf{d \cdot g}} = \frac{1}{\sum\limits_j d_j g_j}
\begin{pmatrix}
(d_1)^2 & d_1 d_2 & \cdots & d_1 d_n \\
d_2 d_1 & (d_2)^2 & \cdots & d_2 d_n \\
\vdots & \vdots & \ddots & \vdots \\
d_n d_1 & d_n d_2 & \cdots & (d_n)^2
\end{pmatrix}
$$

$$\tag{13.27}$$

$$
\mathbf{C}_2 \triangleq \frac{\mathbf{gd}^{\mathrm{T}}}{\mathbf{d \cdot g}} = \frac{1}{\sum\limits_j d_j g_j}
\begin{pmatrix}
g_1 d_1 & g_1 d_2 & \cdots & g_1 d_n \\
g_2 d_1 & g_2 d_2 & \cdots & g_2 d_n \\
\vdots & \vdots & \ddots & \vdots \\
g_n d_1 & g_n d_2 & \cdots & g_n d_n
\end{pmatrix}
$$

Notice that both $\mathbf{C}_1$ and $\mathbf{C}_2$ are **rank one** with every row a multiple of every other.

BFGS Search of Custom Computer Example

Algorithm 13F details a search algorithm based on BFGS update formula $\boxed{13.46}$. Table 13.7 and Figure 13.18 then track its application to our Custom Computer curve-fitting model (13.7).

The initial iteration of Algorithm 13F for this minimize problem employs identity deflection matrix $\mathbf{D}_0 = \mathbf{I}$, which leaves the first direction

$$\Delta \mathbf{x}^{(1)} = -\mathbf{D}_0 \nabla f(\mathbf{x}^{(0)}) = -\mathbf{I}\nabla f(\mathbf{x}^{(0)}) = -\nabla f(\mathbf{x}^{(0)})$$

Thus our BFGS procedure follows the same first direction as gradient search Algorithm 13D.

Like gradient Algorithm 13D, and unlike Newton Algorithm 13E, quasi-Newton methods require line search. Table 13.7 shows that the first such search produces step $\lambda = 0.0001$. The result is

$$
\begin{aligned}
\mathbf{x}^{(1)} &\leftarrow \mathbf{x}^{(0)} + \lambda_1 \Delta \mathbf{x}_1 \\
&= (32.00, -0.4000) + (0.0001)(6.240, -1053) \\
&= (32.00, -0.4685)
\end{aligned}
$$

with gradient $\nabla f(\mathbf{x}^{(1)}) = (-23.02, 2.514)$. Thus

$$
\mathbf{d} \triangleq \mathbf{x}^{(t+1)} - \mathbf{x}^{(t)} \qquad = \begin{pmatrix} 32.0004 \\ -0.4685 \end{pmatrix} - \begin{pmatrix} 32 \\ -0.4 \end{pmatrix} = \begin{pmatrix} 0.0004 \\ -0.0685 \end{pmatrix}
$$

$$
\mathbf{g} \triangleq \nabla f(\mathbf{x}^{(t+1)}) - \nabla f(\mathbf{x}^{(t)}) = \begin{pmatrix} -23.02 \\ 2.514 \end{pmatrix} - \begin{pmatrix} -6.240 \\ 1053.4 \end{pmatrix} = \begin{pmatrix} -16.78 \\ -1050.9 \end{pmatrix}
$$

and

$$\mathbf{d \cdot g} = (0.0004, -0.0685) \cdot (-16.78, -1050.9) = 71.9$$

ALGORITHM 13F: BFGS QUASI-NEWTON SEARCH

Step 0: Initialization. Choose any starting solution $\mathbf{x}^{(0)}$, compute gradient $\nabla f(\mathbf{x}^{((0))})$ and pick stopping tolerance $\epsilon > 0$. Also initialize deflection matrix

$$\mathbf{D}_0 = \mp\mathbf{I}$$

($-$ for a maximize, $+$ for a minimize) and set solution index $t \leftarrow 0$.

Step 1: Stationary Point. If norm $||\nabla f(\mathbf{x}^{(t)})|| < \epsilon$, stop. Point $\mathbf{x}^{(t)}$ is sufficiently close to a stationary point.

Step 2: Direction. Use the current deflection matrix $\mathbf{D}_t$ to compute the move direction

$$\Delta\mathbf{x}^{(t+1)} \leftarrow -\mathbf{D}_t\nabla f(\mathbf{x}^{(t)})$$

Step 3: Line Search. Solve (at least approximately) 1-dimensional line search

$$\text{max or min } f(\mathbf{x}^{(t)} + \lambda\Delta\mathbf{x}^{(t+1)})$$

to compute step size λ_{t+1}.

Step 4: New Point. Update

$$\mathbf{x}^{(t+1)} \leftarrow \mathbf{x}^{(t)} + \lambda_{t+1}\Delta\mathbf{x}^{(t+1)}$$

and compute new gradient $\nabla f(\mathbf{x}^{(t_1)})$.

Step 5: Deflection Matrix. Revise the deflection matrix as

$$\mathbf{D}_{t+1} \leftarrow \mathbf{D}_t + \left(1 + \frac{\mathbf{g}\mathbf{D}_t\mathbf{g}}{\mathbf{d}\cdot\mathbf{g}}\right)\frac{\mathbf{d}\mathbf{d}^{\mathrm{T}}}{\mathbf{d}\cdot\mathbf{g}} - \frac{\mathbf{D}_t\mathbf{g}\mathbf{d}^{\mathrm{T}} + \mathbf{d}\mathbf{g}\mathbf{D}_t}{\mathbf{d}\cdot\mathbf{g}}$$

where $\mathbf{d} \triangleq (\mathbf{x}^{(t+1)} - \mathbf{x}^{(t)})$ and $\mathbf{g} \triangleq (\nabla f(\mathbf{x}^{(t+1)}) - \nabla f(\mathbf{x}^{(t)}))$.

Step 6: Advance. Increment $t \leftarrow t + 1$, and return to Step 1.

TABLE 13.7   BFGS Search of Custom Computer Example

t	$\mathbf{x}^{(t)}$	$f(\mathbf{x}^{(t)})$	$\nabla f(\mathbf{x}^{(t)})$	$\|\nabla f(\mathbf{x}^{(t)})\|$	$\mathbf{D}_t$	$\Delta\mathbf{x}^{(t+1)}$	λ_{t+1}
0	(32.00, −0.4000)	174.746	(−6.240, 1053)	1053.4	$\begin{pmatrix} 1.0000 & 0.0000 \\ 0.0000 & 1.0000 \end{pmatrix}$	(6.240, −1053)	0.0001
1	(32.00, −0.4685)	141.139	(−23.02, 2.514)	23.15	$\begin{pmatrix} 1.0002 & -0.0160 \\ -0.0160 & 0.0003 \end{pmatrix}$	(23.06, −0.3684)	0.3755
2	(40.66, −0.6068)	43.258	(−0.823, −51.54)	51.54	$\begin{pmatrix} 0.3759 & -0.0059 \\ -0.0059 & 0.0002 \end{pmatrix}$	(0.0079, 0.0032)	1.4361
3	(4.067, −0.6021)	43.133	(−0.041, 0.100)	0.11	$\begin{pmatrix} 0.3775 & -0.0055 \\ -0.0055 & 0.0002 \end{pmatrix}$	(0.0160, −0.0002)	1.0604
4	(40.69, −0.6024)	43.132	(0.000, −0.008)	0.01	Stop		

Next we compute main update matrices

$$\frac{\mathbf{d}\mathbf{d}^{\mathrm{T}}}{\mathbf{d}\cdot\mathbf{g}} = \frac{1}{71.9}\begin{pmatrix} 0.0004 \\ -0.4685 \end{pmatrix}(0.0004, -0.4685) = \begin{pmatrix} 0.0000000 & -0.0000004 \\ -0.0000004 & 0.0000652 \end{pmatrix}$$

$$\frac{\mathbf{D}_0\mathbf{g}\mathbf{d}^{\mathrm{T}}}{\mathbf{d}\cdot\mathbf{g}} = \frac{1}{71.9}\begin{pmatrix} 1 & 0 \\ 0 & 1 \end{pmatrix}\begin{pmatrix} -16.78 \\ -1050.9 \end{pmatrix}(0.0004, -0.4685) = \begin{pmatrix} -0.000095 & 0.015975 \\ -0.005927 & 1.00062 \end{pmatrix}$$

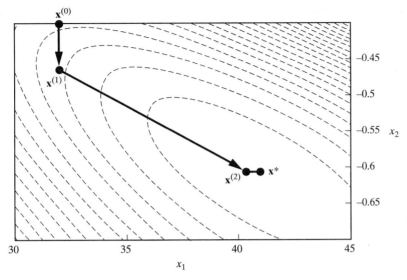

FIGURE 13.18   BFGS Search of Custom Computer Example

Then with

$$\mathbf{gD_0g}^T = (-16.78, -1050.9) \begin{pmatrix} 1 & 0 \\ 0 & 1 \end{pmatrix} \begin{pmatrix} -16.78 \\ -1050.9 \end{pmatrix} = 1,104,583$$

the new deflection matrix is

$$\mathbf{D_1} \leftarrow \mathbf{D_0} + \left(1 + \frac{\mathbf{gD_0g}}{\mathbf{d} \cdot \mathbf{g}}\right) \frac{\mathbf{dd}^T}{\mathbf{d} \cdot \mathbf{g}} - \frac{\mathbf{D_0gd}^T + \mathbf{dg}^T\mathbf{D_0}}{\mathbf{d} \cdot \mathbf{g}}$$

$$= \begin{pmatrix} 1 & 0 \\ 0 & 1 \end{pmatrix} + \left(1 + \frac{1,104,583}{71.9}\right) \begin{pmatrix} 0.0000000 & -0.0000004 \\ -0.0000004 & 0.0000652 \end{pmatrix}$$

$$- \left[\begin{pmatrix} -0.000095 & 0.015975 \\ -0.005927 & 1.00062 \end{pmatrix} + \begin{pmatrix} -0.000095 & -0.005927 \\ 0.015975 & 1.00062 \end{pmatrix} \right]$$

$$= \begin{pmatrix} 1.0002 & -0.0160 \\ -0.0160 & 0.0003 \end{pmatrix}$$

This revised deflection matrix produces the next move direction,

$$\Delta\mathbf{x}^{(1)} \leftarrow -\mathbf{D_1}\nabla f(\mathbf{x}^{(1)}) = -\begin{pmatrix} 1.0002 & -0.0160 \\ -0.0160 & 0.0003 \end{pmatrix} \begin{pmatrix} -23.02 \\ 2.514 \end{pmatrix} = \begin{pmatrix} 23.06 \\ -0.3684 \end{pmatrix}$$

and the search continues.

Algorithm 13F stops when $||\nabla f(\mathbf{x}^{(t)})|| < \epsilon$, indicating that we have reached an approximately stationary point. Using $\epsilon = 0.1$, this occurs at iteration $t = 4$ of Table 13.7.

SAMPLE EXERCISE 13.19: EXECUTING BFGS SEARCH

Suppose that BFGS Algorithm 13F reaches iteration $t = 5$ with

$$\mathbf{x}^{(5)} = (10, 16), \quad \nabla f(\mathbf{x}^{(5)}) = (-1, 1), \quad \mathbf{D_5} = \begin{pmatrix} -10 & 2 \\ 2 & -4 \end{pmatrix}$$

of a maximizing search, and then takes a step of $\lambda_6 = \frac{1}{2}$ in the BFGS direction to reach a new $\mathbf{x}^{(6)}$ with $\nabla f(\mathbf{x}^{(6)}) = (5, -3)$.

(a) Determine the direction $\Delta \mathbf{x}^{(6)}$ that was employed and the new solution $\mathbf{x}^{(6)}$.

(b) Compute the revised deflection matrix $\mathbf{D}_6$ needed for the next iteration.

Analysis:

(a) Following gradient deflection computation $\boxed{13.42}$

$$\Delta \mathbf{x}^{(6)} \leftarrow -\mathbf{D}_5 \nabla f(\mathbf{x}^{(5)}) = -\begin{pmatrix} -10 & 2 \\ 2 & -4 \end{pmatrix}\begin{pmatrix} -1 \\ 1 \end{pmatrix} = \begin{pmatrix} -12 \\ 6 \end{pmatrix}$$

Thus the new solution

$$\mathbf{x}^{(6)} = \mathbf{x}^{(5)} + \lambda_6 \Delta \mathbf{x}^{(6)} = \begin{pmatrix} 10 \\ 16 \end{pmatrix} + \frac{1}{2}\begin{pmatrix} -12 \\ 6 \end{pmatrix} = \begin{pmatrix} 4 \\ 19 \end{pmatrix}$$

(b) We apply BFGS formula $\boxed{13.46}$. First, the difference vectors are

$$\mathbf{d} \triangleq \mathbf{x}^{(6)} - \mathbf{x}^{(5)} \qquad = \begin{pmatrix} 4 \\ 19 \end{pmatrix} - \begin{pmatrix} 10 \\ 16 \end{pmatrix} = \begin{pmatrix} -6 \\ 3 \end{pmatrix}$$

$$\mathbf{g} \triangleq \nabla f(\mathbf{x}^{(6)}) - \nabla f(\mathbf{x}^{(5)}) = \begin{pmatrix} 5 \\ -3 \end{pmatrix} - \begin{pmatrix} -1 \\ 1 \end{pmatrix} = \begin{pmatrix} 6 \\ -4 \end{pmatrix}$$

with $\mathbf{d} \cdot \mathbf{g} = -48$. The update matrices are then

$$\frac{\mathbf{d}\mathbf{d}^{\mathrm{T}}}{\mathbf{d} \cdot \mathbf{g}} = \frac{1}{-48}\begin{pmatrix} -6 \\ 3 \end{pmatrix}(-6, 3) = \begin{pmatrix} -0.75 & 0.375 \\ 0.375 & -0.1875 \end{pmatrix}$$

$$\frac{\mathbf{D}_5\mathbf{g}\mathbf{d}^{\mathrm{T}}}{\mathbf{d} \cdot \mathbf{g}} = \frac{1}{-48}\begin{pmatrix} -10 & 2 \\ 2 & -4 \end{pmatrix}\begin{pmatrix} 6 \\ -4 \end{pmatrix}(-6, 3) = \begin{pmatrix} -8.5 & 4.25 \\ 3.5 & -1.75 \end{pmatrix}$$

Also,

$$\mathbf{g}\mathbf{D}_5\mathbf{g} = (6, -4)\begin{pmatrix} -10 & 2 \\ 2 & -4 \end{pmatrix}\begin{pmatrix} 6 \\ -4 \end{pmatrix} = -520$$

Now substituting in formula $\boxed{13.46}$ gives

$$\mathbf{D}_6 \leftarrow \mathbf{D}_5 + \left(1 + \frac{\mathbf{g}\mathbf{D}_5\mathbf{g}}{\mathbf{d} \cdot \mathbf{g}}\right)\frac{\mathbf{d}\mathbf{d}^{\mathrm{T}}}{\mathbf{d} \cdot \mathbf{g}} - \frac{\mathbf{D}_5\mathbf{g}\mathbf{d}^{\mathrm{T}} + \mathbf{d}\mathbf{g}^{\mathrm{T}}\mathbf{D}_5}{\mathbf{d} \cdot \mathbf{g}}$$

$$= \begin{pmatrix} -10 & 2 \\ 2 & -4 \end{pmatrix} + \left(1 + \frac{-520}{-48}\right)\begin{pmatrix} -0.75 & 0.375 \\ 0.375 & -0.1875 \end{pmatrix}$$

$$- \left[\begin{pmatrix} -8.5 & 4.25 \\ 3.5 & -1.75 \end{pmatrix} + \begin{pmatrix} -8.5 & 3.5 \\ 4.25 & -1.75 \end{pmatrix}\right]$$

$$= \begin{pmatrix} -1.8750 & -1.3125 \\ -1.3125 & -2.7188 \end{pmatrix}$$

Verifying Quasi-Newton Requirements

Although proving most of them is beyond the scope of this book, BFGS update 13.46 can be shown to fulfill all our quasi-Newton requirements.

> **13.47**   BFGS update formula 13.46 produces deflection matrices satisfying quasi-Newton condition 13.43 at every iteration, as well as symmetry requirement 13.44 and improving direction specification 13.45 .

To illustrate, focus on $t = 1$ in Table 13.7. There

$$\mathbf{d} = \mathbf{x}^{(2)} - \mathbf{x}^{(1)} \qquad = \begin{pmatrix} 40.66 \\ -0.6068 \end{pmatrix} - \begin{pmatrix} 32.00 \\ -0.4685 \end{pmatrix} = \begin{pmatrix} 8.66 \\ -0.1383 \end{pmatrix}$$

$$\mathbf{g} = \nabla f(\mathbf{x}^{(2)}) - \nabla f(\mathbf{x}^{(1)}) = \begin{pmatrix} -0.823 \\ -51.54 \end{pmatrix} - \begin{pmatrix} -23.02 \\ 2.514 \end{pmatrix} = \begin{pmatrix} 22.193 \\ -54.05 \end{pmatrix}$$

Thus quasi-Newton condition 13.43 $\mathbf{D}_2\mathbf{g} = \mathbf{d}$ checks

$$\mathbf{D}_2\mathbf{g} = \begin{pmatrix} 0.37594 & -0.00585 \\ -0.00585 & 0.00016 \end{pmatrix} \begin{pmatrix} 22.193 \\ -54.05 \end{pmatrix} \approx \begin{pmatrix} 8.66 \\ -0.1383 \end{pmatrix} = \mathbf{d}$$

It is easy to see that all deflection matrices $\mathbf{D}_t$ in Table 13.7 are also symmetric and positive definite (as required for a minimize problem). For instance, with $\mathbf{D}_3$ the principal minor determinants are

$$\det(0.3775) = 0.3775 > 0 \quad \text{and} \quad \det\begin{pmatrix} 0.3775 & -0.0055 \\ -0.0055 & 0.0002 \end{pmatrix} = 0.00004 > 0$$

SAMPLE EXERCISE 13.20: VERIFYING QUASI-NEWTON REQUIREMENTS

Return to the maximize model of Sample Exercise 13.19, and demonstrate that the computed $\mathbf{D}_6$ satisfies quasi-Newton algorithm requirements 13.43 to 13.45 .

Analysis: Quasi-Newton condition 13.43 is $\mathbf{D}_6\mathbf{g} = \mathbf{d}$. Checking gives

$$\mathbf{D}_6\mathbf{g} = \begin{pmatrix} -1.8750 & -1.3125 \\ -1.3125 & -2.7188 \end{pmatrix} \begin{pmatrix} 6 \\ -4 \end{pmatrix} = \begin{pmatrix} -6 \\ 3 \end{pmatrix} = \mathbf{d}$$

as required. Also, matrix $\mathbf{D}_6$ is symmetric (requirement 13.44) because $d_{1,2}^{(6)} = d_{2,1}^{(6)} = -1.3125$.

To guarantee an improving direction for a maximize problem, $\mathbf{D}_6$ should also be negative definite (requirement 13.45). This too is true because principal minor determinants

$$\det(-1.8750) = -1.8750 \quad \text{and} \quad \det\begin{pmatrix} -1.8750 & -1.3125 \\ -1.3125 & -2.7188 \end{pmatrix} = 3.375$$

alternate in sign and the first is negative.

Approximating the Hessian Inverse with BFGS

Quasi-Newton properties $\boxed{13.43}$ to $\boxed{13.45}$ were motivated to mimic Newton's method's use of Hessian inverse deflection matrices $\mathbf{D}_t = \mathbf{H}^{-1}(\mathbf{x}^{(t)})$. It should not surprise that BFGS and many other quasi-Newton deflection matrices tend to this Newton case.

> $\boxed{13.48}$ As BFGS Algorithm 13F nears a local optimum, deflection matrices $\mathbf{D}_t$ approach the inverse Hessian matrix at that optimum.

Once again, we may illustrate with results for our Custom Computer example. The final deflection matrix of BFGS Table 13.7 is

$$\mathbf{D}_3 = \begin{pmatrix} 0.3775 & -0.0055 \\ -0.0055 & 0.0002 \end{pmatrix}$$

The corresponding Hessian matrix is

$$\mathbf{H}(\mathbf{x}^{(3)}) = \begin{pmatrix} 4.811 & 158.8 \\ 158.8 & 10{,}903 \end{pmatrix}$$

with inverse

$$\mathbf{H}^{-1}(\mathbf{x}^{(3)}) = \begin{pmatrix} 0.4004 & -0.0058 \\ -0.0058 & 0.0002 \end{pmatrix}$$

In accord with principle $\boxed{13.48}$, this $\mathbf{D}_3$ closely approximates $\mathbf{H}^{-1}(\mathbf{x}^{(3)})$.

13.8 OPTIMIZATION WITHOUT DERIVATIVES AND NELDER–MEAD

Sometimes nonlinear programs must be addressed over objective functions that are not differentiable, or at least do not have readily computable derivatives. In such cases, improving search must rely entirely on functional evaluations.

How do algorithms choose search directions without derivatives? Numerous schemes have been proposed. Some simply use the coordinate directions—searching each in turn. Other seek to align the search with the trend of recent progress. We develop here only the method due to Nelder and Mead, which constructs directions by maintaining an ensemble of current points.

Nelder–Mead Strategy

One of the most popular schemes for unconstrained search without derivatives is the Nelder–Mead procedure detailed in Algorithm 13G. Table 13.8 traces its application to our PERT maximum likelihood model (13.10).

In contrast to other improving search methods, which keep only one current point, Nelder-Mead Algorithm 13G maintains a set of $n + 1$.

> $\boxed{13.49}$ In an optimization over n decision variables, the Nelder–Mead algorithm maintains an ensemble of $n+1$ distinct solutions $\mathbf{y}^{(1)}, \ldots, \mathbf{y}^{(n+1)}$, with $\mathbf{y}^{(1)}$ having the best objective function value, $\mathbf{y}^{(2)}$ the second best, and so on.

ALGORITHM 13G: NELDER–MEAD DERIVATIVE-FREE SEARCH

Step 0: Initialization. Choose $(n + 1)$ distinct solutions $\mathbf{x}^{(j)}$ as starting set $\{\mathbf{y}^{(1)}, \ldots, \mathbf{y}^{(n+1)}\}$, evaluate $f(\mathbf{y}^{(1)}), \ldots, f(\mathbf{y}^{(n+1)})$, and initialize iteration index $t \leftarrow 0$.

Step 1: Centroid. Renumber as necessary to arrange the $\mathbf{y}^{(i)}$ in nonimproving sequence by solution value. Then compute best-n centroid

$$\mathbf{x}^{(t)} = \frac{1}{n} \sum_{i=1}^{n} \mathbf{y}^{(i)}$$

Step 2: Stopping. If all solution values $f(\mathbf{y}^{(1)}), \ldots, f(\mathbf{y}^{(n)})$ are sufficiently close to centroid objective value $f(\mathbf{x}^{(t)})$, stop and report the best of $\mathbf{y}^{(1)}$ and $\mathbf{x}^{(t)}$.

Step 3: Direction. Use centroid $\mathbf{x}^{(t)}$ to compute away-from-worst move direction

$$\Delta\mathbf{x}^{(t+1)} \leftarrow \mathbf{x}^{(t)} - \mathbf{y}^{(n+1)}$$

Step 4: Reflection. Try $\lambda = 1$ by computing $f(\mathbf{x}^{(t)} + 1\Delta\mathbf{x}^{(t+1)})$. If this new value is at least as good as current best $f(\mathbf{y}^{(1)})$, go to Step 5 and expand. If it is no better than second-worst value $f(\mathbf{y}^{(n)})$, go to Step 6 and contract. Otherwise, accept $\lambda \leftarrow 1$, and proceed to Step 8.

Step 5: Expansion. Try $\lambda = 2$ by computing $f(\mathbf{x}^{(t)} + 2\Delta\mathbf{x}^{(t+1)})$. If this value is no worse than $f(\mathbf{x}^{(t)} + 1\Delta\mathbf{x}^{(t+1)})$ fix $\lambda \leftarrow 2$, and otherwise set $\lambda \leftarrow 1$. Then proceed to Step 8.

Step 6: Contraction. If reflection value $f(\mathbf{x}^{(t)} + 1\Delta\mathbf{x}^{(t+1)})$ is better than worst current $f(\mathbf{y}^{(n+1)})$, try $\lambda = \frac{1}{2}$ by computing $f(\mathbf{x}^{(t)} + \frac{1}{2}\Delta\mathbf{x}^{(t+1)})$. If not, try $\lambda = -\frac{1}{2}$ by evaluating $f(\mathbf{x}^{(t)} - \frac{1}{2}\Delta\mathbf{x}^{(t+1)})$. Either way, if the result improves on worst current $f(\mathbf{y}^{(n+1)})$, fix λ at the $\pm\frac{1}{2}$ tried and proceed to Step 8. Otherwise, go to Step 7 to shrink.

Step 7: Shrinking. Shrink the current solution set toward best $\mathbf{y}^{(1)}$ by

$$\mathbf{y}^{(i)} \leftarrow \frac{1}{2}\left(\mathbf{y}^{(1)} + \mathbf{y}^{(i)}\right) \qquad \text{for all } i = 2, \ldots, n+1$$

Then compute new $f(\mathbf{y}^{(2)}), \ldots, f(\mathbf{y}^{(n+1)})$, advance $t \leftarrow t + 1$, and return to Step 1.

Step 8: Replacement. Replace worst $\mathbf{y}^{(n+1)}$ in the solution set by

$$\mathbf{x}^{(t)} + \lambda\Delta\mathbf{x}^{(t+1)}$$

Then advance $t \leftarrow t + 1$ and return to Step 1.

Each iteration of the search tries to replace the worst solution $\mathbf{y}^{(n+1)}$ with a better one.

The maximization of Table 13.8 illustrates for $n = 2$. Search begins with the ensemble

$$\mathbf{y}^{(1)} = (5, 3), \qquad \mathbf{y}^{(2)} = (6, 3), \qquad \mathbf{y}^{(3)} = (6, 4)$$
$$f(\mathbf{y}^{(1)}) = 12.425, \quad f(\mathbf{y}^{(2)}) = 11.429, \quad f(\mathbf{y}^{(3)}) = 2.663$$

Notice that the solutions are numbered from best to worst.

TABLE 13.8   Nelder–Mead Search of PERT Example

t	$y^{(1)}$	$y^{(2)}$	$y^{(3)}$	$y^{(t)}$	$\Delta x^{(t+1)}$	Reflect	Second
0	(5.000, 3.000)	(6.000, 3.000)	(6.000, 4.000)	(5.500, 3.000)	(−0.500, −1.000)	$\lambda = 1.0$	$\boxed{\lambda = 0.5}$
	$f = 12.425$	$f = 11.429$	$f = 2.663$	$f = 13.084$		$f = 11.415$	$f = 24.915$
1	(5.250, 2.500)	(5.000, 3.000)	(6.000, 3.000)	(5.125, 2.750)	(−0.875, −0.250)	$\boxed{\lambda = 1.0}$	$\lambda = 2.0$
	$f = 24.915$	$f = 12.425$	$f = 11.429$	$f = 13.074$		$f = 30.005$	$f = 19.654$
2	(4.250, 2.500)	(5.250, 2.500)	(5.000, 3.000)	(4.750, 2.500)	(−0.250, −0.500)	$\lambda = 1.0$	$\boxed{\lambda = 0.5}$
	$f = 30.005$	$f = 24.915$	$f = 12.425$	$f = 30.482$		$f = 18.229$	$f = 17.354$
3	(4.250, 2.500)	(5.250, 2.500)	(4.625, 2.250)	(4.750, 2.500)	(0.125, 0.250)	$\lambda = 1.0$	$\lambda = 0.5$
	$f = 30.005$	$f = 24.915$	$f = 17.354$	$f = 30.482$		$f = 21.434$	$f = 11.231$
4	(4.438, 2.375)	(4.750, 2.500)	(4.250, 2.500)	(4.594, 2.438)	(0.344, −0.063)	$\boxed{\lambda = 1.0}$	
	$f = 35.513$	$f = 30.482$	$f = 30.005$	$f = 16.119$		$f = 31.642$	
5	(4.438, 2.375)	(4.938, 2.375)	(4.750, 2.500)	(4.688, 2.375)	(−0.063, −0.125)	$\lambda = 1.0$	$\lambda = −0.5$
	$f = 35.513$	$f = 31.642$	$f = 30.482$	$f = 15.893$		$f = 17.354$	$f = 26.577$
6	(4.438, 2.375)	(4.594, 2.438)	(4.688, 2.375)	(4.516, 2.406)	(−0.172, 0.031)	$\boxed{\lambda = 1.0}$	
	$f = 35.513$	$f = 16.119$	$f = 15.893$	$f = 25.980$		$f = 31.062$	
7	(4.438, 2.375)	(4.344, 2.438)	(4.594, 2.438)	(4.391, 2.406)	(−0.203, −0.031)	$\lambda = 1.0$	$\boxed{\lambda = 0.5}$
	$f = 35.513$	$f = 31.062$	$f = 16.119$	$f = 32.845$		$f = 17.253$	$f = 28.811$
8	(4.438, 2.375)	(4.344, 2.438)	(4.289, 2.391)	(4.391, 2.406)	(0.102, 0.016)	$\lambda = 1.0$	$\boxed{\lambda = 0.5}$
	$f = 35.513$	$f = 31.062$	$f = 28.811$	$f = 32.845$		$f = 29.758$	$f = 31.979$
9	(4.438, 2.375)	(4.441, 2.414)	(4.344, 2.438)	(4.439, 2.395)	(0.096, −0.043)	$\lambda = 1.0$	$\boxed{\lambda = −0.5}$
	$f = 35.513$	$f = 31.979$	$f = 31.062$	$f = 33.857$		$f = 26.094$	$f = 32.250$
10	(4.438, 2.375)	(4.392, 2.416)	(4.441, 2.414)	(4.415, 2,396)	(−0.027, −0.019)	$\boxed{\lambda = 1.0}$	
	$f = 35.513$	$f = 32.250$	$f = 31.979$	$f = 33.726$		$f = 34.140$	
11	(4.438, 2.375)	(4.388, 2.377)	(4.392, 2.416)	(4.413, 2.376)	(0.021, −0.040)	$\lambda = 1.0$	$\boxed{\lambda = 2.0}$
	$f = 35.513$	$f = 34.140$	$f = 32.250$	$f = 34.957$		$f = 37.893$	$f = 41.050$
12	(4.455, 2.296)	(4.438, 2.375)	(4.388, 2.377)	(4.446, 2.335)	(0.058, −0.042)	$\boxed{\lambda = 1.0}$	$\lambda = 2.0$
	$f = 41.050$	$f = 35.513$	$f = 34.140$	$f = 38.569$		$f = 42.813$	$f = 27.929$
13	(4.504, 2.294)	(4.455, 2.296)	(4.438, 2.375)	(4.479, 2.295)	(0.042, −0.080)	$\boxed{\lambda = 1.0}$	
	$f = 42.813$	$f = 41.050$	$f = 35.513$	$f = 43.202$		$f = 41.193$	
14	(4.504, 2.294)	(4.521, 2.215)	(4.455, 2.296)	(4.513, 2.254)	(0.058, −0.042)	$\lambda = 1.0$	$\boxed{\lambda = −0.5}$
	$f = 42.813$	$f = 41.193$	$f = 41.050$	$f = 42.388$		$f = 28.723$	$f = 44.727$
15	(4.484, 2.275)	(4.504, 2.294)	(4.521, 2.215)	(4.494, 2.285)	(−0.027, 0.070)	$\lambda = 1.0$	$\boxed{\lambda = −0.5}$
	$f = 44.727$	$f = 42.813$	$f = 41.193$	$f = 45.087$		$f = 37.793$	$f = 44.627$
16	(4.484, 2.275)	(4.508, 2.250)	(4.504, 2.294)	(4.496, 2.262)	(−0.009, −0.032)	$\boxed{\lambda = 1.0}$	$\lambda = 2.0$
	$f = 44.727$	$f = 44.627$	$f = 42.813$	$f = 46.671$		$f = 46.356$	$f = 43.832$
17	(4.487, 2.231)	(4.484, 2.275)	(4.508, 2.250)	(4.485, 2.253)	(−0.022, 0.003)	$\lambda = 1.0$	$\boxed{\lambda = −0.5}$
	$f = 46.356$	$f = 44.727$	$f = 44.627$	$f = 45.790$		$f = 42.708$	$f = 47.301$
18	(4.497, 2.251)	(4.487, 2.231)	(4.484, 2.275)	(4.492, 2.241)	(0.008, −0.034)	$\boxed{\lambda = 1.0}$	$\lambda = 2.0$
	$f = 47.301$	$f = 46.356$	$f = 44.727$	$f = 46.977$		$f = 48.520$	$f = 44.839$
19	(4.500, 2.207)	(4.497, 2.251)	(4.487, 2.231)	(4.498, 2.229)	(0.011, −0.002)	$\lambda = 1.0$	$\boxed{\lambda = −0.5}$
	$f = 48.520$	$f = 47.301$	$f = 46.356$	$f = 48.192$		$f = 44.882$	$f = 47.275$
20	(4.500, 2.207)	(4.497, 2.251)	(4.493, 2.230)	(4.498, 2.229)	Stop		
	$f = 48.520$	$f = 47.301$	$f = 47.275$	$f = 48.192$			

It is somewhat arbitrary that exactly $n + 1$ solutions are maintained in the ensemble. Still, too many would complicate computation, and too few would not adequately surround an emerging optimum. Over n decision variables, $n+1$ solutions is just enough to define vertices of a polytope surrounding a point.

SAMPLE EXERCISE 13.21: ARRANGING A NELDER–MEAD ENSEMBLE

Consider the unconstrained nonlinear program

$$\min \ f(x_1, x_2, x_3) \triangleq (x_1)^2 + (x_2 - 1)^2 + (x_3 + 4)^2$$

Choose an initial ensemble of solutions for Nelder–Mead search and arrange them with the notation of $\boxed{13.49}$.

Analysis: With $n = 3$ we need 4 distinct solutions that "bracket" a region likely to produce good solutions. One possibility is

$$\begin{pmatrix} 3 \\ 3 \\ 3 \end{pmatrix}, \quad \begin{pmatrix} -3 \\ 3 \\ -3 \end{pmatrix}, \quad \begin{pmatrix} 3 \\ 0 \\ 3 \end{pmatrix}, \quad \text{and} \quad \begin{pmatrix} -3 \\ 0 \\ -3 \end{pmatrix}$$

Evaluating the objective gives

$$f(3, 3, 3) = 62, \quad f(-3, 3, -3) = 14, \quad f(3, 0, 3) = 59, \quad \text{and} \quad f(-3, 0, -3) = 11$$

Thus, for this minimizing model, we have initial ensemble

$$\mathbf{y}^{(1)} = \begin{pmatrix} -3 \\ 0 \\ -3 \end{pmatrix}, \quad \mathbf{y}^{(2)} = \begin{pmatrix} -3 \\ 3 \\ -3 \end{pmatrix}, \quad \mathbf{y}^{(3)} = \begin{pmatrix} 3 \\ 0 \\ 3 \end{pmatrix}, \quad \text{and} \ \mathbf{y}^{(4)} = \begin{pmatrix} 3 \\ 3 \\ 3 \end{pmatrix}$$

Nelder–Mead Direction

The critical issue in any derivative-free optimization method is how to construct search directions without the aid of partial derivatives. Nelder-Mead Algorithm 13G adopts an away-from-worst approach.

$\boxed{13.50}$ At iteration t, The Nelder–Mead algorithm employs search direction

$$\Delta \mathbf{x} \triangleq \mathbf{x}^{(t)} - \mathbf{y}^{(n+1)}$$

which moves away from worst current solution $\mathbf{y}^{(n+1)}$ through the best-n centroid

$$\mathbf{x}^{(t)} \triangleq \frac{1}{n} \sum_{i=1}^{n} \mathbf{y}^{(i)}$$

The idea is to move away from the worst solution in the ensemble and toward the rest.

Figure 13.19 illustrates for $t = 0$ of the 2-variable search in Table 13.8. The centroid of the best 2 solutions is

$$\mathbf{x}^{(0)} \triangleq \frac{1}{2} \left[\begin{pmatrix} 5 \\ 3 \end{pmatrix} + \begin{pmatrix} 6 \\ 3 \end{pmatrix} \right] = \begin{pmatrix} 5.5 \\ 3 \end{pmatrix}$$

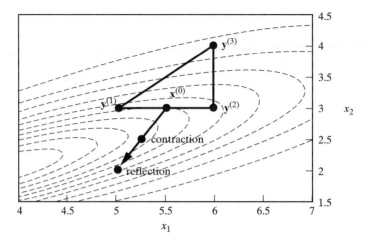

FIGURE 13.19   Nelder-Mead Direction for PERT Example

Thus the search direction of 13.50 is

$$\boldsymbol{\Delta x} \triangleq \mathbf{x}^{(0)} - \mathbf{y}^{(3)} = \begin{pmatrix} 5.5 \\ 3 \end{pmatrix} - \begin{pmatrix} 6 \\ 4 \end{pmatrix} = \begin{pmatrix} -0.5 \\ -1.0 \end{pmatrix}$$

SAMPLE EXERCISE 13.22: COMPUTING NELDER–MEAD DIRECTIONS

Return to Sample Exercise 13.21, and compute the Nelder–Mead direction corresponding to the initial ensemble constructed there.

Analysis: The centroid of the best 3 solutions in that 3-variable model is

$$\begin{aligned} \mathbf{x}^{(0)} &= \tfrac{1}{3}\left(\mathbf{y}^{(1)} + \mathbf{y}^{(2)} + \mathbf{y}^{(3)}\right) \\ &= \tfrac{1}{3}\left[(-3, 0, -3) + (-3, 3, -3) + (3, 0, 3)\right] \\ &= (-1, 1, -1) \end{aligned}$$

Thus direction 13.50 becomes

$$\boldsymbol{\Delta x} = \mathbf{x}^{(0)} - \mathbf{y}^{(4)} = (-1, 1, -1) - (3, 3, 3) = (-4, -2, -4)$$

Nelder–Mead Limited Step Sizes

Centroid $\mathbf{x}^{(t)}$ plays the role of a current solution in Nelder-Mead Algorithm 13G. But what step size λ should be applied to direction 13.50 ?

One strategy would undertake a full line search over

$$\mathbf{x}^{(t)} + \lambda \boldsymbol{\Delta x}$$

However, the required computation is usually not justified for such a crudely derived direction $\boldsymbol{\Delta x}$. Also, too large a step would destroy the "spread" pattern of the ensemble.

Algorithm 13G confronts these concerns by trying only one or two λ's drawn from possibilities $\{+1, +2, +\tfrac{1}{2}, -\tfrac{1}{2}\}$.

13.51 | The Nelder–Mead algorithm explores new points $\mathbf{x}^{(t)} + \lambda \Delta \mathbf{x}$ by first **reflecting** about centroid $\mathbf{x}^{(t)}$ with $\lambda = 1$. If replacing $\mathbf{y}^{(n+1)}$ with this new point would leave it neither best nor worst in the ensemble, it is adopted without further trials. If the reflection point is a new best, the algorithm **expands** to try $\lambda = 2$. If the point would be worst, the procedure **contracts** to try either $\lambda = +\frac{1}{2}$ or $\lambda = -\frac{1}{2}$.

Figure 13.19 illustrates for $t = 0$ in Table 13.8. Trying $\lambda = 1$ produces reflection point

$$\mathbf{x}^{(0)} + \Delta \mathbf{x} = (5.5, 3) + (-0.5, -1.0) = (5, 2)$$

with $f(5, 2) = 11.415$. The term reflection comes from the fact that this new point mirrors worst ensemble $\mathbf{y}^{(3)} = (6, 4)$ on the opposite side of centroid $\mathbf{x}^{(0)} = (5.5, 3)$.

Immediately replacing $\mathbf{y}^{(3)}$ with this reflection point would leave it worst in the new ensemble. Thus we try to do better by contracting with $\lambda = +\frac{1}{2}$. (Contraction would use $\lambda = -\frac{1}{2}$ if the reflection point had been no better than worst $\mathbf{y}^{(3)}$.)

The resulting $(5.25, 2.5)$ yields far superior $f(5.25, 2.5) = 24.915$. Substituting this contraction point for the worst in the current ensemble produces the improved set

$$\mathbf{y}^{(1)} = (5.25, 2.5), \quad \mathbf{y}^{(2)} = (5, 3), \qquad \mathbf{y}^{(3)} = (6, 3)$$

$$f(\mathbf{y}^{(1)}) = 24.915, \quad f(\mathbf{y}^{(2)}) = 12.425, \quad f(\mathbf{y}^{(3)}) = 11.429$$

of $t = 1$. Notice that points have been renumbered to keep them in objective function value sequence.

Table 13.8 shows the direction | 13.50 | for this new ensemble is $\Delta \mathbf{x} = (-0.875, -0.250)$, and new centroid $\mathbf{x}^{(1)} = (5.125, 2.75)$. Now reflection with $\lambda = 1$ produces new best $(4.25, 2.5)$ with $f(4.25, 2.5) = 30.005$. This leads to trying expansion with $\lambda = 2$. However, the result is not as good, so the reflection point joins the ensemble.

SAMPLE EXERCISE 13.23: EXECUTING NELDER–MEAD SEARCH

Suppose that a minimizing Nelder–Mead search has current ensemble objective values $f(\mathbf{y}^{(1)}) = 13$, $f(\mathbf{y}^{(2)}) = 21$, $f(\mathbf{y}^{(3)}) = 25$, and $f(\mathbf{y}^{(4)}) = 50$. For cases (a) to (d) in the following table, describe which step sizes λ would be tried at this iteration, and which (if any) chosen.

	$f(\mathbf{x}^{(t)} - \frac{1}{2}\mathbf{x}\Delta)$	$f(\mathbf{x}^{(t)} + \frac{1}{2}\mathbf{x}\Delta)$	$f(\mathbf{x}^{(t)} + 1\Delta\mathbf{x})$	$f(\mathbf{x}^{(t)} + 2\Delta\mathbf{x})$
(a)	12	15	18	30
(b)	40	22	12	10
(c)	43	51	60	25
(d)	60	49	44	70

Analysis: We follow details of Algorithm 13G.

(a) Reflection ($\lambda = 1$) objective value 18 is neither better than best ensemble value $f(\mathbf{y}^{(1)}) = 13$ nor worse than second worst $f(\mathbf{y}^{(3)}) = 25$. Thus we replace worst current point $\mathbf{y}^{(4)}$ with the one for $\lambda = 1$.

(b) Reflection ($\lambda = 1$) objective value 12 is better than ensemble best $f(\mathbf{y}^{(1)}) = 13$. Thus we expand and try $\lambda = 2$. The resulting value of 10 improves, so we replace worst current point $\mathbf{y}^{(4)}$ with the one for $\lambda = 2$.

(c) Reflection ($\lambda = 1$) objective value 60 is worse than ensemble worst $f(\mathbf{y}^{(4)}) = 50$. Thus we contract and try $\lambda = -\frac{1}{2}$. The resulting value of 43 is better than worst, so we replace $\mathbf{y}^{(4)}$ with the one for $\lambda = -\frac{1}{2}$.

(d) Reflection point ($\lambda = 1$) objective value 44 is better than worst $f(\mathbf{y}^{(4)}) = 50$, but not second worst $f(\mathbf{y}^{(3)}) = 25$. Thus we contract and try $\lambda = \frac{1}{2}$. The resulting value of 49 is better than worst, so we replace $\mathbf{y}^{(4)}$ with the one for $\lambda = \frac{1}{2}$.

Nelder–Mead Shrinking

Sometimes neither the reflection point nor a contraction alternative yields a solution that would much improve the ensemble. This suggests the need for rescaling the entire array of ensemble points.

> 13.52 | When reflection and subsequent contraction fails to improve the Nelder–Mead algorithm's current ensemble, the procedure **shrinks** the whole array toward best point $\mathbf{y}^{(1)}$ by
> $$\mathbf{y}^{(i)} \leftarrow \tfrac{1}{2}(\mathbf{y}^{(1)} + \mathbf{y}^{(i)}) \qquad \text{for all } i = 2, \ldots, n+1$$

Figure 13.20 depicts such a shrinking step at $t = 3$ of the PERT maximum likelihood search in Table 13.8. Ensemble points $\mathbf{y}^{(2)}$ and $\mathbf{y}^{(3)}$ are moved halfway to best $\mathbf{y}^{(1)}$ as

$$\text{new } \mathbf{y}^{(2)} \leftarrow \tfrac{1}{2}(\mathbf{y}^{(1)} + \mathbf{y}^{(2)}) = \tfrac{1}{2}\left[\begin{pmatrix} 4.25 \\ 2.5 \end{pmatrix} + \begin{pmatrix} 5.25 \\ 2.5 \end{pmatrix}\right] = \begin{pmatrix} 4.75 \\ 2.5 \end{pmatrix}$$

$$\text{new } \mathbf{y}^{(3)} \leftarrow \tfrac{1}{2}(\mathbf{y}^{(1)} + \mathbf{y}^{(3)}) = \tfrac{1}{2}\left[\begin{pmatrix} 4.25 \\ 2.5 \end{pmatrix} + \begin{pmatrix} 4.625 \\ 2.25 \end{pmatrix}\right] = \begin{pmatrix} 4.438 \\ 2.375 \end{pmatrix}$$

After renumbering to reflect new objective value sequence, the search proceeds with this more compact ensemble.

SAMPLE EXERCISE 13.24: SHRINKING IN NELDER–MEAD SEARCH

Suppose that it becomes necessary to apply shrinking Step 7 of Algorithm 13G with current ensemble

$$\mathbf{y}^{(1)} = (2, 3), \quad \mathbf{y}^{(2)} = (-4, 5), \quad \text{and} \quad \mathbf{y}^{(3)} = (8, -1)$$

Compute the new ensemble.

Analysis: We apply computation 13.52 .

$$\text{new } \mathbf{y}^{(1)} = \mathbf{y}^{(1)} = (2, 3)$$
$$\text{new } \mathbf{y}^{(2)} = \tfrac{1}{2}(\mathbf{y}^{(1)} + \mathbf{y}^{(2)}) = \tfrac{1}{2}[(2, 3) + (-4, 5)] = (-1, 4)$$
$$\text{new } \mathbf{y}^{(3)} = \tfrac{1}{2}(\mathbf{y}^{(1)} + \mathbf{y}^{(3)}) = \tfrac{1}{2}([2, 3) + (8, -1)] = (5, 1)$$

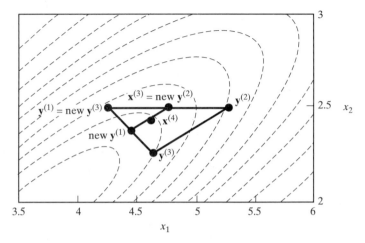

FIGURE 13.20   Nelder–Mead Shrinking in PERT Example

Nelder–Mead Search of PERT Example

Table 13.8 detailed all 20 iterations required in Algorithm 13G search of our maximum likelihood PERT model (13.10). When one of the trial λ values was adopted, it was shown boxed in the table. Otherwise, shrinking procedure $\boxed{13.52}$ was invoked.

Nelder–Mead search stops when objective function values for points in the ensemble become essentially equal. In Table 13.8 this condition was reached when

$$\sqrt{\frac{1}{n+1} \sum_{i=1}^{n+1} \left[f(\mathbf{y}^{(i)}) - f(\mathbf{x}^{(t)}) \right]^2} < \epsilon = 0.5$$

Then the best of the ensemble and centroid points is selected as an approximate optimum $\mathbf{x}^* = (4.5, 2.207)$ with $f(\mathbf{x}^*) = 48{,}520$.

EXERCISES

13-1 A biomedical intrumentation company sells its main product at the rate of 5 units per day. The instrument is manufactured in lots run every few days. It costs the company \$2000 to setup for production of a lot and \$40 per unit per day to hold finished instruments in inventory between runs. The company would like to choose a lot size that minimizes average inventory and setup cost per day assuming that demand occurs smoothly at the given rate.

(a) ⊘ Formulate a 1-variable unconstrained NLP to choose an optimum lot size.

(b) ⊘ ⊞ Plot the objective function of your model and compute an optimum lot size graphically.

13-2 As part of a study of 911 emergency calls, an analyst wishes to choose the value of parameter α in **exponential** probability density function $d(t) \triangleq \alpha e^{-\alpha t}$

that best fits call interarrival times 80, 10, 14, 26, 40, and 22 minutes.

(a) Formulate a 1-variable unconstrained NLP to choose a maximum likelihood estimate of α.

(b) ⊘ ⊞ Plot the objective function of your model and compute an optimal estimate graphically.

13-3 An oil drilling company wishes to locate a supply base somewhere in the jungle area where it is presently exploring for oil. The base will service drilling sites at map coordinates (0,-30), (50,-10), (70,20), and (30,50) with helicopter supply runs. The company wishes to choose a location that minimizes the sum of flying distances to the four sites.

(a) ⊘ Formulate an unconstrained NLP to choose an optimum base location.

(b) ◎ ▢ Use class optimization software to compute at least a local optimum starting from co-ordinates (10,10).

13-4 ◎ Repeat Exercise 13-3, this time minimizing the maximum distance to any drilling site.

13-5 An electronics assembly firm is planning its production staff needs to make a new modem. It has measured one test worker assembling the unit and observed the following data:

Through unit	2	6	20	25	40
Average time	8.4	5.5	4.2	3.7	3.1

Experience shows that **learning curves**, which describe the ability of a worker to improve his or her productivity as more an more units are produced, often take the form

$$\text{average time} = a \, (\text{no. units})^b$$

where a and b are empirical constants.

(a) ◎ Formulate an unconstrained NLP to fit this form to the given data so as to minimize the sum of squared errors.

(b) ◎ ▢ Use class optimization software to compute at least a local optimum for your curve fit-ting model starting with $a = 15$, $b = -0.5$.

13-6 The following shows a series of measurements of the height (in inches) of a new genetically engi-neered tomato plant versus the number of weeks af-ter the plant was replanted outdoors.

Week	1	2	4	6	8	10
Height	9	15	22	33	44	52

Researchers wish to fit this experience with an S-shaped **logistics curve**

$$\text{size} = \frac{k}{1 + e^{a+b(\text{weeks})}}$$

where k, a, and b are empirical parameters.

(a) Formulate an unconstrained NLP to fit this form to the given data so as to minimize the sum of squared errors.

(b) ◎ ▢ Use class optimization software to compute at least a local optimum for your curve fitting model starting with $k = 50$, $a = 3$, $b = -0.3$.

13-7 The university motor pool[3] provides a large number of cars n for faculty and staff traveling on university business. Motor pool cars have an aver-age annual cost of f dollars per car for fixed expenses such as depreciation, insurance, and licensing, plus a variable operating cost of v_m cents per mile driven. Those travelers who cannot be accommodated by the pool must drive their personal cars and be reim-bursed at v_p cents per mile. Demand varies greatly among times of the year, but an extensive analysis of past travel records has fitted regression equations

$$m(n) \triangleq a_m + b_m n + c_m/n$$
$$p(n) \triangleq a_p + b_p n + c_p/n$$

to the annual numbers of miles that would be driven in motor pool and personal cars as a function of the size of the motor pool available. Formulate a 1-variable unconstrained NLP to compute a minimum total cost motor pool size. Ignore integrality.

13-8 Once a site for a new service facility has been chosen, the limits of its market area must be deter-mined,[4] along with the corresponding facility size. Assume (i) that the facility is to be located at the center of a circular market area and sized to cover uniform density d calls per unit area out to a radius of r; (ii) that the cost of operating the facility is $[f + c(\text{size})]$, where f is a fixed cost and c a variable cost per unit size; and (iii) transportation costs per call are propor-tional to (straight-line, one-way) distance from the facility to the customer at t per unit distance. For-mulate a 1-variable unconstrained NLP to choose a market area radius that minimizes the facility's aver-age total cost per call. Evaluate any calculus integrals in your objective.

13-9 Renewing highway pavement markings[5] costs c dollars per mile but reduces social costs from delays, accidents, and other effects of declining

[3]Based on W. W. Williams and O. S. Fowler (1980), "Minimum Cost Fleet Sizing for a University Motor Pool," *Interfaces 10:3*, 21–27.

[4]Based on D. Erlenkotter (1989), "The General Optimal Market Area Model," *Annals of Operations Research, 18*, 45–70.

[5]Based on V. Kouskoulas (1988), "An Optimization Model for Pavement Marking Systems," *Euro-pean Journal of Operational Research, 33*, 298–303.

marking performance over time. Suppose that new markings yield maximum performance p_{max} and that performance t days after renewal can be expressed $p(t) \triangleq p_{max}e^{-\alpha t}$, where α is a constant depending on the durability of the markings. Also assume that each unit of lost performance costs d dollars per day per mile. Formulate a 1-variable unconstrained NLP to choose a time between pavement renewals to minimize average total daily cost per mile. Evaluate any calculus integrals in your objective.

13-10 The number of potential patrons p_i of a new movie theater complex has been estimated from census data for each of the surrounding counties $i = 1, \ldots, 15$. However, the fraction of potential patrons from any i who will actually use the complex varies inversely with its (straight line) distance from the county centroid at coordinates (x_i, y_i). Formulate a 2-variable NLP to choose a maximum total realized patronage location for the complex.

13-11 Denoting by n_t the number of universities using a textbook through semester t of its availability ($n_0 = 0$), the number of new adoptions in any single semester t can be estimated $(a + bn_{t-1})(m - n_{t-1})$, where a and b are parameters relating to the rapidity of success, and m is the maximum number of universities who will ever adopt.

(a) Given values of n_t for $t = 1, \ldots, 10$, formulate an unconstrained NLP to make a nonlinear least squares fit of the foregoing approximation to these data.

(b) ◫ Show that an appropriate change of parameters can covert your model into a linear least squares problem.

13-12 Major aircraft parts undergo inspection and overhaul[6] every t_1 flying hours, and replacement every t_2. Experience shows the cost of overhauling a particular model of jet engine can be expressed as increasing nonlinear function $a(t_1)^b$ of the overhaul cycle, and operating costs per hour are an increasing nonlinear function $c(t_2)^d$ of the replacement time. Each replacement costs a fixed part cost f plus labor cost $g(t_1)^h$ to identify and fix associated problems. Formulate an unconstrained NLP over t_1 and t_2 to find an overhaul and replacement policy that minimizes total cost per flying hour.

13-13 Determine whether each of the following functions is smooth on the specified domain.

(a) ◫ $f(x) \triangleq x^4 + 3x - 19$ for all x
(b) $f(x) \triangleq \max\{2x - 1, 2 - x\}$ for all x
(c) ◫ $f(x) \triangleq |x - 5|$ for $x > 0$
(d) $f(x) \triangleq 3x + \ln(x)$ for $x > 0$
(e) ◫ $f(x_1, x_2) \triangleq x_1 e^{x_2}$ for all x_1, x_2
(f) $f(x_1, x_2) \triangleq |x_1 - 1| + |x_2 + 3|$ for $x_1, x_2 \geq 0$.

13-14 Each of the following plots shows a function $f(x)$. Determine graphically whether each indicated point is an unconstrained local maximum, an unconstrained global maximum, an unconstrained local minimum, an unconstrained global minimum, or none of the above over the domain depicted.

(a) ◫

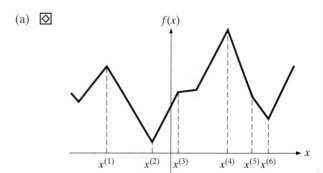

(b)

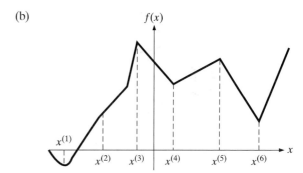

13-15 Each of the following plots shows contours of a smooth function $f(x_1, x_2)$. Determine graphically whether each indicated point is an unconstrained local maximum, an unconstrained global maximum, an unconstrained local minimum, an unconstrained global minimum, or none of the above over the domain depicted.

(a) ◫ Points $\mathbf{x}^{(1)} = (1, 3)$, $\mathbf{x}^{(2)} = (4, 4)$, $\mathbf{x}^{(3)} = (3, 3)$, $\mathbf{x}^{(4)} = (4, 1)$, $\mathbf{x}^{(5)} = (2, 1)$

---

[6]Based on T. C. E. Cheng (1992), "Optimal Replacement of Ageing Equipment Using Geometric Programming," *International Journal of Production Research, 30*, 2151–2158.

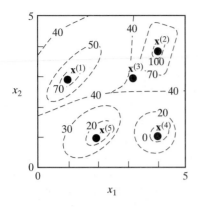

(b) Points $\mathbf{x}^{(1)} = (6, 6)$, $\mathbf{x}^{(2)} = (2, 8)$, $\mathbf{x}^{(3)} = (8, 2)$, $\mathbf{x}^{(4)} = (4, 4)$, $\mathbf{x}^{(5)} = (7, 6)$

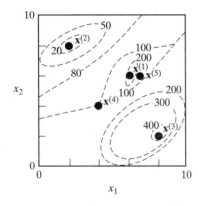

13-16 Determine graphically whether each of the following functions is unimodal for a maximize problem and/or for a minimize problem over the range depicted. If not, identify a violating pair of points.

(a) ⊠ The function of Exercise 13-14(a)
(b) The function of Exercise 13-14(b)
(c) ⊠ The function of Exercise 13-15(a)
(d) The function of Exercise 13-15(b)
(e) ⊠ The function

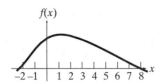

(f) The function

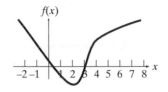

13-17 ⊠ Use golden section Algorithm 13A to find an optimum of the NLP

$$\min \quad 10x + \frac{70}{x}$$
$$\text{s.t.} \quad 1 \le x \le 10$$

to within an error of ± 1.

13-18 Use golden section Algorithm 13A to find an optimum of the NLP

$$\max \quad 500 - x(x - 20)^3$$
$$\text{s.t.} \quad 0 \le x \le 20$$

to within an error of ± 1.

13-19 Suppose that we were given only the lower limit of 1 in the NLP of Exercise 13-17. Apply 3-point pattern Algorithm 13B to compute a corresponding upper limit with which golden section search could begin using each of the following initial step sizes δ.

(a) ⊠ $\delta = 0.5$
(b) ⊠ $\delta = 16$

13-20 Do Exercise 13-19 for the NLP of Exercise 13-18 using $\delta = 2$ and $\delta = 20$.

13-21 ⊠ Use quadratic fit Algorithm 13C to compute an optimum for the NLP of Exercise 13-17 within an error tolerance of 2. Start with the 3-point pattern $\{1, 2, 10\}$.

13-22 Use quadratic fit Algorithm 13C to compute an optimum for the NLP of Exercise 13-18 within an error tolerance of 4. Start with 3-point pattern $\{0, 3, 20\}$.

13-23 Consider the 1-variable function $f(x) \triangleq x^3 - 3x^2 + 11x$ at current point $x = 3$.

(a) ⊠ Derive the first-order Taylor approximation to $f(x + \lambda)$.
(b) ⊠ Derive the second-order Taylor approximation to $f(x + \lambda)$.
(c) ▦ Plot the original function and both Taylor series approximations in the vicinity of x. How accurate do the approximations appear to be? Which is better?

13-24 Do Exercise 13-23 for function $f(x) \triangleq 18x - 50\ln(x)$ at $x = 16$.

13-25 Consider the 2-variable function $f(x_1, x_2) \triangleq (x_1)^3 - 5x_1x_2 + 6(x_2)^2$ with current point $\mathbf{x} = (0, 2)$ and move direction $\Delta\mathbf{x} = (1, -1)$.

(a) ☒ Derive the first-order Taylor approximation to $f(\mathbf{x} + \lambda\Delta\mathbf{x})$.

(b) ☒ Derive the second-order Taylor approximation to $f(\mathbf{x} + \lambda\Delta\mathbf{x})$.

(c) ⊞ Plot the original function and both Taylor series approximations as functions of λ. How accurate do the approximations appear to be? Which is better?

13-26 Do Exercise 13-25 for function $f(x_1, x_2) \triangleq 13x_1 - 6x_1x_2 + 8/x_2$, $\mathbf{x} = (2, 1)$ and $\Delta\mathbf{x} = (3, 1)$.

13-27 For each of the following unconstrained NLPs, either verify that the given $\mathbf{x}$ is a stationary point of the objective function or give a direction $\Delta\mathbf{x}$ that improves at $\mathbf{x}$.

(a) ☒ min $(x_1)^2 + x_1x_2 - 6x_1 - 8x_2$, $\mathbf{x} = (8, -10)$
(b) max $10(x_1)^2 + 12\ln(x_2)$, $\mathbf{x} = (1, 2)$
(c) ☒ min $16x_1 - x_1x_2 + 2(x_2)^2$, $\mathbf{x} = (3, 0)$
(d) max $x_1x_2 - 10x_1 + 4x_2$, $\mathbf{x} = (-4, 10)$

13-28 For each of the following functions f, use conditions $\boxed{13.19}$ to $\boxed{13.22}$ to classify the specified $\mathbf{x}$ as definitely local maximum, possibly local maximum, definitely local minimum, possibly local minimum, and/or definitely neither.

(a) ☒ $f(x_1, x_2) \triangleq 3(x_1)^2 - x_1x_2 + (x_2)^2 - 11x_1$, $\mathbf{x} = (2, 1)$
(b) $f(x_1, x_2) \triangleq -(x_1)^2 - 6x_1x_2 - 9(x_2)^2$, $\mathbf{x} = (-3, 1)$
(c) ☒ $f(x_1, x_2) \triangleq x_1 - x_1x_2 + (x_2)^2$, $\mathbf{x} = (2, 1)$
(d) $f(x_1, x_2) \triangleq 12x_2 - (x_1)^2 + 3x_1x_2 - 3(x_2)^2$, $\mathbf{x} = (12, 8)$
(e) ☒ $f(x_1, x_2) \triangleq x_1(x_2)^3$, $\mathbf{x} = (0, 0)$
(f) $f(x_1, x_2) \triangleq 6x_1 + \ln(x_1) + (x_2)^2$, $\mathbf{x} = (1, 3)$
(g) ☒ $f(x_1, x_2) \triangleq 2(x_1)^2 + 8x_1x_2 + 8(x_2)^2 - 12x_1 - 24x_2$, $\mathbf{x} = (1, 1)$
(h) $f(x_1, x_2) \triangleq 4(x_1)^2 + 3/x_2 - 8x_1 + 3x_2$, $\mathbf{x} = (1, 1)$

13-29 Determine whether each of the following functions is convex, concave, both, or neither over the domain specified.

(a) ☒ $f(x_1, x_2) \triangleq \ln(x_1) + 20\ln(x_2)$ over $x_1, x_2 > 0$
(b) $f(x) \triangleq x\sin(x)$ over $x \in [0, 2\pi)$
(c) ☒ $f(x) \triangleq x(x - 2)^2$ over all $x \geq 0$

(d) $f(x) \triangleq (x - 8)^4 + 132x$ over all x
(e) ☒ $f(x_1, \ldots, x_5) \triangleq 3x_1 + 11x_2 - x_3 - 8x_5$ over all $(x_1, \ldots, x_5)$
(f) $f(x_1, x_2) \triangleq 21x_1 + 63x_2$ over $x_1, x_2 \geq 0$
(g) ☒ $f(x_1, x_2) \triangleq \max\{x_1, x_2\}$ over all x_1, x_2
(h) $f(x_1, x_2) \triangleq \min\{13x_1 - 2x_2, -(x_2)^2\}$ over all x_1, x_2
(i) ☒ $f(x_1, x_2) \triangleq 10/\sqrt{x_1} - 40\sqrt{x_2}$ over $x_1, x_2 \geq 0$
(j) $f(x_1, x_2) \triangleq \ln[-3(x_1)^2 - 9(x_2)^2] - 10/x_2$ over $x_1, x_2 > 0$

13-30 Use convexity/concavity to establish that each of the following solutions $\mathbf{x}$ is either an unconstrained global maximum or an unconstrained global minimum of the f indicated, and explain which.

(a) ☒ $f(x_1, x_2) \triangleq (x_1 - 5)^2 + x_1x_2 + (x_2 - 7)^2$ at $\mathbf{x} = (2, 6)$
(b) $f(x_1, x_2) \triangleq 500 - 8(x_1 + 1)^2 - 2(x_2 - 1)^2 + 4x_1x_2$ at $\mathbf{x} = (-1, 0)$

13-31 Consider the unconstrained NLP

$$\max \quad x_1x_2 - 5(x_1 - 2)^4 - 3(x_2 - 5)^4$$

(a) ▭ Use graphing software to produce a contour map of the objective function for $x_1 \in [1, 4]$, $x_2 \in [2, 8]$.
(b) ☒ Compute the move direction that would be pursued by gradient search Algorithm 13D at $\mathbf{x}^{(0)} = (1, 3)$.
(c) ☒ State the line search problem implied by your direction.
(d) ☒ ⊞ Solve your line search problem graphically and compute the next search point $\mathbf{x}^{(1)}$.
(e) ☒ ⊞ Do two additional iterations of Algorithm 13D to compute $\mathbf{x}^{(2)}$ and $\mathbf{x}^{(3)}$.
(f) ▭ Plot progress of the search on the contour map of part (a).

13-32 Do Exercise 13-31 for the unconstrained NLP

$$\min \quad \frac{1000}{x_1 + x_2} + (x_1 - 4)^2 + (x_2 - 10)^2$$

starting from $\mathbf{x}^{(0)} = (10, 1)$, and plotting $x_1 \in [2, 11]$, $x_2 \in [0, 15]$.

13-33 Return to the unconstrained optimization of Exercise 13-31 starting from $\mathbf{x}^{(0)} = (3, 7)$.

(a) ☒ Write the second-order Taylor approximation to the objective function at $\mathbf{x}^{(0)}$ for unknown $\Delta\mathbf{x}$ and $\lambda = 1$.
(b) ☒ Compute the Newton direction $\Delta\mathbf{x}$ at $\mathbf{x}^{(0)}$ and verify that it is a stationary point of your second-order Taylor approximation.

(c) ⊘ Beginning with your Newton direction, complete 2 iterations of Newton's method Algorithm 13E.

(d) ▭ Plot progress of your search on a contour map like the one of Exercise 13-31(a).

13-34 Do Exercise 13-33 on the NLP of Exercise 13-32 starting from $\mathbf{x}^{(0)} = (10, 1)$.

13-35 Return to the unconstrained optimization of Exercise 13-31 and consider BFGS Algorithm 13F starting at $\mathbf{x}^{(0)} = (2, 3)$.

(a) ⊘ Compute the first direction that would be pursued by Algorithm 13F.

(b) ⊘ Assuming that the optimal step is $\lambda = 0.026$ in that direction, compute the new solution $\mathbf{x}^{(1)}$, the next deflection matrix $\mathbf{D}$, and the next BFGS search direction $\Delta\mathbf{x}$.

(c) Verify that your deflection matrix of part (b) satisfies quasi-Newton conditions $\boxed{13.43}$ (within roundoff error) and is symmetric.

(d) Verify algebraically that your direction of part (b) is improving.

(e) ▭ Plot your first move and second direction on a contour map like that of Exercise 13-31(a).

13-36 Do Exercise 13-35 on the NLP of Exercise 13-32 starting from $\mathbf{x}^{(0)} = (8, 1)$ and using $\lambda = 0.32$ in part (b).

13-37 Consider the unconstrained NLP

$$\min \quad \max\{10 - x_1 - x_2,$$
$$6 + 6x_1 - 3x_2,$$
$$6 - 3x_1 + 6x_2\}$$

(a) ⊘ Explain why Nelder–Mead search is appropriate for solving this unconstrained optimization.

(b) ⊘ Do 3 iterations (moves) of Nelder–Mead Algorithm 13G, starting from initial ensemble (5,0), (10,5), (5,5).

(c) Plot the progress of your search through centroids $\mathbf{x}^{(t)}$ and connecting the 3 points of each ensemble with dashed lines.

13-38 Do Exercise 13-37 for the NLP

$$\max \quad \min\{20 - x_1 - x_2, \quad 6 + 3x_1 - x_2, \quad 6 - x_1 + 3x_2\}$$

starting with ensemble (0,0), (0,2), (2,2).

13-39 Compute the Nelder–Mead Algorithm 13G ensemble that would result from applying the shrinking step to each of the following ($\mathbf{y}^{(1)}$ best objective value, etc.).

(a) ⊘ $\mathbf{y}^{(1)} = (1, 2, 1)$, $\mathbf{y}^{(2)} = (5, 4, 5)$, $\mathbf{y}^{(3)} = (3, 2, 7)$, $\mathbf{y}^{(4)} = (7, 2, 7)$

(b) $\mathbf{y}^{(1)} = (10, 8, 10)$, $\mathbf{y}^{(2)} = (4, 6, 2)$, $\mathbf{y}^{(3)} = (0, 0, 0)$, $\mathbf{y}^{(4)} = (0, 10, 6)$

SUGGESTED READING

Bazarra, Mokhtar S., Hanif D. Sherali, and C. M. Shetty (1993), *Nonlinear Programming Theory and Algorithms*, Wiley, New York.

Luenberger, David G. (1984), *Linear and Nonlinear Programming*, Addison-Wesley, Reading, Mass.

CHAPTER 14

Constrained Nonlinear Programming

· ·

In this chapter we introduce models and methods for constrained nonlinear optimization. We begin by modeling a series of real applications, emphasizing differences between linear and nonlinear constraints, convex and nonconvex feasible sets, separable and nonseparable objective functions, and so on. Then a variety of solution methods are developed, some fairly general purpose and others restricted to special classes on NLPs. Theoretical development draws on Chapters 3, 5 and 13.

14.1 CONSTRAINED NONLINEAR PROGRAMMING MODELS

As usual, we begin our investigation of constrained nonlinear programming with some examples. In this section we formulate three cases, illustrating the broad range on constrained NLP, and in Section 14.2 we present three more, representing some classic special cases. All are based on real application contexts drawn from published reports.

EXAMPLE 14.1: BEER BELGE LOCATION ALLOCATION

One common class of nonlinear programs arises from the **location** and customer **allocation** of distribution and service facilities. The task confronted by a large Belgian brewery we will call Beer Belge illustrates.[1] Beer Belge wishes to realign its 17 depots to more efficiently solve its 24,000 customers throughout Belgium. For purposes of this analysis the customers can be aggregated in 650 regions.

Assigning index dimensions

$$i \triangleq \text{depot number } (i = 1, \ldots, 17)$$
$$j \triangleq \text{customer region number } (j = 1, \ldots, 650)$$

company analysts can determine from maps and past experience

---

[1]Based in part on L. F. Gelders, L. M. Pintelon, and L. N. Van Wassenhove (1987), "A Location-Allocation Problem in a Large Belgian Brewery," *European Journal of Operational Research, 28*, 196–206.

$h_j \triangleq$ x-coordinate of the center of customer region j

$k_j \triangleq$ y-coordinate of the center of customer region j

$d_j \triangleq$ number of delivery trips required to region j per year

We wish to choose locations for the depots and allocate customer region demand to minimize total travel cost.

Beer Belge Location-Allocation Model

As with all such **location-allocation** problems, decision variables in a model of Beer Belge's case must settle two distinct but related questions. First, we need to know where the depots will be located. Then we must allocate the trips required in various regions among the depots. The following address both:

$x_i \triangleq$ x-coordinate of depot i's location

$y_i \triangleq$ y-coordinate of depot i's location

$w_{i,j} \triangleq$ number of trips per year from depot i to customer region j

To obtain an objective function, we adopt the common assumption that round-trip travel cost from depot i to customer region j is proportional to the straight-line (Euclidean) distance between their locations. Then minimizing total transportation cost of the location-allocation amounts to

$$\min \sum_i \sum_j (\text{number of trips from } i \text{ to } j)(\text{distance from } i \text{ to } j)$$

Using symbolic constants and decision variables defined above, we obtain

$$\min \sum_{i=1}^{17} \sum_{j=1}^{650} w_{i,j}\sqrt{(x_i - h_j)^2 + (y_i - k_j)^2}$$

To complete a model, we must add constraints assuring that all required trips to each region are made. The result is

$$\min \quad \sum_{i=1}^{17} \sum_{j=1}^{650} w_{i,j}\sqrt{(x_i - h_j)^2 + (y_i - k_j)^2} \qquad \text{(total travel)}$$

$$\text{s.t.} \quad \sum_{i=1}^{17} w_{i,j} = d_j \quad j = 1, \dots, 650 \qquad \text{(allocate trips)} \tag{14.1}$$

$$w_{i,j} \geq 0 \qquad i = 1, \dots, 17; \quad j = 1, \dots, 650$$

SAMPLE EXERCISE 14.1: FORMULATING LOCATION-ALLOCATION MODELS

Two airstrips are to be constructed in the jungle to service 3 remote oil fields. The first oil field requires 25 tons of supplies per month. The second, which is 75 miles east and 330 miles north of the first, requires 14 tons. The third, which is 225 miles west and 40 miles south of the first, needs 34 tons per month. Formulate a location-allocation model to locate and operate the airstrips to minimize the ton-miles flown per month.

Modeling: Coordinates in this problem may be introduced by taking the first oil field as $(0, 0)$. Then the second is $(-75, 330)$ and the third is $(225, -40)$.

Decisions involve both the location of the airstrips and allocation of workload between the strips. Thus we employ decision variables $(x_1, y_1) \triangleq$ coordinates of the first airstrip, $(x_2, y_2) \triangleq$ coordinates of the second, and $w_{i,j} \triangleq$ tons flown from airstrip i to oil field j. The resulting model is

$$\min \quad w_{1,1}\sqrt{(x_1 - 0)^2 + (y_1 - 0)^2} + w_{1,2}\sqrt{(x_1 + 75)^2 + (y_1 - 330)^2} \quad \text{(ton-miles)}$$
$$+ w_{1,3}\sqrt{(x_1 - 225)^2 + (y_1 + 40)^2} + w_{2,1}\sqrt{(x_2 - 0)^2 + (y_2 - 0)^2}$$
$$+ w_{2,2}\sqrt{(x_2 + 75)^2 + (y_2 - 330)^2} + w_{2,3}\sqrt{(x_2 - 225)^2 + (y_2 + 40)^2}$$

$$\text{s.t.} \quad \begin{aligned} w_{1,1} + w_{2,1} &= 25 && \text{(field 1)} \\ w_{1,2} + w_{2,2} &= 14 && \text{(field 2)} \\ w_{1,3} + w_{2,3} &= 34 && \text{(field 3)} \\ w_{i,j} &\geq 0 \ i = 1, 2; && j = 1, 2, 3 \end{aligned}$$

Here the objective function totals flight distances times tons shipped, and constraints allocate needed shipments among airstrips.

Linearly Constrained Nonlinear Programs

Location-allocation model (14.1) illustrates the many large-scale nonlinear programs having all constraints linear. Only its objective function renders the model an NLP.

Linearly constrained nonlinear programs form an important special class because much of the tractability of linear programs extends to cases with only the objective function nonlinear. Fortunately, linearly constrained NLPs are also very common.

> **14.1** Most large-scale nonlinear programs that can be solved effectively have all or nearly all their constraints linear.

EXAMPLE 14.2: TEXACO GASOLINE BLENDING

Beginning as early as the 1950s, oil companies have used mathematical programming to help plan gasoline blending at refineries. Texaco is no exception.[2]

Figure 14.1 sketches gasoline processing in refineries. Crude oil is first distilled into a range of materials from light ones such as gasoline to much heavier ones such as fuel oil. The "straight-run" gasoline produced in distilling is not nearly enough to meet market demands economically. Thus most of the distillates that are heavier or lighter than gasoline are re-formed and cracked to produce other gasoline forms. A final processing stage combines some additives with these various stocks of gasoline to produce market blends (e.g., premium unleaded, regular unleaded) with suitable quality indexes (e.g., octane, lead content, sulfur content, volatility).

Typical operations plan blending for a month at a time. Using index dimensions

$$i \triangleq \text{gasoline or additive stock number } (i = 1, \ldots, m)$$
$$j \triangleq \text{market blend number } (j = 1, \ldots, n)$$
$$k \triangleq \text{quality index number } (k = 1, \ldots, h)$$

[2]Based on C. W. DeWitt, L. S. Lasdon, A. D. Waren, D. A. Brenner and S. A. Melhem (1989), "OMEGA: An Improved Gasoline Blending System for Texaco," *Interfaces, 19:1*, 85–101.

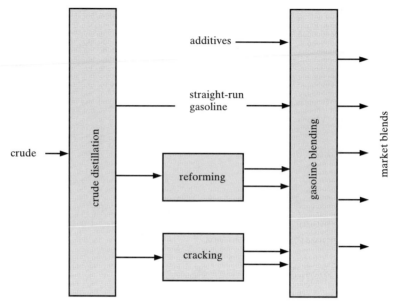

FIGURE 14.1   Gasoline Refinery Flow

We will assume that the following constants are known as each optimization begins:

$p_j \triangleq$ estimated selling price per unit of output blend j

$r_j \triangleq$ quantity of output blend j required

$s_i \triangleq$ quantity of input stock i available

$v_i \triangleq$ estimated cost per unit of input stock i

$a_{i,k} \triangleq k$th quality index of input stock i

$\ell_{j,k} \triangleq$ lowest acceptable level of quality index k in output blend j

$u_{j,k} \triangleq$ highest acceptable level of quality index k in output blend j

We want to find a maximum profit (sales income minus stock costs) plan to meet all blending requirements with available stocks.

Texaco Gasoline Blending Model

Clearly, the decisions in gasoline blending involve how much of which stocks to use. Thus we will develop a model with decision variables

$$x_{i,j} \triangleq \text{quantity of input stock } i \text{ used in output blend } j$$

In accord with principle $\boxed{14.1}$, much of the model will be linear. We begin with the objective to maximize sales income minus stock cost:

$$\max \quad \sum_{i=1}^{m} \sum_{j=1}^{n} (p_j - v_i) x_{i,j} \qquad \text{(profit)}$$

Linear constraints enforce stock availabilities and output demands:

$$\sum_{j=1}^{n} x_{i,j} \leq s_i \qquad i = 1, \ldots, m$$

$$\sum_{i=1}^{m} x_{i,j} \geq r_j \qquad j = 1, \ldots, n$$

Most of the blending constraints also take the linear form of the blending LPs in Section 4.2. That is, we require the average level of each quality index to fall between upper and lower limits for each output blend with constraints

$$\ell_{j,k} \leq \frac{\sum_{i=1}^{m} a_{i,k} x_{i,j}}{\sum_{i=1}^{m} x_{i,j}} \leq u_{j,k} \qquad j = 1, \ldots, n; \quad \text{linear } k$$

What introduces nonlinearity in gasoline blending are two classes of quality indexes that do not combine linearly. The volatility quality measures perform logarithmically to produce constraints

$$\ell_{j,k} \leq \ln\left(\frac{\sum_{i=1}^{m} a_{i,k} x_{i,j}}{\sum_{i=1}^{m} x_{i,j}}\right) \leq u_{j,k} \qquad j = 1, \ldots, n; \quad \text{volatility } k$$

Octane measures are even more complex. Typical schemes use known constants $b_{i,k}$, $c_{i,k}$, $d_{i,k}$, and $e_{i,k}$ to specify octane limits with fourth-order expressions

$$\ell_{j,k} \leq \frac{\sum_{i=1}^{m} b_{i,k} x_{i,j}}{\sum_{i=1}^{m} x_{i,j}} + \frac{\sum_{i=1}^{m} c_{i,k} (x_{i,j})^2}{\left(\sum_{i=1}^{m} x_{i,j}\right)^2} + \frac{\sum_{i=1}^{m} d_{i,k} (x_{i,j})^3}{\left(\sum_{i=1}^{m} x_{i,j}\right)^3}$$

$$+ \frac{\sum_{i=1}^{m} e_{i,k} (x_{i,j})^4}{\left(\sum_{i=1}^{m} x_{i,j}\right)^4} \leq u_{j,k} \qquad j = 1, \ldots, n; \quad \text{octane } k$$

Collecting all the above and adding variable-type constraints completes the full gasoline blending model:

$$\max \sum_{i=1}^{m} \sum_{j=1}^{n} (p_j - v_i) x_{i,j} \qquad \text{(profit)}$$

$$\text{s.t.} \quad \sum_{j=1}^{n} x_{i,j} \leq s_i \qquad\qquad i = 1, \ldots, m \qquad \text{(availability)}$$

$$\sum_{i=1}^{m} x_{i,j} \geq r_j \qquad\qquad j = 1, \ldots, n \qquad \text{(demand)}$$

$$\ell_{j,k} \leq \frac{\sum_{i=1}^{m} a_{i,k} x_{i,j}}{\sum_{i=1}^{m} x_{i,j}} \leq u_{j,k} \qquad j = 1, \ldots, n; \text{ linear } k \quad \text{(blend 1)}$$

$$\ell_{j,k} \leq \ln\left(\frac{\sum_{i=1}^{m} a_{i,k} x_{i,j}}{\sum_{i=1}^{m} x_{i,j}}\right) \leq u_{j,k} \qquad j = 1, \ldots, n; \text{ volatility } k \quad \text{(blend 2)}$$

$$\ell_{j,k} \leq \frac{\sum_{i=1}^{m} b_{i,k} x_{i,j}}{\sum_{i=1}^{m} x_{i,j}} + \frac{\sum_{i=1}^{m} c_{i,k} (x_{i,j})^2}{\left(\sum_{i=1}^{m} x_{i,j}\right)^2} + \frac{\sum_{i=1}^{m} d_{i,k} (x_{i,j})^3}{\left(\sum_{i=1}^{m} x_{i,j}\right)^3}$$

$$+ \frac{\sum_{i=1}^{m} e_{i,k} (x_{i,j})^4}{\left(\sum_{i=1}^{m} x_{i,j}\right)^4} \leq u_{j,k} \qquad j = 1, \ldots, n; \text{ octane } k \quad \text{(blend 3)}$$

$$x_{i,j} \geq 0 \qquad\qquad i = 1, \ldots, m; \quad j = 1, \ldots, n$$

(14.2)

This time it is the last two systems of main constraints that make the model a nonlinear program. All other elements are linear.

Engineering Design Models

The mostly linear location-allocation [model (14.1)] and gasoline blending [model (14.2)] nonlinear programs treated so far are typical of large-scale applications. Still, the NLPs arising in **engineering design** often have quite a different character.

> 14.2 | Optimal engineering design of structures and processes frequently leads to constrained nonlinear programs with relatively few variables but highly nonlinear constraints and objective functions.

EXAMPLE 14.3: OXYGEN SYSTEM ENGINEERING DESIGN

To illustrate the smaller but more nonlinear models arising in engineering, we consider the design of the oxygen production system for a basic oxygen furnace in the steel industry.[3] Figure 14.2(a) shows the main components of the system. The oxygen plant produces at a constant rate, with output compressed and stored in a tank. The furnace then uses the oxygen in a cycle like the one depicted in Figure 14.2(b). Relatively low demand level d_1 applies during the first t_1 minutes of the cycle, followed by a burst to demand d_2 from time t_1 through t_2.

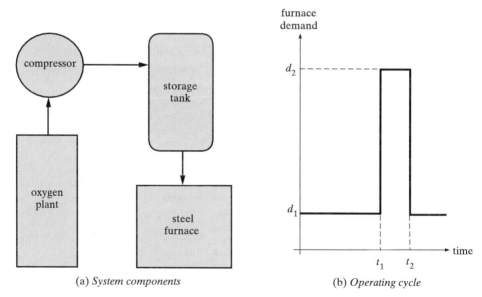

(a) *System components*   (b) *Operating cycle*

FIGURE 14.2   Oxygen System Engineering Design Example

We seek a minimum cost sizing of the system components. In particular, we wish to choose values for 4 decision variables:

---

[3]Based on F. C. Jen, C. C. Pegels, and T. M. DuPuis (1967), "Optimal Capacities of Production Facilities," *Management Science, 14*, B573–B580.

$x_1 \triangleq$ production rate of the oxygen plant

$x_2 \triangleq$ pressure in the storage tank

$x_3 \triangleq$ compressor power

$x_4 \triangleq$ storage tank volume

Also, physical limitations require that the storage tank pressure be at least p_0.

Oxygen System Engineering Design Model

The procurement and operation costs of this oxygen system has 4 parts:

$$\begin{pmatrix} \text{total} \\ \text{cost} \end{pmatrix} = \begin{pmatrix} \text{oxygen} \\ \text{plant cost} \end{pmatrix} + \begin{pmatrix} \text{compressor} \\ \text{cost} \end{pmatrix} + \begin{pmatrix} \text{storage tank} \\ \text{cost} \end{pmatrix} + \begin{pmatrix} \text{electrical} \\ \text{power cost} \end{pmatrix}$$

Using prior experience with similar systems, engineers in this application estimate oxygen plant cost to be linear in the production rate as

$$\text{oxygen plant cost} = 61.8 + 5.72 \,(\text{production rate})$$

Compressor costs grow with compressor power in the nonlinear function

$$\text{compressor cost} = 0.0175 \,(\text{power})^{0.85}$$

Storage vessel costs perform similarly in the required volume

$$\text{storage tank costs} = 0.0094 \,(\text{volume})^{0.75}$$

Finally electricity for the compressor is proportional to the power and time of operation as

$$\text{electrical power cost} = 0.006 \,(\text{time of operation})(\text{compressor power})$$

One constraint of the model comes from the requirement that the oxygen production rate outputs at least what is required over the cycle, that is,

$$t_2 x_1 \geq d_1 t_1 + d_2 (t_2 - t_1)$$

Also, the minimum pressure p_0 imposes bound

$$x_2 \geq p_0$$

The remainder of the constraints enforce the physical relations among the decision variables. First, the power of the compressor is related to the pressure that it must generate in the storage tank. In particular, the compressor power must address the maximum stored inventory $(d_2 - x_1)(t_2 - t_1)$ just before the burst part of the operating cycle. Including standard temperature and gas constants gives the constraint

$$x_3 = 36.25 \frac{(d_2 - x_1)(t_2 - t_1)}{t_1} \ln\left(\frac{x_2}{p_0}\right)$$

Finally, we must relate storage volume to the pressure needed to keep the maximum required inventory. Using appropriate temperature and gas constants gives

$$x_4 = 348{,}300 \frac{(d_2 - x_1)(t_2 - t_1)}{x_2}$$

Collecting all the above and adding variable-type constraints, we obtain the following NLP model of our oxygen system design problem:

$$\min \quad 61.8 + 5.72x_1 + 0.0175(x_3)^{0.85} + 0.0094(x_4)^{0.75}$$
$$+ 0.006t_1x_3 \qquad \text{(total cost)}$$

$$\text{s.t.} \quad t_2x_1 \geq d_1t_1 + d_2(t_2 - t_1) \qquad \text{(demand)}$$

$$x_2 \geq p_0 \qquad \text{(pressure)}$$

(14.3)

$$x_3 = 36.25\frac{(d_2 - x_1)(t_2 - t_1)}{t_1} \ln\left(\frac{x_2}{p_0}\right) \qquad \begin{pmatrix} \text{power vs.} \\ \text{pressure} \end{pmatrix}$$

$$x_4 = 348,300\frac{(d_2 - x_1)(t_2 - t_1)}{x_2} \qquad \begin{pmatrix} \text{volume vs.} \\ \text{pressure} \end{pmatrix}$$

$$x_1, x_2, x_3, x_4 \geq 0$$

With constants

$$d_1 = 2.5, \quad d_2 = 40, \quad t_1 = 0.6, \quad t_2 = 1.0, \quad p_0 = 200$$

an optimal design uses production rate $x_1^* = 17.5$, storage pressure $x_2^* = 473.7$, compressor power $x_3^* = 468.8$, and storage volume $x_4^* = 6618$ for total cost approximately 173.7.

SAMPLE EXERCISE 14.2: FORMULATING ENGINEERING DESIGN MODELS

A closed cylindrical tank is being designed to carry at least 20 cubic feet of chemicals. Metal for the top and sides costs $2 per square foot, but the heavier metal of the base costs $8 per square foot. Also, the height of the tank can be no more than twice its diameter to keep it from being top heavy. Formulate a constrained nonlinear program to find a design of minimum cost.

Modeling: The decision variables the designer must choose are

$$x_1 \triangleq \text{diameter of the tank}$$

$$x_2 \triangleq \text{height of the tank}$$

Then we obtain the NLP

$$\min \quad 2\left(\pi x_1 x_2 + \pi\frac{(x_1)^2}{4}\right) + 8\left(\pi\frac{(x_1)^2}{4}\right) \qquad \text{(metal cost)}$$

$$\text{s.t.} \quad \pi\frac{(x_1)^2}{4}x_2 \geq 20 \qquad \text{(volume)}$$

$$x_2 \leq 2x_1 \qquad \text{(height-to-diameter ratio)}$$

$$x_1 \geq 0, \ x_2 \geq 0$$

The objective function minimizes metal cost for the sides and top at $2 per square foot, plus cost for the base at $8. One main constraint makes the tank have the required volume, and the other enforces the height-to-diameter ratio limit.

14.2 CONVEX, SEPARABLE, QUADRATIC AND POSYNOMIAL GEOMETRIC PROGRAMMING SPECIAL NLP FORMS

Linear programming is one especially tractable special case of constrained nonlinear programming, but it is not the only one. Several other common forms have special characteristics that can be exploited in search algorithms.

In this section we define and illustrate models in four of these unusually tractable classes: **convex programs**, **separable programs**, **quadratic programs**, and **posynomial geometric programs**. Then in Sections 14.7 to 14.9 we illustrate how their special properties can be exploited. As usual, the models presented are drawn from real application contexts, usually cases described in published accounts.

EXAMPLE 14.4: PFIZER OPTIMAL LOT SIZING

One common application context yielding NLPs with special structure arises in managing **inventories** and choosing manufacturing **lot sizes**. Work at Pfizer, a large manufacturer of pharmaceuticals,[4] offers a good example.

The production of pharmaceuticals involves a series of fermentation and organic synthesis steps done in large tanks holding several thousand gallons. Each "campaign" or **lot** of a product consists of one or more "batches" processed serially.

The main issue to be decided is how many batches should be included in a campaign or lot of each product. Large changeover times and costs are incurred each time a new campaign starts because of the need to clean and reconfigure equipment carefully for the next product. These changeover burdens must be balanced against the cost of holding inventories built up during campaigns. In particular, rigorous quality control requires that all batches of a campaign be held at the end of each manufacturing step until every one is ready for the next step.

Our specific (fictitious) version of this lot sizing problem will consider products $j = 1, \ldots, 4$, each having steps $k = 1, 2$ of production ($k = 0$ refers to raw materials). Table 14.1 details values of the associated input parameters

$d_j \triangleq$ number of batches of product j demanded annually

$t_{j,k} \triangleq$ changeover time per campaign of product j at step k (in weeks)

$p_{j,k} \triangleq$ process time per batch of product j in step k (in weeks)

$v_{j,k} \triangleq$ value per batch of product j at the end of step k (in thousands of dollars)

TABLE 14.1   Pfizer Lot Sizing Example Data

Product Number, j	Annual Demand, d_j	Changeover Time (weeks)		Process Time (weeks)		Value ($000)		
		$t_{j,1}$	$t_{j,2}$	$p_{j,1}$	$p_{j,2}$	$v_{j,0}$	$v_{j,1}$	$v_{j,2}$
1	150	0.5	0.7	1.5	3.2	10	14	27
2	220	1.3	2.0	4.0	1.5	50	70	110
3	55	0.3	0.2	2.5	4.2	18	29	40
4	90	0.9	1.8	2.0	3.5	43	69	178

---

[4]Based on P. P. Kleutghen and J. C. McGee (1985), "Development and Implementation of an Integrated Inventory Management Program at Pfizer Pharmaceuticals," *Interfaces, 15:1*, 69–87.

We will also assume that 3000 total production weeks are available to manufacture the 4 products, that holding inventory cost $\frac{1}{2}$% per week of the product value, and that changeover activities cost $c = \$12,000$ thousand per week.

Pfizer Optimal Lot Sizing Model

To model our version of the Pfizer application, introduce decision variables

$$x_j \triangleq \text{number of batches of product } j \text{ in each campaign or lot}$$

To begin, notice that the implied number of campaigns per year is

$$\frac{\text{annual demand}}{\text{campaign size}} = \frac{d_j}{x_j}$$

Then the total annual cost objective function can be expressed as

$$\min \quad \sum_{j=1}^{4} \frac{d_j}{x_j} \left[\begin{pmatrix} \text{changeover costs} \\ \text{per campaign of } j \end{pmatrix} + \begin{pmatrix} \text{holding costs per} \\ \text{campaign of } j \end{pmatrix} \right] \qquad (14.4)$$

Changeover costs for product j are easily expressed as

$$\begin{pmatrix} \text{changeover costs} \\ \text{per campaign of } j \end{pmatrix} = c \sum_{k=1}^{2} t_{j,k}$$

To compute the corresponding inventory holding costs, consider the cycle illustrated in Figure 14.3. Each campaign of product j begins with x_j batches valued at raw material cost $v_{j,0}$. Over the $x_j p_{j,1}$ weeks it takes all batches to complete the first step of processing, their value increases to $x_j v_{j,1}$. Then the next $x_j p_{j,2}$ weeks raise the value to $x_j v_{j,2}$. Finally, the finished pharmaceutical is distributed to customers over the $52/(d_j/x_j)$ weeks until the next campaign is completed. Inventory cost per campaign is then $\frac{1}{2}$% of the area under the implied inventory value curve.

Substituting in expression (14.4) gives the complete objective function

$$\min \quad \sum_{j=1}^{4} \frac{d_j}{x_j} \left[c \sum_{k=1}^{2} t_{j,k} + 0.005 \left(\sum_{k=1}^{2} \tfrac{1}{2} p_{j,k}(v_{j,k-1} + v_{j,k})(x_j)^2 + \frac{52 v_{j,2}}{2 d_j}(x_j)^2 \right) \right]$$

The only main constraint in this model will enforce the limit on production time available. Again summing per campaign, we have

$$\sum_{j=1}^{4} \frac{d_j}{x_j} \left[\begin{pmatrix} \text{changeover time} \\ \text{per campaign of } i \end{pmatrix} + \begin{pmatrix} \text{production time} \\ \text{per campaign of } j \end{pmatrix} \right] \le \begin{pmatrix} \text{time} \\ \text{available} \end{pmatrix}$$

or

$$\sum_{j=1}^{4} \frac{d_j}{x_j} \left(\sum_{k=1}^{2} t_{j,k} + \sum_{k=1}^{2} p_{j,k} x_j \right) \le 3000 \qquad (14.5)$$

Substituting parameter values of Table 14.1 and simplifying produces the complete model

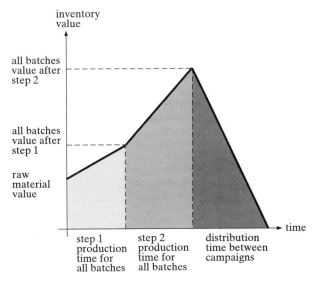

all batches value after step 2

all batches value after step 1

raw material value

inventory value

time

step 1 production time for all batches

step 2 production time for all batches

distribution time between campaigns

FIGURE 14.3   Pfizer Lot Sizing Example Inventory Cycle

$$\min \quad 66.21x_1 + \frac{2160}{x_1} + 426.8x_2 + \frac{8712}{x_2} \qquad \text{(total cost)}$$

$$+ 61.20x_3 + \frac{330}{x_3} + 268.1x_4 + \frac{2916}{x_4} \qquad (14.6)$$

$$\text{s.t.} \quad \frac{180}{x_1} + \frac{726}{x_2} + \frac{27.5}{x_3} + \frac{243}{x_4} \le 221.5 \qquad \text{(production time)}$$

$$x_1, \dots, x_4 \ge 0$$

An optimal plan runs

$$x_1^* = 7.161, \quad x_2^* = 5.665, \quad x_3^* = 2.911, \quad x_4^* = 4.135 \qquad (14.7)$$

campaigns per year of the 4 products. Total annual cost will be approximately $6,837,000.

Convex Programs

Lot sizing model (14.6) is special, in part, because it forms a convex program.

14.3   A constrained nonlinear program in functional form

$$\text{max or min } f(\mathbf{x})$$

$$\text{s.t.} \quad g_i(\mathbf{x}) \quad \left\{ \begin{array}{c} \ge \\ \le \\ = \end{array} \right\} \quad b_i \qquad i = 1, \dots, m$$

is a **convex program** if f is concave for a maximize or convex for a minimize, each g_i of a $\ge$ constraint is concave, each g_i of a $\le$ constraint is convex, and each g_i of an $=$ constraint is linear.

In Section 13.4 we defined convex and concave functions (definition 13.23) and showed their importance in characterizing the most tractable objective functions. Convex programs extend these ideas to constraints. Every $\geq$ constraint should be concave (after collecting terms involving the decision variables on the left-hand side), every $\leq$ constraint convex, and every $=$ constraint linear.

To see that lot sizing model (14.6) satisfies these requirements, we first consider its objective function:

$$\min \quad f(\mathbf{x}) \triangleq 66.21x_1 + \frac{2160}{x_1} + 426.8x_2 + \frac{8712}{x_2}$$
$$+ 61.20x_3 + \frac{330}{x_3} + 268.1x_4 + \frac{2916}{x_4} \tag{14.8}$$

Notice that it is a sum of terms

$$\alpha_j x_j + \frac{\beta_j}{x_2}$$

with positive constants α_j and β_j. The first such terms are linear and so convex (principle 13.28). But linear functions are also concave, which makes reciprocal terms convex for $x_j \geq 0$ (principle 13.32). It follows that the objective function is a sum of convex functions, and so itself convex (principle 13.29). This is just what is needed in definition 14.3 .

Now consider constraints

$$g_1(\mathbf{x}) \triangleq \frac{180}{x_1} + \frac{726}{x_2} + \frac{27.5}{x_3} + \frac{243}{x_4} \leq 221.5$$
$$g_2(\mathbf{x}) \triangleq x_1 \geq 0$$
$$g_3(\mathbf{x}) \triangleq x_2 \geq 0 \tag{14.9}$$
$$g_4(\mathbf{x}) \triangleq x_3 \geq 0$$
$$g_5(\mathbf{x}) \triangleq x_4 \geq 0$$

Definition 14.3 requires g_1 convex and $g_2, \ldots, g_5$ concave. Function g_1 conforms for much the same reason as the objective: It is the positive-weighted sum of reciprocals of nonnegative linear functions. Other functions $g_2, \ldots, g_5$ are concave because they are linear.

SAMPLE EXERCISE 14.3: RECOGNIZING CONVEX PROGRAMS

Determine whether each of the following mathematical programs is a convex program.

(a) max $3w_1 - w_2 + 8\ln(w_1)$
    s.t. $4(w_1)^2 - w_1w_2 + (w_2)^2 \leq 100$
    $w_1 + w_2 = 4$
    $w_1, w_2 \geq 0$

(b) min $3w_1 - w_2 + 8\ln(w_1)$
     s.t. $(w_1)^2 + (w_2)^2 \geq 10$
          $w_1 + w_2 = 4$
          $w_1, w_2 \geq 0$

(c) max $w_1 + 7w_2$
     s.t. $w_1 w_2 \leq 14$
          $(w_1)^2 + (w_2)^2 = 40$
          $w_1, w_2 \geq 0$

(d) min $w_1 + 7w_2$
     s.t. $w_1 + w_2 \leq 14$
          $w_1 - w_2 \geq 0$
          $2w_1 + 5w_2 = 18$
          $w_1, w_2 \geq 0$

Analysis: We apply definition 14.3 .

(a) The objective function of this model is concave because it is the sum of a linear function and $8\ln(w_1)$, which has negative second derivative (principle 13.27). Also, the first main constraint function is convex because the Hessian matrix

$$\mathbf{H(w)} = \begin{pmatrix} 8 & -1 \\ -1 & 2 \end{pmatrix}$$

is positive definite (principle 13.27). Since all 3 other constraints are linear, the model is a convex program. It maximizes a concave objective, subject to a convex $\leq$ constraints, a linear $=$ constraint, and concave (because linear) $\geq$ constraints.

(b) This NLP is not a convex program because its objective function, which is the concave one of part (a), is inappropriate for a minimize. Also, the first main constraint

$$(w_1)^2 + (w_2)^2 \geq 10$$

is convex in a $\geq$ form.

(c) This model is also not a convex program. Its first main constraint involves function $g_1(w_1, w_2) = w_1 w_2$, which is neither convex nor concave. Its Hessian matrix is

$$\mathbf{H_1(w)} = \begin{pmatrix} 0 & 1 \\ 1 & 0 \end{pmatrix}$$

Furthermore, the second main constraint is a nonlinear equality.

(d) This model is a linear program and so a convex program. Linear objectives and constraints, which are both convex and concave (principle 13.28), fulfill all requirements of definition 14.3 .

Special Tractability of Convex Programs

To appreciate the special tractability of convex programs, reach back to our characterization in Section 3.4 of the models most convenient for improving search. The best objective functions are those that are unimodal (definition 3.26), and principle

13.33 has already established that maximizing a concave function, or minimizing a convex, produces an objective with that desirable property.

For constraints we would like the implied feasible set to be convex (definition 3.29). Requirements for a convex program assure exactly that.

14.4 | The feasible set defined by constraints

$$g_i(\mathbf{x}) \quad \left\{ \begin{matrix} \geq \\ \leq \\ = \end{matrix} \right\} \quad b_i \qquad i = 1, \ldots, m$$

is convex if each g_i of a $\geq$ constraint is concave, each g_i of a $\leq$ constraint is convex, and each g_i of an $=$ constraint is linear.

We can see why property 14.4 is true by considering a single convex constraint

$$g(\mathbf{x}) \leq b$$

and two points $\mathbf{x}^{(1)}$ and $\mathbf{x}^{(2)}$ that satisfy it. If the corresponding feasible set is to be convex, every point along the line segment between $\mathbf{x}^{(1)}$ and $\mathbf{x}^{(2)}$ must also satisfy the constraint, that is (principle 3.31), every point representable as

$$\mathbf{x}^{(1)} + \lambda(\mathbf{x}^{(2)} - \mathbf{x}^{(1)})$$

for some $\lambda \in [0, 1]$. Using

$$g(\mathbf{x}^{(1)}) \leq b$$
$$g(\mathbf{x}^{(2)}) \leq b$$

we may multiply the first by $(1 - \lambda)$, the second by λ, and sum to conclude that

$$(1 - \lambda)g(\mathbf{x}^{(1)}) + (\lambda)g(\mathbf{x}^{(2)}) \leq (1 - \lambda)b + (\lambda)b$$

which simplifies to

$$g(\mathbf{x}^{(1)}) + \lambda\left(g(\mathbf{x}^{(2)}) - g(\mathbf{x}^{(1)})\right) \leq b$$

With g convex, the left side of this expression is at most $g(\mathbf{x}^{(1)} + \lambda(\mathbf{x}^{(2)} - \mathbf{x}^{(1)}))$ (definition 13.23). Thus

$$g(\mathbf{x}^{(1)} + \lambda(\mathbf{x}^{(2)} - \mathbf{x}^{(1)})) \leq b$$

which shows that the line segment point for λ is feasible are required. A similar argument demonstrates that feasible sets for concave $\geq$ constraints are convex, and we already know that linear constraints yield convex feasible sets (principle 3.32).

Combining the unimodality of objective functions for convex programs with the convexity of their feasible sets leads (via principle 3.34) to their main convenience for improving search.

14.5 | Every local optimum of a convex program is a global optimum.

Improving search procedures that produce locally optimal solutions automatically yield global optima.

Separable Programs

Besides being a convex program, the objective and constraint functions of Pfizer lot sizing model (14.6) also have a special separable property.

14.6   Function $s(\mathbf{x})$ is **separable** if it can be expressed as the sum

$$s(x_1, \ldots, x_n) \triangleq \sum_{j=1}^{n} s_j(x_j)$$

of single-variable functions $s_1(x_1), \ldots, s_n(x_n)$.

That is, a function is separable if it is the sum of 1-variable functions of its arguments.
    To see that Pfizer objective function (14.8) is separable, consider the 1-variable functions

$$f_1(x_1) \triangleq 66.21x_1 + \frac{2160}{x_1}$$

$$f_2(x_2) \triangleq 426.8x_1 + \frac{8712}{x_2}$$

$$f_3(x_3) \triangleq 61.20x_1 + \frac{330}{x_3}$$

$$f_4(x_4) \triangleq 268.1x_1 + \frac{2916}{x_4}$$

Clearly, the full objective $f(\mathbf{x})$ separates into a sum of these functions of single decision variables.
    Similar definitions establish that constraint functions $g_1(\mathbf{x})$ through $g_5(\mathbf{x})$ are separable. For example,

$$g_1(\mathbf{x}) = g_{1,1}(x_1) + g_{1,2}(x_2) + g_{1,3}(x_3) + g_{1,4}(x_4)$$

where $g_{1,1}(x_1) = 180/x_1, g_{1,2}(x_2) = 726/x_2, g_{1,3}(x_3) = 27.5/x_3$, and $g_{1,4}(x_4) = 243/x_4$.
    When both the objective function and all constraint functions are separable in this way, we term an NLP a separable program.

14.7   A constrained nonlinear program in functional form

$$\max \text{ or } \min f(\mathbf{x})$$

$$\text{s.t.} \quad g_i(\mathbf{x}) \quad \begin{Bmatrix} \geq \\ \leq \\ = \end{Bmatrix} \quad b_i \qquad i = 1, \ldots, m$$

is a **separable program** if f and every g_i is separable.

Pfizer model (14.6) provides an example.

SAMPLE EXERCISE 14.4: RECOGNIZING SEPARABLE PROGRAMS

Return to the NLPs of Sample Exercise 14.3, and determine whether each is a separable program.

Analysis: We apply definition $\boxed{14.7}$.

(a) The objective function and most constraints of this model are separable. However, it fails the definition of a separable program because the first main constraint contains the term $-w_1 w_2$, which involves both variables.

(b) This model is a separable program. Each objective and constraint function can be expressed as a sum of separate functions, one involving only w_1 and the other only w_2.

(c) Again, the objective and most constraint functions of this NLP are separable. Still, the first main constraint is not, because it includes the term $w_1 w_2$. Thus the model is not a separable program.

(d) This model is a linear program. By definition, linear functions involve sums of constant multiples of the decision variables. Thus all are separable, and every linear program is a separable program.

Special Tractability of Separable Programs

Separable programs, where objective and constraint functions are sums of 1-variable functions, have much in common with linear programs where the corresponding functions are limited to scalar multiples of the decision variables. This relationship can be exploited by **piecewise-linear approximation** of each component function as suggested in Figure 14.4. That is, each of the single-variable components of the objective and constraint functions is approximated by a series of linear (straight-

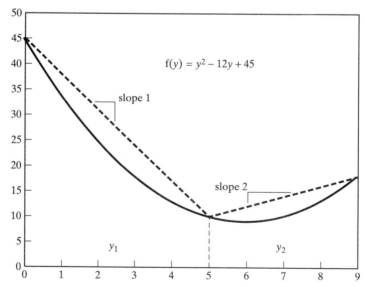

FIGURE 14.4 Piecewise-Linear Approximation of a One-Variable Function

line) segments. Then, if the separable program also conforms to convex program definition 14.3 , the approximation can be solved by linear programming.

> 14.8 | Separable convex nonlinear programs can be addressed by linear programming through piecewise-linear approximation of the objective and constraint functions.

Details are provided in Section 14.8.

EXAMPLE 14.5: QUADRATIC PORTFOLIO MANAGEMENT

We may illustrate another important class of constrained nonlinear programs by returning to an application setting that we encountered earlier in Chapter 8: **finance**. Financial managers are constantly planning and controlling market decisions, and a wide array of nonlinear programs have been employed to help.

We will consider a simple case of **portfolio management**—dividing investment funds to maximize return and minimize risk. Our manager, whom we will call Barney Backroom, must decide how to split the available funds among three classes of investments: common stocks, money markets, and corporate bonds. Table 14.2 shows the experience of the past 6 years on which Barney will base his decisions. He would like to average an 11% return on investments while accepting minimum risk.

TABLE 14.2   Return Experience for Portfolio Example

Category	Percent Returns for Year:						Average
	1	2	3	4	5	6	
Stocks	22.24	16.16	5.27	15.46	20.62	–0.42	13.22
Money market	9.64	7.06	7.68	8.26	8.55	8.26	8.24
Bonds	10.08	8.16	8.46	9.18	9.26	9.06	9.03

Quadratic Portfolio Management Model

The obvious decision variables for a model of this problem are

$$x_1 \triangleq \text{fraction of the portfolio invested in common stocks}$$
$$x_2 \triangleq \text{fraction of the portfolio invested in money markets}$$
$$x_3 \triangleq \text{fraction of the portfolio invested in corporate bonds}$$

Then one main constraint forces all funds to be invested:

$$x_1 + x_2 + x_3 = 1$$

One reasonable assumption about returns is that each class will achieve the average experienced in the years of Table 14.2. Then the goal of returning 11% can be expressed as

$$13.22x_1 + 8.24x_2 + 9.03x_3 \geq 11$$

The more difficult issue is how to model variability of return. One measure is the **variance**—the average squared deviation from the mean. If the three classes of investments varied independently, the variance of the overall return would be simply the sum of the variances of each class. However, financial markets in various commodities tend to interact. Thus a workable model should include the **covariances** relating movement in the different categories of investment.

Given a series of n observations like those in Table 14.2, covariances can be estimated as

$v_{i,j} \triangleq$ estimated covariance between categories i and j

$$= \frac{1}{n}\sum_{t=1}^{n}\left(\begin{array}{c}i \text{ value} \\ \text{in period } t\end{array}\right)\left(\begin{array}{c}j \text{ value} \\ \text{in period } t\end{array}\right) - \frac{1}{n^2}\left[\sum_{t=1}^{n}\left(\begin{array}{c}i \text{ value} \\ \text{in period } t\end{array}\right)\right]\left[\sum_{t=1}^{n}\left(\begin{array}{c}j \text{ value} \\ \text{in period } t\end{array}\right)\right]$$

Then the variance of the total return is approximately

$$\text{variance of return} = \sum_{i=1}^{n}\sum_{j=1}^{n} v_{i,j}x_i x_j$$

$$= \mathbf{x}\mathbf{V}\mathbf{x}$$

for $\mathbf{V}$ is the matrix of $v_{i,j}$.

Table 14.2's data for our example yields

$$\mathbf{V} = \begin{pmatrix} 66.51 & 2.61 & 2.18 \\ 2.61 & 0.63 & 0.48 \\ 2.18 & 0.48 & 0.38 \end{pmatrix}$$

Thus we may combine with the constraints above to obtain the model

$$\begin{array}{lll} \min & 66.51(x_1)^2 + 2(2.61)x_1 x_2 + 2(2.18)x_1 x_3 + 0.63(x_2)^2 & \text{(variance)} \\ & + 2(0.48)x_2 x_3 + 0.38(x_3)^2 & \\ \text{s.t.} & x_1 + x_2 + x_3 = 1 & \text{(invest 100\%)} \quad (14.10) \\ & 13.22x_1 + 8.24x_2 + 9.03x_3 \geq 11 & \text{(return)} \\ & x_1, x_2, x_3 \geq 0 & \end{array}$$

An optimal portfolio invests a fraction $x_1^* = 0.47$ of funds in common stocks, $x_2^* = 0$ of funds in money markets, and $x_3^* = 0.53$ in corporate bonds to produce a minimum variance of 15.895.

Quadratic Programs Defined

Model (14.10) illustrates the special case of a quadratic program.

> **14.9** A constrained nonlinear program is a **quadratic program** if its objective function is quadratic, that is,
>
> $$f(\mathbf{x}) \triangleq \sum_j c_j x_j + \sum_i \sum_j q_{i,j} x_i x_j = \mathbf{c} \cdot \mathbf{x} + \mathbf{x}\mathbf{Q}\mathbf{x}$$
>
> and all constraints are linear.

Portfolio model (14.10) clearly qualifies. All 5 constraints are linear, and the objective function involves only squares and products of 2 variables. In the matrix format of definition 14.9, its objective function has

$$\mathbf{c} = \mathbf{0}, \qquad \mathbf{Q} = \mathbf{V}$$

SAMPLE EXERCISE 14.5: RECOGNIZING QUADRATIC PROGRAMS

Determine whether each of the following NLPs is a quadratic program. For those that are, also place the objective function in matrix format $\mathbf{c} \cdot \mathbf{w} + \mathbf{w}\mathbf{Q}\mathbf{w}$.

(a) max $3w_1 - 5w_2 + 12(w_1)^2 + 8w_1w_2 + (w_2)^2$
s.t. $w_1 + w_2 = 9$
          $w_1, w_2 \geq 0$

(b) min $w_1w_2w_3$
s.t. $(w_1)^2 + (w_2)^2 \leq 25$

(c) min $5w_1 + 19w_2$
s.t. $w_1 + w_2 = 9$
          $w_1, w_2 \geq 0$

Analysis: We apply definition $\boxed{14.9}$.

(a) This model is a quadratic program because its objective function involves only second-order terms, and its constraints are linear. Here

$$\mathbf{c} = \begin{pmatrix} 3 \\ -5 \end{pmatrix}, \qquad \mathbf{Q} = \begin{pmatrix} 12 & 4 \\ 4 & 1 \end{pmatrix}$$

Notice that the coefficient 8 on cross-product w_1w_2 is split so that

$$q_{1,2}w_1w_2 + q_{2,1}w_1w_2 = 4w_1w_2 + 4w_1w_2 = 8w_1w_2$$

(b) This model is not a quadratic program. Its objective involves a 3-way product, which is not quadratic, and its constraint is not linear.

(c) This model is a linear program and so quadratic. Objective function elements

$$\mathbf{c} = \begin{pmatrix} 5 \\ 19 \end{pmatrix}, \qquad \mathbf{Q} = \mathbf{0}$$

Special Tractability of Quadratic Programs

Being nonlinear in only the objective function, and only quadratic there, quadratic programs have much in common with LPs. In particular, powerful dual and complementary slackness properties make possible efficient specialized algorithms for many cases. Section 14.7 provides details.

EXAMPLE 14.6: COFFERDAM DESIGN

To illustrate another special type of constrained NLP arising often in engineering design applications, we consider the planning of a cofferdam.[5] Cofferdams are used to block streams temporarily while construction is in progress. Figure 14.5 illustrates a common design. Each "cycle" of the dam consists of a large steel cylinder filled with soil and joined to the next with curved connecting plates.

---

[5]Based on F. Neghabat and R. M. Stark (1972), "A Cofferdam Design Optimization," *Mathematical Programming, 3*, 263–275.

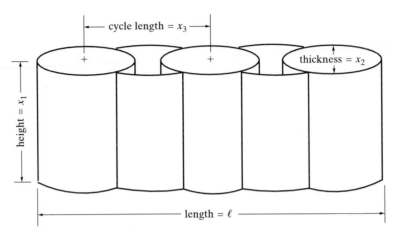

FIGURE 14.5   Cofferdam Example Structure

Given

$$\ell \triangleq \text{total dam length (in feet)}$$
$$t \triangleq \text{design life of the dam (in days)}$$

and other characteristics of site and materials, we wish to determine a minimum cost design. Main decision variables are

$$x_1 \triangleq \text{height of the dam (in feet)}$$
$$x_2 \triangleq \text{average thickness of the dam (in feet)}$$
$$x_3 \triangleq \text{length of a cycle of the dam (in feet)}$$

Cofferdam Example Model

To develop a model, we begin with the various elements of cost. Filling cost is roughly proportional to the dam volume. Using a cost of $0.21 per cubic foot, this produces

$$\text{fill cost} \approx 0.21(\text{length})(\text{height})(\text{thickness})$$
$$= 0.21\ell x_1 x_2$$

Similarly, steel cost depends on the area of the dam's front and back, plus that of the two cylinder sides passing through the dam in each cycle. Pricing steel at $2.28 per square foot, we have

$$\text{steel cost} \approx 2.28 \left[2(\text{length})(\text{height}) + 2 \left(\frac{\text{length}}{\text{cycle length}} \right) (\text{height})(\text{thickness}) \right]$$
$$= 4.56\ell x_1 + 4.56\ell \frac{x_1 x_2}{x_3}$$

A very low dam would minimize construction cost, but flood risk must also be considered. Analysis of prior experience suggests that flood cost can be approximated as

$$\text{flood cost} \approx (\text{cost per flood}) \left(\frac{\text{design life}}{\text{empirical}} \right)$$
$$= 40,000 \frac{t}{x_4}$$

where each flood costs \$40,000, and x_4 is an intermediate decision variable related to dam height by

$$x_4 + 33.3 \leq 0.8x_1$$

Other constraints of the model come from possible failure modes. One possibility is slipping on the river bottom. This can be prevented if

$$1.0425(\text{height}) \leq \text{thickness} \quad \text{or} \quad 1.0425x_1 \leq x_2$$

The other main consideration is tension stresses at cycle joints. These require that

$$(\text{height})(\text{cycle length}) \leq 2857 \quad \text{or} \quad x_1 x_3 \leq 2857$$

Collecting all the above, simplifying to make all constraint right-hand sides $= 1$, and assuming a dam of length $\ell = 800$ feet that should last $t = 365$ days, our cofferdam design task reduces to the following constrained NLP model:

$$\min \quad 168x_1x_2 + 3648x_1 + 3648\frac{x_1x_2}{x_3} + \frac{1.46 \times 10^7}{x_4} \quad \text{(cost)}$$

$$\text{s.t.} \quad \frac{1.25x_4}{x_1} + \frac{41.625}{x_1} \leq 1 \quad \text{(empirical)}$$

$$\frac{1.0425x_1}{x_2} \leq 1 \quad \text{(slipping)} \tag{14.11}$$

$$0.00035x_1x_3 \leq 1 \quad \text{(tension)}$$

$$x_1, x_2, x_3, x_4 > 0$$

An optimal design has height $x_1^* = 62.65$ feet, average thickness $x_2^* = 65.32$ feet, cycle length $x_3^* = 45.60$ feet, intermediate variable $x_4^* = 16.82$, and total cost \$2.111 million.

Posynomial Geometric Programs

The objective and constraint functions of cofferdam model (14.11) have a special posynomial form.

14.10 Function $p(\mathbf{x})$ is a **posynomial** if it can be expressed

$$p(x_1, \ldots, x_n) \triangleq \sum_k d_k \left(\prod_{j=1}^{n} (x_j)^{a_{k,j}} \right)$$

for given $d_k > 0$ and exponents $a_{k,j}$ of arbitrary sign.

For example, objective function

$$f(x_1, x_2, x_3, x_4) \triangleq 168x_1x_2 + 3648x_1 + 3648\frac{x_1x_2}{x_3} + \frac{1.46 \times 10^7}{x_4}$$

of model (14.11) is a posynomial with

$$\begin{array}{llll}
d_1 = 168, & d_2 = 3648, & d_3 = 3648, & d_4 = 1.46 \times 10^7 \\
a_{1,1} = 1, & a_{1,2} = 1, & a_{1,3} = 0, & a_{1,4} = 0 \\
a_{2,1} = 1, & a_{2,2} = 0, & a_{2,3} = 0, & a_{2,4} = 0 \\
a_{3,1} = 1, & a_{3,2} = 1, & a_{3,3} = -1, & a_{3,4} = 0 \\
a_{4,1} = 0, & a_{4,2} = 0, & a_{4,3} = 0, & a_{4,4} = -1
\end{array} \tag{14.12}$$

Notice that powers $a_{k,j}$ can have any sign, but coefficients d_k must be positive. Thus

$$h(x_1, x_2, x_3) \triangleq 13(x_1)^2(x_3) + 29(x_2)^{0.534}(x_3)^{-0.451}$$

is a posynomial, but the variation

$$h(x_1, x_2, x_3) \triangleq 13(x_1)^2(x_3) - 29(x_2)^{0.534}(x_3)^{-0.451}$$

is not because $d_2 = -29 < 0$.

Posynomial geometric programs are NLPs over posynomial functions and positive variables.

> **14.11** An NLP is a **posynomial geometric program** if it can be expressed in the form
>
> $$\begin{aligned} \min \quad & f(\mathbf{x}) \\ \text{s.t.} \quad & g_i(\mathbf{x}) \leq 1 \qquad i = 1, \dots, m \\ & \mathbf{x} > \mathbf{0} \end{aligned}$$
>
> where f and all g_i are posynomial functions of $\mathbf{x}$.

With posynomial terms written out, the format is

$$\begin{aligned} \min \quad & \sum_{k \in K_0} d_k \prod_{j=1}^{n} (x_j)^{a_{k,j}} \\ \text{s.t.} \quad & \sum_{k \in K_i} d_k \prod_{j=1}^{n} (x_j)^{a_{k,j}} \leq 1 \qquad i = 1, \dots, m \\ & x_j > 0 \qquad\qquad\qquad\quad j = 1, \dots, n \end{aligned} \qquad (14.13)$$

where nonoverlapping the K_i index terms k in various posynomials.

Notice that we allow only a minimize objective function form and $\leq$ constraints. Also, we have only positive-valued variables, and positive right-hand sides have been divided through to produces 1's. Usual conversions to reverse directions for other cases fail in the geometric programming case because they change the sign of some d_k and thus destroy the posynomial property.

Cofferdam model (14.11) satisfies all these conditions. In detail format (14.13), model (14.12) shows coefficients for objective function terms in $K_0 = \{1, 2, 3, 4\}$. Those for constraint sets $K_1 = \{5, 6\}$, $K_2 = \{7\}$, and $K_3 = \{8\}$ are

$$\begin{array}{llll} d_5 = 1.25, & d_6 = 41.625, & d_7 = 1.0425, & d_8 = 0.00035 \\ a_{5,1} = -1, & a_{5,2} = 0, & a_{5,3} = 0, & a_{5,4} = 1 \\ a_{6,1} = -1, & a_{6,2} = 0, & a_{6,3} = 0, & a_{6,4} = 0 \\ a_{7,1} = 1, & a_{7,2} = -1, & a_{7,3} = 0, & a_{7,4} = 0 \\ a_{8,1} = 1, & a_{8,2} = 0, & a_{8,3} = 1, & a_{8,4} = 0 \end{array}$$

SAMPLE EXERCISE 14.6: RECOGNIZING POSYNOMIAL GEOMETRIC PROGRAMS

Determine whether each of the following NLPs is a posynomial geometric program. For those that are, also detail coefficients d_k and $a_{k,j}$ in standard form (14.13).

(a) min $144\dfrac{w_1}{\sqrt{w_2}} + 6w_3$

$19w_1 + (w_2)^2 \le w_3$

$w_1 w_2 w_3 \le 44$

$w_1, w_2, w_3 > 0$

(b) max $144\dfrac{w_1}{\sqrt{w_2}} + 6w_3$

$19w_1 - (w_2)^2 \le w_3$

$w_1 w_2 w_3 \ge 44$

$w_1, w_2, w_3 > 0$

Analysis: We apply definitions $\boxed{14.10}$ and $\boxed{14.11}$.

(a) Dividing both constraints by their right-hand sides places this model in posynomial geometric program format (14.13). Coefficients are

$$K_0 = \{1, 2\}, \quad K_1 = \{3, 4\}, \quad K_2 = \{5\}$$

$$d_1 = 144, \quad d_2 = 6, \quad d_3 = 19, \quad d_4 = 1, \quad d_5 = \tfrac{1}{44}$$

$$a_{1,1} = 1, \quad a_{1,2} = -0.5, \quad a_{1,3} = 0$$

$$a_{2,1} = 0, \quad a_{2,2} = 0, \quad a_{2,3} = 1$$

$$a_{3,1} = 1, \quad a_{3,2} = 0, \quad a_{3,3} = -1$$

$$a_{4,1} = 0, \quad a_{4,2} = 2, \quad a_{4,3} = -1$$

$$a_{5,1} = 1, \quad a_{5,2} = 1, \quad a_{5,3} = 1$$

(b) This model is not a posynomial geometric program for several reasons. First, its objective maximizes a posynomial, and definition $\boxed{14.11}$ requires a minimize. Similarly, the second main constraint has $\ge$ form, not the $\le$ as appropriate for geometric programs. Finally, the first main constraint function has a negative coefficient, so it is not a posynomial.

Special Tractability of Posynomial Geometric Programs

Posynomial functions need not be convex, and thus geometric programs $\boxed{14.11}$ are often not convex programs. For example,

$$h(x_1, x_2) \triangleq (x_1)^2 x_2 + 7x_2 \qquad (14.14)$$

has Hessian matrix at $\mathbf{x} = (1, 1)$

$$\mathbf{H}(1, 1) = \begin{pmatrix} 2 & 2 \\ 2 & 0 \end{pmatrix}$$

This matrix is not positive semidefinite, and thus (principle $\boxed{13.27}$) f is not convex over even $\mathbf{x} > \mathbf{0}$.

Still, posynomial geometric programs can be made convex with a suitable change of variables.

$\boxed{14.12}$ Posynomial geometric programs covert to convex programs when original variables x_j are replaced by $z_j \triangleq \ln(x_j)$.

For example, in the case of function (14.14), substituting $z_j \triangleq \ln(x_j)$ or $x_j = e^{z_j}$ produces

$$h(z_1, z_2) \triangleq (e^{z_1})^2(e^{z_2}) + 7(e^{z_2}) = e^{2z_1 + z_2} + 7(e^{z_2})$$

Both exponential terms are convex under composition rule $\boxed{13.31}$, so their sum is also convex.

Notice the role that details of definition $\boxed{14.11}$ play in making this transformation work. First, a logarithmic transformation would be impossible if any x_j could be zero or negative. Also, the transformation could produce a mixed-sign sum of convex terms if all coefficients c_k were not positive in definition $\boxed{14.10}$. Finally, convex constraint functions would not be appropriate for a convex program (definition $\boxed{14.3}$) unless all constraints were of $\leq$ form.

Further transformations of posynomial geometric programs can lead to even greater tractability. Section 14.9 provides details.

14.3 LAGRANGE MULTIPLIER METHODS

If we can see how to view a constrained nonlinear program as one with only $=$ constraints (i.e., no inequalities of any form), calculus methods predating most of the numerical search techniques of this book can sometimes be applied to find an optimal solution. We briefly explore such **Lagrange multiplier techniques** in this section. Also, Lagrangian ideas will be seen to motivate more general approaches in later sections of the chapter.

Reducing to Equality Form

Lagrange multiplier solution techniques are most easily applied to models in equality constrained format.

$\boxed{14.13}$ Lagrange multiplier solution techniques address NLPs in pure equality form

$$\begin{array}{ll} \min \text{ or max} & f(\mathbf{x}) \\ \text{s.t.} & g_i(\mathbf{x}) = b_i \qquad \text{for all } i = 1, \dots, m \end{array}$$

That is, they consider only a set of constraints assumed active, which can be taken as $=$'s.

There is one equality for each original equality and one for each active inequality. Notice that not even variable-type inequalities such as nonnegativity constraints are allowed.

To see how such models can arise, consider the Pfizer lot sizing model of Section 14.2:

$$\min \quad f(\mathbf{x}) \triangleq 66.21x_1 + \frac{2160}{x_1} + 426.8x_2 + \frac{8712}{x_2} \qquad \text{(total cost)} \qquad (14.15)$$

$$+ 61.20x_3 + \frac{330}{x_3} + 268.1x_4 + \frac{2916}{x_4}$$

s.t. $\dfrac{180}{x_1} + \dfrac{726}{x_2} + \dfrac{27.5}{x_3} + \dfrac{243}{x_4} \le 221.5$ (production time)

$x_1 \ldots, x_4 \ge 0$

If none of the constraints are active at an optimal $\mathbf{x}^*$, that solution will be an unconstrained optimum of the objective function alone. Since we have shown in Section 14.2 that the objective function is convex, we may compute $\mathbf{x}^*$ by finding a stationary point (principle $\boxed{13.22}$) i.e., a point where all partial derivatives $= 0$. There x_1^* solves

$$\frac{\partial f}{\partial x_1} = 66.21 - \frac{2160}{(x_1)^2} = 0 \quad \text{so that} \quad x_1^* = \sqrt{\frac{2160}{66.21}} = 5.712$$

Similarly, unconstrained $x_2^* = 4.518$, $x_3^* = 2.322$, and $x_4^* = 3.298$.

Checking the main constraint of (14.15) gives

$$\frac{180}{5.712} + \frac{726}{4.518} + \frac{27.5}{2.322} + \frac{243}{3.298} = 277.7 \not\le 221.5$$

Thus our unconstrained solution is infeasible, and the main constraint must be treated as active at a true optimum. Assuming that nonnegativity constraints will remain inactive, we may deal with the model in pure equality form

min $f(\mathbf{x}) \triangleq 66.21x_1 + \dfrac{2160}{x_1} + 426.8x_2 + \dfrac{8712}{x_2}$ (total cost)

$+\, 61.20x_3 + \dfrac{330}{x_3} + 268.1x_4 + \dfrac{2916}{x_4}$ (14.16)

s.t. $\dfrac{180}{x_1} + \dfrac{726}{x_2} + \dfrac{27.5}{x_3} + \dfrac{243}{x_4} = 221.5$ (production time)

Lagrangian Function and Lagrange Multipliers

Lagrangian techniques begin with conversion of equality constrained model $\boxed{14.13}$ to an unconstrained form by weighting constraints in the objective function with **Lagrange multipliers**, v_i. The result is a **Lagrangian function** of both $\mathbf{x}$ and $\mathbf{v}$.

$\boxed{14.14}$ The Lagrangian function associated with a nonlinear program over equality constraints $g_1(\mathbf{x}) = b_1, \ldots, g_m(\mathbf{x}) = b_m$ is

$$L(\mathbf{x}, \mathbf{v}) \triangleq f(\mathbf{x}) + \sum_{i=1}^{m} v_i[b_i - g_i(\mathbf{x})]$$

where v_i is the Lagrange multiplier for constraint i.

For example, the Lagrangian function for Pfizer equality form (14.16) is

$$L(x_1, x_2, x_3, x_4, v) \triangleq 66.21x_1 + \frac{2160}{x_1} + 426.8x_2 + \frac{8712}{x_2} + 61.20x_3 + \frac{330}{x_3}$$

$$+\, 268.1x_4 + \frac{2916}{x_4} + v\left(221.5 - \frac{180}{x_1} - \frac{726}{x_2} - \frac{27.5}{x_3} - \frac{243}{x_4}\right)$$

(14.17)

Notice what happens when we form the Lagrangian. We have relaxed the given constrained model into unconstrained form. Still, the Lagrangian function value coincides with the original objective function value at every feasible point because

$$L(\mathbf{x}, \mathbf{v}) = f(\mathbf{x}) + \sum_{i=1}^{m} v_i [b_i - g_i(\mathbf{x})] = f(\mathbf{x}) + \sum_{i=1}^{m} v_i(0) = f(\mathbf{x}) \qquad (14.18)$$

SAMPLE EXERCISE 14.7: FORMING THE LAGRANGIAN FUNCTION

Consider the equality-constrained nonlinear program

$$\begin{aligned} \min \quad & 6(x_1)^2 + 4(x_2)^2 + (x_3)^2 \\ \text{s.t.} \quad & 24x_1 + 24x_2 = 360 \\ & x_3 = 1 \end{aligned}$$

Form the corresponding Lagrangian function.

Analysis: Here $m = 2$. Following format $\boxed{14.14}$, we form the Lagrangian by rolling the 2 equality constraints into the objective function with Lagrange multiplier v_1 and v_2. The result is

$$L(x_1, x_2, x_3, v_1, v_2) \triangleq 6(x_1)^2 + 4(x_2)^2 + (x_3)^2 + v_1 (360 - 24x_1 - 24x_2) + v_2 (1 - x_3)$$

Stationary Points of the Lagrangian Function

Think now about stationary points of the Lagrangian [i.e., points where gradient $\nabla L(\mathbf{x}^*, \mathbf{v}^*) = \mathbf{0}$]. Components of gradient $\nabla L(\mathbf{x}, \mathbf{v})$ are partial derivatives with respect to the $\mathbf{x}$ and $\mathbf{v}$ parts of a solution, respectively. Setting both $= \mathbf{0}$ produces the key **stationary-point conditions**.

$\boxed{14.15}$ Solution $(\mathbf{x}^*, \mathbf{v}^*)$ is a stationary point of Lagrangian function $L(\mathbf{x}, \mathbf{v})$ if it satisfies

$$\sum_i \nabla g_i(\mathbf{x}^*) v_i^* = \nabla f(\mathbf{x}^*) \quad \text{or} \quad \sum_i \frac{\partial g_i}{\partial x_j} v_i = \frac{\partial f}{\partial x_j} \qquad \text{for all } j$$

and

$$g_i(\mathbf{x}^*) = b_i \qquad \text{for all } i$$

SAMPLE EXERCISE 14.8: FORMING STATIONARY CONDITIONS FOR LAGRANGIANS

Return to the model and Lagrangian function of Sample Exercise 14.7. State the corresponding stationary-point conditions on $x_1^*, x_2^*, x_3^*, v_1^*,$ and v_2^*.

Analysis: We construct conditions $\boxed{14.15}$ for Lagrangian function

$$L(x_1, x_2, x_3, v_1, v_2) \triangleq 6(x_1)^2 + 4(x_2)^2 + (x_3)^2 + v_1 (360 - 24x_1 - 24x_2) + v_2 (1 - x_3)$$

Gradients

$$\nabla f(\mathbf{x}) = (12x_1, 8x_2, 2x_3)$$
$$\nabla g_1(\mathbf{x}) = (24, 24, 0)$$
$$\nabla g_2(\mathbf{x}) = (0, 0, 1)$$

Thus the required conditions are

$$
\begin{aligned}
24v_1^* &&&= 12x_1^* \\
24v_1^* &&&= 8x_2^* \\
&&+1v_2^* &= 2x_3^* \\
24x_1^* &+24x_2^* &&= 360 \\
1x_3^* &&&= 1
\end{aligned}
$$

Lagrangian Stationary Points and the Original Model

The value of Lagrangian functions 14.14 in solving equality-constrained NLPs 14.13 becomes apparent upon careful examination of Lagrangian stationary conditions 14.15 . We originally formed the Lagrangian to obtain a relaxed model that matches the original objective function at feasible points. For any fixed choice $\mathbf{v}$ of Lagrange multipliers, an unconstrained optimum $\mathbf{x}^*$ of that relaxed model must be a stationary point (principle 13.19). This is exactly what the first part of conditions 14.15 demand.

But the second part of 14.15 requires any stationary point of the Lagrangian to satisfy all constraints of the original model. Thus if the $\mathbf{x}^*$ components of a stationary point can be shown to optimize the relaxation with multipliers fixed at the $\mathbf{v}^*$ of a stationary point, that $\mathbf{x}^*$ solves a relaxed form, achieves feasibility, and has the same objective value as the original $f(\mathbf{x}^*)$. It must solve the constrained model.

14.16 If $(\mathbf{x}^*, \mathbf{v}^*)$ is a stationary point of Lagrangian function $L(\mathbf{x}, \mathbf{v})$ and $\mathbf{x}^*$ is an unconstrained optimum of $L(\mathbf{x}, \mathbf{v}^*)$, then $\mathbf{x}^*$ is an optimum of the corresponding equality-constrained NLP 14.13 .

(Readers may wish to compare with principle 12.10 of Section 12.2 for a similar result in the context of integer programming.)

Lagrange Multiplier Procedure

The Lagrangian approach to solving equality-constrained NLPs exploits sufficient optimality condition 14.16 . Specifically, we:

1. Reduce the given model to pure equality form 14.13 .
2. Form Lagrangian function 14.14 .
3. Solve conditions 14.15 for a stationary point of the Lagrangian function.
4. Try to establish that the $\mathbf{x}$ part of the stationary point is optimal for the Lagrangian with $\mathbf{v} = \mathbf{v}^*$ and thus (principle 14.16) optimal for the original model.

Lagrangian (14.17) shows the result of steps 1 and 2 for our Pfizer lot sizing example. Solving for a stationary point, we first compute

$$\frac{\partial L}{\partial x_1} = 66.21 - \frac{2160}{(x_1)^2} + \frac{180v}{(x_1)^2} = 0 \quad \text{or} \quad x_1^* = \sqrt{\frac{2160 - 180v^*}{66.21}}$$

$$\frac{\partial L}{\partial x_2} = 426.8 - \frac{8712}{(x_2)^2} + \frac{726v}{(x_2)^2} = 0 \quad \text{or} \quad x_2^* = \sqrt{\frac{8712 - 726v^*}{426.8}}$$

$$\frac{\partial L}{\partial x_3} = 61.20 - \frac{330}{(x_3)^2} + \frac{27.5v}{(x_3)^2} = 0 \quad \text{or} \quad x_3^* = \sqrt{\frac{330 - 27.5v^*}{61.20}} \qquad (14.19)$$

$$\frac{\partial L}{\partial x_4} = 268.1 - \frac{2916}{(x_4)^2} + \frac{243v}{(x_4)^2} = 0 \quad \text{or} \quad x_4^* = \sqrt{\frac{2916 - 243v^*}{268.1}}$$

Then setting

$$\frac{\partial L}{\partial v} = 221.5 - \frac{180}{x_1} - \frac{726}{x_2} - \frac{27.5}{x_3} - \frac{243}{x_4} = 0$$

and substituting gives

$$221.5 = \frac{180}{\sqrt{\dfrac{2160 - 180v^*}{66.21}}} + \frac{726}{\sqrt{\dfrac{8712 - 726v^*}{426.8}}} + \frac{27.5}{\sqrt{\dfrac{330 - 27.5v^*}{61.20}}} + \frac{243}{\sqrt{\dfrac{2916 - 243v^*}{268.1}}}$$

Now taking advantage of the fact that changeover costs in the objective function of model (14.16) are multiples (by \$12,000 per week) of changeover times, we may factor

$$221.5 = \frac{1}{\sqrt{12 - v^*}} \left(\frac{180}{\sqrt{\dfrac{180}{66.21}}} + \frac{726}{\sqrt{\dfrac{726}{426.8}}} + \frac{27.5}{\sqrt{\dfrac{27.5}{61.20}}} + \frac{243}{\sqrt{\dfrac{243}{268.1}}} \right)$$

and solve

$$v^* = 12 - \frac{1}{221.5} \left[\sqrt{180(66.21)} + \sqrt{726(426.8)} + \sqrt{27.5(61.20)} + \sqrt{243(268.1)} \right]^2$$

$$= -6.865$$

Finally, substitution is (14.19) gives $x_1^* = 7.161$, $x_2^* = 5.665$, $x_3^* = 2.911$, and $x_4^* = 4.135$.

We know that this $\mathbf{x}^*$ part of the stationary point solution yields an optimum because it agrees with values reported in Section 14.2. To complete step 4 of the Lagrangian approach and apply principle $\boxed{14.16}$, however, we must examine the Lagrangian objective with $v = v^* = -6.865$:

$$L(x_1, x_2, x_3, x_4, -6.865) \triangleq 66.21x_1 + \frac{2160}{x_1} + 426.8x_2 + \frac{8712}{x_2} + 61.20x_3 + \frac{330}{x_3}$$

$$+ 268.1x_4 + \frac{2916}{x_4} - 6.865 \left(221.5 - \frac{180}{x_1} - \frac{726}{x_2} - \frac{27.5}{x_3} - \frac{243}{x_4} \right)$$

$$= 66.21x_1 + \frac{3395.7}{x_1} + 426.8x_2 + \frac{13{,}696}{x_2} + 61.20x_3 + \frac{518.79}{x_3}$$

$$+ 268.1x_4 + \frac{4584.2}{x_4} - 1520.6$$

This function is convex for the same reasons that the original objective was in Section 14.2. Thus the computed stationary point is indeed a global minimum (principle $\boxed{13.22}$).

SAMPLE EXERCISE 14.9: OPTIMIZING WITH LAGRANGIAN METHODS

Use Lagrange multiplier methods to compute an optimal solution to the model of Sample Exercises 14.7 and 14.8.

Analysis: Stationary-point conditions from Sample Exercise 14.8 are

$$
\begin{aligned}
24v_1^* &= 12x_1^* \\
24v_1^* &= 8x_2^* \\
+ 1v_2^* &= 2x_3^* \\
24x_1^* + 24x_2^* &= 360 \\
1x_3^* &= 1
\end{aligned}
$$

Using the first 3 to substitute for $x_1, \ldots, x_3$ in the last 2 of these constraints yields

$$24(2v_1^*) + 24(3v_1^*) = 360 \quad \text{or} \quad v_1^* = 3$$
$$1(\tfrac{1}{2}v_2^*) = 1 \quad \text{or} \quad v_2^* = 2$$

Then solving for corresponding x_j^* produces the unique stationary point

$$(x_1^*, x_2^*, x_3^*, v_1^*, v_2^*) = (6, 9, 1, 3, 2)$$

It remains to be sure that the computed point is a minimum of the Lagrangian function

$$L(x_1, x_2, x_3, v_1^*, v_2^*)$$
$$\triangleq 6(x_1)^2 + 4(x_2)^2 + (x_3)^2 + 3(360 - 24x_1 - 24x_2) + 2(1 - x_3)$$
$$= 6(x_1)^2 + 4(x_2)^2 + (x_3)^2 - 72x_1 - 72x_2 - 2x_3 + 1082$$

But this function is obviously convex (squares plus linear), so that the computed stationary point does indeed solve the original constrained model (principle $\boxed{14.16}$).

Interpretation of Lagrange Multipliers

Those readers who have followed large parts of this book will recall that we employed **dual variables** v_i on constraints of linear programs (definition $\boxed{7.20}$, Section 7.3) to analyze the sensitivity of results to changes in the right-hand-side coefficients of the model. It is no accident that the same notation is used here for Lagrange multipliers.

$\boxed{14.17}$ The optimal Lagrange multiplier, v_i^* associated with constraint $g_i(\mathbf{x}) = b_i$ can be interpreted as the rate of change in optimal value per unit increase in right-hand side b_i.

To see that this interpretation applies, we need only examine the Lagrangian function at an optimal $(\mathbf{x}^*, \mathbf{v}^*)$:

$$L(\mathbf{x}^*, \mathbf{v}^*) = f(\mathbf{x}^*) + \sum_{i=1}^{m} v_i^* [b_i - g_i(\mathbf{x}^*)]$$

The rate of change with RHS b_i is

$$\frac{\partial L}{\partial b_i} = v_i^*$$

For a specific example, return to the $v^* = -6.865$ on the sole constraint of Pfizer equality model (14.16). This quantity is the partial derivative of Lagrangian function (14.17) with respect to the right-hand side, which represents a production capacity in weeks. Thus because the Lagrangian function and the true objective function coincide at stationary points [expression (14.18)], this v^* tells us that small increases in the RHS would decrease (improve) the optimal objective value at the rate of $6865 per week.

SAMPLE EXERCISE 14.10: INTERPRETING OPTIMAL LAGRANGE MULTIPLIERS

Use the optimal Lagrange multipliers of Sample Exercise 14.9 to analyze sensitivity of results to changes in constraint right-hand sides.

Analysis: From Sample Exercise 14.7, model constraints are

$$24x_1 + 24x_2 = 360$$
$$x_3 = 1$$

Corresponding optimal Lagrange multipliers are $v_1^* = 3$ and $v_2^* = 2$. We see that an increase in either right-hand side would increase (degrade) the optimal value of 541. At least for small changes, every unit increase in RHS 360 adds $v_1^* = 3$, and each unit increase in RHS 1 costs $v_2^* = 2$.

Limitations of the Lagrangian Approach

Although Lagrange multiplier methods work well for some models, it should be apparent that they have serious limitations:

- Stationary-point conditions 14.15 can be solved only if they are linear or very simple nonlinear functions. In other cases, solving those conditions may be more difficult than directly searching for an optimal solution to the original model.
- If a given model has many inequality constraints, it may be an explosively combinatorial task to determine which will be active at an optimal solution, so that equality-constrained Lagrangian methods may be applied.
- We can be certain that the stationary point computed from system 14.15 is a global optimum only if the original model functions were tractable enough to apply principle 14.16 . Other cases may produce ambiguous results.

There is also another, more subtle difficulty with applying Lagrange multiplier techniques to some NLPs. Principle 14.16 tells us when the $\mathbf{x}^*$ part of an optimum for the Lagrangian function must be optimal in the original model. But the converse is not always true. That is, an optimal solution in the original model need not correspond to a stationary point of the associated Lagrangian. Although most models occurring in application do have optima satisfying conditions 14.15 , we will present an example in the next section where the property fails.

SAMPLE EXERCISE 14.11: UNDERSTANDING LIMITS OF LAGRANGIAN METHODS

Consider the equality-constrained nonlinear program

$$\max \quad w_1 w_2$$
$$\text{s.t.} \quad 9(w_1)^4 - 17(w_1)^3 + 6(w_1)^2 + 3w_1 + 11e^{w_2} = 100$$

Describe the difficulties that would be encountered in trying to address this model by Lagrange multiplier methods.

Analysis: Using Lagrange multiplier v on the single constraint yields Lagrangian

$$L(w_1, w_2, v) \triangleq w_1 w_2 + v\left[100 - 9(w_1)^4 + 17(w_1)^3 - 6(w_1)^2 - 3w_1 - 11e^{w_2}\right]$$

Corresponding stationary point conditions 14.15 are

$$\frac{\partial L}{\partial w_1} = w_2 - 36v(w_1)^3 + 51v(w_1)^2 - 12vw_1 - 3v \quad = 0$$

$$\frac{\partial L}{\partial w_1} = w_1 - 11ve^{w_2} \quad = 0$$

$$\frac{\partial L}{\partial w_1} = 100 - 9(w_1)^4 + 17(w_1)^3 - 6(w_1)^2 - 3w_1 - 11e^{w_2} = 0$$

Solving these highly nonlinear conditions would probably be as difficult as solving the original NLP. Furthermore, the original objective function is neither convex nor concave, and the constraint function is still less tractable. Even if a stationary point could be computed, it would probably be impossible to argue that the w_1^*, w_2^* part represented an optimal solution.

14.4 KARUSH–KUHN–TUCKER OPTIMALITY CONDITIONS

The Lagrangian stationary-point conditions 14.15 are often difficult to solve directly for an optimum, but they do provide useful conditions that optima must (usually) satisfy. In this section we develop the elaborated form known as **Karush–Kuhn–Tucker (KKT) conditions**, which we will see are intimately related to whether a point is a local optimum of the given NLP.

Full Differentiable NLP Model

The Lagrangian discussion of Section 14.3 deals only with equality constraints. KKT conditions address the full differentiable nonlinear program.

14.18 | Differentiable nonlinear programs have the general form

$$\text{max or min}\quad f(\mathbf{x})$$
$$\text{s.t.}\qquad g_i(\mathbf{x}) \geq b_i \qquad \text{for all } i \in G$$
$$g_i(\mathbf{x}) \leq b_i \qquad \text{for all } i \in L$$
$$g_i(\mathbf{x}) = b_i \qquad \text{for all } i \in E$$

where f and all g_i are differentiable functions, and sets G, L, and E index the $\geq$, $\leq$, and $=$ constraints, respectively.

Complementary Slackness Conditions

The difficulty in extending Lagrangian stationarity conditions | 14.15 | to inequality cases arises in knowing what inequalities will be active at a local optimum $\mathbf{x}$ (i.e., hold as equality). When we know that an inequality will be active, we may treat it as an equality and include it in the Lagrangian. If it will be inactive, we want it left out.

One way to formalize such requirements is to assign a Lagrange variable v_i to every constraint but require those for inactive inequalities to = 0. That is, we enforce **complementary slackness** constraints like those of | 7.26 | (Section 7.3) for linear programs.

14.19 | Either inequality constraints should be active at a local optimum or the corresponding Lagrange variable should = 0, that is,

$$v_i[b_i - g_i(\mathbf{x})] = 0 \qquad \text{for all inequalities } i$$

We may illustrate with quadratic portfolio example (14.10).

$$\min\quad 66.51(x_1)^2 + 2(2.61)x_1x_2 + 2(2.18)x_1x_3$$
$$+ 0.63(x_2)^2 \qquad\qquad \text{(variance)}$$
$$+ 2(0.48)x_2x_3 + 0.38(x_3)^2$$
$$\text{s.t.}\quad x_1 + x_2 + x_3 = 1 \qquad\qquad \text{(invest 100\%)} \qquad (14.20)$$
$$13.22x_1 + 8.24x_2 + 9.03x_3 \geq 11 \qquad \text{(return)}$$
$$x_1, x_2, x_3 \geq 0$$

Numbering constraints in the order given, the corresponding complementary slackness conditions are

$$v_2(11 - 13.22x_1 - 8.24x_2 - 9.03x_3) = 0$$
$$v_3(-x_1) = 0$$
$$v_4(-x_2) = 0 \qquad\qquad (14.21)$$
$$v_5(-x_3) = 0$$

Notice that none is needed on equality constraint $i = 1$.

Lagrange Multiplier Sign Restrictions

We saw with interpretation | 14.17 | that Lagrange multipliers should reflect the rate of change in optimal value per unit increase in right-hand side b_i. Just as with linear programming results $\boxed{7.21}$, this interpretation implies Lagrange multiplier sign restrictions when the constraints are inequalities. For example, we know that increasing RHS b_i relaxes a $\leq$ inequality, so that it can only increase the optimal value in a maximize problem or decrease it in a minimize problem. Other cases are similar.

$\boxed{14.20}$ Lagrange multipliers v_i on constraints i of $\boxed{14.18}$ should satisfy the following sign restrictions:

Objective	i is $\leq$	i is $\geq$	i is $=$
Minimize	$v_i \leq 0$	$v_i \geq 0$	Unrestricted
Maximize	$v_i \geq 0$	$v_i \leq 0$	Unrestricted

Again illustrating with minimizing portfolio management model (14.20), the needed sign restrictions are

$$v_1 \text{ URS}; \quad v_2, v_3, v_4, v_5 \geq 0 \tag{14.22}$$

because the first constraint is an equality, and the rest are $\geq$'s of a minimize model.

KKT Conditions and KKT Points

We are now ready to state Karush–Kuhn–Tucker conditions for general (differentiable) model $\boxed{14.18}$.

$\boxed{14.21}$ Solutions $\mathbf{x}$ and $\mathbf{v}$ satisfy the **Karush–Kuhn–Tucker conditions** for differentiable nonlinear program $\boxed{14.18}$ if they fulfill complementary slackness conditions $\boxed{14.19}$, sign restrictions $\boxed{14.20}$, gradient equation

$$\sum_i \nabla g_i(\mathbf{x}) v_i = \nabla f(\mathbf{x})$$

and primal constraints

$$g_i(\mathbf{x}) \geq b_i \qquad \text{for all } i \in G$$
$$g_i(\mathbf{x}) \leq b_i \qquad \text{for all } i \in L$$
$$g_i(\mathbf{x}) = b_i \qquad \text{for all } i \in E$$

Any $\mathbf{x}$ for which there exist a corresponding $\mathbf{v}$ satisfying these conditions is called a **KKT point**.

Our portfolio model (14.20) has objective function gradient

$$\nabla f(x_1, x_2, x_3) = \begin{pmatrix} 133.02x_1 + 5.22x_2 + 4.36x_3 \\ 5.22x_1 + 1.26x_2 + 0.96x_3 \\ 4.36x_1 + 0.96x_2 + 0.76x_3 \end{pmatrix}$$

and those of the 5 linear constraints are

$$\nabla g_1(x_1, x_2, x_3) = (1, 1, 1)$$
$$\nabla g_2(x_1, x_2, x_3) = (13.22, 8.24, 9.03)$$
$$\nabla g_3(x_1, x_2, x_3) = (1, 0, 0)$$
$$\nabla g_4(x_1, x_2, x_3) = (0, 1, 0)$$
$$\nabla g_5(x_1, x_2, x_3) = (0, 0, 1)$$

Thus the gradient equation part of KKT conditions $\boxed{14.21}$ is

$$1v_1 + 13.22v_2 + v_3 = 133.02x_1 + 5.22x_2 + 4.36x_3$$
$$1v_1 + 8.24v_2 + v_4 = 5.22x_1 + 1.26x_2 + 0.96x_3 \qquad (14.23)$$
$$1v_1 + 9.03v_2 + v_5 = 4.36x_1 + 0.96x_2 + 0.76x_3$$

The rest of the conditions are primal constraints

$$x_1 + x_2 + x_3 = 1$$
$$13.22x_1 + 8.24x_2 + 9.03x_3 \geq 11 \qquad (14.24)$$
$$x_1, x_2, x_3 \geq 0$$

complementary slackness (14.21), and sign restrictions (14.22).

Notice the direct parallel to Lagrangian stationary-point conditions $\boxed{14.15}$. Both sets of conditions require the objective function gradient to be expressible as a multiplier-weighted combination of constraint gradients, while primal constraints are also satisfied. The new elements are complementary slackness conditions and sign restrictions arising from inequalities.

SAMPLE EXERCISE 14.12: FORMULATING KKT CONDITIONS

Consider the nonlinear program

$$\begin{array}{ll} \max & 2w_1 + 7w_2 \\ \text{s.t.} & (w_1 - 2)^2 + (w_2 - 2)^2 = 1 \\ & w_1 \leq 2 \\ & w_2 \leq 2 \\ & w_1 \geq 0 \\ & w_2 \geq 0 \end{array}$$

State the Karush–Kuhn–Tucker conditions for this model.

Analysis: We apply definition $\boxed{14.21}$. Numbering constraints in the order given,

$$\nabla f(w_1, w_2) = (2, 7)$$
$$\nabla g_1(w_1, w_2) = (2w_1, 2w_2)$$
$$\nabla g_2(w_1, w_2) = (1, 0)$$
$$\nabla g_3(w_1, w_2) = (0, 1)$$
$$\nabla g_4(w_1, w_2) = (1, 0)$$
$$\nabla g_5(w_1, w_2) = (0, 1)$$

Thus the KKT conditions consist of primal constraints

$$(w_1 - 2)^2 + (w_2 - 2)^2 = 1$$
$$w_1 \leq 2$$
$$w_2 \leq 2$$
$$w_1 \geq 0$$
$$w_2 \geq 0$$

gradient equation

$$\begin{pmatrix} 2w_1 \\ 2w_2 \end{pmatrix} v_1 + \begin{pmatrix} 1 \\ 0 \end{pmatrix} v_2 + \begin{pmatrix} 0 \\ 1 \end{pmatrix} v_3 + \begin{pmatrix} 1 \\ 0 \end{pmatrix} v_4 + \begin{pmatrix} 0 \\ 1 \end{pmatrix} v_5 = \begin{pmatrix} 2 \\ 7 \end{pmatrix}$$

complementary slackness

$$v_2(2 - w_1) = 0$$
$$v_3(2 - w_2) = 0$$
$$v_4(0 - w_1) = 0$$
$$v_5(0 - w_2) = 0$$

and sign restrictions

$$v_2, v_3 \geq 0$$
$$v_3, v_4 \leq 0$$

Improving Feasible Directions and Local Optima Revisited

To see the importance of KKT conditions 14.21 in constrained nonlinear programming, we must return to the elementary improving search notions of Sections 3.2 and 3.3. Move directions Δx pursued by implementations of improving search should be both improving and feasible. That is (definitions 3.13 and 3.14), they should improve the objective and maintain feasibility for sufficiently small steps λ.

   If there is such an **improving feasible direction** available at a current solution in the search, the point cannot be even locally optimal (principle 3.16). Progress is still possible in every neighborhood of the current point by advancing in the available direction.

   When no improving feasible direction exists, the current solution is under mild assumptions, at least a local optimum (principle 3.17). Still, cases like Figure 3.8(b) show that the absence of improving feasible directions does not always imply local optimality. We can only be certain the search will stop.

> 14.22   Absence of an improving feasible direction at the current point of an improving search, which causes an improving search to stop, provides a working definition of when a local optimum has been reached.

   When does a direction Δx improve at current x? First-order Taylor series approximation (principle 13.17)

$$f(x + \Delta x) \approx f(x) + \nabla f(x) \cdot \Delta x$$

suggests that improvement depends on the sign of $\nabla f(\mathbf{x}) \cdot \Delta \mathbf{x}$. This yields conditions
$\boxed{3.21}$ and $\boxed{3.22}$:

$\boxed{14.23}$ The linear Taylor approximation to smooth objective function $f(\mathbf{x})$ shows the following about move direction $\Delta \mathbf{x}$:

Objective	$\nabla f(\mathbf{x}) \cdot \Delta \mathbf{x} > 0$	$\nabla f(\mathbf{x}) \cdot \Delta \mathbf{x} < 0$
Maximize	Improving	Nonimproving
Minimize	Nonimproving	Improving

If $\nabla f(\mathbf{x}) \cdot \Delta \mathbf{x} = 0$, more information is required to classify $\Delta \mathbf{x}$.

A direction can still improve if $\nabla f(\mathbf{x}) \cdot \Delta \mathbf{x} = 0$, but the absence of any $\Delta \mathbf{x}$ satisfying first-order condition $\boxed{14.23}$ is a strong indication that no improving directions exist.

The corresponding feasibility conditions $\boxed{3.25}$ are exact for linear constraints. Direction $\Delta \mathbf{x}$ is feasible if

$$\mathbf{a} \cdot \Delta \mathbf{x} \begin{cases} \leq 0 & \text{for active constraints} \quad \mathbf{a} \cdot \mathbf{x} \leq b \\ = 0 & \text{for all constraints} \quad \mathbf{a} \cdot \mathbf{x} = b \\ \geq 0 & \text{for active constraints} \quad \mathbf{a} \cdot \mathbf{x} \geq b \end{cases}$$

Inactive constraints need not be considered because they have no immediate impact on feasibility.

To generalize for nonlinear constraints we may again employ first-order Taylor series approximations (principle $\boxed{13.17}$):

$$g_i(\mathbf{x} + \Delta \mathbf{x}) \approx g_i(\mathbf{x}) + \nabla g_i(\mathbf{x}) \cdot \Delta \mathbf{x} \tag{14.25}$$

If g_i is active at $\mathbf{x}$, $g_i(\mathbf{x}) = b_i$. Thus feasibility in (14.25) depends on the sign of the term $\nabla g_i(\mathbf{x}) \cdot \Delta \mathbf{x}$.

$\boxed{14.24}$ Direction $\Delta \mathbf{x}$ is feasible at $\mathbf{x}$ for the linear Taylor approximation to constrained nonlinear program $\boxed{14.18}$ if

$$\nabla g_i(\mathbf{x}) \cdot \Delta \mathbf{x} \begin{cases} \geq 0 & \text{for active} \geq \text{constraints} \\ \leq 0 & \text{for active} \leq \text{constraints} \\ = 0 & \text{for all} = \text{constraints} \end{cases}$$

KKT Conditions and Existence of Improving Feasible Directions

We are now in a position to link Karush–Kuhn–Tucker conditions $\boxed{14.21}$ with the existence of improving feasible directions at a current search point $\mathbf{x}$.

$\boxed{14.25}$ Karush–Kuhn–Tucker conditions provide a first-order, working test of the absence of improving feasible directions. More specifically, $\mathbf{x}$ is a KKT point if and only if no direction of movement from $\mathbf{x}$ fulfills first-order tests $\boxed{14.23}$ and $\boxed{14.24}$ for an improving feasible direction.

To illustrate, recall that an optimal solution for portfolio management model (14.20) is

$$x_1^* = 0.43, \quad x_2^* = 0, \quad x_3^* = 0.57$$

Certainly, there are no improving feasible directions at this global optimum, and we may demonstrate that fact by finding corresponding v_i^* to satisfy KKT conditions (14.21) to (14.24). Values that will do the job are

$$v_1^* = -132.026, \quad v_2^* = 14.892, \quad v_3^* = 0.0, \quad v_4^* = 12.275, \quad v_5^* = 0.0 \qquad (14.26)$$

It is easy to check that these primal and multiplier values satisfy complementary slackness conditions (14.21), sign restrictions (14.22), and primal constraints (14.24). In gradient equation (14.23)

$$1v_1 + 13.22v_2 + v_3 = 1(-132.026) + 13.22(14.892) + (0.0) = 64.8$$
$$133.02x_1 + 5.22x_2 + 4.36x_3 = 133.02(0.47) + 5.22(0) + 4.36(0.53) = 64.8$$
$$1v_1 + 8.24v_2 + v_4 = 1(-132.026) + 8.24(14.892) + (12.275) = 2.96$$
$$5.22x_1 + 1.26x_2 + 0.96x_3 = 5.22(0.47)(+1.26(0.0) + 0.96(0.53) = 2.96$$
$$1v_1 + 9.03v_2 + v_5 = 1(-132.026) + 9.03(14.892) + (0.0) = 2.45$$
$$4.36x_1 + 0.96x_2 + 0.76x_3 = 4.36(0.47) + 0.96(0.0) + 0.76(0.53) = 2.45$$

$$\tag{14.27}$$

Contrast this optimal point with $\mathbf{x} = (1, 0, 0)$, where direction $\Delta\mathbf{x} = (-1, 0, 1)$ satisfies first-order conditions 14.23 and 14.24 because

$$\nabla f(1, 0, 0) \cdot \Delta\mathbf{x} = (133.02, 5.22, 4.36) \cdot (-1, 0, 1) = -128.66 < 0$$

and active constraints

$$\nabla g_1(1, 0, 0) \cdot \Delta\mathbf{x} = (1, 1, 1) \cdot (-1, 0, 1) = 0$$
$$\nabla g_4(1, 0, 0) \cdot \Delta\mathbf{x} = (0, 1, 0) \cdot (-1, 0, 1) \geq 0$$
$$\nabla g_5(1, 0, 0) \cdot \Delta\mathbf{x} = (0, 0, 1) \cdot (-1, 0, 1) \geq 0$$

(Note that $v_2 = v_3 = 0$ for complementary slackness.) But now the unique solution to gradient equation

$$\begin{pmatrix} 1 \\ 1 \\ 1 \end{pmatrix} v_1 + \begin{pmatrix} 0 \\ 1 \\ 0 \end{pmatrix} v_4 + \begin{pmatrix} 0 \\ 0 \\ 1 \end{pmatrix} v_5 = (\; 133.02, 5.22, 4.36 \;)$$

is

$$v_1 = 133.02, \quad v_2 = -127.8, \quad v_3 = -128.66$$

which violates sign restrictions $v_4, v_5 \geq 0$. KKT conditions cannot be satisfied.

To see why principle 14.25 must always be true, we may think of improving feasible conditions 14.23 and 14.24 as a linear program in decision variables $\Delta\mathbf{x}$. Taking the minimize case, and we want to

$$\begin{aligned} \min \quad & \nabla f(\mathbf{x}) \cdot \Delta\mathbf{x} \\ \text{s.t.} \quad & \nabla g_i(\mathbf{x}) \cdot \Delta\mathbf{x} \geq 0 \quad && \text{for active } \geq\text{'s} \\ & \nabla g_i(\mathbf{x}) \cdot \Delta\mathbf{x} \leq 0 \quad && \text{for active } \leq\text{'s} \\ & \nabla g_i(\mathbf{x}) \cdot \Delta\mathbf{x} = 0 \quad && \text{for all } =\text{'s} \end{aligned} \qquad (14.28)$$

Now we apply some linear programming duality from Section 7.4. Over multipliers v_i, the dual of (14.28) is

$$\max \quad \sum_{i \text{ active}} (0) v_i = 0$$

$$\text{s.t.} \quad \sum_{i \text{ active}} \nabla g_i(\mathbf{x}) v_i = \nabla f(\mathbf{x}) \tag{14.29}$$

$$v_i \geq 0 \quad \text{for active} \geq\text{'s}$$

$$v_i \leq 0 \quad \text{for active} \leq\text{'s}$$

Notice that the feasibility requirements of dual (14.29) are identical to the gradient equation and sign restriction part of KKT conditions $\boxed{14.21}$ at $\mathbf{x}$ (assuming that $v_i = 0$ on inactive constraints to satisfy complementary slackness). If any v_i fulfill these conditions, the dual is feasible and its objective value is constant zero. It follows (principle $\boxed{7.31}$) that the optimal $\Delta \mathbf{x}$ in primal (14.28) has the same optimal value $\nabla f(\mathbf{x}) \cdot \Delta \mathbf{x} = 0$. If KKT conditions are fulfilled, no $\Delta \mathbf{x}$ can fulfill all the conditions of $\boxed{14.23}$ and $\boxed{14.24}$.

On the other hand, if some $\Delta \mathbf{x}$ meets all the conditions of $\boxed{14.23}$ and $\boxed{14.24}$, there is a feasible solution to the primal with $\nabla f(\mathbf{x}) \cdot \Delta \mathbf{x} < 0$. Then the dual must be infeasible, because every dual solution bounds the primal optimum (principle $\boxed{7.30}$) and any would have objective value $= 0$. It follows that KKT conditions cannot be fulfilled at $\mathbf{x}$.

SAMPLE EXERCISE 14.13: VERIFYING KKT AS A NO-DIRECTION CHECK

Consider the constrained NLP

$$\min \quad (w_1)^2 + (w_2)^2$$
$$\text{s.t.} \quad w_1 + w_2 = 1$$
$$w_1, w_2 \geq 0$$

A global optimum is $w_1^* = w_2^* = \frac{1}{2}$.

(a) State the KKT conditions for this problem.

(b) Verify that $\Delta \mathbf{w} = (1, -1)$ satisfies the first-order conditions for an improving feasible direction at $\mathbf{w} = (0, 1)$, and that the corresponding KKT conditions have no solution.

(c) Verify that KKT conditions hold at the optimal $\mathbf{w}^*$, so that no direction can meet first-order tests for being improving and feasible.

Analysis:

(a) Following $\boxed{14.21}$ with Lagrange multipliers v_1, v_2, v_3 on the three constraints, conditions are

$$w_1 + w_2 = 1 \quad \text{(primal constraints)}$$
$$w_1, w_2 \geq 0$$
$$v_2(-w_1) = 0 \quad \text{(complementary slackness)}$$
$$v_3(-w_2) = 0$$

$$\begin{pmatrix} 1 \\ 1 \end{pmatrix} v_1 + \begin{pmatrix} 1 \\ 0 \end{pmatrix} v_2 + \begin{pmatrix} 0 \\ 1 \end{pmatrix} v_3 = \begin{pmatrix} 2w_1 \\ 2w_2 \end{pmatrix} \qquad \text{(gradient equation)}$$

$$v_2, v_3 \geq 0 \qquad \text{(sign restrictions)}$$

(b) Direction $\Delta \mathbf{w} = (1, -1)$ meets improving test $\boxed{14.23}$ at $\mathbf{w} = (0, 1)$ because

$$\nabla f(0, 1) \cdot \Delta \mathbf{w} = (0, 2) \cdot (1, -1) < 0$$

It is also feasible because active constraints have

$$\nabla g_1(0, 1) \cdot \Delta \mathbf{w} = (1, 1) \cdot (1, -1) = 0$$
$$\nabla g_2(0, 1) \cdot \Delta \mathbf{w} = (1, 0) \cdot (1, -1) \geq 0$$

Solution $\mathbf{w} = (0, 1)$ satisfies the primal constraint part of KKT conditions in part (a), and making $v_2 = 0$ will assure complementary slackness. Solving gradient equation

$$\begin{pmatrix} 1 \\ 1 \end{pmatrix} v_1 + \begin{pmatrix} 1 \\ 0 \end{pmatrix} v_2 = \begin{pmatrix} 0 \\ 2 \end{pmatrix}$$

yields unique solution $v_1 = 2$, $v_2 = -2$, which violates the sign restriction on v_2. KKT conditions cannot be satisfied.

(c) Optimum $w_1^* = w_2^* = \frac{1}{2}$ satisfies primal constraints and is active only in the first, thus corresponding $v_2 = v_3 = 0$ to conform to complementary slackness. This leaves gradient equation

$$\begin{pmatrix} 1 \\ 1 \end{pmatrix} v_1 = \begin{pmatrix} 1 \\ 1 \end{pmatrix}$$

which has solution $v_1 = 1$ satisfying all sign restrictions. KKT conditions do hold.

Sufficiency of KKT Conditions for Optimality

Since principle $\boxed{14.25}$ shows that a KKT point is one that admits no direction satisfying first-order conditions for improving feasibility, it follows that KKT conditions are sufficient to establish optimality whenever the absence of improving feasible directions is sufficient. The most common case is **convex programs** (definition $\boxed{14.3}$).

$\boxed{14.26}$ If $\mathbf{x}$ is a KKT point of a convex program, $\mathbf{x}$ is a global optimum.

For example, our portfolio example (14.20) is a convex program because all its constraints are linear, and its objective function is convex because it has positive-definite Hessian matrix

$$\mathbf{Hx} = \begin{pmatrix} 133.02 & 5.22 & 4.36 \\ 5.22 & 1.26 & 0.96 \\ 4.36 & 0.96 & 0.76 \end{pmatrix} \tag{14.30}$$

Thus, when we verified that $x_1^* = 0.47$, $x_2^* = 0.0$, $x_3^* = 0.53$ is a KKT point in computation (14.27), we proved that the solution was optimal (principle $\boxed{14.26}$).

Necessity of KKT Conditions for Optimality

A much more subtle issue than when KKT conditions are sufficient to establish a point's optimality is when they are necessary. That is, when must optimal points satisfy KKT conditions?

To see the issue, consider the NLP

$$\begin{array}{rl} \min & (y_1)^2 + 4y_2 \\ \text{s.t.} & (y_1 - 1)^2 + (y_2)^2 \leq 1 \\ & (y_1 + 1)^2 + (y_2)^2 \leq 1 \end{array} \qquad (14.31)$$

It is easy to check that this model is a convex program because the objective and both constraints are convex. Also, the only feasible solution $y_1 = y_2 = 0$ has to be optimal.

For KKT conditions $\boxed{14.21}$,

$$\nabla f(y_1, y_2) = (2y_1, 4), \quad \nabla g_1(y_1, y_2) = (2y_1 - 2, 2y_2), \quad \nabla g_2(y_1, y_2) = (2y_1 + 2, 2y_2)$$

so that the gradient equation part at $\mathbf{y} = (0, 0)$ becomes

$$\begin{pmatrix} -2 \\ 0 \end{pmatrix} v_1 + \begin{pmatrix} 2 \\ 0 \end{pmatrix} v_2 = \begin{pmatrix} 0 \\ 4 \end{pmatrix}$$

Clearly, there is no solution v_1, v_2. Even though $y_1 = y_2 = 0$ is a global optimum of a convex program, it is not a KKT point.

Fortunately, such cases where KKT conditions are not necessary are rare in common models. Also, a variety of **constraint qualifications** have been derived to characterize models where every local or global optimum is a KKT point. We present here only the easiest to apply.

$\boxed{14.27}$ A local optimal solution of a constrained differentiable NLP must be a KKT point if (1) all constraints are linear; or (2) the gradients of all constraints active at the local optimum are linearly independent.

SAMPLE EXERCISE 14.14: VERIFYING NECESSITY OF KKT CONDITIONS

Without actually solving, verify that every local optimum of the following models must be a KKT point.

(a) $\max \quad (w_1)^2 + e^{w_2} + w_1 w_2$
    s.t. $\quad 3w_1 + w_2 \leq 9$
    $\quad w_1, w_2 \geq 0$

(b) $\max \quad (w_1)^2 + e^{w_2} + w_1 w_2$
    s.t. $\quad (w_1 - 1)^2 \leq 1$
    $\quad (w_2 - 2)^2 \leq 4$

Analysis: We apply constraint qualifications $\boxed{14.27}$, which depend only on the constraints of the models.

(a) All constraints of this model are linear, so every local optimum must be a KKT point.

(b) Constraint gradients are

$$\nabla g_1(w_1, w_2) = (2w_1 - 2, 0) \quad \text{and} \quad \nabla g_2(w_1, w_2) = (0, 2w_2 - 4)$$

These constraints are linearly independent except at $\mathbf{w} = (1, 2)$, which is not feasible. Thus the active constraints at any local optimum will be linearly independent, and all such solutions must satisfy KKT conditions.

14.5 PENALTY AND BARRIER METHODS

One approach to solving constrained nonlinear programs is to convert them to a series of unconstrained ones. In this section we investigate such **sequential unconstrained min/maximization techniques** (**SUMT**), also known as **penalty** and **barrier** methods.

Penalty Methods

One scheme for transforming constrained into unconstrained NLPs uses penalty methods.

> **14.28** **Penalty methods** drop constraints of nonlinear programs and substitute new terms in the objective function penalizing infeasibility in the form
>
> $$\max \text{ or } \min F(\mathbf{x}) \triangleq f(\mathbf{x}) \pm \mu \sum_i p_i(\mathbf{x})$$
>
> (+ for minimize problems and − for maximize problems), where μ is a positive **penalty multiplier** and the p_i are functions satisfying
>
> $$p_i(\mathbf{x}) \begin{cases} = 0 & \text{if } \mathbf{x} \text{ satisfies constraint } i \\ > 0 & \text{otherwise} \end{cases}$$

Many alternatives are available for the **penalty functions** $p_i(\mathbf{x})$ associated with particular constraints.

> **14.29** Among the common penalty functions employed for constrained NLPs are
>
> $$\max\{0, b_i - g_i(\mathbf{x})\} \quad \text{and} \quad \max^2\{0, b_i - g_i(\mathbf{x})\} \qquad \text{for } \geq\text{'s}$$
>
> $$\max\{0, g_i(\mathbf{x}) - b_i\} \quad \text{and} \quad \max^2\{0, g_i(\mathbf{x}) - b_i\} \qquad \text{for } \leq\text{'s}$$
>
> $$|g_i(\mathbf{x}) - b_i| \quad \text{and} \quad |g_i(\mathbf{x}) - b_i|^2 \qquad \text{for } =\text{'s}$$

Each imposes no penalty when the corresponding constraint is satisfied, but adds a growing cost if it is violated.

EXAMPLE 14.7: SERVICE DESK DESIGN

Penalty methods are most often used in engineering design applications where many of the constraints are nonlinear. We will illustrate with a contrived example to design the service desk of a catalog order company.

Figure 14.6 displays the problem. A service desk $\frac{1}{2}$ meter in width is to be centered around two 1-meter conveyors bringing orders from warehouse storage.

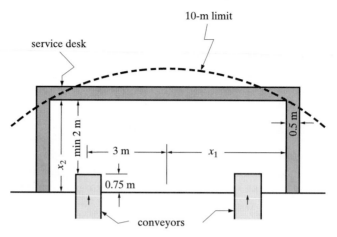

FIGURE 14.6   Service Desk Design Example

The conveyors are 6 meters apart (center-to-center) and protrude 0.75 meter into the work area. For employees to work efficiently behind the counter, there should be at least 2 meters clearance in front of the conveyors, and no part of the inside counter perimeter should total more than 10 meters from the conveyors. Within these limits we wish to maximize the customer room provided by the outside perimeter of the counter.

To model this simple example, introduce an origin halfway between the conveyors, and define decision variables

$$x_1 \triangleq \text{half-length of the work area inside the counter}$$

$$x_2 \triangleq \text{width of the work area inside the counter}$$

Then the problem can be modeled:

$$\max \quad 2x_1 + 2x_2 + 2 \qquad \text{(outer perimeter)}$$

$$\text{s.t.} \quad \frac{(x_1)^2}{(5)^2} + \frac{(x_2)^2}{(4)^2} \le 1 \qquad \text{(10-m distance limit)} \qquad (14.32)$$

$$x_1 \ge 3.5 \qquad \text{(outside conveyors)}$$

$$x_2 \ge 2.75 \qquad \text{(2-m inside space)}$$

The objective function maximizes the outer perimeter. The first (ellipse) constraint keeps the most distant point inside the counter at most a total of 10 meters from the conveyors. The lower bound on x_1 ensures that the counter falls outside the conveyors, and that of x_2 enforces the 2-meter inside clearance. An optimal design uses a desk with inside dimensions $2x_1^* = 2(3.63)$ meters by $x_2^* = 2.75$ meters and outside perimeter 14.76 meters.

Penalty Treatment of the Service Desk Example

Any of the penalty function alternatives in construction 14.29 might be used to deal with the $\le$ constraint and two $\ge$ constraints of service desk model (14.32). We will choose the second, squared penalty forms. For example, in the first constraint

$$p_1(x_1, x_2) \triangleq \max{}^2 \left\{ 0, \frac{(x_1)^2}{25} + \frac{(x_2)^2}{16} - 1 \right\}$$

When the constraint is satisfied $[(x_1)^2/25 + (x_2)^2/16 - 1] \leq 0$ and $p_1(x_1, x_2) = 0$. However, violations of the constraint make $[(x_1)^2/25 + (x_2)^2/16 - 1] > 0$ and impose a penalty equal to the square of the violation. Proceeding in this manner with all constraints yields the unconstrained penalty model

$$\max \quad 2x_1 + 2x_2 + 2 - \mu \left(\max{}^2 \left\{ 0, \frac{(x_1)^2}{25} + \frac{(x_2)^2}{16} - 1 \right\} \right.$$

$$\left. + \max{}^2\{0, 3.5 - x_1\} + \max{}^2\{0, 2.75 - x_2\} \right) \tag{14.33}$$

Infeasible solutions in the constrained model (14.32) are now allowed, but they are discouraged by subtracting a penalty in the objective function.

SAMPLE EXERCISE 14.15: FORMING PENALTY MODELS

Use absolute value (unsquared) penalty functions to reduce the following constrained NLP to an unconstrained penalty model.

$$\begin{aligned}
\min \quad & (w_1)^4 - w_1 w_2 w_3 \\
\text{s.t.} \quad & w_1 + w_2 + w_3 = 5 \\
& (w_1)^2 + (w_2)^2 \leq 9 \\
& w_3 w_2 \geq 1
\end{aligned}$$

Analysis: Using the first, unsquared penalty alternatives of $\boxed{14.29}$, the corresponding unconstrained model $\boxed{14.28}$ is

$$\begin{aligned}
\min \quad & (w_1)^4 - w_1 w_2 w_3 + \mu \left(|w_1 + w_2 + w_3 - 5| \right. \\
& \left. + \max\{0, (w_1)^2 + (w_2)^2 - 9\} + \max\{0, 1 - w_3 w_2\} \right)
\end{aligned}$$

where μ is a positive penalty multiplier.

Concluding Constrained Optimality with Penalties

By definition, the penalty terms of constructions $\boxed{14.28}$ must $= 0$ at any $\mathbf{x}$ feasible in the given constrained NLP. This provides a way to know when unconstrained optimization of the penalty problem yields an optimal solution for the original model.

$\boxed{14.30}$ If an optimal $\mathbf{x}^*$ in unconstrained penalty problem $\boxed{14.28}$ is feasible in the original constrained model, it is optimal in that NLP.

Any better solution to the constrained model would also have all penalty terms $= 0$, so it would have to best $\mathbf{x}^*$ in penalty model objective value.

Differentiability of Penalty Functions

One consideration in choosing among the penalty options in $\boxed{14.29}$ is differentiability. Most of Chapter 13's unconstrained methods that might be employed to optimize penalty model $\boxed{14.28}$ assume that the function is smooth. None of the first options listed in principle $\boxed{14.29}$ meets this differentiability requirement, but the second, squared options do.

$\boxed{14.31}$ Squared penalty options of construction $\boxed{14.29}$ are differentiable whenever the underlying g_i are differentiable.

We can see why by examining a $\leq$ inequality $g_i(\mathbf{x}) \leq b_i$ with g_i smooth. The corresponding squared penalty term can be expressed

$$p_i(\mathbf{x}) = \begin{cases} 0 & \text{if } \mathbf{x} \text{ satisfies } g_i(\mathbf{x}) \leq b_i \\ [g_i(\mathbf{x}) - b_i]^2 & \text{otherwise} \end{cases}$$

Associated partial derivatives are

$$\frac{\partial p_i}{\partial x_j} = \begin{cases} 0 & \text{if } \mathbf{x} \text{ satisfies } g_i(\mathbf{x}) \leq b_i \\ 2[g_i(\mathbf{x}) - b_i]\dfrac{\partial g_i}{\partial x_j} & \text{otherwise} \end{cases}$$

Notice that these two expressions match at the boundary where $g_i(\mathbf{x}) = b_i$. Thus partial derivatives are well defined and continuous.

Exact Penalty Functions

We would also like penalty functions to be **exact** penalty functions. That is, we would like there to be a large enough $\mu > 0$ that the unconstrained penalty model $F(\mathbf{x})$ yields an optimal solution in the original constrained form under principle $\boxed{14.30}$ by driving out all infeasibility.

A trivial example shows that the squared alternatives of $\boxed{14.29}$ may not be exact. Consider

$$\begin{array}{ll} \min & y \\ \text{s.t.} & y \geq 0 \end{array} \tag{14.34}$$

The squared choice in $\boxed{14.29}$ produces penalty model

$$\min \; F(y) \triangleq y + \mu \max^2\{0, -y\}$$

$$= \begin{cases} y & \text{if } y \geq 0 \\ y + \mu y^2 & \text{if } y < 0 \end{cases}$$

Differentiating

$$\frac{dF}{dy} = \begin{cases} 1 & \text{if } y \geq 0 \\ 1 + 2\mu y & \text{if } y < 0 \end{cases}$$

shows that the only stationary point is

$$y^* = -\frac{1}{2\mu}$$

But this unconstrained minimum is negative and infeasible, regardless of the magnitude of μ.

14.32 Squared penalty alternatives 14.29 are usually not exact; that is, there will often be no choice of penalty multiplier μ for which the corresponding unconstrained optimum of penalty model F is optimal in the original NLP.

Suppose that we had used nondifferentiable penalty function $\max\{0, b_i - g_i(\mathbf{x})\}$ on example (14.34). The corresponding penalty model is

$$\min \ F(y) \triangleq y + \mu \max\{0, -y\}$$

Now for any $\mu > 1$, $F(y)$ is minimized at $y = 0$. That is, a finite μ is large enough to make the penalty optimum feasible in the original NLP.

All of the nonsquared penalty forms in 14.29 are exact in this way.

14.33 If nonsquared penalty forms of 14.29 are applied to a constrained non-linear program having an optimal solution, and mild assumptions hold, there exists a finite multiplier μ sufficiently large that an optimum in unconstrained penalty problem 14.28 is optimal in the given NLP.

Managing the Penalty Multiplier

In squared cases such as (14.33), a constrained optimum can be obtained with penalty methods only by letting $\mu \to \infty$. With exact methods of principle 14.33 there is a large enough finite μ to do the job. Either way, μ needs to grow large, and we cannot know how large when we begin.

Why not just use a very large μ from the start? Figure 14.7 shows the risk with trivial model

$$\begin{aligned} \min \quad & w \\ \text{s.t.} \quad & 3 \le w \le 5 \end{aligned} \qquad (14.35)$$

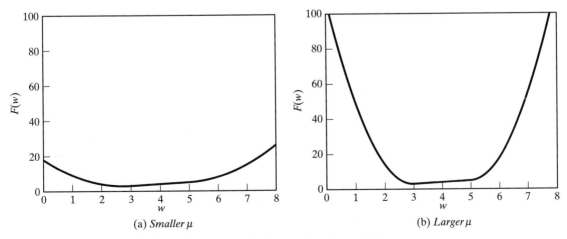

(a) Smaller μ (b) Larger μ

FIGURE 14.7 Effect of Penalty Multiplier on Unconstrained Tractability

When μ is comparitively large, the corresponding penalty objective function F becomes very steep. Small moves have dramatic impacts on its value. The result is an unconstrained model that is difficult to solve with any of the methods of Chapter 13.

These competing demands on the penalty multiplier μ motivate a sequential strategy that slowly increases the multiplier.

> $\boxed{14.34}$ When addressing a constrained nonlinear program by penalty methods, the multiplier μ should be started at a relatively low value > 0 and increased as computation proceeds.

Sequential Unconstrained Penalty Technique

Formalization of principle $\boxed{14.34}$'s strategy for slowly increasing the penalty multiplier μ produces the **sequential unconstrained penalty technique** of Algorithm 14A. Multiplier μ begins relatively small and grows with each search. For each value of μ, unconstrained penalty problem $\boxed{14.28}$ is solved beginning with the optimum of the preceding search. If the result is ever feasible in the original model, we stop with an optimum (principle $\boxed{14.30}$). Otherwise, we continue until the unconstrained optimum is sufficiently close to feasible.

ALGORITHM 14A: SEQUENTIAL UNCONSTRAINED PENALTY TECHNIQUE

Step 0: Initialization. Form penalty model $\boxed{14.28}$, and choose initial penalty multipier $\mu_0 > 0$ relatively small and starting solution $\mathbf{x}^{(0)}$. Also, initialize solution index $t \leftarrow 0$, and pick an escalation factor $\beta > 1$.

Step 1: Unconstrained Optimization. Beginning from $\mathbf{x}^{(t)}$, solve penalty optimization problem $\boxed{14.28}$ with $\mu = \mu_t$ to produce optimum $\mathbf{x}^{(t+1)}$.

Step 2: Stopping. If $\mathbf{x}^{(t+1)}$ is feasible or sufficiently close to feasible in the constrained model given, stop and output $\mathbf{x}^{(t+1)}$.

Step 3: Increase. Enlarge the penalty multiplier as

$$\mu_{t+1} \leftarrow \beta\mu_t$$

Then advance $t \leftarrow t + 1$, and return to Step 1.

Table 14.3 illustrates Algorithm 14A for service desk design model (14.32) and corresponding penalty form (14.33). With initial multiplier $\mu = \frac{1}{4}$ a first search produces the unconstrained optimum

$$\mathbf{x}^{(1)} = (9.690, 6.202)$$

which violates the first constraint of the original model by 5.160. Then the multiplier is increased by factor $\beta = 4$ and a new unconstrained search initiated from $\mathbf{x}^{(1)}$. The resulting optimum $\mathbf{x}^{(2)}$ starts a third search after μ is quadrupled again. The process continues until μ is large enough that the unconstrained optimum approaches feasibility. There we stop to obtain constrained optimal solution $\mathbf{x}^* = (3.63, 2.75)$.

TABLE 14.3 Sequential Penalty Solution of
Service Desk Example

t	μ	Optimal $\mathbf{x}^{(t+1)}$	Constraint Violation, i		
			1	**2**	**3**
0	$\frac{1}{4}$	(9.690, 6.202)	5.160	0.000	0.000
1	1	(6.632, 4.244)	1.885	0.000	0.000
2	4	(4.981, 3.188)	0.627	0.000	0.000
3	16	(4.221, 2.749)	0.185	0.000	0.001
4	64	(3.806, 2.748)	0.051	0.000	0.002
5	256	(3.677, 2.749)	0.013	0.000	0.001
6	1024	(3.643, 2.750)	0.003	0.000	0.000
7	4096	(3.634, 2.750)	0.001	0.000	0.000

Barrier Methods

The penalty methods above begin anywhere and try to force the unconstrained optimum into the feasible set of the given NLP. Barrier methods adopt the alternative of beginning with a feasible solution and trying to prevent the unconstrained search from leaving the feasible region.

> **14.35** | **Barrier methods** drop constraints of nonlinear programs and substitute new terms in the objective function, discouraging any approach to the boundary of the feasible region in the form
>
> $$\text{max or min } F(\mathbf{x}) \triangleq f(\mathbf{x}) \pm \mu \sum_i q_i(\mathbf{x})$$
>
> (+ for minimize problems and − for maximize problems), where μ is a positive **barrier multiplier** and the q_i are functions with
>
> $$q_i(\mathbf{x}) \to \infty$$
>
> as constraint i approaches being active.

Since the boundary cannot be avoided with equality constraints, barrier methods are applicable only when constraints are all inequalities. Many alternatives are available for the $q_i(\mathbf{x})$ associated with such constraints.

> **14.36** | Among the common barrier functions associated with inequality constrained NLPs are
>
> $$-\ln[g_i(\mathbf{x}) - b_i] \quad \text{and} \quad \frac{1}{g_i(\mathbf{x}) - b_i} \qquad \text{for } \geq\text{'s}$$
>
> $$-\ln[b_i - g_i(\mathbf{x})] \quad \text{and} \quad \frac{1}{b_i - g_i(\mathbf{x})} \qquad \text{for } \leq\text{'s}$$

Each explodes toward $+\infty$ as the corresponding inequality approaches being satisfied as an equality.

Barrier Treatment of Service Desk Example

All constraints of service desk design model (14.32) are inequalities, so barrier methods could be applied. We will adapt the more common logarithmic forms. For example, the first constraint produces barrier term

$$q_1(x_1, x_2) \triangleq -\ln\left(1 - \frac{(x_1)^2}{25} - \frac{(x_2)^2}{16}\right)$$

When $\mathbf{x}$ is well inside the feasible region, this barrier function affects the objective only modestly. But as $((x_1)^2/25 + (x_2)^2/16) \to 1$, the negative of the logarithm goes to $+\infty$. Similar treatment of all constraints yields the unconstrained barrier model

$$\max \quad 2x_1 + 2x_2 + 2 + \mu\left[\ln\left(1 - \frac{(x_1)^2}{25} - \frac{(x_2)^2}{16}\right)\right.$$

$$\left. + \ln(x_1 - 3.5) + \ln(x_2 - 2.75)\right]$$

(14.36)

With $\mu > 0$, an approach to any part of the boundary is discouraged.

SAMPLE EXERCISE 14.16: FORMING BARRIER MODELS

Use reciprocal barrier functions to reduce the following constrained NLP to an unconstrained barrier model.

$$\min \quad (w_1)^4 - w_1 w_2 w_3$$
$$\text{s.t.} \quad (w_1)^2 + (w_2)^2 \leq 9$$
$$w_3 w_2 \geq 1$$

Analysis: Using second, reciprocal barrier alternatives of $\boxed{14.36}$, the corresponding unconstrained model $\boxed{14.35}$ is

$$\min \quad (w_1)^4 - w_1 w_2 w_3 + \mu\left(\frac{1}{9 - (w_1)^2 - (w_2)^2} + \frac{1}{w_3 w_2 - 1}\right)$$

where μ is a positive barrier multiplier.

Converging to Optimality with Barrier Methods

Unlike penalty methods, barrier functions affect the objective function at feasible points. We can illustrate the difficulty this causes by returning to the trivial min y, s.t. $y \geq 0$ example of (14.34). Using, say, the logarithmic barrier alternative of $\boxed{14.36}$, the corresponding barrier problem is

$$\min \ F(y) \triangleq y - \mu \ln(y)$$

Differentiating (with $y > 0$) yields

$$\frac{dF}{dy} = 1 - \frac{\mu}{y}$$

which has its only stationary point at

$$y^* = \mu$$

This unconstrained optimum never reaches the true optimum of $y = 0$ for any $\mu > 0$.

Similar behavior occurs for all barrier versions of constrained NLPs with an optimum on the boundary of the feasible set.

> __14.37__ The optimum of barrier function __14.35__ can never equal the true optimum of the given constrained NLP if $\mu > 0$ and the optimum lies on the boundary of the feasible set.

As with penalty methods, however, there is a pattern to the unconstrained optima. As $\mu \to 0$, the unconstrained optimum comes closer and closer to the constrained solution.

> __14.38__ Although none may actually solve the given NLP, if mild assumptions hold, the sequence of unconstrained barrier function optima converges to an optimal solution to the given constrained NLP as multiplier $\mu \to 0$.

Managing the Barrier Multiplier

Property __14.38__ makes it clear that we have to let barrier multipliers μ approach zero if we expect to obtain a constrained optimum. Why not simply start close to zero? Figure 14.8 illustrates the difficulty with the trivial model (14.35).

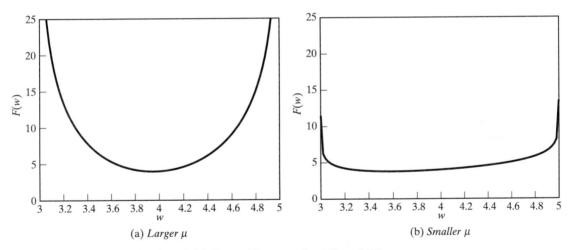

(a) *Larger μ*   (b) *Smaller μ*

FIGURE 14.8   Effect of Barrier Multiplier on Unconstrained Tractability

When μ is comparatively large, a strong barrier keeps the search far from boundary areas, where it might bog down. But small μ's allow the search to approach the boundary.

The best strategy is to start big and reduce the multiplier slowly.

> __14.39__ When addressing a constrained nonlinear program by barrier methods, the multiplier μ should be started at a relatively high value > 0 and decreased as the search proceeds.

Sequential Unconstrained Barrier Technique

Formalization of principle $\boxed{14.39}$'s strategy for slowly decreasing the barrier multiplier μ produces the **sequential unconstrained barrier technique** of Algorithm 14B. Processing begins at an interior feasible $\mathbf{x}^{(0)}$ where none of the constraints are active. Multiplier μ starts relatively large and becomes smaller with each search. For each value μ, unconstrained barrier problem $\boxed{14.35}$ is solved beginning with the optimum of the preceding search. We stop when μ is sufficiently close to zero. (Readers may wish to compare with barrier methods for linear programming in Section 6.4.)

ALGORITHM 14B: SEQUENTIAL UNCONSTRAINED BARRIER TECHNIQUE

Step 0: Initialization. Form barrier function $\boxed{14.35}$, and choose initial barrier multipier $\mu_0 > 0$ relatively large and feasible interior starting solution $\mathbf{x}^{(0)}$. Also initialize solution index $t \leftarrow 0$, and pick a reduction factor $\beta < 1$.

Step 1: Unconstrained Optimization. Beginning from $\mathbf{x}^{(t)}$, solve barrier optimization problem $\boxed{14.35}$ with $\mu = \mu_t$ to produce optimum $\mathbf{x}^{(t+1)}$.

Step 2: Stopping. If μ is sufficiently small, stop and output $\mathbf{x}^{(t+1)}$.

Step 3: Reduce. Decrease the penalty multiplier as

$$\mu_{t+1} \leftarrow \beta\mu_t$$

Then advance $t \leftarrow t + 1$, and return to Step 1.

Table 14.4 illustrates this for service desk design model (14.32) and corresponding barrier form (14.36). Unlike penalty methods, which can start anywhere, a barrier search must begin at an interior feasible point of the constrained NLP. Here we employ

$$\mathbf{x}^{(0)} = (3.52, 2.77)$$

The bulk of the computation involves a sequence of unconstrained barrier model searches with decreasing multipliers μ. Table 14.4 shows the first used $\mu_0 = 4$ to obtain the unconstrained optimum

$$\mathbf{x}^{(1)} = (3.573, 2.794)$$

Then the multiplier was decreased by a factor of $\beta = \frac{1}{2}$ and a new search initiated from $\mathbf{x}^{(1)}$. The resulting $\mathbf{x}^{(2)}$ starts a third search after μ is halved again. Searches continue until μ is close enough to zero that an approximate optimum is at hand. Then we stop with constrained optimum $\mathbf{x}^* = (3.63, 2.75)$.

TABLE 14.4   Sequential Barrier Solution of Service Desk Example

t	μ	$\mathbf{x}^{(t+1)}$	t	μ	$\mathbf{x}^{(t+1)}$
0	4	$(3.573, 2.794)$	2	1	$(3.608, 2.769)$
1	2	$(3.588, 2.786)$	3	$\frac{1}{2}$	$(3.629, 2.752)$

14.6 REDUCED GRADIENT ALGORITHMS

Chapter 5's simplex algorithm for linear programming is the most widely employed of all optimization procedures. In this section we develop natural extensions to the nonlinear case known as **reduced gradient** algorithms, or more generally as the **variable elimination method**.

Standard Form for NLPs with Linear Constraints

Most of our development of reduced gradient methods will assume a linearly constrained nonlinear program. In particular, we assume that constraints have the standard form like Section 5.1's definition 5.12 .

14.40 | **Standard form** for linearly constrained nonlinear programs is

$$\text{min or max} \quad f(\mathbf{x})$$
$$\text{s.t.} \quad \mathbf{A}\mathbf{x} = \mathbf{b}$$
$$\mathbf{x} \geq \mathbf{0}$$

We will also assume for simplicity that the rows of matrix $\mathbf{A}$ are linearly independent, which can always be achieved by dropping redundant constraints.

EXAMPLE 14.8: FILTER TUNING

For a tiny example on which to illustrate reduced gradient notions, consider the tuning of an electronic filter. Two parameters

$$x_1 \triangleq \text{value of the first tuning parameter}$$
$$x_2 \triangleq \text{value of the second tuning parameter}$$

must be chosen to minimize distortion

$$f(x_1, x_2) \triangleq (x_1 - 5)^2 - 2x_1x_2 + (x_2 - 10)^2$$

with x_1 in the range $[0,3]$, x_2 in the range $[0,5]$, and their total at most 6. The result is the linearly constrained nonlinear program

$$\begin{aligned}
\min \quad & f(x_1, x_2) \triangleq (x_1 - 5)^2 - 2x_1x_2 + (x_2 - 10)^2 \\
\text{s.t.} \quad & x_1 + x_2 \leq 6 \\
& 0 \leq x_1 \leq 3 \\
& 0 \leq x_2 \leq 5
\end{aligned} \tag{14.37}$$

Figure 14.9 graphs this tuning model. A global optimum occurs at $\mathbf{x}^{(3)} = (1.75, 4.25)$.

We may convert to standard form 14.40 by adding slacks $x_3, x_4,$ and x_5:

$$\begin{aligned}
\min \quad & f(x_1, x_2) \triangleq (x_1 - 5)^2 - 2x_1x_2 + (x_2 - 10)^2 \\
\text{s.t.} \quad & +x_1 + x_2 + x_3 = 6 \\
& +x_1 + x_4 = 3 \\
& + x_2 + x_5 = 5 \\
& x_1, x_2, x_3, x_4, x_5 \geq 0
\end{aligned} \tag{14.38}$$

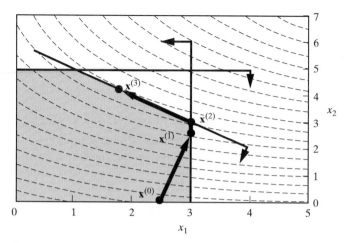

FIGURE 14.9   Reduced Gradient Search of Filter Tuning Example

Conditions for Feasible Directions with Linear Constraints

Any improving search algorithm attempts to move along improving feasible directions. Principle $\boxed{3.25}$ of Section 3.3 has already detailed the requirements for maintaining feasibility with standard-form linear constraints.

> $\boxed{14.41}$ At $\mathbf{x}$ feasible for standard-form linear constraints $\boxed{14.40}$, $\Delta\mathbf{x}$ is a feasible direction if
> $$\mathbf{A}\Delta\mathbf{x} = \mathbf{0}$$
> $$\Delta x_j \geq 0 \qquad \text{for all } j \text{ with } x_j = 0$$

To illustrate, consider point $\mathbf{x}^{(0)}$ at $(2.5, 0)$ of tuning example Figure 14.9. Including corresponding values for slacks, the full standard-form solution is

$$\mathbf{x}^{(0)} = (2.5, 0, 3.5, 0.5, 5)$$

The only active inequality is the nonnegativity constraint on x_2. Thus the corresponding conditions $\boxed{14.41}$ for a feasible direction $\Delta\mathbf{x}$ are

$$
\begin{aligned}
+\ \Delta x_1 \ +\ \Delta x_2 \ +\ \Delta x_3 \ & \qquad\qquad\qquad\quad = 0 \\
+\ \Delta x_1 \ & \qquad\ +\ \Delta x_4 \qquad\quad\ = 0 \\
+\ \Delta x_2 \ & \qquad\qquad\qquad +\ \Delta x_5 = 0 \\
\Delta x_2 \geq 0 \ &
\end{aligned}
\tag{14.39}
$$

Bases of the Main Linear Equalities

Section 5.2 developed the idea of **bases** or basic column sets of the matrix $\mathbf{A}$. Bases are maximal sets of linearly independent columns.

The important characteristic of a basis, and the corresponding **basic variables**, is that we may solve for the values of basic variables once all other, **nonbasic variables** have been fixed. That is, we may view the basics as functions of the nonbasics.

| 14.42 | Identification of a basic set of variables in a system of linear equations partitions solutions into independent, nonbasic versus dependent, basic components. |

Basic, Nonbasic, and Superbasic Variables

The simplex algorithms of Chapter 5 proceed through **basic solutions** in which all nonbasic variables take on lower bound value zero. We saw in Section 5.2 how this restriction can lead to a search through extreme points of the feasible set.

　　With nonlinear models, we know that an optimal solution may very well fall in the interior of the feasible set, or at a nonextreme point of the boundary. This does not change the fact that basic variables are dependent on our choice of nonbasics. It only means that nonbasics cannot be restricted to $= 0$. A new **superbasic** category of variables arises, which are nonbasics at positive value.

| 14.43 | Reduced gradient algorithms classify variables as basics, nonbasics at bound value zero, and superbasics nonbasic at values > 0. |

　　To illustrate, return to tuning example standard form (14.38), and choose x_1, x_3, and x_5 basic. Initial solution

$$\mathbf{x}^{(0)} = (2.5, 0, 3.5, 0.5, 5)$$

of Figure 14.9 is implied by nonbasic values $x_2^{(0)} = 0.0$ and $x_4^{(0)} = 0.5$. Setting all nonbasics $= 0$ would produce an extreme point of the feasible set. But with $x_4^{(0)} > 0$, and thus superbasic, nonextreme $\mathbf{x}^{(0)}$ can be represented.

SAMPLE EXERCISE 14.17: DISTINGUISHING BASIC, NONBASIC, AND SUPERBASIC

Consider the standard-form nonlinear program

$$\begin{aligned}
\max \quad & f(\mathbf{w}) \triangleq 50 - (w_1)^2 + 6w_1 - (w_2)^2 + 6w_2 + w_3 \\
\text{s.t.} \quad & + w_1 \quad - w_2 \quad + 3w_3 \qquad\qquad = 1 \\
& + 3w_1 \quad + 2w_2 \qquad\qquad + 2w_4 = 6 \\
& w_1, w_2, w_3, w_4 \geq 0
\end{aligned}$$

(a) Show that w_3 and w_4 form a basic set of variables.

(b) Assuming the basis of part (a), classify remaining variables as nonbasic at lower bound or superbasic for the solution $\mathbf{w} = (0, 2, 1, 1)$.

Analysis:

(a) A basis is a maximal set of linearly independent vectors. The columns of w_3 and w_4 are linearly independent because the corresponding matrix

$$\mathbf{B} = \begin{pmatrix} 3 & 0 \\ 0 & 2 \end{pmatrix}$$

has nonzero determinant $= 6$, and two is the maximum number of linearly independent 2-vectors.

(b) The specified solution has w_1 nonbasic at value zero, and w_2 superbasic because its value is positive.

Maintaining Equalities by Solving Main Constraints for Basic Variables

Return now to the feasibility requirements of $\boxed{14.41}$. Let $\mathbf{B}$ be a submatrix of $\mathbf{A}$ formed by a basic set of columns, and $\mathbf{N}$ the submatrix of all other columns. Then divide direction vector $\Delta\mathbf{x}$ into corresponding parts denoted $\Delta\mathbf{x}^{(B)}$ for components on columns in $\mathbf{B}$ and $\Delta\mathbf{x}^{(N)}$ for those in $\mathbf{N}$.

Conditions $\boxed{14.41}$ require that

$$\mathbf{A}\Delta\mathbf{x} = \mathbf{B}\Delta\mathbf{x}^{(B)} + \mathbf{N}\Delta\mathbf{x}^{(N)} = \mathbf{0}$$

Solving for basic components as a function of nonbasic variables,

$$\Delta\mathbf{x}^{(B)} = -\mathbf{B}^{-1}\mathbf{N}\Delta\mathbf{x}^{(N)} \tag{14.40}$$

Thus we can find a direction maintaining the equalities by choosing directional components for nonbasic variables and solving for basics as in (14.40).

$\boxed{14.44}$ Direction $\Delta\mathbf{x} \triangleq (\Delta\mathbf{x}^{(B)}, \Delta\mathbf{x}^{(N)})$ maintains feasibility in standard-form equality constraints $\mathbf{A}\mathbf{x} = \mathbf{b}$ if

$$\Delta\mathbf{x}^{(B)} = -\mathbf{B}^{-1}\mathbf{N}\Delta\mathbf{x}^{(N)}$$

where $\mathbf{B}$ is a basis submatrix of $\mathbf{A} \triangleq (\mathbf{B}, \mathbf{N})$.

For example, solving feasibility conditions (14.39) of our tuning example for basic variable components Δx_1, Δx_2, and Δx_3 produces

$$\begin{pmatrix} \Delta x_1 \\ \Delta x_3 \\ \Delta x_5 \end{pmatrix} = -\begin{pmatrix} 0 & 1 \\ 1 & -1 \\ 1 & 0 \end{pmatrix}\begin{pmatrix} \Delta x_2 \\ \Delta x_4 \end{pmatrix} \tag{14.41}$$

SAMPLE EXERCISE 14.18: SOLVING FOR BASIC DIRECTION COMPONENTS

Return to the maximizing nonlinear program of Sample Exercise 14.17 and express basic components of a move direction at $\mathbf{w} = (0, 2, 1, 1)$ as a function of the nonbasic components so that the resulting direction is feasible for the main constraints.

Analysis: We apply principle $\boxed{14.44}$ in solving $\mathbf{A}\Delta\mathbf{w} = \mathbf{0}$ conditions for basic components. The result is

$$\Delta w_3 = -(\tfrac{1}{3}\Delta w_1 - \tfrac{1}{3}\Delta w_2)$$
$$\Delta w_4 = -(\tfrac{3}{2}\Delta w_1 + 1\Delta w_2)$$

Active Nonnegativities and Degeneracy

The second part of feasible direction conditions $\boxed{14.41}$ requires that $\Delta x_j \geq 0$ whenever the corresponding nonnegativity constraint $x_j \geq 0$ is active. As with the simplex algorithm (see Section 5.6), we will not strictly enforce these requirements on basic j. Algorithms developed below endeavor to keep basic variables strictly positive, that is,

$$x_j > 0 \quad \text{for all } j \in B \tag{14.42}$$

For example, basic components $j = 1, 3, 5$ of tuning example solution $\mathbf{x}^{(0)} = (2.5, 0, 3.5, 0.5, 5)$ are all positive.

Under this **nondegeneracy assumption**, only nonnegativity constraints for nonbasics can be active, and we enforce

$$\Delta x_j \geq 0 \quad \text{for all } j \in N \quad \text{with} \quad x_j = 0 \tag{14.43}$$

Reduced Gradients

We now know from construction $\boxed{14.44}$ how to produce a feasible direction by limiting independent choices to components for a nonbasic set of variables. It remains to construct a feasible direction that improves the objective.

For small steps the change in objective along direction $\Delta\mathbf{x}$ is the first-order Taylor approximation term

$$\nabla f(\mathbf{x}) \cdot \Delta\mathbf{x}$$

Subdividing the gradient $\nabla f(\mathbf{x})$ into basic and nonbasic parts $(\nabla f(\mathbf{x})^{(B)}, \nabla f(\mathbf{x})^{(N)})$, we can eliminate the basic components to see fully the impact of choices for the nonbasics:

$$\begin{aligned} \nabla f(\mathbf{x}) \cdot \Delta\mathbf{x} &= \nabla f(\mathbf{x})^{(B)} \cdot \Delta\mathbf{x}^{(B)} + \nabla f(\mathbf{x})^{(N)} \cdot \Delta\mathbf{x}^{(N)} \\ &= \nabla f(\mathbf{x})^{(B)} \cdot \left(-\mathbf{B}^{-1}\mathbf{N}\Delta\mathbf{x}^{(N)}\right) + \nabla f(\mathbf{x})^{(N)} \cdot \Delta\mathbf{x}^{(N)} \\ &= \left(\nabla f(\mathbf{x})^{(N)} - \nabla f(\mathbf{x})^{(B)}\mathbf{B}^{-1}\mathbf{N}\right)\Delta\mathbf{x}^{(N)} \end{aligned} \tag{14.44}$$

These derived coefficients on directional components are called the **reduced gradient**.

$\boxed{14.45}$ The reduced gradient associated with basis matrix $\mathbf{B}$ at current solution $\mathbf{x}$ is $\mathbf{r} \triangleq (\mathbf{r}^{(B)}, \mathbf{r}^{(N)})$ with

$$\mathbf{r}^{(B)} \triangleq \mathbf{0}$$

$$\mathbf{r}^{(N)} \triangleq \nabla f(\mathbf{x})^{(N)} - \nabla f(\mathbf{x})^{(B)}\mathbf{B}^{-1}\mathbf{N}$$

To illustrate, return to tuning model (14.38) at $\mathbf{x}^{(0)} = (2.5, 0, 3.5, 0.5, 5)$. The corresponding gradient is

$$\nabla f(\mathbf{x}^{(0)}) = \begin{pmatrix} 2(x_1 - 5) - 2x_2 \\ -2x_1 + 2(x_2 - 10) \\ 0 \\ 0 \\ 0 \end{pmatrix} = \begin{pmatrix} -5 \\ -25 \\ 0 \\ 0 \\ 0 \end{pmatrix}$$

Now using system (14.41), which derives $\mathbf{B}^{-1}\mathbf{N}$ to express feasible direction conditions with x_1, x_3, and x_5 basic,

$$\mathbf{r}^{(B)} = (r_1, r_3, r_5) = (0, 0, 0) \tag{14.45}$$

and

$$\begin{aligned}
\mathbf{r}^{(N)} &= (r_2, r_4) \\
&= \nabla f^{(N)} - \nabla f^{(B)} \left(\mathbf{B}^{-1}\mathbf{N}\right) \\
&= (-25, 0) - (-5, 0, 0) \begin{pmatrix} 0 & 1 \\ 1 & -1 \\ 1 & 0 \end{pmatrix} \\
&= (-25, 5)
\end{aligned} \tag{14.46}$$

SAMPLE EXERCISE 14.19: COMPUTING REDUCED GRADIENTS

Return to the nonlinear program of Sample Exercises 14.17 and 14.18 with basis $B = \{3, 4\}$. Compute the corresponding reduced gradient at solution $\mathbf{w} = (0, 2, 1, 1)$.

Analysis: With objective function

$$f(\mathbf{w}) \triangleq 50 - (w_1)^2 + 6w_1 - (w_2)^2 + 6w_2 + w_3$$

the gradient at the $\mathbf{w}$ specified is

$$\nabla f(\mathbf{w}) = \begin{pmatrix} -2w_1 + 6 \\ -2w_2 + 6 \\ 1 \\ 0 \end{pmatrix} = \begin{pmatrix} 6 \\ 2 \\ 1 \\ 0 \end{pmatrix}$$

Now applying definition $\boxed{14.45}$, basic components of the reduced gradient become

$$r_3 = r_4 = 0$$

Corresponding nonbasic components are

$$\begin{aligned}
r_1 &= 6 \quad -(1, 0) \cdot (\tfrac{1}{3}, \tfrac{3}{2}) \quad = \tfrac{17}{3} \\
r_2 &= 2 \quad -(1, 0) \cdot (-\tfrac{1}{3}, 1) = \tfrac{7}{3}
\end{aligned}$$

Reduced Gradient Move Direction

Reduced gradient algorithms seek to move nonbasics in reduced gradient direction $\Delta \mathbf{x}^{(N)} = \pm \mathbf{r}^{(N)}$ (+ for maximize models, − for minimize models). However, some adjustment must be made to avoid decreasing any x_j already $= 0$ [i.e., to enforce feasibility requirement (14.43)].

$\boxed{14.46}$ The reduced gradient algorithm moves from feasible point $\mathbf{x}$ in direction $\Delta \mathbf{x}$ derived from reduced gradient $\boxed{14.45}$ as (+ to maximize, − to minimize)

$$\Delta x_j \leftarrow \begin{cases} \pm r_j & \text{if } \pm r_j > 0 \text{ or } x_j > 0 \\ 0 & \text{otherwise} \end{cases}$$

on nonbasic components $j \in N$ and

$$\Delta x^{(B)} \leftarrow -B^{-1}N\Delta x^{(N)}$$

for basics.

In the minimizing tuning example of (14.45)–(14.46), construction $\boxed{14.46}$ makes nonbasic components

$$\Delta x_2 = -r_2 = 25 \quad \text{and} \quad \Delta x_4 = -r_4 = -5 \tag{14.47}$$

Corresponding basic components are derived from expression (14.41) as

$$\begin{pmatrix} \Delta x_1 \\ \Delta x_3 \\ \Delta x_5 \end{pmatrix} = - \begin{pmatrix} 0 & 1 \\ 1 & -1 \\ 1 & 0 \end{pmatrix} \begin{pmatrix} 25 \\ -5 \end{pmatrix} = \begin{pmatrix} 5 \\ -30 \\ -25 \end{pmatrix} \tag{14.48}$$

We have constructed direction Δx of $\boxed{14.46}$ to be feasible (assuming all basics positive). If $\Delta x = 0$, the current x can be shown to be a KKT point, and the algorithm stops. Otherwise [using (14.44) and $\boxed{14.45}$],

$$\nabla f(x) \cdot \Delta x = r^{(N)} \cdot \Delta x^{(N)}$$

$$= \sum_{\Delta x_j = \pm r_j \neq 0} (r_j)(\pm r_j)$$

shows that Δx has the proper sign to be an improving direction.

SAMPLE EXERCISE 14.20: CONSTRUCTING REDUCED GRADIENT DIRECTIONS

Return to the nonlinear program of Sample Exercises 14.17 to 14.19. Construct the move direction that would be pursued by the reduced gradient algorithm at solution $w = (0, 2, 1, 1)$.

Analysis: Using reduced gradient results of Sample Exercise 14.19 and construction $\boxed{14.46}$, nonbasic components for this maximizing model are

$$\Delta w_1 = +r_1 = \tfrac{17}{3} \quad \text{and} \quad \Delta w_2 = +r_2 = \tfrac{7}{3}$$

Then basic components derive from the representation of Sample Exercise 14.18 as

$$\Delta w_3 = -(\tfrac{1}{3} \Delta w_1 - \tfrac{1}{3} \Delta w_2) = -\tfrac{10}{9}$$

$$\Delta w_4 = -(\tfrac{3}{2} \Delta w_1 + 1 \Delta w_2) = -\tfrac{65}{6}$$

Line Search in Reduced Gradient Methods

Having specified reduced gradient direction $\boxed{14.46}$, we next need to decide how far to follow it. As with most nonlinear methods, a line search will be required to determine the largest λ for which the direction improves. However, constraints add feasibility considerations. Our direction Δx will satisfy equality constraints $A(x + \lambda \Delta x) = b$ for arbitrarily large λ, but nonnegativity constraints cannot be ignored. The line search must be limited by the same sort of "minimum ratio" check employed in LP algorithms of Chapters 5 and 6.

> **14.47** The step size λ applied at each step of the reduced gradient algorithm is determined by the 1-dimensional optimization
>
> $$\text{min or max} \quad f(\mathbf{x} + \lambda \mathbf{\Delta x})$$
> $$\text{s.t.} \qquad\qquad 0 \le \lambda \le \lambda_{max}$$
>
> where $\mathbf{x}$ is the current point, $\mathbf{\Delta x}$ is the move direction, and λ_{max} is the maximum feasible step
>
> $$\lambda_{max} = \min \left\{ \frac{x_j}{-\Delta x_j} : \Delta x_j < 0 \right\}$$

For example, tuning example direction (14.47)–(14.48) is negative for components $j = 3, 4, 5$. Thus the maximum feasible step at $\mathbf{x}^{(0)} = (2.5, 0, 3.5, 0.5, 5)$ is

$$\lambda_{max} = \min \left\{ \tfrac{3.5}{30}, \tfrac{.5}{5}, \tfrac{5}{25} \right\} = 0.1 \tag{14.49}$$

Distortion function $f(\mathbf{x}^{(0)} + \lambda \mathbf{\Delta x})$ decreases for all $\lambda \in [0, 0.1]$, so that the step size chosen will be the full $\lambda = 0.1$.

SAMPLE EXERCISE 14.21: COMPUTING MAXIMUM FEASIBLE STEPS

Sample Exercise 14.20 computed reduced gradient move direction

$$\mathbf{\Delta w} = (\tfrac{17}{3}, \tfrac{7}{3}, -\tfrac{10}{9}, -\tfrac{65}{6}) \quad \text{at solution} \quad \mathbf{w} = (0, 2, 1, 1)$$

of a standard-form, linearly constrained NLP. Determine the maximum feasible step in this direction.

Analysis: For a standard-form model with linear equality main constraints, the only possible loss of feasibility occurs when some variable drops to its lower bound of 0. This occurs here at (principle $\boxed{14.47}$)

$$\lambda_{max} = \min \left\{ \frac{1}{10/9}, \frac{1}{65/6} \right\} = \frac{6}{65}$$

Basis Changes in Reduced Gradient Methods

One final issue relates to nondegeneracy assumption (14.42). All our analysis has been based on basic variables always having positive values. As with the simplex algorithms of Chapter 5 (see Section 5.6), such nondegeneracy cannot always be guaranteed. Still, it is sufficient for functioning of the reduced gradient algorithm that we replace a variable in the basis if the most recent move forced it to $= 0$.

With many nonbasics (and superbasics) changing during the move, it is not as easy as with simplex to decide which nonbasic should enter the basis. To be assured of keeping a basis, we need only be careful to select a nonbasic actually affecting the blocking basic in computation $\boxed{14.44}$.

> **14.48** When movement in reduced gradient direction **14.46** is blocked by a nonnegativity constraint on a basic variable x_i, that variable should be replaced in the basis by a nonbasic x_j, preferably superbasic, such that the coefficient of $-\mathbf{B}^{-1}\mathbf{N}$ relating i and j in **14.44** is nonzero.

Superbasics are preferred because they have the positive value desired for a basic.

In the move along tuning example direction (14.47)–(14.48), the blocking variable of step computation (14.49) was nonbasic x_4. No basis adjustment is required.

SAMPLE EXERCISE 14.22: CHANGING THE BASIS IN REDUCED GRADIENT

The nonlinear program of Sample Exercises 14.17 to 14.21 computed maximum feasible step size $\lambda_{max} = \frac{65}{6}$ in the direction

$$\Delta \mathbf{w} = (\tfrac{17}{3}, \tfrac{7}{3}, -\tfrac{10}{9}, -\tfrac{65}{6}) \quad \text{at solution} \quad \mathbf{w} = (0, 2, 1, 1)$$

with w_3 and w_4 nonbasic. Assume that a full step $\lambda = \frac{65}{6}$ is chosen by the line search.

(a) Determine whether a basis change is now needed.

(b) If a change is required, select a new basis.

Analysis: We apply principle **14.48**.

(a) After a full step, the new solution will be

$$\mathbf{w} + \lambda \Delta \mathbf{w} = (\tfrac{34}{65}, \tfrac{144}{65}, \tfrac{35}{39}, 0)$$

Since basic variable w_4 drops to 0, a basis change is required.

(b) We must replace w_4 in the basis with a nonbasic variable that influenced its value on this move—preferably one that is now superbasic. Reference back to Sample Exercise 14.20 show that Δw_4 was affected (had nonzero coefficients) by both nonbasics. With both nonbasics now superbasic, we arbitrarily choose w_2 to produce new basis $\{w_2, w_3\}$.

Reduced Gradient Algorithm

All the building blocks of a reduced gradient search are now in place. Algorithm 14C provides details.

Computation begins at any feasible point and a corresponding basis. Each iteration follows reduced gradient direction **14.46** until either objective progress stops or the feasibility limit is reached. Bases are changes as in **14.48** whenever a basic variable dropping to zero blocks progress. Termination occurs when the computed direction is sufficiently close to the zero vector.

Reduced Gradient Search of Filter Tuning Example

Figure 14.9 has already displayed the sequence of points visited by reduced gradient Algorithm 14C in solving our tuning example from initial point $(2.5, 0)$. Table 14.5 provides details.

ALGORITHM 14C: REDUCED GRADIENT SEARCH

Step 0: Initialization. Choose stopping tolerance $\epsilon > 0$ and any starting feasible solution $\mathbf{x}^{(0)}$. Then construct a corresponding basis B with as many basic $x_j^0 > 0$ as possible, and set solution index $t \leftarrow 0$.

Step 1: Reduced Gradient Direction. Compute reduced gradient $\mathbf{r}$ at $\mathbf{x}^{(t)}$ as in $\boxed{14.45}$, and use $\mathbf{r}$ to generate move direction $\Delta\mathbf{x}^{t+1}$ per $\boxed{14.46}$.

Step 2: Stopping. If $\|\Delta\mathbf{x}^{t+1}\| \leq \epsilon$, stop and output local optimum $\mathbf{x}^{(t)}$.

Step 3: Feasibility Limit. Compute feasiblity limiting step λ_{max} according to $\boxed{14.47}$ ($\lambda_{max} = \infty$ if $\Delta\mathbf{x}^{t+1} \geq \mathbf{0}$).

Step 4: Line Search. Perform a 1-dimensional optimization to determine λ_{t+1} solving

$$\begin{aligned} \min \text{ or max} \quad & f(\mathbf{x} + \lambda\Delta\mathbf{x}^{t+1}) \\ \text{s.t.} \quad & 0 \leq \lambda \leq \lambda_{max} \end{aligned}$$

Step 5: New Point. Advance

$$\mathbf{x}^{(t+1)} \leftarrow \mathbf{x}^{(t)} + \lambda_{t+1}\Delta\mathbf{x}^{t+1}$$

Step 6: Basis Change. If any basic $x_j^{(t+1)} = 0$, replace one such j in the basis with some superbasic j'.

Step 7: Advance. Increment $t \leftarrow t + 1$, and return to Step 1.

The first move of the search follows the direction (14.47)–(14.48) for a full step $\lambda_{max} = 0.1$. No basis change is required because the blocking variable is nonbasic. Thus gradient computations are simply repeated to produce new direction

$$\Delta\mathbf{x} = (0, 21, -21, 0, -21)$$

Once again the direction improves all the way to maximum feasible step $\lambda_{max} = 0.0238$. This time, however, the blocking variable is basic x_3. Replacing x_3 in the basis with superbasic x_2 keeps basic variables strictly positive without losing linear independence of basic columns.

Recomputation produces the next move direction,

$$\Delta\mathbf{x} = (-10, 10, 0, 10, -10)$$

Notice that $\Delta x_3 = 0$ even though $-r_3 = -20$, because decreasing x_3 would produce immediate infeasibility.

The maximum feasible step in the chosen $\Delta\mathbf{x}$ is $\lambda_{max} = 0.2$. Still, a line search over $\lambda \in (0, 0.2]$ discovers a minimum at $\lambda = 0.125$. Thus the search advances only to

$$\mathbf{x}^{(3)} = (1.75, 4.25, 0, 1.25, 0.75)$$

This point proves (at least locally) optimal when $\Delta\mathbf{x} = \mathbf{0}$ computes as the next search direction.

TABLE 14.5 Reduced Gradient Search of Filter Tuning Example

	x_1	x_2	x_3	x_4	x_5	
min $f(\mathbf{x})$		$(x_1-5)^2 - 2x_1x_2 + (x_2-10)^2$				**b**
	1	1	1	0	0	6
A	1	0	0	1	0	3
	0	1	0	0	1	5
$t=0$	B	N	B	N	B	
$\mathbf{x}^{(0)}$	2.5	0.0	3.5	0.5	5.0	$f(\mathbf{x}^{(0)}) = 106.25$
$\nabla f(\mathbf{x}^{(0)})$	−5.0	−25.0	0.0	0.0	0.0	
$\mathbf{r}$	0.0	−25.0	0.0	5.0	0.0	
$\Delta\mathbf{x}$	5.0	25.0	−30.0	−5.0	−25.0	$\lambda_{max} = 0.1, \lambda = 0.1$
$t=1$	B	N	B	N	B	
$\mathbf{x}^{(1)}$	3.0	2.5	0.5	0.0	2.5	$f(\mathbf{x}^{(1)}) = 42.25$
$\nabla f(\mathbf{x}^{(1)})$	−9.0	−21.0	0.0	0.0	0.0	
$\mathbf{r}$	0.0	−21.0	0.0	9.0	0.0	
$\Delta\mathbf{x}$	0.0	21.0	−21.0	0.0	−21.0	$\lambda_{max} = 0.0238, \lambda = 0.0238$
$t=2$	B	B	N	N	B	
$\mathbf{x}^{(2)}$	3.0	3.0	0.0	0.0	2.0	$f(\mathbf{x}^{(2)}) = 35.00$
$\nabla f(\mathbf{x}^{(2)})$	−10.0	−20.0	0.0	0.0	0.0	
$\mathbf{r}$	0.0	0.0	20.0	−10.0	0.0	
$\Delta\mathbf{x}$	−10.0	10.0	0.0	10.0	−10.0	$\lambda_{max} = 0.2, \lambda = 0.125$
$t=3$	B	B	N	N	B	
$\mathbf{x}^{(3)}$	1.75	4.25	0.0	1.25	0.75	$f(\mathbf{x}^{(3)}) = 28.75$
$\nabla f(\mathbf{x}^{(3)})$	−15.0	−15.0	0.0	0.0	0.0	
$\mathbf{r}$	0.0	0.0	15.0	0.0	0.0	
$\Delta\mathbf{x}$	0.0	0.0	0.0	0.0	0.0	Stop

Major and Minor Iterations in Reduced Gradient

At any point in a reduced gradient search, the superbasic variables represent a "free" set in that they can increase or decrease without losing feasibility. A refinement that has proved useful exploits this relative ease of movement by dividing the search into major iterations and minor iterations.

14.49 **Minor iterations** of reduced gradient procedures change only superbasic and basic variable values by adopting at $\Delta\mathbf{x}$ the move direction $\Delta\mathbf{x}$ with ($+$ for maximize, $-$ for minimize)

$$\Delta x_j \leftarrow \begin{cases} \pm r_j & \text{if } x_j > 0 \\ 0 & \text{otherwise} \end{cases}$$

for nonbasic components $j \in N$ and

$$\Delta\mathbf{x}^{(B)} \leftarrow -\mathbf{B}^{-1}\mathbf{N}\Delta\mathbf{x}^{(N)}$$

for basics. Major iterations follow construction 14.46 by also allowing changes in nonbasics = 0.

A minor iteration leaves nonbasics at bound zero fixed, changing only superbasics.

When progress slows, we undertake a major iteration changing more nonbasics as in Algorithm 14C.

Second-Order Extensions of Reduced Gradient

Major/minor direction procedure $\boxed{14.49}$ can be productively extended even further by employing second-order information on the objective function. Thinking of objective function $f(\mathbf{x})$ as a function of the superbasics alone, with other nonbasics fixed $= 0$, and basics taking implied values, we are left with an unconstrained optimization in the superbasics. Quasi-Newton methods of Section 13.7 can then be employed to quickly find a good choice of superbasic values. Afterward, having completed several minor iterations, we consider making other nonbasics positive.

Generalized Reduced Gradient Procedures for Nonlinear Constraints

To this point we have assumed that all constraints of the given nonlinear program are linear. **Generalized reduced gradient** algorithms extend to nonlinear constraints.

Suppose that we are given the nonlinear equality constrained standard form

$$\text{min or max} \quad f(\mathbf{x})$$
$$\text{s.t.} \quad g_i(\mathbf{x}) = b_i \quad i \in E \tag{14.50}$$
$$\mathbf{x} \geq \mathbf{0}$$

Using first-order Taylor approximations, it is natural to consider linearizing constraints around a current $\mathbf{x}^{(t)}$ as

$$b_i = g_i(\mathbf{x}) \approx g_i(\mathbf{x}^{(t)}) + \nabla g_i(\mathbf{x}^{(t)}) \cdot (\mathbf{x} - \mathbf{x}^{(t)}) \tag{14.51}$$

Noting that feasibility implies $g_i(\mathbf{x}^{(t)}) = b_i$, this linearization simplifies to

$$\nabla g_i(\mathbf{x}^{(t)}) \cdot \mathbf{x} = \nabla g_i(\mathbf{x}^{(t)}) \cdot \mathbf{x}^{(t)} \quad \text{for all } i \in E \tag{14.52}$$

Including nonnegativity constraints with system (14.52) yields linear-constrained format

$$\mathbf{A}^{(t)}\mathbf{x} = \mathbf{b}^{(t)}$$
$$\mathbf{x} \geq \mathbf{0}$$

with rows of $\mathbf{A}^{(t)}$ being $\nabla g_i(\mathbf{x}^{(t)})$ and components of $\mathbf{b}^{(t)}$ equaling $\nabla g_i(\mathbf{x}^{(t)}) \cdot \mathbf{x}^{(t)}$. We are now in a position to employ linear-constrained reduced gradient Algorithm 14C (or its second-order extensions).

Dealing with successive systems (14.52) is essentially the strategy of generalized reduced gradient algorithms. Still, there is a difficulty. Approximation (14.51) is not exact for nonlinear constraints. Thus enforcement of (14.52) is not guaranteed to keep $\mathbf{x}$ feasible.

Generalized reduced gradient algorithms address this difficulty by following each reduced gradient move with **corrector steps** to restore feasibility. In essence, a penalty function is introduced (see Section 14.5) and a new move is chosen to minimized the penalized objective function. Once feasibility is restored, a new move can be computed using equations (14.52).

14.7 QUADRATIC PROGRAMMING METHODS

A constrained nonlinear program is a **quadratic program** or **QP** if its objective function is quadratic and all its constraints are linear (definition $\boxed{14.9}$). In this section we investigate special methods adapted to this class of NLPs.

General Symmetric Form of Quadratic Programs

It will be useful to express quadratic programs in general symmetric form.

$\boxed{14.50}$ Quadratic programs can be placed in the **general symmetric form**

$$\text{max or min} \quad f(\mathbf{x}) \triangleq c_0 + \mathbf{c} \cdot \mathbf{x} + \mathbf{x} \mathbf{Q} \mathbf{x}$$

$$\text{s.t.} \qquad \mathbf{a}^{(i)} \mathbf{x} \geq b_i \qquad \text{for all } i \in G$$
$$\mathbf{a}^{(i)} \mathbf{x} \leq b_i \qquad \text{for all } i \in L$$
$$\mathbf{a}^{(i)} \mathbf{x} = b_i \qquad \text{for all } i \in E$$

where $\mathbf{Q}$ is a symmetric matrix, and sets G, L, and E index the $\geq$, $\leq$, and $=$ constraints, respectively.

Notice that nonnegativity and other variable-type restrictions are treated as main constraints.

The assumption that Q is symmetric $(= \mathbf{Q}^T)$ merely simplifies notation. There is no loss of generality because a model with asymmetric $\bar{\mathbf{Q}}$ has the same objective value as one with symmetric

$$\mathbf{Q} = \tfrac{1}{2}(\bar{\mathbf{Q}} + \bar{\mathbf{Q}}^T)$$

Quadratic Program Form of the Filter Tuning Example

We illustrate quadratic programming methods with the tiny distortion tuning example model (14.37) (Section 14.6). In vector format $\boxed{14.50}$, the model is

$$\text{min} \quad 125 + (-10, -20) \cdot \mathbf{x} + \mathbf{x} \begin{pmatrix} 1 & -1 \\ -1 & 1 \end{pmatrix} \mathbf{x}$$

$$\text{s.t.} \quad (1, 0) \cdot \mathbf{x} \geq 0$$
$$(0, 1) \cdot \mathbf{x} \geq 0 \qquad\qquad\qquad (14.53)$$
$$(1, 1) \cdot \mathbf{x} \leq 6$$
$$(1, 0) \cdot \mathbf{x} \leq 3$$
$$(0, 1) \cdot \mathbf{x} \leq 5$$

with $G = \{1, 2\}$, $L = \{3, 4, 5\}$, and $E = \emptyset$.

SAMPLE EXERCISE 14.23: UNDERSTANDING STANDARD QP NOTATION

Return to the quadratic program used in Sample Exercises of Section 14.6:

$$\text{max} \quad f(\mathbf{w}) \triangleq 50 - (w_1)^2 + 6w_1 - (w_2)^2 + 6w_2 + w_3$$
$$\text{s.t.} \quad + w_1 \quad - w_2 \quad + 3w_3 \qquad\qquad = 1$$
$$+ 3w_1 \quad + 2w_2 \qquad\qquad + 2w_4 = 6$$
$$w_1, w_2, w_3, w_4 \geq 0$$

Identify elements c_0, $\mathbf{c}$, $\mathbf{Q}$, G, L, E, $\mathbf{a}^{(i)}$, and b_i of general form $\boxed{14.50}$.

Analysis: Arranging objective function elements in matrix form 14.50 yields $c_0 = 50$,

$$\mathbf{c} = \begin{pmatrix} 6 \\ 6 \\ 1 \\ 0 \end{pmatrix} \text{ and } \mathbf{Q} = \begin{pmatrix} -1 & 0 & 0 & 0 \\ 0 & -1 & 0 & 0 \\ 0 & 0 & 0 & 0 \\ 0 & 0 & 0 & 0 \end{pmatrix}$$

With $E = \{1, 2\}$, $G = \{3, 4, 5, 6\}$, and $L = \emptyset$, corresponding constraint coefficients are

$$\begin{aligned}
\mathbf{a}^{(1)} &= (1, -1, 3, 0), & b_1 &= 1 \\
\mathbf{a}^{(2)} &= (3, 2, 0, 2), & b_2 &= 6 \\
\mathbf{a}^{(3)} &= (1, 0, 0, 0), & b_3 &= 0 \\
\mathbf{a}^{(4)} &= (0, 1, 0, 0), & b_4 &= 0 \\
\mathbf{a}^{(5)} &= (0, 0, 1, 0), & b_5 &= 0 \\
\mathbf{a}^{(6)} &= (0, 0, 0, 1), & b_6 &= 0
\end{aligned}$$

Equality-Constrained Quadratic Programs and KKT Conditions

It is instructive to begin our investigation of quadratic programming with the pure equality case:

$$\text{max or min} \quad f(\mathbf{x}) \triangleq \mathbf{cx} + \mathbf{xQx}$$
$$\text{s.t.} \qquad \mathbf{Ax} = \mathbf{b} \tag{14.54}$$

Here $G = L = \emptyset$ in general form 14.50, and coefficient vectors $\mathbf{a}^{(i)}$ for equalities $i \in E$ have been collected as rows of a matrix $\mathbf{A}$.

With all constraints equalities, Karush–Kuhn–Tucker conditions (principle 14.21) for model (14.54) require no sign restrictions or complementary slackness constraints. Furthermore, the objective function gradient ($\mathbf{Q}$ symmetric) is

$$\nabla f(\mathbf{x}) = \mathbf{c} + 2\mathbf{Qx}$$

and constraint gradients $\nabla g_i(\mathbf{x})$ are the rows of $\mathbf{A}$. Thus KKT conditions for model (14.54) reduce to

$$\sum_i \mathbf{a}^{(i)} v_i = \mathbf{c} + 2\mathbf{Qx}$$
$$\mathbf{Ax} = \mathbf{b}$$

What makes pure-equality quadratic programs special is that these conditions can be rearranged into a square system of linear equations.

14.51 Karush–Kuhn–Tucker optimality conditions for pure equality quadratic programs (14.54) are the linear equations

$$\begin{pmatrix} -2\mathbf{Q} & \mathbf{A}^\mathsf{T} \\ \mathbf{A} & \mathbf{0} \end{pmatrix} \begin{pmatrix} \mathbf{x} \\ \mathbf{v} \end{pmatrix} = \begin{pmatrix} \mathbf{c} \\ \mathbf{b} \end{pmatrix}$$

SAMPLE EXERCISE 14.24: FORMING KKT CONDITIONS FOR EQUALITY QPS

Form Karush–Kuhn–Tucker optimality conditions for the equality-constrained quadratic program

$$\min \quad 4(y_1)^2 - 6y_1 y_2 + 5(y_2)^2 + y_3$$
$$\text{s.t.} \quad +y_1 - 3y_2 - 9y_3 = 11$$
$$-y_1 + 7y_2 + 7y_3 = -9$$

at $\mathbf{y} = (2, 0, -1)$.

Analysis: Here

$$\mathbf{A} = \begin{pmatrix} 1 & -3 & -9 \\ -1 & 7 & 7 \end{pmatrix}, \quad \mathbf{c} = \begin{pmatrix} 0 \\ 0 \\ 1 \end{pmatrix}, \quad \text{and} \quad \mathbf{Q} = \begin{pmatrix} 4 & -3 & 0 \\ -3 & 5 & 0 \\ 0 & 0 & 0 \end{pmatrix}$$

Thus KKT conditions $\boxed{14.51}$ are

$$\begin{pmatrix} -8 & 6 & 0 & 1 & -1 \\ 6 & -10 & 0 & -3 & 7 \\ 0 & 0 & 0 & -9 & 7 \\ 1 & -3 & -9 & 0 & 0 \\ -1 & 7 & 7 & 0 & 0 \end{pmatrix} \begin{pmatrix} y_1 \\ y_2 \\ y_3 \\ v_1 \\ v_2 \end{pmatrix} = \begin{pmatrix} 0 \\ 0 \\ 1 \\ 11 \\ -9 \end{pmatrix}$$

Direct Solution of KKT Conditions for Quadratic Programs

The unusually simple form of KKT conditions $\boxed{14.51}$ for equality-constrained quadratic programs suggests an approach to solution. We could simply form the KKT system of linear equations and solve for KKT point $\mathbf{x}$ and corresponding Lagrange multipliers $\mathbf{v}$.

This is the approach taken in many methods.

$\boxed{14.52}$ Equality-constrained quadratic programs are often approached by direct solution of (linear) Karush–Kuhn–Tucker conditions $\boxed{14.51}$.

Sophisticated methods of linear algebra may be used to compute answers, but the process remains essentially one solving the KKT system.

Unique solvability of system $\boxed{14.51}$ would mean that model (14.54) has a unique KKT point. Since equality constraints assure that every local optimum is a KKT point (principle $\boxed{14.27}$), a unique $\boxed{14.51}$ solution must correspond to a unique local (and thus global) maximum or minimum unless the model has no extrema at all. Other cases may have multiple KKT points, or none at all. Still, any local optimum must be a solution to system $\boxed{14.51}$.

SAMPLE EXERCISE 14.25: SOLVING KKT CONDITIONS FOR EQUALITY QPS

Solve the KKT conditions of the equality-constrained quadratic program in Sample Exercise 14.24 to find a KKT point of the model.

Analysis: The unique solution to this KKT system has primal solution

$$y_1 = -0.0834, \quad y_2 = -0.0992, \quad y_3 = -1.1984$$

and corresponding Lagrange multipliers

$$v_1 = -0.2486, \quad v_2 = -0.1768$$

Further analysis would be required to determine whether this KKT solution is a global maximum, a global minimum, or a saddle point.

Active Set Strategies for Quadratic Programming

Active set methods exploit the linear equation form of KKT conditions for equality-constrained QPs by reducing general quadratic programs $\boxed{14.50}$ to a sequence of equality cases. To see how, define

$$S \triangleq \text{set of indices of active constraints at current feasible solution}$$
$$\mathbf{x}^{(t)} \text{ in general QP model } \boxed{14.50}$$

$$\mathbf{A}_S \triangleq \text{matrix with rows formed by the coefficient vectors}$$
$$\mathbf{a}^{(i)} \text{ of } i \in S$$

Every equality constraint of E belongs to S, along with active inequalities of G and L.

Suppose that we require all active constraints $i \in S$ to continue being satisfied as equalities during our next move. Then an optimal move $\Delta\mathbf{x}$ from $\mathbf{x}^{(t)}$ should solve

$$\begin{array}{ll} \text{max or min} & f(\mathbf{x}^{(t)} + \Delta\mathbf{x}) = f(\mathbf{x}^{(t)}) + \nabla f(\mathbf{x}^{(t)}) \cdot \Delta\mathbf{x} + \Delta\mathbf{x}\mathbf{Q}\Delta\mathbf{x} \\ \text{s.t.} & \mathbf{A}_t \Delta\mathbf{x} = \mathbf{0} \end{array} \tag{14.55}$$

The (14.55) objective merely rewrites $f(\mathbf{x}^{(t)} + \Delta\mathbf{x})$ in terms of the second-order Taylor representation (principle $\boxed{13.17}$), which is exact for quadratic functions. Constraints enforce the familiar requirements $\sum a_{i,j}\Delta x_k = 0$ (principle $\boxed{3.25}$) for a move to preserve linear equality constraints.

Notice that subproblem (14.55) is now in equality-constrained format (14.54). Thus we can compute a move $\Delta\mathbf{x}$ by solving the corresponding KKT linear equations $\boxed{14.51}$.

$\boxed{14.53}$ Active set methods for general quadratic programs compute the move $\Delta\mathbf{x}$ at current solution $\mathbf{x}^{(t)}$ by solving Karush–Kuhn–Tucker conditions

$$\begin{pmatrix} -2\mathbf{Q} & \mathbf{A}_S^{\mathsf{T}} \\ \mathbf{A}_S & \mathbf{0} \end{pmatrix} \begin{pmatrix} \mathbf{x} \\ \mathbf{v}^{(S)} \end{pmatrix} = \begin{pmatrix} \nabla f(\mathbf{x}^{(t)}) \\ \mathbf{0} \end{pmatrix}$$

where $\mathbf{A}_S$ is the coefficient matrix of active constraints and $\mathbf{v}^{(S)}$ is the corresponding Lagrange multiplier vector. All v_i for $i \notin S$ are fixed $= 0$.

As with models having only equality constraints, a KKT solution to system $\boxed{14.53}$ may not exist, or not correspond to the desired minimum for a minimize problem or maximum for a maximize problem over the active constraints. Still, construction $\boxed{14.53}$ provides good results when the objective function is reasonably well behaved.

To illustrate, return to tuning model (14.53), which is a convex program. At $\mathbf{x}^{(0)} = (2.5, 0)$, only nonnegativity constraint $i = 2$ is active, so

$$S = \{2\} \quad \text{and} \quad \mathbf{A}_S = (0, 1)$$

With $\nabla f(\mathbf{x}^{(t)}) = (-5, -25)$, the corresponding move-finding KKT system $\boxed{14.53}$ is

$$\begin{pmatrix} -2 & 2 & 0 \\ 2 & -2 & 1 \\ 0 & 1 & 0 \end{pmatrix} \begin{pmatrix} \Delta x_1 \\ \Delta x_2 \\ v_2 \end{pmatrix} \begin{pmatrix} -5 \\ -25 \\ 0 \end{pmatrix}$$

This system has unique solution

$$\Delta x_1 = 2.5, \quad \Delta x_2 = 0, \quad v_2 = -30 \tag{14.56}$$

Step Size with Active Set Methods

If a move $\Delta \mathbf{x} \neq 0$ results from subproblem (14.55), the usual update

$$\mathbf{x}^{(t+1)} \leftarrow \mathbf{x}^{(t)} + \Delta \mathbf{x}$$

optimizes the objective over the active constraints. However, we have ignored inactive constraints in forming (14.55). A full step in direction $\Delta \mathbf{x}$ may cause some such constraint to be violated.

To account for this possibility, we introduce a now-familiar maximum step rule:

$\boxed{14.54}$ If the $\Delta \mathbf{x}$ computed from the active constraints at solution $\mathbf{x}^{(t)}$ is nonzero, active set algorithms adopt step λ in direction $\Delta \mathbf{x}$, where

$$\lambda_G \quad \leftarrow \quad \min\left\{ \frac{\mathbf{a}^{(i)}\mathbf{x}^{(t)} - b_i}{-\mathbf{a}^{(i)}\Delta\mathbf{x}} : \mathbf{a}^{(i)}\Delta\mathbf{x} < 0,\ i \in G \right\}$$

$$\lambda_L \quad \leftarrow \quad \min\left\{ \frac{b_i - \mathbf{a}^{(i)}\mathbf{x}^{(t)}}{\mathbf{a}^{(i)}\Delta\mathbf{x}} : \mathbf{a}^{(i)}\Delta\mathbf{x} > 0,\ i \in L \right\}$$

$$\lambda \quad \leftarrow \quad \min\{1, \lambda_G, \lambda_L\}$$

The first two possibilities for λ check inactive $\geq$ and $\leq$ constraints, respectively, and the 1 in the last step provides for the possibility that the full move is feasible. For example, we would compute the appropriate step in tuning example direction $\Delta \mathbf{x} = (2.5, 0)$ of (14.56) from $\mathbf{x}^{(0)} = (2.5, 0)$ as

$$\lambda_G = +\infty$$
$$\lambda_L = \min\left\{\tfrac{3.5}{2.5}, \tfrac{0.5}{2.5}\right\} = 0.2 \tag{14.57}$$
$$\lambda = \min\{1, +\infty, 0.2\} = 0.2$$

SAMPLE EXERCISE 14.26: CHOOSING STEP SIZE IN ACTIVE SET QP

Suppose that an active set search of a quadratic program with constraints

$$2y_1 + 3y_2 \geq 10$$
$$1y_1 + 7y_2 \leq 40$$
$$1y_1 + 3y_2 = 17$$

has reached $\mathbf{y} = (2, 5)$ and computed (construction $\boxed{14.53}$) move $\Delta \mathbf{y} = (-3, 1)$.

(a) Determine the appropriate step size λ to apply.

(b) How would the λ change if the second constraint were $1y_1 + 7y_2 \leq 80$?

Analysis: We apply rule $\boxed{14.54}$.

(a) Only the last, equality constraint is active. Changes in the other constraints per unit step in direction $\Delta \mathbf{y}$ are

$$\mathbf{a}^{(1)} \cdot \Delta \mathbf{y} = (2, 3) \cdot (-3, 1) = -3$$
$$\mathbf{a}^{(2)} \cdot \Delta \mathbf{y} = (1, 7) \cdot (-3, 1) = 4$$

Thus

$$\lambda_G = \frac{19 - 10}{3}, \quad \lambda_L = \frac{40 - 37}{4}, \quad \lambda = \min\left\{1, 3, \tfrac{3}{4}\right\} = \tfrac{3}{4}$$

(b) With this revised right-hand side,

$$\lambda_G = 3, \quad \lambda_L = \frac{80 - 37}{4}, \quad \lambda = \min\left\{1, 3, \tfrac{43}{4}\right\} = 1$$

Although step sizes up to $\lambda = \min\left\{3, \tfrac{43}{4}\right\}$ are feasible, the optimal one of KKT computation $\boxed{14.53}$ occurs at $\lambda = 1$.

Stopping at a KKT Point with Active Set Methods

Update $\mathbf{x}^{(t+1)} \leftarrow \mathbf{x}^{(t)} + \lambda \Delta \mathbf{x}$ advances us toward an optimum to (a reasonable well-behaved) general quadratic program $\boxed{14.50}$ as long as move direction $\Delta \mathbf{x} \neq \mathbf{0}$. Should we stop when $\Delta \mathbf{x} = \mathbf{0}$? It depends on whether Lagrange multipliers $\mathbf{v}$ computed from linear system $\boxed{14.53}$ for active constraints complete a KKT solution for the full model.

$\boxed{14.55}$ If $\Delta \mathbf{x} = \mathbf{0}$ in a solution to conditions $\boxed{14.53}$, active set methods will stop at a KKT point for the full model $\boxed{14.50}$ if all corresponding Lagrange multipliers $\mathbf{v}$ satisfy sign restrictions

Objective	Active $i \in G$	Active $i \in L$
Minimize	$v_i \geq 0$	$v_i \leq 0$
Maximize	$v_i \leq 0$	$v_i \geq 0$

Lagrange multipliers satisfying conditions $\boxed{14.55}$ suffice for a KKT point in the full model $\boxed{14.50}$ because the corresponding optimality conditions are

$$\mathbf{c} + 2\mathbf{Q}\mathbf{x}^{(t)} = \sum_i \mathbf{a}^{(i)} v_i = \mathbf{A}_S^{\mathsf{T}} \mathbf{v}^{(S)} \tag{14.58}$$

together with sign restrictions $\boxed{14.55}$, complementary slackness on all inequalities, and feasibility in primal constraints. But (14.58) is the first part of the linear system solved in $\boxed{14.53}$; complementary slackness is automatic because only active constraints are allowed to have $v_i \neq 0$; primal feasibility is enforced by the second part of equation system $\boxed{14.53}$ and step size rule $\boxed{14.54}$. Thus the only additional requirements for a KKT point are the sign restrictions of principle $\boxed{14.55}$.

SAMPLE EXERCISE 14.27: STOPPING IN ACTIVE SET SEARCH OF QPs

Consider an active set search of a maximizing quadratic program with currently active constraints

$$w_1 + 2w_2 \geq 4$$
$$w_3 - 8w_4 + w_5 \leq 2$$
$$3w_1 + 2w_2 + 2w_3 + 2w_4 + 2w_5 = 16$$

Determine whether the procedure would stop if solution of linear equation system $\boxed{14.53}$ produces:

(a) $\Delta \mathbf{w} = (0, 0, 0, 0, 0)$, $\mathbf{v} = (-33, 10, 14)$

(b) $\Delta \mathbf{w} = (0, 0, 0, 0, 0)$, $\mathbf{v} = (33, 10, 14)$

(c) $\Delta \mathbf{w} = (2, -1, -1, 0, 1)$, $\mathbf{v} = (33, 10, 14)$

Analysis: We apply principle $\boxed{14.55}$.

(a) For this maximize model, $v_1 = -33$ is appropriate for a $\geq$ constraint, and $v_2 = 10$ is suitable for a $\leq$. The search would terminate with the current solution a KKT point.

(b) For this maximize model, $v_1 = 33$ violates sign restrictions of $\boxed{14.55}$. The search would not terminate.

(c) Here the move direction $\Delta \mathbf{w} \neq \mathbf{0}$. The search would continue.

Dropping a Constraint from the Active Set

Clearly, the active set must change if further progress is to be achieved when sign restriction $\boxed{14.55}$ are not fulfilled even though system $\boxed{14.53}$ produced a move $\Delta \mathbf{x} = \mathbf{0}$. In particular, one or more now active constraint $i \in S$ must be allowed to become a strict inequality.

　　To see which active constraint to drop from S, focus again on the Lagrange multipliers computed in solving system $\boxed{14.53}$. We know these multipliers can be interpreted (principle $\boxed{14.17}$) as the change in (14.55) optimal value with its constraint right-hand sides. For example, $v_i > 0$ for a $\leq$ inequality i of a minimize subproblem indicates that allowing inequality i to become strict (i.e., relaxing to $\mathbf{a}^{(i)} \cdot \Delta \mathbf{x} \leq 0$) will help the objective function. That is, a violation of sign conditions $\boxed{14.55}$ indicates an active constraint that could be productively dropped.

> 14.56   When solution of system 14.53 produces a $\Delta x = 0$ but v_i not all satisfying the sign restrictions of 14.55, active set algorithms drop from S some i with a violating v_i.

SAMPLE EXERCISE 14.28: DROPPING CONSTRAINTS IN ACTIVE SET QP

For each of the cases in Sample Exercise 14.27 where the procedure did not terminate, determine how the active set S should be modified.

Analysis: We apply principle 14.56.

(a) The procedure stops in this case and no modification of S is required.

(b) For this case we drop constraint $i = 1$ from S and re-solve linear system 14.53 because Lagrange multiplier $v_1 = -33$ violates sign restrictions 14.55.

(c) No change is needed in S for this case because the move direction $\Delta w \neq 0$.

Active Set Solution of the Filter Tuning Example

Algorithm 14D collects principles 14.53 to 14.56 in an active set procedure for quadratic programs. Figure 14.10 and Table 14.6 detail the application of Algorithm 14D to tuning model (14.53), beginning at $x^{(0)} = (2.5, 0.0)$.

ALGORITHM 14D: ACTIVE SET METHOD FOR QUADRATIC PROGRAMS

Step 0: Initialization. Pick starting feasible solution $x^{(0)}$, and initialize working active set S with indices of all constraints active at $x^{(0)}$. Also, choose stopping tolerance $\epsilon > 0$, and initialize iteration index $t \leftarrow 0$.

Step 1: Subproblem. With $A_S \triangleq$ coefficient matrix of active constraints in S, solve the Karush–Kuhn–Tucker conditions

$$\begin{pmatrix} -2Q & A_S^T \\ A_S & 0 \end{pmatrix} \begin{pmatrix} \Delta x^{t+1} \\ v_S^{(t+1)} \end{pmatrix} = \begin{pmatrix} \nabla f(x^{(t)}) \\ 0 \end{pmatrix}$$

of direction problem (14.55) for move direction Δx^{t+1} and active constraint Lagrange multipliers $v_S^{(t+1)}$. Lagrange multipliers for nonactive $i \notin S$ are fixed $v_i^{(t+1)} \leftarrow 0$.

Step 2: KKT Point. If $||\Delta x^{(t+1)}|| \leq \epsilon$ and $v^{(t+1)}$ satsifies sign restrictions of 14.55, stop; the current $x^{(t)}$ is a Karush–Kuhn–Tucker point of the given symmetric quadratic program 14.50. Otherwise, if $||\Delta x^{(t+1)}|| \leq \epsilon$, go to Step 3, and if not, proceed to Step 4.

Step 3: Active Dropping. Choose some $i \in S$ with Lagrange multiplier $v_i^{(t+1)}$ violating sign restrictions 14.55, and remove it from S. Then go to Step 6.

Step 4: Step Size. Compute the maximum appropriate step λ in direction $\Delta x^{(t+1)}$ via

$$\lambda_G \quad \leftarrow \quad \min\left\{ \frac{\mathbf{a}^{(i)}\mathbf{x}^{(t)} - b_i}{-\mathbf{a}^{(i)}\Delta\mathbf{x}^{t+1}} : \mathbf{a}^{(i)}\Delta\mathbf{x}^{t+1} < 0, \; i \in G \right\}$$

$$\lambda_L \quad \leftarrow \quad \min\left\{ \frac{b_i - \mathbf{a}^{(i)}\mathbf{x}^{(t)}}{\mathbf{a}^{(i)}\Delta\mathbf{x}^{t+1}} : \mathbf{a}^{(i)}\Delta\mathbf{x}^{t+1} > 0, \; i \in L \right\}$$

$$\lambda \quad \leftarrow \quad \min\{1, \lambda_G, \lambda_L\}$$

Step 5: Move. Step

$$\mathbf{x}^{(t+1)} \leftarrow \mathbf{x}^{(t)} + \lambda\Delta\mathbf{x}^{(t+1)}$$

and update S with the indices of any newly active constraints.
Step 6: Advance. Increment $t \leftarrow t + 1$, and return to Step 1.

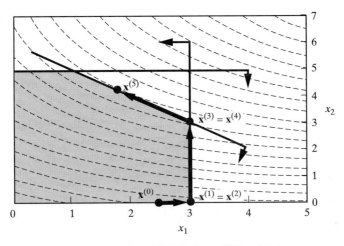

FIGURE 14.10   Active Set QP Solution of Filter Tuning Example

TABLE 14.6   Active Set QP Solution of Filter Tuning Example

		Variables		Constraints					
		x_1	x_2	$i = 1$	$i = 2$	$i = 3$	$i = 4$	$i = 5$	
$t = 0$	$\mathbf{x}^{(0)}$	2.5	0.0	S no	yes	no	no	no	
	$\Delta\mathbf{x}$	2.5	0.0	$\mathbf{v}$ 0.0	−30.0	0.0	0.0	0.0	$\lambda = 0.2000$
$t = 1$	$\mathbf{x}^{(1)}$	3.0	0.0	S no	yes	no	yes	no	
	$\Delta\mathbf{x}$	0.0	0.0	$\mathbf{v}$ 0.0	−26.0	0.0	−4.0	0.0	drop $i = 2$
$t = 2$	$\mathbf{x}^{(2)}$	3.00	0.00	S no	no	no	yes	no	
	$\Delta\mathbf{x}$	0.0	13.0	$\mathbf{v}$ 0.0	0.0	0.0	−30.0	0.0	$\lambda = 0.2308$
$t = 3$	$\mathbf{x}^{(3)}$	3.00	3.00	S no	no	yes	yes	no	
	$\Delta\mathbf{x}$	0.0	0.0	$\mathbf{v}$ 0.0	0.0	−20.0	10.0	0.0	drop $i = 4$
$t = 4$	$\mathbf{x}^{(4)}$	3.00	3.00	S no	no	yes	no	no	
	$\Delta\mathbf{x}$	−1.25	1.25	$\mathbf{v}$ 0.0	0.0	−15.0	0.0	0.0	$\lambda = 1.0000$
$t = 5$	$\mathbf{x}^{(5)}$	1.75	4.25	S no	no	yes	no	no	
	$\Delta\mathbf{x}$	0.00	0.00	$\mathbf{v}$ 0.0	0.0	−15.0	0.0	0.0	KKT point

Computations (14.56) and (14.57) have already established the first move direction $\Delta \mathbf{x} = (2.5, 0.0)$ and step size $\lambda = 0.2$, which lead to

$$\mathbf{x}^{(1)} = \mathbf{x}^{(0)} + \lambda \Delta \mathbf{x} = (2.5, 0.0) + 0.2(2.5, 0.0) = (3.0, 0.0)$$

The active set $S = \{2, 4\}$ at $\mathbf{x}^{(1)}$ includes the previous $x_2 \geq 0$ and newly active $x_1 \leq 3$. Solution of the corresponding linear system $\boxed{14.53}$ produces null direction $\Delta \mathbf{x} = \mathbf{0}$. We have not yet reached a KKT point for the full model because the sign of $v_2 = -26$ is wrong for a $\geq$ constraint in a minimize problem (principle $\boxed{14.55}$). Thus we drop $i = 2$ from S and re-solve equations $\boxed{14.53}$.

The new direction produces a nonzero move, and the search continues. At $t = 5$ the computed direction is again $\Delta \mathbf{x} = \mathbf{0}$. This time, however, Lagrange multipliers satisfy sign restrictions of principle $\boxed{14.55}$. The search terminates with KKT point (here global optimum) $\mathbf{x}^* = (1.75, 4.25)$.

14.8 SEPARABLE PROGRAMMING METHODS

Separable functions decompose into sums of functions of single decision variables (definition $\boxed{14.6}$), and **separable programs** are NLPs over separable objective functions and constraints (definition $\boxed{14.7}$). That is, they take the general form

$$\text{max or min} \quad f(\mathbf{x}) \triangleq \sum_j f_j(x_j)$$

$$\text{s.t.} \quad g_i(\mathbf{x}) \triangleq \sum_j g_{i,j}(x_j) \geq b_i \qquad \text{for all } i \in G$$

$$g_i(\mathbf{x}) \triangleq \sum_j g_{i,j}(x_j) \leq b_i \qquad \text{for all } i \in L \qquad (14.59)$$

$$g_i(\mathbf{x}) \triangleq \sum_j g_{i,j}(x_j) = b_i \qquad \text{for all } i \in E$$

$$x_j \geq 0 \qquad \text{for all } j$$

Here G, L, and E index the $\geq$, $\leq$, and $=$ constraints, respectively. For notational convenience, we also assume that nonnegativity constraints apply to all variables.

Pfizer Example Revisited

Section 14.2 developed Pfizer pharmaceutical manufacturing lot size model

$$\min \quad 66.21x_1 + \frac{2160}{x_1} + 426.8x_2 + \frac{8712}{x_2} \qquad \text{(total cost)}$$

$$+ 61.20x_3 + \frac{330}{x_3} + 268.1x_4 + \frac{2916}{x_4} \qquad (14.60)$$

$$\text{s.t.} \quad \frac{180}{x_1} + \frac{726}{x_2} + \frac{27.5}{x_3} + \frac{243}{x_4} \leq 221.5 \qquad \text{(production time)}$$

$$x_1, \ldots, x_4 \geq 0$$

Variables in this example are

$$x_j \triangleq \text{number of batches in each run or lot of product } j$$

It is easy to see that this model is separable because its objective function can be written

$$f(x_1, x_2, x_3, x_4) \triangleq f_1(x_1) + f_2(x_2) + f_3(x_3) + f_4(x_4)$$

where

$$f_1(x_1) \triangleq 66.21x_1 + \frac{2160}{x_1}$$

$$f_2(x_2) \triangleq 426.8x_2 + \frac{8712}{x_2}$$

$$f_3(x_2) \triangleq 61.20x_3 + \frac{330}{x_3}$$

$$f_4(x_2) \triangleq 268.1x_4 + \frac{2916}{x_4}$$

and the main production time constraint decomposes similarly with

$$g_1(x_1) \triangleq \frac{180}{x_1}$$

$$g_2(x_2) \triangleq \frac{726}{x_2}$$

$$g_3(x_3) \triangleq \frac{27.5}{x_3}$$

$$g_4(x_4) \triangleq \frac{243}{x_4}$$

The remaining four (nonnegativity) constraints are linear and thus automatically separable.

14.57 Linear functions are always separable.

By definition, linear functions consist of a sum of terms $a_j x_j$ involving single decision variables.

SAMPLE EXERCISE 14.29: RECOGNIZING SEPARABLE FUNCTIONS

Determine whether each of the following functions is separable.

(a) $f(w_1, w_2) \triangleq (w_1)^{3.5} + \ln(w_2)$

(b) $g_1(w_1, w_2) \triangleq 14w_1 - 26w_2$

(c) $g_2(w_1, w_2) \triangleq 14w_1 + w_1 w_2 - 26w_2$

Analysis: We apply definition 14.6 .

(a) This f is separable because it decomposes into the sum of $f_1(w_1) \triangleq (w_1)^{3.5}$ and $f_2(w_2) \triangleq \ln(w_2)$.

(b) This g_1 is separable because it is linear (principle $\boxed{14.57}$).

(c) This g_2 is not separable because the term $w_1 w_2$ involves both variables.

Piecewise Linear Approximation to Separable Functions

The main special convenience of separable programs is that they can sometimes be approximated closely by LPs (principle $\boxed{14.8}$). The transformation begins with piecewise linear approximation of the 1-variable functions f_j and $g_{i,j}$.

$\boxed{14.58}$ **Piecewise linear approximation** of separable programs divides the domain of each decision variable x_j into a series of intervals k and interpolates linearly to approximate corresponding $f_j(x_j)$ and $g_{i,j}(x_j)$ as

$$f_j(x_j) \approx c_{j,0} + \sum_k c_{j,k} x_{j,k}$$

$$g_{i,j}(x_j) \approx a_{i,j,0} + \sum_k a_{i,j,k} x_{j,k}$$

New variables $x_{j,k}$ represent x_j within interval k, and coefficients $c_{j,k}$ and $a_{i,j,k}$ express interpolation intercepts and slopes.

Figure 14.11 illustrates for x_1 in our Pfizer model (14.60) over domain $[0, 8]$. Both objective function term $f_1(x_1)$ and main constraint function $g_1(x_1)$ have been approximated with three linear segments. One covers $x_1 \in [0, u_{1,1}] = [0, 1]$; a second treats $x_1 \in [u_{1,1}, u_{1,2}] = [1, 5.7]$; and the third handles $x_1 \in [u_{1,2}, u_{1,3}] = [5.7, 8]$. First breakpoint $u_{1,1} = 1$ was chosen near the "knee" of the two functions, $u_{1,2} = 5.7$ approximates the minimum of $f_1(x_1)$, and $u_{1,3} = 8$ is a practical upper bound on x_1.

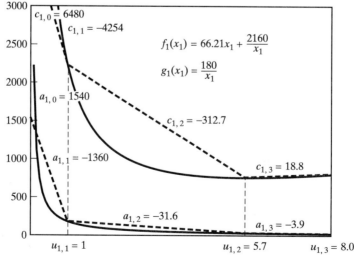

FIGURE 14.11 Piecewise Linear Approximation of Pfizer Example

Introducing new variables $x_{1,1}, x_{1,2}$, and $x_{1,3}$ with coefficients as in Figure 14.11, we have

$$f_1(x_1) \triangleq 66.21x_1 + \frac{2160}{x_1} \approx 6480 - 4254x_{1,1} - 312.7x_{1,2} + 18.8x_{1,3}$$

$$g_1(x_1) \triangleq \frac{180}{x_1} \approx 1540 - 1360x_{1,1} - 31.6x_{1,2} - 3.9x_{1,3}$$

Notice that the same interval limits must be used to approximate x_1 in the objective and all constraints. Upper bounds on the new variables are derived from interval limits

$$0 \le x_{1,1} \le u_{1,1} = 1$$
$$0 \le x_{1,2} \le u_{1,2} - u_{1,1} = 5.7 - 1.0 = 4.7$$
$$0 \le x_{1,3} \le u_{1,3} - u_{1,2} = 8.0 - 5.7 = 2.3$$

SAMPLE EXERCISE 14.30: FORMING PIECEWISE LINEAR APPROXIMATIONS

Consider a separable nonlinear program with objective function and constraint component functions for nonnegative decision variable w_1 given by

$$f_1(w_1) \triangleq (w_1)^2 - 4w_1 + 22$$
$$g_{1,1}(w_1) \triangleq \sqrt{w_1 + 9}$$
$$g_{2,1}(w_1) \triangleq 14w_1$$

Form corresponding piecewise linear approximations using breakpoints $u_{1,1} = 2$ and $u_{1,2} = 5$.

Analysis: We must estimate the interpolation coefficients of definition $\boxed{14.58}$. Intercepts are

$$c_{1,0} = f_1(0) = 22, \quad a_{1,1,0} = g_{1,1}(0) = 3, \quad a_{2,1,0} = g_{2,1}(0) = 0$$

Slopes for the interval $[0, u_{1,1}] = [0, 2]$ are derived as

$$c_{1,1} = \frac{f_1(2) - f_1(0)}{2 - 0} = -2$$

$$a_{1,1,1} = \frac{g_{1,1}(2) - g_{1,1}(0)}{2 - 0} = 0.158$$

$$a_{2,1,1} = \frac{g_{2,1}(2) - g_{2,1}(0)}{2 - 0} = 14$$

Corresponding slopes for interval $[u_{1,1}, u_{2,1}] = [2, 5]$ are

$$c_{1,2} = \frac{f_1(5) - f_1(2)}{5 - 2} = 3$$

$$a_{1,1,2} = \frac{g_{1,1}(5) - g_{1,1}(2)}{5 - 2} = 0.142$$

$$a_{2,1,2} = \frac{g_{2,1}(5) - g_{2,1}(2)}{5 - 2} = 14$$

Thus piecewise linear approximations are

$$f_1(w_1) \approx 22 - 2w_{1,1} + 3w_{1,2}$$
$$g_{1,1}(w_1) \approx 3 + 0.158w_{1,1} + 0.142w_{1,2}$$
$$g_{2,1}(w_1) \approx 0 + 14w_{1,1} + 14w_{1,2}$$

Linear Program Representation of Separable Programs

Applying piecewise linear approximation $\boxed{14.58}$ to every variable of separable program format (14.59) yields an LP approximation.

$\boxed{14.59}$ The **linear programming approximation** to a separable nonlinear program over nonnegative variables can be expressed as

$$\text{max or min} \sum_j \left(c_{j,0} + \sum_k c_{j,k} x_{j,k} \right)$$

$$\text{s.t.} \quad \sum_j \left(a_{i,j,0} + \sum_k a_{i,j,k} x_{j,k} \right) \geq b_i \qquad \text{for all } i \in G$$

$$\sum_j \left(a_{i,j,0} + \sum_k a_{i,j,k} x_{j,k} \right) \leq b_i \qquad \text{for all } i \in L$$

$$\sum_j \left(a_{i,j,0} + \sum_k a_{i,j,k} x_{j,k} \right) = b_i \qquad \text{for all } i \in E$$

$$0 \leq x_{j,k} \leq u_{j,k} - u_{j,k_1} \qquad \text{for all } j, k$$

where $u_{j,k}$ are the interval breakpoints for variable x_j ($u_{j,0} \triangleq 0$), and the coefficients $c_{j,k}$ and $a_{i,j,k}$ express interpolation intercepts and slopes.

This model is a linear program solvable by methods of Chapters 5 and 6.

Full application of construction $\boxed{14.59}$ to our Pfizer lot sizing model produces the linear approximation

$$
\begin{aligned}
\text{min} \quad & 42,354 - 4254x_{1,1} - 312.7x_{1,2} + 18.8x_{1,3} \\
& - 16997x_{2,1} - 1509x_{2,2} + 184.8x_{2,3} \\
& - 598.8x_{3,1} - 82.3x_{3,2} + 43.3x_{3,3} \\
& - 5564x_{4,1} - 615.5x_{4,2} + 157.6x_{4,3} \\
\text{s.t.} \quad & 7030 - 1360x_{1,1} - 31.6x_{1,2} - 3.9x_{1,3} \\
& - 2452x_{2,1} - 161.3x_{2,2} - 20.2x_{2,3} \\
& - 555x_{3,1} - 12x_{3,2} - 1.5x_{3,3} \\
& - 1486x_{4,1} - 73.6x_{4,2} - 9.2x_{4,3} \leq 221.5 \\
& 0 \leq x_{1,1} \leq 1, \ 0 \leq x_{1,2} \leq 4.7, \ 0 \leq x_{1,3} \leq 2.3 \\
& 0 \leq x_{2,1} \leq 1, \ 0 \leq x_{2,2} \leq 3.5, \ 0 \leq x_{2,3} \leq 3.5 \\
& 0 \leq x_{3,1} \leq 1, \ 0 \leq x_{3,2} \leq 1.3, \ 0 \leq x_{3,3} \leq 5.7 \\
& 0 \leq x_{4,1} \leq 1, \ 0 \leq x_{4,2} \leq 2.3, \ 0 \leq x_{4,3} \leq 4.7
\end{aligned}
$$

$$(14.61)$$

An optimal solution is

$$
\begin{array}{llll}
x^*_{1,1} = 1.0, & x^*_{1,2} = 4.7, & x^*_{1,3} = 2.3, & \text{or} \quad x^*_{1,1} + x^*_{1,2} + x^*_{1,3} = x^*_1 = 8.0 \\
x^*_{2,1} = 1.0, & x^*_{2,2} = 3.5, & x^*_{2,3} = 2.4, & \text{or} \quad x^*_{2,1} + x^*_{2,2} + x^*_{2,3} = x^*_2 = 6.9 \\
x^*_{3,1} = 1.0, & x^*_{3,2} = 1.3, & x^*_{3,3} = 0.0, & \text{or} \quad x^*_{3,1} + x^*_{3,2} + x^*_{3,3} = x^*_3 = 2.3 \\
x^*_{4,1} = 1.0, & x^*_{4,2} = 2.3, & x^*_{4,3} = 0.0, & \text{or} \quad x^*_{4,1} + x^*_{4,2} + x^*_{4,3} = x^*_4 = 3.3
\end{array}
\tag{14.62}
$$

which corresponds fairly well to the nonlinear optimum of (14.7).

Correctness of the LP Approximation to Separable Programs

Does linear program $\boxed{14.59}$ correctly model (14.59) (except for interpolation error)? Sometimes yes, sometimes no. To see the potential difficulty, suppose that we attempt to

$$
\begin{array}{ll}
\max & f(y) \triangleq y^2 - 12y + 45 \\
\text{s.t.} & 0 \le y \le 9
\end{array}
$$

This f is the function in Figure 14.4 (Section 14.2). Obviously, the optimal $y^* = 0$, because $f(0) = 45, f(9) = 18$, and every y in between has a lower function value.

Forming linear program representation $\boxed{14.59}$ for this simple example yields

$$
\begin{array}{ll}
\max & 45 - 7y_1 + 2y_2 \\
\text{s.t.} & 0 \le y_1 \le 5 \\
& 0 \le y_2 \le 4
\end{array}
$$

An optimal solution makes $y^*_1 = 0, y^*_2 = 4$, which implies that $y^* = y^*_1 + y^*_2 = 4$. What went wrong? The approximation

$$
f(y) \approx c_0 + \sum_k c_k y_k
$$

corresponding to $\boxed{14.58}$ is correct only if we assume that segment variables y_k satisfy a certain sequence at optimality. For $y \in [0, 5]$ we want first segment y_1 to represent y. If $y \in [5, 9]$, we want segment 1 to run to its upper limit and segment 2 to do the rest.

In general, each segment of a piecewise linear approximation must reach its upper bound before the next is available for use.

$\boxed{14.60}$ Linear program representation $\boxed{14.59}$ gives a correct approximation to separable program (14.59) whenever optimal values for segment variables satisfy

$$
x^*_{j,k+1} > 0 \quad \text{only if} \quad x^*_{j,k} = (u_{j,k} - u_{j,k-1}) \quad \text{for all } j \text{ and } k
$$

with $u_{j,0} \triangleq 0$.

Pfizer example results (14.62) illustrate a case where conditions $\boxed{14.60}$ are satisfied. For example, at $j = 2$, segment upper bounds in formulation (14.61) are 1.0, 3.5, and 3.5, while

$$
x^*_{2,1} = 1.0, \quad x^*_{2,2} = 3.5, \quad x^*_{2,3} = 2.4
$$

Each positive segment has the preceding one at its upper bound.

SAMPLE EXERCISE 14.31: CHECKING PIECEWISE LINEAR APPROXIMATIONS

A piecewise linear approximation to a separable nonlinear program in nonnegative variables w_1 and w_2 uses breakpoints $u_{1,1} = 2$ and $u_{1,2} = 6$ for the first variable, together with $u_{2,1} = 7$ and $u_{2,2} = 20$ for the second. Determine whether each of the following LP approximation solutions provides a correct answer (except for interpolation error) to the original nonlinear program.

(a) $w_{1,1}^* = 2$, $w_{1,2}^* = 3$, $w_{2,1}^* = 6$, $w_{2,2}^* = 0$

(b) $w_{1,1}^* = 0$, $w_{1,2}^* = 3$, $w_{2,1}^* = 1$, $w_{2,2}^* = 13$

Analysis: Following formulation $\boxed{14.59}$, bounds on the segment variables will be

$$0 \le w_{1,1} \le 2 - 0 = 2, \quad 0 \le w_{1,2} = 6 - 2 = 4$$
$$0 \le w_{2,1} \le 7 - 0 = 7, \quad 0 \le w_{2,2} = 20 - 7 = 13$$

(a) This optimal solution does give a correct approximation because it satisfies sequencing conditions $\boxed{14.60}$. Variable $w_{1,1}^*$ equals its upper bound, so $w_{1,2}^*$ can be positive.

(b) This solution yields an incorrect approximation because it violates sequencing conditions $\boxed{14.60}$. Variable $w_{1,2}^* > 0$ and $w_{1,1}^*$ does not equal its upper bound.

Convex Separable Programs

Suppose now that the given separable program (14.59) is also a convex program. That is, we are minimizing a convex objective or maximizing a concave one, subject to concave $\ge$ constraints, convex $\le$ ones, and linear equalities (definition $\boxed{14.3}$).

These requirements relate to the entire objective of constraint functions. Still, it is easy to see that all component functions for each variable in separable form (14.59) must have similar properties.

$\boxed{14.61}$ Separable function

$$s(x_1, \ldots, x_n) \triangleq \sum_{j=1}^{n} s_j(x_j)$$

is convex if and only if each component s_j is convex. It is concave if and only if each s_j is concave.

We already know from principle $\boxed{13.29}$ that sums of convex (or concave) functions are convex (concave). But $\boxed{14.61}$ asserts that the converse is also true for separable functions. To see why, we need only choose points $\mathbf{x}^{(1)}$ and $\mathbf{x}^{(2)}$ with all components equal except the jth. Then a step λ from $\mathbf{x}^{(1)}$ toward $\mathbf{x}^{(2)}$ changes just component $s_j(x_j^{(1)} + \lambda(x_j^{(2)} - x_j^{(1)}))$ of s. Convex function definition $\boxed{13.23}$ can hold for s only if it holds for s_j.

Given their many convenient properties, it should not surprise us that separable convex programs satisfy requirement $\boxed{14.60}$ for good approximation by linear programming.

$\boxed{14.62}$ Linear approximations $\boxed{14.59}$ to a separable convex programs have an optimal solution satisfying sequencing conditions $\boxed{14.60}$ if they have any optimum at all.

Again, Pfizer example model (14.61) illustrates. We have seen in Section 14.2 that the nonlinear version (14.60) is a convex program. Thus it was no accident that segment optima (14.62) satisfy sequencing condition $\boxed{14.60}$.

To understand why separable convex programs have the required property for effective LP approximation, recall (principle $\boxed{13.27}$) that second derivatives of convex functions are nonnegative and those of concave functions are nonpositive. It follows that first derivatives, or slopes, are nondecreasing and nonincreasing, respectively.

For a minimizing convex objective, this implies that the least cost slope $c_{j,k}$ of each approximation occurs at $k = 1$, and for a maximizing concave objective, the first segment will also be the most preferred. In a similar way, coefficients in constraint rows also exhibit a preference for lower-numbered segments. For example, if constraint i is $\leq$, and thus convex, $a_{i,j,1}$ is the smallest slope in the approximation of $g_{i,j}$. Thus it does least damage to feasibility. If the constraint is $\geq$, and thus concave, $a_{i,j,1}$ is largest and advances feasibility most rapidly.

Combining these observations about objective and constraint approximations, we see that the first segment $x_{j,k}$ of each piecewise linear approximation gives the greatest objective function payoff with the least burden on constraints. An optimal LP solution must choose it first, making the second segment positive only when the first has reached its upper bound. The pattern continues through each interval k, so that property $\boxed{14.60}$ is satisfied.

Difficulties with Nonconvex Separable Programs

When the given separable program is not convex, property $\boxed{14.60}$ may not hold automatically. However, it can be enforced artificially.

Suppose that we apply upper and lower bounded simplex Algorithm 5E to linear approximation $\boxed{14.59}$. Each iteration chooses either a nonbasic lower-bounded variable to increase or a nonbasic upper-bounded variable to decrease.

General separable programming searches simply restrict the choices even further to maintain sequencing property $\boxed{14.60}$. A nonbasic lower-bounded segment variable cannot increase unless the preceding segment is already at its upper bound. Similarly, a nonbasic upper-bounded segment variable cannot decrease unless the succeeding segment has value 0.

Of course, these extra limitations may prevent the simplex from computing an optimal solution to linear program $\boxed{14.59}$. However, they do assure that property

$\boxed{14.60}$ is enforced. When no allowable pivots remain, we stop with a heuristic optimum.

An alternative approach producing globally optimal solutions can be derived via integer linear programming methods of Section 11.1. Binary variables $y_{j,k}$ are introduced that parallel each $x_{j,k}$, with

$$y_{j,k} \triangleq \begin{cases} 1 & \text{if } x_{j,k} > 0 \\ 0 & \text{otherwise} \end{cases}$$

Then the switching constraints

$$\left(u_{j,k} - u_{j,k-1}\right) y_{j,k+1} \leq x_{j,k} \leq \left(u_{j,k} - u_{j,k-1}\right) y_{j,k} \qquad \text{for all } j, k$$

enforce sequencing conditions $\boxed{14.60}$ by pushing each segment to its upper bound if the next is positive.

14.9 POSYNOMIAL GEOMETRIC PROGRAMMING METHODS

As we have seen in Sections 14.1 and 14.2, many important applications of nonlinear programming arise in engineering design, where decision variables are physical dimensions, pressures, and so on. Such models are often highly nonlinear and have many locally optimal solutions. This section deals with special cases called **posynomial geometric programs**, which address the "variables to powers" form of many engineering design models and constitute the only broad class of nonconvex NLPs readily solved to global optimality.

Posynomial Geometric Program Form

In Section 14.2 we introduced (definition $\boxed{14.10}$) **posynomial** functions, which are positive-weighted sums of products of decision variables raised to arbitrary powers. A **posynomial geometric program** (**GP**) is an NLP minimizing a posynomial objective function over positive variables and $\leq$ posynomial main constraints (definition $\boxed{14.11}$). The general form is

$$
\begin{aligned}
\min \quad & \sum_{k \in K_0} d_k \prod_{j=1}^{n} (x_j)^{a_{k,j}} \\
\text{s.t.} \quad & \sum_{k \in K_i} d_k \prod_{j=1}^{n} (x_j)^{a_{k,j}} \leq 1 \qquad i = 1, \ldots, m \\
& x_j > 0 \qquad\qquad\qquad\quad j = 1, \ldots, n
\end{aligned}
\tag{14.63}
$$

where nonoverlapping sets K_i index the posynomial terms in the objective and constraints, values d_k are the corresponding weights, and the $a_{k,j}$ are exponents of variables x_j in terms k.

Several details of the format will prove critical to its tractability:

- The objective must minimize a posynomial. Maximizations are not allowed.
- Coefficients d_k must all be positive.
- Constraints must enforce $\leq$ requirements on a posynomial. Any $=$ and $\geq$ forms destroy the structure.
- Decision variables must be limited to positive values. We will want to take their logarithms.

Cofferdam Example Revisited

We illustrate GP methods with Section 14.2's model (14.11) to optimize a cofferdam design:

$$\min \quad 168x_1x_2 + 3648x_1 + 3648\frac{x_1x_2}{x_3} + \frac{1.46 \times 10^7}{x_4} \quad \text{(cost)}$$

$$\text{s.t.} \quad \frac{1.25x_4}{x_1} + \frac{41.625}{x_1} \leq 1 \quad \text{(empirical)}$$

$$\frac{1.0425x_1}{x_2} \leq 1 \quad \text{(slipping)}$$

$$0.00035x_1x_3 \leq 1 \quad \text{(tension)}$$

$$x_1, x_2, x_3, x_4 > 0$$

(14.64)

In the notation of (14.63), $K_0 = \{1, 2, 3, 4\}$, $K_1 = \{5, 6\}$, $K_2 = \{7\}$, and $K_3 = \{8\}$. Corresponding coefficients are

$$d_1 = 168, \quad d_2 = 3648, \quad d_3 = 3648, \quad d_4 = 1.46 \times 10^7$$
$$d_5 = 1.25, \quad d_6 = 41.625, \quad d_7 = 1.0425, \quad d_8 = .00035$$

$$
\begin{array}{llll}
a_{1,1} = 1, & a_{1,2} = 1, & a_{1,3} = 0, & a_{1,4} = 0 \\
a_{2,1} = 1, & a_{2,2} = 0, & a_{2,3} = 0, & a_{2,4} = 0 \\
a_{3,1} = 1, & a_{3,2} = 1, & a_{3,3} = -1, & a_{3,4} = 0 \\
a_{4,1} = 0, & a_{4,2} = 0, & a_{4,3} = 0, & a_{4,4} = -1 \\
a_{5,1} = -1, & a_{5,2} = 0, & a_{5,3} = 0, & a_{5,4} = 1 \\
a_{6,1} = -1, & a_{6,2} = 0, & a_{6,3} = 0, & a_{6,4} = 0 \\
a_{7,1} = 1, & a_{7,2} = -1, & a_{7,3} = 0, & a_{7,4} = 0 \\
a_{8,1} = 1, & a_{8,2} = 0, & a_{8,3} = 1, & a_{8,4} = 0
\end{array}
$$

SAMPLE EXERCISE 14.32: PLACING GEOMETRIC PROGRAMS IN STANDARD FORM

Identify the constants and index sets of standard form (14.63) for the following posynomial geometric program:

$$\min \quad 3\frac{w_1^{.43}}{w_2} + 14w_2w_3$$

$$\text{s.t.} \quad w_1\sqrt{w_3} + w_2\sqrt{w_3} \leq 20$$

$$\frac{w_1}{w_2} \leq 1$$

$$w_1, w_2, w_3 > 0$$

Analysis: Begin by dividing through the main constraint by 20 to obtain standard right-hand side 1. Then the objective function has terms $k \in K_0 \triangleq \{1, 2\}$, the first constraint involves $k \in K_1 \triangleq \{3, 4\}$, and the second has only the one $k \in K_2 \triangleq \{5\}$. The corresponding standard-form coefficients are

$$d_1 = 3, \quad d_2 = 14, \quad d_3 = 0.05, \quad d_4 = 0.05, \quad d_5 = 1$$

and

$$
\begin{aligned}
a_{1,1} &= 0.43, & a_{1,2} &= -1, & a_{1,3} &= 0 \\
a_{2,1} &= 0, & a_{2,2} &= 1, & a_{2,3} &= 1 \\
a_{3,1} &= 1, & a_{3,2} &= 0, & a_{3,3} &= .5 \\
a_{4,1} &= 0, & a_{4,2} &= 1, & a_{4,3} &= .5 \\
a_{5,1} &= 1, & a_{5,2} &= -1, & a_{5,3} &= 0
\end{aligned}
$$

Logarithmic Change of Variables in GPs

Posynomial functions need not be convex [see example (14.14)], and thus geometric programs (14.63) are often not convex programs. However, a change of variables can make them convex. In particular (principle $\boxed{14.12}$), we consider substituting

$$z_j \triangleq \ln(x_j) \tag{14.65}$$

or

$$x_j \triangleq e^{z_j} \tag{14.66}$$

Under transformation (14.66), terms k of posynomials in (14.63) simplify as

$$d_k \prod_j (x_j)^{a_{k,j}} = d_k \prod_j \left(e^{z_j}\right)^{a_{k,j}} = d_k \prod_j e^{a_{k,j} z_j} = d_k e^{\sum_j a_{k,j} z_j} = d_k e^{\mathbf{a}^{(k)} \cdot \mathbf{z}}$$

where $\mathbf{a}^{(k)} \triangleq (a_{k,1}, \ldots, a_{k,n})$ and $\mathbf{z} \triangleq (z_1, \ldots, z_n)$. Then original geometric program form (14.63) becomes

$$
\begin{aligned}
\min \quad & f(\mathbf{z}) \triangleq \sum_{k \in K_0} d_k e^{\mathbf{a}^{(k)} \cdot \mathbf{z}} \\
\text{s.t.} \quad & g_i(\mathbf{z}) \triangleq \sum_{k \in K_i} d_k e^{\mathbf{a}^{(k)} \cdot \mathbf{z}} \leq 1 \qquad i = 1, \ldots, m \\
& z_j \text{ URS} \qquad\qquad\qquad\qquad j = 1, \ldots, n
\end{aligned}
\tag{14.67}
$$

For example, cofferdam model (14.64) transforms to

$$
\begin{aligned}
\min \quad & 168 e^{\mathbf{a}^{(1)} \cdot \mathbf{z}} + 3648 e^{\mathbf{a}^{(2)} \cdot \mathbf{z}} + 3648 e^{\mathbf{a}^{(3)} \cdot \mathbf{z}} + (1.46 \times 10^7) e^{\mathbf{a}^{(4)} \cdot \mathbf{z}} \\
\text{s.t.} \quad & 1.25 e^{\mathbf{a}^{(5)} \cdot \mathbf{z}} + 41.625 e^{\mathbf{a}^{(6)} \cdot \mathbf{z}} \leq 1 \\
& 1.0425^{\mathbf{a}^{(7)} \cdot \mathbf{z}} \qquad\qquad\qquad \leq 1 \\
& 0.00035 e^{\mathbf{a}^{(8)} \cdot \mathbf{z}} \qquad\qquad \leq 1 \\
& \mathbf{z} \text{ URS}
\end{aligned}
\tag{14.68}
$$

with

$$
\begin{aligned}
\mathbf{a}^{(1)} &= (1, 1, 0, 0), & \mathbf{a}^{(2)} &= (1, 0, 0, 0) \\
\mathbf{a}^{(3)} &= (1, 1, -1, 0), & \mathbf{a}^{(4)} &= (0, 0, 0, -1) \\
\mathbf{a}^{(5)} &= (-1, 0, 0, 1), & \mathbf{a}^{(6)} &= (-1, 0, 0, 0) \\
\mathbf{a}^{(7)} &= (1, -1, 0, 0), & \mathbf{a}^{(8)} &= (1, 0, 1, 0)
\end{aligned}
$$

SAMPLE EXERCISE 14.33: CHANGING VARIABLES IN GEOMETRIC PROGRAMS

Return to the posynomial geometric program of Sample Exercise 14.32. Change variables via (14.66) to produce a convex program in format (14.67).

Analysis: Using the coefficients of Sample Exercise 14.32, the transformed model is

$$\min \quad 3e^{0.43z_1 - 1z_2} + 14e^{1z_2 + 1z_3}$$
$$\text{s.t.} \quad 0.05e^{1z_1 + 0.5z_3} + 0.05e^{1z_2 + 0.5z_3} \leq 1$$
$$1e^{1z_1 - 1z_2} \leq 1$$
$$z_1, z_2, z_3 \text{ URS}$$

Convex Transformed GP Model

The power of this simple change of variables (14.66) is to convert a posynomial geometric program to a convex program (definition 14.3).

14.63   The transformed model obtained from a geometric program by substituting $x_j = e^{z_j}$ is a convex program in new variables z_j.

To see why the transformed model is convex in $\mathbf{z}$, observe that the objective and constraint functions are positive-weighted sums of terms

$$p_k(\mathbf{z}) \triangleq e^{\mathbf{a}^{(k)} \cdot \mathbf{z}}$$

Linear exponent $\mathbf{a}^{(k)} \cdot \mathbf{z}$ is convex in $\mathbf{z}$ (principle 13.28), and $h(y) \triangleq e^y$ is nondecreasing and convex. Composition principle 13.31 then implies that each p_k is convex, so that their transformed sum must be, too. It follows that format (14.67) minimizes a convex objective, subject to convex $\leq$ constraints, which makes it a convex program.

Notice how this analysis depends on details of the posynomial GP format. Coefficients $d_{j,k}$ must be positive for weighted sums of convex terms to be guaranteed convex. Also, a minimizing objective and $\leq$ nonlinear constraints are essential if a model over convex functions is to be a convex program.

Direct Solution of the Transformed Primal GP

Principle 14.63 provides a direct avenue to computing global optimal solutions to posynomial geometric programs (14.63). We need only substitute $x_j = e^{z_j}$, solve the resulting convex program in $\mathbf{z}$ by methods of Section 14.6, and transform back by

$$x_j^* \leftarrow e^{z_j^*} \quad \text{for all } j \tag{14.69}$$

For example, application of reduced gradient Algorithm 14C to transformed model (14.68) produces optimal solution

$$z_1^* = 4.138, \quad z_2^* = 4.179, \quad z_3^* = 3.820, \quad z_4^* = 2.823$$

Then inverse transformation (14.69) gives the following optimal solution in original variables:

$$x_1^* = e^{4.138} = 62.65, \quad x_2^* = e^{4.179} = 65.32,$$
$$x_3^* = e^{3.820} = 45.60, \quad x_4^* = e^{2.823} = 16.82$$

Dual of a Geometric Program

Sometimes even more efficient methods than solving convex program (14.67) can be used to optimize geometric programs. The process begins with still another transformation of the given model termed its dual.

In addition to the usual Lagrange multiplier for each constraint, geometric programming duals introduce variables for every term k of the objective and constraint posynomials:

$$v_i \triangleq \text{Lagrange multiplier for constraint } i$$

$$\delta_k \triangleq \text{dual variable for posynomial term } k$$

14.64 The **GP dual** corresponding to posynomial geometric program (14.63) can be expressed as

$$\max \quad \sum_{\text{all } k} \delta_k \ln\left(\frac{d_k}{\delta_k}\right) - \sum_{i=1}^{m} v_i \ln(-v_i)$$

$$\text{s.t.} \quad \sum_{\text{all } k} \mathbf{a}^{(k)} \delta_k = 0$$

$$\sum_{k \in K_0} \delta_k = 1$$

$$\sum_{k \in K_i} \delta_k = -v_i \qquad i = 1, \ldots, m$$

$$\delta_k \geq 0 \qquad\qquad \text{for all } k$$

$$v_i \leq 0 \qquad\qquad i = 1, \ldots, m$$

where v_i is the Lagrange multiplier on main primal constraint i and δ_k is the dual variable for posynomial term k.

Cofferdam model (14.64) illustrates. Dual form **14.64** is

$$\max \quad \delta_1 \ln\left(\frac{168}{\delta_1}\right) + \delta_2 \ln\left(\frac{3648}{\delta_2}\right) + \delta_3 \ln\left(\frac{3648}{\delta_3}\right)$$

$$+ \delta_4 \ln\left(\frac{1.46 \times 10^7}{\delta_4}\right) + \delta_5 \ln\left(\frac{1.25}{\delta_5}\right) + \delta_6 \ln\left(\frac{41.625}{\delta_6}\right) \tag{14.70}$$

$$+ \delta_7 \ln\left(\frac{1.0425}{\delta_7}\right) + \delta_8 \ln\left(\frac{0.00035}{\delta_8}\right)$$

$$- v_1 \ln(-v_1) - v_2 \ln(-v_2) - v_3 \ln(-v_3)$$

$$\text{s.t.} \quad +\delta_1 \quad +\delta_2 \quad +\delta_3 \qquad -\delta_5 \quad -\delta_6 \quad +\delta_7 \quad +\delta_8 \ = \ 0$$

$$+\delta_1 \qquad +\delta_3 \qquad -\delta_7 \qquad = \ 0$$

$$-\delta_3 \qquad +\delta_8 \ = \ 0$$

$$-\delta_4 \quad +\delta_5 \qquad = \ 0$$

$$+\delta_1 \quad +\delta_2 \quad +\delta_3 \quad +\delta_4 \qquad = \ 1$$

$$+\delta_5 \quad +\delta_6 \qquad = \ -v_1$$

$$+\delta_7 \qquad = \ -v_2$$

$$+\delta_8 \ = \ -v_3$$

$$\delta_1, \ldots, \delta_8 \geq 0$$

$$v_1, v_2, v_3 \leq 0$$

It first 4 constraints weight δ_k with exponents $a_{j,k}$ for primal variables j. The fifth normalizes the δ total of variables relating to the objective function. The remaining main constraints simply recover (negatives of) Lagrange multipliers for the 3 primal constraints as sums of associated δ_k. An optimal solution is

$$v_1^* = -1.225, \quad v_2^* = -0.481, \quad v_3^* = -0.155,$$
$$\delta_1^* = 0.326, \quad \delta_2^* = 0.108, \quad \delta_3^* = 0.155, \quad \delta_4^* = 0.411 \tag{14.71}$$
$$\delta_5^* = 0.411, \quad \delta_6^* = 0.814, \quad \delta_7^* = 0.481, \quad \delta_8^* = 0.155$$

with optimal value 14.563.

SAMPLE EXERCISE 14.34: FORMULATING GEOMETRIC PROGRAM DUALS

Form the dual of the posynomial geometric program in Sample Exercise 14.33.

Analysis: Following format $\boxed{14.64}$, we introduce Lagrange multipliers v_1 and v_2 for the main constraints, and variables $\delta_1, \ldots, \delta_5$ for the five posynomial terms. Then the dual becomes

$$\max \quad \delta_1 \ln\left(\frac{3}{\delta_1}\right) + \delta_2 \ln\left(\frac{14}{\delta_2}\right) + \delta_3 \ln\left(\frac{0.05}{\delta_3}\right) + \delta_4 \ln\left(\frac{0.05}{\delta_4}\right) + \delta_5 \ln\left(\frac{1}{\delta_5}\right)$$

$$- v_1 \ln(-v_1) - v_2 \ln(-v_2)$$

$$\text{s.t.} \quad 0.43\delta_1 + 1\delta_3 + 1\delta_5 = 0$$

$$-1\delta_1 + 1\delta_2 + 1\delta_4 - 1\delta_5 = 0$$

$$1\delta_2 + .5\delta_3 + .5\delta_4 = 0$$

$$1\delta_1 + 1\delta_2 = 1$$

$$1\delta_3 + 1\delta_4 = -v_1$$

$$1\delta_5 = -v_2$$

$$\delta_1, \ldots, \delta_5 \geq 0$$

$$v_1, v_2 \leq 0$$

Degrees of Difficulty and Solving the GP Dual

Dual problem 14.64 is a separable program (definition 14.7) with linear constraints. Furthermore, its objective function can be shown to be concave over feasible $(\delta, \mathbf{v})$, even though it is not concave over all choices of the decision variables.

> **14.65** Posynomial geometric program dual 14.64 is a separable convex program over linear constraints.

Thus either the separable programming methods of Section 14.8 or the reduced gradient algorithms of Section 14.6 can be employed to compute a global optimum.

Sometimes the task is even easier. Noting that the last main system of constraints entirely determines the v_i in terms of the δ_k, the degree of difficulty of a model depends on the number of truly independent variables δ_k in other main constraints.

> **14.66** The **degree of difficulty** of a geometric program is
>
> (number of posynomial terms k) $-$ (number of variables j) $- 1$

The first two sets of main constraints in dual 14.64 have one variable for each k and $(n + 1)$ constraints for the n primal decision variables. Thus the degree of difficulty bounds the number of δ-variables that must be fixed in value to determine the rest uniquely. Some models even have degree of difficulty zero, meaning that the dual can be optimized by solving a system of linear equations. Cofferdam example dual (14.70) has degree of difficulty

$$\text{terms} - \text{variables} - 1 = 8 - 4 - 1 = 3$$

> **SAMPLE EXERCISE 14.35: DETERMINING GP DEGREES OF DIFFICULTY**
>
> Determine the degrees of difficulty in posynomial geometric program of Sample Exercises 14.32 to 14.34.
>
> **Analysis:** The model has 5 posynomial terms and 3 variables. Thus its degree of difficulty is $5 - 3 - 1 = 1$.

Recovering a Primal GP Solution

We have seen that the dual problem 14.64 may be convenient to solve, but primal 14.11 is the model of true interest. How can we retrieve a primal optimum $\mathbf{x}^*$?

An elegant dual relationship makes recovery straightforward when optimal Lagrange multipliers are known for the dual.

14.67 Suppose that $\mathbf{z}^*$ are the optimal Lagrange multipliers on constraints $\sum_k \mathbf{a}^{(k)} \delta_k = \mathbf{0}$ in geometric programming dual **14.64**. Then

$$x_j^* \leftarrow e^{-z_j^*}$$

yields a global optimum in the corresponding primal.

For example, optimal Lagrange multipliers for the first 4 constraints of coffer-dam model dual (14.70) are

$$z_1^* = -4.138, \quad z_2^* = -4.179, \, z_3^* = -3.820, \quad z_3^* = -2.823$$

Application of transformation **14.67** recovers the same primal optimum that we have seen before:

$$x_1^* = e^{4.138} = 62.65, \quad x_2^* = e^{4.179} = 65.32$$
$$x_3^* = e^{3.820} = 45.60, \quad x_4^* = e^{2.823} = 16.82$$

Derivation of the GP Dual

Why is problem **14.64** termed a dual, and why can primal optima be recovered by construction **14.67**? Begin by taking logarithms of both sides in constraints and the objective function of transformed model (14.67).

$$
\begin{aligned}
\min \quad & \ln(f(\mathbf{z})) \triangleq \ln\left(\sum_{k \in K_0} d_k e^{\mathbf{a}^{(k)} \cdot \mathbf{z}}\right) \\
\text{s.t.} \quad & \ln(g_i(\mathbf{z})) \triangleq \ln\left(\sum_{k \in K_i} d_k e^{\mathbf{a}^{(k)} \cdot \mathbf{z}}\right) \le 0 \quad i = 1, \ldots, m \\
& z_j \text{ URS} \qquad\qquad\qquad\qquad\qquad j = 1, \ldots, n
\end{aligned}
\tag{14.72}
$$

Minimizing $\ln(f)$ is equivalent to minimizing f, and $g_i(\mathbf{z}) > 0$, so that logarithms always exist.

Noting the convexity of transformed model (14.67), this logarithmically retransformed form (14.72) is also a convex program. Karush–Kuhn–Tucker conditions will be sufficient for an optimal solution (principle **14.26**).

With Lagrange multipliers, v_i, the main rows of KKT conditions are

$$\frac{1}{f(\mathbf{z})} \sum_{k \in K_0} d_k a_{k,j} e^{\mathbf{a}^{(k)} \cdot \mathbf{z}} - \sum_{i=1}^{m} \frac{v_i}{g_i(\mathbf{z})} \sum_{k \in K_i} d_k a_{k,j} e^{\mathbf{a}^{(k)} \cdot \mathbf{z}} = 0 \quad \text{for all } j \tag{14.73}$$

Now substituting

$$
\begin{aligned}
\delta_k &\triangleq \frac{1}{f(\mathbf{z})}\left(d_k e^{\mathbf{a}^{(k)} \cdot \mathbf{z}}\right) \qquad k \in K_0 \\
\delta_k &\triangleq \frac{-v_i}{g_i(\mathbf{z})}\left(d_k e^{\mathbf{a}^{(k)} \cdot \mathbf{z}}\right) \qquad k \in K_i, \quad i = 1, \ldots, m
\end{aligned}
\tag{14.74}
$$

equations (14.73) become

$$\sum_{\text{al } k} a_{k,j}\delta_k = 0 \quad \text{for all } j = 1, \ldots, n$$

These are exactly the first main constraints of dual $\boxed{14.64}$. Sign restrictions on the v_i and δ_k also correspond.

To make new variables δ_k perform according to their definitions (14.74), we must also enforce

$$f(\mathbf{z}) = \sum_{k \in K_0} d_k e^{\mathbf{a}^{(k)} \cdot \mathbf{z}}$$

or, dividing by $f(\mathbf{z})$,

$$1 = \sum_{k \in K_0} \delta_k \tag{14.75}$$

Similarly,

$$g_i(\mathbf{z}) = \sum_{k \in K_i} d_k e^{\mathbf{a}^{(k)} \cdot \mathbf{z}} \quad \text{for all } i = 1, \ldots, m$$

becomes upon multiplication by $-v_i/g_i(\mathbf{z})$

$$-v_i = \sum_{k \in K_i} \delta_k \text{ for all } i = 1, \ldots, m \tag{14.76}$$

Expressions (14.75) and (14.76) complete the constraints of dual $\boxed{14.64}$.

Only the complementary slackness part of KKT conditions now remain to be formulated. Instead of explicitly including such conditions, dual formulation $\boxed{14.64}$ maximizes objective function

$$\sum_{\text{all } k} \delta_k \ln\left(\frac{d_k}{\delta_k}\right) - \sum_i v_i \ln(-v_i) \tag{14.77}$$

over $(\delta, \mathbf{v})$, fulfilling the other constraints. A long but tedious derivation can show that this maximizes the Lagrangian of (14.72) over stationary points $\mathbf{z}$, which has the same effect as enforcing complementary slackness.

Signomial Extension of GPs

Often, models that cannot be formulated as a posynomial geometric program do fit a less restrictive signomial form.

$\boxed{14.68}$ Function $s(\mathbf{x})$ is a **signomial** if it can be expressed

$$s(x_1, \ldots, x_n) \triangleq \sum_k d_k \left(\prod_{j=1}^n (x_j)^{a_{k,j}}\right)$$

for given d_k and $a_{j,k}$ of arbitrary sign.

Note that term weights d_k are not required to be positive.

Obviously, the easy convexity of NLPs over transformed posynomials (principle $\boxed{14.63}$) is lost when weights may be negative. Still, considerable tractability is preserved. The reader is referred to more advanced books on nonlinear programming for details.

EXERCISES

14-1 A lidless, rectangular box is to be manufactured from 30- by 40-inch cardboard stock sheets by cutting squares from the four corners, folding up ends and sides, and joining with heavy tape. The designer wishes to choose box dimensions that maximize volume.

(a) ⬦ Formulate this design problem as a constrained NLP.

(b) ⬦ ▭ Use class optimization software to start from a feasible solution and compute at least a local optimum.

14-2 A partially buried, rectangular office building is to be constructed with a volume of at least 50,000 cubic meters. To minimize energy for heating and cooling, the exterior roof and sidewall surface exposed above ground should not exceed 2250 square meters. Within these limits, the designer wishes to choose dimensions that minimize the volume excavated for the buried part of the building.

(a) Formulate this design problem as a constrained NLP.

(b) ⬦ ▭ Use class optimization software to start from a feasible solution and compute at least a local optimum.

14-3 A company maintains inventories of its 5 products, replenishing the stock of an item whenever it reaches zero by manufacturing a fixed lot size of new units. The following table shows the setup cost for manufacturing, the unit volume, the unit annual inventory holding cost, and the estimated annual demand for each item.

	Product				
	1	**2**	**3**	**4**	**5**
Setup	300	120	440	190	80
Volume	33	10	12	15	26
Holding	87	95	27	36	135
Demand	800	2000	250	900	1350

Managers wish to choose lot sizes for each item that minimize total average annual setup and holding costs while assuring that the maximum combined stored volume will not exceed the 4000 cubic meters available. Assume that lots arrive the instant they are ordered.

(a) ⬦ Formulate this operations problem as a constrained NLP. (*Hint:* What is the average on hand inventory of item j as a function of the lot size for j?)

(b) ⬦ ▭ Use class optimization software to start from a feasible solution and compute at least a local optimum.

14-4 A print shop plans to maintain 5 different presses, replacing each every few years on a regular cycle. The following table shows the replacement cost (in thousands of dollars) of each press, and the estimated annual income (in thousands of dollars) that each can generate when new. However, as the presses grow older, their productivity declines; final values in the table show the (simple, not compound) percent income loss each year of life.

	Press				
	1	**2**	**3**	**4**	**5**
Replace	110	450	150	675	320
Income	90	110	55	220	250
Decline	5%	20%	30%	20%	40%

The owner wishes to choose a replacement (cycle) time for each press that minimizes total replacement and lost income costs within the $250,000 she can average annually for purchasing new presses.

(a) Formulate this replacement problem as a constrained NLP. (*Hint:* What is the average income loss on press j as a function of the replacement time for j?)

(b) ⬦ ▭ Use class optimization software to start from a feasible solution and compute at least a local optimum.

14-5 A machinist will remove excess metal from a rotary (round) machine part by passing the cutting tool of a lathe along 42 inches of the part length. For a lathe turning at N revolutions per minute and advancing the tool at a feed rate of f inches per revolution, classic empirical relationships project the effective life of a cutting tool (in minutes) at

$$\text{tool life} = \left(\frac{5}{Nf^{0.60}}\right)^{6.667}$$

Each time a tool wears out the operator must install a new one and spend 0.1 hour realigning the machine. Engineers wish to choose the machining plan that minimizes total cost at $52 per hour for machinist time and $87 each for new tools. Speed N must be in the interval [200, 600] and feed rate f in the interval [0.001, 0.005].

(a) ⊠ Formulate this machining problem as a constrained NLP.

(b) ⊠ ▢ Use class optimization software to start from a feasible solution and compute at least a local optimum.

14-6 A warehousing firm services orders for its 5 products from an automatic storage and retrieval (ASAR) system, refilling storage from backup areas whenever the ASAR stock of any item reaches zero. The following table shows the weekly demand and the unit volume (cubic feet) for each product.

	Product				
	1	**2**	**3**	**4**	**5**
Demand	100	25	30	50	200
Volume	2	5	3	7	5

Managers wish to decide how many of each item to accommodate within the 1000 cubic feet of available storage to minimize the total number of refilling operations per week.

(a) Formulate this operations problem as a constrained NLP.

(b) ⊠ ▢ Use class optimization software to start from a feasible solution and compute at least a local optimum.

14-7 A solid waste company must locate 2 disposal sites to service the demand (tons per day) of the 5 communities detailed in the following table.

	Community				
	1	**2**	**3**	**4**	**5**
Demand	60	90	35	85	70
E-W coordinate	0	4	30	20	16
N-S coordinate	0	30	8	17	15

Each site will be able to handle up to 200 tons per day, and planners want to select the site locations to minimize to total ton-miles of hauling from community (mile) coordinates shown in the table to disposal sites. Assume that hauling distance is proportional to straight-line distance.

(a) ⊠ Formulate this location-allocation problem as a constrained NLP.

(b) ⊠ ▢ Use class optimization software to start from reasonable locations and compute at least a local optimum.

14-8 A light manufacturing firm is planning a new factory in a rural part of the western United States. A total of 100 employees are to be hired from the 5 surrounding communities. The following table shows the number of (equally) qualified workers available in each community and community location coordinates (in miles).

	Community				
	1	**2**	**3**	**4**	**5**
Available	70	15	20	40	30
E-W coordinate	0	10	6	1	2
N-S coordinate	0	1	8	9	3

Planners want to choose a factory site that minimizes total employee travel distance. Assume that travel distance is proportional to straight-line distance from community to factory site.

(a) Formulate this location-allocation problem as a constrained NLP.

(b) ⊠ ▢ Use class optimization software to start from reasonable locations and compute at least a local optimum.

14-9 An investor has decided to divide his $1.5 million portfolio among government bonds, interest-sensitive stocks, and technology stocks because some of these categories tend to increase return in periods when the others decrease. Specifically, his analysis of recent experience produced the following average returns and covariances among categories:

	Bonds	Interest-Sensitive Stocks	Technology Stocks
Mean return	5.81%	10.97%	13.02%
	Covariance		
Bonds	1.09	−1.12	−3.15
Interest-sensitive stocks	−1.12	1.52	4.38
Technology stocks	−3.15	4.38	12.95

The investor wants to find the least variance way to divide his portfolio among the three categories while maintaining a 10% average return.

(a) ⊘ Formulate this portfolio problem as a constrained NLP.

(b) ⊘ ▭ Use class optimization software to start from a feasible solution and compute at least a local optimum.

14-10 A farmer wants to allocate between 10 and 60% of his available acreage to each of corn, soybeans, and sunflowers. With markets varying wildly from year to year, he has done some research on past performance to guide his decisions. The following table shows the average return per acre and the covariances among categories that he has computed.

	Corn	Soybeans	Sunflow
Dollar return	77.38	88.38	107.50
	Covariance		
Corn	1.09	−1.12	−3.15
Soybeans	−1.12	1.52	4.38
Sunflow	−3.15	4.38	12.95

The farmer wants the least risk plan that will average at least $90 per acre.

(a) Formulate this portfolio problem as a constrained NLP.

(b) ⊘ ▭ Use class optimization software to start from a feasible solution and compute at least a local optimum.

14-11 A new premium whiskey will be produced by blending up to 5 different distilling products, and the quality of the results will be measured by 3 performance indices. The following table shows the value of each index for the 5 ingredients, along with lower and upper limits for the index of the blend and costs per unit volume.

	Ingredient				
Index	1	2	3	4	5
1	12.6	15.8	17.2	10.1	11.7
2	31.4	30.2	29.6	40.4	28.9
3	115	202	184	143	169
Cost	125	154	116	189	132

	Blend	
Index	Low	High
1	12	16
2	31	36
3	121	164

The index of the blend in the first two cases will be just a volume-weighted sum of those of the chosen ingredients. However, the third index is logarithmic (i.e., the natural logarithm of the blend value will be the logarithm of the volume-weighted ingredient sum). Production planners want to choose a blend that meets upper and lower index requirements at minimum cost.

(a) ⊘ Formulate this blending problem as a constrained NLP.

(b) ⊘ ▭ Use class optimization software to start from a feasible solution and compute at least a local optimum.

14-12 A chemical manufacturer needs to produce 1250 barrels of a special industrial cleaning fluid by blending 5 available ingredients. The quality of the result is measured by 3 quantitative indices. The following table show the index values for each ingredient, along with the minimum and maximum required in the blend and the cost per barrel of ingredients.

	Ingredient				
Index	1	2	3	4	5
1	50.4	45.2	33.1	29.9	44.9
2	13.9	19.2	18.6	25.5	10.9
3	89.2	75.4	99.8	84.3	68.8
Cost	531	339	128	414	307

	Blend	
Index	Minimum	Maximum
1	33	43
2	17	20
3	81	99

The index of the blend in the first case will be just a volume-weighted average of those of the chosen ingredients. However, the square of the second blend index will be the square of the volume-weighted ingredient average, and the logarithm of the third blend index will be the logarithm of the volume-weighted ingredient average. Production planners want to choose a blend that meets upper and lower index requirements at minimum cost.

(a) Formulate this blending problem as a constrained NLP.
(b) Use class optimization software to start from a feasible solution and compute at least a local optimum.

14-13 A laser printer manufacturer can make models $i = 1, \ldots, 6$ at any of plants $j = 1, \ldots, 4$. The fraction of plant j capacity required per unit of printer i has been estimated for each combination at $f_{i,j}$. The laser printer market is very competitive, so the price that can be charged for any model is a decreasing nonlinear (demand) function p_i of the total number of model i printers that are sold. Assuming these demand functions are know, formulate an NLP to determine a maximum total revenue production plan.

14-14 A new automatic storage and retrieval (ASAR)[6] area is being added to an existing warehouse on land already owned by the company. It will have $n \geq 1$ aisles, each with pallet storage cells on both sides and a stacker crane moving in the middle which can carry a pallet to/from any location in the aisle. Storage must be provided for a total of at least p pallet cells of width w, depth d, and height h feet. All racks will be $m \geq 1$ cells high, and k cells from one end of the aisle to the other on each side. The total building height should not be more that t feet, including a clearance of u feet between the top of the racks and the ceiling. Aisle width will be 150% of pallet depth to allow for clear passage of pallets carried by the cranes, and their length should accommodate the k storage cells plus one extra pallet width to provide for an input/output station at the end. The ASAR area and its input/output stations will be enclosed with a roof and walls on three sides, being open only on the end (perpendicular to the aisles) where it adjoins the existing warehouse. Engineers want to find a minimum total cost design for the new facility using $c_1 \triangleq$ unit cost of cranes, $c_2 \triangleq$ unit cost of steel racks per pallet storage cell, together with c_3, c_4, and c_5 being the construction cost per square of foundation/floors, ceilings, and sidewalls, respectively. Formulate an NLP model to choose an optimal design in terms of decision variables n, m, k, and building exterior dimensions x (perpendicular to the aisles) y (parallel to the aisles) and z (height). Take all other symbols as constant.

14-15 Assume that Syntex Laboratories[7] is reexamining the distribution of its sales force across major pharmaceutical products $j = 1, \ldots, 7$. Present force levels e_j are expected to produce s_j units of product sales per month at a profit margin of p_j per unit. However, extensive discussion and surveying has quantified impacts of changing the effort dedicated to different products. Nonlinear functions $r_j(x_j)$ predict the ratio of future to current product sales quantities as a function of $x_j \triangleq$ ratio of future to current sales effort. Formulate an NLP model in terms of decision variables x_j to determine a maximum total profit realignment of salesforce distribution keeping the total force unchanged and the force devoted to each product within $\pm 50\%$ of the current levels.

14-16 A major oil company[8] manufactures petroleum lubricants at sites $j = 1, \ldots, 10$ using a critical additive purchased from suppliers $i = 1, \ldots, 15$ all over the world. Manufacturing site j requires d_j metric tons per month of the additive, and suppliers i can provide up to s_i metric tons. The transportation cost for shipping additive from supplier i to manufacturing site j is a constant $t_{i,j}$ per ton. However, the purchase cost from the supplier varies with volume. Starting with a base price c_i, supplier i reduces the unit price for all purchases from i each month by a fraction $\alpha_i > 0$ for each multiple of quantity q_i ordered. For example, the reduction might be fraction $0.5\% = 0.005$ off the base price for each 1000 metric tons ordered per month. Formulate an NLP model to find a minimum total cost procurement and shipping plan using the decision variables $(i = 1, \ldots, 15; j = 1, \ldots, 10)$

$x_{i,j} \triangleq$ metric tons shipped from supplier i to site j

$y_i \triangleq$ total metric tons purchased from supplier i

14-17 A stirred tank reactor[9] is a tank equipped with a large stirring device that is used in the chemical and

[6]Based on J. Ashayeri, L. Gelders, and L. Van Wassenhove (1985), "A Microcomputer-Based Optimisation Model for Design of Automated Warehouses," *International Journal of Production Research, 23*, 825–839.

[7]Based on L. M. Lodish, E. Curtis, M. Ness, and M. K. Simpson (1988), "Sales Force Sizing and Deployment Using a Decision Calculus Model at Syntex Laboratories," *Interfaces, 18:1*, 5–20.

[8]Based on P. Ghandforoush and J. C. Loo (1992), "A Non-linear Procurement Model with Quantity Discounts," *Journal of the Operational Research Society, 43*, 1087–1093.

[9]Based on L. Ong (1988), "Hueristic Approach for Optimizing Continuous Stirred Tank Reactors in Series Using Michaelis–Menten Kinetics," *Engineering Optimization, 14*, 93–99.

biochemical industry to produce chemical reactions. A series of 5 such tanks will be used to lower the concentration of toxic chemical from c_0 at input for tank 1, to no more that $\bar{c}$ on exit from tank 5. A flow rate of q liters per minute will be maintained through the tank sequence, but the effect of each tank depends on its volume v_i. In particular, the units of toxic chemical removed per minute will be approximately

$$\gamma \frac{\text{output concentration}}{1 + \text{output concentration}}$$

times the tank volume. Stirred tank cost can also be estimated from the volume at $\alpha(\text{volume})^\beta$. Formulate an NLP model to find a minimum total cost sequence of tanks using the decision variables ($i = 1, \ldots, 5$)

$$c_i \triangleq \text{output concentration of tank } i$$
$$v_i \triangleq \text{size of tank } i$$

Assume that all other symbols are constants.

14-18 Each day q_i tons of freight arrive by sea[10] in Japan bound for in-country regions $i = 1, \ldots, 50$. These goods may arrive at any of the major ports $j = 1, \ldots, 17$, but the internal transportation cost per ton $c_{i,j}$ varies by port and destination. The government plans a capital investment program in port facilities to pick daily tonnage processing capacities at each port j that minimize these internal transportation costs plus associated port maintenance costs, plus delay costs from freight passing through each port. Port j maintenance costs can be expressed $a_j(\text{capacity } j)^{b_j}$, where a_j and b_j and known constants. Delay cost at j can be estimated $d/[(\text{capacity } j) - (\text{traffic through } j)]$, where d is the delay cost per ton per day. Formulate an NLP model to optimize the ports using the decision variables ($i = 1, \ldots, 50; j = 1, \ldots, 17$)

$$x_{i,j} \triangleq \text{tons shipped through port } j \text{ for } i$$
$$x_j \triangleq \text{total tons shipped through port } j$$
$$y_j \triangleq \text{capacity of port } j$$

14-19 Three urban neighborhoods are mutually connected by freeways admitting traffic in both directions. Net output $b_{i,k}$ (per hour) at each neighborhood k of vehicles originating at i can be estimated

from known patterns ($b_{i,i} = -\sum_{k \neq i} b_{i,k}$). The delay vehicles experience on any arc (i, j) is an increasing nonlinear function $d_{i,j}$ of the total flow on that arc reflecting the number of lanes and other characteristics of the road link. Formulate an NLP to compute a "system optimal" traffic flow (i.e., one that minimizes the total delay experienced in carrying the required traffic), using the decision variables ($i, j, k = 1, \ldots, 3$)

$$x_{i,j,k} \triangleq \text{flow on arc } (i, j) \text{ bound for node } k$$

(Observe that this may not be the same as a "user optimal" flow where each driver tries to minimize his or her own delay.)

14-20 The commander of a battlefront[11] must plan how to employ his f frontline and r reserve firepower to minimize the advance achieved over days $t = 1, \ldots, 14$ by an attack of opposing forces with firepower a. Intelligence and battle simulations predict that each surviving unit of firepower in the attacking force will kill p units of defender firepower per day, and each unit of defender firepower committed to the battle will kill q units of attacker per day. Also, the kilometer advance on day t can be estimated in terms of the ratio of forces fighting that day as

$$\exp\left[-4\left(\frac{\text{defender firepower}}{\text{attacker firepower}}\right)^2\right]$$

Reserves will not be in the battle on day 1, but they may be committed as desired over the ensuing days. Once committed, reserves cannot be withdrawn from the battle. Formulate an NLP model to choose an optimal plan for the defender using the decision variables ($t = 1, \ldots, 14$)

$$x_t \triangleq \text{attacker firepower fighting on day } t$$
$$y_t \triangleq \text{defender firepower fighting on day } t$$
$$z_t \triangleq \text{defender reserve firepower newly committed on day } t$$

Assume that forces are only lost to enemy fire.

14-21 Chilled-water building cooling systems[12] operate as indicated in the following sketch.

---

[10]Based on M. Noritake and S. Kimura (1990), "Optimum Allocation and Size of Seaports," *Journal of Waterway, Port, Coastal and Ocean Engineering, 116*, 287–299.

[11]Based on K. Y. K. Ng and M. N. Lam (1995), "Force Deployment in a Conventional Theatre-Level Military Engagement," *Journal of the Operational Research Society, 46*, 1063–1072.

[12]Based on R. T. Olson and J. S. Liebman (1990), "Optimization of a Chilled Water Plant Using Sequential Quadratic Programming," *Engineering Optimization, 15*, 171–191.

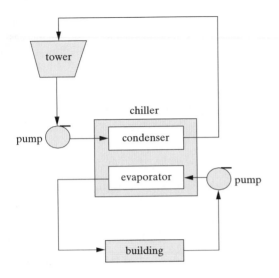

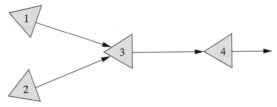

Water flows at a rate of F_1 gallons per minute around the lower loop, entering the chiller at temperature $T_{1,1}$ and being cooled to temperature $T_{1,2}$ before passing through the building. An upper loop flowing at rate F_2 gallons per minute absorbs heat within the chiller in a separate water stream, and passes it through an outdoor cooling tower to reduce its temperature from $T_{2,1}$ to $T_{2,2}$. Tower output temperature $T_{2,2}$ is a nonlinear function f_1 of $T_{2,1}$, F_2, the tower fan speed S, and the ambient air temperature T_0. The lower loop must absorb a load H of heat within the building (measured in water temperature difference times flow). Similarly, the heat exchanged between the two loops should balance within the chiller after adjusting for extra heat generated by the unit and its pumps, which is a constant multiple k of the electrical energy they consume. That energy is, in turn, a nonlinear function f_2 of all four temperatures and both flow rates. Energy consumed in the cooling tower is another nonlinear function f_3 of fan speed. On any given day, temperature T_0, heat load H, constant k, and all 3 functions will be known, but other system variables can be controlled within ranges $[\underline{T}_{i,j}, \overline{T}_{i,j}]$, $[\underline{F}_i, \overline{F}_i]$, and $[\underline{S}, \overline{S}]$, respectively. Formulate an NLP model to decide how to operate such a system at minimum total electrical consumption.

14-22 The figure below shows a system of reservoirs and hydroelectric dams of the sort operated by large utilities such as California's PG&E.[13]

Each node is a reservoir with a power plant releasing water on the downstream side. The water then requires one month to reach the reservoir of the next dam on the river system. Acre-feet inflows $b_{i,t}$ from streams feeding each reservoir i (other than the indicated rivers) can be estimated for coming months $t = 1, \ldots, 3$, and reservoirs begin month 1 with storage $s_{i,0}$. Lower and upper bounds $\underline{s}_i$ and $\overline{s}_i$ restrict the number of acre-feet of water that should be stored in each reservoir i at all times, and similar bounds $\underline{f}_i$ and $\overline{f}_i$ limit the release flow through the dams. The hydropower obtained from any dam i in any month can be estimated by nonlinear function $h_i(s, f)$ reflecting both the flow f through the dam that month and the water pressure, which is a consequence of the amount of water s stored in the corresponding reservoir at the end of the month. Managers would like to operate the system to maximize the total power produced.

(a) Sketch a time-expanded flow network for this problem with arcs $f_{i,t}$ reflecting monthly release flows at i, and other arcs $s_{i,t}$ reflecting ending storage there at month t. Label arcs with lower and upper bounds, and nodes with net inflows. Assume that all releases $f_{i,0} = 0$.

(b) Formulate an NLP to maximize the total power produced in your network of part (a).

14-23 Determine whether each of the following NLP's is a convex program.

(a) ⊗   max $\ln(x_1) + 3x_2$
       s.t. $x_1 \geq 1$
           $2x_1 + 3x_2 = 1$
           $(x_1)^2 + (x_2)^2 \leq 9$

(b)   min $x_1 + x_2$
     s.t. $x_1 x_2 \leq 9$
         $-5 \leq x_1 \leq 5$
         $-5 \leq x_2 \leq 5$

[13]Based on Y. Ikura, G. Gross, and G. S. Hall (1986), "PG&E's State-of-the-Art Scheduling Tool for Hydro Systems," *Interfaces 16:1*, 65-82.

(c) ◇ max $x_1 + 6/x_1 + 5(x_2)^2$
  s.t. $4x_1 + 6x_2 \leq 35$
      $x_1 \geq 5, x_2 \geq 0$

(d) min $14x_1 + 9x_2 - 7x_3$
  s.t. $6x_1 + 2x_2 \leq 20$
      $3x_2 + 11x_3 \leq 25$
      $x_1, x_2, x_3 \geq 0$

(e) ◇ min $e^{x_1+x_2} - 28x_2$
  s.t. $(x_1 - 3)^2 + (x_2 - 5)^2 \leq 4$
      $14x_1 - 6x_2 = 12$
      $-2(x_1)^2 + 2x_1x_2 - (x_2)^2 \geq 0$

(f) max $62x_1 + 123x_2$
  s.t. $\ln(x_1) + \ln(x_2) = 8$
      $7x_1 + 2x_2 \leq 900$
      $x_1, x_2 \geq 1$

14-24 ◇ Determine which of the NLPs in Exercise 14-23 are separable programs.

14-25 Determine whether each of the following NLPs is a quadratic program, and if so, identify the **c** and **Q** of matrix objective function form $\mathbf{c} \cdot \mathbf{x} + \mathbf{x}\mathbf{Q}\mathbf{x}$.

(a) ◇ min $x_1x_2 + 134/x_3 + \ln(x_1)$
  s.t. $x_1 + 4x_2 - x_3 \leq 7$
      $14x_1 + 2x_3 = 16$
      $x_1, x_2, x_3 \geq 0$

(b) min $12(x_1)^2 + 34x_1x_2 + 5(x_2)^2$
      $-12x_1 + 19x_2$
  s.t. $7x_1 + 3x_2 \geq 15$
      $93x_1 + 27x_2 + 11x_3 \leq 300$
      $x_3 \geq 0$

(c) ◇ max $2x_1x_2 + (x_2)^2 + 9x_3$
  s.t. $x_1 + x_2 + x_3 \geq 6$
      $x_j \leq 5 , j = 1, \ldots, 3$

(d) max $(x_1)^2 + (x_2)^2$
  s.t. $(x_1 - 10)^2 + (x_2 - 4)^2 \leq 9$
      $x_1, x_2 \geq 0$

14-26 Determine whether each of the following is a posynomial.

(a) ◇ $23x_1 - 34x_2 + 60x_3$
(b) $54x_1 + 89x_2 + 52x_3$
(c) ◇ $7x_1x_2/(x_3)^{2.3} + 4\sqrt{x_1}$
(d) $44x_1/\ln(x_2) + e^{-x_3}$

14-27 Demonstrate that each of the following NLPs is a posynomial geometric program by placing the model in standard form and detailing the sets K_i, and associated coefficients d_k and $a_{k,j}$.

(a) ◇ min $13x_1x_2/x_3 + 9\sqrt{x_1x_3}$
  s.t. $3x_1 + 8x_2 \leq x_3$
      $20/(x_3)^4 \leq 4$
      $x_1, x_2, x_3 > 0$

(b) min $40/x_1 + x_2/\sqrt{x_3}$
  s.t. $x_1x_2 \leq (x_3)^2$
      $18x_1 + 14x_2 \leq 2$
      $x_1, x_2, x_3 > 0$

14-28 Consider the nonlinear program

  min $8(x_1 - 2)^2 + 2(x_2 - 1)^2$
  s.t. $32x_1 + 12x_2 = 126$

(a) ◇ Form the Lagrangian function for this model.
(b) ◇ Write stationary conditions for the Lagrangian.
(c) ◇ Solve your stationary conditions for x_1 and x_2, and explain why your answers are optimal in the original model.
(d) Explain why a constraint $x_1 \leq 2$ would be active if added to the original model.
(e) ◇ Use the Lagrangian approach to compute an optimal solution to the model with the added constraint of part (d).

14-29 Do Exercise 14-28 for the NLP

  max $300 - 5(x_1 - 20)^2 - 4(x_2 - 6)^2$
  s.t. $x_1 + x_2 = 8$

and part (d) extra constraint $x_2 \geq 0$.

14-30 State the Karush–Kuhn–Tucker optimality conditions for each of the following mathematical programs.

(a) ◇ min $14(x_1 - 9)^2 + 3(x_2 - 5)^2 + (x_3 - 11)^2$
  s.t. $2x_1 + 18x_2 - x_3 = 19$
      $6x_1 + 8x_2 + 3x_3 \leq 20$
      $x_1, x_2 \geq 0$

(b) max $6x_1 + 40x_2 + 5x_3$
  s.t. $x_1 \sin(x_2) + 9x_3 \leq 2$
      $e^{18x_1+3x_2} + 14x_3 \geq 50$
      $x_2, x_3 \geq 0$

(c) ◇ min $100 - (x_1 - 3)^2 - (x_2)^4 + 19x_3$
  s.t. $5(x_1 - 1)^2 + 30(x_2 - 2)^2 \geq 35$
      $60x_2 + 39x_3 = 159$
      $x_1, x_2, x_3 \geq 0$

(d) max $7\ln(x_1) + 4\ln(x_2) + 11\ln(x_3)$
  s.t. $(x_1 + 2)^2 - x_1x_2 + (x_2 - 7)^2 \leq 80$
      $5x_1 + 7x_3 = 22$
      $x_1, x_2, x_3 \geq 1$

14-31 ◇ For each mathematical program in Exercise 14-30, determine whether principle 14.26 assures that a KKT point is a global optimum.

14-32 Consider the NLP

$$\min \quad 15(x_1)^2 + 4(x_2)^2$$
$$\text{s.t.} \quad 3x_1 + 2x_2 = 8$$
$$x_1, x_2 \geq 0$$

(a) ⊘ State the KKT optimality conditions for this model.
(b) Verify that at solution $\mathbf{x} = (0, 4)$ there exists an improving feasible direction $\Delta \mathbf{x} = (2, -3)$.
(c) Confirm that KKT conditions have no solution for the nonoptimal $\mathbf{x}$ of part (b).
(d) Explain why every local optimum of the model must be a KKT point.
(e) ⊘ Show that global optimal solution $\mathbf{x}^* = (1, \frac{5}{2})$ is a KKT point.

14-33 Do Exercise 14-32 for NLP

$$\max \quad 2\ln(x_1) + 8\ln(x_2)$$
$$\text{s.t.} \quad 4x_1 + x_2 = 8$$
$$x_1, x_2 \geq 1$$

with nonoptimal point $\mathbf{x} = (\frac{3}{2}, 2)$, improving feasible direction $\Delta \mathbf{x} = (-1, 4)$, and global optimum $\mathbf{x}^* = (1, 4)$.

14-34 ⊘ Use absolute value (unsquared) penalty functions to reduce each NLP of Exercise 14-30 to an unconstrained penalty model.

14-35 ⊘ Do Exercise 14-34 using squared penalty functions.

14-36 Consider the NLP

$$\min \quad 2(x_1 - 3)^2 - x_1 x_2 + (x_2 - 5)^2$$
$$\text{s.t.} \quad (x_1)^2 + (x_2)^2 \leq 4$$
$$0 \leq x_1 \leq 2, \; x_2 \geq 0$$

with optimal solution $\mathbf{x}^* = (1.088, 1.678)$.

(a) ⊘ Use unsquared penalty functions to reduce this problem to an unconstrained penalty model.
(b) ⊘ Explain why local minima of the unconstrained model in part (a) must be global minima for all $\mu \geq 0$.
(c) ⊘ Determine whether the penalty objective of part (a) is differentiable. Explain.
(d) ⊘ Determine whether there will be a penalty multiplier μ large enough that the unconstrained optimum in part (a) is optimal in the original model. Explain.
(e) ⊘ Suppose that we are solving the given constrained NLP by the sequential unconstrained penalty Algorithm 14A. Explain why it is reasonable to begin with multiplier $\mu = 0.5$ and

increase it by a factor $\beta = 2$ after each unconstrained optimization.
(f) ⊘ 💬 Use class optimization software to apply Algorithm 14A, starting at $\mathbf{x}^{(0)} = (3, 5)$ and managing the penalty multiplier as in part (e).

14-37 ⊘ Do Exercise 14-36 using squared penalty functions. Stop the search in part (f) when the total constraint violation is ≤ 0.2.

14-38 Do Exercise 14-36 for the NLP

$$\max \quad 100 - 8(x_1)^2 - 3(x_2 - 3)^2$$
$$\text{s.t.} \quad x_2 \geq 2/x_1$$
$$0 \leq x_1 \leq 2$$
$$0 \leq x_2 \leq 2$$

Start at $\mathbf{x}^{(0)} = (2, 2)$ with multiplier $\mu = 0.5$, and increase by the factor $\beta = 4$.

14-39 Do Exercise 14-38 using squared penalty functions. Stop the search in part (f) when total constraint violation ≤ 0.2.

14-40 ⊘ Determine whether barrier methods can be applied to each of the NLPs in Exercise 14-23, and if so, use log barrier functions to reduce the constrained optimization model to an unconstrained barrier model.

14-41 ⊘ Do Exercise 14-40 using reciprocal barrier functions.

14-42 Consider solving the NLP of Exercise 14-36 by barrier methods.

(a) ⊘ Use logarithmic barrier functions to reduce this problem to an unconstrained barrier model.
(b) ⊘ Explain why local minima of the unconstrained model in part (a) for all $\mu \geq 0$ must be global minima.
(c) ⊘ Determine whether the barrier objective of part (a) is differentiable. Explain.
(d) ⊘ Determine whether there will be a barrier multiplier $\mu > 0$ small enough that the unconstrained optimum in part (a) is optimal in the original model. Explain.
(e) ⊘ Suppose that we are solving the given constrained NLP by the sequential unconstrained barrier Algorithm 14B. Explain why it is reasonable to begin with multiplier $\mu = 2$ and decrease it by a factor $\beta = \frac{1}{4}$ after each unconstrained optimization.
(f) ⊘ 💬 Use class optimization software to apply Algorithm 14B, starting at $\mathbf{x}^{(0)} = (3, 5)$ and managing the barrier multiplier as in part (e). Proceed while $\mu \geq \frac{1}{32}$.

14-43 ⊘ Do Exercise 14-42 using reciprocal barrier functions.

14-44 Do Exercise 14-42 for the NLP of Exercise 14-38. Start at $\mathbf{x}^{(0)} = (1.8, 1.8)$ with multiplier $\mu = 8$, and decrease with factor $\beta = \frac{1}{4}$.

14-45 Do Exercise 14-44 using reciprocal barrier functions.

14-46 Consider the nonlinear program

$$\min \quad (x_1 - 8)^2 + 2(x_2 - 4)^2$$
$$\text{s.t.} \quad 2x_1 + 8x_2 \leq 16$$
$$x_1 \leq 7$$
$$x_1, x_2 \geq 0$$

(a) ⊘ Introduce slack variables x_3 and x_4 to place the model in standard form for reduced gradient Algorithm 14C.

(b) ⊘ Show that x_2 and x_4 form a basic set of variables for your standard form.

(c) ⊘ Assuming the basis of part (b), classify variables as basic, nonbasic, or superbasic at initial solution $\mathbf{x}^{(0)} = (0, 1, 8, 7)$.

(d) ⊘ Compute the reduced gradient corresponding to the basis and $\mathbf{x}^{(0)}$ of part (c).

(e) ⊘ Construct the move direction that would be pursued by Algorithm 14C at the basis and $\mathbf{x}^{(0)}$ of part (c).

(f) ⊘ Compute the maximum feasible step λ in the direction of part (e). Then, assuming (correctly) that the direction of part (e) remains improving all the way to the maximum λ, compute the resulting new solution $\mathbf{x}^{(1)}$.

(g) ⊘ Explain why a basis change would by required by Algorithm 14C at the $\mathbf{x}^{(1)}$ of part (f), and choose an appropriate new basis.

14-47 ⊞ Return to the standard form NLP of Exercise 14-46(a).

(a) ⊘ Apply reduced gradient Algorithm 14C to compute an optimal solution starting from the $\mathbf{x}^{(0)} = (0, 1, 8, 7)$.

(b) Graph your progress in a plot of the feasible (x_1, x_2).

14-48 Do Exercise 14-46 for nonlinear program

$$\max \quad 500 - 3(x_1 + 1)^2 + 2x_1x_2 - (x_2 - 10)^2$$
$$\text{s.t.} \quad x_1 - x_2 \leq 1$$
$$x_2 \leq 5$$
$$x_1, x_2 \geq 0$$

using basis $\{x_1, x_4\}$ and standard-form starting solution $\mathbf{x}^{(0)} = (2, 1, 0, 4)$.

14-49 ⊞ Do Exercise 14-47 on the standard-form NLP of Exercise 14-48(a).

14-50 Consider the equality-constrained quadratic program

$$\min \quad 6(x_1)^2 + 2(x_2)^2 - 6x_1x_2 + 4(x_3)^2$$
$$+ 5x_1 + 15x_2 - 16x_3$$
$$\text{s.t.} \quad x_1 + 3x_2 - 2x_3 = 2$$
$$3x_1 - x_2 + x_3 = 3$$

(a) ⊘ Identify the $\mathbf{Q}$, $\mathbf{c}$, $\mathbf{A}$, and $\mathbf{b}$ of (symmetric) quadratic program standard form.

(b) ⊘ State Karush–Kuhn–Tucker optimality conditions for the model as a system of linear equalities.

(c) ⊘ ⊞ Solve your system of part (b) for the unique KKT point of the model.

14-51 Do Exercise 14-50 for the equality-constrained quadratic program

$$\max \quad -(x_1)^2 - 8(x_2)^2 - 2(x_3)^2 + 10x_2x_3$$
$$+ 8x_1 - 14x_2 + 20x_3$$
$$\text{s.t.} \quad x_1 + 4x_3 = 2$$
$$-x_2 + 3x_3 = 1$$

14-52 ⊞ Return to the NLP of Exercise 14-46, and consider solving by active set Algorithm 14D starting from solution $\mathbf{x}^{(0)} = (0, 1)$.

(a) ⊘ Demonstrate that the model is a quadratic program by deriving the c_0, $\mathbf{c}$, $\mathbf{Q}$, $\mathbf{a}^{(1)}, \ldots, \mathbf{a}^{(4)}$, $b_1, \ldots, b_4$, G, L, and E of general symmetric form $\boxed{14.50}$.

(b) ⊘ State and solve as a system of linear equalities the active set optimality conditions $\boxed{14.53}$ corresponding to initial solution $\mathbf{x}^{(0)}$.

(c) ⊘ Determine the step λ that would be applied to the direction resulting from part (b), and compute the new point $\mathbf{x}^{(1)}$.

(d) ⊘ Verify by forming and solving the optimality conditions $\boxed{14.53}$ corresponding to $\mathbf{x}^{(1)}$ that no further progress can be made if all inequalities active at $\mathbf{x}^{(1)}$ of part (c) are included in the active set S.

(e) ⊘ Use the results of part (d) to show which active constraint should be dropped from S.

(f) ⊘ Begin from part (e) and complete the solution of this quadratic program.

(g) Graph your progress in a plot of the feasible (x_1, x_2).

(h) Compare the evolution of active set Algorithm 14D in part (g) with corresponding reduced gradient Algorithm 14C computations in Exercise 14-47.

14-53 ⊞ Do Exercise 14-52 for the NLP of Exercise 14-48 starting from solution $\mathbf{x}^{(0)} = (2, 1)$.

14-54 Form linear programming approximations 14.59 to each of the following separable programs using breakpoints $u_{1,0} = 0$, $u_{1,1} = 1$, $u_{1,2} = 3$, $u_{2,0} = 0$, $u_{2,1} = 2$, $u_{2,2} = 4$.

(a) ◇ $\min \quad x_1/(4 - x_1) + (x_2 - 1)^2$
$\text{s.t.} \quad 2x_1 + x_2 \geq 2$
$\qquad 4(x_1 + 1)^3 - 9(x_2)^2 \leq 25$
$\qquad 0 \leq x_1 \leq 3$
$\qquad 0 \leq x_2 \leq 4$

(b) $\max \quad 500 - (x_1 - 1)^2 - 25/(x_2 + 1)$
$\text{s.t.} \quad \sqrt{x_1} - (x_2 + 1)^2 \geq -3$
$\qquad 6x_1 + 2x_2 \leq 10$
$\qquad 0 \leq x_1 \leq 3$
$\qquad 0 \leq x_2 \leq 4$

14-55 Consider the trivial separable program

$$\min \quad 2(x - 3)^2$$
$$\text{s.t.} \quad 0 \leq x \leq 6$$

(a) Verify that the model is a convex program.
(b) ◇ Verify by inspection that an optimal solution occurs at $x^* = 3$.
(c) ◇ Form a linear programming approximation 14.59 using $u_0 = 0$, $u_1 = 2$, $u_2 = 6$.
(d) ◇ Solve your LP approximation of part (c) by inspection, and determine whether correctness condition 14.60 is satisfied by the approximate optimum.
(e) Discuss how convexity of part (a) relates to correct sequencing 14.60 in part (d).
(f) Verify by inspection that $x^* = 0$ and $x^* = 6$ are alternative optima in the original NLP when the objective is maximized instead of minimized.
(g) ◇ Repeat part (d), this time maximizing the objective function.
(h) Comment on the errors introduced in objective function and other values when sequencing condition 14.60 is violated in part (g).

14-56 Consider the standard-form posynomial geometric program

$$\min \quad 3/\sqrt{x_1} + x_1 x_2 + 10/(x_3)^3$$
$$\text{s.t.} \quad 0.5 x_1 x_2/(x_3)^2 \leq 1$$
$$0.167 x_1 + 0.25(x_1)^{0.4} x_2 + 0.0833 x_3 \leq 1$$
$$x_1, x_2, x_3 > 0$$

(a) ⊘ Change variables to convert this geometric program into a convex program.
(b) ⊘ ▱ Use class optimization software to solve your convex program of part (a) and transform optimal variable values back to obtain an optimal solution for the original NLP.
(c) ⊘ Form the geometric programming dual of the original NLP.
(d) ⊘ Determine the degree of difficulty of the original NLP.
(e) ⊘ ▱ Use class optimization software to solve the dual of part (c) and retrieve an optimal primal solution from the corresponding Lagrange multipliers.

14-57 Do Exercise 14-56 for the posynomial geometric program

$$\min \quad 10/(x_1 x_2 x_3)^2$$
$$\text{s.t.} \quad 12(x_1)^2 x_2 + 4x_3 \leq 1$$
$$0.1 x_2 \sqrt{x_1} + x_2 x_3 \leq 1$$
$$(x_1 x_2)^{0.333} \leq 1$$
$$x_1, x_2, x_3 > 0$$

14-58 A water distribution system[14] is a network with (positive = forward or negative = reverse) flows $x_{i,j}$ in pipes between nodes $i, j = 0, \ldots, m$ representing storage tanks and pipe intersections. Pressures at the nodes i can be measured in hydraulic "head," which is the height to which water will rise in an open-ended vertical pipe installed at the node, relative to the "ground node" 0. Heads have assigned values s_i for storage tank nodes i, and net outflows r_i are established for all nodes ($\sum_{i=0}^{m} r_i = 0$). The ground node 0 is connected to each storage node i by an arc $(0, i)$, and to no others. To determine how the system will perform at steady state, engineers need to find flows $f_{i,j}$ and heads h_i that (i) maintain net flow balance at every node, (ii) achieve assigned heads s_i at storage nodes, and (iii) satisfy nonlinear head-to-flow equations

$$h_j - h_i = \phi_{i,j}(f_{i,j}) \qquad \text{nonground } (i, j)$$

---

[14]Based on M. Collins, L. Cooper, R. Helgason, J. Kennington, and L. LeBlanc (1978), "Solving the Pipe Network Analysis Problem Using Optimization Techniques," *Management Science, 24*, 747–760.

where functions $\phi_{i,j}(x_{i,j})$ are known relations between head difference and the flow on particular arcs (i, j) that reflect length, size, pumping, grade, and other characteristics.

(a) Formulate a related NLP over unrestricted flows $x_{i,j}$ having only flow balance constraints at all nodes and a minimizing objective function summing terms

$$f_{i,j}(x_{i,j}) \begin{cases} s_j x_{0,j} & \text{arcs } (0, j) \\ \int_0^{x_{i,j}} \phi_{i,j}(z)\, dz & \text{other } (i, j) \end{cases}$$

(b) Explain why your NLP of part (a) is a separable program. (With mild assumptions on the $\phi_{i,j}$ it can also be shown to be convex.)

(c) State Karush–Kuhn–Tucker conditions for the primal model of part (a) and explain why they must be satisfied by a locally optimal $\mathbf{x}^*$.

(d) Interpret conditions of part (c) to show that a solution to steady-state equation system (i)–(iii) above can be obtained from locally optimal flows $\mathbf{x}^*$ in part (a) and corresponding KKT multipliers $\mathbf{v}^*$.

SUGGESTED READING

Bazarra, Mokhtar S., Hanif D. Sherali, and C. M. Shetty (1993), *Nonlinear Programming Theory and Algorithms*, Wiley, New York.

Luenberger, David G. (1984), *Linear and Nonlinear Programming*, Addison-Wesley, Reading, Mass.

Selected Answers

• • • • • • • • • • • • • • • • • • • •

Chapter 1

1-1 **(a)** s **(b)** d, p, and b **(c)** $\min(d/s)^2$ **(d)** $ps \leq b, s$ nonnegative and integer

1-2 **(a)** feasible and optimal **(b)** neither because infeasible **(c)** feasible not optimal

1-5 **(a)** exact numerical optimization **(c)** closed-form optimization

1-8 **(b)** 16.4 hours; 166.1 days; 110.6 years; 6.5 million years

1-9 **(a)** random variable **(c)** deterministic **(e)** deterministic **(g)** random variable **(i)** deterministic

Chapter 2

2-1 **(a)** $\max 200x_1 + 350x_2$, s.t. $5x_1 + 5x_2 \leq 300$, $0.6x_1 + 1.5x_2 \leq 63$, $x_1 \leq 50$, $x_2 \leq 35$, $x_1 \geq 0$, $x_2 \geq 0$ **(b)** $x_1^* = $ basic $= 30$, $x_2^* = $ deluxe $= 30$ **(d)** all optimal from $\mathbf{x} = (30, 30)$ to $\mathbf{x} = (17.5, 35)$

2-2 **(b)** $x_1^* = $ domestic $= \$5$ million, $x_2^* = $ foreign $= \$7$ million

2-3 **(b)** $x_1^* = $ Squawking Eagle $= 40$ thousand, $x_2^* = $ Crooked Creek $= 10$ thousand

2-4 **(b)** $x_1^* = $ beef $= 25$g, $x_2^* = $ chicken $= 100$ g

2-5 **(b)** $v^* = 7000$, $c^* = 0$

2-6 **(a)** $\min x_1 + x_2$, s.t. $5x_1 + 3x_2 \geq 15$, $2x_1 + 5x_2 \geq 10$, $0 \leq x_1 \leq 4$, $0 \leq x_2 \leq 4$, x_1, x_2 integer **(b)** partial patterns make no physical sense **(c)** Either $x_1^* = x_2^* = 2$, or $x_1^* = 3$, $x_2^* = 1$

2-7 **(a)** $\min 16x_1 + 16x_2$, s.t. $x_1 x_2 = 500$, $x_1 \geq 2x_2$, $x_2 \leq 15$, $x_1 \geq 0$, $x_2 \geq 0$ **(b)** $x_1^* = $ length $= 33\frac{1}{3}$ feet, $x_2^* = $ width $= 15$ feet **(d)** $x_1 \leq 25$ leaves no feasible

2-8 **(b)** $x_1^* = $ diameter $= 78.16$ feet, $x_2^* = $ floors $= 31.26$

2-9 **(a)** $\max w_1$ **(b)** $\max 5w_1 + 2w_2$ **(c)** $\max w_2$

2-11 **(a)** $\min \sum_{i=3}^{4} i \sum_{j=1}^{2} y_{i,j}$ **(c)** $\max \sum_{i=1}^{p} \alpha_i y_{i,4}$ **(e)** $\sum_{j=1}^{4} y_{i,j} = s_i$, $i = 1, \ldots, 3$

2-12 **(a)** $\sum_{i=1}^{17} x_{i,j,t} \leq 200$, $j = 1, \ldots, 5$, $t = \ldots, 7$; 35 constraints **(b)** $\sum_{j=1}^{5} \sum_{t=1}^{7} x_{5,j,t} \leq 4000$; 1 constraint **(c)** $\sum_{j=1}^{5} x_{i,j,t} \geq 100$, $i = 1, \ldots, 17$; $t = 1, \ldots, 7$; 119 constraints

2-16 **(a)** $f(y_1, y_2, y_3) \triangleq (y_1)^2 y_2/y_3$, $g_1(y_1, y_2, y_3) \triangleq y_1 + y_2 + y_3$, $b_1 = 13$, $g_2(y_1, y_2, y_3) \triangleq 2y_1 - y_2 + 9y_3$, $b_2 = 0$, $g_3(y_1, y_2, y_3) \triangleq y_1$, $b_3 = 0$, $g_4(y_1, y_2, y_3) \triangleq y_3$, $b_4 = 0$

2-17 **(a)** linear **(c)** nonlinear **(e)** nonlinear **(g)** nonlinear

2-18 **(a)** LP **(c)** NLP

2-19 **(a)** continuous **(c)** discrete

2-20 **(a)** $\sum_{j=1}^{8} x_j = 3$ **(c)** $x_3 + x_8 \leq 1$

2-21 **(a)** $\max 85x_1 + 70x_2 + 62x_3 + 93x_4$, s.t. $700x_1 + 400x_2 + 300x_3 + 600x_4 \leq 1000$, $x_j = 0$ or 1, $j = 1, \ldots, 4$ **(b)** fund 2 and 4 (i.e., $x_1^* = x_3^* = 0$, $x_2^* = x_4^* = 1$)

2-22 **(b)** build 2 and 4 (i.e., $y_1^* = y_3^* = 0$, $y_2^* = y_4^* = 1$)

2-23 **(a)** ILP **(c)** INLP **(e)** INLP

2-24 **(a)** model (b) **(c)** model (d)

2-26 **(b)** nonzeros: $x_5^* = 1000$, $x_{12}^* = 15,000$

2-27 **(i)** $x_1^* = x_2^* = x_3^* = 1100$, $x_4^* = x_6^* = 1500$, $x_5^* = 1400$, $x_7^* = 400$, $x_8^* = x_{10}^* = 0$, $x_9^* = 1900$

2-28 **(h)** $x_1^* = x_3^* = x_6^* = x_7^* = 1$

2-29 **(h)** nonzeros: $x_{1,1}^* = 81$, $x_{1,2}^* = 93$, $x_{1,3}^* = 166$, $x_{1,5}^* = 90$, $x_{1,6}^* = 85$, $x_{1,7}^* = 145$, $x_{2,2}^* = 301$, $x_{3,1}^* = 166$, $x_{3,4}^* = 105$, $x_{4,3}^* = 99$

2-32 **(g)** $x_2^* = x_3^* = x_4^* = 1$, others $= 0$

2-34 **(h)** nonzero values: $x_{4,2}^* = 115$, $x_{4,3}^* = 165$, $x_{5,1}^* = 85$, $x_{5,3}^* = 225$

2-36 (i) $x_{10}^* = 50, x_{15}^* = 25, x_{20}^* = 5, y_1^* = 5000,$
$y_2^* = 8500$

2-38 (i) nonzeros:
$x_{2,2}^* = x_{2,4}^* = x_{3,1}^* = x_{3,3}^* = x_{5,5}^* = 1, y_2^* = y_3^* = y_5^* = 1$

Chapter 3

3-1 (a) feasible and local max; infeasible; feasible; feasible, local and global max

3-2 (a) $\mathbf{y}^{(1)} = (8, -2, 5), \mathbf{y}^{(2)} = (3, 8, 10),$
$\mathbf{y}^{(3)} = (3, 11, 10)$

3-3 (a) $\Delta\mathbf{w}^{(1)} = (4, -2, 6), \Delta\mathbf{w}^{(2)} = (0, -2, 8),$
$\Delta\mathbf{w}^{(3)} = (-1, 0, 3)$

3-4 (a) nonimproving **(c)** improving
(e) nonimproving

3-5 (a) feasible **(c)** feasible **(e)** infeasible

3-6 (a) $\lambda = 3$; not unbounded **(c)** $\lambda = +\infty$; unbounded

3-7 (a) improving **(c)** need more information
(e) improving

3-8 (a) $\Delta\mathbf{w} = (3, -2, 0, 1)$ **(c)** $\Delta\mathbf{w} = (-8, 3)$

3-9 (a) [ii], [iv]

3-10 (a) feasible **(c)** infeasible

3-11 (a) $2\,\Delta w_1 + 3\,\Delta w_3 = 0,$
$1\,\Delta w_1 + 1\,\Delta w_2 + 2\,\Delta w_3 = 0, \ \Delta w_1 \geq 0$
(c) $1\,\Delta w_1 + 1\,\Delta w_2 = 0, 2\,\Delta w_1 - 1\,\Delta w_2 \geq 0$

3-12 (a) $(4, 7) \cdot (2, 0) > 0, (4, 7) \cdot (-2, 4) > 0$
(b) $\Delta\mathbf{z}^{(1)}$ for $\lambda = 2$ to $\mathbf{z}^{(1)} = (4, 0), \Delta\mathbf{z}^{(2)}$ for $\lambda = \frac{3}{4}$ to
$\mathbf{z}^{(2)} = (\frac{5}{2}, 3), \Delta\mathbf{z}^{(1)}$ for $\lambda = \frac{1}{4}$ to $\mathbf{z}^{(3)} = (3, 3)$

3-14 (a) unimodal **(c)** unimodal **(e)** not unimodal,
$\mathbf{x}^{(1)} = (1, 3), \mathbf{x}^{(2)} = (2, 1)$ **(g)** unimodal

3-15 (a) $(3, 1, 0) + \lambda(-3, 3, 9), \lambda \in [0, 1]; \lambda = \frac{1}{3}$ for
$\mathbf{z}^{(3)}$; no λ gives $\mathbf{z}^{(4)}$

3-16 (a) not convex, $\mathbf{x}^{(1)} = (0, 3), \mathbf{x}^{(2)} = (3, 0)$
(c) convex **(e)** not convex, $\mathbf{x}^{(1)} = (0, 0, 0, 4),$
$\mathbf{x}^{(2)} = (0, 0, 0, 5)$

3-17 (a) min $w_4 + w_5$, s.t.
$40w_1 + 30w_2 + 10w_3 + w_4 = 150, w_1 - w_2 \leq 0,$
$4w_2 + w_3 + w_5 \geq 10, w_1, w_2, w_3, w_4, w_5 \geq 0;$
$w_4 = 150, w_5 = 10$ **(c)** min $w_3 + w_4 + w_5$, s.t.
$(w_1 - 3)^2 + (w_2 - 3)^2 - w_3 \leq 4, 2w_1 + 2w_2 + w_4 = 5,$
$w_1 + w_5 \geq 3, w_3, w_4, w_5 \geq 0; w_3 = 14, w_4 = 5,$
$w_5 = 3$

3-18 (a) stop and conclude model is infeasible
(c) proceed with Phase II from initial solution
$\mathbf{y} = (1, 3, 1)$

3-19 (a) max $22w_1 - w_2 + 15w_3 - M(w_4 + w_5)$, s.t.
$40w_1 + 30w_2 + 10w_3 + w_4 = 150, w_1 - w_2 \leq 0,$
$4w_2 + w_3 + w_5 \geq 10, w_1, w_2, w_3, w_4, w_5 \geq 0;$
$w_4 = 150, w_5 = 10$ **(c)** min
$2w_1 + 3w_2 + M(w_3 + w_4 + w_5)$, s.t.
$(w_1 - 3)^2 + (w_2 - 3)^2 - w_3 \leq 4, 2w_1 + 2w_2 + w_4 = 5,$
$w_1 + w_5 \geq 3, w_3, w_4, w_5 \geq 0; w_3 = 14, w_4 = 5,$
$w_5 = 3$

3-20 (a) stop and conclude model is infeasible if M is big enough, else increase M and repeat **(c)** stop and conclude $\mathbf{y} = (1, 3, 1)$ is a local optimum for the original model

Chapter 4

4-1 (a) $x_j \triangleq$ cases shipped to region j, max
$1.60x_1 + 1.40x_2 + 1.90x_3 + 1.20x_4$, s.t.
$\sum_{j=1}^4 x_j = 1200, 310 \leq x_1 \leq 434, 245 \leq x_2 \leq 343,$
$255 \leq x_3 \leq 357, 190 \leq x_4 \leq 266$ **(b)** $x_1^* = \text{NE} = 408,$
$x_2^* = \text{SE} = 245, x_3^* = \text{MW} = 357, x_4^* = \text{W} = 190$

4-2 (b) nonzero values are $x_{1,1}^* = 70, x_{1,2}^* = 10,$
$x_{2,2}^* = 40, x_{2,3}^* = 5, x_{2,4}^* = 35, x_{3,4}^* = 80$

4-3 (a) $x_j \triangleq$ fraction of ingredient j; min
$200x_1 + 150x_2 + 100x_3 + 75x_4$, s.t. $\sum_{j=1}^4 x_j = 1,$
$60x_1 + 80x_2 + 55x_3 + 40x_4 \geq 60,$
$50x_1 + 70x_2 + 40x_3 + 100x_4 \leq 60,$
$90x_1 + 30x_2 + 60x_3 + 80x_4 \geq 60$ **(b)** the 3 main
inequalities **(c)** $x_1^* = \text{oats} = 0.157, x_2^* = \text{corn}$
$= 0.271, x_3^* = \text{alfalfa} = 0.401, x_4^* = \text{hulls} = 0.171$

4-4 (c) $x_1^* = 0.176, x_2^* = 0.353, x_3^* = 0.000,$
$x_4^* = 0.471$

4-5 (a) $45 \sum_{i=1}^m x_{i,j} \leq \sum_{i=1}^m a_{i,11}x_{i,j} \leq 48 \sum_{i=1}^m x_{i,j},$
$j = 1, \ldots, n$ **(c)** $\sum_{i=1}^m a_{i,15}x_{i,15} \geq 116 \sum_{i=1}^m x_{i,15}$
(e) $7x_{1,j} \leq 3 \sum_{i=2}^m x_{i,j}, j = 6, \ldots, 11$
(g) $3 \sum_{i=3}^6 \sum_{j=1}^n x_{i,j} \geq \sum_{i=1}^m \sum_{j=1}^n x_{i,j}$

4-7 (a) $x_j \triangleq$ number of cuts with pattern j; min
$0.34x_1 + 0.22x_2 + 0.27x_3$, s.t. $2x_1 + 1x_3 \geq 37,$
$5x_2 + 3x_3 \geq 211, x_j \geq 0, j = 1, \ldots, 4$ **(b)** $x_1^* = 147.4,$
$x_2^* = 0.0, x_3^* = 21.1$

4-8 (b) $x_1^* = \text{santas} = 147.4, x_2^* = \text{trees} = 0.0,$
$x_3^* = \text{houses} = 21.1$

4-9 (a) max $30x_1 + 45x_2$, s.t.
$0.30x_1 + 0.30x_2 + 0.10x_3 + 0.15x_4 + 0.50x_5 \leq 80,$
$1.5x_3 + 2.5x_4 \leq 500, x_3 = 4x_1, x_4 = 4x_2, x_5 = x_1 + x_2,$
$x_j \geq 0, j = 1, \ldots, 5$ **(b)** the 3 equalities **(c)** $x_1^* = 27.8,$
$x_2^* = 33.3, x_3^* = 111.1, x_4^* = 133.3, x_5^* = 61.1$

4-10 (c) $x_1^* = 0.0$, $x_2^* = 666.7$, $x_3^* = 0.0$, $x_4^* = 2000$, $x_5^* = 12{,}000$

4-11 (a) $\sum_{p=1}^{n} x_{i,p} = \sum_{p=1}^{n} d_{i,p} + \sum_{k=1}^{m} \sum_{p=1}^{n} a_{i,k} x_{k,p}$, $i = 1, \ldots, m$

4-12 (a) $x_1 \triangleq$ number with 5 days starting Sunday,..., $x_7 \triangleq$ number with 5 days starting Saturday;

min $\sum_{j=1}^{7} x_j$,

s.t. $x_1 + x_4 + x_5 + x_6 + x_7 \geq 8$,

$x_1 + x_2 + x_5 + x_6 + x_7 \geq 6$,

$x_1 + x_2 + x_3 + x_6 + x_7 \geq 6$,

$x_1 + x_2 + x_3 + x_4 + x_7 \geq 6$,

$x_1 + x_2 + x_3 + x_4 + x_5 \geq 6$,

$x_2 + x_3 + x_4 + x_5 + x_6 \geq 10$,

$x_3 + x_4 + x_5 + x_6 + x_7 \geq 10$,

$x_j \geq 0$, $j = 1, \ldots, 7$

(b) all main constraints **(c)** $x_1^* = 0.0$, $x_2^* = 0.67$, $x_3^* = 2.0$, $x_4^* = 2.67$, $x_5^* = 2.0$, $x_6^* = 2.67$, $x_7^* = 0.67$

4-13 (c) $x_1^* = 5$, $x_2^* = 0$, $x_3^* = 0$, $x_4^* = 0$, $x_5^* = 3$, $x_6^* = 3$

4-14 (a) $x_{j,t} \triangleq$ investment in option j, year t; max $1.05x_{1,4} + 1.12x_{2,3} + 1.21x_{3,1}$, s.t.

$10 = x_{1,1} + x_{2,1} + x_{3,1}$,

$1.05x_{1,1} + 10 = x_{1,2} + x_{2,2}$,

$1.05x_{1,2} + 1.12x_{2,1} + 10 = x_{1,3} + x_{2,3}$,

$1.05x_{1,3} + 1.12x_{2,2} = x_{1,4}$, all variables nonnegative

(b) all main constraints **(c)** 4 years **(d)** nonzero values are $x_{2,1}^* = x_{2,2}^* = 10$, $x_{2,3}^* = x_{1,4}^* = 21.2$

4-15 (d) $x_1^* = x_3^* = x_4^* = 1200$, $x_2^* = 650$, $z_1^* = 0$, $z_2^* = 150$, $z_3^* = 1250$, $z_4^* = 1600$

4-16 (a) $\sum_{i=1}^{m} a_{i,k} x_{i,t} \leq b_k$, $k = 1, \ldots, q$, $t = 1, \ldots, n$ **(c)** $x_{i,1} = d_{i,1} + z_{i,1}$, $i = 1, \ldots, m$

4-17 (a) $w_j^+ \triangleq$ overestimation on point j, $w_j^- \triangleq$ underestimation on point j;

min $\sum_{j=1}^{4} \left(w_j^+ + w_j^- \right)$,

s.t. $\beta_0 + 2\beta_1 = 1 + w_1^+ - w_1^-$,

$\beta_0 + 3\beta_1 = 3 + w_2^+ - w_2^-$,

$\beta_0 + 5\beta_1 = 3 + w_3^+ - w_3^-$,

$\beta_0 + 7\beta_1 = 5 + w_4^+ - w_4^-$,

all variables nonnegative **(b)** $\beta_0^* = 0.000$, $\beta_1^* = 0.714$, nonzero over $w_1^{+*} = 0.429$, $w_3^{+*} = 0.571$, nonzero under $w_2^{-*} = 0.857$ **(c)** $0.429, 0.857, 0.571, 0.000$

4-18 (b) $x^* = 10$, nonzero runs $w_3^{+*} = 15$, $w_1^{-*} = 5$

4-19 (a) $z \triangleq$ largest deviation, $w_j^+ \triangleq$ overestimation on point j, $w_j^- \triangleq$ underestimation on point j; min z, s.t.

$\beta_0 + 2\beta_1 = 1 + w_1^+ - w_1^-$,

$\beta_0 + 3\beta_1 = 3 + w_2^+ - w_2^-$,

$\beta_0 + 5\beta_1 = 3 + w_3^+ - w_3^-$,

$\beta_0 + 7\beta_1 = 5 + w_4^+ - w_4^-$,

$z \geq w_j^+$, $j = 1, \ldots, 4$, $z \geq w_j^-$, $j = 1, \ldots, 4$, all variables nonnegative

(b) $\beta_0^* = 0.333$, $\beta_1^* = 0.667$, $z^* = 0.667$, nonzero over $w_1^{+*} = w_3^{+*} = 0.667$, nonzero under $w_2^{-*} = 0.667$ **(c)** $0.667, 0.667, 0.667, 0.000$

4-21 (a) min $\sum_{j=1}^{5} c_j x_j$, s.t. $\sum_{j=1}^{5} a_{i,j} x_j \geq r_i$, $i = 1, \ldots, 7$; $0 \leq x_j \leq u_j$, $j = 1, \ldots, 5$; where the $a_{i,j}$ are the yield fractions in the table, c_j the costs, and u_j the availabilites; the r_i are the given requirements. **(b)** $x_1^* = 18.75$, $x_2^* = 125.00$, $x_3^* = 150.00$, $x_4^* = 650.00$, $x_5^* = 0.00$

4-22 (b) Nonzeros

$x_{1,1}^* = x_{1,2}^* = x_{1,3}^* = 833.3$, $x_{1,4}^* = 500.0$,

$x_{2,1}^* = x_{2,2}^* = x_{2,3}^* = 10{,}000.0$, $x_{2,4}^* = 6000.0$,

$x_{3,1}^* = x_{3,2}^* = x_{3,3}^* = 833.3$, $x_{3,4}^* = 500.0$,

$x_{4,1}^* = x_{4,2}^* = x_{4,3}^* = 833.3$, $x_{4,4}^* = 500.0$,

$x_{5,1}^* = x_{5,2}^* = x_{5,3}^* = 833.3$, $x_{5,4}^* = 500.0$,

$h_{5,1}^* = 533.3$, $h_{5,2}^* = 166.7$, $h_{5,4}^* = 200.0$

4-23 (b) $s_1^* = s_2^* = 0$, $s_3^* = 8$, $s_4^* = 12$, $s_5^* = s_6^* = 24$, $s_7^* = 32$, $t_1^* = 12$, $t_2^* = t_7^* = 8$, $t_3^* = t_5^* = t_6^* = 16$, $t_4^* = 20$

Chapter 5

5-1 (b) boundary and extreme, infeasible, interior, boundary not extreme **(c)** $-w_1 + w_2 \leq 1$ and $w_2 \leq 3$ active; no active; $w_2 \leq 3$ active **(d)** optimal or unique, neither, neither, optimal not unique

5-3 (a) $\mathbf{A} = \begin{pmatrix} 1 & -4 & 1 & 1 & 0 \\ 9 & 0 & 6 & 0 & 0 \\ -5 & 9 & 0 & 0 & -1 \end{pmatrix}$,

$\mathbf{b} = (12, 15, 3)$, $\mathbf{c} = (4, 2, -33, 0, 0)$ **(c)**

$\mathbf{A} = \begin{pmatrix} 2 & -1 & -1 & -1 & 0 & 0 & 0 \\ 1 & 0 & 0 & 0 & 1 & 0 & 0 \\ 0 & 1 & 0 & 0 & 0 & 1 & 0 \\ 0 & 0 & 1 & 0 & 0 & 0 & 1 \end{pmatrix}$,

$\mathbf{b} = (0, 3, 3, 3)$, $\mathbf{c} = (15, 41, -11, 0, 0, 0, 0)$

(e) A = $\begin{pmatrix} 1 & -1 & 1 & 5 & 1 \\ 0 & 3 & -3 & -9 & 0 \end{pmatrix}$, **b** = $(10, -6)$,

c = $(2, 1, -1, 4, 0)$

5-4 (b) A = $\begin{pmatrix} -1 & 1 & 1 & 0 \\ 5 & 0 & 0 & 1 \end{pmatrix}$, **b** = $(2, 10)$

(c) yes, no, yes, yes, no, no **(d) y** = $(2, 4, 0, 0)$
feasible, **y** = $(0, 0, 2, 10)$ feasible, **y** = $(-2, 0, 0, 20)$
infeasible **(e) y** = $(2, 4)$, **y** = $(0, 0)$, **y** = $(-2, 0)$

5-6 (a) $5\,\Delta w_1 + 1\,\Delta w_2 - 1\,\Delta w_3 = 0$,
$3\,\Delta w_1 - 4\,\Delta w_2 + 8\,\Delta w_3 = 0$, $\Delta w_2 \geq 0$

5-7 (a) x = $(1, 0, 3, 0)$ **(b)** $\Delta x = (-3, 1, -1, 0)$ for
x_2, $\Delta x = (1, 0, -5, 1)$ for x_4 **(c)** each has $A\Delta x = 0$,
$\Delta x_2 \geq 0$ and $\Delta x_4 \geq 0$ **(d)** no, yes **(e)** $\lambda = \frac{1}{3}$, $\{x_2, x_3\}$,
$\lambda = \frac{3}{5}$, $\{x_1, x_4\}$

5-9 (a) $z_1^* = 4$, $z_2^* = 2$ **(b)**

A = $\begin{pmatrix} -2 & 1 & 1 & 0 & 0 \\ 1 & 1 & 0 & 1 & 0 \\ 1 & 0 & 0 & 0 & 1 \end{pmatrix}$,

b = $(2, 6, 4)$, **c** = $(3, 1, 0, 0, 0)$ **(c)** bases $\{z_3, z_4, z_5\}$,
then either $\{z_1, z_3, z_4\}$, $\{z_1, z_2, z_3\}$ or $\{z_2, z_4, z_5\}$,
$\{z_1, z_2, z_5\}$, $\{z_1, z_2, z_3\}$ **(d)** $z^{(0)} = (0, 0)$, then either
$z^{(1)} = (4, 0)$, $z^{(2)} = (4, 2)$ or $z^{(1)} = (0, 2)$,
$z^{(2)} = (\frac{10}{3}, \frac{8}{3})$, $z^{(3)} = (4, 2)$

5-11 (a) yes **(c)** no **(e)** no

5-12 (a) $z = 10 - 29x_2 + 10x_3$, $x_1 = 1 - (3x_2 - 1x_4)$,
$x_2 = 3 - (1x_2 + 5x_4)$

5-13 (a) no **(c)** yes

5-14 (b) A = $\begin{pmatrix} -1 & 1 & 1 & 0 \\ 1 & -1 & 0 & 1 \end{pmatrix}$,

b = $(4, 10)$, **c** = $(4, 5, 0, 0)$ **(c)** $y^{(0)} = (0, 0, 4, 10)$,
then either $y^{(1)} = (10, 0, 14, 0)$ and $\Delta y = (1, 1, 0, 0)$
for y_2 improves without limit, or $y^{(1)} = (0, 4, 0, 14)$
and $\Delta y = (1, 1, 0, 0)$ for y_1 improves without limit

5-16 (a) min $w_6 + w_7$, s.t. $w_1 + w_2 + w_4 = 18$,
$-2w_1 + w_3 - w_6 = -2$, $3w_2 + 5w_3 - w_5 + w_7 = 15$,
$w_1, \ldots, w_7 \geq 0$; $\{w_4, w_6, w_7\}$

5-17 (a) max $2w_1 + w_2 + 9w_3 - M(w_6 + w_7)$ s.t. same
constraints and starting basis as in Exercise 5-16

5-18 (b) A = $\begin{pmatrix} -2 & 1 & -1 & 0 & 1 \\ 0 & 1 & 0 & 1 & 0 \end{pmatrix}$,

b = $(2, 1)$, min **d** = $(0, 0, 0, 0, 1)$
(c) $y^{(0)} = (0, 0, 0, 1, 2)$, $y^{(1)} = (0, 1, 0, 0, 1)$,
optimal value positive

5-20 (a) $(4!)/(2!\,2!) = 6$ **(c)** $(2340!)/(1150!\,1190!)$

5-21 (a) yes **(c)** no

5-22 (a) alternative optima **x** = $(3, 6)$ through
x = $(6, 3)$ **(b)**

A = $\begin{pmatrix} 1 & 1 & 1 & 0 & 0 \\ -2 & 1 & 0 & 1 & 0 \\ 1 & -2 & 0 & 0 & 1 \end{pmatrix}$, **b** = $(9, 0, 0)$,

c = $(1, 1, 0, 0, 0)$ **(c)** bases $\{x_3, x_4, x_5\}$, then either
$\{x_1, x_3, x_4\}$ and $\{x_1, x_2, x_4\}$, or $\{x_2, x_3, x_5\}$ and
$\{x_1, x_2, x_5\}$ **(d)** $x^{(0)} = x^{(1)} = (0, 0)$, then either
$x^{(2)} = (3, 6)$ or $x^{(2)} = (6, 3)$

5-24 (a) B^{-1} = $\begin{pmatrix} 0 & 0.500 \\ 0.250 & 0.125 \end{pmatrix}$ **(b) v** = $(0, 0.5)$

(c) no for x_2, yes for x_4 **(d) E** = $\begin{pmatrix} 1 & 0.2 \\ 0 & 0.2 \end{pmatrix}$,

B^{-1} = $\begin{pmatrix} 0.050 & 0.525 \\ 0.050 & 0.025 \end{pmatrix}$

5-26 same as Exercise 5-9

5-28 (a) nonimproving, $\lambda = 2$, (L, B, B, L, B)
(c) improving, $\lambda = 5$, (L, B, U, B, B)

5-29 (a) $z_1^* = 3$, $z_2^* = 0$

(b) A = $\begin{pmatrix} 1 & 1 & -1 & 0 \\ 3 & 2 & 0 & -1 \end{pmatrix}$, **b** = $(3, 8)$,

c = $(5, 6, 0, 0)$ **(c)** basis statuses $(U, U, \text{1st}, \text{2nd})$,
then either $(L, U, \text{1st}, \text{2nd})$, $(L, \text{2nd}, \text{1st}, L)$,
$(\text{1st}, \text{2nd}, L, L)$ and $(\text{1st}, L, L, \text{2nd})$, or
$(U, L, \text{1st}, \text{2nd})$ and $(\text{1st}, L, L, \text{2nd})$ **(d)** $z^{(0)} = (6, 5)$,
then either $z^{(1)} = (0, 5)$, $z^{(2)} = (0, 4)$ $z^{(3)} = (2, 1)$ and
$z^{(3)} = (3, 0)$, or $z^{(1)} = (6, 0)$ and $z^{(2)} = (3, 0)$

Chapter 6

6-1 (a) $w_1^* = \frac{3}{2}$, $w_2^* = 2$ **(b)** $\Delta w = (2, 3)$ **(d)** check
constraints strictly satisfied
(e) $\lambda_{\max} = \frac{5}{17}$ **(f)** $w^{(1)} = (\frac{27}{17}, \frac{32}{17})$

6-3 (a) no **(c)** no **(e)** yes

6-4 (a) no **(c)** yes **(e)** no

6-5 (a) $2\,\Delta w_1 + 3\,\Delta w_2 - 3\,\Delta w_3 = 0$,
$4\,\Delta w_1 - 1\,\Delta w_2 + 1\,\Delta w_3 = 0$

6-6 (a) $\Delta x = (-0.8, -1.6, 4)$,

$\begin{pmatrix} 1 & 2 & 1 \\ -2 & 1 & 0 \end{pmatrix} \Delta x = 0$

6-7 (a) d = $(-14, -3, -5)$

(b) $P = \begin{pmatrix} \frac{1}{6} & -\frac{1}{6} & \frac{1}{3} \\ -\frac{1}{6} & \frac{1}{6} & -\frac{1}{3} \\ \frac{1}{3} & -\frac{1}{3} & \frac{2}{3} \end{pmatrix}$ **(c)** $\Delta z = (-\frac{7}{2}, \frac{7}{2}, -7)$

(d) $\begin{pmatrix} 2 & 0 & -1 \\ 1 & 1 & 0 \end{pmatrix} \Delta z = 0$,

$(14, 3, 5) \cdot (-\frac{7}{2}, \frac{7}{2}, -7) < 0$

6-9 (a) $(\frac{1}{2}, \frac{1}{5}, 1, \frac{1}{9})$ **(c)** $(\frac{3}{2}, 1, 1, \frac{2}{3})$

6-10 (a) $(2, 5, 1, 9)$ **(c)** $(6, 25, 1, 54)$

6-11 (a) $x^* = (0, \frac{12}{5}, 0)$ **(b)** $y^{(3)} = (1, 1, 1)$,
$y^* = (0, \frac{12}{5}, 0)$ **(c)** min $4y_1 + 3y_2 + 5y_3$, s.t.
$4y_1 + 5y_2 + 3y_3 = 12, y_1, y_2, y_3 \geq 0$

6-13 (a) $\Delta x = (-0.64, 1.6, -2.24)$,
$\Delta y = (-0.32, 1.6, -2.24)$ **(b)** $(2, 3, 5) \cdot \Delta x < 0$,
$(2, 5, 3) \cdot \Delta x = 0$ **(c)** $\lambda = 0.36084$
(d) $x^{(4)} = (1.769, 1.577, 0.192)$,
$y^{(4)} = (0.884, 1.577, 0.192)$

6-15 (a) check feasible and strictly positive **(b)** min
$40y_1 + 3y_2$, s.t. $4y_1 - 3y_2 + 2y_3 = 3, 3y_2 - y_3 = 2$,
$y_1, y_2, y_3 \geq 0$ **(c)** $\Delta x = (-7.669, 7.669, 7.669)$

(d) $(10, 1, 0) \cdot \Delta x < 0, \begin{pmatrix} 1 & -1 & 2 \\ 0 & 1 & -1 \end{pmatrix} \Delta x = 0$

(e) $\lambda = 0.12037, x^{(1)} = (3.077, 3.923, 1.923)$

6-17 (a) yes, unbounded **(c)** no,
$x^{(12)} = (3.949, 0.514, 9)$

6-18 (a) $w^* = (4, 0, 0)$ **(b)** max
$13w_1 - 2w_2 + w_3 + \mu(\ln(w_1) + \ln(w_2) + \ln(w_3))$, s.t.
$3w_1 + 6w_2 + 4w_3 = 12, w_1, w_2, w_3 \geq 0$ **(c)** 17.7 and
16.45, 3.088 and -78.10, moderate bonus in middle
vs. major penalty near boundary
(d) $w^* = (3.850, 0.035, 0.060)$,
$w^* = (2.799, 0.285, 0.474)$,
$w^* = (3.850, 0.350, 0.600)$

6-20 (a) yes **(c)** no

6-21 (a) curve II

6-22 (a) check feasible and strictly positive **(b)** min
$4x_1 - x_2 + 2x_3 - 10[\ln(x_1) + \ln(x_2) + \ln(x_3)]$, s.t.
$4x_1 - 3x_2 + 2x_3 = 13, 3x_2 - x_3 = 1, x_1, x_2, x_3 \geq 0$
(c) $\Delta x = (-4.6415, 6.1887, 18.566)$
(d) $(0.667, -11, -3) \cdot \Delta x < 0$,
$\begin{pmatrix} 4 & -3 & 2 \\ 0 & 3 & -1 \end{pmatrix} \Delta x = 0$ **(e)** $\lambda_{max} = 0.64634$,
$\lambda = 0.10$ **(f)** decrease, then increase

6-24 (a) check feasible and strictly positive
(b) $\Delta x = (-16.899, 16.899, 16.899)$

(c) $(7.5, -2.333, -10) \cdot \Delta x < 0$,
$\begin{pmatrix} 1 & -1 & 2 \\ 0 & 1 & -1 \end{pmatrix} \Delta x = 0$

(d) $\lambda_{max} = 0.2367$ **(e)** $\lambda = 0.1$,
$x^{(1)} = (2.310, 4.690, 2.690)$

Chapter 7

7-1 (b) dollars of profit, thousand beta zappers,
thousand freeze phasers, extrusion hours, trimming
hours, assembly hours **(c)** make thousand beta
zappers, make thousand freeze phasers **(d)** input1:
5 hours extrusion, 1 hour trimming, 12 hours
assembly; output1: thousand beta zappers, $2500
profit; input2: 9 hours extrusion, 2 hours trimming,
15 hours assembly; output2: thousand freeze
phasers, $1600 profit

7-3 (b) $ cost, professional-equivalent hours
production, Proof hours supervision, grad
maximum hours **(c)** ugrad hour programming, grad
hour programming, professional hour programming
(d) input1: 0.2 hour Proof supervision, $4 cost;
output1: 0.2 professional-equivalent hour
programming; input2: 0.15 hour Proof supervision, 1
hour grad maximum, $10 cost; output2: 0.3
professional-equivalent hour programming; input3:
0.15 hour Proof supervision, $25 cost; output3: 1
professional-equivalent hour programming

7-5 (a) tighten, decrease, more steep **(c)** relax,
decrease, less steep **(e)** relax, increase, less steep
(g) tighten, increase, more steep

7-6 (a) tighten, decrease **(c)** tighten, increase

7-7 (a) increase, more steep **(c)** increase, less steep

7-8 (a) $v_1 \triangleq$ dollar change in optimal profit per
thousand increase in beta zapper demand, $v_2 \triangleq$
dollar change in optimal profit per thousand
increase in freeze phaser demand, $v_3 \triangleq$ dollar
change in optimal profit per hour increase in
extrusion capacity, $v_4 \triangleq$ dollar change in optimal
profit per hour increase in trimming capacity, $v_5 \triangleq$
dollar change in optimal profit per hour increase in
assembly capacity **(b)** $v_1, v_2 \leq 0; v_3, v_4, v_5 \geq 0$; RHS
increase tightens $\geq$ and relaxes $\leq$
(c) $v_1 + 5v_3 + v_4 + 12v_5 \geq 2500$,
$v_2 + 9v_3 + 2v_4 + 15v_5 \geq 1600$, implicit cost of
activities should equal or exceed objective function
return **(d)** min $10v_1 + 15v_2 + 320v_3 + 300v_4 + 480v_5$,

minimize total implicit cost (f) $x_1 = 10$ or $v_1 = 0$, $x_2 = 15$ or $v_2 = 0$, $5x_1 + 9x_2 = 320$ or $v_3 = 0$, $x_1 + 2x_2 = 300$ or $v_4 = 0$, $12x_1 + 15x_2 = 480$ or $v_5 = 0$, either a primal constraint is active or small RHS change has no objective function impact (g) $v_1 + 5v_3 + v_4 + 12v_5 = 2500$ or $x_1 = 0$, $v_2 + 9v_3 + 2v_4 + 15v_5 = 1600$ or $x_2 = 0$, a primal variable should be used only if its implicit cost matches its objective function return

7-10 (a) $v_1 \triangleq$ dollar change in optimal cost per professional-equivalent hour increase in required production, $v_2 \triangleq$ dollar change in optimal cost per hour increase in Proof supervision, $v_3 \triangleq$ dollar change in optimal cost per hour increase in grad availability **(b)** $v_1 \geq 0$; $v_2, v_3 \leq 0$; RHS increase tightens $\geq$ and relaxes $\leq$ **(c)** $0.2v_1 + 0.2v_2 \leq 4$, $0.3v_1 + 0.15v_2 + v_3 \leq 10$, $v_1 + 0.15v_2 \leq 25$, implicit value of activities should not exceed objective function cost **(d)** max $1000v_1 + 164v_2 + 500v_3$, maximize total implicit value **(f)** $0.2x_1 + 0.3x_2 + x_3 = 1000$ or $v_1 = 0$, $0.2x_1 + 0.15x_2 + 0.15x_3 = 164$ or $v_2 = 0$, $x_2 = 500$ or $v_3 = 0$, either a primal constraint is active or small RHS change has no objective function impact **(g)** $0.2v_1 + 0.2v_2 = 4$ or $x_1 = 0$, $0.3v_1 + 0.15v_2 + v_3 = 10$ or $x_2 = 0$, $v_1 + 0.15v_2 = 25$ or $x_3 = 0$, a primal variable should be used only if its implicit value matches its objective function cost

7-12 (a) max $40v_1 + 10v_2$, s.t. $2v_1 + 4v_2 \leq 17$, $3v_1 + 4v_2 \leq 29$, $2v_1 + 3v_3 \leq 0$, $3v_1 + v_2 - v_3 \leq 1$, $v_1 \leq 0$, $v_2 \geq 0$, v_3 URS **(c)** min $10v_1 + 19v_2 + 5v_3$, s.t. $2v_1 + v_3 \geq 30$, $-3v_1 + 4v_2 + v_3 = 0$, $-v_2 + v_3 \leq -2$, $9v_1 = 10$, $v_1 \geq 0$, $v_2 \leq 0$, v_3 URS **(e)** min $10v_1 + 11v_3$, s.t. $3v_1 + v_2 = 0$, $2v_1 + v_3 \geq 2$, $-v_1 + 3v_3 \geq 9$, $-v_2 + v_3 = 0$, $v_1 \leq 0$, $v_2 \geq 0$, v_3 URS **(g)** max $40v_1 + 18v_2 + 11v_3$, s.t. $15v_1 + 12v_2 \leq 0$, $15v_1 - 90v_2 \leq 32$, $15v_1 \leq 50$, $14v_2 + v_3 \leq 0$, $v_1 \leq -19$, v_1 URS, $v_2 \geq 0$, $v_3 \leq 0$

7-13 (a) $2x_1 + 3x_2 + 2x_3 + 3x_4 = 40$ or $v_1 = 0$, $4x_1 + 4x_2 + x_4 = 10$ or $v_2 = 0$, $2v_1 + 4v_2 = 17$ or $x_1 = 0$, $3v_1 + 4v_2 = 29$ or $x_2 = 0$, $2v_1 + 3v_2 = 0$ or $x_3 = 0$, $3v_1 + v_2 - v_3 = 1$ or $x_4 = 0$ **(c)** $2x_1 - 3x_2 + 9x_4 = 10$ or $v_1 = 0$, $4x_2 - x_3 = 19$ or $v_2 = 0$, $2v_1 + v_3 = 30$ or $x_1 = 0$, $-v_2 + v_3 = -2$ or $x_3 = 0$ **(e)** $3w + 2x_1 - x_2 = 10$ or $v_1 = 0$, $w - y = 0$ or $v_2 = 0$, $2v_1 + v_3 = 2$ or $x_1 = 0$, $-v_1 + 3v_3 = 9$ or

$x_2 = 0$ **(g)** $12x_1 - 90x_2 + 14x_4 = 18$ or $v_2 = 0$, $x_4 = 11$ or $v_3 = 0$, $15v_1 + 12v_2 = 0$ or $x_1 = 0$, $15v_1 - 90v_2 = 32$ or $x_2 = 0$, $15v_1 = 50$ or $x_3 = 0$, $14v_2 + v_3 = 0$ or $x_4 = 0$ $v_1 = -19$ or $x_5 = 0$

7-14 (a) min $14v_1 + 14v_2$, s.t. $2v_1 + 5v_2 \geq 14$, $5v_1 + 2v_2 \geq 7$, $v_1, v_2 \geq 0$; $\mathbf{x}^* = (2, 2)$, $\mathbf{v}^* = (\frac{1}{3}, \frac{8}{3})$ **(c)** max $24v_1 + 11v_2$, s.t. $2v_1 + 3v_2 \leq 8$, $9v_1 + v_2 \leq 11$, $v_1, v_2 \geq 0$; $\mathbf{x}^* = (3, 2)$, $\mathbf{v}^* = (1, 2)$

7-15 (a) $v_1 = 3$, $v_2 = \frac{1}{3}$ **(c)** $v_1 = \frac{1}{3}$, $v_2 = 4$

7-16 (a) min $4v_1 + 12v_2$, s.t. $2v_1 \geq 4$, $v_1 + 3v_2 \geq 1$, $v_1 \leq 0$, $v_2 \geq 0$; unbounded, infeasible **(c)** max $2v_1 + 5v_2$, s.t. $v_1 \geq 10$, $v_1 - v_2 \geq 3$, $v_1, v_2 \geq 0$, infeasible, unbounded **(e)** max $2v_1 + 5v_2$, s.t. $-v_1 + v_2 \leq -3$, $2v_1 - 2v_2 \leq 4$, $v_1, v_2 \geq 0$; infeasible, infeasible

7-17 (a) dual optimal value ≤ 70

7-19 (a) no; 51.25 **(b)** \$0; \$208.33 **(c)** increase \$20,833; increase at least \$39,375 and at most \$41,667 **(d)** increase \$31,875; decrease at least \$25,925 and at most \$31,875 **(e)** 87.5 hours **(f)** \$624.99

7-21 (a) $v_1^* = \$25.882$ **(b)** \$1294.10, at least \$2588.20 **(c)** yes, increase \$82.35; increase at least \$376.45 **(d)** \$3.118 **(e)** \$4917.65; at least \$17,376.01 and at most \$18,670.58 **(f)** yes **(g)** no **(h)** \$20.12

Chapter 8

8-1 (b) (166,85.2) to (0,184.8) **(c)** first: $\mathbf{x}^* = (24, 166, 50)$; second: $\mathbf{x}^* = (24, 0, 216)$

8-3 (b) (2.4,19.8) to (2.8,12.5) to (3.2,12.1) to (4.2,11.3) **(c)** first: $\mathbf{x}^* = (0, 1, 0, 0, 1)$; second: $\mathbf{x}^* = (1, 0, 1, 1, 0)$

8-4 (b) approximately (9.05,4202) to (12,2760) to (20,1809) to (30,1365) to (43.29,1110) **(c)** first: $\mathbf{x}^* = (6.65, 5.56, 4.46, 3.33)$; second: $\mathbf{x}^* = (1, 1, 1, 17)$

8-5 (a) first: $\mathbf{x}^* = (0, 6)$, second: $\mathbf{x}^* = (4, 1)$ **(b)** no, no, yes, yes, no, yes **(c)** (9,4) to (27,2) to (30,0)

8-7 (a) first: first objective, s.t. original constraints, points $(1, \frac{9}{2})$ through $(3, \frac{3}{2})$ alternative optima; second: second objective, s.t. original constraints and $6x_1 + 4x_2 \geq 26$, $\mathbf{x}^* = (1, \frac{9}{2})$ **(b)** first: second objective, s.t. original constraints, $\mathbf{x}^* = (0, 5)$;

second: first objective, s.t. original constraints and $x_2 \geq 5$, $\mathbf{x}^* = (0, 5)$

8-9 **(a)** min $15x_1 + 110x_2 + 92x_3 + 123x_4$

8-10 **(a)** $\mathbf{x}^* = (1, \frac{9}{2})$ **(b)** $\mathbf{x}^* = (0, 5)$

8-12 **(a)** min $d_1 + d_2$, s.t. $3x_1 + 5x_2 - x_3 - d_1 \leq 20$, $11x_2 + 23x_3 + d_2 \geq 100$, $d_1, d_2 \geq 0$, and all original constraints **(c)** min $d_1 + d_2 + d_3$, s.t. $40x_1 + 23x_2 + d_1 \geq 700$, $20x_1 - 20x_2 - d_2 \leq 25$, $5x_2 + x_3 - d_3 \leq 65$, $d_1, d_2, d_3 \geq 0$, and all original constraints **(e)** min $d_1 + d_2$, s.t. $22x_1 + 8x_2 + 13x_3 - d_1 \leq 20$, $3x_1 + 6x_2 + 4x_3 + d_2 \geq 12$, $d_1, d_2 \geq 0$, and all original constraints

8-13 **(b)** min $d_1 + d_2$, s.t. $x_1 + d_1 \geq 3$, $2x_1 + 2x_2 + d_2 \geq 14$, $2x_1 + x_2 \leq 9$, $x_1' \leq 4$, $x_2 \leq 7$, $x_1, x_2, d_1, d_2 \geq 0$ **(c)** least total distance to the two contours

8-15 **(a)** min d_1, s.t. $x_1 + d_1 \geq 3$, $2x_1 + 2x_2 + d_2 \geq 14$, $2x_1 + x_2 \leq 9$, $x_1 \leq 4$, $x_2 \leq 7$, $x_1, x_2, d_1, d_2 \geq 0$; any feasible solution with $x_1 \geq 3$ is alternative optimal; min d_2, s.t. $d_1 \leq 0$ and other constraints of the first LP; $\mathbf{x}^* = (3, 3)$ **(b)** min $100d_1 + d_2$, s.t. same constraints as first LP in part (a) **(c)** yes; yes

8-17 **(a)** first: first objective s.t. original constraints, $\mathbf{x}^* = (24, 166, 50)$; second: second objective s.t. original constraints and $x_2 \geq 166$, $\mathbf{x}^* = (24, 166, 50)$ **(b)** max $0.50x_1 + 2.20x_2 + 0.80x_3$, s.t. original constraints, $\mathbf{x}^* = (24, 166, 50)$ **(c)** min $d_1 + d_2$, s.t. $x_2 + d_1 \geq 100$, $0.50x_1 + 0.20x_2 + 0.80x_3 + d_2 \geq 144$, $d_1, d_2 \geq 0$, and all original constraints **(e)** $\mathbf{x}^* = (24, 100, 116)$ **(f)** first: min d_1, s.t. constraints of part (c), any feasible solution with $x_2 \geq 100$ is optimal; second: min d_2, s.t. $d_1 \leq 0$ and constraints of part (c), $\mathbf{x}^* = (24, 100, 116)$ **(g)** min $10,000d_1 + d_2$

8-19 **(a)** first: first objective s.t. original constraints, $\mathbf{x}^* = (0, 1, 0, 0, 1)$; second: second objective s.t. original constraints and $1.0x_1 + 0.4x_2 + 1.4x_3 + 1.8x_4 + 2.0x_5 \leq 2.4$, $\mathbf{x}^* = (0, 10, 0, 0, 1)$ **(b)** min $4.5x_1 + 6.2x_2 + 7.4x_3 + 7.8x_4 + 18.4x_5$, s.t. original constraints, $\mathbf{x}^* = (1, 0, 1, 1, 0)$ **(c)** min $d_1 + d_2$, s.t. $1.0x_1 + 0.4x_2 + 1.4x_3 + 1.8x_4 + 2.0x_5 - d_1 \leq 3.0$, $2.5x_1 + 5.4x_2 + 4.6x_3 + 4.2x_4 + 14.4x_5 - d_2 \leq 12$, $d_1, d_2 \geq 0$, and all original constraints **(d)** part (c)

optimal value $\neq 0$ **(e)** $\mathbf{x}^* = (1, 1, 0, 1, 0)$, new intermediate **(f)** first: min d_1, s.t. constraints of part (c), $\mathbf{x}^* = (1, 1, 1, 0, 0)$; second: min d_2, s.t. $d_1 \leq 0$ and constraints of part (c), same $\mathbf{x}^*$ **(g)** min $10,000d_1 + d_2$

8-21 **(a)** $x_1 \triangleq$ singles, $x_2 \triangleq$ doubles, $x_3 \triangleq$ luxuries; min $2d_1^- + d_2^-$, s.t. $40x_1 + 60x_2 + 120x_3 \leq 10,000$; $0.7x_1 + 0.4x_2 + 0.9x_3 + d_1^- \geq 100$; $0.3x_1 + 0.6x_2 + 0.1x_3 + d_2^- \geq 120$; all variables nonnegative **(b)** $x_1^* = 76.92$, $x_2^* = 115.38$, $x_3^* = 0.00$, $d_1^{-*} = 0.00$, $d_2^{-*} = 27.69$

8-22 **(b)** nonzeros: $x_1^* = 4.4$, $x_2^* = 10.0$, $x_3^* = 12.3$, $x_4^* = 16.5$, $r_1^* = 0.6$, $r_4^* = 1.5$

8-28 **(b)** $v^* = 285$, $f^* = 0.007$, $d^* = 0.040$, $g_1^{+*} = g_1^{-*} = 0$, $g_2^* = 0.637$

8-32 **(c)** $x_1^* = 53.33$, $x_2^* = 500$, $x_3^* = 166.67$, $y_1^* = 100$, $y_2^* = 1150$, $y_3 = 250$, all $d_k^* = 0$ except $d_1^* = 466.67$

Chapter 9

9-1 **(a)** nodes: 1,2,3,4,5; arcs: (3,2), (3,4), (3,5); edges: (1,2), (1,3), (1,4), (2,5), (4,5) **(b)** yes, no, yes, no **(c)** replace each edge with two opposed arcs of the same length as the edge

9-3 **(a)** 1–3–5–2, length 6, 1–3, length 3, 1–4, length 7, 1–3–5, length 4 **(b)** 1–3 best for 1 to 3, 1–3–5 best for 1 to 5, 3–5 best for 3 to 5, 3–5–2 best for 3 to 2, 5–2 best for 5 to 2 **(c)** $v[1] = 0$, $v[2] = 6$, $x_{1,3}[2] = x_{3,5}[2] = x_{2,5}[2] = 1$, $v[3] = 3$, $x_{1,3}[3] = 1$, $v[4] = 7$, $x_{1,4}[4] = 1$, $v[5] = 4$, $x_{1,3}[5] = x_{3,5}[5] = 1$ **(d)** $v[1] = 0$, $v[2] = \min\{v[1] + 10, v[3] + 5, v[5] + 2\}$, $v[3] = \min\{v[1] + 3\}$, $v[4] = \min\{v[1] + 7, v[3] + 9, v[5] + 4\}$, $v[5] = \min\{v[2] + 2, v[3] + 1, v[4] + 4\}$ **(f)** positive lengths preclude negative dicycles

9-5 **(a)** 1–3–2, length 4, 1–3, length 3, 1–3–2–4, length 10, 2–1, length 8, 2–4–3, length 10, 2–4, length 6, 3–1, length 3, 3–2, length 1, 3–2–4, length 7, 4–3–1, length 7, 4–3–2, length 5, 4–3 length 3 **(b)** 1–3 best for 1 to 3, 1–3–2 best for 1 to 2, 3–2 best for 3 to 2, 3–2–4 best for 3 to 4, 2–4 best for 2 to 4 **(c)** $v[k][k] = 0$, $k = 1, \ldots, 4$, $v[1][2] = 4$, $x_{1,3}[1][2] = x_{3,2}[1][2] = 1$, $v[1][3] = 3$, $x_{1,3}[1][3] = 1$, $v[1][4] = 10$, $x_{1,3}[1][4] = x_{3,2}[1][4] = x_{2,4}[1][4] = 1$, $v[2][1] = 8$, $x_{1,2}[2][1] = 1$, $v[2][3] = 10$,

$x_{2,4}[2][3] = x_{4,3}[2][3] = 1$, $v[2][4] = 6$, $x_{2,4}[2][4] = 1$,
$v[3][1] = 3$, $x_{1,3}[3][1] = 1$, $v[3][2] = 1$, $x_{3,2}[3][2] = 1$,
$v[3][4] = 7$, $x_{3,2}[3][4] = x_{2,4}[3][4] = 1$,
$v[4][1] = 7$, $x_{4,3}[4][1] = x_{1,3}[4][1] = 1$,
$v[4][2] = 5$, $x_{4,3}[4][2] = x_{3,2}[4][2] = 1$, $v[4][3] = 3$,
$x_{4,3}[4][3] = 1$ **(d)** $v[k][k] = 0, k = 1, \ldots, 4$,
$v[1][2] = \min\{8, v[1][3] + v[3][2], v[1][4] + v[4][1]\}$,
$v[1][3] = \min\{3, v[1][2] + v[2][3], v[1][4] + v[4][2]\}$,
$v[1][4] = \min\{v[1][2] + v[2][4], v[1][3] + v[3][4]\}$,
$v[2][1] = \min\{8, v[2][3] + v[3][1], v[2][4] + v[4][1]\}$,
$v[2][3] = \min\{v[2][1] + v[1][3], v[2][4] + v[4][3]\}$,
$v[2][4] = \min\{6, v[2][1] + v[1][4], v[2][3] + v[3][4]\}$,
$v[3][1] = \min\{3, v[3][2] + v[2][1], v[3][4] + v[4][1]\}$,
$v[3][2] = \min\{1, v[3][1] + v[1][2], v[3][4] + v[4][2]\}$,
$v[3][4] = \min\{v[3][1] + v[1][4], v[3][2] + v[2][4]\}$,
$v[4][1] = \min\{v[4][2] + v[2][1], v[4][3] + v[3][1]\}$,
$v[4][2] = \min\{6, v[4][1] + v[1][2], v[4][3] + v[3][2]\}$,
$v[4][3] = \min\{4, v[4][1] + v[1][3], v[4][2] + v[2][3]\}$
(f) positive lengths preclude negative dicycles

9-7 (a) 1–2, length -5, 1–3, length 10, 1–3–4, length
12 **(b)** 1–2–4–1, 1–3–4–1 **(c)** no, yes **(d)** functional
equations nolonger sufficient

9-9 (a) one to all, no negative dicycles **(b)** see
Exercise 9-3(a) **(c)** see Exercise 9-3(a) **(d)** 1–3–2,
length 8, 1–3, length 3, 1–4, length 7, 1–3–5, length 4
(e) 4, labels correct after $t = 4$

9-11 (a) 1–3–4–1

9-12 (a) all to all, no negative dicycles **(b)** see
Exercise 9-5(a) **(c)** see Exercise 9-5(a) **(d)** 1–2,
length 8, 1–3, length 3, 1–2–4, length 14, 2–1, length
8, 2–1–3, length 11, 2–4, length 6, 3–1, length 3, 3–2,
length 1, 3–2–4, length 7, 4–2–1, length 14, 4–2,
length 6, 4–3 length 3

9-14 (a) 1–3–4–1

9-15 (a) one to all, lengths nonnegative **(b)** see
Exercise 9-3(a) **(c)** see Exercise 9-3(a) **(d)** 1–3–2,
length 8, 1–3, length 3, 1–4, length 7, 1–3–5,
length 4

9-17 (a) Bellman–Ford, Floyd–Warshall, Dijkstra
(best), 1–3–2, length 5, 1–3, length 2, no path to 4
(c) none apply

9-18 (a) acyclic; one numbering: $a = 1, b = 3$,
$c = 4, d = 2, e = 6, f = 5$ **(c)** not acyclic; dicycle
a–d–e–c–b–a

9-19 (a) acyclic digraph **(b)** $v[1] = 0$, $v[2] = \infty$,
$v[3] = 2$, $v[4] = 12$, $v[5] = -1$, $v[6] = 11$ **(c)** 1–3,
1–3–4, 1–3–5, 1–3–5–6

9-22 (a)

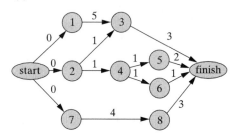

(b) check arcs (i, j) have $i < j$ **(c)** $v[1] = 0$, $v[2] = 0$,
$v[3] = 5$, $v[4] = 1$, $v[5] = 2$, $v[6] = 2$, $v[7] = 0$,
$v[8] = 4$, $finish = 8$ **(d)** $start$–1–3–$finish$ **(e)** late
starts: $1 = 2, 2 = 6, 3 = 7, 4 = 7, 5 = 8, 6 = 9$,
$7 = 3, 8 = 7$; slacks: $1 = 2, 2 = 6, 3 = 2, 4 = 6$,
$5 = 6, 6 = 7, 7 = 3, 8 = 3$

9-24 (a)

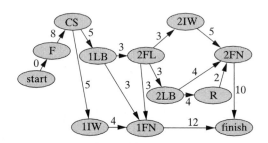

(b) one numbering: F = 1, CS = 2, 1LB = 3, 1IW =
4, 2FL = 5, 1FN = 6, 2LB = 7, 2IW = 8, R = 9, 2FN
= 10 **(c)** $v[F] = 0$, $v[CS] = 8$, $v[1LB] = 13$,
$v[1IW] = 13$, $v[2FL] = 16$, $v[1FN] = 19$,
$v[2LB] = 19$, $v[2IW] = 19$, $v[R] = 23$, $v[2FN] = 25$,
$finish = 35$

(d) $start$–F–CS–1LB–2FL–2LB–R–2FN–$finish$
(e) late starts: F = 0, CS = 8, 1LB = 13, 1IW = 19,
2FL = 16, 1FN = 23, 2LB = 19, 2IW = 20, R = 23,
2FN = 25; slacks: F = 0, CS = 0, 1LB = 0, 1IW = 6,
2FL = 0, 1FN = 4, 2LB = 0, 2IW = 1, R = 0. 2FN
= 0

9-26 (b)

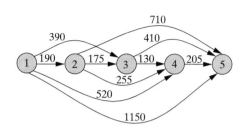

(d) produce 30, 35, 0, 35 **(e)** produce 30, 25

9-28 (b)

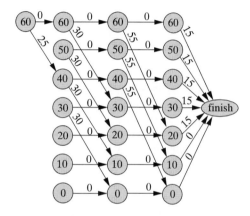

(d) take 1 and 3

9-30 (c) for 8: 1–2–13–3–8, length 26; for 10: 1–2–13–3–8–10, length 41; for 11: 1–2–13–4–11, length 19; for 12: 1–2–13–3–8–6–12, length 35

9-31 (c) 1–2, length 6; 1–2–7–3, length 42; 1–2–7–4, length 30; 1–2–7–5, length 20; 1–6, length 11; 1–2–7, length 12; 1–2–7–5–8, length 34; 1–2–7–5–8–9, length 57; 1–2–7–5–8–10, length 59

9-32 (b) yes **(d)** 7–11–2 with one 4-hour and one 3-hour, cost \$90

9-33 (d) 1–3–6–7 (or 1–3–5–7), length 9 gridsize

9-34 (d) using S for shed and I for the unlabeled intersection: S–5–I–1, length 24; S–5–3–2, length 26; S–5–3, length 14; S–5–4, length 25; S–5, length 5

Chapter 10

10-1 (a) $V = \{1, 2, 3, 4, 5\}$,
$A = \{(1, 2), (1, 4), (2, 5), (3, 1), (3, 4),$

$(3, 5), (4, 2), (4, 5), (5, 2)\}$ **(b)** source: 1,3; sink: 2,5; transshipment: 4 **(d)** $\min 5x_{1,2} + 10x_{1,4}$
$-6x_{3,1} + 2x_{3,4} + 8x_{4,2} + 6x_{4,5} + 1x_{5,2} + 3x_{5,3}$,
s.t. $-x_{1,2} - x_{1,4} + x_{3,1} = -50$,
$x_{1,2} - x_{2,5} + x_{4,2} + x_{5,2} = 20$,
$-x_{3,1} - x_{3,4} + x_{5,3} = -70$,
$x_{1,4} + x_{3,4} - x_{4,2} - x_{4,5} = 0$,
$x_{2,5} + x_{4,5} - x_{5,2} - x_{5,3} = 100$,
$x_{1,4} \le 40, x_{2,5} \le 10, x_{3,1} \le 4, x_{4,2} \le 20$, all $x_{i,j} \ge 0$
(e) A =

$$\begin{pmatrix} -1 & -1 & 0 & 1 & 0 & 0 & 0 & 0 & 0 \\ 1 & 0 & -1 & 0 & 0 & 1 & 0 & 1 & 0 \\ 0 & 0 & 0 & -1 & -1 & 0 & 0 & 0 & 1 \\ 0 & 1 & 0 & 0 & 1 & -1 & -1 & 0 & 0 \\ 0 & 0 & 1 & 0 & 0 & 0 & 1 & -1 & -1 \end{pmatrix}$$

10-3 (a) check every column has at most a -1 and a $+1$ **(b)**

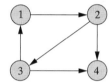

10-5 (a) $150 > 80$ **(b)** node 4, demand 70; zero cost arcs (1,4) and (2,4)

10-7 (a) $\min 25x_{P1,C1} + 30x_{P1,C2} + 15x_{P1,W} + 45x_{P2,C1} + 23x_{P2,C2} + 15x_{P2,W} + 11x_{W,C1} + 14x_{W,C2}$,
s.t. $x_{P1,C1} + x_{P1,C2} + x_{P1,W} \le 400$,
$x_{P2,C1} + x_{P2,C2} + x_{P2,W} \le 600$,
$x_{P1,W} + x_{P2,W} - x_{W,C1} - x_{W,C2} = 0$,
$x_{P1,C1} + x_{P2,C1} + x_{W,C1} = 160$,
$x_{P1,C2} + x_{P2,C2} + x_{W,C2} = 700$, all $x_{i,j} \ge 0$
(b) $x^*_{P1,C1} = 160, x^*_{P1,W} = 100, x^*_{P2,C2} = 600$,
$x^*_{W,C2} = 100$
(c)

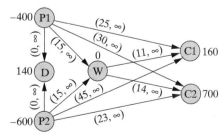

(d) source: P1,P2; sink: C1,C2,D; transshipment: W

10-9 (a) cycle **(b)** chain, path **(c)** chain **(d)** cycle, dicycle

10-11 (a) check flow balance and capacities
(b) $\Delta x =$
$\pm(1, 0, 1, -1, 0), \pm(0, 1, 0, 1, -1), \pm(1, 1, 1, 0, -1)$
(d) no, yes, yes, no, no, no **(e)** no, yes, yes, no, no, yes
(f) 20, 20, 10

10-13 (a) $x^* = (0, 35, 60, 40, 0)$

10-15 (a) add node 0; (1,0) flow 50, (3,0) flow 70, (0,2) flow 20, (0,5) flow 100

10-16 (a) yes **(c)** no

10-17 (a) min
$23x_{N,B} + 77x_{N,W} + 8x_{S,B} + 94x_{S,W} + 53x_{T,B} + 41x_{T,W}$,
s.t. $x_{N,B} + x_{N,W} = 50, x_{S,B} + x_{S,W} = 50$,
$x_{T,B} + x_{T,W} = 50, x_{N,B} + x_{S,B} + x_{T,B} = 60$,
$x_{N,W} + x_{S,W} + x_{T,W} = 90$, all $x_{i,j} \geq 0$ **(b)** $x^*_{N,B} = 10$,
$x^*_{N,W} = 40, x^*_{S,B} = 50, x^*_{T,W} = 50$
(c)

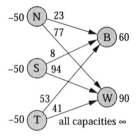

(d)

		B		W	
		23		77	
N		10		40	50
		8		94	
S		50		0	50
		8		94	
T		0		50	50
		60		90	

10-19 (a) max $\sum_{i=1}^{4} \sum_{j=1}^{4} r_{i,j}x_{i,j}$, s.t. $\sum_{j=1}^{4} x_{i,j} = 1$,
$i = 1, \ldots, 4, \sum_{i=1}^{4} x_{i,j} = 1, j = 1, \ldots, 4$, all
$x_{i,j} = 0$ or 1, where $r_{i,j} \triangleq$ rating of member i for task
j **(b)** $x^*_{1,1} = x^*_{2,4} = x^*_{3,3} = x^*_{4,2} = 1$ **(c)** 4 member
nodes with supply 1, 4 task nodes with demand 1,
arcs from every member to every task with cost the

negative of the corresponding rating **(d)** network
flow with integer supplies and demands

10-21 (a) supply 1 at 3, demand 1 at 2, all other net
demands 0, $x^* = (0, 0, 0, 1, 0)$ **(c)** supply 1 at 1,
demand 1 at 4, all other net demands 0,
$x^* = (1, 0, 1, 0, 1)$

10-23 (a)

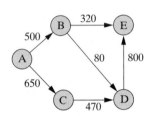

(b) $x^*_{A,B} = 400, x^*_{A,C} = 470, x^*_{B,D} = 80, x^*_{B,E} = 320$,
$x^*_{C,D} = 470, x^*_{D,E} = 550$ **(c)** add infinite capacity, cost
$= -1$ arc from E to A; all other cost $= 0$; all net
demands $= 0$

10-25 (a) add infinite capacity, cost -1 return arc
(2,3), all other costs and all net demands 0;
$x^* = (30, 0, 30, 100, 0)$ **(c)** add infinite capacity, cost
-1 return arc (4,1), all other costs and all net
demands 0; $x^* = (50, 40, 0, 50, 40)$

10-26 (a) min $25x_{P1,1,C1,1} + 30x_{P1,1,C2,1} +$
$15x_{P1,1,W,1} + 45x_{P2,1,C1,1} + 23x_{P2,1,C2,1} + 15x_{P2,1,W,1} +$
$11x_{W,1,C1,1} + 14x_{W,1,C2,1} + 25x_{P1,2,C1,2} + 30x_{P1,2,C2,2} +$
$15x_{P1,2,W,2} + 45x_{P2,2,C1,2} + 23x_{P2,2,C2,2} + 15x_{P2,2,W,2} +$
$11x_{W,2,C1,2} + 14x_{W,2,C2,2} + 10x_{W,1,W,2}$, s.t.
$x_{P1,1,C1,1} + x_{P1,1,C2,1} + x_{P1,1,W,1} \leq 400$,
$x_{P2,1,C1,1} + x_{P2,1,C2,1} + x_{P2,1,W,1} \leq 600$,
$x_{P1,1,W,1} + x_{P2,1,W,1} - x_{W,1,C1,1} - x_{W,1,C2,1} - x_{W,1,W,2} = 0$,
$x_{P1,1,C1,1} + x_{P2,1,C1,1} + x_{W,1,C1,1} = 150$,
$x_{P1,1,C2,1} + x_{P2,1,C2,1} + x_{W,1,C2,1} = 700$,
$x_{P1,2,C1,2} + x_{P1,2,C2,2} + x_{P1,2,W,2} \leq 400$,
$x_{P2,2,C1,2} + x_{P2,2,C2,2} + x_{P2,2,W,2} \leq 600$,
$x_{P1,2,W,2} + x_{P2,2,W,2} - x_{W,2,C1,2} - x_{W,2,C2,2} + x_{W,1,W,2} = 0$,
$x_{P1,2,C1,2} + x_{P2,2,C1,2} + x_{W,2,C1,2} = 300$,
$x_{P1,2,C2,2} + x_{P2,2,C2,2} + x_{W,2,C2,2} = 810$, all $x_{i,k,j,l} \geq 0$
(b) $x^*_{P1,1,C1,1} = 160, x^*_{P1,1,W,1} = 210, x^*_{P2,1,C2,1} = 600$,
$x^*_{W,1,C2,1} = 100, x^*_{P1,2,C1,2} = 300, x^*_{P1,2,W,2} = 100$,
$x^*_{P2,2,C2,2} = 600, x^*_{W,2,C2,2} = 210, x^*_{W,1,W,2} = 110$

(c)

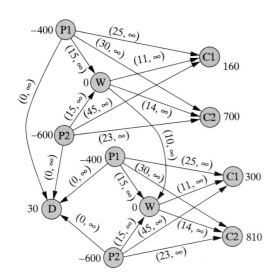

10-28 **(a)** apply weights $+1, +1$ **(c)** apply weights $+1, -1, -1$

10-29 **(a)** $x_{1,3} = 50, x_{1,4} = 10, x_{2,3} = 25, x_{2,5} = 5,$
$x_{3,5} = 65, x_{4,5} = 10$; feasible **(c)** $x_{1,2} = 25, x_{1,4} = 35,$
$x_{2,3} = 55, x_{3,5} = 45, x_{4,5} = 35$; infeasible **(e)** not
basis, cycle 1–3–2–1 **(g)** not basis, not connected

10-30 **(a)** forms spanning tree, all off-tree at bounds
(b) increase (2,3), $\Delta x = (1, -1, 1, 0, 0)$; increase
(3,4), $\Delta x = (-1, 1, 0, -1, 1)$ **(c)** yes,yes **(d)** 15,10

10-32 **(a)** $x_{2,4}^* = 35, x_{3,1}^* = 60, x_{3,2}^* = 40$ **(c)** $x_{1,2}^* = 15,$
$x_{1,3}^* = 35, x_{2,3}^* = 15, x_{3,4}^* = 10$

10-33 **(a)** check flow balance and capacities
(b)

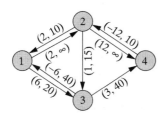

(c) direction of 1–2–3–1 or 2–3–4–2 **(d)** 15 or 10

10-35 **(a)** $x_{2,4}^* = 35, x_{3,1}^* = 60, x_{3,2}^* = 40$ **(c)** $x_{1,2}^* = 15,$
$x_{1,3}^* = 35, x_{2,3}^* = 15, x_{3,4}^* = 10$

10-36 **(a)** feasible, spanning tree, all off-tree at
bounds **(b)** 2–3–4–2 **(c)** 4–2–1–3–4

10-37 **(a)** min $\sum_{k=1}^{3}(x_{k,1,3} + x_{k,2,1} + x_{k,3,2})$,
s.t. $x_{1,3,1} + x_{1,2,1} - x_{1,1,2} - x_{1,1,3} = -7,$
$x_{1,1,2} + x_{1,3,2} - x_{1,2,3} - x_{1,2,1} = 0,$
$x_{1,2,3} + x_{1,1,3} - x_{1,3,1} - x_{1,3,2} = 7,$
$x_{2,3,1} + x_{2,2,1} - x_{2,1,2} - x_{2,1,3} = 7,$
$x_{2,1,2} + x_{2,3,2} - x_{2,2,3} - x_{2,2,1} = -7,$
$x_{2,2,3} + x_{2,1,3} - x_{2,3,1} - x_{2,3,2} = 0,$
$x_{3,3,1} + x_{3,2,1} - x_{3,1,2} - x_{3,1,3} = 0,$
$x_{3,1,2} + x_{3,3,2} - x_{3,2,3} - x_{3,2,1} = 7,$
$x_{3,2,3} + x_{3,1,3} - x_{3,3,1} - x_{3,3,2} = -7,$
$\sum_{k=1}^{3} x_{k,1,2} \le 11, \sum_{k=1}^{3} x_{k,2,3} \le 11,$
$\sum_{k=1}^{3} x_{k,3,1} \le 11$, all $x_{k,i,j} \ge 0$ **(b)**
$x_{1,1,2}^* = x_{3,1,2}^* = x_{1,2,3}^* = x_{2,2,3}^* = x_{2,3,1}^* = x_{3,3,1}^* = 5.5,$
$x_{1,1,3}^* = x_{2,2,1}^* = x_{3,3,2}^* = 1.5$

(c)

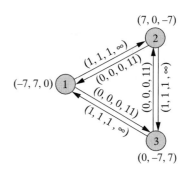

(d) an optimal solution ships each 7 units to itself at
zero cost **(e)** integer flows not guaranteed in
multicommodity flows

10-39 **(a)** min $7(x_{1,1} + x_{1,2}) + 8(x_{2,1} + x_{2,2})$
$+5(x_{3,1} + x_{3,2})$, s.t. $\sum_{j=1}^{2} x_{i,j} \le 300, i = 1, \ldots, 3,$
$0.7x_{1,1} + 0.8x_{2,1} + 0.6x_{3,1} = 350,$
$0.6x_{1,2} + 0.8x_{2,2} + 0.7x_{3,2} = 275$, all $x_{i,j} \ge 0$
(b) $x_{1,1}^* = 300, x_{2,1}^* = 175, x_{2,2}^* = 81.25, x_{3,2}^* = 300$

(c)

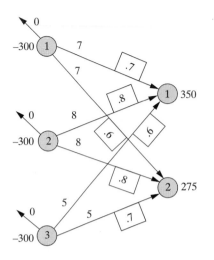

(d) integer flows not guaranteed in flow with losses

Chapter 11

11-1 (a) proportional to variable magnitude
(b) $x^* = (96.67, 5, 48.33)$ **(c)** min
$2250y_1 + 1650y_2 + 2700y_3$, s.t.
$150y_1 + 150y_2 + 150y_3 = 150$,
$300y_1 + 600y_2 + 300y_3 \le 310$,
$600y_1 + 450y_2 + 150y_3 \le 450$, $y_j = 0$ or 1,
$j = 1, \dots, 3$ **(d)** $x^* = (0, 0, 150)$, $y^* = (0, 0, 1)$
(e) min $15x_1 + 11x_2 + 18x_3 + 400(y_1 + y_2 + y_3)$, s.t.
$x_j \le 150y_j$, $y_j = 0$ or $1, j = 1, \dots, 3$, and all original
constraints **(f)** $x^* = (0, 0, 150)$, $y^* = (0, 0, 1)$
(g) add variables y_1, y_2, y_3, and constraints
$50y_j \le x_j \le 150y_j$, $y_j = 0$ or $1, j = 1, \dots, 3$
(h) $x^* = (100, 0, 50)$, $y^* = (1, 0, 1)$

11-2 (h) $x^* = (175, 0, 125)$, $y^* = (1, 0, 1)$

11-3 (a) max $4.5x_1 + 4.1x_2 + 8x_3 + 7x_4$, s.t.
$4x_1 + 3.8x_2 + 6x_3 + 7.2x_4 \le 8$ $x_j = 0$ or 1,
$j = 1, \dots, 4$ **(b)** $x^* = (1, 1, 0, 0)$

11-5 (a) budget: $100 total; mutual exclusiveness:
alternatives for NW and SE parcels; dependency:
IE tunnel upon IE lab **(b)** max
$9x_1 + 2x_2 + 10x_3 + 2x_4 + 5x_5 + 8x_6 + 10x_7 + 1x_8$,
s.t. $48x_1 + 20.8x_2 + 32x_3 + 28x_4 + 44x_5 + 17.2x_6 +$
$36.8x_7 + 1.2x_8 \le 100$, $x_1 + x_4 + x_5 + x_6 \le 1$,
$x_2 + x_7 \le 1$, $x_8 \le x_7, x_1, \dots, x_8 = 0$ or 1
(c) $x^* = (0, 0, 1, 0, 0, 1, 1, 1)$

11-7 (a) max $45x_1 + 30x_2 + 84x_3 + 73x_4 + 80x_5$
$+70x_6 + 61x_7 + 91x_8$, s.t. $x_1 + x_2 \le 1, x_2 + x_5 \le 1$,
$x_4 + x_7 \le 1, x_5 + x_6 + x_7 \le 1, x_1, \dots, x_8 = 0$ or 1
(b) $x^* = (1, 0, 1, 1, 0, 0, 0, 1)$

11-9 (a) min $40x_1 + 65x_2 + 43x_3 + 48x_4$
$+72x_5 + 36x_6$, s.t. $x_1 + x_2 \ge 1, x_1 + x_4 \ge 1$,
$x_2 + x_3 \ge 1, x_2 + x_5 \ge 1, x_3 + x_5 \ge 1, x_3 + x_6 \ge 1$,
$x_4 + x_5 \ge 1, x_5 + x_6 \ge 1, x_1, \dots, x_6 = 0$ or 1 **(b)**
$x^* = (0, 1, 1, 1, 0, 1)$ **(c)** min $\sum_{i=1}^{9} y_i$, s.t.
$x_1 + x_2 + y_1 \ge 1$,
$x_1 + x_4 + y_2 \ge 1, x_2 + x_3 + y_3 \ge 1$,
$x_2 + x_4 + y_4 \ge 1, x_2 + x_5 + y_5 \ge 1$,
$x_3 + x_5 + y_6 \ge 1, x_3 + x_6 + y_7 \ge 1$,
$x_4 + x_5 + y_8 \ge 1, x_5 + x_6 + y_9 \ge 1$,
$\sum_{j=1}^{6} x_j \le 2, x_1, \dots, x_6 = 0$ or 1,
$y_1, \dots, y_9 = 0$ or 1 **(d)** $x^* = (0, 1, 0, 0, 1, 0)$,
$y^* = (0, 1, 0, 0, 0, 0, 1, 0, 0)$

11-11 (a) min $1.40x_1 + 0.96x_2 + 1.52x_3 + 1.60x_4 +$
$1.32x_5 + 1.12x_6 + 0.84x_7 + 1.54x_8$,
s.t. $x_2 + x_5 + x_8 = 1, x_1 + x_3 + x_4 = 1$,
$x_5 + x_6 + x_8 = 1, x_1 + x_3 + x_7 + x_8 = 1$,
$x_3 + x_4 + x_6 = 1, x_2 + x_4 + x_5 + x_7 = 1$,
$x_1, \dots, x_8 = 0$ or 1
(b) $x^* = (0, 0, 1, 0, 1, 0, 0, 0)$

11-13 (a) min $18x_{1,1} + 26x_{1,2} + 31x_{1,4} + 50x_{2,2}$
$+22x_{2,3} + 40x_{3,1} + 29x_{3,2} + 52x_{3,3} + 39x_{3,4} + 43x_{4,3} +$
$46x_{4,4}$, s.t. $x_{1,1} + x_{1,2} + x_{1,4} = 1, x_{2,2} + x_{2,3} = 1$,
$x_{3,1} + x_{3,2} + x_{3,3} + x_{3,4} = 1, x_{4,3} + x_{4,4} = 1$,
$x_{1,1} + x_{3,1} = 1, x_{1,2} + x_{2,2} + x_{3,2} = 1$,
$x_{2,3} + x_{3,3} + x_{4,3} = 1, x_{1,4} + x_{3,4} + x_{4,4} = 1$,
all $x_{i,j} \ge 0$ **(b)** can be viewed as a network flow with
unit supplies and demands
(c) $x_{1,1}^* = x_{2,3}^* = x_{3,2}^* = x_{4,4}^* = 1$

11-15 (a) min $60x_{H,1}x_{E,2} + 120x_{H,1}x_{E,3} +$
$36x_{H,1}x_{M,2} + 72x_{H,1}x_{M,3} + 60x_{H,2}x_{E,1} + 20x_{H,2}x_{E,3} +$
$36x_{H,2}x_{M,1} + 12x_{H,2}x_{M,3} + 120x_{H,3}x_{E,1} + 20x_{H,3}x_{E,2} +$
$72x_{H,3}x_{M,1} + 12x_{H,3}x_{M,2} + 42x_{E,1}x_{M,2} + 84x_{E,1}x_{M,3} +$
$42x_{E,2}x_{M,1} + 14x_{E,2}x_{M,3} + 84x_{E,3}x_{M,1} + 14x_{E,3}x_{M,2}$,
s.t. $x_{H,1} + x_{H,2} + x_{H,3} = 1, x_{E,1} + x_{E,2} + x_{E,3} = 1$,
$x_{M,1} + x_{M,2} + x_{M,3} = 1, x_{H,1} + x_{E,1} + x_{M,1} = 1$,
$x_{H,2} + x_{E,2} + x_{M,2} = 1, x_{H,3} + x_{E,3} + x_{M,3} = 1$, all
$x_{i,j} = 0$ or 1 **(b)** can assess cost only after pairs of
assignments **(c)** $x_{H,3}^* = x_{E,2}^* = x_{M,1}^* = 1$

11-17 (a) min $\sum_{i=1}^{6}(c_{i,F}x_{i,F} + c_{i,B}x_{i,B})$, s.t.
$x_{i,F} + x_{i,B} = 1, i = 1, \dots, 6, \sum_{i=1}^{6} t_i x_{i,F} \le 200$,

$\sum_{i=1}^{6} t_i x_{i,B} \leq 190$, all $x_{i,j} = 0$ or 1, where $c_{i,j} \triangleq$ materials handling to move i to j and $t_i \triangleq$ time for i **(b)** more than one unit of time allocated by each decision **(c)** $x_{1,B}^* = x_{2,B}^* = x_{3,F}^* = x_{4,F}^* = x_{5,B}^* = x_{6,F}^* = 1$

11-19 (a) min $\sum_{i=1}^{5} \sum_{j=i+1}^{6} c_{i,j} x_{i,j}$, s.t.

$x_{1,2} + x_{1,3} + x_{1,4} + x_{1,5} + x_{1,6} = 1$,
$x_{1,2} + x_{2,3} + x_{2,4} + x_{2,5} + x_{2,6} = 1$,
$x_{1,3} + x_{2,3} + x_{3,4} + x_{3,5} + x_{3,6} = 1$,
$x_{1,4} + x_{2,4} + x_{3,4} + x_{4,5} + x_{4,6} = 1$,
$x_{1,5} + x_{2,5} + x_{3,5} + x_{4,5} + x_{5,6} = 1$,
$x_{1,6} + x_{2,6} + x_{3,6} + x_{4,6} + x_{5,6} = 1$,

all $x_{i,j} = 0$ or 1, where $c_{i,j} \triangleq$ cost of pairing i with j **(b)** objects to be paired do not come from distinct sets **(c)** $x_{1,5}^* = x_{2,3}^* = x_{4,6}^* = 1$

11-21 (a) seeks a minimum total length closed route visiting every point
(b)

	1	2	3	4	5	6
1	—	40	230	160	220	40
2	40	—	190	120	180	20
3	230	190	—	70	30	190
4	160	120	70	—	60	120
5	220	180	30	60	—	180
6	40	20	190	120	180	—

(c) min $\sum_{i=1}^{5} \sum_{j=i+1}^{6} d_{i,j} x_{i,j}$, s.t.

$x_{1,2} + x_{1,3} + x_{1,4} + x_{1,5} + x_{1,6} = 2$,
$x_{1,2} + x_{2,3} + x_{2,4} + x_{2,5} + x_{2,6} = 2$,
$x_{1,3} + x_{2,3} + x_{3,4} + x_{3,5} + x_{3,6} = 2$,
$x_{1,4} + x_{2,4} + x_{3,4} + x_{4,5} + x_{4,6} = 2$,
$x_{1,5} + x_{2,5} + x_{3,5} + x_{4,5} + x_{5,6} = 2$,
$x_{1,6} + x_{2,6} + x_{3,6} + x_{4,6} + x_{5,6} = 2$, all $x_{i,j} = 0$ or 1,

where $d_{i,j} \triangleq$ i-to-j distance of part (b)
(d) $x_{1,2}^* = x_{1,6}^* = x_{2,6}^* = x_{3,4}^* = x_{3,5}^* = x_{4,5}^* = 1$
(e) $x_{1,3} + x_{1,4} + x_{1,5} + x_{2,3} + x_{2,4}$
$+ x_{2,5} + x_{3,6} + x_{4,6} + x_{5,6} \geq 2$
(f) $x_{1,2}^* = x_{1,6}^* = x_{2,3}^* = x_{3,5}^* = x_{4,5}^* = x_{4,6}^* = 1$
(g) model $\boxed{11.27}$ with $d_{i,j}$ as in part (b)

11-23 (a) seeks a minimum total setup time closed tour of the four products, and times are not symmetric **(b)** min $\sum_{i=1}^{4} \sum_{j \neq i} c_{i,j} x_{i,j}$, s.t. $\sum_{j \neq i} x_{i,j} = 1$, $i = 1, \ldots, 4$, $\sum_{i \neq j} x_{i,j} = 1$ $j = 1, \ldots, 4$, all $x_{i,j} \geq 0$, where $c_{i,j} \triangleq$ given i to j setup time.
(c) $x_{1,3}^* = x_{3,1}^* = x_{2,4}^* = x_{4,2}^* = 1$
(d) $x_{1,2} + x_{1,4} + x_{3,2} + x_{3,4} \geq 1$

(e) $x_{1,2}^* = x_{2,4}^* = x_{3,1}^* = x_{4,3}^* = 1$ **(f)** model $\boxed{11.27}$ with $d_{i,j}$ the given setup times

11-25 (a) min $\sum_{i=1}^{3} (f_i y_i + \sum_{j=1}^{4} d_j c_{i,j} x_{i,j})$, s.t. $\sum_{j=1}^{4} x_{i,j} \leq 4 y_i$, $i = 1, \ldots, 3$, $\sum_{i=1}^{3} x_{i,j} = 1$, $j = 1, \ldots, 4$, $y_1, \ldots, y_3 = 0$ or 1, all $x_{i,j} \geq 0$, where $f_i \triangleq$ given fixed cost for i, $c_{i,j} \triangleq$ given transportation cost from i to j, and $d_j \triangleq$ given demand at j.
(b) $y_3^* = x_{3,1}^* = x_{3,2}^* = x_{3,3}^* = 1$

11-27 (a) min $2x_{F1,S1} + 8000y_{F1,S1} + 2x_{F2,F1} + 6000y_{F2,F1} + 2x_{F2,S1} + 10,000y_{F2,S1} + 2x_{F2,S2} + 14,000y_{F2,S2} + 2x_{S1,S2} + 2000y_{S1,S2} + 2x_{S1,T} + 2x_{S2,T}$,
s.t. $x_{F2,F1} - x_{F1,S1} = -800$,
$-x_{F2,F1} - x_{F2,S1} - x_{F2,S2} = -600$,
$x_{F1,S1} + x_{F2,S1} - x_{S1,S2} - x_{S1,T} = 0$,
$x_{F2,S2} + x_{S1,S2} - x_{S2,T} = 0$,
$x_{S1,T} + x_{S2,T} = 1400$, $x_{F1,S1} \leq 1000y_{F1,S1}$,
$x_{F2,F1} \leq 1000y_{F2,F1}$, $x_{F2,S1} \leq 1000y_{F2,S1}$,
$x_{F2,S2} \leq 1000y_{F2,S2}$, $x_{S1,S2} \leq 1000y_{S1,S2}$,
all $x_{i,j} \geq 0$, all $y_{i,j} = 0$ or 1 **(b)** $y_{F1,S1}^* = y_{F2,S2}^* = 1$, $x_{F1,S1}^* = x_{S1,T}^* = 800$, $x_{F2,S2}^* = x_{S2,T}^* = 600$

11-29 (a) $x_1 + 10 \leq x_2 + M(1 - y_{1,2})$,
$x_2 + 2 \leq x_1 + My_{1,2}$,
$x_1 + 10 \leq x_3 + M(1 - y_{1,3})$,
$x_3 + 16 \leq x_1 + My_{1,3}$,
$x_1 + 10 \leq x_4 + M(1 - y_{1,4})$,
$x_4 + 8 \leq x_1 + My_{1,4}$,
$x_2 + 3 \leq x_3 + M(1 - y_{2,3})$,
$x_3 + 16 \leq x_2 + My_{2,3}$,
$x_2 + 3 \leq x_4 + M(1 - y_{2,4})$,
$x_4 + 8 \leq x_2 + My_{2,4}$,
$x_3 + 16 \leq x_4 + M(1 - y_{3,4})$,
$x_4 + 8 \leq x_3 + My_{3,4}$, $x_1 \geq 0$, $x_2 \geq 20$, $x_3 \geq 1$,
$x_4 \geq 12$, all $y_{i,j} = 0$ or 1 **(b)** 39, 23.5, 38, 15.25, 19, 2.75, 19, 4.75 **(c)** min
$\frac{1}{4}(x_1 + 10 + x_2 + 3 + x_3 + 16 + x_4 + 8)$, s.t.
constraints of part (a) **(d)** $x_1^* = 0$, $x_2^* = 20$, $x_3^* = 23$, $x_4^* = 2$ **(e)** mean flow time and mean lateness
(f) min z, s.t. $z \geq x_1 - 2$, $z \geq x_2 - 27$, $z \geq x_3 - 4$, $z \geq x_4 - 13$, plus all constraints of part (a)
(g) $x_1^* = 0$, $x_2^* = 34$, $x_3^* = 10$, $x_4^* = 26$ **(h)** maximum tardiness

11-31 (a) $x_{1,1} + 10 \leq x_{1,2}$, $x_{1,2} + 3 \leq x_{1,3}$,
$x_{2,1} + 2 \leq x_{2,3}$, $x_{2,3} + 1 \leq x_{2,2}$,
$x_{3,2} + 6 \leq x_{3,1}$, $x_{3,1} + 12 \leq x_{3,3}$,
$x_{1,1} + 10 \leq x_{2,1} + M(1 - y_{1,2,1})$,

$x_{2,1} + 2 \leq x_{1,1} + My_{1,2,1},$

$x_{1,1} + 10 \leq x_{3,1} + M(1 - y_{1,3,1}),$

$x_{3,1} + 12 \leq x_{1,1} + My_{1,3,1},$

$x_{2,1} + 2 \leq x_{3,1} + M(1 - y_{2,3,1}),$

$x_{3,1} + 12 \leq x_{2,1} + My_{2,3,1},$

$x_{1,2} + 10 \leq x_{2,2} + M(1 - y_{1,2,2}),$

$x_{2,2} + 2 \leq x_{1,2} + My_{1,2,2},$

$x_{1,2} + 10 \leq x_{3,2} + M(1 - y_{1,3,2}),$

$x_{3,2} + 12 \leq x_{1,2} + My_{1,3,2},$

$x_{2,2} + 2 \leq x_{3,2} + M(1 - y_{2,3,2}),$

$x_{3,2} + 12 \leq x_{2,2} + My_{2,3,2},$

$x_{1,3} + 10 \leq x_{2,3} + M(1 - y_{1,2,3}),$

$x_{2,3} + 2 \leq x_{1,3} + My_{1,2,3},$

$x_{1,3} + 10 \leq x_{3,3} + M(1 - y_{1,3,3}),$

$x_{3,3} + 12 \leq x_{1,3} + My_{1,3,3},$

$x_{2,3} + 2 \leq x_{3,3} + M(1 - y_{2,3,3}),$

$x_{3,3} + 12 \leq x_{2,3} + My_{2,3,3},$ all $x_{j,k} \geq 0,$ all $y_{i,j,k} = 0$ or 1 **(b)** mean flow time **(c)** min $\frac{1}{3}(x_{1,3} + 14 + x_{2,2} + 4 + x_{3,3} + 8),$ s.t. all constraints of part (a) **(d)** $x_{1,1}^* = 2, x_{1,2}^* = 12, x_{1,3}^* = 15, x_{2,1}^* = 0,$ $x_{2,2}^* = 6, x_{2,3}^* = 5, x_{3,1}^* = 12, x_{3,2}^* = 0, x_{3,3}^* = 29$

Chapter 12

12-1 (a) $\mathbf{x}^* = (0, 1, 1, 0)$

12-2 (a) 16, 21

12-3 (a) yes **(c)** no

12-4 (a) same except last two constraints replaced by $0 \leq x_1 \leq 1$

12-5 (a) yes **(c)** no

12-6 (a) optimal ILP value ≤ 54.5 **(c)** optimal ILP value ≥ 19

12-7 (a) no **(c)** yes

12-8 (a) ILP optimal value ≤ 18.5 **(b)** not optimal; $\hat{\mathbf{x}} = (0, 0, 1, 1, 0)$ **(c)** $14 \leq$ optimal value ≤ 18.5 **(d)** $\mathbf{x}^* = (0, 0, 1, 1, 0),$ value 14

12-10 (a) ILP optimal value ≥ 26.5 **(b)** optimal **(c)** $26.5 = $ optimal value **(d)** $\mathbf{x}^* = (0.5, 1, 0, 1, 0.5),$ value 26.5

12-12 (a) $\mathbf{x}^* = (0, 0, 0, 0), y^* = 0$ **(c)** same integer-feasible solutions of all $= 0$ or all $= 1$ **(d)** solution value 0 vs. solution value -42

12-14 (a) valid; yes **(b)** valid; no **(c)** not valid **(d)** valid, yes

12-16 (a) valid; yes **(c)** not valid

12-17 (a) 4, 6 **(b)** original LP optimum violates $x_2 \leq 6y_2$ **(c)** new $\tilde{\mathbf{x}} = (0, 3.2), \tilde{\mathbf{y}} = (0, 0.533)$

12-19 (a) max $30x_1 + 55x_2 + 20x_3 + v_1(55 - 40x_1 + 12x_2 - 11x_3) + v_2(20 - 19x_1 - 60x_2 - 3x_3),$ s.t. all undualized constraints; $v_1 \geq 0, v_2 \leq 0$

12-20 (a) $x_{1,1}^* = x_{1,2}^* = y_1^* = 1$ **(b)** min $3x_{1,1} + 6x_{1,2} + 5x_{2,1} + 2x_{2,2} + 250y_1 + 300y_2 + v_1(1 - x_{1,1} - x_{2,1}) + v_2(1 - x_{1,2} + x_{2,2}),$ s.t. all given constraints except the two relaxed **(c)** optimizations are now independent for facilities 1 and 2 **(d)** all $\tilde{x}_{i,j}$ and $\tilde{y}_i = 0,$ relaxation value $0 \leq 259$ **(e)** all $\tilde{x}_{i,j}$ and $\tilde{y}_i = 0,$ relaxation value $200 \leq 259$ **(f)** all $\tilde{x}_{i,j}$ and $\tilde{y}_i = 1,$ relaxation value $-934 \leq 259$

12-22 (a) $(0, 0, 0), (0, 0, 1), (1, 0, 0), (1, 0, 1)$

12-23 (a) (#, #, #, #), (#, #, 1, #), (#, 1, 1, #), (#, 0, 1, #), (#, #, 0, #) **(b)** branched: 0, 1; terminated: 2, 3, 4 **(c)** 0, 3

12-25 (a) given ILP plus constraints $x_2 = 1, x_5 = 0$

12-26 (a) branch on $x_1 = 1$ vs. $x_1 = 0$ **(c)** terminate by solving after saving $\tilde{\mathbf{x}}$ as a new incumbent solution **(e)** terminate by bound

12-27 partial solutions: (#, #, #), (#, #, 0), (0, #, 0), (0, 1, 0), (0, 0, 0), (1, #, 0), (1, 0, 0), (1, 1, 0), (#, #, 1), (1, #, 1), (0, #, 1); $\mathbf{x}^* = (1, 0, 1, 0)$

12-31 0: branch on fractional x_3; 1: branch on fractional x_2; 2: branch on fractional x_1; 3: terminate by solving after saving incumbent solution $\hat{\mathbf{x}} = (1, 0, 0), \hat{v} = 90$; 4: terminate infeasible; 5: terminate by solving after saving incumbent solution $\hat{\mathbf{x}} = (0, 1, 0), \hat{v} = 50$; 6: terminate by bound $54 \geq 50$; $\mathbf{x}^* = (0, 1, 0)$

12-33 (b) 0: round for incumbent solution $\hat{\mathbf{x}} = (0, 0, 1), \hat{v} = 54,$ branch on fractional x_3; 1: round for new incumbent solution $\hat{\mathbf{x}} = (0, 1, 0),$ $\hat{v} = 50,$ branch on fractional x_2; 2: terminate by bound $75 \geq 50$; 3: terminate by bound $90 \geq 50$; 4: terminate by bound $54 \geq 50$; $\mathbf{x}^* = (0, 1, 0)$

12-35 (a) 18 to ∞, 18 to ∞, 18 to ∞, 18 to 90, 18 to 90, 18 to 50, 50 to 50

12-36 (a) $a, b; a, b, c$ **(b)** 212 **(c)** 17; 8.7 percent

12-38 (a) partial solutions: (#, #, #), (#, #, 0), (#, #, 1), (0, #, 0), (1, #, 0), (1, 0, 0), (1, 1, 0), (1, #, 1), (0, #, 1); (0, 1, 0), (0, 0, 0); $\mathbf{x}^* = (1, 0, 1, 0)$ **(b)** partial solutions: (#, #, #), (#, #, 0), (0, #, 0), (0, 1, 0),

(#, #, 1), (1, #, 1), (1, #, 0), (1, 0, 0), (1, 1, 0), (0, #, 1); (0, 0, 0); $\mathbf{x}^* = (1, 0, 1, 0)$

12-41 0: fractional so introduce valid inequality $4x_1 + 3x_2 \geq 3$, which cuts off the current relaxation; 1: lacking further valid inequalties, branch on fractional x_1; 2: fractional so introduce valid inequality $x_2 + x_3 \leq 1$ which cuts off the current relaxation; 3: terminate infeasible; 4: lacking further valid inequalities, branch on fractional x_3; 5: terminate by solving after saving incumbent solution $\hat{\mathbf{x}} = (0, 1, 1)$, $\hat{v} = 75$; 6: terminate infeasible; $\mathbf{x}^* = (0, 1, 1)$

12-43 **(a)** $\mathbf{x}^* = (1, 1, 0)$ **(b)** $(0,0,1)$, $(1,0,1)$, $(0,1,1)$, $(0,0,2)$ **(c)** $\hat{\mathbf{x}} = (1, 0, 1)$ **(d)** $\hat{\mathbf{x}} = (1, 1, 0)$ **(f)** $\hat{\mathbf{x}} = (1, 1, 0)$

12-45 **(a)** $\mathbf{x}^* = (0, 1, 1, 0)$ **(b)** $\hat{\mathbf{x}} = (1, 0, 0, 0)$ **(c)** $\hat{\mathbf{x}} = (0, 1, 1, 0)$

12-49 $\hat{\mathbf{x}} = (0, 1, 1, 0)$

12-51 $\hat{\mathbf{x}} = (0, 1, 1, 0)$

12-53 **(a)** check both feasible **(b)** after 1 or 2: $\mathbf{x}^{(3)} = \mathbf{x}^{(1)}$, $\mathbf{x}^{(4)} = \mathbf{x}^{(2)}$; after 3: $\mathbf{x}^{(3)} = (0, 0, 1, 1)$, $\mathbf{x}^{(4)} = (0, 0, 0, 0)$ **(c)** all feasible except $\mathbf{x}^{(3)}$ cutting after 3; infeasibles must either be excluded from the population or included with a large negative objective value

12-55 $(0, 1, 1, 0)$, value 16; $(0, 0, 1, 0)$, value 9; $(0, 1, 0, 1)$, value 15; and any feasible immigrant such as $(0, 1, 0, 0)$, value 7

12-57 **(a)** $\mathbf{x}^* = (1, 0, 1, 0)$ **(b)** most payoff per unit constraint usage **(c)** $\hat{\mathbf{x}} = (0, 1, 0, 1)$

Chapter 13

13-1 **(a)** min $40(x/2) + 2000/(x/5)$ **(b)** $x^* = 22.4$

13-2 **(b)** $\alpha^* = 0.031$

13-3 **(a)** min $\sqrt{(x_1)^2 + (x_2 + 30)^2}$ $+\sqrt{(x_1 - 50)^2 + (x_2 + 10)^2}$ $+\sqrt{(x_1 - 70)^2 + (x_2 - 20)^2}$ $+\sqrt{(x_1 - 30)^2 + (x_2 - 50)^2}$ **(b)** $\mathbf{x}^* = (45.8, 2.7)$

13-4 **(b)** $\mathbf{x}^* = (27.7, 5.2)$

13-5 **(a)** min $\sum_{i=1}^{5}[t_i - a(u_i)^b]^2$, where u_i and t_i are the given units and average time values **(b)** $a^* = 10.36$, $b^* = -0.322$

13-6 **(b)** $k^* = 62.79$, $a^* = 2.011$, $b^* = -0.363$

13-11 **(b)** Defining $\alpha = am$, $\beta = (bm - a)$, $\gamma = -b$, the fitted equation has the linear form $\alpha + \beta n_{t-1} + \gamma(n_t)^2$

13-13 **(a)** yes **(c)** no **(e)** yes

13-14 **(a)** local maximum; local and global minimum; nothing; local and global maximum; nothing; local minimum

13-15 **(a)** local maximum; local and global maximum; nothing; local and global minimum; local minimum

13-16 **(a)** no max, $x^{(1)}$, $x^{(4)}$; no min, $x^{(6)}$, $x^{(2)}$ **(c)** no max, $\mathbf{x}^{(1)}$, $\mathbf{x}^{(2)}$; no min, $\mathbf{x}^{(5)}$, $\mathbf{x}^{(4)}$ **(e)** yes max; no min $x = 3, -2$

13-17 at $t = 4$, $x^* \approx 2.47$

13-19 **(a)** $x^{(\text{hi})} = 4.5$ **(b)** $x^{(\text{hi})} = 9$

13-21 at $t = 3$, $x^* \approx 2.79$

13-23 **(a)** $f_1(3 + \lambda) = 33 + 20\lambda$ **(b)** $f_2(3 + \lambda) = 33 + 20\lambda + 6\lambda^2$

13-25 **(a)** $f_1((0, 2) + \lambda(1, -1)) = 24 - 34\lambda$ **(b)** $f_2((0, 2) + \lambda(1, -1)) = 24 - 34\lambda + 11\lambda^2$

13-27 **(a)** stationary **(c)** $\Delta\mathbf{x} = (-16, 3)$

13-28 **(a)** definitely local minimum **(c)** definitely neither **(e)** possibly local maximum and possibly local minimum **(g)** possibly local minimum

13-29 **(a)** concave **(c)** neither **(e)** both **(g)** convex **(i)** convex

13-30 **(a)** global minimum

13-31 **(b)** $\Delta\mathbf{x} = (23, 97)$ **(c)** max $100 - 5[(1 + 23\lambda) - 2]^4 - 3[(3 + 97\lambda) - 5]^4 + (1 + 23\lambda)(3 + 97\lambda)$, s.t. $\lambda > 0$ **(d)** $\lambda = 0.027$, $\mathbf{x}^{(1)} = (1.621, 5.619)$ **(e)** $\mathbf{x}^{(2)} = (2.627, 5.435)$, $\mathbf{x}^{(3)} = (2.675, 5.592)$

13-33 **(a)** $f_2((3, 7) + \Delta\mathbf{x}) \triangleq 68 + (-13, -93)\Delta\mathbf{x} + \frac{1}{2}\Delta\mathbf{x}\begin{pmatrix} -60 & 1 \\ 1 & -144 \end{pmatrix}\Delta\mathbf{x}$ **(b)** $\Delta\mathbf{x} = (-0.2275, -0.6474)$ **(c)** $\mathbf{x}^{(1)} = (2.773, 6.353)$, $\mathbf{x}^{(2)} = (2.681, 5.942)$

13-35 **(a)** $\Delta\mathbf{x} = (3, 98)$ **(b)** $\mathbf{x}^{(1)} = (2.078, 5.548)$, $\mathbf{D} = \begin{pmatrix} -1.003 & -0.0268 \\ -0.0268 & -0.0267 \end{pmatrix}$, $\Delta\mathbf{x} = (5.555, 0.151)$

13-37 **(a)** the function is not differentiable everywhere **(b)** $\mathbf{x}^{(0)} = (5, 2.5)$, $\mathbf{x}^{(1)} = (0, 1.25)$, $\mathbf{x}^{(2)} = (-1.25, -0.9375)$, $\mathbf{x}^{(3)} = (-1.5625, -0.2344)$

13-39 (a) $\mathbf{y}^{(1)} = (1, 2, 1)$, $\mathbf{y}^{(2)} = (3, 3, 3)$,
$\mathbf{y}^{(3)} = (2, 2, 4)$, $\mathbf{y}^{(4)} = (4, 2, 4)$

Chapter 14

14-1 (a) max ℓwh, s.t. $w + 2h \le 30$, $\ell + 2h \le 40$,
$\ell, w, h \ge 0$ (b) $\ell^* = 28.685$, $w^* = 18.685$, $h^* = 5.657$

14-2 (b) $\ell^* = 27.386$, $w^* = 27.386$, $h^* = 66.667$,
$d^* = 52.974$

14-3 (a) min $\sum_{j=1}^{5}(h_j x_j/2 + s_j d_j/x_j)$, s.t.
$\sum_{j=1}^{5} v_j x_j \le 4000$, $x_1, \ldots, x_5 \ge 0$, where s_j, v_j, h_j, and
d_j are the values in the table.
(b) $\mathbf{x}^* = (41.423, 55.937, 47.602, 52.600, 27.441)$

14-4 (b) $\mathbf{x}^* = (10.338, 9.456, 6.304, 8.189, 3.740)$

14-5 (a) min
$52[42/(60Nf) + 0.1(42/(5^{6.667}N^{-5.667}f^{-3}))] +$
$87[42/(5^{6.667}N^{-5.667}f^{-3})]$, s.t. $200 \le N \le 600$,
$0.001 \le f \le 0.005$ (b) $N^* = 200$, $f^* = 0.001$

14-6 (b) $\mathbf{x}^* = (83.052, 26.263,$
$37.142, 31.391, 74, 284)$

14-7 (a) min $\sum_{i=1}^{5} \sum_{j=1}^{2} w_{i,j}\sqrt{(e_i - x_j)^2 + (n_i - y_j)^2}$,
s.t. $\sum_{j=1}^{2} w_{i,j} = d_i$, $i = 1, \ldots, 5$, $\sum_{i=1}^{5} w_{i,j} \le 200$,
$j = 1, 2$, $w_{i,j} \ge 0$, $i = 1, \ldots, 5$, $j = 1, 2$, where d_i, e_i
and n_i are the values in the table. (b) $x_1^* = 20$,
$y_1^* = 17$, $x_2^* = 4$, $y_2^* = 30$, $w_{1,1}^* = 10$, $w_{1,2}^* = 50$,
$w_{2,2}^* = 90$, $w_{3,1}^* = 35$, $w_{4,1}^* = 85$, $w_{5,1}^* = 80$

14-8 (b) $x^* = 1.897$, $y^* = 7.196$,
$\mathbf{w}^* = (10, 0, 20, 40, 30)$

14-9 (a) min $\sum_{i=1}^{3} \sum_{j=1}^{3} v_{i,j} x_i x_j$, s.t. $\sum_{i=1}^{3} x_i = 1.5$,
$\sum_{i=1}^{3} m_i x_i \ge (10)(1.5)$, $x_1, x_2, x_3 \ge 0$, where the m_i
are the mean return rates and $v_{i,j}$ the covariances in
the given table. (b) $x_1^* = 0.282$, $x_2^* = 1.218$, $x_3^* = 0.0$

14-10 (b) $\mathbf{x}^* = (0.2, 0.6, 0.2)$

14-11 (a) min $\sum_{j=1}^{5} c_j x_j$, s.t. $\sum_{j=1}^{5} x_j = 1$,
$12 \le \sum_{j=1}^{5} a_{1,j} x_j \le 16$, $31 \le \sum_{j=1}^{5} a_{2,j} x_j \le 36$,
$\ln(121) \le \ln\left(\sum_{j=1}^{5} a_{3,j} x_j\right) \le \ln(164)$, $x_1, \ldots, x_5 \ge 0$,
where the c_j are the costs, and $a_{i,j}$ the ingredient
index values in the table.
(b) $\mathbf{x}^* = (0.778, 0, 0.222, 0, 0)$

14-12 (b) $\mathbf{x}^* = (0, 0, 1217.742, 0, 32.288)$

14-15 max $\sum_{j=1}^{7} p_j s_j r_j(x_j)$, s.t. $\sum_{j=1}^{7} e_j x_j = \sum_{j=1}^{7} e_j$;
$0.5 \le x_j \le 1.5$, $j = 1, \ldots, 7$

14-21 min $f_2(T_{1,1}, T_{1,2}, T_{2,1}, T_{2,2}, F_1, F_2) + f_3(S)$, s.t.
$T_{2,2} = f_1(T_{2,1}, F_2, S, T_0)$; $(T_{1,1} - T_{1,2})F_1 = H$;
$(T_{1,1} - T_{1,2})F_1 + k f_2(T_{1,1}, T_{1,2}, T_{2,1}, T_{2,2}, F_1, F_2) =$
$(T_{2,1} - T_{2,2})F_2$; $\underline{T}_{i,j} \le T_{i,j} \le \overline{T}_{i,j}$, $i, j = 1, 2$;
$\underline{F}_i \le F_i \le \overline{F}_i$, $i = 1, 2$; $\underline{S} \le S \le \overline{S}$

14-23 (a) yes (c) no (e) yes

14-24 (a) yes (c) yes (e) no

14-25 (a) not QP (c) QP: $\mathbf{c} = (0, 0, 9)$,
$$\mathbf{Q} = \begin{pmatrix} 0 & 1 & 0 \\ 1 & 1 & 0 \\ 0 & 0 & 0 \end{pmatrix}$$

14-26 (a) no (c) yes

14-27 (a) min $13x_1 x_2 (x_3)^{-1} + 9(x_1)^{1/2}(x_3)^{1/2}$, s.t.
$3x_1(x_3)^{-1} + 8x_2(x_3)^{-1} \le 1$, $5(x_3)^{-4} \le 1$, $x_1, x_2, x_3 > 0$;
$K_0 = \{1, 2\}$, $K_1 = \{3, 4\}$, $K_2 = \{5\}$, $d_1 = 13$, $d_2 = 9$,
$d_3 = 3$, $d_4 = 8$, $d_5 = 5$, $a_{1,1} = 1$, $a_{1,2} = 1$, $a_{1,3} = -1$,
$a_{2,1} = \frac{1}{2}$, $a_{2,2} = 0$, $a_{2,3} = \frac{1}{2}$, $a_{3,1} = 1$, $a_{3,2} = 0$,
$a_{3,3} = -1$, $a_{4,1} = 0$, $a_{4,2} = 1$, $a_{4,3} = -1$, $a_{5,1} = 0$,
$a_{5,2} = 0$, $a_{5,3} = -4$

14-28 (a) $8(x_1 - 2)^2 + 2(x_2 - 1)^2$
$+v(126 - 32x_1 - 12x_2)$ (b) $16(x_1 - 2) - 32v = 0$,
$4(x_2 - 1) - 12v = 0$, $32x_1 + 12x_2 = 126$ (c) $x_1^* = 3$,
$x_2^* = \frac{5}{2}$ (e) $x_1^* = 2$, $x_2^* = 5.167$

14-30 (a) all given primal constraints, plus
$2v_1 + 6v_2 + v_3 = 28(x_1 - 9)$,
$18v_1 + 8v_2 + v_4 = 6(x_2 - 5)$,
$-v_1 + 3v_2 = 2(x_3 - 11)$, $v_2 \le 0$, $v_3, v_4 \ge 0$,
$v_2(20 - 6x_1 - 8x_2 - 3x_3)$, $v_3(-x_1) = 0$,
$v_4(-x_2) = 0$ (c) all given primal constraints, plus
$[10(x_1 - 1)]v_1 + v_3 = -2(x_1 - 3)$,
$[60(x_2 - 2)]v_1 + 60v_2 + v_4 = -4(x_2)^3$,
$39v_2 + v_5 = 19$, $v_1, v_3, v_4, v_5 \ge 0$,
$v_1[35 - 5(x_1 - 1)^2 - 30(x_2 - 2)^2] = 0$, $v_3(-x_1) = 0$,
$v_4(-x_2) = 0$, $v_5(-x_3) = 0$

14-31 (a) yes (c) no

14-32 (a) all given primal constraints, plus
$3v_1 + v_2 = 30x_1$, $2v_1 + v_3 = 8x_2$, $v_2, v_3 \ge 0$,
$v_2(-x_1) = 0$, $v_3(-x_2) = 0$ (e) $v_1^* = 10$, $v_2^* = v_3^* = 0$

14-34 (a) min $14(x_1 - 9)^2 + 3(x_2 - 5)^2 + (x_3 -$
$11)^2 + \mu(|2x_1 + 18x_2 - x_3 - 19| + \max\{0, 6x_1 +$
$8x_2 + 3x_3 - 20\} + \max\{0, -x_1\} + \max\{0, -x_2\})$
(c) min $100 - (x_1 - 3)^2 - (x_2)^4 + 19x_3$
$+\mu(\max\{0, 35 - 5(x_1 - 1)^2 - 30(x_2 - 2)^2\} + |60x_2 +$
$39x_3 - 159| + \max\{0, -x_1\} + \max\{0, -x_2\} +$
$\max\{0, -x_3\})$

14-35 **(a)** min $14(x_1 - 9)^2 + 3(x_2 - 5)^2 + (x_3 - 11)^2 + \mu(|2x_1 + 18x_2 - x_3 - 19|^2 + \max^2\{0, 6x_1 + 8x_2 + 3x_3 - 20\} + \max^2\{0, -x_1\} + \max^2\{0, -x_2\})$ **(c)** min $100 - (x_1 - 3)^2 - (x_2)^4 + 19x_3 + \mu(\max^2\{0, 35 - 5(x_1 - 1)^2 - 30(x_2 - 2)^2\} + |60x_2 + 39x_3 - 159|^2 + \max^2\{0, -x_1\} + \max^2\{0, -x_2\} + \max^2\{0, -x_3\})$

14-36 **(a)** min $2(x_1 - 3)^2 - x_1x_2 + (x_2 - 5)^2 + \mu(\max\{0, (x_1)^2 + (x_2)^2 - 4\} + \max\{0, x_1 - 2\} + \max\{0, -x_1\} + \max\{0, -x_2\})$ **(b)** original model was convex program and convexity is preserved by the chosen penalty functions **(c)** no **(d)** yes **(e)** penalty multipliers should start relatively low and increase slowly **(f)** $x^{(0)} = (3, 5)$; with $\mu = .5$, $x^{(1)} = (3.229, 4.646)$; with $\mu = 1$, $x^{(2)} = (2.457, 3.743)$; with $\mu = 2$, $x^{(3)} = (1.841, 2.730)$; with $\mu = 4$, $x^{(4)} = (1.147, 1.762)$; with $\mu = 8$, $x^{(5)} = (1.088, 1.678)$

14-37 **(a)** min $2(x_1 - 3)^2 - x_1x_2 + (x_2 - 5)^2 + \mu(\max^2\{0, (x_1)^2 + (x_2)^2 - 4\} + \max^2\{0, x_1 - 2\} + \max^2\{0, -x_1\} + \max^2\{0, -x_2\})$ **(b)** original model was convex program and convexity preserved by the chosen penalty functions **(c)** yes **(d)** no **(e)** penalty multipliers should start relatively low and increase slowly **(f)** $x^{(0)} = (3, 5)$; with $\mu = 0.5$, $x^{(1)} = (1.449, 2.190)$; with $\mu = 1$, $x^{(2)} = (1.308, 1.991)$; with $\mu = 2$, $x^{(3)} = (1.214, 1.858)$; with $\mu = 4$, $x^{(4)} = (1.157, 1.777)$; with $\mu = 8$, $x^{(5)} = (1.124, 1.730)$

14-40 **(a)** not applicable due to equality constraint **(c)** max $x_1 + 6/x_1 + 5(x_2)^2 + \mu(\ln(35 - 4x_1 - 6x_2) + \ln(x_1 - 5) + \ln(x_2))$ **(e)** not applicable due to equality constraint **(f)** not applicable due to equality constraint

14-41 **(a)** not applicable due to equality constraint **(c)** max $x_1 + 6/x_1 + 5(x_2)^2 - \mu[1/(35 - 4x_1 - 6x_2) + 1/(x_1 - 5) + 1/(x_2)]$ **(e)** not applicable due to equality constraint

14-42 **(a)** min $2(x_1 - 3)^2 - x_1x_2 + (x_2 - 5)^2 - \mu[\ln(4 - (x_1)^2 - (x_2)^2] + \ln(2 - x_1) + \ln(x_1) + \ln(x_2)$ **(b)** original model was convex program and convexity preserved by the chosen barrier functions

for $x > 0$ **(c)** yes **(d)** no **(e)** barrier multipliers should start relatively high and decrease to 0 **(f)** $x^{(0)} = (3, 5)$; with $\mu = 2$, $x^{(1)} = (0.991, 1.615)$; with $\mu = \frac{1}{2}$, $x^{(2)} = (1.059, 1.663)$; with $\mu = \frac{1}{8}$, $x^{(3)} = (1.081, 1.674)$; with $\mu = \frac{1}{32}$, $x^{(4)} = (1.086, 1.677)$

14-43 **(a)** min $2(x_1 - 3)^2 - x_1x_2 + (x_2 - 5)^2 + \mu[1/(4 - (x_1)^2 - (x_2)^2] + 1/(2 - x_1) + 1/x_1 + 1/x_2$ **(b)** original model was convex program and convexity preserved by the chosen barrier functions for $x > 0$ **(c)** yes **(d)** no **(e)** barrier multipliers should start relatively high and decrease to 0 **(f)** $x^{(0)} = (3, 5)$; with $\mu = 2$, $x^{(1)} = (0.978, 1.553)$; with $\mu = \frac{1}{2}$, $x^{(2)} = (1.031, 1.614)$; with $\mu = \frac{1}{8}$, $x^{(3)} = (1.061, 1.645)$; with $\mu = \frac{1}{32}$, $x^{(4)} = (1.075, 1.661)$

14-46 **(a)** min $(x_1 - 8)^2 + 2(x_2 - 4)^2$, s.t. $2x_1 + 8x_2 + x_3 = 16, x_1 + x_4 = 7, x_1, x_2, x_3, x_4 \geq 0$ **(b)** columns linearly independent **(c)** N,B,S,B **(d)** $r = (-13, 0, 1.5, 0)$ **(e)** $\Delta x = (13, -3.0625, -1.5, -13)$ **(f)** $\lambda_{max} = 0.32653$, $x^{(1)} = (4.2449, 0, 7.5102, 2.7551)$ **(g)** basic x_2 dropped to 0; new basis $\{x_3, x_4\}$ or $\{x_1, x_4\}$

14-47 **(a)** $x^{(0)} = (0, 1, 8, 7)$, $x^{(1)} = (4.2449, 0, 7.5102, 2.7551)$; then either $x^{(2)} = (4.6393, 0.8402, 0, 2.3607)$, $x^* = x^{(3)} = (6.2222, 0.4444, 0, 0.7778)$, or $x^{(2)} = (7, 0, 2, 0)$, $x^{(3)} = (7, 0.25, 0, 0)$, $x^* = x^{(4)} = (6.2222, 0.4444, 0, 0.7778)$

14-50 **(a)** $Q = \begin{pmatrix} 6 & -3 & 0 \\ -3 & 2 & 0 \\ 0 & 0 & 4 \end{pmatrix}$, $c = \begin{pmatrix} 5 \\ 15 \\ -16 \end{pmatrix}$, $A = \begin{pmatrix} 1 & 3 & -2 \\ 3 & -1 & 1 \end{pmatrix}$, $b = \begin{pmatrix} 2 \\ 3 \end{pmatrix}$ **(b)** $-12x_1 + 6x_2 + 1v_1 + 3v_2 = 5$, $6x_1 - 4x_2 + 3v_1 - 1v_2 = 15, -8x_3 - 2v_1 + 1v_2 = -16$, $1x_1 + 3x_2 - 2x_3 = 2, 3x_1 - 1x_2 + 1x_3 = 3$ **(c)** $x_1^* = x_2^* = x_3^* = 1, v_1^* = 5, v_2^* = 2$

14-52 **(a)** $c_0 = 96, c = (-16, -16), Q = \begin{pmatrix} 1 & 0 \\ 0 & 2 \end{pmatrix}$, $a^{(1)} = (2, 8), a^{(2)} = (1, 0), a^{(3)} = (1, 0), a^{(4)} = (0, 1)$, $b_1 = 16, b_2 = 7, b_3 = 0, b_4 = 0, G = \{3, 4\}$, $L = \{1, 2\}, E = \emptyset$ **(b)** $-2\Delta x_1 + 1v_3 = -16$,

$-4\,\Delta x_2 = -12, 1\,\Delta x_1 = 0$; solution $\Delta x_1 = 0$, $\Delta x_2 = 3, v_3 = -16$ **(c)** $\lambda = \frac{1}{3}$, $\mathbf{x}^{(1)} = (0,2)$ **(d)** $-2\,\Delta x_1 + 2v_1 + 1v_3 = -16, -4\,\Delta x_2 + 8v_1 = -8$, $2\,\Delta x_1 + 8\,\Delta x_2 = 0$, $\Delta x_1 = 0$; solution $\Delta x_1 = 0$, $\Delta x_2 = 0, v_1 = -1, v_3 = -14; \Delta\mathbf{x} = \mathbf{0}$ implies no further progress **(e)** drop $i = 3, x_1 \geq 0$ **(f)** $\mathbf{x}^{(2)} = (0,2)$, $\mathbf{x}^{(3)} = \mathbf{x}^* = (6.222, 0.444)$

14-54 (a) min $1 + x_{1,1}/3 + 4x_{1,2}/3 + 4x_{2,2}$, s.t. $2x_{1,1} + 2x_{1,2} + 1x_{2,1} + 1x_{2,2} \geq 2$, $4 + 28x_{1,1} + 112x_{1,2} - 18x_{2,1} - 54x_{2,2} \leq 25$, $0 \leq x_{1,1} \leq 1, 0 \leq x_{1,2} \leq 2, 0 \leq x_{2,1} \leq 2$, $0 \leq x_{2,2} \leq 2$

14-55 (b) unconstrained minimum is feasible **(c)** min $18 - 8x_1 + 4x_2$, s.t. $0 \leq x_1 + x_2 \leq 6$,

$0 \leq x_1 \leq 2, 0 \leq x_2 \leq 4$ **(d)** $x_1^* = 2, x_2^* = 0$; sequence correct **(g)** $x_1^* = 0, x_2^* = 4$; sequence incorrect

14-56 (a) min $3\exp(-0.5z_1) + \exp(z_1 + z_2)$ $+10\exp(-3z_3)$, s.t. $0.5\exp(z_1 + z_2 - 2z_3) \leq 1$, $0.167\exp(z_1) + 0.25\exp(.4z_1 + z_2) + 0.0833\exp(z_3) \leq 1$, $\mathbf{z}$ URS **(b)** $\mathbf{z}^* = (1.360, -31.1, 1.433)$, $\mathbf{x}^* = (3.898, 3 \times 10^{-14}, 4.191)$ **(c)** max $\delta_1\ln(3/\delta_1) + \delta_2\ln(1/\delta_2) + \delta_3\ln(10/\delta_3) + \delta_4\ln(0.5/\delta_4) + \delta_5\ln(0.167/\delta_5) + \delta_6\ln(0.25/\delta_6) + \delta_7\ln(0.0833/\delta_7) - v_1\ln(-v_1) - v_2\ln(-v_2)$, s.t. $-0.5\delta_1 + \delta_2 + \delta_4 + \delta_5 + 0.4\delta_6 = 0, \delta_2 + \delta_4 + \delta_6 = 0$, $-3\delta_3 - 2\delta_4 + \delta_7 = 0, \delta_1 + \delta_2 + \delta_3 = 1, \delta_4 = -v_1$, $\delta_5 + \delta_6 + \delta_7 = -v_2, \delta_1, \ldots, \delta_7 \geq 0, v_1, v_2 \leq 0$ **(d)** 4 **(e)** $\delta^* = (0, 0, 1, 0, 0, 0, 3)$, $\mathbf{v}^* = (0, -3)$, $\mathbf{x}^*$ as in part (b)

Index